Table 1.3: SI units and Prefixes.

(a) SI units

Quantity	Unit	SI symbol	Formula
SI base units			
Length	meter	m	-
Mass	kilogram	kg	-
Time	second	s	-
Temperature	kelvin	K	-
SI supplementary unit			
Plane angle	radian	rad	-
SI derived units			
Energy	joule	J	N-m
Force	newton	N	kg-m/s^2
Power	watt	W	J/s
Pressure	pascal	Pa	N/m^2
Work	joule	J	N-m

(b) SI prefixes

Multiplication factor	Prefix	SI symbol for prefix
$1{,}000{,}000{,}000{,}000 = 10^{12}$	tera	T
$1{,}000{,}000{,}000 = 10^{9}$	giga	G
$1{,}000{,}000 = 10^{6}$	mega	M
$1000 = 10^{3}$	kilo	k
$100 = 10^{2}$	hecto	h
$10 = 10^{1}$	deka	da
$0.1 = 10^{-1}$	deci	d
$0.01 = 10^{-2}$	centi	c
$0.001 = 10^{-3}$	milli	m
$0.000\,001 = 10^{-6}$	micro	μ
$0.000\,000\,001 = 10^{-9}$	nano	n
$0.000\,000\,000\,001 = 10^{-12}$	pico	p

Table 1.

Defin	
Accel	
Ener	
	Kilocal... required to raise 1 kg of water ... (1 kcal = 4187 J)
Length	1 mile = 5280 ft
	1 nautical mile = 6076.1 ft
Power	1 horsepower = 550 ft-lb/s
Pressure	1 bar = 10^5 Pa
Temperature	Fahrenheit: $t_F = \frac{9}{5}t_C + 32$
	Rankine: $t_R = t_F + 459.67$
	Kelvin: $t_K = t_C + 273.15$ (exact)
Kinematic viscosity	1 poise = 0.1 kg/m-s
	1 stoke = 0.0001 m^2/s
Volume	1 cubic foot = 7.48 gal
Useful conversion factors	
	1 in = 0.254 m = 25.4 mm
	1 lbm = 0.4536 kg
	1 °R = $\frac{5}{9}$K
	1 ft = 0.3048 m
	1 lb = 4.448 N
	1 lb = 386.1 lbm-in./s^2
	1 ton = 2000 lb (shortton)
	or 2240 lb (long ton)
	1 tonne = 1000 kg (metric ton)
	1 kgf = 9.807 N
	1 lb/in.^{2} = 6895 Pa
	1 ksi = 6.895 MPa
	1 Btu = 1055 J
	1 ft-lb = 1.356 J
	1 hp = 746 W = 2545 Btu/hr [a]
	1 kW = 3413 Btu/hr
	1 quart = 0.000946 m^3 = 0.946 liter
	1 kcal =3.968 Btu

[a] Note that in countries using the metric system, a horsepower is defined as 75 kpm/s, or 736 W.

Table 5.1: Deflection for common cantilever and simply-supported beam conditions.

Type of Loading	Deflection
Cantilever, point load P at distance a from wall (b to free end), length l	$y = -\frac{P}{6EI}\left(\langle x-a\rangle^3 - x^3 + 3x^2a\right)$ When $b=0$, $y = -\frac{P}{6EI}(3lx^2 - x^3)$ and $y_{max} = y(l) = -\frac{PL^3}{3EI}$
Cantilever, distributed load w_o over length b starting at a, length l	$y = -\frac{w_o}{24EI}\left[4bx^3 - 12bx^2\left(a+\frac{b}{2}\right) - \langle x-a\rangle^4\right]$ When $a=0$ and $b=l$, $y = \frac{w_o}{24EI}(6l^2x^2 - 4lx^3 + x^4)$ and $y_{max} = -\frac{w_o l^4}{8EI}$
Cantilever, end moment M, length l	$y = -\frac{Mx^2}{2EI}$, $y_{max} = -\frac{Ml^2}{2EI}$
Simply supported, point load P at a (b to right support), reactions $\frac{Pb}{l}$, $\frac{Pa}{l}$	$y = \frac{P}{6EI}\left(\frac{b}{l}\langle x\rangle^3 - \langle x-a\rangle^3 + 3a^2x - 2alx - \frac{a^3x}{l}\right)$
Simply supported, distributed load w_o over b between a and c, reactions $\frac{w_ob}{l}\left(c+\frac{b}{2}\right)$, $\frac{w_ob}{l}\left(a+\frac{b}{2}\right)$	$y = \frac{w_o b}{24lEI}\left\{4\left(c+\frac{b}{2}\right)x^3 - \frac{l}{b}\left(\langle x-a\rangle^4 - \langle x-a-b\rangle^4\right) + x\left[b^3 + 6bc^2 + 4b^2c + 4c^3 - 4l^2\left(c+\frac{b}{2}\right)\right]\right\}$

Fundamentals of

Machine Elements

Third Edition

Fundamentals of

Machine Elements

Third Edition

Steven R. Schmid
Bernard J. Hamrock
Bo O. Jacobson

CRC Press is an imprint of the
Taylor & Francis Group, an **informa** business

CRC Press
Taylor & Francis Group
6000 Broken Sound Parkway NW, Suite 300
Boca Raton, FL 33487-2742

Printed on acid-free paper
Version Date: 20130827

International Standard Book Number-13: 978-1-4398-9132-2 (Hardback)

Visit the Taylor & Francis Web site at
http://www.taylorandfrancis.com

and the CRC Press Web site at
http://www.crcpress.com

Dedication

This book is dedicated to Professor Bernard J. Hamrock, a great friend and mentor. Those who have had the pleasure of knowing him understand that his is a rare intellect: a world-class researcher who fundamentally changed machine design with his contributions to contact mechanics and lubrication theory; a gifted instructor and research advisor; a prolific author of exceptional papers and books; and a valuable colleague to all who have come to know him.

Professor Hamrock's professional accomplishments are exceeded only by his personal ones: A beloved husband, his love for his wife, Rosemary, is unwavering, as is his dedication as a father and grandfather; friendly to all, and a trusted friend when needed. He is by no means the stereotypical bookish professor. A football athlete in his youth, he maintains a love of the Buckeyes, of his world travels and his wine sommeliering. Those who know Bernie are grateful for the experience.

Steven R. Schmid
Notre Dame, Indiana

Contents

Part I — Fundamentals

Preface

The nature of the engineering profession is changing. It was once commonplace that students had significant machinery exposure before studying mechanical engineering, and it always was assumed that students would receive practical experience in internships or some form of co-operative employment during their college years, if not sooner. Students were historically drawn from much less diverse groups than today; students from a few decades ago (such as the authors) naturally gained experience with machinery from working on their car or tractor, and this experience was especially helpful for courses in design of machine elements. The demographics have changed, permanently and irrevocably, and the characteristics of incoming students have also changed. This has been exacerbated by the advances in technology that make maintenance of most machinery a discipline for only the specially trained. However, with a broad perspective, it has become clear that the demographics change has been an extremely positive development for the profession.

Design presents a number of challenges and opportunities to instructors. As a topic of study it is exciting because of its breadth and unending ability to provide fascinating opportunities for research, analysis, and creativity. Literally every discipline and sub-discipline in engineering has strong ties to design, and most universities have used design and manufacturing as the basis of a capstone course that culminates a mechanical engineering bachelor's degree. To students of engineering, it is, at first, an intimidating field so enormous that any semester or academic year sequence in machine design can do nothing but scratch the surface of the subject. This perception is absolutely true; like so many other areas of specialization within engineering, design truly is an area where lifelong learning is necessary.

Machine design is a challenge to both instructors and students. There are a number of courses, such as statics, dynamics, solid and fluid mechanics, etc., where topics for study are broken down into small portions and where closed-form, quantitative problems are routinely solved by students and by faculty during lectures. Such problems are important for learning concepts, and they give students a sense of security in that absolute answers can be determined. Too often, machine design is presented in a similar fashion. While, in practice, such closed-form solutions do exist, they are relatively rare. Usually, multiple disciplines are blended, and the information available is insufficient to truly optimize a desired outcome. In practice, engineers need to apply good judgment after they have researched a problem as best they can, given budgetary and time restrictions. They must then state or decide upon a solution, if not an answer. These difficult open-ended problems are much more demanding than closed-form solutions, and require a different mindset. Instead of considering a number as valid or invalid (usually by checking against the answer provided in the book or by the instructor), an open-ended problem can be evaluated only with respect to whether the result is reasonable and if good scientific methods were used. As experimental philosophers, design engineers should not hesitate proving their designs with prototypes or demonstrations. Of course, many students are taught that three weeks of modeling can save a day in the laboratory. (Sadly, this statement is not always recognized as ironic.)

This book is intended to provide the undergraduate student with a clear and thorough understanding of both the theory and application of the fundamentals of machine elements. It is expected that this book will also be used as a reference by practicing engineers. The book is not directed toward lower level undergraduate students — familiarity with differential and integral calculus is often needed to comprehend the material presented. The design of machine elements involves a great deal of geometry as well. Therefore, the ability to sketch the various configurations that arise, as well as to draw a free-body diagram of the loads acting on a component, are also needed. The material covered in this text is appropriate as a third- or fourth-year engineering course for students who have studied basic engineering sciences, including physics, engineering mechanics, and materials and manufacturing processes.

The book is divided into two parts. Part I (Chapters 1 to 8) presents the fundamentals, and Part II (Chapters 9 to 19) uses the fundamentals in considering the design of various machine elements. The material in Part I is sequential; material presented in early chapters is needed in subsequent chapters. This building-block approach provides the foundation necessary to design the various machine elements considered in Part II.

Learning Tools

The following pedagogical devices are used in each chapter to improve understanding and motivate the student:

- Each chapter will open with a photograph that clearly depicts the machine elements or topics covered in the chapter. Chapters will also have an opening quotation that is related to the chapter; the goal is to pique the reader's interest in the subject matter and start each chapter with a positive and entertaining feature to draw the students into the topic.

- In the margin to the side of the illustration, the contents, examples, case studies, and design procedures present in the chapter are listed.

- After the illustration, each chapter has a brief abstract that indicates the contents at a very high level. Part of this abstract will include a list of machine elements covered in the chapter, the typical applications of the machine elements in the chapter, and the alternate machine elements that can be considered by designers.

- A list of symbols and subscripts is then presented to help students with nomenclature as they read the chapter.

- Figures and tables have been redrawn in this edition to use modern graphical procedures of three-dimensional sketches, thick boundary lines, and sans-serif fonts in illustrations.

- Examples are printed with a light gray background to differentiate them from the text. Examples demonstrate the mathematical procedures covered and are useful for students performing quantitative problems.

- Design procedures are printed with a light color background to differentiate them from the text and examples. The design procedures are useful guides to common design problems and aide students with all levels of Bloom's taxonomy of learning.

- Case studies are printed with a light color background and are placed just before the chapter-ending summary. Case studies are mostly qualitative descriptions of important modern applications of the chapter's machine elements, but at a depth that requires an understanding of the chapter material. Case studies are intended to reference the chapter's subject matter and place it in the proper design framework so that students have no doubt that the chapter is relevant and important.

- After the summary, the chapter has a list of key words that the student can use for study or to help with jargon when necessary.

- A summary of equations is contained after the key words, and is intended to help students as they work on chapter-ending problems. The summary of equations is also a useful handout for instructors to copy and give to the students for exams.

- Every chapter includes lists of recommended readings consisting of modern as well as classic books and other resources that are especially timeless and relevant.

- The styles of the chapter-ending problems have been designed to cover every stage in modern learning taxonomies. Chapter-ending problems are organized as:

    1. Questions. These address the "remembering" task of learning taxonomies.
    2. Qualitative Problems. These are carefully designed to take an understanding of machine elements gleaned from the book and lecture and applying them to a new situation.
    3. Quantitative Problems. These problems focus on numerical analysis, with some extension to evaluating designs and results. Historically, machine element texts have provided only such analysis problems. Answers to the majority of quantitative problems are given. Solutions to the homework problems can be found in the Instructor's Solutions Manual, available to instructors who adopt the text. In addition, most problems have worksheets, where a partial solution is provided.
    4. Synthesis, Design, and Projects. These are open-ended, often team-based exercises that require creation of new designs or principles and that go beyond normal analysis problems.

Engineering educators will recognize that the end-of-chapter problems are designed to accommodate taxonomies of learning, allowing students of all backgrounds to develop an understanding, familiarity, and mastery of the subject matter.

The qualitative problems and synthesis, design, and projects class of problems also promote a useful method of active learning. In addition to conventional lecture format classes, an instructor can incorporate these problems in "seminar" sessions, active learning, or else for group projects. The authors have found this approach to be very useful and appreciated by students.

Certain users will recognize a consistent approach and pedagogy as the textbook *Manufacturing Engineering and Technology*, and will find that the texts complement each other. This is by intent, and it is hoped that the engineering student will realize quickly that *to do manufacturing or design, one needs to know both.*

Web Site

A web site containing other book-related resources can be found at www.crcpress.com/product/isbn/9781439891322. The web site provides reported errata, web links to related sites of interest, password-protected solutions to homework problems for instructors, a bulletin board, and information about ordering books and supplements. The web site also contains presentation files for instructors and students, using full-color graphics whenever possible.

Contents

Chapter 1 introduces machine design and machine elements and covers a number of topics, such as safety factors, statistics, units, unit checks, and significant figures. In designing a machine element it is important to evaluate the kinematics, loads, and stresses at the critical section. Chapter 2 describes the applied loads (normal, torsional, bending, and transverse shear) acting on a machine element with respect to time, the area over which the load is applied, and the location and method of application. The importance of support reaction, application of static force and moment equilibrium, and proper use of free-body diagrams is highlighted. Shear and moment diagrams applied to beams for various types of singularity function are also considered. Chapter 2 then describes stress and strain separately.

Chapter 3 focuses on the properties of solid engineering materials, such as the modulus of elasticity. (Appendix A gives properties of ferrous and nonferrous metals, ceramics, polymers, and natural rubbers. Appendix B explores the stress-strain relationships for uniaxial, biaxial, and triaxial stress states.) Chapter 4 describes the stresses and strains that result from the types of load described in Chapter 2, while making use of the general Hooke's law relationship developed in Appendix B. Chapter 4 also considers straight and curved members under these four types of load.

Certainly, ensuring that the design stress is less than the yield stress for ductile materials and less than the ultimate stress for brittle materials is important for a safe design. However, attention must also be paid to displacement (deformation) since a machine element can fail by excessive elastic deformation. Chapter 5 attempts to quantify the deformation that might occur in a variety of machine elements. Some approaches investigated are the integral method, the singularity function, the method of superposition, and Castigliano's theorem. These methods are applicable for distributed loads.

Stress raisers, stress concentrations, and stress concentration factors are investigated in Chapter 6. An important cause of machine element failure is cracks within the microstructure. Therefore, Chapter 6 covers stress levels, crack-producing flaws, and crack propagation mechanisms and also presents failure prediction theories for both uniaxial and multiaxial stress states. The loading throughout Chapter 6 is assumed to be static (i.e., load is gradually applied and equilibrium is reached in a relatively short time). However, most machine element failures involve loading conditions that fluctuate with time. Fluctuating loads induce fluctuating stresses that often result in failure by means of cumulative damage. These topics, along with impact loading, are considered in Chapter 7.

Chapter 8 covers lubrication, friction, and wear. Not only must the design stress be less than the allowable stress and the deformation not exceed some maximum value, but also lubrication, friction, and wear (tribological considerations) must be properly understood for machine elements to be successfully designed. Stresses and deformations for concentrated loads, such as those that occur in rolling-element bearings and gears, are also determined in Chapter 8. Simple expressions are developed for the deformation at the center of the contact as well as for the maximum stress. Chapter 8 also describes the properties of fluid film lubricants used in a number of machine elements. Viscosity is an important parameter for establishing the load-carrying capacity and performance of fluid-film lubricated machine elements. Fluid viscosity is greatly affected by temperature, pressure, and shear rate. Chapter 8 considers not only lubricant viscosity, but also pour point and oxidation stability, greases and gases, and oils.

Part II (Chapters 9 to 20) relates the fundamentals to various machine elements. Chapter 9 deals with columns, which receive special consideration because yielding and excessive deformation do not accurately predict the failure of long columns. Because of their shape (length much larger than radius) columns tend to deform laterally upon loading, and if deflection becomes critical, they fail catastrophically. Chapter 9 establishes failure criteria for concentrically and eccentrically loaded columns.

Chapter 10 considers cylinders, which are used in many engineering applications. The chapter covers tolerancing of cylinders; stresses and deformations of thin-walled, thick-walled, internally pressurized, externally pressurized, and rotating cylinders; and press and shrink fits.

Chapter 11 considers shafting and associated parts, such as keys, snap rings, flywheels, and couplings. A shaft design procedure is applied to static and cyclic loading; thus, the material presented in Chapters 6 and 7 is directly applied to shafting. Chapter 11 also considers critical speeds of rotating shafts.

Chapter 12 presents the design of hydrodynamic bearings — both thrust and journal configurations — as well as design procedures for the two most commonly used slider bearings. The procedures provide an optimum pad configuration and describe performance parameters, such as normal applied load, coefficient of friction, power loss, and lubricant flow through the bearing. Similar design information is given for plain and nonplain journal bearings. The chapter also considers squeeze film and hydrostatic bearings, which use different pressure-generating mechanisms.

Rolling-element bearings are presented in Chapter 13. Statically loaded radial, thrust, and preloaded bearings are considered, as well as loaded and lubricated rolling-element bearings, fatigue life, and dynamic analysis. The use of the elastohydrodynamic lubrication film thickness is integrated with the rolling-element bearing ideas developed in this chapter.

Chapter 14 covers general gear theory and the design of spur gears. Stress failures are also considered. The transmitted load is used to establish the design bending stress in a gear tooth, which is then compared with an allowable stress to establish whether failure will occur. Chapter 14 also considers fatigue failures. The Hertzian contact stress with modification factors is used to establish the design stress, which is then compared with an allowable stress to determine whether fatigue failure will occur. If an adequate protective elastohydrodynamic lubrication film exists, gear life is greatly extended.

Chapter 15 extends the discussion of gears beyond spur gears as addressed in Chapter 14 to include helical, bevel, and worm gears. Advantages and disadvantages of the various types of gears are presented.

Chapter 16 covers threaded, riveted, welded, and adhesive joining of members, as well as power screws. Riveted and threaded fasteners in shear are treated alike in design and failure analysis. Four failure modes are presented: bending of member, shear of rivet, tensile failure of member, and compressive bearing failure. Fillet welds are highlighted, since they are the most frequently used type of weld. A brief stress analysis for lap and scarf adhesively bonded joints is also given.

Chapter 17 treats the design of springs, especially helical compression springs. Because spring loading is most often continuously fluctuating, Chapter 16 considers the design allowance that must be made for fatigue and stress concentration. Helical extension springs are also covered in Chapter 16. The chapter ends with a discussion of torsional and leaf springs.

Brakes and clutches are covered in Chapter 18. The brake analysis focuses on the actuating force, the torque transmitted, and the reaction forces in the hinge pin. Two theories relating to clutches are studied: the uniform pressure model and the uniform wear model.

Chapter 19 deals with flexible machine elements. Flat belts and V-belts, ropes, and chains are covered. Methods of effectively transferring power from one shaft to another while using belts, ropes, and chains are also presented. Failure modes of these flexible machine elements are considered.

What's New in This Edition

This third edition represents a major revision from the second edition. In addition to the pedagogy enhancements mentioned above, the contents have been greatly expanded and organized to aide students of all levels in design synthesis and analysis approaches. Design synthesis is generally taught or expected of students only after a machine elements course in most college curricula. This book attempts to provide guidance through design procedures for synthesis issues, but it also exposes the reader to a wide variety of machine elements.

Users of the second edition will immediately recognize that this third edition has been completely re-typeset using a space-saving, two-column approach, and all figures redrawn to match the new column widths. The space-saving typesetting format has saved over 300 pages from the previous edition, while the content has been expanded considerably. This was, in fact, a goal: too many textbooks are difficult to use because they give the impression of completeness, but this is often illusory. Large margins and gaps between topics artificially produce heavy tomes. Our goal was to create a book with good coverage that can be more easily carried by students.

In every chapter opening box, the reader is directed toward other machine elements that can serve the same purpose, which can also help in synthesis. As an example, a student designing a gear set for power transmission between two shafts may thus be reminded that a belt drive is perhaps an alternative worthy of consideration.

The book has been designed to compliment the well-known manufacturing textbooks *Manufacturing Processes for Engineering Materials* and *Manufacturing Engineering and Technology* by Kalpakjian and Schmid. Students who use both texts in their engineering studies will recognize similarities in organization, graphical styles, and, it is hoped, clarity.

The classes of chapter-ending problems have been introduced above, but they have been carefully designed to aid students to develop a deep understanding of each chapter's subject matter. They have been developed using learning taxonomies that require ever-sophisticated cognitive effort. That is, students are required to remember (Questions), apply knowledge to fairly simple and straightforward questions (Qualitative Problems), extend the knowledge to analytical problems (Quantitative Problems), and finally asked to extend their analytical abilities to open-ended and synthesis problems requiring creativity in their solution (Synthesis and Design Problems).

A major effort has been made to expand coverage in all areas. Specific changes to this edition include:

- In Chapter 1, additional design considerations have been listed in Section 1.4, additional examples and case studies have been added, and life cycle engineering has been included.
- Chapter 3 now includes a description of hardness and common hardness tests used for metals; this clarifies the use of these concepts in gear design. In addition, the manufacturing discussion has been expanded.
- The use of retaining rings in Chapter 11 necessitated the inclusion of flat groove stress concentration factors in Chapter 6.
- Chapter 7, on fatigue design, has been significantly expanded. The staircase method for determining endurance limits has been added in Design Procedure 7.2, the fatigue strength concentration factor descriptions are longer with more mathematical models, and Haigh diagrams are included to show the effects of mean stress. Additional material data has been included for the fracture mechanics approaches to fatigue design.
- In Chapter 8, a streamlined discussion of typical surface finishes in machine elements, and manufacturing processes used to produce them, has been prepared. In addition, a discussion of the commonly used bearing materials has been added.
- Chapter 11 has been expanded considerably. In addition to an expanded discussion of keys and set screws, the chapter presents new treatment of spline, pin, and retaining ring design, and has a new section on the design of shaft couplings.
- Hydrodynamic bearings are increasingly important because of their widespread use in transportation and power industries; while the discussion of thrust and journal bearings has been retained, the analysis is simplified and more straightforward. The discussion of squeeze film and hydrostatic bearings has been expanded.
- Chapter 13 has been extensively rewritten to reflect the latest International Standards Organization standards that unify the approach used to design rolling element bearings. This has allowed a simplification of bearing selection and analysis, as will be readily apparent. Further, this remains the only machine element book that accurately depicts the wide variety of bearings available. This treatment now includes the topic of toroidal bearings, a novel design that is now widely available, and leads to compact and high load carrying designs. Life adjustment factors and effects of variable loading have been expanded, and an industrially relevant case study on windmill bearings has been exhaustively researched and included in the chapter.
- The treatment of spur gear design in Chapter 14 has been modified to reflect the latest advances in materials, including powder metal materials that have become extremely popular for automotive applications. The importance of lubrication in gears has been emphasized. Further, a design synthesis approach for spur gear design has been included in Section 14.14.
- Geometry factors for bevel gears in Chapter 15 have been simplified without loss in accuracy. Also, a design synthesis approach for worm gears has been included.
- The discussion of fasteners and welds in Chapter 16 has been expanded considerably. The importance of the heat affected zone for weld quality is discussed, and the classes of welds and their analysis methods are described. This includes the treatment of modern welding approaches such as friction stir welding as well as laser and electron beam welding.
- Gas springs and wave springs have been added to the discussion of Chapter 17.
- Chapter 18 has been reorganized, starting with fundamental principles that apply to all brake and clutch systems, especially thermal effects. Additional automotive examples have been added.
- Chapter 19 has been essentially rewritten to reflect the latest standards and manufacturer's recommendations on belt design, chains, and wire ropes. In addition, silent chains have been included into the chain discussion.
- The appendices have been expanded to provide the student with a wide variety of material properties, geometry factors for fracture analysis, and new summaries of beam deflection. While it is recognized that modern students have such information readily available via the Internet, making such material available in the textbook is useful for reference purposes.

This text has been under preparation for over four years, and required meticulous efforts at maintaining a consistent approach, careful statement of design procedures wherever they were useful, and expansion of chapter-ending problems. We hope the student of machine element design will enjoy and benefit from this text.

Steven R. Schmid
The University of Notre Dame

Bernard J. Hamrock
The Ohio State University

Bo O. Jacobson
Lund University

Acknowledgments

Many people helped to produce this textbook. Jonathan Plant, who was the editor of the first edition, was extremely supportive during the preparation of this edition. Professor Serope Kalpakjian (ret.) of the Illinois Institute of Technology made numerous comments and helpful suggestions, and also assisted in review of the manuscript. Professor William Dornfeld kindly used draft versions of this textbook in a number of semesters and made very constructive comments and suggestions. Dr. Michael Kotsalas of the Timken company was of great assistance in modernizing the chapters in bearing design. A number of faculty members from the mechanical engineering departments at Notre Dame and Ohio State University provided comments about the text. In particular, the authors would like to thank Professors John Renaud, Anthony Luscher, Si Lee, Necip Berme, and Bharat Bhushan. Special thanks are also due to Triodyne, Inc. personnel, especially Ralph L. Barnett, whose insights were invaluable. A number of resources from Triodyne were kindly made available in the preparation of the text.

Many reviewers have been associated with this project. The authors would like to acknowledge them for their time, expertise, and advice. The list of reviewers are Timothy Rodts, Michael Giordano, Matthew Prygoski, and Amy Libardi, University of Notre Dame; Miguel Angel Sellés Cantó, Escola Politècnica Superior d' Alcoi, Universitat Politècnica de València; Peder Klit, Technical University of Denmark; Robert W. Ellis, Lawrence Technological University; James Adams, Metal Powder Industries Federation; Thomas Kurfess, Georgia Institute of Technology; Paul K. Wright, University of California at Berkeley; Jian Cao, Northwestern University; K. Scott Smith, University of North Carolina at Charlotte; William G. Ovens, Rose-Hulman Institute of Technology; John D. Reid, University of Nebraska–Lincoln; Steven Y. Liang, Georgia Institute of Technology; Steven A. Velinsky, University of California–Davis; Thierry Blanchet, Rensselaer Polytechnic Institute; K.V.C. Rao, Michigan Technological University; Alexander G. Liniecki, San Jose State University; Jesa Kreiner, California State University–Fullerton; Clarence Maday, North Carolina State University; J. Darrell Gibson, Rose-Hulman Institute of Technology; Terry F. Lehnhoff, University of Missouri–Rolla; Yu Michael Wang, University of Maryland; B.K. Rao, Idaho State University; W. Brent Hall, University of Illinois–Urbana-Champaign; N. Duke Perreira, Lehigh University; Gordon Pennock, Purdue University; Richard E. Dippery, Kettering University; Anthony Luscher, Ohio State University; Michael Peterson, Colorado State University; Gary McDonald, The University of Tennessee at Chattanooga; Wayne D. Milestone, University of Wisconsin–Madison; and Dean Taylor, Cornell University.

The authors also acknowledge the use of tables and illustrations from the following publishers: American Gear Manufacturers Association, American Society of Mechanical Engineers, American Society for Testing and Materials, BHRA Fluid Engineering, Buttersworths, Elsevier Science Publishing Company, Engineering Sciences Data Unit, Ltd., Heinemann (London), Hemisphere Publishing Corporation, Macmillan Publishing Company, Inc., Mechanical Technology Incorporated, Metal Powder Industries Federation, Non-Ferrous Founders Society, Oxford University Press, Inc., Penton Publishing Inc., Society of Automotive Engineers, Society of Tribologists and Lubrication Engineers, VCH Publishers, John Wiley & Sons, and Wykeham Publications (London), Ltd. The specific sources are identified in the text.

About the Authors

Steven R. Schmid received his B.S. degree in Mechanical Engineering from the Illinois Institute of Technology in 1986. He then joined Triodyne, Inc., where his duties included investigation of machinery failures and consultation in machine design. He earned his Master's degree from Northwestern University in 1989 and his Ph.D. in 1993, both in mechanical engineering. In 1993 he joined the faculty at the University of Notre Dame, where he is currently a Professor of Aerospace and Mechanical Engineering and teaches and conducts research in the fields of design and manufacturing. Dr. Schmid received the American Society of Mechanical Engineers Newkirk Award and the Society of Manufacturing Engineers Parsons Award in 2000. He was also awarded the Kaneb Center Teaching Award in 2000, 2003, and 2010, and served as a Kaneb fellow in 2003. Dr. Schmid holds professional engineering (P.E.) and certified manufacturing engineer (C.Mfg.E.) licenses. He is co-author (with S. Kalpakjian) of *Manufacturing Engineering and Technology* (2014) and *Manufacturing Processes for Engineering Materials* (2008), both published by Prentice Hall. He was awarded the ASME Foundation Swanson Fellowship in 2012, and is a Fellow of the American Society of Mechanical Engineers.

Bernard J. Hamrock joined the staff of Ohio State University as a professor of Mechanical Engineering in 1985 and is now Professor Emeritus. Prior to joining Ohio State University he spent 18 years as a research consultant in the Tribology Branch of the NASA Lewis Research Center in Cleveland, Ohio. He received his Ph.D. and Doctor of Engineering degrees from the University of Leeds, Leeds, England. Professor Hamrock's research has resulted in a book with Duncan Dowson, *Ball Bearing Lubrication*, published in 1982 by Wiley Interscience, three separate chapters for handbooks, and over 150 archival publications. His second book, *Fundamentals of Fluid Film Lubrication* was published in 1993 by McGraw-Hill, and a second edition, with co-authors Steven Schmid and Bo Jacobson, in 2004. His awards include the 1976 Melville Medal from the American Society of Mechanical Engineers, the NASA Exceptional Achievement Medal in 1984, the 1998 Jacob Wallenberg Award given by The Royal Swedish Academy of Engineering Sciences, and the 2000 Mayo D. Hersey Award from the American Society of Mechanical Engineers.

Bo O. Jacobson received his Ph.D. and D.Sc. degrees from Lund University in Sweden. From 1973 until 1987 he was Professor of Machine Elements at Luleå Technology University in Sweden. In 1987 he joined SKF Engineering & Research Centre in the Netherlands, while retaining a professorship at Chalmers University from 1987 to 1991 and at Luleå Technical University from 1992 to 1997. In 1997 he was appointed Professor of Machine Elements at Lund University, Sweden. Professor Jacobson was a NRC Research Fellow at NASA Lewis Research Center from 1981 to 1982. He has published four compendia used at Swedish universities. His text *Rheology and Elastohydrodynamic Lubrication* was published by Elsevier in 1991, and his book *Rolling Contact Phenomena* (with J. Kalker) was published in 2000 by Springer. Professor Jacobson has more than 100 archival publications. His awards include the prestigious Gold Medal given by the Institution of Mechanical Engineers, England, and the Wallenberg Award in 1984.

Part I — Fundamentals

Outline

Chapter 1

Introduction

The i8 concept car, a hybrid sports car requiring three liters per 100 km (94 mpg) and acceleration from 0 to 100 km/hr in under five seconds. *Source:* Courtesy of BMW.

The invention all admir'd, and each, how he
To be th' inventor miss'd; so easy it seem'd,
Once found, which yet unfound most would have thought
Impossible

John Milton

Contents

Design is arguably the most important specialization in modern industrial society. Integral to most engineering curricula, and also making up its own specialization in many schools, design is critical for wealth generation, development of economic activity, and the creation of jobs. This chapter introduces design synthesis, where a new machine or system is produced to address a need or to satisfy customer requirements. Design analysis is also discussed, which involves the use of engineering disciplines to determine critical dimensions, select acceptable materials, and even optimize designs. To bring high-quality products to market quickly, it is important to integrate multiple disciplines early in the design process, including solid and fluid mechanics, materials selection, marketing, manufacturing, safety, and environmental concerns. Many times, constraints are applied to mechanical designs, such as those mandated by governmental codes or industrial standards. It has been observed that simultaneous higher quality and lower costs arise when manufacturing can exploit automation, but automation can only be justified for larger production runs than most products can justify. One method of achieving the benefits of large-scale manufacture is to use standard sizes and types of machine elements in design, but this requires some sophistication with respect to significant figures, measurement units, and specification of dimensions.

Symbols

n_s	safety factor
n_{sx}	safety factor involving quality of materials, control over applied load, and accuracy of stress analysis
n_{sy}	safety factor involving danger to personnel and economic impact
σ_{all}	allowable normal stress, Pa
σ_d	design normal stress, Pa

1.1 What is Design?

Design means different things to different people. A clothing manufacturer believes that incorporating different materials or colors into a new dress constitutes design. A potter paints designs onto china to complement its surroundings. An architect designs ornamental facades for residences. An engineer chooses a bearing from a catalog and incorporates it into a speed-reducer assembly. These design activities, although they appear to be fundamentally different, share a common thread: they all require significant creativity, practice, and vision to be done well.

"Engineering," wrote Thomas Tredgold, "is the art of directing the great sources of power in nature for the use and convenience of man" [Florman 1987]. It is indeed significant that this definition of engineering is more than 60 years old — few people now use the words "engineering" and "art" in the same sentence, let alone in a definition. However, many products are successful for nontechnical reasons, reasons that cannot be proved mathematically. On the other hand, many problems are mathematically tractable, but usually because they have been inherently overconstrained. Design problems are, almost without exception, open-ended problems combining hard science and creativity. Engineering is indeed an art, even though *parts* of engineering problems lend themselves well to analysis.

For the purposes of this textbook, **design** is the transformation of concepts and ideas into useful machinery. A **machine** is a combination of mechanisms and other components that transforms, transmits, or uses energy, load, or motion for a specific purpose. If Tredgold's definition of engineering is accepted, design of machinery is the fundamental practice in engineering.

A machine comprises several different machine elements properly designed and arranged to work together as a whole. Fundamental decisions regarding loading, kinematics, and the choice of materials must be made during the design of a machine. Other factors, such as strength, reliability, deformation, tribology (friction, wear, and lubrication), cost, and space requirements also need to be considered. The objective is to produce a machine that not only is sufficiently rugged to function properly for a reasonable time, but also is economically feasible. Further, nonengineering decisions regarding marketability, product liability, ethics, politics, etc. must be integrated early in the design process. Since few people have the necessary tools to make all these decisions, machine design in practice is a discipline-blending human endeavor.

This textbook emphasizes one of the disciplines necessary in design — mechanical engineering. It therefore involves calculation and consideration of forces, energies, temperatures, etc., — concepts instilled into an engineer's psyche.

To "direct the great sources of power in nature" in machine design, the engineer must recognize the functions of the various machine elements and the types of load they transmit. A **machine element** may **function** as a normal load transmitter, a torque transmitter, an energy absorber, or a seal. Some common load transmitters are rolling-element bearings, hydrodynamic bearings, and rubbing bearings. Some torque transmitters are gears, shafts, chains, and belts. Brakes and dampers are energy absorbers. All the machine elements in Part II can be grouped into one of these classifications.

Engineers must produce safe, workable, good designs, as stated in the first fundamental canon in the *Code of Ethics for Engineers* [ASME 2012]:

> Engineers shall hold paramount the safety, health, and welfare of the public in the performance of their professional duties.

Designing reasonably safe products involves many design challenges to ensure that components are large enough, strong enough, or tough enough to survive the loading environment. One subtle concept, but of huge importance, is that the engineer has a duty to protect the welfare of the general public. Welfare includes economic well-being, and it is well known that successful engineering innovations lead to wealth and job creation. However, products that are too expensive are certain to fail in a competitive marketplace. Similarly, products that do not perform their function well will fail. Economics and functionality are always pressing concerns, and good design inherently means safe, economical, and functional design.

1.2 Design of Mechanical Systems

A **mechanical system** is a synergistic collection of machine elements. It is synergistic because as a design it represents an idea or concept greater than the sum of the individual parts. For example, a mechanical clock, although merely a collection of gears, springs, and cams, also represents the physical realization of a time-measuring device. Mechanical system design requires considerable flexibility and creativity to obtain good solutions. Creativity seems to be aided by familiarity with known successful designs, and mechanical systems are often collections of well-designed components from a finite number of proven classes.

Designing a mechanical system is a different type of problem than selecting a component. Often, the demands of the system make evident the functional requirements of a component. However, designing a large mechanical system, potentially comprising thousands or even millions of machine elements, is a much more open, unconstrained problem.

To design superior mechanical systems, an engineer must have a certain sophistication and experience regarding machine elements. Studying the design and selection of machine elements affords an appreciation for the strengths and limitations of classes of components. They can then be more easily and appropriately incorporated into a system. For example, a mechanical system cannot incorporate a worm gear or a Belleville spring if the designer does not realize that these devices exist.

A toolbox analogy of problem solving can be succinctly stated as, "If your only tool is a hammer, then every problem is a nail." The purpose of studying machine element design is to fill the toolbox so that problem solving and design synthesis activities can be flexible and unconstrained.

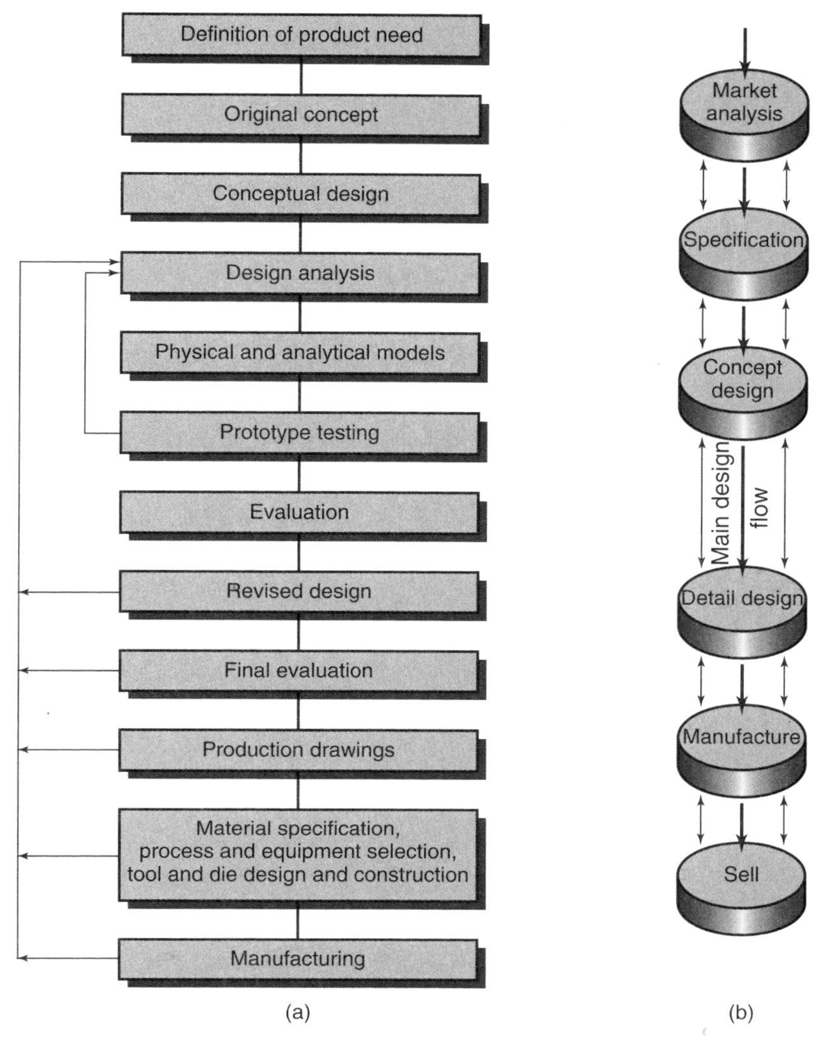

Figure 1.1: Approaches to product development. (a) Classic approach, with large design iterations typical of the over-the-wall engineering approach. *Source:* Adapted from Kalpakjian and Schmid [2003]. (b) A more modern approach, showing a main design flow with minor iterations representing concurrent engineering inputs. *Source:* Adapted from Pugh [1996].

1.3 Design as a Multidisciplinary Endeavor

The quality revolution that transformed the manufacturing sector in the early 1980s has forever changed the approach companies and engineers take toward product development. A typical design process of the recent past (Fig. 1.1a) shows that the skills involved in machine element design played an essential role in the process. This approach was commonly used in the United States in the post-World War II era.

The term *over-the-wall engineering* (OTW) has been used to describe this design approach. Basically, someone would apply a particular skill and then send the product "over the wall" to the next step in development. A product design could sometimes flow smoothly from one step to the next and into the marketplace within weeks or months. This was rarely the case, however, as usually a problem would be discovered. For example, a manufacturing engineer might ask that workpieces be more easily clamped into a milling machine fixture. The design engineer would then alter the design and send the product back downstream. A materials scientist might then point out that the material chosen had drawbacks and suggest a different choice. The design engineer would make the alteration and resubmit the design. This process could continue *ad infinitum*, with the result that the product would take a long time to develop. It is not surprising that modifications to this approach started being developed in the 1970s and 1980s.

Figure 1.1b shows a more modern design approach. Here, there is still the recognized general flow of information from product conception through introduction into the marketplace, but there is immediate involvement of many disciplines in the design stage. Different disciplines are involved simultaneously instead of sequentially as with the OTW approach. Some tasks are extremely technical, such as design analysis (the main focus of this book) or manufacturing. Others are nontechnical, such as market analysis. **Concurrent engineering** is the philosophy of involving many disciplines from the beginning of a design effort and keeping them involved throughout product development. Thus, redundant efforts are minimized and higher quality products are developed more quickly. Although design iterations still occur, the iteration loops are smaller and incur much less wasted time, effort, and expense. Also, design shortcomings can be corrected before they are incorporated. For example, service personnel can inform design engineers of excessive component failures in previous designs during the conceptual design stage for a new model, and shortcomings can be corrected instead of persisting. No such mechanism for correcting design shortcomings ever existed in conventional design approaches or management structures.

Another important concept in design is the time to market. Bringing high-quality products to market quickly is normal practice in the consumer electronics industry, where rapid change shortened useful market life to a few months. Minimizing time to market is now recognized as essential for controlling development costs. Further, new products introduced before their competitors' products usually enjoy a larger share of the market and profits. Thus, manufacturers who could ship products weeks or even days faster than their competitors had a distinct sales advantage. Saving development time through concurrent engineering made companies much more competitive in the global marketplace.

Concurrent engineering has profoundly affected design engineers. They can no longer work alone and must participate in group discussions and design reviews. They need good communications skills. Designing machinery has become a cooperative endeavor.

Clearly, many disciplines now play a role in product development, but design engineers cannot merely focus on their discipline and rely on experts for the rest. They need familiarity with other disciplines, at least from a linguistics standpoint, to integrate them into the design process. Thus, modern engineers may need to speak the language of materials science, law, marketing, etc., even if they are not experts in these fields.

1.4 Design of Machine Elements

Specifying a mechanical system is only the beginning of the design synthesis process. Particular machine element classes need to be chosen, possibly leading to further design iterations. Designing a proper machine element usually involves the following steps:

1. Selecting a suitable type of machine element from consideration of its function
2. Estimating the size of the machine element that is likely to be satisfactory
3. Evaluating the machine element's performance against design requirements or constraints
4. Modifying the design and dimensions until the performance is near to whichever optimum is considered most important

The last two steps in the process can be handled fairly easily by someone who is trained in analytical methods and understands the fundamental principles of the subject. The first two steps, however, require some creative decisions and, for many, represent the most difficult part of design.

After a suitable type of machine element has been selected for the required function, the specific machine element is designed by analyzing kinematics, load, and stress. These analyses, coupled with proper material selection, will enable a stress-strain-strength evaluation in terms of a safety factor (as discussed in Section 1.4.1). A primary question in designing any machine element is whether it will fail in service. Most people, including engineers, commonly associate **failure** with the actual breaking of a machine element. Although breaking is one type of failure, a design engineer must have a broader understanding of what really determines whether a part has failed.

A machine element is considered to have failed:

1. When it becomes completely inoperable
2. When it is still operable but is unable to perform its intended function satisfactorily
3. When serious deterioration has made it unreliable or unsafe for continued use, necessitating its immediate removal from service for repair or replacement

The role of the design engineer is to predict the circumstances under which failure is likely to occur. These circumstances are stress-strain-strength relationships involving the bulk of the solid members and such surface phenomena as friction, wear, lubrication, and environmental deterioration.

The principles of design are universal. An analysis is equally valid regardless of the size, material, and loading. However, an analysis by itself should not be looked on as an absolute and final truth. An analysis is limited by the assumptions imposed and by its range of applicability. Thus, designers often must check and verify if they have addressed considerations such as:

- Have all alternative designs been thoroughly investigated?
- Can the design be simplified and the number of its components minimized without adversely affecting its intended functions and performance?
- Can the design be made smaller and lighter?
- Are there unnecessary features to the product or some of its components, and if so, can they be eliminated or combined with other features?
- Have modular design and building-block concepts been considered for a family of similar products and for servicing and repair, upgrading, and installing options?
- Are the specified dimensional tolerances and surface finish unnecessarily tight, thereby significantly increasing product cost, and can they be relaxed without any adverse effects?
- Will the product be difficult or excessively time consuming to assemble and disassemble for maintenance, servicing, or recycling of some or all of its components?
- Is the use of fasteners minimized, including their quantity and variety?
- Have environmental considerations been taken into account and incorporated into product design, as well as material and process selection?
- Have green design and life-cycle engineering principles been applied, including recycling considerations?

Design analysis attempts to predict the strength or deformation of a machine element so that it can safely carry the imposed loads for as long as required. Assumptions have to be made about the material properties under different loading types (axial, bending, torsion, and transverse shear, as well as various combinations) and classes (static, sustained, impact, or cyclic). These loading constraints may vary throughout the machine as they relate to different machine elements, an important factor for the design engineer to keep in mind.

1.4.1 Safety in Mechanical Design

The code of Hammurabi, a Babylonian doctrine over 3000 years old, had this requirement:

> If a builder build a house for a man and do not make its construction firm, and the house which he has built collapse and cause the death of the owner of the house, that builder shall be put to death.

It could be argued that engineers are getting off a lot easier these days. Modern legal doctrines do not call for the death of manufacturers of unsafe products or of the engineers who designed them. Regardless of the penalty, however, engineers have a moral and legal obligation to produce reasonably safe products. A number of fundamental concepts and tools are available to assist them in meeting this challenge.

Safety Factor

If 500 tension tests are performed on a specimen of one material, 500 different yield strengths will be obtained if the precision and accuracy of measurement are high enough. With some materials, a wide range of strengths can be achieved; in others, a reasonable guaranteed minimum strength can be found. However, this strength does not usually represent the stress that engineers apply in design.

Using results from small-scale tension tests, a design engineer prescribes a stress somewhat less than the semi-empirical strength of a material. The **safety factor** can be expressed as

$$n_s = \frac{\sigma_{\text{all}}}{\sigma_d} \tag{1.1}$$

where σ_{all} is the allowable normal stress and σ_d is the design normal stress. If $n_s > 1$, the design is adequate. The larger n_s, the safer the design. If $n_s < 1$, the design may be inadequate and redesign may be necessary. In later chapters, especially Chapter 6, more will be said about σ_{all} and σ_d. The rest of this section focuses on the left side of Eq. (1.1).

It is difficult to accurately evaluate the various factors involved in engineering design problems. One factor is the shape of a part. For an irregularly shaped part, there may be no design equations available for accurate stress computation. Sometimes the load is uncertain. For example, the loading applied to a bicycle seat and frame depends on the size of the rider, speed, and size of bumps encountered. Another factor is the consequences of part failure; life-threatening consequences require more consideration than non-life-threatening consequences.

Engineers use a safety factor to ensure against such uncertain or unknown conditions. The engineering student is often asked, What safety factor was used in the design, and which value should be used? Safety factors are sometimes prescribed by code, but usually they are rooted in design experience. That is, design engineers have established through a product's performance that a safety factor is sufficient. Future designs are often based on safety factors found adequate in previous products for similar applications.

Particular design experience for specific applications does not form a basis for the rational discussion of illustrative examples or for the guidance of engineering students. The Pugsley [1966] method for determining the safety factor is a potential approach for obtaining safety factors in design, although the reader should again be warned that safety factor selection is somewhat nebulous in the real world and the Puglsey method can be unconservative; that is, it predicts safety factors that are too low for real applications. Pugsley systematically determined the safety factor from

Table 1.1: Safety factor characteristics A, B, and C.

Characteristic[a]		B			
A	C	vg	g	f	p
vg	vg	1.1	1.3	1.5	1.7
	g	1.2	1.45	1.7	1.95
	f	1.3	1.6	1.9	2.2
	p	1.4	1.75	2.1	2.45
g	vg	1.3	1.55	1.8	2.05
	g	1.45	1.75	2.05	2.35
	f	1.6	1.95	2.3	2.65
	p	1.75	2.15	2.55	2.95
f	vg	1.5	1.8	2.1	2.4
	g	1.7	2.05	2.4	2.75
	f	1.9	2.3	2.7	3.1
	p	2.1	2.55	3.0	3.45
p	vg	1.7	2.15	2.4	2.75
	g	1.95	2.35	2.75	3.15
	f	2.2	2.65	3.1	3.55
	p	2.45	2.95	3.45	3.95

[a] vg = very good, g = good, f = fair, and p = poor.
A = quality of materials, workmanship, maintenance, and inspection.
B = control over load applied to part.
C = accuracy of stress analysis, experimental data or experience with similar parts.

Table 1.2: Safety factor characteristics D and E.

Characteristic E[a]	D		
	ns	s	vs
ns	1.0	1.2	1.4
s	1.0	1.3	1.5
vs	1.2	1.4	1.6

[a] vs = very serious, s = serious, and ns = not serious
D = danger to personnel
E = economic impact

$$n_s = n_{sx} n_{sy} \tag{1.2}$$

where

- n_{sx} = safety factor involving characteristics A, B, and C
- A = quality of materials, workmanship, maintenance, and inspection
- B = control over load applied to part
- C = accuracy of stress analysis, experimental data, or experience with similar devices
- n_{sy} = safety factor involving characteristics D and E
- D = danger to personnel
- E = economic impact

Table 1.1 gives n_{sx} values for various A, B, and C conditions. To use this table, estimate each characteristic for a particular application as being very good (vg), good (g), fair (f), or poor (p). Table 1.2 gives n_{sy} values for various D and E conditions. To use this table, estimate each characteristic for a particular application as being very serious (vs), serious (s), or not serious (ns). Substituting the values of n_{sx} and n_{sy} into Eq. (1.2) yields a proposed safety factor.

Although a simple procedure to obtain safety factors, the Pugsley method illustrates the concerns present in safety factor selection. Many parameters, such as material strength and applied loads, may not be well known, and confidence in the engineering analysis may be suspect. For these reasons the safety factor has sometimes been called an "ignorance factor," as it compensates for ignorance of the total environment, a situation all design engineers encounter to some extent. Also, the Pugsley method is merely a guideline and is not especially conservative; most engineering safety factors are much higher than those resulting from Eq. (1.2), as illustrated in Example 1.1.

Example 1.1: Safety Factor of Wire Rope in an Elevator

Given: A wire rope is used on an elevator transporting people to the 20th floor of a building. The design of the elevator can be 50% overloaded before the safety switch shuts off the motor.

Find: What safety factor should be used?

Solution: The following values are assigned:

A = vg, because life threatening
B = f to p, since large overloads are possible
C = vg, due to being highly regulated
D = vs, people could die if the elevator fell from the 20th floor
E = vs, possible lawsuits

From Tables 1.1 and 1.2 the safety factor is

$$n_s = n_{sx} n_{sy} = (1.6)(1.6) = 2.56$$

Note that the value of $n_{sx} = 1.6$ was obtained by interpolation from values in Table 1.1. By improving factors over which there is some control, n_{sx} can be reduced from 1.6 to 1.0 according to the Pugsley method, thus reducing the required safety factor to 1.6.

Just for illustrative purposes, the safety factor for this situation is prescribed by an industry standard [ANSI 2010] and cannot be lower than 7.6 and may need to be as high as 11.9. The importance of industry standards is discussed in Section 1.4.2, but it is clear that the Pugsley method should be used only with great caution.

Product Liability

When bringing a product to the market, it is probable that safety will be a primary consideration. A design engineer must consider the **hazards**, or injury producers, and the **risk**, or likelihood of obtaining an injury from a hazard, when evaluating the safety of a system. Unfortunately, this is mostly a qualitative evaluation, and combinations of hazard and risk can be judged acceptable or unacceptable.

The ethical responsibilities of engineers to provide safe products are clear, but the legal system also enforces societal expectations through a number of legal theories that apply to designers and manufacturers of products. Some of the more common legal theories are the following:

- **Caveat Emptor**. Translated as "Let the buyer beware," this is a doctrine founded on Roman laws. In the case of a defective product or dangerous design, the purchaser or user of the product has no legal recourse to recover losses. In a modern society, such a philosophy is incompatible with global trade and high-quality products, and is mentioned here only for historical significance.

- **Negligence.** In negligence, a party is liable for damages if they failed to act as a reasonable and prudent party would have done under like or similar circumstances. For negligence theory to apply, the injured party, or *plaintiff*, must demonstrate:

  1. That a standard of care was violated by the accused party, or *defendant*.
  2. That this violation was the *proximate cause* of the accident.
  3. That no contributory negligence of the plaintiff caused the misfortune.

- **Strict liability**. Under the strict liability doctrine, the actions of the plaintiff are not an issue; the emphasis is placed on the machine. To recover damages under the strict liability legal doctrine, the plaintiff must prove that:

  1. The product contained a defect that rendered it unreasonably dangerous. (For example, an inadequately sized or cracked bolt fastening a brake stud to a machine frame.)
  2. The defect existed at the time the machine left the control of the manufacturer. (The manufacturer used the cracked bolt.)
  3. The defect was a proximate cause of the accident. (The bolt broke, the brake stud fell off the machine, the machine's brake didn't stop the machine, resulting in an accident.) Note that the plaintiff does not need to demonstrate that the defect was the proximate cause; the actions of the plaintiff that contribute to his or her own accident are not considered under strict liability.

- **Comparative fault.** Used increasingly in courts throughout the United States, juries are asked to assess the relative contributions that different parties had in relation to an accident. For example, a jury may decide that a plaintiff was 75% responsible for an accident, and reduce the monetary award by that amount.

- **Assumption of risk.** Although rarely recognized, the *assumption of risk* doctrine states that a plaintiff has limited recourse for recovery of loss if they purposefully, knowingly, and intentionally conducted an unsafe act.

One important requirement for engineers is that their products must be reasonably safe for their intended uses as well as their *reasonably foreseeable misuses*. For example, a chair must be made structurally sound and stable enough for people to sit on (this is the intended use). In addition, a chair should be stable enough so that someone can stand on the chair to change a light bulb, for example. It could be argued that chairs are designed to be sat upon, and that standing on a chair is a misuse. This may be true, but represents a reasonably foreseeable misuse of the chair, and must therefore be considered by designers. In the vast majority of states, misuses of a product that are not reasonably foreseeable do not have to be considered by the manufacturers.

The legal doctrines and ethical requirements that designers produce safe products are usually consistent. Sometimes, the legal system does result in requirements that engineers cannot meet. For example, in the famous *Barker vs. Lull* case in New Jersey, the court ruled that product manufacturers have a nondelegable duty to warn of the unknowable.

Liability proofing is the practice of incorporating design features with the intent of limiting product liability exposure without other benefits. This can reduce the safety of machinery. For example, one approach to liability proofing is to place a very large number of warnings onto a machine, with the unfortunate result that all of the warnings are ignored by machine operators. The few hazards that are not obvious and can be effectively warned against are then "lost in the noise" and a compromise of machine safety can occur.

Case Study 1.1: *Mason v. Caterpillar Tractor Co.*

Wilma Mason brought action under negligence theory against Caterpillar Tractor Company and Patton Industries for damages after her husband received fatal injuries while trying to repair a track shoe on a Caterpillar tractor. Mr. Mason was repairing the track shoe with a large sledgehammer, when a small piece of metal from the track shoe shot out, striking him, and causing fatal injuries. The plaintiff alleged that the tractor track was defective because the defendants failed to use reasonable methods of heat treatment, failed to use a sufficient amount of carbon in the steel, and failed to warn the decedent of "impending danger."

The Trial and Appellate courts both granted summary judgements in favor of the defendants. They ruled that the plaintiff failed to show evidence of a product defect that existed when the machine left the control of the manufacturer. Mr. Mason used a large, 20-lb sledgehammer with a full swing, striking a raised portion of the track shoe. There was no evidence that the defendants were even aware that the track shoes were being repaired or reassembled by sledgehammers. It was also noted by the court that the decedent wore safety glasses, indicating his awareness of the risk of injury.

Safety Hierarchy

A design rule that is widely accepted in general is the **safety hierarchy**, which describes the steps that a manufacturer or designer should use when addressing hazards. The safety hierarchy is given in Design Procedure 1.1. Eliminating hazards through design can imply a number of different approaches. For example, a mechanical part that is designed so that its failure is not reasonably foreseeable is one method of eliminating a hazard or risk of injury. However, design of a system that eliminates injury producers or moves them away from people also represents a reasonable approach.

This book emphasizes mechanical analysis and design of parts to reduce or eliminate the likelihood of failure. As such, it should be recognized that this approach is one of the fundamental, necessary skills required by engineers to provide reasonably safe products.

Design Procedure 1.1: The Safety Hierarchy

A designer should attempt the following, in order, in attempting to achieve reasonable levels of safety:

1. Eliminate hazards through design.
2. Reduce the risk or eliminate the hazard through safeguarding technology.
3. Provide warnings.
4. Train and instruct.
5. Provide personal protective equipment.

There is a general understanding that primary steps are more efficient in improving safety than later steps. That is, it is more effective to eliminate hazards through design than to use guards, which are more effective than warnings, etc. Clearly, the importance of effective design cannot be overstated.

Failure Mode and Effects Analysis and Fault Trees

Some common tools available to design engineers are **failure mode and effects analysis** (FMEA) and **fault tree analysis**. FMEA addresses component failure effects on the entire system. It forces the design engineer to exhaustively consider reasonably foreseeable failure modes for every component and its alternatives.

FMEA is flexible, allowing spreadsheets to be tailored for particular applications. For example, an FMEA can also be performed on the steps taken in assembling components to identify critical needs for training and/or warning.

In fault tree analysis, statistical data are incorporated into the failure mode analysis to help identify the most likely (as opposed to possible) failure modes. Often, hard data are not available, and the engineer's judgment qualitatively identifies likely failure modes.

As discussed above, machine designers are legally required to provide reasonably safe products and to consider the product's intended uses as well as foreseeable misuses. FMEA and fault tree analysis help identify unforeseeable misuses as well. For example, an aircraft designer may identify aircraft-meteorite collision as a possible loading of the structure. However, because no aircraft accidents have resulted from meteorite collisions and the probability of such occurrences is extremely low, the design engineer ignores such hypotheses, recognizing they are not reasonably foreseeable.

Load Redistribution, Redundancy, Fail Safe, and the Doctrine of Manifest Danger

One potential benefit of failure mode and effects analysis and fault tree analysis is that they force the design engineer to think of minimizing the effects of individual component failures. A common goal is that the failure of a single component should not result in a catastrophic accident. The design engineer can ensure this by designing the system so that, upon a component failure, loads are redistributed to other components without exceeding their nominal strengths — a philosophy known as **redundancy** in design. For example, a goose or other large bird sucked into an aircraft engine may cause several components to fail and shut down the engine. This type of accident is not unheard of and is certainly reasonably foreseeable. Thus, modern aircraft are designed with sufficient redundancy to allow a plane to fly and land safely with one or more engines shut down.

Many designs incorporate redundancy. Redundant designs can be *active* (where two or more components are in use but only one is needed) or *passive* (where one component is inactive until the first component fails). An example of an active redundant design is the use of two deadbolt locks on a door: both bolts serve to keep the door locked. A passive redundant design example would entail adding a chain lock on a door having a deadbolt lock: if the deadbolt lock fails, the chain will keep the door closed.

An often-used philosophy is to design machinery with **fail-safe** features. For example, a brake system (see Chapter 18) can be designed so that a pneumatic cylinder pushes the brake pads or shoes against a disk or drum, respectively. Alternatively, a spring could maintain pressure against the disk or drum and a pneumatic system could work against the spring to release the brake. If the pressurized air supply were interrupted, such a design would force brake actuation and prevent machinery motion. This alternative design is fail safe as long as the spring is far more reliable than the pneumatic system.

The **doctrine of manifest danger** is a powerful tool used by machinery designers to prevent catastrophic losses. If danger becomes manifest, troubleshooting is straightforward and repairs can be quickly made. Thus, if a system can be designed so that imminent failure is detectable or so that single-component failure is detectable before other elements fail in turn, a safer design results. A classic application of the doctrine of manifest danger is in the design of automotive braking systems, where the brake shoe consists of a friction material held onto a metal backing plate by rivets. By making the rivets long enough, an audible and tactile indication is given to the car driver when the brake system needs service. That is, if the friction material has worn, the rivets will contact the disk or drum, indicating through noise and vibration that maintenance is required, and this occurs long before braking performance is compromised.

Reliability

Safety factors are a way of compensating for variations in loading and material properties. Another approach that can be extremely successful in certain circumstances is the application of **reliability** methods.

As an example, consider the process of characterizing a material's strength through tension tests (see Section 3.4). Manufacturing multiple tension test specimens from the same extruded billet of aluminum would result in little difference in measured strength from one test specimen to another. Thus, aluminum in general (as well as most metals) is a *deterministic* material, and deterministic methods can be used in designing aluminum structures if the load is known. For example, in a few hundred tensile tests, a guaranteed minimum strength can be defined that is below the strength of any test specimen and that would not vary much from one test population to another. This guaranteed minimum strength is then used as *the* strength for design analysis. Such deterministic methods are used in most solid mechanics and mechanics courses. That is, all specimens of a given material have a single strength and the loading is always well defined.

Most ceramics, however, would have a significant range of any given material property, including strength. Thus, ceramics are *probabilistic*, and an attempt to define a minimum strength for a population of ceramic test specimens would be an exercise in futility. There would not necessarily be a guaranteed minimum strength. One can only treat ceramics in terms of a likelihood or probability of strength exceeding a given value. There are many such probabilistic materials in engineering practice.

Some loadings, on the other hand, are well known and never vary much. Examples are the stresses inside intravenous (IV) bags during sterilization, the load supported by counterweight springs, and the load on bearings supporting centrifugal fans. Other loads can vary significantly, such as the force exerted on automotive shock absorbers (depends on the size of the pothole and the speed at impact) or on wooden pins holding a chair together (depends on the weight of the seated person or persons) or the impact force on the head of a golf club.

For situations where a reasonable worst-case scenario cannot be defined, reliability methods are sometimes a reasonable design approach. In reliability design methods, the goal is to achieve a reasonable likelihood of survival under the loading conditions during the intended design life. This approach has its difficulties as well, including the following:

1. To use statistical methods, a reasonable approximation of an infinite test population must be defined. That is, mean values and standard deviations about the mean, and even the nature of the distribution about the mean, must be known. However, they are not usually very well characterized after only a few tests. After all, if only a few tests were needed to quantify a distribution, deterministic methods would be a reasonable, proper, and less mathematically intensive approach. Thus, characterization can be expensive and time consuming, since many experiments are needed.

2. Even if strengths and loadings are known well enough to quantify their statistical distributions, defining a desired reliability is as nebulous a problem as defining a desired safety factor. A reliability of 99% might seem acceptable, unless that were the reliability of an elevator you happened to be occupying. A reliability of 100% is not achievable, or else deterministic methods would be used. A reliability of 99.9999. . .% should be recognized as an extremely expensive affair, and as indicative of overdesign as a safety factor of 2000.

3. The mathematical description of the data has an effect on reliability calculations. A quantity may be best described by a Gaussian or normal distribution, a lognormal distribution, a binary distribution, a Weibull distribution, etc. Often, one cannot know beforehand which distribution is best. Some statisticians recommend using a normal distribution until it is proved ineffective.

The implications are obvious: Reliability design is a complicated matter and even when applied does not necessarily result in the desired reliability if calculated from insufficient or improperly reduced data.

This textbook will emphasize deterministic methods for the most part. The exceptions are the treatments of rolling-element bearings and gears and reliability in fatigue design. For more information on reliability design, refer to the excellent text by Lewis [1995] among others.

1.4.2 Government Codes and Industry Standards

In many cases, engineers must rely on government codes and industry-promulgated standards for design criteria. Some of the most common sources for industry standards are:

1. ANSI, the American National Standards Institute
2. ASME, the American Society of Mechanical Engineers
3. ASTM, the American Society for Testing and Materials
4. AGMA, the American Gear Manufacturers Association
5. AISI, the American Iron and Steel Institute
6. AISC, the American Institute of Steel Construction
7. ISO, the International Standards Organization
8. NFPA, the National Fire Protection Association
9. UL, Underwriters Laboratories

Government codes are published annually in the Code of Federal Regulations (CFR) and periodically in the Federal Register (FR) at the national level. States and local municipalities have codes as well, although most relate to building standards and fire prevention.

Code compliance is important for many reasons, some of which have already been stated. However, one essential goal of industry standards is conformability. For example, bolt geometries are defined in ANSI standards so that bolts have fixed thread dimensions and bolt diameters. Therefore, bolts can be mass produced; resulting in inexpensive, high-quality threaded fasteners. Also, maintenance is simplified in that standard bolts can be purchased anywhere, making replacement parts readily available.

1.4.3 Manufacturing

Design and manufacturing are difficult to consider apart from one another. The tenet of "form follows function" suggests that shapes are derived only from applied loads in the design environment. However, this is not always the case, and the shapes of products are often natural progressions from arbitrary beginnings.

Design for manufacturability (DFM) is a well-established and important tool for design engineers. Manufacturability plays a huge role in the success of commercial products. After all, a brilliant concept that cannot be manufactured cannot be a successful design (per the definition in Section 1.1). Also, because most manufacturing costs are determined by decisions made early in the design process, market success depends on early consideration of a complete product lifecycle, including manufacturing. Individual components should be designed to be easily fabricated, assembled, and constructed (*design for assembly*, DFA). Although manufacturing and assembly are outside the scope of this text, Fig. 1.2 shows their effect on design.

Engineers must wear many hats. Some predominant concerns of a design engineer have been discussed, but many more exist, including:

1. *Environmental or sustainable design:* This issue addresses whether products can be produced that are less harmful to the environment. Biodegradable or easily recycled materials may need to be selected to satisfy this concern.

2. *Economics:* Deciding whether a product will be profitable is of utmost concern.

3. *Legal considerations* - Violating patents and placing unreasonably dangerous products into the marketplace are not only ethically wrong but have legal ramifications as well.

4. *Marketing:* The features of a product that attract consumers and the product's presentation to the marketplace play a significant role in a product's success.

5. *Serviceability:* If a part breaks, can repairs be done in the field, or must customers send the product back to the manufacturer at excessive expense? Unless such concerns are incorporated into design, long-term customer loyalty is compromised.

6. *Quality:* Approaches such as *total quality engineering* and *Taguchi methods* have been successfully applied to make certain that no defects are shipped.

These are merely a few of the concerns faced by design engineers.

The design process may appear so elaborate and involved that no one can master it. In actuality, one important skill makes the design process flow smoothly: effective communication. Communication between diverse disciplines involved in product design ensures that all voices are heard and all design constraints are satisfied early, before significant costs are incurred. Effective communication skills, written and oral, are the most important trait of a good engineer. Although this text emphasizes the more analytical and technical sides of design, it is important to remember that design is not merely an analytical effort but one of human interaction.

1.4.4 Life Cycle Engineering

Life cycle engineering (LCE) involves consecutive and interlinked stages of a product or a service, from the very beginning to its disposal or recycling; it includes the following:

1. Extraction of natural resources

2. Processing of raw materials

3. Manufacturing of products

4. Transportation and distribution of the product to the customer

5. Use, maintenance, and reuse of the product

6. Recovery, recycling, reuse, or disposal of its components

All of these factors are applicable to any type of product. Each product can have its own metallic and nonmetallic materials, processed into individual components and assembled; thus, each product has its own life cycle. Moreover, (a) some products are intentionally made to be disposable, particularly those made of paper, cardboard, inexpensive plastic, and glass, but nonetheless are all recyclable, and (b) numerous other products are completely reusable.

A major aim of LCE is to consider reusing and recycling the components of a product, beginning with the earliest stage of product design. This is also called **green design** or **green engineering**. These considerations also include environmental factors, optimization, and numerous technical factors regarding each component of a product.

As is now universally acknowledged, the natural resources on Earth are limited, thus clearly necessitating the need and urgency to conserve materials and energy. The concept of **sustainable design** emphasizes the need for conserving resources, particularly through proper maintenance and reuse. While profitability is important to an organization, sustainable design is meant to meet purposes such as (a) increase the life cycle of products, (b) eliminate harm to the environment and the ecosystem, and (c) ensure our collective well-being, especially for the benefit of future generations.

Case Study 1.2: Sustainable Manufacturing in the Production of Nike Athletic Shoes

Among numerous examples from industry, the production of Nike shoes clearly has indicated the benefits of sustainable manufacturing. The athletic shoes are assembled using adhesives. Up to around 1990, the adhesives used contained petroleum-based solvents, which pose health hazards

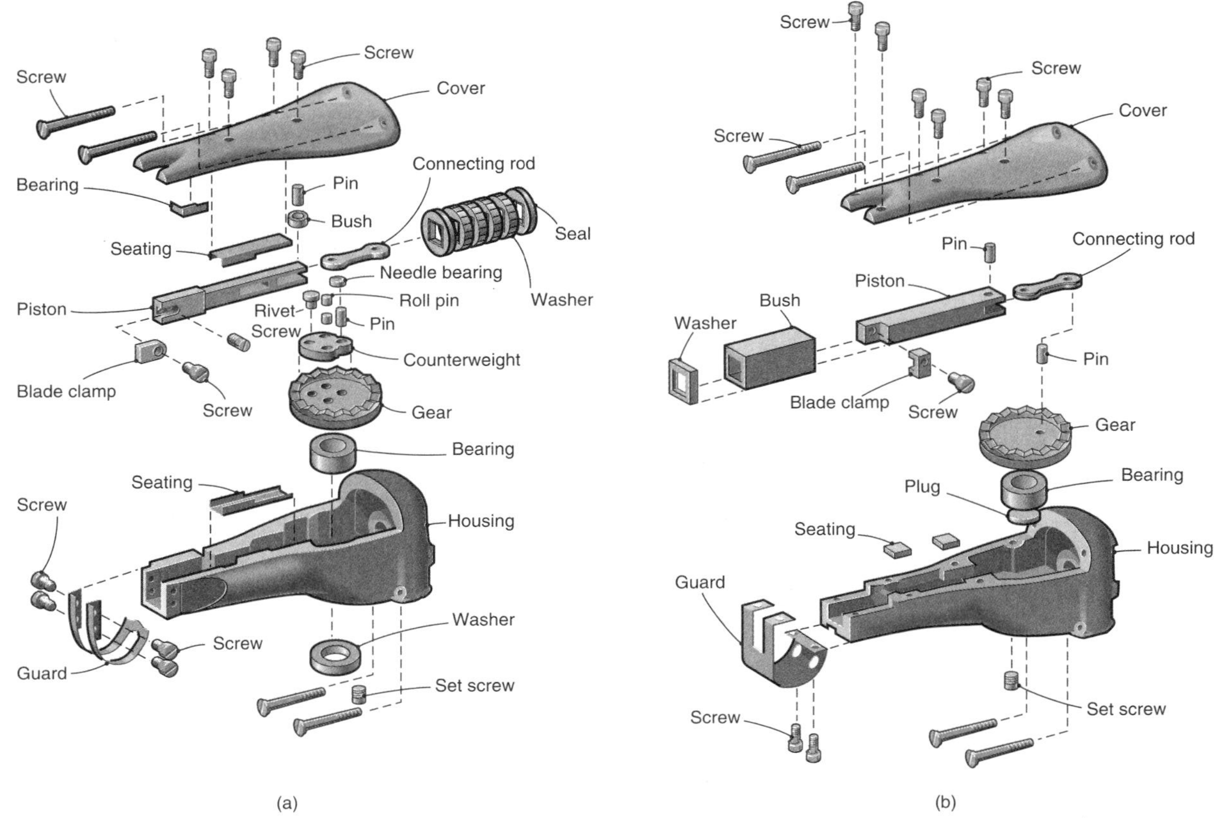

Figure 1.2: Effect of manufacturing and assembly considerations on the design of a reciprocating power saw. (a) Original design, with 41 parts and 6.37-min assembly time; (b) modified design, with 29 parts and 2.58-min assembly time. *Source:* Adapted from Boothroyd [1992].

and contribute to petrochemical smog. The company cooperated with chemical suppliers to successfully develop a water-based adhesive technology, now used in the majority of shoe-assembly operations. As a result, solvent use in all manufacturing processes in the subcontracted facilities in Asia has greatly been reduced.

Regarding another component of the shoe, the rubber outsoles are made by a process that results in significant amounts of extra rubber around the periphery of the sole (called *flashing*). With about 40 factories using thousands of molds and producing over a million outsoles a day, the flashing constitutes the largest chunk of waste in manufacturing the shoes. In order to reduce this waste, the company developed a technology that grinds the flashing into 500-μm rubber powder, which is then added back into the rubber mixture needed to make the outsole. With this approach, waste was reduced by 40%. Moreover, it was found that the mixed rubber had better abrasion resistance and durability, and its overall performance was higher than the best premium rubber.

Source: Adapted from Kalpakjian and Schmid [2010].

1.5 Computers in Design

Computer-aided design (CAD) also means different things to different people, but in this text it is the application of computer technology to planning, performing, and implementing the design process. Computers allow virtual integration of all phases of the design process, whether technical or managerial activities. With sophisticated hardware and software, manufacturers can now minimize design costs, maximize efficiency, improve quality, reduce development time, and maintain an edge in domestic and international markets.

CAD allows the design engineer to visualize geometries without making costly models, iterations, or prototypes. These systems can now analyze designs from simple brackets to complex structures quickly and easily. Designs can be optimized and modified directly and easily at any time. Information stored by computer can be accessed and retrieved from anywhere within the organization.

Whereas some restrict the term "CAD" to drafting activities, others will generically group all computer-assisted functions arbitrarily as CAD. **Artificial intelligence** (AI) attempts to duplicate how the human mind works and apply it to processes on the computer. Sometimes, AI is used to describe the cases where computers are used as more than mere drafting tools and actually help in the intellectual design tasks. **Expert systems** are rule-based computer programs that solve specialized problems and provide problem-solving skills to the design engineer. For example, an expert system could analyze a part drafted on a computer system for ease in manufacturing. If an excessively small tolerance is found, the expert system warns the engineer that manufacturing difficulties will ensue and suggests easing the tolerance. Similarly, an expert system could analyze a design to standardize parts (e.g., to make sure that an assembly uses only one bolt size instead of the optimum size for each location, thereby easing inventory and maintenance difficulties). Artificial intelligence is a more elaborate form of an expert system; AI is sometimes restricted to systems that can learn new information.

Rapid prototyping, also called *3D printing* or *additive manufacturing*, is another computer-driven technology that produces parts from geometry data files in hours or even minutes. Rapid prototyping has been especially helpful in design visualization and rapid detection of design errors. For example, a casting with an excessively thin wall is easily detected when a solid model is held in the hand, a subtlety that is difficult to discern when viewing a part drawing on a computer screen. Significant developments have occurred in rapid prototyping in recent years. Currently, a wide variety of polymers can be used, as well as metals and ceramics.

Finite element analysis (FEA) is the most prevalent computational method for solid and fluid mechanics analysis, as well as heat transfer. The finite element computational method solves complex shapes, such as those found in machinery, and replaces the complex shape with a set of simple elements interconnected at a finite set of node points. In FEA, a part geometry is sectioned into many subsections or *elements*. The stiffness of each element is known and is expressed in terms of a stiffness matrix for that element. By combining all the stiffness matrices, applying kinematic and stress boundary conditions, and solving for unknown stresses or displacements, complicated geometries and loading conditions can be easily analyzed.

1.6 Catalogs and Vendors

Manufacturing concerns are inseparable from design. Clearly, many machine elements are mass produced because there is an economic justification for large production runs using hard automation. Hard automation generally results in higher quality, tighter tolerance parts than soft automation or hand manufacture, and usually results in less expensive parts as well. In fact, many industry standards mentioned in Section 1.4.2 exist to prescribe geometries that can be mass produced in order to achieve quality and cost benefits. For example, a centerless grinder can produce many high-quality 15-mm-diameter bushings, whereas a single 15-mm bushing is difficult to manufacture and would be very expensive by comparison. Therefore, the practice of machine design often involves selecting mass-produced components from suppliers, often as summarized in catalogs or web sites.

Mechanical designers know the importance of good vendor identification and readily available and up-to-date catalog information. The Internet has brought a huge variety of machine element catalogs to every designer's desktop, and it will be assumed that students are well-aware of Internet search tools and can quickly retrieve product catalogs if desired. Often in this textbook, portions of a manufacturer's catalog will be provided so that data are convenient for problem solving, but it should be recognized that the complete product portfolios are usually much larger than the abstracted data presented.

1.7 Units

The solutions to engineering problems must be given in specific and consistent units that correspond to the specific parameter being evaluated. Two systems of units are mainly used in this text:

1. Systéme International d'Unites (SI units): Force is measured in newtons, length in meters (sometimes millimeters are more convenient for certain applications), time in seconds, mass in kilograms, and temperature in degrees Celsius. In addition, absolute temperature is measured in degrees Kelvin, where the temperature in Kelvin is the temperature in Celsius plus 273.15°.

2. English units: Force is measured in pounds force, length in inches, time in seconds, mass in pounds mass, and temperature in degrees Fahrenheit.

In Chapter 8 an additional measure, viscosity, is given in the centimeter-gram-seconds (cgs) system.

The data needed to solve problems are not always in the same system of units. It is therefore necessary to convert from one system to another. The SI units, prefixes, and symbols used throughout the text are shown in Table 1.3 as well as inside the front cover. The primary units of the text are SI, but problems are given in English units as well as in SI units to enable the student to handle either system.

Basic SI units, some definitions, and fundamental and other useful conversion factors are given in Table 1.4, which is also inside the front cover. Note that many units can be quite confusing. For example, a ton in the United States and Canada refers to a weight of 2000 lb, while in the United Kingdom it is a term for weight or mass equivalent to 2240 lb. Sometimes, a weight is reported in short tons (2000 lb) or long tons (2240 lb). The metric equivalent, called a metric ton or a tonne, is 1000 kg. As another example, a horsepower in English units is 550 ft-lb/s or 746 W. However, in metric countries, a horsepower is defined as 736 W. Keeping track of units is a necessary task for design analysis, as illustrated in Case Study 1.3.

Example 1.2: Length of Electrical Connections in a Supercomputer

Given: A supercomputer has a calculation speed of 1 gigaflop $= 10^9$ floating-point operations per second. Performance can be limited if the electrical connections within the supercomputer are so long that electron travel times are greater than the calculation's speed.

Find: Determine the critical length of electrical wire for such connections if the electron speed for coaxial cables is 0.9 times the speed of light (3×10^8 m/s).

Solution: If the speed is determined only by the cable length,

$$l = \frac{(0.9)\left(3 \times 10^8\right)}{10^9} = 0.27 \text{ m} = 27 \text{ cm}$$

The mean cable length must be less than 27 cm.

Example 1.3: Astronomical Distances

Given: The distance from Earth to α-Centauri is 4 light-years.

Find: How many terameters away is α-Centauri? Note that the speed of light is 3×10^8 m/s.

Solution: Note that $1 \text{ year} = (365)(24)(3600) \text{ s} = \left(3.1536 \times 10^7\right) \text{ s}$. The distance from Earth to α-Centauri is

$$(4)(3.1536 \times 10^7 \text{ s})(3 \times 10^8) \text{ m/s} = 3.784 \times 10^{16} \text{ m}$$

From Table 1.3b, 1T = 10^{12}. Therefore, the distance is 37,840 Tm.

Table 1.3: SI units and prefixes.

(a) SI units

Quantity	Unit	SI symbol	Formula
SI base units			
Length	meter	m	-
Mass	kilogram	kg	-
Time	second	s	-
Temperature	kelvin	K	-
SI supplementary unit			
Plane angle	radian	rad	-
SI derived units			
Energy	joule	J	N-m
Force	newton	N	kg-m/s^2
Power	watt	W	J/s
Pressure	pascal	Pa	N/m^2
Work	joule	J	N-m

(b) SI prefixes

Multiplication factor	Prefix	SI symbol for prefix
1,000,000,000,000 = 10^{12}	tera	T
1,000,000,000 = 10^9	giga	G
1,000,000 = 10^6	mega	M
1000 = 10^3	kilo	k
100 = 10^2	hecto	h
10 = 10^1	deka	da
0.1 = 10^{-1}	deci	d
0.01 = 10^{-2}	centi	c
0.001 = 10^{-3}	milli	m
0.000 001 = 10^{-6}	micro	μ
0.000 000 001 = 10^{-9}	nano	n
0.000 000 000 001 = 10^{-12}	pico	p

Table 1.4: Conversion factors and definitions.

Definitions	
Acceleration of gravity	1 g = 9.8066 m/s^2 (32.174 ft/s^2)
Energy	Btu (British thermal unit) = amount of energy required to raise 1 lbm of water 1°F (1 Btu = 778.2 ft-lb)
	kilocalorie = amount of energy required to raise 1 kg of water 1K (1 kcal = 4187 J)
Length	1 mile = 5280 ft
	1 nautical mile = 6076.1 ft
Power	1 horsepower = 550 ft-lb/s
Pressure	1 bar = 10^5 Pa
Temperature	Fahrenheit: $t_F = \frac{9}{5}t_C + 32$
	Rankine: $t_R = t_F + 459.67$
	Kelvin: $t_K = t_C + 273.15$ (exact)
Kinematic viscosity	1 poise = 0.1 kg/m-s
	1 stoke = 0.0001 m^2/s
Volume	1 cubic foot = 7.48 gal
Useful conversion factors	
	1 in. = 0.254 m = 25.4 mm
	1 lbm = 0.4536 kg
	1° R = $\frac{5}{9}$K
	1 ft = 0.3048 m
	1 lb = 4.448 N
	1 lb = 386.1 lbm-in./s^2
	1 ton = 2000 lb (shortton) or 2240 lb (long ton)
	1 tonne = 1000 kg (metric ton)
	1 kgf = 9.807 N
	1 lb/in.^{2} = 6895 Pa
	1 ksi = 6.895 MPa
	1 Btu = 1055 J
	1 ft-lb = 1.356 J
	1 hp = 746 W = 2545 Btu/hr[a]
	1 kW = 3413 Btu/hr
	1 quart = 0.000946 m^3 = 0.946 liter
	1 kcal = 3.968 Btu

[a] Note that in countries using the metric system, a horsepower is defined as 75 kpm/s or 736 W.

Case Study 1.3: Loss of the Mars Climate Orbiter

On December 11, 1998, the Mars Climate Orbiter was launched to start its nearly 10-month journey to Mars. The Mars Climate Orbiter was a $125 million satellite intended to orbit Mars and measure the atmospheric conditions on that planet over a planetary year. It was also intended to serve as a communications relay for the Mars Climate Lander, which was due to reach Mars in December 1999. The Mars Climate Orbiter was destroyed on September 23, 1999 as it was maneuvering into orbit.

The cause for the failure was quickly determined: the manufacturer, Lockheed Martin, programmed the entry software in English measurements. However, the navigation team at NASA's Jet Propulsion Laboratory in Pasadena, California assumed the readings were in metric units. As a result, trajectory errors were magnified instead of corrected by mid-course thruster firings. This painful lesson demonstrated the importance of maintaining and reporting units with all calculations.

1.8 Unit Checks

Unit checks should always be performed during engineering calculations to make sure that each term of an equation is in the same system of units. The importance of knowing the units of the various parameters used in an equation cannot be overemphasized. In this text, a symbol list giving the units of each parameter is provided at the beginning of each chapter. If no units are given for a particular phenomenon, it is dimensionless. This symbol list can be used as a partial check during algebraic manipulations of an equation.

Design Procedure 1.2: Procedure for Unit Checks

It is generally advisable to carry units throughout calculations. However, an expression can generally be evaluated by:

1. Establish units of specific terms of an equation while making use of Table 1.3a.

2. Place units of terms into both sides of an equation and reduce.

3. The unit check is complete if both sides of an equation have the same units.

Example 1.4: Unit Checks

Given: The centrifugal force, P, acting on a car going through a curve with a radius, r, at a velocity, v, is $m_a v^2/r$, where m_a is the mass of the car. Assume a 1.3-tonne car drives at 100 km/hr through a 100-m-radius bend.

Find: Calculate the centrifugal force.

Solution: Rewriting using metric units gives

$$m_a = 1.3 \text{ tonne} = 1300 \text{ kg}$$

$$v = 100 \text{ km/hr} = \frac{(100 \text{ m})(1000)}{3600 \text{ s}} = 27.78 \text{ m/s}$$

The centrifugal force is

$$P = \frac{m_a v^2}{r} = \frac{(1300 \text{ kg})(27.78 \text{ m/s})^2}{100 \text{ m}}$$

This results in $P = 10,030$ kg-m/s^2 or 10,030 N.

1.9 Significant Figures

The accuracy of a number is specified by how many significant figures it contains. Throughout this text, four significant figures will be used unless otherwise limited. For example, 8201 and 30.51 each have four significant figures. When numbers begin or end with a zero, however, it is difficult to tell how many significant figures there are. To clarify this situation, the number should be reported by using *scientific notation* involving powers of 10. Thus, the number 8200 can be expressed as 8.200×10^3 to represent four significant figures. Also, 0.005012 can be expressed as 5.012×10^{-3} to represent four significant figures.

Example 1.5: Significant Figures

Given: A car with a mass of 1502 kg is accelerated by a force of 14.0 N.

Find: Calculate the acceleration with the proper number of significant figures.

Solution: Newton's equation gives that acceleration equals the force divided by the mass

$$a = \frac{P}{m_a} = \frac{14.0}{1502} = 0.00932091 \text{ m/s}^2$$

Since force is accurate to three figures, the acceleration can only be calculated with the accuracy of

$$\pm\frac{0.5}{140} = \pm 0.004 = \pm 0.4\%$$

Therefore, the acceleration is 0.00932 m/s^2.

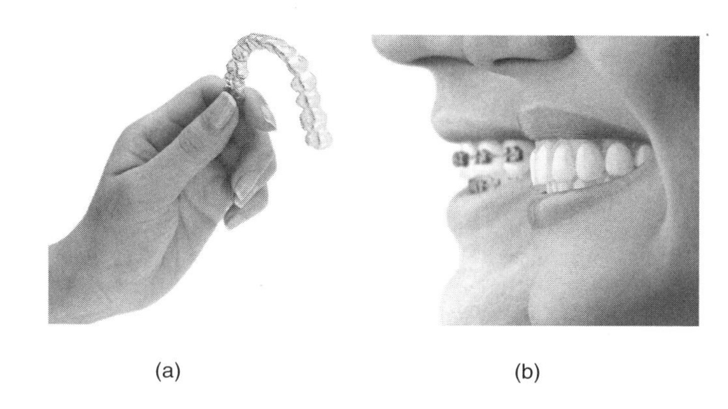

(a) (b)

Figure 1.3: The Invisalign® product. (a) An example of an Aligner; (b) a comparison of conventional orthodontic braces and a transparent Aligner. *Source:* Courtesy of Align Technology, Inc.

Case Study 1.4: Design and Manufacture of the Invisalign Orthodontic Product

Widespread healthcare and improved diet and living habits have greatly extended the expected lifetime of people within the last century. Modern expectations are not only that life will be extended, but also that the *quality* of life will be maintained late in life. One important area where this concern manifests itself is with teeth; straight teeth lead to a healthy bite with low tooth stresses, and they also lend themselves to easier cleaning and therefore are more resistant to decay. Thus, straight teeth, in general, last longer with less pain. Of course, there are aesthetic reasons that people wish to have straight teeth as well.

Orthodontic braces have been available to straighten teeth for over 50 years. These involve metal, ceramic, or plastic brackets that are adhesively bonded to teeth, with fixtures for attachment to a wire that then forces compliance on the teeth and straightens them to the desired shape within a few years. Conventional orthodontic braces are a well-known and wholly successful approach to long-term dental health. However, there are many drawbacks to conventional braces, including:

- They are aesthetically unappealing.
- The sharp wires and brackets can cause painful oral irritation to the teeth and gums.
- They trap food, leading to premature tooth decay.
- Brushing and flossing of teeth are far more difficult with braces in place, and therefore they are less effective for most individuals.
- Certain foods must be avoided because they will damage the braces.

One innovative solution is the Invisalign product produced by Align Technology. Invisalign consists of a series of Aligners, each of which the patient wears for approximately two weeks. Each Aligner (see Fig. 1.3) consists of a precise geometry which incrementally moves teeth to their desired positions. Because they are inserts that can be removed for eating, brushing, and flossing, most of the drawbacks of conventional braces are eliminated. Further, since they are produced from transparent plastic, they do not seriously affect the patient's appearance.

The Invisalign product uses an impressive combination of advanced technologies, and the production process is shown in Fig. 1.4. The treatment begins with an orthodontist creating a polymer impression of the patient's teeth or a direct digital image of the teeth using a 3D intra oral scanner (Fig. 1.4a). In case of physical impressions, the impressions are then used to create a three-dimensional CAD representation of the patient's teeth, as shown in Fig. 1.4b. Proprietary computer-aided design software then assists in the development of a treatment strategy for moving the teeth in optimal fashion.

Specially produced software called ClinCheck then produces a digital video of the incremental movements which can be reviewed by the treating orthodontist and modified if necessary.

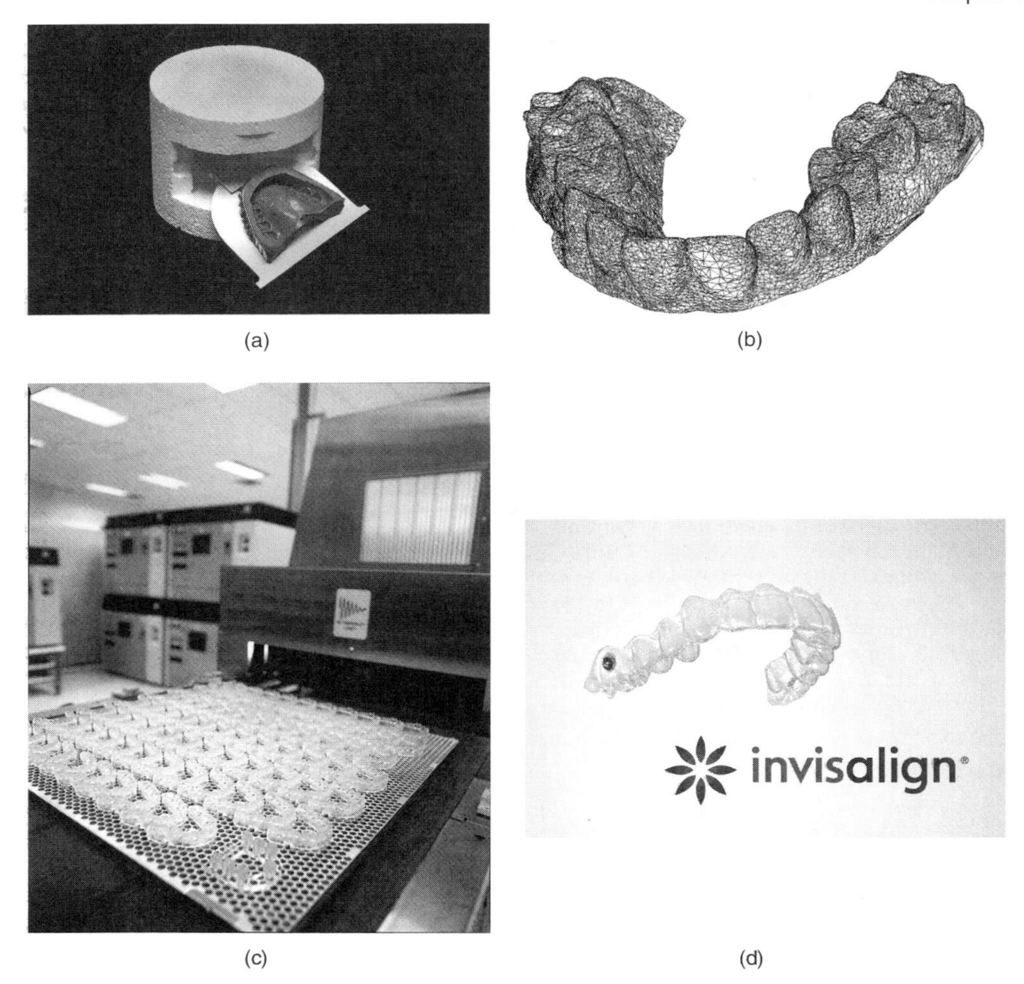

Figure 1.4: The process used in application of Invisalign orthodontic treatment. (a) Impressions are made of the patient's teeth by the orthodontist and shipped to Align Technology, Inc. These are used to make plaster models of the patient's teeth. (b) High-resolution, three-dimensional representations of the teeth are produced from the plaster models. The correction plan is then developed using computer tools. (c) Rapid-prototyped molds of the teeth at incremental positions are produced through stereolithography. (d) An Aligner is produced by molding a transparent plastic over the stereolithography part. Each Aligner is used for approximately two weeks. The patient is left with a beautiful smile. *Source:* Courtesy of Align Technology, Inc.

Once a treatment plan has been designed, the computer-based information needs to be used to produce the Aligners. This is done through a novel application of rapid prototyping technology. Stereolithography is a process that uses a focused laser to cure a liquid photopolymer. The laser only cures a small depth of the polymer, so a part can be built on a tray that is progressively lowered into a vat of photopolymer as layers or slices of the desired geometry are traced and rastered by the laser.

A number of materials are available for stereolithography, but these have a characteristic yellow-brown shade to them and are therefore unsuitable for direct application as an orthodontic product. Instead, the stereolithography machine produces patterns of the desired incremental positions of the teeth (Fig. 1.4c). A sheet of clear polymer is then molded over these patterns to produce the Aligners. These are sent to the treating orthodontist and new Aligners are given to the patient as needed, usually every two weeks or so.

The Invisalign product has proven to be very popular for patients who wish to have straight teeth without almost anyone knowing they are in treatment. It depends on advanced engineering technologies, precise force delivery through custom engineered shapes, CAD and manufacturing, and rapid prototyping and advanced polymer manufacturing processes.

Summary

This chapter introduced the concept of design as it applies to machines and machine elements. The most important goal of the design process is to ensure that the design does not fail. To avoid failure, the design engineer must predict the circumstances under which failure is first likely to occur. These circumstances or criteria can involve the material properties and applied loads, as well as surface phenomena, including friction, wear, lubrication, and environmental deterioration.

The concept of failure was quantified by using a safety factor, which is the ratio of the allowable stress established for the material to the maximum design stress that will occur. Besides the simple safety factor, other failure models, such as failure mode and effects analysis and fault tree analysis, among many others, were presented. The ethical requirements in producing safe designs were stated, along with strategies for achieving this constraint, including the Safety Hierarchy and Doctrine of Manifest Danger. Design was found to be a cooperative endeavor where multidisciplinary approaches are invaluable.

Key Words

artificial intelligence (AI) attempts to duplicate how the human mind works in computer processes

computer-aided design (CAD) application of computer technology to planning, performing, and implementing the design process

concurrent engineering design approach wherein all disciplines involved with a product are in the development process from beginning to end

design transformation of concepts and ideas into useful machinery

English units system of units where:

force is measured in pounds force (lbf)

length in inches (in.)

time in seconds (s)

mass in pounds mass (lbm)

temperature in degrees Fahrenheit (°F)

expert systems computer programs that solve specialized problems on an expert level

fail-safe design approach where no catastrophic loss can occur as a result of a component failure

failure the condition of a machine element when it is completely inoperable, cannot perform its intended function adequately, or is unreliable for continued safe use

failure mode and effects analysis (FMEA) systematic consideration of component failure effects on the entire system

fault tree analysis statistical data used to identify the most likely failure modes

finite element analysis (FEA) computational method used for solving for stress, strain, temperature, etc. in complex shapes, such as those found in machinery; replaces the complex shape with a set of simple elements interconnected at a finite set of node points

machine combination of mechanisms and other components that transform, transmit, or use energy, load, or motion for a specific purpose

machine element function normal load transmitter, torque transmitter, energy absorber, or seal

manifest danger design approach where needed service is made apparent before catastrophic failure

mechanical system synergistic collection of machine elements

rapid prototyping parts produced quickly from computer geometry description files

redundancy additional capacity or incorporation of backup systems so that a component failure does not lead to catastrophic loss

safety factor ratio of allowable stress to design stress

SI units system of units where:

force is measured in newtons (N)

length in meters (m)

time in seconds (s)

mass in kilograms (kg)

temperature in degrees Kelvin (K)

Recommended Readings

General Engineering

Florman, S.C. (1976) *The Existential Pleasures of Engineering*, St. Martin's Press.
Petroski, H. (1992) *To Engineer Is Human*, Vintage Books.

General Design

Haik, Y., (2003) *Engineering Design Process*, Thomson.
Hyman, B. (2003) *Fundamentals of Engineering Design*, 2nd ed., Prentice-Hall.
Lindbeck, J.R., (1995) *Product Design and Manufacture*, Prentice-Hall.
Otto, K., and Wood, K., (2000) *Product Design: Techniques in Reverse Engineering and New Product Development*, Prentice-Hall.
Ullman, D.G., (2009) *The Mechanical Design Process*, McGraw-Hill.
Vogel, C.M., and Cagan, J. (2012) *Creating Breakthrough Products*, 2nd ed., Prentice-Hall.

Manufacturing/Design for Manufacture

Boothroyd, G., Dewhurst, P., and Knight, W. (2010) *Product Design for Manufacture and Assembly*, 3rd ed., Taylor & Francis.
Boothroyd, G. (2005) *Assembly Automation and Product Design*, 2nd ed., Taylor & Francis.
DeGarmo, E.P., Black, J.T., and Kohser, R.A. (2011) *DeGarmo's Materials and Processes in Manufacturing*, 9th ed., Prentice-Hall.
Dieter, G.E. and Schmidt, L. (2008) *Engineering Design*, 4th ed., McGraw-Hill.
Kalpakjian, S., and Schmid, S.R. (2008) *Manufacturing Processes for Engineering Materials*, 5th ed., Prentice-Hall.
Kalpakjian, S., and Schmid, S.R. (2010) *Manufacturing Engineering and Technology*, 6th ed., Pearson.
Wright, P.K. (2001) *21st Century Manufacturing*, Prentice-Hall.

Concurrent Engineering

Anderson, D.M. (2010) *Design for Manufacturability & Concurrent Engineering*, CIM Press.
Nevins, J.L., and Whitney, D.E. (Eds.) (1989) *Concurrent Design of Products and Processes*, McGraw-Hill.
Prasad, B. (1996) *Concurrent Engineering Fundamentals*, Prentice-Hall.
Pugh, S. (1996) *Creating Innovative Products Using Total Design*, Addison-Wesley.
Pugh, S. (1991) *Total Design*, Addison-Wesley.

References

ANSI (2010) A17.1 "Minimum Safety Requirements for Passenger Elevators," American National Standards Institute.

ASME (2012) *Code of Ethics for Engineers*, Board on Professional Practice and Ethics, American Society of Mechanical Engineers.

Boothroyd, G. (1992) *Assembly Automation and Product Design*, Marcel Dekker.

Florman, S.C. (1987) *The Civilized Engineer*, St. Martin's Press.

Kalpakjian, S. and Schmid, S.R. (2003) *Manufacturing Processes for Engineering Materials*, 4th ed., Prentice-Hall.

Kalpakjian, S., and Schmid, S.R. (2010) *Manufacturing Engineering and Technology*, 6th ed., Pearson.

Lewis, E.E. (1995) *Introduction to Reliability Engineering*, 2nd ed., Wiley.

Petroski, H. (1992) *To Engineer Is Human*, Vintage Books.

Pugh, S. (1996) *Creating Innovative Products Using Total Design*, Addison-Wesley.

Pugsley, A.G. (1966) *The Safety of Structures*, Edward Arnold.

Questions

1.1 What is design?

1.2 What is *over-the-wall engineering*?

1.3 What is failure?

1.4 Define *safety factor*.

1.5 Explain the terms "product liability," "negligence," and "strict liability."

1.6 What is the Safety Hierarchy?

1.7 Give two examples of standards promulgating bodies.

1.8 What is a life cycle?

1.9 How do you define a product's life cycle?

1.10 Name two unit systems.

Qualitative Problems

1.11 Describe the differences between a safety factor and reliability.

1.12 Explain why it is said that design casts the largest shadow.

1.13 List factors that you feel should be considered when selecting a safety factor.

1.14 List some of the concerns that must be considered by a product designer.

1.15 What are the advantages and disadvantages of the Pugsley method for estimating safety factor?

1.16 Journal bearings on train boxcars in the early 19th century used a "stink additive" in their lubricant. If the bearing got too hot, it would attain a noticeable odor, and an oiler would give the bearing a squirt of lubricant at the next train stop. What design philosophy does this illustrate? Explain.

1.17 Explain why engineers must work with other disciplines, using specific product examples.

1.18 A car is being driven at 150 km/hr on a mountain road where the posted speed limit is 100 km/hr. At a tight turn, one of the tires fails (a blowout) causing the driver to lose control and results in an accident involving property losses and injuries but no loss of life. Afterward, the driver decides to file a lawsuit against the tire manufacturer. Explain which legal theories give him a viable argument to make a claim.

1.19 Give three examples of fail-safe and three examples of fail-unsafe products.

1.20 List three measures that are known within (*a*) one; (*b*) two; (*c*) three; and (*d*) more than three significant figures.

Quantitative Problems

1.21 A hand-held drilling machine has a bearing to take up radial and thrust load from the drill. Depending on the number of hours the drill is expected to be used before it is scrapped, different bearing arrangements will be chosen. A rubbing bushing has a 50-hr life. A small ball bearing has a 300-hr life. A two-bearing combination of a ball bearing and a cylindrical roller bearing has a 10,000-hr life. The cost ratios for the bearing arrangements are 1:5:20. What is the optimum bearing type for a simple drill, a semiprofessional drill, and a professional drill?

1.22 Using the hand-held drill described in Problem 1.21, if the solution with the small ball bearing was chosen for a semiprofessional drill, the bearing life could be estimated to be 300 hr until the first spall forms in the race. The time from first spall to when the whole rolling-contact surface is covered with spalls is 200 hr, and the time from then until a ball cracks is 100 hr. What is the bearing life

(a) If high precision is required?

(b) If vibrations are irrelevant?

(c) If an accident can happen when a ball breaks?

1.23 The dimensions of skis used for downhill competition need to be determined. The maximum force transmitted from one foot to the ski is 2500 N, but the snow conditions are not known in advance, so the bending moment acting on the skis is not known. Estimate the safety factor needed.

1.24 A crane has a loading hook that is hanging in a steel wire. The allowable normal tensile stress in the wire gives an allowable force of 100,000 N. Estimate the safety factor that should be used.

(a) If the wire material is not controlled, the load can cause impact, and fastening the hook in the wire causes stress concentrations. (If the wire breaks, people can be seriously hurt and expensive equipment can be destroyed.)

(b) If the wire material is extremely well controlled, no impact loads are applied and the hook is fastened in the wire without stress concentrations. (If the wire breaks, no people or expensive equipment can be damaged.)

1.25 Calculate the following:

(a) The velocity of hair growth in meters per second, assuming hair grows 0.75 in. in one month.

(b) The weight of a 1-in. diameter steel ball bearing in meganewtons.

(c) The mass of a 1-kg object on the surface of the moon.

(d) The equivalent rate of work in watts of 4 horsepower.

1.26 The unit for dynamic viscosity in the SI system is newton-seconds per square meter, or pascal-seconds (N-s/m^2 = Pa-s). How can that unit be rewritten using the basic relationships described by Newton's law for force and acceleration?

1.27 The unit for dynamic viscosity in Problem 1.26 is newton-seconds per square meter (N-s/m^2) and the kinematic viscosity is defined as the dynamic viscosity divided by the fluid density. Find at least one unit for kinematic viscosity.

1.28 A square surface has sides 1 m long. The sides can be split into decimeters, centimeters, or millimeters, where 1 m = 10 dm, 1 dm = 10 cm, and 1 cm = 10 mm. How many millimeters, centimeters, and decimeters equal 1 m? Also, how many square millimeters, square centimeters , and square decimeters equal a square meter?

1.29 A volume is 1 tera (mm^3) large. Calculate how long the sides of a cube must be to contain that volume.

1.30 A ray of light travels at a speed of 300,000 km/s = 3×10^8 m/s. How far will it travel in 1 ps, 1 ns, and 1 μs?

1.31 Two smooth flat surfaces are separated by a 10-μm-thick lubricant film. The viscosity of the lubricant is 0.100 Pa-s. One surface has an area of 1 dm^2 and slides over the plane surface with a velocity of 1 km/hr. Determine the friction force due to shearing of the lubricant film. Assume the friction force is the viscosity times the surface area times the velocity of the moving surface and divided by the lubricant film thickness.

1.32 A firefighter sprays water on a house. The nozzle diameter is small relative to the hose diameter, so the force on the nozzle from the water is

$$F = v\frac{dm_a}{dt}$$

where v is the water velocity and dm_a/dt is the water mass flow per unit time. Calculate the force the firefighter needs to hold the nozzle if the water mass flow is 3 tons/hr and the water velocity is 100 km/hr.

1.33 The mass of a car is 1346 kg. The four passengers in the car weigh 643 N, 738 N, 870 N, and 896 N. It is raining and the additional mass due to the water on the car is 1.349 kg. Calculate the total weight and mass of the car, including the passengers and water, using four significant figures.

1.34 During an acceleration test of a car the acceleration was measured to be 1.4363 m/s^2. Because slush and mud adhered to the bottom of the car, the mass was estimated to be 1400 $\pm$ 100 kg. Calculate the force driving the car and indicate the accuracy.

Design and Projects

1.35 Design transport containers for milk in 1 gallon and 1 liter sizes.

1.36 Design a kit of tools for campers so they can prepare and eat meals. The kit should have all of the implements needed, and be lightweight and compact.

1.37 An acid container will damage the environment and people around it if it leaks. The cost of the container is proportional to the container wall thickness. The safety can be increased either by making the container wall thicker or by mounting a reserve tray under the container to collect the leaking acid. The reserve tray costs 10% of the thick-walled container cost. Which is less costly, to increase the wall thickness or to mount a reserve tray under the container?

Chapter 2

Load, Stress, and Strain

Collapse of the Tacoma Narrows bridge in 1940. *Source:* AP Photos.

The careful text-books measure
(Let all who build beware!)
The load, the shock, the pressure
Material can bear.
So when the buckled girder
Lets down the grinding span,
The blame of loss, or murder,
Is laid upon the man.
Not on the stuff - The Man!

Rudyard Kipling, *Hymn of Breaking Strain*

This chapter addresses fundamental problems essential to design: determining the location in a part that is likely to fail, and how to analyze stresses and strains that occur at the critical location. The concept of the critical section is discussed, and the terminology of different loads is defined. The concepts of equilibrium and free-body diagrams are then presented, leading to the production of shear and bending moment diagrams for beams. There are numerous methods of producing such diagrams, and three of the most common and powerful techniques are presented. Stress and strain are discussed next, with an emphasis that they are tensors. The common circumstances of plane stress and plane strain are defined. The ability to determine stress states based on orientation is demonstrated through stress transformation equations and Mohr's circle diagrams, and the procedure for finding principal stresses for a generalized three-dimensional stress state is given. The useful concept of octahedral stresses is presented, and the chapter ends by briefly describing the use of strain gages and rosettes to experimentally determine strains.

Contents

Examples

Design Procedures

Symbols

A	area, m^2
d	diameter, m
g	gravitational acceleration, 9.807 m/s^2
l	length, m
M	moment, N-m
m_a	mass, kg
n	any integer
P	force, N
q	load intensity function, N/m
R	reaction force, N
r	radius of Mohr's circle, m
$\mathbf{S}$	stress tensor
$\mathbf{S}'$	principal stress tensor
$\mathbf{T}$	strain tensor
T	torque, N-m
V	transverse shear force, N
W	normal applied load, N
w_o	load per unit length, N/m
x, y, z	Cartesian coordinate system, m
x', y', z'	rotated Cartesian coordinate system, m
γ	shear strain
δ	elongation, m
ϵ	normal strain
θ	deviation from initial right angle or angle of force application, deg
μ	coefficient of friction
σ	normal stress, Pa
τ	shear stress, Pa
$\tau_{1/2}$, $\tau_{2/3}$, $\tau_{1/3}$	principal shear stresses in triaxial stress state, Pa
ϕ	angle of oblique plane, deg

Subscripts

a	axial
b	biaxial stress
c	center
e	von Mises
r	roller
t	triaxial stress; transverse
x, y, z	Cartesian coordinates
x', y', z'	rotated Cartesian coordinates
θ	angle representing deviation from initial right angle
σ	normal stress
τ	shear stress
ϕ	angle of oblique plane
1, 2, 3	principal axes

2.1 Introduction

The focus of this text is the design and analysis of machines and machine elements. Since machine elements carry **loads**, it follows that an analysis of loads is essential in machine element design. Proper selection of a machine element often is a simple matter of calculating the stresses or deformations expected in service and then choosing a proper size so that critical stresses or deformations are not exceeded. The first step in calculating the stress or deformation of a machine element is to accurately determine the load. Load, stress, and strain in all its forms are the foci of this chapter, and the information developed here is used throughout the text.

2.2 Critical Section

To determine when a machine element will fail, the designer evaluates the stress, strain, and strength at the critical section. The **critical section**, or the location in the design where the largest internal load is developed and failure is most likely, is often not intuitively known beforehand. Design Procedure 2.1 lists the common steps in determining the critical section and loading. The first and second steps arise from system design. The third step is quite challenging and may require analysis of a number of locations or failure modes before the most critical is found. For example, a beam subjected to a distributed load might conceivably exceed the maximum deflection at a number of locations; thus, the beam deflection would need to be calculated at more than one position.

In general, the critical section will often occur at locations of geometric nonuniformity, such as where a shaft changes its diameter along a fillet, or at an interface between two different materials. Also, locations where load is applied or transferred are often critical locations. Finally, areas where the geometry is most critical are candidates for analysis. This topic will be expanded upon in Chapter 6.

Design Procedure 2.1: Critical Section and Loading

To establish the critical section and the critical loading, the designer:

1. Considers the external loads applied to a machine (e.g., a gyroscope)
2. Considers the external loads applied to an element within the machine (e.g., a ball bearing)
3. Locates the critical section within the machine element (e.g., the inner race)
4. Determines the loading at the critical section (e.g., contact stresses)

Example 2.1: Critical Section of a Simple Crane

Given: A simple crane, shown in Fig. 2.1a, consists of a horizontal beam loaded vertically at one end with a load of 10 kN. The beam is pinned at the other end. The force at the pin and roller must not be larger than 30 kN to satisfy other design constraints.

Find: The location of the critical section and also whether the load of 10 kN can be applied without damage to the crane.

Solution: The forces acting on the horizontal beam are shown in Fig. 2.1b. Summation of moments about the pin (at $x = 0$) gives

$$(1.0)P = (0.25)W_r,$$

so that W_r is found to be 40 kN. Summation of vertical forces gives

$$-W_p + W_r - 10 \text{ kN} = 0,$$

which results in $W_p = 30$ kN. The critical section is at the roller, since $W_r > W_p$. Also, since $W_r > W_{\text{all}}$, failure will occur. To avoid failure, the load at the end of the horizontal beam must be reduced.

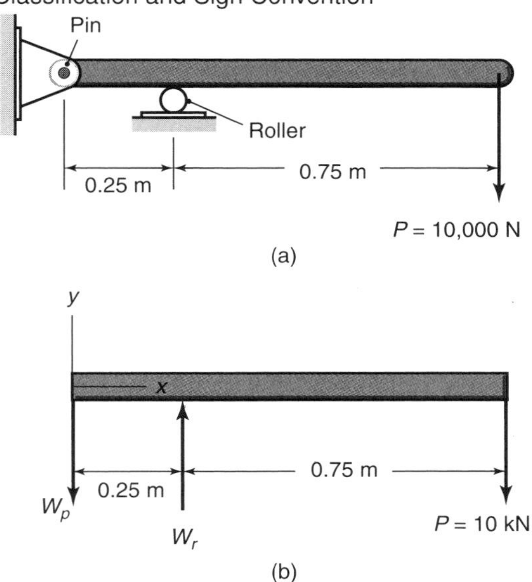

Figure 2.1: A schematic of a simple crane and applied forces considered in Example 2.1. (a) Assembly drawing; (b) free-body diagram of forces acting on the beam.

2.3 Load Classification and Sign Convention

Any applied load can be classified with respect to time in the following ways:

1. *Static load* — Load is gradually applied and equilibrium is reached in a relatively short time. The structure experiences no dynamic effects.

2. *Sustained load* — Load, such as the weight of a structure, is constant over a long time.

3. *Impact load* — Load is rapidly applied. An impact load is usually attributed to an energy imparted to a system.

4. *Cyclic load* — Load can vary and even reverse its direction and has a characteristic period with respect to time.

The load can also be classified with respect to the area over which it is applied:

1. *Concentrated load* — Load is applied to an area much smaller than the loaded member, such as presented for nonconformal surfaces in Section 8.4. An example would be the contact between a caster and a support beam on a mechanical crane, where the contact area is around 100 times smaller than the surface area of the caster. For these cases, the applied force can be considered to act at a point on the surface.

2. *Distributed load* — Load is spread along a large area. An example would be the weight of books on a bookshelf.

Loads can be further classified with respect to location and method of application. Also, the coordinate direction must be determined before the sign of the loading can be established:

1. *Normal load* — The load passes through the centroid of the resisting section. Normal loads may be tensile (Fig. 2.2a) or compressive (Fig. 2.2b). The established sign convention has tensile loads being positive and compressive loads being negative.

2. *Shear load* — The separated bar in Fig. 2.2c illustrates the action of positive shearing. The figure has been redrawn to show the surface of interest on the right side. A shear force is positive if the force direction and the normal direction are both positive or both negative. The shear force, V, shown on the left surface of Fig. 2.2c is in the positive y-direction, which is upward, and the normal to the surface is in the positive x-direction. Thus, the shear force is positive. On the right surface of Fig. 2.2c the shear force is also positive, since the direction of the shear force and the normal to the surface are both negative. A shear force is negative if the force direction and the normal direction have different signs. If the positive y-coordinate had been chosen to be upward (negative) rather than downward (positive) in Fig. 2.2c, the shear force would be negative rather than positive. Thus, to establish whether a shear force is positive or negative, the positive x- and y-coordinates must be designated.

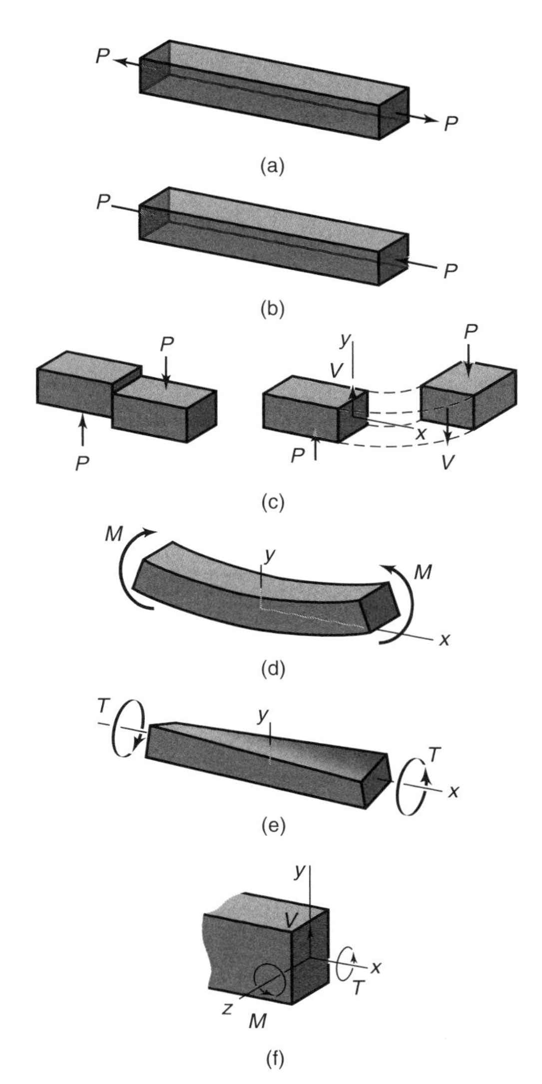

Figure 2.2: Load classified as to location and method of application. (a) Normal, tensile; (b) normal, compressive; (c) shear; (d) bending; (e) torsion; (f) combined.

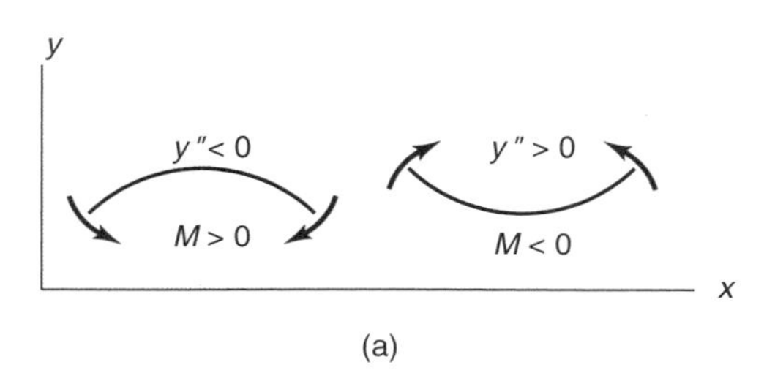

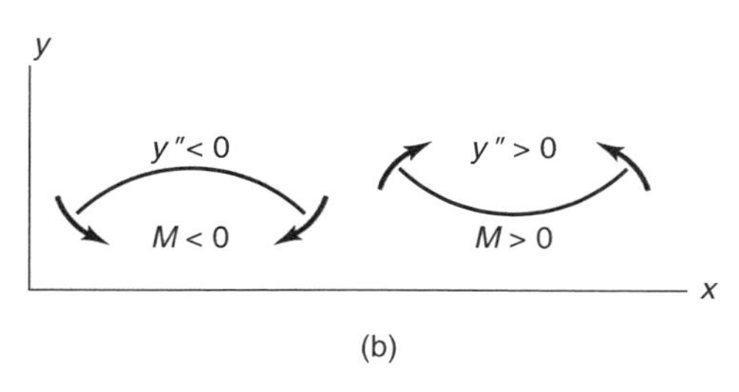

Figure 2.3: Sign conventions used in bending. (a) Positive moment leads to a tensile stress in the positive y-direction; (b) positive moment acts in a positive direction on a positive face. The sign convention shown in (b) will be used in this book.

3. *Bending load* — This commonly occurs when load is applied transversely to the longitudinal axis of the member. Figure 2.2d shows a member that is subject to equal and opposite moments applied at its ends. The moment results in normal stresses in a cross section transverse to the normal axis of the member, as described further in Section 4.5.2.

   The sign convention used in bending stress analysis should be briefly discussed. Two common sign conventions are used in engineering practice, as illustrated in Fig. 2.3. The difference between these two sign conventions is in the sign of the moment applied, and each sign convention has its proponents and critics. The proponents of the sign convention shown in Fig. 2.3a prefer that the stresses that arise in the beam follow the rule that, for a positive moment, a positive distance from the neutral axis results in a positive (tensile) stress. On the other hand, the sign convention shown in Fig. 2.3b allows certain mnemonic methods for its memorization, such as a positive moment results in a deformed shape that "holds water" or has a positive second derivative. Perhaps the best reason for using the sign convention in Fig. 2.3b is that the convention for bending moments is the same as for applied shear forces — that a positive force or moment acting on a face with a positive outward pointing normal acts in a positive direction when using a right-handed coordinate system.

   It should be recognized that sign conventions are arbitrary, and correct answers can be obtained for problems using any sign convention, as long as the sign convention is applied consistently within a problem. In this book, the bending sign convention of Fig. 2.3b will be used, but this should not be interpreted as mandatory for solution of problems.

4. *Torsion load* — Such a load subjects a member to twisting motion, as shown in Fig. 2.2e. The twist results in a distribution of shear stresses on the transverse cross section of the member. Positive torsion occurs in Fig. 2.2e. The right-hand rule is applicable here.

5. *Combined load* — Figure 2.2f shows a combination of two or more of the previously defined loads (e.g., shear, bending, and torsion acting on a member). Note that positive shear, bending, and torsion occur in this figure.

Example 2.2: Classification of Load Types

Given: A diver jumping on a diving board.

Find:
a) The load type when the diver lands on the diving board
b) The load type when the diver stands motionless waiting for the signal to jump
c) The load type on the diving board just as the diver jumps
d) The load type of the diving board assembly against the ground when no dynamic loads are acting

Solution:
a) Impact load — as the diver makes contact with the diving board.
b) Static load — when the diver is motionless.
c) Cyclic load — when the diving board swings up and down just after the dive
d) Sustained load — when gravity acts on the diving board structure, pressing it against the ground

Example 2.3: Loads on a Lever Assembly

Given: The lever assembly shown in Fig 2.4a.

Find: The normal, shear, bending, and torsional loads acting at section B.

Solution: Figure 2.4b shows the various loads acting on the lever, all in the positive direction. To the right of the figure, expressions are given for the loading at section B of the lever shown in Fig. 2.4a.

2.4 Support Reactions

Reactions are forces developed at supports. For two-dimensional problems (i.e., bodies subjected to coplanar force systems), the types of support most commonly encountered, along with the corresponding reactions, are shown in Table 2.1. (Note the direction of the forces on each type of support and the reaction they exert on the attached member.) One way to determine the support reaction is to imagine the attached member as being translated or rotated in a particular direction. If the support prevents translation in a given direction, a force is developed on the member in that direction. Likewise, if the support prevents rotation, a moment is applied to the member. For example, a roller prevents translation only in the contact direction, perpendicular (or normal) to the surface; thus, the roller cannot develop a coupled moment on the member at the point of contact.

2.5 Static Equilibrium

Equilibrium of a body requires both a balance of forces, to prevent the body from translating (moving) along a straight or curved path, and a balance of moments, to prevent the body from rotating. From statics, it is customary to present

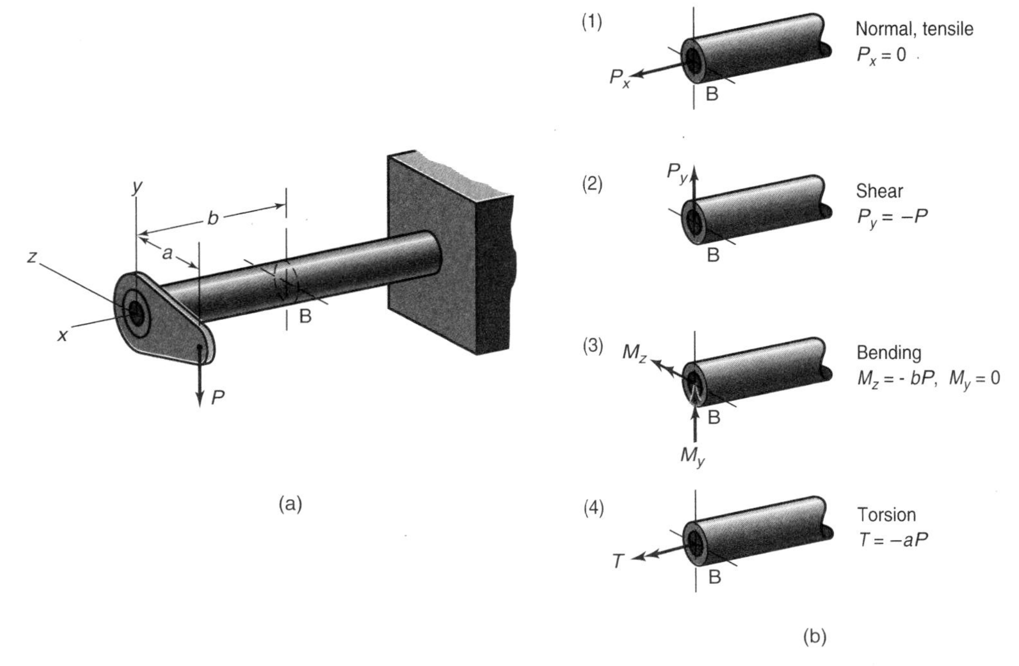

Figure 2.4: Lever assembly and results. (a) Lever assembly; (b) results showing (1) normal, tensile, (2) shear, (3) bending, and (4) torsion on section B of lever assembly.

Table 2.1: Four types of support with their corresponding reactions.

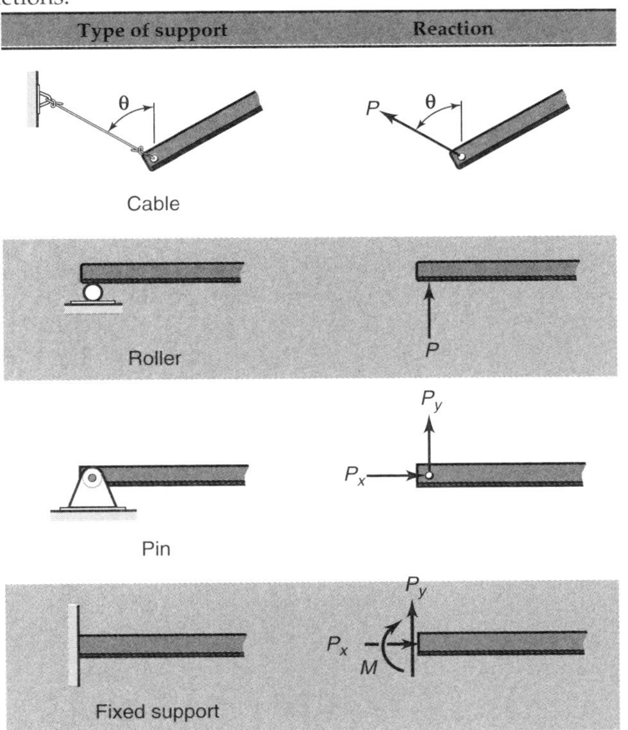

Type of support	Reaction
Cable (θ)	P, θ
Roller	P
Pin	P_x, P_y
Fixed support	P_x, P_y, M

these equations as

$$\sum P_x = 0, \qquad \sum P_y = 0, \qquad \sum P_z = 0, \tag{2.1}$$

$$\sum M_x = 0, \qquad \sum M_y = 0, \qquad \sum M_z = 0. \tag{2.2}$$

Often, in engineering practice, the loading on a body can be represented as a system of coplanar forces. If this is the case, and the forces lie in the x-y plane, the equilibrium conditions of the body can be specified by only three equations:

$$\sum P_x = 0, \qquad \sum P_y = 0, \qquad \sum M_z = 0. \tag{2.3}$$

Note that the moment, M_z, is a vector perpendicular to the plane that contains the forces. Successful application of the equilibrium equations requires complete specification of all the known and unknown forces acting on the body.

Example 2.4: Static Equilibrium of a Ladder

Given: A painter stands on a ladder that leans against the wall of a house. Assume the painter is at the midheight of the ladder. The ladder stands on a horizontal surface with a coefficient of friction of 0.3 and leans at an angle of 20° against the house, which also has a coefficient of friction of 0.3.

Find: Whether the painter and ladder are in static equilibrium and what critical coefficient of friction, μ_{cr}, will not provide static equilibrium.

Solution: Figure 2.5 shows a diagram of the forces acting on the ladder due to the weight of the painter as well as the weight of the ladder. The mass of the ladder is m_l and the mass of the painter is m_p. If the ladder starts to slide, the friction force will counteract the motion.

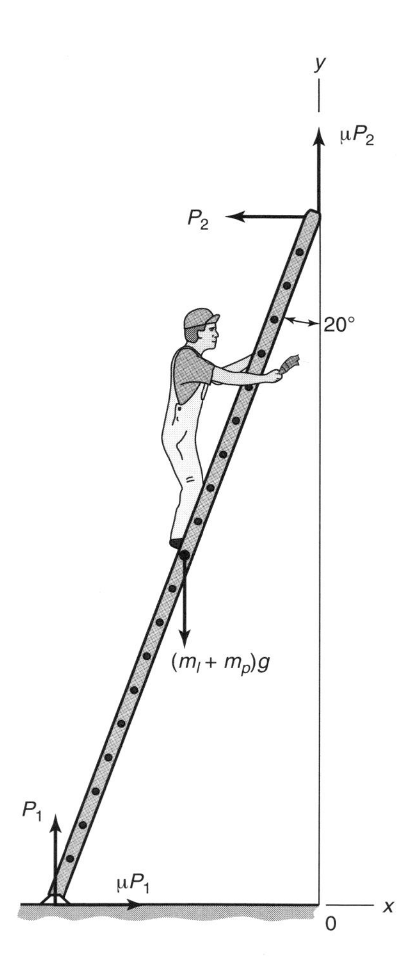

Figure 2.5: Ladder in contact with a house and the ground while having a painter on the ladder.

Summation of horizontal forces gives

$$\sum P_x = \mu_{\text{cr}} P_1 - P_2 = 0, \qquad (a)$$

or $P_2 = \mu_{\text{cr}} P_1$. Summation of vertical forces results in

$$P_1\left(1 + \mu_{\text{cr}}^2\right) = (m_l + m_p)\, g.$$

Therefore,

$$P_1 = \frac{(m_l + m_p)\, g}{1 + \mu_{cr}^2}. \qquad (b)$$

Making use of Eq. (a) gives

$$P_2 = \frac{\mu_{cr}\,(m_l + m_p)\, g}{1 + \mu_{cr}^2}. \qquad (c)$$

Applying moment equilibrium about point 0 results in

$$P_1 l \sin 20^\circ - P_2 l \cos 20^\circ - (m_l + m_p)\, g \frac{l}{2} \sin 20^\circ = 0, \qquad (d)$$

where l is the ladder length. Substituting Eqs. (b) and (c) into Eq. (d) gives

$$0 = \frac{(m_l + m_p)\, gl \sin 20^\circ}{1 + \mu_{cr}^2} - \frac{\mu_{cr}\,(m_l + m_p)\, gl \cos 20^\circ}{1 + \mu_{cr}^2}$$

$$- (m_l + m_p)\, g \frac{l}{2} \sin 20^\circ,$$

or

$$0 = \frac{1}{1 + \mu_{cr}^2} - \frac{\mu_{cr}}{\tan 20^\circ\,(1 + \mu_{cr}^2)} - \frac{1}{2}.$$

Through algebraic manipulation,

$$0.5 = \frac{\tan 20^\circ - \mu_{cr}}{\tan 20^\circ\,(1 + \mu_{cr}^2)}$$

$$0.5 \tan 20^\circ + 0.5\mu_{cr}^2 \tan 20^\circ = \tan 20^\circ - \mu_{cr}$$

$$\mu_{cr}^2 + \frac{\mu_{cr}}{0.5 \tan 20^\circ} - 1 = 0$$

so that

$$\mu_{cr} = 0.1763.$$

Since μ is given as 0.3, the ladder will not move, so that the painter and ladder are in static equilibrium. The critical coefficient where the ladder starts to slide is 0.1763.

2.6 Free-Body Diagram

An entire machine, any individual machine element, or any part of a machine element can be represented as a free body. Static equilibrium is assumed at each level. The best way to account for the forces and moments in the equilibrium equations is to draw the free-body diagram. For the equilibrium equations to be correctly applied, the effects of all the applied forces and moments must be represented in the free-body diagram.

A **free-body diagram** is a sketch of a machine, a machine element, or part of a machine element that shows all acting forces, such as applied loads and gravity forces, and all reactive forces. The reactive forces are supplied by the ground, walls, pins, rollers, cables, or other means. The sign of the reaction may not be known, but it can be assigned arbitrarily or guessed. If, after the static equilibrium analysis, the sign of the reactive force is positive, the initial direction is correct; if it is negative, the direction is opposite to that initially guessed.

Example 2.5: Equilibrium of a Suspended Sphere

Given: A steel sphere, shown in Fig. 2.6a, has a mass of 10 kg and hangs from two wires. A spring attached to the bottom of the sphere applies a downward force of 150 N.

Find: The forces acting on the two wires. Also, draw a free-body diagram showing the forces acting on the sphere.

Solution: Figure 2.6b shows the free-body diagram of the forces acting on the sphere. Summation of the vertical forces gives

$$2P \cos 60^\circ - m_a g - 150 = 0.$$

or

$$P = \frac{(10)(9.807) + 150}{2 \cos 60^\circ} = 248.1 \text{ N}.$$

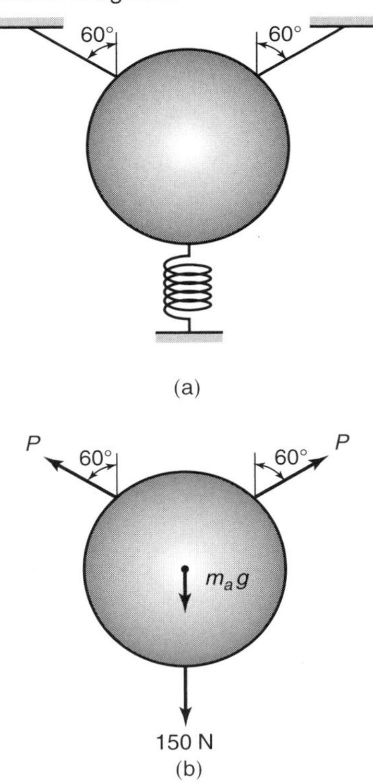

Figure 2.6: Sphere and applied forces. (a) Sphere supported with wires from top and spring at bottom; (b) free-body diagram of forces acting on sphere.

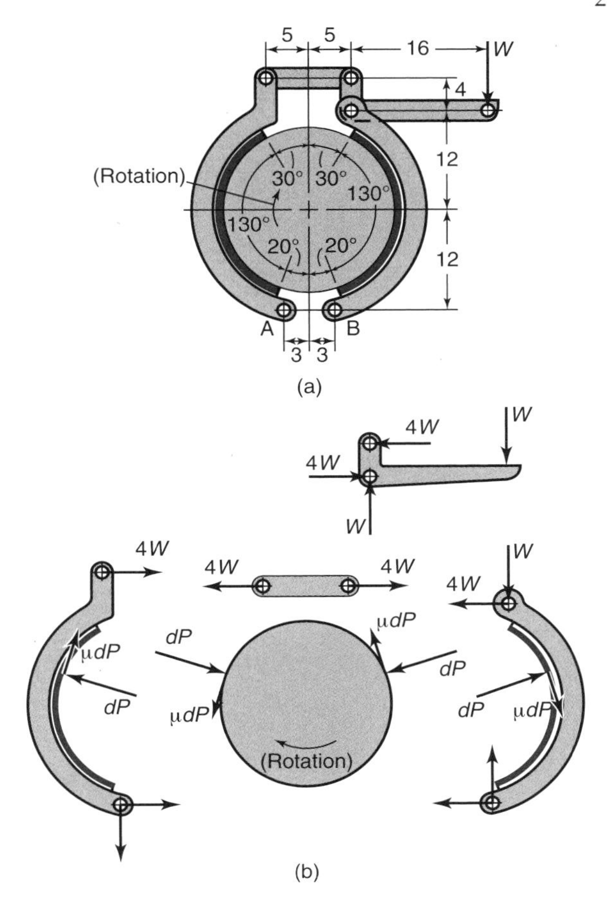

Figure 2.7: External rim brake and applied forces, considered in Example 2.6. (a) External rim brake; (b) external rim brake with forces acting on each part. (Linear dimensions are in millimeters.)

Example 2.6: Free-Body Diagram of an External Rim Brake

Given: The external rim brake shown in Fig. 2.7a.

Find: Draw a free-body diagram of each component of the system.

Solution: Figure 2.7b shows each brake component as well as the forces acting on them. The static equilibrium of each component must be preserved, and the friction force acts opposite to the direction of motion on the drum and in the direction of motion on both shoes. The $4W$ value in Fig. 2.7b was obtained from the moment equilibrium of the lever. Details of brakes are considered in Chapter 18, but in this chapter it is important to be able to draw the free-body diagram of each component.

2.7 Supported Beams

A **beam** is a structural member designed to support loading applied perpendicular to its longitudinal axis. In general, beams are long, often straight bars having a constant cross-section. Often, they are classified by how they are supported. Three major types of support are shown in Fig. 2.8:

1. A **simply supported beam** (Fig. 2.8a) is pinned at one end and roller-supported at the other.
2. A **cantilevered beam** or **cantilever** (Fig. 2.8b) is fixed at one end and free at the other.
3. An **overhanging beam** (Fig. 2.8c) has one or both of its ends freely extending past its supports.

Two major parameters used in evaluating beams are strength and deflection, as discussed in Chapter 5. Shear and bending are the two primary modes of beam loading. However, if the height of the beam is large relative to its width, elastic instability can become important and the beam can twist under loading (see *unstable equilibrium* in Section 9.2.3).

2.8 Shear and Moment Diagrams

Designing a beam on the basis of strength requires first finding its maximum shear and moment. This section describes three common and powerful approaches for developing shear and moment diagrams. Usually, any of these methods will be sufficient to analyze any statically determinate beam, so the casual reader may wish to emphasize one method and then continue to the remaining sections.

2.8.1 Method of Sections

One way to obtain shear and moment diagrams is to apply equilibrium to sections of the beam taken at convenient locations. This allows expression of the transverse shear force, V, and the moment, M, as functions of an arbitrary position, x, along the beam's axis. These shear and moment functions can then be plotted as shear and moment diagrams from which the maximum values of V and M can be obtained.

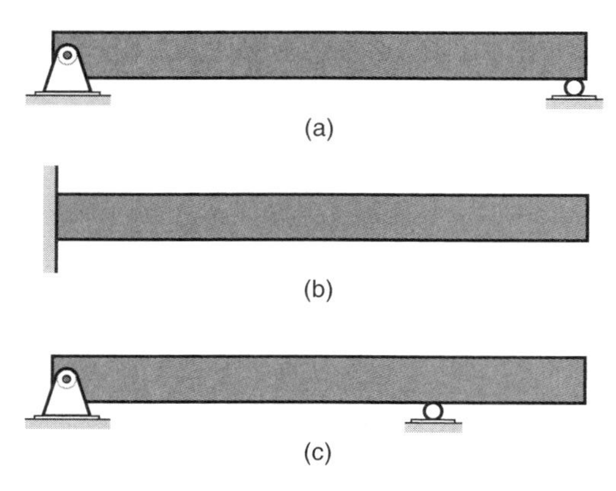

Figure 2.8: Three types of beam support. (a) Simply supported; (b) cantilevered; (c) overhanging.

Design Procedure 2.2: Drawing Shear and Moment Diagrams by the Method of Sections

The procedure for drawing shear and moment diagrams by the method of sections is as follows:

1. Draw a free-body diagram and determine all the support reactions. Resolve the forces into components acting perpendicular and parallel to the beam's axis.
2. Choose a position, x, between the origin and the length of the beam, l, thus dividing the beam into two segments. The origin is chosen at the beam's left end to ensure that any x chosen will be positive.
3. Draw a free-body diagram of the two segments and use the equilibrium equations to determine the transverse shear force, V, and the moment, M.
4. Plot the shear and moment functions versus x. Note the location of the maximum moment. Generally, it is convenient to show the shear and moment diagrams directly below the free-body diagram of the beam.
5. Additional sections can be taken as necessary to fully quantify the shear and moment diagrams.

Example 2.7: Shear and Moment Diagrams by Method of Sections

Given: The bar shown in Fig. 2.9a.

Find: Draw the shear and moment diagrams.

Solution: For $0 \le x < l/2$, the free-body diagram of the bar section is as shown in Fig. 2.9b. The unknowns V and M are positive. Applying the equilibrium equations gives

$$\sum P_y = 0 \rightarrow V = -\frac{P}{2}, \qquad (a)$$

$$\sum M_z = 0 \rightarrow M = \frac{P}{2}x. \qquad (b)$$

For $l/2 \le x < l$, the free-body diagram is shown in Fig. 2.9c. Again, V and M are shown in the positive direction.

$$\sum P_y = 0 \rightarrow \frac{P}{2} - P + V = 0, \quad \text{or} \quad V = P/2. \qquad (c)$$

$$\sum M_z = 0 \rightarrow M + P\left(x - \frac{l}{2}\right) - \frac{P}{2}x = 0.$$

Therefore,

$$M = \frac{P}{2}(l - x). \qquad (d)$$

The shear and moment diagrams in Fig. 2.9d can be obtained directly from Eqs. (a) to (d).

2.8.2 Direct Integration

Note that if $q(x)$ is the load intensity function in the y-direction, the transverse shear force is

$$V(x) = -\int_{-\infty}^{x} q(x)\,dx, \qquad (2.4)$$

and the bending moment is

$$M(x) = -\int_{-\infty}^{x} V(x)\,dx = \int_{-\infty}^{x}\int_{-\infty}^{x} q(x)\,dx\,dx. \qquad (2.5)$$

For simple loading cases, direct integration is often the most straightforward method of producing shear and moment diagrams. Since the integral of a curve is its area, graphically producing a shear or moment diagram follows directly from the loading. The only complication arises from point loadings and their use in developing a shear diagram. With concentrated loadings, the shear diagram will take a "jump" equal in magnitude to the applied load. The sign convention used for moment diagrams is important; recall that the sign convention described in Fig. 2.3b is used in this textbook.

Example 2.8: Shear and Moment Diagrams by Direct Integration

Given: The beam shown in Fig. 2.10a. From static equilibrium, it can be shown that R_A = 12 kN and R_B = 4 kN in the directions shown.

Find: The shear and moment diagrams by direct integration. Determine the location and magnitude of the largest shear force and moment.

Solution: The shear diagram will be constructed first. Consider the loads on the beam and work from left to right to construct the shear diagram. The following steps are followed to construct the shear diagram:

1. At the left end (at $x = 0$), there is a downward acting force. As discussed above, this means that the shear diagram will see a jump in its value at $x = 0$. From Eq. (2.4), a downward acting load leads to an upward acting shear force (that is, its sign is opposite to the loading). Thus, the diagram jumps upward by a magnitude of 4 kN.
2. Moving to the right, this value is unchanged until $x = 2$ m, where a 12 kN concentrated load acts upward. This results in a downward jump as shown.

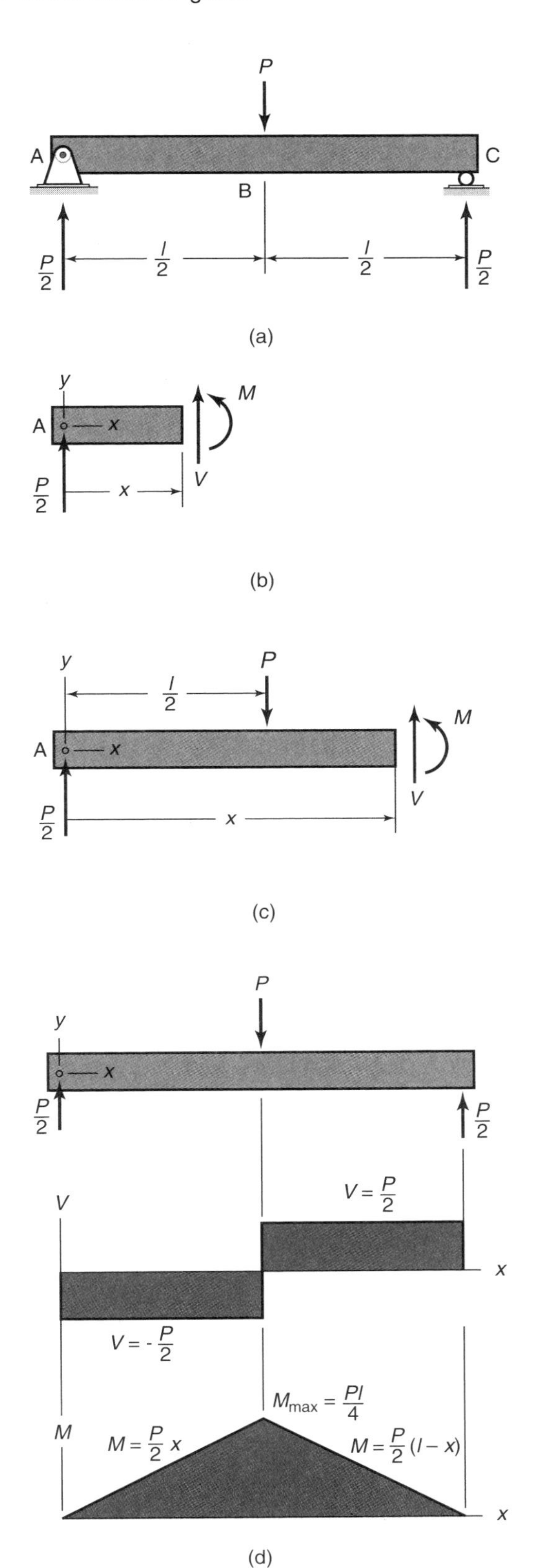

Figure 2.9: Simply supported beam. (a) Midlength load and reactions; (b) free-body diagram for $0 < x < l/2$; (c) free-body diagram for $l/2 \leq x < l$; (d) shear and moment diagrams.

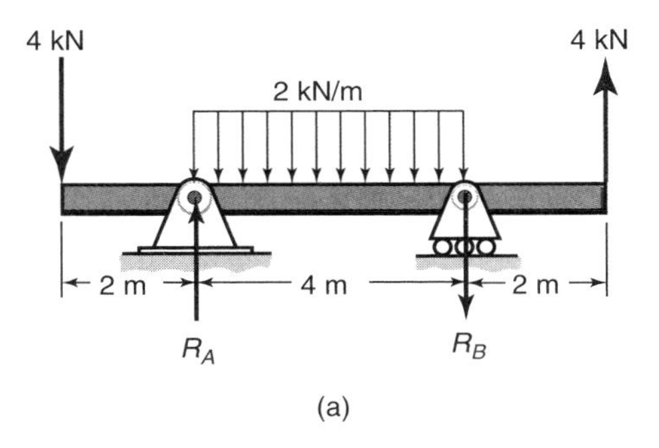

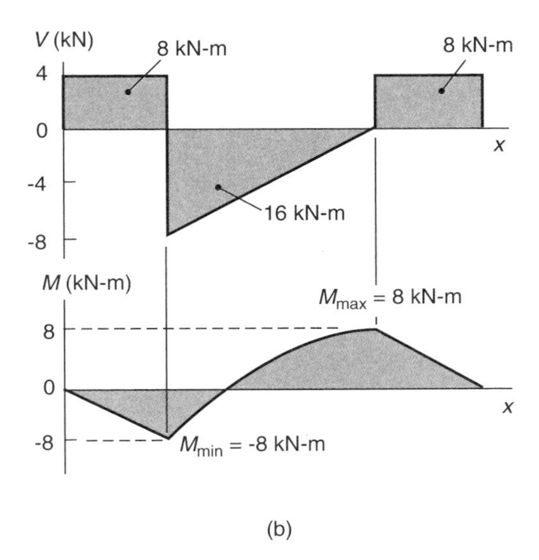

Figure 2.10: Beam for Example 2.8. (a) Applied loads and reactions; (b) shear diagram with areas indicated, and moment diagram with maximum and minimum values indicated.

3. The constant distributed loading to the right of $x = 2$ m will result in a shear force that changes linearly with respect to x. From Eq. (2.4), the magnitude of the total change is the integral of the applied load, or just its area. Thus, the total change due to the 2kN/m distributed load from $x = 2$ to $x = 6$ is 8 kN, and since the distributed load acts downward, this change is upward in the shear diagram because of the sign convention used in Eq. (2.4). Therefore, the value of the shear force at $x = 6$ is $(-8 \text{ kN}) + 8 \text{ kN} = 0$. The line from $x = 2$ to $x = 6$ is shown.

4. At $x = 6$, there is a concentrated force associated with the downwards acting force R_B, so there is an upward jump of 4 kN.

5. At $x = 8$, the upward acting force leads to a downward jump of 4 kN, returning the shear to zero.

The bending moment is obtained from repeated application of Eq. (2.5). However, note that the integral of the shear force is the area under the shear force curve. The shear diagram just developed consists of rectangles and triangles, where the area is calculated from geometry. The areas have been indicated in the shear diagram. For example, the shear diagram up to $x = 2$ consists of a rectangle with a height of $V = 4$ kN and a base of $x = 2$ m. Thus, its area is 8 kN-m.

The moment diagram is then constructed using the following steps.

1. At a starting value of $M = 0$ at $x = 0$, the diagram will be constructed from left to right. From $x = 0$ to $x = 2$ m, the value of the shear diagram is positive and constant. Integrating this curve results in a linear profile. Since the shear diagram is positive, the moment that results must be negative according to Eq. (2.5), and at $x = 2$ m, the value is 8 kN-m. This linear profile is shown in the figure.

2. From $x = 2$ m to $x = 6$ m, the shear diagram is linear with respect to x, so that the moment diagram will be quadratic. At $x = 6$ m, it is known that the moment will have a value of 8 kN-m by summing the areas of the shear diagram segments. The slope of the moment curve is equal to the value of the shear curve, as seen by taking the derivative of Eq. (2.5). Thus, the slope is initially large and at $x = 6$ it is zero.

3. From $x = 6$ m to $x = 8$ m, the moment diagram has a linear profile and ends at $M = 0$. This can be seen by summing the areas in the shear diagram, remembering that areas below the abscissa are considered negative.

The shear and moment diagrams are shown in Fig. 2.10b. It can be seen that the largest magnitude of shear stress is at $x = 2$ m and has a value of $|V|_{\text{max}} = 8$ kN. The largest magnitude of bending moment is $|M|_{\text{max}} = 8$ kN-m.

2.8.3 Singularity Functions

If the loading is simple, the method for obtaining shear and moment diagrams described in Sections 2.8.1 or 2.8.2 can be used. Often, however, this is not the situation. For more complex loading, methods such as **singularity functions** can be used. A singularity function in terms of a variable, x, is written as

$$f_n(x) = \langle x - a \rangle^n. \tag{2.6}$$

where n is any integer (positive or negative) including zero, and a is a reference location on a beam. Singularity functions are denoted by using angular brackets. The advantage of using a singularity function is that it permits writing an analytical expression directly for the transverse shear and moment over a range of discontinuities.

Table 2.2 shows six singularity and load intensity functions along with corresponding graphs and expressions. Note in particular the inverse ramp example. A unit step is constructed beginning at $x = a$, and the ramp beginning at $x = a$ is subtracted. To have the negative ramp discontinued at $x = a + b$, a positive ramp beginning at this point is constructed; the summation results in the desired loading.

Design Procedure 2.3: Singularity Functions

Some general rules relating to singularity functions are:

1. If $n > 0$ and the expression inside the angular brackets is positive (i.e., $x \geq a$), then $f_n(x) = (x - a)^n$. Note that the angular brackets to the right of the equal sign in Eq. (2.6) are now parentheses.

2. If $n > 0$ and the expression inside the angular brackets is negative (i.e., $x < a$), then $f_n(x) = 0$.

3. If $n < 0$, then $f_n(x) = 0$.

4. If $n = 0$, then $f_n(x) = 1$ when $x \geq a$ and $f_n(x) = 0$ when $x < a$.

5. If $n \geq 0$, the integration rule is

$$\int_{-\infty}^{x} \langle x - a \rangle^n = \frac{\langle x - a \rangle^{n+1}}{n+1}.$$

Note that this is the same as if there were parentheses instead of angular brackets.

6. If $n < 0$, the integration rule is

$$\int_{-\infty}^{x} \langle x - a \rangle^n dx = \langle x - a \rangle^{n+1}.$$

7. When $n \geq 1$, then

$$\frac{d}{dx}\langle x - a \rangle^n = n\langle x - a \rangle^{n-1}.$$

Design Procedure 2.4: Shear and Moment Diagrams by Singularity Functions

The procedure for drawing the shear and moment diagrams by making use of singularity functions is as follows:

1. Draw a free-body diagram with all the applied distributed and concentrated loads acting on the beam, and determine all support reactions. Resolve the forces into components acting perpendicular and parallel to the beam's axis.

2. Write an expression for the load intensity function $q(x)$ that describes all the singularities acting on the beam. Use Table 2.2 as a reference, and make sure to "turn off" singularity functions for distributed loads and the like that do not extend across the full length of the beam.

3. Integrate the negative load intensity function over the beam length to get the shear force. Integrate the negative shear force distribution over the beam length to get the moment, in accordance with Eqs. (2.4) and (2.5).

4. Draw shear and moment diagrams from the expressions developed.

Example 2.9: Shear and Moment Diagrams Using Singularity Functions

Given: The same conditions as in Example 2.7.

Find: Draw the shear and moment diagrams by using a singularity function for a concentrated force located midway on the beam.

Table 2.2: Singularity and load intensity functions with corresponding graphs and expressions.

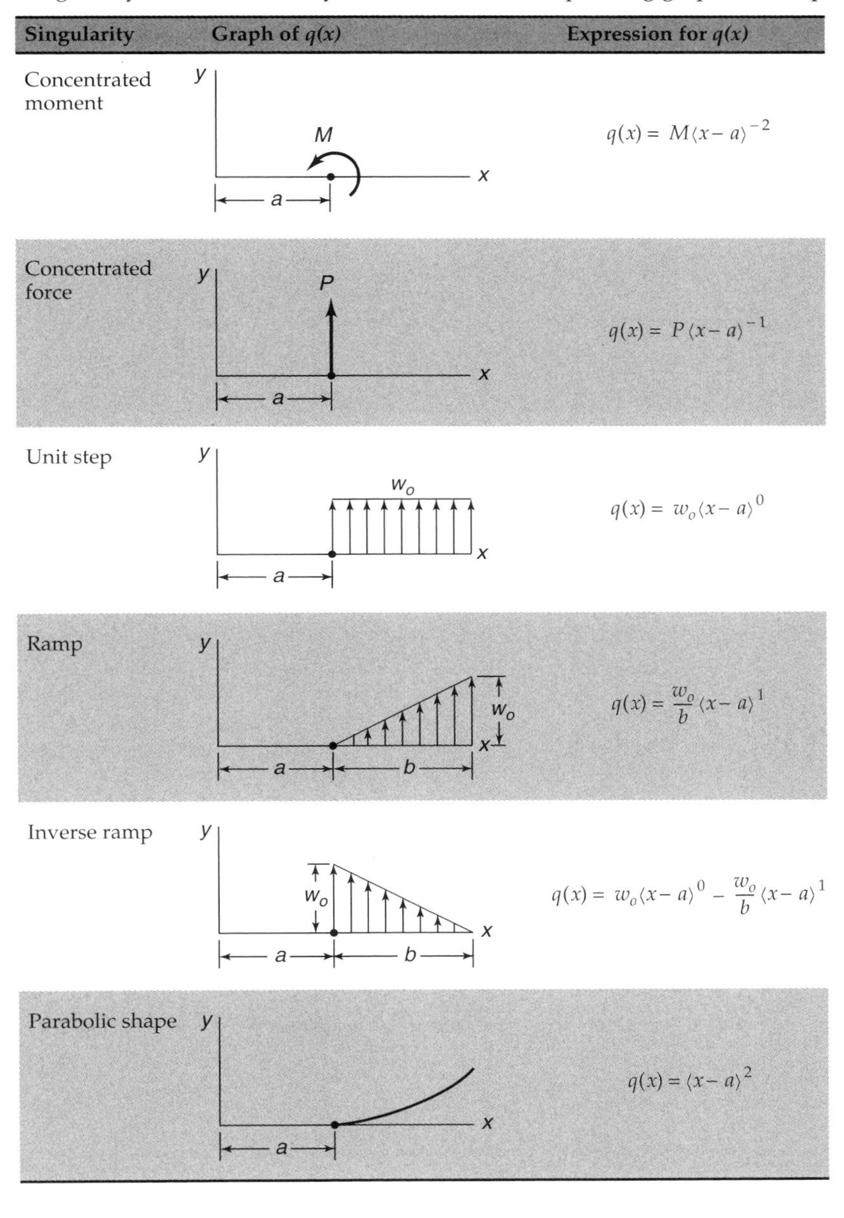

Singularity	Graph of $q(x)$	Expression for $q(x)$
Concentrated moment		$q(x) = M\langle x - a\rangle^{-2}$
Concentrated force		$q(x) = P\langle x - a\rangle^{-1}$
Unit step		$q(x) = w_o\langle x - a\rangle^{0}$
Ramp		$q(x) = \frac{w_o}{b}\langle x - a\rangle^{1}$
Inverse ramp		$q(x) = w_o\langle x - a\rangle^{0} - \frac{w_o}{b}\langle x - a\rangle^{1}$
Parabolic shape		$q(x) = \langle x - a\rangle^{2}$

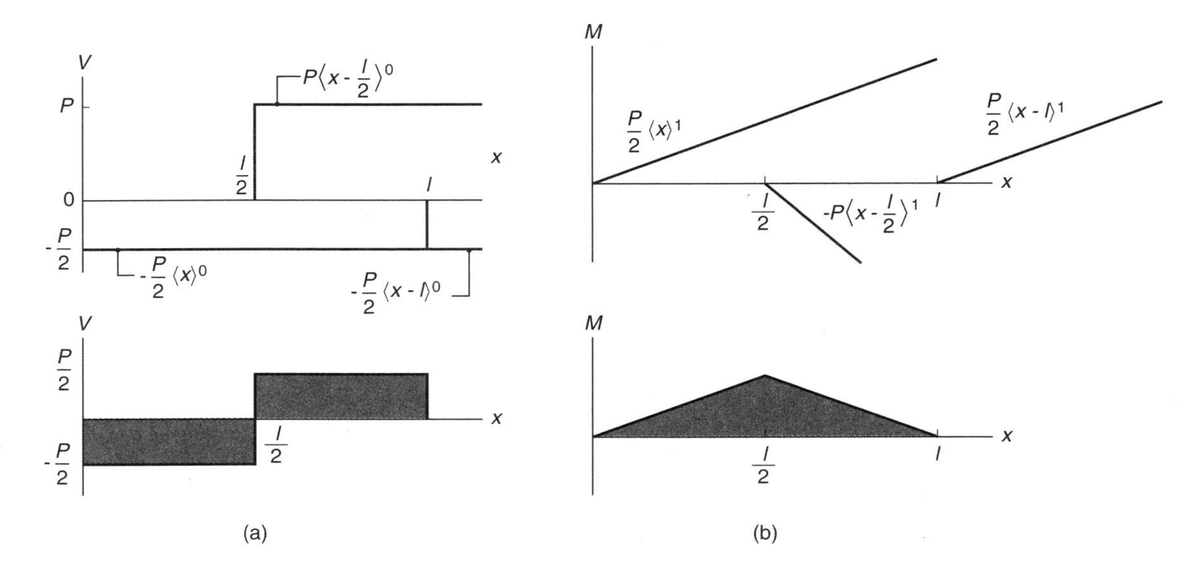

Figure 2.11: (a) Shear and (b) moment diagrams for Example 2.9.

Solution: The load intensity function for the simply supported beam shown in Fig. 2.9a is

$$q(x) = \frac{P}{2}\langle x\rangle^{-1} - P\left\langle x - \frac{l}{2}\right\rangle^{-1} + \frac{P}{2}\langle x - l\rangle^{-1}$$

The shear expression is

$$V(x) = -\int_{-\infty}^{x}\left[\frac{P}{2}\langle x\rangle^{-1} - P\left\langle x - \frac{l}{2}\right\rangle^{-1} + \frac{P}{2}\langle x - l\rangle^{-1}\right]dx$$

or

$$V(x) = -\frac{P}{2}\langle x\rangle^{0} + P\left\langle x - \frac{l}{2}\right\rangle^{0} - \frac{P}{2}\langle x - l\rangle^{0}$$

Figure 2.11a shows the resulting shear diagrams. The diagram at the top shows individual shear, and the diagram below shows the composite of these shear components. The moment expression is

$$M(x) = -\int_{-\infty}^{x}\left[-\frac{P}{2}\langle x\rangle^{0} + P\left\langle x - \frac{l}{2}\right\rangle^{0} - \frac{P}{2}\langle x - l\rangle^{0}\right]dx$$

or

$$M(x) = \frac{P}{2}\langle x\rangle^{1} - P\left\langle x - \frac{l}{2}\right\rangle^{1} + \frac{P}{2}\langle x - l\rangle^{1}$$

Figure 2.11b shows the moment diagrams. The diagram at the top shows individual moments; the diagram at the bottom is the composite moment diagram. The slope of M_2 is twice that of M_1 and M_3, which are equal. The resulting shear and moment diagrams are the same as those found in Example 2.7.

Example 2.10: Shear and Moment Expressions Using Singularity Functions

Given: A simply supported beam shown in Fig. 2.12a where $P_1 = 8$ kN, $P_2 = 5$ kN, $w_o = 4$ kN/m, and $l = 12$ m.

Find: The shear and moment expressions as well as their corresponding diagrams while using singularity functions.

Solution: The first task is to solve for the reactions at $x = 0$ and $x = l$. The force representation is shown in Fig. 2.12b. Note that w_o is defined as the load per unit length for the central part of the beam. In Fig. 2.12b it can be seen that the unit step w_o over a length of $l/2$ produces a resultant force of $w_o l/2$ and that the positive ramp over the length of $l/4$ can be represented by a resultant vector of

$$w_o\left(\frac{l}{4}\right)\left(\frac{1}{2}\right) \quad \text{or} \quad \frac{w_o l}{8}$$

Also, note that the resultant vector acts at

$$x = \left(\frac{2}{3}\right)\left(\frac{l}{4}\right) = \frac{l}{6}$$

From force equilibrium

$$0 = R_1 + P_1 + P_2 + R_2 - \frac{w_o l}{2} - \frac{w_o l}{8} \qquad (a)$$

$$R_1 + R_2 = -P_1 - P_2 + \frac{5w_o l}{8} \qquad (b)$$

Making use of moment equilibrium and the moment of the triangular section load gives

$$\frac{(P_1 + 2P_2)l}{4} - \frac{w_o l^2}{4} - \frac{w_o l}{8}\left(\frac{l}{6}\right) + R_2 l = 0$$

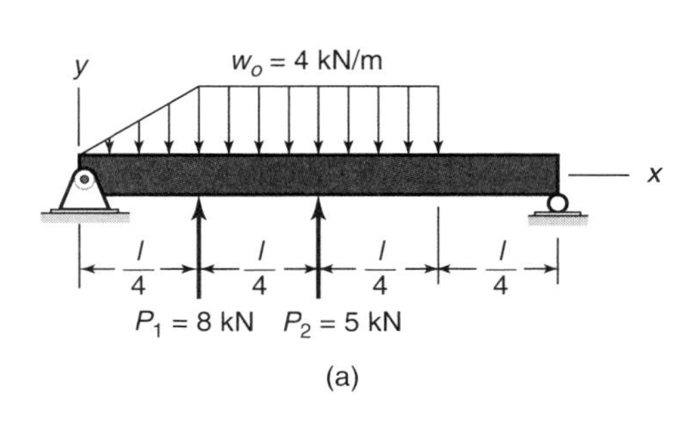

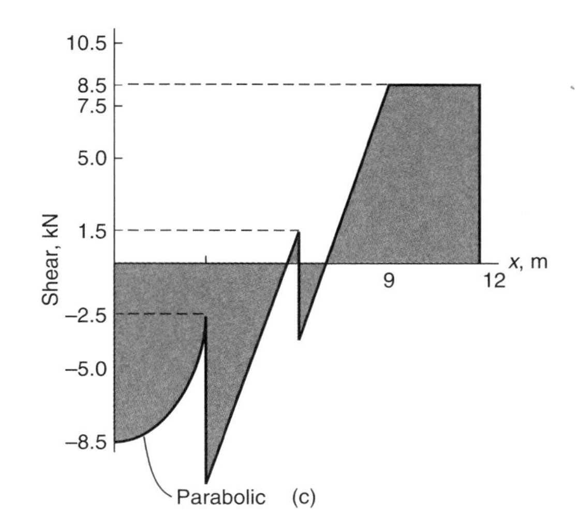

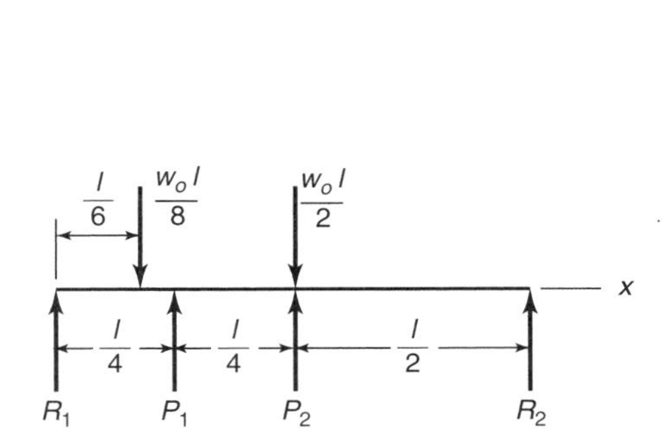

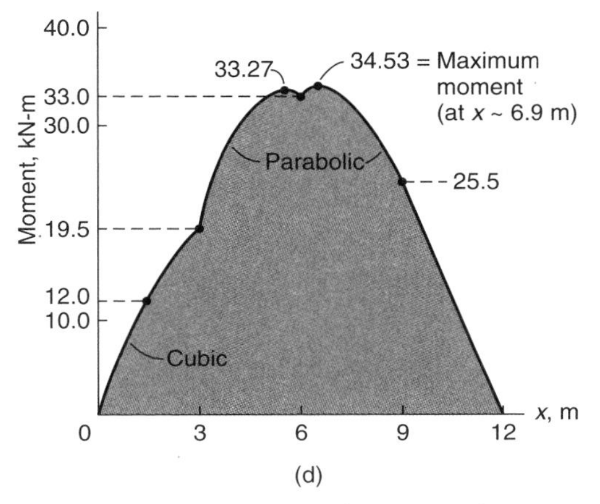

Figure 2.12: Simply supported beam examined in Example 2.10. (a) Forces acting on beam when P_1 = 8 kN, P_2 = 5 kN; w_o = 4 kN/m; l = 12 m; (b) free-body diagram showing resulting forces; (c) shear and (d) moment diagrams.

or

$$R_2 = \frac{13w_o l}{48} - \frac{P_1 + 2P_2}{4} \qquad (c)$$

Substituting Eq. (*c*) into Eq. (*b*) gives

$$R_1 = -\frac{3P_1}{4} - \frac{P_2}{2} + \frac{17w_o l}{48} \qquad (d)$$

Substituting the given values for P_1, P_2, w_o, and l gives

$$R_1 = 8.5\text{ kN} \qquad \text{and} \qquad R_2 = 8.5\text{ kN} \qquad (e)$$

The load intensity function can be written as

$$\begin{aligned} q(x) \;=\; & R_1\langle x\rangle^{-1} - \frac{w_o}{l/4}\langle x\rangle^1 + \frac{w_o}{l/4}\left\langle x - \frac{l}{4}\right\rangle^1 \\ & + P_1\left\langle x - \frac{l}{4}\right\rangle^{-1} + P_2\left\langle x - \frac{l}{2}\right\rangle^{-1} \\ & + w_o\left\langle x - \frac{3l}{4}\right\rangle^0 + R_2\langle x - l\rangle^{-1} \end{aligned}$$

Note that a unit step beginning at $l/4$ is created by initiating a ramp at $x = 0$ acting in the negative direction and summing it with another ramp starting at $x = l/4$ acting in the positive direction, since the slopes of the ramps are the same. The second and third terms on the right side of the load intensity function produce this effect. The sixth term on the right side of the equation turns off the unit step. Integrating the load intensity function gives the shear force as

$$\begin{aligned} V(x) \;=\; & -R_1\langle x\rangle^0 + \frac{2w_o}{l}\langle x\rangle^2 - \frac{2w_o}{l}\left\langle x - \frac{l}{4}\right\rangle^2 \\ & - P_1\left\langle x - \frac{l}{4}\right\rangle^0 - P_2\left\langle x - \frac{l}{2}\right\rangle^0 \\ & - w_o\left\langle x - \frac{3l}{4}\right\rangle^1 - R_2\langle x - l\rangle^0 \end{aligned}$$

Integrating the shear force gives the moment, and substituting the values for w_o and l gives

$$\begin{aligned} M(x) \;=\; & 8.5\langle x\rangle^1 + \frac{2}{9}\langle x\rangle^3 - \frac{2}{9}\langle x - 3\rangle^3 + 8\langle x - 3\rangle^1 \\ & + 5\langle x - 6\rangle^1 + 2\langle x - 9\rangle^2 + 8.5\langle x - 12\rangle^1 \end{aligned}$$

The shear and moment diagrams are shown in Fig. 2.12c and d, respectively.

2.9 Stress

One of the fundamental problems in engineering is determining the effect of a loading environment on a part. This determination is an essential part of the design process; one cannot choose a dimension or a material without first understanding the intensity of force inside the component being analyzed. **Stress** is the term used to define the intensity and direction of the internal forces acting at a given point. Strength, on the other hand, is a property of a material and will be covered in later chapters.

For normal loading on a load-carrying member in which the external load is uniformly distributed over a cross-section, the magnitude of the average normal stress can be calculated from

$$\sigma_{\text{avg}} = \frac{\text{Average force}}{\text{Cross-sectional area}} = \frac{P}{A}. \tag{2.7}$$

Thus, the unit of stress is force per unit area. Consider a small area ΔA on the cross section, and let ΔP represent the internal forces acting on this small area. The average intensity of the internal forces transmitted by the area ΔA is obtained by dividing ΔP by ΔA. If the internal forces transmitted across the section are assumed to be continuously distributed, the area ΔA can be made increasingly smaller and will approach a point on the surface in the limit. The corresponding force ΔP will also become increasingly smaller. The stress at the point on the cross section to which ΔA converges is

$$\sigma = \lim_{\Delta A \to 0} \frac{\Delta P}{\Delta A} = \frac{dP}{dA}. \tag{2.8}$$

The stress at a point acting on a specific plane is a vector and thus has a magnitude and a direction. Its direction is the limiting direction ΔP as area ΔA approaches zero. Similarly, the shear stress can be defined in a specific plane. Thus, a stress must be defined with respect to a direction.

Example 2.11: Stress in Beam Supports

Given: As shown in Fig. 2.13a, a 3-m-long beam is supported at the left end by a 6-mm-diameter steel wire and at the right end by a 10-mm-diameter steel cylinder. The bar carries a mass $m_{a1} = 200$ kg and the bar's mass is $m_{a2} = 50$ kg.

Find: Determine the stresses in the wire and in the cylinder.

Solution: The wire and cylinder areas are $A_B = 28.27$ mm^2 and $A_C = 78.54$ mm^2. Figure 2.13b shows a free-body diagram of the forces acting on the bar. Moment equilibrium about point C gives

$$3R_{\text{B}} = 2(200)(9.81) + 1.5(50)(9.81) = 4660 \text{ N}$$

or $R_{\text{B}} = 1553$ N. From force equilibrium,

$$R_{\text{B}} - m_{a1}g - m_{a2}g + R_{\text{C}} = 0$$

$$R_{\text{C}} = g\left(m_{a1} + m_{a2}\right) - R_{\text{B}} = 9.81(200 + 50) - 1553 = 900 \text{ N}$$

The stresses at points B and C are, from Eq. (2.7),

$$\sigma_{\text{B}} = \frac{R_{\text{B}}}{A_{\text{B}}} = \frac{1553}{28.27} = 54.93 \text{ N/mm}^2 = 54.93 \text{ MPa}$$

$$\sigma_{\text{C}} = -\frac{R_{\text{C}}}{A_{\text{C}}} = -\frac{900}{78.54} = -11.46 \text{ N/mm}^2 = -11.46 \text{ MPa}$$

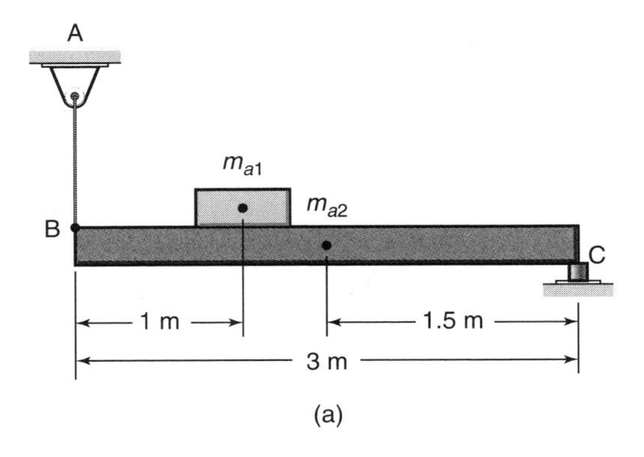

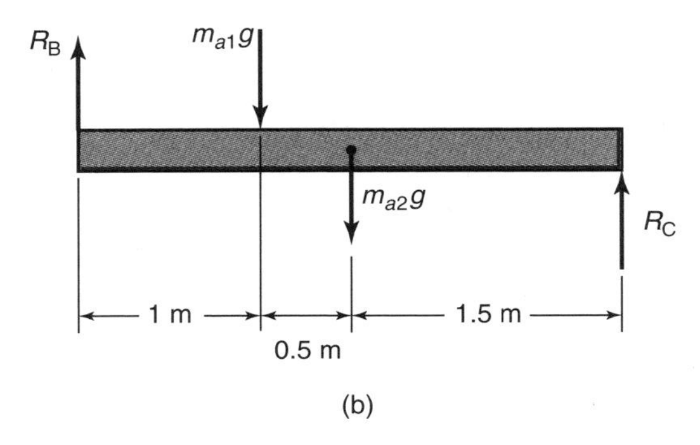

Figure 2.13: Figures used in Example 2.11. (a) Load assembly drawing; (b) free-body diagram.

2.10 Stress Element

Figure 2.14 shows a stress element with the origin of stress placed inside the element. Across each of the mutually perpendicular surfaces there are three stresses, yielding a total of nine stress components. Of the three stresses acting on a given surface, the normal stress is denoted by σ and the shear stress by τ. A normal stress will receive a subscript indicating the direction in which the stress acts (e.g., σ_x). A shear stress requires two subscripts, the first to indicate the plane of the stress and the second to indicate its direction (e.g., τ_{xy}). The **sign convention for normal stress** distinguishes positive for tension and negative for compression. A positive shear stress points in the positive direction of the coordinate axis denoted by the second subscript if it acts on a surface with an outward normal in the positive direction. The **sign convention for shear stress** is directly associated with the coordinate directions. If both the normal from the surface and the shear are in the positive direction or if both are in the negative direction, the shear stress is positive. Any other combinations of the normal and the direction of shear will produce a negative shear stress. The surface stresses of an element have the following relationships:

1. The normal and shear stress components acting on opposite sides of an element must be equal in magnitude but opposite in direction.

2. Moment equilibrium requires that the shear stresses be symmetric, implying that the subscripts can be reversed in order, or

$$\tau_{xy} = \tau_{yx}, \qquad \tau_{xz} = \tau_{zx}, \qquad \tau_{yz} = \tau_{zy}, \tag{2.9}$$

thus reducing the nine different stresses acting on the

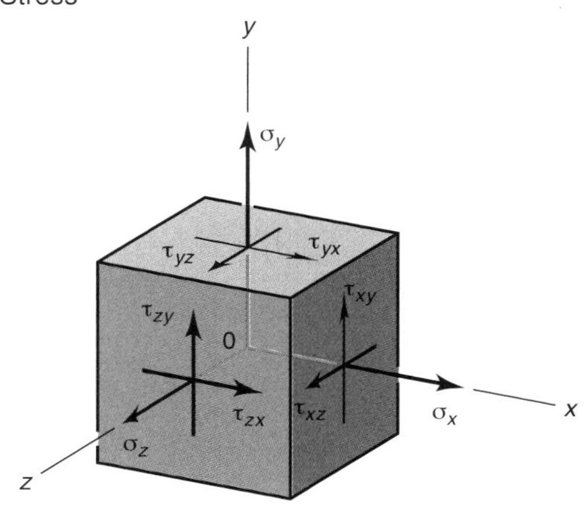

Figure 2.14: Stress element showing general state of three-dimensional stress with origin placed in center of element.

element to six: three normal stresses σ_x, σ_y, σ_z and three shear stresses τ_{xy}, τ_{yz}, τ_{xz}.

The general laws of stress transformation, given in Appendix B (Section B.1), enable the determination of stresses acting on any new orthogonal coordinate system.

Example 2.12: Stresses in Stress Element

Given: The stress element shown in Fig. 2.14 is put into a pressure vessel and pressurized to 10 MPa. An additional shear stress of 5 MPa acting on the bottom surface is directed in the positive x-direction.

Find: Are the stresses positive or negative?

Solution: The normal stress here is thus $\sigma = -10$ MPa. A positive pressure results in a negative normal stress by definition; since the element is loaded in compression by the pressure, it has a negative value. For the shear stress, τ_{zx}, a positive shear stress is directed in the positive coordinate direction when the normal to the surface is directed in the positive coordinate direction. A shear stress acting on a surface with the normal in the negative coordinate direction is positive when the stress is directed in the negative coordinate direction. The shear stress in this problem acts on a surface with the normal in the negative y-direction, but the stress is directed in the positive x-direction. Thus, the shear is negative.

2.11 Stress Tensor

In engineering practice, it is common to encounter scalar quantities, those that have numerical value. Vectors, such as force, have a magnitude as well as a direction. Stress requires six quantities for its definition; thus, stress is a *tensor*. From the stress element of Fig. 2.14 and Eq. (2.9) the stress tensor is

$$\mathbf{S} = \begin{pmatrix} \sigma_x & \tau_{xy} & \tau_{xz} \\ \tau_{xy} & \sigma_y & \tau_{yz} \\ \tau_{xz} & \tau_{yz} & \sigma_z \end{pmatrix}, \tag{2.10}$$

which is a symmetrical tensor. A property of a symmetrical tensor is that there exists an equivalent tensor with an orthogonal set of axes 1, 2, and 3 (called *principal axes*) with respect to which the tensor elements are all zero except for those in the principal diagonal; thus,

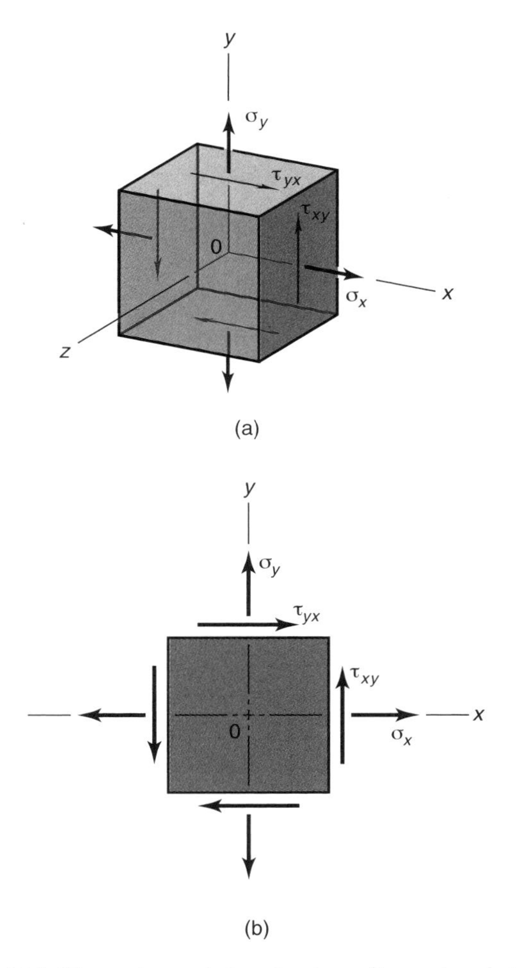

Figure 2.15: Stress element showing two-dimensional state of stress. (a) Three-dimensional view; (b) plane view.

$$\mathbf{S}' = \begin{pmatrix} \sigma_1 & 0 & 0 \\ 0 & \sigma_2 & 0 \\ 0 & 0 & \sigma_3 \end{pmatrix}, \tag{2.11}$$

where σ_1, σ_2, and σ_3 are principal stresses and will be discussed further below. Note that no shear stresses occur in Eq. (2.11).

2.12 Plane Stress

Many cases of stress analysis can be simplified to the case of **plane stress**, where the stresses all occur inside one plane. This is a common and valuable simplification, as the third direction can thus be neglected, and all stresses on the stress element act on two pairs of faces rather than three, as shown in Fig. 2.15. This two-dimensional stress state is sometimes called **biaxial** or **plane stress**.

In comparing the two views of the plane stress element shown in Fig. 2.15, note that all stresses shown in Fig. 2.15b act on surfaces perpendicular to the paper, with the paper being designated as either the x-y plane or the z plane. The stresses shown in Fig. 2.15 all have positive values in accordance with the conventions presented in Section 2.10.

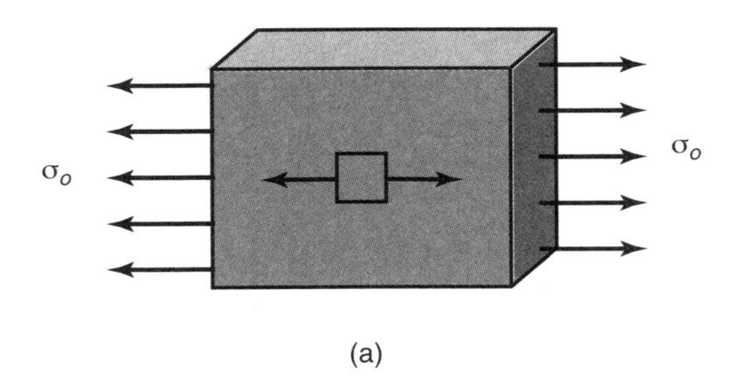

(a)

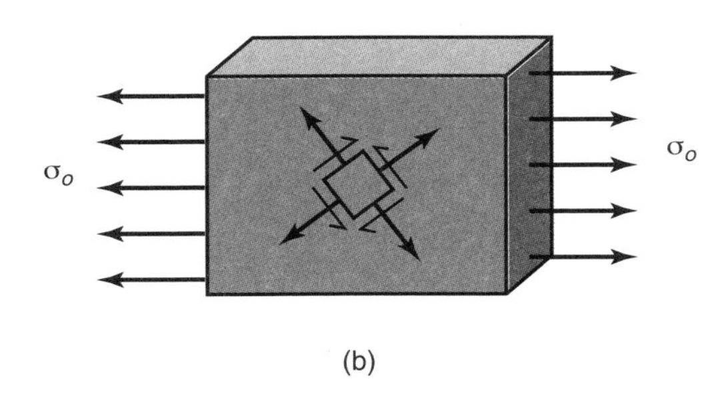

(b)

Figure 2.16: Illustration of equivalent stress states. (a) Stress element oriented in the direction of applied stress; (b) stress element oriented in different (arbitrary) direction.

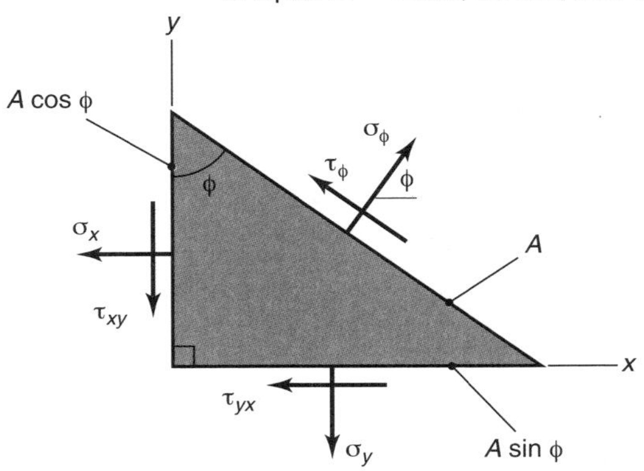

Figure 2.17: Stresses in an oblique plane at an angle ϕ.

The magnitude of stress depends greatly on the coordinate system orientation. For example, consider the stress element shown in Fig. 2.16a. When a uniform stress is applied to the element, the stress state is clearly $\sigma_x = \sigma_o$, $\sigma_y = 0$, and $\tau_{xy} = 0$. However, if the original orientation of the element were as shown in Fig. 2.16b, this would no longer be the case, and all stress components in the plane would be nonzero. A profound question can be raised at this point: How does the material know the difference between these stress states? The answer is that there is no difference between the stress states of Fig. 2.16, so that they are *equivalent*. Obviously, it is of great importance to be able to transform stresses from one orientation to another, and the resultant stress transformation equations will be of great use throughout the remainder of the text.

Consider if, instead of the stresses acting as shown in Fig. 2.15b, they act in an oblique plane at angle ϕ as shown in Fig. 2.17. The stresses σ_x, σ_y, and τ_{xy} can then be determined in terms of the stresses on an inclined surface whose normal stress makes an angle ϕ with the x-axis.

Note from Fig. 2.17 that if the area of the inclined surface is A (length of the surface times the thickness into the paper), the area of the horizontal side of the triangular element will be $A\sin\phi$, and the area of the vertical side, $A\cos\phi$. From force equilibrium

$$\begin{aligned}\sigma_\phi A &= \tau_{xy}\sin\phi A\cos\phi + \tau_{yx}\cos\phi A\sin\phi \\ &\quad +\sigma_x\cos\phi A\cos\phi + \sigma_y\sin\phi A\sin\phi.\end{aligned}$$

This reduces to

$$\sigma_\phi = 2\tau_{xy}\sin\phi\cos\phi + \sigma_x\cos^2\phi + \sigma_y\sin^2\phi. \tag{2.12}$$

By using trigonometric identities for the double angle, Eq. (2.12) can be written as

$$\sigma_\phi = \frac{\sigma_x+\sigma_y}{2} + \frac{\sigma_x-\sigma_y}{2}\cos 2\phi + \tau_{xy}\sin 2\phi. \tag{2.13}$$

Similarly, from force equilibrium, the shear stress in the oblique plane can be expressed as

$$\tau_\phi = \tau_{xy}\cos 2\phi - \frac{\sigma_x-\sigma_y}{2}\sin 2\phi. \tag{2.14}$$

Equations (2.13) and (2.14) have maximum and minimum values that are of particular interest in stress analysis. The angle ϕ_σ, which gives the extreme value of σ_ϕ, can be determined by differentiating σ_ϕ with respect to ϕ and setting the result equal to zero, giving

$$\frac{d\sigma_\phi}{d\phi} = -(\sigma_x-\sigma_y)\sin 2\phi_\sigma + 2\tau_{xy}\cos 2\phi_\sigma = 0$$

or

$$\tan 2\phi_\sigma = \frac{2\tau_{xy}}{\sigma_x-\sigma_y}. \tag{2.15}$$

where ϕ_σ is the angle where normal stress is extreme. Equation (2.15) has two roots, 180° apart, and for the double-angle nature of the left side of Eq. (2.15) this suggests roots of ϕ_σ being 90° apart. One of these roots corresponds to the maximum value of normal stress, the other to the minimum value.

Substituting Eq. (2.15) into Eqs. (2.13) and (2.14) gives the following after some algebraic manipulation:

$$\sigma_1, \sigma_2 = \frac{\sigma_x+\sigma_y}{2} \pm \sqrt{\tau_{xy}^2 + \frac{(\sigma_x-\sigma_y)^2}{4}}, \tag{2.16}$$

$$\tau_{\phi_\sigma} = 0. \tag{2.17}$$

At this stress element orientation, where the normal stresses are extreme, the shear stress is zero. The axes that define this orientation are called the **principal axes**, and the normal stresses from Eq. (2.16) are called the **principal normal stresses**. Principal stresses are given numerical subscripts to differentiate them from stresses at any other orientation. A common convention is to order the principal stresses according to

$$\sigma_1 \geq \sigma_2 \geq \sigma_3. \tag{2.18}$$

In plane stress, one of the principal stresses is always zero.

Another orientation of interest is the one where the shear stress takes an extreme value. Differentiating Eq. (2.14) with respect to ϕ and solving for τ gives the orientation ϕ_τ, with resulting extreme shear stress of

$$\tau_{\max}, \tau_{\min} = \tau_1, \tau_2 = \pm\sqrt{\left(\frac{\sigma_x-\sigma_y}{2}\right)^2 + \tau_{xy}^2} \tag{2.19}$$

and

$$\sigma_{\phi_\tau} = \frac{\sigma_x+\sigma_y}{2}. \tag{2.20}$$

The shear stresses from Eq. (2.19) are called **principal shear stresses**. Thus, on the stress element oriented to achieve a maximum shear stress, the normal stresses on the two faces are equal. Also, it can be shown that

$$|\phi_\tau - \phi_\sigma| = \frac{\pi}{4}. \tag{2.21}$$

In summary, for a plane stress situation where σ_x, σ_y, and τ_{xy} are known, the normal and shear stresses σ_ϕ and τ_ϕ can be determined for any oblique plane at angle ϕ from Eqs. (2.13) and (2.14). Also, the principal normal and shear stresses σ_1, σ_2, τ_1, and τ_2 can be determined from Eqs. (2.16) and (2.19).

If the principal normal stresses σ_1 and σ_2 are known, the normal and shear stresses at any oblique plane at angle ϕ can be determined from the following equations:

$$\sigma_\phi = \frac{\sigma_1 + \sigma_2}{2} + \frac{\sigma_1 - \sigma_2}{2} \cos 2\phi \tag{2.22}$$

$$\tau_\phi = \frac{\sigma_1 - \sigma_2}{2} \sin 2\phi. \tag{2.23}$$

In Eq. (2.23) a second subscript is not needed because τ_ϕ represents a shear stress acting on any oblique plane at angle ϕ as shown in Fig. 2.17.

Example 2.13: Stress Transformation

Given: A thin, square steel plate is oriented in the x- and y-directions. A tensile stress, σ, acts on the four sides. Thus, $\sigma_x = \sigma_y = \sigma$.

Find: The normal and shear stresses acting on the diagonal of the plate.

Solution: From Eq. (2.13),

$$\sigma_\phi = \frac{\sigma_x + \sigma_y}{2} + \frac{\sigma_x - \sigma_y}{2} \cos 2\phi + \tau_{xy} \sin 2\phi$$

Thus,

$$\sigma_{45^\circ} = \frac{\sigma + \sigma}{2} + \frac{\sigma - \sigma}{2} \cos 90^\circ + \tau_{xy} \sin 90^\circ = \sigma$$

Similarly, from Eq. (2.14)

$$\tau_\phi = \tau_{xy} \cos 2\phi - \frac{\sigma_x - \sigma_y}{2} \sin 2\phi$$

Therefore,

$$\tau_{45^\circ} = \tau_{xy} \cos 90^\circ - \frac{\sigma - \sigma}{2} \sin 90^\circ = 0$$

2.13 Mohr's Circle

Mohr's circle for a triaxial state of stress at a point was first constructed in 1914 by a German engineer, Otto Mohr, who noted that Eqs. (2.13) and (2.14) define a circle in a σ-τ plane. This circle is used extensively as a convenient method of graphically visualizing the state of stress acting in different planes passing through a given point. The approach used in this text is first to apply Mohr's circle to a two-dimensional stress state; a three-dimensional stress state is discussed in Section 2.14. Indeed, Mohr's circle is most useful for stress visualization in plane stress situations.

Figure 2.18 shows a typical Mohr's circle diagram. A number of observations can be made:

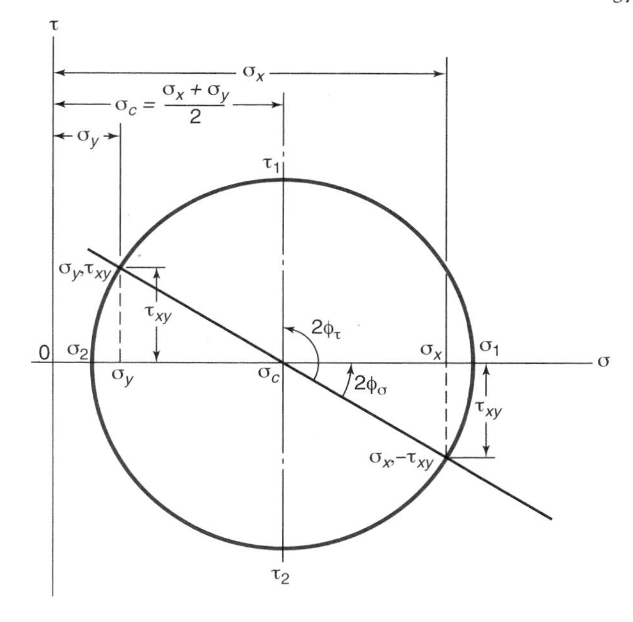

Figure 2.18: Mohr's circle diagram of Eqs. (2.13) and (2.14).

1. Normal stresses are plotted along the abscissa (x-axis), and shear stresses are plotted on the ordinate (y-axis).
2. The circle defines all stress states that are equivalent.
3. The biaxial stress state for any direction can be scaled directly from the circle.
4. The principal normal stresses (i.e., the extreme values of normal stress) are at the locations where the circle intercepts the x-axis.
5. The maximum shear stress equals the radius of the circle.
6. A rotation from a reference stress state in the real plane of ϕ corresponds to a rotation of 2ϕ from the reference points in the Mohr's circle plane.

Design Procedure 2.5: Mohr's Circle

The steps in constructing and using Mohr's circle in two dimensions are as follows:

1. Calculate the plane stress state for any x-y coordinate system so that σ_x, σ_y, and τ_{xy} are known.
2. The center of the Mohr's circle can be placed at

$$\left(\frac{\sigma_x + \sigma_y}{2}, 0\right). \tag{2.24}$$

3. Two points diametrically opposite to each other on the circle correspond to the points $(\sigma_x, -\tau_{xy})$ and (σ_y, τ_{xy}). Using the center and either point allows one to draw the circle.
4. The radius of the circle can be calculated from stress transformation equations or through geometry by using the center and one point on the circle. For example, the radius is the distance between points $(\sigma_x, -\tau_{xy})$ and the center, which directly leads to

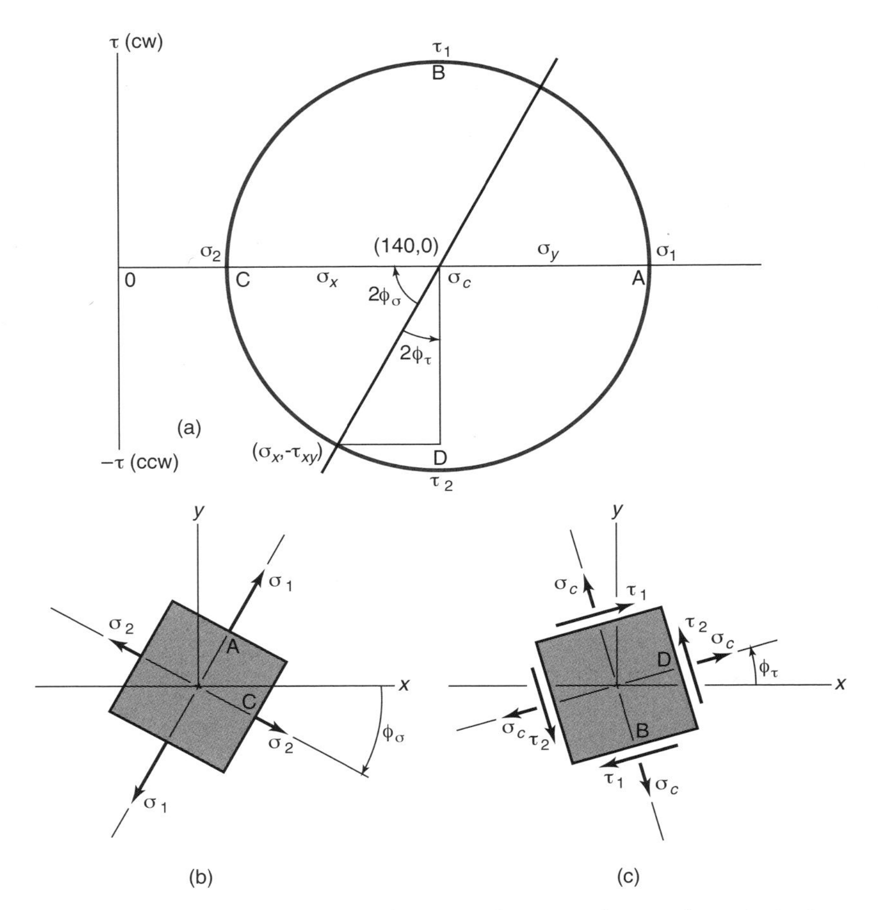

Figure 2.19: Results from Example 2.14. (a) Mohr's circle diagram; (b) stress element for principal normal stress shown in x-y coordinates; (c) stress element for principal shear stresses shown in x-y coordinates.

$$r = \sqrt{\left(\frac{\sigma_x - \sigma_y}{2}\right)^2 + \tau_{xy}^2}. \tag{2.25}$$

5. The principal stresses have the values $\sigma_{1,2}$ = center ± radius.

6. The maximum shear stress equals the radius.

7. The principal axes can be found by calculating the angle between the x-axis in the Mohr's circle plane and the point $(\sigma_x, -\tau_{xy})$. The principal axes in the real plane are rotated one-half this angle in the same direction relative to the x-axis in the real plane.

8. The stresses in an orientation rotated ϕ from the x-axis in the real plane can be read by traversing an arc of 2ϕ in the same direction on the Mohr's circle from the reference points $(\sigma_x, -\tau_{xy})$ and (σ_y, τ_{xy}). The new points on the circle correspond to the new stresses $(\sigma_{x'}, -\tau_{x'y'})$ and $(\sigma_{y'}, \tau_{x'y'})$, respectively.

Example 2.14: Mohr's Circle

Given: The plane stresses σ_x = 90 MPa, σ_y = 190 MPa, and τ_{xy} = 80 MPa.

Find: Draw the Mohr's circle and find the principal normal and shear stresses in the x-y plane. Determine the stress state when the axes are rotated 15° counterclockwise.

Solution: This solution will demonstrate the eight-step approach given in Design Procedure 2.5, with the first step already done in the problem statement. Step 2 advises to calculate the center of the circle and place it at $(\sigma_c, 0)$, where

$$\sigma_c = \frac{\sigma_x + \sigma_y}{2} = \frac{(90 + 190)}{2} = 140 \text{ MPa}.$$

According to Step 3, either point $(\sigma_x, -\tau_{xy})$ or (σ_y, τ_{xy}) can be used to draw the circle. This has been done with the point $(\sigma_x, -\tau_{xy})$ = (90 MPa, −80 MPa) to draw the circle as shown in Fig. 2.19. From Step 4 and from the triangle defined by the x-axis and the point $(\sigma_x, -\tau_{xy})$, the radius can calculated as

$$r = \sqrt{(90 - 140)^2 + (-80)^2} = 94.3 \text{ MPa}.$$

From Step 5 the principal stresses have the values $\sigma_{1,2} = 140 \pm 94.3$, or $\sigma_1 = 234.3$ MPa and $\sigma_2 = 45.7$ MPa. From Step 6, the maximum shear stress equals the radius, or $\tau_{max} = 94.3$ MPa. The principal stress orientation can be determined, if desired, from trigonometry. In the Mohr's circle plane (Fig. 2.19a), the point $(\sigma_x, -\tau_{xy})$ makes an angle of $2\phi = \tan^{-1}(80/50) = 58°$ with the x-axis. To reach the point on the x-axis, an arc of this angle is needed in the clockwise direction on the Mohr's circle. Thus, the principal plane is $\phi = 29°$ clockwise from the x-axis. Finally, the stresses at an angle of 15° can be obtained From Eqs. (2.13) and (2.14) using

$$\sigma_{y'} = 140 + (94.3)\cos 28° = 223.2 \text{ MPa},$$

$$\tau_{x'y'} = (94.3)\sin 28° = 44.3 \text{ MPa}.$$

Figure 2.19b shows an element of the principal normal stresses as well as the appropriate value of ϕ_σ. Figure 2.19c shows an element of the principal shear stresses as well as the appropriate value of ϕ_τ. The stress at the center of the Mohr's circle diagram is also represented in Fig. 2.19c along with the principal shear stresses.

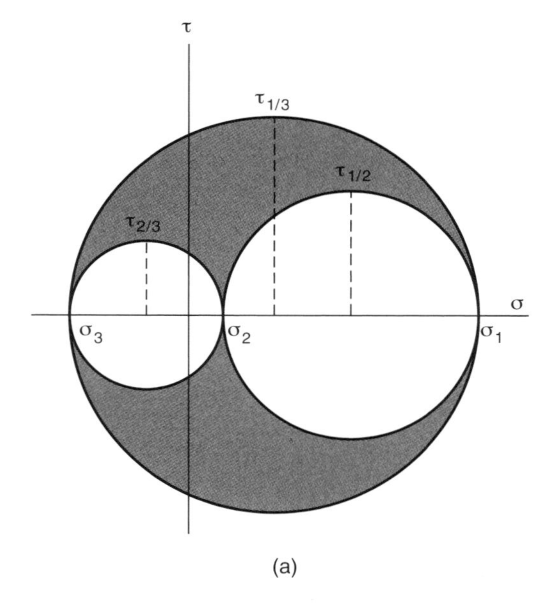

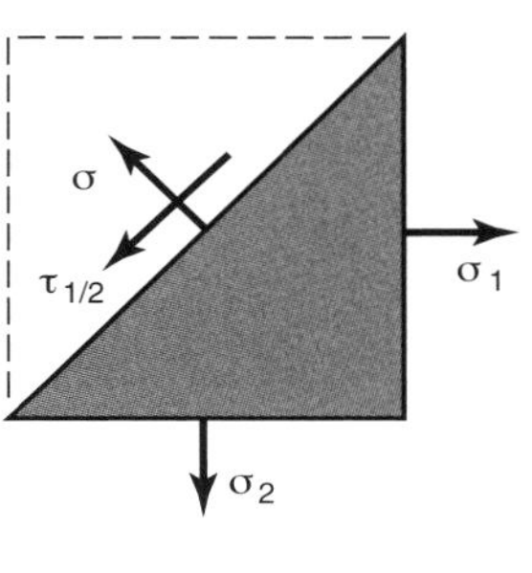

Figure 2.20: Mohr's circle for triaxial stress state. (a) Mohr's circle representation; (b) principal stresses on two planes.

2.14 Three-Dimensional Stresses

The general laws of strain transformation, given in Appendix B, enable the determination of strains acting on any orthogonal coordinate system. Considering the general situation shown in Fig. 2.14, the stress element has six faces, implying that there are three principal directions and three principal stresses σ_1, σ_2, and σ_3. Six stress components (σ_x, σ_y, σ_z, τ_{xy}, τ_{xz}, and τ_{yz}) are required to specify a general state of stress in three dimensions, in contrast to the three stress components (σ_x, σ_y, and τ_{xy}) that were used for two-dimensional (plane or biaxial) stress. Determining the principal stresses for a three-dimensional situation is much more difficult. The process involves finding the three roots to the cubic equation

$$\begin{aligned} 0 \;=\; & \sigma^3 - (\sigma_x + \sigma_y + \sigma_z)\sigma^2 + \\ & \left(\sigma_x\sigma_y + \sigma_x\sigma_z + \sigma_y\sigma_z - \tau_{xy}^2 - \tau_{yz}^2 - \tau_{zx}^2\right)\sigma \\ & - \left(\sigma_x\sigma_y\sigma_z + 2\tau_{xy}\tau_{yz}\tau_{zx} - \sigma_x\tau_{yz}^2 - \sigma_y\tau_{zx}^2 - \sigma_z\tau_{xy}^2\right). \end{aligned} \quad (2.26)$$

In most design situations many of the stress components are zero, greatly simplifying evaluation of this equation.

If the principal orientation of an element associated with a three-dimensional stress state, as well as the principal stresses, is known, this condition is called **triaxial stress**. Figure 2.20 shows a Mohr's circle for a triaxial stress state. It consists of three circles, two externally tangent and inscribed within the third circle. The principal shear stresses shown in Fig. 2.20 are determined from

$$\tau_{1/2} = \frac{\sigma_1 - \sigma_2}{2}, \quad \tau_{2/3} = \frac{\sigma_2 - \sigma_3}{2}, \quad \tau_{1/3} = \frac{\sigma_1 - \sigma_3}{2}. \quad (2.27)$$

The principal normal stresses must be ordered as described in Eq. (2.18). From Eq. (2.27), the maximum principal shear stress is $\tau_{1/3}$.

A Mohr's circle can be generated for triaxial stress states, but this is often unnecessary. In most circumstances it is not necessary to know the orientations of the principal stresses; it is sufficient to know their values. Thus, Eq. (2.26) is usually all that is needed.

Example 2.15: Three-Dimensional Mohr's Circle

Given: Assume that the principal normal stresses obtained in Example 2.14 are the same for triaxial consideration with $\sigma_3 = 0$. That is, $\sigma_1 = 234.3$ MPa, $\sigma_2 = 45.7$ MPa, and $\sigma_3 = 0$.

Find:

(*a*) Determine the principal shear stresses for a triaxial stress state and draw the appropriate Mohr's circle diagram.

(*b*) If the shear stress τ_{xy} is changed from 80 to 160 MPa, show how the Mohr's circles for the biaxial and triaxial stress states change.

Solution:

(*a*) From Eq. (2.27), the principal shear stresses in a triaxial stress state are

$$\tau_{1/2} = \frac{\sigma_1 - \sigma_2}{2} = \frac{234.3 - 45.7}{2} = 94.3 \text{ MPa},$$

$$\tau_{2/3} = \frac{\sigma_2 - \sigma_3}{2} = \frac{45.7}{2} = 22.85 \text{ MPa},$$

$$\tau_{1/3} = \frac{\sigma_1 - \sigma_3}{2} = \frac{(23.43 - 0)}{2} = 117.15 \text{ MPa}.$$

Figure 2.21a shows the appropriate Mohr's circle diagram for the triaxial stress state.

(*b*) If the shear stress in Example 2.14 is doubled (τ_{xy} = 160 MPa instead of 80 MPa), Eq. (2.19) gives

$$\begin{aligned}\tau_1, \tau_2 &= \pm\sqrt{\tau_{xy}^2 + \left(\frac{\sigma_x - \sigma_y}{2}\right)^2}\\ &= \pm\sqrt{160^2 + \left(\frac{90-190}{2}\right)^2} \text{ MPa}\\ &= \pm 167.6 \text{ MPa}.\end{aligned}$$

The principal normal stresses for the biaxial stress state are

$$\sigma_1 = \sigma_c + \tau_1 = 140 + 167.6 = 307.6 \text{ MPa},$$

$$\sigma_2 = \sigma_c - \tau_2 = 140 - 167.6 = -27.6 \text{ MPa}.$$

Figure 2.21b shows the resultant Mohr's circle diagram for the biaxial stress state. In a triaxial stress state that is ordered σ_1 = 307.6 MPa, σ_2 = 0, and σ_3 = -27.6 MPa, from Eq. (2.27) the principal shear stresses can be written as

$$\tau_{1/2} = \frac{\sigma_1 - \sigma_2}{2} = \frac{307.6 - 0}{2} \text{ MPa} = 153.8 \text{ MPa},$$

$$\tau_{2/3} = \frac{\sigma_2 - \sigma_3}{2} = \frac{0 + 27.6}{2} \text{ MPa} = 13.8 \text{ MPa},$$

$$\tau_{1/3} = \frac{\sigma_1 - \sigma_3}{2} = \frac{307.6 + 27.6)}{2} \text{ MPa} = 167.6 \text{ MPa}.$$

Figure 2.21c shows the Mohr's circle diagram for the triaxial stress state. From Fig. 2.21b and c, the maximum shear stress in the biaxial stress state, τ_1, is equivalent to $\tau_{1/3}$, the maximum shear stress in the triaxial stress state. However, comparing Figs. 2.19a and 2.21a shows that the maximum shear stress in the plane (or biaxial) stress state is not equal to that in the triaxial stress state. Furthermore, the maximum triaxial stress is larger than the maximum biaxial stress. Thus, if σ_1 and σ_2 have the same sign in the biaxial stress state, the triaxial maximum stress $\tau_{1/3}$ must be used for design considerations. However, if σ_1 and σ_2 have opposite signs in the biaxial stress state, the maximum biaxial and triaxial shear stresses will be the same and either one can be used in the analysis.

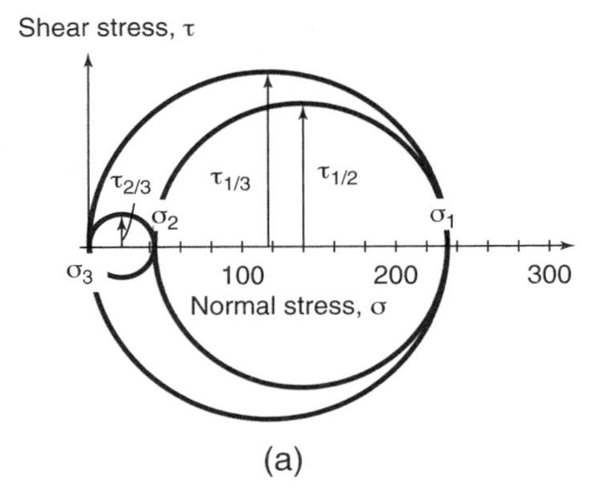

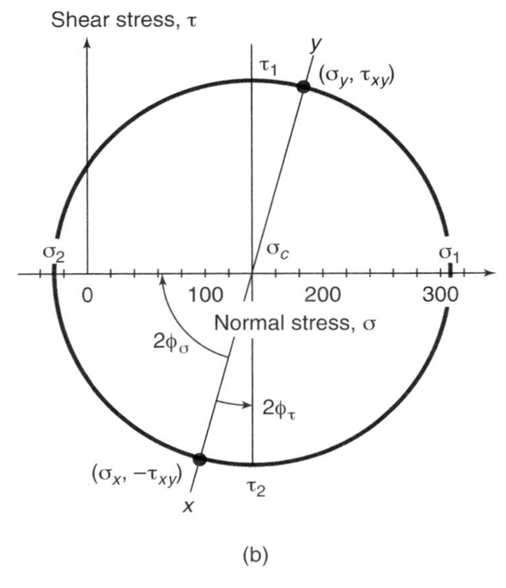

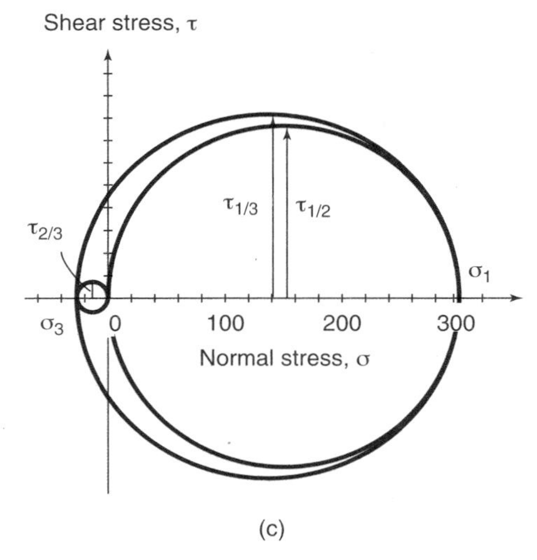

Figure 2.21: Mohr's circle diagrams for Example 2.15. (a) Triaxial stress state when σ_1 = 234.3 MPa, σ_2 = 457 MPa and σ_3 = 0; (b) biaxial stress state when σ_1 = 307.6 MPa and $\sigma_2 = -27.6$ MPa; (c) triaxial stress state when σ_1 = 307.6 MPa, σ_2 = 0, and $\sigma_3 = -27.6$ MPa.

2.15 Octahedral Stresses

Sometimes it is advantageous to represent the stresses on an octahedral stress element rather than on a conventional cubic element of principal stresses. Figure 2.22 shows the orientation of one of the eight octahedral planes that are associated with a given stress state. Each **octahedral plane** cuts across a corner of a principal element, so that the eight planes together form an octahedron (Fig. 2.22). The following characteristics of the stresses on a octahedral plane should be noted:

1. Identical normal stresses act on all eight planes. Thus, the normal stresses tend to compress or enlarge the octahedron but do not distort it.
2. Identical shear stresses act on all eight planes. Thus, the shear stresses tend to distort the octahedron without changing its volume.

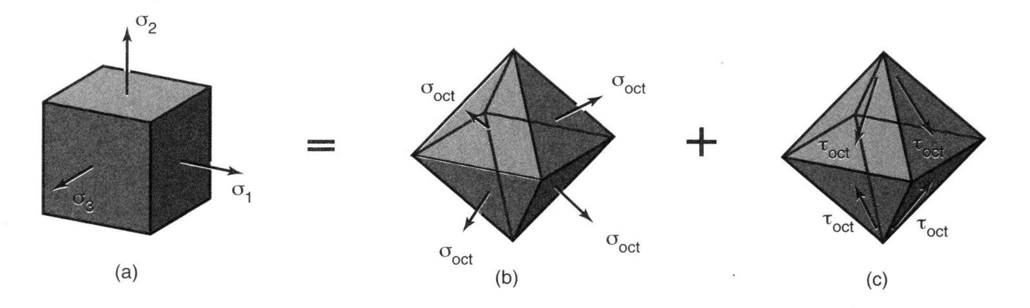

Figure 2.22: Stresses acting on octahedral planes. (a) General state of stress; (b) normal stress; (c) octahedral shear stress.

The fact that the normal and shear stresses are the same for the eight planes is a powerful tool in failure analysis.

The normal octahedral stress can be expressed in terms of the principal normal stresses, or the stresses in the x, y, z coordinates, as

$$\sigma_{\text{oct}} = \frac{\sigma_1 + \sigma_2 + \sigma_3}{3} = \frac{\sigma_x + \sigma_y + \sigma_z}{3}, \tag{2.28}$$

$$\begin{aligned} \tau_{\text{oct}} &= \frac{1}{3}\left[(\sigma_1 - \sigma_2)^2 + (\sigma_2 - \sigma_3)^2 + (\sigma_3 - \sigma_1)^2\right]^{1/2} \\ &= \frac{2}{3}\left[\tau_{1/2}^2 + \tau_{2/3}^2 + \tau_{1/3}^2\right]^{1/2}. \end{aligned} \tag{2.29}$$

In terms of octahedral normal stresses,

$$\begin{aligned} 9\tau_{\text{oct}}^2 &= (\sigma_x - \sigma_y)^2 + (\sigma_y - \sigma_z)^2 + (\sigma_z - \sigma_x)^2 \\ &\quad + 6\left(\tau_{xy}^2 + \tau_{yz}^2 + \tau_{xz}^2\right). \end{aligned} \tag{2.30}$$

Example 2.16: Octahedral Stresses

Given: Consider the stress state from Example 2.15, where $\sigma_1 = 234.5$ MPa, $\sigma_2 = 45.7$ MPa, and $\sigma_3 = 0$.

Find: Determine the octahedral stresses.

Solution: The normal and octahedral stress can be written as

$$\sigma_{\text{oct}} = \frac{\sigma_1 + \sigma_2 + \sigma_3}{3} = \frac{(234.3 + 45.7 + 0)}{3} = 93.3 \text{ MPa}.$$

The shear octahedral stress from Eq. (2.29) can be written as

$$\begin{aligned} \tau_{\text{oct}} &= \frac{2}{3}\left[\tau_{1/2}^2 + \tau_{2/3}^2 + \tau_{1/3}^2\right]^{1/2} \\ &= \frac{2}{3}\left[94.3^2 + 22.9^2 + 117.2^2\right]^{1/2} \\ &= 101.4 \text{ MPa}. \end{aligned}$$

2.16 Strain

Strain is defined as the displacement per length produced in a solid as the result of stress. In designing a machine element, not only must the design be adequate when considering the stress relative to the strength, but it must also be ensured that the displacements and/or deformations are not excessive and are within design constraints. Depending on the application, these deformations may be either highly visible or practically unnoticeable.

Just as the direction and intensity of the stress at any given point are important with respect to a specific plane passing through that point, the same is true for strain. Thus, just as for stress, strain is a tensor. Also, just as there are normal and shear stresses, so too there are normal and shear strains. **Normal strain**, designated by the symbol ϵ, is used to describe a measure of the elongation or contraction of a linear segment of an element in which stress is applied. The average normal strain is

$$\epsilon_{\text{avg}} = \frac{\delta}{l} = \frac{\text{Average elongation}}{\text{Original length}}. \tag{2.31}$$

Note that strain is dimensionless. Furthermore, the strain at a point is

$$\epsilon = \lim_{\Delta l \to 0} \frac{\Delta\delta_{\text{avg}}}{\Delta l} = \frac{d\delta_{\text{avg}}}{dl}. \tag{2.32}$$

Figure 2.23 shows the strain on a cubic element subjected to uniform tension in the x-direction. The element elongates in the x-direction while simultaneously contracting in the y- and z-directions, a phenomenon known as the **Poisson effect**, and discussed further in Section 3.5.2. From Eq. (2.32), the normal strain components can be written as

$$\epsilon_x = \lim_{x \to 0} \frac{\delta_x}{x}, \quad \epsilon_y = \lim_{y \to 0} \frac{\delta_y}{y}, \quad \epsilon_z = \lim_{z \to 0} \frac{\delta_z}{z}. \tag{2.33}$$

Figure 2.24 shows the shear strain of a cubic element due to shear stress in both a three-dimensional view and a two-dimensional (or plane) view. The **shear strain**, designated by γ, is used to measure angular distortion (the change in angle between two lines that are orthogonal in the undeformed state). The shear strain as shown in Fig. 2.24 is defined as

$$\gamma_{yx} = \lim_{y \to 0} \frac{\delta_x}{y} = \tan\theta_{yx} \approx \theta_{yx}, \tag{2.34}$$

where θ_{yx} is the angle representing deviation from initial right angle. Note that a small angle approximation has been used for θ_{yx} in Eq. (2.34).

The subscripts used to define the shear strains are like those used to define the shear stresses in Section 2.10. The first subscript designates the coordinate direction perpendicular to the plane in which the strain acts, and the second subscript designates the coordinate direction in which the strain acts. For example, γ_{yx} is the strain resulting from taking adjacent planes perpendicular to the y-axis and displacing them relative to each other in the x-direction. The sign conventions for strain follow directly from those developed for stress. A positive stress produces a positive strain and a negative stress produces a negative strain. The shear strain shown

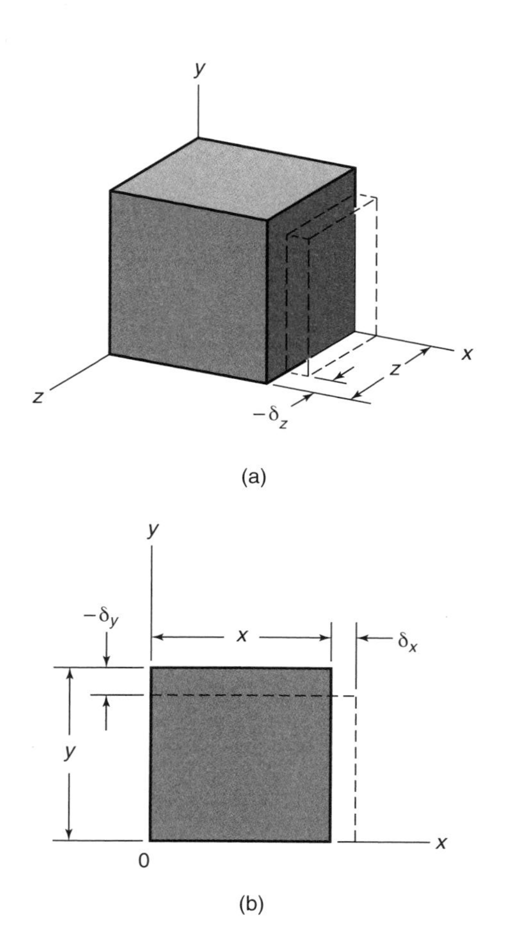

Figure 2.23: Normal strain of cubic element subjected to uniform tension in x-direction. (a) Three-dimensional view; (b) two-dimensional (or plane) view.

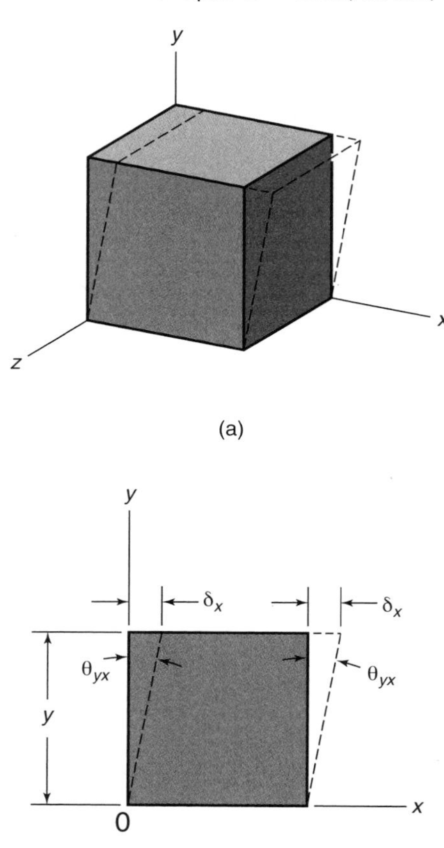

Figure 2.24: Shear strain of cubic element subjected to shear stress. (a) Three-dimensional view; (b) two-dimensional (or plane) view.

in Fig. 2.24 and described in Eq. (2.34) is positive. The strain of the cubic element thus contains three normal strains and six shear strains, just as found for stresses. Similarly, symmetry reduces the number of shear strain elements from six to three.

Example 2.17: Calculation of Strain

Given: A 300-mm-long circular aluminum bar with a 50-mm diameter is subjected to a 125-kN axial load. The axial elongation is 0.2768 mm and the diameter is decreased by 0.01522 mm.

Find: The transverse and axial strains in the bar.

Solution: The axial strain is

$$\epsilon_a = \frac{\delta}{l} = \frac{0.2768}{300} = 9.227 \times 10^{-4}$$

The transverse strain is

$$\epsilon_t = \frac{\delta_t}{d} = \frac{-0.01522}{50} = -3.044 \times 10^{-4}$$

The sign for the transverse strain is negative because the diameter decreased after the bar was loaded. The axial strain is positive because the axial length increased after loading. Note that strain has no dimension, although it is commonly reported in units of m/m or μm/m.

2.17 Strain Tensor

For strains within the elastic range, the equations relating normal and shear strains with the orientation of the cutting plane are analogous to the corresponding equations for stress given in Eq. (2.10). Thus, the state of strain can be written as a tensor:

$$\mathbf{T} = \begin{pmatrix} \epsilon_x & \frac{1}{2}\gamma_{xy} & \frac{1}{2}\gamma_{xz} \\ \frac{1}{2}\gamma_{xy} & \epsilon_y & \frac{1}{2}\gamma_{yz} \\ \frac{1}{2}\gamma_{xz} & \frac{1}{2}\gamma_{yz} & \epsilon_z \end{pmatrix} \tag{2.35}$$

In comparing Eq. (2.35) with Eq. (2.10) note that ϵ_x, ϵ_y, and ϵ_z are analogous to σ_x, σ_y, and σ_z, respectively, but it is half of the shear strain, $\gamma_{xy}/2$, $\gamma_{yz}/2$, $\gamma_{zx}/2$ that is analogous to τ_{xy}, τ_{yz}, and τ_{zx}, respectively.

2.18 Plane Strain

Instead of the six strains for the complete strain tensor, in plane strain the components ϵ_z, γ_{xz}, and γ_{yz} are zero. Thus, only two normal strain components, ϵ_x and ϵ_y, and one shear strain component, γ_{xy}, are considered. Figure 2.25 shows the deformation of an element caused by each of the three strains considered in plane strain. The normal strain components ϵ_x and ϵ_y, shown in Fig. 2.25a and b, are produced by changes in element length in the x- and y-directions, respectively. The shear strain γ_{xy}, shown in Fig. 2.25c, is produced by the relative rotation of two adjacent sides of the element. Figure 2.25c also helps to explain the physical significance that τ is analogous to $\gamma/2$ rather than to γ. Each side of an element changes in slope by an angle $\gamma/2$ when subjected to pure shear.

The following sign convention is to be used for strains:

1. Normal strains ϵ_x and ϵ_y are positive if they cause elongation along the x- and y-axes, respectively. In Fig. 2.25a and b, ϵ_x and ϵ_y are positive.

2. Shear strain γ_{xy} is positive when the interior angle of a strain element (A0B in Fig. 2.25c) becomes smaller than $90°$.

The principal strains, planes, and directions are directly analogous to those found earlier for principal stresses. The principal normal strains in the x-y plane, the maximum shear strain in the x-y plane, and the orientation of the principal axes relative to the x- and y-axes are

$$\epsilon_1, \epsilon_2 = \frac{\epsilon_x + \epsilon_y}{2} \pm \sqrt{\left(\frac{1}{2}\gamma_{xy}\right)^2 + \left(\frac{\epsilon_x - \epsilon_y}{2}\right)^2}, \tag{2.36}$$

$$\gamma_{\max} = \pm 2\sqrt{\left(\frac{1}{2}\gamma_{xy}\right)^2 + \left(\frac{\epsilon_x - \epsilon_y}{2}\right)^2}, \tag{2.37}$$

$$2\phi = \tan^{-1}\left(\frac{\gamma_{xy}}{\epsilon_x - \epsilon_y}\right). \tag{2.38}$$

From here there are two important problem classes:

1. If the principal strains are known and it is desired to find the strains acting at a plane oriented at angle ϕ from the principal direction, the equations are

$$\epsilon_\phi = \frac{\epsilon_1 + \epsilon_2}{2} + \frac{\epsilon_1 - \epsilon_2}{2}\cos 2\phi, \tag{2.39}$$

$$\gamma_\phi = (\epsilon_1 - \epsilon_2)\sin 2\phi. \tag{2.40}$$

In Eq. (2.40), γ_ϕ represents a shear strain acting on the ϕ plane and directed $90°$ from the ϕ-axis. Just as for stress, the second subscript is omitted for convenience and no ambiguity results. A Mohr's circle diagram can also be used to represent the state of strain.

2. The second problem of interest is the case where a normal strain component has been measured in three different but specified directions and it is desired to obtain the strains ϵ_x, ϵ_y, and γ_{xy} from these readings. In this case the equation

$$\epsilon_\theta = \epsilon_x \cos^2\theta + \epsilon_y \sin^2\theta + \gamma_{xy}\sin\theta\cos\theta \tag{2.41}$$

is of great assistance. Here, ϵ_θ is the measured strain in the direction rotated θ counterclockwise from the x-axis, and ϵ_x, ϵ_y, and γ_{xy} are the desired strains. Thus, measuring a strain in three different directions gives three equations for the three unknown strains and is sufficient for their quantification. Strain gages are often provided in groups of three, called *rosettes*, for such purposes.

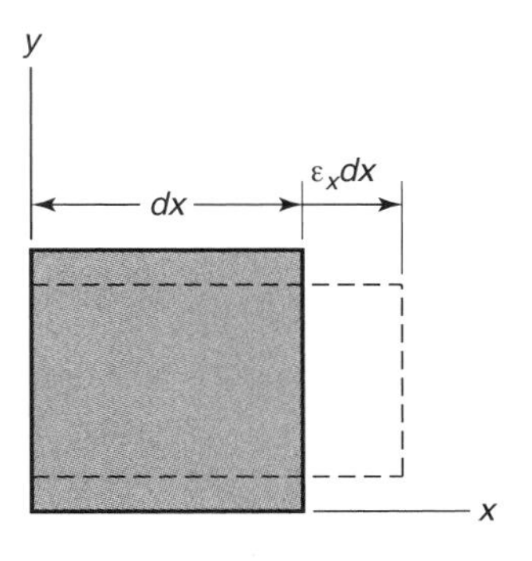

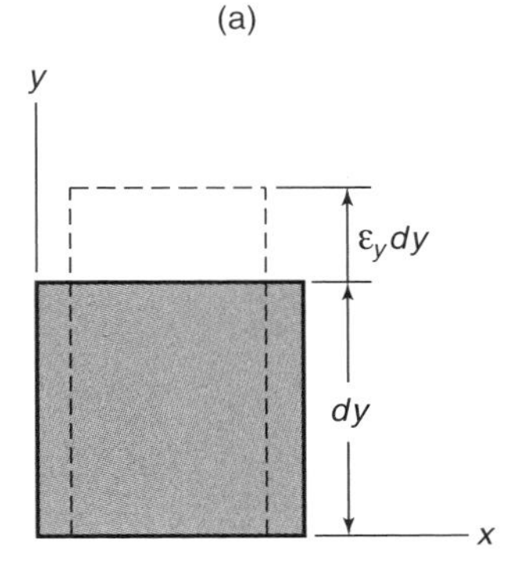

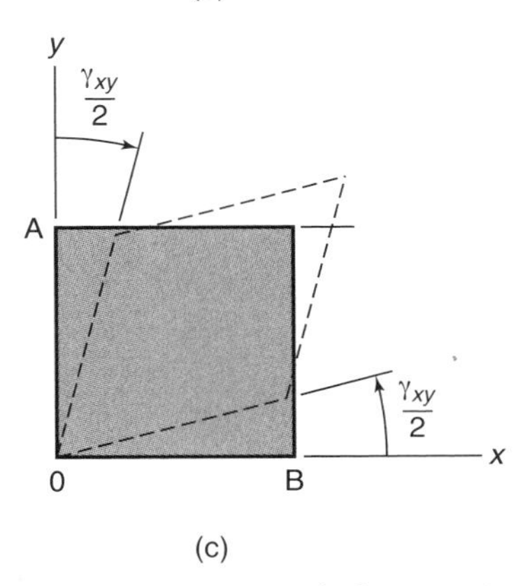

Figure 2.25: Graphical depiction of plane strain element. (a) Normal strain ϵ_x; (b) normal strain ϵ_y; and (c) shear strain γ_{xy}.

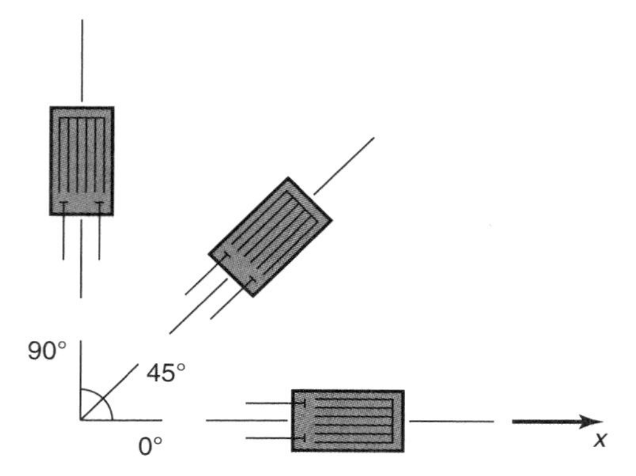

Figure 2.26: Strain gage rosette used in Example 2.18.

Example 2.18: Strain Gage Rosette

Given: A 0° − 45° − 90° strain gage rosette as shown in Fig. 2.25 is attached to a structure with the 0° gage placed along a reference (x) axis. Upon loading, the strain in the 0° direction reads +50 μm/m, the strain gage in the 45° direction reads −27 μm/m, and the gage in the 90° direction reads 0.

Find: The strains ϵ_x, ϵ_y, and γ_{xy}.

Solution: Equation (2.41) can be applied three times to obtain three equations:

$$\epsilon_{0^\circ} = 50\ \mu\text{m/m} = \epsilon_x \cos^2(0^\circ) + \epsilon_y \sin^2(0^\circ) + \gamma_{xy} \sin 0^\circ \cos 0^\circ$$

or $\epsilon_x = 50\ \mu$m/m,

$$\epsilon_{90^\circ} = 0 = \epsilon_x \cos^2(90^\circ) + \epsilon_y \sin^2(90^\circ) + \gamma_{xy} \sin 90^\circ \cos 90^\circ$$

or $\epsilon_y = 0$, and

$$\epsilon_{45^\circ} = \epsilon_x \cos^2(45^\circ) + \epsilon_y \sin^2(45^\circ) + \gamma_{xy} \sin 45^\circ \cos 45^\circ$$

or $\gamma_{xy} = -27 - 50 - 0 = -77\ \mu$m/m.

2.19 Summary

This chapter described how load, stress, and strain affect the design of machine elements. If the proper type of machine element has been selected, a potential cause of failure is the design stress exceeding the strength of the machine element. Therefore, it is important to evaluate the stress, strain, and strength of the machine element at the critical section. To do so first requires a determination of load in all its forms. The applied load on a machine element was described with respect to time, the area over which load is applied, and the location and method of application. Furthermore, the importance of support reactions, application of static force and moment equilibrium, and proper use of free-body diagrams were investigated.

The chapter then focused on shear and moment diagrams applied to a beam. Singularity functions introduced by concentrated moment, concentrated force, unit step, ramp, inverse ramp, and parabolic shape were considered. Various combinations of these singularity functions can exist within a beam. Integrating the load intensity function for the various beam singularity functions over the beam length establishes the shear force. Integrating the shear force over the beam length determines the moment. From these analytical expressions the shear and moment diagrams can be readily constructed.

Stress defines the intensity and direction of the internal forces at a particular point and acting on a given plane. The stresses acting on an element have normal and shear components. Across each mutually perpendicular surface there are two shear stresses and one normal stress, yielding a total of nine stresses (three normal stresses and six shear stresses). The sign conventions for both the normal and shear stresses were presented. The nine stress components may be regarded as the components of a second-order Cartesian tensor. It was found that the stress tensor is symmetrical, implying that the tensor can be written with zero shear stress and the principal normal stresses along its diagonal.

In many engineering applications, stress analysis assumes that a surface is free of stress or that the stress in one plane is small relative to the stresses in the other two planes. The two-dimensional stress situation is called the biaxial (or plane) stress state and can be expressed in terms of two normal stresses and one shear stress, for example σ_x, σ_y, and τ_{xy}. That the stresses can be expressed in any oblique plane is important in deriving and applying Mohr's circle for a biaxial stress.

The concepts of strain and deflection were also investigated, since these are often design constraints. Just as with stress, strain is a tensor, and transformation equations and Mohr's circle are equally applicable to strain analysis. The concept of strain gage rosettes was introduced as a method to obtain plane strains.

Key Words

beam structural member designed to support loads perpendicular to its longitudinal axis

bending load load applied transversely to longitudinal axis of member

biaxial or plane stress condition where one surface is comparatively free of stress

cantilevered beam support where one end is fixed and the other end is free

combined load combination of two or more previously defined loads

concentrated load load applied to small nonconformal area

critical section section where largest internal stress occurs

cyclic load load varying throughout a cycle

distributed load load distributed over entire area

free-body diagram sketch of part showing all acting forces

impact load load rapidly applied

loads force, moment, or torque applied to a mechanism or structure

Mohr's circle method used to graphically visualize state of stress acting in different planes passing through a given point

normal load load passing through centroid of resisting section

normal strain elongation or contraction of linear segment of element in which stress is applied

overhanging beam support where one or both ends freely extend past support

principal normal stresses combination of applied normal and shear stresses that produces maximum principal normal stress or minimum principal normal stress, with a third principal stress between or equivalent to the extremes

principal shear stresses combination of applied normal and shear stresses that produces maximum principal shear stress or minimum principal shear stress

shear load load collinear with transverse shear force

shear strain measure of angular distortion in which shear stress is applied

sign convention for normal strain positive if elongation is in direction of positive axes

sign conversion for normal stress positive for tension and negative for compression

sign convention for shear strain positive if interior angle becomes smaller after shear stress is applied

sign convention for shear stress positive if both normal from surface and shear are in positive or negative direction; negative for any other combination

simply supported beam support where one end is pinned and the other is roller-supported

singularity functions functions used to evaluate shear and moment diagrams, especially when discontinuities, such as concentrated load or moment, exist

static load load gradually applied and equilibrium reached in a short time

strain dimensionless displacement produced in solid as a result of stress

stress intensity and direction of internal force acting at a given point on a particular plane

sustained load a load that is constant over a long time

symmetrical tensor condition where principal normal stresses exist while all other tensor elements are zero

torsion load a load that results in twisting deformation

triaxial stress stress where all surfaces are considered

uniaxial stress condition where two perpendicular surfaces are comparatively free of stress

Summary of Equations

Force equilibrium: $\sum P_x = 0, \sum P_y = 0, P_z = 0$

Moment equilibrium: $\sum M_x = 0, \sum M_y = 0, \sum M_z = 0$

Transverse shear in beams: $V(x) = -\int_{-\infty}^{x} q(x)dx$

Bending moment in beams: $M(x) = -\int_{-\infty}^{x} V(x)dx$

Principal stresses in plane stress:

$$\sigma_1, \sigma_2 = \frac{\sigma_x + \sigma_y}{2} \pm \sqrt{\tau_{xy}^2 + \frac{(\sigma_x - \sigma_y)^2}{4}}$$

Mohr's circle

Center: $\left(\frac{\sigma_x + \sigma_y}{2}, 0\right)$

Radius: $r = \sqrt{\left(\frac{\sigma_x - \sigma_y}{2}\right)^2 + \tau_{xy}^2}$

Octahedral Stresses:

Normal: $\sigma_{\text{oct}} = \frac{\sigma_1 + \sigma_2 + \sigma_3}{3} = \frac{\sigma_x + \sigma_y + \sigma_z}{3}$

Shear:

$$\begin{aligned}\tau_{\text{oct}} &= \frac{1}{3}\left[(\sigma_1 - \sigma_2)^2 + (\sigma_2 - \sigma_3)^2 + (\sigma_3 - \sigma_1)^2\right]^{1/2} \\ &= \frac{2}{3}\left[\tau_{1/2}^2 + \tau_{2/3}^2 + \tau_{1/3}^2\right]^{1/2}\end{aligned}$$

Principal strains in plane strain:

$$\epsilon_1, \epsilon_2 = \frac{\epsilon_x + \epsilon_y}{2} \pm \sqrt{\left(\frac{1}{2}\gamma_{xy}\right)^2 + \left(\frac{\epsilon_x - \epsilon_y}{2}\right)^2}$$

Recommended Readings

Beer, F.P., Johnson, E.R., DeWolf, J., and Mazurek, D. (2011) *Mechanics of Materials*, 6th ed., McGraw-Hill.
Craig, R.R. (2011) *Mechanics of Materials*, 3rd ed., Wiley.
Hibbeler, R.C. (2010) *Mechanics of Materials*, 8th ed. Prentice-Hall, Upper Saddle River.
Popov, E.P. (1968) *Introduction to Mechanics of Solids*, Prentice-Hall.
Popov, E.P. (1999) *Engineering Mechanics of Solids*, 2nd ed., Prentice-Hall.
Riley, W.F., Sturges, L.D., and Morris, D.H. (2006) *Mechanics of Materials*, 6th ed., Wiley.
Shames, I.H., and Pitarresi, J.M. (2000) *Introduction to Solid Mechanics*, 3rd ed., Prentice-Hall.
Ugural, A.C. (2007) *Mechanics of Materials*, Wiley.

Questions

2.1 What is a concentrated load? What is a distributed load?

2.2 What kind of reaction occurs with a roller support? What occurs with a pin?

2.3 Define *static equilibrium*.

2.4 What is a simply supported beam? What is a cantilever?

2.5 Why are singularity functions useful?

2.6 Under what conditions does a singularity function *not* equal zero?

2.7 Define the terms *stress* and *strain*.

2.8 What is a tensor?

2.9 Define *normal stress* and *shear stress*.

2.10 What is Mohr's circle?

2.11 What is a principal stress?

2.12 What are the units for stress? What are the units for strain?

2.13 What are octahedral stresses?

2.14 What is elongation?

2.15 What is a rosette?

Qualitative Problems

2.16 Give three examples of (a) static loads; (b) sustained loads; (c) impact loads; and (d) cyclic loads.

2.17 Explain the sign convention for shear forces.

2.18 Explain the common sign conventions for bending moments. Which is used in this book?

2.19 Without the use of equations, explain a methodology for producing shear and moment diagrams.

2.20 Give two examples of scalars, vectors, and tensors.

2.21 Explain the difference between plane stress and plain strain. Give an example of each.

2.22 Without the use of equations, qualitatively determine the bending moment diagram for a bookshelf.

2.23 Explain why $\tau_{xy} = \tau_{yx}$.

2.24 Define and give two examples of (a) uniaxial stress state; (b) biaxial stress state; and (c) triaxial stress state.

2.25 Sketch and describe the characteristics of a three-dimensional Mohr's circle.

2.26 What are the similarities and differences between deformation and strain?

2.27 The text stated that 0°–45°–90° strain gage rosettes are common. Explain why.

2.28 Draw a free body diagram of a book on a table.

2.29 If the three principal stresses are determined to be 100 MPa, −50 MPa and 75 MPa, which is σ_2?

2.30 Derive Eq. (2.16).

Quantitative Problems

2.31 The stepped shaft A-B-C shown in Sketch *a* is loaded with the forces P_1 and/or P_2. Note that P_1 gives a tensile stress σ in B-C and $\sigma/4$ in A-B and that P_2 gives a bending stress σ at B and 1.5σ at A. What is the critical section

(a) If only P_1 is applied?

(b) If only P_2 is applied?

(c) If both P_1 and P_2 are applied?

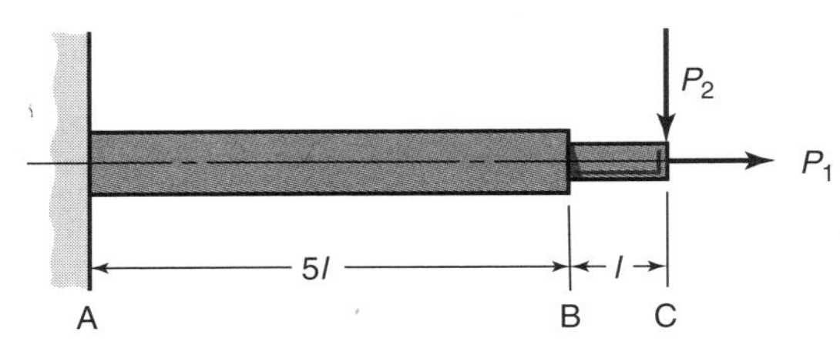

Sketch *a*, used in Problems 2.31 and 2.32.

2.32 The stepped shaft in Sketch *a* has loads P_1 and P_2. Find the load classification if P_1's variation is sinusoidal and P_2 is the load from a weight

(a) If only P_1 is applied

(b) If only P_2 is applied

(c) If both P_1 and P_2 are applied

2.33 A bar hangs freely from a frictionless hinge. A horizontal force P is applied at the bottom of the bar until it inclines 45° from the vertical direction. Calculate the horizontal and vertical components of the force on the hinge if the acceleration due to gravity is g, the bar has a constant cross section along its length, and the total mass is m_a. *Ans.* $R_x = \frac{1}{2}m_a g$, $R_y = m_a g$.

2.34 Sketch *b* shows the forces acting on a rectangle. Is the rectangle in equilibrium? *Ans.* No.

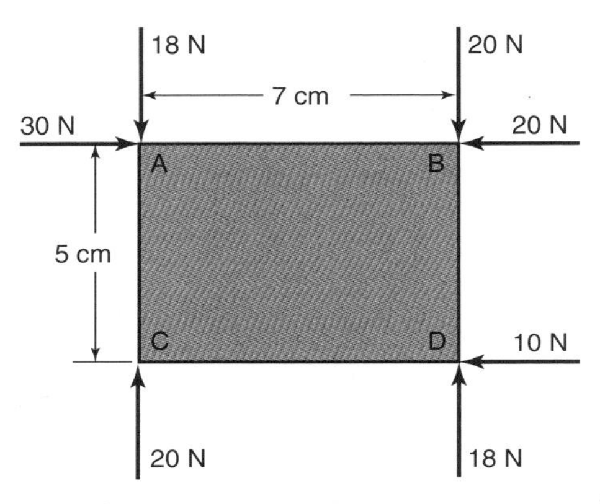

Sketch *b*, used in Problem 2.34

2.35 Sketch *c* shows the forces acting on a triangle. Is the triangle in equilibrium? *Ans.* Yes.

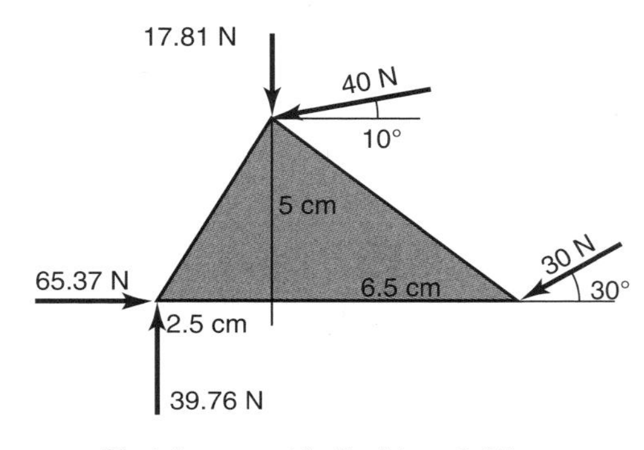

Sketch *c*, used in Problem 2.35

2.36 Given the components shown in Sketches d and e, draw the free-body diagram of each component and calculate the forces.

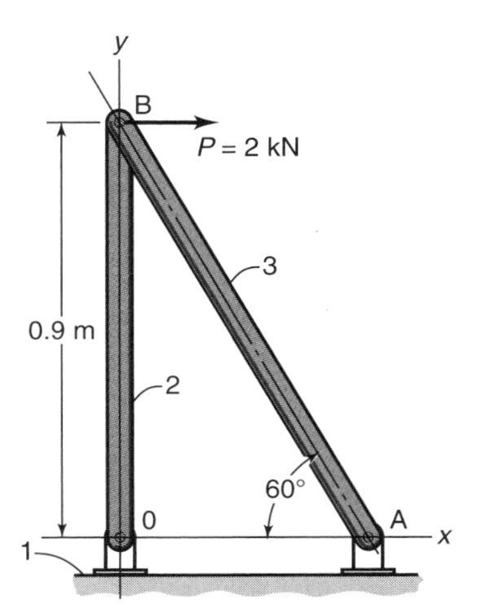

Sketch d, used in Problem 2.36

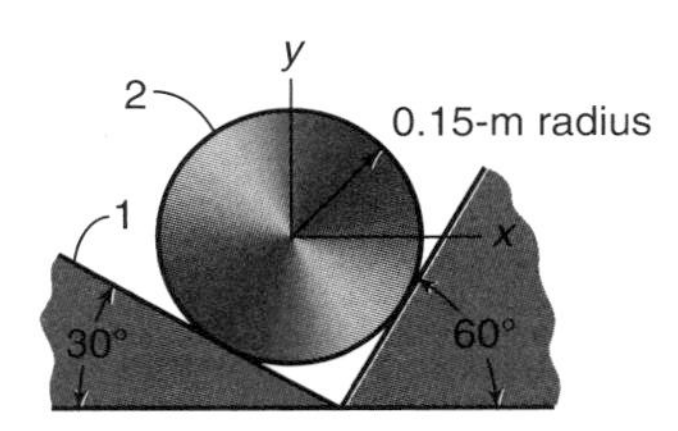

Sketch e, used in Problem 2.36

2.37 Sketch f shows a cube with side lengths a and eight forces acting at the corners. Is the cube in equilibrium? *Ans.* Yes.

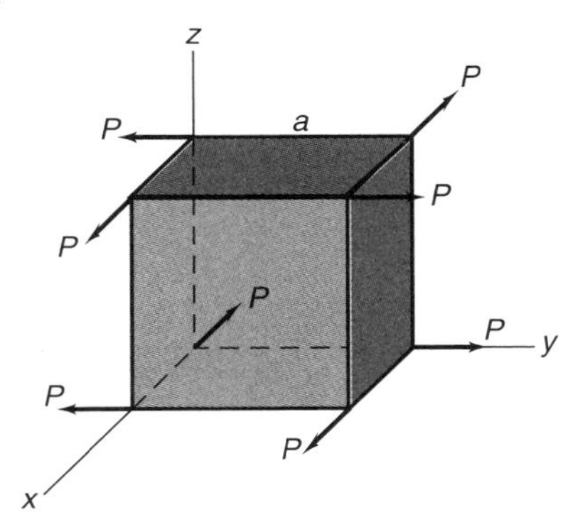

Sketch f, used in Problem 2.37

2.38 A 5-m-long beam is loaded as shown in Sketch g. The beam cross section is constant along its length. Draw the shear and moment diagrams and locate the critical section. *Ans.* $|V_{\max}| = 6.8$ kN, $|M_{\max}| = 6.4$ kN-m.

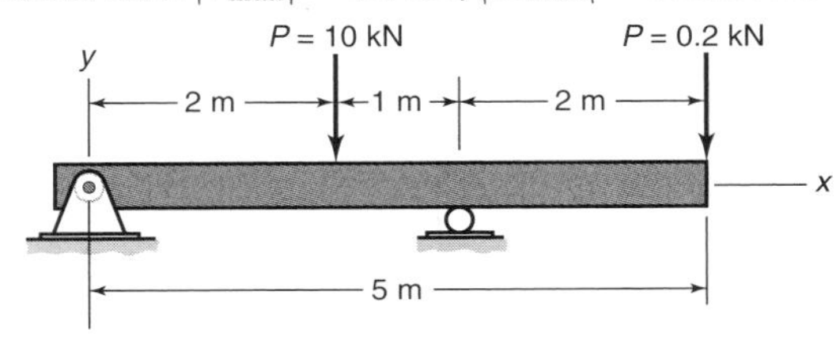

Sketch g, used in Problem 2.38

2.39 Sketch h shows a 0.06-m-diameter steel shaft supported by self-aligning bearings at A and B (which can provide radial but not bending loads on the shaft). Two gears attached to the shaft cause applied forces as shown. The shaft weight can be neglected. Determine the forces at A and B and the maximum bending moment. Draw shear and moment diagrams. *Ans.* $|M_{\max}| = 296$ Nm.

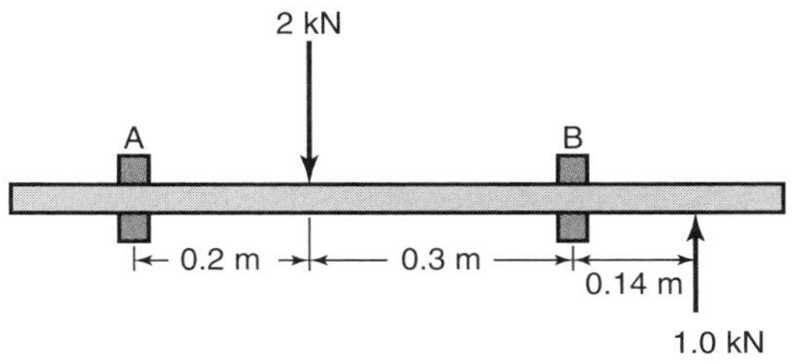

Sketch h, used in Problem 2.39

2.40 A beam is loaded as shown in Sketch i. Determine the reactions and draw the shear and moment diagrams for $P = 500$ lb. *Ans.* $A_y = B_y = 1000$ lb, $|M_{\max}| = 2000$ ft-lb.

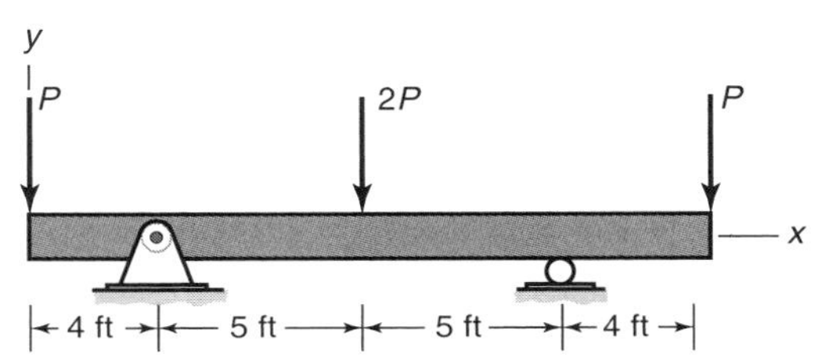

Sketch i, used in Problem 2.40

2.41 Sketch j shows a simply supported beam loaded with a force P at a position one-third of the length from one of the supports. Determine the largest shear force and bending moment in the beam. Also, draw the shear and moment diagrams. *Ans.* $|M_{\max}| = \frac{2}{9}Pl$.

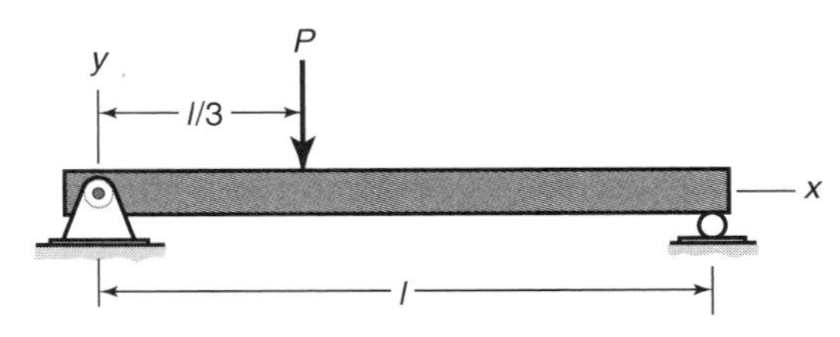

Sketch j, used in Problem 2.41

2.42 Sketch k shows a simply supported beam loaded by two equally large forces P at a distance $l/4$ from its ends. Determine the largest shear force and bending moment in the beam, and find the critical location with respect to bending. Also, draw the shear and moment diagrams. *Ans.* $|M_{\max}| = \frac{1}{4}Pl$.

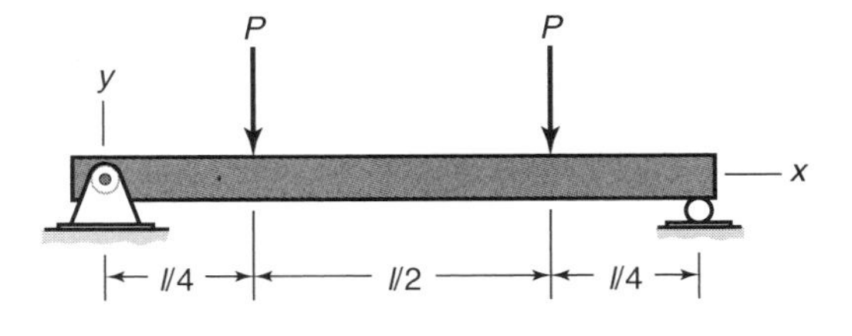

Sketch k, used in Problem 2.42

2.43 The beam shown in Sketch l is loaded by the force P. Draw the shear and moment diagrams for the beam, indicating maximum values. *Ans.* $|M|_{\max} = Pl$.

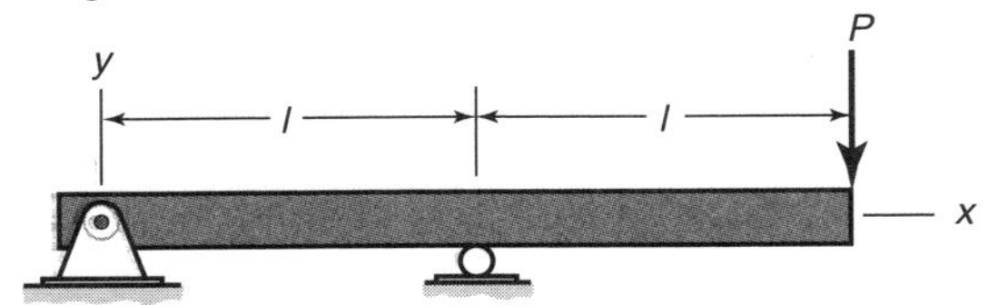

Sketch l, used in Problem 2.43

2.44 Sketch m shows a simply supported beam with a constant load per unit length, w_o, imposed over its entire length. Determine the shear force and bending moment as functions of x. Draw a graph of these functions. Also, find the critical section with the largest bending moment. *Ans.* $M_{\max} = \frac{1}{8}w_o l^2$.

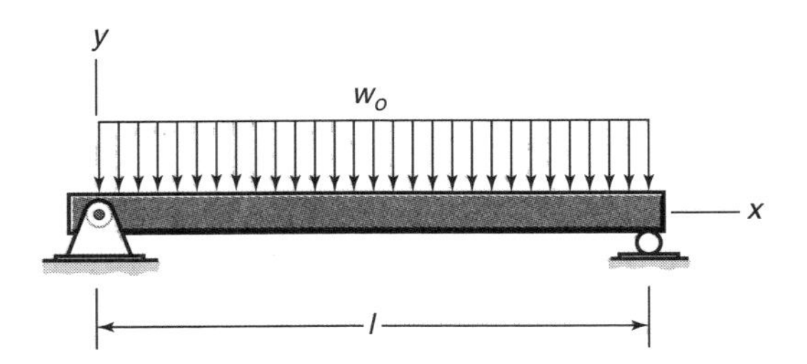

Sketch m, used in Problem 2.44

2.45 Sketch n shows a simply supported beam loaded with a ramp function over its entire length, the largest value being P/l. Determine the shear force and the bending moment and the critical section with the largest bending moment. Also, draw the shear and moment diagrams. *Ans.* $|M_{\max}| = \frac{2}{9\sqrt{3}}Pl$.

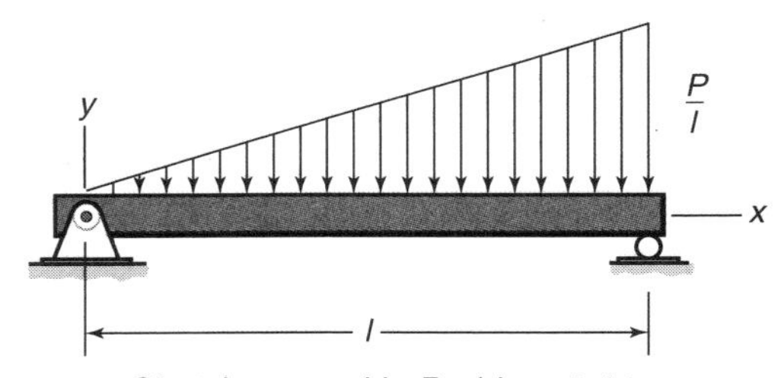

Sketch n, used in Problem 2.44

2.46 The simply supported beam shown in Sketch o has P_1 = 5 kN, P_2 = 10 kN, w_o = 5 kN/m, and l = 12 m. Use singularity functions to determine the shear force and bending moment as functions of x. Also, draw the shear force and bending moment diagrams. *Ans.* $|M_{\max}| =$ 52.5 kN-m.

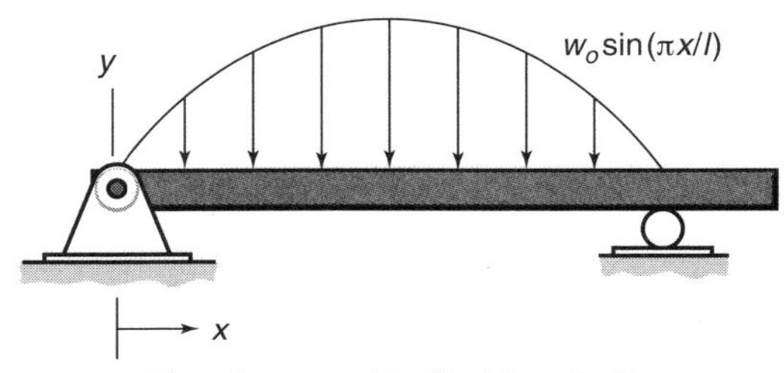

Sketch o, used in Problem 2.46

2.47 Sketch p shows a sinusoidal distributed force applied to a beam. Determine the reactions and largest shear force and bending moment for each section of the beam. *Ans.* Reactions $= \frac{lw_o}{\pi}$, $|M_{\max}| = \frac{l^2 w_o}{2\pi}$.

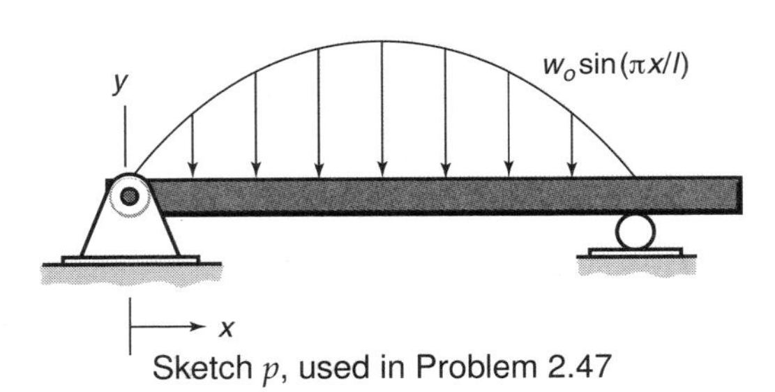

Sketch p, used in Problem 2.47

2.48 Find the length c that gives the smallest maximum bending moment for the load distribution shown in Sketch q. *Ans.* $x = 0.207l$.

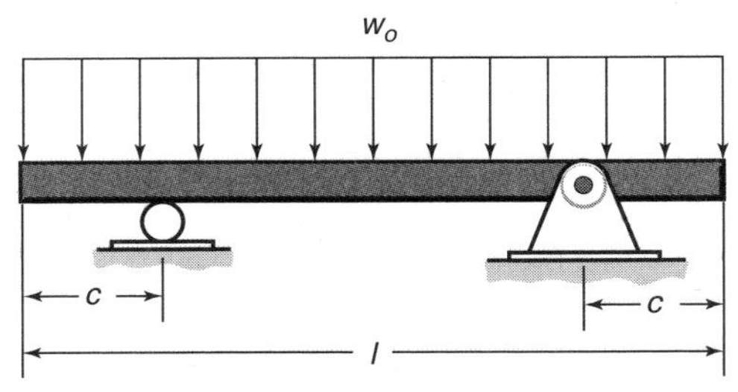

Sketch q, used in Problem 2.48

2.49 Draw the shear and moment diagrams and give the reaction forces for the load distribution shown in Sketch r. *Ans.* $R = w_o a$, $M_{\max} = \frac{1}{2}w_o a^2$.

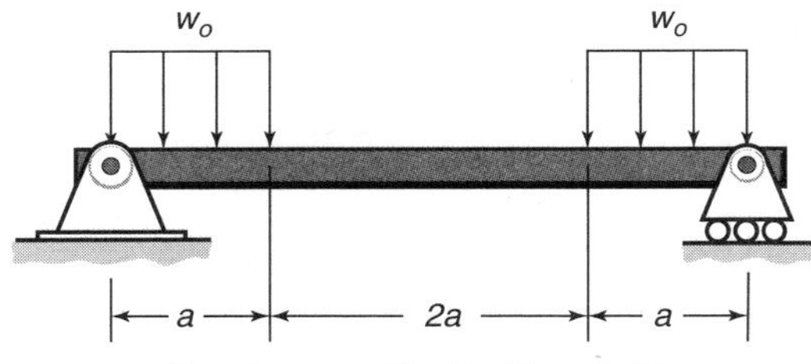

Sketch r, used in Problem 2.49

2.50 Use singularity functions for the force system shown in Sketch s to determine the load intensity, the shear force, and the bending moment in the beam. From a force analysis determine the reaction forces R_1 and R_2. Also, draw the shear and moment diagrams. *Ans.* $R_1 = -14.29$ lb, $R_2 = 54.29$ lb, $|M|_{\max} = 274.3$ in-lb.

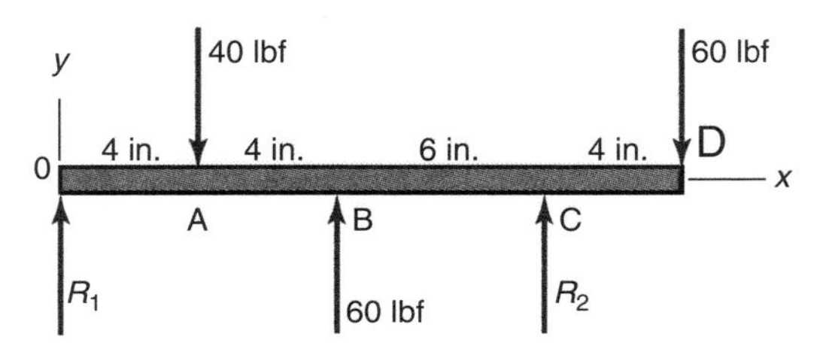

Sketch s, used in Problem 2.50

2.51 Use singularity functions for the force system shown in Sketch t to determine the load intensity, the shear force, and the bending moment. Draw the shear and moment diagrams. Also, from a force analysis determine the reaction forces R_1 and R_2. *Ans.* $R_1 = 92.22$ lb, $R_2 = 47.78$ lb, $M_{\max} = 485.6$ in-lb.

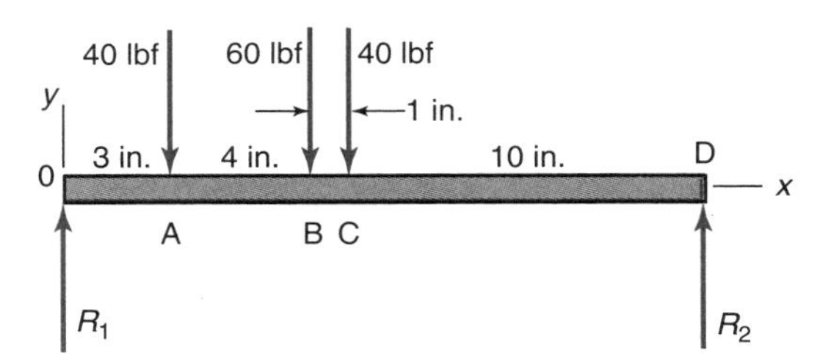

Sketch t, used in Problem 2.51

2.52 Draw a free-body diagram of the forces acting on the simply supported beam shown in Sketch u, with $w_o = 6$ kN/m and $l = 10$ m. Use singularity functions to draw the shear force and bending moment diagrams. *Ans.* $M_{\max} = 63.89$ kN-m.

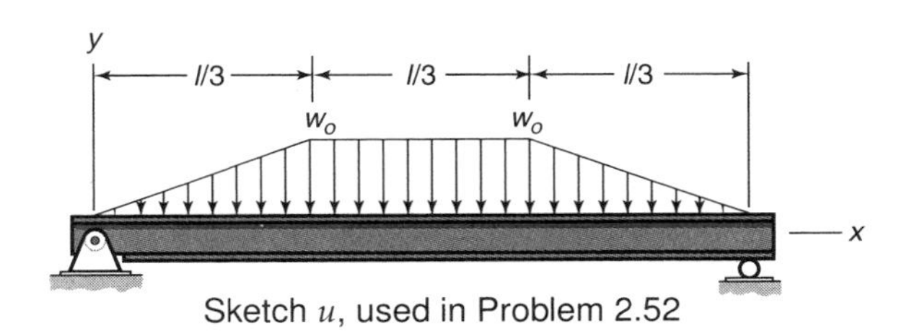

Sketch u, used in Problem 2.52

2.53 Repeat Example 2.8 using singularity functions.

2.54 Sketch v shows a simply supported beam with $w_o = 6$ kN/m and $l = 10$ m. Draw a free-body diagram of the forces acting along the beam as well as the coordinates used. Use singularity functions to determine the shear force and the bending moment. *Ans.* $M_{\max} = 25$ kN-m.

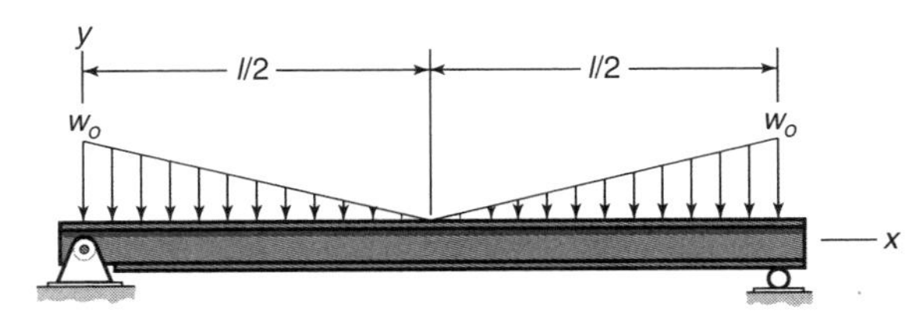

Sketch v, used in Problem 2.54

2.55 An additional concentrated force with an intensity of 20 kN is applied downward at the center of the simply supported beam shown in Sketch v. Draw a free-body diagram of the forces acting on the beam. Assume $l = 10$ m and $w_o = 5$ kN/m. Use singularity functions to determine the shear force and bending moments and draw the diagrams. *Ans.* $M_{\max} = 75$ kN-m.

2.56 Draw a free-body diagram of the beam shown in Sketch w and use singularity functions to determine the shear force and the bending moment diagrams. Determine the maximum moment. *Ans.* $|M_{|\max}| = 16.19$ kN-m.

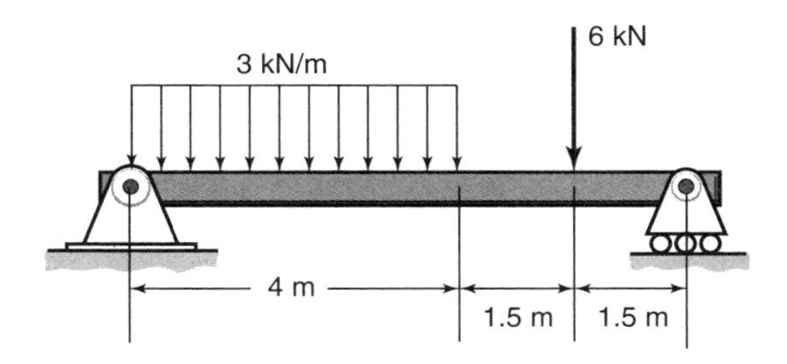

Sketch w, used in Problems 2.56 and 2.57

2.57 Use direct integration to determine the shear force and bending moment diagrams for the beam shown in Sketch w. Determine the maximum moment. *Ans.* $|M_{|\max}| = 16.19$ kN-m.

2.58 Draw the shear and bending moment diagrams for the beam shown in Sketch x. Determine the magnitude and location of the maximum moment. *Ans.* $|M_{\max}| = 2.5$ kN-m.

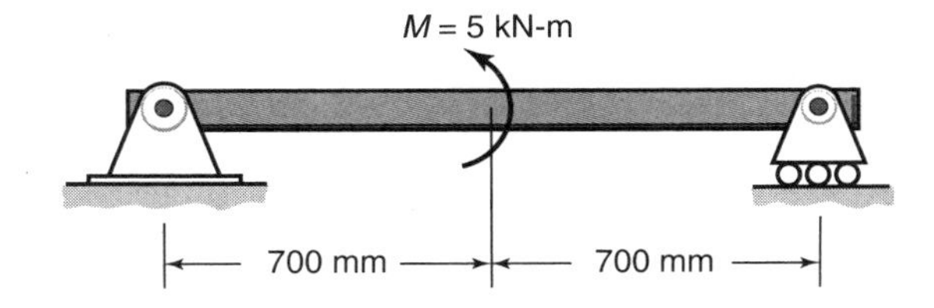

Sketch x, used in Problem 2.58

2.59 Using singularity functions, draw the shear and moment diagrams for the beam shown in Sketch y. Use $P = 20$ kN and $l = 4$ m. *Ans.* $|M_{\max}| = 40$ kN-m.

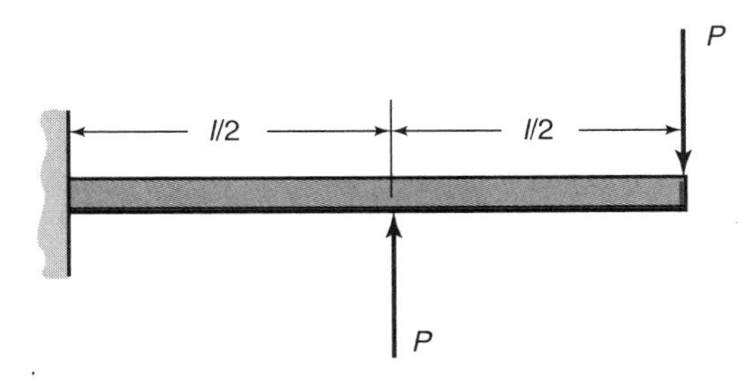

Sketch y, used in Problem 2.59

2.60 Determine the location and magnitude of maximum shear stress and bending moment for the beam shown in Sketch z. Use $w_o = 10$ kN/m and $l = 5$ m. *Ans.* $|V_{\max}| = 25$ kN, $|M_{\max}| = 95.75$ kN-m.

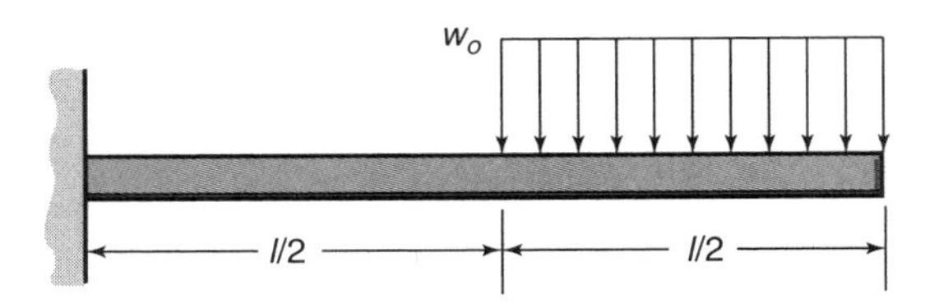

Sketch z, used in Problem 2.60

2.61 Sketch the shear and bending moment diagrams for the beam shown in Sketch *aa*. Determine the maximum shear force and bending moment. *Ans.* $|V_{\max}| = 18$ kN, $|M_{\max}| = 55$ kN-m.

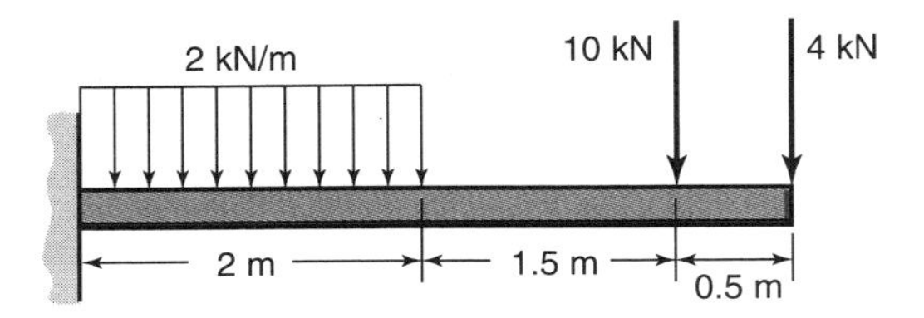

Sketch *aa*, used in Problem 2.61

2.62 A steel bar is loaded by a tensile force P = 20 kN. The cross section of the bar is circular with a radius of 10 mm. What is the normal tensile stress in the bar? *Ans.* $\sigma = 63.66$ MPa.

2.63 A stainless-steel bar of square cross section is subjected to a tensile force of $P = 10$ kN. Calculate the required cross section to provide a tensile stress in the bar of 90 MPa. *Ans.* $l = 10.54$ mm.

2.64 What is the maximum length, $l_{\max}$, of a copper wire if its weight should not produce a stress higher than 70 MPa when it is hanging vertically? The density of copper is 8900 kg/m^3, and the density of air is so small relative to that of copper that it may be neglected. The acceleration of gravity is 9.81m/s^2. *Ans.* $l_{\max} = 801$ m.

2.65 A machine with a mass of 5000 kg will be lifted by a steel rod with an ultimate tensile strength of 860 MPa. A safety factor of 4 is to be used. Determine the diameter needed for the steel rod. *Ans.* $d = 17.04$ mm.

2.66 A string on a guitar is made of nylon and has a diameter of 0.5 mm. It is tightened with a force $P = 12$ N. What is the stress in the string? *Ans.* $\sigma = 61.12$ MPa.

2.67 Determine the normal and shear stresses due to axial and shear forces at sections A and B in Sketch *bb*. The cross sectional area of the rod is 0.025 m^2 and $\theta = 30°$. *Ans.* At section AA, $\sigma = 200$ kPa, $\tau = 346.4$ kPa. At section BB, τ =400 kPa.

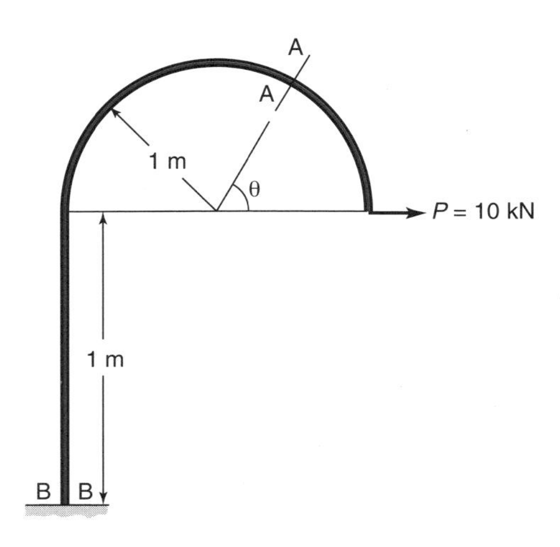

Sketch *bb*, used in Problem 2.67

2.68 Determine the normal and shear stresses in sections A and B of Sketch *cc*. The cross-sectional area of the rod is 0.00250 m^2. Ignore bending and torsional effects. *Ans.* In AA, $\sigma = -3.464$ MPa, $\tau = -2.00$ MPa.

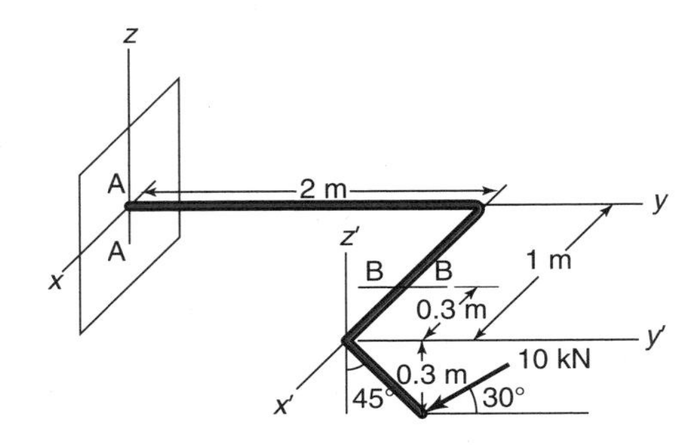

Sketch *cc*, used in Problem 2.68

2.69 Sketch *dd* shows a distributed load on a semi-infinite plane. The stress in polar coordinates based on plane stress is

$$\sigma_r = -\frac{2w_o \cos\theta}{\pi r}$$

$$\sigma_\theta = \tau_{r\theta} = \tau_{\theta r} = 0$$

Determine the expressions σ_x, σ_y, and τ_{xy} in terms of r and θ. *Ans.* σ_x=-$\dfrac{2w_o \cos^3\theta}{\pi r}$.

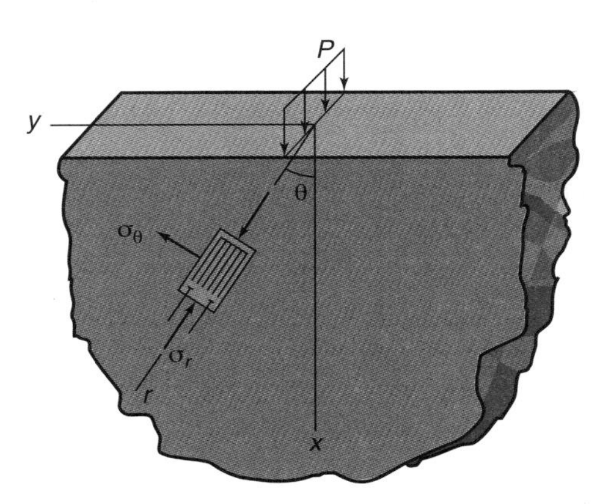

Sketch *dd*, used in Problem 2.69

2.70 Sketch *ee* shows loading of a thin but infinitely wide and long plane. Determine the angle θ needed so that the stress element will have no shear stress. *Ans.* $\theta = 0$.

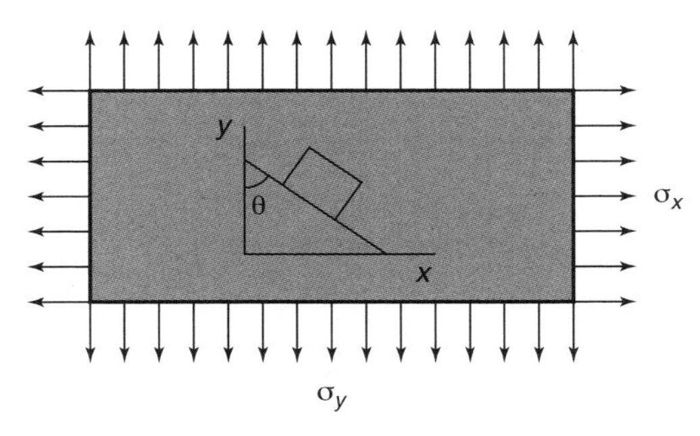

Sketch *ee*, used in Problem 2.70

2.71 A stress tensor is given by

$$\mathbf{S} = \begin{pmatrix} 200 & 40 & 0 \\ 40 & 25 & 0 \\ 0 & 0 & 0 \end{pmatrix}$$

where all values are in megapascals. Calculate the principal normal stresses and the principal shear stresses. *Ans.* $\sigma_1 = 208.7$ MPa, $\sigma_2 = 16.29$ MPa, $\sigma_3 = 0$.

2.72 A thin, square steel plate is oriented with respect to the x- and y-directions. A tensile stress σ acts in the x-direction, and a compressive stress $-\sigma$ acts in the y-direction. Determine the normal and shear stresses on the diagonal of the square. *Ans.* $\sigma_{45°}$=0, $\tau_{45°} = -\sigma$.

2.73 A thin, rectangular brass plate has normals to the sides in the x- and y-directions. A tensile stress σ acts on the four sides. Determine the principal normal and shear stresses. *Ans.* $\sigma_1 = \sigma_2 = \sigma$, τ=0.

2.74 Given the thin, rectangular brass plate in Problem 2.73, but with the stress in the y-direction being $\sigma_y = -\sigma$ instead of $+\sigma$, determine the principal normal and shear stresses and their directions. *Ans.* $\sigma_1 = -\sigma_2 = \sigma$, $\tau = \pm\sigma$.

2.75 For the following stress states, sketch the stress element, draw the appropriate Mohr's circle, determine the principal stresses and their directions, and sketch the principal stress elements:

(a) $\sigma_x = 8$, $\sigma_y = 14$, and $\tau_{xy} = 4$. *Ans.* $\sigma_1 = 16$ MPa, $\sigma_2 = 6$ MPa.

(b) $\sigma_x = -15$, $\sigma_y = 9$, and $\tau_{xy} = 5$. *Ans.* $\sigma_1 = 10$ MPa, $\sigma_2 = -16$ MPa.

(c) $\sigma_x = 12$, $\sigma_y = 28$, and $\tau_{xy} = 15$. *Ans.* $\sigma_1 = 35$ MPa, $\sigma_2 = 5$ MPa.

(d) $\sigma_x = -54$, $\sigma_y = 154$, and $\tau_{xy} = -153$. *Ans.* $\sigma_1 = 235$ MPa, $\sigma_2 = -135$ MPa.

All stresses are in megapascals.

2.76 Repeat Problem 2.75 for

(a) $\sigma_x = \sigma_y = -10$, and $\tau_{xy} = 0$. *Ans.* $\tau_{xy} = 0$ MPa.

(b) $\sigma_x = 0$, $\sigma_y = 30$, and $\tau_{xy} = 20$. *Ans.* $\sigma_1 = 40$ MPa, $\sigma_2 = -10$ MPa.

(c) $\sigma_x = -20$, $\sigma_y = 40$, and $\tau_{xy} = -40$. *Ans.* $\sigma_1 = 60$ MPa, $\sigma_2 = -40$ MPa.

(d) $\sigma_x = 30$, $\sigma_y = 0$, and $\tau_{xy} = -20$. *Ans.* $\sigma_1 = 40$ MPa, $\sigma_2 = -10$ MPa.

All stresses are in megapascals.

2.77 Given the state of stresses shown in the two parts of Sketch *ff* determine the principal stresses and their directions by using Mohr's circle and the stress equations. Show the stress elements. All stresses in Sketch *ff* are in megapascals. *Ans.* (a) $\sigma_1 = 34$ MPa, $\sigma_2 = -38$ MPa.

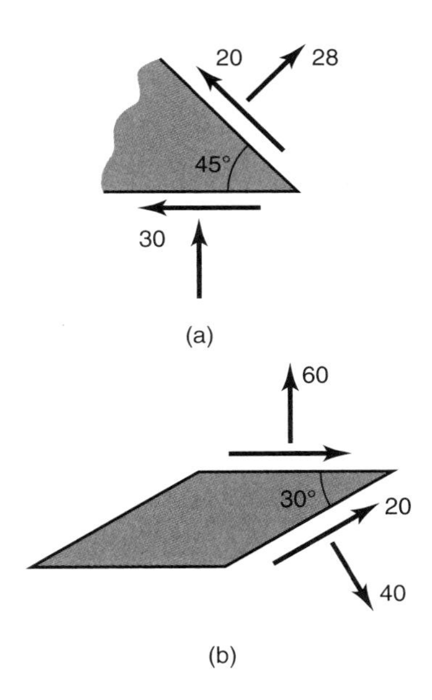

Sketch *ff*, used in Problem 2.77

2.78 A certain loading on a machine element leads to a stress state of $\sigma_x = a$, $\sigma_y = a/2$, and $\tau_{xy} = a/4$. What value of a results in the maximum allowable shear stress of 100 MPa? *Ans.* $a = 282.8$ MPa.

2.79 Given the normal and shear stresses $\sigma_x = 4$ ksi, $\sigma_y = 6$ ksi, and $\tau_{xy} = -5$ ksi, draw the Mohr's circle diagram and the principal normal and shear stresses on the x-y axis. Determine the triaxial stresses and give the corresponding Mohr's circle diagram. *Ans.* $\sigma_1 = 10.10$ ksi, $\sigma_2 = 0$, $\sigma_3 = -0.099$ ksi.

2.80 Given the normal and shear stresses $\sigma_x = 0$, $\sigma_y = 10$ ksi, and $\tau_{xy} = 12$ ksi, draw the Mohr's circle diagram and the principal normal and shear stresses on the x-y axis. Determine the triaxial stresses and give the corresponding Mohr's circle diagram. *Ans.* $\sigma_1 = 18$ ksi, $\sigma_2 = 0$, $\sigma_3 = -8$ ksi.

2.81 Given the normal and shear stresses $\sigma_x = 10$ ksi, $\sigma_y = 24$ ksi, and $\tau_{xy} = -12$ ksi, draw the Mohr's circle diagram and the principal normal and shear stresses on the x-y axis. Determine the triaxial stresses and give the corresponding Mohr's circle diagram. *Ans.* $\sigma_1 = 30.89$ ksi, $\sigma_2 = 3.107$ ksi, $\sigma_3 = 0$.

2.82 A stress element in plane stress encounters $\sigma_x = 20$ MPa, $\sigma_y = -10$ MPa, and $\tau_{xy} = 13$ MPa. (a) Determine the three principal stresses and maximum shear stress. (b) Using a Mohr's circle diagram, explain the effect of superimposing a hydrostatic pressure p on the principal stresses and maximum shear stress. *Ans.* (a) $\sigma_1 = 24.85$ MPa, $\sigma_2 = 0$, $\sigma_3 = -14.85$ MPa.

2.83 In a three-dimensional stress field, the stresses are found to be $\sigma_x = 4$ ksi, $\sigma_y = 2$ ksi, $\sigma_z = 6$ ksi, $\tau_{xy} = -2$ ksi, $\tau_{yz} = 0$, and $\tau_{xz} = 2$ ksi. Draw the stress element for this case. Determine the principal stresses and sketch the corresponding Mohr's circle diagram. *Ans.* $\sigma_1 = 7.064$ ksi, $\sigma_2 = 4.694$ ksi, $\sigma_3 = 0.2412$ ksi.

2.84 Given the normal and shear stresses $\sigma_x = -9$ ksi, $\sigma_y = 15$ ksi, and $\tau_{xy} = 5$ ksi, determine or draw the following:

(a) Two-dimensional Mohr's circle diagram.

(b) Normal principal stress element in the x-y plane.

(c) Shear principal stress.

(d) Three-dimensional Mohr's circle diagram and corresponding principal normal and shear stresses. *Ans.* $\sigma_1 = 16$ ksi, $\sigma_2 = 0$, $\sigma_3 = -10$ ksi.

2.85 The strain tensor in a machine element is

$$\mathbf{T} = \begin{pmatrix} 0.0012 & -0.0001 & 0.0007 \\ -0.0001 & 0.0003 & 0.0002 \\ 0.0007 & 0.0002 & -0.0008 \end{pmatrix}$$

Find the strain in the x-, y-, and z-directions, in the direction of the space diagonal $\left(\frac{1}{\sqrt{3}}; \frac{1}{\sqrt{3}}; \frac{1}{\sqrt{3}}\right)$, and in the direction ϵ_x, ϵ_y, and ϵ_z. *Ans.* In the direction of the diagonal, $\epsilon = 0.0005$.

2.86 A strain tensor is given by

$$\mathbf{T} = \begin{pmatrix} 0.0023 & 0.0006 & 0 \\ 0.0006 & 0.0005 & 0 \\ 0 & 0 & 0 \end{pmatrix}$$

Calculate the maximum shear strain and the principal strains. *Ans.* $\epsilon_1 = 0.00248$, $\epsilon_2 = 0.00032$.

Design and Projects

2.87 Without using the words "stress" or "strain," define *elastic modulus*.

2.88 A bookshelf sees a uniform distributed load across its entire length, and is supported by two brackets. Where should the brackets be located? *Hint:* See Problem 2.48.

2.89 For the beam shown in Sketch *gg*, determine the force P so that the maximum bending moment in the beam is as small as possible. What is the value of $|M|_{\max}$? *Ans.* $|M_{\max}| = \frac{1}{2}w_o l^2$

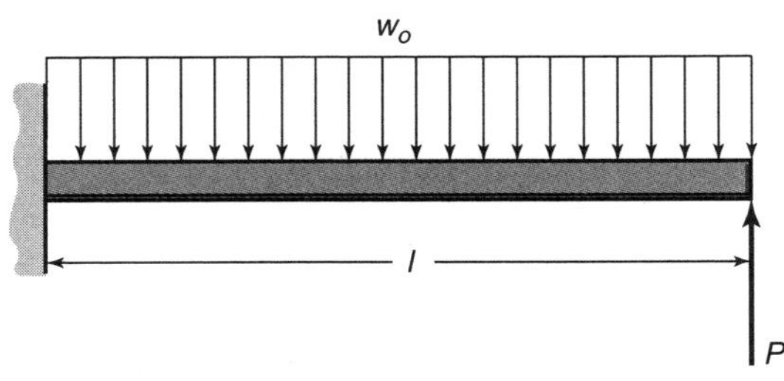

Sketch gg, used in Problem 2.89

2.90 Three- and four-point bending tests are common tests used to evaluate materials.

(a) Sketch the shear and bending moment diagrams for each test.

(b) Is there any difference in the stress state for the two tests?

(c) Which test will cause specimens to fail at a lower bending moment? Why?

2.91 Sketch *hh* shows a beam that is fixed on both ends with a central load. Can you determine the shear and moment diagrams for this case? If not, explain what additional information you would need and how you would go about solving this problem.

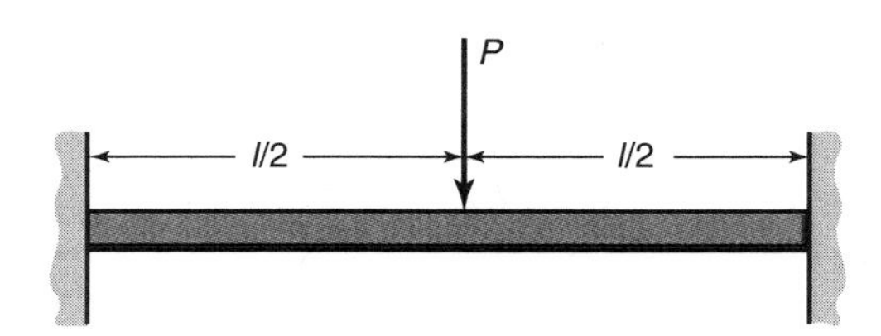

Sketch hh, used in Problem 2.91

2.92 For a 0°–60°–120° strain gage rosette with one base on the x-axis, derive the strains ϵ_x, ϵ_y, and γ_{xy} as a function of ϵ_0, ϵ_{60}, and ϵ_{120}. *Ans.* $\epsilon_x = \epsilon_0$, $\epsilon_y = \frac{2}{3}(\epsilon_{60} + \epsilon_{120}) - \frac{1}{3}\epsilon_0$.

2.93 A stress tensor is given by

$$\mathbf{S} = \begin{pmatrix} 200 & 40 & -30 \\ 40 & 25 & 10 \\ -30 & 10 & -25 \end{pmatrix}$$

where all values are in megapascals. Calculate the principal normal stresses and the principal shear stresses. *Ans.* $\sigma_1 = 211.8$ MPa, $\sigma_2 = 21.46$ MPa, $\sigma_3 = -33.31$ MPa.

Chapter 3

Introduction to Materials and Manufacturing

A hob cuts gear teeth. *Source:* Courtesy of Sandvik Coromant.

Give me matter, and I will construct a world out of it.
Immanuel Kant

Contents

Examples

Case Study

Material and manufacturing process selection are essential for design. Being able to exploit a material's potential and characteristics is necessary to ensure that the best material is used for a particular machine element. This chapter will classify, characterize, and guide selection of solid materials in a general sense. Physical and mechanical properties of engineering materials will be examined and two-parameter charts will be used to suggest materials for specific situations. The hardness of materials will be defined, hardness testing will be described, and the importance of hardness will be discussed. A brief summary of the manufacturing processes that are available for each material class will be discussed, with an introduction to their relationship to mechanical design. Processes can be categorized as casting, where a metal is melted, placed in a mold, and cooled to solidification; bulk forming, where a material is forced to take on a new shape; sheet forming, which uses a rolled metal with small thickness to produce a desired shape; material removal processes such as machining, grinding, and other finishing processes; various methods to produce polymer parts; and methods to produce ceramics and composite materials.

Symbols

A	area, m^2
$\bar{a}$	linear thermal expansion coefficient, $(°C)^{-1}$
C_p	specific heat of material, J/(kg-°C)
C_R	relative cost
d	fiber diameter, m
E	modulus of elasticity, Pa
%EL	elongation, percent
G	shear modulus, Pa
g	gravitational acceleration, 9.807 m/s^2
H	hardness, N/m^2
HB	Brinell hardness number
HK	Knoop hardness number
HR	Rockwell hardness number
HV	Vickers hardness number
K	bulk modulus, Pa
K_t	thermal conductivity, W/(m-°C)
k	spring rate, N/m
k_1	Archard wear constant, $(Pa)^{-1}$
L	sliding distance, m
l	length, m
m_a	mass of body, kg
P	force, N
p	normal pressure, Pa
p_l	limiting pressure, Pa
Q	quantity of heat, J
r	radius, m
r_o	atom size, m
S	strength, Pa
S_u	ultimate strength, Pa
S_y	yield strength, Pa
t_m	temperature, °C
Δt_m	temperature change, °C
t_h	thickness, m
U_r	modulus of resilience, Pa
v	volume fraction
W	weight, N
W_r	wear rate, m^2
γ	shear strain
δ	deformation; deflection, m
ϵ	strain
ν	Poisson's ratio
ρ	density, kg/m^3
σ	normal stress, Pa
τ	shear stress, Pa
τ_f	fiber-matrix bond strength, Pa

Subscripts

a	axial
all	allowable
c	composite; cross sectional
cr	critical
f	fiber
fr	at fracture
g	glass transition
i	inner
m	matrix; mean
t	transverse
o	without load
1,2,3	principal axes

3.1 Introduction

The cost of the design stage of a product lifecycle is usually low, typically less than 5% of the total cost. Much higher costs are associated with the materials and manufacturing processes used in a product's production. However, the materials and manufacturing costs are to a great extent set during the design stage. This recognition has been stated as "Design casts the largest shadow."

Recognizing the impact of a design on product cost and quality, design cannot be done without careful consideration of materials and manufacturing. Intelligent material selection requires knowledge of material capabilities and characteristics. Knowledge of manufacturing leads to design of easy-to-manufacture geometries. Selection of material impacts manufacturing, as some processes are incompatible with certain materials. Thus, design, materials selection, and manufacturing are inseparable, and good design requires sophistication regarding materials and manufacturing.

This chapter introduces materials science, distinguishing between ductile and brittle materials in Section 3.2. This is followed by a discussion of the main classes of solids, namely, metals, polymers, ceramics, and composites. Deformation of materials subjected to stress is then discussed, with the behavior of the different material classes discussed in Section 3.3 and important physical and mechanical properties summarized for a wide variety of engineering materials in Section 3.5. Section 3.7 discusses the use of two-material charts in selecting materials for particular loading types, and limits to the utility of these charts is explained. Finally, the chapter closes with a discussion of the major classes of manufacturing processes and their effects on material properties.

3.2 Ductile and Brittle Materials

3.2.1 Ductile Materials

Ductility is a measure of the degree of plastic deformation attained at fracture. Designers often use ductile materials because they can absorb shock (or energy) and, if they become overloaded, will usually exhibit large deformations before failing. Also, stress concentrations (Sections 6.2 and 7.7) can be partially relieved through the deformations that can be achieved by ductile materials.

One way to quantify ductility is by the percent elongation, %EL, given by

$$\%\text{EL} = \left(\frac{l_{\text{fr}} - l_o}{l_o}\right) \times 100\%, \tag{3.1}$$

where l_{fr} is the length of the specimen at fracture and l_o is the original length of the specimen (without load).

A **ductile material** is one with a large percent elongation before failure, arbitrarily defined as 5% or higher for the purposes of this text. Table A.2 shows that the %ELs for low-carbon (AISI 1020), medium-carbon (AISI 1040), and high-carbon steels (AISI 1080) are 37%, 30%, and 25%, respectively. (Note that selected materials from tables in Appendix A also appear on the inside front cover.) Thus, steel is ductile because it far exceeds the 5% elongation described in Eq. (3.1). Also note from Eq. (3.1) that the original length of the specimen, l_o, is an important value because a significant portion of the plastic deformation at fracture is confined to the neck region. Thus, the magnitude of %EL will depend on the specimen length. The shorter l_o, the greater the fraction of total elongation from the neck and, consequently, the higher the

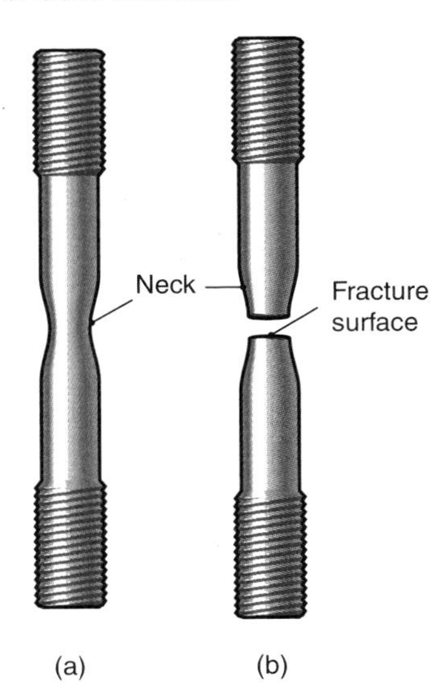

Figure 3.1: Ductile material from a standard tensile test apparatus. (a) Necking; (b) failure.

value of %EL. Therefore, l_o should be specified when percent elongation values are cited.

Figure 3.1a shows a test specimen of a ductile material in which **necking** (decreasing cross-sectional area and localized deformation) is occurring. Figure 3.1b shows the same specimen just after fracture. Note the considerable amount of plastic deformation at fracture.

Example 3.1: Ductility of Materials

Given: A flat plate is formed into a hollow cylinder with an inner radius of 100 mm and a wall thickness of 60 mm.

Find: Determine which temper of AISI 301 stainless steel (see Table A.4) cannot be cold-formed to produce the cylinder. Assume that the midplane of the plate does not experience either tension or compression and will thus not experience any elongation.

Solution: The length without load is at the midplane of the plate, or

$$l_o = 2\pi\left(r_i + \frac{t_h}{2}\right) = 2\pi\left(100 + \frac{60}{2}\right) = 260\pi \text{ mm}.$$

The length at fracture at the outer diameter of the cylinder is

$$l_{\text{fr}} = 2\pi r_o = 2\pi(r_i + t_h) = 2\pi(100 + 60) = 320\pi \text{ mm}.$$

Thus, from Eq. (3.1) the percent elongation is

$$\begin{aligned} \%\text{EL} &= \left(\frac{l_{\text{fr}} - l_o}{l_o}\right) \times 100\% \\ &= \left(\frac{320\pi - 260\pi}{260\pi}\right) \times 100\% \\ &= 23.04\%. \end{aligned}$$

From Table A.4, half-hard, 3/4-hard, and full-hard AISI 301 cannot achieve this strain. However, a lower temper, such as 1/4-, 1/8-, or 1/16-hard, or a fully annealed specimen, would be suitable.

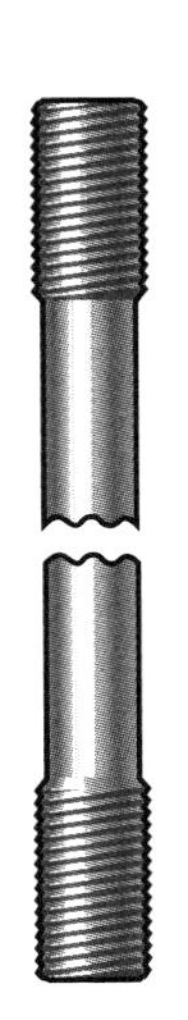

Figure 3.2: Failure of a brittle material from a standard tensile test apparatus.

3.2.2 Brittle Material

A **brittle material** produces little (%EL<5%) or no plastic deformation before failure. Figure 3.2 shows a brittle test specimen at failure. Note that little or no necking occurs prior to failure, in contrast to Fig. 3.1.

3.3 Classification of Solid Materials

Engineering materials fall into four major classes: metals, ceramics (including glasses), polymers (including elastomers), and composites. The members of each class generally have the following common features:

1. Similar properties, chemical makeup, and atomic structure
2. Similar processing routes
3. Similar applications

3.3.1 Metals

Metals are combinations of metallic elements. They have large numbers of nonlocalized electrons (i.e., electrons not bound to particular atoms). Metals are extremely good conductors of electricity and heat and are not transparent to visible light; a polished metal surface has a lustrous appearance. Furthermore, metals are strong and usually deformable, making them extremely important materials in machine design.

Metals are usually ductile and can be made stronger by alloying and by mechanical and heat treatment. High-strength alloys can have a percent elongation as low as 2%, but even this is enough to ensure that the material yields before it fractures. Some cast metals have very low ductility, however. Metals are often used in circumstances where cyclic loading is encountered (see Chapter 7), and they are generally resistant to corrosion. Ductile materials, such as steel, accommodate stress concentrations by deforming in a way that redistributes the load more evenly.

An **isotropic** material has properties that are the same in all directions; a material with directional properties is **anisotropic**. On a microscopic scale, metals form well-defined crystals with ordered packing of atoms. The crystals in a metal are vary small and are randomly oriented. Thus, while a crystal may be anisotropic, a metal should be considered polycrystalline, and the averaged results of many crystals leads to a reasonable assumption of isotropy. However, the crystals, or grains, in a metal may be elongated or oriented, leading to anisotropic behavior, especially with sheet metals. Thus, manufacturing history will determine whether a metal is isotropic or anisotropic.

Most metals are initially cast and then can be further processed to achieve the desired shape. Further processes include secondary casting (such as sand, shell, investment or die casting), bulk forming (forging, extrusion, rolling, drawing), sheet forming (deep drawing, stretch forming, stamping) or machining (milling, turning, grinding, polishing). An additional option for metals is to produce metal powders and form desired shapes through powder metallurgy techniques. Manufacturing processes are addressed in Section 3.8.

3.3.2 Ceramics and Glasses

Ceramics are compounds of metallic and nonmetallic elements, most frequently oxides, nitrides, and carbides. For example, the ceramic material aluminum oxide (also known as alumina, carborundum, or in single crystal form, sapphire), is Al_2O_3. **Glasses** are made up of metallic and nonmetallic elements just as are ceramics, but glasses typically have no clear crystal structure. A typical soda-lime glass consists of approximately 70 wt% silicon dioxide (SiO_2), the balance being mainly soda (Na_2O) and lime (CaO). Both ceramics and glasses typically insulate against the passage of electricity and heat and are more resistant to high temperatures and harsh chemical environments than are metals and polymers.

Ceramics and glasses, like metals, have high density. However, instead of being ductile like metals, ceramics and glasses are brittle at room temperature. They are typically 15 times stronger in compression than in tension. They are stiff, hard, and abrasion resistant (hence their use for bearings and cutting tools). Thus, they must be considered an important class of engineering material for use in machine elements.

Ceramics are usually obtained by forming a ceramic slurry (a suspension of ceramic powders in water) with binders, and then firing the ceramics to develop a strong bond between particles. Some machining operations are possible with ceramics, but they are often too brittle to be successfully machined. However, grinding and polishing are commonly performed successfully with ceramics.

Example 3.2: Thermal Expansion

Given: A piece of stabilized zirconia (ZrO_2) has a thermal expansion coefficient of $12 \times 10^{-6}/^\circ$C, and is to be implanted into a steel ring. The fit between the steel and the zirconia is a medium press fit at room temperature. When the temperature fluctuates from room temperature to 500°C, the zirconia should not loosen.

Find: The correct class of steel to be used from those given in Table 3.1.

Solution: Since the zirconia should not loosen from the steel when the temperature is increased, a slightly smaller coefficient of thermal expansion, but very close to the zirconia value, is desired for the stainless steel. From Table 3.1, a low- or medium alloy has a coefficient of thermal expansion of $11 \times 10^{-6}/^\circ$C and would therefore be the preferred classes of steel.

3.3.3 Polymers and Elastomers

Polymers are organic compounds composed of carbon, hydrogen, and other nonmetallic elements. Polymers have large and complex molecular structures.

Polymers, also called **plastics**, are of two basic types: thermoplastics and thermosets. **Thermoplastics** are long-chain molecules, sometimes with branches, where the strength arises from interference between chains and branches. **Thermosets** have a higher degree of cross-linking so that molecular chains are linked together and cannot slide over each other. In general, thermoplastics are more ductile than thermosets, and at elevated temperatures they soften significantly and melt. Thermosets are more brittle, do not soften as much as thermoplastics, and usually degrade chemically before melting.

Elastomers have a networked cross-linked structure, but not as extensive as that for more rigid thermosets, so they produce large elastic deformations at relatively light loads. A common example of an elastomer is the material in a rubber band, which displays the typical characteristics of large elastic deformation followed by brittle fracture without any plastic deformation. Further, rubber bands are highly nonlinearly elastic, which is typical of elastomers.

Polymers and elastomers can be extremely flexible with large elastic deformations. Polymers are roughly five times less dense than metals but have nearly equivalent strength-to-weight ratios. Because polymers creep (the time-dependent permanent deformation that occurs under static stress) even at room temperature, a polymer machine element under load may, with time, acquire a permanent set. The properties of polymers and elastomers change greatly with variations in temperature. For example, a polymer that is tough and flexible at 20°C may be brittle at the 4°C environment of a household refrigerator and yet creep rapidly at the 100°C of boiling water.

The mechanical properties of polymers are characterized using the same parameters as for metals (i.e., modulus of elasticity and tensile, impact, and fatigue strengths). However, polymers vary much more in strength, stiffness, etc, than do metals. This variation can be explained by the difference in chain lengths and the amount of polymer that is in a crystalline or amorphous state. Further, a thermoplastic that is deformed plastically will have its molecules aligned in the direction of strain, leading to higher strength and anisotropy. Thus, two polymers with the identical chemical constituents can have very different microstructures and associated variation in mechanical and physical properties. In addition, the mechanical characteristics of polymers, for the most part, are highly sensitive to the rate of deformation, temperature, and chemical nature of the environment (the presence of water, oxygen, organic solvents, etc.). Therefore, the particular values given for polymer mechanical properties should be used with caution.

Polymers are easy to shape: complicated parts performing several functions can be molded from a polymer in a single operation (see Section 3.8.2). However, injection molding operations have high tooling costs and can only be justified for large production runs. Large elastic deflections allow the design of polymer components that snap together, making assembly fast and inexpensive. Polymers are corrosion resistant and have low coefficients of friction.

Thermoplastics and thermosets have very different manufacturing options and strategies. Thermoplastics are generally heated to a temperature above their melting point, formed into a desired shape and then cooled. Examples of common manufacturing processes for thermoplastics include extrusion, injection molding, blow molding, and ther-

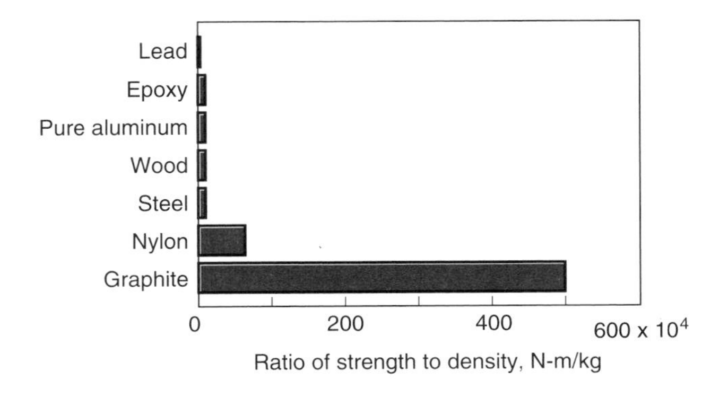

Figure 3.3: Strength/density ratio for various materials.

moforming. Thermosets are blended from their constituents, formed to a desired shape and then cured at elevated temperature and/or pressure to develop cross links. Common manufacturing methods used for thermosets include reaction injection molding, compression molding, and potting (similar to casting).

3.3.4 Composites

Figure 3.3 compares a number of materials from a minimum-weight design standpoint (i.e., a larger strength-to-density ratio leads to a lighter design). Fibers can have much better strength-to-weight ratios than conventional extruded bars, molded plastics, and sintered ceramics. However, fibers are often susceptible to corrosion, even in air. For example, graphite fibers will oxidize readily in air and cannot provide their exceptional strength for long.

Many modern technologies require machine elements with demanding combinations of properties that cannot be met by conventional metal alloys, ceramics, and polymeric materials. Present-day technologies require solid materials that have low density, high strength, stiffness and abrasion resistance, and that are not easily corroded. This combination of characteristics is rather formidable, considering that strong materials are usually relatively dense and that increasing stiffness generally decreases impact strength.

Composite materials combine the attractive properties of two or more material classes while avoiding some of their drawbacks. A composite is designed to display a combination of the best characteristics of each component material. For example, graphite-reinforced epoxy acquires strength from the graphite fibers while the epoxy protects the graphite from oxidation. The epoxy also helps support shear stresses and provides toughness.

The three main types of composite material are:

1. *Particle reinforced*, which contain particles with approximately the same dimensions in all directions distributed in a matrix, such as concrete.

2. *Discontinuous fiber reinforced*, which use fibers of limited length-to-diameter ratio in a matrix, such as fiberglass.

3. *Continuous fiber reinforced*, where continuous fibers are incorporated, such as seen in graphite tennis rackets.

Figure 3.4 shows a cross section of a continuous fiber-reinforced composite material. Most such composites contain glass, polymer or carbon fibers, and a polymer matrix. These composites cannot be used at elevated temperatures because the polymer matrix softens or degrades, but at room temperature their performance can be outstanding. Some disadvantages of composites are that they are expensive and relatively difficult to form and join.

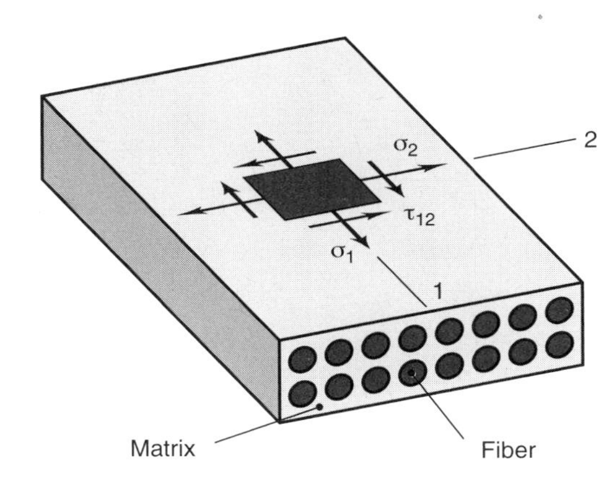

Figure 3.4: Cross section of fiber-reinforced composite material.

Composite materials have many characteristics that are different from those of the other three classes of material considered. Whereas metals, polymers, and ceramics are **homogeneous** (properties are not a function of position in the solid), **isotropic** (properties are the same in all directions at a point in the solid), or **anisotropic** (properties are different in all directions at a point in the solid), composites are **nonhomogeneous** and **orthotropic**. An orthotropic material has properties that are different in three mutually perpendicular directions at a point in the solid but has three mutually perpendicular planes of material symmetry. Consideration in this text is limited to simple, unidirectional, fiber-reinforced orthotropic composite materials, such as shown in Fig. 3.4.

An important parameter in discontinuous fiber-reinforced composites is the fiber length. Some critical fiber length is necessary for effective strengthening and stiffening of the composite material. The critical length, l_{cr}, of the fiber depends on the fiber diameter, d, its ultimate strength, S_u, and the fiber-matrix bond strength, τ_f, according to

$$l_{cr} = \frac{S_u d}{2\tau_f}. \tag{3.2}$$

The two in the denominator of Eq. (3.2) accounts for the fact that the fiber is embedded in the matrix and splits into two parts at failure. For a number of glass- and carbon-fiber-reinforced composites this critical length is about 1 mm, or 20 to 150 times the fiber diameter.

Discontinuous and particle-reinforced composites share many of the manufacturing methods with thermosetting polymers and metals, depending on the matrix materials. Polymer matrix materials are commonly molded or placed onto a form (lay-up) and then cured in an oven. Metal matrix composites use variants of casting or powder metallurgy techniques. Continuous fiber reinforced polymers have unique manufacturing approaches, including tape layup, pultrusion, pulforming, and filament winding.

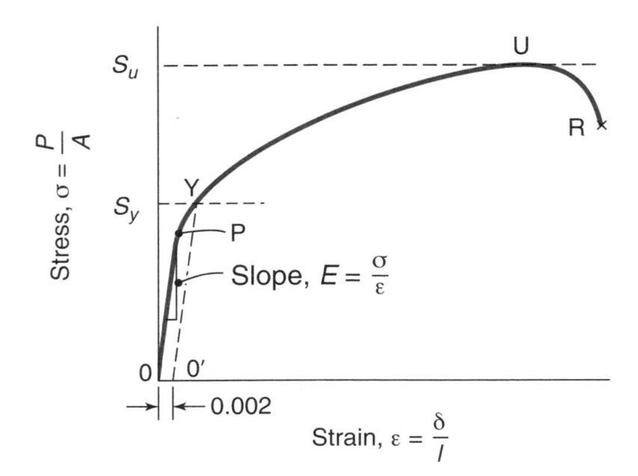

Figure 3.5: Typical stress-strain curve for a ductile material.

Example 3.3: Strength of a Composite Material

Given: A fiber-reinforced plastic contains carbon fibers having an ultimate strength of 1 GPa and a modulus of elasticity of 150 GPa. The fibers are 3 mm long with a diameter of 30 μm.

Find: Determine the required fiber-matrix bond strength in order to fully develop the strength of the fiber reinforcement.

Solution: From Eq. (3.2), the fiber-matrix bond strength can be expressed as

$$\tau_f = \frac{S_u d}{2l_{\text{cr}}} = \frac{(10^9)(30 \times 10^{-6})}{2(0.003)} = 5 \text{ MPa}.$$

3.4 Stress-Strain Diagrams

The stress-strain diagram is important in designing machine elements because it yields data about a material's strength without regard to its size or shape. Because stress-strain diagrams differ considerably for the different classes of material, each will be treated separately. The exception is that a stress-strain diagram for composites will not be presented because of the diverse nature of these materials.

3.4.1 Metals

Figure 3.5 shows the stress-strain diagram for a ductile metal. Although the stress shown in the figure is tensile, the stress-strain diagrams for most metals are essentially the same for compression and tension.

Figure 3.6 better clarifies the mechanisms and behavior near the yield stress. A number of points presented in Figs. 3.5 and 3.6 need to be defined:

1. **Proportional limit** (point P): Stress at which the stress-strain curve first deviates from linear behavior.

2. **Elastic limit** (point E): Highest stress the material can withstand and still return exactly to its original length when unloaded.

3. **Yield strength** (point Y): The elastic limit is difficult to obtain experimentally, so the yield stress as shown is often used. The yield point is determined by starting a line at a deformation of 0.002 (0.2%) on the x-axis of Fig. 3.5, and that has a slope equal to the initial elastic modulus. The yield point is the location where this line intercepts the stress-strain curve.

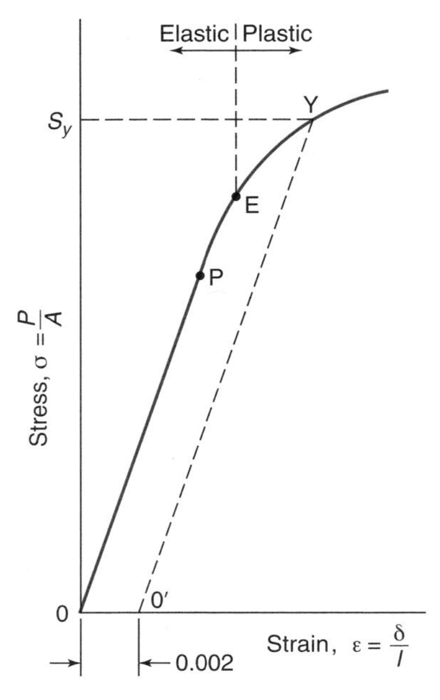

Figure 3.6: Typical stress-strain behavior for ductile metal showing elastic and plastic deformations and yield strength S_y.

4. **Ultimate strength** (point U): Maximum stress reached in the stress-strain diagram

5. **Fracture stress** (point R): Stress at the time of fracture or rupture.

Note that the elastic limit (point E) is not shown in Fig. 3.5 but is shown in Fig. 3.6 and clarifies behavior near the elastic-plastic demarcation point. In practice, the change in slope is not as pronounced as in Fig. 3.6, and the difference between points E and Y can be more subtle. Note that loading occurs along 0PEY; unloading occurs along Y0′ and is assumed to be linear.

Figure 3.6 can be divided into elastic and plastic regions. The demarcation point is point E, the elastic limit, although Y, the yield point, is often used because it is more easily characterized. Three different phenomena occur during plastic behavior:

1. **Yielding**: A slight increase in stress above the elastic limit (point E) will cause the metal to deform permanently (plastic deformation).

2. **Strain hardening**: After yielding and before reaching the ultimate strength, S_u, strain hardening occurs. That is, ductile metals become harder and stronger as they plastically deform.

3. **Necking**: Necking, which is a localized deformation and decrease in cross-sectional area, occurs after the ultimate stress is reached and it continues until fracture occurs, as shown in Fig. 3.1.

Thus far, the discussion of stress-strain diagrams has focused on ductile metals. Figure 3.7 compares typical tensile stress-strain diagrams for a brittle and a ductile metal loaded to

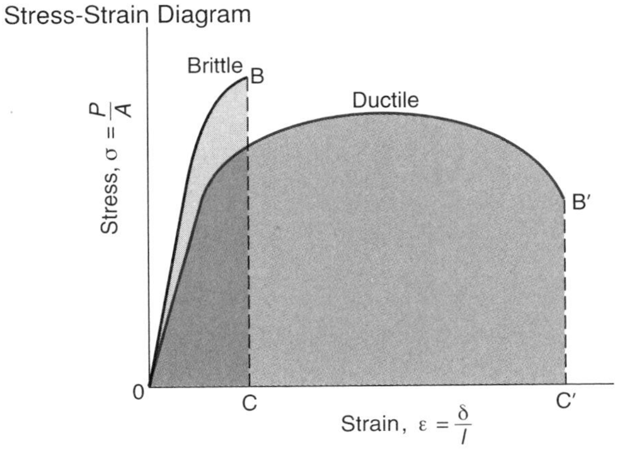

Figure 3.7: Typical tensile stress-strain diagrams for brittle and ductile metals loaded to fracture.

fracture. The brittle metal experiences little or no plastic deformation, whereas the ductile metal attains large strains before fracture occurs. Also, brittle materials have considerably higher (typically 10 times or greater) ultimate strength in compression than in tension. In contrast, ductile materials have essentially the same ultimate strength in compression and in tension.

3.4.2 Ceramics

The stress-strain behavior of ceramics is not usually determined by the tensile test used for metals. The reason for this is twofold. First, it is difficult to prepare and test specimens having the required geometry; and second, there is a significant difference in results obtained from tests conducted in compression and in tension. Therefore, a more suitable transverse bending test is most frequently used, in which a specimen having either a circular or rectangular cross section is bent until it fractures. Stress is computed from the specimen thickness, the bending moment, and the moment of inertia of the cross section. The maximum stress, or the stress at fracture, is sometimes called the **modulus of rupture**, and is an important parameter used in characterizing ceramics.

The stress-strain behavior for ceramic materials obtained in the transverse bending test is similar to the tensile test results for metals. A linear relationship exists between stress and strain. In Fig. 3.8, the stress-strain diagram for a ceramic shows that strength depends on whether the loading is compressive or tensile. This is understandable because ceramics are notoriously brittle and, if loaded in tension, imperfections in the material become fracture initiation and propagation sites. In compression, on the other hand, defects such as microcracks are squeezed, so that they do not compromise the material's strength. This kind of behavior can also be seen with cast metals, which have large numbers of voids in their lattices.

Strength for ceramics refers to fracture strength in tension and crushing strength in compression; typically, the compressive strength is many times larger than the tensile strength. Once the crushing strength is reached, the strain increases significantly but the stress decreases. The x in Fig. 3.8 designates the elastic strain when the fracture strength in tension is reached.

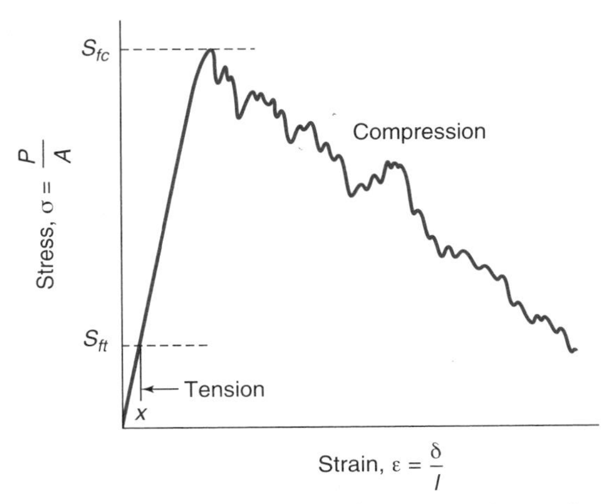

Figure 3.8: Stress-strain diagram for ceramic in tension and in compression.

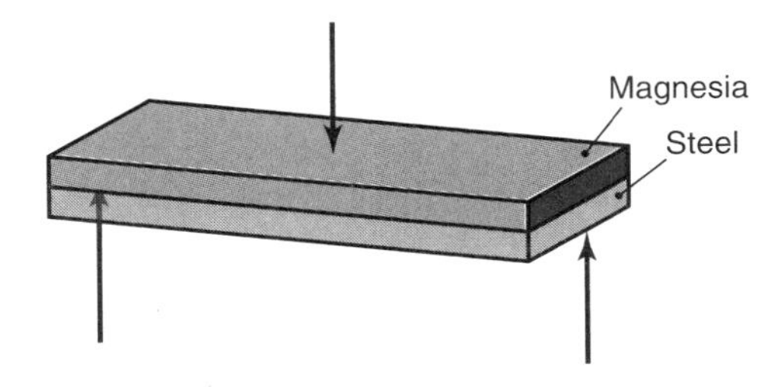

Figure 3.9: Beam loaded in three-point bending used in Example 3.4.

Example 3.4: Design of a Composite Beam

Given: A beam, shown in Fig. 3.9, consists of equal lengths and cross sections of magnesia and AISI 1080 steel. The magnesia and steel sections are bonded together so that they act as a single beam. The beam is simply supported and subjected to three-point bending as shown. Assume that for magnesia, $E_m = 207$ GPa and $S_{um} = 105$ MPa and for 1080 steel, $E_s = 207$ GPa, $S_{ys} = 380$ MPa, and $S_{us} = 615$ MPa.

Find: Is the beam strongest when the steel is at the bottom (as shown in Fig. 3.9) or at the top?

Solution: An important feature of ceramics is that they are typically 15 times stronger in compression than in tension, whereas steels have the same yield stress in compression or tension. In Fig. 3.9, the top member is in compression while the bottom member is in tension. Thus, for the steel at the bottom and magnesia at the top, as shown in Fig. 3.9, the magnesia is in compression with a compressive strength of around $(15)(105\times10^6) = 1575$ MPa and is much stronger than the steel, which is in tension with a strength of 380 MPa. If the magnesia were at the bottom instead of the top, it would be in tension with a strength of 105 MPa and thus would be weaker than the steel, which would be in compression with a strength of 380 MPa. Taking the weakest members of the two beam combinations (since that is where it will fail), we find that the beam is about $380/105 = 3.642$ times stronger when the steel is at the bottom in Fig. 3.9.

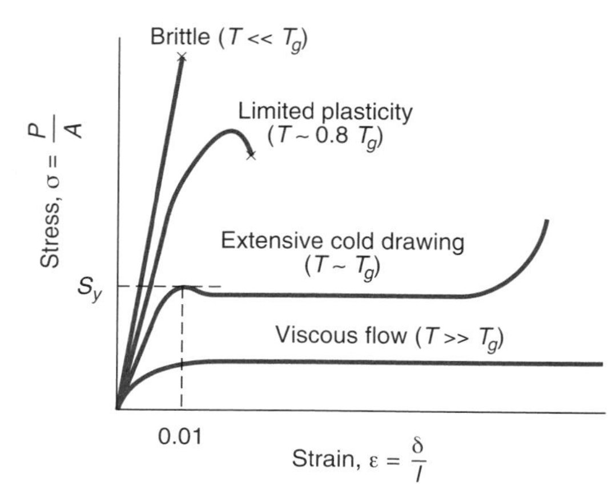

Figure 3.10: Stress-strain diagram for polymer below, at, and above its glass transition temperature, T_g.

Example 3.5: Glass Transition Temperature

Given: A plastic cup is made of polymethylmethacrylate. The room-temperature elongation at fracture is 5% and the glass transition temperature is 90°C. The cup is sterilized with 100°C superheated steam at a high pressure, stressing the plastic to 30 MPa.

Find: Can the cup be expected to maintain its shape during sterilization?

Solution: Since the sterilization temperature is 10°C above the glass transition temperature, the plastic will deform by more than 5% (see Fig. 3.10). The stress is approximately half the ultimate strength at room temperature. It can therefore be concluded that the cup will deform during sterilization.

3.4.3 Polymers

The testing techniques and specimen configurations used to develop stress-strain diagrams for metals can be used with slight modifications for polymers, especially for highly elastic materials such as rubbers. However, the stress-strain behavior of these materials is unique, and is very sensitive to temperature. For this reason, it also helps to distinguish between thermoplastics and thermosets. Thermosets typically have little plastic deformation, so that their stress-strain diagrams in tension are essentially the same as the brittle materials in Fig. 3.7 or the ceramic in Fig. 3.8. Polymers are slightly stronger (~20%) in compression than in tension.

Thermoplastics behave very differently. Figure 3.10 shows stress-strain behavior for a thermoplastic polymer below, at, and above its glass transition temperature. The **glass transition temperature**, T_g, is the temperature at which a polymer transforms from a hard and brittle material to a rubbery or leathery solid. For $T \ll T_g$ the polymer fractures after relatively small strains. For $T \approx T_g$ the plastic can undergo very large strains. Above the melting temperature (used for manufacturing applications), the material flows like a viscous liquid as shown. Finally, the deformation displayed for $T \ll T_g$ is fully elastic.

Elastomers, like thermosetting polymers, are brittle, but elastomers can survive large strains before fracture (as can be seen with a rubber band). Elastomers may be elastic, but they are highly nonlinear. After an initial linear zone, the stress-strain diagram becomes markedly nonlinear, having typically a strain of 0.01, as shown in Fig. 3.10. This nonlinearity may be caused either by shear yielding, the irreversible slipping of molecular chains, or by crazing (the formation of low-density, crack-like volumes that scatter light, making the polymer look white).

These stress-strain diagrams for metals, polymers, and ceramics reveal the following characteristics:

1. For brittle solids (ceramics, glasses, brittle polymers, and brittle metals), a yield strength may be difficult to determine and the fracture strength or ultimate strength is used in design.

2. For metals, ductile polymers, and most composites, the ultimate strength is larger than the yield strength by a factor of 1.1 to 4. The reason for this is mainly work hardening or, in the case of composites, load transfer to the reinforcement. Either yield or ultimate strength may be used in design, depending on the particular application and constraints.

3.5 Properties of Solid Materials

This section defines the various engineering properties of solid materials needed to select the proper materials for machine elements. All these properties may not be important for each machine element considered in Chapters 9 to 19, but they are important for the wide range of applications of the various machine elements. For each property the relative behaviors of the classes of material are presented as well as the relative behaviors of the various materials within a specific class.

3.5.1 Density

Density is the mass per unit volume. The SI unit of density is kilograms per cubic meter. Typical densities of solid materials lie between 10^3 and 10^4 kg/m^3. Figure 3.11 illustrates the density ranking of various metals, polymers, and ceramics. Metals, such as lead, copper, and steel, have the highest mass density. Polymers, such as nylon, natural rubber, and polyethylene, have the lowest mass density. Table 3.1 gives values of density at room temperature (20°C).

Metals are dense because they are made of heavy atoms in an efficient packing. Ceramics, for the most part, are also efficiently packed, but have lower densities than metals because they contain oxygen, nitrogen, and carbon atoms. Polymers have low densities because they are mainly made of carbon and hydrogen, and they are never completely crystalline, so that the atoms are not efficiently packed.

Alloying changes the density of metals only slightly. To a first approximation, the highest density of an alloy (metallic solid resulting from combining two or more metals) is given by a linear interpolation between the densities of the alloy concentrations.

3.5.2 Modulus of Elasticity, Poisson's Ratio, and Shear Modulus

The **modulus of elasticity** (or **Young's modulus**) is defined as the slope of the linear-elastic part of the stress-strain curve. In Fig. 3.6, the linear portion of the stress-strain curve is between the origin and point P, or at stresses lower than the proportional limit stress. The modulus of elasticity can be written as

$$E = \frac{\sigma}{\epsilon}. \tag{3.3}$$

Since strain has no dimension, the modulus of elasticity has the same units as stress, or N/m^2. Figure 3.12 and Table 3.1

Table 3.1: Typical physical properties of common engineering materials.

Material	Elastic modulus, E, (GPa)	Poisson's ratio, ν	Density, ρ, (kg/m^3)	Thermal conductivity, K_t (W/m° C)	Thermal expansion coefficient, $\bar{\alpha}$ (μm/m° C)	Specific heat, C_p (J/kg° C)
Metals						
Aluminum	62	0.33	2700	209	23	900
Aluminum alloys	70	0.33	2630-2820	221-239	24	900
Aluminum tin	63	0.33	2700	180	24	960
Babbitt, lead-based	29	0.33	7530	24	20	150
Babbitt, tin-based	52	0.33	7340	56	23	210
Brasses	100	0.33	7470-8940	120	19	390
Bronze, aluminum	117	0.33	8940	50	18	380
Bronze, leaded	97	0.33	9100	47	18	380
Bronze, phosphor	110	0.33	8500	50	18	380
Bronze, porous	60	0.22	8040	30	18	380
Copper	124	0.33	8970	170	18	380
Iron, gray cast	109	0.26	7860	50	11	420
Iron, malleable cast	170	0.26	7860		11	420
Iron, spheroidal graphite	159	0.26	7860	30	11	420
Iron, porous	80	0.20	7460	28	12	460
Iron, wrought	170	0.30	7860	70	12	460
Magnesium alloys	41	0.33	1770	110	27	1000
Nickel alloys	221	0.31	8850	52-63	15	440
Steel, low alloys	196	0.30	7800	35	11	450
Steel, medium and high alloys	200	0.30	7850	30	11	450
Steel, stainless	193	0.30	8030	15	17	500
Steel, high speed	212	0.30	7860	30	11	450
Titanium alloys	110	0.32	4510	8-12	8.4	520
Zinc alloys	50	0.27	7135	110	27	400
Polymers						
Acetal (polyformaldehyde)	2.7	0.35	1400	0.24	90	1460
Nylons (polyamides)	1.9	0.40	1140	0.25	100	1700
Polyethylene, high density	0.9	0.35	940	0.5	126	1800
Phenol formaldehyde	7.0	0.35	1362	0.17	25-40	1600
Rubber, natural	0.004	0.50	930	1.6	80-120	2000
Ceramics						
Alumina (Al_2O_3)	390	0.28	961	25	5.0	880
Graphite	27	0.31	2400	125	1.4-4.0	840
Silicon carbide (SiC)	450	0.19	3210	15	4.3	750
Silicon nitride (Si_2N_4)	314	0.26	3290	30	3.2	710

give data for the elastic modulus for various metals, polymers, and ceramics at room temperature. The elastic moduli for metals and ceramics are high and quite similar, but those for polymers are considerably lower.

The elastic moduli of most materials depend on two factors: bond strength and bond density per unit area. A bond is like a spring: it has a spring rate, k (in newtons per meter). The modulus of elasticity, E, is roughly

$$E = \frac{k}{r_o}. \tag{3.4}$$

where r_o is the atom size (this can be obtained from the mean atomic volume $4\pi r_o^3/3$, which is generally known). The wide range of modulus of elasticity in Fig. 3.12 and Table 3.1 is largely caused by the range of k in materials. The covalent bond is stiff (k = 20 to 200 N/m), while the metallic and ionic bonds are somewhat less stiff (k = 15 to 100 N/m). Diamond, although not shown in Fig. 3.12 or Table 3.1, has a very high modulus of elasticity because the carbon atom is small (giving a high bond density) and its atoms are linked by extremely strong bonds (200 N/m). Metals have a high modulus of elasticity because close packing gives a high bond density and the bonds are strong, although not as strong as those of diamond. Polymers contain both strong diamond-like covalent bonds and weak hydrogen (or van der Waals) bonds (k = 0.5 to 2 N/m); these weak bonds stretch when a polymer is deformed, giving a lower modulus of elasticity. Elastomers have a low modulus of elasticity because they have only an extremely weak restoring force, which is associated with tangled, long-chain molecules when the material is loaded.

In a tension test, there will be axial deformation, but there will also be dimensional changes in the transverse direction, for as a bar extends axially, it contracts transversely. The transverse strain, ϵ_t, is related to the axial strain, ϵ_a, by **Poisson's ratio**, ν, such that

$$\epsilon_t = -\nu\epsilon_a. \tag{3.5}$$

The negative sign simply means that the transverse deformation will be in the opposite sense to the axial deformation. Poisson's ratio is dimensionless. Table 3.1 gives quantitative values of Poisson's ratio for various metals, polymers, and ceramics at room temperature. The highest Poisson's ratio approaches 0.5 for rubber, and the lowest is 0.19 for silicon carbide and cemented carbides (although it approaches zero for cork or foams in compression). Poisson's ratio cannot be less than zero (or else the second law of thermodynamics would be violated), nor can it exceed 0.5 (or else a material's volume would increase when compressed).

Shear stress and strain are proportional to each other; that is

$$\tau = G\gamma, \tag{3.6}$$

where G is the **shear modulus** or **modulus of rigidity**. This relation is only true for the linear-elastic region of the shear stress-strain curve (from 0 to P in Fig. 3.6). G has units of stress (N/m^2).

The three material properties E, G, and ν are related by the following equation:

$$G = \frac{E}{2(1+\nu)}. \tag{3.7}$$

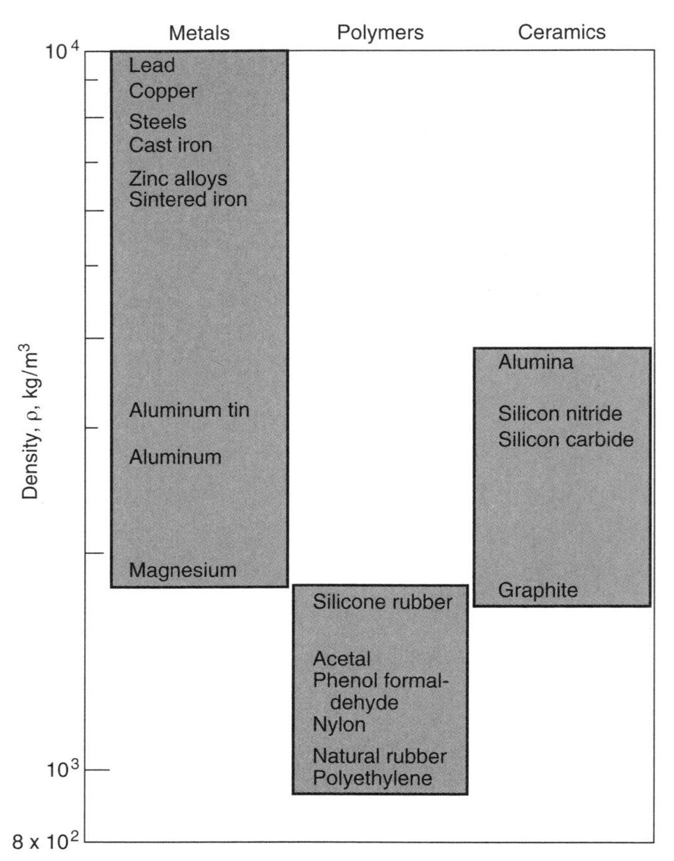

Figure 3.11: Density for various metals, polymers, and ceramics at room temperature (20°C).

Thus, when two parameters are known, the third can easily be determined from Eq. (3.7). That is, if for a particular material, the modulus of elasticity and Poisson's ratio are obtained from Table 3.1, then the shear modulus can be obtained from Eq. (3.7).

The material presented thus far is valid for metals, polymers, or ceramics. To establish the modulus of elasticity for a unidirectional fiber-reinforced composite in the direction of the fibers, it is assumed that the fiber-matrix interfacial bond is good, so that deformation of both matrix and fibers is the same. Under these conditions, the total load sustained by the composite, P_c, is equal to the loads carried by the matrix, P_m, and the fiber, P_f, or

$$P_c = P_m + P_f, \tag{3.8}$$

where subscripts c, m, and f refer to composite, matrix, and fiber, respectively. Substituting Eq. (3.8) into the definition of stress given by Eq. (2.7) results in

$$\sigma_c = \sigma_m \frac{A_m}{A_c} + \sigma_f \frac{A_f}{A_c}. \tag{3.9}$$

If the composite, matrix, and fiber lengths are equal, Eq. (3.9) becomes

$$\sigma_c = \sigma_m v_m + \sigma_f v_f, \tag{3.10}$$

where v_m is the volume fraction of the matrix and v_f is the volume fraction of the fiber in the composite material. Because the same deformation of matrix and fibers was assumed and the composite consists of only matrix and fibers

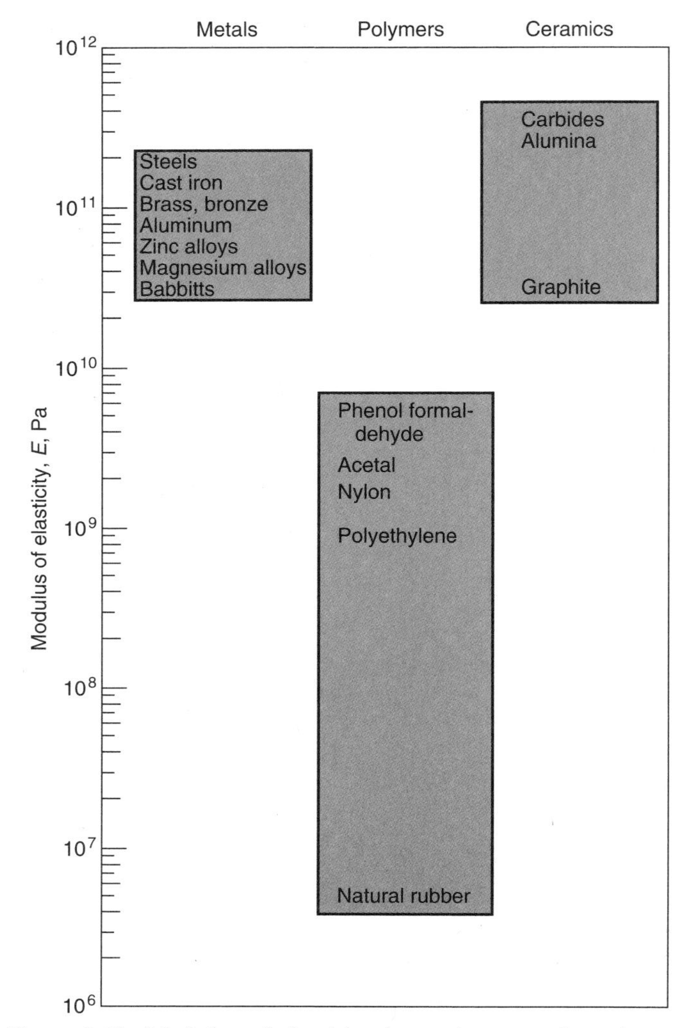

Figure 3.12: Modulus of elasticity for various metals, polymers, and ceramics at room temperature (20°C).

(i.e., $v_m + v_f = 1$), Eq. (3.10) can be rewritten in terms of the moduli of elasticity as

$$E_c = E_m v_m + E_f v_f \tag{3.11}$$

or

$$E_c = E_m(1 - v_f) + E_f v_f. \tag{3.12}$$

Thus, Eq. (3.12) enables the composite's modulus of elasticity to be determined when the elastic moduli of the matrix and the fiber and the volume fractions of each are known. It can also be shown that the ratio of the load carried by the fibers to that carried by the matrix is

$$\frac{P_f}{P_m} = \frac{E_f v_f}{E_m v_m}. \tag{3.13}$$

Recall that Eqs. (3.8) to (3.13) are only applicable to unidirectional, fiber-reinforced composites.

Example 3.6: Stiffness of a Fiber Reinforced Polymer

Given: A fiber-reinforced plastic contains 10 vol% glass fibers (E = 70 GPa, S_u = 0.7 GPa).

Find: Calculate how this fiber percentage has to be changed to give the same elastic properties if the glass fibers are changed to carbon fibers (E = 150 GPa, S_u = 1 GPa). Assume the matrix material has E_m = 2 GPa.

Solution: According to Eq. (3.11), the modulus of elasticity for a fiber composite is

$$E_c = E_m v_m + E_f v_f.$$

Therefore, for the glass fiber-reinforced material,

$$E_c = \left(2 \times 10^9\right)(0.9) + \left(70 \times 10^9\right)(0.1) = 8.8 \text{ GPa}.$$

This composite modulus of elasticity should be maintained for the carbon-reinforced plastic:

$$8.8 \times 10^9 = \left(2 \times 10^9\right) x + \left(150 \times 10^9\right)(1 - x),$$

which is solved as $x = 0.954$ or $1 - x = 0.046$. Thus, the plastic should contain 4.6 vol% carbon fibers to get the same elastic properties as the glass-fiber-reinforced plastic with 10% glass fibers.

3.5.3 Strength

The strength of a machine element depends on the class, treatment, and geometry of the specimen as well as the type of loading that the machine element will experience. This section focuses on the various classes of material and their strength characteristics.

Metals

Metals can be divided into ferrous and nonferrous alloys. Ferrous alloys are those in which iron is the primary component, but carbon as well as other alloying elements may be present. Nonferrous alloys are all those alloys that are not iron-based. Recall that the strength of metals is essentially the same in compression or in tension. Tables A.1 through A.12 show the yield strengths for various metals.

Polymers

The strength of polymers is determined by a strain of 0.010, as opposed to the nonrecoverable plastic strain of 0.002 used to define yield strength for metals. Polymers are somewhat stronger ($\sim$20%) in compression than in tension. Also, when dealing with polymers the strength of interest is the ultimate strength at fracture rather than the yield strength as for metals. The tensile strengths at fracture for selected thermoplastic and thermosetting polymers are given in Table A.13.

Ceramics

The strength of interest for ceramics is the fracture strength. Ceramics, being brittle materials, are much stronger in compression (typically 15 times stronger) than in tension. Table A.14 gives fracture strength in tension for selected ceramic materials.

3.5.4 Resilience and Toughness

Resilience is a material's capacity to absorb energy when it is deformed elastically and then, upon unloading, to release this energy. The **modulus of resilience**, U_r, is the strain energy per unit volume required to stress a material from an unloaded state to the point of yielding. Mathematically, this is expressed as

$$U_r = \int_0^{\epsilon_y} \sigma \, d\epsilon, \tag{3.14}$$

where ϵ_y is the yield strain, or the strain when the stress is the yield strength, S_y. For the linear-elastic region, the area below the stress-strain diagram is the modulus of resilience or

$$U_r = \frac{S_y \epsilon_y}{2}. \tag{3.15}$$

Making use of Eq. (3.3) gives

$$U_r = \frac{S_y^2}{2E}. \tag{3.16}$$

An example of a material with high modulus of resilience is high-carbon steel. Resilience is extremely useful in selecting a material for springs (see Chapter 16), or for energy storage, making high-carbon steel alloys primary candidate materials for such applications.

Toughness is a material's ability to absorb energy up to fracture. For the static (low-strain rate) situation, toughness can be obtained from the stress-strain curve (e.g., see Fig. 3.5) up to the point of fracture or rupture. **Resilience** is the strain energy per unit volume up to the yield strength of the material (point Y in Fig. 3.5); whereas toughness is energy per unit volume to rupture (point R). While resilience represents energy that can be recovered, toughness is associated with energy that is absorbed by the material during deformation, only some of which may be recovered upon unloading.

For a material to be tough, it must display both strength and ductility; and often, ductile materials are tougher than brittle materials, as demonstrated in Fig. 3.7.

Example 3.7: Resilience

Given: In a mining operation, iron ore is dumped into a funnel-shaped hopper that fills box cars for transport by train. The inside of the hopper wears rapidly because the impact of the ore produces plastic deformation in the hopper surface. A change of surface material is considered.

Find: Which is the better choice of hopper surface material, hard steel (AISI 1080) or rubber?

Solution: A key parameter to be used in this evaluation is the resilience of the two materials. For AISI 1080 high-carbon steel from Table A.2, $S_y = 380$ MPa and $E = 207$ GPa. From Eq. (3.16) the modulus of resilience for AISI 1080 steel is

$$(U_r)_{\text{steel}} = \frac{S_y^2}{2E} = \frac{\left(380 \times 10^6\right)^2}{2\left(207 \times 10^9\right)} = 348,800 = 0.3488 \text{ MPa}.$$

For natural rubber, from Tables A.13, $S_u = 30$ MPa and $E =$ 0.004 GPa. Therefore, from Eq. (3.16)

$$(U_r)_{\text{rubber}} = \frac{\left(30 \times 10^6\right)^2}{2\left(4 \times 10^6\right)} = 112.5 \times 10^6 = 112.5 \text{ MPa}.$$

Rubber is over two orders of magnitude more resilient than steel. Based on this parameter only, it would appear that the inside of the hopper should have a rubber lining.

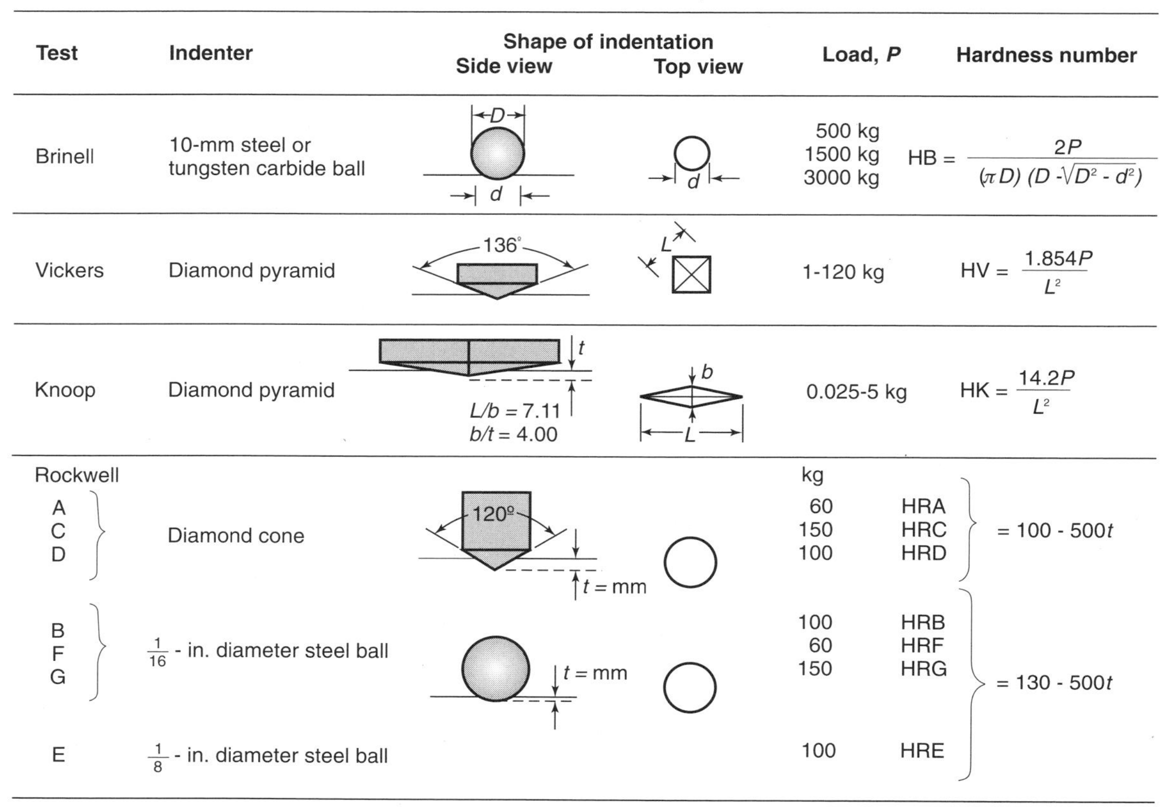

Test	Indenter	Shape of indentation: Side view	Shape of indentation: Top view	Load, P	Hardness number
Brinell	10-mm steel or tungsten carbide ball	D, d	d	500 kg 1500 kg 3000 kg	$HB = \dfrac{2P}{(\pi D)(D - \sqrt{D^2 - d^2})}$
Vickers	Diamond pyramid	136°	L	1-120 kg	$HV = \dfrac{1.854P}{L^2}$
Knoop	Diamond pyramid	t $L/b = 7.11$ $b/t = 4.00$	b, L	0.025-5 kg	$HK = \dfrac{14.2P}{L^2}$
Rockwell A C D	Diamond cone	120º t = mm		kg 60 150 100	HRA HRC HRD = 100 - 500t
B F G	$\frac{1}{16}$ - in. diameter steel ball	t = mm		100 60 150	HRB HRF HRG = 130 - 500t
E	$\frac{1}{8}$ - in. diameter steel ball			100	HRE = 130 - 500t

Figure 3.13: General characteristics of selected hardness tests, with equations for calculating hardness. *Source:* Kalpakjian and Schmid [2010].

3.5.5 Hardness

Hardness is a commonly used property, since it correlates well to material strength and wear resistance. **Hardness** is usually defined as the resistance of a surface to plastic indentation. There are many particular geometries and loads that are useful for defining hardness. Figure 3.13 summarizes some of the popular hardness tests. Some of these are described below, and further information is contained in Kalpakjian and Schmid [2010].

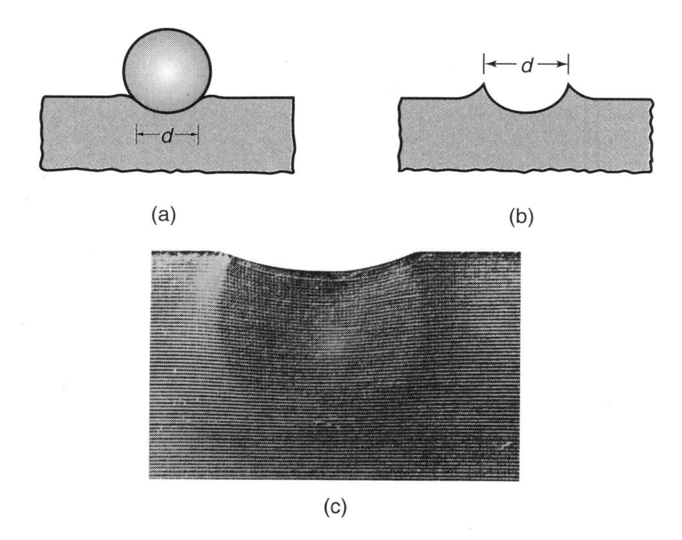

Figure 3.14: Indentation geometry in Brinell hardness testing: (a) annealed metal; (b) work-hardened metal; (c) deformation of mild steel under a spherical indenter. Note that the depth of the permanently deformed zone is about one order of magnitude larger than the depth of indentation. For a hardness test to be valid, this zone should be fully developed in the material. *Source:* Kalpakjian and Schmid [2010].

1. A **Brinell hardness test** involves pressing a steel or tungsten-carbide ball 10 mm in diameter against a surface, with a load of 500, 1500, or 3000 kg (Fig. 3.13). The Brinell hardness number (HB) is defined as the ratio of the applied load in kilograms force to the curved surface area of the indentation given in square millimeters. The harder the material to be tested, the smaller the impression, hence a 1500-kg or 3000-kg load is usually recommended in order to obtain impressions sufficiently large for accurate measurement. Depending on the condition of the material, one of two types of impression develops on the surface after the test (Fig. 3.14) or of any of the other tests described in this section. The impressions in annealed metals generally have a rounded profile (Fig. 3.14a); in cold-worked metals they usually have a sharp profile (Fig. 3.14b). The correct method of measuring the indentation diameter, d, is shown in the figure.

   The indenter, which has a finite elastic modulus, also undergoes elastic deformation under the applied load; as a result, hardness measurements may not be as accurate as expected. One method for minimizing this effect is to use tungsten-carbide balls because of their higher modulus of elasticity. Carbide balls are usually recommended for Brinell hardness numbers greater than 500.

2. A **Rockwell hardness test**, of which there are many variations, measures the depth of penetration instead of the diameter of the indentation. The indenter is pressed onto the surface, first with a minor load and then with a major load; the difference in the depth of penetration is a measure of the hardness of the material. Some of the more common Rockwell hardness scales and the indenters used are shown in Fig. 3.13, but many more forms have been standardized for particular applications. Rockwell superficial hardness tests have also been developed using the same type of indenters but at lighter loads.

3. The **Vickers test** uses a pyramid-shaped diamond indenter and a load that ranges from 1 kg to 120 kg. The Vickers hardness number is indicated by HV. The impressions obtained are typically less than 0.5 mm (0.020 in.) on the diagonal. The Vickers test gives essentially the same hardness number regardless of the load, and is suitable for testing materials with a wide range of hardness, including heat-treated steels. More recently, test procedures have been developed to perform Vickers-type tests in atomic force microscopes and nanoindenters, to estimate hardness at penetration depths as low as 20 nm.

4. The **Knoop test** (Fig. 3.13) uses a diamond indenter in the shape of an elongated pyramid, with applied loads ranging generally from 25 g to 5 kg. The Knoop hardness number is indicated by HK. Because of the light loads that are applied, it is a microhardness test; therefore, it is suitable for very small or very thin specimens, and for brittle materials such as carbides, ceramics, and glass. This test is also used for measuring the hardness of the individual grains and components in a metal alloy. The size of the indentation is generally in the range from 0.01 to 0.10 mm (0.0004 to 0.004 in.); consequently, surface preparation is very important. Because the hardness number obtained depends on the applied load, Knoop test results should always cite the load used.

5. Rebound tests such as the **Scleroscope** and **Leeb tests** are commonly performed. The scleroscope is an instrument in which a diamond-tipped indenter (hammer) enclosed in a glass tube is dropped onto the specimen from a certain height. The hardness is related to the rebound of the indenter: the higher the rebound, the harder the material. The impression made by a scleroscope is very small. Because obtaining reliable results with a scleroscope is difficult, a modern electronic version, called a *Leeb*, or *Equotip*, test, has been developed. In this test, a carbide hammer impacts the surface, and incident and rebound velocities are electronically measured. A Leeb number is then calculated and usually converted to Rockwell or Vickers hardness.

6. The hardness of materials such as rubbers, plastics, and similar soft and elastic nonmetallic materials is generally measured using a **Shore test** with an instrument called a **durometer** (from the Latin *durus*, meaning "hard"). An indenter is pressed against the surface and then a constant load is rapidly applied. The depth of penetration is measured after one second; the hardness is inversely related to the penetration. The hardness numbers in these tests range from 0 to 100.

Because hardness is the resistance to permanent indentation, it can be likened to performing a compression test on a small volume of a material's surface (Fig. 3.14c). Studies have shown that (in the same units) the hardness of a cold-worked metal is about three times its yield stress, S_y; for annealed metals, it is about five times S_y. An additional relationship has been established between the ultimate tensile strength, S_u, and the Brinell hardness, HB, for steels as measured for a load of 3000 kg:

$$S_u = 3.5\text{HB}. \tag{3.17}$$

Note that the units in Eq. (3.17) are important; for the equation to be valid, S_u must be expressed in MPa and the Brinell hardness measured in kilograms force per square millimeter.

3.5.6 Thermal Conductivity

The rate at which heat is conducted through a solid at steady state (meaning that temperature does not vary with time) is a measure of the **thermal conductivity**, K_t. When two bodies at different temperatures are brought together, the faster-moving molecules of the warmer body collide with the slower-moving molecules of the cooler body and transfer some of their motion to the latter. The warmer body loses energy (drops in temperature) while the cooler one gains energy (rises in temperature). The transfer process stops when the two bodies reach the same temperature. This transfer of molecular motion through a material is called **heat conduction**. Materials differ in how fast they conduct heat. The SI unit of thermal conductivity, K_t, is watts per meter-Celsius, and the English unit is British thermal units per foot-hour-Fahrenheit.

Figure 3.15 ranks the thermal conductivity for various metals, polymers, and ceramics. Metals and ceramics in general are good conductors (high K_t) and polymers are good insulators (low K_t). Table 3.1 quantifies the thermal conductivity values given in Fig. 3.15. In Fig. 3.15 and Table 3.1, unless otherwise stated, the temperature is assumed to be room temperature (20°C; 68°F).

3.5.7 Linear Thermal Expansion Coefficient

Different materials may expand more or less when heated. Thermal strain is proportional to the temperature change, and their ratio is constant over a fairly large temperature range. This observation can be expressed by

$$\epsilon = \bar{a}\Delta t_m, \tag{3.18}$$

where $\bar{a}$ is called the *linear expansivity* or **linear thermal expansion coefficient**. The SI unit of $\bar{a}$ is $(°\text{C})^{-1}$; the English unit is $(°\text{F})^{-1}$. Figure 3.16 ranks the linear thermal expansion coefficient for various metals, polymers, and ceramics applied over the temperature range 20 to 200°C (68 to 400°F). Polymers have the highest thermal expansion coefficient, followed by metals and then ceramics. Table 3.1 gives quantitative values of $\bar{a}$ for various metals, polymers, and ceramics from 20 to 200°C (68 to 392°F).

3.5.8 Specific Heat Capacity

The nature of a material determines the amount of heat required to change its temperature by a given amount. Imagine an experiment in which a cast iron ball and a babbitt (lead-based white metal) ball of the same size are heated to the temperature of boiling water and then laid on a block of wax. The cast iron ball would melt a considerable amount of wax, but the babbitt ball, in spite of its greater mass, would melt

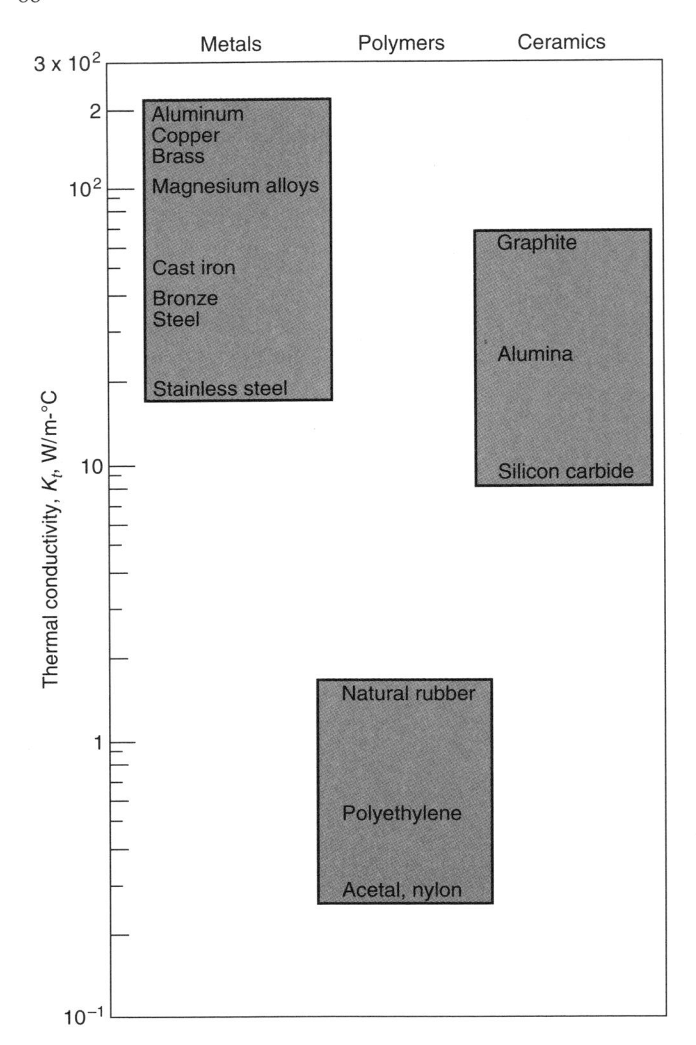

Figure 3.15: Thermal conductivity for various metals, polymers, and ceramics at room temperature (20°C).

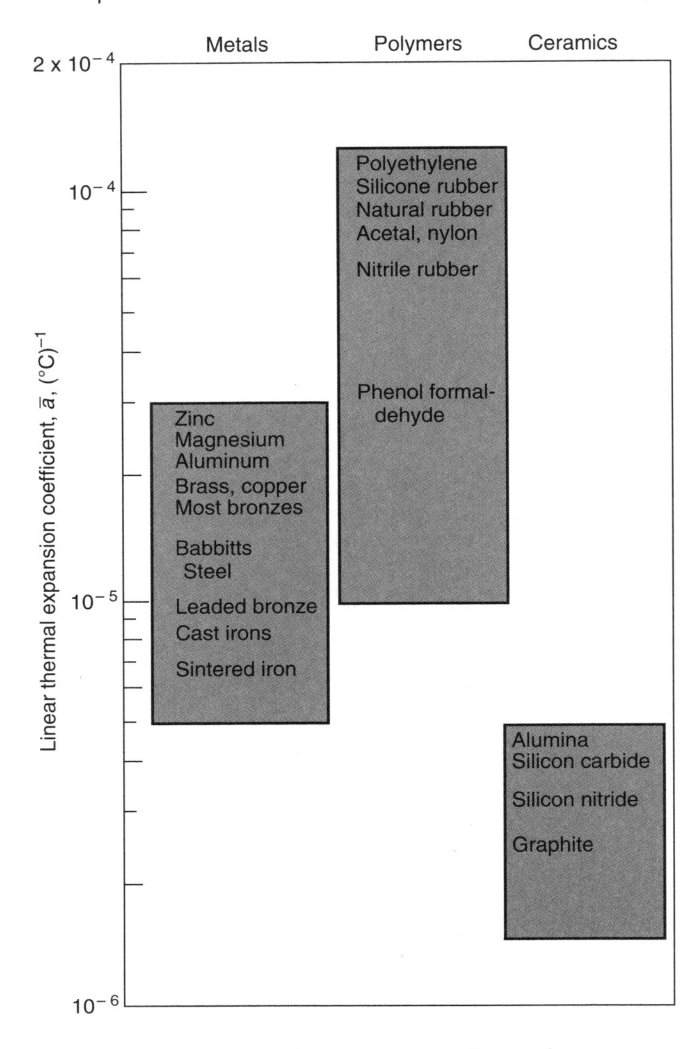

Figure 3.16: Linear thermal expansion coefficient for various metals, polymers, and ceramics at room temperature (20°C).

hardly any. It therefore would seem that different materials, in cooling through the same temperature range, release different amounts of heat.

The quantity of heat energy given up or taken on when a body changes its temperature is given by

$$Q = C_p m_a (\Delta T), \tag{3.19}$$

where
Q = quantity of heat, J
C_p = the **specific heat** of the material, J/(kg-°C)
m_a = mass of body, kg
ΔT= temperature change, °C

Figure 3.17 illustrates the specific heat capacity of various metals, polymers, and ceramics at room temperature (20°C), while Table 3.1 provides quantitative values. Polymers have considerably higher specific heat than metals or ceramics.

Example 3.8: Heat Content of Materials

Given: A thermos is made of two steel bottles, separated by a vacuum, where the inner bottle weighs 200 g and is filled with 500 g of boiling water. The initial temperature of the thermos is 20°C. The specific heat capacity for water C_p = 4180 J/kg-°C.

Find:

(*a*) The maximum temperature of the water in the thermos when the heat has spread to the thermos walls

(*b*) The maximum temperature if the thermos is preheated by hot water, then emptied and refilled.

Solution: From Table 3.1, note that C_p for steel is 450 J/kg-°C.

(*a*) Assuming that no heat is dissipated to the surrounding area,

$$(m_a C_p)_{\text{steel}} (T - 20^\circ\text{C}) = (m_a C_p)_{\text{water}} (100^\circ\text{C} - T),$$

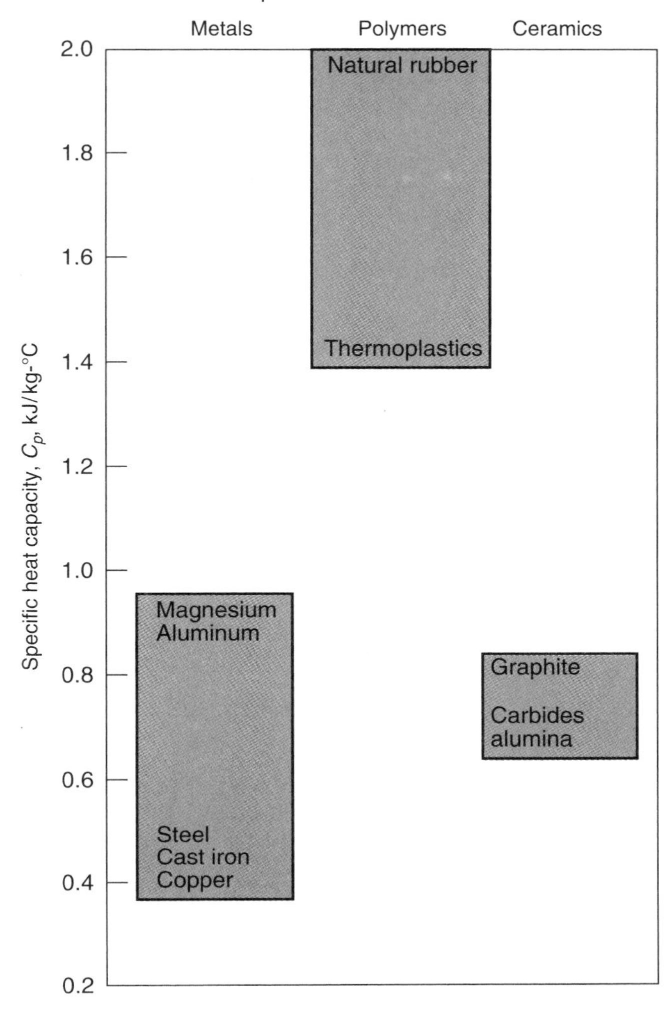

Figure 3.17: Specific heat capacity for various metals, polymers, and ceramics at room temperature (20°C).

or

$$\begin{aligned} T &= \frac{100\,(m_a C_p)_{\text{water}} + 20\,(m_a C_p)_{\text{steel}}}{(m_a C_p)_{\text{water}} + (m_a C_p)_{\text{steel}}} \\ &= \frac{100(0.5)(4.180) + 20(0.2)(0.45)}{(0.5)(4.180) + (0.2)(0.45)} \\ &= 96.7^\circ\text{C}. \end{aligned}$$

(*b*) The only change if the thermos is preheated is that 20°C is replaced with 96.7°C.

$$T = \frac{100(0.5)(4.180) + 96.7(0.2)(0.45)}{(0.5)(4.180) + (0.2)(0.45)} = 99.9^\circ\text{C}.$$

3.5.9 Archard Wear Constant

Wear is more difficult than the other properties of solid materials to quantify, partly because it is a surface phenomenon and not a bulk phenomenon, and also because wear involves interactions between two materials. When solids slide against each other under a normal load, the volume of material lost from the softer surface is characterized by the **Archard wear equation** as:

$$v = k_1 \frac{WL}{3H}, \tag{3.20}$$

where v is the wear volume, W is the normal load, L is the sliding distance, and H is the hardness. k_1 is the **Archard wear constant**, and is dimensionless if H is in N/m^2 or psi. The equations of wear are covered in greater detail in Section 8.9.

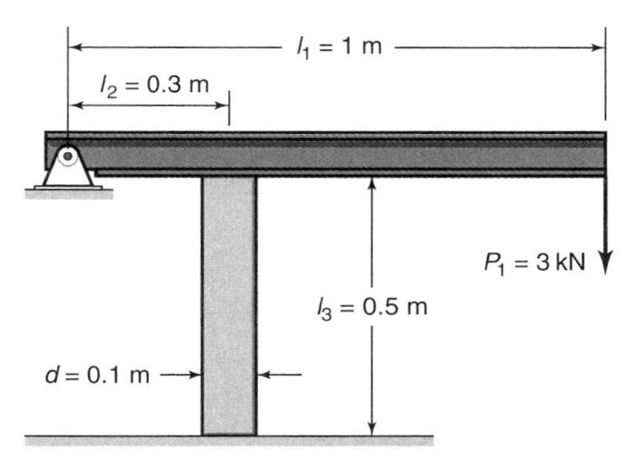

Figure 3.18: Rigid beam assembly use in Example 3.9.

3.6 Stress-Strain Relationships

As discussed earlier in this chapter, the stress-strain diagrams for most engineering materials exhibit a linear relationship between stress and strain within the elastic limit. Therefore, an increase in stress causes a proportionate increase in strain. First discovered by Robert Hooke in 1678, this linear relationship between stress and strain in the elastic range is known as **Hooke's law**. Thus, for a uniaxial stress state, Hooke's law can be expressed as

$$\sigma = E\epsilon. \tag{3.21}$$

For isotropic materials in a triaxial stress state, the stress and strain are related by the generalized Hooke's law given by:

$$\epsilon_1 = \frac{\sigma_1}{E} - \frac{\nu\sigma_2}{E} - \frac{\nu\sigma_3}{E}, \tag{3.22}$$

$$\epsilon_2 = \frac{\sigma_2}{E} - \frac{\nu\sigma_1}{E} - \frac{\nu\sigma_3}{E}, \tag{3.23}$$

$$\epsilon_3 = \frac{\sigma_3}{E} - \frac{\nu\sigma_1}{E} - \frac{\nu\sigma_2}{E}. \tag{3.24}$$

These equations, and versions for some other common conditions, are derived in Appendix B.

Example 3.9: Deformation

Given: A 1-m-long rigid beam shown in Fig. 3.18 is pinned at its left end, carries a 3-kN vertical load at its right end, and is kept horizontal by a vertical pillar located 0.3 m from the left end. The pillar is a 0.5-m-long steel tube with an outer diameter of 0.1 m and a wall thickness of 5 mm. The modulus of elasticity of the steel is 205 GPa.

Find: How much does the right end of the beam deflect due to the 3-kN force?

Solution: Moment equilibrium about the hinge pin gives

$$P_1 l_1 - P_2 l_2 = 0,$$

where P_2 is the force in the pillar. Solving for P_2,

$$P_2 = \frac{P_1 l_1}{l_2} = \frac{(3000)(1)}{0.3} = 10,000 \text{ N}.$$

Because the beam can rotate only about the hinge pin, the deflection δ_1 at P_1 can be described by the angle of rotation, α, and this angle is the same when describing the deflection δ_2 at P_2. Thus, for small rotations, $\delta_1 = l_1\alpha$ and $\delta_2 = l_2\alpha$. This implies that

$$\delta_1 = \frac{l_1 \delta_2}{l_2}. \qquad (a)$$

The compression of the vertical pillar, δ_2, can be obtained by using Hooke's law or

$$\delta_2 = \frac{l_3 P_2}{EA} = \frac{l_3 P_2}{\pi E \left(r_o^2 - r_i^2\right)}. \qquad (b)$$

Substituting Eq. (*b*) into Eq. (*a*) gives

$$\begin{aligned}\delta_1 &= \frac{l_1 l_3 P_2}{\pi l_2 E \left(r_o^2 - r_i^2\right)} \\ &= \frac{1(0.5)(10,000)}{\pi(0.3)\left(205 \times 10^9\right)\left(0.05^2 - 0.045^2\right)} \\ &= 5.45 \times 10^{-5} \text{ m} \\ &= 54.5\,\mu\text{m}.\end{aligned}$$

3.7 Two-Parameter Materials Charts

Material properties limit the performance and life of machine elements. Robust designs seldom derive from consideration of just one property of the solid materials. Thus, the information presented in Section 3.5 on the individual properties of solid materials is not adequate for selecting a material for a particular application. Instead, one or several combinations of properties are needed. Some important property combinations are

- Stiffness versus density (E versus ρ)
- Strength versus density (S versus ρ)
- Stiffness versus strength (E versus S)
- Wear constant versus limiting pressure (k_1 versus p_l)

A number of other combinations might be useful in material selection, but these are the primary considerations in designing machine elements. Further information using the approach in this section, including additional material parameters and applications, are contained in Ashby [2010].

3.7.1 Stiffness versus Density

The modulus of elasticity and density are familiar properties in selecting solid materials. Figure 3.19 shows the full range of elastic modulus, E, and density, ρ, for engineering materials. Data for a particular class of material cluster together and are enclosed by shaded domains. The same class cluster appears on all the diagrams.

Figure 3.19 shows that the moduli of elasticity for engineering materials span five decades from 0.01 to 1000 GPa (1.47 to 147×10^3 ksi); the density spans a factor of 200, from less than 100 to 20,000 kg/m^3.

The chart helps in common problems of material selection for applications in which weight must be minimized. For example, consider a simple tension member where the weight is to be minimized and where the strain cannot exceed a given value, ϵ_{cr}. If the material is loaded up to its yield point, the stress is given by $\sigma = P/A$. Also, from Hooke's law,

$$\sigma = E\epsilon_{cr}.$$

Equating the stresses gives

$$A = \frac{P}{E\epsilon_{cr}}. \qquad (3.25)$$

The weight of the member is

$$W = Agl\rho. \qquad (3.26)$$

Substituting Eq. (3.25) into Eq. (3.26) gives

$$\frac{W}{g} = \left(\frac{Pl}{\epsilon_{cr}}\right)\left(\frac{1}{E/\rho}\right). \qquad (3.27)$$

Note that the first fraction contains design constraints and the second fraction contains all of the relevant material properties. Thus, the optimum material for minimizing Eq. (3.27) is one that maximizes the quantity E/ρ. In Fig. 3.19 this is accomplished by considering lines parallel to the $E/\rho = C$ reference. Those materials with the greatest value in the direction normal to these lines are the optimum materials (i.e., those farthest toward the top and left of the charts). In Fig. 3.19, the reference lines refer to the minimum-weight design subjected to strain requirements under the following conditions:

$E/\rho = C$	Minimum-weight design of stiff tension members
$E^{1/2}/\rho = C$	Minimum-weight design of stiff beams and columns
$E^{1/3}/\rho = C$	Minimum-weight design of stiff plates

Example 3.10: Material for a Solid Fishing Rod

Given: A fishing rod is to be made of a material that gives low weight and high stiffness.

Find: From Fig. 3.19 determine which is better, a rod made of plastic (without fiber reinforcement) or a split-cane rod (bamboo fibers glued together).

Solution: Figure 3.19 shows that only very special polymers have moduli of elasticity as high as the best wooden fibers. The polymers are also two to three times more dense than wood. A split-cane rod will therefore give a lower weight for a given stiffness than any plastic.

3.7.2 Strength versus Density

The weight of an object depends on its volume and its density. Strength, on the other hand, means different things for different classes of solid material. For metals it is the yield

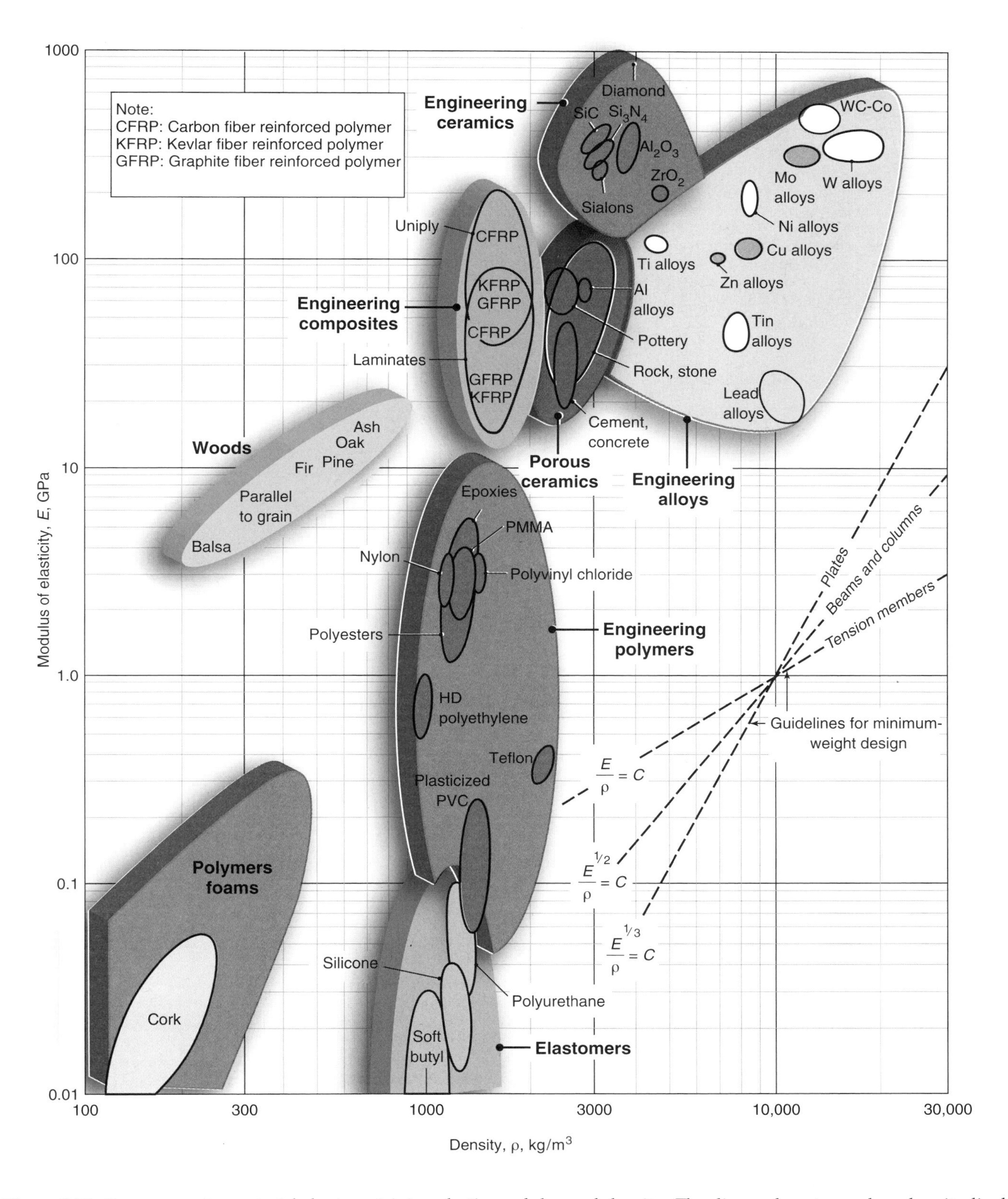

Figure 3.19: Two-parameter material chart containing elastic modulus and density. The diagonal contours show longitudinal wave velocity. The guidelines of constant E/ρ, $E^{1/2}/\rho$ and $E^{1/3}/\rho$ allow selection of materials for minimum-weight, deflection-limited design. *Source:* Adapted from Ashby [2010].

strength, which is the same in tension and compression. For brittle ceramics it is the crushing strength in compression, not that in tension, which is about 15 times smaller. For elastomers strength means the fracture strength. For composites it is the tensile failure strength (compressive strength can be lower because of fiber buckling).

Figure 3.20 shows these strengths, for which the symbol S is used (despite the different failure mechanisms involved), plotted against density, ρ. The considerable vertical extension of the strength domain for an individual material reflects its wide range of available properties, caused by the degree of alloying, work hardening, grain size, porosity, etc.

Figure 3.20 is useful for determining optimum materials based on strength where deformation under loading is not an issue. Just as before, one chooses a reference line, and materials located at the greatest distance from this line (up and to the left) are superior. The following circumstances correspond to the reference lines in Fig. 3.20:

$\sigma/\rho = C$	Tension members
$\sigma^{2/3}/\rho = C$	Beams and shafts
$\sigma^{1/2}/\rho = C$	Plates

The range of strength for engineering materials spans five decades, from 0.1 MPa (foams used in packaging and energy-absorbing systems) to 10^4 MPa (diamond). The range of density is the same as in Fig. 3.19.

Example 3.11: Material for a Tubular Fishing Rod

Given: The fishing rod given in Example 3.10 is manufactured in the form of a tapered tube with a given wall thickness distributed along its length.

Find: The material that makes the rod as strong as possible for a given weight.

Solution: Figure 3.20 shows that the strongest materials for a given density are diamond and silicon carbide and other ceramics. It is difficult and expensive to use these as fishing rod materials. The best choice is carbon-fiber-reinforced plastic or glass-fiber-reinforced plastic that has 800- to 1000-MPa strength for a density of 1500 kg/m^3.

3.7.3 Stiffness versus Strength

Figure 3.21 plots modulus of elasticity versus strength. Note that contours of normalized strength, S/E, etc., appear as a family of straight parallel lines.

The reference lines in Fig. 3.21 are useful for the following circumstances:

$S/E = C$	Design of seals and hinges
$S^{3/2}/E = C$	Elastic components such as knife-edges and diaphragms
$S^2/E = C$	Elastic energy storage per volume (for compact energy adsorption)

Example 3.12: Elastic Strain

Given: The springs in a car suspension can be made of rubber, steel, or a uniply carbon-fiber-reinforced plastic. The geometries of the different suspension springs are quite different, depending on allowable elastic deformations.

Find: The maximum elastic strains in the three types of spring if the rubber is polyurethane (PU) and the steel has a strength of 1 GPa.

Solution: From Fig. 3.21, PU rubber has a strength of around 30 MPa and a modulus of elasticity of 0.05 GPa. The maximum elastic strain is

$$\left(\frac{S}{E}\right)_{\text{rubber}} = \frac{30}{50} = 0.60$$

Likewise, for steel and carbon-fiber-reinforced plastic

$$\left(\frac{S}{E}\right)_{\text{steel}} = \frac{1}{205} = 0.005$$

$$\left(\frac{S}{E}\right)_{\text{plastic}} = \frac{1}{200} = 0.005$$

The rubber has a maximum elastic strain of 60%, whereas the steel and carbon-fiber-reinforced plastic springs have a maximum elastic strain of 0.5%. Also, from Fig. 3.20 the steel spring will be five times heavier than the carbon-fiber-reinforced plastic spring. It will be seen in Chapter 17 that steel can be made into an effective spring by utilizing a helical shape, or coil.

3.7.4 Wear Rate versus Limiting Pressure

Wear presents a new set of problems in attempting to choose a solid material. If the materials are unlubricated, sliding motion is occurring, and if one of the surfaces is steel, the wear rate is defined as

$$W_r = \frac{\text{Volume of material removed}}{\text{Sliding distance}} = \frac{v}{L}. \tag{3.28}$$

The wear rate W_r thus has the SI unit of square meters. At low limiting pressure, p_l,

$$W_r = k_1 A p_l, \tag{3.29}$$

where k_1 is the Archard wear constant, A is the contact area, and p_l is the limiting pressure. Figure 3.22 shows the constant k_1 as a function of limiting pressure p_l. Each class cluster shows the constant value of k_1 at low p_l and the steep rise as p_l is approached. Materials cannot be used above their limiting pressure.

Example 3.13: Design for Wear

Given: A polytetrafluoroethylene (PTFE, or Teflon) slider is in contact with high-carbon steel. The sliding distance is 300 m, and the thickness of the Teflon layer allowed to be worn away is 3 mm.

Find: How large does the PTFE slider surface have to be so that it will not have excessive wear and the limiting pressure will not be exceeded if the load carried is 10 MN?

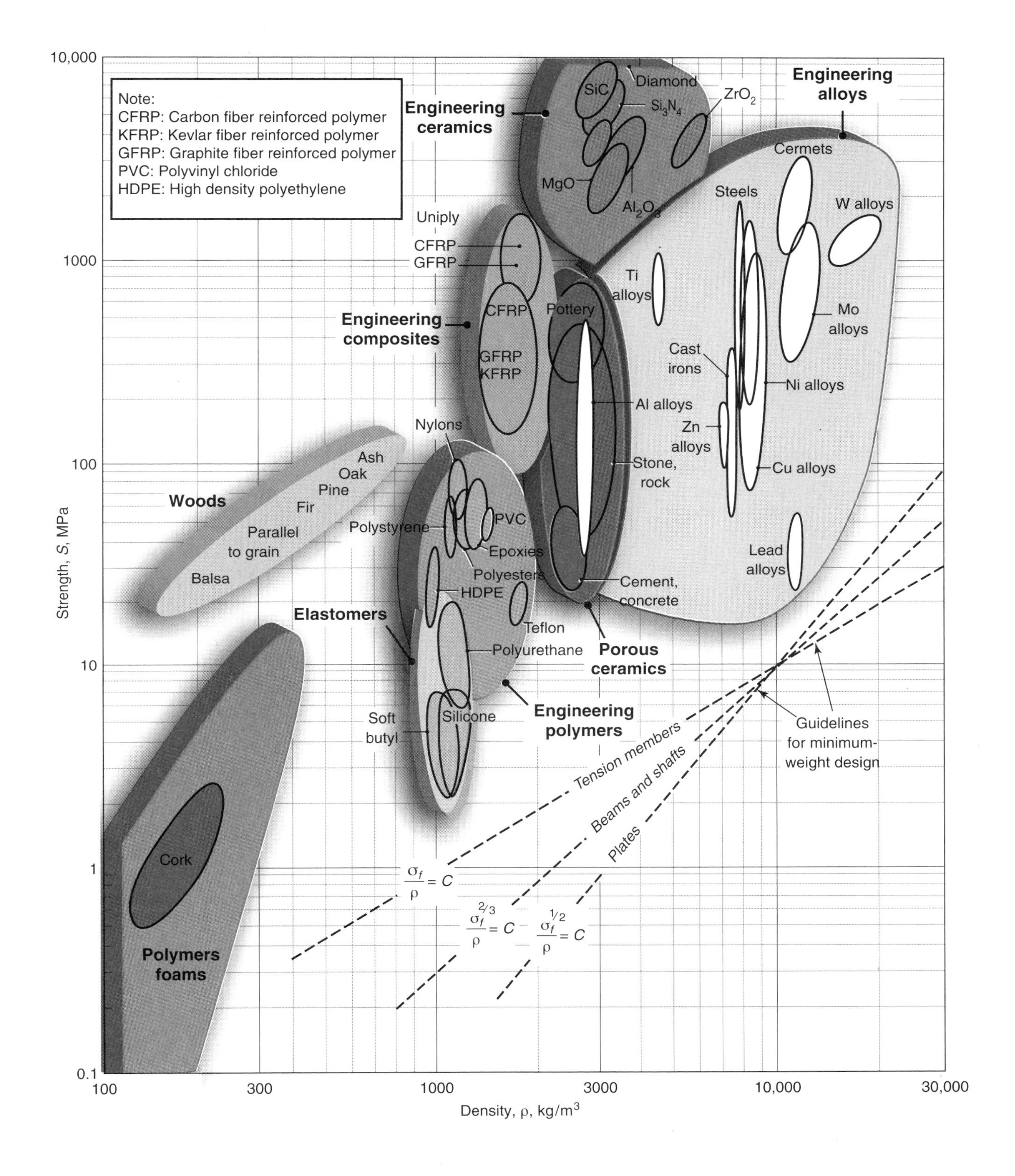

Figure 3.20: Strength versus density. The guidelines of constant S/ρ, $S^{2/3}/\rho$, and $S^{1/2}/\rho$ allow selection of materials for minimum-weight, yield-limited design. *Source:* Adapted from Ashby [2010].

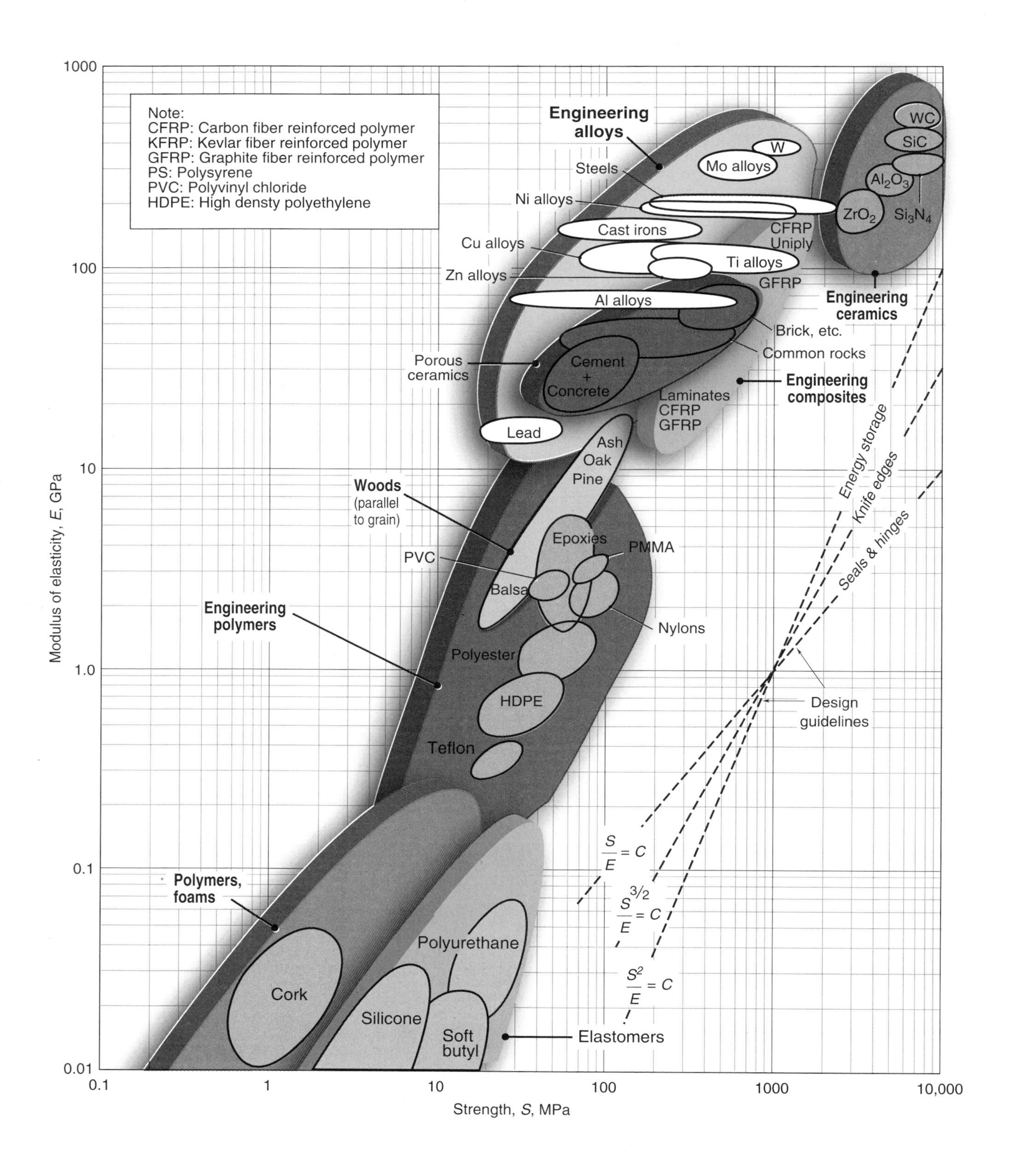

Figure 3.21: A plot of elastic modulus versus strength for engineering materials. The design guidelines help with the selection of materials for such machine elements as springs, knife-edges, diaphragms, and hinges. *Source:* Adapted from Ashby [2010].

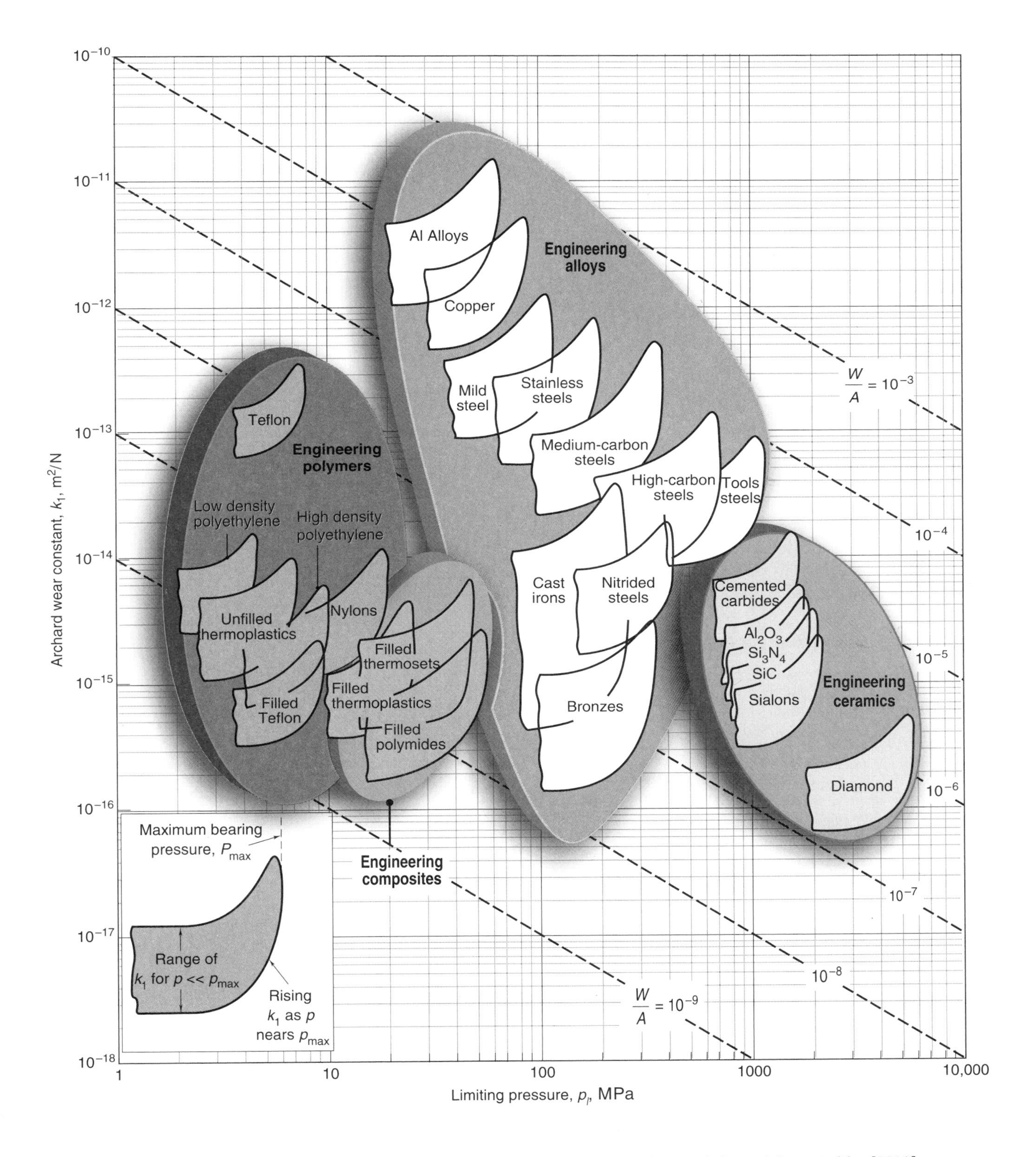

Figure 3.22: Archard wear constant as a function of limiting pressure. *Source:* Adapted from Ashby [2010].

Solution: From Fig. 3.22, the limiting pressure for PTFE on steel is p_l = 8 MPa, and the Archard wear constant is $k_1 = 2 \times 10^{-13}$ m²/N. From Eq. (3.29),

$$\frac{W_r}{A} = k_1 p_l = \left(2 \times 10^{-13}\right)\left(8 \times 10^6\right) = 1.6 \times 10^{-6}.$$

The worn volume of the material is

$$At_h = W_r L,$$

where L is the sliding distance and t_h is the wear depth. Therefore,

$$\frac{W_r}{A} = \frac{t_h}{L} = \frac{0.003}{300} = 10^{-5}.$$

The pressure can be written as

$$p = \left(\frac{W_r}{A}\right)\frac{1}{k_1} = \frac{10^{-5}}{2 \times 10^{-13}} = 0.5 \times 10^8 \text{ Pa} = 50 \text{ MPa}.$$

Since $p \gg p_l$, the limiting pressure is needed to determine the size of the slider. Thus,

$$p_l A = \left(10 \times 10^6\right) \text{ N} = 10^7 \text{ N}.$$

Solving for the area,

$$A = \frac{10^7}{p_l} = \frac{10^7}{8 \times 10^6} = 1.25 \text{ m}^2.$$

The surface area has to be 1.25 m² in order to avoid excessive compressive stress. For these conditions, the wear depth will be only 0.48 mm.

3.7.5 Young's Modulus versus Relative Cost

In practice, design engineers consider cost much more than it has been considered thus far in this text. Figure 3.23 shows the stiffness of a material versus the relative cost (i.e., the cost per weight of the material divided by the cost per weight of mild steel). The reference lines are useful for the following:

$E/C_R\rho = C$	Minimum-cost design of stiff tension members
$E^{1/2}/C_R\rho = C$	Minimum-cost design of stiff beams and columns
$E^{1/3}/C_R\rho = C$	Minimum-cost design of stiff plates

Figure 3.23 does much to explain why steel and concrete are so valuable as building materials for public works projects where cost is to be minimized. Although a bridge manufactured from teflon is certainly possible, it would be far more costly than a steel-and-concrete bridge.

3.8 Effects of Manufacturing

The proper selection of an engineering material is a critical task for successful design. Equally as important for its performance and economic impact is the selection of a manufacturing process or processes for each component. Selection of a manufacturing process has a large effect on the material's microstructure and can dramatically affect the strength, ductility, and other material properties.

This section is a brief introduction to manufacturing process effects on machine element design. Much more information is contained in Kalpakjian and Schmid [2010], or Schey [2000].

3.8.1 Manufacture of Metals

Casting

A wide variety of casting processes are available, all of which involve molten (liquid) metal solidifying within a mold. Castings in general have a microstructure that contains a large number of micropores. In tension, these pores act as stress risers, while in compression, the pores close upon themselves; the main result is a much higher strength in compression than in tension. Castings can have limited ductility because of this microstructure, but ductility can be improved by annealing. The main advantages to castings are low cost, especially for moderate production runs, and highly intricate shapes.

Casting processes are usually classified as expendable mold-expendable pattern; expendable mold-permanent pattern; or permanent mold processes. Some of the most common casting processes are:

- **Expendable mold, expendable pattern.** Investment casting and evaporative pattern casting are common examples of this class of casting process. In investment casting, a pattern is created from a low melting point solid such as a thermoplastic or wax. This pattern is then coated by successive dips into a slurry (ceramic particles suspended in water). Once a desired coating thickness has been developed, the coated pattern is placed in an oven, melting the wax and leaving a cavity. Molten metal is then poured into the cavity, which solidifies in the shape of the original pattern.

  In evaporative pattern casting, a polystyrene foam pattern is produced with the desired shape of the metal part; this foam is then buried in sand. Molten metal is poured onto the foam, evaporating the polystyrene and displacing it in the sand cavity. After solidification, the mold is broken and the part is removed.

- **Expendable mold, permanent pattern.** This class of operations includes sand casting (shown in Fig. 3.24), shell casting, plaster-mold, and ceramic-mold casting. Sand casting is the most prevalent form of casting, with typical applications including machine tool bases, engine blocks, and machine housings. In sand casting and other such processes, a mold is created from a pattern, but the pattern can be reused for many parts.

- **Permanent mold.** Including processes such as die casting, pressure casting, and centrifugal casting, permanent mold processes have high tooling costs and are therefore limited to large production runs. In these operations, the desired part shape is produced into a mold of a metal with a higher melting temperature than the workpiece, or graphite in some applications. Molten metal is injected under high pressure into the mold cavity, so that it fills the mold completely before solidification. Examples of parts produced in permanent mold operations are transmission housings, valve bodies, hand tools, computer housings, toys, and camera frames. Permanent mold processes generally have better surface finish and tolerances than other casting operations.

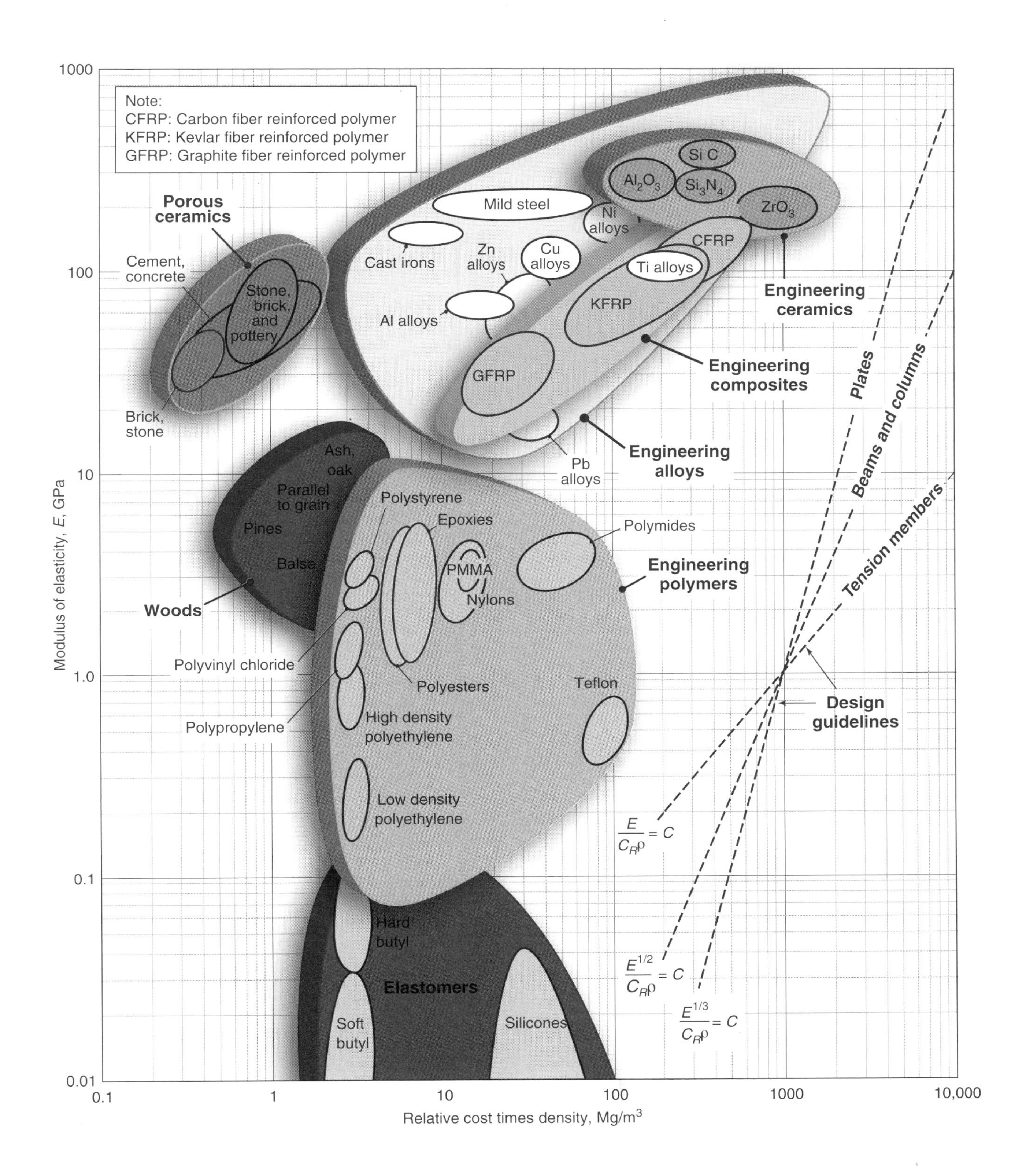

Figure 3.23: Modulus of elasticity as a function of the product of cost and density. The reference lines help with selection of materials for machine elements. *Source:* After Ashby [2010].

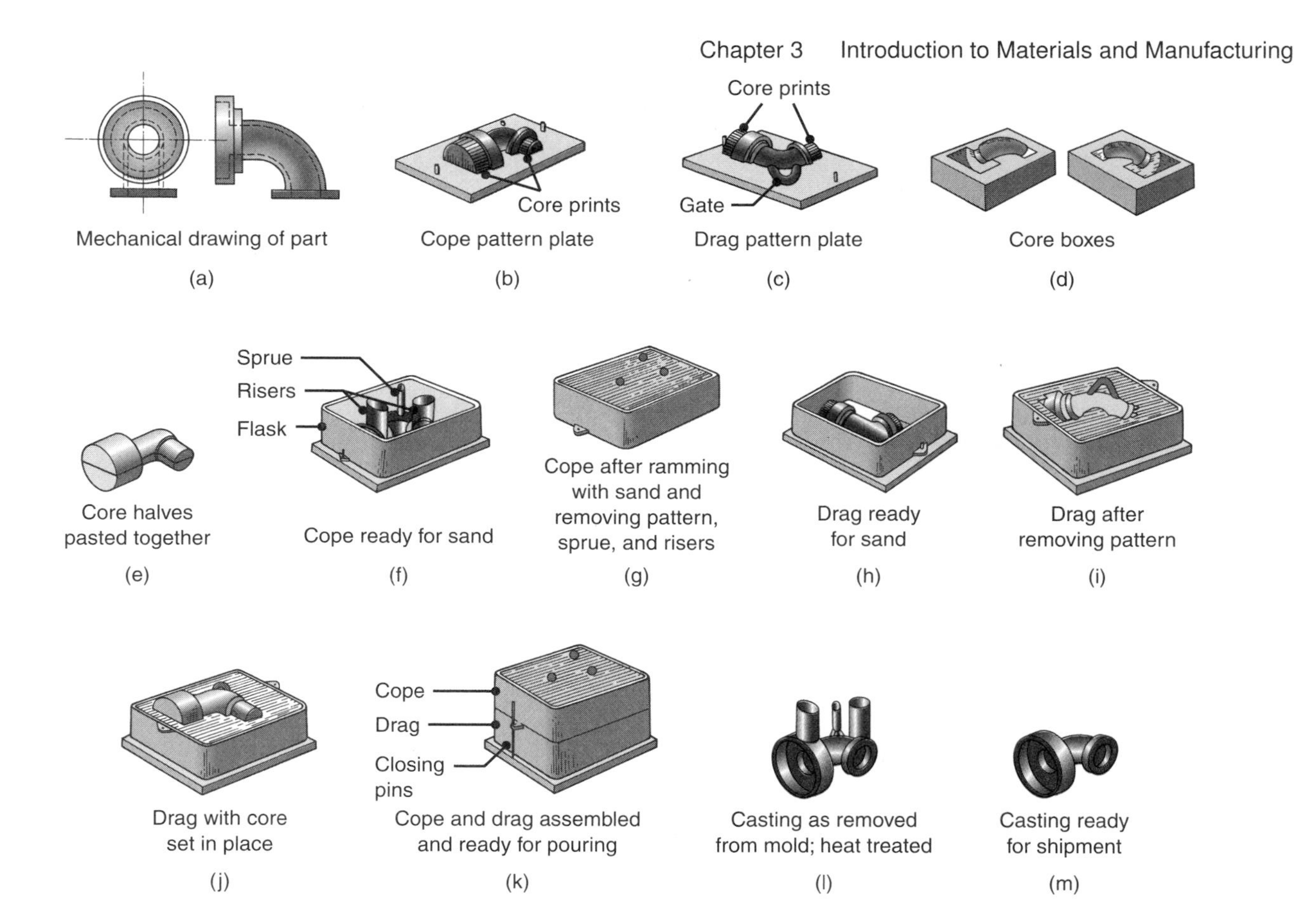

Figure 3.24: Schematic illustration of the sequence of operations for sand casting. (a) A mechanical drawing of the part is used to generate a design for the pattern. Considerations such as part shrinkage and tapers must be built into the drawing. (b, c) Patterns have been mounted on plates equipped with pins for alignment. Note the presence of core prints designed to hold the core in place. (d, e) Core boxes produce core halves, which are pasted together. The cores will be used to produce the hollow area of the part shown in (a). (f) The cope half of the mold is assembled by securing the cope pattern plate to the flask with aligning pins and attaching inserts to form the sprue and risers. (g) The flask is rammed with sand and the plate and inserts are removed. (h) The drag half is produced in a similar manner with the pattern inserted. A bottom board is placed below the drag and aligned with pins. (i) The pattern, flask, and bottom board are inverted; and the pattern is withdrawn, leaving the appropriate imprint. (j) The core is set in place within the drag cavity. (k) The mold is closed by placing the cope on top of the drag and securing the assembly with pins. The flasks then are subjected to pressure to counteract buoyant forces in the liquid, which might lift the cope. (l) After the metal solidifies, the casting is removed from the mold. (m) The sprue and risers are cut off and recycled, and the casting is cleaned, inspected, and heat treated (when necessary). *Source:* Courtesy of the Steel Founders' Society of America.

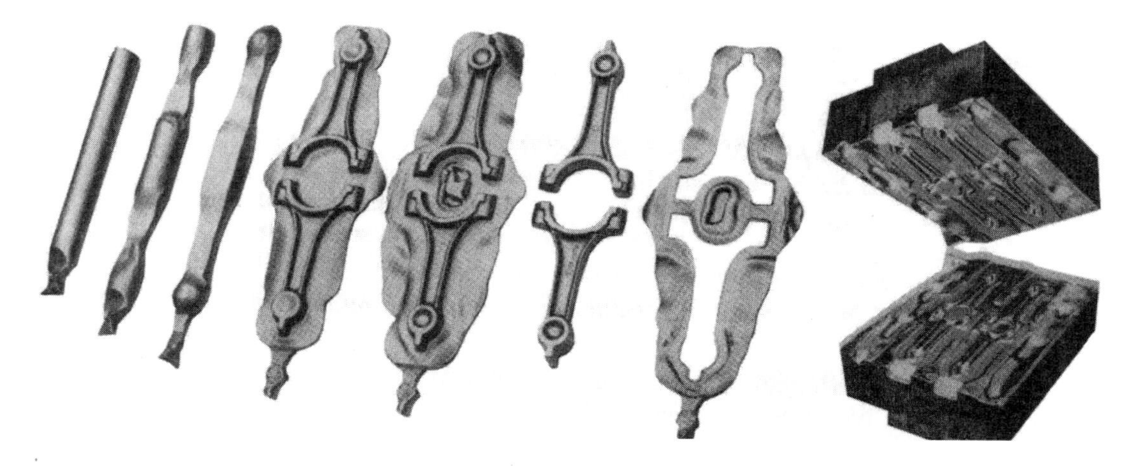

Figure 3.25: An example of the steps in forging a connecting rod for an internal combustion engine, and the die used. *Source:* Schey [2000].

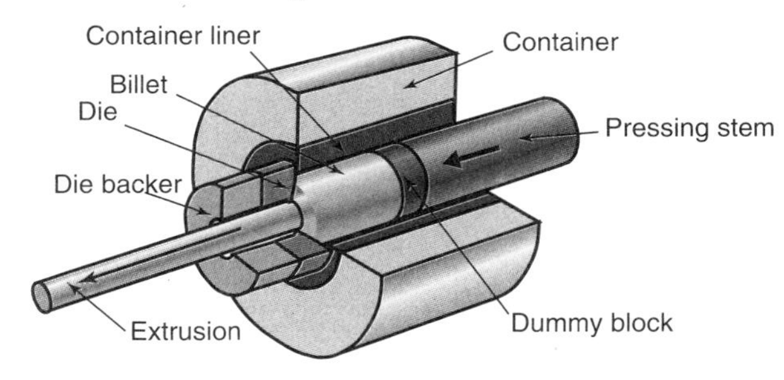

(a)

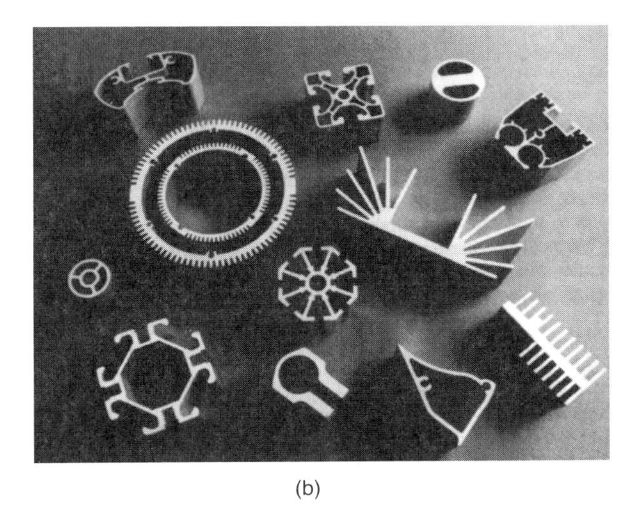

(b)

Figure 3.26: The extrusion process. (a) Schematic illustration of the forward or direct extrusion process *Source:* Kalpakjian and Schmid [2010]. (b) Examples of extruded cross sections. *Source:* Schey [2000].

Bulk Forming

In bulk forming processes, materials are subjected to large strains in order to achieve the desired shape. Thus, the material must be quite ductile in order to be formed without fracturing. Some of the common bulk forming operations are:

- **Forging**. Forging is controlled deformation of metal through the application of compressive stresses. In *open die forging*, simple die shapes such as flats and rounds are used to obtain a rough shape in the workpiece. In *closed die forging*, as shown in Fig. 3.25, a cavity is carefully prepared in a die, so that the metal will conform to and acquire the cavity shape during forging. Note that the workpiece in forging will develop *flash*, which must be trimmed off; the flash is necessary to ensure that the metal completely fills the die.

- **Extrusion**. In extrusion, shown in Fig. 3.26a, a *billet* is pushed through a die to produce a product with a constant cross-section. Structural shapes are commonly produced through extrusion, and smaller parts can be produced by cutting extrusions to desired lengths.

- **Rolling**. Arguably the most common bulk forming operation, rolling is performed on approximately 90% of metals. In *flat rolling*, a billet is reduced in thickness through the compressive stress applied by two rollers. Rolling can also be used to produce structural shapes, tubes, rings (such as bearing races), spheres (such as balls in rolling element bearings), and screw threads.

Bulk deformation can take place at elevated temperatures (hot working) to exploit increases in material ductility and decreases in strength and stiffness at elevated temperatures. Cold working has associated with it superior surface finish, improved mechanical properties due to strain hardening, and a more refined microstructure.

Sheet Forming

One of the main advantages of metals is that they have sufficient ductility to be rolled into thin sheets. These sheets are then used for further processing, allowing the economic production of high-quality parts with large aspect ratios. Since sheet metals are cold-rolled, they will usually have small, elongated grains, some anisotropy, and higher strength than bulk forms of the same material.

The most common sheet forming operations are bending, stretch forming, and deep drawing. In bending, a sheet or tube is forced around a mandrel to a desired shape, or sheet metal is plastically deformed in a die with the desired bend shape. In stretch forming, a sheet is forced between two dies with the desired profile; the sheet-metal plastically deforms to match the die profile. In deep drawing, a cup-shaped part is produced by forcing a sheet or blank into a die cavity; the punch and die have a clearance that is slightly larger than the sheet thickness to avoid shearing the blank. Typical deep drawn parts include cookware, oil pans, and beverage containers.

Powder Metallurgy (PM)

Metals can be produced in powder form through various approaches. These powders can be further processed by a number of methods, including:

- **Pressing and sintering**. The powder is placed inside a die cavity and compressed under high pressure; the part that is ejected is called a *green compact* and has a strength comparable to chalk. These parts are then *sintered*, or heated in a controlled atmosphere furnace at up to 90% of their absolute melting temperature for up to four hours. In the sintering furnace, the powder particles fuse and develop a strong bond.

- **Metal injection molding**. The powder is mixed with a polymer binder and the mixture is processed by injection molding, as described below. After molding, the part is sintered and perhaps infiltrated by a metal with a lower melting point.

- **Cold and hot isostatic pressing**. The powder is placed in a compliant mold and then placed into a pressurized chamber. The high pressures that result yield a PM part that is strong and has tight tolerances.

PM parts are very porous; they can have up to 20% porosity. As opposed to castings, the pores in PM are not isolated; they are all interconnected. As such, PM parts can be thought of as a sponge, in that once infiltrated by a lubricant, the lubricant is always present. This is one of the main reasons that PM parts are very popular for tribological applications such as gears, cams, bearings, and sleeves.

Machining

Material removal processes such as metal cutting, grinding, or electrical discharge machining are used to remove material from a bulk form (or near-net shaped form) to achieve desired surface finishes, tolerances, or shapes that are difficult to obtain otherwise. The machining process affects part design in a number of ways, including:

- All machining operations result in feed marks on the workpiece, which can limit the fatigue life. This topic will be discussed in detail in Chapter 7.
- Machining is comparatively expensive and slow.
- Very smooth surfaces or very low tolerances can be produced only through expensive manufacturing operations. Thus, a goal of designers is to specify rough surfaces whenever possible in order to have maximum economy in their designs.
- Machining does not affect the material microstructure. Thus, a brittle material will not become ductile because its surface was machined.
- Machining requires parts to be held in fixtures during machining. Therefore, designers need to incorporate clamping and fixturing locations in their parts to allow machining.
- It is difficult to produce sharp external corners in casting or bulk forming operations, but it is easier to produce sharp external corners in machining.

3.8.2 Manufacture of Polymers

Polymers are produced by a wide variety of manufacturing operations; only a few very popular approaches are described here. There are two basic forms of polymers, and these determine the manufacturing strategy that will be used. Thermoplastics are polymers that have a defined melting point; heating them greatly reduces their strength and allows them to flow into desired shapes. Once they are cooled, they regain their strength and hardness. Thermosets chemically degrade when heated long before they melt. The basic strategy used with thermosets is to form the polymer constituents into the desired shape, then through the application of heat and/or pressure, cause these polymers to cure, or set. After the cure cycle, the thermoset part is ready for use.

An innovative set of processes, called *rapid prototyping*, are described in the Case Study at the end of this chapter. Rapid prototyping is experiencing rapid development; some of the latest advanced materials in rapid prototyping are described in Appendix A.15.

Thermoplastic Manufacture

Thermoplastics are very common, versatile materials, available in a wide variety of shapes and colors. Thermoplastics are produced from their chemical constituents and are available in bulk, pellet, or powder form for processing into their final desired shape. It should be noted that a "melted" polymer is quite viscous; its consistency can be closer to bread dough or soft taffy than a liquid. This should be understood when trying to visualize the processes described here.

Among the common methods of processing thermoplastics are:

- **Extrusion**. In extrusion, a polymer in pellet or powder form is heated in an extruder, shown in Fig. 3.27. The polymer is melted in the extruder barrel by the heat input from heating elements as well as the friction between pellets and between pellets and screw. As a result, a liquid polymer is forced out through a die with a desired cavity, resulting in a part with a constant cross section. The polymer is then cooled, usually by forced cool air convection. Polymer tubes, structural members, and rods are produced through extrusion.
- **Injection molding**. Instead of forcing the polymer through a die opening, the polymer is injected into a die cavity. Usually, metal dies are used that incorporate channels for coolant, so that the heat from the polymer is quickly removed. Once solid, the polymer part is ejected and another cycle can begin. Injection molding is extremely popular; a wide variety of automotive, consumer electronic, and home use products are produced by injection molding.
- **Thermoforming**. Thermoforming involves the use of an extruded film that is heated and draped over a (usually) metal die with an intended shape. The soft thermoplastic complies with and cools against the die, and hardens. Plastic packaging, advertising signs, refrigerator liners, and the Invisalign product in Case Study 1.4 are examples of thermoformed products.
- **Blow molding**. In extrusion blow molding, an extruded tube is clamped and expanded by internal pressure against a die-defined cavity. In injection blow molding, a short tubular piece called a *parison* is first injection molded, and then transferred to a blow molding die. Blow molding produces hollow containers such as plastic beverage bottles.

Manufacture of Thermosets

Thermosets are also very common. When compared to thermoplastics, thermosets in general have better strength (especially at elevated temperatures) and strength-to-weight ratios. As discussed above, thermosets are manufactured by blending polymer constituents, forming them into the desired shape, and then curing the plastic. Some of the important thermoset manufacturing operations are:

1. **Reaction injection molding**. Similar to injection molding described above, reaction injection molding involves injecting liquid thermoset into a heated die; the elevated temperature then allows curing of the thermoset.
2. **Compression molding**. Compression molding shares many similarities with forging. A powder or clay-like consistency polymer is placed into a heated mold and is compressed to fill the cavity. The polymer then cures into the shape defined by the mold. Compression molded parts have good tolerances and surface finish, and yield a polymer with high molecular weight and crystallinity (and therefore strength) compared to other proceses.

3.8.3 Manufacture of Ceramics

Ceramics share many processing similarities to PM techniques. Ceramics are processed from powder form, usually by mixing with water and binders. Once formed, a ceramic part will be placed into a furnace, or kiln, to fuse the particles. In all processing techniques, the ceramics will be somewhat porous, and the mechanical properties will depend on this porosity. The most common processes for producing ceramic parts are:

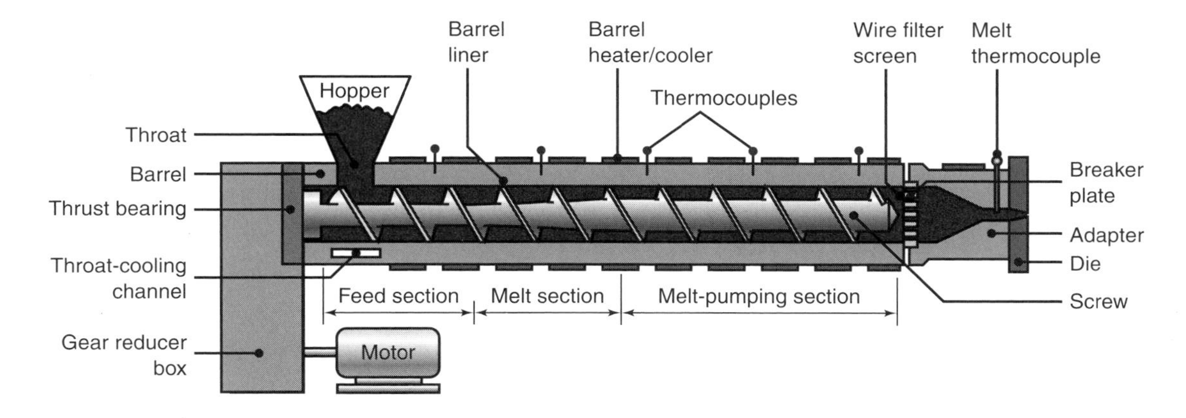

Figure 3.27: Schematic illustration of a typical extruder. *Source:* Kalpakjian and Schmid [2010].

- **Slip casting**. In slip casting, a ceramic slurry is poured into a permeable mold. Some of the water in the slurry diffuses into the mold, leaving a locally high slurry concentration next to the mold wall. After a few minutes, the slurry is poured from the mold, leaving a coating adhering to the mold. After a few hours, more of the water has been extracted from the slurry and it is removed from the mold. It is very fragile at this step and is fired in a furnace or kiln to fuse the ceramic particles.

- **Dry or wet pressing**. This approach is similar to pressing of metal powders; the resultant compact needs to be fired in a furnace or kiln to develop strength.

- **Doctor-blade process**. In this operation, a claylike consistency of ceramic is spread by a blade. The ceramic can be spread into a sheet or can be formed into a desired shape such as plates or bowls (*jiggering*).

- **Pressing and injection molding**. These are similar to the PM and polymer processes of the same name, respectively. The part from these operations needs to be fired to develop strength.

Ceramics and PM parts share many design considerations. Since there is significant shrinkage during sintering or firing, there is a possibility for warpage in large parts, and thin cross sections are likely to fracture. Sharp corners will crumble while the part is in the compact stage, and chamfers are preferable to radii because of tooling design issues.

3.8.4 Selection of Manufacturing Processes

Manufacturing process selection is a difficult task and must incorporate design parameters as well as economic considerations. A full discussion of manufacturing process selection is beyond the scope of this text, and the interested reader is again directed to the text by Kalpakjian and Schmid [2010]. However, it should be recognized in the design of machine elements that there are certain trends in processes used. For example:

- Processes that are advantageous for small production runs usually have low capital equipment costs, but high labor costs, whereas hard automation is expensive but has low labor costs. Therefore, the cost per part can be greatly reduced if the parts are produced in quantities large enough to justify purchase of hard automation. In practice, this means that bolts, gears, bearings, etc, are mass produced in standard sizes at far greater economy than if produced to order. From a design standpoint, it is therefore important to specify dimensions as standard sizes, such as 1 in. instead of 0.9456 in., or 50 mm instead of 43.6 mm.

- It should be recognized that product quality and robustness also increase with large production runs.

- Certain materials are commercially available in only some forms (see Table 3.2). The desire to use materials in other forms is not impossible, but is probably not economically advisable.

- Manufacturing processes are determined, to a great extent, by the design requirements of size, strength, tolerance, and surface finish. Figure 3.28 can be used as a guide to help select manufacturing processes based on tolerance and roughness.

Table 3.2: Commercially available forms of materials.

Material	Available forms[a]
Aluminum	B, F, I, P, S, T, W
Ceramics	B, p, s, T
Copper and brass	B, f, I, P, s, T, W
Elastomers	b, P, T
Glass	B, P, s, T, W
Graphite	B, P, s, T, W
Magnesium	B, I, P, S, T, w
Plastics	B, f, P, T, w
Precious metals	B, F, I P, t, W
Steels and stainless steels	B, I, P, S, T, W
Zinc	F, I, P, W

[a] B = bar and rod; F = foil; I = ingot; P = plate and sheet; S = structural shapes; T = tubing; W = wire. Lowercase letters indicate limited availability. Most of the metals are also available in powder form, including prealloyed powders.

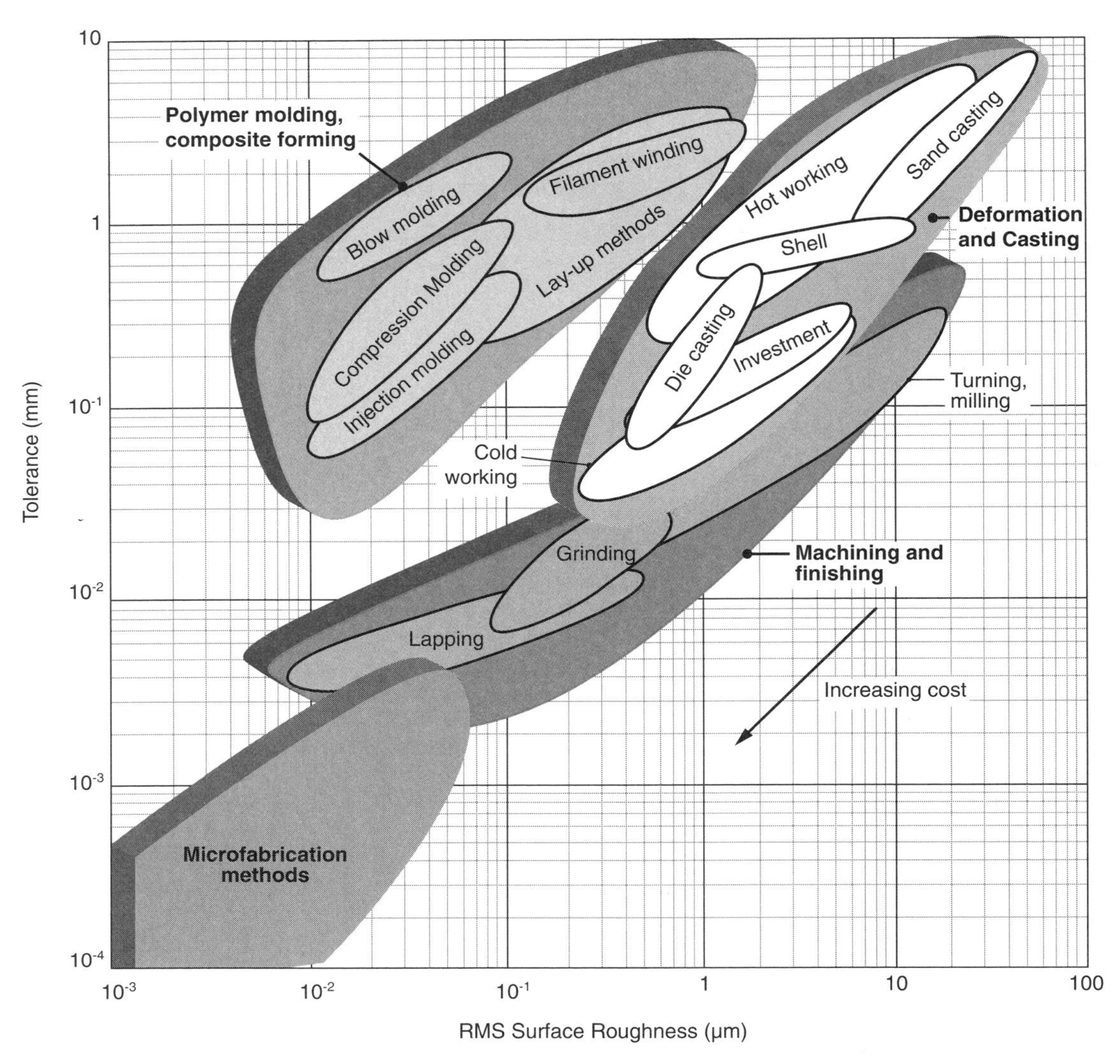

Figure 3.28: A plot of achievable tolerance versus surface roughness for assorted manufacturing operations. The dashed lines indicate cost factors; an increase in precision corresponding to the separation of two neighboring lines gives an increase in cost for a given process, or a factor of two. *Source:* Adapted from Ashby [2010].

Case Study: The Maker Movement

Since the 1980s, a large number of manufacturing processes have been developed that use computer representations of parts and computer control to precisely deposit material where desired. These processes are unique in that they create parts layer by layer, using thin slices to get very good part definition. Figure 3.29 summarizes some of the most popular **rapid prototyping** operations, which are also referred to as *3D printing* or *additive manufacturing*. These approaches have the common ability to create individual parts within a few minutes or hours without molds or tools, a capability that is extremely valuable in many circumstances.

Early rapid prototyping machines had significant drawbacks: they were very expensive, materials were initially limited to polymers (and, in some processes, paper), and the mechanical properties that were available were inferior to those that resulted from other manufacturing processes. Also, materials were very expensive; the control of the deposition process placed large demands on the chemical composition and purity of the polymers used.

In the 1990s, a number of technical advances occurred that improved the usefulness of rapid prototyping:

- Materials advances led to the development of polymers with mechanical properties that were comparable to those obtained from other manufacturing processes. While these materials are still relatively expensive, their cost is lower, so that rapid prototyping is more competitive.
- New processes were developed that allowed rapid prototyping to be applied to metals and ceramics, including sand. This led to the direct application of rapid prototyped parts into applications requiring the strength of metals, and also allowed the development of rapid tooling, where tools and dies for other manufacturing processes are produced with the assistance of rapid prototyping operations.
- Computer processors became faster, and CAD software incorporated greater 3D-modeling capability.

Process	Notes

Fused Deposition Modeling (FDM)

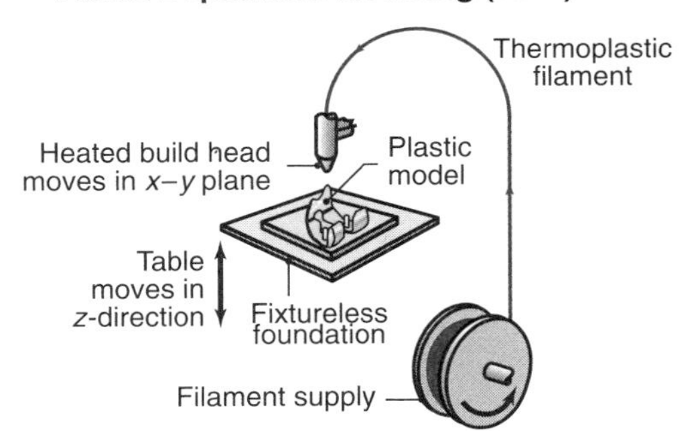

Process involves extrusion of thermoplastic filament through an extruder head; position of the extruder head is computer controlled. Materials used include ABS, polycarbonate, and polysulfone. ABS available in many colors, but is most commonly white. Strength of polymers can be as high as 72 MPa for Ultem 9085 (a trade name). FDM is the main process used in the Maker movement.

Selective Laser Sintering (SLS)

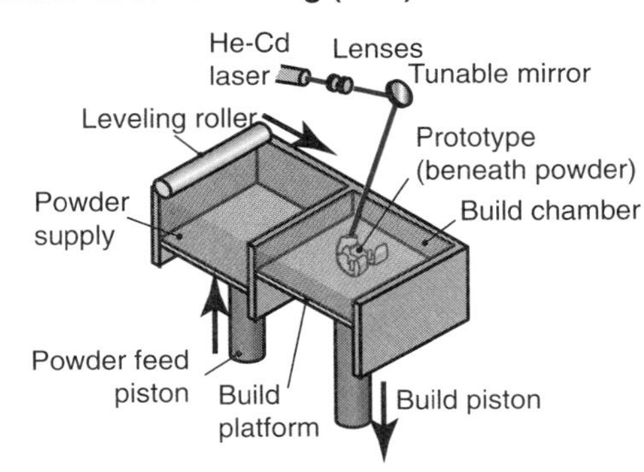

Layers of loose powder are applied over the build platform, and a laser slectively fuses (sinters) them. Successive layers build the part. An electron beam can be used (electron beam melting, or EBM). Materials can be polymers, metals, or polymer-coated ceramic or sand particles. Strengths can range from very low for elastomer-mimicing compliant materials to 78 MPa for Windform XT (a trade name). Strength in SLS can be over 300 MPa for bronze-infiltrated stainless steel powder and over 900 MPa for Ti-6Al-4V in EBM.

Stereolithography

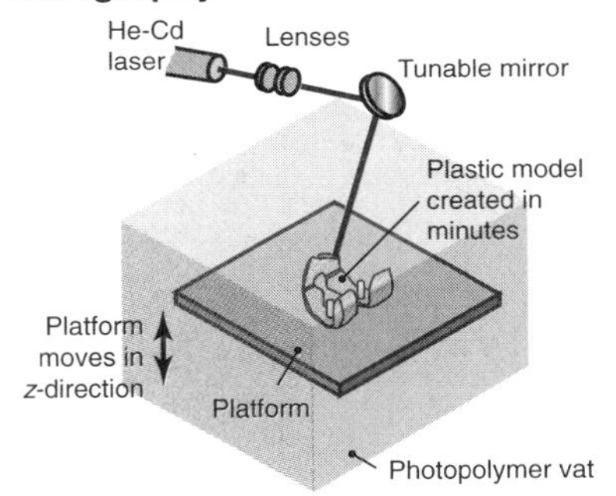

A platform is located in a vat of liquid photopolymer, and lowered to a desired depth; a computer controlled laser then cures the photopolymer, producing parts layer by layer. Can produce transparent parts. Materials range from compliant parts similar to elastomers to rigid plastics with strength up to 68 MPa (Accura 60, a trade name).

Three Dimensional Printing (3DP)

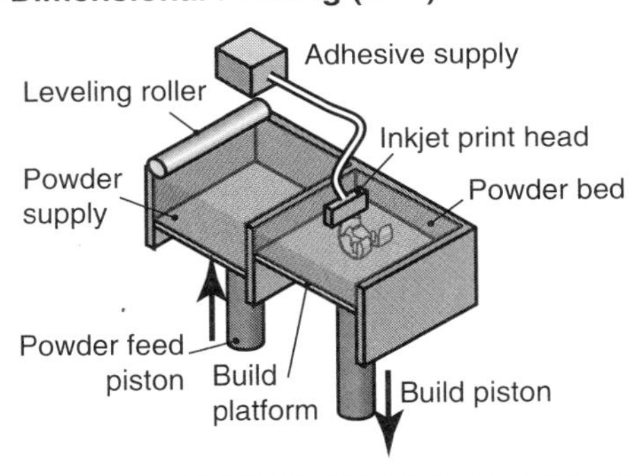

Layers of powder are selectively fused by printing an adhesive binder in similar fashion as inkjet printers; the binder cures in seconds. Can utilize multi-colored print heads to produce full-color parts. Mechanical properties are similar to selective-laser sintering.

Figure 3.29: Selected popular rapid prototyping operations. These processes allow the production of parts based on computer representations of geometry without the use of tools. Fused deposition modeling is the most commonly used process by the Maker Movement.

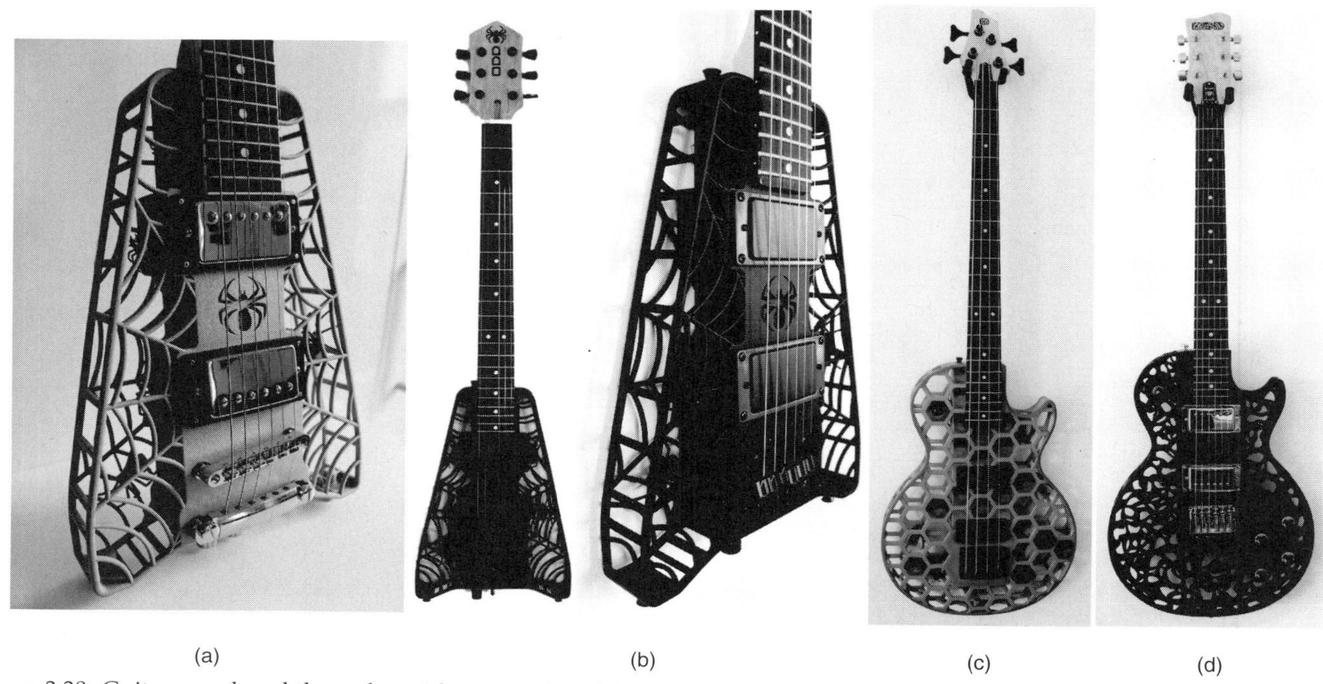

(a) (b) (c) (d)

Figure 3.30: Guitars produced through rapid prototyping. (a) Spider design body detail; (b) finished Spider guitars; (c) Hive-b and (d) Atom designs. *Source:* Courtesy of O. Diegel, Massey University.

- New colors and levels of transparency became available, and even multi-colored components could be produced. Variants of three-dimensional printing were developed that used multi-colored binders applied in similar fashion as inkjet printers to achieve full-color prototypes.

- Three-dimensional scanners became more popular and reasonably priced, which allowed the capture of an objects geometry and its import into CAD packages, where the geometry could be modified and then sent directly to a 3D printer. There are currently apps for smart phones that capture the 3D geometry of an object.

These developments are all impressive and led to the great current interest in rapid prototyping operations, and new business models have developed as a result. For example, some businesses will accept a 3D graphics file format by web submission, print the part and mail it to the purchaser by overnight delivery. A clever approach has been used with toys: With care, a 3D scan can be taken of an individuals face or head; a miniaturized model of the head can be added to a toy to produce a superhero doll with an individuals face attached. Creative approaches have been used to design home decorations, clothing, phone and tablet computer cases, musical instruments, etc. Figure 3.30 illustrates some extremely innovative guitar designs that are made possible by the flexible manufacturing offered by rapid prototyping.

While these are interesting and exciting capabilities, much greater interest has developed in rapid prototyping machinery in recent years. The original patents on the fused deposition modeling process have lapsed, and many companies and individuals can now practice this process.

A crowdsourcing community, known as The Maker Movement, has organized, is linked with Internet communication tools, and has developed a number of so-called *Makerbots*. Such machines are freely available as plans that can be downloaded from the Internet and used to build fully functional 3D printers for only a few hundred dollars. Alternatively, some very inexpensive machines have been marketed based on these crowd-sourced designs, such as the system shown in Fig. 3.31.

Materials researchers have continually improved the mechanical properties and ease of manufacture of polymers. Appendix A summarizes mechanical properties of selected common materials used in rapid prototyping. Also, researchers are able to apply new and innovative materials to rapid prototyping machines, including novel approaches such as printing of food or biological material for producing medical implants.

It has been suggested that the Maker Movement will revolutionize modern society, that design and manufacturing will be an activity performed by anyone with a computer and a 3D printer. This may be true for parts or products that do not need to support significant forces or transmit torque or power. However, as will be seen in Chapters 6 and 7, the design of components for robust performance in highly stressed environments requires a good understanding of engineering principles. Many machine elements have outstanding performance and economy because of mass production; rolling element bearings exemplify this category. And many materials depend on manufacturing processes to develop their full strength; even if a fastener could be produced, it is not necessarily an SAE Grade 8 bolt unless it is properly heat treated, for example.

Regardless, this is an exciting time for the Maker Movement, and this industry is experiencing rapid change and frequent developments that expand the applications and utility of rapid prototyping.

Figure 3.31: A Replicator 2 by MakerBot. This inexpensive 3D printer has a 410 in^3 (6400 cm^3) build volume and can produce parts from a variety of polymers. Shown in the foreground is a model of a tractor produced on a Replicator 2. *Source:* Courtesy MakerBot Industries, LLC.

3.9 Summary

Eight important mechanical properties of solid materials were discussed in this chapter. The differences between ductile and brittle materials were presented. It was found that, at fracture, ductile materials exhibit considerable plastic deformation, whereas brittle materials exhibit little or no yielding before failure.

Four major classes of solid material were described: metals, ceramics and glasses, polymers and elastomers, and composites. The members of each class have common features, such as similar chemical makeup and atomic structure, similar processing routes, and similar applications.

A stress-strain diagram was presented for each class of solid material because they differ significantly. The stress-strain diagrams for metals are essentially the same for compression and tension. This feature is not true for polymers or ceramics. Results of transverse bending tests for ceramics were found to be similar to tensile test results for metals.

A number of solid material properties used in choosing the correct material for a particular application were presented: mass density, modulus of elasticity, hardness, Poisson's ratio, shear modulus, strength, resilience, toughness, thermal conductivity, linear thermal expansion coefficient, specific heat capacity, and Archard wear constant. These parameters were presented for a large selection of materials.

Two-parameter materials charts were also presented. Stiffness versus weight, strength versus weight, stiffness versus strength, and wear constant versus limiting pressure for the various classes of material can give a better idea of the best material for a particular machine element.

An introduction to manufacturing processes and their effects on materials was also presented, with the most common approaches presented for each class of material. These include casting, bulk and sheet forming for metals, many types of molding for polymers, and casting and molding operations for ceramics. Each material can be machined to some extent as well.

The Case Study described the emerging Maker Movement, wherein designers can produce their desired products in a CAD program and have them manufactured directly on a 3D printer. A large number of applications and materials have been used by the Maker Movement.

Key Words

anisotropic material having different properties in all directions at a point in a solid

Archard wear constant wear property of a material

Brinell hardness a hardness measure that results from a Brinell test, where a steel or tungsten carbide ball is impressed onto a material.

brittle material material that fractures at strain below 5%

ceramics compounds of metallic and nonmetallic elements

composite materials combinations of two or more materials, usually consisting of fiber and thermosetting polymer

density mass per unit volume

ductile material material that can sustain elongation greater than 5% before fracture

ductility degree of plastic deformation sustained at fracture

elastic limit stress above which material acquires permanent deformation

elastomers polymers with intermediate amount of cross-linking

fracture stress stress at time of fracture or rupture

glasses compounds of metallic and nonmetallic elements with no crystal structure

hardness resistance to surface penetration

homogeneous material having properties not a function of position in a solid

isotropic material having the same properties in all directions at a point in a solid

Knoop hardness a hardness measure that results from a Knoop test, also known as a microhardness test

metals combinations of metallic elements

modulus of elasticity proportionality constant between stress and strain

modulus of rupture stress at rupture from bending test, used to determine strength of ceramics

necking decreasing cross-sectional area that occurs after ultimate stress is reached and before fracture

orthotropic material having different properties in three mutually perpendicular directions at a point in a solid and having three mutually perpendicular planes of material symmetry

Poisson's ratio absolute value of ratio of transverse to axial strain

polymers compounds of carbon and other elements forming long-chain molecules

proportional limit stress above which stress is no longer linearly proportional to strain

resilience capacity of material to release absorbed energy

Rockwell hardness a hardness measure that results from a Rockwell test, where the penetration of a cone or ball into a material is measured

rule of mixtures linear interpolation between densities of alloy concentration

specific heat capacity ratio of heat stored per mass to change in temperature of material

strain hardening increase in hardness and strength of ductile material as it is plastically deformed

thermal conductivity ability of material to transmit heat

thermal expansion coefficient ratio of elongation in material to temperature rise

thermoplastics polymers without cross-links

thermosets polymers with highly cross-linked structure

toughness ability to absorb energy up to fracture

ultimate strength maximum stress achieved in stress-strain diagram

Vickers hardness a hardness measure that results from a Vickers test, where a diamond pyramid is impressed onto a material

yield strength stress level defined by intersection of reference line (with slope equal to initial material elastic modulus and x-intercept of 0.2%) and material stress-strain curve

yielding onset of plastic deformation

Young's modulus (see modulus of elasticity)

Summary of Equations

Material Properties:

Percent elongation: $\%\text{EL} = \left(\dfrac{l_{\text{fr}} - l_o}{l_o}\right) \times 100\%$

Elastic modulus: $E = \dfrac{\sigma_{\text{avg}}}{\epsilon_{\text{avg}}}$

Poisson's ratio: $\nu = -\dfrac{\text{Transverse strain}}{\text{Axial strain}}$

Shear modulus: $G = \dfrac{E}{2(1+\nu)}$

Modulus of resilience: $U_r = \dfrac{S_y^2}{2E}$

Heat capacity: $Q = C_p m_a (\Delta T)$

Composite Materials:

Elastic modulus: $E_c = E_m(1 - v_f) + E_f v_f$

Load sharing: $\dfrac{P_f}{P_m} = \dfrac{E_f v_f}{E_m v_m}$

Hardness:

Brinell hardness: $\text{HB} = \dfrac{2P}{(\pi D)\left(D - \sqrt{D^2 - d^2}\right)}$

$S_u = 3.5\text{HB}$

Vickers hardness: $\text{HV} = \dfrac{1.854P}{L^2}$

Knoop hardness: $\text{HK} = \dfrac{14.2P}{L^2}$

Rockwell hardness: $\text{HRA} = 100 - 500t$, $\text{HRB} = 130 - 500t$, etc.

Archard Wear Law: $v = k_1 \dfrac{WL}{3H}$

Recommended Readings

Ashby, M.F. (2010) *Materials Selection in Mechanical Design*, 4th ed., Butterworth-Heinemann.

ASM Metals Handbook, 8th ed. (2009) American Society for Metals.

Brandt, D.A., and Warner, J.C. (2009) *Metallurgy Fundamentals: Ferrous and Nonferrous*, 5th ed. Goodheart-Wilcox.

Budinski, K., and Budinski, M. (2009) *Engineering Materials, Properties and Selection*, 9th ed., Prentice-Hall.

Callister, W.D., and Rethwisch, D.G. (2011) *Fundamentals of Materials Science and Engineering: An Integrated Approach*, 4th ed. Wiley.

Farag, M.M. (2007) *Selection of Materials and Manufacturing Processes for Engineering Materials*, 2nd ed. Prentice-Hall.

Flinn, R. A., and Trojan, P. K. (1986) *Engineering Materials and Their Applications*, Houghton Mifflin.

Kalpakjian, S., and Schmid, S.R. (2010) *Manufacturing Engineering and Technology*. 6th ed., Pearson.

Raman, A. (2006) *Materials Selection and Applications in Mechanical Engineering*, Industrial Press.

Schey, J.A. (2000) *Introduction to Manufacturing Processes*, 3rd ed., McGraw-Hill.

References

Ashby, M.J. (2010) *Materials Selection in Mechanical Design*, 4th ed., Butterworth-Heinemann.

Kalpakjian, S., and Schmid, S.R. (2010) *Manufacturing Engineering and Technology*, 6th. ed., Pearson.

Schey, J.A. (2000) *Introduction to Manufacturing Processes*, 3rd ed., McGraw-Hill.

Questions

3.1 Define the terms ductile and brittle.

3.2 What are the three basic classifications of solids?

3.3 What is the difference between a thermoset and a thermoplastic?

3.4 What is a composite material?

3.5 Define proportional limit. How is this different from elastic limit? How are these different from the yield strength?

3.6 How is the yield strength defined?

3.7 What is strain hardening?

3.8 What is the glass transition temperature for polymers? Does this behavior occur with other materials? Explain.

3.9 What is the elastic modulus of a material?

3.10 What are the advantageous properties of glass?

3.11 What is hardness?

3.12 What is ductility, and how is it measured?

3.13 What is the difference between thermal expansion and specific heat?

3.14 What is casting? What materials can be cast?

3.15 Give three examples of bulk metal forming operations.

3.16 What is PM?

3.17 What is sintering?

3.18 What is a thermoplastic? A thermoset?

3.19 What are the main manufacturing processes applicable to ceramics?

3.20 Why are the molds for producing PM parts larger than the desired shape?

Qualitative Problems

3.21 What are the primary functions of the reinforcement in a fiber reinforced polymer? What are the primary functions of the matrix?

3.22 Explain why ceramics and cast metals are much stronger in compression than in tension.

3.23 Without using the words *stress* or *strain*, define *elastic modulus*.

3.24 Sketch typical stress-strain diagrams for metals, ceramics, and polymers.

3.25 What is the main difference between resilience and toughness? What is the main similarity between resilience and toughness?

3.26 Describe the events that occur when a specimen undergoes a tension test. Sketch a plausible stress-strain curve, and identify all significant regions and points between them. Assume that loading continues up to fracture.

3.27 Which hardness tests and scales would you use for very thin strips of metal, such as aluminum foil? Explain.

3.28 List the factors that you would consider in selecting a hardness test. Explain your answer.

3.29 A statistical sampling of Rockwell C hardness tests is conducted on a material, and it is determined that the material is defective because of insufficient hardness. The supplier claims that the tests are flawed because the diamond-cone indenter was probably dull. Is this a valid claim? Explain.

3.30 In a Brinell hardness test, the resulting impression is found to be an ellipse. Give possible explanations for this result.

3.31 Which of the properties described in this chapter are important for (a) pots and pans, (b) gears, (c) clothing, (d) paper clips, (e) music wire, (f) beverage cans? Explain your answers.

3.32 Identify products that cannot be made of steel, and explain why this is so. (For example, electrical contacts commonly are made of gold or copper because their softness results in low contact resistance, while for steel the contact resistance would be very high.)

3.33 What characteristics make polymers advantageous for applications such as gears? What characteristics are drawbacks for such applications?

3.34 Review Fig. 3.25 and list reasons why the connecting rod is manufactured in multiple steps, instead of one cavity with one press stroke.

3.35 Review Fig. 3.26b and list potential applications for extruded products.

Quantitative Problems

3.36 The design specification for a metal requires a minimum hardness of 80 HRA. If a Rockwell test is performed and the depth of penetration is 60 μm, is the material acceptable? *Ans.* No.

3.37 It can be shown that thermal distortion in precision devices is low for high values of thermal conductivity divided by thermal expansion coefficient. Rank the materials in Table 3.1 according to their suitability to resist thermal distortion.

3.38 If a material has a target hardness of 300 HB, what is the expected indentation diameter? Assume the applied load is 3000 kg. *Ans.* $d = 2.95$ mm.

3.39 For a material that follows a power law curve for stress-strain behavior, that is, $\sigma = K\epsilon^n$, where K is the strength coefficient and n is the strain hardening exponent, find the strain at which necking occurs. *Ans.* $\epsilon = n$.

3.40 A 2-m-long polycarbonate tensile rod has a cross-sectional diameter of 150 mm. It is used to lift a tank weighing 45 tons (45,000 kg) from a 1.8-m-deep ditch onto a road. The vertical motion of the crane's arc is limited to 4.2 m. Will it be possible to lift the tank onto the road? *Ans.* Yes.

3.41 Materials are normally classified according to their properties, processing routes, and applications. Give examples of common metal alloys that do not show some of the typical metal features in their applications.

3.42 Equation (B.56) in Appendix B gives the relationship between stresses and strains in isotropic materials. For a polyurethane rubber, the elastic modulus at 100% elongation is 7 MPa. When the rubber is exposed to a hydrostatic pressure of 10 MPa, the volume shrinks 0.5%. Calculate Poisson's ratio for the rubber. *Ans.* $\nu = 0.499$.

3.43 A fiber-reinforced plastic has fiber-matrix bond strength $\tau_f = 15$ MPa and fiber ultimate strength $S_u = 1$ GPa. The fiber length is constant for all fibers at $l = 1.25$ mm. The fiber diameter $d = 30$ μm. Find whether the fiber strength or the fiber-matrix bond will determine the strength of the composite. *Ans.* The fiber determines the composite strength.

3.44 Using the same material as in Problem 3.43 but with fiber length $l = 0.75$ mm, calculate if it is possible to increase the fiber stress to $S_u = 1$ GPa by making the fiber rectangular instead of circular, maintaining the same cross-sectional area for each fiber, and if so, give the cross-section dimensions. *Ans.* $h_t = 8.22$ μm, $w_t = 86.0$ μm.

3.45 A copper bar is stressed to its ultimate strength, $S_u = 150$ MPa. The cross-sectional area of the bar before stressing is 120 mm^2, and the area at the deformed cross section where the bar starts to break at the ultimate strength is 70 mm^2. How large a force is needed to reach the ultimate strength? *Ans.* 18 kN.

3.46 AISI 440C stainless steel has ultimate strength $S_u = 807$ MPa and fracture strength $S_{fr} = 750$ MPa. At the ultimate strength the cross-sectional area of a tension bar made of AISI 440C is 80% of its undeformed value. At the fracture point the minimum cross-sectional area has shrunk to 70%. Calculate the real stresses at the point of ultimate strength and at fracture. *Ans.* At fracture, $\sigma = 1071$ MPa; at ultimate strength, $\sigma = 1009$ MPa.

3.47 According to Sketch a, a beam is supported at point A and at either B or C. At C the silicon nitride tensile rod is lifting the beam end with force $P = S_{fr}^t$, where A_c is the cross-sectional area of the rod. Find the distance A-B such that the silicon nitride rod would be crushed if it took up a compressive force at B instead of a tensile force at C. Note that the strength in compression is fifteen times larger than the strength in tension for silicon nitride. Also, find the reaction forces at A for the two load cases. *Ans.* $l' = \frac{1}{15}l$, $A_y = \frac{120}{7}P$.

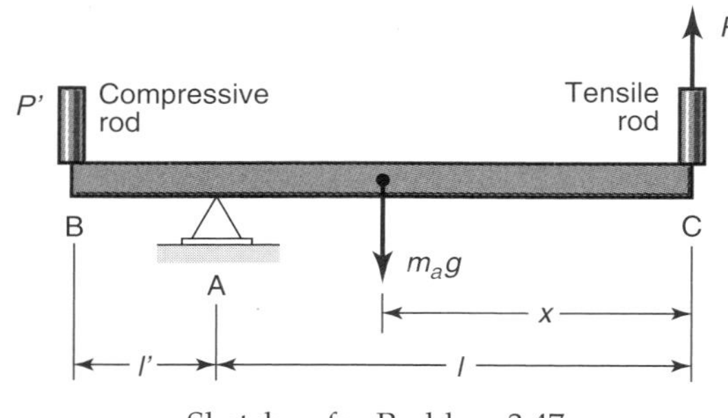

Sketch a, for Problem 3.47

3.48 Polymers have different properties depending on the relationship between the local temperature and the polymer's glass transition temperature T_g. The rubber in a bicycle tire has $T_g = -12$°C. Could this rubber be used in tires for an Antarctic expedition at temperatures down to −70°C? *Ans.* No.

3.49 What is the modulus of resilience of a highly cold-worked steel with a hardness of 300 HB? Of a highly cold-worked aluminum with a hardness of 120 HB? *Ans.* $U_{rs} = 1.957$ MPa.

3.50 Given an aluminum bronze with 25 wt% aluminum and 75 wt% copper, find the density of the aluminum bronze. *Ans.* $\rho_{bronze} = 7382$ kg/m^3

3.51 The glass-fiber-reinforced plastic in Example 3.6 (Section 3.5.2) is used in an application where the bending deformations, caused by the applied static load, will crack the plastic by overstressing the fibers. Will a carbon-fiber-reinforced plastic also crack if it has the same elastic properties as the glass-fiber-reinforced plastic? *Ans.* Yes.

3.52 In Problem 3.51, carbon fibers were used to reinforce a polymer matrix. The concentration of fibers was decreased in Example 3.6 (Section 3.5.2) to give the same elastic properties for the carbon-fiber-reinforced polymer as for the glass-fiber-reinforced polymer. If instead, the fiber concentration were kept constant at 10% when the glass fibers were changed to carbon fibers, how much smaller would the deformation be for the same load, and would the fibers be overstressed? The material properties are the same as in Example 3.6.

3.53 A bent beam, shown in Sketch b, is loaded with force $P = 125,000$ N. The beam has a square cross section a^2. The length of a side $a = 30$ mm. The length $l_1 = 50$ mm and $l_2 = 100$ mm. The yield strength $S_y = 350$ MPa (medium-carbon steel). Find whether the stresses in tension and shear are below the allowable stresses. Neglect bending.

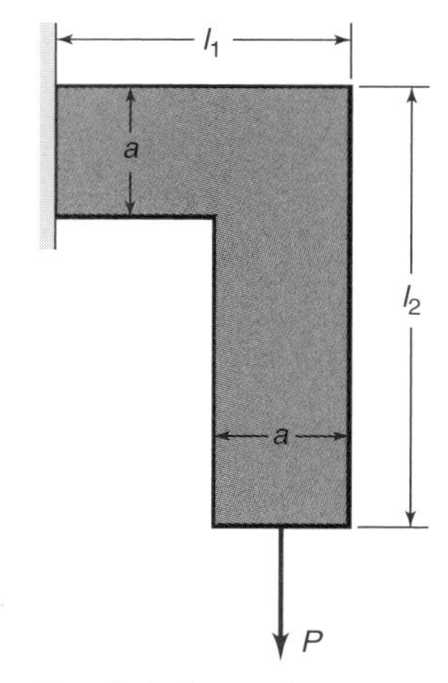

Sketch b, for Problem 3.53

3.54 A steel cube has side length $l = 0.1$ m, modulus of elasticity $E = 206$ GPa, and Poisson's ratio $\nu = 0.3$. A stress σ is applied to one direction on the cube. Find the compressive stresses needed on the four remaining cube faces to give the same elongation that results from σ. *Ans.* $\sigma_c = -1.67\sigma$.

3.55 For the stressed steel cube in Problem 3.54, calculate the volume ratio (v_t/v_c) when $\sigma = 500$ MPa. *Ans.* $v_t/v_c = 1.0042$.

3.56 A tough material, such as soft stainless steel, has yield strength $S_y = 200$ MPa, ultimate strength $S_u = 500$ MPa, and 200% elongation. Find the ratio of the material toughness to the resilience at fracture (the *hyperstatic resilience*), assuming that the stress-strain curve consists of two straight lines according to Sketch c. *Ans.* Toughness/resilience = 1080.

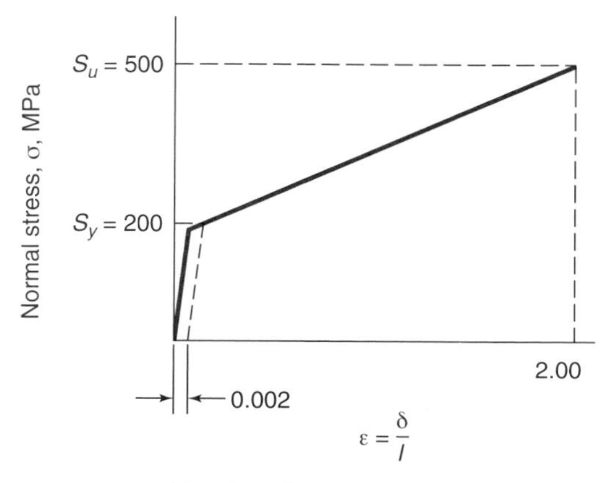

Sketch c, for Problem 3.56

3.57 Hooke's law describes the relationship between uniaxial stress and uniaxial strain. What is the ratio of the strain encountered by the most compliant material mentioned in this chapter compared to the stiffest? *Ans.* 1.125×10^5.

3.58 According to Archard's wear equation, the wear depth is proportional to the sliding distance and the contact pressure. How will the contact pressure be distributed radially for a disk brake if the wear rate is the same for all radii? *Ans.* Pressure is inversely proportional to radius.

3.59 Given a brake pad for a disk brake on a car, and using Archard's wear constant, determine how the wear is distributed over the brake pad if the brake pressure is constant over the pad. *Ans.* Wear rate is proportional to radius.

3.60 Derive an expression for the toughness of a material represented by the stress-strain law $\sigma = K(\epsilon + 0.2)^n$ and whose fracture strain is denoted by ϵ_f.

Design and Projects

3.61 List and explain the desirable mechanical properties for (a) an elevator cable, (b) a paper clip, (c) a leaf spring for a truck, (d) a bracket for a bookshelf, (e) piano wire, (f) a wire coat hanger, (g) the clip for a pen, and (h) a staple.

3.62 List applications where the following properties would be desirable: (a) high density, (b) low density, (c) high stiffness, (d) low stiffness, (e) high thermal conductivity, and (f) low thermal conductivity.

3.63 Give several applications in which both specific strength and specific stiffness are important.

3.64 Conduct a literature search and add the following materials to Table 3.1: cork, concrete, ice, sugar, lithium, chromium, and platinum.

3.65 A recent development in the automotive industry is to use steel alloys with a high manganese content, called TRIP, TWIP, and martensitic steels. Conduct an Internet search and literature review and write a one-page summary of these materials and their mechanical properties.

3.66 Design an actuator to turn on a switch when temperature drops below a certain level. Use two materials with different coefficients of thermal expansion in your design.

3.67 Assume that you are in charge of public relations for a large steel-producing company. Outline all of the attractive characteristics of steels that you would like your customers to be informed about.

3.68 Assume that you are in competition with the steel industry and are asked to list all of the characteristics of steels that are not attractive. Make a list of these characteristics and explain their relevance to engineering applications.

3.69 Aluminum is being used as a substitute material for steel in automobiles. Describe your concerns, if any, in purchasing an aluminum automobile.

3.70 Add a column to Table 3.1 and add values of electrical conductivity for the materials given.

3.71 Review the technical literature, and produce a figure similar to Fig. 3.24 for investment casting.

3.72 Perform an Internet search, and produce a Powerpoint presentation that summarizes applications of rapid prototyping. Whenever possible, indicate the material and machinery used.

Chapter 4

Stresses and Strains

The failed Hyatt Regency Hotel walkway (Kansas City, 1981) that was directly attributable to changes in design that over-stressed structural members. *Source:* AP Photos.

Contents

Examples

Case Study

I am never content until I have constructed a mechanical model of the subject I am studying. If I succeed in making one, I understand; otherwise I do not.

William Thomson (Lord Kelvin)

This chapter discusses stresses and strains that arise from common loading conditions discussed in Chapter 2, using the constitutive rules described in Chapter 3. The approaches in this chapter are mainly developed for the analysis of beams and shafts, but are easily adaptable to any machine element, and are usually sufficient for stress analysis and design. The chapter begins with a review of cross-sectional properties such as the centroid of an area, moment of inertia (and the use of the parallel-axis theorem to obtain the moment of inertia), the radius of gyration, section modulus, and mass moment of inertia. After introducing normal stresses and strains due to uniaxial loadings, the chapter considers torsion stresses and strains that are essential for shaft design. Bending stress and strain are then considered for both straight and curved beams, as well as the associated transverse shear stresses. The combination of these stresses due to complicated loadings is then examined.

Symbols

A	cross sectional area, m^2
A'	partial cross sectional area, m^2
a	width, m
b	height, m
c	distance from neutral axis to outer fiber of solid, m
d_x, d_y	distance between two parallel axes, one of which contains centroid of area, m
E	modulus of elasticity, Pa
e	eccentricity, distance separating centroidal and neutral radii of curved member, m
G	shear modulus of elasticity, Pa
h	height of triangular cross-sectional area, m
h_p	power, W
I	area moment of inertia, m^4
I_m	mass moment of inertia, kg-m^2
J	polar area moment of inertia, m^4
$\bar{J}$	polar area moment of inertia about centroidal coordinates, m^4
k	spring rate, N/m
k_a	angular spring rate, N-m
l	length, m
M	bending moment, N-m
m_a	mass, kg
P	force, N
Q	first moment about neutral axis, m^3
r	radius, m
$\bar{r}$	centroidal radius, m
r_g	radius of gyration, m
s	length of a line segment, m
T	torque, N-m
u	velocity, m/s
V	transverse shear force, N
w_t	width, m
x, y, z	Cartesian coordinate system, m
$\bar{x}, \bar{y}, \bar{z}$	centroidal coordinate system, m
x', y'	coordinates parallel to x- and y-axes
Z_m	section modulus, I/c, m^3
γ	shear strain
δ	deformation, m
ϵ	normal strain
ρ	density, kg/m^3
θ	angle of twist, rad
σ	normal stress, Pa
τ	shear stress, Pa
ω	angular velocity, rad/s

Subscripts

i	inner
o	outer
x, y, z	Cartesian coordinates
$\bar{x}, \bar{y}, \bar{z}$	centroidal coordinates
x', y'	coordinates parallel to x- and y-axes

4.1 Introduction

Normal, torsional, bending, and transverse shear loadings were described in Section 2.3. This chapter describes the stresses and strains resulting from these types of loading while making use of the general Hooke's law relation developed in Chapter 3. The theory developed in this chapter is applicable to any machine element. For the purposes of this chapter, however, the member is assumed to have a symmetrical cross section and to be made of an isotropic, homogeneous, linear-elastic material. Later in the chapter, a curved member is also considered for bending. More complicated geometries are considered in Chapter 6.

This chapter provides a quick review of stress analysis and more in-depth coverage can be found in the Recommended Readings at the end of the chapter. Section 4.2 describes the approach for calculating important properties of a beam's cross-section, namely its centroid, moment of inertia, section modulus, radius of gyration, and mass moment of inertia. Readers familiar with these topics may wish to proceed to Section 4.3, which introduces the concepts of normal stress and strain, or Section 4.4 which deals with torsion. Bending is discussed in Section 4.5 and the effect of transverse shear in bending is presented in Section 4.6.

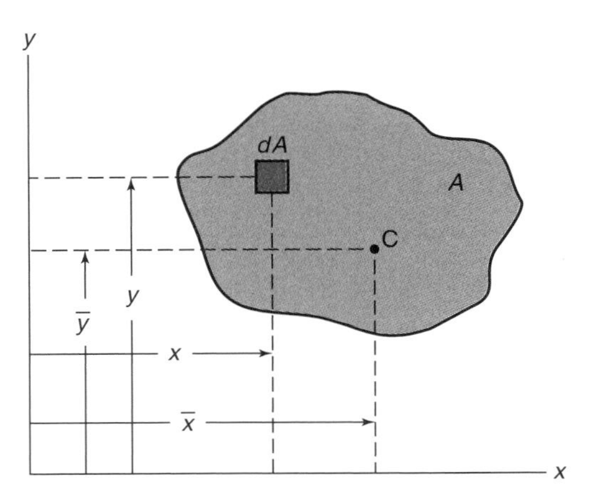

Figure 4.1: Centroid of area. The centroid is at point C with coordinates $(\bar{x}, \bar{y})$.

4.2 Properties of Beam Cross Sections

The centroid of an area, moment of inertia of an area, parallel-axis theorem, radius of gyration, section modulus, and mass moment of inertia are important concepts that need to be defined before proceeding with the remainder of this chapter. These concepts will be used throughout the text.

4.2.1 Centroid of Area

The **centroid** of an area (Fig. 4.1), or the *center of gravity* of an area, refers to the point that defines the area's geometric center. Conceptually, this is the point of balance where equal areas exist on each side of the point across arbitrary sections of the body. A finite thickness version of the area would balance on a pin placed at the centroid. Mathematically, it is that point at which the sum of the first moments of area about an axis through it is zero, or

$$\int_A (y - \bar{y}) \, dA = 0, \tag{4.1}$$

$$\int_A (x - \bar{x}) \, dA = 0. \tag{4.2}$$

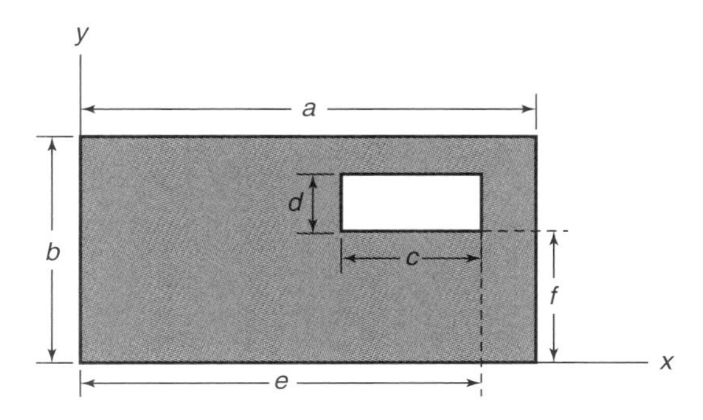

Figure 4.2: Rectangular hole within a rectangular section used in Example 4.1.

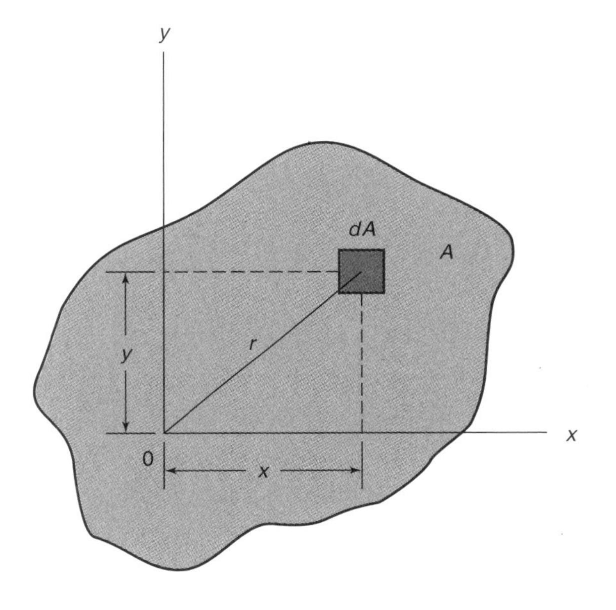

Figure 4.3: Area with coordinates used in describing area moment of inertia.

Solving for $\bar{x}$ and $\bar{y}$ gives

$$\bar{y} = \frac{\int_A y\, dA}{\int_A dA} = \frac{\int_A y\, dA}{A}, \tag{4.3}$$

$$\bar{x} = \frac{\int_A x\, dA}{\int_A dA} = \frac{\int_A x\, dA}{A}. \tag{4.4}$$

A complicated area can usually be divided into simple subareas and Eqs. (4.3) and (4.4) can be applied by making the numerator equal to the sum of the first moment integrals of the separate areas. The denominator is the total area. Thus, the centroid of the composite area is given by

$$\bar{y} = \frac{A_1\bar{y}_1 + A_2\bar{y}_2 + \ldots}{A_1 + A_2 + \ldots}, \tag{4.5}$$

$$\bar{x} = \frac{A_1\bar{x}_1 + A_2\bar{x}_2 + \ldots}{A_1 + A_2 + \ldots}. \tag{4.6}$$

Example 4.1: Determining the Centroid of an Area

Given: Figure 4.2 shows a rectangular part having dimensions $a \times b$ with a rectangular hole $c \times d$. It also shows the coordinates and location of this hole within the rectangular cross section. The dimensions are $a = 10$ cm, $b = 5$ cm, $c = 3$ cm, $d = 1$ cm, $e = 9$ cm, and $f = 3$ cm.

Find: The centroid of this part.

Solution: In this problem, one could break the section into a number of rectangles to obtain the correct solution. However, a useful approach is demonstrated here, namely treating the hole as a "negative" area. Then, making use of Eqs. (4.5) and (4.6) gives

$$\begin{aligned}
\bar{y} &= \frac{A_1\bar{y}_1 - A_2\bar{y}_2}{A_1 - A_2} = \frac{ab(b/2) - cd(f + d/2)}{ab - cd} \\
&= \frac{(10)(5)(5/2) - (3)(1)(3 + 1/2)}{(10)(5) - (3)(1)} = 2.436 \text{ cm}, \\
\bar{x} &= \frac{A_1\bar{x}_1 - A_2\bar{x}_2}{A_1 - A_2} = \frac{ab(a/2) - cd(e - c/2)}{ab - cd} \\
&= \frac{(10)(5)(10/2) - (3)(1)(9 - 3/2)}{(10)(5) - (3)(1)} = 4.840 \text{ cm}.
\end{aligned}$$

4.2.2 Area Moment of Inertia

The terms **area moment of inertia** and *second moment of area* are used interchangeably. In Section 4.2.1, the first moment of area, $\int_A y\, dA$, was associated with the centroid and in this section the second moment of area, $\int_A y^2\, dA$, is used to define the moment of inertia.

Figure 4.3 shows the coordinates that describe the area moments of inertia, which are designated by the symbol I. The moments of inertia with respect to the x- and y-axes, respectively, can be expressed as

$$I_x = \int_A y^2\, dA \quad \text{and} \quad I_y = \int_A x^2\, dA. \tag{4.7}$$

When the reference axis is normal to the plane of the area, through 0 in Fig. 4.3, the integral is called the **polar moment of inertia**, J, and can be written as

$$J = \int_A r^2\, dA = \int_A \left(x^2 + y^2\right)\, dA = \int_A x^2\, dA + \int_A y^2\, dA. \tag{4.8}$$

Note from Eq. (4.7) that

$$J = I_x + I_y. \tag{4.9}$$

In defining the polar moment of inertia, the x- and y-axes can be any two mutually perpendicular axes intersecting at 0. The unit of moment of inertia and polar moment of inertia is length raised to the fourth power.

Example 4.2: Area and Polar Moment of Inertia

Given: Figure 4.4 shows a circular cross section with radius r and x-y coordinates.

Find: The area moment of inertia of the circular area about the x- and y-axes and the polar moment of inertia about the centroid.

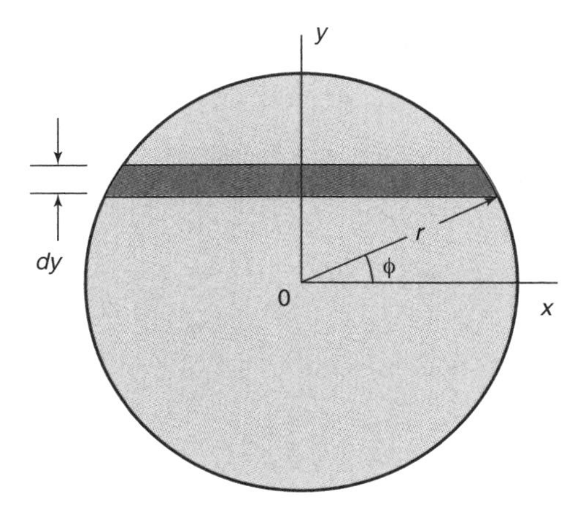

Figure 4.4: Circular cross section used in Example 4.2.

Solution: Consider the shaded slice of the circle shown in Fig. 4.4. For any position defined by ϕ between $\pi/2$ and $-\pi/2$, the incremental area is

$$dA = 2r\cos\phi\,dy.$$

From Eq. (4.7), the area moment of inertia of the circular area about the x-axis is

$$I_x = \int y^2\,dA = \int y^2 2r\cos\phi\,dy.$$

But because $y = r\sin\phi$ and $dy = r\cos\phi\,d\phi$,

$$\begin{aligned} I_x &= 2r^4\int_{-\pi/2}^{\pi/2}\sin^2\phi\cos^2\phi\,d\phi \\ &= \frac{r^4}{2}\int_{-\pi/2}^{\pi/2}\frac{1-\cos 4\phi}{2}\,d\phi \\ &= \frac{\pi r^4}{4}. \end{aligned}$$

Because of symmetry, the moment of inertia with respect to the y-axis is

$$I_y = \frac{\pi r^4}{4}.$$

Thus, the area moments of inertia about the x- and y-axes are identical and equivalent to $\pi r^4/4$. From Eq. (4.9) the polar moment of inertia about the centroid is

$$J_z = I_x + I_y = 2I_x = \frac{\pi r^4}{2}.$$

4.2.3 Parallel-Axis Theorem

When an area's moment of inertia has been determined with respect to a given axis through its centroid, the moment of inertia with respect to any parallel axis can be obtained by means of the parallel-axis theorem. Figure 4.5 shows the coordinates and distances to be used in deriving the parallel-axis theorem.

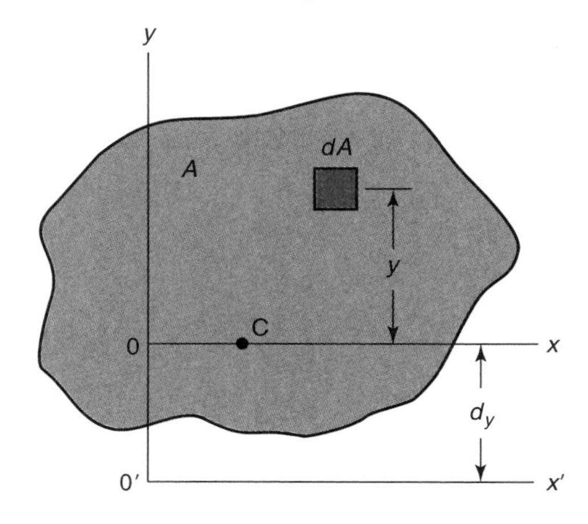

Figure 4.5: Coordinates and distance used in describing the parallel-axis theorem.

The moment of inertia of the area A about the axis x' is

$$\begin{aligned} I_{x'} &= \int_A (y+d_y)^2\,dA \\ &= \int_A y^2\,dA + 2d_y\int_A y\,dA + d_y^2\int_A dA, \end{aligned}$$

or

$$I_{x'} = I_x + 2d_y\int_A y\,dA + Ad_y^2. \tag{4.10}$$

If the x-axis passes through the centroid of the area, $\int_A y\,dA$ is zero and Eq. (4.10) reduces to

$$I_{x'} = I_x + Ad_y^2. \tag{4.11}$$

Similarly, for an axis y' parallel to the y-axis that goes through the centroid and is separated by a distance d_x,

$$I_{y'} = I_y + Ad_x^2. \tag{4.12}$$

Equations (4.11) and (4.12) are known as the **parallel-axis theorem**, and allow calculation of an area's moment of inertia with respect to any axis.

Example 4.3: Moment of Inertia and Centroid from Parallel Axis Theorem

Given: Figure 4.6 shows a triangular cross section with a circular hole.

Find: The area moment of inertia and the centroid.

Solution: Assume that the y-axis starts at the midwidth of the base and is positive in the upward direction. The height of the triangle is

$$h = \frac{6}{2}\tan 60^\circ = 5.196\text{ cm}.$$

The triangle and circle are defined with an a and b subscript, respectively. The centroids and areas of the triangle and the circle can be expressed separately as

$$\bar{y}_a = \frac{h}{3} = \frac{5.196}{3} = 1.732\text{ cm},$$

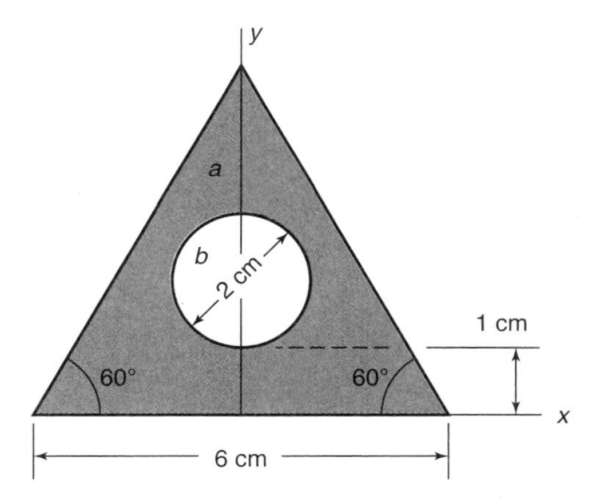

Figure 4.6: Triangular cross section with circular hole, used in Example 4.3.

$$A_a = \frac{1}{2}bh = \frac{1}{2}(6)(5.196) = 15.59 \text{ cm}^2,$$

$$A_b = \frac{\pi d^2}{4} = \frac{\pi (2)^2}{4} = \pi = 3.142 \text{ cm}^2.$$

The centroid of the composite figure is

$$\bar{y} = \frac{\bar{y}_a A_a - \bar{y}_b A_b}{A_a - A_b} = \frac{(1.732)(15.59) - (2)(3.142)}{15.59 - 3.142} = 1.664 \text{ cm}.$$

The moments of inertia of the triangle and circle are

$$I_a = \frac{bh^3}{36} = \frac{6(5.196)^3}{36} = 23.38 \text{ cm}^4,$$

$$I_b = \frac{\pi d^4}{64} = \frac{\pi (2)^4}{64} = 0.7854 \text{ cm}^4.$$

From the parallel-axis theorem, the moment of inertia of the composite area about the centroidal axis is

$$\begin{aligned} I_x &= I_a + (\bar{y} - \bar{y}_a)^2 A_a - I_b - (\bar{y} - \bar{y}_b)^2 A_b \\ &= 23.38 + (1.664 - 1.732)^2 (15.59) - 0.7854 \\ &\quad -(1.664 - 2)^2 (3.142) \\ &= 2.231 \times 10^{-7} \text{ m}^4. \end{aligned}$$

Example 4.4: Parallel Axis Theorem

Given: Figure 4.7 shows a circular cross section of radius r, and x'-y' coordinates.

Find: The area moments of inertia $I_{x'}$, $I_{y'}$ and the polar moment of inertia J_z', relative to the x', y', and z' axes.

Solution: Using Eqs. (4.11) and (4.12) and the results from Example 4.2 gives

$$I_{x'} = I_x + Ad_y^2 = \frac{\pi r^4}{4} + \pi r^2 (4r)^2 = 16.25\pi r^4,$$

$$I_{y'} = I_y + Ad_x^2 = \frac{\pi r^4}{4} + \pi r^2 (3r)^2 = 9.25\pi r^4$$

$$J_{z'} = I_{x'} + I_{y'} = 25.5\pi r^4.$$

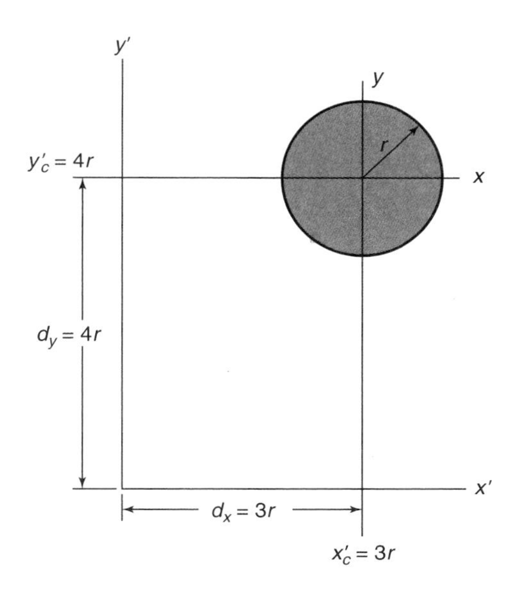

Figure 4.7: Circular cross-sectional area relative to x'-y' coordinates, used in Example 4.4.

4.2.4 Radius of Gyration

An area's **radius of gyration** is the length that, when squared and multiplied by the area, will give the area's moment of inertia with respect to the specific axis, or

$$I = r_g^2 A. \tag{4.13}$$

This is another way of expressing the area's moment of inertia. The radius of gyration can be written as

$$r_g = \sqrt{\frac{I}{A}}. \tag{4.14}$$

The radius of gyration is not the distance from the reference axis to a fixed point in the area (such as the centroid), but it is a useful property of the area and the specified axis.

Example 4.5: Radius of Gyration

Given: The same circular area given in Example 4.4.

Find: The radius of gyration with respect to the x' and y' axes.

Solution: Using Eq. (4.14) and the results from Example 4.4 give

$$r_{gx'} = \sqrt{\frac{I_{x'}}{A}} = \sqrt{\frac{16.25\pi r^4}{\pi r^2}} = 4.03r,$$

$$r_{gy'} = \sqrt{\frac{I_{y'}}{A}} = \sqrt{\frac{9.25\pi r^4}{\pi r^2}} = 3.04r.$$

When the circle is small relative to the distances d_x and d_y, the radii of gyration are just a little larger than the centroidal distances $3r$ and $4r$. This is a useful observation that will be applied later in the text, for example in the analysis of spot welds in Section 16.6.

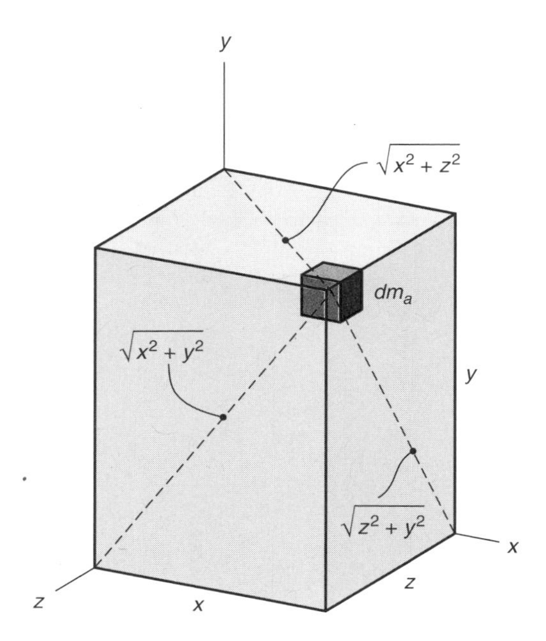

Figure 4.8: Mass element with three-dimensional coordinates.

4.2.5 Section Modulus

The **section modulus** is an area's moment of inertia divided by the farthest distance from the centroidal axis to the outer fiber of the solid, c, or

$$Z_m = \frac{I}{c}. \tag{4.15}$$

The unit of section modulus is length to the third power. The area moment of inertia, I in Eqs. (4.13) to (4.15), holds for I_x and I_y with appropriate changes in r_g and Z_m. Thus, r_{gx} and Z_{mx} would correspond to the use of I_x, and r_{gy}, and Z_{my} would correspond to the use of I_y.

Table 4.1 gives the centroid, area, and moment of inertia for seven different cross sections. Note that data is presented for moments of inertia about the centroid (using $\bar{x}$ and $\bar{y}$ as subscripts) as well as the exterior reference axes (using x and y subscripts). Also, $\bar{J}$ implies that the area polar moment of inertia was taken with the centroidal coordinates; J (without an overbar) indicates that the area polar moment of inertia was taken with respect to coordinates x and y.

4.2.6 Mass Moment of Inertia

The **mass moment of inertia** of an element is the product of the element's mass and the square of the element's distance from the axis. Figure 4.8 shows a mass element with three-dimensional coordinates. From this figure, the mass moments of inertia can be expressed with respect to the x-, y-, and z-axes as

$$I_{mx} = \int \left(y^2 + z^2\right)\, dm_a, \tag{4.16}$$

$$I_{my} = \int \left(x^2 + z^2\right)\, dm_a, \tag{4.17}$$

$$I_{mz} = \int \left(x^2 + y^2\right)\, dm_a. \tag{4.18}$$

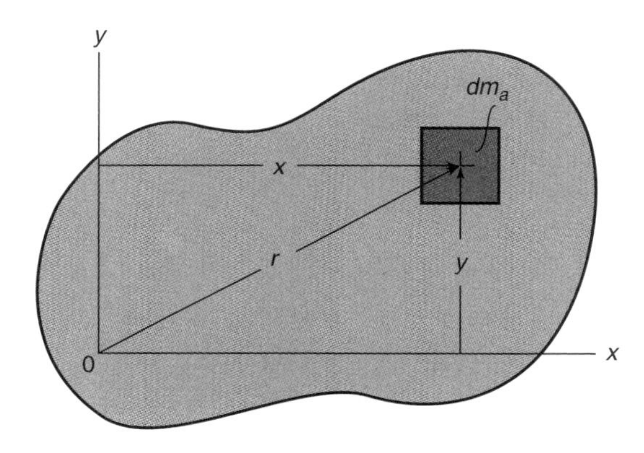

Figure 4.9: Mass element in two-dimensional coordinates and distance from the two axes.

The SI unit of mass moment of inertia is kg-m^2. If, instead of the three-dimensional coordinates given in Fig. 4.8, there is a thin plate (see Fig. 4.9) so that only the x-y plane needs to be considered, the mass moments of inertia become

$$I_{mx} = \int y^2\, dm_a, \tag{4.19}$$

$$I_{my} = \int x^2\, dm_a, \tag{4.20}$$

$$J_o = \int r^2\, dm_a = \int \left(x^2 + y^2\right)\, dm_a = I_{mx} + I_{my}. \tag{4.21}$$

The polar mass moment of inertia is given in Eq. (4.21) as J_o. Table 4.2 gives the mass and mass moment of inertia of six commonly used shapes. Just as with Table 4.1, the origin and coordinates are important for the equations relative to a specific shape.

4.3 Normal Stress and Strain

The bar shown in Fig. 4.10 supports a force, P, and is in tension. Such a force increases the length of the bar. For a section some distance away from the ends, an average intensity of the normal force on the cross section, or **average normal stress**, σ_{avg}, can be written as

$$\sigma_{avg} = \frac{P}{A}. \tag{4.22}$$

Had the section been cut near the ends, the situation would be more complicated. The stress system would no longer be simple tension uniformly distributed over the cross section because any grips or tooling used to apply the loads invariably result in uneven load distributions. Fortunately, the stresses farther from the load application point are fairly uniform (i.e., stress concentrations, covered in detail in Chapter 6, are reduced as the distance from them increases). This tendency was first noted by Saint-Venant, so it is known as **Saint-Venant's principle**. Its importance should not be underestimated; it makes the application of this chapter's equations possible.[1]

[1]The exact distance that one must travel from a stress concentration before stresses can be considered uniform varies and can range from one characteristic length (diameter, grip length, etc.) to over ten characteristic lengths for some composite materials.

Table 4.1: Centroid, area moment of inertia, and area for common cross sections.

Cross section	Centroid	Area moment of inertia	Area
Circle 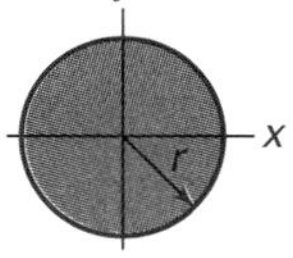	$\bar{x} = 0$ $\bar{y} = 0$	$I_x = I_{\bar{x}} = \frac{\pi}{4} r^4$ $I_y = I_{\bar{y}} = \frac{\pi}{4} r^4$ $J = \frac{\pi}{2} r^4$	$A = \pi r^2$
Hollow circle 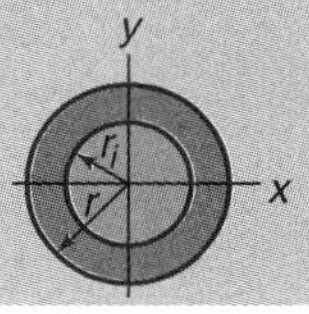	$\bar{x} = 0$ $\bar{y} = 0$	$I_x = I_{\bar{x}} = \frac{\pi}{4}\left(r^4 - r_i^4\right)$ $I_y = I_{\bar{y}} = \frac{\pi}{4}\left(r^4 - r_i^4\right)$ $J = \frac{\pi}{2}\left(r^4 - r_i^4\right)$	$A = \pi\left(r^2 - r_i^2\right)$
Triangle 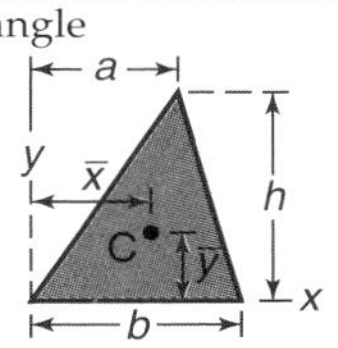	$\bar{x} = \frac{a+b}{3}$ $\bar{y} = \frac{h}{3}$	$I_x = \frac{bh^3}{12}, I_{\bar{x}} = \frac{bh^3}{36}$ $I_y = \frac{bh\left(b^2 + ab + a^2\right)}{12}$ $I_{\bar{y}} = \frac{bh\left(b^2 - ab + a^2\right)}{36}$ $\bar{J} = \frac{bh}{36}\left(b^2 + h^2 + a^2 - ab\right)$	$A = \frac{bh}{2}$
Rectangle 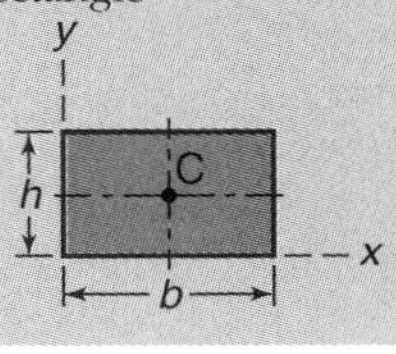	$\bar{x} = \frac{b}{2}$ $\bar{y} = \frac{h}{2}$	$I_x = \frac{bh^3}{3}, I_{\bar{x}} = \frac{bh^3}{12}$ $I_y = \frac{b^3 h}{3}, I_{\bar{y}} = \frac{b^3 h}{12}$ $\bar{J} = \frac{bh}{12}\left(b^2 + h^2\right)$	$A = bh$
Circular sector 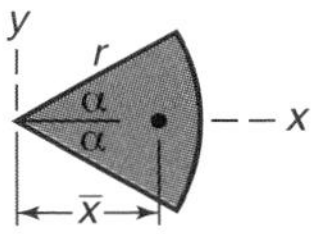	$\bar{x} = \frac{2}{3}\frac{r \sin\alpha}{\alpha}$	$I_x = \frac{r^4}{4}\left(\alpha - \frac{1}{2}\sin 2\alpha\right)$ $I_y = \frac{r^4}{4}\left(\alpha + \frac{1}{2}\sin 2\alpha\right)$ $J = \frac{1}{2} r^4 \alpha$	$A = r^2\alpha$
Quarter-circle 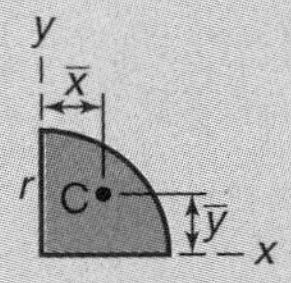	$\bar{x} = \bar{y} = \frac{4r}{3\pi}$	$I_x = I_y \frac{\pi r^4}{16}$ $I_{\bar{x}} = I_{\bar{y}} = \left(\frac{\pi}{16} - \frac{4}{9\pi}\right) r^4$ $J = \frac{\pi r^4}{8}$	$A = \frac{\pi r^2}{4}$
Elliptical quadrant 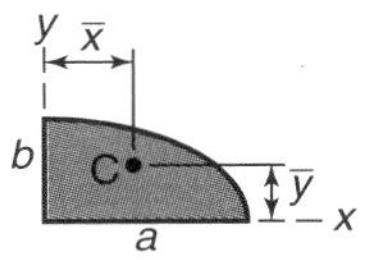	$\bar{x} = \frac{4a}{3\pi}$ $\bar{y} = \frac{4b}{3\pi}$	$I_x = \frac{\pi a b^3}{16}, I_{\bar{x}} = \left(\frac{\pi}{16} - \frac{4}{9\pi}\right) ab^3$ $I_y = \frac{\pi a^3 b}{16}, I_{\bar{y}} = \left(\frac{\pi}{16} - \frac{4}{9\pi}\right) a^3 b$ $J = \frac{\pi ab}{16}\left(a^2 + b^2\right)$	$A = \frac{\pi ab}{4}$

Table 4.2: Mass and mass moment of inertia of six solids.

Shape	Equations
Rod 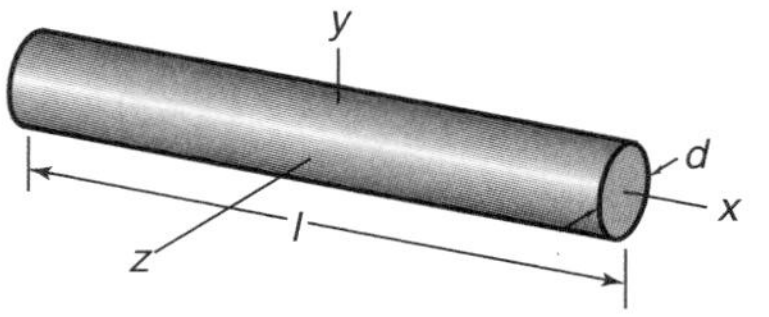	$m_a = \frac{d^2 l \rho}{4}$ $I_{my} = I_{mz} = \frac{m_a l^2}{12}$
Disk 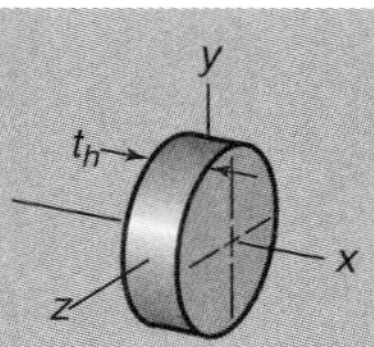	$m_a = \frac{d^2 t_h \rho}{4}$ $I_{mx} = \frac{m_a d^2}{8}$ $I_{my} = I_{mz} = \frac{m_a d^2}{16}$
Rectangular prism 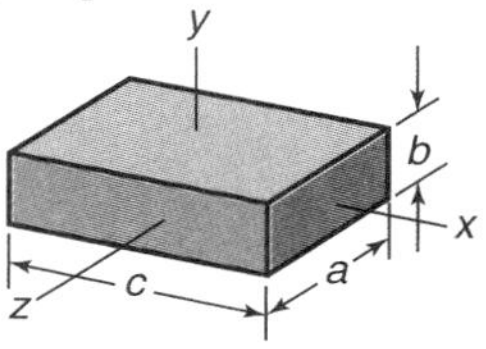	$m_a = abc\rho$ $I_{mx} = \frac{m_a (a^2 + b^2)}{12}$ $I_{my} = \frac{m_a (a^2 + c^2)}{12}$ $I_{mz} = \frac{m_a (b^2 + c^2)}{12}$
Cylinder 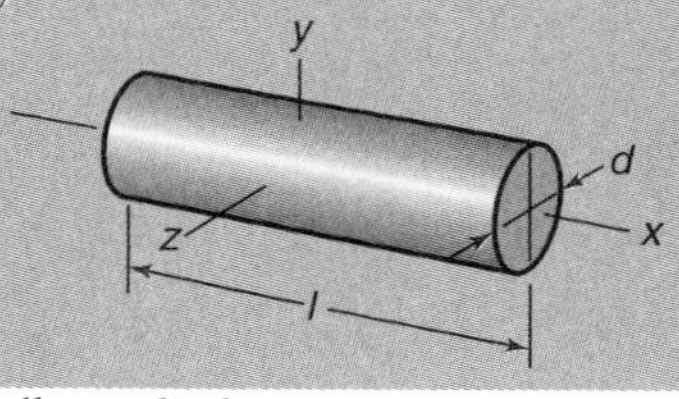	$m_a = \frac{d^2 l \rho}{4}$ $I_{mx} = \frac{m_a d^2}{8}$ $I_{my} = I_{mz} = \frac{m_a (3d^2 + 4l^2)}{48}$
Hollow cylinder 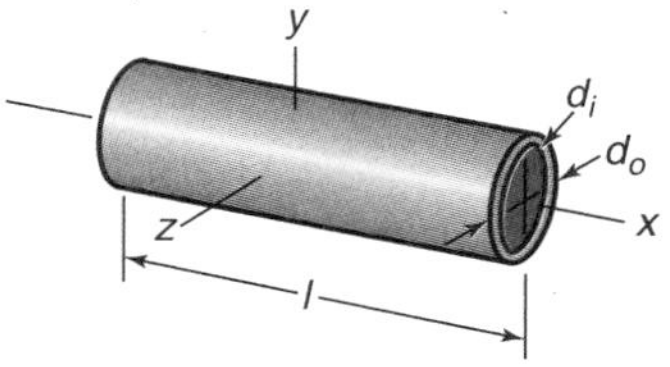	$m_a = \frac{l \rho (d_o^2 - d_i^2)}{4}$ $I_{mx} = \frac{m_a (d_o^2 - d_i^2)}{8}$ $I_{my} = I_{mz} = \frac{m_a (3d_o^2 + 3d_i^2 + 4l^2)}{48}$
Sphere 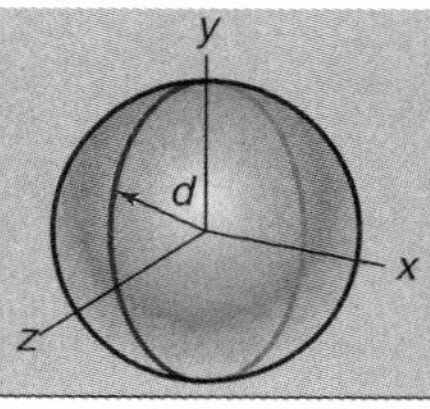	$m_a = \frac{d^3 \rho}{6}$ $I_{mx} = I_{my} = I_{mz} = \frac{m_a d^2}{10}$

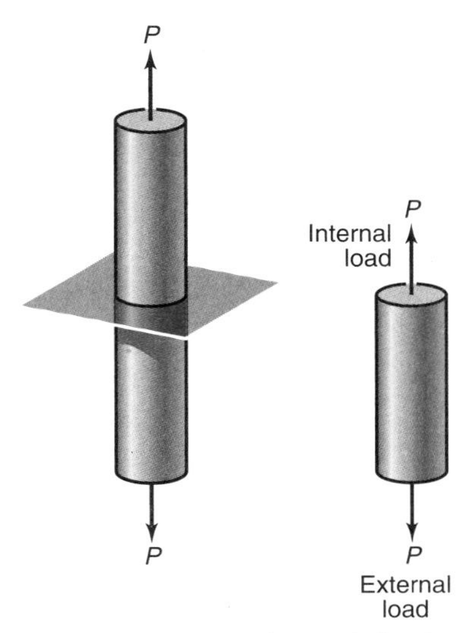

Figure 4.10: Circular bar with tensile load applied.

Consistent with the sign convention presented in Chapter 2, a tensile normal stress is taken as positive and a compressive normal stress is negative. The total length change in a uniform bar caused by an axial load is called the **elastic deformation**, δ. The **normal strain** is

$$\epsilon = \frac{\delta}{l} = \frac{\text{Elastic deformation}}{\text{Original length}}. \tag{4.23}$$

Although the strain is dimensionless, it is often expressed in terms of meter per meter, or often μm/m. From Hooke's law for a uniaxial normal loading,

$$\sigma = \epsilon E \qquad \text{or} \qquad \epsilon = \frac{\sigma}{E} \tag{4.24}$$

where E is the modulus of elasticity described in Section 3.5.2 and is a constant for a given material. Substituting Eqs. (4.22) and (4.24) into Eq. (4.23) gives

$$\delta = \epsilon l = \frac{\sigma l}{E} = \frac{Pl}{AE}. \tag{4.25}$$

Equations (4.24) and (4.25) are valid for either tension or compression. The **spring rate** is the ratio of normal to elastic deflection or, for axial loading,

$$k = \frac{P}{\delta} = \frac{AE}{l}. \tag{4.26}$$

Equations (4.22) to (4.26) hold for any cross section that is constant over the length l.

Example 4.6: Normal Stress, Deformation, and Spring Rate

Given: A hollow carbon steel shaft 50 mm long must carry a normal force of 5000 N at a normal stress of 100 MPa. The inside diameter is 65% of the outside diameter.

Find: The outside diameter, the axial deformation, and the spring rate.

Solution: The cross-sectional area can be expressed as

$$\begin{aligned} A &= \frac{\pi}{4}\left(d_o^2 - d_i^2\right) = \frac{\pi d_o^2}{4}\left[1 - \left(\frac{d_i}{d_o}\right)^2\right] \\ &= \frac{\pi d_o^2}{4}\left[1 - \left(\frac{0.65 d_o}{d_o}\right)^2\right] = 0.4536 d_o^2. \end{aligned}$$

From Eq. (4.22),

$$A = \frac{P}{\sigma} = \frac{5000}{100 \times 10^6} = 5 \times 10^{-5}\ \text{m}^2.$$

Therefore,

$$0.4536 d_o^2 = 5 \times 10^{-5}\ \text{m}^2$$

or

$$d_o = 1.05 \times 10^{-2}\ \text{m} = 10.5\ \text{mm}.$$

For carbon steel, the modulus of elasticity is 207 GPa (see Table 3.1). From Eq. (4.25), the elastic deformation is

$$\delta = \frac{Pl}{AE} = \frac{(5000)(0.05)}{(5 \times 10^{-5})(207 \times 10^9)} = 24.15 \times 10^{-6}\ \text{m},$$

or $\delta = 24.15\ \mu$m. From Eq. (4.26), the spring rate is

$$k = \frac{P}{\delta} = \frac{5000}{24.15 \times 10^{-6}} = 2.07 \times 10^8\ \text{N/m}.$$

Example 4.7: Elongation and Spring Rate in Tension

Given: A fisherman catches a salmon with a lure fastened to his 0.45-mm-diameter nylon line. When the fish bites, it is 46 m from the reel. The modulus of elasticity of the line material is 4 GPa, and the salmon pulls with a force of 50 N.

Find: The elastic elongation of the line, the spring rate, and the tensile stress in the line.

Solution: The cross-sectional area of the line is

$$A = \frac{\pi d^2}{4} = \frac{\pi (0.45)^2}{4} = 0.1590\ \text{mm}^2.$$

The stress being exerted on the line by the salmon is

$$\sigma = \frac{P}{A} = \frac{50}{0.159 \times 10^{-6}} = 0.3144\ \text{GPa}.$$

The elongation of the line is

$$\delta = \epsilon l = \frac{\sigma l}{E} = \frac{(0.3144)(46)}{4} = 3.615\ \text{m}.$$

The spring rate of the line is $k = P/\delta = 13.83$ N/m.

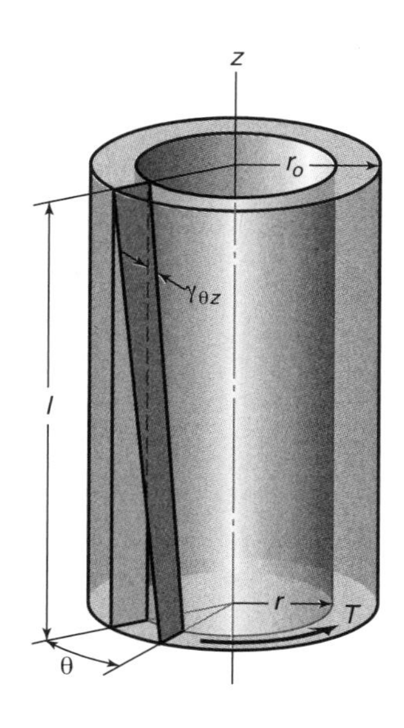

Figure 4.11: Twisting of member due to applied torque.

4.4 Torsion

A **shaft** is a slender element that is mainly loaded by axial moment or twist (torque), which causes torsional deformation and associated shear stresses. Thus, **torsion** is a loading that results in twisting of the shaft. Design of shafts will be investigated in detail in Chapter 11; the focus here is on the twisting, or torsion, to which shafts are subjected and the resulting stresses. A major use of the shaft is to transfer or transmit mechanical power from one point to another. Designers are interested primarily in the twisting moment that can be transmitted by the shaft without damaging the material or exceeding deformation constraints. Hence, it is important to be able to calculate the stresses in the shaft and the angle of twist. Solid circular members are of primary concern because most torque-transmitting shafts are of this shape. For cross sections that are not circular, the interested reader is referred to the classic text by Timoshenko and Goodier [1970].

4.4.1 Stress and Strain

Figure 4.11 shows the twisting of a member due to an applied torque. The circular shaft deforms such that each plane cross section originally normal to the axis remains plane and normal and does not distort within its own plane. While Fig. 4.11 shows a hollow shaft, a solid shaft is simply a special case where the inner radius equals zero. The shaft is fixed at the top and a torque is applied to the bottom end. The angle of twist is θ and the extension strains are assumed to be zero. Thus, $\epsilon_r = \epsilon_\theta = \epsilon_z = \gamma_{r\theta} = \gamma_{rz} = 0$, and the only nonzero strain is

$$\gamma_{\theta z} = r\frac{d\theta}{dz} \approx \frac{r\theta}{l}. \tag{4.27}$$

Equation (4.27) can also be obtained from Fig. 4.11 by observing the common surface, as it relates $l\gamma_{\theta z}$ and $r\theta$.

From Hooke's law, the stress is related to the strain by

$$\tau_{\theta z} = G\gamma_{\theta z} = Gr\frac{d\theta}{dz} \approx \frac{Gr\theta}{l}, \tag{4.28}$$

where G is the shear modulus of elasticity given in Eq. (3.7). In Eq. (4.28), the shear strain, $\gamma_{\theta z}$, and the shear stress, $\tau_{\theta z}$, vary linearly with respect to the rate of twist $d\theta/dz \approx \theta/l$. Also, the shear stress does not change in the θ-direction (because of symmetry) or in the z-direction (because the deformation and the stress pattern are uniform along the shaft's length).

The rate of twist $d\theta/dz$, or θ/l in Eq. (4.28), still must be determined. To do so, the stresses must satisfy equilibrium. The applied twisting moment, or torque, is

$$T = \int_A r\left(\tau_{\theta z}\, dA\right) = \frac{G\theta}{l}\int_A r^2\, dA. \tag{4.29}$$

The polar moment of inertia is

$$J = \int_A r^2\, dA. \tag{4.30}$$

Substituting Eq. (4.30) into Eq. (4.29) gives the torque and the angle of twist as

$$T = \frac{G\theta J}{l} \quad \text{and} \quad \theta = \frac{Tl}{GJ}. \tag{4.31}$$

Also, substituting Eqs. (4.29) and (4.30) into Eq. (4.28) gives

$$\tau_{\theta z} = \frac{Tr}{J}. \tag{4.32}$$

The maximum stress is

$$\tau_{\max} = \frac{Tc}{J}. \tag{4.33}$$

where c is the distance from the neutral axis to the outer fiber. The angular spring rate can be expressed as

$$k_a = \frac{T}{\theta} = \frac{JG}{l}. \tag{4.34}$$

Example 4.8: Angle of Twist and Spring Rate in Torsion

Given: A 50-mm-long, hollow circular shaft made of carbon steel must carry a torque of 5000 N-m at a maximum shear stress of 70 MPa. The inside diameter is one-half the outside diameter.

Find: The outside diameter, the angle of twist, and the angular spring rate.

Solution: From Table 4.1 for a hollow cylinder,

$$J = \frac{\pi}{2}\left(r^4 - r_i^4\right) = \frac{\pi r^4}{2}\left[1 - \left(\frac{r_i}{r}\right)^4\right]. \tag{a}$$

From Eq. (4.15), the section modulus is

$$Z_m = \frac{J}{c} = \frac{T}{\tau_{\max}} = \frac{5000}{70 \times 10^6} = 71.43 \times 10^{-6}\ \text{m}^3. \tag{b}$$

Note that

$$\frac{r_i}{r} = \frac{r_i}{r_o} = 0.5 \quad \text{and} \quad c = \frac{d_o}{2}. \tag{c}$$

Therefore, by making use of Eqs. (*a*) and (*c*), Eq. (*b*) becomes

$$\frac{\pi r_o^3}{2}\left[1 - (0.5)^4\right] = 71.43 \times 10^{-6}\ \text{m}^3.$$

Solving for r_o,

$$r_o^3 = 48.51 \times 10^{-6}\ \text{m}^3,$$

so that $r_o = 0.03647$ m $= 36.47$ mm and $d_o = 72.94$ mm. Note from Eqs. (*b*) and (*c*) that

$$J = cZ_m = \frac{d_o}{2}Z_m = \left(\frac{0.07294}{2}\right)(71.43 \times 10^{-6}),$$

or $J = 2.605 \times 10^{-6}$ m^4. The shear modulus of elasticity for carbon steel is 80 GPa. Making use of Eq. (4.31) gives the angular twist due to torsion as

$$\theta = \frac{Tl}{GJ} = \frac{(5000)(0.050)}{(80 \times 10^9)(2.605 \times 10^{-6})} = 0.00120 \text{ rad}.$$

From Eq. (4.34) the angular spring rate is

$$k_a = \frac{T}{\theta} = \frac{5000}{0.00120} = 4.167 \times 10^6 \text{ N-m/rad}.$$

4.4.2 Power Transfer

One of the most common uses of a circular shaft is in power transmission; therefore, no discussion of torsion can exclude this important topic. **Power** is the rate of doing work or

$$h_p = Pu. \tag{4.35}$$

For rotating systems,

$$P = \frac{T}{r}. \tag{4.36}$$

Since $u = \omega r$,

$$h_p = Pu = T\omega, \tag{4.37}$$

where
- P = force, N
- u =velocity, m/s
- T =torque, N-m
- ω = angular velocity, rad/s

The SI unit of work is a joule, or a newton-meter. The SI unit of power is a watt, or a joule per second. Solving Eq. (4.37) for torque yields

$$T = \frac{h_p}{\omega}. \tag{4.38}$$

Example 4.9: Power Transmitted by a Shaft

Given: A shaft carries a torque of 1000 N-m and turns at 900 rpm.

Find: The power transmitted in kilowatts.

Solution: The angular velocity is

$$\omega = (900 \text{ rpm})\frac{2\pi}{60} = 94.25 \text{ rad/s}$$

From Eq. (4.37),

$$h_p = T\omega = (1000)(94.25) = 94.25 \text{ kW}$$

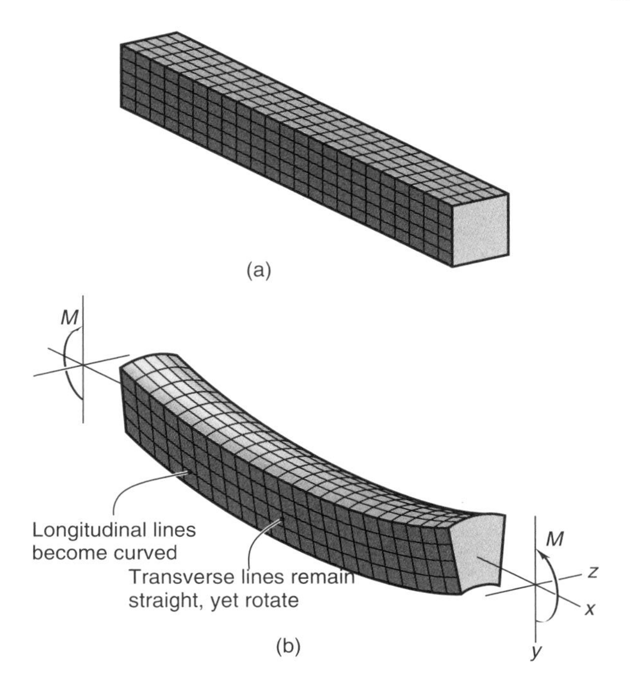

Figure 4.12: Bar made of elastic material to illustrate effect of applied bending moment. (a) Undeformed bar; (b) deformed bar.

4.5 Bending Stress and Strain

4.5.1 Straight Member

A **beam** is usually a long and slender member that supports predominantly bending loads. Analysis of beam bending is a common concern to engineers. Bending occurs in horizontal members of buildings (joists) subjected to vertical floor loading, in leaf springs on a truck, and in wings of an airplane supporting the weight of the fuselage. In each of these applications the stress and deformation are important design considerations.

Throughout this section the following assumptions are made:

1. The material is Hookian (see Section 3.6).
2. Deformations are small.

Figure 4.12 shows bending occurring in a highly deformable material, such as rubber, which is well-suited for demonstration purposes. The behavior of this bar can be extended to traditional beam materials such as steel, but with much smaller deformations. Figure 4.12a shows an undeformed bar with a square cross section marked by longitudinal and transverse grid lines. In Fig. 4.12b, a moment is applied. The longitudinal lines become curved while the transverse lines remain straight and undergo a rotation. The longitudinal lines have a radius when the bar is deformed, even though initially they were straight. Inspection of Fig. 4.12b suggests that the neutral axis will shift upward when the moment is applied. For elastic bending of stiffer beam materials, strains are low and warping of cross sections and shifting of the neutral axis will be ignored. Figure 4.13 further demonstrates the effect of bending. The bending moment causes the material in the bottom portion of the beam to stretch, or be

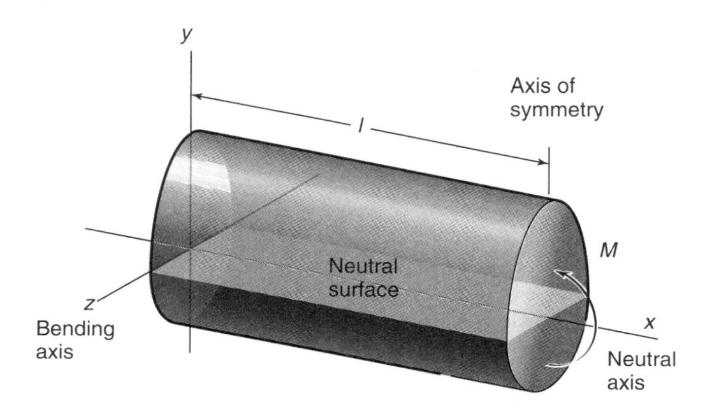

Figure 4.13: Bending occurring in a cantilevered bar, showing neutral surface.

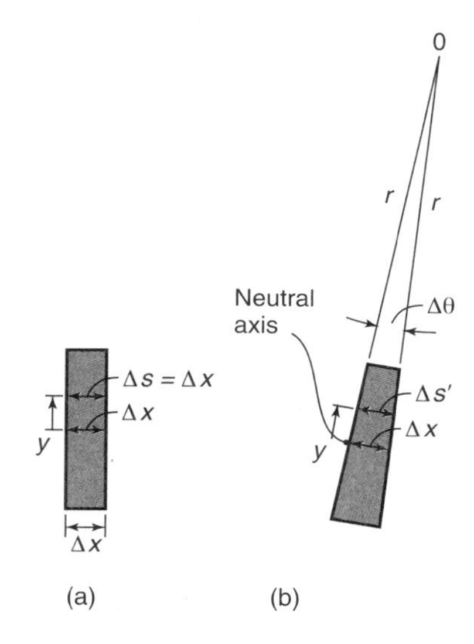

Figure 4.14: Undeformed and deformed elements in bending.

in tension, and that in the top to be in compression. Consequently, between these two regions there must be a surface, called the *neutral surface*, in which the longitudinal fibers of the material will not undergo a change in length. On this neutral surface no bending stress is occurring, and it is neither in tension nor in compression.

Figure 4.14 shows undeformed and deformed elements when bending occurs. The normal strain along line segment Δs is

$$\epsilon = \lim_{\Delta s \to 0} \frac{\Delta s' - \Delta s}{\Delta s}. \tag{4.39}$$

From Fig. 4.14, where r is the radius of curvature of the element's longitudinal axis,

$$\Delta x = \Delta s = r\Delta\theta,$$

$$\Delta s' = (r - y)\Delta\theta.$$

Substituting these equations into Eq. (4.39) gives

$$\epsilon = \lim_{\Delta s \to 0} \frac{(r - y)\Delta\theta - r\Delta\theta}{r\Delta\theta} = -\frac{y}{r}. \tag{4.40}$$

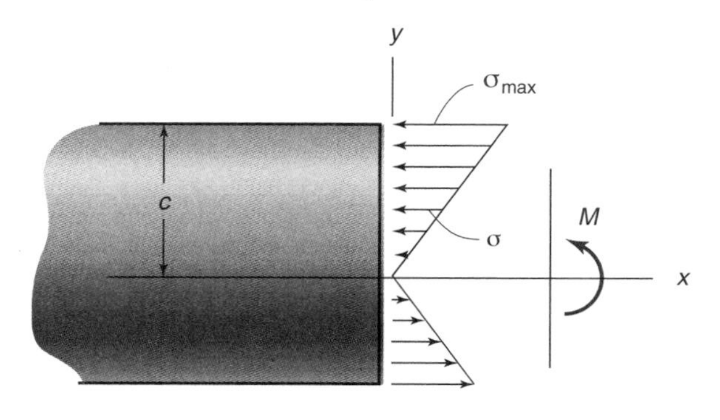

Figure 4.15: Profile view of bending stress variation.

The longitudinal normal strain will vary linearly with distance from the neutral axis. The maximum strain occurs at the outermost fiber, located at a distance, c, from the neutral axis. Therefore,

$$\frac{\epsilon}{\epsilon_{\max}} = -\frac{y/r}{c/r}$$

or

$$\epsilon = -\frac{y}{c}\epsilon_{\max}. \tag{4.41}$$

Similarly, a linear variation of normal stress over the cross-sectional area occurs, or

$$\sigma = -\frac{y}{c}\sigma_{\max}. \tag{4.42}$$

Figure 4.15 shows a profile view of the normal stress. For positive y, the normal stress is compressive and for negative y, the normal stress is tensile. The normal stress is zero at the neutral axis.

4.5.2 Bending Stress

From force equilibrium,

$$0 = \int_A dP = \int_A \sigma\, dA = \int_A -\frac{y}{c}\sigma_{\max}\, dA = -\frac{\sigma_{\max}}{c}\int_A y\, dA.$$

Because $\sigma_{\max}$ is not equal to zero, this can only be assured if

$$\int_A y\, dA = 0. \tag{4.43}$$

Since the first moment of the member's cross-sectional area about the neutral axis must be zero, this requires that the neutral axis pass through the centroid. The moment may be expressed as

$$M = \int_A y\, dP = \int_A y\sigma\, dA = -\frac{\sigma_{\max}}{c}\int_A y^2\, dA. \tag{4.44}$$

The area moment of inertia is

$$I = \int_A y^2\, dA.$$

Substituting this relationship into Eq. (4.44) and solving for stress yields

$$\sigma_{\max} = -\frac{Mc}{I}. \tag{4.45}$$

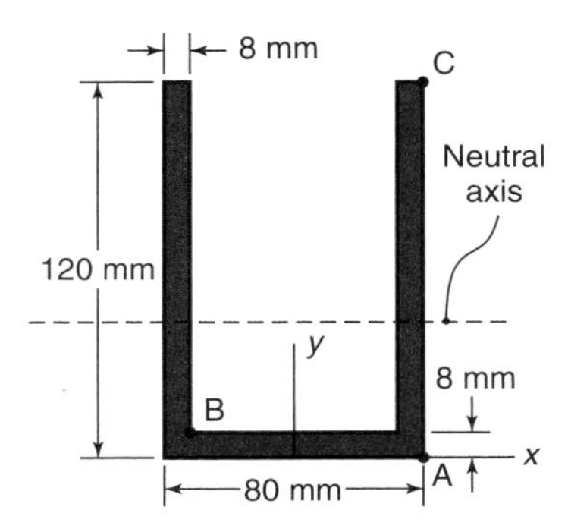

Figure 4.16: U-shaped cross section used in Example 4.10.

The stress at any intermediate distance y is

$$\sigma = -\frac{My}{I}. \tag{4.46}$$

Making use of Eqs. (3.22), (4.40), and (4.46) yields

$$\frac{1}{r} = \frac{M}{EI}. \tag{4.47}$$

From Eq. (4.47), when the bending moment is positive, the curvature is positive (i.e., concave in the y-direction as shown in Fig. 2.3).

Example 4.10: Stress in Bending

Given: An aluminum alloy beam with the cross section shown in Fig. 4.16 experiences positive bending from an applied moment, M. The allowable stress is 150 MPa.

Find: (*a*) The maximum moment that can be applied to the beam. (*b*) The stresses at points A, B, and C when the maximum moment is applied.

Solution: (*a*) The cross-sectional area, if subscript 1 refers to a horizontal section and subscript 2 refers to a vertical section, is (see Fig. 4.16):

$$A = A_1 + 2A_2 = 8(80-16) + 2(120)(8) = 2432 \text{ mm}^2.$$

The centroid of the cross section is

$$\bar{y} = \frac{\bar{y}_1 A_1 + 2\bar{y}_2 A_2}{A} = \frac{4(64)(8) + 2(60)(8)(120)}{2432} = 48.21 \text{ mm}.$$

The distances from the neutral axis to the centroids of the horizontal and vertical bars are

$$d_{n1} = 48.21 - 4 = 44.21 \text{ mm},$$

$$d_{n2} = 60 - 48.21 = 11.79 \text{ mm}.$$

The area moment of inertia of the composite structure is

$$I = I_1 + A_1 d_{n1}^2 + 2(I_2 + A_2 d_{n2}^2)$$

or

$$I = \frac{64(8)^3}{12} + (64)(8)(44.2)^2 + 2\left[\frac{8(120)^3}{12} + 8(120)(11.79)^2\right]$$

or $I = 3.574 \times 10^6 \text{ mm}^4$. The distances from the neutral axis to the points where the stress is to be evaluated are

$$d_{nA} = 48.21 \text{ mm},$$

$$d_{nB} = 48.21 - 8 = 40.21 \text{ mm},$$

$$d_{nC} = 120 - 48.21 = 71.79 \text{ mm}.$$

Point C is the farthest from the neutral axis and is thus the location where the stress is the largest. Furthermore, as discussed in Section 2.3, positive bending implies that the portion of the composite structure above the neutral axis is in compression while the portion below it is in tension. From Eq. (4.45), the maximum moment is

$$M_{\max} = \frac{\sigma_{\text{all}} I}{d_{nC}} = \frac{(150 \times 10^6)(3.574 \times 10^{-6})}{71.79 \times 10^{-3}} = 7468 \text{ Nm}.$$

(*b*) The stresses at the various points of interest are

$$\sigma_A = \frac{M_{\max} d_{nA}}{I} = \frac{(7468)(48.21 \times 10^{-3})}{3.574 \times 10^{-6}} = 100.7 \text{ MPa},$$

$$\sigma_B = \frac{M_{\max} d_{nB}}{I} = \frac{(7468)(40.21 \times 10^{-3})}{3.574 \times 10^{-6}} = 84.0 \text{ MPa},$$

$$\sigma_C = \sigma_{\text{all}} = -150 \text{ MPa}.$$

4.5.3 Curved Member

In a straight member, the normal stress and strain vary linearly with distance from the neutral axis, as shown in Eqs. (4.45) and (4.46). In a curved member, the stress and strain are not linearly related. Examples of curved members are hooks and chain links, which are not slender but have a high degree of curvature.

Figure 4.17 shows a curved member in bending. Positive bending causes surface dc to rotate through $d\phi$ to $d'c'$. Note that y is negative from the neutral axis to the outer radius r_o, the region where the member is experiencing tension, and positive from the neutral axis to the inner radius r_i, the region where the member is experiencing compression. No bending occurs at the neutral axis; thus, the member is neither in compression nor in tension. The neutral axis occurs at a radius of r_n. From Fig. 4.17, note that the neutral radius r_n and the centroidal radius $\bar{r}$ are not the same, although they are the same for a straight member. The difference between r_n and $\bar{r}$ is the eccentricity, e. Also, the radius r locates an arbitrary area element dA, as shown in the cross-sectional view of Fig. 4.17b.

The strain for an arbitrary radius, r, can be expressed as

$$\epsilon = \frac{(r - r_n)\, d\phi}{r\phi}. \tag{4.48}$$

The strain is zero when r is the neutral radius, r_n, and is largest at the outer fiber, or $r = r_o$. The normal stress can be written as

$$\sigma = \epsilon E = \frac{E(r - r_n)\, d\phi}{r\phi}. \tag{4.49}$$

For r less than r_n the stress is compressive and for r greater than r_n the stress is tensile.

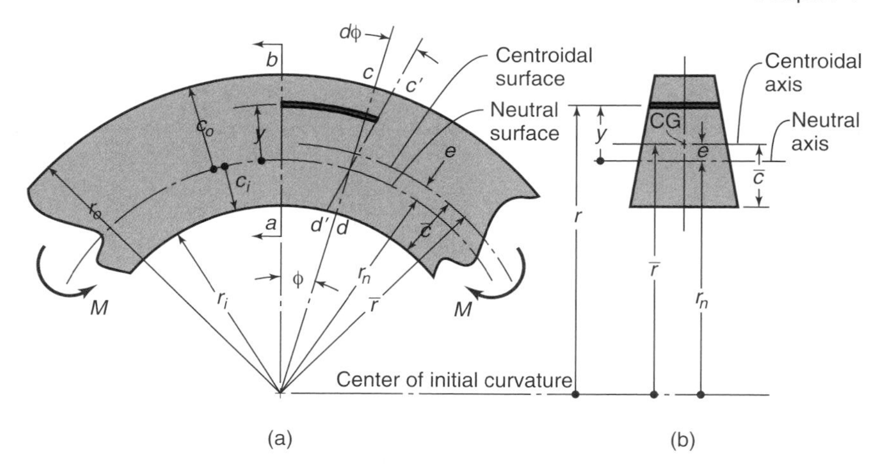

Figure 4.17: Curved member in bending. (a) Circumferential view; (b) cross-sectional view.

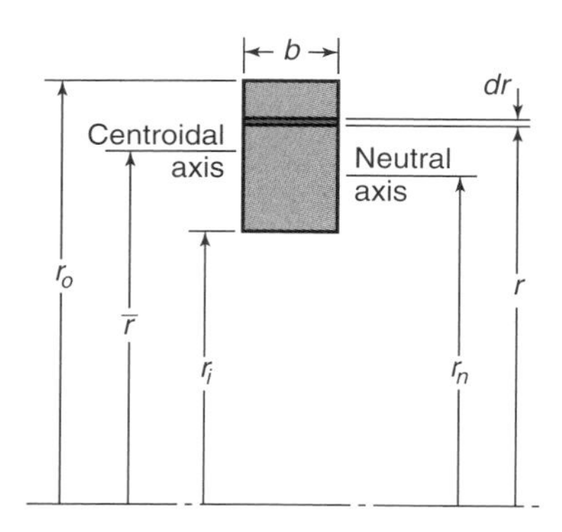

Figure 4.18: Rectangular cross section of curved member.

Solving for the strain and stress in Eqs. (4.48) and (4.49) requires the location of the neutral axis. This location is obtained by taking the sum of the normal stresses acting on the section and setting it to zero. Thus, making use of Eq. (4.49) gives

$$\int_A \sigma\, dA = \frac{E\, d\phi}{\phi} \int_A \frac{r - r_n}{r}\, dA = 0. \tag{4.50}$$

This equation reduces to

$$A - r_n \int_A \frac{dA}{r} = 0,$$

or

$$r_n = \frac{A}{\int_A \frac{dA}{r}}. \tag{4.51}$$

Equation (4.51) clearly indicates that the neutral radius is a function of the cross-sectional area.

Rectangular Cross Sections

Figure 4.18 shows a rectangular cross section of a curved member with its centroidal and neutral axes. From Eq. (4.51), the neutral radius for a rectangular cross section is

$$r_n = \frac{b(r_o - r_i)}{\int_{r_i}^{r_o} \frac{b\, dr}{r}} = \frac{r_o - r_i}{\ln\left(\frac{r_o}{r_i}\right)}. \tag{4.52}$$

The centroidal radius is

$$\bar{r} = \frac{r_i + r_o}{2}. \tag{4.53}$$

The eccentricity is

$$e = \bar{r} - r_n = \frac{r_i + r_o}{2} - \frac{r_o - r_i}{\ln\left(\frac{r_o}{r_i}\right)}. \tag{4.54}$$

Circular Cross Sections

The neutral radius for a circular cross section is

$$r_n = \frac{\bar{r} + \sqrt{\bar{r}^2 - c^2}}{2}, \tag{4.55}$$

where c is the radius of the cross section and $\bar{r}$ is the location of the centroid. Having established the location of the neutral radius for two different cross sections of a curved member, consider again the stress in Eq. (4.49). The bending moment is the moment arm $(r - r_n)$ multiplied by the force $\sigma\, dA$ integrated over the cross-sectional area, or

$$M = \int (r - r_n)(\sigma\, dA).$$

Making use of Eq. (4.49) gives

$$\begin{aligned} M &= \frac{E\, d\phi}{\phi} \int \frac{(r - r_n)^2\, dA}{r} \\ &= \frac{E\, d\phi}{\phi} \left(\int r\, dA - r_n A - r_n A + r_n^2 \int \frac{dA}{r} \right). \end{aligned}$$

From Eq. (4.51), this equation reduces to

$$M = \frac{E\, d\phi}{\phi} \left(\int r\, dA - r_n A \right). \tag{4.56}$$

From the definition of a centroid,

$$\bar{r} = \frac{1}{A} \int r\, dA.$$

Equation (4.56) then becomes

$$M = E\frac{d\phi}{\phi}Ae, \tag{4.57}$$

where

$$e = \bar{r} - r_n. \tag{4.58}$$

Substituting Eq. (4.49) into Eq. (4.57),

$$M = \frac{r\sigma Ae}{r - r_n},$$

or

$$\sigma = \frac{M(r - r_n)}{Aer} = \frac{My}{Ae(r_n + y)}, \tag{4.59}$$

where

$$y = r - r_n. \tag{4.60}$$

The stress distribution is hyperbolic in shape. The maximum stress occurs at either the inner or outer surface:

$$\sigma_i = -\frac{Mc_i}{Aer_i}, \tag{4.61}$$

$$\sigma_o = \frac{Mc_o}{Aer_o}. \tag{4.62}$$

Example 4.11: Stress in Curved Member

Given: A curved member with a rectangular cross section shown in Fig. 4.18 has the dimensions b = 25 mm and $h = r_o - r_i$ = 75 mm, and is subjected to a bending moment of 2000 N-m. No other type of loading acts on the member. Positive bending occurs.

Find: The maximum stress for the following geometries:

(*a*) A straight member

(*b*) A member whose centroidal axis has a radius of $\bar{r}$ = 375 mm.

(*c*) A member whose centroidal axis has a radius of $\bar{r}$ = 75 mm.

Solution:

(*a*) For a straight member,

$$I = \frac{bh^3}{12} \quad \text{and} \quad c = \frac{h}{2}.$$

Therefore,

$$\frac{I}{c} = \frac{bh^2}{6},$$

$$|\sigma| = \frac{Mc}{I} = \frac{(2000)6}{(0.025)(0.075)^2} = 85.33 \text{ MPa}.$$

Therefore,

$$\sigma_i = -85.33 \text{ MPa},$$

$$\sigma_o = 85.33 \text{ MPa}.$$

(*b*) The outer and inner radii relative to the centroidal radius of 375 mm are

$$r_o = \bar{r} + \frac{h}{2} = 375 + \frac{75}{2} = 412.5 \text{ mm},$$

$$r_i = \bar{r} - \frac{h}{2} = 375 - \frac{75}{2} = 337.5 \text{ mm}.$$

From Eq. (4.52), the neutral radius is

$$r_n = \frac{r_o - r_i}{\ln\left(\frac{r_o}{r_i}\right)} = \frac{412.5 - 337.5}{\ln\left(\frac{412.5}{337.5}\right)} = 373.7 \text{ mm}.$$

Therefore,

$$e = \bar{r} - r_n = 375 - 373.7 = 1.3 \text{ mm}.$$

The distances from the neutral axis to the inner and outer fibers are

$$c_o = \frac{h}{2} + e = 37.5 + 1.3 = 38.8 \text{ mm},$$

$$c_i = \frac{h}{2} - e = 37.5 - 1.3 = 36.2 \text{ mm}.$$

The corresponding normal stresses are

$$\begin{aligned}
\sigma_i &= -\frac{Mc_i}{Aer_i} = -\frac{(2000)(0.0362)}{(0.075)(0.025)(0.0013)(0.3375)} \\
&= -88.0 \text{ MPa}, \\
\sigma_o &= \frac{Mc_o}{Aer_o} = -\frac{(2000)(0.0388)}{(0.075)(0.025)(0.0013)(0.4125)} \\
&= 77.12 \text{ MPa}.
\end{aligned}$$

Because the radius of curvature is large, there is little difference from the stresses found for the straight beam in part a.

(*c*) The outer and inner radii relative to the centroidal radius of 75 mm are

$$r_o = \bar{r} + \frac{h}{2} = 0.075 + \frac{0.075}{2} = 112.5 \text{ mm},$$

$$r_i = \bar{r} - \frac{h}{2} = 0.075 - \frac{0.075}{2} = 37.5 \text{ mm},$$

$$r_n = \frac{r_o - r_i}{\ln\left(\frac{r_o}{r_i}\right)} = \frac{0.1125 - 0.0375}{\ln\left(\frac{112.5}{37.5}\right)} = 68.27 \text{ mm},$$

$$e = \bar{r} - r_n = 75 - 68.27 = 6.73 \text{ mm}.$$

Therefore,

$$c_o = \frac{h}{2} + e = 37.5 + 6.73 = 44.23 \text{ mm},$$

$$c_i = \frac{h}{2} - e = 37.5 - 6.73 = 30.77 \text{ mm},$$

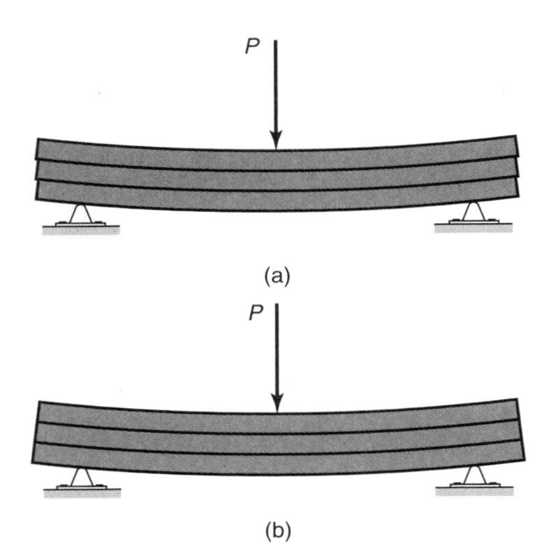

Figure 4.19: Development of transverse shear. (a) Boards not bonded together; (b) boards bonded together.

$$\begin{aligned}\sigma_i &= -\frac{Mc_i}{Aer_i} = -\frac{(2000)(0.03077)}{(0.075)(0.025)(0.00673)(0.0375)} \\ &= -130.0 \text{ MPa}, \\ \sigma_o &= -\frac{Mc_o}{Aer_o} = -\frac{(2000)(0.04423)}{(0.075)(0.025)(0.00673)(0.1125)} \\ &= 62.31 \text{ MPa}.\end{aligned}$$

Thus, the more curved the surface is, the greater the departure from the straight-beam results.

4.6 Transverse Shear Stress and Strain

In addition to the bending stresses considered in Section 4.5, moments can also cause shear stresses within the member. Figure 4.19 illustrates how transverse shear is developed. Figure 4.19a shows three boards that are not bonded together. The application of a force, P, will cause the boards to slide relative to one another and the stack will deflect as shown, with the ends no longer flush as they were when no load was applied. On the other hand, if the boards are bonded together, they will act as a single unit, as shown in Fig. 4.19b. In a solid beam the same shear stress is present. As a result of the internal shear stress distribution, shear strains will be developed and these will tend to distort the cross section in a rather complex manner. An undeformed bar, shown in Fig. 4.20a and made of a highly deformable material and marked with horizontal and vertical grid lines, tends to deform when a shear force is applied. The deformed pattern is shown in Fig. 4.20b. The squares near the top and bottom of the bar retain their original shapes. The strain on the center square of the bar will cause it to have the greatest deformation. The transverse shear stress causes the cross section to warp. Figure 4.20b shows that the deformation caused by transverse shear is much more complex than that caused by another type of loading (axial, torsion, or bending).

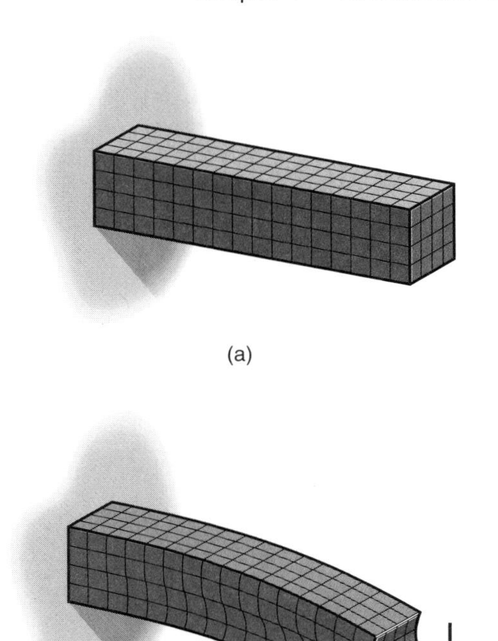

Figure 4.20: Cantilevered bar made of highly deformable material and marked with horizontal and vertical grid lines to show deformation due to transverse shear. (a) Undeformed; (b) deformed.

The transverse shear formulas also apply to Fig. 4.21. The shaded top segment of the element has been sectioned at y' from the neutral axis. This section has a width, w_t, at the section and has cross-sectional sides, each having an area A'. Because the resultant moments on each side of the element differ by dM, force equilibrium will only be satisfied from Fig. 4.21 if a longitudinal shear stress, τ, acts over the bottom face of the segment. This longitudinal shear stress is also responsible for the results shown in Fig. 4.19a. The expression for the shear stress in the member at the point located a distance y' from the neutral axis is

$$\tau = \frac{VQ}{Iw_t}, \tag{4.63}$$

where

V = transverse shear force, N
I = moment of inertia of entire cross section computed about neutral axis, m^4
w_t = width at point where τ is determined, m
Q = first moment about neutral axis of shaded portion of Fig. 4.21 given by

$$Q = \int_{A'} y\, dA = \bar{y}' A', \tag{4.64}$$

and where A' is the cross-sectional area of the top portion shown in Fig. 4.21, and $\bar{y}'$ is the distance to the centroid of A', measured from the neutral axis. Applying the shear formula given in Eq. (4.63) for a rectangular cross section gives

$$\tau = \frac{6V}{bh^3}\left(\frac{h^2}{4} - y^2\right), \tag{4.65}$$

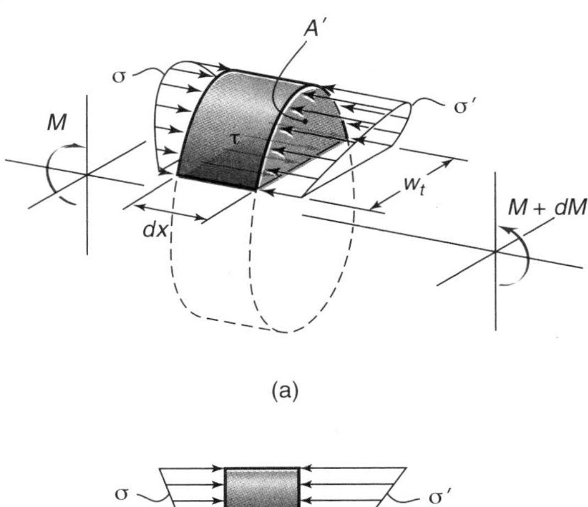

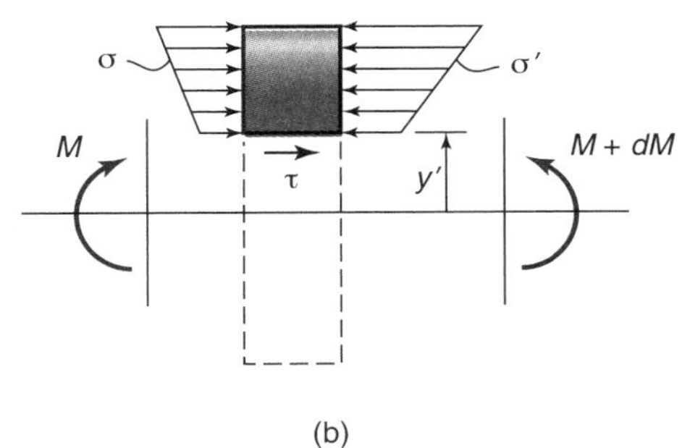

Figure 4.21: Three-dimensional and profile views of moments and stresses associated with shaded top segment of element that has been sectioned at y' about neutral axis. (a) Three-dimensional view; (b) profile view.

where b and h are the base and height of a rectangular section, respectively. From Eq. (4.65), the shear stress intensity for a rectangular cross section varies from zero at the top and bottom ($y = \pm h/2$) to a maximum at the neutral axis ($y = 0$). At other intermediate values it is parabolic in shape. For $A = bh$ and $y = 0$, Eq. (4.65) becomes

$$\tau_{\max} = 1.5\frac{V}{A}. \tag{4.66}$$

The maximum shear stress depends on the shape of the cross section, but this approach can be used for any arbitrary beam profile. Table 4.3 summarizes maximum values for common cross-sectional shapes. In all cases, shear stress is zero on extreme fibers, is maximum on the neutral axis, and has a parabolic distribution through the thickness.

Table 4.3: Maximum shear stress for different beam cross sections.

Cross section	Maximum shear stress
Rectangle	$\tau_{\max} = \dfrac{3V}{2A}$
Circle	$\tau_{\max} = \dfrac{4V}{3A}$
Round tube	$\tau_{\max} = \dfrac{2V}{A}$
I-beam (Web, Flange)	$\tau_{\max} = \dfrac{V}{A_{web}}$

Example 4.12: Stress Due to Transverse Shear

Given: A cantilever having a square cross section is loaded by a shear force perpendicular to the beam centerline at the free end. The sides of the square cross section are 50 mm, and the shear force is 10,000 N.

Find: Calculate the transverse shear stress at the beam centerline:

(*a*) If the shear force is parallel with two sides of the square cross section

(*b*) If the shear force is parallel with one of the diagonals of the square cross section

Solution: The area moment of inertia is

$$I = \frac{bh^3}{12} = \frac{(50)(50)^3}{12} = 520,800 \text{ mm}^4.$$

This moment is valid in all directions due to symmetry. Thus, the shear force and the area moment of inertia are the same for both parts of this example.

(*a*) The width of the point where τ is calculated is 50 mm. The evaluation of Q is

$$Q = \int_A y\,dA = \int_0^{25} y(50)\,dy = \frac{50(25)^2}{2} = 15,625 \text{ mm}^3.$$

From Eq. (4.63) the bending shear stress is

$$\tau = \frac{VQ}{Iw_t} = \frac{(10,000)(15,625)}{(520,800)(50)} = 6.0 \text{ MPa}.$$

(*b*) The width at the point where τ is applied is

$$w_t = \sqrt{2(50)^2} = 50\sqrt{2} = 70.71 \text{ mm}.$$

Q is evaluated as

$$\begin{aligned} Q &= \int_0^{25\sqrt{2}} 2y\left(25\sqrt{2} - y\right) dy \\ &= \left[25\sqrt{2}y^2 - 2\left(25\sqrt{2}\right)^3/3\right]_{y=0}^{y=25\sqrt{2}} \\ &= 14,730 \text{ mm}^3. \end{aligned}$$

From Eq. (4.63) the bending shear stress is

$$\tau = \frac{VQ}{Iw_t} = \frac{(10,000)(14,730)}{(520,800)(70.71)} = 4.0 \text{ N/mm}^2 = 4.0 \text{ MPa}.$$

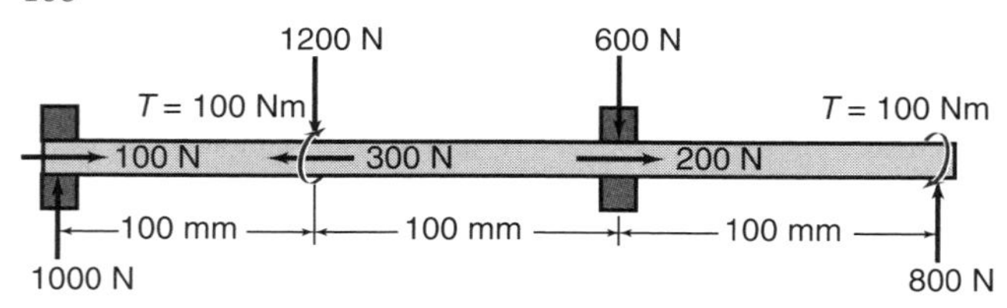

Figure 4.22: Shaft with loading considered in Example 4.13.

Example 4.13: Critical Location in a Beam

Given: A shaft is loaded by the forces and torques shown in Fig. 4.22, which result from the actions of helical gears and the shaft's rolling element bearing supports. The shaft has a diameter of 25 mm.

Find: Determine the location in the shaft where the stresses are highest. What are the principal stresses at this location?

Solution: First of all, it should be noted that with a diameter of 25 mm, the following can be calculated:

$$A = \frac{\pi d^2}{4} = \frac{\pi (0.025)^2}{4} = 4.909 \times 10^{-4}\ \text{m}^2,$$

$$I = \frac{\pi d^4}{64} = \frac{\pi (0.025)^4}{64} = 1.917 \times 10^{-8}\ \text{m}^4,$$

$$J = \frac{\pi d^4}{32} = \frac{\pi (0.025)^4}{32} = 3.835 \times 10^{-8}\ \text{m}^4.$$

The shear and moment diagrams are shown in Fig. 4.23. At first, it is not clear where the critical location is; the absolute shear force is highest between 0 and 100 mm, and the moment is highest at $x = 100$ mm. However, the torque is highest between 100 and 300 mm, and the axial load is tensile between 100 and 200 mm but compressive between 0 and 100 mm.

In practice, a designer must analyze *all* potential critical locations to determine the most critical. This example will analyze the location just to the right of the gear that acts at $x = 100$ mm, where $V = 200$ N, $P = 300$ N, $T = 100$ Nm, and $M = 100$ Nm. These result in the stresses in Table 4.4.

Consider the cross section of the shaft shown in Fig. 4.24, with the stress element locations shown. At location A, the axial stress is tensile, but the bending stress is compressive; the resultant normal stress is $\sigma_x = 0.611 - 65.11 = -64.60$ MPa. The shear stress distributions for torsion and shear are shown in Fig. 4.25; clearly at the top of the shaft, the shear stress due to the vertical shear force can be ignored, while the shear stress due to torsion is at its maximum value. Thus, $\tau_{xz} = 32.59$ MPa.

At location B, the bending stress is zero (see Fig. 4.25). Thus, the normal stress is due to axial force only, or $\sigma_x = 0.611$ MPa. The shear stresses due to torsion and vertical shear are additive; thus, $\tau_{xy} = 0.543 + 32.59 = 33.13$ MPa.

At location C, the normal stresses are both tensile; thus, $\sigma_x = 0.611 + 65.21 = 65.82$ MPa. The shear stress is the same as at location A, but it is now negative, so that $\tau_{xz} = -32.59$ MPa.

At location D, the bending stress is zero, so the normal stress is the same as at location B, or $\sigma_x = 0.611$ MPa. The shear stress due to torsion is zero, but the shear stress due to the vertical shear force has its maximum value of 0.543 MPa.

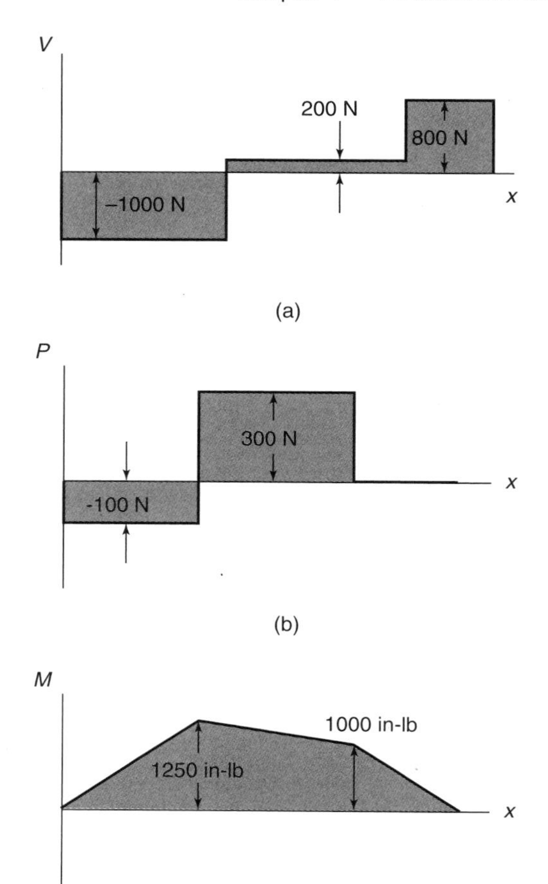

Figure 4.23: (a) Shear force; (b) normal force and (c) bending moment diagrams for the shaft in Fig. 4.22.

From consideration of these stress elements, we conclude that the element at location C is critical; it has the largest normal and shear stresses. It has the stress state $\sigma_x = 65.82$ MPa, $\sigma_y = \sigma_z = 0$, $\tau_{xz} = -32.59$ MPa, and $\tau_{xy} = \tau_{yz} = 0$. This is a two-dimensional stress state; therefore, one of the principal stresses is zero. The other two principal stresses can be obtained from Mohr's circle or from Eq. (2.16), using proper subscripts, as

$$\begin{aligned}\sigma &= \frac{\sigma_x + \sigma_z}{2} \pm \sqrt{\tau_{xz}^2 + \frac{(\sigma_x - \sigma_z)^2}{4}} \\ &= \frac{65.82}{2} \pm \sqrt{32.59^2 + \frac{65.82^2}{4}} \\ &= 32.91 \pm 73.45 \text{ MPa}.\end{aligned}$$

Therefore, the principal stresses are, in their proper order according to Eq. (2.18), $\sigma_1 = 106.4$ MPa, $\sigma_2 = 0$, and $\sigma_3 = -40.54$ MPa.

Table 4.4: Stresses obtained in Example 4.13.

Loading	Resultant maximum stress	Reference	Value
Axial	$\sigma_x = \dfrac{P}{A} = \dfrac{300}{4.909 \times 10^{-4}}$	Eq. (2.7)	0.611 MPa
Bending	$\sigma_x = \dfrac{Mc}{I} = \dfrac{(100)(0.0125)}{1.917 \times 10^{-8}}$	Eq. (4.45)	65.21 MPa
Vertical shear	$\tau_{xy} = \dfrac{4V}{3A} = \dfrac{4(200)}{3(4.909 \times 10^{-4})}$	Table 4.3	0.543 MPa
Torsion	$\tau = \dfrac{rT}{J} = \dfrac{(0.0125)(100)}{3.835 \times 10^{-8}}$	Eq. (4.32)	32.59 MPa

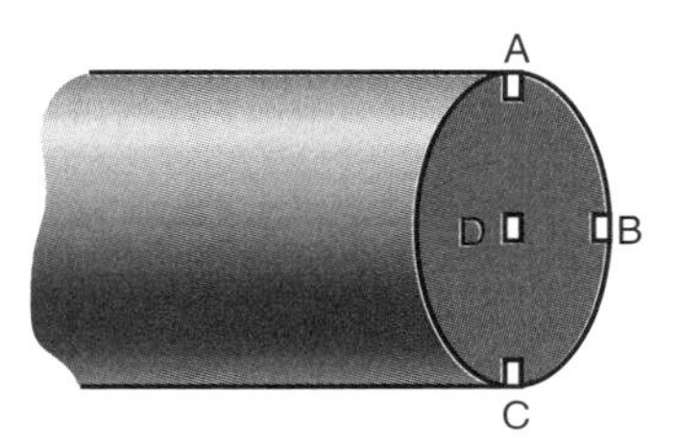

Figure 4.24: Cross section of shaft at $x = 100$ mm, with identification of stress elements considered in Example 4.13.

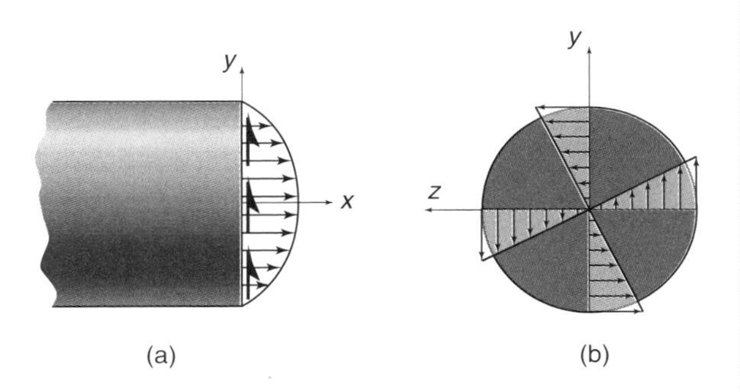

Figure 4.25: Shear stress distributions. (a) Shear stress due to a vertical shear force; (b) shear stress due to torsion.

Case Study: Design of a Shaft for a Coil Slitter

Given: Flat rolled sheets are produced in wide rolling mills, but many products are manufactured from strip stock. Figure 4.26a depicts a coil slitting line, where large sheets are cut into ribbons or strips. Figure 4.26b shows a shaft supporting the cutting blades. The rubber rollers support the sheet during cutting and prevent wrinkling. For such slitting lines the shafts that support the slitting knives are a highly stressed and critical component. Figure 4.26c is a free-body diagram of a shaft for a short slitting line where a single blade is placed in the center of the shaft and a motor drives the shaft through a pulley at the far right end.

Find: If the maximum shear stress is 40 MPa and the largest gage sheet causes a blade force of 2000 N, what shaft diameter is needed?

Solution: The reaction forces are found through statics to be R_A = 1720 N and R_B = 2600 N and are shown in Fig. 4.26c. Figure 4.27 shows the shear and bending moment diagrams. The maximum shear occurs just to the left of the pulley and equals 2880 N. The maximum bending moment is 430 Nm. In addition, there is a torque of 216 Nm between the pulley and the knife blade. Two locations must be analyzed: the location in the shaft where the moment is largest, and the location where the shear is largest.

(a) Moment. The magnitude of the normal stress in the x-direction at the location of maximum moment is given by Eq. (4.45) as

$$\sigma_x = \frac{Mc}{I} = \frac{(430)\left(\dfrac{d}{2}\right)}{\dfrac{\pi d^4}{64}} = \frac{4380}{d^3}.$$

The shear stress due to the torque exerted on the pulley is, from Eq. (4.33),

$$\tau_{xy} = \frac{Tc}{J} = \frac{1100}{d^3}.$$

A Mohr's circle can be constructed as discussed in Design Procedure 2.5. The Mohr's circle for this case is shown in Fig. 4.28 and has a radius of 2450 Nm/d^3. Setting this equal to the maximum allowable shear stress of 40 MPa yields

$$\frac{2450}{d^3} = 40 \times 10^6.$$

Therefore,

$$d = 0.0394 \text{ m} = 39.4 \text{ mm}.$$

(b) Shear. The maximum shear stress at the location of maximum shear is, from Table 4.1,

$$\tau_{\max} = \frac{4V}{3A} = \frac{4(2880)}{3\left(\dfrac{\pi d^2}{4}\right)} = \frac{4890}{d^2}.$$

At one end of the shaft the torsion-induced shear stress is subtracted from this shear stress; at the other end, the effects are cumulative. Thus, the total shear is

$$\tau_{\text{tot}} = \frac{4890}{d^2} + \frac{1100}{d^3} = 40 \text{ MPa}.$$

Solving numerically gives $d = 31.5$ mm.

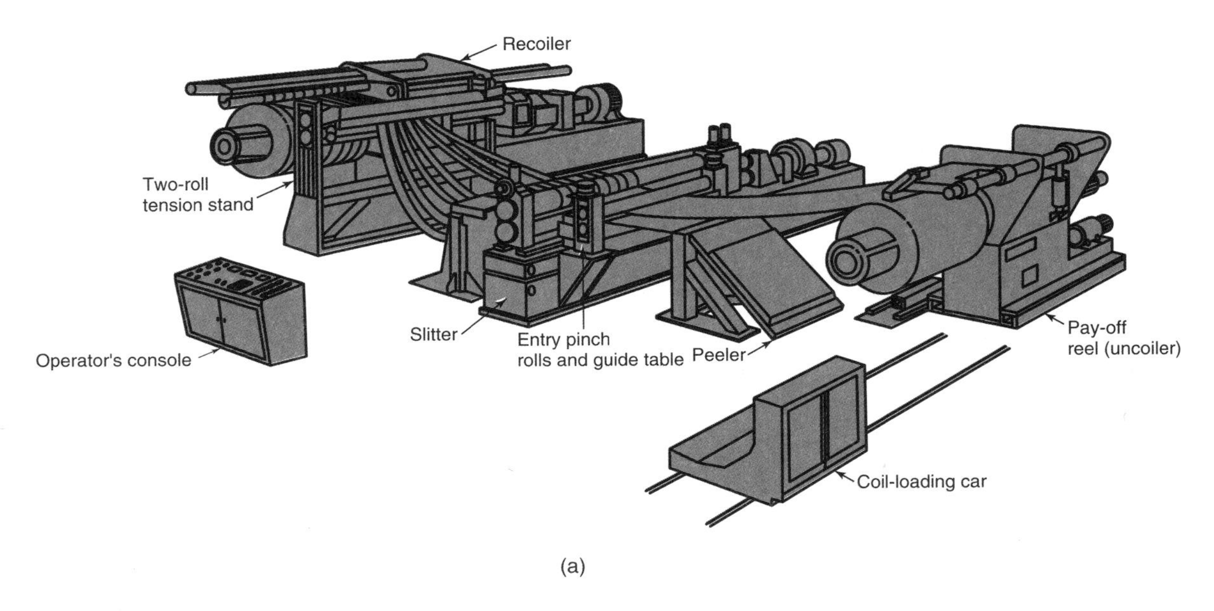

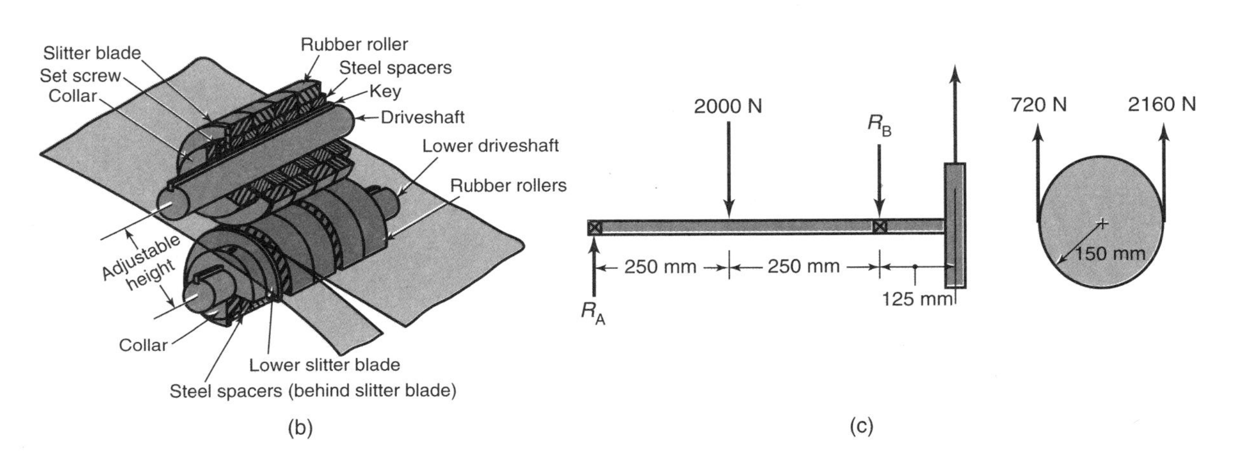

Figure 4.26: Design of shaft for coil slitting line. (a) Illustration of coil slitting line; (b) knife and shaft detail; (c) free-body diagram of simplified shaft for case study. Illustrations (a) and (b) are adapted from *Tool and Manufacturing Engineers Handbook, Fourth Edition, Volume 2 Forming.* Reprinted with permission of the Society of Manufacturing Engineers, ©1984.

(c) Discussion. As in most shaft applications the normal stresses due to bending determine the shaft diameter, so that a shaft with a diameter not less than $39.4 \approx 40$ mm should be used. A number of points should be made regarding this analysis:

- When the normal stress due to bending was calculated, the shear stress due to shear was neglected, even though there was shear in the shaft at that location. However, the distribution of normal stress is such that it is extreme at the top and the bottom, where the shear stress is zero.
- When the maximum shear stress due to vertical shear was calculated, the effects of bending were ignored.
- The bending stress is zero at the neutral axis, the location of the maximum shear stress.
- There are two general shaft applications. Some shafts are extremely long, as in this problem, whereas others are made much shorter to obtain compact designs. A coil slitter can have shafting more than 10 m long, but more supporting bearings would be needed for stiffness. This shaft was used only as an illustrative example; in actuality, the supporting bearings would be placed much closer to the load application and more than two bearing packs would probably be appropriate.

4.7 Summary

The first four chapters have provided the essentials needed to describe the stress and strain for the four types of loading (normal, torsional, bending, and transverse shear) that might occur. These stresses and strains will be used in designing machine elements later in the text. Each chapter attempted to build from the knowledge learned in previous chapters. It was assumed throughout the first four chapters that the member experiencing one of the four types of loading had a symmetric cross section and was made of a Hookian material. Chapter 4 started by defining centroid, moment of inertia, the parallel-axis theorem, radius of gyration, and section modulus. These concepts needed to be understood before stresses and strains resulting from normal, torsional, bending, or transverse shear could be explored. These definitions, which are used throughout the text, were followed by the stresses and strains found in the four types of loading. While evaluating the stress and strain due to normal loading, it was convenient to include the axial displacement and the spring rate. For torsional loading, the stresses and strains as well as the angular twist and angular spring rate were presented. Also, in discussing torsion, the relevant equations associated with power transfer were given. The stresses and strains associated with bending were explained, as well as the importance of the neutral axis. Both a straight and a curved member were analyzed. Section 4.6 defined the stresses associated with transverse shear. Deformations associated with bending and combined transverse loading were not covered in this chapter but are considered more fully in Chapter 5.

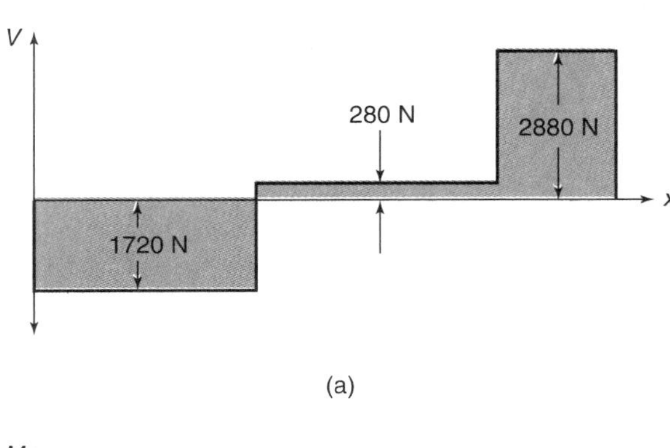

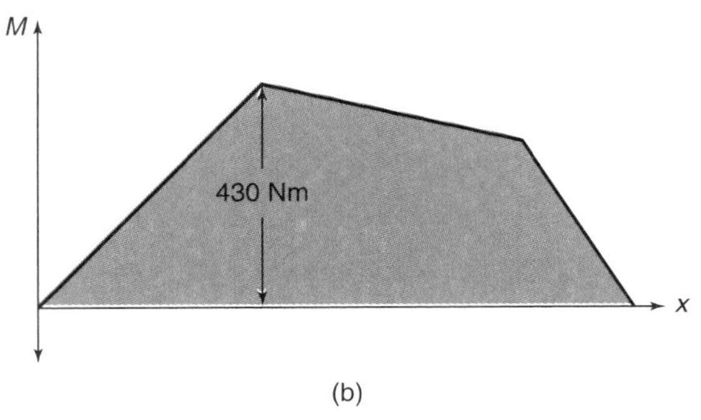

Figure 4.27: (a) Shear diagram and (b) moment diagram for idealized coil slitter shaft.

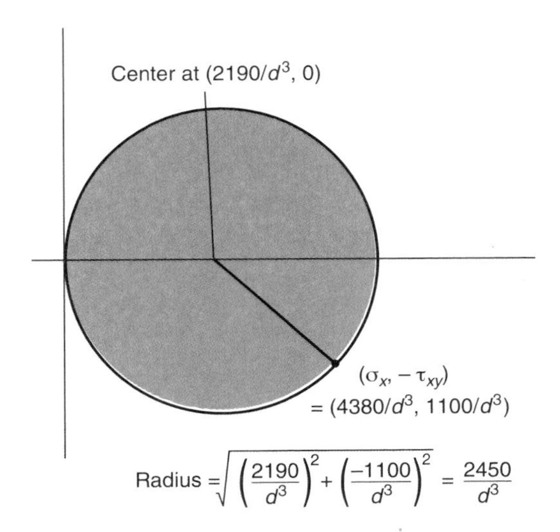

Figure 4.28: Mohr's circle at the location of maximum bending stress.

Key Words

area moment of inertia also called *second moment of area*, a property of a cross section that relates bending stress, applied moment, and distance from the neutral axis, m^4

average normal stress average normal load divided by cross-sectional area, Pa

centroid of area geometric center of an area, m

mass moment of inertia product of element's mass and square of element's distance from the axis, kg-m^2

normal strain elastic deformation divided by original length

parallel-axis theorem a theorem that allows calculation of the moment of inertia about any axis

power rate of doing work, or the product of force and velocity or torque and angular velocity, Nm/s

radius of gyration radius that when squared and multiplied by area gives area moment of inertia, m

section modulus moment of inertia divided by farthest distance from centroidal axis to outer fiber of solid, m^3

spring rate normal load divided by elastic deformation, N/m

torsion loading resulting in twisting of shaft

Summary of Equations

Beam cross sections:

Centroid of area: $\bar{x} = \dfrac{A_1\bar{x}_1 + A_2\bar{x}_2 + \ldots}{A_1 + A_2 + \ldots}$ and $\bar{y} = \dfrac{A_1\bar{y}_1 + A_2\bar{y}_2 + \ldots}{A_1 + A_2 + \ldots}$

Moment of inertia: $I_x = \int_A y^2\, dA$ and $I_y = \int_A x^2\, dA$

Polar moment of inertia: $J = I_x + I_y$

Parallel-axis theorem: $I_{x'} = I_x + Ad_y^2$

Radius of gyration: $r_g = \sqrt{\dfrac{I}{A}}$

Section modulus: $Z_m = I/c$

Mass moment of inertia: $I_{mx} = \int \left(y^2 + z^2\right) dm_a$, etc.

Normal stress: $\sigma = \dfrac{P}{A}$

Normal strain: $\epsilon = \dfrac{\delta}{l} = \dfrac{\sigma}{E}$

Shear stress in torsion: $\tau = \dfrac{Tr}{J}$

Deflection in torsion: $\theta = \dfrac{Tl}{GJ}$

Power: $h_p = Pu = T\omega$

Bending stress: $\sigma = -\dfrac{My}{I}$

Radius of curvature in bending: $\dfrac{1}{r} = \dfrac{M}{EI}$

Internal shear stress due to transverse shear: $\tau = \dfrac{VQ}{Iw_t}$

Spring rate:

Tension member: $k = \dfrac{AE}{l}$

Torsion: $k = \dfrac{T}{\theta} = \dfrac{JG}{l}$

Recommended Readings

Beer, F.P., Johnson, E.R., DeWolf, J., and Mazurek, D. (2011) *Mechanics of Materials*, 6th ed., McGraw-Hill.

Craig, R.R. (2011) *Mechanics of Materials*, 3rd ed., Wiley.

Gere, J.M, and Goodno, B.J. (2012) *Mechanics of Materials*, 8th ed., CL Engineering.

Hibbeler, R.C. (2010) *Mechanics of Materials*, 8th ed., Prentice-Hall.

Popov, E.P. (1968) *Introduction to Mechanics of Solids*, Prentice-Hall.

Popov, E.P. (1999) *Engineering Mechanics of Solids*, 2nd ed., Prentice-Hall.

Riley, W.F., Sturges, L.D., and Morris, D.H. (2006) *Mechanics of Materials*, 6th ed., Wiley.

Shames, I.H., and Pitarresi, J.M. (2000) *Introduction to Solid Mechanics*, 3rd ed., Prentice-Hall.

Ugural, A.C. (2007) *Mechanics of Materials*, Wiley.

Reference

Timoshenko and Goodier (1970) *Theory of Elasticity*, 3rd ed., McGraw-Hill.

Questions

4.1 What is the centroid of an area?

4.2 What is the difference between the moment of inertia and the second moment of area?

4.3 Describe the importance of the parallel-axis theorem.

4.4 How is the radius of gyration used?

4.5 Explain the difference between a beam and a shaft.

4.6 What is the equation for normal stress for a bar loaded in uniaxial tension?

4.7 Does a torque acting on a shaft lead to shear or normal stresses?

4.8 In the equation $\tau = \dfrac{VQ}{Iw_t}$, what is Q?

4.9 What is the equation for normal stress in a beam that is loaded by a bending moment?

4.10 What is the equation for vertical shear stress in a beam that is loaded in internal shear?

4.11 What is Saint-Venant's principle?

4.12 Explain how a shaft transmits power.

4.13 What is the section modulus of an area?

4.14 What is the difference between a moment and a torque, with respect to stresses they cause?

4.15 What is the polar moment of inertia?

Qualitative Problems

4.16 Explain why an I-beam can support a higher bending moment than an H-Beam. That is, why is orientation important for an I-beam?

4.17 Why are tubes more efficient at supporting bending moments than cylinders? Is this also true for torsion?

4.18 Is it possible for the centroid of an area to be outside of the area? Explain.

4.19 List the functions of the web and flange in an I-beam.

4.20 It is desired to drill a hole though a beam with a rectangular cross-section, transverse to its axis. Where would you recommend that the hole be placed?

4.21 Does the stiffness of a balloon increase when it is inflated? Why or why not?

4.22 Explain why the tension test specimen shown in Fig. 3.1 uses a smaller test section than the cross section at the grips. Often, such a specimen has a parabolic profile with a minimum value at its center. Explain why.

4.23 Construct a Design Procedure for determining the critical location in a beam. *Hint:* Make extensive use of Example 4.13.

4.24 Explain why a beam is much stiffer in axial loading than in torsion or in bending.

4.25 Refer to Table 4.2 and explain the conditions under which you would consider a cylinder to be a rod.

Quantitative Problems

4.26 An area in the x-y coordinate system, as shown in Sketch a, consists of a large circle having radius r out of which are cut three smaller circles having radii $r/3$. Find the x- and y-coordinates for the centroid. The radius r is 10 cm. *Ans.* $\bar{x} = 10$ cm, $\bar{y} = 8.889$ cm.

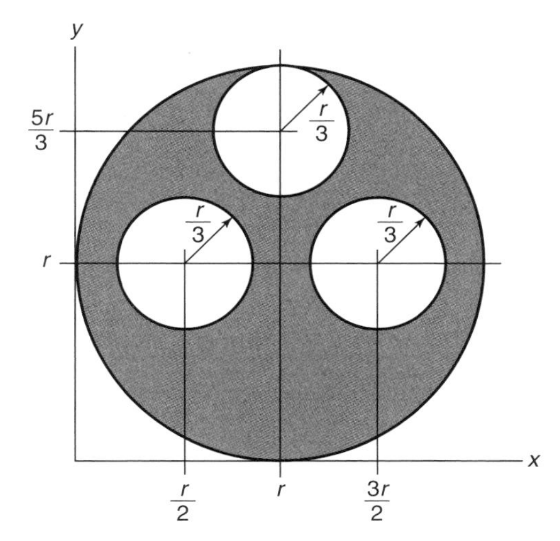

Sketch a, for Problems 4.26 and 4.27.

4.27 The circular surface in Sketch a has circular cutouts glued onto it below the top cutout such that the centroids of the three cutouts are at $\bar{x} = r$ and $\bar{y} = r/3$. Find the x- and y- coordinates for the centroid ($r = 10$ cm). *Ans.* $\bar{x} = 10$ cm, $\bar{y} = 7.037$ cm.

4.28 The rectangular area shown in Sketch b has side lengths of a in the x-direction and b in the y-direction. Find the moments of inertia I_x and I_y and the polar moment of inertia J for the rectangular surface. *Ans.* $I_x = \frac{1}{12}ab^3$.

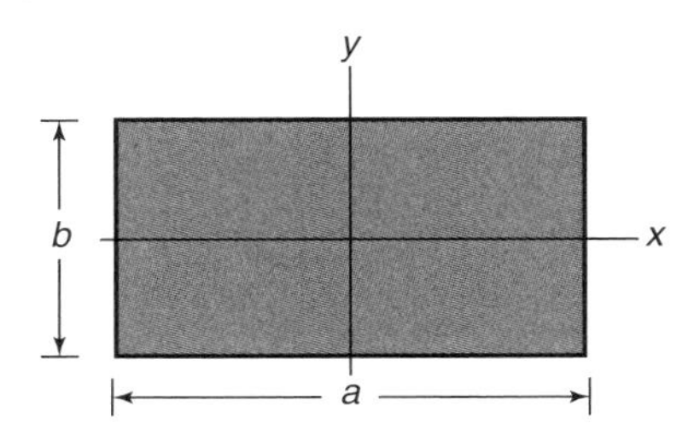

Sketch b, for Problem 4.28.

4.29 Derive the area moment of inertia for the hollow circular area shown in Table 4.1.

4.30 Derive the area moment of inertia for the elliptical quadrant shown in Table 4.1.

4.31 Derive the area moment of inertia for the triangular section shown in Table 4.1.

4.32 Derive the area moment of inertia for a rectangular section with a cutout as shown in Sketch c. *Ans.* $I_y = 4.141 \times 10^{-6}$ m^4, $I_x = 2.202 \times 10^{-5}$ m^4.

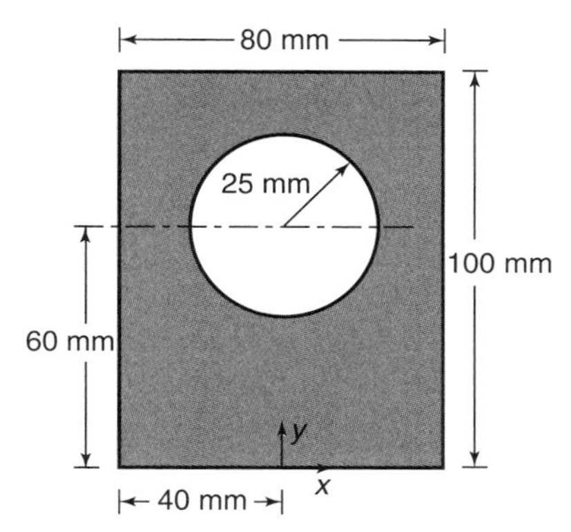

Sketch c, for Problem 4.32.

4.33 Evaluate the moment of inertia of the shape bounded by the curves shown in Sketch d, up to a value of $x = 3$. *Ans.* $I_x = 6.646$ m^4, $I_y = 15.39$ m^4, $J = 22.04$ m^4.

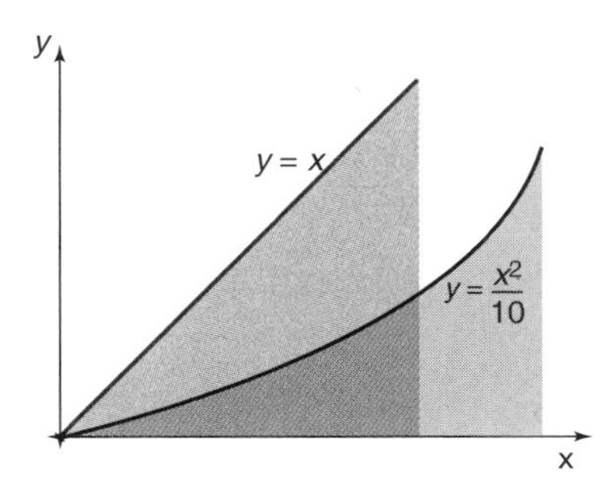

Sketch d, for Problem 4.33.

4.34 An elevator is hung by a steel rope. The rope has a cross-sectional area of 250 mm^2 and a modulus of elasticity of 70 GPa. The upward acceleration when the elevator starts is 4 m/s^2. The rope is 100 m long and the elevator weighs 1000 kg. Determine the stress in the rope, the elongation of the rope due to the elevator's weight, and the extra elongation due to the acceleration. *Ans.* $\sigma = 55.2$ MPa, $\delta_{wt} = 56.06$ mm, $\delta_{acc} = 22.86$ mm.

4.35 The 700-m-long cables in a suspension bridge are stressed to a 200-MPa tensile stress. The total force in each cable is 10 MN. Calculate the cross-sectional area, the total elongation of each cable, and the spring rate when the modulus of elasticity is 70 GPa. *Ans.* $A = 0.05$ m^2, $\delta = 2$ m, $k = 5$ MN/m.

4.36 A 1.00-m-long steel piston in a hydraulic cylinder exerts a force of 40 kN. The piston is made of AISI 1080 steel and has a diameter of 50 mm. Calculate the stress in the piston, the elongation, and the spring rate. *Ans.* $\sigma = -20.41$ MPa, $\delta = 9.859 \times 10^{-5}$ m, $k = 405.7$ MN/m.

4.37 A steel pillar supporting a highway bridge is 14 m high and made of steel tubing having an outer diameter of 1.5 m and a wall thickness of 30 mm. The weight carried by the pillar is 12 MN. Calculate the deformation of the pillar, the spring rate, and the stress in the pillar. *Ans.* $\sigma = -86.6$ MPa, $\delta = -5.86$ mm, $k = 2.048$ GN/m.

4.38 The foundation of a bronze statue is made of a 3-m-high conical tube of constant wall thickness (8 mm). The tube's outer diameter is 200 mm at the top (just under the statue) and 400 mm at the ground. The tube material is AISI 316 stainless steel. The statue weighs 16,000 N. Calculate the deformation of the tube, the spring rate, and the maximum and minimum compressive stresses in the tube. *Ans.* $\delta = -35.48\ \mu$m, $k = 450.9$ MN/m, $\sigma_{\text{max}} = -3.33$ MPa, $\sigma_{\text{min}} = -1.63$ MPa.

4.39 Calculate the deflection at point A of the hanging cone shown in Sketch *e* if the cone is constructed of aluminum. *Ans.* $\delta = 1.369$ mm.

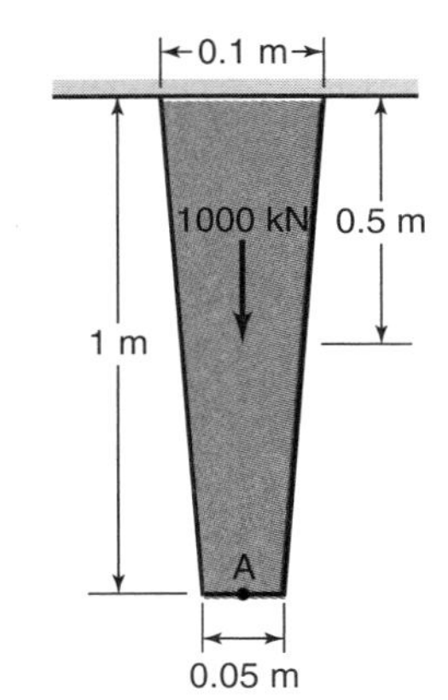

Sketch *e*, for Problem 4.39.

4.40 An electric motor transmits 100 kW to a gearbox through a 50-mm-diameter solid steel shaft that rotates at 1000 rpm. Find the torque transmitted through the shaft and the angular torsion of the 1-m-long shaft. *Ans.* $T = 955$ Nm, $\theta = 1.12°$.

4.41 The torque-transmitting shaft in Problem 4.40 is too heavy for the application, so it is exchanged for a circular tube having a 50-mm outside diameter and a 40-mm inside diameter. Find the angular torsion of the tube-formed shaft, which is 1 m long, when 100 kW is transmitted at 1000 rpm. The shear modulus is 80,000 N/mm^2. Also find the maximum shear stress in the tube and the percentage of weight decrease from the solid shaft. *Ans.* $\theta = 1.888°$, 64% weight savings.

4.42 A torque-transmitting, hollow steel shaft with a circular cross section has an outer diameter of 50 mm and an inner diameter of 40 mm. Find the maximum length possible for the shaft if the torsion should be below 5° at a torque of 2000 N-m. *Ans.* $l = 1.258$ m.

4.43 An aluminum core having a diameter d_i of 30 mm is placed within a tubular steel shaft having a diameter d_o of 50 mm, as shown in Sketch *f*. A flange is welded to the end of the shaft, and a pulley force of 200 kN is applied. The shaft is 100 mm long. Find the deflection at the end of the shaft and the stresses induced in the aluminum and steel sections of the shaft. Assume that the moduli of elasticity are 207 GPa for steel and 70 GPa for aluminum. *Ans.* $\delta = 0.6647$ mm, $\sigma_{\text{al}} = 46.5$ MPa, $\sigma_s = 132.9$ MPa.

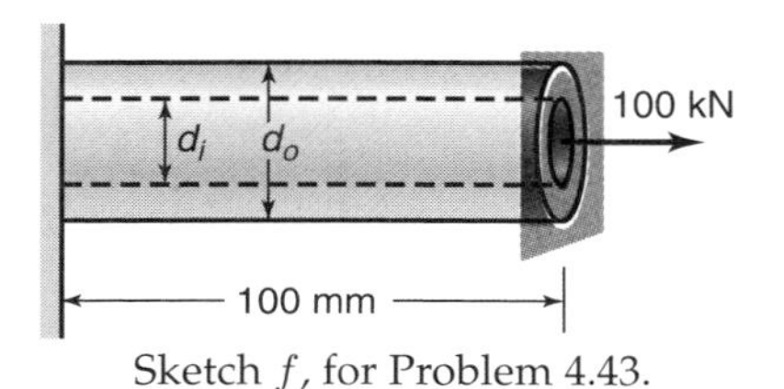

Sketch *f*, for Problem 4.43.

4.44 A bar of weight W is supported horizontally by three weightless rods as shown in Sketch g. Assume that the cross-sectional areas, A, the moduli of elasticity, E, and the yield stresses, S_y, are the same for the three rods. What is the maximum weight that can be supported? *Ans.* $W = 2.5 S_y A$.

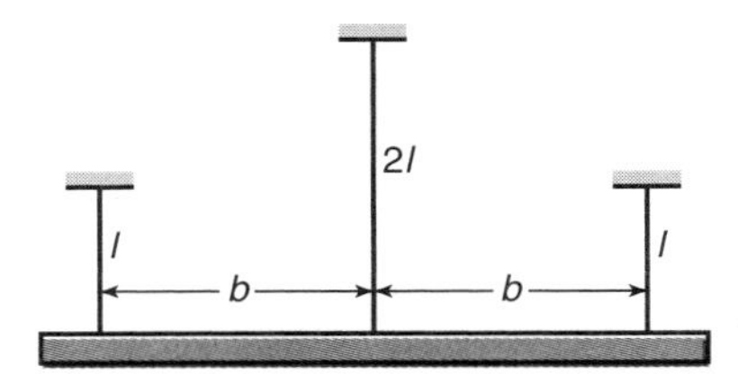

Sketch g for Problem 4.44.

4.45 The bronze statue described in Problem 4.38 is asymmetrical, so that when a gale force wind blows against it, a twisting torque of 800 N-m is applied to the tube. Calculate how much the statue twists. The tube's wall thickness, 6 mm at the top and 12 mm at the ground, is assumed to be proportional to its diameter. *Ans.* $\theta = 0.01569°$.

4.46 Determine the minimum diameter of a solid shaft used to transmit 500 kW of power from a 3000-rpm motor so that the shear stress does not exceed 50 MPa. *Ans.* $d = 54.48$ mm.

4.47 A steel coupling is used to transmit a torque of 30,000 N-m. The coupling is connected to the shaft by a number of 10-mm-diameter bolts placed equidistant on a pitch circle of 0.4-m diameter. The inner diameter of the coupling is 0.1 m. The allowable shear stress on the bolts is 500 MPa. Find the minimum number of bolts needed. *Ans.* 6.

4.48 A shaft and a coupling are to transmit 50 kW of power at an angular speed of 1000 rpm. The coupling is connected to the shaft by 10 bolts, 20 mm in diameter, placed on a pitch circle of 200 mm. For an allowable stress on the bolts of 100 MPa, are the bolts able to transfer this power? *Ans.* Yes.

4.49 A beam transmitting a bending moment M of 5000 N-m has a square cross section with sides of 100 mm. The weight is decreased by making either one or two axial circular holes along the beam, as shown in Sketch h. Determine whether one or two holes gives the lowest weight for the beam at a given bending stress. Neglect the transverse shear stress. *Ans.* Two holes gives lower weight.

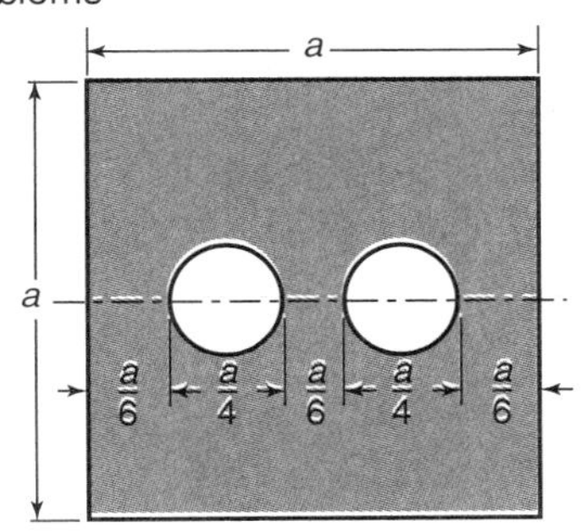

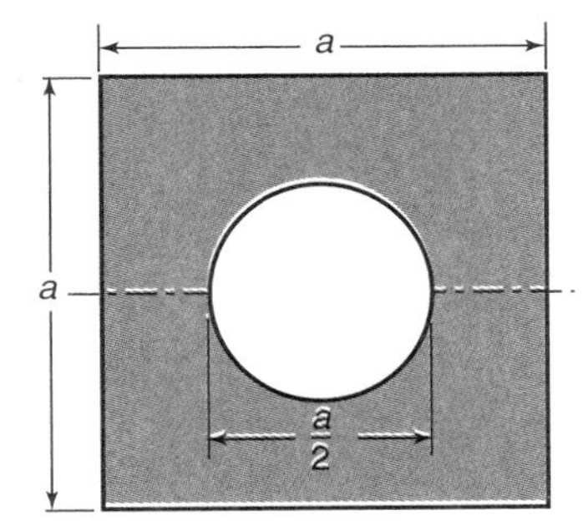

Sketch h, for Problem 4.49

4.50 A straight beam is loaded at the ends by moments M. The area moment of inertia for the beam is $I = a^3b/12$. Find the bending stress distribution in the beam when M = 1000 N-m, a = 3 cm, and b = 6 cm. Also find the radius of curvature to which the beam is bent. The beam's modulus of elasticity is 207 GPa. *Ans.* σ = (7.407 GN/m^3)y, $r = 27.95$ m.

4.51 The beam in Problem 4.50 is bent in a perpendicular direction so that $I = \frac{1}{12}ab^3$. Find the bending stress distribution and the radius of curvature to which the beam is bent. *Ans.* r = 111.8 m.

4.52 A curved bar has a rectangular cross section with height $h = r_o - r_i = 50$ mm and width $b = 100$ mm. Its inner radius is 200 mm. Find the distance between the neutral axis and the centroid. *Ans.* $e = 0.9290$ mm.

4.53 The curved bar in Problem 4.52 is loaded with a bending moment of 3000 N-m. Find the stress at the innermost and outermost radii. *Ans.* $\sigma_i = -77.73$ MPa, $\sigma_o = 6.699$ MPa.

4.54 Two beams with rectangular cross sections $a \times b$ are placed on top of each other to form a beam having height $2a$ and width b. Find the area moments of inertia I_x and I_y for the two beams:

(a) When they are welded together along the length. *Ans.* $I_{\bar{x}} = \frac{2}{3}ba^3$, $I_{\bar{y}} = \frac{1}{6}b^3a$.

(b) When they are not welded together.

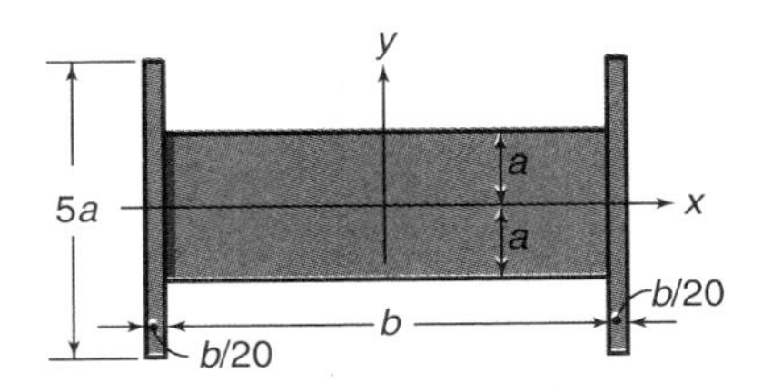

Sketch i, for Problem 4.55

4.55 Two thin steel plates having width $b/20$ and height $5a$ are placed one at each side of the two beams in Problem 4.54, as shown in Sketch i. Find the moments of inertia around the x- and y-axes:

(a) When the plates are not welded together. *Ans.* $I_x = 1.21ba^3$, $I_y = 0.304ab^3$.

(b) When they are welded together as in Sketch i.

(c) When they are welded together to form a closed tube $5a \times 1.1b$. *Ans.* $I_x = 9.21ba^3$.

4.56 A cantilevered beam with a rectangular cross section is loaded by a force perpendicular to the beam centerline at the free end. The cross section is 100 mm high and 25 mm wide. The vertical load at the beam end is 40,000 N. Calculate how long the beam should be to give tensile and compressive stresses 10 times higher than the maximum shear stress. Also, calculate these stresses. *Ans.* l = 250.0 mm.

4.57 A cantilevered beam with a circular-tube cross section has an outer diameter of 100 mm and a wall thickness of 10 mm. The load perpendicular to the beam is 15 000 N, and the beam is 1.2 m long from the point of force to the wall where the beam is fastened. Calculate the maximum bending and shear stresses. *Ans.* $\sigma_{\max} = 171.2$ MPa, $\tau_{\max} = 7.958$ MPa.

4.58 A fishing rod is made of glass-fiber-reinforced plastic in the form of a tube having an outer diameter of 10 mm and a wall thickness of 1.5 mm. The glass fibers are parallel to the tube axis, so that the bending shear stress is carried by the plastic and the bending stresses are carried by the fibers. The fishing rod is 2 m long. Determine whether the rod fails from tensile stresses in the fibers or from shear overstressing in the plastic. The bending strength of the fiber-reinforced plastic is 800 MPa, and its shear strength is 3.2 MPa.

4.59 Two aluminum beams, like the one shown in Fig. 4.16, are welded together to form a closed cross section with dimensions of 240 × 80 mm. The weld was badly done, so that the wall is only 2 mm thick instead of 8 mm. The allowable shear stress in the weld is 50 MPa, whereas the maximum allowable bending stress is 150 MPa. Find whether the 2-m-long beam fails first at the outermost fibers or at the welds.

4.60 Obtain the largest principal stresses at $x = 10$ in. for the shaft considered in Example 4.13. *Ans.* $\sigma_1 = 12,350$ psi.

4.61 To decrease as much as possible the weight of beams subjected to bending, the center of gravity of the cross section is placed as far away from the beam center of gravity as possible. A beam supporting a floor has a cross section as shown in Sketch j. The bending moment acting on the beam is 3000 N-m and the shear force is 1000 N. Calculate the maximum bending stress and maximum shear stress. *Ans.* $\sigma_{\max}$ = 19.7 MPa, $\tau_{\max}$ = 1.74 MPa.

Sketch j, for Problem 4.61

Design and Projects

4.62 Using a deck of playing cards and two small C-clamps, demonstrate the importance of internal shear in the ability of a beam to support bending moments.

4.63 Design a minimum weight cantilever with a rectangular cross-section that supports a load at its free end and does not exceed a specified stress anywhere in the beam. Consider (a) a constant-thickness and (b) a constant width cross-section.

4.64 Using a single sheet of foam board and glue, design and construct a beam that supports the largest possible load across a span.

4.65 For a beam with the cross-section shown in Sketch i, select a location for a strain gage rosette if it is to detect complicated loadings.

4.66 With very old trees, it can be seen that the inside is old and rotten, but an exterior ring remains vibrant and strong. Explain why this allows trees to live longer than if the central region lived longer than the outside.

Chapter 5

Deformation

Testing of 787 Dreamliner wings. *Source:* Courtesy of Boeing Corp.

Let me tell you the secret that has led me to my goal. My strength lies in my tenacity.

Louis Pasteur

Often, deflection is as important as stress in determining the acceptability of a design, and machine components are often sized in order to achieve acceptable deflections. This chapter presents a number of important approaches for calculating the deflection of beams and other structures. The topics discussed in Chapters 2 though 4 are built upon to demonstrate the methods used to calculate deflection of beams, using both direct integration of moment diagrams or using singularity functions. When complicated loadings exist on a body that encounters small strains (i.e., a linear elastic situation), the method of superposition is a powerful approach to calculate the deflection of the body by considering all loads in tractable combinations. Strain energy approaches, including Castigliano's Theorem, are then presented. These approaches are especially valuable for impact or energy loadings.

Contents

Examples

Design Procedures

Symbols

A	area, m^2
a	length dimension, m
b	length dimension, m
C_1, C_2	constants
c	distance from neutral axis to outer fiber, m
E	modulus of elasticity, Pa
G	shear modulus of elasticity, Pa
h	height, m
I	area moment of inertia, m^4
J	polar moment of inertia, m^4
l	length, m
M	moment, N-m
P	force, N
Q	load applied at point of deformation, N
q	load intensity, N/m
R	reaction force, N
T	torque, N-m
U	strain energy, N-m
V	shear force, N
v	volume, m^3
w_o	unit step load distribution, N/m
x, y, z	Cartesian coordinates, m
γ	shear strain
δ	deformation (deflection), m
$\delta_{\max}$	maximum deformation (deflection), m
δ_P	deflection at location of applied load, m
ϵ	strain
θ	slope
ν	Poisson's ratio
σ	normal stress, Pa
τ	shear stress, Pa

Subscripts

a, b	solids a and b
H	horizontal
V	vertical

5.1 Introduction

The focus of Chapters 2 through 4 has been on describing load, stress, and strain for various conditions that may occur in machine elements. Knowing the design stress and making sure it is less than the yield strength for ductile materials and less than the ultimate strength for brittle materials are important for a safe design. However, attention also needs to be paid to strain limitation and displacement, since a machine element can fail if a part deforms excessively. For example, in high-speed machinery with close tolerances, excessive deflections can cause interference between moving parts. Excessive deflection of a spindle on a milling machine would compromise the tolerances that could be achieved; many other situations exist where deflection is of great importance. This chapter attempts to quantify the deformation that may occur in a great variety of machine elements.

Chapter 4 described the deformation for normal stresses (Section 4.3) and defined the angle of twist for torsional stress (Section 4.4.1) as well as the spring rate and the angular spring rate for normal and torsional stresses. This chapter focuses on describing the deformation for distributed loading, as occurs in a beam. Some approaches described are the integral method, the singularity function, the method of superposition, and Castigliano's theorem.

5.2 Moment-Curvature Relation

Figure 4.14b shows a deformed element of a straight beam in pure bending. The **radius of curvature**, r, can be expressed in Cartesian coordinates as

$$\frac{1}{r} = \frac{\dfrac{d^2y}{dx^2}}{\left[1 + \left(\dfrac{dy}{dx}\right)^2\right]^{3/2}}. \tag{5.1}$$

However, dy/dx is much less than unity, so that

$$\frac{1}{r} = \frac{d^2y}{dx^2}. \tag{5.2}$$

Substituting Eq. (5.2) into Eq. (4.47) gives

$$\frac{d^2y}{dx^2} = \frac{M}{EI}. \tag{5.3}$$

This equation relates the transverse displacement to a bending moment. Even though an approximation to the curvature was used in reducing Eq. (5.1) to Eq. (5.2) that is valid only for small bending angles, this is a reasonable approximation for most beams. Equation (5.3) is the **moment-curvature relation** and is sometimes referred to as the *equation of the elastic line*.

It is convenient to summarize the load intensity, shear force, moment, slope, and deformation in the following group of ordered derivatives:

$$\frac{q}{EI} = \frac{d^4y}{dx^4} \tag{5.4}$$

$$-\frac{V}{EI} = \frac{d^3y}{dx^3} \tag{5.5}$$

$$\frac{M}{EI} = \frac{d^2y}{dx^2} \tag{5.3}$$

$$\theta = \frac{dy}{dx} \tag{5.6}$$

$$y = f(x) \tag{5.7}$$

If SI units are used in these equations, the appropriate units are newtons per meter for load intensity, newtons for shear force, newton-meters for moment, and meters for deformation. Slope is dimensionless, meaning that it is given in radians.

Integrating Eq. (5.3) gives the dimensionless slope at any point x as

$$EI\frac{dy}{dx} = EI\theta = -Mx + C_1. \tag{5.8}$$

Integrating Eq. (5.8) gives the deflection at any point x as

$$EIy(x) = -M\frac{x^2}{2} + C_1x + C_2. \tag{5.9}$$

Note that integrating the load intensity function will produce the negative of the shear force distribution, or

$$-V(x) = \int_0^x q(x)\,dx. \tag{5.10}$$

Furthermore, integrating the shear force gives the negative of the moment, or

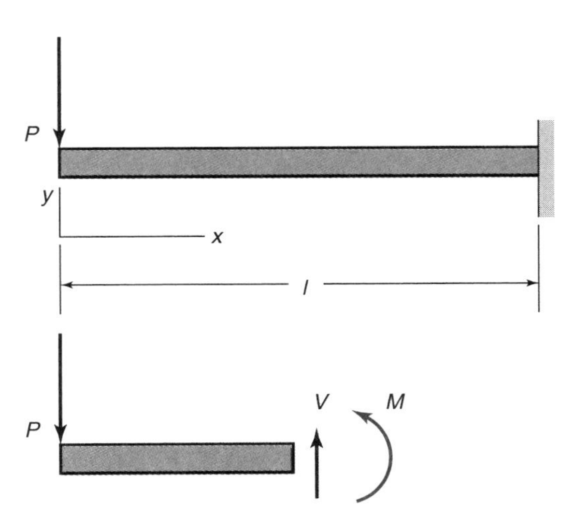

Figure 5.1: Cantilevered beam with concentrated force applied at free end.

$$M(x) = -\int_0^x V(x)\,dx. \tag{5.11}$$

Equations (5.8) to (5.11) are used in directly obtaining the deflection due to any type of loading.

Example 5.1: Deflection Obtained by Direct Integration

Given: A perpendicular force, P, acts at the end of a cantilevered beam with length l, as shown in Fig. 5.1. Assume that the cross section is constant along the beam and that the material is the same throughout, thus implying that the area moment of inertia, I, and the modulus of elasticity, E, are constant.

Find: The slope and deformation at any x, and the location and value of the maximum slope.

Solution: The moment is $M = -Px$. From Eq. (5.3),

$$\frac{d^2y}{dx^2} = \frac{M}{EI} = -\frac{Px}{EI}.$$

Integrating once gives

$$\frac{dy}{dx} = -\frac{Px^2}{2EI} + C_1.$$

Integrating again gives

$$y = -\frac{Px^3}{6EI} + C_1x + C_2.$$

The boundary conditions are

1. At $x = l, y = 0$
2. At $x = l, dy/dx = 0$

From boundary condition 2,

$$C_1 = \frac{Pl^2}{2EI}.$$

Therefore,

$$y = -\frac{Px^3}{6EI} + \frac{Pl^2x}{2EI} + C_2.$$

Making use of boundary condition 1 gives

$$C_2 = \frac{Pl^3}{6EI} - \frac{Pl^3}{2EI} = -\frac{Pl^3}{3EI}.$$

Therefore,

$$y = \frac{P}{6EI}\left(-x^3 + 3l^2x - 2l^3\right),$$

and thus

$$\frac{dy}{dx} = \frac{P\left(l^2 - x^2\right)}{2EI}.$$

The maximum slope occurs at $x = 0$ and is

$$\left(\frac{dy}{dx}\right)_{x=0} = \frac{Pl^2}{2EI}.$$

5.3 Singularity Functions

A **singularity function** permits expressing in one equation what would normally be expressed in several separate equations with boundary conditions. Singularity functions were introduced in Section 2.8.3, and their application to beam deflection is straightforward using Eqs. (5.4) through (5.7).

Design Procedure 5.1: Deflection by Singularity Functions

1. Draw a free-body diagram showing the forces acting on the system.
2. Use force and moment equilibria to establish reaction forces acting on the system.
3. Obtain an expression for the load intensity function for all the loads acting on the system while making use of Table 2.2.
4. Integrate the negative load intensity function to give the shear force and then integrate the negative shear force to give the moment.
5. Make use of Eq. (5.9) to describe the deflection at any location.
6. Plot the following as a function of x:
   (a) Shear
   (b) Moment
   (c) Slope
   (d) Deflection

Example 5.2: Deflection of a Beam by Singularity Functions

Given: A beam with a load applied between simply supported ends.

Find: The deflection for any x by using singularity functions.

Solution: Figure 5.2a shows a free-body diagram of the simply supported beam, while Fig. 5.2b has a section of the beam. From Table 2.2, for concentrated forces, the load intensity equation for the forces shown in Fig. 5.2b can be written as

$$q(x) = \frac{Pb}{l}\langle x\rangle^{-1} - P\langle x-a\rangle^{-1}. \qquad (a)$$

Integrating twice gives the moment as

$$M(x) = \frac{Pb}{l}\langle x\rangle^{1} - P\langle x-a\rangle^{1}. \qquad (b)$$

Making use of Eq. (5.3) gives

$$EI\frac{d^2y}{dx^2} = M(x) = \frac{Pb}{l}\langle x\rangle^{1} - P\langle x-a\rangle^{1}. \qquad (c)$$

Because EI is constant along the beam, integrating Eq. (c) gives

$$EI\frac{dy}{dx} = \frac{Pb}{2l}\langle x\rangle^{2} - \frac{P}{2}\langle x-a\rangle^{2} + C_1. \qquad (d)$$

Integrating again gives

$$EIy = \frac{Pb}{6l}\langle x\rangle^{3} - \frac{P}{6}\langle x-a\rangle^{3} + C_1x + C_2. \qquad (e)$$

The boundary conditions are

1. $y=0$ at $x=0$, which results in $C_2=0$.
2. $y=0$ at $x=l$, so that

$$C_1 = -\frac{Pbl}{6} + \frac{Pb^3}{6l} = \frac{Pb}{6l}\left(b^2-l^2\right) = -\frac{Pb}{6l}\left(l^2-b^2\right). \qquad (f)$$

or

$$y(x) = -\frac{P}{6EI}\left[\frac{xb}{l}\left(l^2-x^2-b^2\right) + \langle x-a\rangle^3\right]. \qquad (g)$$

Note from Eq. (g) that when $x \le a$ the last term on the right side of the equal sign is zero, and when $x > a$ the angular brackets become round brackets. (For more information about angular brackets see Section 2.8.3. Also note that from Fig. 5.2 that positive y is upward; in Eq. (g), y is negative, meaning deflection is downward. This can be intuitively confirmed by inspecting Fig. 5.2.)

Example 5.3: Deflection of a Cantilever Using Singularity Functions

Given: A cantilevered beam has a unit step distribution over a part of the beam as shown in Fig. 5.3. The beam is clamped or built into the structure at A and free at C. The unit step distribution begins at the free end.

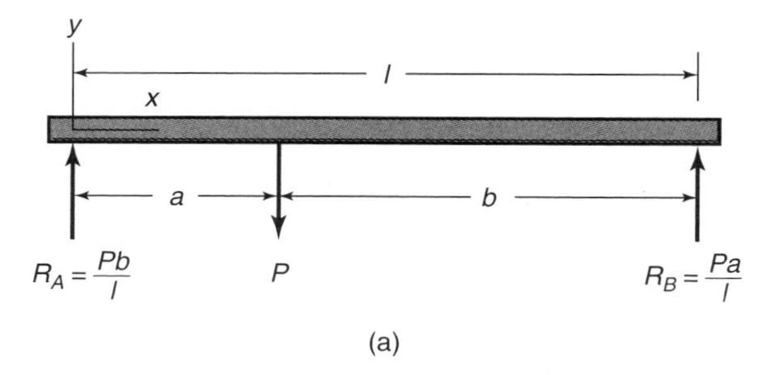

(a)

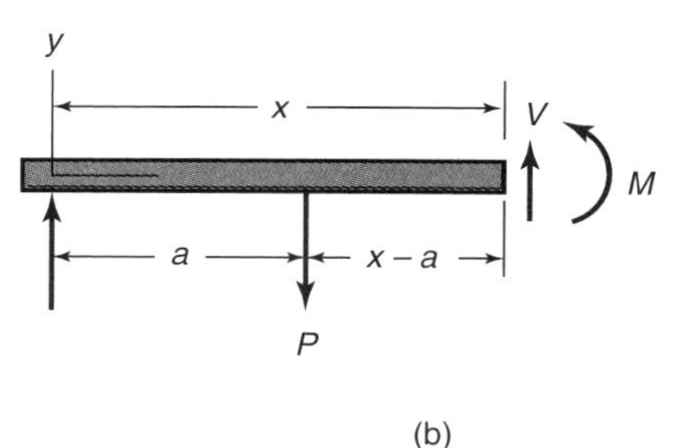

(b)

Figure 5.2: Free-body diagram of force anywhere between simply supported ends. (a) Complete beam; (b) portion of beam.

Find: Derive a general expression for the deflection at any x and at the free end by using singularity functions. Also, describe the deflection for two special cases: (a) When no unit step load is applied, and (b) When the unit step load is applied over the entire length.

Solution: The load intensity equation for the forces and moments shown in Fig. 5.3c can be expressed by using Table 2.2 as

$$q(x) = w_o b\langle x\rangle^{-1} - w_o b\left(a+\frac{b}{2}\right)\langle x\rangle^{-2} - w_o\langle x-a\rangle^{0}.$$

Integrating twice gives

$$M(x) = w_o b\langle x\rangle^{1} - w_o b\left(a+\frac{b}{2}\right)\langle x\rangle^{0} - \frac{w_o}{2}\langle x-a\rangle^{2}.$$

Making use of Eq. (5.3) gives

$$EI\frac{d^2y}{dx^2} = w_o b\langle x\rangle^{1} - w_o b\left(a+\frac{b}{2}\right)\langle x\rangle^{0} - \frac{w_o}{2}\langle x-a\rangle^{2}.$$

Integrating once gives

$$EI\frac{dy}{dx} = \frac{w_o b}{2}\langle x\rangle^{2} - w_o b\left(a+\frac{b}{2}\right)\langle x\rangle^{1} - \frac{w_o}{6}\langle x-a\rangle^{3} + C_1.$$

Integrating again gives

$$\begin{aligned} EIy(x) &= \frac{w_o b}{6}\langle x\rangle^{3} - \frac{w_o b}{2}\left(a+\frac{b}{2}\right)\langle x\rangle^{2} \\ &\quad -\frac{w_o}{24}\langle x-a\rangle^{4} + C_1x + C_2. \end{aligned}$$

The boundary conditions are:

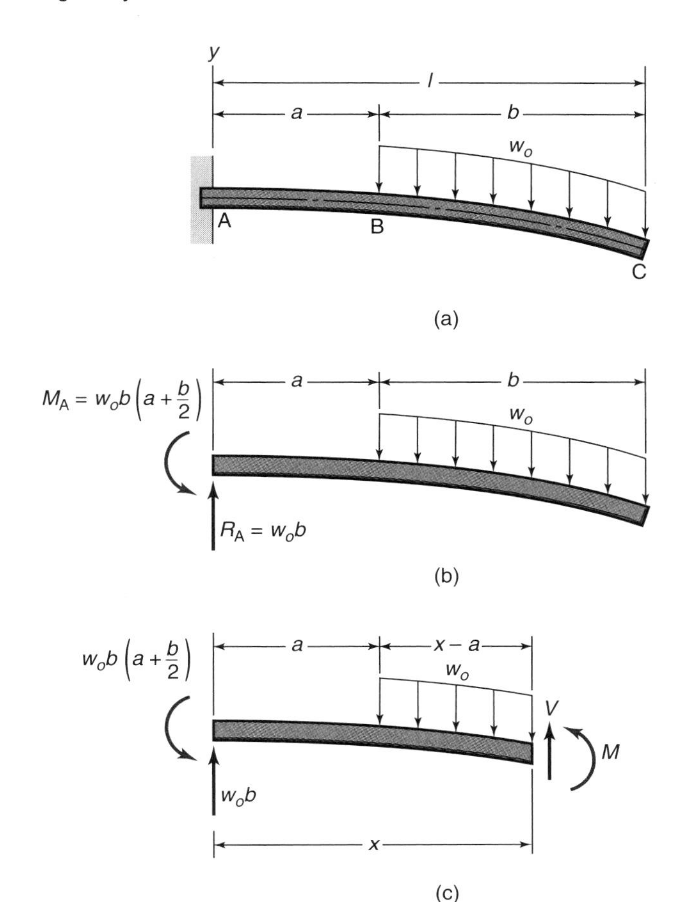

Figure 5.3: Cantilevered beam with unit step distribution over part of beam. (a) Loads and deflection acting on cantilevered beam; (b) free-body diagram of forces and moments acting on entire beam; (c) free-body diagram of forces and moments acting on portion of beam.

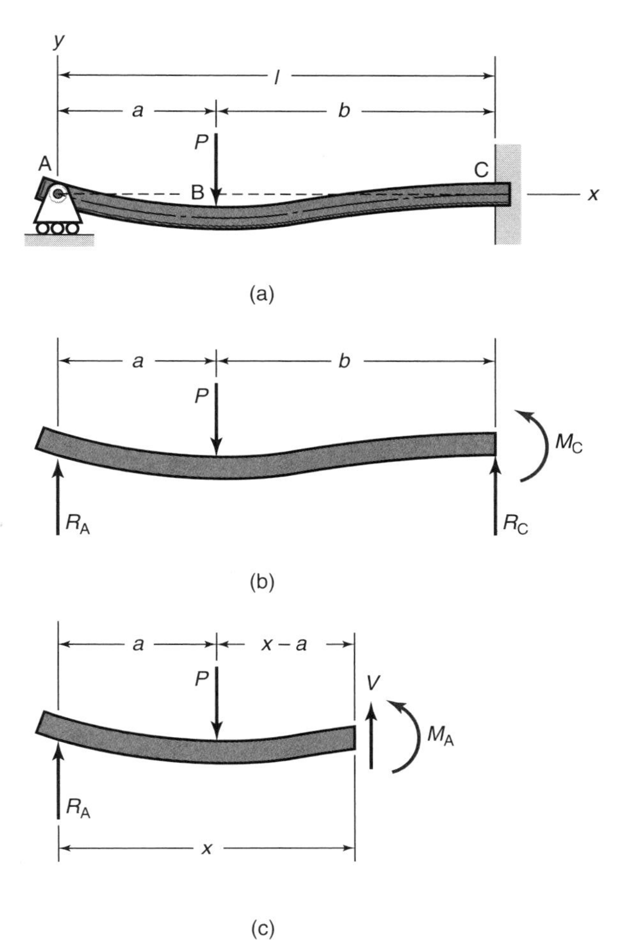

Figure 5.4: Pinned-fixed beam with concentrated force acting anywhere along beam. (a) Sketch of assembly; (b) free-body diagram of entire beam; (c) free-body diagram of part of beam.

1. $\frac{dy}{dx} = 0$ at $x = 0$, or $C_1 = 0$
2. $y = 0$ at $x = 0$, or $C_2 = 0$

Therefore,

$$y = \frac{w_o}{EI}\left[\frac{bx^3}{6} - \frac{bx^2}{2}\left(a + \frac{b}{2}\right) - \frac{1}{24}\langle x - a\rangle^4\right].$$

The deflection at the free end, or where $x = a + b$, is

$$\begin{aligned} y_l &= -\frac{w_o}{EI}\left[\frac{b^4}{24} + \frac{b(a+b)^2}{2}\left(a + \frac{b}{2}\right) - \frac{b(a+b)^3}{6}\right] \\ &= -\frac{w_o b}{EI}\left[\frac{a^3}{3} + \frac{3}{4}a^2 b + \frac{ab^2}{2} + \frac{b^3}{8}\right]. \end{aligned}$$

(a) For the special case of $b = 0$, $y_{\max} = 0$ (no deflection occurs).

(b) For the special case of $a = 0$,

$$y_l = -\frac{w_o b^4}{8EI}.$$

In this situation the unit step extends completely across the length l.

Example 5.4: Deflection of Statically Indeterminate Beam Using Singularity Functions

Given: A beam is fixed at one end and has its other end simply supported. A concentrated force acts at any point along the beam, as shown in Fig. 5.4.

Find: Determine a general expression for the deflection of the beam by using singularity functions.

Solution: Note from Fig. 5.4b that there are three unknowns, R_A, R_C, and M_C. The force and moment equilibrium conditions produce only two equations. A solution is to take one of the reaction forces as an unknown and express the other two in terms of that unknown. Thus,

$$R_C = P - R_A, \qquad (a)$$

$$M_C = -Pb + R_A l. \qquad (b)$$

The load intensity equation for the forces shown in Fig. 5.4a can be expressed as

$$q(x) = R_A \langle x\rangle^{-1} - P\langle x - a\rangle^{-1}. \qquad (c)$$

Integrating twice gives the moment as

$$M_B(x) = R_A \langle x \rangle^1 - P \langle x - a \rangle^1. \quad (d)$$

Making use of Eq. (5.3) gives

$$EI\frac{d^2y}{dx^2} = M(x) = R_A x - P\langle x - a\rangle^1. \quad (e)$$

Integrating once gives

$$EI\frac{dy}{dx} = R_A\frac{x^2}{2} - \frac{P}{2}\langle x - a\rangle^2 + C_1. \quad (f)$$

Integrating again gives the deflection as

$$EIy = R_A\frac{x^3}{6} - \frac{P}{6}\langle x - a\rangle^3 + C_1 x + C_2. \quad (g)$$

The boundary conditions are:

1. $y = 0$ at $x = 0$, which results in $C_2 = 0$
2. $\frac{dy}{dx} = 0$ at $x = l$ gives

$$C_1 = -R_A\frac{l^2}{2} + \frac{P}{2}(l-a)^2. \quad (h)$$

Therefore,

$$EIy = R_A\frac{x^3}{6} - \frac{P}{6}\langle x-a\rangle^3 - \frac{R_A l^2 x}{2} + \frac{Px}{2}(l-a)^2. \quad (i)$$

3. $y = 0$ at $x = l$:

$$R_A = \frac{Pb^2}{2l^3}(3l - b). \quad (j)$$

The general expression for the deformation is

$$y = \frac{P}{6EI}\left[\frac{xb^2}{2l^3}\left(-3l^3 + 3lx^2 + 3bl^2 - bx^2\right) - \langle x-a\rangle^3\right]. \quad (k)$$

Substituting Eq. (j) into Eq. (d) gives

$$M_B(x) = \frac{Pxb^2}{2l^3}(3l-b) - P\langle x-a\rangle^1. \quad (l)$$

Note that

$$M_B(x=0) = 0, \quad (m)$$

$$M_B(x=a) = \frac{Pb^2a}{2l^3}(3l-b), \quad (n)$$

$$M_B(x=l) = \frac{Pb^2l}{2l^2}(3l-b) - Pb = -\frac{Pab}{2l^2}(2l-b). \quad (o)$$

From Eqs. (n) and (o), the moment is negative at $x = a$ and positive at $x = l$, but is not clear whether the maximum magnitude exists at $x = a$ or at $x = l$. When $a = \left(\sqrt{2}-1\right)l = 0.414l$, the magnitude of the bending moment at B equals that at C. When $a < 0.414l$, the greater moment occurs at B; and when $a > 0.414l$, the greater moment occurs at C.

5.4 Method of Superposition

The **method of superposition** uses the principle that the deflection *at any point* in a body is equal to the sum of the deflections caused by each load acting separately. Thus, if a beam is loaded by n separate forces, the deflection at a particular point is the sum of the n deflections, one for each force. This method depends on the linearity of the governing relations between the load and the deflection, and it involves reducing complex conditions of load and support into a combination of simple loading conditions for which solutions are available. The solution of the original problem then takes the form of a superposition of these solutions. The solution assumes that the deflection of the beam is linearly proportional to the applied load. Thus, for n different loads, Eq. (5.3) can be written as

$$EI\frac{d^2y}{dx^2} = EI\frac{d^2}{dx^2}(y_1 + y_2 + \cdots + y_n) = M_1 + M_2 + \cdots + M_n. \quad (5.12)$$

Table 5.1 gives solutions for some simple beam deflection situations that may be combined to produce the deflection for a more complex situation. Additional cases are summarized in Appendix D.

Example 5.5: Deflection of a Cantilever Through Superposition

Given: Figure 5.5 shows a cantilevered beam (i.e., fixed at one end and free at the other end). A moment is applied at the free end, and a concentrated force is applied at a distance, a, from the fixed end.

Find: Use the method of superposition to determine the deflection at the free end.

Solution: Figure 5.5b depicts the deflection with one end fixed and one end free for a concentrated force at any point within the length of the beam. Figure 5.5c shows a moment applied to the free end and the deformation. From Table 5.1 the individual deflections can be obtained directly as

$$y_{l,1} = -\frac{Pa^2}{6EI}(3l-a) \quad (a)$$

and

$$y_{l,2} = \frac{Ml^2}{2EI}. \quad (b)$$

The resultant deflection from the method of superposition is

$$y_l = y_{l,1} + y_{l,2} = \frac{-Pa^2(3l-a) + 3Ml^2}{6EI}. \quad (c)$$

The deflection at any point on the beam is

$$y = -\frac{P}{6EI}\left[\langle x-a\rangle^3 - x^3 + 3x^2a\right] + \frac{Mx^2}{2EI}. \quad (d)$$

5.5 Strain Energy

Statically indeterminate beams and beams of varying material properties or cross sections cannot be successfully analyzed by using the methods discussed thus far. Also, when a loading is energy related (such as an object striking a beam with a given initial velocity), the exact forces are not readily apparent. For this reason *energy methods* are often very useful.

When loads are applied to a machine element, it will deform. In the process, the external work done by the loads will be converted into internal work (called **strain energy**), provided that no energy is lost in the form of heat. This strain

Table 5.1: Deflection for common cantilever and simply-supported beam conditions. See also Appendix D.

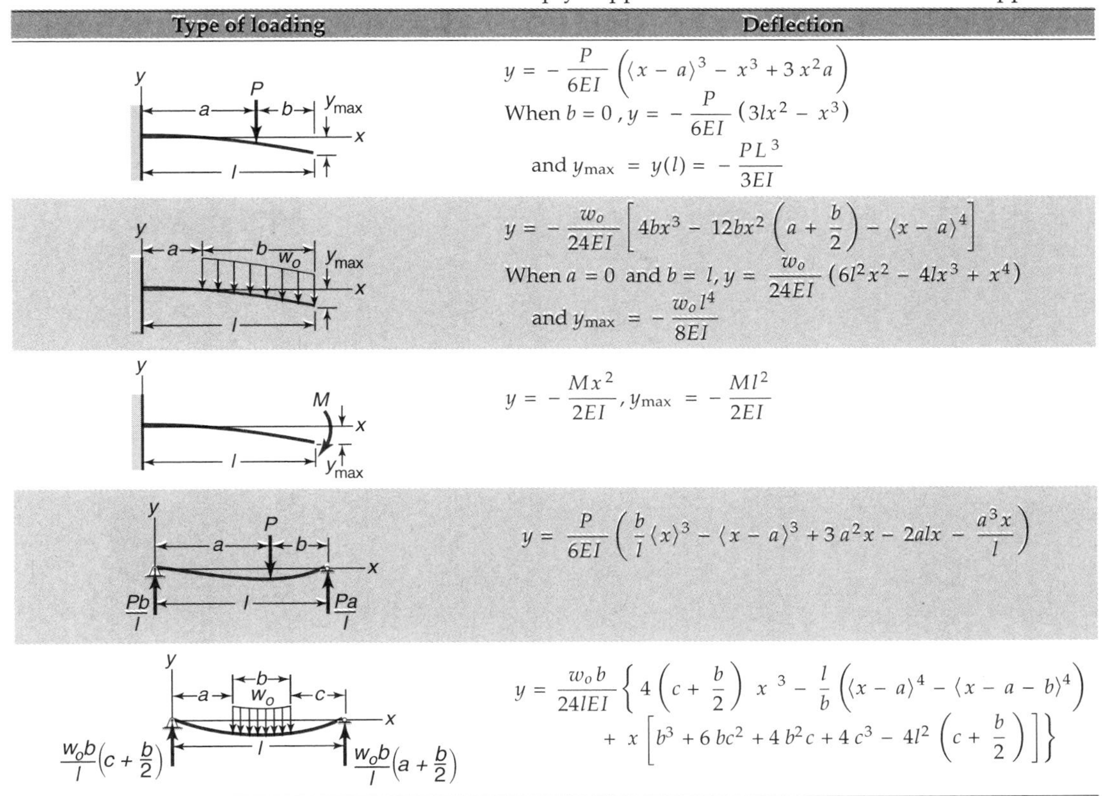

Type of loading	Deflection
	$y = -\frac{P}{6EI}\left(\langle x-a\rangle^3 - x^3 + 3x^2a\right)$ When $b = 0$, $y = -\frac{P}{6EI}(3lx^2 - x^3)$ and $y_{\max} = y(l) = -\frac{PL^3}{3EI}$
	$y = -\frac{w_o}{24EI}\left[4bx^3 - 12bx^2\left(a + \frac{b}{2}\right) - \langle x-a\rangle^4\right]$ When $a = 0$ and $b = l$, $y = \frac{w_o}{24EI}(6l^2x^2 - 4lx^3 + x^4)$ and $y_{\max} = -\frac{w_o l^4}{8EI}$
	$y = -\frac{Mx^2}{2EI}, y_{\max} = -\frac{Ml^2}{2EI}$
	$y = \frac{P}{6EI}\left(\frac{b}{l}\langle x\rangle^3 - \langle x-a\rangle^3 + 3a^2x - 2alx - \frac{a^3x}{l}\right)$
	$y = \frac{w_o b}{24lEI}\left\{4\left(c + \frac{b}{2}\right)x^3 - \frac{l}{b}\left(\langle x-a\rangle^4 - \langle x-a-b\rangle^4\right) + x\left[b^3 + 6bc^2 + 4b^2c + 4c^3 - 4l^2\left(c + \frac{b}{2}\right)\right]\right\}$

energy is stored in the body. The unit of strain energy is newton-meters. Strain energy is always positive even if the stress is compressive because stress and strain are always in the same direction. The symbol U is used to designate strain energy.

5.5.1 Normal Stress

When a tension test specimen is subjected to an axial load, a volume element (shown in Fig. 5.6) is subjected to an axial force and the force develops a stress

$$dP = \sigma_z\, dA = \sigma_z\, dx\, dy \tag{5.13}$$

on the top and bottom faces of the element after it undergoes a vertical elongation of $\epsilon_z dz$.

Because the force, ΔP, is increased uniformly from zero to its final magnitude dP when the displacement $\epsilon_z\, dz$ is attained, the average force magnitude $dP/2$ times the displacement $\epsilon_z\, dz$ is the strain energy or

$$dU = \left(\frac{1}{2}dP\right)\epsilon_z\, dz.$$

Making use of Eq. (5.13) and the fact that $dv = dx\, dy\, dz$ gives

$$dU = \frac{1}{2}\sigma_z\epsilon_z\, dv. \tag{5.14}$$

In general, if the member is subjected only to a uniaxial normal stress, the strain energy is

$$U = \int_v \frac{\sigma\epsilon}{2}\, dv. \tag{5.15}$$

Also, if the material acts in a linear-elastic manner, Hooke's law can be applied and Eq. (3.3) can be substituted into Eq. (5.15) giving

$$U = \int_v \frac{\sigma^2}{2E}\, dv. \tag{5.16}$$

For axial loading of a beam of length l, making use of Eq. (2.7) gives

$$U = \int_v \frac{P^2}{2EA^2}\, dv, \tag{5.17}$$

but $dv = A\, dx$, so that

$$U = \int_0^l \frac{P^2}{2AE}\, dx = \frac{P^2 l}{2AE}. \tag{5.18}$$

If the loading is from bending moments, making use of Eq. (4.46) gives

$$U = \int_v \frac{\sigma^2}{2E}\, dv = \int \frac{M^2y^2}{2EI^2}\, dv.$$

Since $dv = dA\, dx$, where dA represents an element of the cross-sectional area. Also, recall that $M^2/2EI^2$ is a function of x alone; then,

$$U = \int_0^l \frac{M^2}{2EI^2}\left(\int y^2\, dA\right) dx.$$

Equation (5.7) then gives

$$U = \int_0^l \frac{M^2}{2EI}\, dx. \tag{5.19}$$

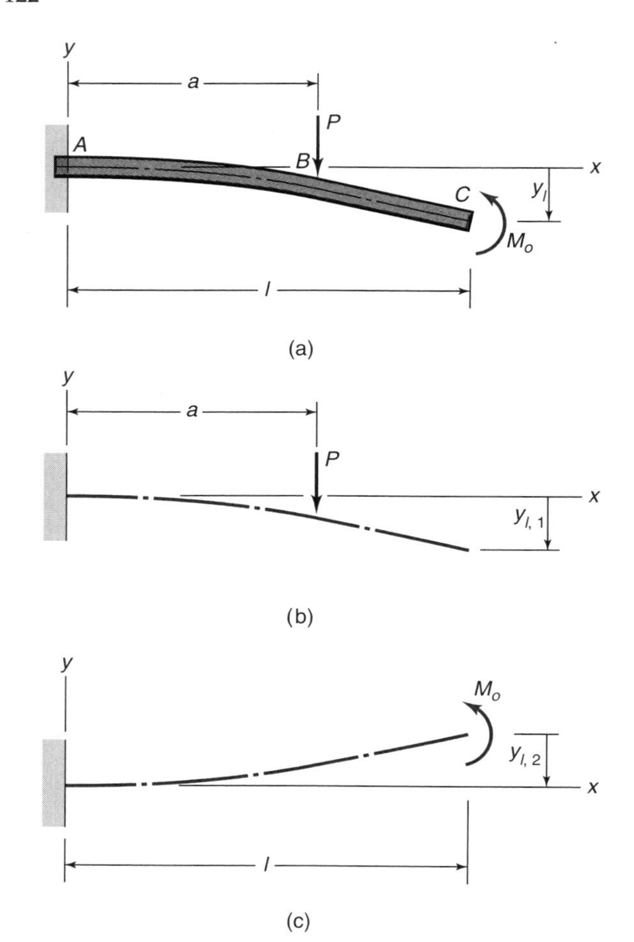

Figure 5.5: Beam fixed at one end and free at the other with moment applied to free end and concentrated force at any distance from free end. (a) Complete assembly; (b) free-body diagram showing effect of concentrated force; (c) free-body diagram showing effect of moment.

5.5.2 Shear Stress

Consider the volume element shown in Fig. 5.7. The shear stress causes the element to deform such that only the shear force $dV = \tau\, dx\, dy$, acting on the top of the element, is displaced $\gamma\, dz$ relative to the bottom surface. Only the vertical surfaces rotate, and therefore the shear forces on these surfaces do not contribute to the strain energy. Thus, the strain energy stored in the element due to a shear stress, τ, is

$$dU = \frac{1}{2}(\tau\, dx\, dy)\,\gamma\, dz,$$

or

$$dU = \frac{1}{2}\tau\gamma\, dv. \tag{5.20}$$

Integrating over the entire volume gives the strain energy stored in the member due to shear stress as

$$U = \int_v \frac{\tau\gamma}{2}\, dv. \tag{5.21}$$

Making use of Hooke's law for shear stress, expressed in Eq. (4.28), gives

$$U = \int_v \frac{\tau^2}{2G}\, dv. \tag{5.22}$$

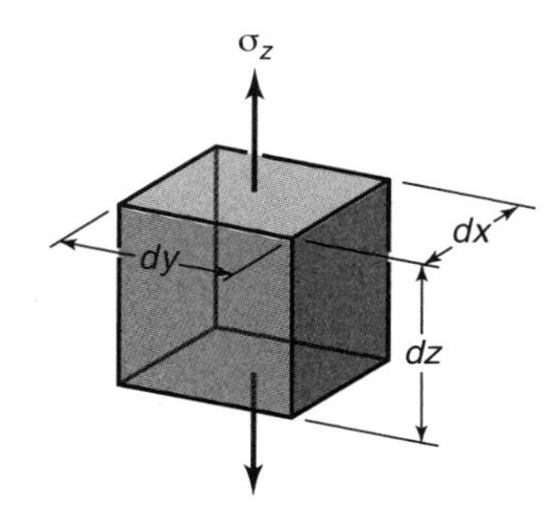

Figure 5.6: Element subjected to normal stress.

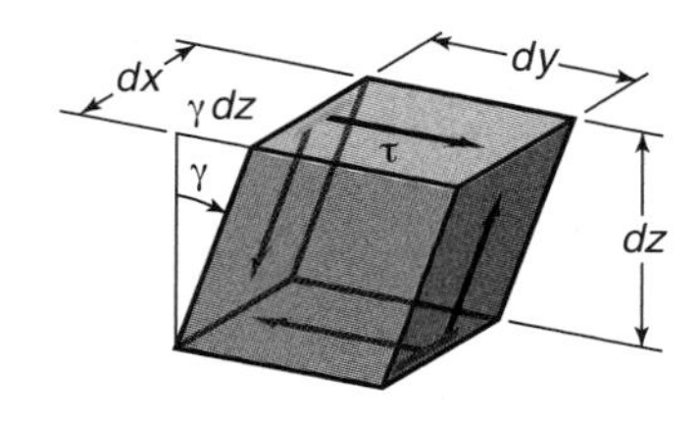

Figure 5.7: Element subjected to shear stress.

For strain energy in torsion of circular shafts, making use of Eqs. (4.27) to (4.33) gives

$$U = \int_v \frac{T^2 r^2}{2GJ^2}\, dv,$$

but $dv = dA\, dx$, and thus $T^2/2GJ^2$ is a function of x only, so that

$$U = \int_0^l \frac{T^2}{2GJ^2}\left(\int r^2\, dA\right) dx.$$

Making use of Eq. (4.30) gives

$$U = \int_0^l \frac{T^2}{2GJ}\, dx. \tag{5.23}$$

If the shaft has a uniform cross section,

$$U = \frac{T^2 l}{2GJ}. \tag{5.24}$$

Recall that Table 4.1 contains values of J for a number of different cross sections, but Eq. (5.24) can be used only for cross sections with circular symmetry.

Example 5.6: Strain Energy in a Shaft

Given: A 1-m-long solid shaft with circular cross section has a diameter of 40 mm over 0.5 m of the length and 30 mm over the rest of the length. The shaft is subjected to a torque of 1100 N-m. The shaft material is AISI 1080 high-carbon steel.

Find: Calculate the strain energy in the shaft; also calculate the ratio of the strain energies stored in the thinner and thicker parts of the shaft.

Solution: From Eq. (5.23),

$$U = \int_0^l \frac{T^2}{2GJ}\,dx = \frac{T^2}{2G}\left(\frac{1}{J_1}\int_0^{0.5} dx + \frac{1}{J_2}\int_{0.5}^{1.0} dx\right).$$

From Table 3.1 for steel, $E = 207$ GPa and $\nu = 0.3$. From Eq. (3.7), the shear modulus is

$$G = \frac{E}{2(1+\nu)} = \frac{(207 \times 10^9)}{2(1+0.3)} = 79.6 \text{ GPa}.$$

For a solid circular cross section,

$$J_1 = \frac{\pi r^4}{2} = \frac{\pi}{2}(0.02)^4 = 25.13 \times 10^{-8}\text{ m}^4,$$

$$J_2 = \frac{\pi}{2}(0.015)^4 = 7.952 \times 10^{-8}\text{ m}^4.$$

The strain energy becomes

$$U = \frac{1100^2}{2\,(79.6 \times 10^9)}\left(\frac{10^8}{25.13} + \frac{10^8}{7.952}\right)\left(\frac{1}{2}\right) = 62.91 \text{ N-m}.$$

The ratio of the energy stored in the two parts can be expressed as

$$\frac{U_1}{U_2} = \frac{J_2}{J_1} = \frac{(7.952 \times 10^{-8})}{(25.13 \times 10^{-8})} = 0.3164.$$

5.5.3 Transverse Shear Stress

The strain energy due to shear stress can be obtained from Eq. (4.65). For a rectangular cross section with width b and height h,

$$\tau = \frac{3V}{2A}\left(1 - \frac{y^2}{c^2}\right) = \frac{3V}{2bh}\left(1 - \frac{y^2}{c^2}\right). \qquad (5.25)$$

Substituting this equation into Eq. (5.22) and integrating gives

$$U = \frac{1}{2G}\left(\frac{3V}{2bh}\right)^2 \int \left(1 - \frac{y^2}{c^2}\right)^2 dv.$$

Setting $dv = b\,dx\,dy$ results in

$$U = \frac{9V^2}{8Gbh^2}\int_{-c}^{c}\left(1 - \frac{2y^2}{c^2} + \frac{y^4}{c^4}\right) dy \int_0^l dx.$$

Integrating yields

$$U = \frac{9V^2 l}{8Gbh^2}\left(y - \frac{2y^3}{3c^2} + \frac{y^5}{5c^4}\right)\Bigg|_{y=-c}^{y=c}.$$

Substituting the limits gives

$$U = \frac{6V^2 lc}{5Gbh^2}.$$

Recalling that $c = h/2$,

$$U = \frac{3V^2 l}{5Gbh}. \qquad (5.26)$$

This is the general strain energy due to transverse shear stress for a rectangular cross section. Table 5.1 summarizes the strain energy for four types of loading. Recall that the transverse shear is valid only for a rectangular cross section. For torsion, Table 4.1 should be used for J, the polar area moment of inertia for a circular cross section. For bending, I corresponds to I_x in Table 4.1, since the beam axis is in the x-direction.

5.5.4 General State of Stress

The total strain energy due to a general state of stress can be expressed as

$$\begin{aligned} U &= \int_v \left(\frac{\sigma_x \epsilon_x}{2} + \frac{\sigma_y \epsilon_y}{2} + \frac{\sigma_z \epsilon_z}{2}\right) dv \\ &\quad + \int_v \left(\frac{\tau_{xy}\gamma_{xy}}{2} + \frac{\tau_{yz}\gamma_{yz}}{2} + \frac{\tau_{xz}\gamma_{xz}}{2}\right) dv. \end{aligned} \qquad (5.27)$$

Making use of Eq. (B.54) gives

$$\begin{aligned} U &= \int_v \frac{1}{2E}\left(\sigma_x^2 + \sigma_y^2 + \sigma_z^2\right) dv \\ &\quad - \int_v \frac{\nu}{E}\left(\sigma_x\sigma_y + \sigma_y\sigma_z + \sigma_z\sigma_x\right) \\ &\quad + \int_v \frac{1}{G}\left(\tau_{xy}^2 + \tau_{yz}^2 + \tau_{zx}^2\right) dv. \end{aligned} \qquad (5.28)$$

If only the principal stresses σ_1, σ_2, and σ_3 act on the elements, Eq. (5.28) reduces to

$$U = \int_v \left[\frac{1}{2E}\left(\sigma_1^2 + \sigma_2^2\sigma_3^2\right) - \frac{\nu}{E}\left(\sigma_1\sigma_2 + \sigma_2\sigma_3\sigma_3\sigma_1\right)\right] dv. \qquad (5.29)$$

5.6 Castigliano's Theorem

It is often necessary to calculate the elastic deformation due to distributed loads that are not as simple as those already presented. Castigliano's theorem can be applied to a wide range of deflection problems. Extensive use is made here of the strain energy material presented in Section 5.5.

Castigliano's theorem states that when a body is elastically deformed by a system of loads, the deflection at any point, p, in any direction, a, is equal to the partial derivative of the strain energy (with the system of loads acting) with respect to a load at p in the direction a, or

$$y_i = \frac{\partial U}{\partial Q_i}. \qquad (5.30)$$

The load Q_i is applied to a particular point of deformation and therefore is not a function of x. Thus, it is permissible to take the derivative with respect to Q_i before integrating for the general expressions for the strain energy. Also, the load may be any of the loads presented in Section 2.3 and throughout the text: normal, shear, bending, and transverse shear. Table 5.2 shows the strain energy for the various types of loading.

The best way to understand how to apply Castigliano's theorem is to observe how it is used in a number of different examples. Examples 5.7 through 5.10 demonstrate various features of Castigliano's approach.

Table 5.2: Strain energy for four types of loading.

Loading type	Factors involved	Strain energy for special case where all three factors are constant with x	General expression for strain energy
Axial	P, E, A	$U = \dfrac{P^2 l}{2EA}$	$U = \int_o^l \dfrac{P^2}{2EA}\,dx$
Bending	M, E, I	$U = \dfrac{M^2 l}{2EI}$	$U = \int_o^l \dfrac{M^2}{2EI}\,dx$
Torsion	T, G, J	$U = \dfrac{T^2 l}{2GJ}$	$U = \int_o^l \dfrac{T^2}{2GJ}\,dx$
Transverse shear (rectangular section)	V, G, A	$U = \dfrac{3V^2 l}{5GA}$	$U = \int_o^l \dfrac{3V^2}{5GA}\,dx$

Design Procedure 5.2: Procedure for Using Castigliano's Theorem

The following procedure is to be employed in using Castigliano's theorem:

1. Obtain an expression for the total strain energy, including:
   (a) Loads (P, M, T, V) acting on the element (use Table 5.2)
   (b) A fictitious force Q acting at the location and in the direction of the desired deflection
2. Obtain deflection from $y = \partial U/\partial Q$.
3. If Q is fictitious, set $Q = 0$ and solve the resulting equation.

Example 5.7: Deflection of a Beam From Castigliano's Theorem

Given: The simply supported beam shown in Fig. 5.2 with the force P applied at $x = l/2$.

Find: Determine the deflection, δ_P, at the location of the applied force by using Castigliano's theorem. Consider both bending and transverse shear.

Solution: Because of symmetry, the deflection at the point of an applied force can be obtained by doubling the solution from zero to $l/2$. The two types of loading being applied to the beam are:

(*a*) Bending

$$M = \frac{Px}{2} \quad \text{and} \quad \frac{dM}{dP} = \frac{x}{2}. \qquad (a)$$

(*b*) Transverse shear

$$V = \frac{P}{2} \quad \text{and} \quad \frac{dV}{dP} = \frac{1}{2}. \qquad (b)$$

By making use of Table 5.2 the total strain energy can be expressed as

$$U = 2\int_0^{l/2} \frac{M^2}{2EI}\,dx + 2\int_0^{l/2} \frac{3V^2}{5GA}\,dx. \qquad (c)$$

Using Eqs. (*a*) and (*b*) this equation becomes

$$U = \int_0^{l/2} \frac{P^2x^2}{4EI}\,dx + \int_0^{l/2} \frac{3P^2}{10GA}\,dx.$$

From Castigliano's theorem

$$-\delta_P = \frac{\partial U}{\partial P} = +\int_0^{l/2} \frac{Px^2}{2EI}\,dx + \int_0^{l/2} \frac{3P}{5GA}\,dx.$$

Because P, E, I, G, and A are not functions of x, this equation becomes

$$\delta_P = -\frac{Pl^3}{48EI} - \frac{3Pl}{10GA}. \qquad (d)$$

The first term on the right side is due to bending and the second is due to transverse shear.

To better understand the bending and transverse shear contributions to the total deflection at the applied force, assume that the beam has a rectangular cross section. From Table 4.1 for a rectangular shape, $I = bh^3/12$ and $A = bh$. Substituting these expressions into Eq. (*d*) gives

$$\delta_P = -\frac{Pl^3}{4Ebh^3} - \frac{3Pl}{10Gbh} \qquad (e)$$

which can be rewritten as

$$\delta_P = -\frac{3Pl}{10Gbh}\left[\left(\frac{5}{6}\right)\left(\frac{G}{E}\right)\left(\frac{l}{h}\right)^2 + 1\right]. \qquad (f)$$

For carbon steel $G/E = 0.383$. The length-to-height ratio, l/h, of a beam is typically at least 10. Assume that the smallest value is $l/h = 10$. After substituting the values for G/E and l/h into Eq. (*f*), the first term within the brackets is 32 times the second term. Thus, in most applications, the transverse shear term will be considerably smaller than the bending moment term (typically, less than 3%). This is an important finding that will be used often; for beams, deflections and energies can be assumed to be due to bending only.

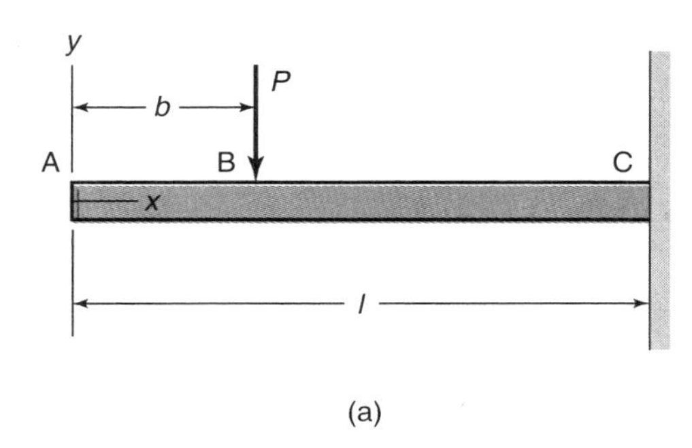

(a)

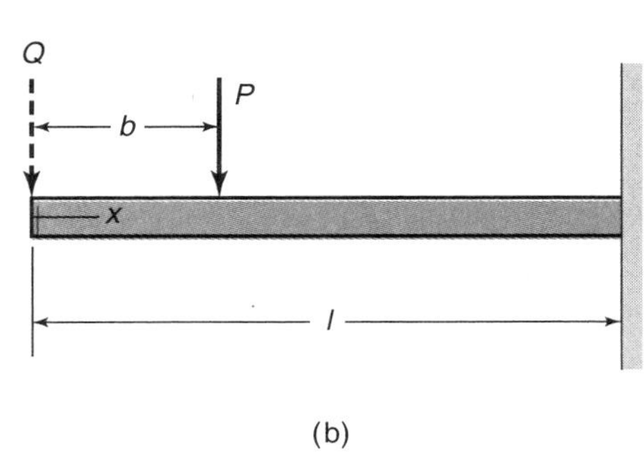

(b)

Figure 5.8: Cantilevered beam with concentrated force acting at a distance b from free end. (a) Coordinate system and significant points shown; (b) fictitious force, Q, shown along with concentrated force, P.

Example 5.8: Deflection Using Castigliano's Theorem

Given: A cantilevered beam has a concentrated force acting at a distance, b, from the free end, as shown in Fig. 5.8.

Find: Determine the deflection at the free end using Castigliano's theorem. Transverse shear can be neglected.

Solution: Note from Fig. 5.8 that, since no force is acting at the free end, a fictitious force is created. The moment at any x can be expressed as

$$M = -Qx - P\langle x - b\rangle .$$

The only force contributing to the total strain rate is the bending moment; from Table 5.2:

$$U = \frac{1}{2EI}\int_0^l \left[Q^2x^2 + 2Q_xP\langle x-b\rangle + (P\langle x-b\rangle)^2\right] dx.$$

From Castigliano's theorem

$$\delta_A = \frac{\partial U}{\partial Q} = \frac{1}{2EI}\left[\frac{2Ql^3}{3} + 2P\int_0^l x\langle x-b\rangle\, dx\right]. \qquad (a)$$

Thus, setting $Q = 0$ and integrating gives

$$\delta_A = \frac{P}{6EI}\left[l^2(2l - 3b) + b^3\right]. \qquad (b)$$

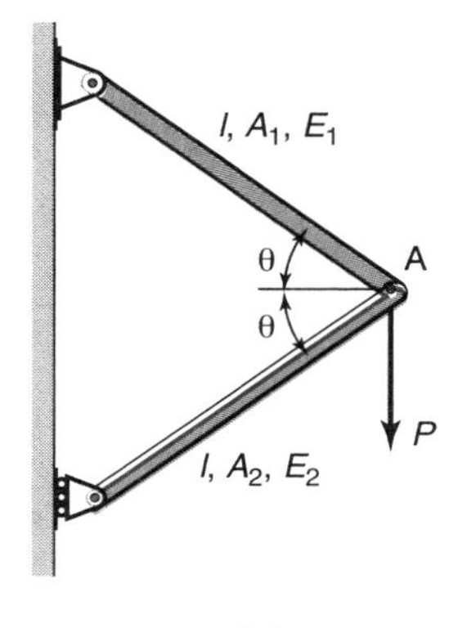

(a)

(b)

Figure 5.9: System arrangement. (a) Entire assembly; (b) free-body diagram of forces acting at point A.

Example 5.9: Castigliano's Theorem Applied to a Pinned Structure

Given: The assembly shown in Fig. 5.9a can be made of different materials and have different cross-sectional areas in its two equal parts, denoted by the subscripts 1 and 2.

Find: Calculate the horizontal displacement at the point of vertical force application by using Castigliano's theorem.

Solution: Note from the free-body diagram in Fig. 5.9b that, since no force is acting horizontally, a fictitious force Q has to be created. From force equilibrium of the vertical and horizontal forces

$$\sum P_v = 0 \rightarrow -P + P_1\sin\theta + P_2\sin\theta = 0,$$

$$\sum P_H = 0 \rightarrow Q - P_1\cos\theta + P_2\cos\theta = 0.$$

Solving for P_1 and P_2 gives

$$P_1 = \frac{1}{2}\left(\frac{P}{\sin\theta} + \frac{Q}{\cos\theta}\right) \quad \text{and} \quad \frac{\partial P_i}{\partial Q} = \frac{1}{2\cos\theta}. \qquad (a)$$

$$P_2 = \frac{1}{2}\left(\frac{P}{\sin\theta} - \frac{Q}{\cos\theta}\right) \quad \text{and} \quad \frac{\partial P_i}{\partial Q} = -\frac{1}{2\cos\theta}. \qquad (b)$$

From Table 5.2 the total strain energy for axial loading can be written as

$$U = \frac{P_1^2 l}{2A_1E_1} + \frac{P_2^2 l}{2A_2E_2}.$$

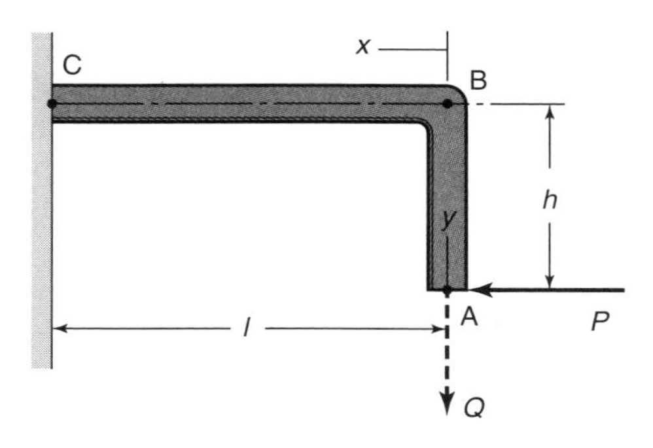

Figure 5.10: Cantilevered beam with 90° bend acted upon by horizontal force, P, at free end.

From Castigliano's theorem, the horizontal displacement at point A is

$$\delta_{A,H} = \frac{\partial U}{\partial Q} = \frac{1}{2}\left(\frac{2P_1\dfrac{\partial P_1}{\partial Q}}{A_1E_1} + \frac{2P_2\dfrac{\partial P_2}{\partial Q}}{A_2E_2}\right).$$

Substituting Eqs. (a) and (b) into the above equation gives

$$\begin{aligned}\delta_{A,H} &= \frac{l}{4A_1E_1}\left(\frac{P}{\sin\theta} + \frac{Q}{\cos\theta}\right)\frac{1}{\cos\theta} \\ &\quad - \frac{l}{4A_2E_2}\left(\frac{P}{\sin\theta} - \frac{Q}{\cos\theta}\right)\frac{1}{\cos\theta}.\end{aligned}$$

Setting $Q = 0$ gives

$$\delta_{A,H} = \frac{lP}{4\sin\theta\cos\theta}\left(\frac{1}{A_1E_1} - \frac{1}{A_2E_2}\right). \qquad (c)$$

Note that if the beam sections 1 and 2 have the same cross-sectional area and are made of the same material (i.e., the modulus of elasticity is the same), Eq. (c) gives $\delta_{A,H} = 0$, and thus there would not be any horizontal displacement.

Example 5.10: Deflection of a Bent Beam from Castigliano's Theorem

Given: Figure 5.10 shows a cantilevered beam with a 90° bend acted on by a horizontal force P at the free end.

Find: Calculate the vertical deflection at the free end if transverse shear is neglected. Use Castigliano's theorem.

Solution: Note from Fig. 5.10 that, since no vertical force exists at the free end, a fictitious force was created. Thus, the four components used to define the total strain energy are:

(a) Bending in AB, where $M_{AB} = Py$

(b) Bending in BC, where $M_{BC} = Qx + Ph$

(c) Axial load in AB of magnitude Q

(d) Axial load in BC of magnitude P

From Table 5.2 the total strain energy can be written as

$$\begin{aligned}U &= \int_0^h \frac{P^2y^2}{2EI}\,dy + \int_0^l \frac{(Qx+Ph)^2}{2EI}\,dx \\ &\quad + \int_0^h \frac{Q^2\,dy}{2EA} + \int_0^l \frac{P^2\,dx}{2EA}\end{aligned}$$

Since the beam's material and cross sections are the same in sections AB and BC, from Castigliano's theorem

$$\delta_{A,V} = \int_0^l \frac{(Qx+Ph)x\,dx}{EI} + \int_0^h \frac{Q\,dy}{EA} = \frac{Ql^3}{3EI} + \frac{Phl^2}{2EI} + \frac{Qh}{EA}$$

Setting $Q = 0$ gives

$$\delta_{A,V} = \frac{Phl^2}{2EI}.$$

5.7 Summary

The three main failure modes for machine elements are (a) from being overstressed, (b) from excessive elastic deformations, and (c) from the lack of a tribological film. This chapter described the deformations that machine elements, especially beams, may experience. Deformations due to distributed and concentrated loads were both considered. For a distributed load four major approaches to describing the deformations were presented: the moment-curvature relation, singularity functions, the method of superposition, and Castigliano's theorem. Each has its particular strengths and limitations. The type of load being applied (normal, bending, shear, or transverse shear) determines the approach. Castigliano's theorem is the most versatile of the four approaches considered, since it can be applied to a wide range of deflection problems.

Key Words

Castigliano's theorem theorem that when a body is elastically deformed by a system of loads, deflection at any point in any direction is equal to the partial derivative of strain energy (for the system of loads) with respect to load in the direction of interest

method of superposition principle that deflection at any point in beam is equal to sum of deflections caused by each load acting separately

moment-curvature relation relationship between beam curvature and bending moment

radius of curvature distance from center to inside edge of beam in bending

singularity function function that permits expressing in one equation what would normally be expressed in several separate equations with boundary conditions

strain energy internal work that was converted from external work done by applying load

Summary of Equations

Beam equations:

Radius of curvature: $\frac{d^2y}{dx^2} = \frac{M}{EI}$

Load intensity: $\frac{q}{EI} = \frac{d^4y}{dx^4}$

Shear force: $-\frac{V}{EI} = \frac{d^3y}{dx^3} = \int_0^x q(x)\,dx$

Moment: $\frac{M}{EI} = \frac{d^2y}{dx^2} = -\int_0^x V(x)\,dx$

Slope: $\theta = \frac{dy}{dx}$

Strain energy:

Normal stress: $U = \frac{P^2l}{2AE}$

Bending stress: $U = \int_0^l \frac{M^2}{2EI}\,dx$

Torque: $U = \frac{T^2l}{2GJ}$

Transverse shear: $U = \frac{3V^2l}{5Gbh}$ (rectangular cross section)

Castigliano's Theorem: $y_i = \frac{\partial U}{\partial Q_i}$

Recommended Readings

Beer, F.P., Johnson, E.R., DeWolf, J., and Mazurek, D. (2011) *Mechanics of Materials*, 6th ed., McGraw-Hill.
Budynas, R.G., and Nisbett, J.K. (2011), *Shigley's Mechanical Engineering Design*, 9th ed., McGraw-Hill.
Craig, R.R. (2001) *Mechanics of Materials*, 2nd ed., Wiley.
Hibbeler, R.C. (2010) *Mechanics of Materials*, 8th ed. Prentice-Hall.
Juvinall, R.C., and Marshek, K.M. (2012) *Fundamentals of Machine Component Design*, 5th ed., Wiley.
Norton, R.L. (2011) *Machine Design*, 4th ed., Prentice-Hall.
Popov, E.P. (1999) *Engineering Mechanics of Solids*, 2nd ed., Prentice-Hall.
Riley, W.F., Sturges, L.D., and Morris, D.H. (2006) *Mechanics of Materials*, 6th ed., Wiley.
Ugural, A.C. (2007) *Mechanics of Materials*, Wiley.

Questions

5.1 How are bending moment and deflection related in beams?

5.2 What is the moment-curvature relation?

5.3 How can one obtain deflection in a beam?

5.4 How does stress depend on the radius of curvature in a beam?

5.5 What are singularity functions?

5.6 What is the Method of Superposition?

5.7 What is the difference between strain and strain energy?

5.8 What is Castigliano's Theorem?

5.9 Why was a fictitious load used in Example 5.9?

5.10 What are the units of slope in Eq. (5.6)?

Qualitative Problems

5.11 It was mentioned in the text that the radius of curvature in a beam is measured from the center of curvature to the inside surface of the beam. Can the radius of curvature ever equal zero? Explain.

5.12 Design Procedure 2.1 discussed singularity functions. Which of the rules are useful for the material presented in this chapter?

5.13 In general, what method for calculating beam deflection would you use for an impact loading?

5.14 Can the Method of Superposition be used for impact loads? Explain.

5.15 Could you use Castigliano's Theorem in Example 5.9 if the bars are replaced by cables? Why or why not?

5.16 List the strengths and weaknesses of singularity functions compared to direct integration in order to obtain beam deflection.

5.17 How can Castigliano's Theorem be used for statically indeterminate beams?

5.18 Define Castigliano's Theorem without the use of equations.

5.19 Can Castigliano's Theorem be used for viscoelastic materials? Explain.

5.20 Assume that the summary of the chapter is not present and write a suitable one- or two-page summary.

Quantitative Problems

5.21 A beam is loaded by a concentrated bending moment M at the free end. Find the vertical and angular deformations along the beam by using the equation of the elastic line, Eq. (5.3). *Ans.* $y = -\frac{Mx^2}{2EI}$.

5.22 A simply supported beam of length l carries a force P. Find the ratio between the bending stresses in the beam when P is concentrated in the middle of the beam and evenly distributed along it. Use the moment-curvature relation given in Eq. (5.3). Also, calculate the ratio of the deformations at the middle of the beam. *Ans.* $\frac{y_{\text{conc}}}{y_{\text{dist}}} = 1.6$.

5.23 A simply supported beam with length l is centrally loaded with a force P. How large a moment needs to be applied at the ends of the beam

(a) To maintain the slope angle of zero at the supports? *Ans.* $M_o = \frac{1}{8}Pl$.

(b) To maintain the midpoint of the beam without deformation when the load is applied? Use the equation of the elastic line, Eq. (5.3). *Ans.* $M_o = \frac{1}{6}Pl$

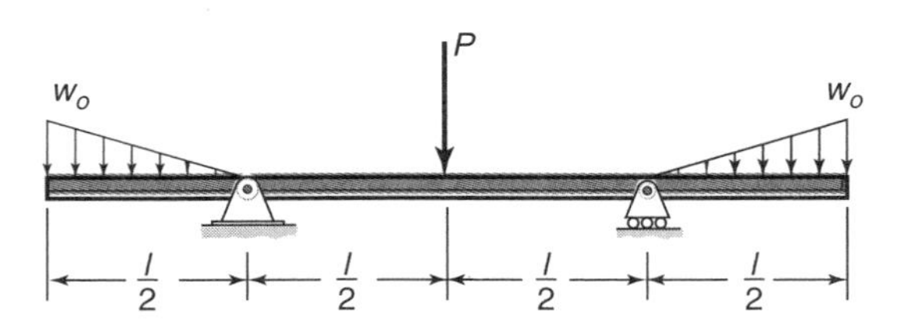

Sketch a, for Problem 5.24

5.24 Find the relation between P and w_o so that the slope of the deflected beam is zero at the supports for the loading conditions shown in Sketch a. Assume that E and A are constant. *Ans.* $P = \frac{2}{3}w_o l$.

5.25 Given a simply supported beam with two concentrated forces acting on it as shown in Sketch b, determine the expression for the elastic deformation of the beam for any x by using singularity functions. Assume that E and I are constant. Also determine the location of maximum deflection and derive an expression for it.

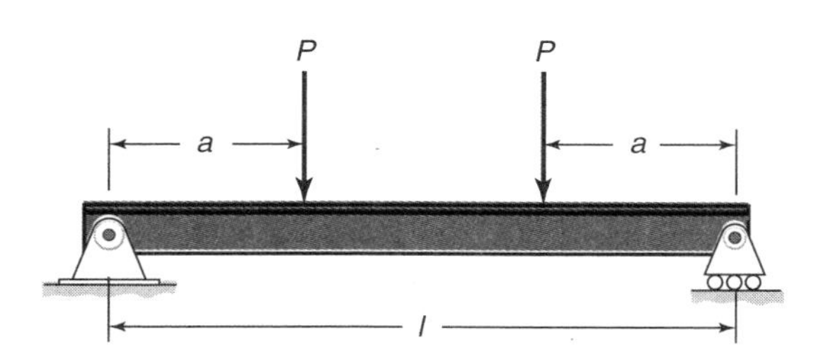

Sketch b, for Problem 5.25

5.26 For the loading condition described in Sketch c obtain the internal shear force $V(x)$ and the internal moment $M(x)$ by using singularity functions. Draw $V(x)$, $M(x)$, $q(x)$, and $y(x)$ as a function of x. Assume that $w_o = 9$ kN/m and $l = 3$ m.

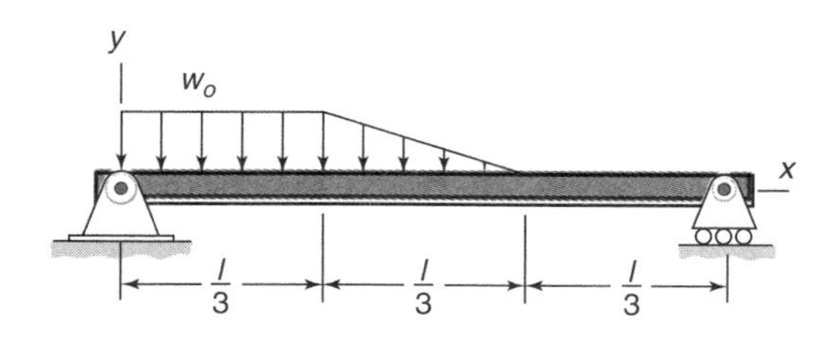

Sketch c, for Problem 5.26

5.27 A simply supported beam is shown in Sketch d with $w_o = 4$ kN/m and $l = 12$ m.

(a) Draw the free-body diagram of the beam.

(b) Use singularity functions to determine shear force, bending moment, slope, and deflection.

(c) Construct diagrams of shear force, bending moment, slope, and deflection.

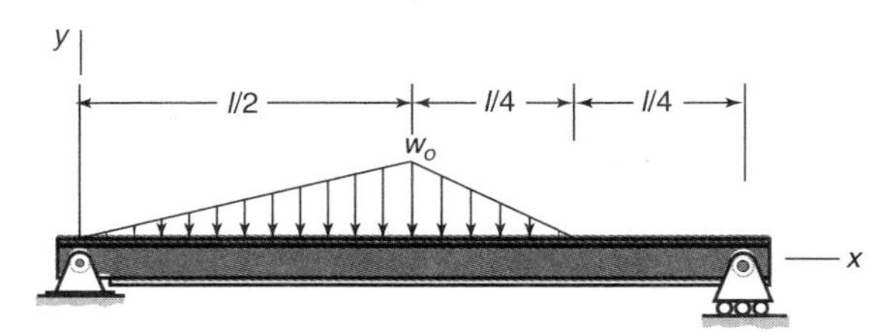

Sketch d, for Problem 5.27

5.28 The simply supported beam in Problem 5.25 is altered so that instead of a concentrated force P, a concentrated moment M is applied at the same location. The moments are positive and act parallel with each other. Determine the deformation of the beam for any position x along it by using singularity functions. Assume that E and I are constant. Also, determine the location of the maximum deflection.

5.29 The simply supported beam considered in Problem 5.28 has moments applied in opposite directions so that the moment at $x = a$ is M_o and at $x = l - a$ the moment is $-M_o$. Find the elastic deformation of the beam by using singularity functions. Also, determine the location and size of the maximum deflection.

5.30 Given the loading condition shown in Sketch e, find the deflection at the center and ends of the beam. Assume that EI = 750 kNm^2. *Ans.* $y_{\text{end}} = -0.05477$ m, $y_{\text{mid}} = 0.01359$ m.

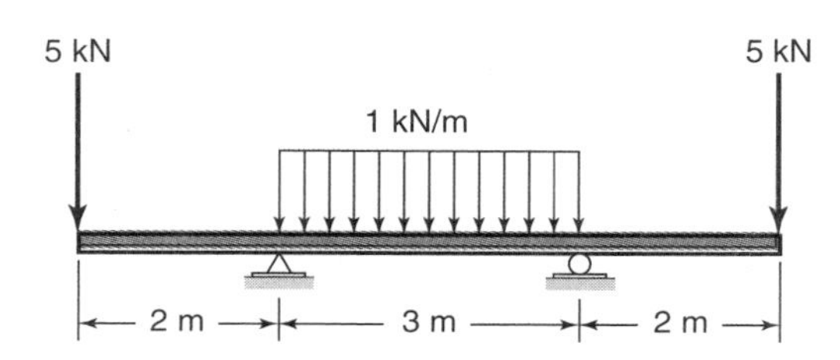

Sketch e, for Problem 5.30

5.31 Given the loading condition shown in Sketch f obtain an expression for the deflection at any location on the beam. Assume that EI is constant.

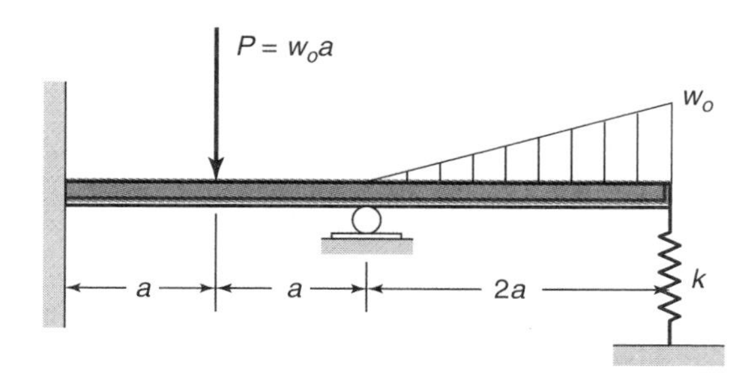

Sketch f, for Problem 5.31

5.32 Given the loading condition and spring shown in Sketch g determine the stiffness of the spring so that the bending moment at point B is zero. Assume that EI is constant. *Ans.* $k = \frac{2EI}{l^3}$.

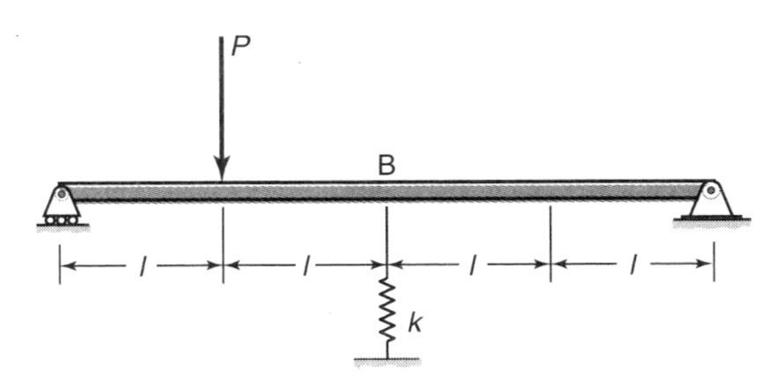

Sketch g, for Problem 5.32

5.33 When there is no load acting on the cantilevered beam shown in Sketch h, the spring has zero deflection. When there is a spring and a force of 20 kN is applied at point C, a deflection of 25 mm occurs at the spring. If a 50-kN load is applied at the location shown in Sketch h, what will be the deflection of the beam? Assume that the stiffness of the spring is 450 kN/m.

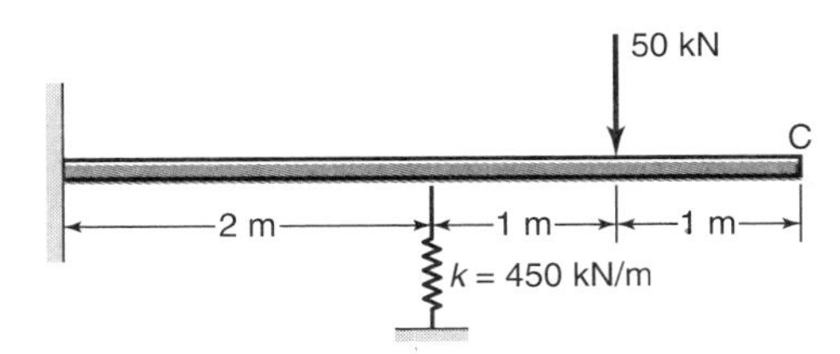

Sketch h, for Problem 5.33

5.34 Determine the deflection at point A and the maximum moment for the loading shown in Sketch i. Consider only bending effects and assume that EI is constant. *Ans.* $\delta_H = \dfrac{Pr^3\pi}{8EI}$.

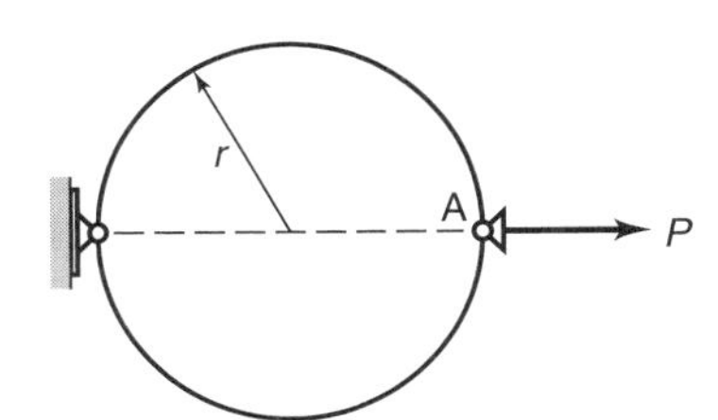

Sketch i, for Problem 5.34

5.35 Determine the maximum deflection of the beam shown in Sketch j. *Ans.* $y_{\max} = \dfrac{w_o l^2}{24EI}(9l^2 + 20al + 12a^2)$.

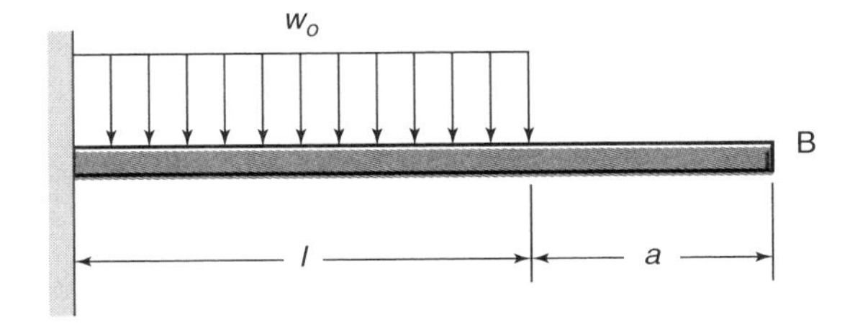

Sketch j, for Problem 5.35

5.36 Determine the deflection at any point in the beam shown in Sketch k using singularity functions.

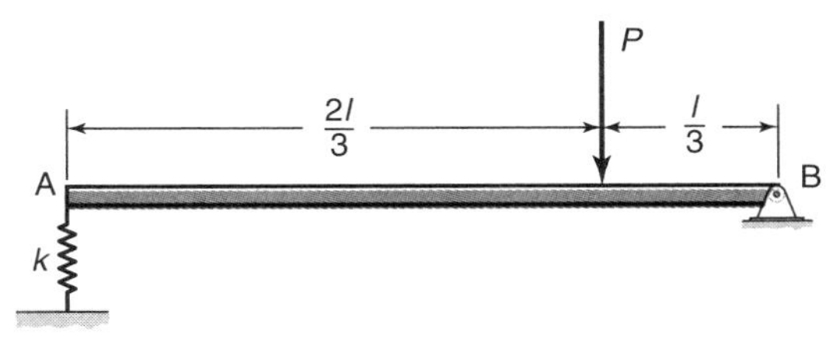

Sketch k, for Problem 5.36

5.37 Determine the deformation of a cantilevered beam with loading shown in Sketch l as a function of x. Also determine the maximum bending stress in the beam and the maximum deflection. Assume that E = 207 GPa, $I = 2.50 \times 10^{-6}$ m^4, P = 1000 N, w_o = 3000 N/m, $a = 0.5$ m, $b = 0.15$ m, $c = 0.45$ m. The distance from the neutral axis to the outermost fiber of the beam is 0.040 m. *Ans.* $y_{\max} = -0.5826$ mm, $\sigma_{\max} = 33.44$ MPa.

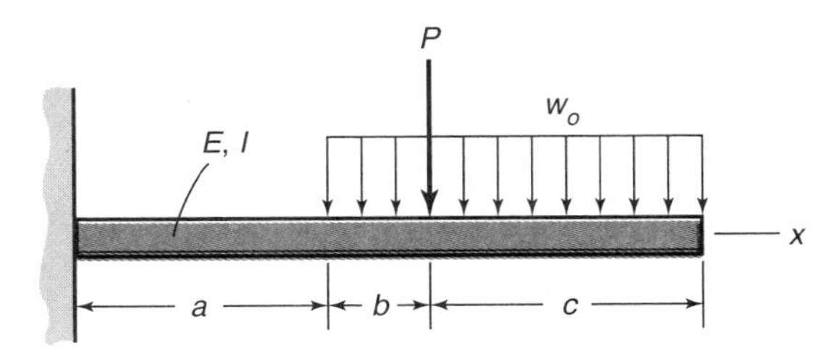

Sketch l, for Problem 5.37

5.38 Given the loading shown in Sketch m, let a = 0.6 m, b = 0.7 m, M = 6500 N-m, and w_o = 20,000 N/m. The beam has a square cross section with sides of 75 mm and the beam material has a modulus of elasticity of 207 GPa. Determine the beam deformation by using the method of superposition. Also, calculate the maximum bending stress and maximum beam deformation. *Ans.* $y_{\max} = -3.988$ mm, $\sigma_{\max} = 96.71$ MPa.

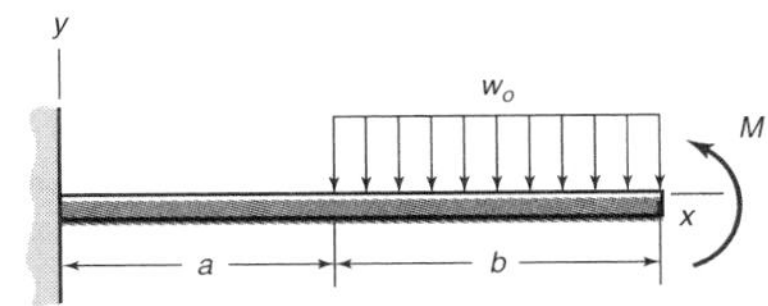

Sketch m, for Problem 5.38

5.39 The cantilevered beam shown in Sketch n has both a concentrated force and a moment acting on it. Let a = 1 m, b = 0.7 m, P = 8700 N, and M = 4000 N-m. The beam cross section is rectangular with a height of 80 mm and a width of 35 mm. Also, E = 207 GPa. Calculate the beam deformation by using the method of superposition. Find how large M has to be to give zero deformation at $x = a$. *Ans.* M = 5800 Nm.

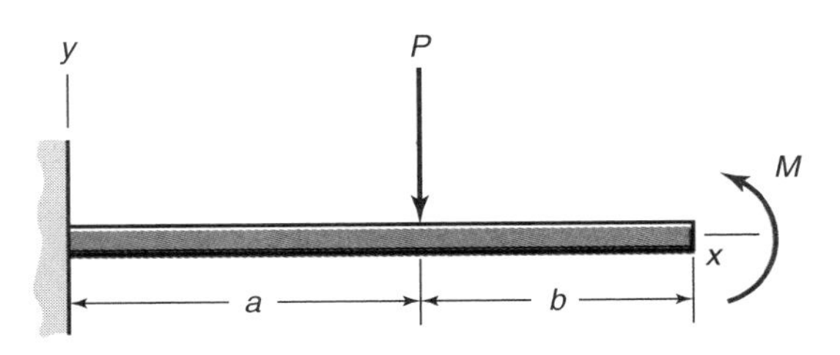

Sketch n, for Problem 5.39

5.40 A simply supported beam has loads as shown in Sketch *o*. Calculate the beam deflection by using the method of superposition. Also, calculate the maximum bending stress and the deflection at mid-span. Assume that $E = 207$ GPa and that the beam has a rectangular cross section with a height of 30 mm and a width of 100 mm. Also, P = 1200 N, $w_o = 10,000$ N/m, $a = 0.2$ m, $b = 0.1$ m, $c = 0.4$ m, and $d = 0.2$ m. *Ans.* $\sigma_{\max} = 74.47$ MPa, $y_{\max} = -0.7277$ mm.

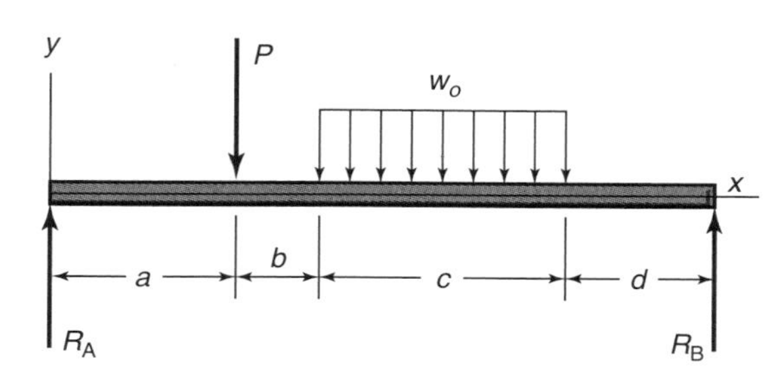

Sketch *o*, for Problem 5.40

5.41 The beam shown in Sketch *p* is fixed at both ends and center loaded with a force of 2300 N. The beam is 3.2 m long and has a square tubular cross section with an outside width of 130 mm and a wall thickness of 10 mm. The tube material is AISI 1080 high-carbon steel. Calculate the deformation along the beam by using the method of superposition. What is the deformation at mid-span? *Ans.* $y(x = 1.6 \text{ m}) = 0.1593$ mm.

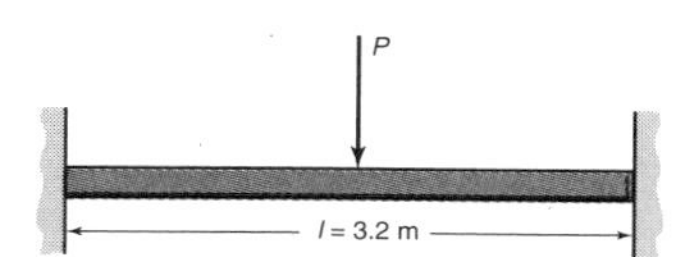

Sketch *p*, for Problem 5.41

5.42 Beam A shown in Sketch *q* is a 13-mm-diameter aluminum beam; beam B is an 8-mm-diameter steel beam. The lower member is of uniform cross section and is assumed to be rigid. Find the distance x if the lower member is to remain horizontal. Assume that the modulus of elasticity for steel is three times that for aluminum. *Ans.* $x = 0.564$ m.

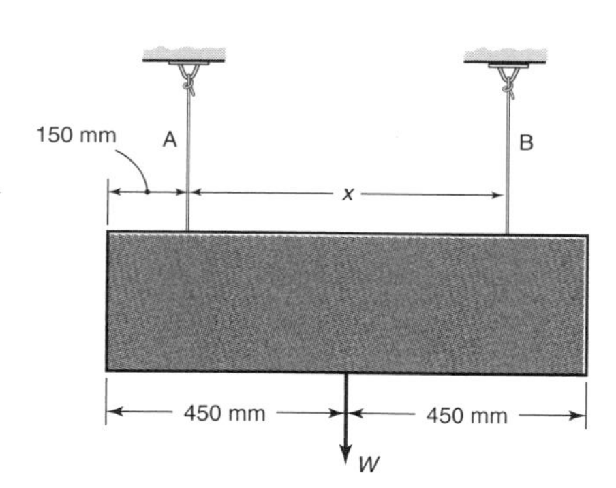

Sketch *q*, for Problem 5.42

5.43 An aluminum rod 3/4 in. in diameter and 48 in. long and a nickel steel rod 1/2 in. in diameter and 32 in. long are spaced 60 in. apart and fastened to a horizontal beam that carries a 2000-lbf load, as shown in Sketch *r*. The beam is to remain horizontal after load is applied. Assume that the beam is weightless and absolutely rigid. Find the location x of the load and determine the stresses in each rod. *Ans.* $x = 40.0$ in., $\sigma_A = 1510$ psi, $\sigma_B = 6741$ psi.

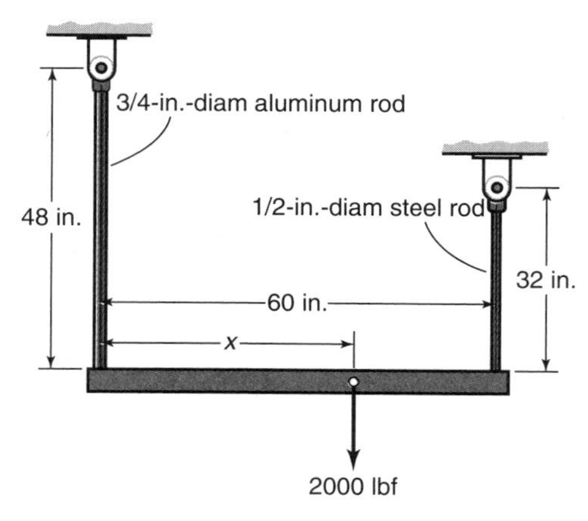

Sketch *r*, for Problem 5.43

5.44 Find the force on each of the vertical wires shown in Sketch *s*. The weight is assumed to be rigid and horizontal, implying that the three vertical bars are connected to the weight in a straight line. Also, assume that the support at the top of the bars is rigid. The bar materials and its circular cross-sectional area are given in the sketch. *Ans.* $P_s = 1785$ lb, $P_B = 1430$ lb.

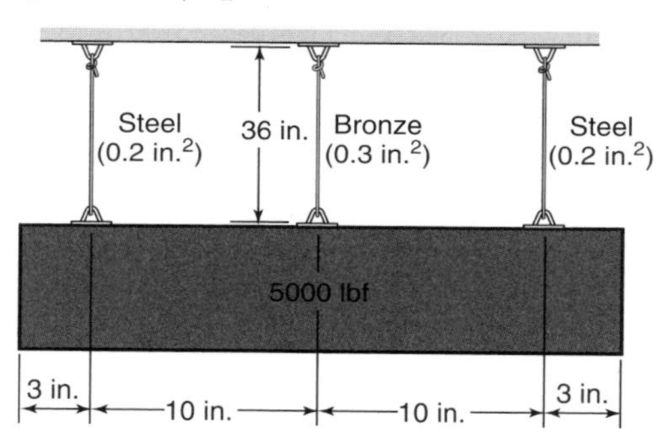

Sketch *s*, for Problem 5.44

5.45 Two solid spheres, one made of aluminum alloy 2014 and the other made of AISI 1040 medium-carbon steel, are lowered to the bottom of the sea at a depth of 10,000 m. Both spheres have a diameter of 0.3 m. Calculate the elastic energy stored in the two spheres when they are at the bottom of the sea if the density of water is 1000 kg/m^3 and the acceleration of gravity is 9.807 m/s^2. Also, calculate how large the steel sphere has to be to have the same elastic energy as the 0.3-m-diameter aluminum sphere. *Ans.* $U_{al} = 960.6$ Nm, $U_s = 393.1$ Nm.

5.46 Use Castigliano's approach instead of singularity functions to determine the maximum deflection of the beam considered in Problem 5.25. Assume that transverse shear is negligible.

5.47 Using Castigliano's Theorem, find the maximum deflection of the two-diameter cantilevered beam shown in Sketch *t*. Neglect transverse shear. *Ans.* $y = \dfrac{3Pl^3}{16EI}$.

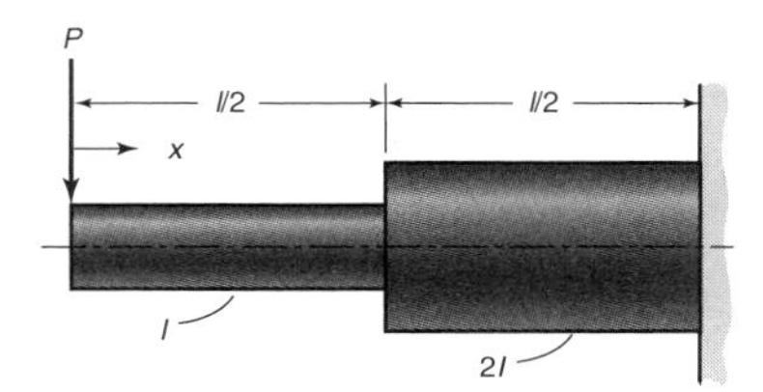

Sketch t for Problem 5.47

5.48 The right-angle-cantilevered bracket shown in Sketch u is loaded with force P in the z-direction. Derive an expression for the deflection of the free end in the z-direction by using Castigliano's Theorem. Neglect transverse shear effects.

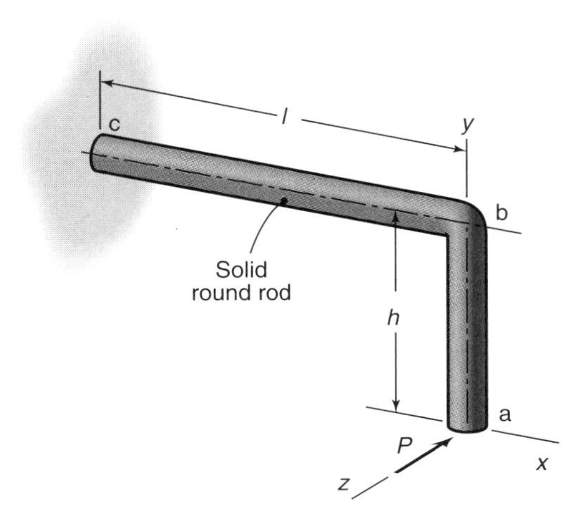

Sketch u for Problem 5.48

5.49 A triangular cantilevered plate is shown in Sketch v. Use Castigliano's Theorem to derive an expression for the deflection at the free end, assuming that transverse shear is neglected. *Ans.* $\delta = \dfrac{6Pl^3}{Ebh^3}$.

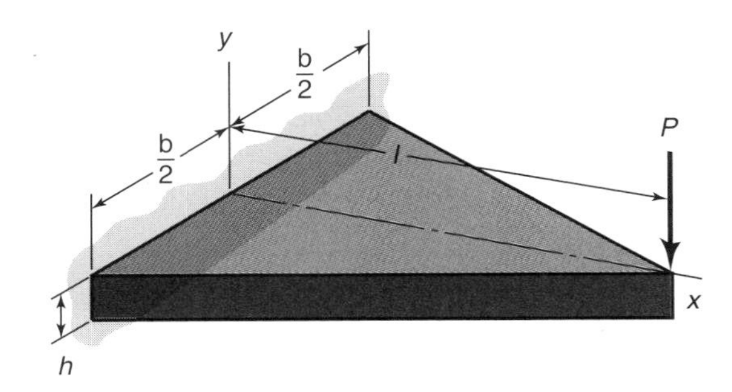

Sketch v for Problem 5.49

5.50 A right-angle-cantilevered bracket with concentrated load and torsional loading at the free end is shown in Sketch w. Using Castigliano's Theorem, find the deflection at the free end in the z-direction. Neglect transverse shear effects.

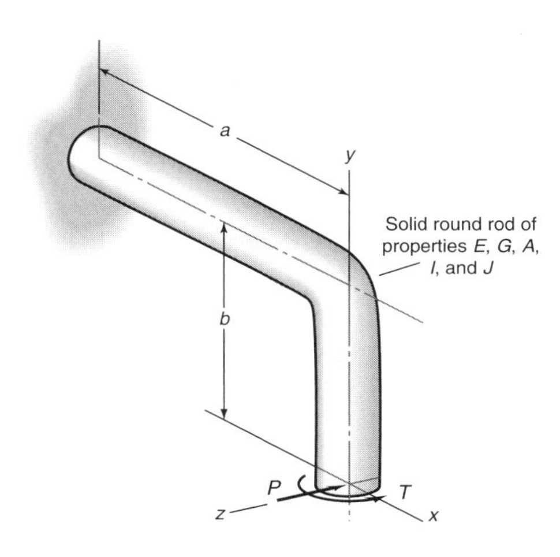

Sketch w for Problem 5.50

5.51 A cantilevered I-beam has a concentrated load applied to the free end as shown in Sketch x. What upward force at point S is needed to reduce the deflection at S to zero? Use Castigliano's Theorem. Transverse shear can be neglected. *Ans.* $S_y = 10$ kN.

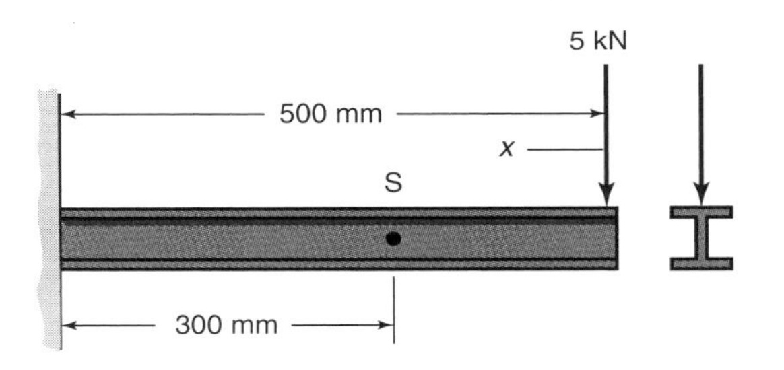

Sketch x for Problem 5.51

5.52 Using Castigliano's Theorem calculate the horizontal and vertical deflections at point A shown in Sketch y. Assume that E and A are constant.

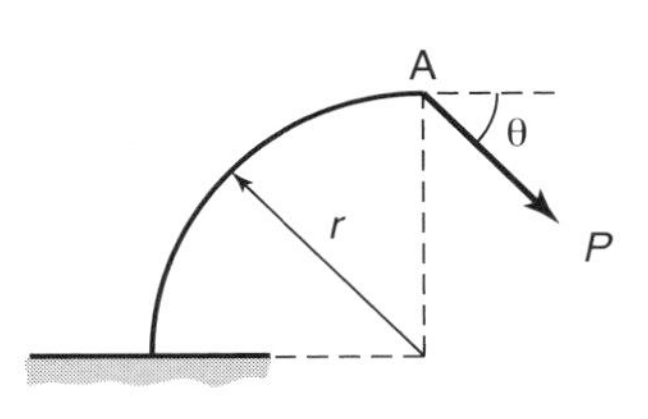

Sketch y for Problem 5.52

5.53 Calculate the deflection at the point of load application and in the load direction for a load applied as shown in Sketch z. Assume that E and I are constant.

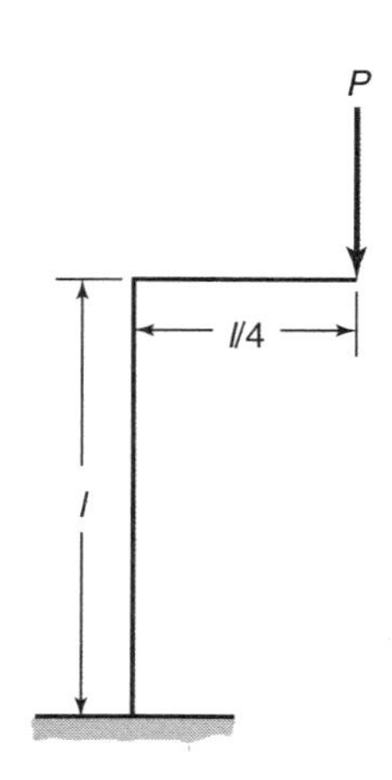

Sketch z, for Problem 5.53

5.54 Using Castigliano's Theorem determine the horizontal and vertical deflections at point A of Sketch aa. Assume that E and I are constant.

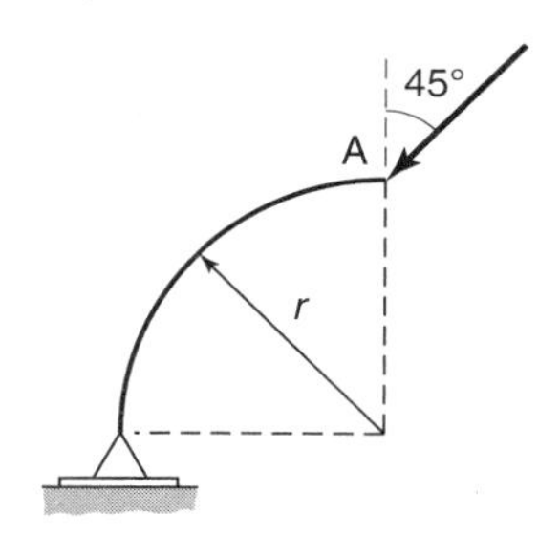

Sketch aa for Problem 5.54

5.55 For the structure shown in Sketch bb find the force in each member and determine the deflection at point A. Assume that E and A are the same in each member.

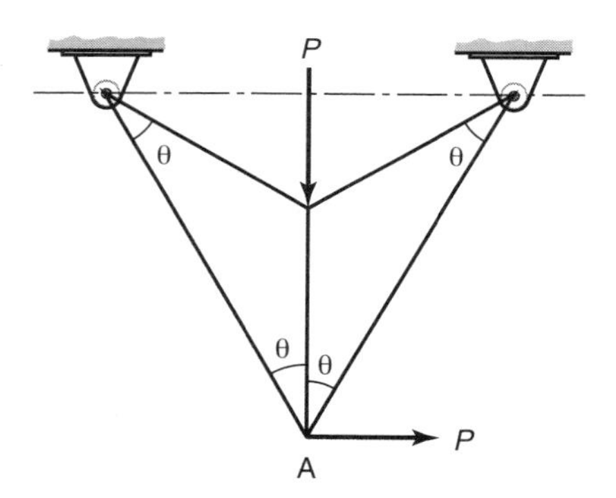

Sketch bb for Problem 5.55

5.56 Obtain the maximum deflection of the beam given in Problem 2.59 as a function of the load P, span l, moment of inertia, I, and elastic modulus, E. *Ans.* $y_{\max} = -0.2292Pl^3$.

Design and Projects

5.57 Construct a Design Procedure for the Method of Superposition.

5.58 Design an experiment to verify the energy stored in a beam is as given in Table 5.1.

5.59 Sketch cc shows a split ring used as a compression ring on an automotive piston (see also Fig. 12.31). To install the ring, it is necessary to open a gap δ as shown. If EI of the cross section is constant, derive the required force P as a function of ring radius. *Ans.* $P = \dfrac{\delta EI}{3\pi r^3}$.

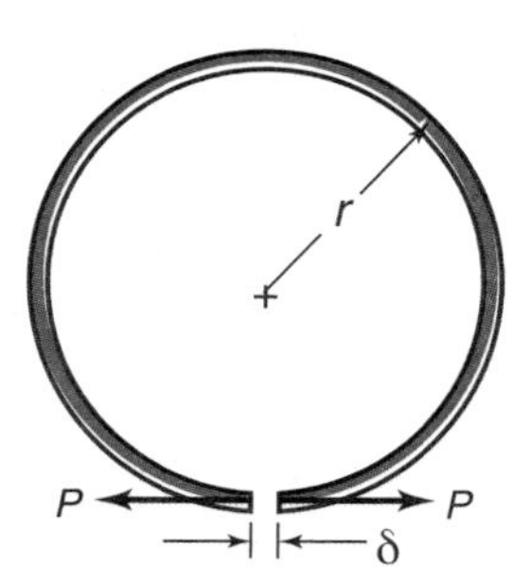

Sketch cc, for Problem 5.59

5.60 Consider the situation of a golf club striking a golf ball. Set up a system of equations to describe the deformation and motion of the golf ball.

5.61 Table 5.1 has an entry for the strain energy associated with transverse shear for a rectangular beam. Why is there no entry for circular cross-sections? Explain.

5.62 Assume you are the instructor of a course covering the subject matter in this chapter. Prepare two qualitative and two quantitative problems for the chapter and provide solutions.

Chapter 6

Failure Prediction for Static Loading

The liberty bell, a classic case of brittle fracture. *Source:* Shutterstock.

The concept of failure is central to the design process, and it is by thinking in terms of obviating failure that successful designs are achieved.

Henry Petroski, *Design Paradigms*

Contents

Examples

Design Procedures

Case Study

Static failure involves excessive deformation or fracture under relatively constant loads, often complicated by the presence of geometric or material discontinuities called stress concentrations. Stress concentrations commonly occur when the cross section of a part changes over a short distance, so that the largest stress encountered is higher than the nominal stress. The chapter presents stress concentration factors for the common and important situations encountered in design. The chapter then investigates fracture mechanics, including the modes of crack displacement and the circumstances under which a crack can propagate across a part and cause its failure, including the importance of a material's fracture toughness. Uniaxial and multiaxial failure theories are then presented to provide criteria for failure (yielding, crumbling, or fracture) for different materials, and discusses the conditions when the different failure criterion are used.

Symbols

A	area, m^2
a	half of crack length, m
b	plate width, m
c	distance from neutral axis to outer fiber, m
D	major diameter, m
d	diameter, m
H	major height, m
h	minor height, m
I	area moment of inertia, m^4
J	polar area moment of inertia, m^4
K_c	stress concentration factor
K_{Ic}	fracture toughness, MPa$\sqrt{m}$
K_i	stress intensity factor, MPa$\sqrt{m}$
l	length, m
M	bending moment, N·m
n_s	safety factor
P	force, N
q	volume flow rate, m^3/s
r	radius, m
S_{uc}	ultimate stress in compression, Pa
S_{ut}	ultimate stress in tension, Pa
S_y	yield stress, Pa
S_{yt}	yield stress in tension, Pa
T	torque, N·m
u	velocity, m/s
Y	geometry correction factor
σ	normal stress, Pa
σ_e	von Mises stress, Pa
$\sigma_1, \sigma_2, \sigma_3$	principal normal stresses, Pa
τ	shear stress, Pa

6.1 Introduction

This chapter considers static loads, where the load is gradually applied and equilibrium is reached in a relatively short time. Under static loads, a machine element may often fail at sites of local stress concentration caused by geometrical or microstructural discontinuities. Calculation of stress concentration factors is examined in detail in Section 6.2, and the mechanics are explained through a flow analogy. Another failure mode that is often encountered is fracture, whereby cracks within a microstructure grow or propagate in an uncontrolled fashion, leading to failure. Fracture mechanics is a technique used to determine the stress level at which preexisting cracks of known size will propagate, or the largest allowable crack for a given stress and material can be estimated. Geometric factors and a material property called fracture toughness are used in the theory. The chapter ends with failure criteria for both uniaxial and multiaxial stress states. The most common failure criteria in engineering practice are presented, and their applicability to different materials is discussed.

6.2 Stress Concentration

Stresses at or near a *discontinuity*, such as a hole in a plate, are higher than if the discontinuity did not exist. Figure 6.1 shows a rectangular bar with a hole under an axial load. The stress is largest near the hole; therefore, failure will first occur at the hole. The same can be deduced for any other discontinuity, such as a fillet (a narrowing in the width of a plate), a notch (a sharp groove or cut especially intended to initiate failure), an inclusion (such as a discontinuous fiber in a polymer matrix), or in the vicinity of applied loads.

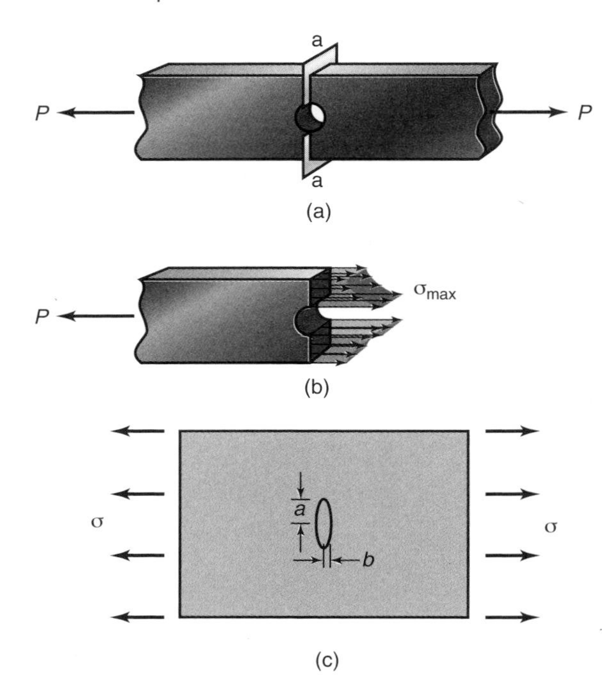

Figure 6.1: Rectangular plate with hole subjected to axial load. (a) Plate with cross-sectional plane. (b) Half of plate showing stress distribution. (c) Plate with elliptical hole subjected to axial load.

A **stress raiser** is any discontinuity in a part that alters the stress distribution so that the elementary stress equations described in Ch. 4 no longer describe the stress in the part. The **stress concentration** is the region where stress raisers are present. The **stress concentration factor**, K_c, is the factor used to relate the actual maximum stress at the discontinuity to the average or nominal stress:

$$K_c = \frac{\sigma_{\text{max}}}{\sigma_{\text{nom}}}. \tag{6.1}$$

The stress concentration factor allows consideration of stress raisers without excessively complicating the mathematics. The value of K_c is usually determined by some experimental technique, such as photoelastic analysis of a plastic model of a part, or by analytical or numerical simulation of the stress field.

An understanding of stress concentrations can be obtained by considering the example of an elliptical hole in a plate loaded in tension, as depicted in Fig. 6.1c. The theoretical stress concentration at the edge of the hole is given by

$$K_c = 1 + 2\left(\frac{a}{b}\right). \tag{6.2}$$

where a is the half-length of the ellipse transverse to the stress direction and b is the half-width in the direction of applied stress. Note that as b approaches zero, the ellipse becomes sharper, and the situation approaches that of a very sharp crack. If the crack is sharp and has a tip radius of ρ, the theoretical stress concentration factor is given by

$$K_c = 1 + 2\left(\sqrt{\frac{a}{\rho}}\right). \tag{6.3}$$

For this case, the stress at the edge of the crack is very large (infinite in the limiting case of $\rho = 0$). As either a approaches zero or b or ρ becomes very large, the effects of the stress concentration become smaller. Thus, the *size* and *orientation* of geometric discontinuities with respect to applied stress play a large role in determining the stress concentration. For more complicated situations, additional factors will play a role, as discussed in the following section.

It should also be noted that stress concentrations can arise from abrupt variations in material properties such as elastic modulus, thermal expansion coefficient, or thermal conductivity. The greater the change in properties in the area of interest, the higher the stress concentration. Thus, the design goal is to try to use constant geometries and use materials that have uniform properties. When this is not practical, the changes should be as gradual as possible.

6.2.1 Stress Concentration Factor Charts

As discussed above, the stress concentration factor is a function of the type and shape of the discontinuity (hole, fillet, or groove), and the type of loading being experienced. Consideration here will be limited to only two geometries, a flat plate and a round bar. Figures 6.2 to 6.4 display the stress concentration factor due to bending and axial load for a flat plate with a hole, fillet, or groove. These figures also give the expressions for nominal stresses. Many of these curves are developed from photoelastic studies. Note from Fig. 6.2a that a small hole in a plate loaded in tension ($d/b \to 0$) leads to $K_c = 3.0$, which is consistent with Eq. (6.2). Figures 6.5 and 6.6 show the stress concentration factor for a round bar with a fillet and a groove, respectively. Figure 6.7 shows the stress concentration factor for a flat groove, such as is used as a seat for retaining rings (see Section 11.7). Figure 6.8 shows the effect of a radial hole in a shaft. These examples are by no means all the possible geometries, but are those most often encountered in practice; for other geometries, refer to Pilkey and Pilkey [2008] or Young and Budynas [2001].

From these figures, the following observations can be made about stress concentration factors:

1. The stress concentration factor, K_c, is *independent* of material properties.

2. K_c is significantly affected by part geometry. Note that as the radius of the discontinuity is decreased, the stress concentration is increased.

3. K_c is also affected by the type of discontinuity; the stress concentration factor is considerably lower for a fillet (Fig. 6.3) than for a hole (Fig. 6.2).

The stress concentration factors given in Figs. 6.2 to 6.8 were determined on the basis of static loading, with the additional assumption that the stress in the material does not exceed its proportional limit. In practice, this is usually approximated by the yield stress. If the material is brittle, the proportional limit is the rupture stress, so failure for this part will begin at the point of stress concentration when the proportional limit is reached. It is thus important to apply stress concentration factors when using brittle materials. On the other hand, if the material is ductile and subjected to a static load, designers often neglect stress concentration factors, since a stress that exceeds the proportional limit will not result in a crack. Instead, the ductile material will flow plastically and can strain harden. Furthermore, as a material yields near a stress concentration, deformation results in blunting of notches, so that the stress concentration is reduced. In applications where stiff designs and tight tolerances are essential, stress concentration should be considered regardless of material ductility.

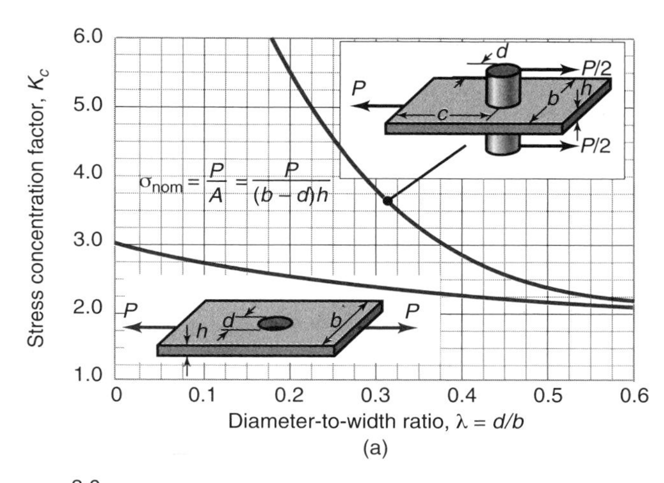

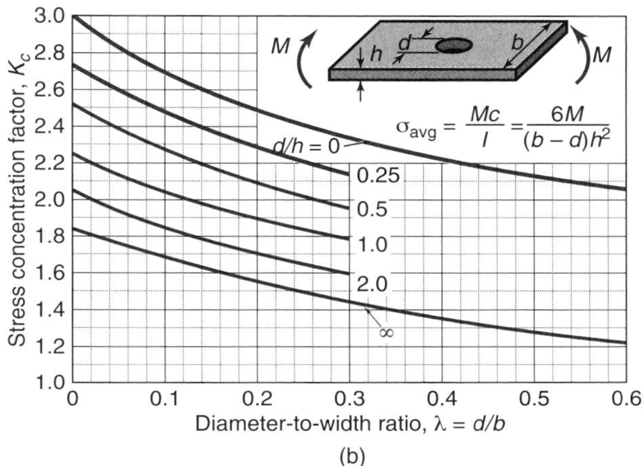

Figure 6.2: Stress concentration factors for rectangular plate with central hole. (a) Axial load and pin-loaded hole; (b) bending.

Example 6.1: Theoretical Stress Concentration Factor

Given: A flat plate made of a brittle material and a width of $b = 20$ mm, a major height of $H = 100$ mm, a minor height of $h = 50$ mm, and a fillet radius of $r = 10$ mm.

Find: The stress concentration factor and the maximum stress for the following conditions:

(*a*) Axial loading with $P = 10,000$ N

(*b*) Pure bending with $M = 100$ Nm

(*c*) Axial loading of $P = 10,000$ N, with fillet radius reduced to 5 mm.

Solution:

(*a*) Axial loading. Note from the geometry that

$$\frac{H}{h} = \frac{100}{50} = 2.0.$$

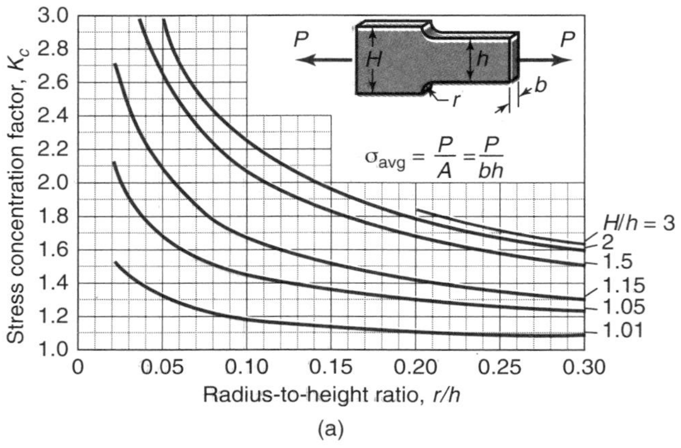

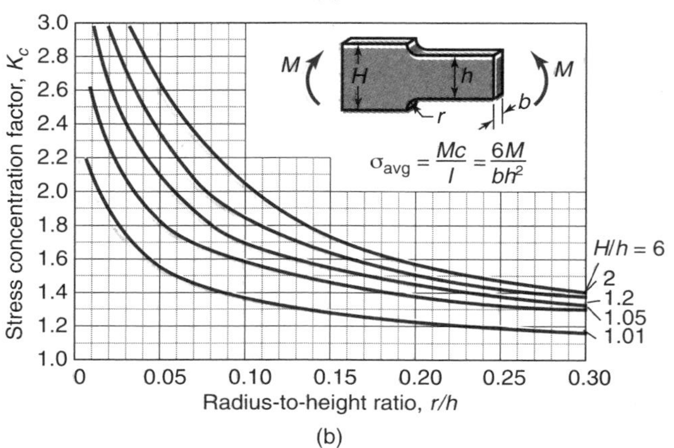

Figure 6.3: Stress concentration factors for rectangular plate with fillet. (a) Axial load; (b) bending.

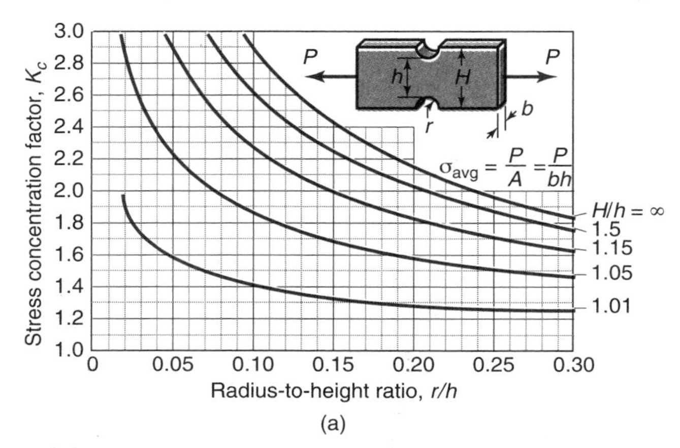

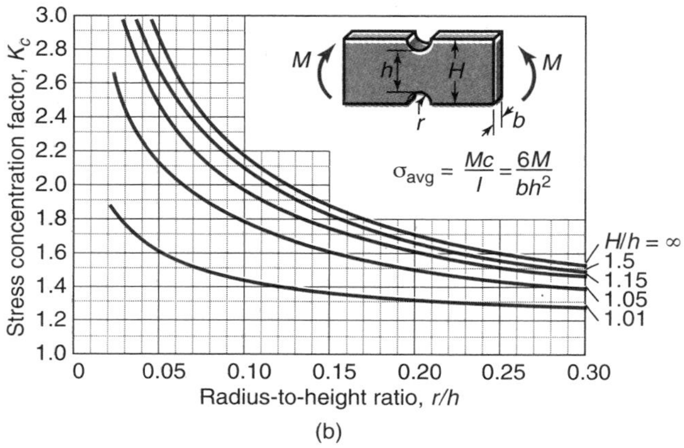

Figure 6.4: Stress concentration factors for rectangular plate with groove. (a) Axial load; (b) bending.

$$\frac{H}{h} = \frac{100}{50} = 2.0.$$

Also,

$$\frac{r}{h} = \frac{10}{50} = 0.2.$$

From Fig. 6.3a, $K_c = 1.8$. From Eq. (6.1) and Fig. 6.3a, the maximum stress is

$$\sigma_{\max} = 1.8\sigma_{\text{avg}} = \frac{1.8(10,000)}{(0.02)(0.05)} = 18 \text{ MPa}.$$

(b) Pure bending. From Fig. 6.3b, $K_c = 1.5$. The maximum stress is

$$\sigma_{\max} = 1.5\frac{6M}{bh^2} = \frac{9(100)}{(0.02)(0.05)^2} = 18 \text{ MPa}.$$

(c) Axial loading but with fillet radius changed to 5 mm. For this case

$$\frac{r}{h} = \frac{5}{50} = 0.1.$$

From Fig. 6.3a, $K_c = 2.1$. The maximum stress is

$$\sigma_{\max} = \frac{2.1P}{bh} = \frac{(2.1)(10,000)}{(0.02)(0.05)} = 21 \text{ MPa}.$$

Thus, reducing the fillet radius by one-half increases the maximum stress by around 17%.

Example 6.2: Allowable Loads in the Presence of a Stress Concentration

Given: A 50-mm-wide, 5-mm-high rectangular plate has a 5-mm-diameter central hole. The allowable tensile stress is 700 MPa.

Find:

(a) The maximum tensile force that can be applied.

(b) The maximum bending moment that can be applied to reach the maximum stress.

(c) The maximum tensile force and the maximum bending moment if the hole is not present. Express the results as a ratio when compared to parts *(a)* and *(b)*.

Solution:

(a) The diameter-to-width ratio is $d/b = 5/50 = 0.1$. The cross-sectional area with the hole is

$$A = (b-d)h = (50-5)5 = 225 \text{ mm}^2 = 0.225\times10^{-3} \text{ m}^2.$$

From Fig. 6.2a for $d/b = 0.1$ the stress concentration factor $K_c = 2.70$ for axial loading. The maximum force is

$$P_{\max} = \frac{(700 \times 10^6)(0.225 \times 10^{-3})}{2.70} = 58,330 \text{ N}.$$

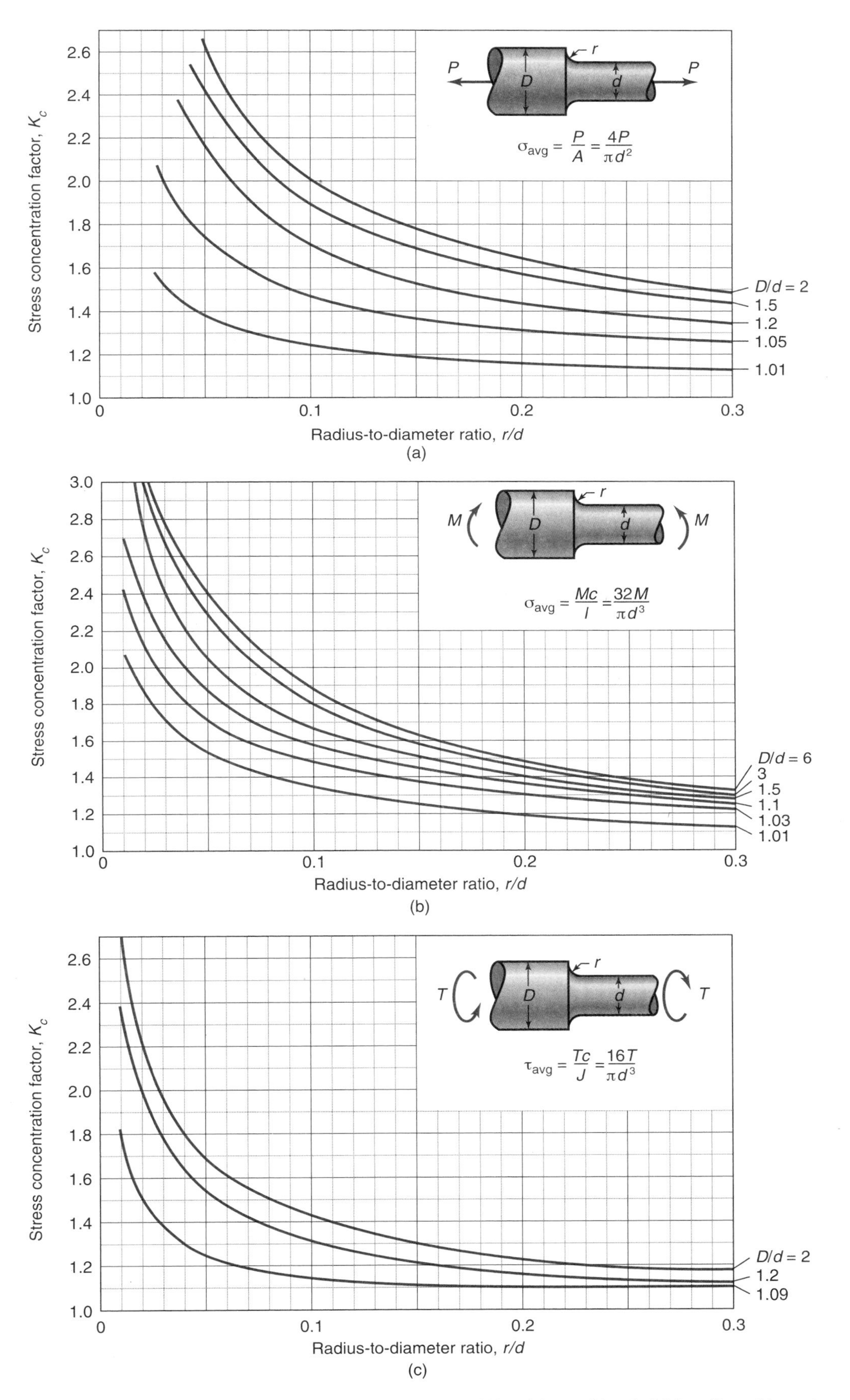

Figure 6.5: Stress concentration factors for round bar with fillet. (a) Axial load; (b) bending; (c) torsion.

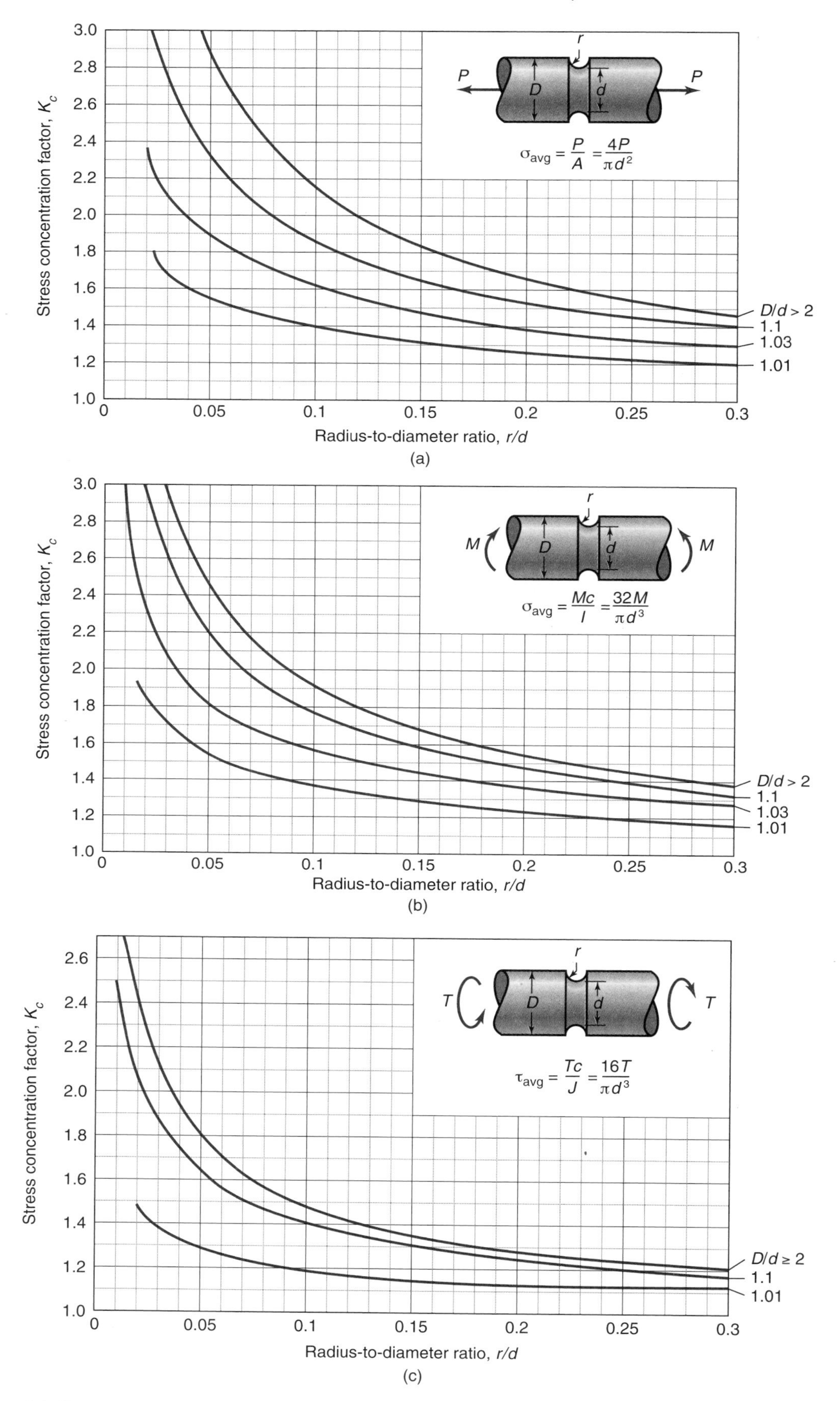

Figure 6.6: Stress concentration factors for round bar with groove. (a) Axial load; (b) bending; (c) torsion.

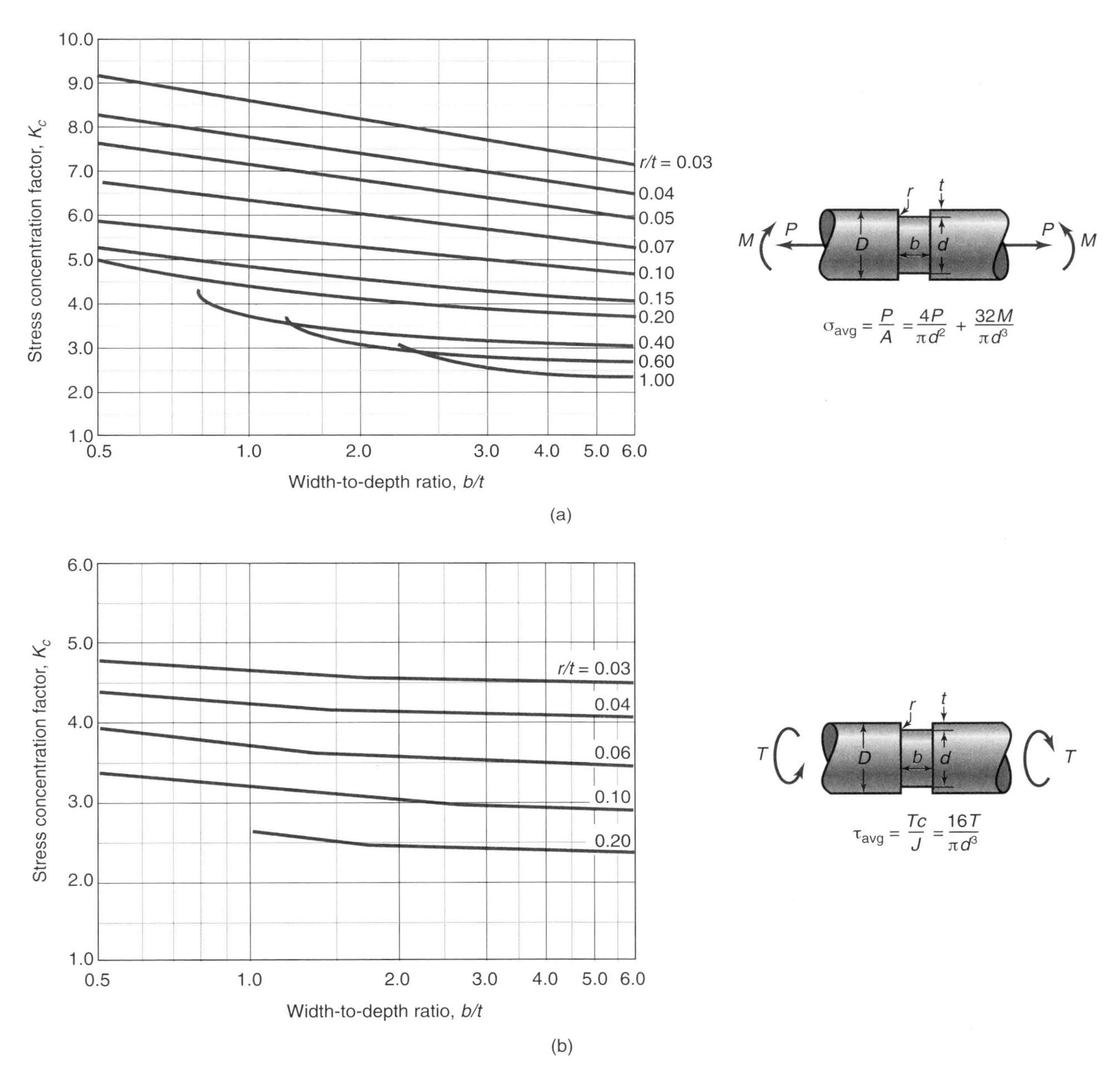

Figure 6.7: Stress concentration factors for round bar with a flat groove.

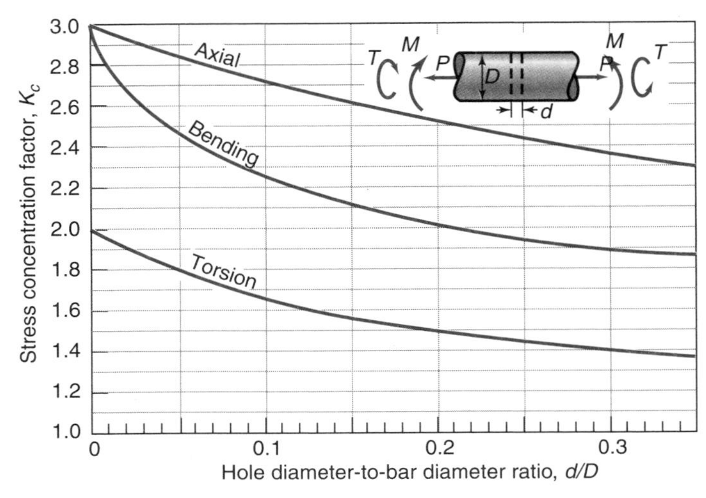

Nominal stresses:

Axial load:

$$\sigma_{avg} = \frac{P}{A} = \frac{P}{(\pi D^2/4) - Dd}$$

Bending (plane shown is critical):

$$\sigma_{nom} = \frac{Mc}{I} = \frac{M}{(\pi D^3/32) - (dD^2/6)}$$

Torsion:

$$\tau_{avg} = \frac{Tc}{J} = \frac{T}{(\pi D^3/16) - (dD^2/6)}$$

Figure 6.8: Stress concentration factors for round bar with hole.

(*b*) From Fig. 6.2b for bending when d/b = 0.1 and d/h = 5/5 = 1, the stress concentration factor is $K_c = 2.04$. The maximum bending moment is

$$\begin{aligned} M_{\max} &= \frac{(b-d)h^2\sigma_{\text{all}}}{6K_c} \\ &= \frac{(0.225\times10^{-3})(5\times10^{-3})(700\times10^{6})}{6(2.04)} \\ &= 64.34 \text{ Nm}. \end{aligned}$$

(*c*) The cross-sectional area without the hole is

$$A = bh = (50)5 = 250 \text{ mm}^2 = 0.250 \times 10^{-3} \text{ m}^2.$$

Therefore,

$$P_{\max} = \sigma_{\text{all}}A = \left(700\times10^{6}\right)\left(0.250\times10^{-3}\right) = 175 \text{ kN}.$$

The force ratio is 175/58.33=3.00. For bending,

$$\begin{aligned} M_{\max} &= \frac{\sigma_{\text{all}}bh^2}{6} = \frac{\sigma_{\text{all}}Ah}{6} \\ &= \frac{(700\times10^{6})(0.25\times10^{-3})(5\times10^{-3})}{6} \\ &= 145.8 \text{ Nm}. \end{aligned}$$

The bending moment ratio is 145.8/64.34 = 2.266.

6.2.2 Flow Analogy

Good practice drives the designer to reduce or eliminate stress concentrations as much as is practical. Recommending methods of reducing the stress concentration requires better understanding of what occurs at the discontinuity to increase the stress. One way of achieving this understanding is to observe similarity between the velocity of fluid flow in a channel and the stress distribution of an axially loaded plate when the channel dimensions are comparable to the size of the plate. The analogy is accurate, since the equations of flow potential in fluid mechanics and stress potential in solid mechanics are of the same form.

If the channel has constant dimensions throughout, the velocities are uniform and the streamlines are equally spaced. Similarly, for a bar of constant dimensions under axial load, the stresses are uniform and stress contours are equally spaced. At any point within the channel the flow must be constant, where the volume flow is

$$q = \int u \, dA. \tag{6.4}$$

From solid mechanics, the force must be constant at any location in the plate,

$$P = \int \sigma \, dA. \tag{6.5}$$

If the channel section changes sharply, the flow velocity increases near the shape change and, in order to maintain equal flow, the streamlines must narrow and crowd together. In a stressed member of the same cross section, the increase in stress is analogous to the increase in fluid velocity, or inversely to the change in the spacing of the streamlines. Figure 6.9a shows the stress distribution around the sharp corners of an axially loaded flat plate. The situation in Fig. 6.9a produces a stress concentration factor greater than 3. Recall from Eq. (6.1) that this implies that the maximum stress is more than three times greater than the average stress. However, this stress concentration factor can be reduced, typically from 3 down to around 1.5, by rounding the corners as shown in Fig. 6.9b. A still further reduction in stress concentration factor can be achieved by introducing small grooves or holes, as shown in Fig. 6.9c and d, respectively. In Fig. 6.9b to d, the design helps to reduce the rigidity of the material at the corners, so that the stress and strain are more evenly spread throughout the flat plate. The improvements in these designs can be gleaned from the flow analogy.

6.3 Introduction to Fracture Mechanics

The science of flaw development and crack extension as a function of applied load is called **fracture mechanics**. A **crack**

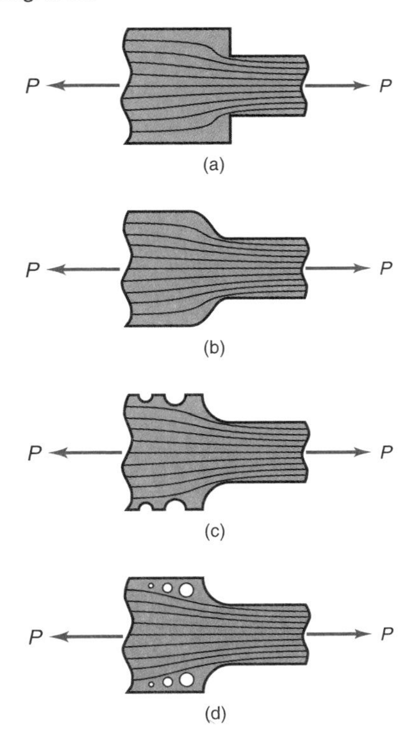

Figure 6.9: Axially loaded flat plate with fillet showing stress contours: (a) square corners, (b) rounded corners, (c) small grooves, and (d) small holes.

is an initially small (perhaps microscopic) flaw that exists under normal conditions on the surface and within the body of the material. No materials or manufacturing processes yield defect-free structures, hence such imperfections are always present. Under an applied stress, the crack can grow, resulting in a small extension; if the stress is high enough, the crack will grow rapidly and cause material failure by propagating across an entire cross section.

Lower stress is required to propagate a crack than to start one. Many consumer product packages are designed with a perforation or edge notch to allow a person to manually open the package; this is a demonstration of crack propagation. Clearly, without a perforation, such packaging would be much more difficult to open. Fracture can occur at stress levels well below the yield stress of a solid material. Fracture mechanics is a highly developed field, and is used extensively for fatigue (Ch. 7) or fatigue wear (Ch. 8). This chapter examines the critical crack length that will make a part fail under static loads. **Fracture control** consists of maintaining nominal stresses and existing cracks below a critical level for the material being used.

6.4 Modes of Crack Growth

There are three fundamental **modes of crack propagation**, as shown in Fig. 6.10:

1. *Mode I – opening.* The opening (or tensile) mode, shown in Fig. 6.10a, is the most often encountered mode of crack propagation. The crack faces separate symmetrically with respect to the crack plane.

2. *Mode II – sliding.* The sliding (or in-plane shearing) mode occurs when the crack faces slide relative to each other symmetrically with respect to the normal to the crack plane but asymmetrically with respect to the crack plane, as shown in Fig. 6.10b.

3. *Mode III – tearing.* The tearing (or antiplane) mode occurs when the crack faces slide asymmetrically with respect to both the crack plane and its normal, as illustrated in Fig. 6.10c.

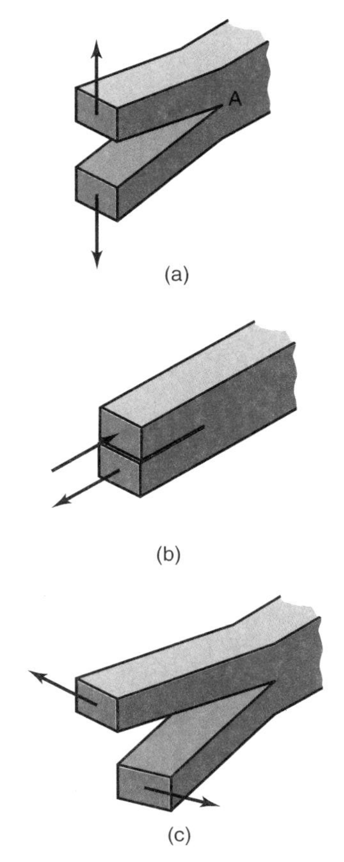

Figure 6.10: Three modes of crack displacement. (a) Mode I, opening; (b) Mode II, sliding, (c) Mode III, tearing.

The crack propagation modes are known by their Roman numeral designations given above (e.g., Mode I). Although Mode I is the easiest to visualize as a crack-propagating mechanism, applying the discussion of stress raisers in Section 6.2 to geometries such as Fig. 6.10 suggests that crack propagation will occur when stresses are higher at the crack tip than elsewhere in the solid.

6.5 Fracture Toughness

In this text, considerations of fracture toughness are restricted to Mode I crack propagation since this is the most commonly encountered failure. An important concept is that of **stress intensity factor**, K_i, which specifies the stress intensity at the crack tip (see point A in Fig. 6.10a). The SI unit of K_i is megapascals times meters$^{1/2}$ (MPa$\sqrt{\text{m}}$). The equation for stress intensity factor is

$$K_i = Y\sigma\sqrt{\pi a}, \tag{6.6}$$

where

Y = dimensionless correction factor that accounts for

geometry of the part containing a crack
σ = nominal stress, MPa
a = one-half of crack length, m

The values of Y for common cases are summarized in Appendix C. Some assumptions imposed in deriving Eq. (6.6) are that the load is applied far from the crack and that the crack length, $2a$, is small relative to the width.

Fracture toughness is the critical value of stress intensity at which crack extension occurs, and is a material property. Fracture toughness is used as a design criterion in fracture prevention for brittle materials, just as yield strength is used as a design criterion to avoid plastic deformation in ductile materials. At fracture, the stress intensity factor, given by Eq. (6.6), will equal the fracture toughness. Table 6.1 shows room-temperature yield stress and fracture toughness data for selected engineering materials. Fracture toughness is denoted as K_{Ic} to indicate that it corresponds to Mode I and that it is a critical stress intensity factor. K_{Ic} depends on many factors, with the most important being temperature, strain rate, and microstructure. The magnitude of K_{Ic} diminishes with increasing strain rate and decreasing temperature. Furthermore, enhancing yield strength by a material treating process, such as strain hardening, produces a corresponding decrease in K_{Ic}.

Design Procedure 6.1: Fracture Mechanics Applied to Design

The use of fracture mechanics in design generally involves the following steps for geometries that lend themselves to simple analysis. For more complicated geometries, fracture modes or loadings, a numerical analysis, and/or experimental design verification are usually required.

1. Given a candidate material, obtain its fracture toughness. See Table 6.1 for selected materials, or else find the value in the technical literature or from experiments.

2. The dimensionless correction factor for the part geometry, Y, can be obtained from Appendix C for common design situations. Y is tabulated for additional geometries by Pilkey and Pilkey [2008] and Young and Budynas [2001].

3. Equation (6.6) allows calculation of allowable stress as a function of semi-crack length, a; similarly, the largest allowable crack (with length $2a$) can be determined from the required stress.

4. If design criteria cannot be met, the following alternatives can be pursued:

   (a) Increasing the part thickness will reduce the nominal stress, σ_{nom}.

   (b) A different material with a higher fracture toughness can be selected.

   (c) Local reinforcement of critical areas can be pursued, such as locally increasing thickness.

   (d) The manufacturing process can have a significant impact on the initial flaw size. The class of operations (casting versus forging, compression molding versus extrusion, etc.), quality control procedures, and quality of incoming material are all important factors. See, for example, Kalpakjian and Schmid [2014] for further information.

Example 6.3: Critical Crack Length

Given: The following two materials:

(*a*) AISI 4340 steel, tempered at 260°C

(*b*) Aluminum alloy 7075–T651

Assume that the applied stress of the material is 0.8 times the yield stress and that the dimensionless correction factor, Y, is unity.

Find: The critical crack length at room temperature.

Solution:

(*a*) From Table 6.1 for AISI 4340, $S_y = 1640$ MPa and $K_{Ic} = 50$ MPa$\sqrt{\text{m}}$. Therefore,

$$\sigma = 0.8S_y = 1310 \text{ MPa}.$$

From Eq. (6.6),

$$a = \frac{1}{\pi}\left(\frac{K_{Ic}}{Y\sigma}\right)^2 = \frac{1}{\pi}\left[\frac{(50.0)}{(1)(1310)}\right]^2 = 0.000464 \text{ m}.$$

Since this is one-half the critical crack length, the critical crack length for AISI 4340 steel is 0.928 mm.

(*b*) From Table 6.1 for aluminum alloy 7075–T651, $S_y = 505$ MPa and $K_{Ic} = 29$ MPa$\sqrt{\text{m}}$. Therefore,

$$\sigma = 0.8S_y = 404 \text{ MPa}.$$

From Eq. (6.6)

$$a = \frac{1}{\pi}\left(\frac{K_{Ic}}{Y\sigma}\right)^2 = \frac{1}{\pi}\left[\frac{29}{(1)(404)}\right]^2 = 0.00164 \text{ m},$$

or a critical crack length of 3.28 mm. Note that the stronger material (the steel) has a smaller critical crack length.

Example 6.4: Allowable Crack Length

Given: A container used for compressed air is made of aluminum alloy 2024–T351. The required safety factor against yielding is 1.6, and the largest crack allowed through the thickness of the material is 6 mm. Assume $Y = 1$.

Find:

(*a*) The stress intensity factor and the safety factor guarding against brittle fracture

(*b*) Whether a higher safety factor will be achieved if the material is changed to a stronger aluminum alloy 7075–T651. Use the same crack dimensions, as this relates to an inspection capability.

Solution:

(*a*) From Table 6.1 for aluminum alloy 2024–T351, S_y = 325 MPa and K_{Ic} = 36 MPa$\sqrt{\text{m}}$. The nominal stress is

$$\sigma_{\text{nom}} = \frac{S_y}{n_s} = \frac{325}{1.6} = 203.1 \text{ MPa}.$$

The crack half-length is a = 3 mm = 0.003 m. The stress intensity factor from Eq. (6.6) is

Table 6.1: Yield stress and fracture toughness data for selected engineering materials at room temperature.

Material	Yield strength, S_y		Fracture toughness, K_{Ic}	
	ksi	MPa	ksi $\sqrt{\text{in}}$	MPa $\sqrt{\text{m}}$
Metals				
Aluminum alloys				
2014-T4	65	450	26	29
2024-T3	57	390	31	34
2024-T351	47	325	33	36
7075-T651	73	505	26	29
7079-T651	68	470	30	33
Steels				
4340 tempered at 260°C	238	1640	45.8	50.0
4340 tempered at 425°C	206	1420	80.0	87.4
D6AC, tempered at 540°C	217	1495	93	102
A538	250	1722	100	111
Titanium alloys				
Ti-6Al-4V	119	820	96	106
Ti-13V-11Cr-3Al	164	1130	25	27
Ti-6Al-6V-2S	157	1080	34	37
Ti-6Al-2Sn-4Z-6Mo	171	1180	24	26
Ceramics				
Aluminum oxide	—	—	2.7–4.8	3.0-5.3
Silicon nitride	—	—	3.5–7	4-8
Silicon carbide	—	—	1.8–4.5	2-5
Soda-lime glass	—	—	0.64–0.73	0.7–0.8
Concrete	—	—	0.18–1.27	0.2–1.4
Polymers				
Polymethyl methacrylate	3–7	20–50	0.9–2.7	1–3
Polystyrene	4.5–11.5	30–80	0.9–1.8	1–2
Polycarbonate	8.5–10	60–70	2.3–2.7	2.5–3
Polyvinyl chloride	5.8–7	40–50	1.8–2.7	2–3

$$\begin{aligned} K_i &= Y\sigma_{\text{nom}}\sqrt{\pi a} \\ &= (1)\left(203.1\times10^6\right)\sqrt{\pi(0.003)} \\ &= 19.72\ \text{MPa}\sqrt{\text{m}}. \end{aligned}$$

The safety factor for brittle fracture is then

$$n_{s,f} = \frac{K_{Ic}}{K_i} = \frac{36}{19.72} = 1.826.$$

(*b*) From Table 6.1 for the stronger aluminum alloy 7075–T651, S_y = 505 MPa and K_{Ic} = 29 MPa$\sqrt{\text{m}}$. The safety factor guarding against yielding is

$$n_{s,y} = 1.6\left(\frac{505}{325}\right) = 2.49.$$

Thus, the increased strength of 7075–T651 results in a higher safety factor guarding against yielding. The safety factor guarding against crack propagation is then

$$\frac{K_{Ic}}{K_i} = \frac{29}{19.72} = 1.47.$$

Note that the stronger material will fail more easily from crack propagation. These examples demonstrate that material strength can be counterintuitive with respect to failure, and that consideration of fracture toughness is important for fault tolerant design (see Section 7.12).

6.6 Failure Prediction for Uniaxial Stress State

Experimental data exist for uniaxial stress states and are widely available. Failure is predicted if the design stress, σ_d, is greater than the known allowable stress, σ_{all}. The safety factor from Eq. (1.1) in the uniaxial stress state can be expressed as

$$n_s = \frac{\sigma_{\text{all}}}{\sigma_d}. \tag{6.7}$$

Of course, the larger n_s is, the safer is the design, and values of $n_s < 1$ mean that a redesign is necessary.

Example 6.5: Uniaxial Failure of a Leaf Spring

Given: The leaf springs of a truck's rear wheels are loaded in pure bending (see Section 17.6). The 8-ton axle load is taken up by the two springs, giving a bending moment of 9800 N-m in each spring at the load application point. The steel used for the spring is AISI 4340, tempered at 260°C. The dimensions of the leaf spring are such that the width is 10 times the thickness. Assume a safety factor of 5.

Find: The cross section of the leaf spring.

Solution: From Table 6.1 for AISI 4340 steel tempered at 260°C, S_y = 1640 MPa. The design stress from Eq. (6.7) is

$$\sigma_d = \frac{\sigma_{\text{all}}}{n_s} = \frac{1640}{5} = 328\ \text{MPa}. \tag{a}$$

From Table 4.1 for a rectangular section,

$$I = \frac{bh^3}{12} \quad \text{and} \quad c = h/2. \tag{b}$$

It is given that $b = 10h$. Substituting Eq. (b) into Eq. (4.45) gives the magnitude of the bending design stress as

$$\sigma_d = \frac{Mc}{I} = \frac{9800\,(h/2)}{(10h)h^3/12} = \frac{5880}{h^3}. \tag{c}$$

Making use of Eq. (a) gives

$$328 \times 10^6 = \frac{5880}{h^3},$$

or

$$h^3 = \frac{5880}{328 \times 10^6} = 17.93 \times 10^{-6}\ \text{m}^3,$$

so that

$$h = 0.026\ \text{m} = 26\ \text{mm}.$$

The cross section of the leaf spring is thus 26×260 mm.

6.7 Failure Prediction for Multiaxial Stress State

A multiaxial stress state can be biaxial or triaxial. In practice, it is difficult to devise experiments to cover every possible combination of critical stresses because tests are expensive and a large number is required to obtain results with good confidence. Therefore, a theory is needed that compares the normal and shear stresses σ_x, σ_y, σ_z, τ_{xy}, τ_{yz}, and τ_{xz} with the uniaxial stress, for which experimental data are relatively easy to obtain. Several failure prediction theories are presented for a multiaxial stress state while under static loading.

6.7.1 Ductile Materials

As discussed in Section 3.2, most metals and thermoplastic polymers are considered to be *ductile*. There are exceptions, however: metal castings are not as ductile as wrought or cold-worked parts and can thus behave in a brittle manner. Ductile materials typically have the same tensile strength as compressive strength and are not as susceptible to stress raisers as are brittle materials. For the purposes of this text, a ductile material is considered to have failed when it yields. Although in some applications a small amount of plastic deformation may be acceptable, this is rarely the case in machinery elements. Two theories of yield criteria are presented next: the maximum-shear-stress theory and the distortion-energy theory.

Maximum-Shear-Stress Theory

The **maximum-shear-stress theory** (MSST) was first proposed by Coulomb [1773] but was independently developed by Tresca [1868] and is therefore often called the **Tresca yield criterion**. Tresca noted that plastically deforming platinum exhibited bright shear bands under small strains, indicating that metals deformed under shear in all circumstances and that the shear was highly localized on well-defined planes. His observations led to the MSST, which states that a part subjected to any combination of loads will fail (by yielding or fracturing) whenever the maximum shear stress exceeds a critical value. The critical value can be determined from standard uniaxial tension tests. Experimental evidence verifies

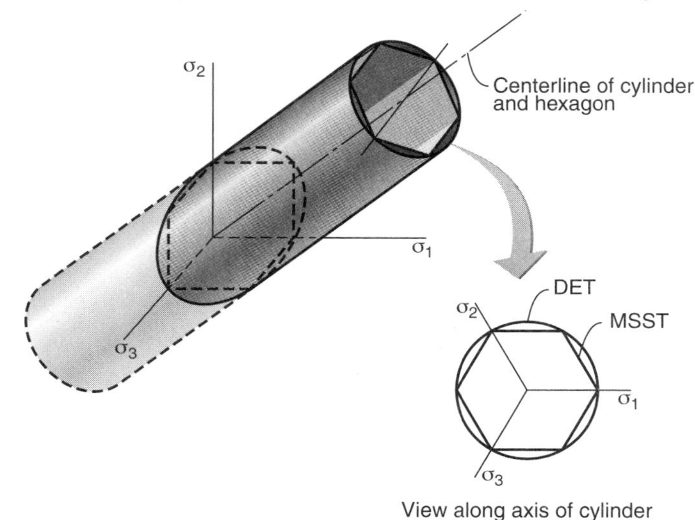

Figure 6.11: Three-dimensional yield locus for MSST and DET.

that the MSST is a good theory for predicting yielding of ductile materials, and it is a common approach in design. If the nomenclature $\sigma_1 \geq \sigma_2 \geq \sigma_3$ is used for the principal stresses in accordance with Eq. (2.18), the maximum shear stress theory predicts yielding when

$$\sigma_1 - \sigma_3 = \frac{S_y}{n_s}, \tag{6.8}$$

where S_y is the yield strength of the material and n_s is the safety factor.

For a three-dimensional stress state, the maximum-shear-stress theory provides an envelope describing the stress combinations that cause yielding, as illustrated in Fig. 6.11. The curve, defined by a yield criterion, is known as a *yield locus*. Any stress state in the interior of the yield locus results in elastic deformation. Points outside the yield locus are not possible because such stress states would cause yielding in the solid before these stresses could be attained. Stress states outside the yield locus cannot be supported by the material. If a situation arises that would increase the material strength (such as strain rate effects or work hardening), the yield locus expands, so that fracture may not necessarily be the result.

It is helpful to present the yield criterion in a plane stress circumstance, for which there will be two principal stresses in the plane as well as a principal stress equal to zero perpendicular to the plane. Figure 6.12 graphically depicts failure prediction in the plane stress state by the maximum-shear-stress theory. The principal stresses used in the figure are labeled as σ_1 and σ_2, but the ordering of normal stresses ($\sigma_1 \geq \sigma_2 \geq \sigma_3$) is not being enforced. Remember that the principal stress outside the page is zero. In the first quadrant, where σ_1 and σ_2 are by definition positive, this means that the value of σ_3 in Eq. (6.8) would be zero and that yielding would occur whenever σ_1 or σ_2 reached the uniaxial yield strength, S_y. In the second quadrant, where σ_1 is negative and σ_2 is positive, Eq. (6.8) would result in a line as shown in Fig. 6.12. The third and fourth quadrants of the curve follow the same reasoning in their development.

Distortion-Energy Theory

The **distortion-energy theory** (DET), also known as the **von Mises criterion**, postulates that failure is caused by the elas-

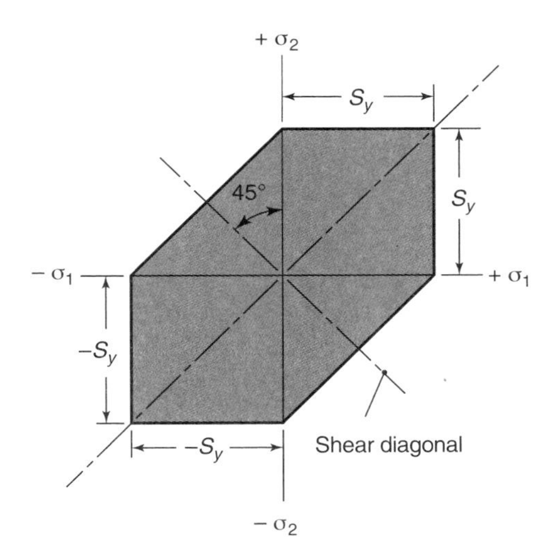

Figure 6.12: Graphical representation of maximum-shear-stress theory (MSST) for biaxial stress state ($\sigma_z = 0$).

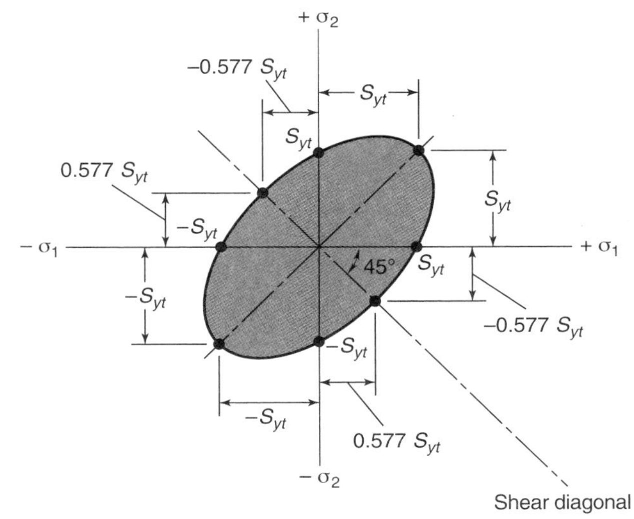

Figure 6.13: Graphical representation of distortion-energy theory (DET) for biaxial stress state ($\sigma_z = 0$).

tic energy associated with shear deformation in the material. This theory is valid for ductile materials and predicts yielding under combined loading and works well (although the differences between the DET and the MSST are small).

The DET can be derived mathematically in a number of ways, but one of the more straightforward is based on the hypothesis that yielding occurs when the root mean shear stress exceeds a critical value. Mathematically, this can be expressed as

$$\left[(\sigma_1-\sigma_2)^2+(\sigma_2-\sigma_3)^2+(\sigma_3-\sigma_1)^2\right]^{1/2}=\text{Constant.} \tag{6.9}$$

Since in uniaxial yielding, $\sigma_1 = S_y$ and $\sigma_2 = \sigma_3 = 0$, the constant in Eq. (6.9) is evaluated as $\sqrt{2}S_y$. Eq. (6.9) then becomes

$$\frac{1}{\sqrt{2}}\left[(\sigma_1-\sigma_2)^2+(\sigma_2-\sigma_3)^2+(\sigma_3-\sigma_1)^2\right]^{1/2}=S_y. \tag{6.10}$$

Incorporating the safety factor, Eq. (6.10) becomes

$$\sigma_e=\frac{1}{\sqrt{2}}\left[(\sigma_1-\sigma_2)^2+(\sigma_2-\sigma_3)^2+(\sigma_3-\sigma_1)^2\right]^{1/2}=\frac{S_y}{n_s}, \tag{6.11}$$

where σ_e is the **von Mises stress**. For a biaxial stress state, assuming $\sigma_3 = 0$, the von Mises stress is given by

$$\sigma_e=\left(\sigma_1^2+\sigma_2^2-\sigma_1\sigma_2\right)^{1/2}=\left(\sigma_x^2+\sigma_y^2-\sigma_x\sigma_y+3\tau_{xy}^2\right)^{1/2}. \tag{6.12}$$

The DET yield locus is shown in Fig. 6.11 for a three-dimensional stress state and in Fig. 6.13 for plane stress loading (biaxial stress state). Compared to the MSST, the DET has the advantage that the yield criterion is continuous in its first derivative, an important consideration for applications in plasticity.

Example 6.6: Determination of von Mises Stress

Given: In the rear-wheel suspension of the Volkswagen Beetle, the spring motion was provided by a torsion bar fastened to an arm on which the wheel was mounted, as depicted in Fig. 6.14. The torque in the torsion bar was created by a 2500-N force acting on the wheel from the ground through a 300-mm lever arm. Because of space limitations, the bearing holding the torsion bar was situated 100 mm from the wheel shaft. The diameter of the torsion bar was 28 mm.

Find: The stresses in the torsion bar at the bearing by using the distortion-energy theory.

Solution: The stresses acting on the torsion bar are a shear stress from torsion and a perpendicular tensile/compressive stress from bending, resulting in a plane stress loading (see Example 4.13). Using Eq. (4.33) gives the shear stress from torsion as

$$\tau=\frac{Tc}{J}=\frac{2500(0.3)(0.014)32}{\pi(0.028)^4}\ \text{Pa}=174.0\ \text{MPa}.$$

Using Eq. (4.45) gives the tensile stress from bending as

$$\sigma=\frac{Mc}{I}=\frac{2500(0.1)(0.014)64}{\pi(0.028)^4}\ \text{Pa}=116.0\ \text{MPa}.$$

From Eq. (2.16), the principal normal stresses then are

$$\begin{aligned}\sigma_1,\sigma_2 &= \frac{\sigma_x+\sigma_y}{2}\pm\sqrt{\tau_{xy}^2+\left(\frac{\sigma_x-\sigma_y}{2}\right)^2}\\ &= \frac{116.0}{2}\pm\sqrt{(174.0)^2+\left(\frac{116.0}{2}\right)^2},\end{aligned}$$

or $\sigma_1 = 241.4$ MPa and $\sigma_2 = -125.4$ MPa. From Eq. (6.12), the von Mises stress is

$$\begin{aligned}\sigma_e &= \left(\sigma_1^2+\sigma_2^2-\sigma_1\sigma_2\right)^{0.5}\\ &= \left[(241.4)^2+(-125.4)^2-241.4(-125.4)\right]^{0.5}\\ &= 322.9\ \text{MPa}.\end{aligned}$$

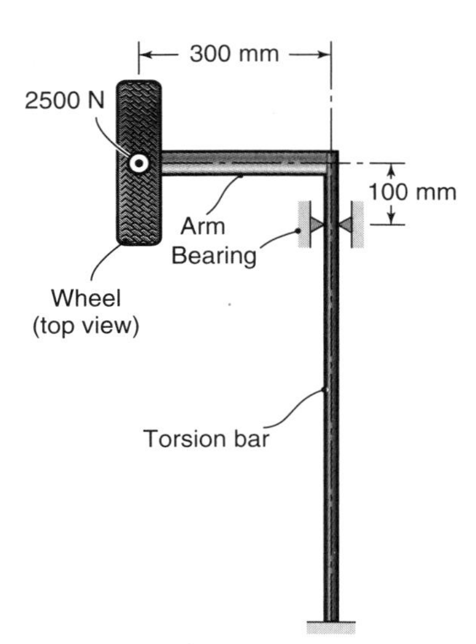

Figure 6.14: Rear wheel suspension used in Example 6.6.

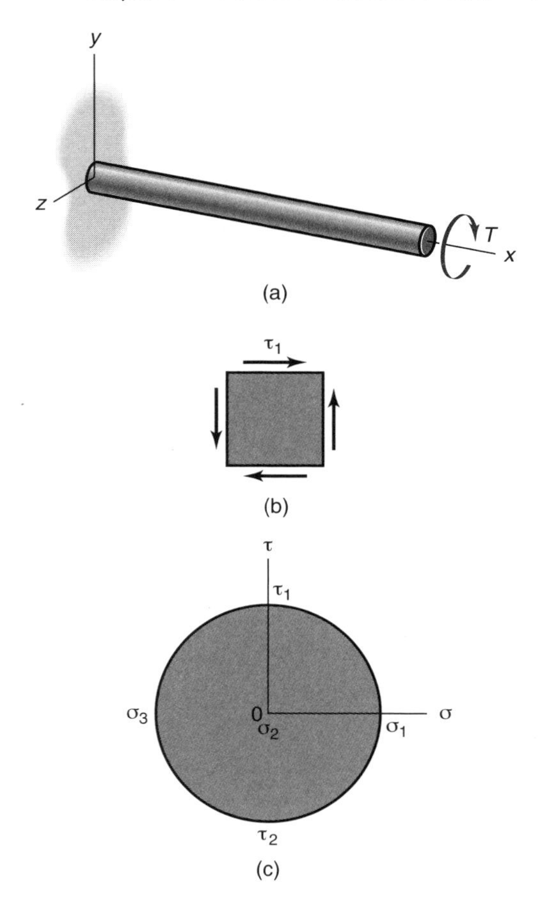

Figure 6.15: Cantilevered, round bar with torsion applied to free end (used in Example 6.7). (a) Bar with coordinates and load; (b) stresses acting on an element; (c) Mohr's circle representation of stresses.

Example 6.7: Yielding of a Ductile Bar

Given: A round, cantilevered bar made of a ductile material is subjected to a torque applied to the free end.

Find: Determine when yielding will occur by using (*a*) the MSST and (*b*) the DET.

Solution: Figure 6.15 shows the cantilevered bar, the stresses acting on an element, and a Mohr's circle representation of the stress state. Since the goal is to determine stresses at yielding, the safety factor will be taken as $n_s = 1$. The principal stresses are $\sigma_1 = \tau_1$ and $\sigma_3 = -\tau_1$.

(*a*) Using Eq. (6.8) the MSST predicts failure if

$$|\sigma_1 - \sigma_3| = 2\tau_{\max} = \frac{S_y}{n_s},$$

or

$$\tau_{\max} = 0.5S_y. \qquad (a)$$

(*b*) Using Eq. (6.11) yields

$$\begin{aligned}\sigma_e &= \left[\sigma_1^2 - \sigma_1\sigma_2 + \sigma_2^2\right]^{1/2} \\ &= \left[\tau_1^2 - \tau_1(-\tau_1) + (-\tau_1)^2\right]^{1/2} = \sqrt{3}\tau_1 \\ &= \sqrt{3}\tau_{\max} = \frac{S_y}{n_s},\end{aligned}$$

or

$$\tau_{\max} = \frac{1}{\sqrt{3}}S_y = 0.577S_y. \qquad (b)$$

Equations (*a*) and (*b*) show that the MSST and the DET are in fairly good agreement. This circumstance, that is, a loading of pure shear, results in the greatest difference between the MSST and the DET, suggesting that both theories will give close to the same results.

Example 6.8: Yield Criteria Applied to Design

Given: A round, cantilevered bar, similar to that considered in Example 6.7, is subjected not only to torsion but also to a transverse load at the free end, as shown in Fig. 6.16a. The bar is made of a ductile material having a yield stress of 350 MPa. The transverse force is 2000 N and the torque is 100 Nm applied to the free end. The bar is 150 mm long and a safety factor of 2 is assumed. Transverse shear can be neglected.

Find: Determine the minimum diameter to prevent yielding by using both (*a*) the MSST and (*b*) the DET.

Solution: Figure 6.16b shows the stress element on the top of the bar at the wall. Note that, in this example, $\sigma_z = 0$, so that the element encounters plane or biaxial stress. The critical section occurs at the wall. By using Eqs. (4.45) and (4.33) the normal and shear stresses can be written as

$$\sigma_x = \frac{Mc}{I} = \frac{Pl(d/2)}{\pi d^4/64} = \frac{32Pl}{\pi d^3},$$

$$\tau_{xy} = \frac{Tc}{J} = \frac{T(d/2)}{\pi d^4/32} = \frac{16T}{\pi d^3}.$$

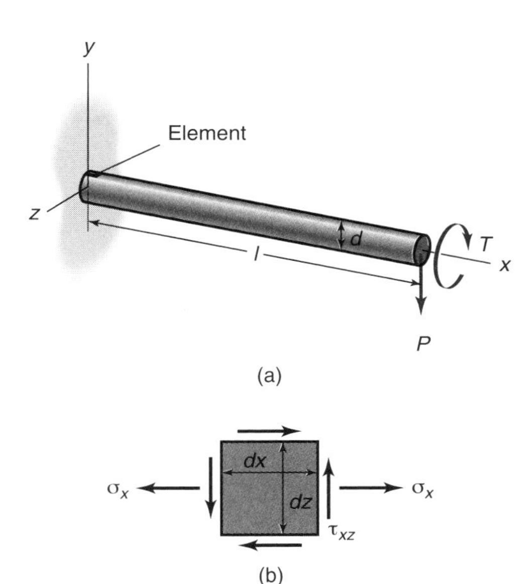

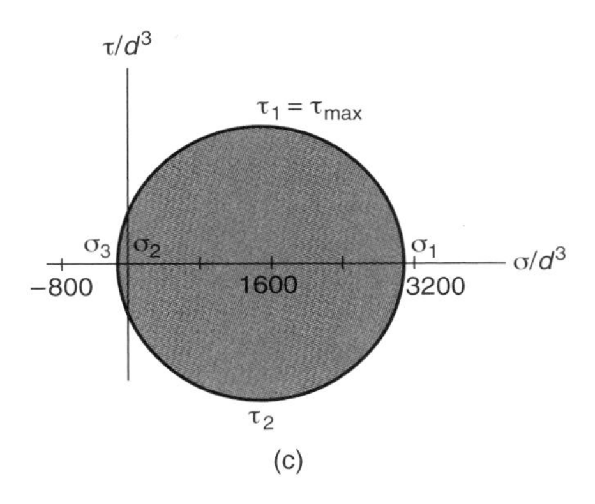

Figure 6.16: Cantilevered, round bar with torsion and transverse force applied to free end (used in Example 6.8). (a) Bar with coordinates and loads; (b) stresses acting on element at top of bar and at wall; (c) Mohr's circle representation of stresses.

From Eq. (2.16), the principal normal stresses in biaxial stress can be written as

$$\begin{aligned}\sigma_1, \sigma_2 &= \frac{\sigma_x}{2} \pm \sqrt{\left(\frac{\sigma_x}{2}\right)^2 + \tau_{xz}^2} \\ &= \frac{16Pl}{\pi d^3} \pm \sqrt{\left(\frac{16Pl}{\pi d^3}\right)^2 + \left(\frac{16T}{\pi d^3}\right)^2}.\end{aligned}$$

Therefore,

$$\sigma_1, \sigma_2 = \frac{16}{\pi d^3}\left(Pl \pm \sqrt{(Pl)^2 + T^2}\right).$$

Substituting for P and l results in

$$\sigma_1, \sigma_2 = \frac{16}{\pi d^3}\left[(2000)(0.150) \pm \sqrt{(2000)^2(0.150)^2 + (100)^2}\right].$$

Therefore,

$$\sigma_1 = \frac{3140}{d^3} \quad \text{and} \quad \sigma_2 = -\frac{82.6}{d^3}.$$

Note that the stresses are in the wrong order; to ensure $\sigma_1 \geq \sigma_2 \geq \sigma_3$, they are rearranged so that $\sigma_1 = 3140/d^3$, $\sigma_2 = 0$, and $\sigma_3 = -82.6/d^3$.

From Eq. (2.19), the maximum and principal shear stresses can be written as

$$\tau_1, \tau_2 = \pm\sqrt{\tau_{xy}^2 + \frac{(\sigma_x - \sigma_z)^2}{4}},$$

or

$$\begin{aligned}\tau_{\max} &= \tau_1 = \frac{16}{\pi d^3}\sqrt{(Pl)^2 + T^2} \\ &= \frac{16}{\pi d^3}\sqrt{(2000)^2(0.150)^2 + (100)^2}.\end{aligned}$$

Therefore,

$$\tau_{\max} = \tau_1 = \frac{1610}{d^3}.$$

(*a*) Using Eq. (6.8), the MSST predicts that failure will be avoided if

$$|\sigma_1 - \sigma_3| = 2\tau_1 = 2\tau_{\max} < \frac{S_y}{n_s}.$$

Therefore,

$$\left|\frac{3140}{d^3} + \frac{82.6}{d^3}\right| < \frac{350 \times 10^6}{2},$$

or $d = 0.0210$ m = 21.0 mm.

(*b*) This is a plane stress state, since one of the principal stresses is zero. Therefore, Eq. (6.12) yields

$$\begin{aligned}\sigma_e &= \left[\left(\frac{3140}{d^3}\right)^2 + \left(\frac{3140}{d^3}\right)\left(\frac{82.6}{d^3}\right) + \left(\frac{82.6}{d^3}\right)^2\right]^{1/2} \\ &= \frac{3180}{d^3}.\end{aligned}$$

Thus, using Eq. (6.11) the DET predicts that failure occurs when

$$\sigma_e = \frac{S_y}{n_s},$$

$$\frac{26,950}{d^3} = \frac{50,000}{2}.$$

Hence $d = 20.9$ mm. Note that both theories give approximately the same solution.

6.7.2 Brittle Materials

As discussed in Chapter 3, brittle materials do not yield, but instead they fracture. Thus, a failure criterion applied to brittle materials addresses the circumstances under which the material will literally break. One important consideration with brittle materials is that their strengths in compression are usually much greater than their strengths in tension. Consequently, the failure criterion will show a difference in tensile and compressive behavior. Three failure criteria are presented next: the maximum-normal-stress theory, the internal friction theory, and the modified Mohr theory.

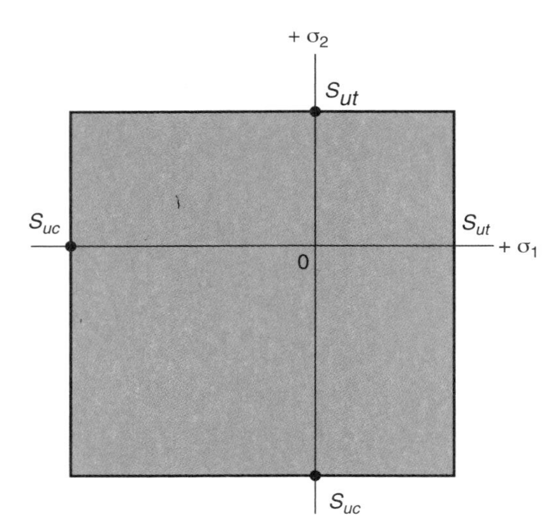

Figure 6.17: Graphical representation of maximum-normal-stress theory (MNST) for a biaxial stress state ($\sigma_z = 0$).

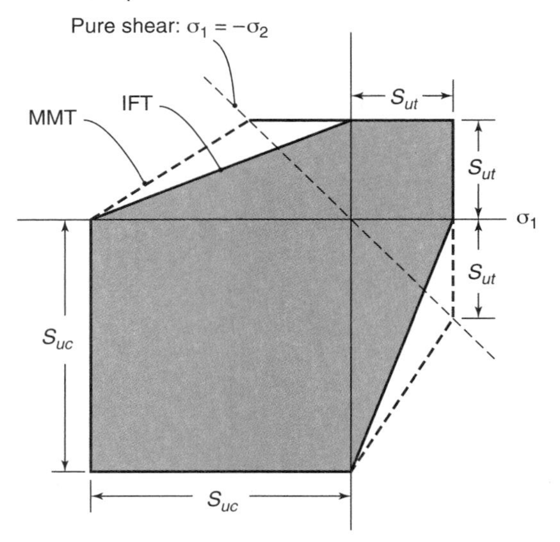

Figure 6.18: Internal friction theory and modified Mohr theory for failure prediction of brittle materials.

Maximum Normal Stress Theory

The **maximum-normal-stress theory** (MNST) states that a part subjected to any combination of loads will fail when the greatest positive principal stress exceeds the material's tensile strength or when the greatest negative principal stress exceeds the compressive strength. This theory matches experiments best for fibrous brittle materials and some glasses, but other failure criteria give better general predictions and the MNST is not usually recommended. It is presented here as an alternative for some special materials and because it is simple mathematically. Failure will occur, using the MNST theory, if

$$\sigma_1 \geq \frac{S_{ut}}{n_s}, \tag{6.13}$$

or

$$\sigma_3 \leq \frac{S_{uc}}{n_s}, \tag{6.14}$$

where

$\sigma_1 \geq \sigma_2 \geq \sigma_3$ = principal normal stresses, Pa
S_{ut} = uniaxial ultimate strength in tension, Pa
S_{uc} = uniaxial ultimate strength in compression, Pa
n_s = safety factor

Note that failure is predicted to occur when either Eq. (6.13) or (6.14) is satisfied.

Figure 6.17 graphically presents failure prediction for brittle materials in the biaxial stress state by the MNST. The $\sigma_1 - \sigma_2$ plot for biaxial stresses (i.e., $\sigma_3 = 0$) shows that the MNST predicts failure for all combinations of σ_1 and σ_2 falling on the boundary of the shaded area in Fig. 6.17; any stress state within the shaded area represents an elastic stress state.

Internal Friction Theory

The MNST is most useful for brittle or fibrous materials and some glasses, and is not generally applicable to other materials. Most brittle materials, such as ceramics and cast metals widely used in machine components, have behavior that departs from the MNST, and a different fracture theory is needed. Furthermore, the maximum shear stress theory is difficult to apply to brittle materials, since their strength in compression is so much higher than the tensile strength. In addition, some materials, such as certain magnesium alloys, are ductile but are stronger in tension than in compression. A logical extension to the MSST is to separate compressive and tensile strengths. In mathematical terms,

If $\sigma_1 > 0$ and $\sigma_3 < 0$,

$$\frac{\sigma_1}{S_{ut}} - \frac{\sigma_3}{S_{uc}} = \frac{1}{n_s}. \tag{6.15}$$

If $\sigma_3 > 0$,

$$\sigma_1 = \frac{S_{ut}}{n_s}; \tag{6.16}$$

and if $\sigma_1 < 0$,

$$\sigma_3 = \frac{S_{uc}}{n_s}, \tag{6.17}$$

where

$\sigma_1 \geq \sigma_2 \geq \sigma_3$ = principal stresses, Pa
S_{ut} = fracture strength in tension, Pa
S_{uc} = fracture strength in compression, Pa
n_s = safety factor

Although this would appear to be an arbitrary extension of the MSST, Eqs. (6.15) to (6.17) can be derived analytically if internal friction is considered. For this reason, this fracture criterion is known as the **internal friction theory** (IFT) and is also known as the **Coulomb-Mohr theory**. The IFT has the advantage of being more accurate than either the MNST or MSST for materials that have a pronounced difference in tensile and compressive strengths. Figure 6.18 depicts the IFT for a two-dimensional stress state ($\sigma_3 = 0$).

Modified Mohr Theory

The IFT has an analytical basis, but the **modified Mohr theory** arose through efforts at fitting test data. The modified Mohr theory (MMT) best predicts brittle material behavior, especially in the fourth quadrant in Fig. 6.17. The MMT can be expressed as follows:

If $\sigma_1 > 0$ and $\sigma_3 < -S_{ut}$,

$$\sigma_1 - \frac{S_{ut}\sigma_3}{S_{uc} - S_{ut}} = \frac{S_{uc}S_{ut}}{n_s S_{uc} - S_{ut}}. \tag{6.18}$$

If $\sigma_3 > -S_{ut}$,

$$\sigma_1 = \frac{S_{ut}}{n_s}. \tag{6.19}$$

If $\sigma_1 < 0$,

$$\sigma_3 = \frac{S_{uc}}{n_s}. \tag{6.20}$$

Figure 6.18 depicts the MMT, along with the IFT, to demonstrate that there is only a slight difference between the two criteria.

Example 6.9: Failure of a Brittle Bar

Given: Repeat Example 6.7 but with the cantilever constructed from a brittle material.

Find: Determine when fracture will occur by using (*a*) the MNST, (*b*) the IFT, and (*c*) the MMT. Assume that the compressive strength of the material is twice its tensile strength.

Solution: Just as in Example 6.7, the stress state is $\sigma_1 = \tau_1$, $\sigma_2 = 0$, and $\sigma_3 = -\tau_1$.

(*a*) Using Eq. (6.13) the MNST predicts failure if

$$\sigma_1 > \frac{S_{ut}}{n_s} \quad \text{or} \quad \tau_1 = \tau_{\max} \geq \frac{S_{ut}}{n_s}. \tag{a}$$

Tensile fracture stress was used in Eq. (a) because the material will fail in tension before it fails in compression.

(*b*) Recall that, when using the IFT, σ_1 is positive and σ_3 is negative. Thus, Eq. (6.14) yields

$$\frac{\sigma_1}{S_{ut}} - \frac{\sigma_3}{S_{uc}} = \frac{1}{n_s},$$

or

$$\frac{\tau_1}{S_{ut}} - \frac{(-\tau_1)}{2S_{ut}} = \frac{1}{n_s}.$$

Solving for τ_1,

$$\tau_1 = \frac{2}{3}\frac{S_{ut}}{n_s}. \tag{b}$$

(*c*) An examination of Figures 6.17 and 6.18 shows that for pure shear there is no difference between the MMT and the MNST.

6.7.3 Selecting Failure Criteria

Selecting a failure criterion to use in design is somewhat of an art. Figure 6.19 shows some test data on biaxial failures or yielding for several materials. The DET fits ductile materials slightly better than does the MSST, but most of the data fall between the two curves. Although the DET fits the data well for ductile solids, the MSST is often applied. The MSST is mathematically simple and conservative, meaning that for a given circumstance the MSST will predict yielding at lower loads than does the DET. Thus, because there is an added safety factor in using the MSST, it is often used in design. On the other hand, many commercial stress analyses and finite element codes use a von Mises stress to graphically present results, and thus there is a natural tendency to use a DET in such circumstances.

Ductile materials are produced with high repeatability in terms of strength, so designers can use either failure criterion with good confidence. Brittle materials are somewhat more difficult to analyze because their material properties generally vary much more than those of ductile materials. Owing to the statistical nature of brittle material properties, experimental strength verification is usually preferred over theoretical predictions for most applications.

When a ductile material is tested, as in tension, measured yield stresses deviate little among specimens, especially if the specimens have been obtained from the same batch in its production. Thus, relatively few experiments are needed to specify the strength of a ductile material with good certainty. For most brittle materials, however, many more tests must be performed to accurately assess the strength distribution. Test specimens do not fail at the same stress, even if they were manufactured in the same batch with the same processes. It is difficult to give a strength value for a brittle material with any great certainty.

A student of engineering design often wonders why the IFT theory is presented at all if the MMT better fits brittle material data, as is suggested by Fig. 6.19. Given the variation in strengths between brittle test specimens, the difference in the three failure criteria presented here is insignificant. To

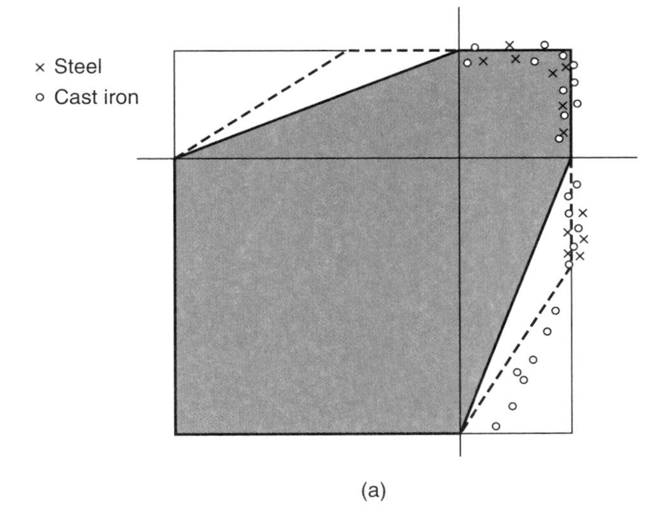

(a)

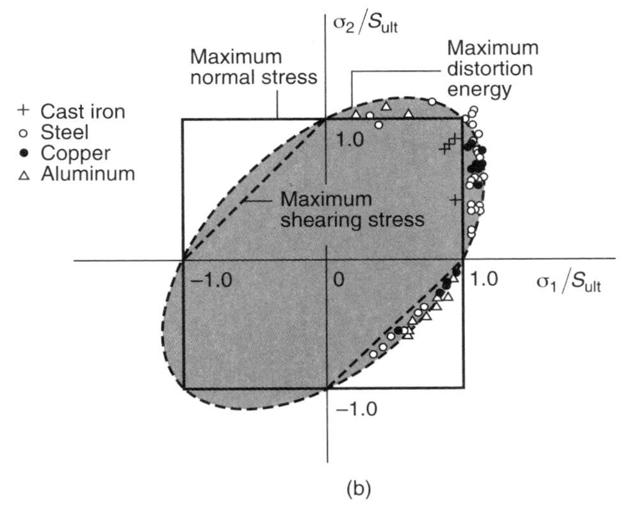

(b)

Figure 6.19: Experimental verification of yield and fracture criteria for several materials. (a) Brittle fracture. (b) Ductile yielding. *Source:* Data from Dowling [1993] and Murphy [1964].

the question, "Which theory is better or best?" no absolute answer can be given, and a designer or organization should use the one with which it has the most experience and history.

The IFT and the MSST are identical for most metals because the yield strengths of metals in compression and tension are approximately equal. Thus, applying the IFT theory at all times would ensure a conservative solution. Designers are often reluctant to follow this procedure, since the IFT is the most complex criterion from a mathematical standpoint.

The failure criteria given in this chapter are difficult to apply to composite materials and polymers. The behavior of polymers is complex, and can include viscoplastic behavior, where a yield point is difficult to define. Composite materials require more complex failure theories to account for fiber length and orientation with respect to load; the interested reader is referred to Daniels and Ishai [2005], Reddy [1996], and Kaw [1997].

For the purposes of this text, ductile materials can be analyzed using the DET or the MSST. Brittle materials should be analyzed by using the MNST, the IFT, or the MMT. A wide variety of additional failure and yield criteria is given in the technical literature. However, those presented in this chapter are by far the most commonly applied, and a suitable yield criterion can usually be chosen from those presented.

Design Procedure 6.2 summarizes some of the main concerns and gives general suggestions on selection of a failure criterion.

Design Procedure 6.2: Selection of a Failure Criterion

Given a material, where the tensile and compressive yield, ultimate and/or fracture stresses are known, the following steps can be used to help select a failure criterion:

1. For a ductile metal, where the strength is the same in tension and compression, use either the MSST or DET. These criteria are fairly close, with the largest difference of 15% occurring for pure shear in a plane stress loading. The MSST is more conservative; that is, it predicts yielding at a lower stress level than DET.
2. If a ductile metal has a different strength in compression than in tension, such as with certain magnesium alloys, the IFT or MMT are reasonable options.
3. Brittle materials are difficult to analyze using failure criteria, and confidence in strength values is difficult to obtain. However, the IFT leads to good results without the mathematical complication of the MMT.
4. For circumstances where improved performance is required, MMT may be justified over the IFT.

Regardless of these suggestions, failure criteria are often specified for the designer by their customer, employer, or supplier. Sometimes, more elaborate yield criterion are used, such as those that incorporate viscoelastic behavior or creep. This design procedure is generally applicable, but deviations in industrial practice are not uncommon.

Table 6.2: Safety factors from using different criteria for three different materials used in Example 6.10.

Part	Criterion	Equation used	Safety factor
a)	MSST	(6.7)	1.5
	DET	(6.10)	1.73
b)	MSST	(6.7)	1.28
	DET	(6.10)	1.33
c)	IFT	(6.14)	1.61
	MMT	(6.17)	1.69

Example 6.10: Selection of Yield and Failure Criteria

Given: The following materials and loadings (remaining stresses are zero):

(*a*) Pure aluminum: S_y = 30 MPa, σ_x = 10 MPa, σ_y = -10 MPa, $\tau_{xy} = 0$

(*b*) 0.2% carbon steel: S_y = 295 MPa, σ_x = -0 MPa, σ_y = -200 MPa, τ_{xy} = 75 MPa

(*c*) Gray cast iron: S_{ut} = 125 MPa, S_{uc} = 450 MPa, σ_x = -100 MPa, σ_y = 50 MPa, $\tau_{xy} = 0$

Find: The safety factors for these circumstances.

Solution: The results are summarized in Table 6.2. Note that the yield criteria selected for aluminum and steel are the MSST and the DET, since these are recognized as ductile metals. The loading for the aluminum corresponds to pure shear (see Section 2.14), and this loading will have the largest difference between the MSST and DET as can be seen from the resulting safety factors. The loading for the steel is not pure shear, and the yield criteria are in good agreement.

Cast iron is not a ductile material, so that neither the MSST or the DET should be applied. Design Procedure 6.2 notes that the IFT yields good results without the mathematical complexity of the MMT. As can be seen from Table 6.2, the difference in safety factors that result from the IFT and MMT is fairly small, so that either could be used.

Case Study: Stress Concentration Factors for Complicated Geometries

The stress concentration factors presented in Section 6.2.1 and in Appendix C are applicable to most design circumstances, but it is not unusual to encounter different and often more complex geometries. Departure from the circumstances given in Figs. 6.2 through 6.8 can involve the use of non-linear materials, varying thickness, more complex geometries, closely spaced stress concentrations, or combined loadings. This case study presents the common methods of determining stress concentration factors for more complex situations.

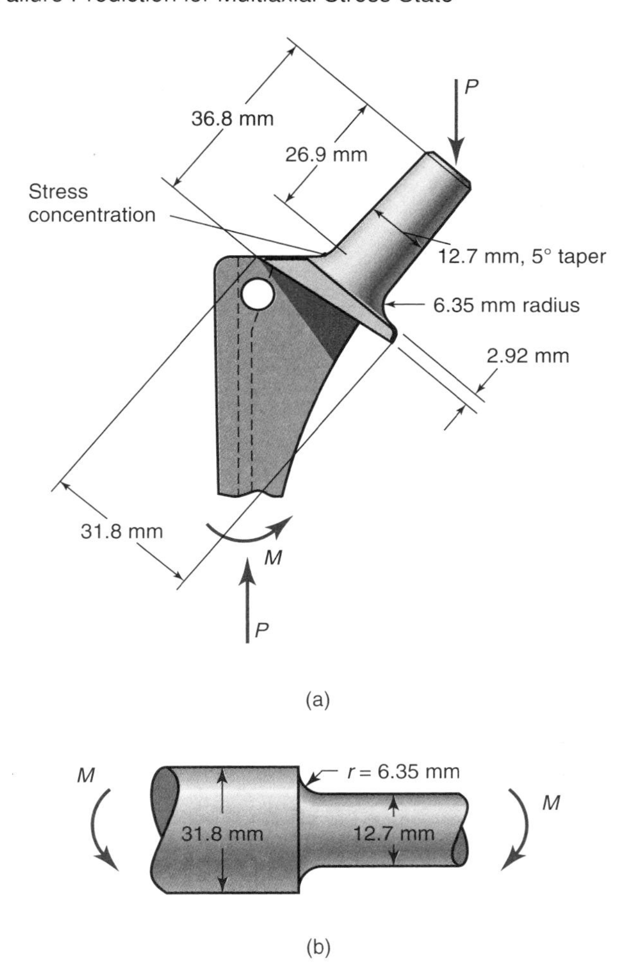

Figure 6.20: (a) Schematic illustration of a portion of a total hip replacement with selected dimensions; (b) idealized geometry used to estimate the stress concentration factor at the fillet.

Stress Concentration Charts

The charts presented in Section 6.2.1 are a small subset of those that have been developed to date. Performing an Internet search or review of standard handbooks of stress concentration factors may result in a chart that best matches a particular design condition. This is often the quickest and best approach to determining a stress concentration factor.

Often, a circumstance that is close to the design problem can be found, but not one that matches it exactly. For example, Fig. 6.20a shows a section of a total hip replacement, with a fillet from the transition of a tapered section and the portion of the stem that is inserted into a femur. The loading encountered is combined compression and bending, and the cross section transitions from circular at the base of the taper to rounded trapezoidal at the stem. An estimate of the stress concentration associated with this geometry can be obtained using the approximated geometry in Fig. 6.20 and Fig. 6.5c, even though the geometry is not exactly the same.

Such estimates are often very useful; they can confirm that designs have high or low safety factors and may justify additional investigations or verify that they are unnecessary. During preliminary design work, this approach can be used to verify that design approaches have merit while avoiding the cost and delay of more elaborate investigations.

Regardless, designs for critical applications (such as the total hip replacement in Fig. 6.20) will require further analysis and verification before they are finalized.

Finite Element Analysis

Finite element analysis (FEA) is a numerical approach that has become very widely used with the proliferation of powerful computers and dedicated software. A detailed treatment of this powerful approach is not within the scope of this book, but the interested reader is directed to the texts by Fish and Belytschko [2007] or Hughes [2000]. This Case Study provides a short outline of the approach.

FEA requires *discretization* of a geometry into a mesh of *elements*; the elements are commonly triangles, quadrilaterals, bar segments, bricks, etc., defined by their *nodes*. It is not uncommon for meshes to consist of thousands of elements and nodes. After discretization, constraints are applied to some nodes and loads are applied to others.

Several FEA programs exist, but the general numerical solution approach is always based on matrix methods. A compliance matrix is assembled for the entire mesh from the compliances of the individual elements; this matrix is then inverted and the displacements of element nodes are determined for the system. Application of Hooke's Law [see Eq. (B.55)] yields the stresses in each element from the displacements at each node.

Depending on the element type, stresses are evaluated at one or more locations within the element. For a simple quadrilateral element, for example, the stress may be evaluated only at its center. For this reason, it must be recognized that highly localized stress concentrations will require small elements near the stress raiser to accurately capture its value. However, the elements cannot be so small that the accuracy of the computer processor limits the analysis. Furthermore, widespread use of small elements results in long computation times.

Experienced analysts will locate many small elements near suspected stress raisers and will place larger elements elsewhere in order to achieve reasonable computation times while capturing close estimates of stress concentrations. Depending on the complexity of the part, it may be necessary to refine meshes to make certain that results are accurate and confirm simulations. If designs are modified, new meshes have to be constructed and the software run to obtain a new result. Thus, finite element analysis should be recognized as a powerful tool, but not without its costs and limitations.

Figure 6.21 depicts a typical mesh for determining the stress associated with a hole in a plate under tension. The figure shows a wireframe representation in order to highlight the elements and their distributions; commonly, the von Mises stresses are plotted as contours. Regardless, one can see the larger elements located away from the stress concentration. The maximum element size is a user-defined parameter in the automatic meshing routines.

Photoelasticity

Certain transparent materials display *birefringence*, wherein a ray of light encounters two refractive indices within the material, and is broken into two rays. Some materials display birefringence only under applied stress; for these materials, the magnitude of the stress determines the extent of birefringence at any point. When a polarized light is passed through the material, the two refractive indices manifest themselves in a phase difference that produces a fringe pattern in a polariscope (Fig. 6.22). Analysis of the refringence pattern allows direct measurement of stresses everywhere in the part geometry.

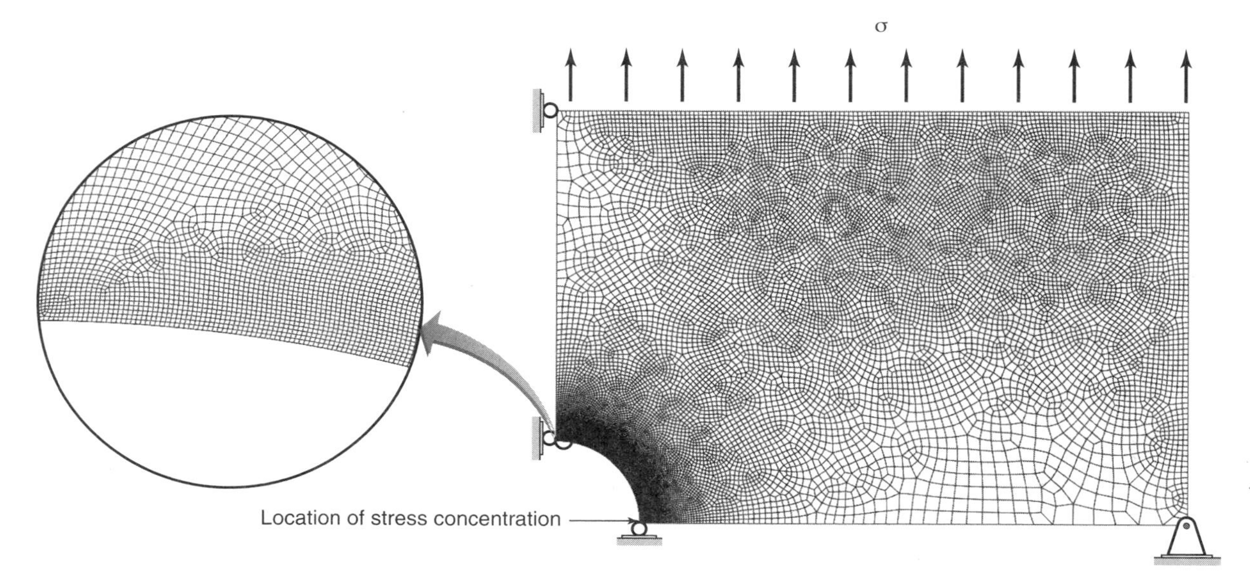

Figure 6.21: Example of a finite element mesh to capture the value of a stress concentration corresponding to Fig. 6.2. Only one-fourth of the problem has been discretized to take advantage of symmetry. Note the large number of elements located near the stress raiser. Boundary conditions and applied loads have been added for image clarity.

Photoelasticity can be applied to three-dimensional stress states and specimens, but the approach is complicated and, as a result, the approach is usually limited to plane stress. In fact, Figs. 6.3 and 6.4 were obtained from photoelastic methods and later confirmed by analytical solutions.

Photoelasticity has become less common since the proliferation of FEA, but it is still used for experimental verification. The cost and time involved in preparing physical specimens can be quite high, and skilled technicians are required to obtain good results. Further details on photoelastic stress analysis can be found in Doyle [2004].

Experimental Confirmation

Critical applications or complex geometries are usually confirmed by mechanical testing to ensure that intended loads can be safely supported. Such tests can be performed at a number of levels. For example, experiments can be conducted on components, systems or assemblies, or entire products to make sure that they support required loads and are robust designs. Further, testing may involve idealized loading, actual field trials, or any level of complexity in between. Rarely is the stress concentration directly measured, but instead failure loads as a whole are determined.

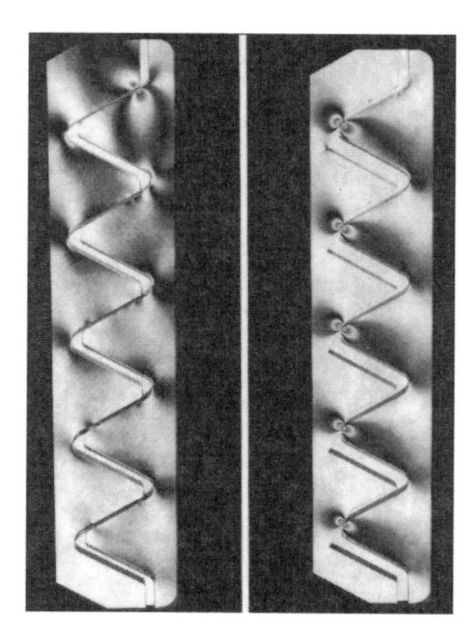

Figure 6.22: Photoelastic comparison of threaded fastener profiles comparing load distribution. The left image shows a conventional profile where load per tooth varies widely, and the right shows a Spiralock© profile with more uniform stresses on each tooth. *Source:* Courtesy of Stanley Engineered Fastening - Spiralock.

6.8 Summary

Of the approaches that can be used to determine stress concentration factors, the use of stress concentration charts from the technical literature (of which Figs. 6.2 through 6.8 are examples) represents the most cost-effective method. Especially in the age of the Internet, it is also usually the fastest approach as well. The finite element method has become an extremely widespread approach as well, but it has drawbacks of longer lead time and therefore higher cost. Experimental confirmation of designs is also relatively expensive and time consuming, but may be necessary to obtain confidence in the safety and suitability of designs.

Key Words

Coulomb-Mohr theory a failure theory for materials with different strengths in tension and compression; identical to the internal friction theory

crack small flaw, always present, that can compromise material strength

distortion-energy theory postulate that failure is caused by elastic energy associated with deformation; identical to von Mises criterion

fracture control maintenance of nominal stress and crack size below critical level

fracture toughness critical value of stress intensity at which crack extension occurs

internal friction theory failure criterion accounting for difference between compressive and tensile strengths of brittle materials; identical to Coulomb-Mohr theory

maximum-normal-stress theory theory that yielding will occur whenever the greatest positive principal stress exceeds the tensile yield strength or whenever the greatest negative principal stress exceeds the compressive yield strength

maximum-shear-stress theory theory that yielding will occur when the largest shear stress exceeds a critical value; identical to the Tresca criterion

modes of crack propagation principal mechanisms for cracks to enlarge: Mode I, opening through tension; Mode II, sliding or in-plane shearing; Mode III, tearing

modified Mohr theory failure postulate similar to Coulomb-Mohr theory, except that curve is altered in quadrants II and IV of plane stress plot of principal stresses

stress concentration region where stress raiser is present

stress concentration factor factor used to relate actual maximum stress at discontinuity to nominal stress

stress intensity factor stress intensity at crack tip

stress raiser discontinuity that alters stress distribution so as to increase maximum stress

Tresca yield criterion theory that yielding will occur when the largest shear stress exceeds a critical value; identical to maximum-shear-stress theory

von Mises criterion postulate that failure is caused by elastic energy associated with deformation; identical to distortion-energy theory

von Mises stress effective stress based on von Mises criterion

Summary of Equations

Stress concentration factor:

Definition: $K_c = \dfrac{\text{Actual maximum stress}}{\text{Average stress}}$

Elliptical hole in plate loaded in tension: $K_c = 1 + 2\left(\dfrac{a}{b}\right)$

Fracture Toughness: $K_{Ic} = Y\sigma_{\text{nom}}\sqrt{\pi a}$

Failure Prediction for Uniaxial Stress State: $n_s = \dfrac{\sigma_{\text{all}}}{\sigma_d}$

Failure Prediction for Multiaxial Stress State:

Maximum shear stress theory (MSST, or Tresca):

$$n_s = \frac{S_y}{\sigma_1 - \sigma_3}$$

von Mises stress:

$$\sigma_e = \frac{1}{\sqrt{2}}\left[(\sigma_1 - \sigma_2)^2 + (\sigma_3 - \sigma_1)^2 + (\sigma_3 - \sigma_2)^2\right]^{1/2}$$

Distortion energy theory (DET, or von Mises): $n_s = \dfrac{S_y}{\sigma_e}$

Maximum normal stress theory (MNST):

$$n_s = \frac{S_{ut}}{\sigma_1} \text{ or } n_s = \frac{S_{uc}}{\sigma_3}, \text{ whichever is lower.}$$

Internal Friction Theory (IFT, or Coulomb-Mohr):

If $\sigma_1 > 0$ and $\sigma_3 < 0$, $\dfrac{\sigma_1}{S_{ut}} - \dfrac{\sigma_3}{S_{uc}} = \dfrac{1}{n_s}$

If $\sigma_3 > 0$, $n_s = \dfrac{S_{ut}}{\sigma_1}$

If $\sigma_1 < 0$, $n_s = \dfrac{S_{uc}}{\sigma_3}$

Modified Mohr Theory (MMT):

If $\sigma_1 > 0$ and $\sigma_3 < -S_{ut}$,

$$\sigma_1 - \frac{S_{ut}\sigma_3}{S_{uc} - S_{ut}} = \frac{S_{uc}S_{ut}}{n_s S_{uc} - S_{ut}}$$

If $\sigma_3 > -S_{ut}$, $n_s = \dfrac{S_{ut}}{\sigma_1}$

If $\sigma_1 < 0$, $\sigma_3 = \dfrac{S_{uc}}{n_s}$

Recommended Readings

Anderson, T.L. (2005), *Fracture Mechanics — Fundamentals and Applications*, 3rd ed., CRC Press.

Budynas, R.G., and Nisbett, J.K. (2011), *Shigley's Mechanical Engineering Design*, 9th ed., McGraw-Hill.

Hill, R. (1950) *The Mathematical Theory of Plasticity*, Oxford.

Dowling, N.E. (1993) *Mechanical Behavior of Materials*, Pearson.

Juvinall, R.C., and Marshek, K.M. (2012) *Fundamentals of Machine Component Design*, 5th ed., Wiley.

Mott, R. L. (2014) *Machine Elements in Mechanical Design*, 4th ed., Pearson.

Norton, R.L. (2011) *Machine Design*, 4th ed., Pearson Education.

Sun, C.T., and Jin, Z.-H. (2012) *Fracture Mechanics*, Elsevier.

References

ASM International (1989) *Guide to Selecting Engineering Materials*, American Society for Metals.

Bowman, K.J. (2004) *Mechanical Behavior of Materials*, Wiley.

Coulomb, C.A. (1773) *Sur une Application des Regles de maximmis et minimus a quelques problemes de statique relatifs a l'architecture.*

Daniels, I.M., and Ishai, O. (2005) *Engineering Mechanics of Composite Materials* 2nd ed., Oxford.

Dowling, N.E. (1993) *Mechanical Behavior of Materials*, Pearson.

Doyle, J.F. (2004) *Experimental Stress Analysis: Completing the Solution of Partially Specified Problems*, Wiley.

Fish, J., and Belytschko, T. (2007) *A First Course in Finite Elements*, Wiley.

Hughes, T.J.R. (2000) *The Finite Element Method*, Dover.
Kalpakjian, S., and Schmid, S.R. (2014) *Manufacturing Engineering and Technology*, 7th ed., Pearson.
Kaw, A.K. (1997) *Mechanics of Composite Materials*, CRC Press.
Murphy, G. (1964) *Advanced Mechanics of Materials*, McGraw-Hill.
Pilkey, W.D., and Pilkey, D.F. (2008) *Peterson's Stress Concentration Factors*, 3rd ed., Wiley.
Reddy, J.N. (1996) *Mechanics of Laminated Composite Plates*, CRC Press.
Tresca, H. (1868) "Mem. prenetes par divers savants," vol. 59, p. 754, *Comptes Rendus Acad. Sci.*.
Young, W.C., and Budynas, R. (2001) *Roark's Formulas for Stress and Strain*, 7th ed., McGraw-Hill.

Questions

6.1 What is a stress concentration?

6.2 What is a stress concentration factor?

6.3 Define the term "crack" as relates to material fracture.

6.4 List the modes of crack growth.

6.5 What is fracture toughness?

6.6 Is a material with a high fracture toughness a ductile material? Explain.

6.7 What is the maximum shear stress theory?

6.8 Do the distortion energy criterion and maximum shear stress criterion give very different results? Explain.

6.9 Without the use of equations, define the von Mises stress.

6.10 What is a yield locus?

6.11 What is the MNST and the IFT?

6.12 For what materials is the IFT most useful?

Qualitative Problems

6.13 A round bar has a fillet with $r/d = 0.15$ and $D/d = 1.5$. The bar transmits both bending moment and torque. A new construction is considered to make the shaft stiffer and stronger by making it equally thick on each side of the fillet or groove. Determine whether that is a good idea.

6.14 Are stress concentrations more severe for tension, bending, or torsion for a bar with a groove?

6.15 A diamond cutting tool is used to make a shallow but sharp groove in a glass plate. The glass is struck with a hard rubber mallet to produce a well-defined cut at the location of the groove. Referring to the discussions in this chapter, explain the phenomena that are important in this application.

6.16 Explain why the approach used to cut glass will not work for a ductile aluminum.

6.17 A plate of titanium is diffusion bonded to a plate of aluminum. There are no holes, grooves, fillets, or notches in either member, and the two plates are then exposed to uniform tension. Sketch the stress distribution through the thickness of the plates. Is there a stress concentration? Explain.

6.18 Give three examples of each mode of crack growth described in Section 6.4.

6.19 List and briefly explain the variables that can influence the fracture toughness of a material.

6.20 Some magnesium alloys have a lower strength in compression than in tension. What failure criterion would you use for such a material? Produce an equivalent sketch as Fig. 6.11 for this failure criterion.

6.21 If a material strain hardens, what effect does plastic deformation have on the yield locus?

6.22 Figs. 6.2 through 6.8 do not show stress concentrations for compressive loadings. Why not?

6.23 Review Fig. 6.4 and plot the stress concentration factor as a function of r/h for the case where the grooves are semicircular.

6.24 Figure 7.3 shows an R.R. Moore test specimen for fatigue tests. Explain why this specimen has its shape, and estimate the stress concentration factor.

Quantitative Problems

6.25 Complete the following table:

Shape	Loading	Geometry (mm)	r (mm)	K_c
Plate w/ hole	Tension	b=50, d=5	—	
Plate w/ hole	Bending	b=50, d=5	—	
Plate w/ fillet	Tension	H=100, h=50		2.0
Bar w/ fillet	Tension	D=50, d=25		2.0
Bar w/ groove	Torsion	D=10, d=9.09	0.9	

6.26 Given that the stress concentration factor is 3.81 for a machine element made of steel with a modulus of elasticity of 207 GPa, find the stress concentration factor for an identical machine element made of aluminum instead of steel. The modulus of elasticity for aluminum is 69 GPa.

6.27 A flat part with constant thickness b is loaded in tension as shown in Fig. 6.3a. The height changes from 50 to 87 mm with a radius $r = 4.0$ mm. Find how much lower a load can be transmitted through the bar if the height increases from 50 to 100 mm and the radius increases from 4.0 to 10 mm. *Ans.* 10% higher load.

6.28 A flat steel plate axially loaded as shown in Sketch a has two holes for electric cables. The holes are situated beside each other and each has a diameter d. To make it possible to draw more cables, the two holes are replaced with one hole having twice the diameter $2d$, as shown in Sketch b. Assume that the ratio of diameter to width is $d/b = 0.2$ for the two-hole plate. Which plate will fail first?

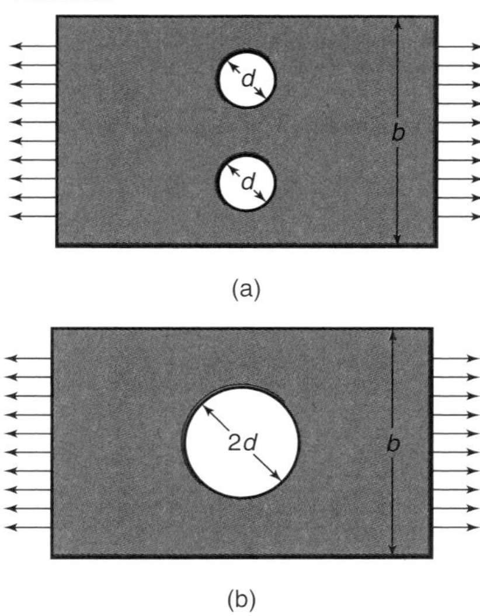

Sketches a and b, for Problem 6.28

6.29 A 5-mm thick, 100-mm wide AISI 1020 steel rectangular plate has a central elliptical hole 8 mm in length transverse to the applied stress and 2 mm in diameter along the stress. Determine the applied load that causes yielding at the edge of the hole. *Ans.* 92.43 kN.

6.30 A 10-in. wide plate loaded in tension contains a 2-in. long, 1/2-in. wide slot. Estimate the stress concentration by:

(a) Approximating the slot as an ellipse that is inscribed within the slot.

(b) Obtaining the stress concentration at the edge of the slot by taking a section through the slot and approximating the geometry as a rectangular plate with a groove.

6.31 A machine has three circular shafts, each with fillets giving stress concentrations. The ratio of fillet radius to shaft diameter is 0.1 for all three shafts. One of the shafts transmits a tensile force, one transmits a bending torque, and one transmits torsion. Because they are stressed exactly to the stress limit ($n_s = 1$), a design change is proposed doubling the notch radii to get a safety factor greater than 1. How large will the safety factors be for the three shafts if the diameter ratio is 2 ($D/d = 2$)? *Ans.* $n_{s,\text{tension}} = 1.21$, $n_{s,\text{bending}} = 1.19$, $n_{s,\text{torsion}} = 1.17$.

6.32 A material with a yield strength of $S_y = 350$ MPa is subjected to the stress state shown in Sketch c. What is the factor of safety based on the maximum shear stress and distortion energy theories? *Ans.* For MSST, $n_s = 11.67$.

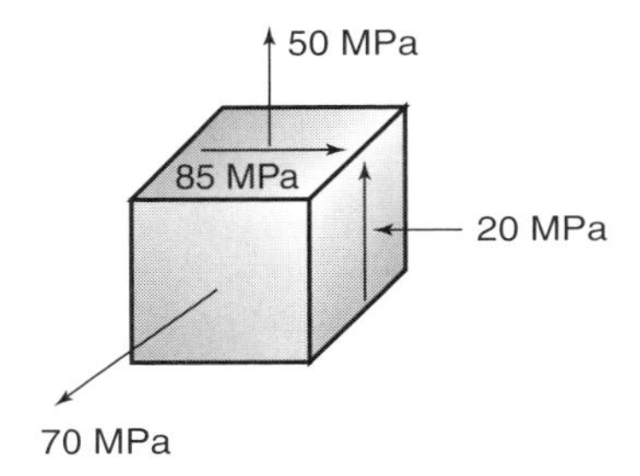

Sketch c, for Problems 6.32 and 6.33

6.33 A material subjected to the stresses shown in Sketch c is known to be yielding. What stress would cause yielding in uniaxial tension? *Ans.* For MSST, $S_y = 30$ MPa.

6.34 The shaft shown in Sketch d is subjected to tensile, torsional, and bending loads. Determine the principal stresses at the location of stress concentration. *Ans.* $\sigma_1 = 53.0$ MPa, $\sigma_2 = 0$, $\sigma_3 = -12.27$ MPa.

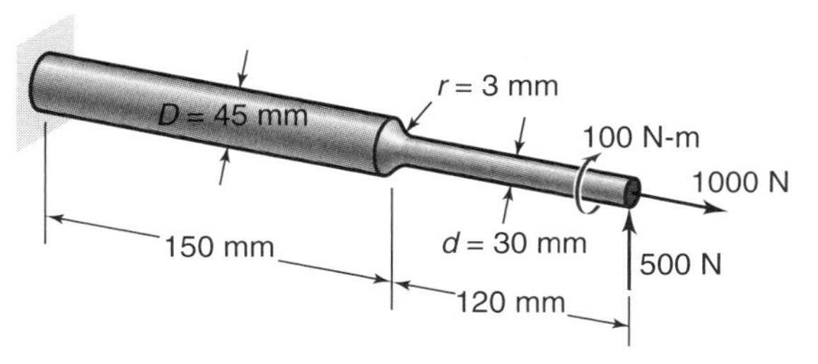

Sketch d, for Problem 6.34

6.35 A steel plate with dimensions shown in Sketch e is subjected to 150-kN tensile force and 300-N-m bending moment. The plate is made of AISI 1080 steel, quenched and tempered at 800°C. A hole is to be punched in the center of the plate. What is the maximum diameter of the hole for a safety factor of 1.5? *Ans.* $d = 170$ mm.

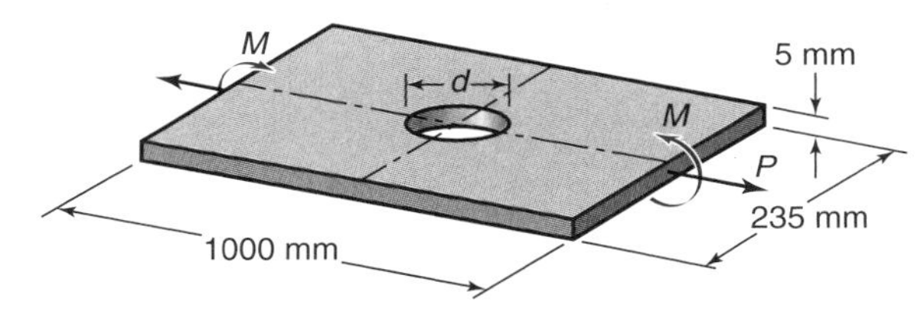

Sketch e, for Problem 6.35

6.36 A Plexiglas plate with dimensions 1 m × 1 m × 1 cm is loaded by a nominal tensile stress of 55 MPa in one direction. The plate contains a small crack perpendicular to the load direction. At this stress level a safety factor of 2 against crack propagation is obtained. Find how much larger the crack can get before it grows catastrophically. *Ans.* $a_2 = 4a_1$.

6.37 A pressurized container is made of AISI 4340 steel. The wall thickness is such that the tensile stress in the material is 1000 MPa. The dimensionless geometry correction factor $Y = 1$ for the given geometry. Find how big the largest crack can be without failure if the steel is tempered

(a) At 260°C. *Ans.* 1.592 mm.

(b) At 425°C. *Ans.* 4.863 mm.

6.38 Two tensile test rods are made of AISI 4340 steel tempered at 260°C and aluminum alloy 2024-T351. The dimensionless geometry correction factor $Y = 1$. Find how high a stress each rod can sustain if there is a crack of 1-mm half-length in each of them. *Ans.* AISI 4340: $\sigma = 892$ MPa.

6.39 A plate made of titanium alloy Ti-6Al-4V has the dimensionless correction factor $Y = 1$. How large can the largest crack in the material be if it still should be possible to plastically deform the plate in tension? *Ans.* 1.488 mm.

6.40 Consider the 2014-T4 aluminum plate shown in Sketch *f*. What is the largest tensile stress this plate can withstand? What is the largest center crack that can exist while ensuring that the material will still yield? *Ans.* $\sigma = 115.7$ MPa, $l_c = 2.644$ mm.

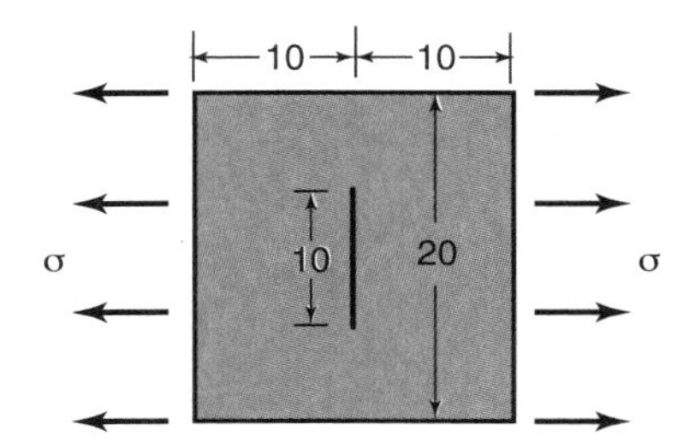

Sketch *f*, for Problem 6.40. All dimensions are in millimeters.

6.41 For a plate with length 50 mm, width 25 mm, and thickness 5 mm, what is the ratio of allowable stress for a center-cracked specimen compared to an edge-cracked specimen? Assume the crack is 10 mm long for both cases. *Ans.* $\sigma_{\text{edge}}/\sigma_{\text{center}} = 0.607$.

6.42 A cylindrical pressure vessel constructed from 4340 steel tempered at 260°C has an outer diameter of 250 mm, a wall thickness of 2.5 mm, and an internal pressure of 4 MPa. What crack can be tolerated in the axial direction before fracture occurs? *Ans.* $l_c = 39.8$ mm.

6.43 A 100-mm wide, 200-mm long plate made of titanium alloy Ti-6Al-4V has a single edge crack. How large can the crack be if it still should be possible to plastically deform the plate in tension? What if the plate is very long? *Ans.* $l_c = 8.79$ mm.

6.44 A polymethylmethacrylate ($K_{Ic} = 3$ MPa$\sqrt{\text{m}}$) model of a gear has a 1-mm half-length crack formed in its fillet curve (where the tensile stress is maximum). The model is loaded until the crack starts to propagate. Assume that $Y = 1.5$. How much higher a load can a gear made of AISI 4340 steel tempered to 425°C carry with the same crack and the same geometry? *Ans.* $\sigma_{\text{steel}}/\sigma_{\text{plexiglass}} = 29.13$.

6.45 A pressure vessel made of aluminum alloy 2024-T351 is manufactured for a safety factor of 3.0 guarding against yielding. The material contains cracks through the wall thickness with a crack half-length less than 3 mm. $Y = 1$. Find the safety factor when considering crack propagation. *Ans.* $n_s = 3.17$.

6.46 The clamping screws holding the top lid of a nuclear reactor are made of AISI 4340 steel tempered at 260°C. They are stressed to a maximum level of 1250 MPa during a pressurization test before starting the reactor. Find the safety factor guarding against yielding and the safety factor guarding against crack propagation if the initial cracks in the material have $Y = 1$ and $a = 1$ mm. Also, find the safety factor if the same material is used, but tempered at 425°C. *Ans.* AISI 4340 tempered at 260°C: $n_s = 0.714$.

6.47 A glass tube used in a pressure vessel is made of aluminum oxide (sapphire) to make it possible to apply 30-MPa pressure and still have a safety factor of 2 guarding against fracture. For a soda-lime glass of the same geometry only 7.5-MPa pressure can be allowed if a safety factor of 2 is to be maintained. Find the size of the cracks the glass tube can tolerate at 7.5 MPa pressure and a safety factor of 2. $Y = 1$ for both tubes. *Ans.* Sapphire: $l_c < 75.2\ \mu$m, glass: $l_c < 65.6\ \mu$m.

6.48 A stress optic model used for demonstrating the stress concentrations at the ends of a crack is made of polymethylmethacrylate. An artificially made crack 50 mm long is perpendicular to the loading direction. $Y = 1$. Calculate the highest tensile stress that can be applied to the model without propagating the crack. *Ans.* $\sigma_{\text{nom}} = 3.57$ MPa.

6.49 A passengerless airplane requires wings that are lightweight and resistant to fracture with cracks up to 2 mm in length. The dimensionless geometry correction factor Y is usually 1.5 for a safety factor of 2.

(a) What is the appropriate alloy for this application? *Ans.* Either Aluminum 2020-T351 or Alloy steel 4340 tempered at 425 °C.

(b) If Y is increased to 4.5, what kind of alloy from Table 6.1 should be used? *Ans.* Al 2020-T351.

6.50 The anchoring of the cables carrying a suspension bridge are made of cylindrical AISI 1080 steel bars, quenched and tempered at 800°C, and are 200 mm in diameter. The force transmitted from the cable to the steel bar is 3.5 MN. Calculate the safety factor against yielding. *Ans.* $n_s = 3.41$.

6.51 The arm of a crane has two steel plates connected with a rivet that transfers the force in pure shear. The rivet is made of annealed AISI 1040 steel and has a circular cross section with a diameter of 25 mm. The load on the rivet is 15 kN. Calculate the safety factor according to the von Mises criterion. *Ans.* $n_s = 5.67$.

6.52 An aluminum alloy yields at a stress of 50 MPa in uniaxial tension. If this material is subjected to the stresses $\sigma_1 = 25$ MPa, $\sigma_2 = 15$ MPa, and $\sigma_3 = -26$ MPa, will it yield? Explain.

6.53 A material with a yield stress of 70 MPa is subjected to principal (normal) stresses of $\sigma_1 = \sigma$, $\sigma_2 = 0$, and $\sigma_3 = -\sigma/2$. What is the value of σ when the metal yields according to the von Mises criterion? What if $\sigma_2 = \sigma/3$? *Ans.* $\sigma = 52.91$ MPa.

6.54 Assume that a material with a uniaxial yield strength S_y yields under a stress state of principal stresses σ_1, σ_2, and σ_3, where $\sigma_1 \geq \sigma_2 \geq \sigma_3$. Show that the superposition of a hydrostatic stress p on this system (such as placing the specimen in a chamber pressurized with a liquid) does not affect yielding. In other words, the material will still yield according to yield criteria.

6.55 A machine element is loaded so that the principal normal stresses at the critical location for a biaxial stress state are $\sigma_1 = 20$ ksi and $\sigma_2 = -10$ ksi. The material is ductile with a yield strength of 60 ksi. Find the safety factor according to

(a) The maximum-shear-stress theory (MSST) *Ans.* $n_s = 2.00$.

(b) The distortion-energy theory (DET) *Ans.* $n_s = 2.27$.

6.56 A bolt is tightened, subjecting its shank to a tensile stress of 80 ksi and a torsional shear stress of 50 ksi at a critical point. All of the other stresses are zero. Find the safety factor at the critical point by the DET and the MSST. The material is high-carbon steel (AISI 1080, quenched and tempered at 800°C). Will the bolt fail because of the static loading? *Ans.* $n_{s,\text{DET}} = 0.47$, $n_{s,\text{MSST}} = 0.43$.

6.57 A torque is applied to a piece of chalk used in a classroom until the chalk cracks. Using the maximum-normal-stress theory (MNST) and assuming the tensile strength of the chalk to be small relative to its compressive strength, determine the angle of the cross section at which the chalk cracks. *Ans.* 45°.

6.58 A cantilevered bar 500 mm long with square cross section has 25-mm sides. Three perpendicular forces are applied to its free end, a 1000 N force is applied in the x-direction, a 100 N force is applied in the y direction, and an equivalent force of 100 N is applied in the z-direction. Calculate the equivalent stress at the clamped end of the bar by using the DET when the sides of the square cross section are parallel with the y- and z-directions. *Ans.* $\sigma_e = 40$ MPa.

6.59 A shaft transmitting torque from a gearbox to the rear axle of a truck is unbalanced, so that a centrifugal load of 500 N acts at the middle of the 3-m-long shaft. The annealed AISI 1040 tubular steel shaft has an outer diameter of 70 mm and an inner diameter of 58 mm. Simultaneously, the shaft transmits a torque of 3000 N-m. Use the DET to determine the safety factor guarding against yielding. *Ans.* $n_s = 2.39$.

6.60 The right-angle-cantilevered bracket used in Problem 5.30, Sketch w, has a concentrated force of 1000 N and a torque of 300 N-m. Calculate the safety factor. Use the DET and neglect transverse shear. Assume that the bracket is made of annealed AISI 1040 steel and use the following values: $a = 0.5$ m, $b = 0.3$ m, $d = 0.035$ m, $E = 205$ GPa, and $\nu = 0.3$. *Ans.* $n_s = 1.76$.

6.61 A 100-mm-diameter shaft is subjected to a 10-kN-m steady bending moment, an 8-kN-m steady torque, and a 120-kN axial force. The yield strength of the shaft material is 600 MPa. Use the MSST and the DET to determine the safety factors for the various types of loading. *Ans.* $n_s = 4.21$.

6.62 Use the MSST and the DET to determine the safety factor for 2024-T4 aluminum alloy for each of the following stress states:

(a) $\sigma_x = 10$ MPa, $\sigma_y = -60$ MPa. *Ans.* $n_{s,\text{MSST}} = 4.64$.

(b) $\sigma_x = \sigma_y = \tau_{xy} = -30$ MPa. *Ans.* $n_{s,\text{DET}} = 5.42$.

(c) $\sigma_x = -\sigma_y = 36$ MPa, and $\tau_{xy} = 65$ MPa. *Ans.* $n_{s,\text{MSST}} = 1.67$.

(d) $\sigma_x = 2\sigma_y = -112$ MPa, and $\tau_{xy} = 33$ MPa. *Ans.* $n_{s,\text{DET}} = 2.31$.

6.63 Four different stress elements, each made of the same material, are loaded as shown in Sketches g, h, i, and j. Use the MSST and the DET to determine which element is the most critical. *Ans.* Sketch g is most critical.

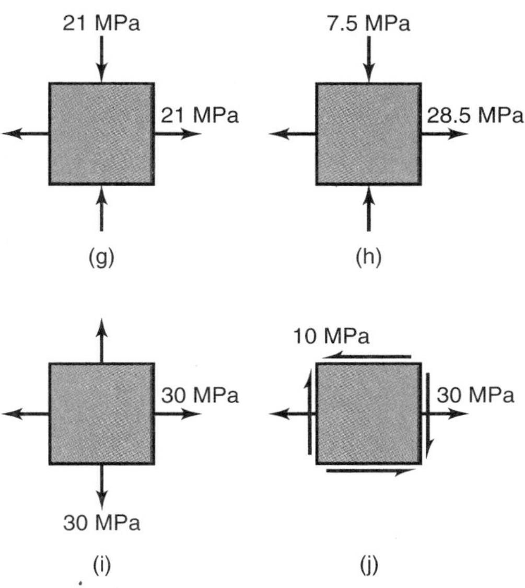

Sketches g, h, i, and j, for Problem 6.63

6.64 The rod shown in Sketch k is made of annealed AISI 1040 steel and has two 90° bends. Use the MSST and the DET to determine the minimum rod diameter for a safety factor of 2 at the most critical section. *Ans.* $d = 35$ mm.

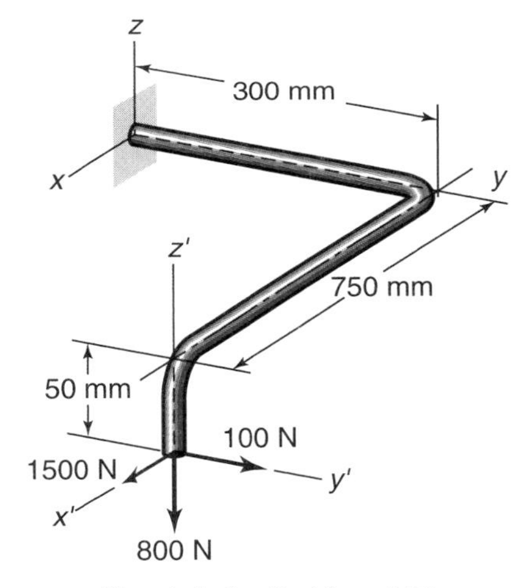

Sketch k, for Problem 6.64

6.65 The shaft shown in Sketch l is made of AISI 1020 steel, quenched and tempered at 870°C. Determine the most critical section by using the MSST and the DET. Dimensions of the various diameters shown in Sketch j are $d = 30$ mm, $D = 45$ mm, and $d_2 = 40$ mm.

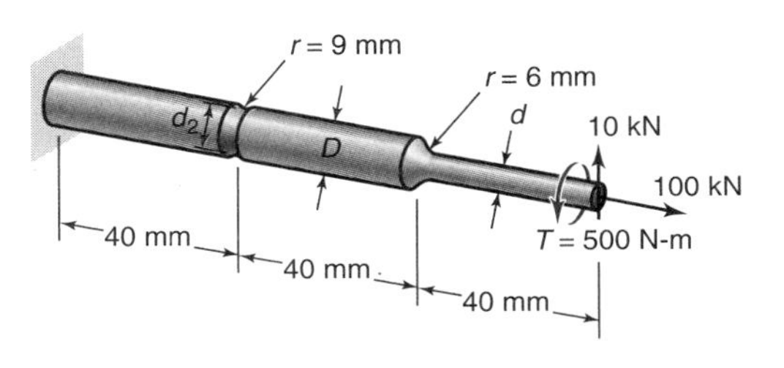

Sketch l, for Problem 6.65

Design and Projects

6.66 Design a test apparatus and procedure that will produce the data depicted in Fig. 6.19.

6.67 A plate of titanium is diffusion bonded to a plate of aluminum. The plates are then loaded in tension. Is there a stress raiser present? A hole is then drilled through the plates, and they are loaded by uniform tension. Sketch the stress distribution through the thickness of the plates at the critical location.

6.68 Perform an Internet search and summarize the important characteristics of Izod and Charpy tests. What properties do these tests measure?

6.69 A steel plate 100 mm in width is to be loaded in tension. If you need to locate a hole with a 25 mm diameter somewhere in the plate, would you place it in the center, at the edge, or an intermediate location? Explain.

6.70 Using a pair of shears, carefully cut the top of a beverage can in half and prepare a drawing of the cross section. What failure mode is used to open the can? Explain.

6.71 Derive the allowable pressure in a pressure vessel as a function of its dimensions and material yield strength for the von Mises and Tresca yield criterion. Consider both a spherical and cylindrical pressure vessel.

6.72 Take a long cylindrical balloon and, with a thin felt-tip pen, mark a small square on it. What will be the shape of this square after you blow up the balloon, (1) a larger square, (2) a rectangle with its long axis in the circumferential directions, (3) a rectangle with its long axis in the longitudinal direction, or (4) an ellipse? Perform this experiment and, based on your observations, explain the results, using appropriate equations. Assume that the material the balloon is made up of is perfectly elastic and isotropic and that this situation represents a thin-walled closed-end cylinder under internal pressure.

6.73 It has been proposed to modify the von Mises yield criterion as

$$(\sigma_1 - \sigma_2)^a + (\sigma_2 - \sigma_3)^a + (\sigma_3 - \sigma_1)^a = \text{Constant}$$

where a is an even integer larger than 2. Plot this yield criterion for $a = 4$ and $a = 12$, along with the Tresca and von Mises criterion, in plane stress. For what types of materials is such a modified yield criterion useful? Explain.

6.74 During plastic deformation, metals act as incompressible solids. Derive the effective Poisson's ratio for plastically deforming metals.

6.75 It is common practice to use a radius on an inner shoulder, and a chamfer at an external corner. (a) Using appropriate sketches and/or equations, explain why this is the case, and why a chamfered inner shoulder is not practical. (b) If a 45° chamfer is produced, it is common to specify a leg, c, of a chamfer to be $c = 0.1\sqrt{d}$, where d is the shaft diameter in millimeters. Replot Fig. 6.5 as a function of diameter using this rule.

6.76 It was mentioned in Section 6.7.2 that some magnesium alloys are well-suited for the internal friction theory because their strength in compression is lower than their strength in tension. Review the technical literature, and write a one-page summary explaining why the behavior of these magnesium alloys is unique.

6.77 Review the technical literature and find the analytical solutions for the case of Fig. 6.3b.

6.78 Sketch m shows an axial plate with a row of circular holes. Qualitatively sketch the stress concentration factor you expect to see as a function of d/b. Then review the technical literature and obtain the stress concentration factor for this situation. A rule of thumb for metals is that rivets should have a spacing not less than three times the diameter. What stress concentration factor results from the associated hole spacing?

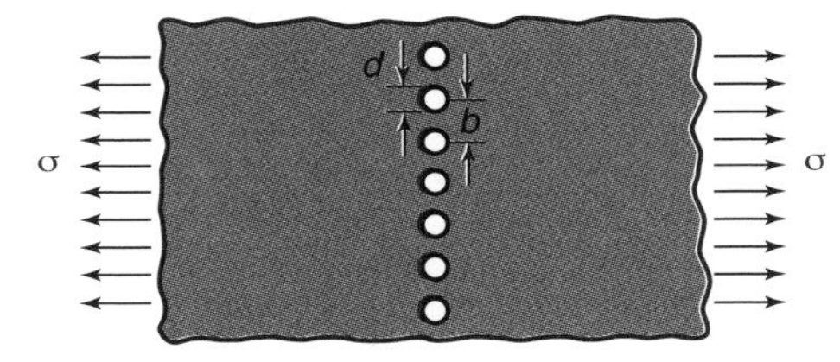

Sketch m, for Problem 6.78.

6.79 For the materials in Table 6.1, calculate the largest allowable center crack that can exist in a long tension member while still yielding. Assume the member has a width of (a) 25 mm, and (b) 1 mm. Include the geometry correction factor, Y, in your analysis.

Chapter 7

Fatigue and Impact

Aloha Airlines Flight 243, a Boeing 737-200, taken April 28, 1988. The midflight fuselage failure was attributed to corrosion-assisted fatigue. *Source:* AP Photos.

All machine and structural designs are problems in fatigue because the forces of Nature are always at work and each object must respond in some fashion.

Carl Osgood, *Fatigue Design*

This chapter introduces the essential concepts of fatigue and impact. Fatigue is a process wherein a material accumulates damage due to a cyclic load; this damage compromises strength and can lead to brittle fracture, even for ductile materials. In fatigue, flaws initially present in the material grow until one dominates; this crack then propagates through the part with every stress cycle. The chapter begins with an analysis of Mode I fracture with uniaxial stresses, and presents S-N diagrams and associated theories for uniaxial fully alternating stresses. Some materials have an endurance limit; if the stresses remain below the endurance limit, then a fatigue failure is not likely. Many factors that affect the endurance limit are discussed, including stress concentrations, surface finish, temperatures, residual stresses, part size, and desired reliability. For materials without an endurance limit, fault tolerant design approaches are needed to avoid fatigue failures. The chapter also describes failure when there is a significant mean stress, as well as when the stress state is more complicated. Fracture mechanics approaches to fatigue are also introduced. The chapter ends with a discussion of impact stresses that result from dynamic loadings.

Contents

Examples

Design Procedures

Case Study

Symbols

A	area, m^2
a	fatigue strength exponent
b	width, m
b_s	slope
C	constant used in Eq. (7.47)
C_1, C_2	integration constants
$\bar{C}$	intercept
c	distance from neutral axis to outer fiber of solid, m
d	diameter, m
E	modulus of elasticity, Pa
e, f	factors used in Eq. (7.19)
g	gravitational acceleration, 9.807 m/s^2 (386.1 in./s^2)
H	height including two notch radii, m
h	height without two notch radii, m
I	area moment of inertia, m^4
I_m	impact factor
K	stress intensity factor, MPa$\sqrt{\text{m}}$
K_c	stress concentration factor
K_f	fatigue stress concentration factor
k	spring rate, N/m
ΔK	stress intensity range, MPa$\sqrt{\text{m}}$
k_f	surface finish factor
k_m	miscellaneous factor
k_r	reliability factor
k_s	size factor
k_t	temperature factor
l_c	crack length, m
M	bending moment, N-m
m	constant used in Eq. (7.47)
N	fatigue life in cycles
N'	number of cycles to failure at a specific stress
N_t'	total number of cycles to failure
n'	number of cycles at a specific stress when $n' < N'$
n_s	safety factor
P	force, N
q_n	notch sensitivity factor
R	stress ratio, $\sigma_{\min}/\sigma_{\max}$
R_a	arithmetic average roughness, m
r	notch radius, m
S_e	modified endurance limit, Pa
S_e'	endurance limit, Pa
S_f	modified fatigue strength, Pa
S_f'	fatigue strength, Pa
S_l'	strength at 10^3 cycles for ductile material, Pa
S_l'	fatigue strength where high-cycle fatigue begins, Pa
S_u	ultimate strength, Pa
S_y	yield strength, Pa
u	sliding velocity, m/s
W	weight, N
Y	plate size correction factor
α	fatigue ductility exponent
α_i	cyclic ratio n_i'/N_i'
δ	deflection, m
δ_{st}	static deflection, m
ϵ_f'	fatigue ductility coefficient
ϕ	phase angle between bending and torsion, rad
$\Delta\epsilon$	total strain
σ	stress, Pa
σ_f'	stress at fracture, Pa
σ_r	stress range, Pa
τ	shear stress, Pa

Subscripts

a	alternating
f	final
i	integer
m	mean
ref	reference

7.1 Introduction

This chapter focuses on understanding and predicting component failure under cyclic and impact loading, conditions that occur in most machine elements. Fluctuating loads induce fluctuating stresses that often result in failure due to cumulative damage. To better understand failures due to fluctuating stresses, consider the back and forth bending of a paper clip. Bending results in compressive and tensile stresses on opposite sides of the paper clip, and these stresses reverse with the bending direction. Thus, the stress at any point around the paper clip will vary as a function of time. Repeated bending of the paper clip will eventually exhaust the material, resulting in failure.

Stresses and deflections in impact loading are in general much greater than those found in static loading; thus, dynamic loading effects are important. A material's physical properties are a function of loading speed. Higher loading rates usually result in increased yield and ultimate strengths. Some examples in which impact loading have to be considered are bumpers and guideways, as well as components of crushing machinery, hammer and hammermills, tampers and the like.

7.2 Fatigue

In the 19th century, the First Industrial Revolution was in full swing. Coal was mined, converted into coke, and used to smelt iron from ore. The iron was used to manufacture bridges, railroads, and trains that brought more coal to the cities, allowing more coke to be produced, etc., in a spiral of ever-increasing production. But then bridges, the monuments to engineering accomplishment, began failing. Even worse, they would fail in extremely confusing ways. A 50-ton locomotive could pass over a bridge without incident; a farmer driving a horse-drawn wagon full of hay would subsequently cause the bridge to collapse. Fear gripped the populace, as people believed death awaited on the bridges (Fig. 7.1).

Today, we no longer fear our "aging infrastructure," certainly not to the extent of the late 19th century. The road construction and improvements that delay traffic from time to time make collapses of civil engineering works, such as bridges, truly rare. But **fatigue** failures, as we now know them to be, are not at all rare in machine elements. Fatigue (and *fatigue wear*, Section 8.9.3) is the single largest cause of failure in metals, estimated to be 90% of all metallic failures. Fatigue failures, especially in structures, are catastrophic and insidious, occurring suddenly and often without warning. For that reason, engineers must apply fatigue considerations in design.

Fatigue has both microscopic and macroscopic aspects. That is, although a rolling-element bearing failing by surface spalling and a shaft breaking in two are quite different events, both the bearing and the shaft have failed due to fatigue. The failure stresses were considerably lower than the yield strengths of the materials. The rolling-element bearing suffered surface fatigue failure; the shaft, structural fatigue

Figure 7.1: "On the Bridge," an illustration from *Punch* magazine in 1891 warning the populace that death was waiting for them on the next bridge. Note the cracks in the iron bridge.

failure. Thus, fatigue failure occurs at relatively low stress levels to a component or structure subjected to fluctuating or cyclic stresses.

Some of the basic concepts associated with fatigue are the following:

1. Fatigue is a complex phenomenon and no universal theories to describe the behavior of materials subjected to cyclic loadings exist; instead, there is a large number of theories, each tailored to particular materials.

2. Fatigue failures act in brittle fashion even in ductile metals; little if any gross plastic deformation is associated with fatigue.

3. Most of the engineering design experience in fatigue is based on an experimental understanding of carbon steel behavior. Much effort has been directed toward extending these semi-empirical rules to other ferrous and nonferrous metals, as well as ceramics, polymers, and composite materials.

4. Fatigue involves the accumulation of damage within a material. Damage usually consists of cracks that grow by a small distance with each stress cycle.

5. Experiments have shown that fatigue cracks generally begin at a surface and propagate through the bulk. Therefore, much attention is paid to the quality of surfaces in fatigue-susceptible machine elements. However, if large subsurface flaws or stress raisers exist in the substrate, fatigue cracks can initiate below the surface.

6. Flaws grow uniformly and simultaneously at several sites until one flaw becomes dominant and grows more rapidly than the others. At this point, the flaw is considered a crack.

7. Fatigue testing is imperative to confirm safe mechanical design.

The last of these points cannot be overemphasized. This book concentrates on theoretical approaches that can guide a designer, but there is much uncertainty in fatigue failures. Guidelines will be presented that are often difficult to directly apply to specific materials, manufacturing processes, or components. Application of these methodologies without experimental verification of designs is acceptable for students, preliminary design or design analysis, but rarely for commercial products. It is also common to use large safety factors whenever possible for components subjected to fatigue.

The total life of a component or structure is the time it takes a crack to initiate plus the time it needs to propagate through the cross-section. Design life can be maximized as summarized in Design Procedure 7.1.

Design Procedure 7.1: Methods to Maximize Design Life

1. *Minimizing initial flaws, especially surface flaws.* Great care is taken to produce fatigue-resistant surfaces through processes such as grinding or polishing that produce exceptionally smooth surfaces. These surfaces are then carefully protected before a product is placed into service.

2. *Maximizing crack initiation time.* Compressive surface residual stresses are imparted (or at least tensile residual stresses are relieved) through manufacturing processes such as shot peening or burnishing, or by a number of surface treatments.

3. *Maximizing crack propagation time.* Substrate properties, especially those that retard crack growth, are also important. For example, in some materials fatigue cracks will propagate more quickly along grain boundaries than through grains. In this case, using a material that has elongated grains transverse to the direction of fatigue crack growth can extend fatigue life (e.g., by using cold-worked components instead of castings).

4. *Maximizing the critical crack length.* Fracture toughness (Section 6.5) is an essential material property, and materials with higher fracture toughnesses are generally better suited for fatigue applications.

Given a finite number of resources, which one of the approaches in Design Procedure 7.1 should the designer emphasize? Ever-smoother surfaces can be manufactured at ever-increasing costs but at ever-decreasing payoff as the surface valleys become smaller. Maximizing initiation time allows parts to function longer with little or no loss in performance. Maximizing propagation time can allow cracks to be detected before they become catastrophic, the approach previously called the "doctrine of manifest danger" (Section 1.4.1). Maximizing the critical crack length can lead to long life with a greater likelihood of recognizing an imminent failure. All of these approaches have merit and are pursued from time to time. The proper emphasis, just as with the safety factor, is product and material specific and is best decided through experience.

The designer's job is to select a material and a processing route that will lead to successful products. However, fatigue is an extremely complex subject. This chapter introduces approaches to design for cyclic loading. In practice, extreme caution should be used in applying any of these approaches without generous safety factors and/or experimental confirmation.

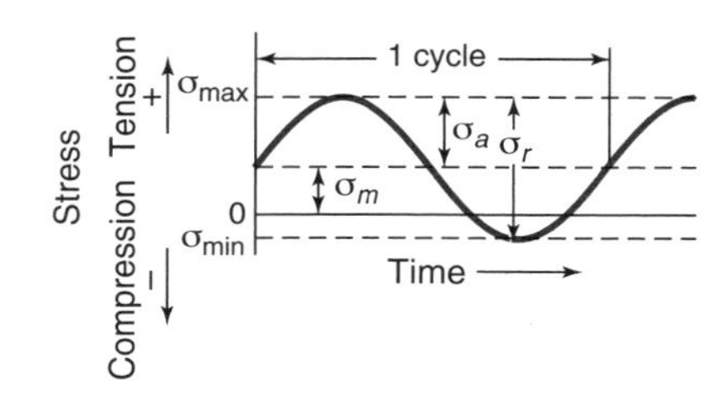

Figure 7.2: Variation in nonzero cyclic mean stress.

7.3 Cyclic Stresses

A **cyclic stress** is time dependent, but the variation is such that the stress sequence repeats itself. The stresses may be axial (tensile or compressive), flexural (bending), or torsional (twisting). Figure 7.2 shows an idealized cyclic variation of a stress that varies with time. Also shown are several parameters used to characterize fluctuating cyclic stress. A stress cycle ($N = 1$) constitutes a single application and removal of a load and then another application and removal of the load in the opposite direction. Thus, $N = 1/2$ means that the load is applied once and then released. The stress amplitude alternates about a **mean stress**, σ_m, defined as

$$\sigma_m = \frac{\sigma_{\max} + \sigma_{\min}}{2}. \tag{7.1}$$

The **stress range**, σ_r, is the difference between $\sigma_{\max}$ and $\sigma_{\min}$, namely,

$$\sigma_r = \sigma_{\max} - \sigma_{\min}. \tag{7.2}$$

The **stress amplitude**, σ_a, is one-half of the stress range, or

$$\sigma_a = \frac{\sigma_r}{2} = \frac{\sigma_{\max} - \sigma_{\min}}{2}. \tag{7.3}$$

The **stress ratio**, R, is the ratio of minimum and maximum stress, or

$$R = \frac{\sigma_{\min}}{\sigma_{\max}}. \tag{7.4}$$

Four frequently encountered patterns of constant-amplitude cyclic stress are:

1. **Completely reversed** ($\sigma_m = 0, R = -1$)
2. **Nonzero mean** (as shown in Fig. 7.2)
3. **Released tension** ($\sigma_{\min} = 0, R = 0, \sigma_m = \sigma_{\max}/2$)
4. **Released compression** ($\sigma_{\max} = 0, R = \infty, \sigma_m = \sigma_{\min}/2$)

Example 7.1: Cyclic Stresses

Given: Breakaway bolts are used to fasten street light poles and fire hydrants to their foundations; these bolts are intended to withstand normal loads such as from the wind, but when struck by a vehicle, their limited ductility causes them to fail. The pole or hydrant is then allowed to move and this reduces the severity of an automobile impact. Consider the case where a bolt is tightened so that it is in tension, but a wind loading applies a compressive stress, so that the repeating stress is $\sigma = \pm 100$ MPa.

Find: The mean stress, the range of stress, the stress amplitude, and the stress ratio for the breakaway bolt.

Solution: Equations (7.1) through (7.4) are needed to solve this problem. The mean stress is

$$\sigma_m = \frac{\sigma_{\max} + \sigma_{\min}}{2} = \frac{100 - 100}{2} = 0.$$

The range of stress is

$$\sigma_r = \sigma_{\max} - \sigma_{\min} = 100 - (-100) = 200 \text{ MPa}.$$

The alternating stress is

$$\sigma_a = \frac{\sigma_r}{2} = \frac{200}{2} = 100 \text{ MPa}.$$

The stress ratio is

$$R = \frac{\sigma_{\min}}{\sigma_{\max}} = \frac{-100}{100} = -1.$$

7.4 Strain Life Theory of Fatigue

As discussed in Section 7.2, fatigue is a damage accumulation process that manifests itself through crack propagation, but no crack propagation is possible without plastic deformation at the crack tip. Although the volume stressed highly enough for plastic deformation can be extremely small, if the stress fields remain elastic, no crack propagation is possible. However, recall from Section 6.2 that the stress concentration associated with a sharp crack is very large, and even low stresses will therefore result in plastic deformation at the crack tip. Further, the use of bulk material properties, such as yield strength or ultimate strength, presents difficulties because cyclic loadings can change these values near a crack tip. They may increase or decrease depending on the material and its manufacturing history. Thus, the material strength at the location where cracks are propagating can differ from the bulk material strength listed in handbooks or obtained from tension tests.

Given these difficulties, several approaches have been suggested for dealing with the strain encountered at a crack tip and predicting fatigue failure. One of the better known is the **Manson-Coffin relationship**, which gives the total strain amplitude as the sum of the elastic and plastic strain amplitudes and relates this to life:

$$\frac{\Delta\epsilon}{2} = \frac{\sigma_f'}{E}\left(2N'\right)^a + \epsilon_f'\left(2N'\right)^\alpha, \tag{7.5}$$

where

$\Delta\epsilon$ = total strain, including both plastic and elastic components
σ_f' = stress at fracture in one stress cycle, Pa
E = elastic modulus of material, Pa
N' = number of cycles that will occur before failure
ϵ_f' = fatigue ductility coefficient (true strain corresponding to fracture in one stress cycle)
a = fatigue strength exponent
α = fatigue ductility exponent

Table 7.1 gives some typical values of the material properties a and α.

The Manson-Coffin relationship is difficult to use in practice because the total strain, $\Delta\epsilon$, is difficult to determine. Stress concentration factors, such as those presented in Chapter 6, are readily available in the technical literature, but strain concentration factors in the plastic range are nowhere to be

Table 7.1: Cyclic properties of some metals. *Source:* After Shigley and Mitchell [1983] and Suresh [1998].

Material	Condition[a]	Yield strength S_y, MPa	Fracture strength σ'_f MPa	Fatigue ductility coefficient, ϵ'_f	Fatigue strength exponent, a	Fatigue ductility exponent, α
Steel						
1015	Normalized	228	827	0.95	–0.110	–0.64
4340	Tempered	1172	1655	0.73	–0.076	–0.62
1045	Q&T[a] 80°F	—	2140	—	–0.065	–1.00
1045	Q&T 306°F	1720	2720	0.07	–0.055	–0.60
1045	Q&T 500°F	1275	2275	0.25	–0.080	–0.68
1045	Q&T 600°F	965	1790	0.35	–0.070	–0.69
4142	Q&T 80°F	2070	2585	—	–0.075	–1.00
4142	Q&T 400°F	1720	2650	0.07	–0.076	–0.76
4142	Q&T 600°F	1340	2170	0.09	–0.081	–0.66
4142	Q&T 700°F	1070	2000	0.40	–0.080	–0.73
4142	Q&T 840°F	900	1550	0.45	–0.080	–0.75
Aluminum						
1100	Annealed	97	193	1.80	–0.106	–0.69
2014	T6	462	848	0.42	–0.106	–0.65
2024	T351	379	1103	0.22	–0.124	–0.59
5456	H311	234	724	0.46	–0.110	–0.67
7075	T6	469	1317	0.19	–0.126	–0.52
Titanium						
Ti-6Al-4V	Solution treated+aged	1185	2030	0.841	–0.104	–0.69
Nickel						
Inconel X	Annealed	700	2255	1.16	–0.117	–0.75

[a] Q&T: Quenched and tempered.

found. The advantage of the Manson-Coffin equation is that it gives insight into important properties in fatigue strength determination. It shows the importance of strength as well as ductility, and it leads to the conclusion that as long as there is a cyclic plastic strain, no matter how small, there will eventually be failure.

7.5 Fatigue Strength

7.5.1 Rotating-Beam Experiments

Fatigue is inherently probabilistic; that is, there is a great range of performance within samples prepared from the same materials. In previous problems and case studies, a valuable approach called the worst-case scenario was described. To apply this approach to fatigue, a designer would select surface finishes, notch sizes, initial flaw size, etc, that minimize the fatigue strength of the candidate specimen. However, this process would result in fatigue specimens with zero strength, a situation that does nothing to aid designers. Thus, data on fatigue often reflect the best-case scenario, and do not reflect actual environments. The designer is strongly cautioned that great care must be taken in applying fatigue design theories based on best case scenarios to critical applications.

Because fatigue is a damage accumulation phenomenon, initial flaws have a large effect on performance. No manufacturing process produces defect-free parts; indeed, it is not uncommon to encounter thousands, even millions, of flaws per cubic millimeter. The flaws are distributed in size, shape, location and orientation, they are often close enough to violate Saint-Venant's principal (see Section 4.3), so that the associated stress concentrations interfere with and compound each other. Analytical approaches that derive fatigue strengths from first principles are thus very difficult, and most knowledge on material fatigue is experiment-based.

Experimental approaches to fatigue use either exemplars or idealized, standard specimens. The former are more reliable and best for critical applications. The latter are often used when a direct simulation of the loading environment is cost prohibitive.

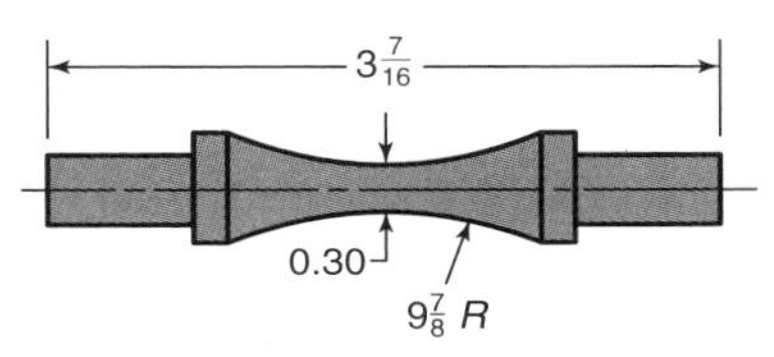

Figure 7.3: R.R. Moore machine fatigue test specimen. Dimensions in inches.

To establish the fatigue strength of an exemplar, a series of tests is performed. The test apparatus duplicates as nearly as possible the stress conditions (stress level, time frequency, stress pattern, etc.) in practice. The exemplar duplicates as nearly as possible any manufacturing and treatment processes. Such experiments give the most direct indication of a component's survivability in the actual loading environment.

To test idealized, standard specimens, a rotating-beam fatigue testing machine is often used, such as the Moore rotating-beam machine. The specimen is subjected to pure bending, and no transverse shear is imposed. The specimen has specific dimensions (Fig. 7.3) and a highly polished surface, with a final polishing in the axial direction to avoid circumferential scratches. If the specimen breaks into two equal pieces, the test is indicative of the material's fatigue strength. If the pieces are unequal, a material or surface flaw has skewed the results. The test specimen is subjected to completely reversed ($\sigma_m = 0$) stress cycling at a relatively large maximum stress amplitude, usually two-thirds of the static ultimate strength, and the cycles to failure are counted. Thus, for each specimen at a specific stress level, the test is conducted until failure occurs. The procedure is repeated on other identical specimens, progressively decreasing the maximum stress amplitude.

7.5.2 Regimes of Fatigue Crack Growth

Figure 7.4a shows the size of a fatigue crack as a function of number of cycles for two stress ratios. Figure 7.4b illustrates the rate of crack growth, and more clearly shows three different regimes of crack growth:

1. *Regime A* is a period of very slow crack growth. Note that the crack growth rate can be even smaller than an atomic spacing of the material per cycle. Regime A should be recognized as a period of non-continuum failure processes. The fracture surfaces are faceted or serrated in this regime, indicating crack growth is primarily due to shear deformations within a grain. The growth rate is so small that crack lengths may be negligible over the life of the component if this regime is dominant. Regime A is strongly affected by material microstructure, environmental effects, and stress ratio, R.

2. *Regime B* is a period of moderate crack growth rate, often referred to as the **Paris regime**. In this regime, the rate of crack growth is influenced by several factors, involving material microstructure, mechanical load variables, and the environment. Thus, it is not surprising that crack propagation rates cannot be determined for a given material or alloy from first principles, and testing is required to quantify the growth rate.

3. *Regime C* is a period of high-growth rate, where the maximum stress intensity factor for the fatigue cycle approaches the fracture toughness of the material. Material microstructural effects and loadings have a large influence on crack growth, and additional static modes such as cleavage and intergranular separation can occur.

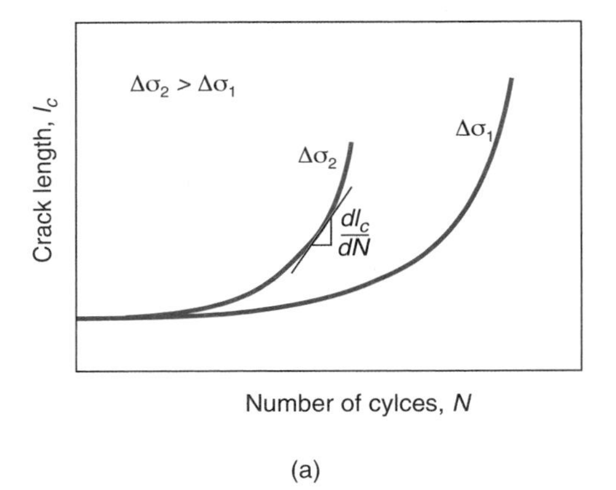

(a)

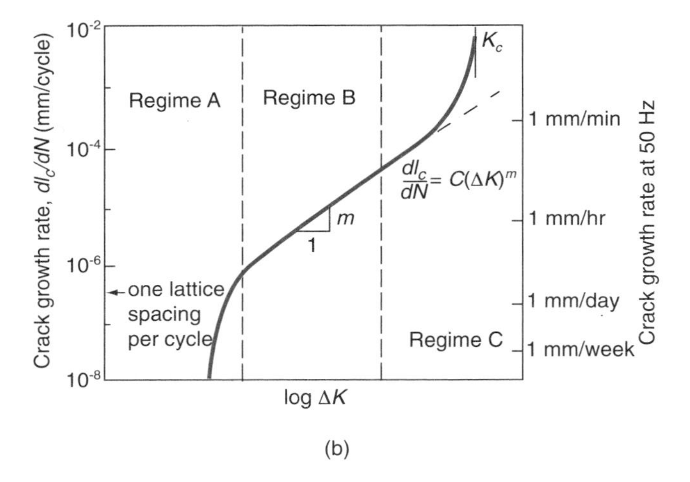

(b)

Figure 7.4: Illustration of fatigue crack growth. (a) Size of a fatigue crack for two different stress ratios as a function of the number of cycles; (b) rate of crack growth, illustrating three regimes.

7.5.3 Microstructure of Fatigue Failures

As discussed above, even the most ductile materials can exhibit brittle behavior in fatigue, and will fracture with little or no plastic deformation. The reasons for this are not at all obvious, but an investigation of fatigue fracture microstructure can help explain this behavior.

A typical fatigue fracture surface is shown in Fig. 7.5, and has the following features:

1. Near the origin of the fatigue crack (Point B in Fig. 7.5), the surface is *burnished*, or very smooth. In the early stages of fatigue, the crack grows slowly and elastic deformations result in microscopic sliding between the two surfaces, resulting in a rubbing of the surfaces and associated mechanical polishing.

2. Near the final fracture location (Point A in Fig. 7.5), *striations* or *beachmarks* are clearly visible to the naked eye. During the last few cycles of a fatigue failure, the crack growth is very rapid, and these striations are indicative of fast growth and growth-arrest processes.

3. Microscopic striations can exist between these two extremes as shown in Fig. 7.5, and are produced by the slower growth of fatigue cracks at this location in the part.

4. The final fracture surface often looks rough and is indicative of brittle fracture, but it can also appear ductile depending on the material.

The actual pattern of striations depends on the particular geometry, material, and loading (Fig. 7.6), and can require experience to evaluate a failure cause.

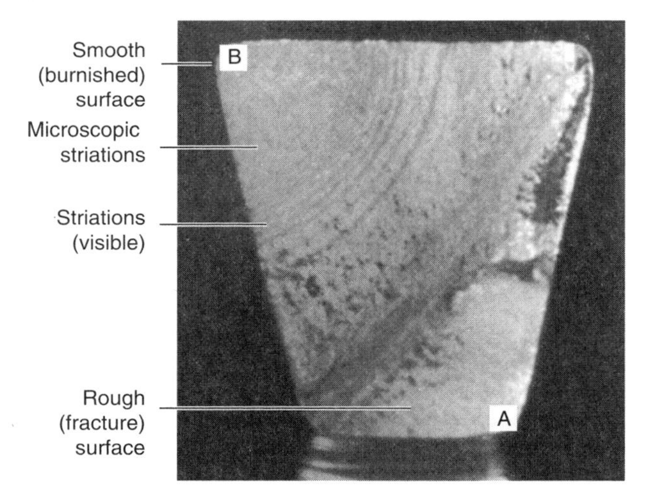

Figure 7.5: Cross section of a fatigued section, showing fatigue striations or beachmarks originating from a fatigue crack at B. *Source:* Rimnac, C., et al., in *ASTM STP 918, Case Histories Involving Fatigue and Fracture*, copyright 1986, ASTM International. Reprinted with permission.

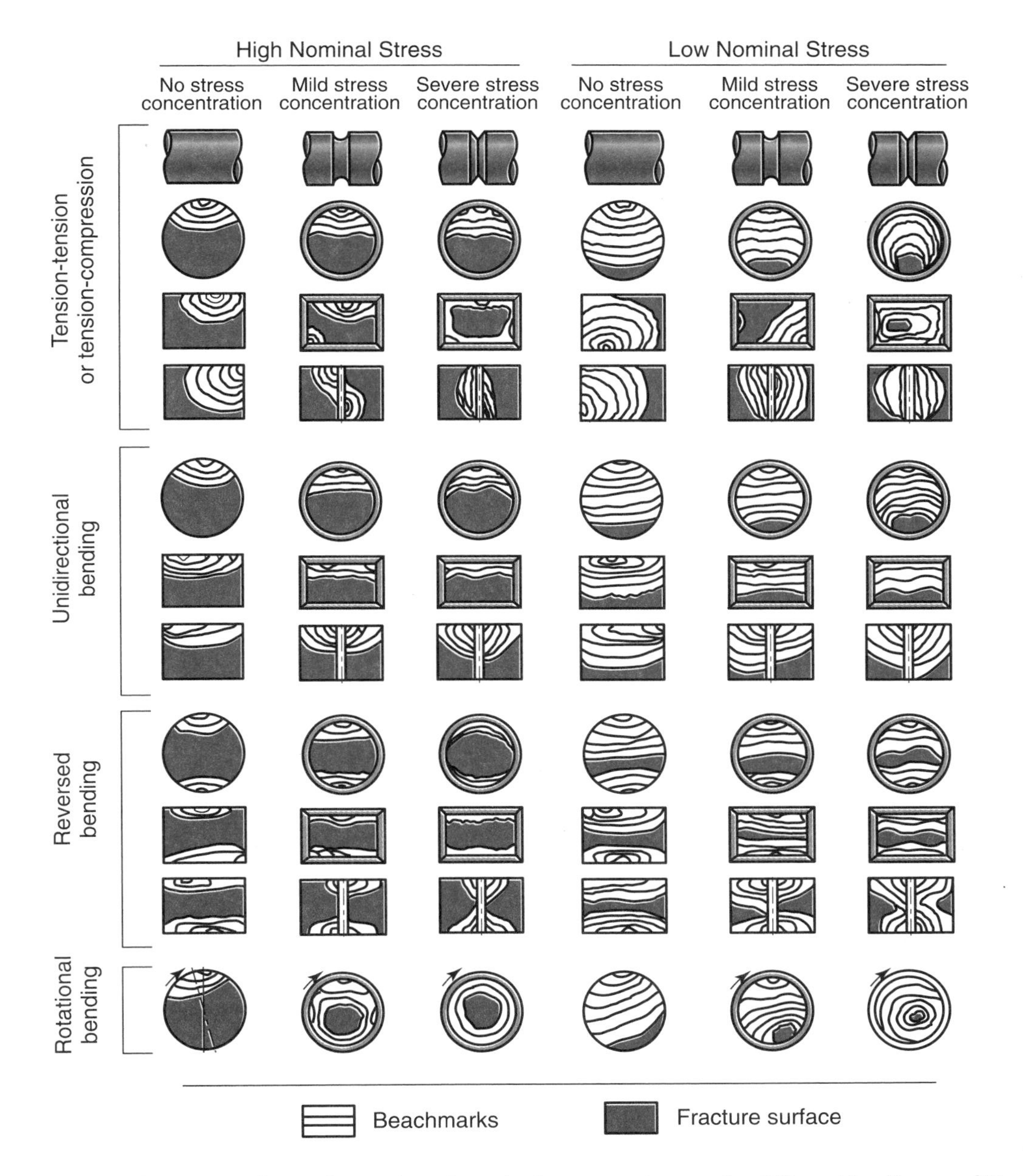

Figure 7.6: Typical fatigue-fracture surfaces of smooth and notched cross-sections under different loading conditions and stress levels. *Source:* Adapted from *Metals Handbook*, American Society for Metals [1975].

7.5.4 S-N Diagrams

Data from reversed bending experiments are plotted as the fatigue strength versus the logarithm of the total number of cycles to failure, N_t', for each specimen. These plots are called **S-N diagrams** or **Wöhler diagrams** after August Wöhler, a German engineer who published his fatigue research in 1870. They are a standard method of presenting fatigue data and are useful and informative. Two general patterns for two classes of material, those with and those without endurance limits, emerge when plotting the fatigue strength versus the logarithm of the number of cycles to failure. Figure 7.7 shows typical results for several materials. Figure 7.7a presents test data for wrought steel. Note the large amount of scatter in the data, even with the great care used to prepare test specimens. Thus, material properties extracted from curves such as those in Fig. 7.7 are all somewhat suspect, and have significant variation between test specimens. Figure 7.7a also shows a common result. For some materials with **endurance limits**, such as ferrous and titanium alloys, a change in slope occurs at low stress levels, called a "knee" in the curve. This implies that an endurance limit S_e' is reached, below which failure will not occur (although this is strictly not true — see Section 7.6.3). This endurance limit S_e' represents the largest fluctuating stress that will *not* cause failure for an infinite number of cycles. For many steels the endurance limit ranges between 35 and 60% of the ultimate strength.

Most nonferrous alloys (e.g., aluminum, copper, and magnesium) *do not* have a significant endurance limit. Their fatigue strength continues to decrease with increasing cycles. Thus, fatigue will occur regardless of the stress amplitude. The fatigue strength for these materials is taken as the stress level at which failure will occur for some specified number of cycles (e.g., 10^6 or 10^7 cycles).

Determining the endurance limit experimentally is lengthy and expensive. The Manson-Coffin relationship given by Eq. (7.5) demonstrates that the fatigue life will depend on the material's fracture strength during a single load cycle, suggesting a possible relationship between static material strength and strength in fatigue. Such a relationship has been noted by several researchers (see Fig. 7.8). The stress endurance limits of steel for three types of loading can be approximated as

$$\begin{aligned} \text{bending}: &\quad S_e' = 0.5S_u \\ \text{axial}: &\quad S_e' = 0.45S_u \\ \text{torsion}: &\quad S_e' = 0.29S_u \end{aligned} \tag{7.6}$$

Equation (7.6) can be used to approximate the endurance limits for other ferrous alloys but it must be recognized that the limits can vary significantly from experimentally determined endurance limits. As depicted by the dashed line in Fig. 7.8, the maximum value of the endurance limit for ferrous alloys is taken as 690 MPa (100 ksi), regardless of the predictions from Eq. (7.6). Even if the ultimate strength and the type of loading are known for other ferrous metals, their endurance limits can be only be approximated from Eq. (7.6).

Other materials, for which there is much less experience, are nevertheless finding increasing uses in fatigue applications. Table 7.2 gives the approximate strengths in fatigue for various material classes. Figure 7.7c gives some stress-life curves for common polymers. Because polymers have a much greater variation in properties than metals, Fig. 7.7c should be viewed as illustrative of fatigue properties and not used for quantitative data.

Given a new material, or when experimental verification of an endurance limit is needed, it is often not required to develop the entire S-N diagram. The designer may only wish to

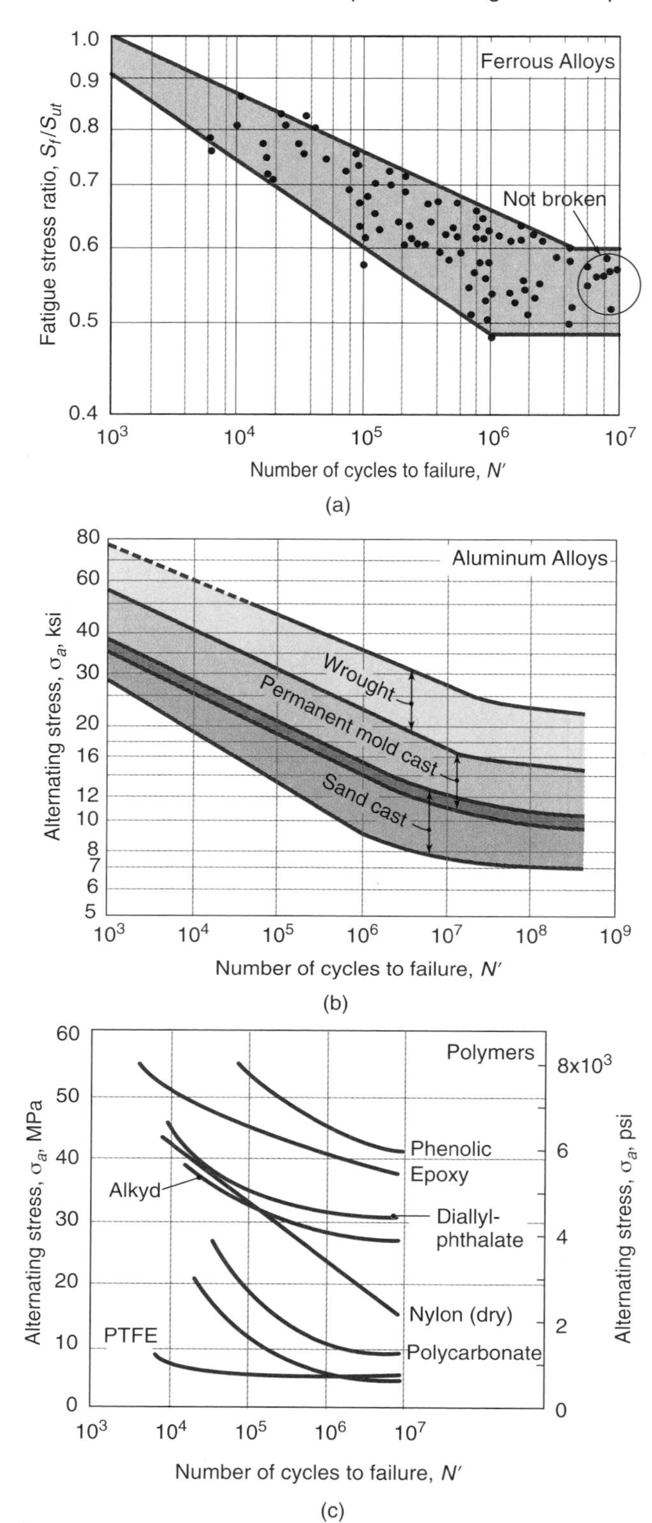

Figure 7.7: Fatigue strength as a function of number of loading cycles. (a) Ferrous alloys, showing clear endurance limit; (b) aluminum alloys, with less pronounced knee and no endurance limit; (c) selected properties of assorted polymer classes. *Source:* (a) Adapted from Lipson and Juvinall [1963], (b) Adapted from Juvinall and Marshek [1991], (c) Adapted from Norton [1996].

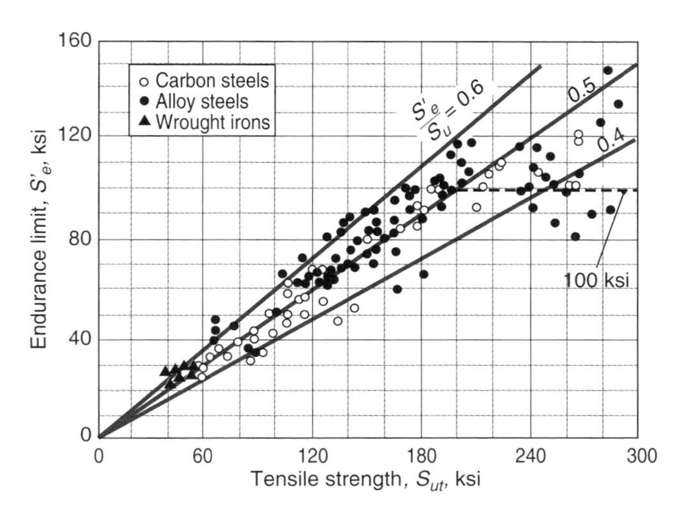

Figure 7.8: Endurance limit as a function of ultimate strength for wrought steels. *Source:* Adapted from Shigley and Mitchell [1983].

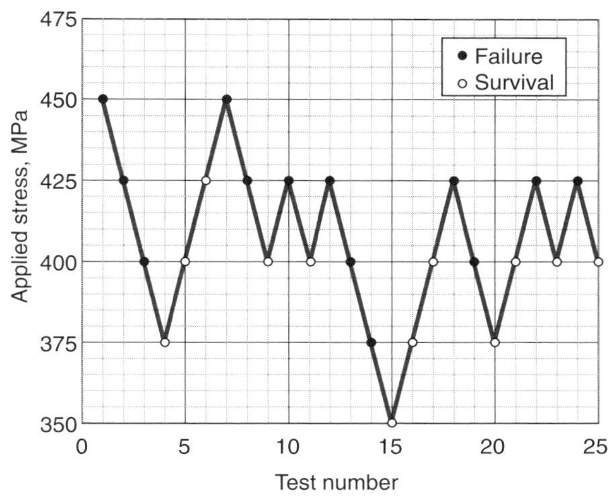

Figure 7.9: Typical results from fatigue tests using the staircase approach, and used in Example 7.2.

experimentally obtain the endurance limit, or the endurance limit at a desired number of stress cycles, using a minimum number of specimens in order to control costs and time required for evaluation. A valuable approach in this case is the **staircase** approach, also known as the *up-and-down* method as outlined in Design Procedure 7.2.

Design Procedure 7.2: Staircase Approach

The staircase approach is useful for determining the endurance limit of a material when the complete S-N curve is not needed. This approach involves the following steps:

1. A designer must first estimate the endurance limit for the material of interest, either with a strength-based approach such as in Eq. (7.6), or through preliminary testing.
2. A test interval is then selected, typically around 10% of the estimated endurance limit.
3. An initial test is performed at a stress level equal to the expected endurance limit.
4. If the specimen breaks, it is recorded as such and the next experiment will be performed at a stress level reduced by the stress interval.
5. At the desired duration (commonly 10^6 or 10^7 cycles), the test is stopped. If the specimen survives, it is recorded as such and the next experiment will be performed at a stress level increased by the stress interval.
6. A plot of typical results is shown in Fig. 7.9.
7. The mean endurance limit can be obtained from the following steps:
   (a) Count the number of failures and survivals in the test results. Proceed with the analysis using the less common test result.
   (b) The number of events (failures or survivals) is assigned to n_i for each stress level σ_i. In this approach, the lowest stress level is denoted as σ_o, the next highest as σ_1, etc.
   (c) Obtain the quantity A_n from

   $$A_n = \sum i n_i.$$

   (d) The endurance limit is then estimated from

   $$S'_e = \sigma_o + d\left(\frac{A_n}{\sum n_i} \pm \frac{1}{2}\right),$$

   where the plus sign is used if the more common experimental result is survival, and the minus sign is used if the more common event is failure.
8. It is recommended that at least 15 experiments be performed, although more can be helpful for more accurate quantification of the endurance limit.

The staircase approach can provide information about the mean endurance limit, and it also can allow characterization of the nature of the strength distribution. Further details about the potential benefits of this approach are contained in Lee, et al. [2005]. This approach is useful for evaluating endurance limit modification factors as well (see Section 7.8).

Example 7.2: Determination of the Endurance Limit using the Staircase Approach

Given: The results of a series of 25 fatigue tests shown in Fig. 7.9. The expected endurance limit was $\sigma_1 = 450$ MPa, with an interval of $d = 25$ MPa.

Find: Determine the endurance limit, S_e, of the material.

Table 7.2: Approximate endurance limit for various materials.

Material	Number of cycles	Relation
Magnesium alloys	10^8	$S_e' = 0.35S_u$
Copper alloys	10^8	$0.25S_u < S_e' < 0.5S_u$
Nickel alloys	10^8	$0.35S_u < S_e' < 0.5S_u$
Titanium	10^7	$0.45S_u < S_e' < 0.65S_u$
Aluminum alloys	5×10^8	$S_e' = 0.40S_u$ (S_u < 48 ksi)
		$S_e' = 19$ ksi ($S_u \geq$ 48 ksi)

Solution: This solution will follow the approach in Design Procedure 7.2, where the first six steps have been completed. From Step 7a, and Fig. 7.9, it can be seen that there are 13 failures and 12 survivals. Therefore, survival is the least common event, and will be used in this analysis. According to Step 7b, the following data are recorded:

i	σ_i (MPa)	n_i
0	350	1
1	375	3
2	400	7
3	425	1
4	450	0

According to Step 7c, A_n is calculated from

$$A_n = \sum in_i = 0(1) + (1)(3) + 2(7) + (3)(1) = 20.$$

Therefore, according to Step 7d, the endurance limit is given by (noting that the more common result is failure):

$$\begin{aligned} S_e' &= \sigma_o + d\left(\frac{A_n}{\sum n_i} - \frac{1}{2}\right) \\ &= 350 + (25)\left(\frac{20}{12} - \frac{1}{2}\right) \\ &= 379 \text{ MPa.} \end{aligned}$$

7.6 Fatigue Regimes

The S-N diagram (Fig. 7.7a) shows different types of behavior as the number of cycles to failure increases. Two basic regimes are **low-cycle fatigue** (generally below 1000 stress cycles) and **high-cycle fatigue** (more than around 1000 but less than one million stress cycles). The slope of the line is much lower in low-cycle fatigue than in high-cycle fatigue.

Another differentiation can be made between **finite life** and **infinite life**. Figure 7.7a shows an endurance limit for ferrous alloys, below which any repeating stress will lead to infinite life in the component. Although a distinction between the finite-life and infinite-life portions of the curve is not always clear, for steels it occurs between 10^6 and 10^7 cycles. Thus, the finite-life classification is considered to hold for any loading below this range.

7.6.1 Low-Cycle Fatigue

Low-cycle fatigue is any loading that causes failure below 1000 cycles. This type of loading is common. A number of devices, such as latches on automotive glove compartments, studs on truck wheels, and setscrews fixing gear locations on shafts, cycle fewer than 1000 times during their service lives. Surviving 1000 cycles means that these devices will last as long as intended.

For components in the low-cycle range, designers either ignore fatigue effects entirely or they reduce the allowable stress level. Ignoring fatigue seems to be a poor approach. However, the low-cycle portion of the curve in Fig. 7.7a has a small slope (i.e., the strength at 1000 cycles has not been reduced a great deal). Further, the y-intercept for the curve is the ultimate strength, not the yield strength. Since static design often uses the yield strength and not the ultimate strength in defining allowable stresses, static approaches are acceptable for designing low-cycle components. In fact, the safety factor compensates for the uncertainty in material strength due to cyclic loading.

Taking low-cycle effects into account allows modifying the material strength based on experimental data. The fatigue strength for steel at which high-cycle fatigue begins can be approximated as

$$\begin{array}{ll} \text{bending:} & S_l' = 0.9S_u \\ \text{axial:} & S_l' = 0.75S_u \\ \text{torsion:} & S_l' = 0.72S_u \end{array} \tag{7.7}$$

7.6.2 High-Cycle, Finite-Life Fatigue

In many applications the number of stress cycles placed on a component during its service life is between 10^3 and 10^7. Examples include car door hinges, aircraft body panels, and aluminum softball bats. Because the strength drops rapidly in this range (Fig. 7.7), strength needs to be expressed as a function of loading cycles. The fatigue strength at any location between S_l' and S_e' can generally be expressed as

$$\log S_f' = b_s \log N_t' + \bar{C}, \tag{7.8}$$

where b_s is the slope and $\bar{C}$ the intercept of the finite-life portion of the S-N diagram. At the end points ($S = S_l'$ for $N = 1000$ and $S = S_e'$ for $N = 10^6$), Eq. (7.8) becomes

$$\log S_l' = b_s \log(10^3) + \bar{C} = 3b_s + \bar{C}, \tag{7.9}$$

$$\log S_e' = b_s \log(10^6) + \bar{C} = 6b_s + \bar{C}. \tag{7.10}$$

Subtracting Eq. (7.10) from Eq. (7.9) gives

$$b_s = -\frac{1}{3}\log\left(\frac{S_l'}{S_e'}\right). \tag{7.11}$$

Substituting Eq. (7.11) into Eq. (7.10) gives

$$\bar{C} = 2\log\left(\frac{S_l'}{S_e'}\right) + \log S_e' = \log\left[\frac{(S_l')^2}{S_e'}\right]. \tag{7.12}$$

Thus, by using Eqs. (7.6) and (7.7), the slope b_s and the intercept $\bar{C}$ can be determined for a specific type of loading. Knowing the slope and the intercept from Eq. (7.8) yields the fatigue strength as

$$S_f' = 10^{\bar{C}}\left(N_t'\right)^{b_s} \qquad \text{for} \qquad 10^3 \leq N_t' \leq 10^6. \tag{7.13}$$

If the fatigue strength is given and the number of cycles until failure is desired,

$$N_t' = \left(S_f' 10^{-\bar{C}}\right)^{1/b_s} \qquad \text{for} \qquad 10^3 \leq N_t' \leq 10^6. \tag{7.14}$$

Thus, by using Eqs. (7.6), (7.7), (7.11), and (7.12), the fatigue strength can be obtained from Eq. (7.13) or the number of cycles to failure can be determined from Eq. (7.14).

Example 7.3: High Cycle, Finite Life Fatigue

Given: The pressure vessel lids of nuclear power plants are bolted down to seal the high pressure exerted by the water vapor (in a boiler reactor) or the pressurized water (in a pressurized water reactor). The ultimate strength of the bolt material is 1080 MPa. In the current design, the bolts are so heavily stressed that they are replaced after the reactors are opened 25 times.

Find: Determine the required stress for a life of 10,000 cycles.

Solution: Equations (7.6) and (7.7) for axial loading give $S_e' = 0.45S_u$ and $S_l' = 0.75S_u$. Note that S_l' is for 1000 cycles and S_e' is for a life of 10^6 cycles. Equation (7.11) gives the slope as

$$b_s = -\frac{1}{3}\log\left(\frac{S_l'}{S_e'}\right) = -\frac{1}{3}\log\left(\frac{0.75S_u}{0.45S_u}\right) = -0.07395.$$

From Eq. (7.12), the intercept is

$$\bar{C} = \log\left[\frac{(S_l')^2}{S_e'}\right] = \log\left[\frac{(0.75)^2(1080)^2}{(0.45)(1080)}\right] = 3.130.$$

Knowing the slope and intercept, Eq. (7.13) gives the fatigue strength as

$$S_f' = 10^{\bar{C}}\left(N_t'\right)^{b_s} = 10^{3.13}\,(10,000)^{-0.07395} = 682.7\text{ MPa}.$$

Thus, the stress has to be decreased to 682.7 MPa to achieve a life of 10,000 cycles.

7.6.3 High-Cycle, Infinite-Life Fatigue

A number of applications call for infinite life, defined for steels as the number of cycles above which an endurance limit can be defined, usually taken as 10^6 cycles. If a material does not have an endurance limit, it cannot be designed for infinite life. Thus, aluminum alloys, for example, will always be designed for finite life (using the approach given in Section 7.6.2) or a fracture mechanics approach will be used (Section 7.12).

It should be recognized that the Manson-Coffin relationship given by Eq. (7.5) suggests that failure will always occur so long as the material encounters strain. Experimental investigations have confirmed this; there is apparently no cyclic stress level that materials can withstand without eventually failing by fatigue. However, the following should be noted:

- With materials such as the carbon steels in Fig. 7.7a, the knee in the curve is pronounced, so that the S-N diagram becomes almost horizontal at around a million stress cycles.

- Note that the x-axis of Fig. 7.7a is on a log scale. Thus, stresses below the endurance limit will cause failure, but only after a very large number of cycles. Often referred to as **gigacycle fatigue** because of the numbers of cycles that are typically involved, this regime has unique behavior. For example, instead of cracks propagating from a surface or stress concentration, gigacycle fatigue failures often initiate at sub-surface material flaws.

- If a part will survive many millions, billions, or even trillions of cycles, this is essentially infinite life. That is, for most components this would translate into a useful life that far exceeds the intended life of the machine itself, so the part would never fail. Another rationale is that machine failures will undoubtedly occur in other components first, and that the machine will certainly be discarded before gigacycle fatigue becomes an issue.

For ferrous and titanium alloys, however, an infinite-life design approach based on an endurance limit can be followed. Basically, the designer determines an endurance limit and uses this strength as the allowable stress. Then, sizing and selection of components can proceed just as in static design. This approach, which is fairly complex, is described in the next sections.

7.7 Stress Concentration Effects

The Manson-Coffin equation [Eq. (7.5)] showed that the life of a component has a direct correlation with the strain to which it is subjected. Because locations of stress concentration are also locations of strain concentration, these locations can be seen as prime candidates for the promotion of fatigue crack initiation and growth. However, the stress concentration factor developed in Chapter 6 cannot be directly applied to fatigue applications since many materials will relieve stresses near a crack tip through plastic flow. That is, because some materials flow plastically near crack tips, fracture is avoided and the crack's growth is retarded.

For *static* loading, the stress concentration factor, K_c, is used, and for fatigue loading the fatigue stress concentration factor, K_f, is used, where

$$K_f = \frac{\text{Endurance limit for notch-free specimen}}{\text{Endurance limit for notched specimen}}. \tag{7.15}$$

A notch or stress concentration may be a hole, fillet, or groove, or a location of abrupt change in material properties. Recall from Section 6.2 that the theoretical stress concentration factor is a function of geometry. The fatigue stress concentration factor is not only a function of geometry but also a function of the material and type of loading. The consideration of the material is often dealt with by using a **notch sensitivity factor**, q_n, defined as

$$q_n = \frac{K_f - 1}{K_c - 1}, \tag{7.16}$$

or

$$K_f = 1 + (K_c - 1)\,q_n. \tag{7.17}$$

Note from Eq. (7.16) that the range of q_n is between zero (when K_f = 1) and unity (when $K_f = K_c$). From Eq. (7.17), observe that obtaining the fatigue stress concentration factor requires knowing the material's notch sensitivity and the type of loading.

Figure 7.10 is a plot of notch sensitivity versus notch radius for some commonly used materials and for various types of loading. For all the materials considered, the notch sensitivity approaches zero as the notch becomes very sharp (that is, as r approaches zero). Also, the harder and stronger steels tend to be more notch sensitive (have a large value of q_n). This is not too surprising, since notch sensitivity is a measure of material ductility and the hardest steels have limited

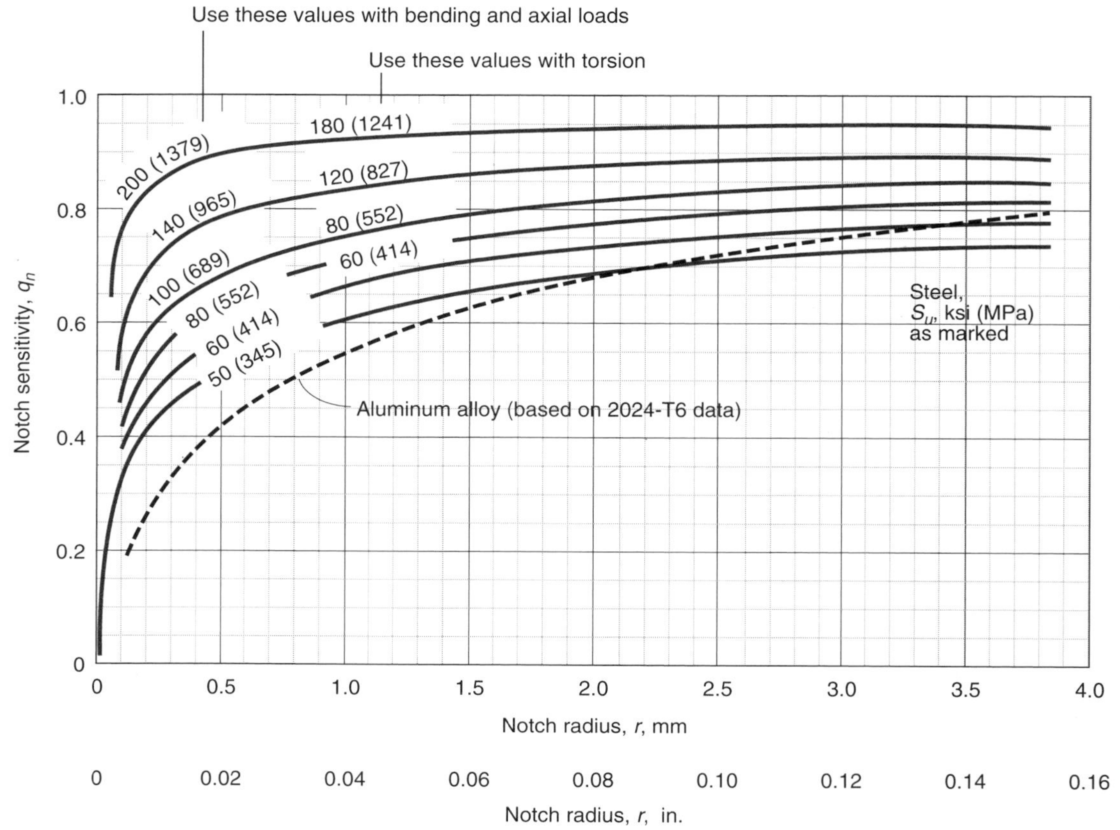

Figure 7.10: Notch sensitivity as a function of notch radius for several materials and types of loading. *Source:* Adapted from Sines and Waisman [1959].

ductility. Figure 7.10 also shows that a given steel is slightly more notch sensitive for torsional loading than for bending and axial loading.

To apply the effects of stress concentrations in fatigue, designers can either reduce the endurance limit or increase the applied stress by K_f. Sometimes designs are based on the ultimate tensile strength, and not the yield strength. In such a case, significant plastic deformation may occur near the stress concentration and reduce its magnitude. It may be reasonable to ignore stress concentrations or to incorporate them within a safety factor. Some design applications are not adversely affected by plastic flow, or else use highly elastic materials that deform sufficiently to relieve stress concentrations. All of these approaches can lead to successful designs. In general, the design constraints will provide guidance on the use of stress concentration factors. In this text, stress concentration factors are used to increase applied stresses, not to reduce allowable strength.

Example 7.4: Fatigue Stress Concentration Factors

Given: The drive shaft for a Formula-1 racing car has a diameter of 30 mm and a half-circular notch with a 1-mm radius. The shaft was dimensioned for a coefficient of friction between the tires and the ground of 1.5 for equal shear and bending stresses. By mounting spoilers and a wing on the car, the load on the tires can be doubled at high speed without increasing the car's mass. Assume from the distortion-energy theory that the equivalent stress is $\sigma_e = \sqrt{\sigma^2 + 3\tau^2}$.

Find: Determine the fatigue stress concentration factors for bending and torsion of the drive shaft if the shaft material has an ultimate tensile strength of 965 MPa. Also, determine if increased acceleration or increased curve handling will give the higher risk of drive shaft failure.

Solution: From Fig. 7.10, for a notch radius of 1 mm and ultimate strength of 965 MPa, the notch sensitivity is 0.82 for bending and 0.85 for torsion. From Fig. 6.6b when $r/d = 1/28 = 0.0357$ and $D/d = 30/28 = 1.0714$, the stress concentration factor is 2.2 for bending, and from Fig. 6.6c the stress concentration factor is 1.8 for torsion. From Eq. (7.17), the fatigue stress concentration factor due to bending is

$$K_f = 1 + (K_c - 1)q_n = 1 + (2.2 - 1)(0.82) = 1.98.$$

The fatigue stress concentration factor due to torsion is

$$K_f = 1 + (1.8 - 1)(0.85) = 1.68.$$

Let σ_{e1} be the equivalent stress for increased curve handling and σ_{e2} be the equivalent stress for increased acceleration. Doubling the load and using the distortion-energy theory results in

$$\frac{\sigma_{e1}}{\sigma_{e2}} = \frac{\sqrt{(2\sigma)^2 + 3\tau^2}}{\sqrt{\sigma^2 + 3(2\tau)^2}} = \frac{2\sqrt{1 + 0.75(\tau/\sigma)^2}}{\sqrt{1 + 12(\tau/\sigma)^2}}.$$

Recall that the shaft was dimensioned such that the shear and bending stresses are equal ($\tau = \sigma$). Thus,

$$\frac{\sigma_{e1}}{\sigma_{e2}} = \frac{2\sqrt{1+0.75}}{\sqrt{1+12}} = 0.7338$$

so that

$$\sigma_{e2} = \frac{\sigma_{e1}}{0.7338} = 1.363\sigma_{e1}.$$

Therefore, increased acceleration gives a higher risk of drive shaft failure compared to increased curve handling.

7.8 The Modified Endurance Limit

As discussed in Section 7.5.1, fatigue experiments use the best possible circumstances for estimating fatigue performance. However, this situation cannot be guaranteed for design applications, so the component's endurance limit must be modified or reduced from the best-case scenario. This is done in practice by using endurance limit modification factors that take important factors into account. The endurance limit modification factors covered in this text are for completely reversed loading ($\sigma_m = 0$). The **modified endurance limit** can be expressed as

$$S_e = k_f k_s k_r k_t k_m S_e', \tag{7.18}$$

where

- S_e' = endurance limit from experimental apparatus under idealized conditions, Pa
- k_f = surface finish factor
- k_s = size factor
- k_r = reliability factor
- k_t = temperature factor
- k_m = miscellaneous factor

Note that the type of loading has already been incorporated into S_e' as presented in Eq. (7.6). As discussed in Section 7.7, the effects of stress concentrations are not included, since these factors are used to increase stress but not to reduce allowable strength.

Equation (7.18) should not be taken as an accurate prediction of endurance limit for complicated situations, but merely a reasonable *approximation* of what should be expected in practice. As will be seen, universally applicable correction factors do not exist, and those that are presented are experimentally based for controlled materials, loadings, and other parameters. This further confirms the observation stated above that experimental confirmation or the use of large safety factors are unavoidable in fatigue design.

7.8.1 Surface Finish Factor

The specimen shown in Fig. 7.3 has a highly polished surface finish with final polishing in the axial direction to smooth any circumferential scratches. Most machine elements do not have such a high-quality finish. The modification factor to incorporate the surface finish effect depends on the process used to generate the surface and on the material. Given a manufacturing process, Fig. 7.11a estimates the surface finish factor when the ultimate strength in tension is known, or else the coefficients from Table 7.3 can be used with the equation

$$k_f = eS_{ut}^f, \tag{7.19}$$

where

- k_f = surface finish factor
- S_{ut} = ultimate tensile strength of material, MPa (ksi)
- e and f = coefficients defined in Table 7.3

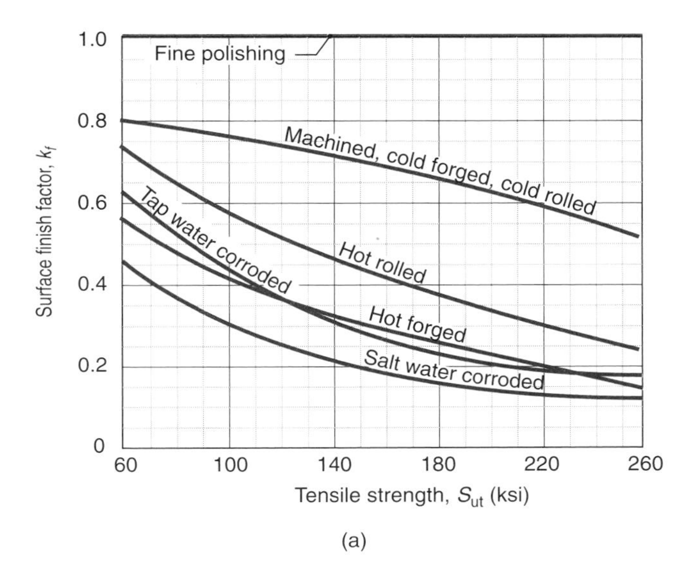

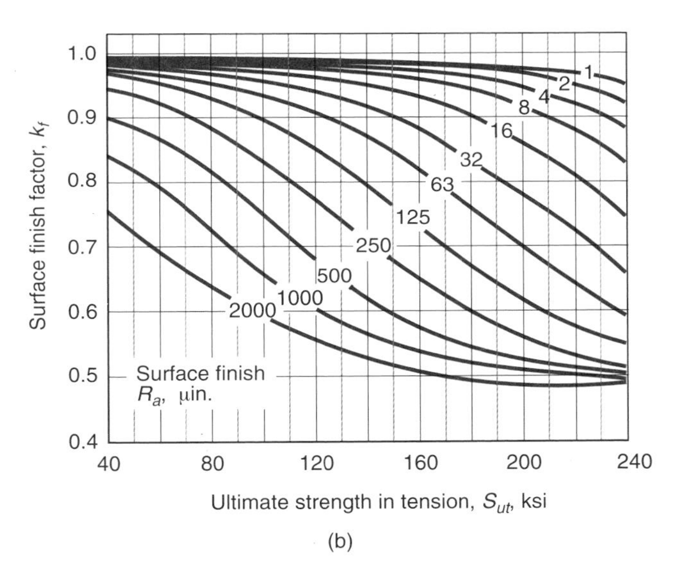

Figure 7.11: Surface finish factors for steel. (a) As a function of ultimate strength in tension for different manufacturing processes; (b) as a function of ultimate strength and surface roughness as measured with a stylus profilometer. *Source:* (a) Adapted from Juvinall and Marshek [1991] and data from the American Iron and Steel Institute; (b) adapted from Johnson [1967].

Table 7.3: Surface finish factor. *Source:* Shigley and Mitchell [1983].

Manufacturing process	Factor e MPa	Factor e ksi	Exponent f
Grinding	1.58	1.34	−0.085
Machining or cold drawing	4.51	2.70	−0.265
Hot rolling	57.7	14.4	−0.718
As forged	272.0	39.9	−0.995

Table 7.4: Reliability factors for six probabilities of survival.

Probability of survival, percent	Reliability factor, k_r
50	1.00
90	0.90
95	0.87
99	0.82
99.9	0.75
99.99	0.70

Note that Eq. (7.19) and Fig. 7.11a can give different results, especially for hot-working processes. This can be understood, recognizing the different sources for the expressions and data, and also because of the wide range of properties that can occur in these processes.

If the process used to obtain the surface finish is not known but the quality of the surface is known from the measured or prescribed arithmetic average surface roughness R_a, the surface finish factor can be obtained from Fig. 7.11b. Note also from the discussion of surface roughness that these values of k_f are approximate; surfaces are not fully characterized by their roughness, and deep and sharp circumferential scratches are the most detrimental to fatigue life, which may not be captured in the R_a roughness value.

These approaches are all approximate and are used only for well-controlled manufacturing processes. It is misleading to apply Table 7.3 for other circumstances or operations. For example, plasma spray operations tend to provide an extremely rough surface, but the fatigue properties are mainly determined by the surface layer beneath the plasma-sprayed coating. Further, the data in Table 7.3 are undoubtedly too stringent. With modern numerically controlled machine tools and improvements in tooling materials, superior finishes are routinely produced that will give slightly better performance from a fatigue standpoint.

7.8.2 Size Factor

The high-cycle fatigue apparatus used to obtain the endurance limit S_e' was for a specific diameter, namely, 0.30 in., and often uses extruded or drawn steel bar stock. For metals, such extrusions have pronounced grain elongation in the direction transverse to fatigue crack growth. Also, the degree of cold work is high and the likelihood of large flaws is low. Similar effects are seen for ceramics and castings but for different reasons (smaller shrinkage pores, etc.). However, it must be noted that the size, shape, and number of flaws in a given cross section are strongly dependent on the manufacturing process.

Many researchers have suggested size factor expressions, but a simple approach suggested by Shigley and Mitchell [1983] is as follows for round bars. For bending or torsion the size factor is

$$k_s = \begin{cases} 0.869d^{-0.112} & 0.3 \text{ in.} < d < 10 \text{ in.} \\ 1 & d < 0.3 \text{ in. or } d \leq 8 \text{ mm} \\ 1.248d^{-0.112} & 8 \text{ mm} < d \leq 250 \text{ mm} \end{cases} \tag{7.20}$$

For axial loading $k_s = 1$.

For components that are not circular in cross-section the size factor is difficult to determine. However, an approach suggested by Kuguel [1969] is often referenced. This approach requires obtaining the cross-sectional area that is loaded above 95% of the maximum stress, denoted as A_{95}. (This is admittedly difficult to obtain for complicated cross sections.) Equating this to the portion of a circular cross section that is loaded above 95% of its maximum stress yields

$$d = \sqrt{\frac{A_{95}}{0.0766}}. \tag{7.21}$$

Equation (7.21) does not differentiate according to processing history, and this is a major shortcoming of the approach. For example, consider a steel bar that has been extruded to a diameter of 50 mm, and is then machined to produce a square cross section with a side length of 25 mm. Machining is a material removal process; it does nothing to improve the substrate microstructure. Any size effect present should be based on the 50 mm initial diameter, not the machined dimension or shape, since this reflects the material state with respect to initial flaws, grain sizes and shapes, etc.

7.8.3 Reliability Factor

Table 7.4 shows the reliability factor for various percentages of survival probability. This table is based on the endurance limit having a standard deviation of 8%, generally the upper limit for steels. The reliability factor for such a case can be expressed as

$$k_r = 0.512 \left[\ln\left(\frac{1}{R}\right) \right]^{0.11} + 0.508, \tag{7.22}$$

where R is the probability of survival. The reliability factor as obtained from Table 7.4 or Eq. (7.22) can be considered only as a guide because the distribution is generally not well characterized for high values of reliability. It can also be assumed that Table 7.4 can be applied to materials other than steel, but care must be taken if the standard deviation is greater than 8%.

7.8.4 Temperature Factor

Many high-cycle fatigue applications take place under extremely high temperatures, such as in aircraft engines, where the material is much weaker than at room temperature. Conversely, in some applications, such as automobile axles in cold climates, the metal is generally less ductile than at room temperature.

In either case, the Manson-Coffin relationship given by Eq. (7.5) would suggest that a major factor affecting fatigue life is the strength in one loading cycle, σ_f'. It is therefore reasonable to follow one of two approaches. The designer can either (a) modify the ultimate strength of the material based on its properties at the temperature of interest before determining a material endurance limit in Eq. (7.18), or (b) use a temperature factor:

$$k_t = \frac{S_{ut}}{S_{ut,\text{ref}}} \tag{7.23}$$

where S_{ut} is the ultimate tensile strength of the material at the desired temperature and $S_{ut,\mathrm{ref}}$ is the ultimate tensile strength at a reference (usually room) temperature. Of course, when fatigue experiments have been conducted at operating temperatures, the resulting endurance limit inherently accounts for the temperature, and k_t can be ignored.

7.8.5 Miscellaneous Effects

Several other phenomena can affect a component's fatigue properties. Whereas the preceding sections have outlined methods for numerically approximating some effects, other considerations defy quantification. Among these are the following:

1. *Manufacturing history*. Manufacturing processes play a major role in determining the fatigue characteristics of engineering materials. This role is partially manifested in the size factor discussed in Section 7.8.2, but there are other effects as well. Because fatigue crack growth is often more rapid along grain boundaries than through grains, any manufacturing process affecting grain size and orientation can affect fatigue. Because some forming operations, such as rolling, extrusion, and drawing, lead to elongated grains, the material's fatigue strength will vary in different directions (anisotropy). With extrusion and drawing this effect is usually beneficial, since the preferred direction of crack propagation becomes the axial direction and crack propagation through the thickness is made more difficult by grain orientation and elongation in metals. Annealing a metal component relieves residual stresses, causes grains to become equiaxed, and may cause increased grain growth rates. Relieving tensile residual stresses at a surface is generally beneficial, but equiaxed or larger grains can be detrimental from a fatigue standpoint.

2. *Residual stresses* can result from manufacturing processes. A **residual stress** is caused by elastic recovery after nonuniform plastic deformation through a component's thickness. Compressive residual stress on a surface retards crack growth; tensile residual stress can encourage crack growth.

   Compressive residual stresses can be imparted through shot or laser peening and roller burnishing and may be obtained in forging, extrusion, or rolling. **Shot peening** is a cold working process in which the surface of a part is impacted with small spherical media called shot. Each impact leads to plastic deformation at the workpiece surface, leading to compressive residual stress after elastic recovery. The layer under compressive residual stress is usually less than 1 mm thick, and the material bulk properties are unaffected. Crack development and propagation are severely retarded by compressive residual stresses; for this reason, shot peening is a common surface treatment for fatigue-susceptible parts such as gears, springs, shafts (especially at stress concentrations), connecting rods, etc.

   The beneficial effect of shot peening on fatigue life can be seen in Fig. 7.12. Similar behavior can be found for other materials. This is an important tool for fatigue design because it represents one of the only strategies that *increases* the fatigue strength of materials, and this increase can be very large. For example, consider an aircraft landing gear, produced from steel with a 2068 MPa (300 ksi) strength. Figure 7.12a shows that shot peening can increase the fatigue strength by a factor of 3 over a polished surface. Similar benefits are possible with other materials, but as seen in the figure, the more typical fatigue strength improvement is 15 to 30%.

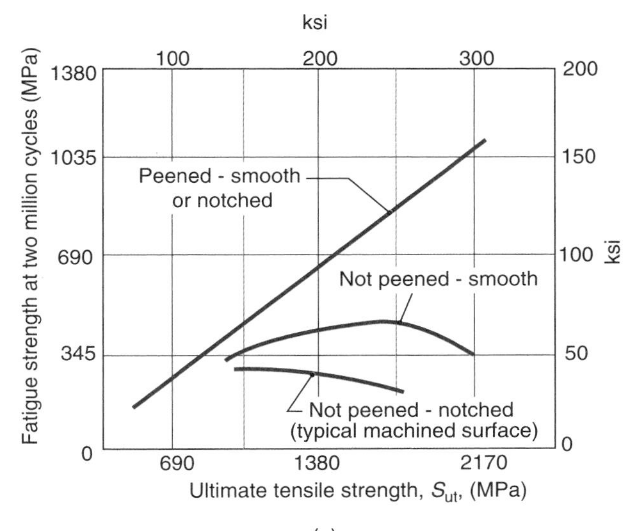

(a)

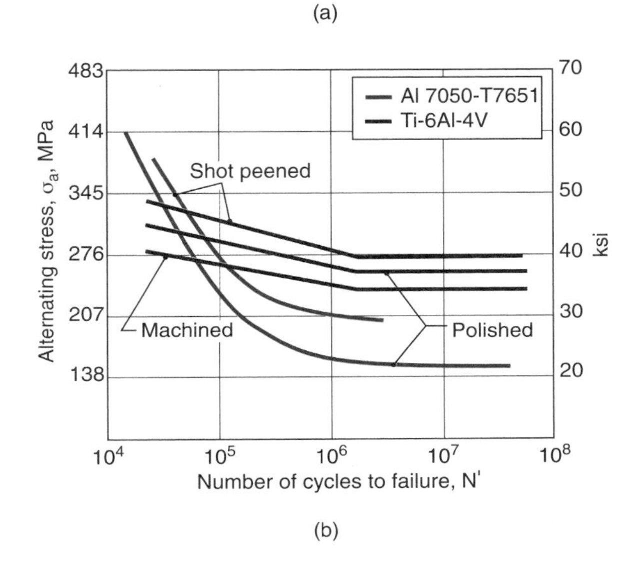

(b)

Figure 7.12: The use of shot peening to improve fatigue properties. (a) Fatigue strength at 2×10^6 cycles for high-strength steel as a function of ultimate strength; (b) typical S-N curves for non-ferrous metals. *Source:* Courtesy of J. Champaigne, Electronics, Inc.

3. *Coatings* can significantly affect fatigue. Some operations, such as carburizing, lead to a high carbon content in steel surface layers (and thus a high fracture strength) and impart a compressive residual stress on the surface. Electroplated surfaces can be porous and promote crack growth, reducing fatigue strengths by as much as 50%. Zinc plating is the main exception where the fatigue strength is not seriously affected. Anodized oxide coatings are also usually porous, reducing fatigue strength. Coatings applied at high temperatures, such as in chemical vapor deposition processes or hot dipping, may induce tensile residual stresses at the surface.

4. *Corrosion*. It is not surprising that materials operating in corrosive environments have lower fatigue strengths. The main adverse reactants in corroding metals are hy-

drogen and oxygen. Hydrogen diffuses into a material near a crack tip, aided by large tensile stresses at the tip, embrittling the material and aiding crack propagation. Oxygen causes coatings to form that are brittle or porous, aiding crack initiation and growth. High temperatures in corrosive environments speed diffusion-based processes.

Design Procedure 7.3: Estimation of Endurance Limit

This procedure is mainly intended for carbon steels, assuming that an endurance limit can be defined. Each of the correction factors are approximate, thus, the endurance limit estimation should be used with generous safety factors or as a basis for experimental design verification, especially if extended to non-ferrous materials. The preferred approach for determining endurance limit is to conduct a series of experiments. Since experiments can be conducted with the loading (bending, tension, shear, etc.) and manufacturing process, part size, etc, that closely match the application, this is the best way to minimize errors.

If an experimental investigation is impractical, the endurance limit can be estimated through the following procedure:

1. The endurance limit for a specimen (S_e') can be estimated for a type of loading from Eq. (7.6). This requires knowledge of the material's ultimate strength, which can be obtained from experiments or from tables of mechanical properties; some steel properties are summarized in Appendix A.
2. Note from Fig. 7.8 that the predicted value should not be assigned a value greater than 690 MPa (100 ksi).
3. The modified endurance limit (S_e) is then obtained from Eq. (7.18), where:
   (a) The surface finish factor, k_f, is obtained from Eq. (7.19) using coefficients from Table 7.3, or else k_f can be estimated from Fig. 7.11.
   (b) The size factor, k_s, can be estimated from Eq. (7.20) for bending or torsion, with k_s=1 for tension. If the part is not round, then an equivalent diameter can be obtained from Eq. (7.21). These equations have high uncertainty, but they do allow size effects to be considered without overly complicating the mathematics.
   (c) The reliability factor, k_r, can be obtained from Table 7.4.
   (d) The effects of temperature, k_t, are best obtained experimentally, but Eq. (7.23) gives a reasonable estimate for this factor.

It should be noted that additional factors can impact the fatigue strength, as discussed in Section 7.8.5.

Example 7.5: Endurance Limits and Modification Factors

Given: Figure 7.13 shows a portion of a round shaft with a flat groove used to seat a retaining ring (see Section 11.7). In order to support the applied loads, a non-standard retaining ring is used so that the depth and width are larger than normal. AISI 1020 steel (quenched and tempered at 870°C) is used for the shaft, which is machined to its final dimensions.

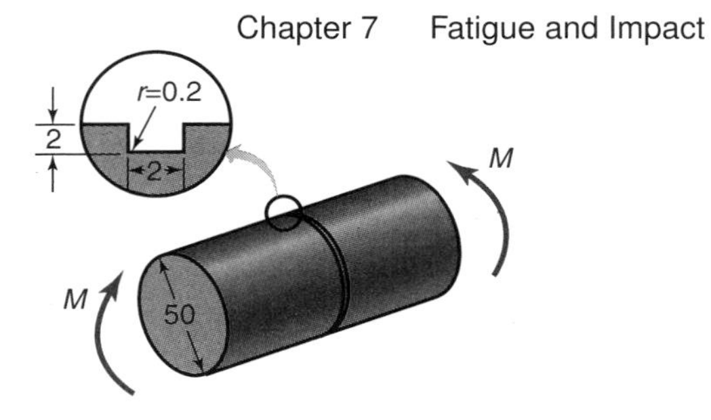

Figure 7.13: Round shaft with a retaining ring groove considered in Example 7.5. All dimensions are in millimeters.

Find: Estimate the modified endurance limit for the shaft and the allowable bending moment using a safety factor of 5.0. Use a reliability of 99% and no thermal or miscellaneous effects.

Solution: The modified endurance limit will be determined using the approach in Design Procedure 7.3, and will preserve its numbering scheme.

1. For this AISI 1020 steel, the ultimate strength is obtained from Table A.2 as 395 MPa. Equation (7.6) then gives the endurance limit for a test specimen constructed from this material as

$$S_e' = 0.5S_u = 0.5(395) = 197.5 \text{ MPa}.$$

Note that the endurance limit was determined for bending, consistent with the loading shown in Fig. 7.13.

2. Equation (7.18) will be used to estimate the modified endurance limit, but first the required correction factors will be obtained.
   (a) The surface finish factor is obtained from Eq. (7.19) using values of $e = 4.51$ and $f = -0.265$ from Table 7.3:

$$k_f = eS_{ut}^f = (4.51)(395)^{-0.265} = 0.92.$$

   By comparison, Fig. 7.11a suggests a value of $k_f = 0.85$. These values are fairly close, and either could be used for the remainder of the problem. The value of $k_f = 0.92$ will be arbitrarily selected for further derivations.

   (b) The size factor is obtained from Eq. (7.20) as

$$k_s = 1.248d^{-0.112} = 1.248(50)^{-0.112} = 0.80.$$

   Note that 50 mm was used as the diameter in this calculation instead of the 46-mm groove dimension. If the groove dimension were used, it would have only minor effect in this case ($k_s = 0.81$), but the reason for using the shaft diameter was because machining the groove does not have an effect on flaw elimination in the shaft material.

   Regardless, it should be recognized that k_s is a rough estimate of size effect.

   (c) The reliability factor is obtained from Table 7.4 as $k_r = 0.82$.
   (d) Per the problem statement, the values of k_t and k_m are set equal to unity. Therefore, from Eq. (7.18),

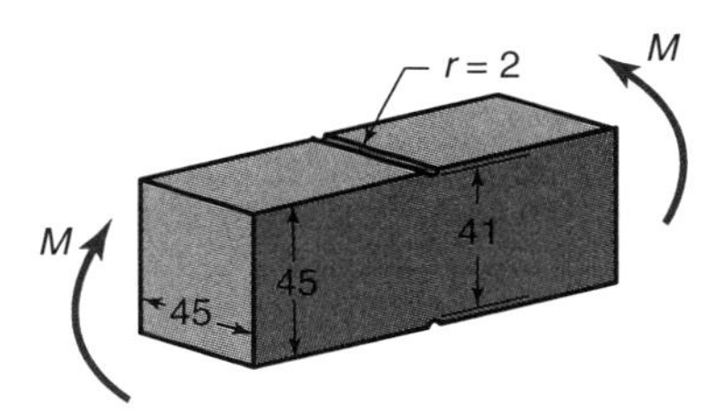

Figure 7.14: Drawn square profile with machined groove considered in Example 7.6. All dimensions are in millimeters.

$$
\begin{aligned}
S_e &= k_f k_s k_r k_t k_m S_e' \\
&= (0.92)(0.80)(0.82)(1)(1)(197.5) \\
&= 119 \text{ MPa}.
\end{aligned}
$$

To determine the allowable bending moment, it is necessary to determine the fatigue stress concentration factor, K_f. The approach is the same as in Example 7.4, where Fig. 6.7a yields $K_c = 5.5$ (using $r/t = 0.10$ and $b/t = 1.0$). From Fig. 7.10 for $S_{ut} = 395$ MPa and $r = 0.2$ mm, $q_n \approx 0.45$. Therefore, from Eq. (7.17),

$$K_f = 1 + (K_c - 1)q_n = 1 + (5.5 - 1)(0.45) = 3.025 \approx 3.0.$$

The nominal stress at the groove root is given in Fig. 6.7a. Therefore,

$$K_f \sigma_{\text{avg}} = K_f \left(\frac{32M}{\pi d^3} \right) = \frac{S_e}{n_s}.$$

Solving for M,

$$
\begin{aligned}
M &= \frac{\pi S_e d^3}{32 K_f n_s} \\
&= \frac{\pi (119 \times 10^6)(0.046)^3}{(32)(3.0)(5.0)} \\
&= 75.8 \text{ Nm}.
\end{aligned}
$$

Example 7.6: Endurance Limits and Modification Factors for Square Cross Section

Given: Instead of the retaining ring for the shaft considered in Example 7.5, an integrated snap fastener (see Section 16.8) is being considered on a drawn square profile with machined groove as shown in Fig. 7.14. Note that the cross sectional areas of the shaft are very close to that in Example 7.5. As in Example 7.5, use AISI 1020 steel with a reliability of 99%, and no thermal or miscellaneous effects.

Find: Estimate the modified endurance limit for the shaft and the allowable bending moment using a safety factor of 5.0.

Solution: The procedure for calculating modified endurance limits is very similar to Example 7.5. The only correction factor that is different in this case is the size factor. Since the cross section is not circular, Eq. (7.21) will be used to obtain the diameter so that Eq. (7.20) can be applied.

As was discussed in Example 7.5, the size factor will be determined for the drawn cross section without the notch. Note that for bending, the maximum stress is at the outer dimension, or

$$\sigma = \frac{Mc}{I} = \frac{M}{I}\frac{0.045}{2} = 0.0245\frac{M}{I}.$$

The location where the stress is 95% of the maximum is:

$$0.95 \left(\frac{M}{I} \right) \left(\frac{0.045}{2} \right) = \frac{My}{I},$$

so that $y = 0.0214$ m. The area where the stress exceeds 95% of the maximum is (noting that there are tensile and compressive areas that contribute):

$$A_{95} = 2(0.045)(0.0225 - 0.0214) = 9.9 \times 10^{-5} \text{ m}^2.$$

From Eq. (7.21),

$$d = \sqrt{\frac{A_{95}}{0.0766}} = \sqrt{\frac{9.9 \times 10^{-5}}{0.0766}} = 0.0359 \text{ m}.$$

Therefore, Eq. (7.20) yields

$$k_s = 1.248(35.9)^{-0.112} = 0.836.$$

Note that this size effect does not differ much from the results of Example 7.5, nor should it. The manufacturing processes are very similar, so the flaws, grain structure, etc, should all be close. As stated above, the cross sections are very close as well. Therefore, the modified endurance limit is, from Eq. (7.18),

$$
\begin{aligned}
S_e &= k_f k_s k_r k_t k_m S_e' \\
&= (0.92)(0.836)(0.82)(1)(1)(197.5) \\
&= 124 \text{ MPa}.
\end{aligned}
$$

For this case, Fig. 6.4b yields $K_c \approx 2.3$ ($H/h = 1.1$, $r/h = 0.048$). From Fig. 7.10, $q_n \approx 0.65$. Therefore, $K_f = 1.845$. Using the revised geometry,

$$K_f \sigma_{\text{avg}} = K_f \left(\frac{6M}{h^3} \right) = \frac{S_e}{n_s}.$$

Solving for M,

$$M = \frac{S_e h^3}{6 K_f n_s} = \frac{(124 \times 10^6)(0.041)^3}{6(1.1845)(5)} = 231 \text{ Nm}.$$

This geometry can support a larger moment, and this is attributable to the lower stress concentration.

7.9 Cumulative Damage

In constructing the S-N curve in Fig. 7.7, it was assumed that the cyclic variation was completely reversed ($\sigma_m = 0$). Furthermore, for any stress level between the strengths S_l' and S_e', say S_1', the maximum stress level in the completely reversed variation was kept constant until failure occurred at N_1' cycles. Operating at stress amplitude S_1' for a number of cycles $n_1' < N_1'$ produced a smaller damage fraction. Because cyclic variations are often not constant in practice, engineers must deal with several different levels of completely reversed stress cycles. Operating over stress levels between S_l' and S_e', say S_i', at a number of cycles $n_i' < N_i'$ results in the damage fraction n_i'/N_i'. When the damage fraction due to different levels of stress exceeds unity, failure is predicted. Thus, failure is predicted if

$$\frac{n_1'}{N_1'} + \frac{n_2'}{N_2'} + \cdots + \frac{n_i'}{N_i'} \geq 1. \quad (7.24)$$

This formulation is frequently called the **linear damage rule** (sometimes called *Miner's rule*), since it states that the damage at any stress level is directly proportional to the number of cycles (assuming that each cycle at a given stress level does the same amount of damage). The rule also assumes that the stress sequence does not matter and that the rate of damage accumulation at a particular stress level is independent of the stress history. This is not strictly true, as it is well-known that more severe loadings cause disproportionate damage, especially if they occur early in the part's life. Despite these shortcomings, the linear damage rule remains popular, largely because it is so simple.

If N_t' is the total number of cycles to failure when there are different cyclic patterns (all of which are completely reversed), the ratio of the number of cycles at a specific stress level to the total number of cycles to failure is

$$\alpha_i = \frac{n_i'}{N_t'} \quad \text{or} \quad n_i' = \alpha_i N_t'. \quad (7.25)$$

Substituting Eq. (7.25) into Eq. (7.24), it is predicted that failure will occur if

$$\sum \frac{\alpha_i}{N_i'} \geq \frac{1}{N_t'}. \quad (7.26)$$

Example 7.7: Cumulative Damage

Given: Consider a steel with an ultimate strength of 440 MPa, so that $S_l' = 330$ MPa and $S_e' = 200$ MPa. A complicated loading cycle is applied, so that the stress is 175 MPa for 20% of the time, 220 MPa for 30%, 250 MPa for 40%, and 275 MPa for 10%.

Find: The number of cycles until cumulative failure.

Solution: First of all, note from Eqs. (7.11) and (7.12) that

$$b_s = -\frac{1}{3}\log\left(\frac{S_l'}{S_e'}\right) = -\frac{1}{3}\log\left(\frac{330}{200}\right) = -0.0725,$$

$$\bar{C} = \log\left[\frac{(S_l')^2}{S_e'}\right] = \log\left(\frac{330^2}{200}\right) = 2.736.$$

Note that $S_1' = 175$ MPa is less than S_e', so that $N_1' = \infty$, implying that at this stress level failure will not occur. From Eq. (7.14) for the other three fatigue strength levels,

$$\begin{aligned} N_2' &= \left(S_f' 10^{-\bar{C}}\right)^{1/b_s} \\ &= \left[(220)(10)^{-2.736}\right]^{-1/0.0725} \\ &= 2.683 \times 10^5 \text{ cycles,} \end{aligned}$$

$$N_3' = \left[(250)(10)^{-2.736}\right]^{-1/0.0725} = 4.601 \times 10^4 \text{ cycles,}$$

$$N_4' = \left[(275)(10)^{-2.736}\right]^{-1/0.0725} = 1.236 \times 10^4 \text{ cycles.}$$

Making use of Eq. (7.26) gives

$$\frac{\alpha_1}{N_1'} + \frac{\alpha_2}{N_2'} + \frac{\alpha_3}{N_3'} + \frac{\alpha_4}{N_4'} = \frac{1}{N_t'};$$

$$\frac{0.2}{\infty} + \frac{0.3}{2.683 \times 10^5} + \frac{0.4}{4.601 \times 10^4} + \frac{0.1}{1.236 \times 10^4} = \frac{1}{N_t'}.$$

This is solved as $N_t' = 55,860$ cycles.

7.10 Influence of Nonzero Mean Stress

Other than in classifying cyclic behavior, completely reversed ($\sigma_m = 0$) stress cycles have been assumed. Many machine elements involve fluctuating stresses about a nonzero mean. The experimental apparatus used to generate the results shown in Fig. 7.7 cannot apply mean and alternating stresses, so other test approaches are needed (such as tension-compression or combined stresses). If a material has been extensively characterized, the effect of a nonzero mean stress can be incorporated through a **Haigh diagram**, also called a *constant life diagram* as shown in Fig. 7.15. Because such data are not generally available, the influence of nonzero mean stress must be estimated by using one of several empirical relationships that determine failure at a given life when alternating and mean stresses are both nonzero.

7.10.1 Ductile Materials

Figure 7.16 shows how four empirical relationships estimate the influence of nonzero mean stress on fatigue life for ductile materials loaded in tension. The ordinate has both the yield strength, S_{yt}, and endurance limit, S_e, indicated, and the abscissa shows both the yield and ultimate strengths. The yield line shown is for reference purposes, as it indicates failure during the first loading cycle. Commonly used criteria for fatigue failure with a non-zero mean stress are shown in the figure.

The effects of stress concentrations on ductile materials (see Section 7.7) require further explanation. If there is a stress concentration and the yield stress is exceeded, then the deformed geometry is difficult to obtain. Recall from Eq. (3.21) that stress is proportional to strain for linearly elastic materials. However, during plastic deformation, stresses are related to strain increment. If the yield stress is exceeded, it is possible that plastic strains are negligible, but it is also possible that they are so large as to effectively eliminate the stress concentration effect. The amount of plastic strain that occurs depends on the nature of a part's supports, its ductility and strain hardening capability, and applied loads.

However, consider the situation where the mean stress is small; in Fig. 7.16 this would suggest that the endurance limit should not be exceeded, consistent with the approach in Section 7.8. Since the endurance limit is much smaller than the yield strength, it is reasonable to apply stress concentration effects to the alternating stress. Applying stress concentration effects to the mean stress has little impact because it was assumed these stresses are small. On the other hand, if the mean stress is large, then plastic deformation will relieve stress concentrations, and applying the fatigue stress concentration factor based on the original geometry to the alternating stress results in a conservative approach. Therefore, a common modeling simplification is to apply stress concentrations to alternating stresses but not mean stresses, and this is the approach followed in this text.

Gerber Line

The **Gerber line** is sometimes called the *Gerber parabolic relationship* because the equation is

$$\frac{K_f n_s \sigma_a}{S_e} + \left(\frac{n_s \sigma_m}{S_{ut}}\right)^2 = 1. \quad (7.27)$$

where

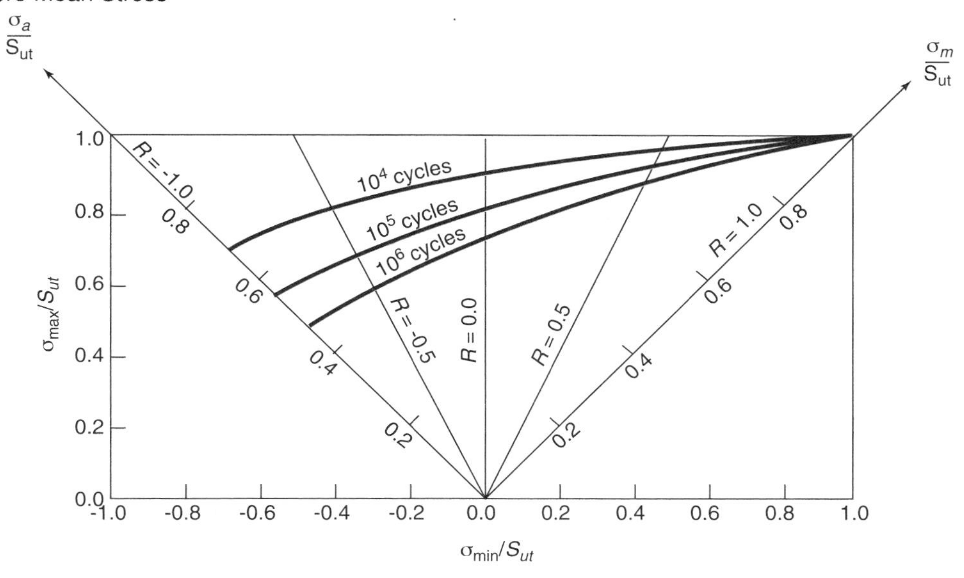

Figure 7.15: A typical Haigh diagram showing constant life curves for different combinations of mean and alternating stresses.

S_e = modified endurance limit, Pa
S_{ut}= ultimate strength in tension, Pa
n_s = safety factor
σ_a = alternating stress, Pa
σ_m = mean stress, Pa
K_f = fatigue stress concentration factor

This line passes through the central portion of the experimental failure points and hence should be the best predictor of failure, but the parabolic nature of the equation complicates mathematics.

Goodman Line

The **Goodman line** proposes connecting the modified endurance limit on the alternating stress axis with the ultimate strength in tension on the mean stress axis in Fig. 7.16 by a straight line, or

$$\frac{K_f\sigma_a}{S_e}+\frac{\sigma_m}{S_{ut}}=\frac{1}{n_s}. \tag{7.28}$$

Note the linearization of Eq. (7.28) relative to Eq. (7.27). Equation (7.28) fits experimental data reasonably well and is simpler to use than Eq. (7.27). The starting and ending points for the Goodman and Gerber lines are the same in Fig. 7.16, but between these points the Goodman line is linear and the Gerber line is parabolic.

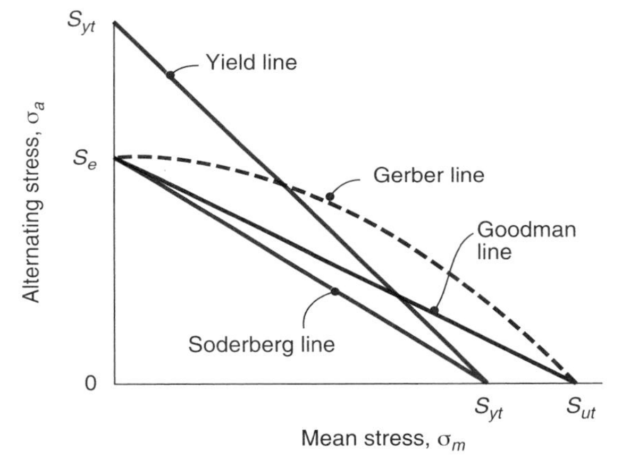

Figure 7.16: Influence of nonzero mean stress on fatigue life for tensile loading as estimated by four empirical relationships.

Example 7.8: Effect of Nonzero Mean Stress

Given: A straight, circular rotating beam with a 30-mm diameter and a 1-m length has an axial load of 30,000 N applied at the end and a stationary radial load of 400 N. The beam is cold-drawn and the material is AISI 1040 steel. Assume that $k_s = k_r = k_t = k_m = 1$.

Find: The safety factor for infinite life by using the Goodman line.

Solution: From Table A.1 for AISI 1040 steel, S_u = 520 MPa. From Fig. 7.11a for the cold-drawn process and S_u = 520 MPa, the surface finish factor is 0.78. From Eq. (7.18), the modified endurance limit while making use of Eq. (7.6) is

$$\begin{aligned} S_e &= k_f k_s k_r k_t k_m S'_e \\ &= (0.78)(1)(1)(1)(1)(0.45)(520) \\ &= 182.5 \text{ MPa.} \end{aligned}$$

The bending stress from Eq. (4.45) gives the alternating stress that the beam experiences as

$$\sigma_a=\frac{Mc}{I}=\frac{(64)(400)(1)(0.03/2)}{\pi(0.03)^4}=150.9 \text{ MPa.}$$

The mean stress due to the axial load is

$$\sigma_m=\frac{P_a}{A}=\frac{(30,000)(4)}{\pi(0.03)^2}=42.44 \text{ MPa.}$$

For an unnotched beam $K_f = 1$. From Eq. (7.28),

$$\frac{K_f \sigma_a}{S_e} + \frac{\sigma_m}{520} = \frac{1}{n_s}.$$

Therefore,

$$\frac{(1)(150.9)}{182.5} + \frac{42.44}{520} = \frac{1}{n_s};$$

$$n_s = \frac{1}{0.9084} = 1.101.$$

Using the Goodman line, the safety factor for infinite life is 1.101.

Soderberg Line

The **Soderberg line** is conservative and is given as

$$\frac{K_f \sigma_a}{S_e} + \frac{\sigma_m}{S_{yt}} = \frac{1}{n_s}. \tag{7.29}$$

Note from Fig. 7.16 and Eqs. (7.28) and (7.29) that the ultimate strength in the Goodman relationship has been replaced with the yield strength in the Soderberg relationship.

Yield Line

To complete the possibilities, the **yield line** is given. It is used to define yielding on the first cycle, or

$$\frac{\sigma_a}{S_{yt}} + \frac{\sigma_m}{S_{yt}} = \frac{1}{n_s}. \tag{7.30}$$

This completes the description of the theories presented in Fig. 7.16.

Modified Goodman Diagram

The Goodman relationship given in Eq. (7.28) is modified by combining fatigue failure with failure by yield. The complete **modified Goodman diagram** is shown in Fig. 7.17.[1] Thus, all points inside a modified Goodman diagram ABCDEFGH correspond to fluctuating stresses that should cause neither fatigue failure nor yielding. The word "complete" is used to indicate that the diagram is valid for both tension and compression. The word "modified" designates that the Goodman line shown in Fig. 7.16 has been modified in Fig. 7.17; that is, in Fig. 7.16 the Goodman line extends from the endurance limit on the alternating stress ordinate to the ultimate strength on the mean stress abscissa. In the modified Goodman diagram in Fig. 7.17, the Goodman line is modified such that for stresses larger than the yield strength the yield line BC is used. Thus, the modified Goodman diagram combines fatigue criteria as represented by the Goodman line and yield criteria as represented by the yield line. Note in Fig. 7.17 that lines AB, DE, EF, and HA are Goodman lines and that lines BC, CD, FG, and GH are yield lines. The static load is represented by line CG. Table 7.5 gives the equations and range of applicability for the construction of the Goodman and yield lines of the complete modified Goodman diagram. Note that when the ultimate and yield strengths are known for a specific material, as well as the corresponding endurance limit for a particular part made of that material, the modified Goodman diagram can be constructed.

[1]Equation (7.28) is referred to in the technical literature as either the *Goodman* or the *modified Goodman* relationship, since it was the combined contribution of multiple researchers. Figure 7.17 is often referred to as the modified Goodman diagram, as is done in this text, but is called an *enhanced modified Goodman diagram* in some sources.

Table 7.5: Equations and range of applicability for construction of complete modified Goodman diagram.

Line	Equation	Range
AB	$\sigma_{max} = \dfrac{S_e}{K_f} + \sigma_m\left(1 - \dfrac{S_e}{S_u K_f}\right)$	$0 \le \sigma_m \le \dfrac{S_y - S_e/K_f}{1 - \dfrac{S_e}{K_f S_u}}$
BC	$\sigma_{max} = S_y$	$\dfrac{S_y - \dfrac{S_e}{K_f}}{1 - \dfrac{S_e}{K_f S_u}} \le \sigma_m \le S_y$
CD	$\sigma_{min} = 2\sigma_m - S_y$	$\dfrac{S_y - \dfrac{S_e}{K_f}}{1 - \dfrac{S_e}{K_f S_u}} \le \sigma_m \le S_y$
DE	$\sigma_{min} = \left(1 + \dfrac{S_e}{K_f S_u}\right)\sigma_m - \dfrac{S_e}{K_f}$	$0 \le \sigma_m \le \dfrac{S_y - \dfrac{S_e}{K_f}}{1 - \dfrac{S_e}{K_f S_u}}$
EF	$\sigma_{min} = \sigma_m - \dfrac{S_e}{K_f}$	$\dfrac{S_e}{K_f} - S_y \le \sigma_m \le 0$
FG	$\sigma_{min} = -S_y$	$-S_y \le \sigma_m \le \dfrac{S_e}{K_f} - S_y$
GH	$\sigma_{max} = 2\sigma_m + S_y$	$-S_y \le \sigma_m \le \dfrac{S_e}{K_f} - S_y$
HA	$\sigma_{max} = \sigma_m + \dfrac{S_e}{K_f}$	$\dfrac{S_e}{K_f} - S_y \le \sigma_m \le 0$

As an example of the way the modified Goodman diagram aids in visualizing the various combinations of fluctuating stress, consider the mean stress indicated by point L in Fig. 7.17. The Goodman criterion indicates that this stress can fluctuate between points M and N. This fluctuation is sketched on the right of the figure.

Also shown in Fig. 7.17 are the four regions of mean stress on the abscissa. Table 7.6 gives the failure equation for each of these regions as well as the validity limits for each equation. Table 7.6 is an extremely valuable guide when applying the modified Goodman diagram.

Example 7.9: Safety Factor Using the Modified Goodman Criterion

Given: For the beam given in Example 7.5 the bending moment varies between 50 and 200 Nm.

Find: Using the modified Goodman relationship, determine the safety factor guarding against fatigue failure.

Solution: From Example 7.5, $S_{ut} = 395$ MPa and $S_y = 295$ MPa. Also, it was determined in Example 7.5 that $S_e = 119$ MPa and $K_f = 3.0$. The nominal stresses at the groove ($d = 0.046$ m) are given by

$$\sigma_{min} = \frac{32M_{min}}{\pi d^3} = \frac{32(50)}{\pi(0.046)^3} = 5.232 \text{ MPa},$$

$$\sigma_{max} = \frac{32M_{min}}{\pi d^3} = \frac{32(200)}{\pi(0.046)^3} = 20.93 \text{ MPa}.$$

Therefore, from Eqs. (7.1) and (7.3),

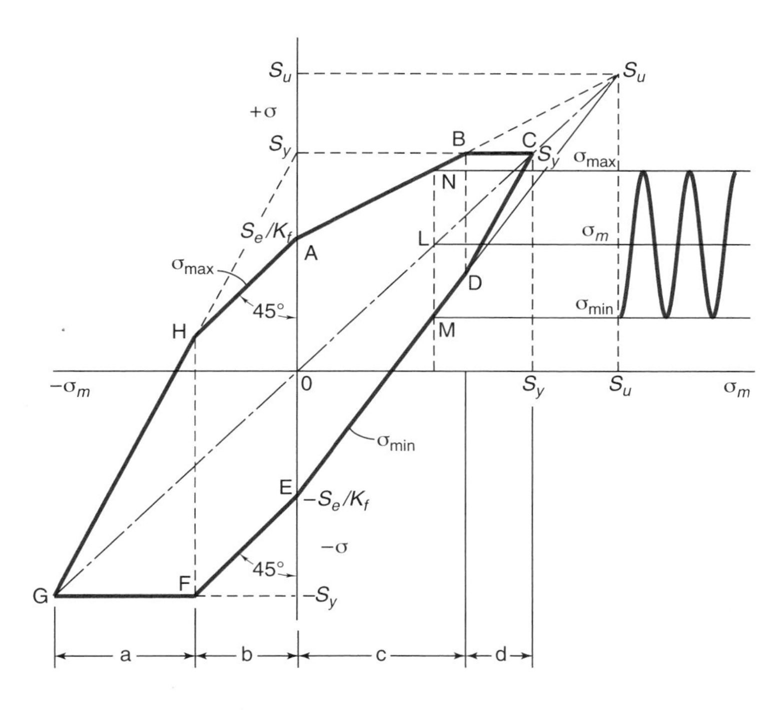

Figure 7.17: Complete modified Goodman diagram, plotting stress as ordinate and mean stress as abscissa.

Table 7.6: Failure equations and validity limits of equations for four regions of complete modified Goodman relationship.

Region in Fig. 7.16	Failure equation	Validity limits of equation
a	$\sigma_{max} - 2\sigma_m = S_y/n_s$	$-S_y \leq \sigma_m \leq \frac{S_e}{K_f} - S_y$
b	$\sigma_{max} - \sigma_m = \frac{S_e}{n_s K_f}$	$\frac{S_e}{K_f} - S_y \leq \sigma_m \leq 0$
c	$\sigma_{max} + \sigma_m \left(\frac{S_e}{K_f S_u} - 1\right) = \frac{S_e}{n_s K_f}$	$0 \leq \sigma_m \leq \frac{S_y - \frac{S_e}{K_f}}{1 - \frac{S_e}{K_f S_u}}$
d	$\sigma_{max} = \frac{S_y}{n_s}$	$\frac{S_y - \frac{S_e}{K_f}}{1 - \frac{S_e}{K_f S_u}} \leq \sigma_m \leq S_y$

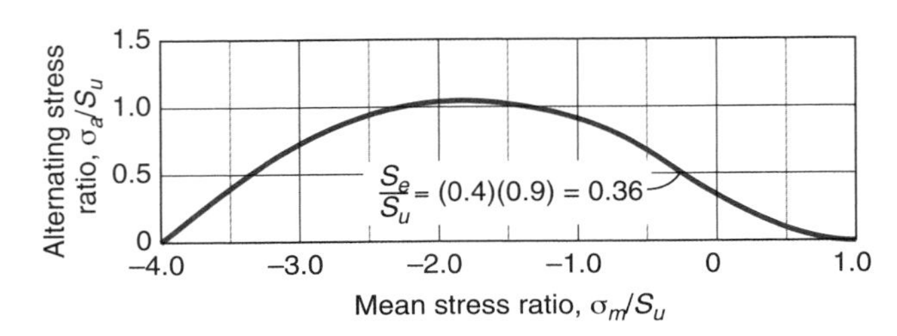

Figure 7.18: Alternating stress ratio as a function of mean stress ratio for axially loaded cast iron.

$$\sigma_a = \frac{\sigma_{\max} - \sigma_{\min}}{2} = \frac{20.93 - 5.232}{2} = 7.849 \text{ MPa},$$

$$\sigma_m = \frac{\sigma_{\max} + \sigma_{\min}}{2} = \frac{20.93 + 5.232}{2} = 13.08 \text{ MPa}.$$

In applying the modified Goodman approach, Table 7.6 is very valuable. Note the requirement for Region a to be valid, then note that

$$\frac{S_e}{K_f} - S_y = \frac{119}{3.0} - 295 = -255 \text{ MPa}.$$

However, since $\sigma_m = 20.93$ MPa, this cannot be true and therefore Region a is not valid. Similarly, Region b cannot be valid since $\sigma_m > 0$. Note that

$$\frac{S_y - \dfrac{S_e}{K_f}}{1 - \dfrac{S_e}{K_f S_u}} = \frac{295 - \dfrac{119}{3.0}}{1 - \dfrac{119}{(3.0)(395)}} = 284 \text{ MPa}.$$

Since $\sigma_m < 284$ MPa, Region c is valid and the failure equation from Table 7.6 is

$$\sigma_{\max} + \sigma_m \left(\frac{S_e}{K_f S_{ut}} - 1 \right) = \frac{S_e}{n_s} K_f;$$

$$20.93 + 13.08 \left(\frac{119}{(3.0)(395)} - 1 \right) = \frac{119}{(3.0) n_s}.$$

This is solved as $n_s = 4.22$.

7.10.2 Brittle Materials

Until recently, the use of brittle materials in a fatigue environment has been limited to gray cast iron in compression. Now, however, carbon fibers and ceramics have had significant use in fatigue environments. Figure 7.18 shows the alternating stress ratio as a function of the mean stress ratio for axially loaded cast iron. The figure is skewed since the compressive strength is typically several times greater than the tensile strength.

In Fig. 7.18, the dimensionalization of the alternating and mean stresses is with respect to the ultimate strength rather than to the yield strength as done for ductile materials. Also, the compressive mean stress permits large increases in alternating stress.

For brittle materials a stress raiser increases the likelihood of failure under either steady or alternating stresses, and it is customary to apply a stress concentration factor to both. Thus, designers apply the fatigue stress concentration factor K_f to the alternating component of stress for ductile materials but apply the stress concentration factor K_c to both the alternating and mean components of stress for brittle materials.

For a single normal stress in brittle materials, the equation for the safety factor with a steady stress σ_m and with the ultimate tensile strength S_{ut} as the basis of failure is

$$n_s = \frac{S_{ut}}{K_c \sigma_m}. \tag{7.31}$$

With an alternating stress, σ_a, and a modified endurance limit, S_e:

$$n_s = \frac{S_e}{K_c \sigma_a}. \tag{7.32}$$

For a single shear stress on a brittle component and with a *steady* shear stress, τ_m, the safety factor is

$$n_s = \frac{S_{ut}}{K_{cs} \tau_m \left[1 + \dfrac{S_{ut}}{S_{uc}} \right]}. \tag{7.33}$$

With an alternating shear stress, τ_a, and a modified endurance limit, S_e:

$$n_s = \frac{S_e}{K_{cs} \tau_a \left[1 + \dfrac{S_{ut}}{S_{uc}} \right]}. \tag{7.34}$$

7.11 Influence of Multi-Axial Stress States

The previous sections have considered fatigue failures for uniaxial stress states. Most machine element applications encounter more complicated loading conditions. Two special cases are important. The first situation is where the applied stresses are in phase, a situation referred to as **simple multiaxial stress**. For example, a cylindrical pressure vessel that is periodically pressurized will have a hoop and axial stress that are both directly related to the pressure, so that they are subjected to their maximum and minimum values at the same time. On the other hand, a shaft with two gears mounted on it will see a periodic variation in normal forces and torques applied by the gears; these may be caused by forcing functions in the driven machinery. Clearly, the normal forces and torques do not have to be in phase in this circumstance; this situation is called **complex multiaxial stress**.

Caution should be given that the current theoretical approaches for multi-axial stress states are even less developed than uniaxial fatigue approaches, so that experimental confirmation of designs is even more imperative.

7.11.1 Simple Multiaxial Stress

Fully Reversing Stresses

For *fully reversing, simple multiaxial stresses*, the mean stress is zero for all applied normal and shear stresses. For such a circumstance, experimental evidence suggests that a combination of an equivalent von Mises effective stress, defined from the alternating principal stress components, can be used in conjunction with uniaxial fatigue failure criteria. That is, an effective stress can be defined from

$$\sigma_e' = \sqrt{\sigma_{a,1}^2 + \sigma_{a,2}^2 + \sigma_{a,3}^2 - \sigma_{a,1}\sigma_{a,2} - \sigma_{a,2}\sigma_{a,3} - \sigma_{a,1}\sigma_{a,3}}. \tag{7.35}$$

If the stress state is two-dimensional (with $\sigma_{a,3} = 0$), this equation can be written as

$$\sigma_e' = \sqrt{\sigma_{a,1}^2 + \sigma_{a,2}^2 - \sigma_{a,1}\sigma_{a,2}}. \tag{7.36}$$

The safety factor can then be calculated from

$$n_s = \frac{S_f}{\sigma_e'} \quad \text{or} \quad n_s = \frac{S_e}{\sigma_e'}. \tag{7.37}$$

S_f is used for finite life applications and is the fatigue strength at the desired life. S_e is used for infinite life applications. Recall that all fatigue strength reduction factors are incorporated in S_f and S_e. Stress concentration effects should be applied in Eqs. (7.35) and (7.36).

Simple Multiaxial Stresses with Non-zero Mean

A number of studies have addressed the situation where the applied stresses are in phase, but the mean stress is non-zero. Two of the more common theories are referred to as the **Sines method** and the **von Mises method**. These theories use the common approach of defining an effective alternating and mean stress, and then inserting these effective stresses in the Modified Goodman failure criterion given in Table 7.5.

The approach of Sines and Waisman [1959] uses the following equivalent stresses:

$$\sigma_a' = \frac{1}{\sqrt{2}}\left\{(\sigma_{a,x}-\sigma_{a,y})^2 + (\sigma_{a,y}-\sigma_{a,z})^2 + (\sigma_{a,z}-\sigma_{a,x})^2 + 6\left(\tau_{a,xy}^2 + \tau_{a,yz}^2 + \tau_{a,zx}^2\right)\right\}^{1/2}, \tag{7.38}$$

$$\sigma_m' = \sigma_{m,x} + \sigma_{m,y} + \sigma_{m,z}. \tag{7.39}$$

For a two-dimensional stress state ($\sigma_z = \tau_{xz} = \tau_{yz} = 0$), these equations can be simplified to

$$\sigma_a' = \sqrt{\sigma_{a,x}^2 + \sigma_{a,y}^2 - \sigma_{a,x}\sigma_{a,y} + 3\tau_{a,xy}^2}, \tag{7.40}$$

$$\sigma_m' = \sigma_{m,x} + \sigma_{m,y}. \tag{7.41}$$

Note that the mean component of shear stress does not appear in these equations. This is acceptable for some circumstances, but is non-conservative for situations where a stress concentration such as a notch or fillet is present. For this reason, another approach using the von Mises effective stresses defines the effective alternating and mean stresses as

$$\sigma_a' = \frac{1}{\sqrt{2}}\left[(\sigma_{a,x}-\sigma_{a,y})^2 + (\sigma_{a,y}-\sigma_{a,z})^2 + (\sigma_{a,z}-\sigma_{a,x})^2 + 6\left(\tau_{a,xy}^2 + \tau_{a,yz}^2 + \tau_{a,zx}^2\right)\right]^{1/2}, \tag{7.42}$$

$$\sigma_m' = \frac{1}{\sqrt{2}}\left[(\sigma_{m,x}-\sigma_{m,y})^2 + (\sigma_{m,y}-\sigma_{m,z})^2 + (\sigma_{m,z}-\sigma_{m,x})^2 + 6\left(\tau_{m,xy}^2 + \tau_{m,yz}^2 + \tau_{m,zx}^2\right)\right]^{1/2}, \tag{7.43}$$

or, for a two dimensional stress state with $\sigma_{a,z} = \sigma_{m,z} = \tau_{a,xz} = \tau_{a,yz} = \tau_{m,xz} = \tau_{m,yz} = 0$,

$$\sigma_a' = \sqrt{\sigma_{a,x}^2 + \sigma_{a,y}^2 - \sigma_{a,x}\sigma_{a,y} + 3\tau_{a,xy}^2}, \tag{7.44}$$

$$\sigma_m' = \sqrt{\sigma_{m,x}^2 + \sigma_{m,y}^2 - \sigma_{m,x}\sigma_{m,y} + 3\tau_{m,xy}^2}. \tag{7.45}$$

One of the difficulties in applying these theories is that conflicts may arise in determining stress concentration factors and fatigue strengths depending on the loading condition selected. Recall from Eq. (7.7) that S_l' will vary depending on whether the loading is bending, axial, or torsional. If the loading is a combination of these loadings, it is not clear how to calculate S_l'. Similarly, stress concentration factors can be defined based on the loading; thus, it is not obvious which value to use in Table 7.5. Recognizing that an experimental verification of a design is imperative for critical applications, it is reasonable to follow any of the following approximations:

1. Perform a worst-case scenario, using the smallest resulting strengths and largest stress concentrations that result from the loading.

2. Since Mode I failure is usually most critical, calculate strengths and stress concentrations based on axial loads when they are present. If normal stresses are present due to bending moments, calculate material strengths and stress concentration effects based on bending.

3. An experienced engineer can evaluate the applied stresses to determine which is the most likely failure mode. That is, if the torsional stresses are dominant, it is reasonable to calculate strengths and concentration factors based on torsion. However, this is a very subjective approach that all but guarantees experimental confirmation of the design will be required.

7.11.2 Complex Multiaxial Stresses

For complex multiaxial stress states, where the normal and shear stress maxima and minima do not occur at the same time, failure theories are not well developed. True asynchronous situations cannot be analyzed with existing failure theories, and an experimental program may be required. A designer's goal in such circumstances is to obtain estimates of machine element dimensions for use in the experimental program. The following approaches have been applied for complex multiaxial stress states:

1. It has been shown that the fatigue strength of some metals is not less than their strengths in a simple multiaxial stress state. Thus, it may be reasonable to approximate the stresses as synchronous, and to analyze the situation as a simple multiaxial stress state using Eqs. (7.38) through (7.39) or (7.42) through (7.43).

2. For situations where the loading consists of bending and torsion (such as is commonly encountered in shafts), an approach in the American Society of Mechanical Engineers Boiler Code can be used. This approach defines an effective stress given by:

$$\sigma_s = \frac{\sigma}{\sqrt{2}}\left[1 + 3\left(\frac{\tau}{\sigma}\right)^2 + \sqrt{1 + \frac{6\tau}{\sigma}\cos 2\phi + 9\left(\frac{\tau}{\sigma}\right)^4}\,\right]^{\frac{1}{2}}, \tag{7.46}$$

where
σ = bending stress amplitude including stress concentration effects
τ = torsional stress amplitude including stress concentration effects
ϕ = phase angle between bending and torsion

Equation (7.46) can be used to obtain both mean and alternating components of stress. It can be shown by comparing Eqs. (7.46) and (7.42) through (7.43) that the von Mises approach is conservative for any phase difference or stress ratio. However, this is true only for high cycle fatigue, and can be non-conservative for finite life applications. Since shafts are usually designed for infinite life, this is rarely a concern.

3. Behavior for a given loading and material combination may be well-quantified within an organization, or can be found in the technical literature.

7.12 Fracture Mechanics Approach to Fatigue

With the increasing interest in materials without clear endurance limits, special attention must be paid to damage accumulation and replacing fatigued components before catastrophic failure can occur. Indeed, this is a main design challenge in the aircraft industry, where aluminum alloys, although they have no endurance limit, are used because of their high strength-to-weight ratios. Routine nondestructive evaluation to determine the size of flaws in stress-bearing members is conducted to identify and remove suspect components. This approach is called **fault-tolerant design** since it recognizes the presence of defects and allows the use of a material as long as the defects remain smaller than a critical size.

Recall the situation depicted in Fig. 7.4b, where Regime B is active. If the slope m is reasonably small, a part can have a long service life before the crack size becomes critical, and this is indeed a common circumstance. Paris et al. [1961] hypothesized that crack growth in such a cyclic loading should follow the rule

$$\frac{dl_c}{dN} = C(\Delta K)^m, \tag{7.47}$$

where dl_c/dN is the change in crack length per load cycle (l_c is the crack length and N is the number of stress cycles), C and m are empirical constants, and ΔK is the stress intensity range defined as

$$\Delta K = K_{\text{max}} - K_{\text{min}}, \tag{7.48}$$

where K_{max} and K_{min} are the maximum and minimum stress intensity factors, respectively, around a crack during a loading cycle. For a center-cracked plate with crack length $l_c = 2a$, recall from Eq. (6.6) that

$$K_{\text{max}} = Y\sigma_{\text{max}}\sqrt{\pi a}, \tag{7.49}$$

where σ_{max} is the maximum far-field stress and Y is a correction factor to account for finite plate sizes. Table 7.7 summarizes conservative values of C and m for steels, while Fig. 7.19 gives fatigue crack propagation data for a variety of materials. Appendix C gives values of Y. Equation (7.47) is known as the **Paris power law** and is the most widely used equation in fracture mechanics approaches to fatigue problems. Suresh [1998] has derived the life of a component based on the Paris power law as

$$N = \frac{2}{(m-2)CY^m(\Delta\sigma)^m\pi^{m/2}} \times \left\{ \frac{1}{(l_{co})^{(m-2)/2}} - \frac{1}{(l_{cf})^{(m-2)/2}} \right\}, \tag{7.50}$$

Table 7.7: Paris law constants for various metals. Data represents worst-case (fastest) crack growth rates reported for the material.

Material	C mm/cycle $(\text{MPa}\sqrt{\text{m}})^m$	C in/cycle $\text{ksi}\sqrt{\text{in}}^m$	m
Steel			
Ferritic-pearlitic	6.89×10^{-9}	3.6×10^{-10}	3.0
Martensitic	1.36×10^{-7}	6.6×10^{-9}	2.25
Austenitic	5.61×10^{-9}	3.0×10^{-10}	3.25
Aluminum			
6061-T6	5.88×10^{-8}	3.1×10^{-9}	3.17
2024-T3	1.6×10^{-11}	8.4×10^{-11}	3.59

unless $m = 2$, when the fatigue life is

$$N = \frac{1}{CY^2(\Delta\sigma)^2\pi}\ln\frac{l_{cf}}{l_{co}}, \tag{7.51}$$

where

- N = fatigue life in cycles
- C and m = material constants
- Y = correction factor to account for finite plate sizes
- $\Delta\sigma$ = range of far-field stresses to which component is subjected
- l_{co} = initial crack size
- l_{cf} = critical crack size based on fracture mechanics

Example 7.10: Fault Tolerant Design

Given: An aluminum alloy aircraft component in the form of a 100-mm-wide plate is subjected to a 100-MPa stress during pressurization of the aircraft cabin. Superimposed on this stress is a fluctuation arising from vibration, with an amplitude of 10 MPa and a frequency of 45 Hz. Nondestructive crack detection techniques do not detect any flaws, but the smallest detectable flaw is 0.2 mm.

Find: Determine the minimum expected life of the component. If upon reinspection a centered 1.1-mm-long crack is found, what is the expected life from the time of inspection? Use a fracture toughness of 29 MPa$\sqrt{\text{m}}$ and a fracture stress of 260 MPa and assume $m = 2.5$ and $C = 6.9\times10^{-12}$ m/cycle for $\Delta\sigma$ in MPa.

Solution: The critical crack length l_{cf} is given by Eq. (6.5) and is found to be 1.60 mm. A worst-case scenario would occur if the largest undetectable flaw resulting in the largest value of Y were located in the geometry. This flaw occurs for the double-edge-cracked tension specimen, where Y equals 2.0. Therefore, the life is found from Eq. (7.50) as

$$\begin{aligned} N &= \frac{2}{(m-2)CY^m(\Delta\sigma)^m\pi^{m/2}} \\ &\quad \times \left\{ \frac{1}{(l_{c0})^{(m-2)/2}} - \frac{1}{(l_{cf})^{(m-2)/2}} \right\} \\ &= 324 \text{ hr}. \end{aligned}$$

If $l_{c0} = 1.1$ mm, the life until fracture is approximately 47 hr.

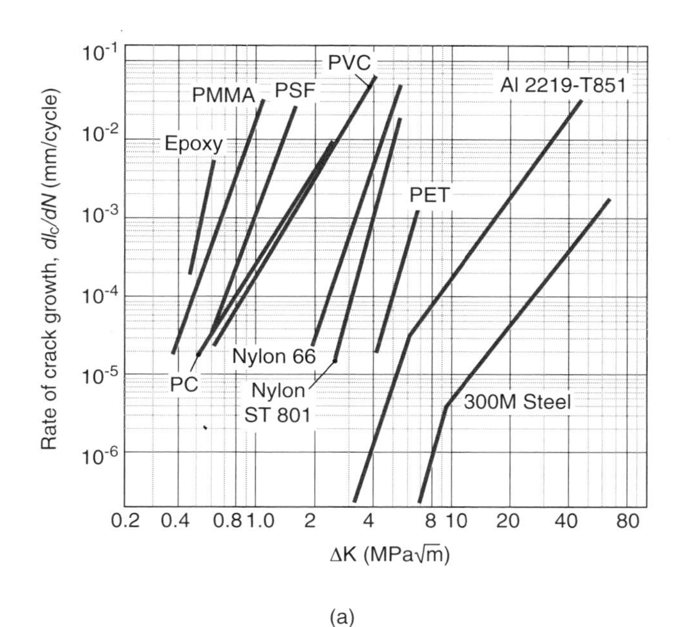

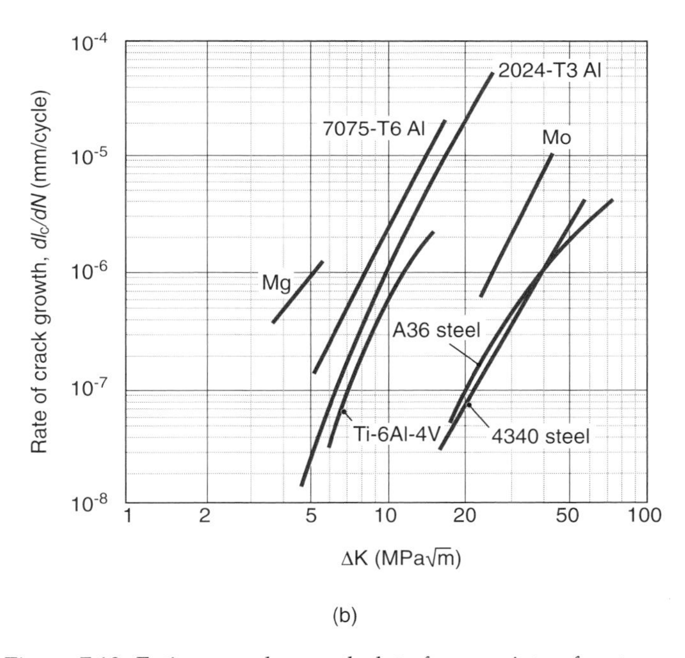

Figure 7.19: Fatigue crack growth data for a variety of materials. (a) Selected polymers in comparison to aluminum and steel; (b) selected metal alloys. *Source:* Adapted from Bowman [2004].

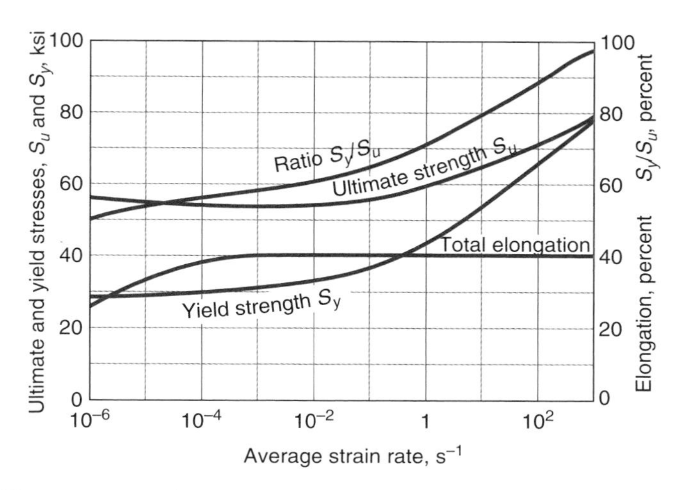

Figure 7.20: Mechanical properties of mild steel at room temperature as function of average strain rate.

7.13 Linear Impact Stresses and Deformations

This chapter thus far has focused on cyclic load variation; this section focuses on **impact loading**. Throughout most of this text a load is applied to a body gradually such that when the load reaches a maximum it remains constant or static. For the material covered thus far in this chapter, the loading is **dynamic** (i.e., it varies with time). When loads are rapidly applied to a body as in impact loading, the stress levels and deformations induced are often much larger than those in static or cyclic loading.

The properties of the material are a function of the loading speed; the more rapid the loading, the higher both the yield and ultimate strengths of the material. Figure 7.20 shows the variation of the mechanical properties with loading speed for a typical mild steel. For average strain rates from 10^{-1} to 10^3 s^{-1} the yield strength increases significantly.

If no energy is lost during impact, the conservation of energy can be applied. Consider a simple block falling a distance h and striking a spring compressed a distance $\delta_{\max}$ before momentarily coming to rest. If the mass of the spring is neglected and it is assumed that the spring responds elastically, conservation of energy requires that the kinetic energy be completely transformed into elastic strain energy:

$$W\left(h+\delta_{\max}\right)=\frac{1}{2}\left(k\delta_{\max}\right)\delta_{\max}, \tag{7.52}$$

where k is the spring constant in N/m and W is the weight of the block. Equation (7.52) can be expressed as a quadratic equation:

$$\delta_{\max}^2-\frac{2W}{k}\delta_{\max}-2\left(\frac{W}{k}\right)h=0. \tag{7.53}$$

Solving for $\delta_{\max}$

$$\delta_{\max}=\frac{W}{k}+\sqrt{\left(\frac{W}{k}\right)^2+2\left(\frac{W}{k}\right)h}. \tag{7.54}$$

If the weight is applied statically (or gradually) to the spring, the static displacement is

$$\delta_{st}=\frac{W}{k}. \tag{7.55}$$

Substituting Eq. (7.55) into Eq. (7.54) gives

$$\delta_{\max} = \delta_{st} + \sqrt{(\delta_{st})^2 + 2\delta_{st}h},$$

or

$$\delta_{\max} = \delta_{st}\left(1 + \sqrt{1 + \frac{2h}{\delta_{st}}}\right). \tag{7.56}$$

From the maximum displacement in Eq. (7.56) the maximum force is

$$P_{\max} = k\delta_{\max}. \tag{7.57}$$

Recall that in dropping the block from some distance, h, the maximum force $P_{\max}$ on impact is essentially instantaneous. The block will continue to oscillate until the motion dampens and the block assumes the static position. This analysis assumes that when the block first makes contact with the spring, the block does not rebound, that is, it does not separate from the spring. Making use of the spring constant k in Eqs. (7.55) and (7.57) relates the static and dynamic effects.

The impact factor can be expressed as

$$I_m = \frac{\delta_{\max}}{\delta_{st}} = \frac{P_{\max}}{W} = 1 + \sqrt{1 + \frac{2h}{\delta_{st}}}. \tag{7.58}$$

Note that once the impact factor is known, the impact load, stresses, and deflections can be calculated. The impact stress is

$$\sigma = \frac{P_{\max}}{A}, \tag{7.59}$$

where A is the area of the spring surface, m^2. If, instead of the block dropping vertically, it slides with a velocity, u, on a surface that provides little frictional resistance (so that it can be neglected) and the block impacts the spring, the block's kinetic energy is transformed into stored energy in the spring, or

$$\frac{1}{2}\left(\frac{W}{g}\right)u^2 = \frac{1}{2}k\delta_{\max}^2, \tag{7.60}$$

where g is gravitational acceleration (9.807 m/s^2). Note that the right side of Eq. (7.52) is identical to the right side of Eq. (7.60). Solving for $\delta_{\max}$ in Eq. (7.60) gives

$$\delta_{\max} = \sqrt{\frac{Wu^2}{gk}}. \tag{7.61}$$

By using Eq. (7.55), Eq. (7.61) becomes

$$\delta_{\max} = \sqrt{\frac{\delta_{st}u^2}{g}}. \tag{7.62}$$

In both situations the moving body (the block) is assumed to be rigid, and the stationary body (the spring) is assumed to be deformable. The material is assumed to behave in a linear-elastic manner. Thus, whether a block falls a distance h and impacts on a spring, or a block moves at a velocity u and strikes a spring, the formulation for the deformation, the impact force, or the impact stress can be determined.

Example 7.11: Impact Stresses

Given: A diver jumps up 0.6 m on the free end of a diving board before diving into the water. Figure 7.21 shows a sketch of the diver and the dimensions of the diving board. The supported end of the diving board is fixed. The modulus of elasticity is 69 GPa and the yield strength is 220 MPa. The weight of the diver is 900 N.

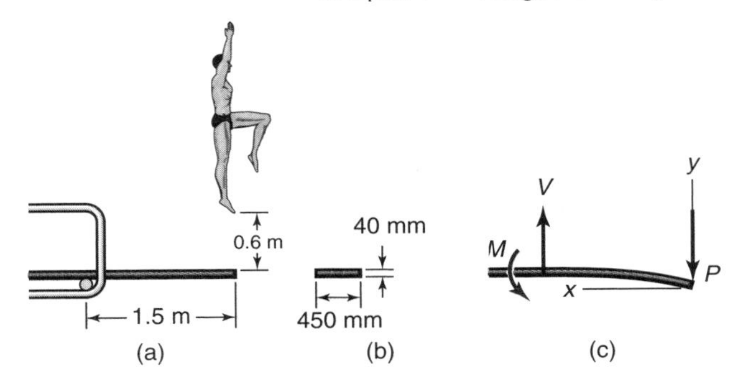

Figure 7.21: Diver impacting diving board, used in Example 7.11. (a) Side view; (b) front view; (c) side view showing forces and coordinates.

Find: The safety factor for impact loading based on yielding.

Solution: Equations (7.54) and (7.55) should be used to determine the maximum deflection at impact at the end of the diving board. The spring rate is not given and will need to be determined. The forces acting are shown in Fig. 7.21c. From Eq. (5.3),

$$\frac{d^2y}{dx^2} = \frac{M}{EI} = -\frac{Px}{EI}.$$

Integrating this equation gives

$$EI\frac{dy}{dx} = -\frac{Px^2}{2} + C_1,$$

where C_1 is an integration constant. Integrating again gives

$$EIy = -\frac{Px^3}{6} + C_1x + C_2,$$

where C_2 is another integration constant. The boundary conditions are

1. $x = l$, $y' = 0$, leading to $C_1 = \frac{Pl^2}{2}$
2. $x = l$, $y = 0$, resulting in $C_2 = -\frac{Pl^3}{3}$

The deflection at the end of the board ($x = 0$ in Fig. 7.21c) is of interest. Therefore,

$$EIy = -\frac{Px^3}{6} + \frac{Pl^2x}{2} - \frac{Pl^3}{3},$$

or

$$\frac{6EIy}{P} = -x^3 + 3l^2x - 2l^3,$$

so that

$$\delta = -\frac{Pl^3}{3EI}.$$

The spring constant is

$$k = -\frac{P}{\delta} = \frac{3EI}{l^3},$$

where I is the area moment of inertia given by

$$I = \frac{bh^3}{12} = \frac{(0.45)(0.04)^3}{12} = 2.40 \times 10^{-6}\ \text{m}^4,$$

l is the length (given as 1.5 m) and E is the modulus of elasticity (69 GPa). Therefore,

$$k = \frac{3(69 \times 10^9)(2.40 \times 10^{-6})}{(1.5)^3} = 147.2 \text{ kN/m}.$$

From Eq. (7.55),

$$\delta_{st} = \frac{W}{k} = \frac{900 \text{ N}}{147.2 \text{ kN/m}} = 0.00611 \text{ m}.$$

From Eq. (7.58), the impact factor is

$$\begin{aligned} I_m &= \frac{\delta_{\max}}{\delta_{st}} \\ &= 1 + \sqrt{1 + \frac{2h}{\delta_{st}}} \\ &= 1 + \sqrt{1 + \frac{2(0.6)}{0.006144}} \\ &= 14.01. \end{aligned}$$

Therefore,

$$\delta_{\max} = I_m \delta_{st} = (14.01)(0.006144) = 0.08566 \text{ m}.$$

From Eq. (7.57),

$$P_{\max} = k\delta_{\max} = \left(147.2 \times 10^3\right)(0.08566) = 12.61 \text{ kN}.$$

The maximum bending stress from Eq. (4.45) is

$$\sigma_{\max} = \frac{Mc}{I} = \frac{(12,610)(1.5)(0.020)}{2.40 \times 10^{-6}} = 157.6 \text{ MPa}.$$

The yield stress is given as 220 MPa. The safety factor for yielding is

$$n_s = \frac{S_y}{\sigma_{\max}} = \frac{\left(220 \times 10^6\right)}{157.6 \times 10^6} = 1.396.$$

Thus, $n_s > 1$ and thus failure should not occur; however, because the safety factor is just above 1, the margin of safety is a minimum.

Case Study 7.1: Fault Tolerant Design in Aircraft

Aircraft are a particularly important application of fatigue design theory, especially fault tolerant design. Since the weight of an aircraft needs to be minimized to control fuel consumption, materials with high strength-to-weight ratios (see Section 3.7.2) are required. Historically, this has necessitated the widespread use of aluminum and titanium alloys, and also composite materials. Unfortunately, these materials often do not display clear endurance limits. A designer can then opt to incorporate very large safety factors, but this would have adverse effects on fuel economy and would make commercial aviation impractical.

Maintenance Intervals

The alternative that is followed in the aviation industry is to pursue periodic inspection and maintenance. These maintenance tasks address a wide variety of concerns, including evaluation of brakes, replacement of lubricants and working fluids, inspection of bearings, etc.This case study emphasizes the structural components of aircraft and the methodologies used to prevent fatigue failure during use.

The four classes of maintenance on an aircraft are, along with their approximate interval:

- **A check.** An A check is performed approximately every 500 flight hours. This check is generally performed overnight at an airport gate, and includes visual inspection for obvious damage, corrosion, and deterioration.

- **B check.** B checks are not generally used with modern aircraft, but in the past consisted of similar tasks as the A check with marginally more detailed procedures. A B check is also generally performed overnight at an airport gate.

(a)

(b)

Figure 7.22: (a) Exterior view of Boeing 747-400 during a D check; (b) inspection of landing gear component for structural integrity. *Source:* Courtesy of Lufthansa Technik.

- **C check.** C checks are generally performed every 12 months of operation or based on flight hours. C checks are high-level tests that are performed in an airplane hangar and use extensive tooling and equipment, and generally take a few days. As related to the aircraft structure, the C check involves a thorough inspection of the structure, and can include non-destructive approaches such as ultrasonic or eddy current inspection. Examination for corrosion or cracks, especially at locations of stress concentration, is carefully performed.

- **D check.** A D Check is a very comprehensive check for an aircraft, and has been characterized as completely disassembling the aircraft and replacing worn or damaged components so that the aircraft is brought to an as-new condition. This characterization is an exaggeration, but a D check does involve extensive evaluation of structural components.

A D check occurs roughly every five years and can cost over $2 million. Figure 7.22a shows a general view of a D check, while Figure 7.22b shows details of non-destructive evaluation of a landing gear structural component during a D check.

Note that the intervals given are only approximate, as each aircraft has a maintenance cycle as defined by the manufacturer and approved by governing bodies such as the Federal Aviation Administration or European Aviation Safety Agency. Modern inspection intervals are determined by damage growth intervals such as predicted by Eq. (7.50), whereas in the past these were based on service experience. There is no limit to the service life of a damage-tolerant aircraft if necessary inspections and corrective actions are carried out.

Design Considerations

Aircraft designers have a very sophisticated understanding of aircraft loads during normal operation and for reasonably foreseeable conditions (wind gusts, hard landings, fuselage pressurization, etc.) that can occur during flight. Computational tools are used in the design stage, and parts are sized so that the largest loads can be safely supported.

It should be noted that many components have complex geometries and loadings, and the loadings can change direction during different maneuvers. Therefore, it has been recognized that multiple site damage is a concern with aircraft structures.

In designing a structural component, fault or damage tolerance can be fundamentally incorporated by selecting proper materials that have a low crack growth rate for the stress intensities encountered in service (see Section 7.12). A given material also defines the maximum damage that can be sustained under extreme loads (see Section 6.5), that is, the allowable crack length (or contributions from multiple cracks) that still allow the plane to fly safely under foreseeable conditions. Stress analysis results in a stress intensity factor that allows prediction of crack growth rates in materials (see Fig. 7.19). Often, initial cracks are smaller than the detection threshold of measurement systems, so designs use the detection threshold as the assumed initial flaw size in these conditions. Therefore, inspection intervals can be defined to ensure that damage is detected before a catastrophic failure occurs. Further, inspection procedures and devices can be developed to capture failure modes of particular concern.

Design for disassembly is not usually a criterion considered by designers, but it certainly is essential for C and D checks on aircraft. Even so, many components are exceedingly difficult to access or test. Recognizing this, larger safety factors or more fatigue resistant materials are used for components that are especially difficult to inspect and repair/replace, even though this may have a cost penalty.

Summary

There is considerable pressure on aircraft manufacturers to extend the interval between service checks. The sophistication with respect to fatigue design has advanced considerably in the past two decades, so that maintenance hours per airplane associated with fatigue and corrosion issues have decreased by an order of magnitude. It has often been noted that flying in commercial aircraft is the safest form of travel, and this is attributable to the widespread use of fault tolerant design.

7.14 Summary

Failures in components or structures are often caused by fluctuating stresses. If the stress variation sequence repeats itself, it is called cyclic. The various cyclic patterns were described. The fatigue strength versus the logarithm of the number of cycles to failure was presented for materials with and without an endurance limit. Ferrous materials tend to have endurance limits, and nonferrous materials tend not to have endurance limits. The endurance limit for ferrous materials is a function of the type of loading the component is subjected to. Also, *low*-cycle fatigue failure has been classified as occurring at less than 10^3 cycles, and *high*-cycle fatigue failure as occurring above 10^3 cycles.

The endurance limit has been experimentally determined for a specimen of specific size with a mirror-like surface finish. The specimen was precisely prepared and tested under controlled conditions. In practice, conditions differ significantly from those in a test situation. To more accurately characterize the conditions that prevail in practice, the modified endurance limit is used. The surface finish and size factors are used along with a fatigue stress concentration factor.

Various approaches for estimating when failure will occur under nonzero mean cyclic stresses were also considered. The Soderberg, Gerber, Goodman, and modified Goodman theories were presented. These theories allow the variation of mean and alternating stresses. The construction of a complete modified Goodman diagram was shown, and fatigue and yield criteria were used. Failure equations were given for specific regions of the modified Goodman diagram.

Most of the results presented were for ductile materials, but the behavior of brittle materials was briefly described. The major difference between brittle and ductile materials is that the compressive and tensile strengths are nearly identical for ductile materials whereas for brittle materials the compressive strength is several times greater than the tensile strength.

Impact loading has to be considered when loads are rapidly applied. The stress levels and deformations induced are much larger than with static or cyclic loading. The more rapid the loading, the higher the yield and ultimate strengths of a material. By equating the kinetic and elastic strain energies, two types of impact loading were considered: a block falling from a given height onto a spring, and a block sliding downward into a spring. An impact factor, maximum and static deformations, and maximum and static loads were also obtained.

Key Words

cyclic stress stress sequence that repeats over time

dynamic adjective indicating variation with time

endurance limit stress level below which infinite life can be realized

fatigue failure, at relatively low stress levels, of structures that are subject to fluctuating and cyclic stresses

finite life life until failure due to fatigue

Gerber line parabolic relationship taking mean and alternating stresses into account

Goodman line theory connecting modified endurance limit and ultimate strength on plot of alternating stress versus mean stress

high-cycle fatigue fatigue failure that occurs above 10^3 cycles but below 10^6 cycles

impact loading load rapidly applied to body

infinite life stress levels that do not cause fatigue failure

low-cycle fatigue fatigue failure that occurs below 10^3 cycles

Manson-Coffin relationship theoretical approach to fatigue based on strain

mean stress average of minimum and maximum stresses in cycle

Miner's rule same as linear damage rule

modified endurance limit corrections for endurance limit based on surface finish, material, specimen size, loading type, temperature, etc.

modified Goodman diagram diagram that defines all stress states not resulting in fatigue failure or yielding

notch sensitivity material property that reflects ability of ductile materials to be less susceptible to stress raisers in fatigue

Paris power law postulate that crack growth in cyclic loading follows power law

residual stress internal stress usually caused by manufacturing process

Soderberg line theory connecting modified endurance limit and yield strength on plot of alternating stress versus mean stress

S-N diagram plot of stress level versus number of cycles before failure

staircase an approach used to determine the endurance limit of a material by varying the applied stress on a modest number of samples.

stress amplitude one-half of stress range

stress range difference between maximum and minimum stresses in cycle

stress ratio ratio of minimum and maximum stresses

Wöhler diagram same as S-N diagram

yield line failure criterion that postulates yielding on first cycle of cyclic loading with nonzero mean

Summary of Equations

Cyclic Stresses:

Mean stress: $\sigma_m = \dfrac{\sigma_{\max} + \sigma_{\min}}{2}$

Stress range: $\sigma_r = \sigma_{\max} - \sigma_{\min}$

Stress amplitude: $\sigma_a = \dfrac{\sigma_r}{2} = \dfrac{\sigma_{\max} - \sigma_{\min}}{2}$

Finite Life Fatigue:

Manson-Coffin Relationship:

$\frac{\Delta\epsilon}{2} = \frac{\sigma_f'}{E}\left(2N'\right)^a + \epsilon_f'\left(2N'\right)^\alpha$

Strength as a function of loading cycles:

$\log S_f' = b_s \log N_t' + \bar{C}$,

$$b_s = -\frac{1}{3}\log\left(\frac{S_l'}{S_e'}\right),\ \bar{C} = \log\left[\frac{\left(S_l'\right)^2}{S_e'}\right]$$

High cycle fatigue limits:

Lower limit ($N \approx 10^3$): for bending, $S_l' = 0.9S_u$;
for axial loads, $S_l' = 0.75S_u$;
for torsion, $S_l' = 0.72S_u$.

Upper limit ($N \approx 10^6$ or 10^7): $S_e' = 0.5S_u$;
for axial loads, $S_e' = 0.45S_u$;
for torsion, $S_e' = 0.29S_u$.

Life as a function of stress: $N_t' = \left(S_f' 10^{-\bar{C}}\right)^{1/b_s}$

Fatigue Stress Concentrations: $K_f = 1 + (K_c - 1)q_n$

Modified Endurance Limit: $S_e = k_f k_s k_r k_m S_e'$

Surface finish factor: $k_f = eS_{ut}^f$

Size factor:

$$k_s = \begin{cases} 0.869d^{-0.112} & 0.3 \text{ in.} < d < 10 \text{ in.} \\ 1 & d < 0.3 \text{ in. or } d \le 8 \text{ mm} \\ 1.189d^{-0.112} & 8 \text{ mm} < d \le 250 \text{ mm} \end{cases}$$

Temperature factor: $k_t = \dfrac{S_{ut}}{S_{ut,\text{ref}}}$

Cumulative Damage:

Miner's rule: $\dfrac{n_1'}{N_1'} + \dfrac{n_2'}{N_2'} + \cdots + \dfrac{n_i'}{N_i'} \ge 1$

Effect of Nonzero Mean Stress:

Gerber parabola: $\dfrac{K_f n_s \sigma_a}{S_e} + \left(\dfrac{n_s \sigma_m}{S_{ut}}\right)^2 = 1$

Goodman line: $\dfrac{K_f \sigma_a}{S_e} + \dfrac{\sigma_m}{S_{ut}} = \dfrac{1}{n_s}$

Soderberg line: $\dfrac{K_f \sigma_a}{S_e} + \dfrac{\sigma_m}{S_{yt}} = \dfrac{1}{n_s}$

Yield line: $\dfrac{\sigma_a}{S_{yt}} + \dfrac{\sigma_m}{S_{yt}} = \dfrac{1}{n_s}$

Multiaxial Stresses:

Simple, fully reversing: $n_s = \dfrac{S_e}{\sigma_e'}$

where

$$\left(\sigma_e'\right)^2 = \sigma_{a,1}^2 + \sigma_{a,2}^2 + \sigma_{a,3}^2 - \sigma_{a,1}\sigma_{a,2} - \sigma_{a,2}\sigma_{a,3} - \sigma_{a,1}\sigma_{a,3}$$

Simple, nonzero mean:

Sines method:

$$2\left(\sigma_a'\right)^2 = \left(\sigma_{a,x} - \sigma_{a,y}\right)^2 + \left(\sigma_{a,y} - \sigma_{a,z}\right)^2 + \left(\sigma_{a,z} - \sigma_{a,x}\right)^2 + 6\left(\tau_{a,xy}^2 + \tau_{a,yz}^2 + \tau_{a,zx}^2\right)$$

$$\sigma'_m = \sigma_{m,x} + \sigma_{m,y} + \sigma_{m,z}$$

von Mises method:

$$2\left(\sigma'_a\right)^2 = \left(\sigma_{a,x} - \sigma_{a,y}\right)^2 + \left(\sigma_{a,y} - \sigma_{a,z}\right)^2 + \left(\sigma_{a,z} - \sigma_{a,x}\right)^2 + 6\left(\tau^2_{a,xy} + \tau^2_{a,yz} + \tau^2_{a,zx}\right)$$

$$2\left(\sigma'_m\right)^2 = \left(\sigma_{m,x} - \sigma_{m,y}\right)^2 + \left(\sigma_{m,y} - \sigma_{m,z}\right)^2 + \left(\sigma_{m,z} - \sigma_{m,x}\right)^2 + 6\left(\tau^2_{m,xy} + \tau^2_{m,yz} + \tau^2_{m,zx}\right)$$

Complex multiaxial stresses:

$$\sigma_s = \frac{\sigma}{\sqrt{2}}\left[1 + 3\left(\frac{\tau}{\sigma}\right)^2 + \sqrt{1 + 6\frac{\tau}{\sigma}\cos 2\phi + 9\left(\frac{\tau}{\sigma}\right)^4}\right]^{1/2}$$

Fracture Mechanics Approach

Paris equation: $\dfrac{dl_c}{dN} = C(\Delta K)^m$

Predicted life:

$$N = \frac{2}{(m-2)CY^m(\Delta\sigma)^m\pi^{m/2}} \times \left\{\frac{1}{(l_{c0})^{(m-2)/2}} - \frac{1}{(l_{cf})^{(m-2)/2}}\right\}$$

$$\text{if } m = 2,\ N = \frac{1}{CY^2\left(\Delta\sigma\right)^2\pi}\ln\frac{l_{cf}}{l_{c0}}$$

Impact

Maximum deformation: $\delta_{\max} = \delta_{st}\left(1 + \sqrt{1 + \dfrac{2h}{\delta_{st}}}\right)$

Impact factor: $I_m = 1 + \sqrt{1 + \dfrac{2h}{\delta_{st}}}$

Recommended Readings

Anderson, T.L. (2004) *Fracture Mechanics: Fundamentals and Applications*, 3rd ed., CRC Press.

Bannantine, J.A., Comer, J.J., and Handrock, J.L. (1989) *Fundamentals of Metal Fatigue Analysis*, Prentice-Hall.

Bathias, C., and Paris, P.C. (2005) *Gigacycle Fatigue in Mechanical Practice*. Wiley.

Bathias, C., and Pineua, A. (2010) *Fatigue of Materials and Structures*. Wiley.

Budynas, R.G., and Nesbitt, J.K. (2011), *Shigley's Mechanical Engineering Design*, 9th ed., McGraw-Hill.

Juvinall, R.C. (1967) *Engineering Considerations of Stress, Strain, and Strengths*, McGraw-Hill, New York.

Madayag, A.F. (1969) *Metal Fatigue: Theory and Design*, Wiley, New York.

Freiman, S., and Mecholsky Jr., J.J. (2012) *The Fracture of Brittle Materials: Testing and Analysis*. Wiley.

Lee, Y.-L., Pan, J., Hathaway, R., and Barkey, M. (2005) *Fatigue Testing and Analysis*. Elsevier.

Manson, S.S., and Halford, G.R. (2006) *Fatigue and Durability of Structural Materials*. American Society for Metals.

Newman, J.C., and Plascik, R.S., eds. (2000) *Fatigue Crack Growth Thresholds, Endurance Limites, and Design*, STP 1372, American Society for Testing and Materials.

Nicholas, T. (2006) *High Cycle Fatigue – A Mechanics of Materials Perspective*. Elsevier.

Norton, R.L. (2011) *Machine Design*, 4th ed., Prentice Hall.

Rice, R.C., ed., (1988) *Fatigue Design Handbook*, Society of Automotive Engineers, Warrendale, PA.

Schijve, J. (2009) *Fatigue of Structures and Materials*, Springer.

Stephens, R.I., Fatemi, A., Stephens, R.R., Fuchs, H.O., and Faterni, A. (2000) *Metal Fatigue in Engineering*, 2nd ed., Wiley.

Suresh, S. (1998) *Fatigue of Materials*, 2nd ed., Cambridge University Press.

Zahavi, E., and Torbilo, V. (1996) *Fatigue Design*, CRC Press.

References

Bowman, K. (2004) *Mechanical Behavior of Materials*, Wiley.

Johnson, R.C. (1967) *Predicting Part Failures Part I* "Machine Design," vol. 45, p. 108.

Juvinall, R.C., and Marshek, K.M. (1991) *Fundamentals of Machine Component Design*, Wiley.

Kuguel, R. (1969), "A Relation Between Theoretical Stress Concentration Factor and Fatigue Notch Factor Deduced from the Concept of Highly Stressed Volume.", Proceedings ASTM, v. 61, pp. 732-748.

Lipson, C., and Juvinall, R.C. (1963) *Handbook of Stress and Strength*, Macmillan.

Metals Handbook, American Society for Metals, 1975.

Norton, R.L. (1996) *Machine Design*, Prentice-Hall.

Paris, P.C., Gomez, M.P., and Anderson, W.P. (1961) *A Rational Analytic Theory of Fatigue*, Trend Engineering, University of Washington, v. 13, no. 1, pp. 9–14.

Shigley, J.E., and Mitchell, L.D. (1983) *Mechanical Engineering Design*, 4th ed. McGraw-Hill.

Sines G., and Waisman, J.L. (1959) *Metal Fatigue*, McGraw-Hill.

Suresh, S. (1998) *Fatigue of Materials*, 2nd ed. Cambridge University Press.

Questions

7.1 Define *fatigue* and *fatigue wear*. Explain how they are related.

7.2 Sketch a completely reversing and a released tension stress cycle.

7.3 How does residual stress affect fatigue?

7.4 What is a cyclic stress?

7.5 What is the difference between a flaw and a crack?

7.6 Define stress range, stress amplitude, and stress ratio.

7.7 What is the difference between fatigue strength and endurance limit?

7.8 What is the Paris regime of crack growth?

7.9 Are there always striations on a fatigue fracture surface?

7.10 What is an S-N diagram? Sketch a typical S-N diagram for steel and one for aluminum.

7.11 What is the notch sensitivity of a material?

7.12 Why does the modified endurance limit depend on the manufacturing process used to produce a part?

7.13 Explain the reasons that there is a size factor required to determine the modified endurance limit.

7.14 Why does shot peening improve fatigue strength?

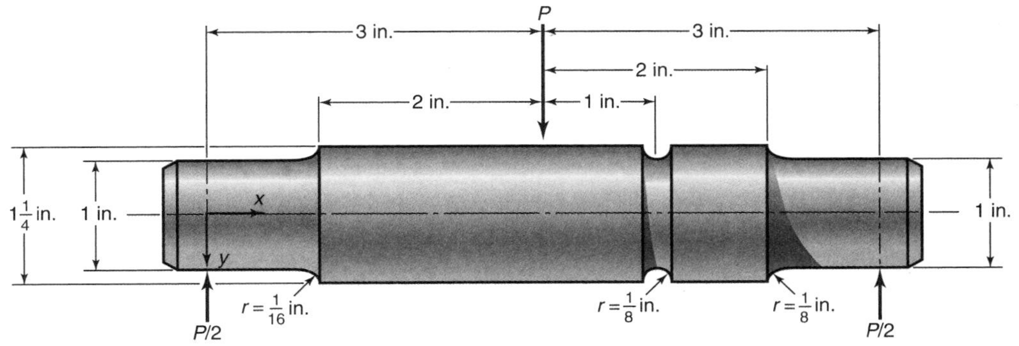

Sketch b, for Problem 7.35

7.15 What are the similarities between the Goodman line and the Gerber parabola?

7.16 What is Miner's rule?

7.17 List the theories that incorporate the influence of mean stress on fatigue failure.

7.18 What is fault tolerant design?

7.19 What is the Paris law?

7.20 Define impact. Why are impact stresses larger than static stresses?

Qualitative Problems

7.21 Review the Manson-Coffin relationship and identify all of the important variables, explaining how each impacts fatigue crack growth.

7.22 List the factors that influence the endurance limit of a carbon steel. Rank them in order of importance.

7.23 In Fig. 7.4, Regime A shows crack growth rates slower than a lattice spacing per cycle. Explain how this is possible.

7.24 Identify the point where fatigue started and the direction of crack growth for the cross-sections shown in Sketch a.

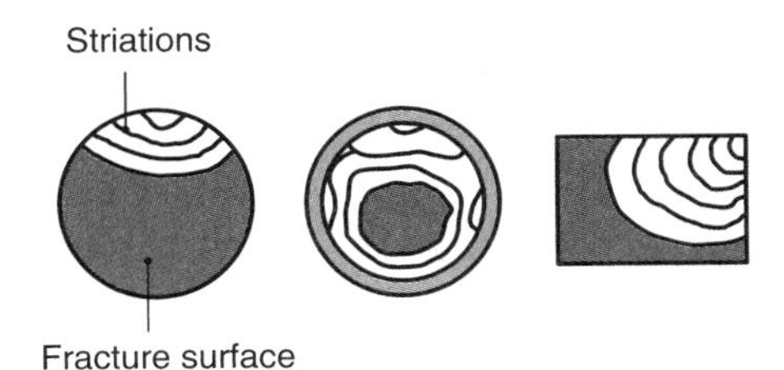

Sketch a, for Problem 7.24

7.25 Is there an endurance limit for aluminum or copper alloys? Explain.

7.26 According to the Manson-Coffin equation, stresses that induce deformation will eventually result in fatigue failure. Does this mean that there cannot be an endurance limit? Explain.

7.27 Demonstrate that Example 7.3 produces the same results regardless of units.

7.28 List factors that affect surface finish that are not included in Eq. (7.19).

7.29 List factors that can be incorporated into the miscellaneous effects modification factor on endurance limit.

7.30 Without using equations, explain the advantages of using the modified Goodman approach.

7.31 Give three examples each of parts that (a) encounter fewer than 1000 cycles; (b) encounter more than 1000 but less than 1 million cycles, and (c) encounter more than 1 million cycles.

Quantitative Problems

7.32 Using the information in Fig. 7.19, estimate the constants C and m from Eq. (7.47) and construct a table of these values for different materials.

7.33 A tuning fork is hit with a pencil and starts to vibrate with a frequency of 440 Hz. The maximum bending stress in the tuning fork is 2 MPa at the end positions. Calculate the mean stress, the range of stress, the stress amplitude, and the stress ratio. Also, calculate how much stress the tuning fork can sustain without being plastically deformed if it is made of AISI 1080 steel, quenched and tempered at 800°C. *Ans.* $\sigma_m = 0$, $\sigma_r = 4$ MPa, $R = -1$.

7.34 The jack for a Volvo consists of a mechanism in which the lift screw extends horizontally through two corners of the mechanism while the other two corners apply a force between the ground and the car to be lifted. Strain gage measurements show that the maximum compressive stress in the jack is 150 MPa when the car is jacked up so high that both wheels on one side of the car are in the air and the load on the jack is 8000 N. How many times can the jack be used for a small truck that weighs 6 tons and loads the jack to 15,000 N before it fails from fatigue? The jack material is AISI 1080 steel, quenched and tempered at 800°C. *Ans.* 807,800 cycles.

7.35 The shaft shown in Sketch b rotates at high speed while the imposed loads remain static. The shaft is machined from ground, high-carbon steel (AISI 1080, quenched and tempered at 800°C). If the loading is sufficiently large to produce a fatigue failure after 1 million cycles, where would the failure most likely occur? Show all necessary computations and reasoning.

7.36 For each of the annealed AISI 1040 steel bars shown in Sketches c and d determine

(a) The static tensile load causing fracture. *Ans.* $P =$ 325 kN.

(b) The alternating (completely reversing) axial load $\pm P$ that would be just on the verge of producing eventual fatigue failure. *Ans.* $P_{\text{notched}} = 62.5$ kN.

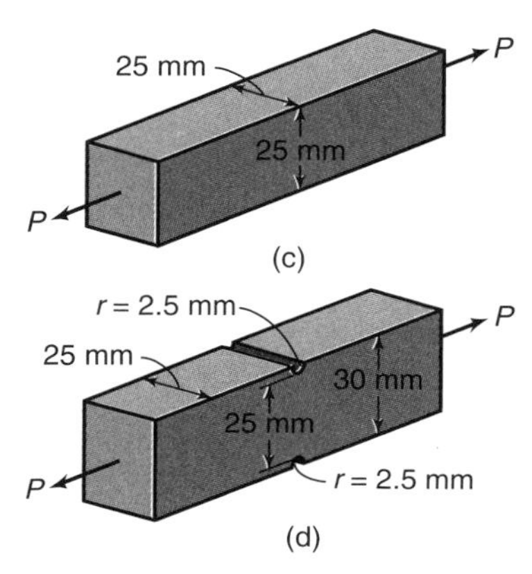

Sketches c and d, for Problem 7.36

7.37 A stepped shaft, as shown in Sketch e, was machined from high-carbon steel (AISI 1080, quenched and tempered at 800°C). The loading is one of completely reversed torsion. During a typical 30 s of operation under overload conditions the nominal stress in the 1-in.-diameter section was calculated to be as shown. Estimate the life of the shaft when it is operating continually under these conditions. *Ans.* 4.79 hr.

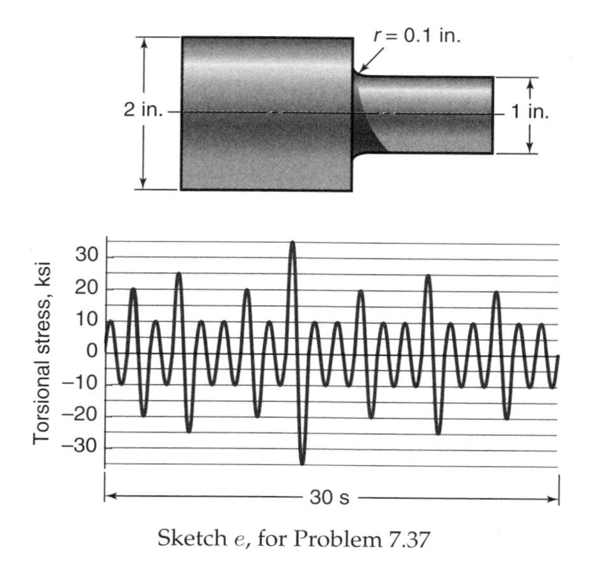

Sketch e, for Problem 7.37

7.38 A flood-protection dam gate is supposed to operate only once per week for 100 years, but after 30 years of use it needs to be operated twice per day (each time the high tide comes in). Find how much lower the bending stress must be from then on to still give a total life of 100 years. The material being fatigued is medium-carbon steel (annealed AISI 1040). *Ans.* 20% lower stress.

7.39 A hydraulic cylinder has a piston diameter $D = 100$ mm and a piston rod diameter $d = 33$ mm. The hydraulic pressure alternately applied to each side of the piston is $p = 35$ MPa. This pressure induces in the rod a compressive force of $\pi p D^2/4$ or a tensile force of $\pi p(D^2 - d^2)/4$, depending on which side of the piston the pressure is applied. The piston rod material is martensitic stainless steel (AISI 410). Find the endurance life of the piston rod. *Ans.* 11,370 cycles.

7.40 A 20-mm-diameter shaft transmits a variable torque of 600 ± 300 N-m. The frequency of the torque variation is 0.1 s^{-1}. The shaft is made of high-carbon steel (AISI 1080, quenched and tempered at 800°C). Find the endurance life of the shaft. *Ans.* 789,200 cycles, or 2192 hr.

7.41 For the shaft in Problem 7.40 determine how large the shaft diameter has to be for infinite life. *Ans.* $d = 22.5$ mm.

7.42 A notched bar has the diameter $D = 25$ mm, notch radius $r = 0.5$ mm, and the bottom diameter of the notch is 24 mm. It experiences rotational bending fatigue. Determine which of the steels AISI 1020, 1040, 1080, or 316 will give the highest allowable bending moment for infinite life. Calculate that moment.

7.43 A straight, circular rotating beam has a diameter of 30 mm and a length of 1 m. At the end of the beam a stationary load of 600 N is applied perpendicularly to the beam, giving a bending deformation. Find which surface machining method can be used to give infinite life to the rotating beam if it is made of

(a) AISI 1020 steel, quenched and tempered at 870°C.

(b) AISI 1080 steel, quenched and tempered at 800°C.

Note that $k_s = k_r = k_m = 1$.

7.44 The pedals on a bicycle are screwed into the crank with opposite threads, one left-hand thread and one right-hand thread, to ensure that the pedals do not accidentally unscrew. Just outside the thread is a 0.75-mm-radius fillet connecting the 12.5-mm-diameter threaded part with an 11-mm-diameter central shaft in the pedal. Make a careful sketch of the geometry. The shaft is made of AISI 4340 alloy steel tempered at a low temperature to give an ultimate stress of 2 GPa. Find the fatigue stress concentration factor for the fillet, and calculate the maximum allowable pedal force for infinite life if the force from the foot is applied 70 mm from the fillet. *Ans.* 1090 N.

7.45 During the development of a new submarine a 1-to-20 model was used. The model was tested to find if there was any risk of fatigue failure in the full-size submarine. To be on the safe side, the stresses in the model were kept 25% higher than the stresses in the full-size submarine. Is it possible to conclude that the full-size submarine will be safe if the model was safe at a 25% higher stress level?

7.46 During the development of a new car it was found that the power from the motor was too high for the gearbox. When maximum torque was transmitted through the gearbox, 50% of the gearboxes failed within the required lifetime of 800 hr. How much stronger should the gearbox be made to ensure that only one gearbox out of 1000 would fail within 800 hr? *Ans.* 33% stronger.

7.47 A tension member in service has been inspected and a fatigue crack has been discovered as shown in Sketch *f*. A proposed solution is to drill a hole at the tip of the crack, the intent being to reduce the stress concentration at the crack tip. The material is 1.25-in.-thick annealed AISI 1040 medium carbon steel, and was produced through forging. If the load is a completely reversing 3500 lb and the original design is as shown,

(a) What was the original factor of safety? *Ans.* $n_s = 3.2$.

(b) What is the smallest drilled hole which restores the safety factor to that of the original design? (Use the nearest 1/32-in. increment that is satisfactory.) Use a reliability of 90%. *Ans.* 3/32 in.

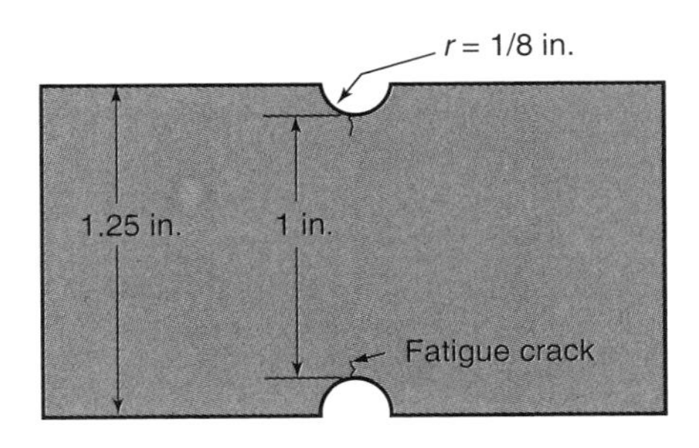

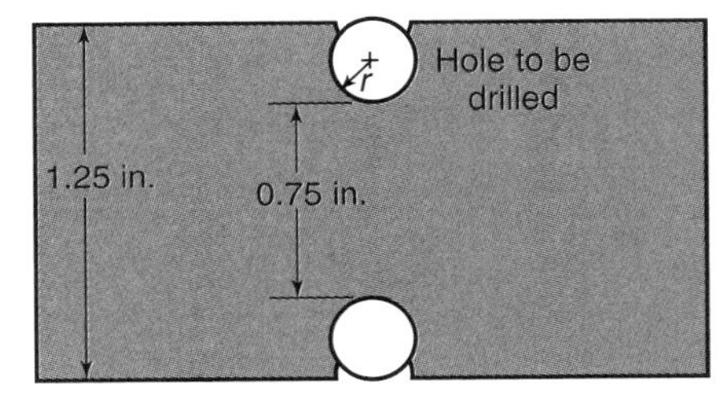

Sketch *f*, for Problem 7.47

7.48 Truck gearboxes are dimensioned for infinite life at maximum torque regarding contact stresses and bending stresses for all gear steps. Car gearboxes are dimensioned for finite life according to typical running conditions. Maximum torque for the first gear can typically be maintained only 3 to 6 s for each acceleration before the maximum speed is reached. If a driver accelerates at full power 20 times per day for 20 years on the first gear, the required life is only 60 to 120 hr. The normal load spectrum gives a life of 200,000 km for 99% of the gearboxes. A driver uses the car twice a year to move his 10-ton boat 50 km (the distance between his home and the harbor) during the 10-year life of the car. Calculate how much of the 200,000-km nominal life of the gearbox is consumed by the boat moving if the life is inversely proportional to the load raised to the 3.2 power. Assume that during the move the gearbox load is four times higher than normal. *Ans.* 42.25% of the gearbox life.

7.49 The hand brakes of a bicycle are often the same type for the front and rear wheels. The high center of gravity (compared with the distance between the wheels) will increase the contact force between the front wheel and the ground and will decrease the contact force between the rear wheel and the ground when the brakes are applied. If equal force is applied to each of the two brakes, the rear wheel will start sliding while the front wheel is still rolling. The manufacturer of the brake wires did not know about this difference between the front and rear wheels, so the wires were dimensioned as if the wheels had equal contact force and thus needed equal force in the two brake wires. The wires were originally dimensioned to withstand 20 years' use with brake applications at a force necessary to lock the two (equally loaded) wheels. How long will the life of the wires be if the friction between the front wheel and the ground is just enough to lift the real wheel from the ground? Assume the endurance life is proportional to the stress raised to the –10 power.

7.50 Ball bearings often run at varying loads and speeds. Bearing life is inversely proportional to the contact stress raised to the 10/3 power for a number of bearing types. For ball bearings the contact stresses are approximately proportional to the third root of the load. A ball bearing in a gearbox is dimensioned to have a life of 1752 million revolutions, half of which are at half the maximum motor torque, one-quarter at full motor torque, and one-quarter at 75% of full motor torque. Calculate the bearing life if the motor torque is kept constant at the maximum level. *Ans.* 681 million revolutions.

7.51 A round cold-drawn steel bar with a solid cross-section is subjected to a cyclic force that ranges from a maximum of 10 kN in tension to a minimum of 5 kN in compression. The ultimate strength of the steel is 600 MPa and the yield strength is 400 MPa. The critical safety factor is 2. Determine the following:

(a) The modified endurance limit as a function of diameter. *Ans.* $S_e = (279 \text{ MPa})d^{-0.112}$.

(b) The cross-sectional area that will produce fatigue failure and the corresponding diameter. *Ans.* $d = 10$ mm.

(c) The region (a, b, c, or d) in Fig. 7.15, assuming modified Goodman criteria, and why.

7.52 Both bars used in Problem 7.36 and shown in sketches *c* and *d* are made from cold-drawn, medium-carbon steel (annealed AISI 1040). Determine the safety factor for each bar. The load varies from 10 to 40 kN. Use the modified Goodman relationship.

7.53 The 7/16-in.-thick component in Sketch *g* is designed with a fillet and a hole. The load varies from 2,000 to 12,000 lb. The following strengths are given: $S_u = 56$ ksi and $S_y = 41$ ksi. Using the modified Goodman failure theory, determine the safety factor for the hole as well as for the fillet. At which location will failure first occur? Will the component fail? Explain. *Ans.* $n_{s,\text{fillet}} = 1.44$.

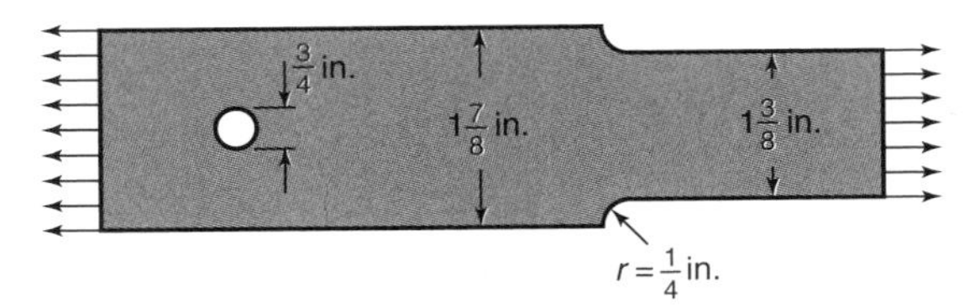

Sketch *g*, for Problem 7.53

7.54 The flat bar shown in Sketch h is made of cold-drawn, high-carbon steel (AISI 1080, quenched and tempered at 800°C). The cyclic, non-zero, mean axial load varies from a minimum of 2 kN to a maximum of 10 kN. Using the Goodman failure theory, determine the safety factors for the hole, the fillet, and the groove. Also, indicate where the flat bar will first fail. *Ans.* $n_{s,\text{hole}} = 2.97$, $n_{s,\text{fillet}} = 3.25$.

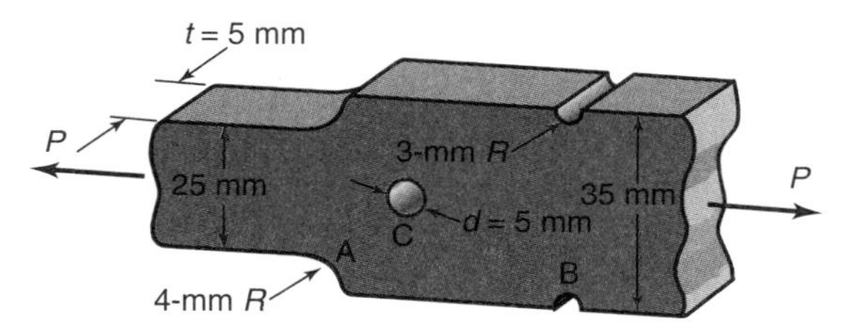

Sketch h, for Problem 7.54

7.55 A straight, circular rotating beam with a diameter of 50 mm and a length of 1 m has an axial load of 3000 N applied at the end and a stationary radial load of 400 N. The beam is cold-drawn and the material is titanium. Note that $k_s = k_r = k_m = 1$. Find the safety factor n_s for infinite life by using the Goodman line. *Ans.* $n_s = 5.269$.

7.56 The annealed AISI 1040 steel bar shown in Sketch i is subjected to a tensile load fluctuating between 1000 and 3000 lb. Estimate the factor of safety using the Goodman criterion. *Ans.* $n_s = 2.889$.

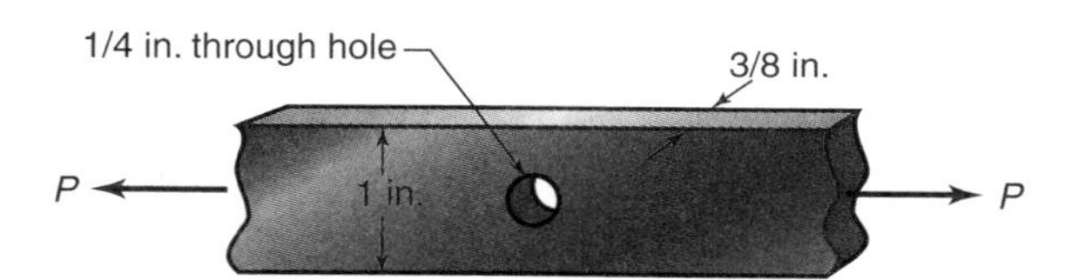

Sketch i, for Problem 7.56

7.57 The cantilever shown in Sketch j carries a downward load F that varies from 500 to 700 lb.

(a) Compute the resulting safety factor for static and fatigue failure if the bar is made from annealed AISI 1040 steel. *Ans.* $n_{s,\text{static}} = 3.491$.

(b) If the notch sensitivity is assumed to remain constant, what fillet radius is needed for a fatigue failure safety factor of 3.0? *Ans.* $r = 0.025$ in.

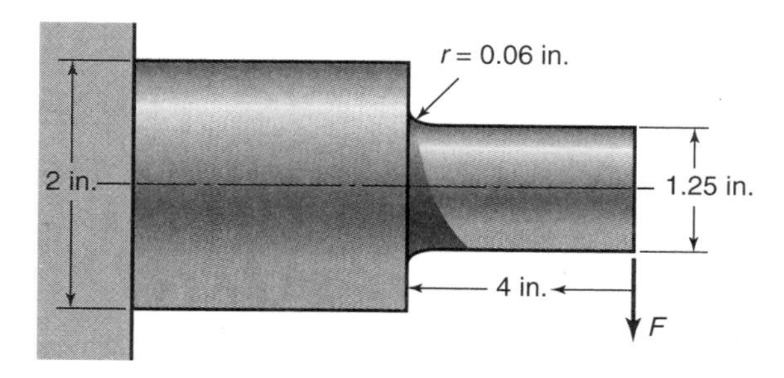

Sketch j, for Problem 7.57

7.58 The 7/16-in. thick component in Problem 7.53 is constructed from AISI 1020 steel (quenched and tempered at 870°C), but is used at elevated temperature so that its yield strength is 70% of its room temperature value, and its ultimate strength is 60% of its room temperature value.

(a) If the desired reliability is 90%, what completely reversing load can be supported without failure? *Ans.* $P_{\max} = 7.81$ kip.

(b) If the load varies from 1000 to 2000 lb, what is the safety factor guarding against fatigue failure? Use the Goodman failure criterion. *Ans.* $n_s = 4.28$.

7.59 The stepped rod shown in Sketch k is subjected to a tensile force that varies between 4000 and 7000 lb. The rod has a machined surface finish everywhere except the shoulder area, where a grinding operation has been performed to improve the fatigue resistance of the rod. Using a 99% probability of survival, determine the safety factor for infinite life if the rod is made of AISI 1080 steel, quenched and tempered at 800°C. Use the Goodman line. Does the part fail at the fillet? Explain.

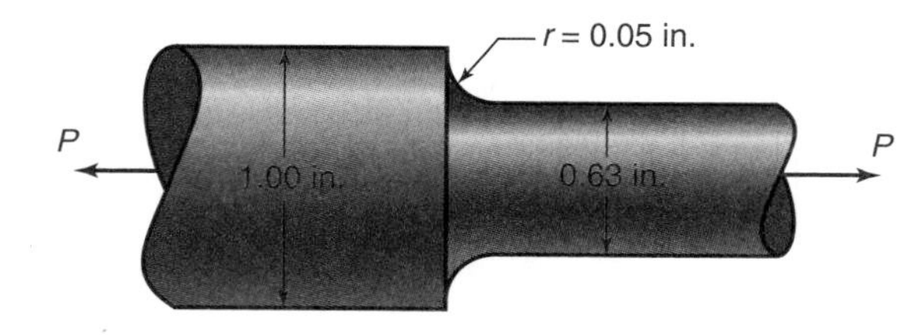

Sketch k, for Problem 7.59

7.60 A section of a 25-mm-diameter shaft shown in Sketch l is drawn from AISI 1080 (quenched and tempered at 800°C) carbon steel, and has a 3-mm-diameter through-hole drilled (machined) into it. During service, the shaft encounters a bending moment that varies from –15 to 45 Nm. For a reliability of 99.5%, determine the safety factor using the Goodman line. *Ans.* $n_s = 3.432$.

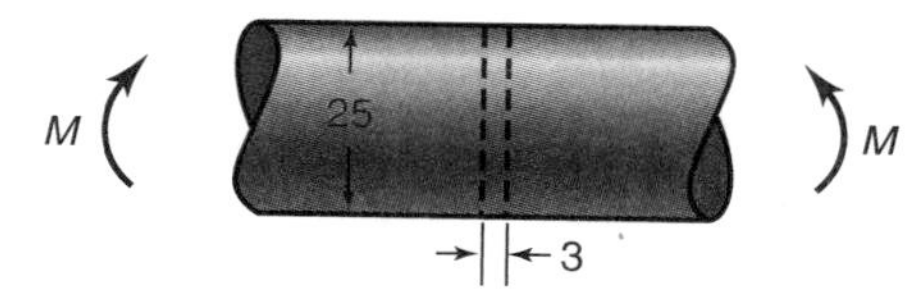

Sketch l, for Problem 7.60

7.61 A 10,000-lb elevator is supported by a stranded steel cable having a 2.5-in.^{2} cross-section and an effective modulus of elasticity of 12×10^6 psi. As the elevator is descending at a constant velocity of 400 ft/min, an accident causes the cable, 60 ft above the elevator, to suddenly stop. Determine the static elongation of the cable, the impact factor, the maximum elongation, and the maximum tensile stress developed in the cable. *Ans.* $\sigma_{\max} = 37.5$ ksi.

7.62 A person is planning a bungee jump from a 40-m-high bridge. Under the bridge is a river with crocodiles, so the person does not want to be submerged into the water. The rubber rope fastened to the ankles has a spring constant of 3600 N divided by the length of the rope. The distance from the ankles to the top of the head is 1.75 m, and the person weighs 80 kg. Calculate how long the rope should be. *Ans.* 21.29 m.

7.63 A toy with a bouncing 50-mm-diameter steel ball has a compression spring with a spring constant of 100,000 N/m. The ball falls from a 3-m height down onto the spring (which can be assumed to be weightless) and bounces away and lands in a hole. Calculate the maximum force on the spring and the maximum deflection during the impact. The steel ball density is 7840 kg/m^3. *Ans.* 1743 N.

7.64 At a building site a 1-ton container hangs from a crane wire and is then placed on the floor so that the wire becomes unloaded. The container is pushed to the elevator shaft where it is to be lowered as shown in Sketch m. By mistake there is a 1-m slack in the wire from the crane when the container falls into the elevator shaft. Calculate the maximum force in the wire if it has a cross-sectional steel surface of 500 mm^2 and an effective modulus of elasticity of 70 GPa and is 25 m long from the crane to the container. *Ans.* 175.8 kN.

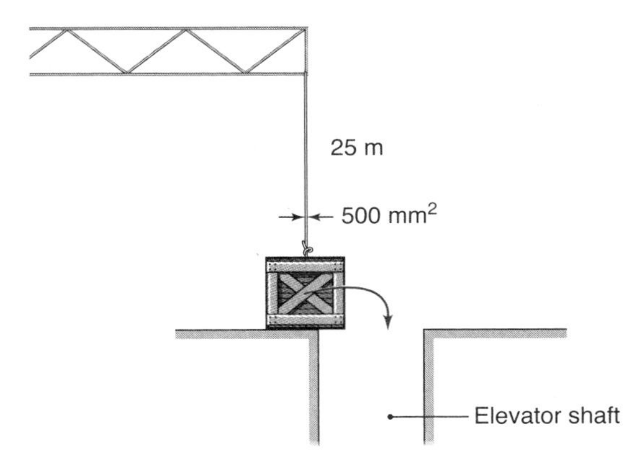

Sketch m, for Problem 7.64

7.65 Modern kitchen drawers have small rubber springs mounted onto the sides of the inside of the front plate to take up the force and stop the drawer when it is being closed. The spring constant for each of the two rubber springs is 400 kN/m. The drawer is full of cutlery, which weighs 5 kg, and is closed with a speed of 0.5 m/s. Calculate the maximum force in each rubber spring if the drawer itself weighs 1 kg and

(a) The cutlery is in a container that is fixed to the drawer so that it moves with the drawer. *Ans.* $P = 548$ N.

(b) The cutlery is in a plastic container that can slide 80 mm with a coefficient of friction of 0.25 inside the drawer. *Ans.* $P = 229.7$ N.

7.66 Car doors are easy to slam shut but difficult to press shut by hand force. The door lock has two latches, the first easily engaged and the second requiring the rubber seal around the door to be quite compressed before it can engage. The rubber seal has a spring constant of 50,000 N/m at the locked position for the door, and the mass moment of inertia for the door around its hinges is 2.5 kg-m^2. The distance between the lock and the hinges is 1 m. Calculate the force needed to press the car door shut at the lock if a speed of 0.8 m/s at the lock slams it shut. *Ans.* 282.8 N.

7.67 Sketch n shows two bar designs. In the left sketch, the bar is cold drawn and has no stress concentrations. In the sketch on the right, a groove has been machined into the bar, with the groove root diameter being the same as the outer diameter as the bar on the left. The bar is produced from low-carbon steel (AISI 1020, quenched and tempered at 870°C). Find the completely reversing force, P, that can be supported based on the endurance limit. *Ans.* $P = 48.21$ kN.

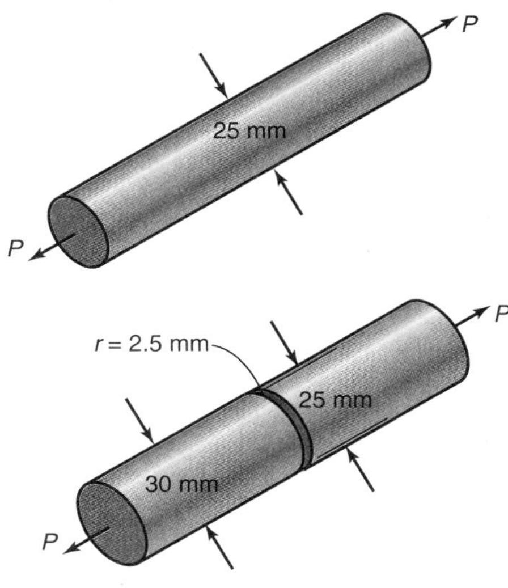

Sketch n, for Problem 7.67

7.68 Repeat Problem 7.67 assuming that the bar is loaded by a completely reversing bending moment, M. *Ans.* $M =$ 20.9 Nm.

7.69 For the two bars given in Problem 7.67, determine the safety factor guarding against fatigue failure if the load varies between 4 and 20 kN. *Ans.* Using Goodman, $n_s =$ 4.39.

7.70 A center-cracked plate made of 2024-T3 aluminum has a width of 100 mm, a thickness of 4 mm, and an initial crack length of $l = 3$ mm. If the applied tensile load varies between 30 and 50 kN, estimate the number of cycles that causes the crack to grow large enough to cause the part to fracture. *Ans.* $N = 119,500$ cycles.

7.71 A fatigue failure of Ti-6Al-4V titanium alloy shows microscopic striations that have an average spacing of 0.5×10^{-7} mm. Determine the average stress intensity factor that would produce such striations. The failed part was a center-cracked tension member, with a thickness of 15 mm and a width of 100 mm and a load of 90 kN applied in released tension. Determine the crack length at fracture. *Ans.* $l_c = 16$ mm.

Synthesis and Design

7.72 It was noted in the chapter that a fatigue crack has a large stress concentration associated with it because of its sharpness. One innovative approach to improving fatigue strength in polymers involves distributing encapsulated uncured polymer within a material; when a fatigue crack encounters and breaks open such as a capsule, the liquid coats the fatigue crack and blunts it. List the parameters that you expect will affect the performance of such a material. Give a short justification for each item in your list.

7.73 Design a test method for determining the rate of fatigue crack growth in a test specimen.

7.74 It is estimated that an average person takes two million steps per year. Conduct a literature search and examine how bone responds to flaw generation and crack growth associated with fatigue.

7.75 Conduct a literature and Internet search and write a one-page paper on the circumstances associated with the Aloha Airlines flight from 1988 depicted in the chapter's opening photograph. Include a failure description and the technical reasons for the failure.

7.76 In rotating beam fatigue tests, specimens are loaded in four-point bending. Would the measured endurance limit be higher or lower if the test was conducted in three-point bending? What about tension? Explain your answer using appropriate sketches and equations from Ch. 4.

7.77 Perform an Internet search to obtain an illustration and description of the *Mannesmann* process for manufacturing tubes. Explain how fatigue plays a role in the Mannesmann process.

7.78 Design a test method to estimate the endurance limit of a material for different mean stresses (see Fig. 7.13).

7.79 It is well-known that fatigue testing of polymers must take place at low speeds so that there are no thermal effects in the material near the crack tip. Explain the reasons why heat is generated in fatigue testing. List the physical and mechanical properties that will influence the maximum allowable test speed and explain why you think they are important.

7.80 An engineer working for a manufacturer of mechanical power presses is assisting in the development of a new size of machine. When starting to design a new brake stud, the engineer finds the information in Sketch *o* from an existing computer-assisted drawing (CAD) of a machine with the closest capacity to the new project. The brake stud supports the brake on the cam shaft of the mechanical power press. The brake actuates with every cycle of the press, stopping the ram at the top-dead-center position so that an operator can remove a workpiece from the dies and insert another workpiece for the next cycle. If the stud fails, the press could continue to coast downward and could result in a serious injury.

Focusing on section A-A the engineer recognizes that no fillet radius had been specified for the brake stud for machines sold previously. Conversations with machinists who routinely worked on the part led to the conclusion that the common practice was to undercut the fillet to make sure assembly was complete; the fillet radius in effect was the radius of the machine tool insert, a value as low as 1/8 in. The immediate concern is whether a product recall is in order, since a brake stud failure could result in a machine operator's hands being in a die while the machine fails to stop after a cycle. Analyze the system and determine whether the machine is safe as manufactured. The carbon steel used for the stud has a minimum ultimate strength of 74.5 ksi, and no yield strength was prescribed in the drawing.

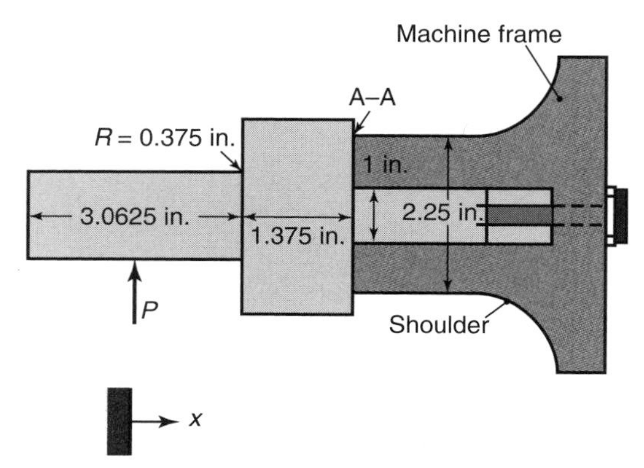

Sketch *o* for Problem 7.81.

Chapter 8

Lubrication, Friction, and Wear

Greases are a necessary lubricant for many applications, including rolling element bearings, for the reduction of friction and wear. *Source:* Courtesy of SKF USA, Inc.

...among all those who have written on the subject of moving forces, probably not a single one has given sufficient attention to the effect of friction in machines..."

Guillaume Amontons

Contents

Examples

Design Procedure

Case Study

This chapter introduces the essential tribological concepts of surfaces, lubrication, friction and wear, each of which plays a critical role in the performance and reliability of machine elements. Surfaces are investigated in detail, including common definitions to quantify roughness and a general discussion of their physical nature. Conformal contacts such as in journal bearings, and non-conformal contacts such as gears and rolling-element bearings are differentiated. The field of Hertzian contact for non-conformal surfaces is presented. Lubrication theory is developed in terms of the film parameter and the regimes of lubrication, including boundary, partial, and hydrodynamic lubrication. Boundary lubrication typically occurs at low speeds, with inviscid lubricants, and/or high loads, and is always a concern during startup. In boundary lubrication, load is transferred between the asperities of surfaces in contact. At the other extreme, hydrodynamic lubrication involves transfer of load across a pressurized lubricant film that completely separates the surfaces. Surfaces that transfer load through a combination of direct asperity contact and pressurized lubricants involve partial lubrication. Elastohydrodynamic lubrication is a condition that is common in machine elements and is therefore also included. Basically, elastohydrodynamic lubrication involves circumstances where the surfaces are separated by a fluid film and where elastic deformation of the surfaces are large compared to the film thickness. Friction and wear theory are also summarized, including the important Archard wear law used to predict the life of machine elements subjected to adhesive or abrasive wear.

Symbols

A	area, m^2
A_o	sum of projected areas in sliding, m^2
A_r	sum of real areas in contact, m^2
b^*	contact semiwidth, m
C_1, C_2	constants used in Eq. (8.26)
D_x	diameter of contact ellipse along x-axis, m
D_y	diameter of contact ellipse along y-axis, m
E	modulus of elasticity, Pa
E'	effective elastic modulus, Pa
$\mathcal{E}$	complete elliptic integral of second kind
F	friction force, N
$\mathcal{F}$	complete elliptic integral of first kind
H	hardness of softer material, Pa
h	film thickness, m
$h_{\min}$	minimum film thickness, m
k_e	ellipticity parameter, D_y/D_x
k_1	adhesive wear constant
k_2	abrasive wear constant
L	sliding distance, m
N	number of measurements of z_i
p	pressure, Pa
p_H	Hertzian contact pressure, Pa
$p_{\max}$	maximum pressure, Pa
R	radius, m; curvature sum, m
R_a	arithmetic average surface roughness, m
R_q	root-mean-square surface roughness, m
r	radius, m
S_y	yield stress in tension, Pa
s	shear strain rate, s^{-1}
u	velocity, m/s
u_b	velocity of upper surface, m/s
v	wear volume, m^3
W	normal load, N
W'	dimensionless load for rectangular contact
w'	load per unit width, N/m
w_a	squeeze velocity, m/s
x, y, z	Cartesian coordinate system, m
z	coordinate in direction of film, m
z_i	height from reference line, m
α_r	radius ratio, R_y/R_x
$\delta_{\max}$	maximum deformation, m
η	absolute viscosity, Pa·s
η_k	kinematic viscosity, m^2/s
η_o	absolute viscosity at $p = 0$ and at constant temperature, Pa·s
θ	cone angle, rad
Λ	dimensionless film parameter
μ	coefficient of sliding friction
ν	Poisson's ratio
ξ	pressure-viscosity coefficient, m^2/N
ρ	density, kg/m^3
τ	shear stress, Pa

Subscripts

a	solid a
b	solid b

8.1 Introduction

Tribology is generally defined as the study of lubrication, friction, and wear, and it plays a significant role in machine element life and performance. The importance of tribology should not be underestimated — the famous Jost Report [1966] attributed tribology-related failures and design shortcomings to be in excess of 10% of gross domestic product in developed countries. For machine elements to be designed properly, not only must the design stress be less than the allowable stress and the deformation not exceed some maximum value, but also tribological considerations must be properly addressed. The interaction of surfaces in relative motion should not be regarded as a special subject; like the strength of materials, tribology is basic to most machine elements.

This chapter presents the lubrication, friction, and wear considerations that are important in the successful design of machine elements. Sections 8.2 and 8.3 introduce the concepts of surface roughness and conformity, which are essential design considerations for most machine elements. The situation of Hertzian contact, involving concentrated stresses in nonconformal applications such as bearings, gears, and cams, is analyzed in Section 8.4. Bearing materials are summarized in Section 8.5. Lubrication is essential for the long life of rolling element bearings, journal bearings, gears of all kinds, and many other applications in machine design and manufacturing. Section 8.6 describes the properties of lubricants, while Section 8.7 introduces the important concept of lubrication regimes that has a dominant effect on component life and performance. Friction is discussed in Section 8.8 and wear in Section 8.9; both can have a large impact on the robustness, life, and performance of machine elements such as brakes and clutches, wire ropes, chains, rubber belts, threaded fasteners, etc.

This chapter is intended to provide tribological background in order to better understand machine element design. Of course, this chapter is highly abstracted, and the interested reader can find much more information in the Recommended Readings.

8.2 Surface Parameters

Designing machine elements is ultimately a problem of two surfaces that are either in direct contact or separated by a thin fluid film. In either case, the surface roughness and texture are important in ensuring long component life. Consider a surface profile as depicted in Fig. 8.1, which has been measured with a *stylus profilometer*. There are numerous other surface measuring instruments available, some of which will measure areas and not just traces across a surface, but Fig. 8.1 displays some typical features that are important to keep in mind. Note that the magnification in the vertical direction is 1000 and that in the horizontal is 20, so that the ratio of vertical to horizontal magnification is 50:1. Directional differences in magnification are often not appreciated, leading to a false impression of the nature of surfaces. Real engineering surfaces are much smoother than profilometer traces would suggest, mainly due to this magnification difference.

The surface profile in Fig. 8.1 shows the surface height variation relative to a mean reference line. By definition, the areas above and below the mean line are equal. **Surface roughness** can then be defined as a measure of surface de-

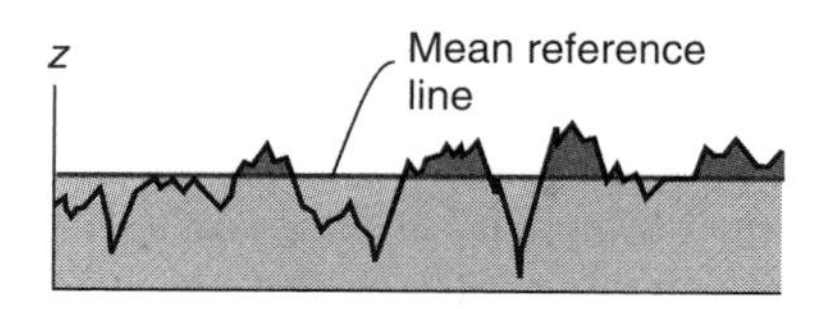

Figure 8.1: Surface profile showing surface height variation relative to mean reference line.

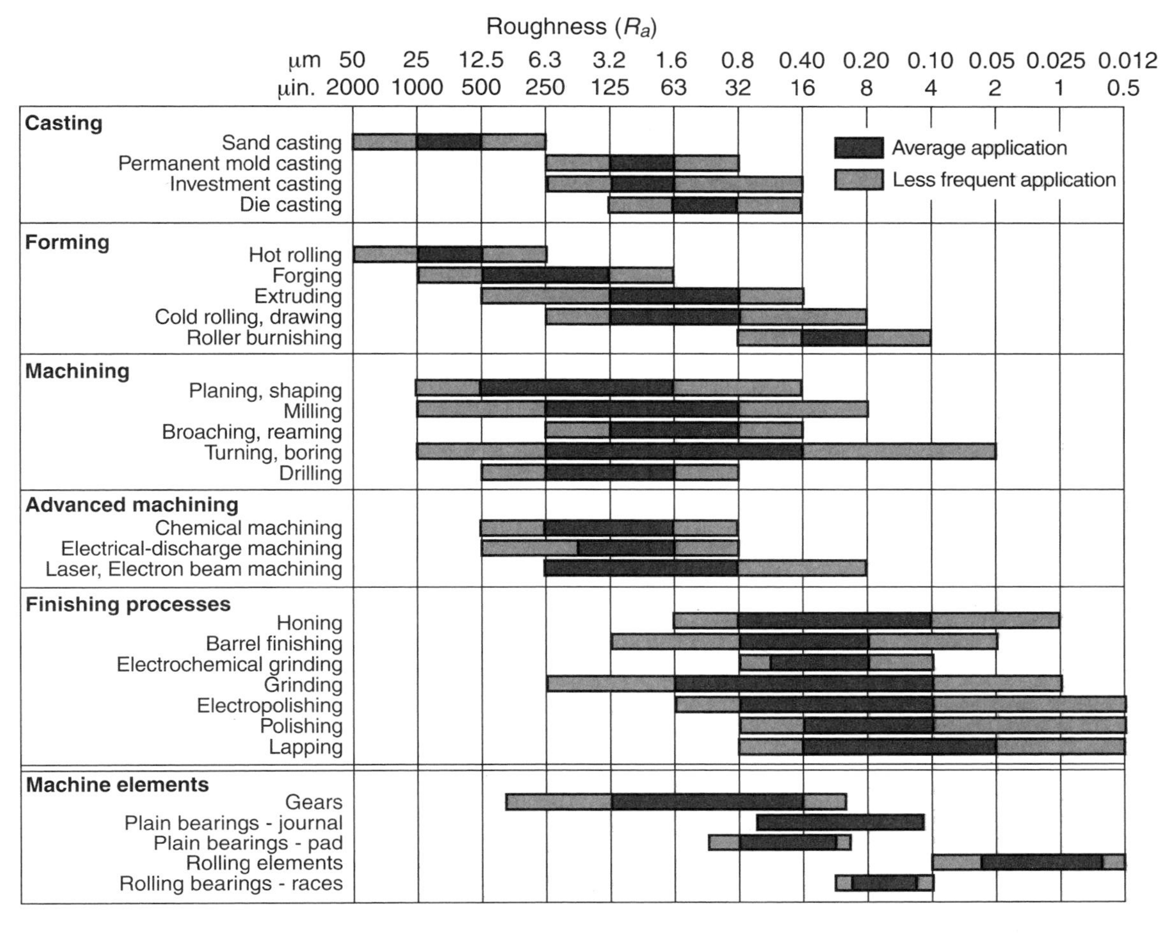

Figure 8.2: Typical arithmetic average surface roughness (R_a) for various manufacturing processes and machine components. *Source:* Adapted from Kalpakjian and Schmid [2010] and Hamrock et al. [2004].

parture from the mean line. Two different surface roughness parameters are in common use:

1. **Centerline average** or **arithmetic average** surface roughness, denoted by R_a,

$$R_a = \frac{1}{N}\sum_{i=1}^{N} |z_i|, \tag{8.1}$$

where z_i is the vertical distance from the reference line and N is the number of height measurements taken.

2. **Root-mean-square** (rms) surface roughness, denoted by R_q,

$$R_q = \left(\frac{1}{N}\sum_{i=1}^{N} z_i^2\right)^{\frac{1}{2}}. \tag{8.2}$$

R_q emphasizes surface peaks and valleys more than R_a and is sometimes preferred for that reason. If a Gaussian height distribution is assumed, R_q has the additional advantage of being the standard deviation of the profile. For a simple sinusoidal distribution, the ratio of R_q to R_a is

$$\frac{R_q}{R_a} = \frac{\pi}{2\sqrt{2}} = 1.11. \tag{8.3}$$

Figure 8.2 gives typical values of the arithmetic average for various processes and components considered in Chapters 9 through 19. Note from this figure that as higher precision processes are applied, the R_a values decrease significantly. The finest process shown is lapping, which produces an R_a between 0.012 and 0.8 μm. A general trend which must be recognized is that the processes that produce smooth surfaces are slower and more expensive than those that produce rougher surfaces. Thus, the specification of a surface roughness for a particular application will have significant cost implications.

8.3 Conformal and Nonconformal Surfaces

Conformal surfaces fit snugly into each other so that the load is carried over a relatively large area. For example, the lubrication area of a journal bearing would be 2π times the radius times the length. The load-carrying surface area remains essentially constant even if the load is increased. Journal bearings (Fig. 8.3) and slider bearings (see Fig. 12.9) have conformal surfaces. In journal bearings, the radial clearance between the journal and the sleeve is typically one-thousandth of the journal diameter; in slider bearings, the inclination of the bearing surface to the runner is typically one part in a thousand.

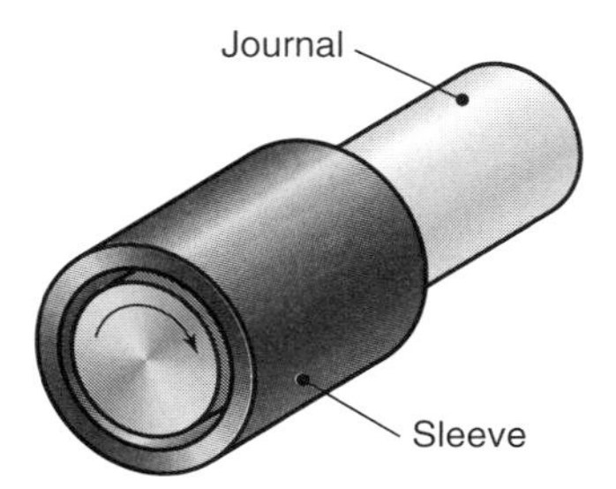

Figure 8.3: Conformal surfaces.

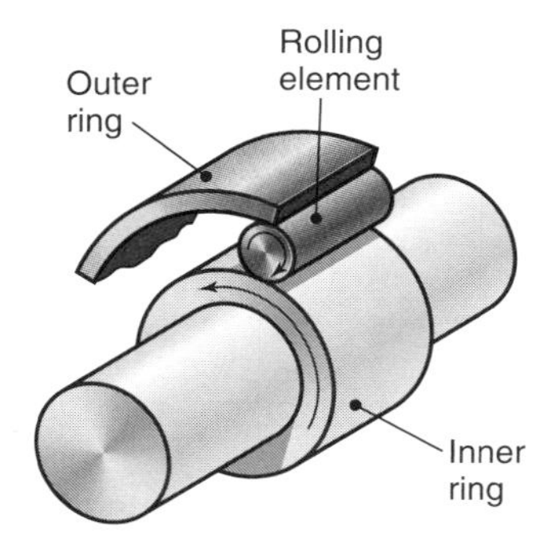

Figure 8.4: Nonconformal surfaces.

Many machine elements have surfaces that do not conform to each other. The full burden of the load must then be carried by a small area. The contact area of a nonconformal conjunction is typically three orders of magnitude smaller than that of a conformal conjunction. In general, the area between **nonconformal surfaces** enlarges considerably with increasing load, but it is still smaller than the contact area between conformal surfaces. Some examples of nonconformal surfaces are mating gear teeth, cams and followers, and rolling-element bearings (Fig. 8.4).

8.4 Hertzian Contact

The text up to this point has mainly been focused on distributed loading. This section describes not only the deformation due to concentrated loading but also the associated surface and subsurface stresses. The focus is mostly on elliptical contacts, but rectangular contact situations are also presented. The theory presented in this section is based on the work of Hertz [1881], and is therefore referred to as **Hertzian contact**.

8.4.1 Elliptical Contacts

The undeformed geometry of nonconformal contacting solids can be represented in general terms by two ellipsoids, as shown in Fig. 8.5. Two solids with different radii of curvature in a pair of principal planes (x and y) passing through the conjunction make contact at a single point under the condition of zero applied load. Such a condition is called **point** or **elliptical contact** because of the shape of the contact patch. It is assumed throughout this book that convex surfaces like those in Fig. 8.5 exhibit positive curvature; concave surfaces exhibit negative curvature. Figure 8.6 shows elements and bearing races and their curvature conventions. The importance of the sign of the radius of curvature will be shown

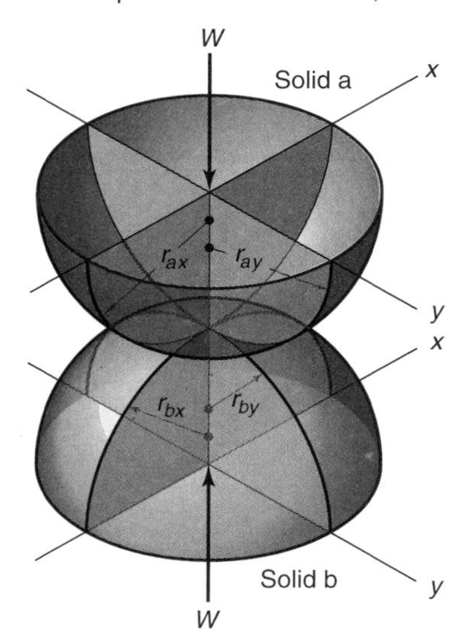

Figure 8.5: Geometry of contacting elastic solids.

to be extremely important, and allows extension of Hertzian contact equations to a wide variety of machine elements.

Note that if coordinates x and y are chosen such that

$$\frac{1}{r_{ax}} + \frac{1}{r_{bx}} \geq \frac{1}{r_{ay}} + \frac{1}{r_{by}}, \tag{8.4}$$

coordinate x then determines the direction of the *minor axis* of the contact area when a load is applied, and y, the direction of the *major axis*. Since the coordinate system at this point is arbitrary, the direction of the rolling or entraining motion is always considered to be along the x-axis. Depending on the machine element geometry, the minor axis can be in either the rolling or transverse direction.

Curvature Sum

The curvature sum, which is important in analyzing contact stresses and deformation, is

$$\frac{1}{R} = \frac{1}{R_x} + \frac{1}{R_y}, \tag{8.5}$$

where

$$\frac{1}{R_x} = \frac{1}{r_{ax}} + \frac{1}{r_{bx}}, \tag{8.6}$$

$$\frac{1}{R_y} = \frac{1}{r_{ay}} + \frac{1}{r_{by}}. \tag{8.7}$$

Equations (8.5) to (8.7) effectively redefine the problem of two ellipsoidal solids in contact in terms of an equivalent solid of radii R_x and R_y in contact with a plane.

The **radius ratio**, α_r, is

$$\alpha_r = \frac{R_y}{R_x}. \tag{8.8}$$

Thus, if Eq. (8.4) is satisfied, $\alpha_r > 1$; and if it is not satisfied, $\alpha_r < 1$.

Machine elements with nonconformal surfaces generally have a range of radius ratios from 0.03 to 100. For example, for a traction drive involving a disk rolling on a plane, α_r is typically 0.03; for a ball-on-plane contact, $\alpha_r = 1.0$, and for a contact so wide it approaches line contact, α_r can be 100 or more, such as in a cylindrical roller bearing against

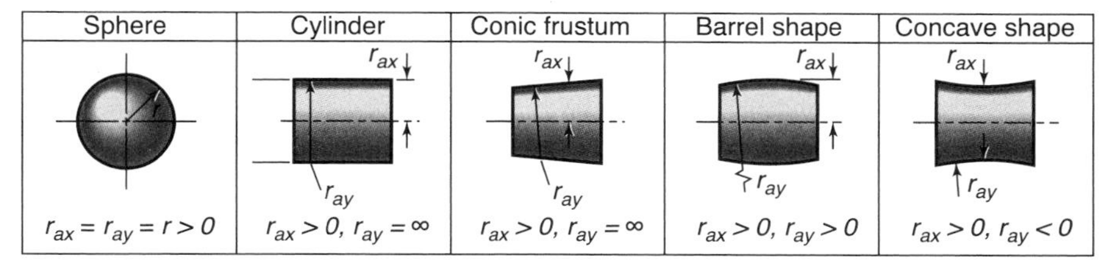

(a)

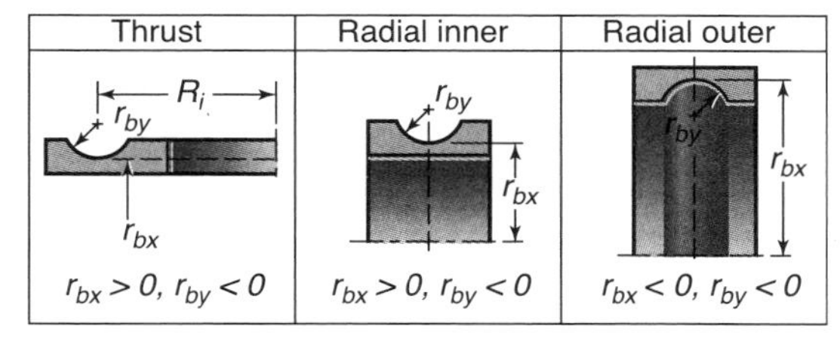

(b)

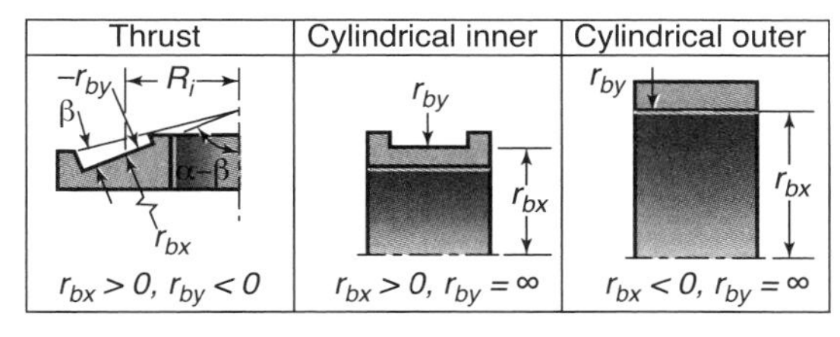

(c)

Figure 8.6: Sign designations for radii of curvature. (a) Rolling elements; (b) ball bearing races; (c) rolling bearing races.

a bearing race (see Figs. 8.6 and 13.1b). Some further examples of machine elements with radius ratios less than one are some (Novikov) gear contacts, locomotive wheel-rail contacts, and roller-flange contacts in a radially loaded roller bearing. Some further examples of machine elements with radius ratios greater than 1 are rolling-element bearings and most gears. These components are considered in more detail later in the text, but the general information on contact geometry of nonconformal surfaces is an important consideration.

Ellipticity Parameter

The **ellipticity parameter**, k_e, is defined as the ratio of elliptical contact diameter in the y-direction (transverse direction) to the elliptical contact diameter in the x-direction (direction of entraining motion), or

$$k_e = \frac{D_y}{D_x}. \tag{8.9}$$

If $\alpha_r \geq 1$, the contact ellipse will be oriented with its major diameter transverse to the direction of motion, and consequently $k_e \geq 1$; otherwise, the major diameter would lie along the direction of motion with both $\alpha_r < 1$ and $k_e < 1$. To avoid confusion, the commonly used solutions to the surface deformation and stresses are presented only for $\alpha_r > 1$.

Note that the ellipticity parameter is a function only of the solids' radii of curvature, but not of load. This is because as the load increases, the semiaxes in the x- and y-directions of the contact ellipse increase proportionately to each other so that the ellipticity parameter remains constant.

Contact Pressure

When an elastic solid is subjected to a load, stresses are produced that increase as the load is increased. These stresses are associated with deformations, which are defined by strains. Unique relationships exist between stresses and their corresponding strains, as shown in Appendix B. For elastic solids, the stresses are linearly related to the strains, with the proportionality constant adopting different values for different materials, as discussed in Section 3.6. The modulus of elasticity, E, and Poisson's ratio, ν, are two important parameters that are used in this chapter to describe contacting solids.

When two elastic nonconformal solids are brought together under a load, a contact area develops whose shape and size depend on the applied load, the elastic properties of the materials, and the curvature of the surfaces. When the two solids shown in Fig. 8.5 have a normal load applied to them, the contact area is elliptical. As stated above, such contacts are therefore referred to as **elliptical** or **point contacts**, although the former is preferred in this text. For the special case where $r_{ax} = r_{ay}$ and $r_{bx} = r_{by}$ the resulting contact is a circle rather than an ellipse, and is sometimes called **circular contact**. When r_{ay} and r_{by} are both infinite, the initial line contact develops into a rectangle when load is applied, so such circumstances are referred to as **line** or **rectangular contacts**.

Hertz [1881] considered the stresses and deformations in two perfectly smooth, ellipsoidal, contacting solids much like those shown in Fig. 8.5. His experiments involved investigations or highly polished glass lenses bearing against an optical flat, and his application of classical elasticity theory to this problem has formed the basis of stress calculations for machine elements such as ball and roller bearings, gears, and cams and followers. Hertz made the following assumptions:

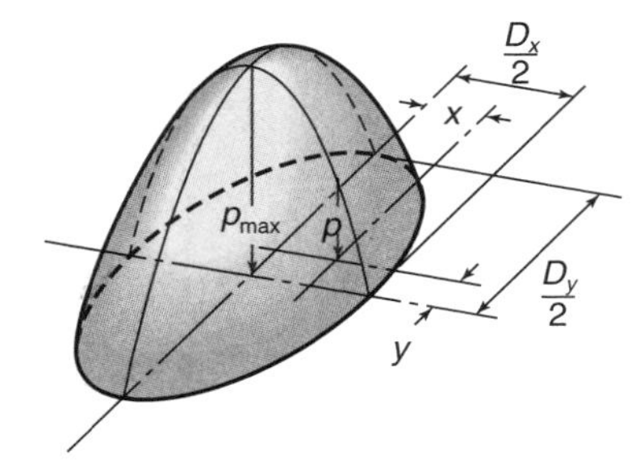

Figure 8.7: Pressure distribution in ellipsoidal contact.

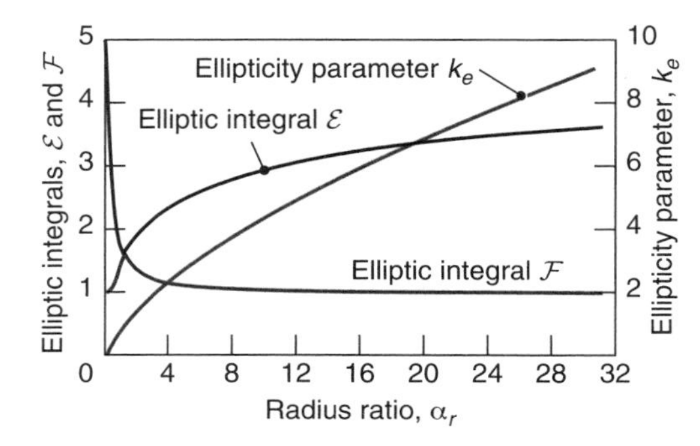

Figure 8.8: Variation of ellipticity parameter and elliptic integrals of first and second kinds as function of radius ratio.

1. The materials are homogeneous and the yield stress is not exceeded anywhere in the contacting bodies. The materials will therefore be approximated as linear elastic solids.
2. No tangential forces are induced between the solids, that is, the contact is frictionless.
3. The solids are continuous, without cracks or discontinuities in their surfaces.
4. Contact is limited to a small portion of the surface such that the dimensions of the contact region are small compared to the radii of the ellipsoids.
5. The solids are at rest and in equilibrium.

Making use of these assumptions, Hertz applied the following expression for the pressure distribution within an ellipsoidal contact (shown in Fig. 8.7):

$$p_H = p_{\max}\left[1 - \left(\frac{2x}{D_x}\right)^2 - \left(\frac{2y}{D_y}\right)^2\right]^{\frac{1}{2}}, \tag{8.10}$$

where D_x and D_y are the diameters of contact ellipse in the x- and y-directions, respectively. The maximum pressure is

$$p_{\max} = \frac{6W}{\pi D_x D_y}, \tag{8.11}$$

where W is the normal applied load. Equation (8.10) determines the distribution of pressure or compressive stress on the common interface, which is clearly a maximum at the contact center and decreases to zero at the periphery.

Simplified Solutions

The classical Hertzian solution requires the calculation of the ellipticity parameter, k_e, as well as the complete elliptic integrals of the first and second kinds, $\mathcal{F}$ and $\mathcal{E}$. This calculation involves finding a solution to a transcendental equation relating k_e, $\mathcal{F}$, and $\mathcal{E}$ to the geometry of the contacting solids and is usually accomplished by some iterative numerical procedure. Hamrock and Brewe [1983] conducted a numerical analysis and curve-fit the data to obtain the simplified equations shown in Table 8.1. From this table, the ellipticity parameter for the complete radius ratio range found in machine elements can be expressed as

$$k_e = \alpha_r^{2/\pi}. \tag{8.12}$$

Figure 8.8 shows the ellipticity parameter and the elliptic integrals of the first and second kinds ($\mathcal{F}$ and $\mathcal{E}$, respectively) for a radius ratio range usually encountered in machine elements. Note from Fig. 8.8 that $\mathcal{F} = \mathcal{E} = \pi/2$ when $\alpha_r = 1$. Also, both $\mathcal{F}$ and $\mathcal{E}$ have discontinuous derivatives at $\alpha_r = 1$, thus the need for two columns in Table 8.1.

When the ellipticity parameter, k_e, the normal applied load, W, Poisson's ratio, ν, and the modulus of elasticity, E, of the contacting solids are known, the major and minor axes of the contact ellipse (D_y and D_x) and the maximum deformation at the contact center can be written as

$$D_y = 2\left(\frac{6k_e^2\mathcal{E}WR}{\pi E'}\right)^{\frac{1}{3}}, \tag{8.13}$$

$$D_x = 2\left(\frac{6\mathcal{E}WR}{\pi k_e E'}\right)^{\frac{1}{3}}, \tag{8.14}$$

$$\delta_{\max} = \mathcal{F}\left[\frac{9}{2\mathcal{E}R}\left(\frac{W}{\pi k_e E'}\right)^2\right]^{\frac{1}{3}}, \tag{8.15}$$

where

$$E' = \frac{2}{\dfrac{(1-\nu_a^2)}{E_a} + \dfrac{(1-\nu_b^2)}{E_b}}. \tag{8.16}$$

Note from these equations that D_y and D_x are proportional to $W^{1/3}$, and $\delta_{\max}$ is proportional to $W^{2/3}$.

Example 8.1: Hertzian Contact I – Ball in Ring

Given: A solid sphere of 20 mm radius ($r_{ax} = r_{ay} = 20$ mm) rolls on the inside of a cylindrical outer race (see Fig. 8.6) with a 100-mm internal radius ($r_{bx} = -100$ mm) and a large width in the axial direction. The sphere is made of silicon nitride and the outer race is made of stainless steel. The normal applied load is 1000 N.

Find:

(*a*) Curvature sum, R

(*b*) Ellipticity parameter, k_e

(*c*) Elliptic integrals of the first and second kinds, $\mathcal{F}$ and $\mathcal{E}$

(*d*) Effective modulus of elasticity, E'

(*e*) Dimensions of elliptical contact, D_x and D_y

Table 8.1: Simplified elliptical contact equations.

Property	Radius ratio range	
	$1 \le \alpha_r \le 100$	$0.01 \le \alpha_r \le 1.0$
Geometry	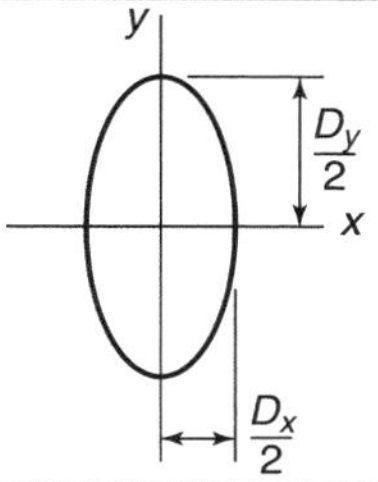	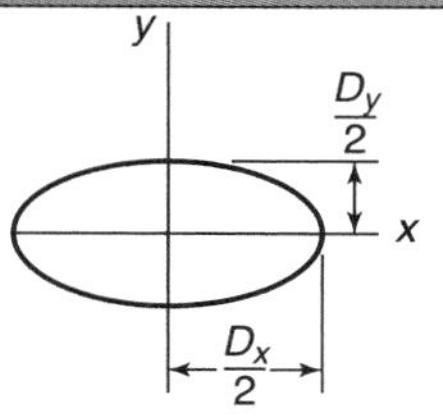
Ellipticity ratio	$k_e = \alpha_r^{2/\pi}$	$k_e = \alpha_r^{2/\pi}$
Elliptic integrals	$\mathcal{F} = \frac{\pi}{2} + \left(\frac{\pi}{2} - 1\right)\ln\alpha_r$	$\mathcal{F} = \frac{\pi}{2} - \left(\frac{\pi}{2} - 1\right)\ln\alpha_r$
	$\mathcal{E} = 1 + \frac{\pi - 2}{2\alpha_r}$	$\mathcal{E} = 1 + \left(\frac{\pi}{2} - 1\right)\alpha_r$

(*f*) Maximum Hertzian pressure or stress on surface, $p_{\max}$

(*g*) Maximum deformation, $\delta_{\max}$

Solution:

(*a*) Note that r_{bx} is negative, so from Eqs. (8.5) to (8.7),

$$\frac{1}{R_x} = \frac{1}{r_{ax}} + \frac{1}{r_{bx}} = \frac{1}{0.02} - \frac{1}{0.10} = 40 \text{ m}^{-1};$$

$$\frac{1}{R_y} = \frac{1}{r_{ay}} + \frac{1}{r_{by}} = \frac{1}{0.02} + \frac{1}{\infty} = 50 \text{ m}^{-1};$$

$$\frac{1}{R} = \frac{1}{R_x} + \frac{1}{R_y} = 40 + 50 = 90 \text{ m}^{-1}.$$

Therefore, $R_x = 0.025$ m, $R_y = 0.02$ m, and the curvature sum $R = 0.0111$ m.

(*b*) α_r is obtained from Eq. (8.8) as

$$\alpha_r = R_y/R_x = 0.020/0.025 = 0.80.$$

Therefore, the ellipticity parameter is given by Eq. (8.12) as

$$k_e = \alpha_r^{2/\pi} = (0.80)^{2/\pi} = 0.8676.$$

(*c*) From Table 8.1, the elliptical integrals of the first and second kinds are (note that $0.01 \le \alpha_r \le 1$):

$$\mathcal{F} = \frac{\pi}{2} - \left(\frac{\pi}{2} - 1\right)\ln(0.8) = 1.69,$$

$$\mathcal{E} = 1 + \left(\frac{\pi}{2} - 1\right)(0.8) = 1.457.$$

(*d*) From Table 3.1, $E_a = 314$ GPa, $E_b = 193$ GPa, $\nu_a = 0.26$, and $\nu_b = 0.30$. The effective modulus of elasticity is obtained from Eq. (8.16) as

$$\begin{aligned} E' &= \frac{2}{\frac{(1-\nu_a^2)}{E_a} + \frac{(1-\nu_b^2)}{E_b}} \\ &= \frac{2}{\frac{(1-0.26^2)}{314 \times 10^9} + \frac{(1-0.32^2)}{193 \times 10^9}} \\ &= 262.5 \text{ GPa}. \end{aligned}$$

(*e*) The dimensions of the elliptical contact, from Eqs. (8.13) and (8.14), are

$$\begin{aligned} D_y &= 2\left(\frac{6k_e^2\mathcal{E}WR}{\pi E'}\right)^{\frac{1}{3}} \\ &= 2\left[\frac{6(0.8676)^2(1.457)(1000)(0.0111)}{\pi(262.5 \times 10^9)}\right]^{\frac{1}{3}} \\ &= 0.8915 \times 10^{-3} \text{ m}. \end{aligned}$$

$$\begin{aligned} D_x &= 2\left(\frac{6\mathcal{E}WR}{\pi k_e E'}\right)^{\frac{1}{3}} \\ &= 2\left[\frac{6(1.457)(1000)(0.0111)}{\pi(0.8676)(262.5 \times 10^9)}\right]^{\frac{1}{3}} \\ &= 1.024 \times 10^{-3} \text{ m}. \end{aligned}$$

(*f*) The maximum pressure, from Eq. (8.11), is

$$\begin{aligned} p_{\max} &= \frac{6W}{\pi D_x D_y} \\ &= \frac{6(1000)}{\pi(0.8915 \times 10^{-3})(1.024 \times 10^{-3})} \\ &= 2.092 \text{ GPa}. \end{aligned}$$

(*g*) The maximum deformation at the center of the conjunction is calculated from Eq. (8.15), noting first that

$$\frac{W}{\pi k_e E'} = \frac{1000}{\pi(0.8676)(262.5 \times 10^9)} = 1.398 \times 10^{-9},$$

so that

$$\delta_{\max} = (1.6981)\left[\frac{4.5\left(1.398 \times 10^{-9}\right)^2}{(1.457)(0.0111)}\right]^{\frac{1}{3}},$$

which is solved as $\delta_{\max} = 13.90\ \mu$m.

Example 8.2: Hertzian Contact II – Ball in Groove

Given: The balls in a deep-groove ball bearing have a 17-mm diameter. The groove in the inner race has an 8.84-mm radius; the radius from the center of the race to the bottom of the groove is 27.5 mm. A 20,000 N is applied. The ball and race are both made of steel.

Find:

(*a*) The dimensions of the contact ellipse, and

(*b*) The maximum deformation.

Solution:

(*a*) Note from Fig. 8.6, and using an a subscript for the ball and a b for the groove, that $r_{ax} = r_{ay} = 8.5$ mm, $r_{bx} = 27.5$ mm, and $r_{by} = -8.84$ mm. From Eqs. (8.6) and (8.7) the curvature sums in the x- and y-directions are

$$\frac{1}{R_x} = \frac{1}{r_{ax}} + \frac{1}{r_{bx}} = \frac{1}{8.5 \times 10^{-3}} + \frac{1}{27.5 \times 10^{-3}},$$

or $R_x = 0.00649$ m $= 6.49$ mm. Similarly,

$$\frac{1}{R_y} = \frac{1}{r_{ay}} + \frac{1}{r_{by}} = \frac{1}{8.5 \times 10^{-3}} - \frac{1}{8.84 \times 10^{-3}},$$

which yields $R_y = 0.221$ m. From Eqs. (8.5),

$$\frac{1}{R} = \frac{1}{R_x} + \frac{1}{R_y} = 158.5\ \text{m}^{-1},$$

or $R = 0.00631$ m $= 6.31$ mm. Using Eq. (8.8) gives the radius ratio as

$$\alpha_r = \frac{R_y}{R_x} = \frac{0.221}{0.00649} = 34.0.$$

From Table 8.1, the ellipticity parameter and the elliptic integrals of the first and second kinds are (note that $1 \le \alpha_r \le 100$):

$$k_e = \alpha_r^{2/\pi} = 34.0^{2/\pi} = 9.446,$$

$$\mathcal{F} = \frac{\pi}{2} + \left(\frac{\pi}{2} - 1\right)\ln\alpha_r = \frac{\pi}{2} + \left(\frac{\pi}{2} - 1\right)\ln 34.0,$$

or $\mathcal{F} = 3.584$.

$$\mathcal{E} = 1 + \frac{\pi - 2}{2\alpha_r} = 1 + \frac{\pi - 2}{2(34.0)} = 1.017.$$

The effective modulus of elasticity is

$$E' = \frac{E}{1 - \nu^2} = \frac{207 \times 10^9}{1 - 0.3^2} = 227.5\ \text{GPa}.$$

From Eqs. (8.13) and (8.9), the dimensions of the contact ellipse are

$$\begin{aligned} D_y &= 2\left(\frac{6k_e^2\mathcal{E}WR}{\pi E'}\right)^{\frac{1}{3}} \\ &= 2\left[\frac{6(9.446)^2(1.017)(20,000)(0.00631)}{\pi\,(227.5 \times 10^9)}\right]^{\frac{1}{3}} \\ &= 9.162\ \text{mm}, \\ D_x &= \frac{D_y}{k_e} = \frac{9.162}{9.446} = 0.9699\ \text{mm}. \end{aligned}$$

(*b*) The maximum deformation at the center of contact is calculated from Eq. (8.15), but note first that

$$\frac{W}{\pi k_e E'} = \frac{20,000}{\pi(9.446)((227.5 \times 10^9)} = 2.962 \times 10^{-9},$$

so that

$$\begin{aligned} \delta_{\max} &= \mathcal{F}\left[\frac{9}{2\mathcal{E}R}\left(\frac{W}{\pi k_e E'}\right)^2\right]^{\frac{1}{3}} \\ &= 3.584\left[\frac{9\left(2.962 \times 10^{-9}\right)^2}{2(1.017)(0.00631)}\right]^{\frac{1}{3}} \\ &= 65.68\ \mu\text{m}. \end{aligned}$$

8.4.2 Rectangular Contacts

For rectangular conjunctions, the contact patch has infinite width. This type of contact is exemplified by a cylinder loaded against a plane, a flat-bottomed groove, or another parallel cylinder, or by a roller loaded against an inner or outer cylindrical race. In these situations, the contact semi-width is given by

$$b^* = R_x\left(\frac{8W'}{\pi}\right)^{\frac{1}{2}}, \tag{8.17}$$

where the dimensionless load is

$$W' = \frac{w'}{E'R_x}, \tag{8.18}$$

and w' is the load per unit length along the contact. The maximum deformation for a rectangular conjunction is

$$\delta_{\max} = \frac{2W'R_x}{\pi}\left[\ln\left(\frac{2\pi}{W'}\right) - 1\right]. \tag{8.19}$$

The maximum Hertzian contact pressure in a rectangular conjunction can be written as

$$p_{\max} = E'\left(\frac{W'}{2\pi}\right)^{\frac{1}{2}}. \tag{8.20}$$

Example 8.3: Hertzian Contact III - Rectangular Contact

Given: A solid cylinder of 20 mm radius (r_{ax} = 20 mm and $r_{ay} = \infty$) rolls around the inside of an outer race with a 100-mm internal radius (r_{bx} = −100 mm) and a large width in the axial (y) direction ($r_{by} = \infty$). The cylinder is made of silicon nitride and the outer race is made of stainless steel. The normal applied load per unit length is 1000 N/m.

Find:

(*a*) Curvature sum R

(*b*) Semiwidth of contact b^*

(*c*) Maximum Hertzian contact pressure $p_{\max}$

(*d*) Maximum deformation $\delta_{\max}$

Also, compare the results with those of Example 8.2.

Solution:

(*a*) From Example 8.1, E' = 262.5 GPa. The curvature sum is

$$\frac{1}{R_x} = \frac{1}{r_{ax}} + \frac{1}{r_{bx}} = \frac{1}{0.02} - \frac{1}{0.10} = 40 \text{ m}^{-1},$$

so that R_x = 0.025 m. The dimensionless load is obtained from Eq. (8.18) as:

$$\begin{aligned} W' &= \frac{w'}{E'R_x} \\ &= \frac{1000}{(2.625 \times 10^{11})(2.5 \times 10^{-2})} \\ &= 1.524 \times 10^{-7}. \end{aligned}$$

(*b*) From Eq. (8.17), the semiwidth of the contact is

$$\begin{aligned} b^* &= R_x \left(\frac{8W'}{\pi}\right)^{\frac{1}{2}} \\ &= (0.025)\left[\frac{8(1.524 \times 10^{-7})}{\pi}\right]^{\frac{1}{2}} \\ &= 15.57\ \mu\text{m}. \end{aligned}$$

(*c*) The maximum contact pressure is given by Eq. (8.20):

$$\begin{aligned} p_{\max} &= E'\left(\frac{W'}{2\pi}\right)^{\frac{1}{2}} \\ &= (2.625 \times 10^{11})\left[\frac{(1.524 \times 10^{-7})}{2\pi}\right]^{\frac{1}{2}} \\ &= 40.88 \text{ MPa}. \end{aligned}$$

(*d*) The maximum deformation at the center of the contact is, from Eq. (8.19),

$$\begin{aligned} \delta_{\max} &= \frac{2W'R}{\pi}\left[\ln\left(\frac{2\pi}{W'}\right) - 1\right] \\ &= \frac{2(1.524 \times 10^{-7})(2.5 \times 10^{-2})}{\pi} \\ &\quad \times \left[\ln\left(\frac{2\pi}{1.524 \times 10^{-7}}\right) - 1\right], \end{aligned}$$

or $\delta_{\max} = 0.0401\ \mu$m.

8.5 Bearing Materials

Materials for conformal surfaces will be discussed here, since these materials apply to hydrodynamic bearings (Chapter 12) as well as more general cases such as guideways or unlubricated sleeves. Materials for rolling-element bearings and gears will be expanded upon in Chapters 13 and 14, respectively. Bearing materials for conformal surfaces fall into two major categories:

1. *Metallics:* babbitts, bronzes, aluminum alloys, porous metals, and metal overlays such as silver, babbitts, and indium.

2. *Nonmetallics:* plastics, rubber, carbon-graphite, wood, ceramics, cemented carbides, metal oxides (e.g., aluminum oxide), and glass.

8.5.1 Metals

Tin- and Lead-Base Alloys

Babbitts, also known as **white metals**, are either tin or lead-base alloys having excellent embeddability and conformability characteristics. The babbitts are among the most widely used materials for lubricated bearings.

Tin- and lead-base babbitts have relatively low load-carrying capacity. This capacity is increased by metallurgically bonding these alloys to stronger backing materials such as steel, cast iron, or bronze. Fatigue strength is increased by decreasing the thickness of the babbitt lining. The optimum thickness of the bearing layer varies with the application, but is generally between 0.02 and 0.12 mm.

Table 8.2 shows the composition and physical properties of some of the most common tin- and lead-base alloys. Note the significant effect of temperature in decreasing the strength of these alloys. The effect of various percentages of alloying elements on the mechanical and physical properties of tin- and lead-base alloys can also be significant. Increasing the copper or the antimony generally increases the hardness and the tensile strength and decreases the ductility.

Copper-Lead Alloys and Bronzes

Copper-lead alloys are commonly used as lining materials on steel-backed bearings. These alloys have high fatigue resistance and can operate at higher temperatures, but they have poor antiseizure properties. They are also used in automotive and aircraft internal combustion engines and in diesel engines. Their high lead content provides a good bearing surface but makes them susceptible to corrosion. Their corrosion resistance and antiseizure properties are improved when they are used as trimetal bearings with a lead-tin or lead-indium overlay electrodeposited onto the copper-lead surface.

Table 8.2: Physical and mechanical properties of selected white metal bearing alloys. *Source:* From Hamrock et al. [2004].

Alloy number	Nominal composition (%)					Specific gravity	Yield strength, MPa		Compressive strength, MPa		Brinell hardness	
	Sn	Sb	Pb	Cu	As		20° C	100° C	20° C	100° C	20° C	100° C
1	91.0	4.5	–	4.5	–	7.34	30.3	18.3	88.6	47.9	17.0	8.0
2	89.0	7.5	–	3.5	–	7.39	42.0	20.6	102.7	60.0	24.5	12.0
3	84.0	8.0	–	8.0	–	7.46	45.4	21.7	121.3	68.3	27.0	14.5
7	10.0	15.0	74.55	–	0.45	9.73	24.5	11.0	107.9	42.4	22.5	10.5
8	5.	15.0	79.55	–	0.45	10.04	23.4	12.1	107.6	42.4	20.0	9.5

Table 8.3: Mechanical properties of selected bronze and copper alloy bearing materials *Source:* Abstracted from Hamrock, et al. [2004].

Material	Designation	Brinell hardness	Tensile strength, MPa	Maximum temperature, °C	Allowable stress, MPa
Copper lead	SAE 480	25	55.2	177	13.8
High-lead tin bronze	AMS 4840	48	172.5	204	20.7
Semiplastic bronze	SAE 67	55	207	232	20.7
Leaded red bronze	SAE 40	60	242	232	24.2
Bronze	SAE 660	60	242	232	27.6
Phosphor bronze	SAE 64	63	242	232	27.6
Gunmetal	SAE 62	65	310	260	27.6
Navy G	SAE 620	68	276	260	27.6
Leaded gunmetal	SAE 63	70	276	260	27.6
Aluminum bronze	ASTM B148-52-9c	195	621	260	31.1

Several bronze alloys, including lead, tin, and aluminum bronzes, are used extensively as bearing materials. Some are described in Table 8.3. Because of their good structural properties, they can be used without a steel backing. Lead bronzes, which contain up to 25% lead, provide higher load-carrying capacity and fatigue resistance and a higher temperature capability than the babbitt alloys. Tin contents up to about 10% are used to improve mechanical properties. Lead bronze bearings are used in pumps, diesel engines, railroad cars, home appliances, and many other applications.

Tin bronzes, which contain 9 to 20% tin and small quantities of lead (usually $< 1\%$), are harder than lead bronzes and are therefore used in heavier-duty applications.

8.5.2 Nonmetallics

Polymers

Although nonmetallic materials such as rubber and graphite have found increasing application, polymeric and plastic materials have had a very large impact in tribological machine elements. The limits of applying nonmetallic materials are shown in Table 8.4. The specific limits shown in this table are load-carrying capacity, maximum temperature, maximum speed, and pu limit, where p is the pressure in Pa and u is the surface speed in m/s.

Of the thermoplastic materials, **nylon** has been recognized as a valuable bearing material, as has the remarkable low-friction polymer polytetrafluoroethylene (PTFE), or **teflon**. The great merit of these materials is that they can operate well without lubricants, although their mechanical properties generally limit their application to lightly loaded conditions and often to low speeds and conforming surfaces.

Nylon has good abrasion resistance, a low wear rate, and good embeddability. Like most plastics, it has good antiseizure properties and softens or chars rather than seizing. Cold flow (creep) under load is one of its main disadvantages. This effect can be minimized by supporting thin nylon liners in metal sleeves. Nylon bearings are used in household applications such as kitchen mixers and blenders and for other lightly loaded applications. Nylon is not affected by petroleum oils and greases, food acids, milk, etc., and thus can be used in applications where these fluids are encountered.

Teflon is resistant to chemical attack by many solvents and chemicals and can be used up to 260°C. Like nylon, it has a tendency to creep. Teflon in its unmodified form also has the disadvantages of low stiffness, a high thermal expansion coefficient, low thermal conductivity, and poor wear resistance. These poor properties are greatly improved by adding fibers such as glass, ceramics, metal powders, metal oxides, graphite, or molybdenum disulfide.

Phenolics are another important class of plastic bearing material. These are in the form of laminates, made by infiltrating sheets of either paper or fabric with phenolic resin, stacking the sheets, and curing with heat and pressure to bond them together. Other filling materials, such as graphite and molybdenum disulfide, are added in powdered form to improve lubrication qualities and strength.

Laminated phenolics operate well with steel or bronze journals when lubricated with oil, water, or other liquids. They have good resistance to seizure. One main disadvantage of these materials is their low thermal conductivity (about 1/150 that of steel), which prevents them from dissipating frictional heat readily and can result in their failure by charring.

Graphite

In addition to its excellent self-lubricating properties, **graphite** has several advantages over conventional materials and lubricants. It can withstand temperatures of approximately 370°C in an oxidizing atmosphere such as air and can be used in inert atmospheres ranging from cryogenic temperatures to 700°C. Graphite is highly resistant to chemical attack and can be used with low-viscosity lubricants, including water, gasoline, or air. A pv value of 15,000 is typically used for graphite when lubricated. In general, low speeds and light loads should be used in nonlubricated applications.

Table 8.4: Limits of application of nonmetallic bearing materials. *Source:* Abstracted from Hamrock, et al. [2004].

Material	Allowable stress, MPa	Maximum temperature, °C	Maximum speed, m/s	pu limit, N/m-s
Carbon graphite	4.1	399	12.7	525×10^3
Phenolics	41.4	93	12.7	525×10^3
Nylon	6.9	93	5.1	105×10^3
PTFE (Teflon)	3.4	260	.51	35×10^3
Reinforced PTFE	17.2	260	5.1	350×10^3
PTFE fabric	414.0	260	.25	875×10^3
Polycarbonate (Lexan)	6.9	104	5.1	105×10^3
Acetal resin (Delrin)	6.9	82	5.1	105×10^3
Rubber	0.34	66	7.6	525×10^3
Wood	13.8	66	10.2	525×10^3

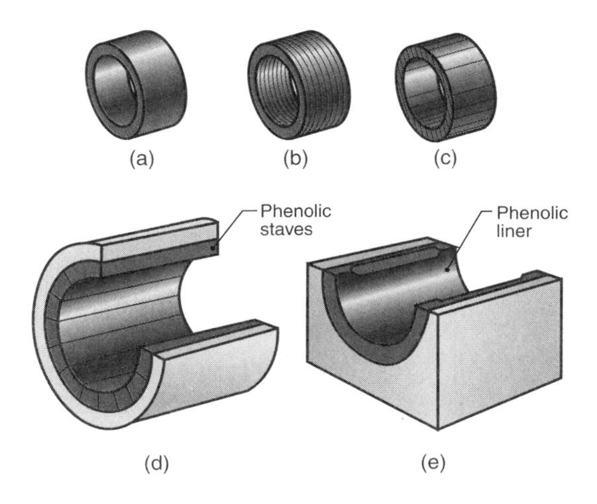

Figure 8.9: Phenolic laminate bearings. (a) Tubular bearing; (b) circumferentially laminated bearing; (c) axially laminated bearing; (d) stave bearing; (e) molded bearing. *Source:* Hamrock et al. [2004].

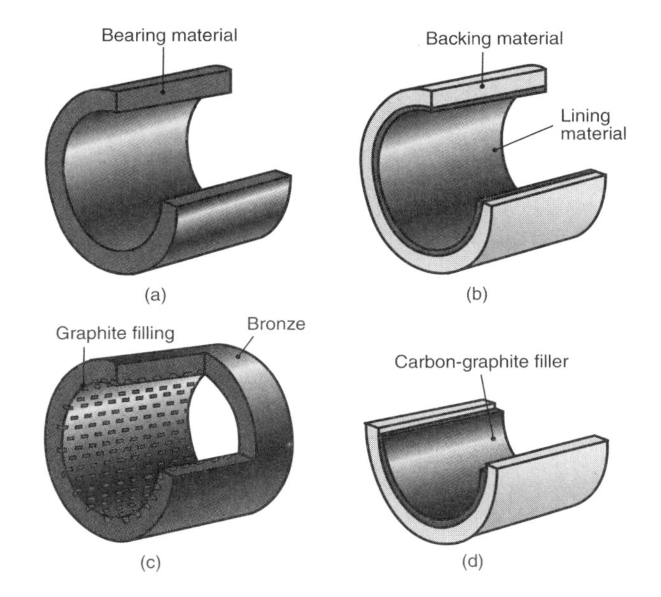

Figure 8.10: Different forms of bearing surfaces. (a) Solid bearing; (b) lined bearing; (c) filled bearing; (d) shrink-fit bearing.

Graphite is commonly used for pump shaft bearings, impeller wear rings in centrifugal pumps, and journal and thrust bearings in covered motor pumps and for many other applications. Because of its low thermal expansion coefficient, graphite liners are often shrink-fit into steel sleeves (see Section 10.6). The steel backing provides mechanical support, improves heat transfer, and helps to maintain shaft clearance.

8.5.3 Bearing Configurations

Figure 8.9 shows the various configurations of phenolic laminates used in bearings. Tubular bearings (Fig. 8.9a) are used where complete bushings are required. Bearings in which the load is taken by the edges of the laminations (Fig. 8.9b and c) are used in light-duty service. Stave bearings (Fig. 8.9d) are used mainly for stern-tube and rudder-stock bearings on ships and for guide bearings on vertical turbines. Molded bearings (Fig. 8.9e) are used for roll-neck bearings in steel mills or for ball-mill bearings.

Other materials may be applied to bearing surfaces in several ways, as shown in Fig. 8.10:

- **Solid bearing** (Fig. 8.10a). Bearings are machined directly from a single material (cast iron, aluminum alloys, bronzes, porous metals, etc.).
- **Lined bearing** (Fig. 8.10b). Bearing material is bonded to a stronger backing material. The thickness of the bearing lining may range from 0.25 mm to as much as 13 mm. Most modern bonding techniques are metallurgical, although chemical and mechanical methods are also used. The lining material may be cast, sprayed, electrodeposited, or chemically applied.
- **Filled bearing** (Fig. 8.10c). A stronger bearing material is impregnated with a bearing material that has better lubricating properties (e.g., graphite impregnated into a bronze backing).
- **Shrink-fit liner bearing** (Fig. 8.10d). Carbon-graphite or plastic liners are shrunk into a metal backing sleeve or held by retaining devices such as setscrews.

8.6 Lubricant Rheology

A **lubricant** is any substance that reduces friction and wear and provides smooth running and a satisfactory life, and **rheology** is the study of the flow and deformation of matter. Most lubricants are liquids (such as mineral oils, synthetic

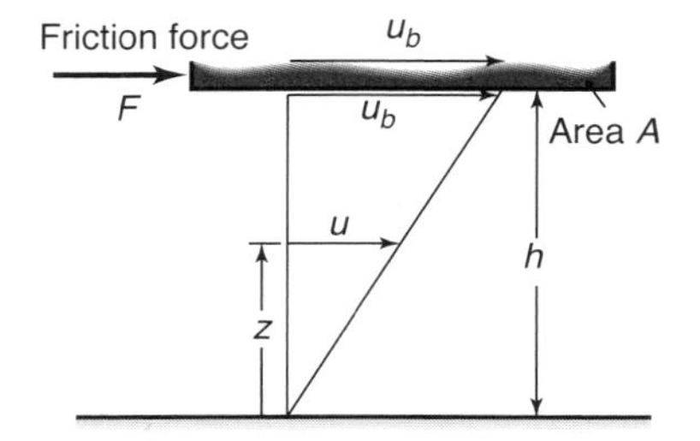

Figure 8.11: Slider bearing illustrating absolute viscosity.

esters, silicone fluids, and water), but they may also be solids (such as graphite or molybdenum disulfide) for use in dry bearings, greases for use in rolling-element bearings, or gases (such as air) for use in gas bearings. The physical and chemical interactions between the lubricant and the lubricating surfaces must be understood in order to provide the machine elements with satisfactory life.

When interposed between solid surfaces, a lubricant facilitates relative sliding or rolling, and controls friction and wear. Separation of the surfaces is, however, not the only function of a lubricant. Liquid lubricants have desirable secondary properties and characteristics:

1. They can be drawn in between moving parts by hydrodynamic action.

2. They have relatively high heat capacity to cool the contacting parts.

3. They are easily blended with chemical additives to give a variety of properties, such as corrosion resistance, detergency, or surface-active layers.

4. They can remove wear particles.

5. They are thermal insulators.

6. They damp vibrations and reduce impact loads.

Liquid lubricants can be divided into those of petroleum origin, known as mineral oils, and those of animal or vegetable origin, known as fatty oils; synthetic oils are often grouped with the latter. For a lubricant to be effective, it must be viscous enough to maintain a lubricant film under operating conditions and to remove heat from machine elements. On the other hand, the lubricant should be as inviscid as possible to minimize power requirements at startup (especially cold starts) and to avoid power loss due to viscous drag. The most important lubricant property, the viscosity, is considered in the following subsections.

8.6.1 Absolute Viscosity

Absolute or **dynamic viscosity** can be defined in terms of the simple model shown in Fig. 8.11, which depicts two parallel flat plates separated by a constant distance, h, with the upper plate moving with velocity u_b and the lower plate stationary. To move the upper plate of area A at a constant velocity across the surface of the oil and cause adjacent layers to flow past each other, a tangential force must be applied. Since the oil will "wet" and cling to the two surfaces, the bottommost layer will not move at all, the topmost layer will move with a velocity equal to the velocity of the upper plate, and the layers between the plates will move with velocities directly proportional to their distances, z, from the stationary plate. This type of orderly movement in parallel layers is known as streamline, laminar, or viscous flow.

The shear stress on the oil causing relative movement of the layers is equal to F/A. The shear strain rate, s, of a particular layer is defined as the ratio of its velocity to its perpendicular distance from the stationary surface, z, and is constant for each layer:

$$s = \frac{u}{z} = \frac{u_b}{h}. \tag{8.21}$$

The shear strain rate has the unit of reciprocal seconds.

Newton correctly deduced that the force required to maintain a constant velocity, u_b, of the upper surface is proportional to the area and the shear strain rate or

$$F = \frac{\eta A u_b}{h}, \tag{8.22}$$

where η is the absolute viscosity. By rearranging Eq. (8.22), the absolute viscosity can be expressed as

$$\eta = \frac{F/A}{u_b/h} = \frac{\text{Shear stress}}{\text{Shear strain rate}}. \tag{8.23}$$

It follows from Eq. (8.23) that the unit of viscosity must be the unit of shear stress divided by the unit of shear strain rate. The units of viscosity for three different systems in general use are:

1. SI units: newton-second per square meter (N-s/m^2) or, since a newton per square meter is also called a pascal, pascal-second (Pa-s).

2. cgs units: dyne-second per square centimeter, or *centipoise*, where 1 cP = 10^{-2} P.

3. English units: pound force-second per square inch (lb-s/in^2), called a *reyn* in honor of Osborne Reynolds

Conversion of absolute viscosity from one system to another can be facilitated by Table 8.5. To convert from a unit in the column on the left side of the table to a unit at the top of the table, multiply by the corresponding value given in the table.

Example 8.4: Units of Viscosity

Given: An absolute viscosity of 0.04 N-s/m^2.

Find: The absolute viscosity in reyn, centipoise, and poise.

Solution: Using Table 8.5 gives

$$\eta = 0.04 \text{ N-s/m}^2 = (0.04) \text{ Pa-s} = 5.8 \times 10^{-6} \text{ lb-s/in.}^2$$

Note also that

$$\eta = 0.04 \text{ N-s/m}^2 = 0.04 \text{ Pa-s} = 5.8 \times 10^{-6} \text{ reyn}$$

and

$$\eta = 0.04 \text{ N-s/m}^2 = 0.004 \times 10^3 \text{ cP} = 40 \text{ cP} = 0.4 \text{ P}.$$

8.6.2 Kinematic Viscosity

In many situations it is convenient to use **kinematic viscosity** rather than absolute viscosity. The kinematic viscosity, η_k, is:

$$\eta_k = \frac{\text{Absolute viscosity}}{\text{Density}} = \frac{\eta}{\rho}. \tag{8.24}$$

The ratio given in Eq. (8.24) is literally kinematic; all terms involving force or mass have canceled out. The units of kinematic viscosity are

Table 8.5: Absolute viscosity conversion factors.

To convert from	To			
	cP	kgf-s/m^2	N-s/m^2	lb-s/in.^{2}
	Multiply by			
cP	1	1.02×10^{-4}	10^{-3}	1.45×10^{-7}
kgf-s/m^2	9.807×10^{3}	1	9.807	1.422×10^{-3}
N-s/m^2	10^{3}	1.02×10^{-1}	1	1.45×10^{-4}
reyn, or lb-s/in.^{2}	6.90×10^{6}	7.03×10^{2}	6.9×10^{3}	1

1. SI units: square meters per second (m^2/s)
2. cgs units: square centimeters per second (cm^2/s), called a *stoke*
3. English units: square inches per second (in^2/s)

Example 8.5: Kinematic Viscosity

Given: Both mercury and water have an absolute viscosity of 1.5 cP at 5°C, but mercury has 13.6 times higher density than water. Assume density changes associated with temperature increases are small.

Find: The kinematic viscosities of mercury and water at 5°C and at 90°C.

Solution: Figure 8.12 gives the absolute viscosities of various fluids as a function of temperature. At 90°C the absolute viscosity of water is 0.32 cP, and for mercury it is 1.2 cP. The kinematic viscosity of mercury at 5°C is

$$\eta_k = \frac{\eta}{\rho} = \frac{1.5 \times 10^{-3}}{13,600} = 0.110 \times 10^{-6} \text{ m}^2/\text{s},$$

and at 90°C it is

$$\eta_k = \frac{\eta}{\rho} = \frac{1.2 \times 10^{-3}}{13,600} = 0.0882 \times 10^{-6} \text{ m}^2/\text{s}.$$

The kinematic viscosity of water at 5°C is

$$\eta_k = \frac{\eta}{\rho} = \frac{1.5 \times 10^{-3}}{1000} = 1.50 \times 10^{-6} \text{ m}^2/\text{s},$$

and at 90°C it is

$$\eta_k = \frac{\eta}{\rho} = \frac{0.32 \times 10^{-3}}{1000} = 0.32 \times 10^{-6} \text{ m}^2/\text{s}.$$

Although mercury has the same absolute viscosity as water at 5°C and 3.75 times higher absolute viscosity at 90°C, the kinematic viscosities for mercury are much lower than those for water because of mercury's high density.

8.6.3 Viscosity-Pressure Effects

In highly loaded contacts such as ball bearings, gears, and cams, the pressure is high enough to increase the lubricant viscosity significantly. The increase of a lubricant's viscosity with pressure is known as a **viscosity-pressure effect** or **piezoviscous effect**, and is especially pronounced in mineral oils and other fluids with a large molecular chain length. The Barus equation relates the viscosity and pressure for isothermal conditions as

$$\eta = \eta_o \exp(\xi p), \tag{8.25}$$

where p is the pressure, η_o is the absolute viscosity at ambient pressure ($p = 0$) and at a constant temperature, and ξ is the **pressure-viscosity coefficient** of the lubricant.

Table 8.6 lists the kinematic viscosities in square meters per second and the absolute viscosities in centipoise of various fluids at zero pressure and different temperatures. These values of the absolute viscosity correspond to η_o in Eq. (8.25) for the particular fluid and temperature used. The pressure-viscosity coefficients, ξ, for these fluids, expressed in square meters per newton, are given in Table 8.7. The values correspond to ξ in Eq. (8.25).

Example 8.6: Piezoviscous Effects

Given: A synthetic paraffin and a fluorinated polyether are used at a temperature of 99°C.

Find:

(*a*) The pressure at which the two oils have the same absolute viscosity

(*b*) The pressure at which the paraffin is 100 times less viscous than the fluorinated polyether

Solution:

(*a*) From Table 8.6, the viscosity of the synthetic paraffin at 99°C is 0.0347 Ns/m^2, and the fluorinated polyether has a viscosity of 0.0202 Ns/m^2. From Table 8.7, the pressure-viscosity coefficient at 99°C for the paraffin is 1.51×10^{-8} m^2/N and for the polyether it is 3.24×10^{-8} m^2/N. Using Eq. (8.25) and equating the viscosity gives

$$0.0347 e^{1.51\times 10^{-8} p} = 0.0202 e^{3.24\times 10^{-8} p}$$

$$1.72 e^{1.51\times 10^{-8} p} = e^{3.24\times 10^{-8} p}$$

$$1.72 = \frac{e^{3.24\times 10^{-8} p}}{e^{1.51\times 10^{-8} p}} = e^{1.73\times 10^{-8} p}.$$

Therefore,

$$p = \frac{\ln 1.72}{1.73 \times 10^{-8}} = 31.3 \text{ MPa}.$$

(*b*) Using Eq. (8.25) while letting the paraffin be 100 times less viscous than the polyether gives

$$1.72(100) e^{1.51\times 10^{-8} p} = e^{3.24\times 10^{-8} p},$$

$$172 = e^{1.73\times 10^{-8} p}.$$

Therefore, $p = 2.97 \times 10^{-8}$ Pa $= 0.297$ GPa. This may seem to be a very high pressure, but it is not an unusual pressure for non-conformal contacts such as rolling element bearings or gears.

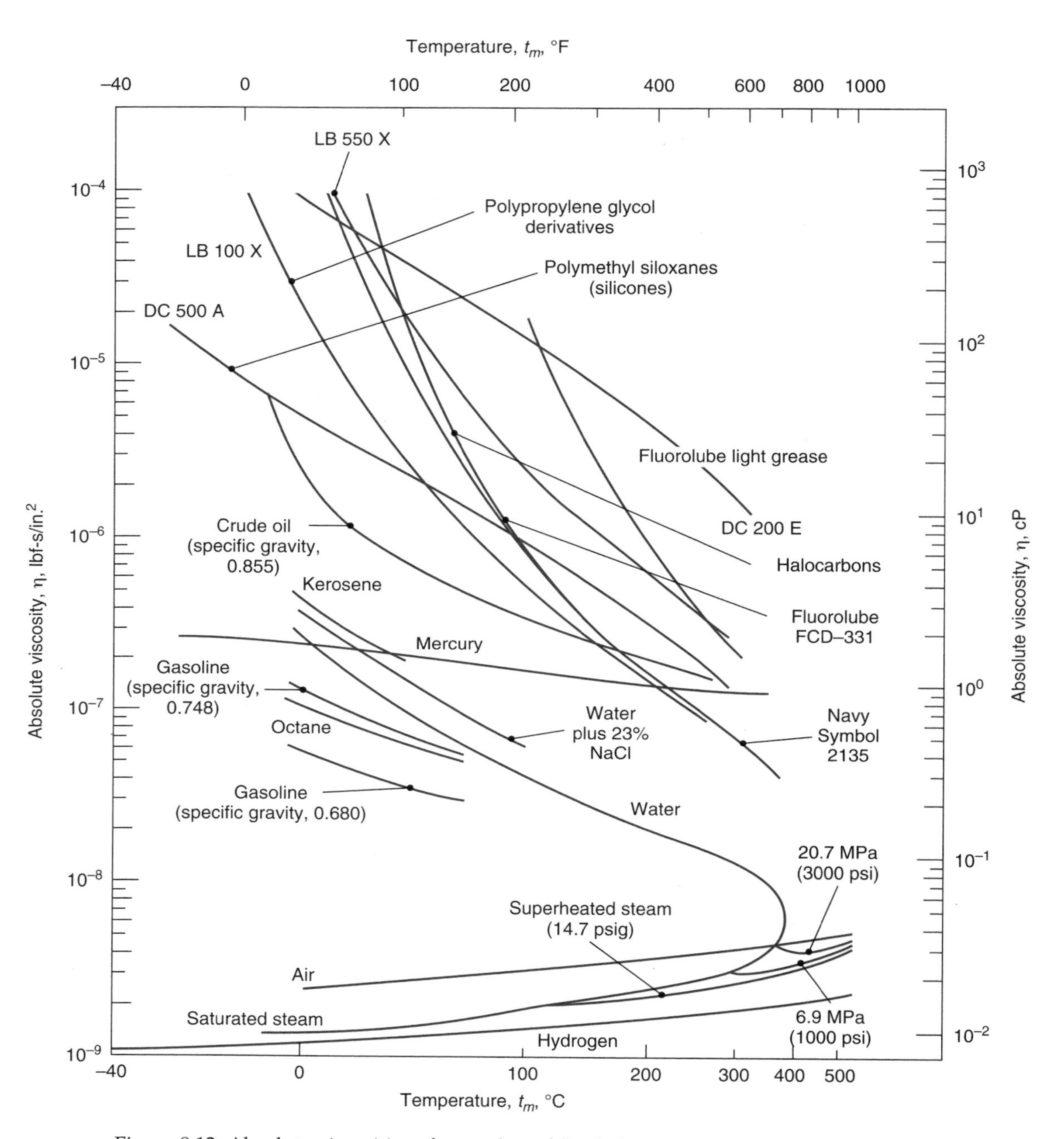

Figure 8.12: Absolute viscosities of a number of fluids for a wide range of temperatures.

Table 8.6: Absolute and kinematic viscosities of various fluids at atmospheric pressure and different temperatures.

Fluid	Absolute viscosity at $p = 0$, η_o, N-s/m²			Kinematic viscosity at $p = 0$, η_k, m²/s		
	Temperature, °C			Temperature, °C		
	38	99	149	38	99	149
Advanced ester	0.0253	0.00475	0.00206	2.58×10^{-5}	0.51×10^{-5}	0.23×10^{-5}
Formulated advanced ester	0.0276	0.00496	0.00215	2.82×10^{-5}	0.53×10^{-5}	0.24×10^{-5}
Polyalkyl aromatic	0.0255	0.00408	0.00180	3.0×10^{-5}	0.50×10^{-5}	0.23×10^{-5}
Synthetic paraffinic oil	0.375	0.0347	0.0101	44.7×10^{-5}	4.04×10^{-5}	1.3×10^{-5}
Synthetic paraffinic oil plus antiwear additive	0.375	0.0347	0.0101	44.7×10^{-5}	4.04×10^{-5}	1.3×10^{-5}
C-ether	0.0295	0.00467	0.00220	2.5×10^{-5}	0.41×10^{-5}	0.20×10^{-5}
Superrefined napthenic mineral oil	0.0681	0.00686	0.002.74	7.8×10^{-5}	0.82×10^{-5}	0.33×10^{-5}
Synthetic hydrocarbon (traction fluid)	0.0343	0.00353	0.00162	3.72×10^{-5}	0.40×10^{-5}	0.19×10^{-5}
Fluorinated polyether	0.181	0.0202	0.00668	9.66×10^{-5}	1.15×10^{-5}	0.4×10^{-5}

Table 8.7: Pressure-viscosity coefficients of various fluids at different temperatures.

Fluid	Temperature, °C		
	38	99	149
	Pressure-viscosity coefficient, ξ, m²/N		
Advanced ester	1.28×10^{-8}	0.987×10^{-8}	0.851×10^{-8}
Formulated advanced ester	1.37×10^{-8}	1.00×10^{-8}	0.874×10^{-8}
Polyalkyl aromatic	1.58×10^{-8}	1.25×10^{-8}	1.01×10^{-8}
Synthetic paraffinic oil	1.99×10^{-8}	1.51×10^{-8}	1.29×10^{-8}
Synthetic paraffinic oil plus antiwear additive	1.96×10^{-8}	1.55×10^{-8}	1.25×10^{-8}
C-ether	1.80×10^{-8}	0.980×10^{-8}	0.795×10^{-8}
Superrefined napthenic mineral oil	2.51×10^{-8}	1.54×10^{-8}	1.27×10^{-8}
Synthetic hydrocarbon (traction fluid)	3.12×10^{-8}	1.71×10^{-8}	0.939×10^{-8}
Fluorinated polyether	4.17×10^{-8}	3.24×10^{-8}	3.02×10^{-8}

8.6.4 Viscosity-Temperature Effects

The viscosity of mineral and synthetic oils decreases with increasing temperature; therefore, the temperature at which a viscosity measurement is taken must be reported. Figure 8.13 shows how absolute viscosity varies with temperature. Figure 8.12 presents the absolute viscosity of several fluids for a wide temperature range. The interesting point of this figure is how drastically the slope and level of viscosity change for different fluids. The viscosity varies by five orders of magnitude, with the slope being highly negative for fluids and positive for gases.

Figure 8.13 gives the viscosity of Society of Automotive Engineers (SAE) oils as a function of temperature. The SAE standards allow for a range of values, and specify a kinematic viscosity, which results in the approximate absolute viscosity curves shown in Fig. 8.13. Real lubricants have viscosities that are functions of strain rate, which serves as another reminder that Fig. 8.13 does not provide exact values. Further, SAE specifies viscosity only at one temperature for the lubricants in Fig. 8.13a, namely 100°C (212°F). Multigrade oils, shown in Fig. 8.13b, have a low temperature viscosity requirement defined at –18°C (0°F), which they must meet in addition to the high temperature viscosity requirement. This demanding set of requirements can be met through the use of viscosity index modifiers. Such complicated formulations require specific characterization of viscosity as functions of temperature and strain rate. However, the data in Fig. 8.13 and Table 8.8 are sufficient for most design and analysis tasks.

The viscosities of SAE oils can also be approximated for the temperatures in Fig. 8.13 using the curve fit

Table 8.8: Curve fit data for SAE single grade oils for use in Eq. (8.26). *Source:* From Seirig and Dandage [1982].

SAE Grade	Constant C_1		Constant C_2
	reyn	N-s/m²	
10	1.58×10^{-8}	1.09×10^{-4}	1157.5
20	1.36×10^{-8}	9.38×10^{-5}	1271.6
30	1.41×10^{-8}	9.73×10^{-5}	1360.0
40	1.21×10^{-8}	8.35×10^{-5}	1474.4
50	1.70×10^{-8}	1.17×10^{-4}	1509.6
60	1.87×10^{-8}	1.29×10^{-4}	1564.0

$$\eta = \begin{cases} C_1 \exp\left(\dfrac{C_2}{t_F + 95}\right) & \text{English units} \\ C_1 \exp\left(\dfrac{C_2}{1.8 t_C + 127}\right) & \text{S.I. units} \end{cases} \tag{8.26}$$

where C_1 and C_2 are constants given in Table 8.8 and t_F and t_C are the lubricant temperature in °F or °C, respectively. Note that Eq. (8.26) can give slightly different results than Fig. 8.13, since the strain rates at evaluation are slightly different.

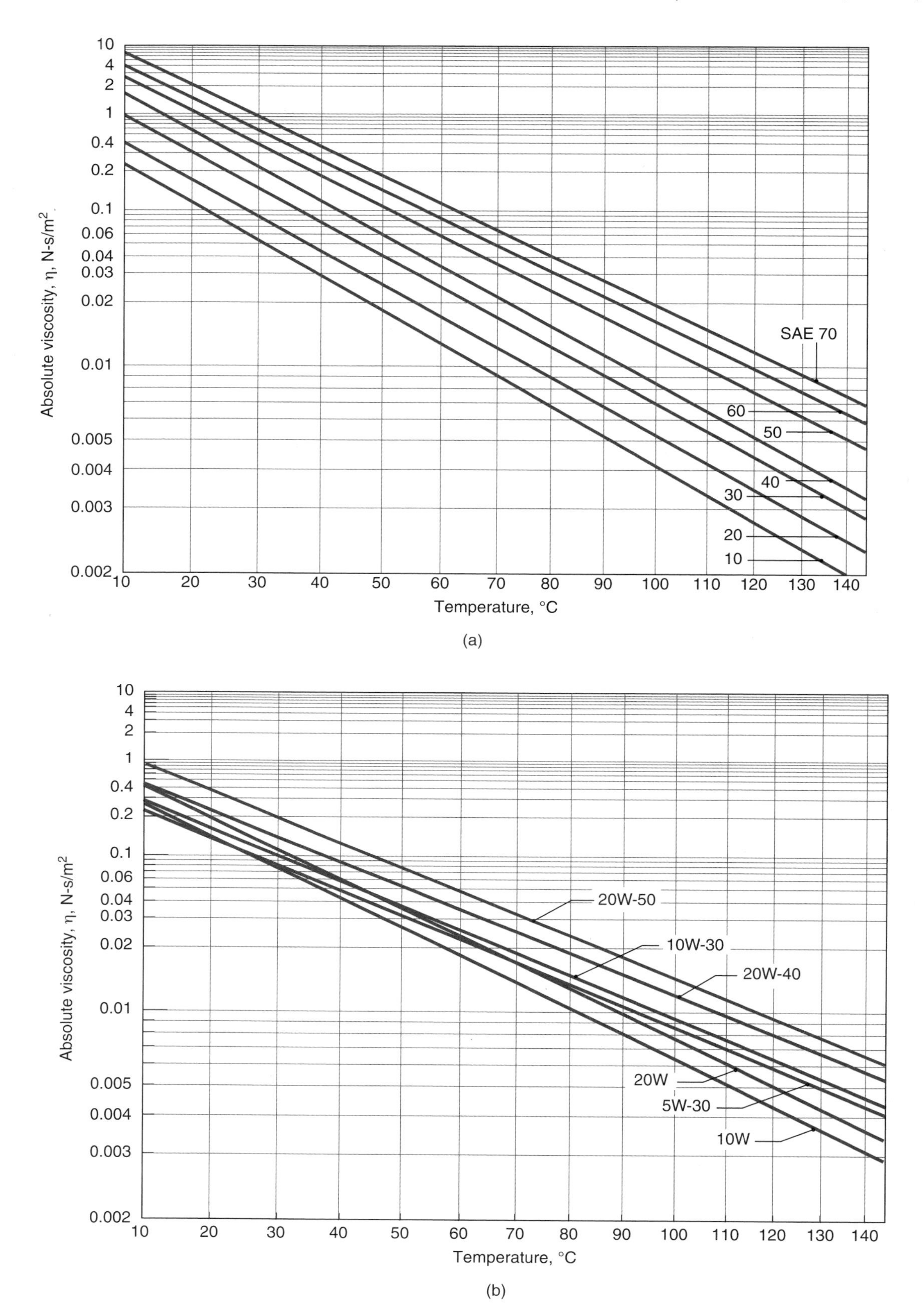

Figure 8.13: Absolute viscosities of SAE lubricating oils at atmospheric pressure. (a) Single grade oils; (b) multigrade oils.

Example 8.7: Viscosity-Temperature Effects

Given: An SAE 10 oil is used in a car motor where the working temperature is 110°C.

Find: How many times more viscous would the oil be at startup when the temperature is 10°C?

Solution: Figure 8.13 gives the viscosity as 0.0033 N-s/m^2 at 110°C and 0.22 N-s/m^2 at 10°. The viscosity ratio is 0.22/0.0033 = 66.67. Alternatively, Eq. (8.26) can be used to obtain the viscosity at 10°C as 0.990 N-s/m^2 and 0.003839 at 110°, resulting in a viscosity ratio of 258. The viscosity at low temperatures is suspect using Eq. (8.26); however, this clearly demonstrates that the startup viscosity is much lower than the viscosity during operation.

8.7 Regimes of Lubrication

Some machine elements such as hydrodynamic bearings (Ch. 12) require lubricants to function; others such as rolling-element bearings (Ch. 13) and gears (Ch. 14-15) would have compromised life and performance in the absence of effective lubrication. Lubrication engineering is a topic worthy of detailed study; this section presents a general overview, emphasizing fundamental topics that have a direct impact on machine elements. The discussion begins with the introduction of the film parameter, and its use to define the regime of lubrication. Each regime is then summarized, using surface roughness (Section 8.2) and lubricant rheology (Section 8.6) to explain performance.

8.7.1 Film Parameter

When machine elements such as rolling-element bearings, gears, hydrodynamic journal and thrust bearings, and seals (all of which are considered later in this text) are adequately designed and lubricated, the surfaces are completely separated by a lubricant film. Endurance testing of ball bearings, for example, has demonstrated that when the lubricant film is thick enough to separate the contacting bodies, bearing fatigue life is greatly extended (see Section 13.9.5 and Fig. 13.24). Conversely, when the film is not thick enough to provide full separation between the asperities in the contact zone, bearing life is adversely affected by the high shear stresses resulting from direct metal-to-metal contact.

An important parameter that indicates the effectiveness of lubrication is the **film parameter**, given by

$$\Lambda = \frac{h_{\min}}{\left(R_{qa}^2 + R_{qb}^2\right)^{\frac{1}{2}}}, \qquad (8.27)$$

where $h_{\min}$ is the minimum film thickness, and R_{qa} and R_{qb} are the rms surface roughnesses of the two surfaces. The film parameter is used to define the four important lubrication regimes. The range for these four regimes is

1. *Boundary lubrication*, $\Lambda < 1$
2. *Partial lubrication*, $1 \leq \Lambda < 3$
3. *Hydrodynamic lubrication*, $3 \leq \Lambda$
4. *Elastohydrodynamic lubrication*, $3 \leq \Lambda < 10$

These values are only approximate, but give useful insight into the importance of lubrication.

Running-in is a process that affects the film parameter. Reviewing Fig. 8.14, it can be seen that surfaces contact at asperity peaks. If microscale wear takes place at these asperity peaks, and material is removed, then the remaining surface is smoother as a result. This process allows wear to occur so that the mating surfaces can adjust to each other to provide for smooth running; this type of wear can be viewed as beneficial. The film parameter will increase with running-in, since the composite surface roughness will decrease. Running-in also has a significant effect on the shape of the asperities that is not captured by the composite surface roughness. With running-in the tips of the asperities in contact become flattened and are less aggressive, also leading to less wear.

Example 8.8: Film Parameter

Given: Gears for an excavator are manufactured by sand casting. The as-cast surface is measured to have a root-mean-square roughness of R_a=18 μm. This high surface roughness makes the gear wear rapidly. The film thickness for the grease-lubricated gears is calculated to be $h_{\min}$=1.6 μm.

Find: To what surface finish should the sand-cast gears be machined to give a film parameter of 1? What manufacturing process would you select for this application?

Solution: Using Eq. (8.27) while assuming that the roughnesses are equal on the two surfaces gives

$$\Lambda = \frac{h_{\min}}{\left(R_{qa}^2 + R_{qb}^2\right)^{1/2}} = \frac{h_{\min}}{R_q\sqrt{2}}.$$

Solving for R_q,

$$R_q = \frac{h_{\min}}{\Lambda\sqrt{2}} = \frac{1.6}{1\sqrt{2}} = 1.131\,\mu\text{m}.$$

Recall from Eq. (8.3) that R_q/R_a is around 1.11 for a sinusoidal distribution, and this will be assumed here to obtain an estimate of required centerline average roughness. Therefore, the surface finish needs to be around R_a = 1.131/1.11 = 1.02 μm. Figure 8.2 shows that for a surface roughness of 1 μm, grinding is the fastest and cheapest method of achieving these surface finishes (honing does not apply to gear geometries). Smoother surfaces can be manufactured by electropolishing, polishing, and lapping, but these processes are considerably more expensive.

8.7.2 Boundary Lubrication

In **boundary lubrication**, the solids are not separated by the lubricant, thus, fluid film effects are negligible and there is significant asperity contact. The nature of surface contact is governed by the physical and chemical properties of thin surface films of molecular proportions. The properties of the bulk lubricant are of minor importance, and the coefficient of friction is essentially independent of fluid viscosity. The frictional characteristics are determined by the properties of the solids and the lubricant film at the common interfaces. The surface films vary in thickness from 1 to 10 nm, depending on the lubricant's molecular size.

Figure 8.14 illustrates the film conditions existing in fluid film and boundary lubrication. The surface slopes and film thicknesses in this figure are greatly distorted for purposes of illustration. To scale, real engineering surfaces would appear

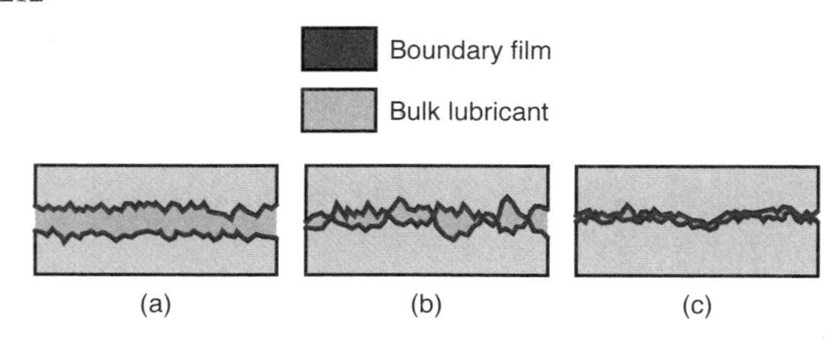

Figure 8.14: Regimes of lubrication. (a) Fluid film lubrication – surfaces completely separated by bulk lubricant film. This regime is sometimes further classified as thick or thin film lubrication. (b) Partial lubrication – both bulk lubricant and boundary film play a role. (c) Boundary lubrication – performance depends essentially on a boundary film.

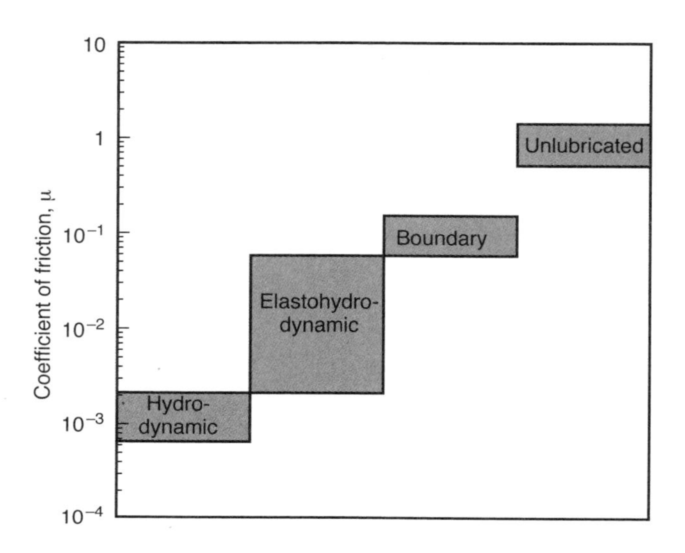

Figure 8.15: Bar diagram showing coefficient of friction for various lubrication conditions. *Source:* Hamrock et al. [2004].

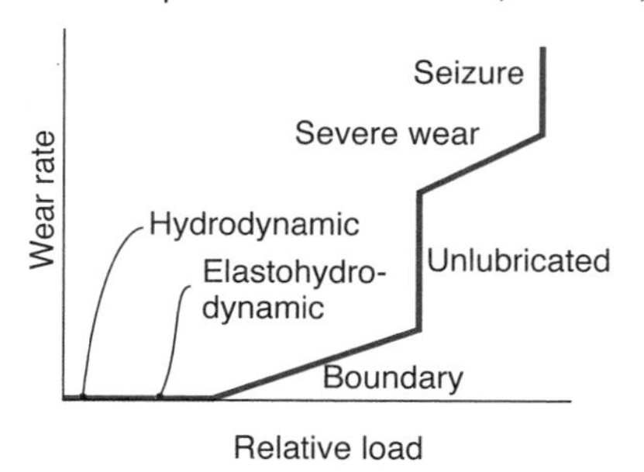

Figure 8.16: Wear rate for various lubrication regimes.

as gently rolling hills with almost imperceptible topography rather than sharp peaks. The surface asperities are not in contact for fluid film lubrication but are in contact for boundary lubrication.

Figure 8.15 shows the behavior of the coefficient of friction in the different lubrication regimes. In boundary lubrication, although the friction is much higher than in the hydrodynamic regime, it is still much lower than for unlubricated surfaces. The mean coefficient of friction increases a total of three orders of magnitude in transitioning from the hydrodynamic to the boundary regime.

Figure 8.16 shows the wear rate in the various lubrication regimes as determined by the operating load. In the hydrodynamic and elastohydrodynamic regimes, there is little or no wear because there is no asperity contact. In the boundary lubrication regime, the degree of asperity interaction and the wear rate increase as the load increases. The transition from boundary lubrication to an unlubricated condition is marked by a drastic change in wear rate. As the relative load is increased in the unlubricated regime, the wear rate increases until scoring or seizure occurs and the machine element can no longer operate successfully. Most machine elements cannot operate long with unlubricated surfaces. Figures 8.15 and 8.16 show that the friction and wear of unlubricated machine element surfaces can be greatly decreased by providing boundary lubrication. Effective boundary lubrication is achieved by lubricant chemistry, usually by including alcohols or fatty acids to the lubricant that attach to and protect metal surfaces, or by providing compounds that develop protective coatings. Boundary lubricants are tailored for particular materials, so that a formulation for steel sliding against steel cannot always be used for bronze sliding against steel. Indeed, additives in such a lubricant may be benign with respect to steel but corrosive to bronze, and would therefore shorten life.

Boundary lubrication is used for machine elements with heavy loads and low running speeds, where it is difficult to achieve fluid film lubrication. In addition, all machine elements encounter boundary lubrication at startup. Mechanisms such as door hinges operate under boundary lubrication conditions; other boundary lubrication applications are those where low cost is of primary importance, such as in rubbing sleeve bearings.

8.7.3 Partial Lubrication

If the pressures in lubricated machine elements are too high or the running speeds are too low, the lubricant film will be penetrated. Some contact will then take place between the asperities, and **partial lubrication** (sometimes called **mixed lubrication**) will occur. The behavior of the conjunction in a partial lubrication regime is governed by a combination of boundary and fluid film effects. Interaction takes place between one or more molecular layers of boundary lubricating films, and, at the same time, a partial fluid film lubrication action develops between the solids. The average film thickness in a partially lubricated conjunction is typically between 0.001 and 1 μm.

Since some load is transferred between contacting asperities, it is essential that a lubricant for partial lubrication incorporate chemical additives just as if it were to be used for a boundary lubrication circumstance. Thus, the concerns mentioned above regarding formulation apply for partial lubrication as well.

It is important to recognize that the transition from boundary to partial lubrication does not take place instantaneously as the severity of loading is increased, but rather a decreasing proportion of the load is carried by pressures within the fluid that fills the space between the opposing solids. As the load increases, a larger part of the load is supported by the contact pressure between the asperities of the solids.

8.7.4 Hydrodynamic Lubrication

Hydrodynamic lubrication (HL) generally occurs with conformal surfaces lubricated by a viscous fluid. A positive

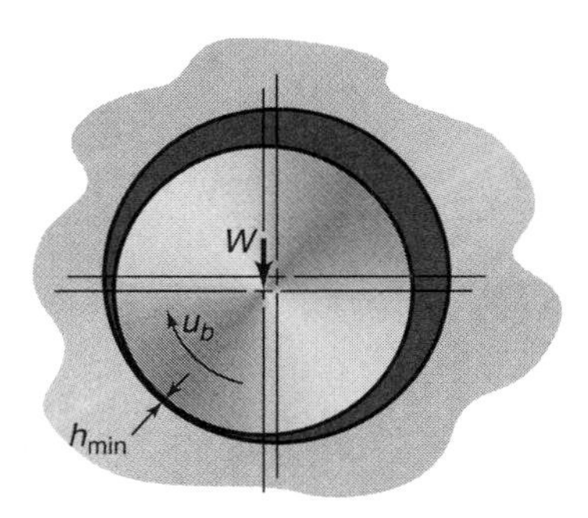

Conformal surfaces
p_{max} ~ 5 MPa
$h_{min} = f(W, u_b, \eta_0, R_x, R_y) > 1$ μm
No elastic effect

Figure 8.17: Characteristics of hydrodynamic lubrication.

pressure develops in a hydrodynamically lubricated journal or thrust bearing because the bearing surfaces converge and their relative motion and the viscosity of the fluid develop a lubricant film that separates the surfaces. The existence of this positive pressure implies that a normal applied load may be supported. The magnitude of the pressure developed (usually less than 5 MPa) is not generally high enough to cause significant elastic deformation of the surfaces.

The minimum film thickness in a hydrodynamically lubricated bearing is a function of normal applied load, W, velocity, u_b, lubricant absolute viscosity, η_o, and geometry (R_x and R_y). Figure 8.17 summarizes some of these characteristics of hydrodynamic lubrication. The minimum film thickness, $h_{\min}$, as a function of entraining speed, u_b, and applied load, W, for sliding motion is given as

$$(h_{\min})_{\text{HL}} \propto \left(\frac{u_b}{W}\right)^{\frac{1}{2}} \tag{8.28}$$

In practice, the minimum film thickness normally exceeds 1 μm.

In hydrodynamic lubrication, the films are generally thick, so that opposing solid surfaces are prevented from coming into contact. This condition is often called the ideal form of lubrication because it provides low friction and very low wear. Lubrication of the solid surfaces is governed by the bulk physical properties of the lubricant, notably the viscosity; the frictional characteristics arise purely from the shearing of the viscous lubricant.

For a normal load to be supported by a bearing, positive pressure profiles must be developed over the bearing length. Figure 8.18 illustrates three ways of developing positive pressure in hydrodynamically lubricated bearings. For positive pressure to be developed in a **slider bearing** (Fig. 8.18a and Section 12.3) the lubricant film thickness must be decreasing in the sliding direction. In a **squeeze film bearing** (Fig. 8.18b and Section 12.5) the bearing surfaces approach each other with velocity, w_a, providing a valuable cushioning effect; positive pressures will be generated only when the film thickness is diminishing. In an **externally pressurized bearing**, sometimes called a *hydrostatic bearing* (Fig. 8.18c and Section 12.6), the pressure inside the bearing supports the load. The load-carrying capacity is independent of bearing motion and lubricant viscosity. A main advantage of these bearings is the absence of wear at startup as there is with a slider bearing.

8.7.5 Elastohydrodynamic Lubrication

Elastohydrodynamic lubrication (EHL) is a form of hydrodynamic lubrication where elastic deformation of the surfaces becomes significant. Often, EHL is called a "condition", and not a "regime" of lubrication, although it is included here because of its importance to machine elements. Historically, EHL may be viewed as one of the major developments in machine design theory of the 20th century. An increased understanding of EHL has led to dramatic performance improvements in machinery of all types. The important features in a hydrodynamically lubricated slider bearing, namely converging film thickness, sliding motion, and a viscous fluid between the surfaces, are also important for EHL. EHL is normally associated with nonconformal surfaces and fluid film lubrication. There are two distinct forms: hard and soft EHL.

Hard EHL relates to materials of high elastic modulus, such as metals and ceramics. In this form of lubrication, elastic deformation and pressure-viscosity effects are both important. Figure 8.19 gives the characteristics of hard elastohydrodynamically lubricated conjunctions. The maximum pressure is typically between 0.5 and 4 GPa; the minimum film thickness normally exceeds 0.1 μm. These conditions are dramatically different from those found in a hydrodynamically lubricated conjunction summarized in Fig. 8.17. At loads normally experienced in nonconformal machine elements, the elastic deformations are two orders of magnitude larger than the minimum film thickness. Furthermore, the lubricant viscosity can vary by as much as 20 orders of magnitude within the lubricated conjunction. The minimum film thickness is a function of the same parameters as for hydrodynamic lubrication (Fig. 8.17) but with the additions of the effective elastic modulus, E', and the pressure-viscosity coefficient, ξ, of the lubricant. Engineering applications where hard EHL is important include gears (Ch. 14 and 15) and rolling-element bearings (Ch. 13).

Soft EHL relates to materials of low elastic modulus, such as rubber and polymers. Figure 8.20 summarizes the characteristics of soft EHL. In soft EHL, the elastic distortions are large, even with light loads. The maximum pressure for soft EHL is 0.5 to 4 MPa (typically 1 MPa), in contrast to 0.5 to 4 GPa (typically 1 GPa) for hard EHL. This low pressure has a negligible effect on the viscosity variation throughout the conjunction. The minimum film thickness is a function of the same parameters as in hydrodynamic lubrication with the addition of the effective elastic modulus. The minimum film thickness for soft EHL is typically 1 μm. Engineering applications where EHL is important for low-elastic-modulus materials include seals, natural human joints, and a number of lubricated machine elements that use rubber as a material. The common features of hard and soft EHL are that the local elastic deformation of the solids provides coherent fluid

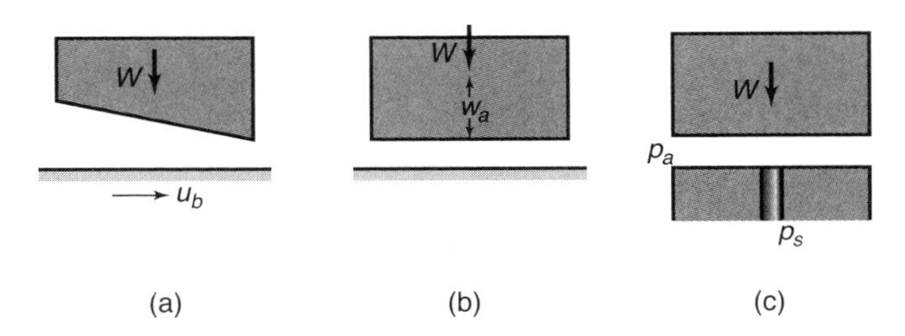

Figure 8.18: Mechanisms of pressure development for hydrodynamic lubrication. (a) Slider bearing; (b) squeeze film bearing; (c) externally pressurized bearing. *Source:* Hamrock et al. [2004].

Nonconformal surfaces
High-elastic-modulus material.
(e.g., steel)
$p_{max} \sim 0.5$ to 4 GPa
$h_{min} = f(W, u_b, \eta_0, R_x, R_y, E', \xi) > 0.1$ μm
Elastic & viscous effects both important

Figure 8.19: Characteristics of hard EHL.

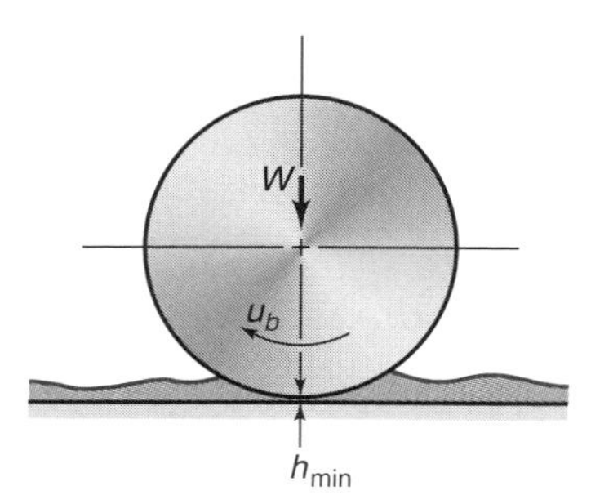

Nonconformal surfaces (e.g., rubber)
$p_{max} \sim 0.5$ to 4 MPa
$h_{min} = f(W, u_b, \eta_0, R_x, R_y, E') \sim 1$ μm
Elastic effects predominate

Figure 8.20: Characteristics of soft EHL.

films and that asperity interaction is largely prevented. Lack of asperity interaction implies that the frictional resistance to motion is due only to lubricant shearing and is therefore relatively low.

8.8 Friction

Friction is the force resisting relative movement between surfaces in contact. The two main classes of friction are sliding and rolling, as shown in Fig. 8.21. Whereas sliding surfaces are conformal, rolling friction involves nonconformal surfaces. However, most rolling contacts do experience some sliding.

In both rolling and sliding contacts, a tangential force, F, in the direction of motion is needed to move the upper body over the stationary lower body. The ratio between the tangential force and the normal applied load, W, is known as the **coefficient of friction**, μ:

$$\mu = \frac{F}{W}. \tag{8.29}$$

Rolling and sliding friction varies from 0.001 in lightly loaded rolling-element bearings to greater than 10 for clean metals sliding against themselves in vacuum. For most common materials sliding in air, the value of μ lies in a narrow

Table 8.9: Typical coefficients of friction for combinations of unlubricated metals in air.

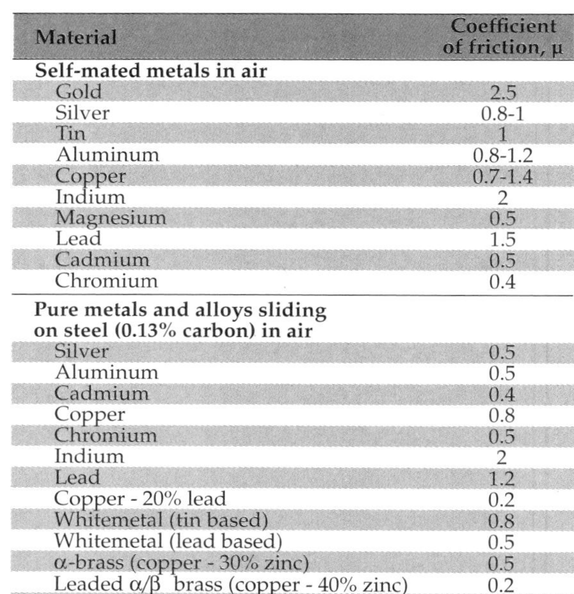
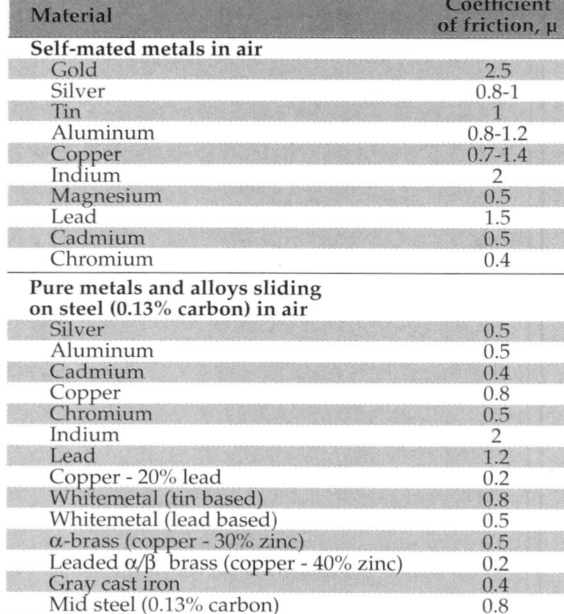

Material	Coefficient of friction, μ
Self-mated metals in air	
Gold	2.5
Silver	0.8-1
Tin	1
Aluminum	0.8-1.2
Copper	0.7-1.4
Indium	2
Magnesium	0.5
Lead	1.5
Cadmium	0.5
Chromium	0.4
Pure metals and alloys sliding on steel (0.13% carbon) in air	
Silver	0.5
Aluminum	0.5
Cadmium	0.4
Copper	0.8
Chromium	0.5
Indium	2
Lead	1.2
Copper - 20% lead	0.2
Whitemetal (tin based)	0.8
Whitemetal (lead based)	0.5
α-brass (copper - 30% zinc)	0.5
Leaded α/β brass (copper - 40% zinc)	0.2
Gray cast iron	0.4
Mid steel (0.13% carbon)	0.8

range from approximately 0.1 to 2.0. Table 8.9 gives typical coefficients of sliding friction for unlubricated metals in air; note that the range is 0.2 to 2.5, one order of magnitude.

8.8.1 Low Friction

In a number of situations low friction is desirable; for example,

1. Turbines and generators in electric or hydroelectric power stations, which use oil-lubricated hydrodynamic bearings

2. Gyroscopes, spinning at very high speeds, which use gas-lubricated hydrodynamic bearings

Friction can be reduced by using special low-friction materials, by lubricating the surface if it is not already being done, or by using clever designs that convert sliding motion into rolling. Reducing the wear and heat produced by friction are important factors in extending machine element life.

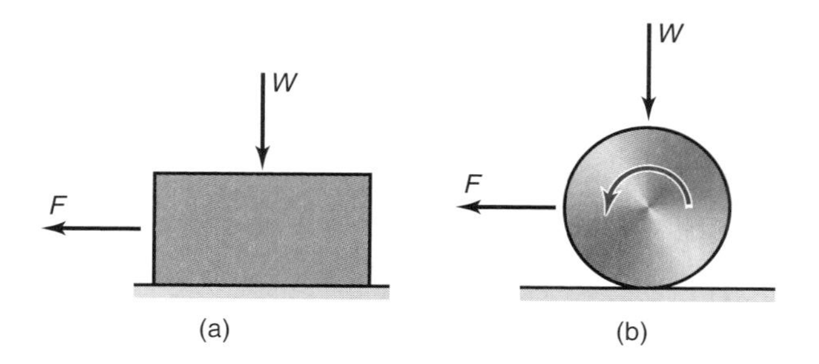

Figure 8.21: Friction force in (a) sliding and (b) rolling.

8.8.2 High Friction

In some situations, high friction is desirable, for example:

1. Brakes
2. Interaction between a shoe and a walking surface
3. Interaction between tire and road
4. Interaction between a nail and the wood into which it is hammered
5. Grip between a nut and a bolt

It must be emphasized that in these applications the friction must be high but controlled.

8.8.3 Laws of Dry Friction

The three **laws of friction** may be stated as follows:

1. The friction force is proportional to the normal load, as given by Eq. (8.29).
2. The friction force is not dependent on the apparent area of the contacting solids; that is, it is independent of the size of the solid bodies.
3. The friction force is independent of the sliding velocity.

These laws are applicable for most sliding conditions in the absence of a lubricant. The first two laws are normally called **Amonton's law**; he rediscovered them in 1699, with Leonardo da Vinci usually given the credit for discovering them some 200 years earlier.

The first two laws are found today to be generally satisfied for metals but are violated when polymers are the solid materials in contact. The third law of friction is less well founded than the first two laws. The friction force needed to begin sliding is usually greater than that necessary to maintain it. However, once sliding is established, the coefficient of friction is often nearly independent of sliding velocity.

A phenomenon important to machine element design is **stick-slip**. When stick-slip occurs, the friction force fluctuates between two extreme values. All stick-slip processes are caused by the fact that the friction force does not remain constant as a function of some other variable, such as distance, time, or velocity.

8.8.4 Sliding Friction of Metals

Adhesive Friction

Bowden and Tabor [1973] recognized that surfaces in contact touch only at points of asperity interaction, and that very high stresses induced in such small areas would lead to local plastic deformation. The penetration of an asperity into the opposing surface can be likened to a miniature hardness test, and the mean normal stress over the real areas of asperity contact can be represented by the hardness, H, of the softer material. Likewise, if τ represents the shear stress of the asperity junctions, the normal applied load is W, and the friction force is F, then the coefficient of friction μ can be expressed as

$$\mu = \frac{F}{W} = \frac{A_r \tau}{A_r H} = \frac{\tau}{H}. \tag{8.30}$$

This expression is an important step in understanding friction, although it is incomplete because it neglects the more complex nature of asperity interactions and deformations and accounts only for the adhesive element of friction. These limitations are apparent when it is recognized that for metals $\tau \approx 0.5S_y$ and $H \approx 3S_y$, where S_y is the yield strength. Thus, from Eq. (8.30), all clean metals should have a coefficient of friction of 1/6, which is not representative of experimental findings.

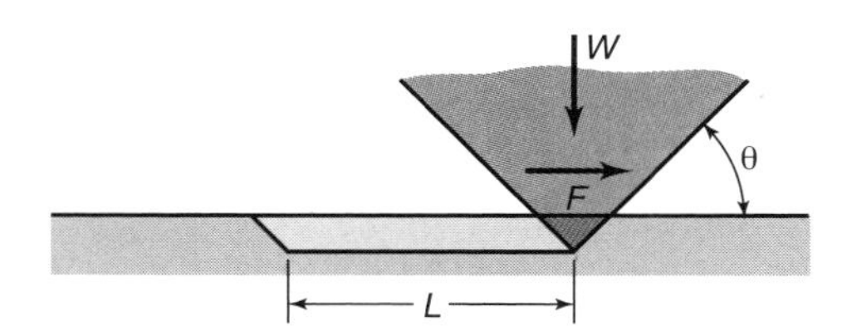

Figure 8.22: Conical asperity having mean angle θ plowing through a softer material. Also simulates abrasive wear.

Abrasive Friction

Another approach to friction is that a force would be required to move hard asperities through or over another surface and that the resulting microcutting motion represents a friction process. If the sum of the projected areas of the indenting asperities perpendicular to the sliding direction is A_o, and if the mean stress resisting plastic deformation of the softer material that is being cut is equal to the hardness, then the friction force, F, can be expressed as

$$F = A_o H. \tag{8.31}$$

The normal applied load, W, carried by a number of asperities can be expressed as

$$W = AH,$$

where A is the contact area. The coefficient of friction then becomes

$$\mu = \frac{F}{W} = \frac{A_o}{A}. \tag{8.32}$$

Figure 8.22 shows a conical asperity having a mean angle θ. The coefficient of friction for this geometry is

$$\mu = \frac{A_o}{A} = \frac{2}{\pi}\tan\theta. \tag{8.33}$$

Note that only the front end of the conical asperity shown in Fig. 8.22 is in contact with the opposing surface.

Combined Adhesion and Abrasion

If both molecular (adhesive) and deformation (plowing) mechanisms occur, the coefficient of friction is a function of both, or

$$\mu = \frac{\tau}{H} + \frac{2}{\pi}\tan\theta. \tag{8.34}$$

Experiments have demonstrated that the adhesion term [first term on the right side of Eq. (8.34)] plays a major role in determining the friction between metals. The abrasive term is dominant when there is a significant hardness mismatch between the materials in contact. In any case, both adhesive and abrasive contributions to friction are always present, and their isolation is difficult, so that a single friction coefficient is usually reported for a material pair.

8.8.5 Sliding Friction of Polymers

The coefficient of friction is usually lower for machine elements made of polymers, such as polyethylene, acrylics, polystyrene, and nylon, than for those made of metals. For clean copper on steel, the coefficient of friction is 1.0, but for most plastics it is typically 0.3 or so. Coefficients of friction in sliding contact, as measured experimentally, vary from as low as 0.02 to 100 or even higher. This is not surprising, in view of the many variables involved in friction.

One major difference between the frictional behavior of polymers and metals is the effect of load and geometry. Geometry refers to the shape of the surfaces, whether they are flat or curved and, if they are curved, how sharp the curvature is. With metals, the area of *true* contact is determined only by the load and the yield pressure and not by the shape of the surfaces. Polymers deform viscoelastically, implying that deformation depends on the normal applied load, W, the geometry, and the time of loading. For a fixed time of loading and fixed geometry, for example, a sphere on a flat surface, the area of true contact is not proportional to the load as it is with metals, but is proportional to W^n, where n is less than 1 and usually near 3/4. Therefore,

$$F \propto W^{3/4} \qquad \text{for polymers} \tag{8.35}$$

$$F \propto W \qquad \text{for metals} \tag{8.36}$$

and the difference in friction behavior is quite significant.

8.8.6 Sliding Friction of Rubber

Rubber is an extreme example of an elastic material that can achieve large deformations. Regardless of the amount that rubber becomes distorted, it will return to its original shape when the deforming force is removed, as long as the rubber is not torn or cut. Not surprisingly, the friction of rubber often deviates significantly from the laws of friction developed for metals. The friction depends on the load and the geometry of the contacting surfaces.

For rubber, the relationship between friction force and the normal applied load is

$$F \propto W^{2/3}. \tag{8.37}$$

Rubber is truly an elastic solid and this relationship is valid over a wide load range.

8.9 Wear

Wear has long been recognized as very important and is usually considered detrimental in machine elements. In a more general sense, practically everything wears out, yet the fundamental actions that govern the process remain elusive. In fact, it remains true that today's engineer is better equipped to design a machine element to withstand known loads than to predict a given life of the machine element.

Wear is the progressive loss of material from the operating surface of a body occurring as a result of loading and relative motion at the surface. Wear can be classified by the physical nature of the underlying process, such as abrasion, adhesion, and fatigue. These three main types of wear are considered in this section. Other types of wear, such as erosion, fretting, and corrosion, are beyond the scope of this text.

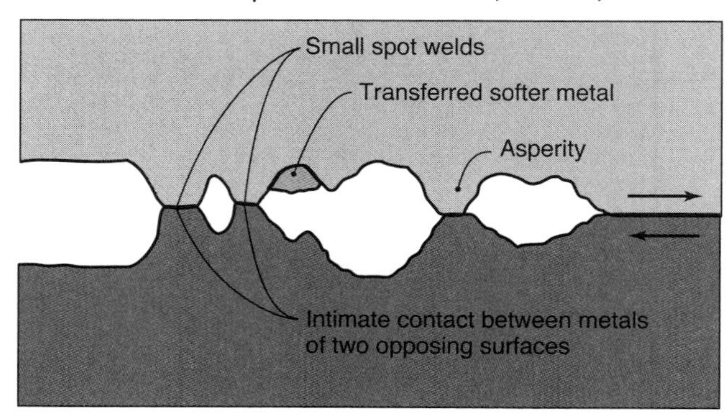

Figure 8.23: Adhesive wear model.

8.9.1 Adhesive Wear

Figure 8.23 shows the mechanism of **adhesive wear**, where material is transferred from one surface to the other by solid-phase welding. Adhesive wear is the most common type of wear and the least preventable. Observe from Fig. 8.23 that, as the moving asperities pass each other, the microscopic weld breaks and, unless the fracture occurs on the original interface, material is removed from one surface or the other. Usually, the surface having the lower strength loses material.

The volume of material removed by wear, v, is directly proportional to the sliding distance, L, and the normal applied load, W, and is inversely proportional to the hardness, H, of the softer of the two materials, or

$$v \propto \frac{WL}{H}. \tag{8.38}$$

After a dimensionless adhesive wear coefficient, k_1, is introduced, Eq. (8.38) becomes

$$v = k_1 \frac{WL}{3H}. \tag{8.39}$$

Equation (8.39) is known as the **Archard wear law**. The dimensionless adhesive wear coefficient, k_1, can be interpreted as a measure of the probability that any interaction between asperities of the two surfaces in contact will produce a wear particle due to any of the wear mechanisms. Thus, if k_1 is 1, every junction involving surface contact will produce a wear fragment; if $k_1 = 0.1$, one-tenth of the contacts between surfaces will produce a wear fragment. Studies have shown that, for unlubricated surfaces, the lowest value of k_1 is obtained for polyethylene sliding on steel. For this situation $k_1 = 1 \times 10^{-7}$, which can be interpreted as 1 in 10 million contacts will produce a wear fragment.

Table 8.10 shows some typical values of k_1 for several materials in contact. As presented in Eq. (8.39), the wear volume is a function not only of k_1 but also of the hardness of the softer of the two materials in contact. For example, although the *wear coefficient* for polyethylene on steel is one-tenth that for tungsten carbide sliding on itself, the *wear volume* is 10 times larger because tungsten carbide wears 10 times less than the polymer. Therefore, the volume of the wear fragments is considerably smaller because tungsten is several orders of magnitude harder than polyethylene.

Table 8.10 also gives the coefficient of friction. The wear constant varies by eight orders of magnitude for the various rubbing materials, but the coefficient of friction varies by only one order of magnitude. To understand why the wear constant varies so much more than the friction coefficient, consider the case of steel sliding against steel. At low loads and

Table 8.10: Coefficients of rubbing friction and adhesive wear constant for nine rubbing materials.

Rubbing materials	Coefficient of friction, μ	Adhesive wear coefficient, k_1
Gold on gold	2.5	0.1-1
Copper on copper	1.2	0.01-0.1
Mild steel on mild steel	0.6	10^{-2}
Brass on hard steel	0.3	10^{-3}
Lead on steel	0.2	2×10^{-5}
Polytetrafluoroethylene (teflon) on steel	0.2	2×10^{-5}
Stainless steel on hard steel	0.5	2×10^{-5}
Tungsten carbide on tungsten carbide	0.35	10^{-6}
Polyethylene on steel	0.5	5×10^{-8}

speeds, wear is low, and contact occurs at oxide-covered asperities. The oxides prevent gross adhesion, and the wear particles are generally observed to be small black particles of iron oxide. At high loads and speeds, the oxide is penetrated or broken, and asperity contact is between the substrate materials. The microwelds at asperity surfaces are much stronger in this case, and this is reflected in the wear particles, which are typically larger and consist of metal flakes. As discussed above, it is very important that wear tests duplicate a machine element's working environment as much as is practical to obtain valuable data.

Example 8.9: Application of the Archard Wear Law

Given: A journal bearing in a dam gate moves slowly and operates under a high load. The entire bearing is made of AISI 1040 steel. To increase the bearing life, a material change is considered for one of the surfaces.

Find: Which of the following materials will give the longest life: brass, lead, polytetrafluoroethylene (PTFE), or polyethylene? The hardness for brass is 225 MPa, for lead it is 30 MPa, for PTFE it is 50 MPa, and for polyethylene it is 70 MPa.

Solution: From Eq. (8.39) the wear volume is

$$v = k_1 \frac{WL}{3H}.$$

The wear volume is a minimum when H/k_1 is a maximum. Making use of Table 8.10 to obtain the wear coefficient for the materials being considered gives

Brass:

$$\frac{H}{k_1} = \frac{225 \times 10^6}{10^{-3}} = 225 \text{ GPa}.$$

Lead:

$$\frac{H}{k_1} = \frac{30 \times 10^6}{(2 \times 10^{-5})} = 1500 \text{ GPa}.$$

PTFE:

$$\frac{H}{k_1} = \frac{50 \times 10^6}{(2 \times 10^{-5})} = 2500 \text{ GPa}.$$

Polyethylene:

$$\frac{H}{k_1} = \frac{70 \times 10^6}{5 \times 10^{-8}} = 1,400,000 \text{ GPa}.$$

Thus, polyethylene will give much longer life than any of the other materials.

8.9.2 Abrasive Wear

Abrasive wear arises when two interacting surfaces are in direct physical contact and one is significantly harder than the other. Under a normal load the asperities of the harder surface penetrate into the softer surface, producing plastic deformation. Figure 8.22 shows a simple abrasive wear model where a hard conical asperity with slope θ under a normal load, W, plows through the softer surface, removing material and producing a groove. The amount of material lost by abrasive wear is

$$v \approx \frac{2k_1}{\pi}\frac{\tan\theta}{H}WL = \frac{k_1 k_2 WL}{H}, \tag{8.40}$$

where k_1 is an adhesive wear coefficient, given in Table 8.10, and k_2 is an abrasive wear coefficient, given by

$$k_2 = \frac{2\tan\theta}{\pi}. \tag{8.41}$$

For most asperities θ is small, typically between 5° and 10°.

A similarity of form is evident between Eqs. (8.39) and (8.40), and it is clear that the general **laws of wear** can be stated as follows:

1. Wear increases with sliding distance.
2. Wear increases with normal applied load.
3. Wear decreases as the hardness of the sliding surface increases.

These laws, along with Eqs. (8.39) and (8.40), reveal the role of the principal variables in abrasive wear, but there are sufficient exceptions to the predicted behavior to justify a measure of caution in their use. Furthermore, wear rates, or the life of rubbing components, cannot be determined without some knowledge of the wear coefficients, which are experimentally determined.

Adhesive and abrasive wear are always present, and are difficult to isolate. For that reason, Eq. (8.40) is simplified to Eq. (8.39) and is used to quantify the combined contributions of both wear modes. Obviously, wear tests need to reproduce the application's conditions as much as is practical to obtain reasonable wear rates. The data presented in the tables in this chapter are good estimates of friction and wear coefficients encountered by machine elements, but critical designs need experimental verification.

Design Procedure 8.1: Wear Avoidance Hierarchy

Wear is a phenomenon that commonly arises in machine elements. It should be noted that wear is not always detrimental - a running-in period, for example, consists of controlled wear that can make surfaces smoother and more conformal, and transfer loads more efficiently. Regardless, when one encounters excessive wear, the following is a reasonable hierarchy of modifications to investigate, ordered roughly in terms of decreasing cost effectiveness.

1. If a lubricant is not being used, use a lubricant if possible.

2. Follow the wear laws; that is, reduce load, decrease the sliding distance, or increase the hardness of the softer surface.

3. Substitute materials that have less propensity for adhesive wear. Consider the use of hard, chemically inert coatings that can reduce wear.

4. If wear is unavoidable, design the system so that wear does not occur on expensive and difficult-to-replace components. Instead, use pads, spacers, or other elements that wear preferentially and can be replaced at lower cost, such as brake pads bearing against a rotor (see Section 18.3).

5. If a lubricant is being used:

   (a) Increase the film thickness by using a lubricant with a higher viscosity and/or viscosity pressure coefficient when possible.

   (b) Make the surfaces smoother so that the film parameter is increased and more load is carried by the lubricant than asperities.

   (c) If the application encounters boundary lubrication, make certain the lubricant contains effective chemical additives, but that their concentration is not so high as to cause corrosion.

   (d) Make certain that excessive temperatures are not encountered; use chillers or heat exchangers to maintain lubricant temperature at reasonable levels.

8.9.3 Fatigue Wear

For nonconformal contacts in such machine elements as rolling-element bearings, gears, friction drives, cams, and tappets, a prevalent form of failure is **fatigue wear**. Fatigue wear will form the basis of mathematical models that describe the useful life of rolling element bearings (see Section 13.9) and gears (see Section 14.12), and will be discussed in greater detail later. In this section, a brief introduction to fatigue wear mechanisms is given.

In **fatigue wear**, the removal of material results from a complicated process driven by a cyclic load. Figure 8.24 illustrates the stresses developed on and below the surface that deform and weaken the metal. Cyclic loading causes defects or cracks to develop below the surface. Eventually, the defects coalesce near the surface. Material at the surface then *spalls*, or is broken away, degrading the component's surface and releasing work-hardened particles that can accelerate abrasive wear. Fatigue wear occurs in nonconformal machine elements, even in well-lubricated situations.

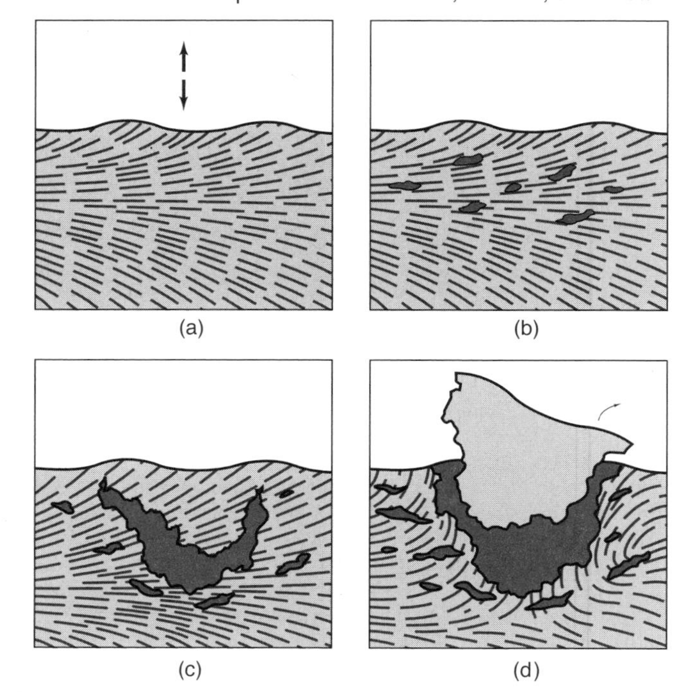

Figure 8.24: Fatigue wear simulation. (a) Machine element surface is subjected to cyclic loading; (b) defects and cracks develop near the surface; (c) the cracks grow and coalesce, eventually extending to the surface until (d) a wear particle is produced, leaving a fatigue spall in the material.

A group of apparently identical nonconformal machine elements subjected to identical loads, speeds, lubrication, and environmental conditions may exhibit wide variation in failure times. For this reason the fatigue wear process must be treated statistically. Thus, the fatigue life of a bearing is normally defined in terms of its statistical ability to survive a certain period of time or by the allowable wear before the component ceases to function adequately. The only requirement for fatigue failure is that the surface material should be loaded. The other wear mechanisms require not only loading but also physical contact between the surfaces. If the surfaces are separated by a lubricant film, adhesive and abrasive wear are virtually eliminated, but fatigue wear can still occur.

Case Study 8.1: Wear in Orthopedic Implants

Total joint arthroplasty has been an extremely successful procedure over the past five decades, and has seen continued technological advances and performance improvements. Examples of very successful implant designs are the total hip replacement (THR) and total knee replacement (TKR) depicted in Fig. 8.25. A wide variety of materials and designs have been used, and designs have evolved and have been discarded as new approaches have come to the fore.

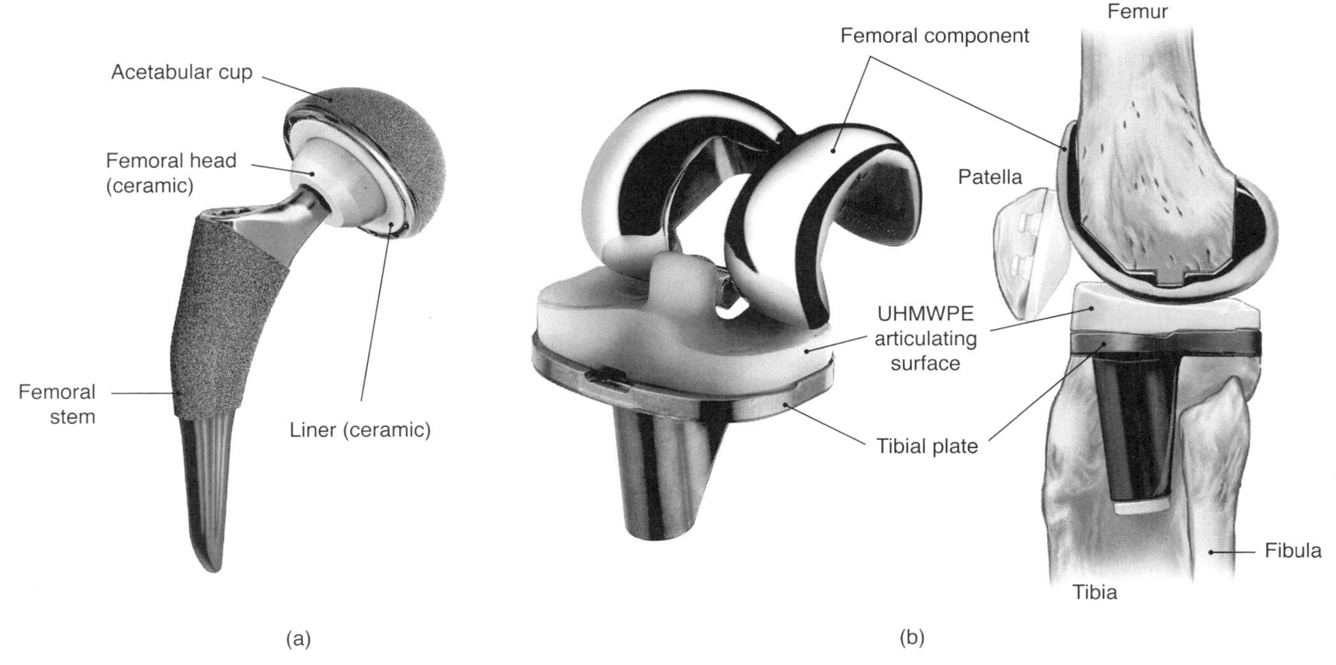

Figure 8.25: Examples of common orthopedic implants. (a) Total hip replacement, using a metal-on-metal interface. *Source:* Courtesy of DePuy Synthes Joint Reconstruction; (b) Total knee replacement using a metal-on-polymer interface. *Source:* Courtesy of Zimmer, Inc.

Selection of materials for orthopedic articulating surfaces is a difficult undertaking. In addition to the need for long life without the need of service, and the absence of effective lubrication, it is essential that the materials be biocompatible.

A number of material combinations satisfy these requirements and are used in artificial joints, with the most common involving metal (usually a cobalt-chrome alloy) on polyethylene, or metal-on-metal. For total hip replacements, ceramic-on-ceramic designs are also available, which are thought to develop significantly higher film parameters (see Section 8.7) because the polished ceramic has a much lower roughness than metal or polymer components.

It should be noted that designs have improved dramatically in the 21st century, and joint replacements that used to last an average of around seven years now routinely record average lives of twenty years or more. Material improvements continue, so that future designs will certainly outperform current technologies.

Conventional Polymers

A large number of polymers were investigated for use in total joint replacements starting in the 1960s, with ultra-high molecular weight polyethylene (UHMWPE) proving to be most successful. This material became preferred because of its low wear rates (see Table 8.10). Not only does this have a direct impact on the life of the polymer components, but also patient comfort, since wear particles can cause inflammation, pain, and *osteolysis*, or a loss of bone. Conventional UHMWPE performed well, but researchers identified a number of concerns, including:

- The manufacture and sterilization of UHMWPE led to the formation of free radicals (unbonded ends of polymer chains). Such free radicals led to oxidation, embrittlement, and higher wear rates.
- Loads from common activities such as walking or climbing stairs causes an unsteady load on the implants. The high-load stance phase of walking occurs in one direction of sliding, and the low-load swing stage occurs over an arc of motion.

It was found that the stance phase of walking led to orientation of UHMWPE crystals in the surface and near-surface, so that the crystals aligned with the sliding direction. Sliding from the swing phase is oblique to this orientation, and the wear associated with the low load swing phase was surprisingly high.

Crosslinked Polyethylene

In the 1990s, it was found that the developments of cross-links in the polyethylene molecule can arrest crystal orientation and therefore significantly reduce the detrimental wear associated with the swing stage of walking. Technologies to develop large numbers of cross-links in UHMWPE were developed, resulting in dramatic increases in implant lives.

Cross-linked UHMWPE could not be introduced into orthopedic implants until approved by regulatory agencies such as the Food and Drug Administration in the United States. Approval of new materials can be a lengthy, involved, and expensive process, but for cross-linked UHMWPE the process required demonstration of benefits through wear testing in joint simulators. Joint simulators are elaborate devices that closely match biomechanical forces and sliding velocities. However, evaluation of a material requires between six and nine months of continuous testing. Since there were a number of manufacturing process variables that could affect the performance of cross-linked UHMWPE, accelerated wear tests were conducted using a pin-on-flat configuration.

These tests served to screen materials so that only the most favorable were tested in the expensive and time-consuming joint simulators. Wear coefficients measured in these experiments are used to estimate the life of implants from the Archard equation (see Section 8.9).

Avoidance of Oxidation

As mentioned above, free radicals produced during sterilization of the UHMWPE led to oxidation and associated embrittlement of the polymer. Once this failure mode was identified, it became necessary to modify implant storage procedures; by packaging and storing the implants in a vacuum, oxygen could not bond at the free radical sites and the UHMWPE did not encounter embrittlement. Note that inside the body, there are no free oxygen molecules in the implant environment, so that embrittlement after placement in the body is not a concern. Recent efforts have focused upon blending or infiltrating UHMWPE with antioxidants, notably vitamin E, to prevent oxidation at free radicals.

Summary

Material advances continue at a rapid pace, and implant designs that exploit new materials continue to be developed. For example, the low wear rates associated with new materials allow implant designers to produce smaller implants that are easier to place in the body. Further, larger radii can be pursued for total joint replacements so that the Hertz contact stresses are lower, dislocation of the implant is less likely, and larger lubricant films can be entrained. Such designs require high wear resistance in the interface materials because the sliding distance is larger. For example, in normal walking, the swing angle of the legs is fixed, so that the sliding distance encountered by a total hip replacement is directly proportional to the head radius. Larger heads result in proportionately larger sliding distances, as shown by Eq. (8.39). However, a decrease in the wear coefficient can still result in longer implant lives. Exciting designs are continually being introduced, but all require tribological sophistication.

8.10 Summary

In this chapter, conformal and nonconformal surfaces were defined. Conformal surfaces fit snugly into each other with a high degree of geometric conformity so that the load is carried over a relatively large area, and the area remains essentially constant as the load is increased. Nonconformal surfaces do not geometrically conform to each other and have small load-carrying areas that enlarge with increasing load but are still small relative to those of conformal surfaces.

A lubricant's physical and chemical actions within a lubricated conjunction were described for the four lubrication regimes: hydrodynamic, elastohydrodynamic, partial, and boundary. Hydrodynamic lubrication is characterized by conformal surfaces, where the lubricating film is thick enough to prevent the opposing solids from coming into contact. Three modes of pressure development within hydrodynamic lubrication were presented: slider, squeeze, and external pressurization.

Elastohydrodynamic lubrication (EHL) is characterized by nonconformal surfaces, without asperity contact of the solid surfaces. Two modes of EHL exist: hard and soft. Hard EHL typically occurs with metals, and soft EHL with elastomeric materials. The pressures developed in hard EHL are high (typically between 0.5 and 4 GPa), so that elastic deformation of the solid surfaces becomes important, as do the pressure-viscosity effects of the lubricant.

In boundary lubrication, considerable asperity contact occurs, and the lubrication mechanism is governed by the physical and chemical properties of thin surface films that are of molecular size (from 1 to 10 nm). Partial or mixed lubrication is governed by a mixture of boundary and fluid film effects. Most of the scientific unknowns lie in this lubrication regime.

Features of rolling and sliding friction were presented, as well as the laws of friction. The coefficient of friction was defined as the ratio of the friction force to the normal applied load. Sliding friction of metals, polymers, plastics, and rubber was also discussed.

The useful life of engineering machine components is limited by breakage, obsolescence, and *wear*. Wear is the progressive loss of substance from the operating surface of a body occurring as a result of loading and relative motion at the surface. The most common forms of wear were discussed:

1. In adhesive wear, material transfers from one surface to another due to solid-phase welding.
2. In abrasive wear, material is displaced by hard particles.
3. In fatigue wear, material is removed by cyclic stress variations even though the surfaces may not be in contact.

Key Words

abrasive wear wear caused by physical damage from penetration of a hard surface into a softer one

absolute viscosity shear stress divided by shear strain rate, having SI units of Pascal-second

adhesive wear wear caused by solid-state weld junctions of two surfaces

boundary lubrication lubrication condition where considerable asperity interaction occurs between solids and lubrication mechanism is governed by properties of thin surface films that are of molecular proportion

centerline average roughness a measure of roughness, denoted by R_a, defined by the mean distance of a surface from its mean line

conformal surfaces surfaces that fit snugly into each other with high degree of conformity as in journal bearings

elastohydrodynamic lubrication a lubrication condition where nonconformal surfaces are completely separated by lubricant film and no asperities are in contact

ellipticity parameter diameter of contact ellipse in y-direction divided by diameter of contact ellipse in x-direction

fatigue wear wear caused by propagation of subsurface damage to surface due to cyclic loading

film parameter minimum film thickness divided by composite surface roughness

fluid film lubrication lubrication condition where lubricated surfaces are completely separated by a lubricant film and no asperities are in contact

friction force resisting relative movement between surfaces in contact

hydrodynamic lubrication fluid film lubrication of conformal surfaces as in journal bearings

kinematic viscosity absolute viscosity divided by density, with SI units of meter squared per second

laws of friction These can be summarized as:

1. Friction force is proportional to normal load.
2. Friction force is not dependent on apparent area of contact solids; that is, it is independent of size of solid bodies.
3. Friction force is independent of sliding velocity.

laws of wear These can be summarized as:

1. Wear increases with sliding distance.
2. Wear increases with normal applied load.
3. Wear decreases as hardness of sliding surface increases.

lubricant any substance that reduces friction and wear and provides smooth running and satisfactory life for machine elements

mixed lubrication same as partial lubrication

nonconformal surfaces surfaces that do not conform to each other very well as in rolling-element bearings

partial lubrication lubrication condition where the load between two surfaces in contact is transmitted partially through lubricant film and partially through asperity contact

root-mean-square roughness a measure of roughness, denoted by R_q, defined by the standard deviation of a surface from its mean line

running-in process through which beneficial wear causes surfaces to adjust to each other and improve performance

tribology study of lubrication, friction, and wear of moving or stationary parts

wear progressive loss of substance from operating surface of body occurring as a result of loading and relative motion of surface

Summary of Equations

Surface Roughness:

$$R_a = \frac{1}{N}\sum_{i=1}^{N}|z_i|$$

$$R_q = \left(\frac{1}{N}\sum_{i=1}^{N} z_i^2\right)^{\frac{1}{2}}$$

Hertzian Contact:

General:

Effective elastic modulus: $E' = \dfrac{2}{\dfrac{(1-\nu_a^2)}{E_a} + \dfrac{(1-\nu_b^2)}{E_b}}$

Elliptical contact:

Curvature sum: $\dfrac{1}{R} = \dfrac{1}{R_x} + \dfrac{1}{R_y}$

Radius ratio: $\alpha_r = \dfrac{R_y}{R_x}$

Ellipticity parameter: $k_e = \dfrac{D_y}{D_x} = \alpha_r^{2/\pi}$

Maximum contact pressure: $p_{\max} = \dfrac{6W}{\pi D_x D_y}$

Contact dimensions: $D_y = 2\left(\dfrac{6k_e^2\mathcal{E}WR}{\pi E'}\right)^{\frac{1}{3}}$,

$$D_x = 2\left(\frac{6\mathcal{E}WR}{\pi k_e E'}\right)^{\frac{1}{3}}$$

Maximum elastic deformation:

$$\delta_{\max} = \mathcal{F}\left[\frac{9}{2\mathcal{E}R}\left(\frac{W}{\pi k_e E'}\right)^2\right]^{\frac{1}{3}}$$

Rectangular contact:

Dimensionless load: $W' = \dfrac{w'}{E'R_x}$

Contact semiwidth: $b^* = R_x\left(\dfrac{8W'}{\pi}\right)^{\frac{1}{2}}$

Maximum contact pressure: $p_{\max} = E'\left(\dfrac{W'}{2\pi}\right)^{\frac{1}{2}}$

Maximum elastic deformation:

$$\delta_{\max} = \frac{2W'R_x}{\pi}\left[\ln\left(\frac{2\pi}{W'}\right) - 1\right]$$

Lubricant Rheology:

Viscosity: $\eta = \dfrac{F/A}{u_b/h} = \dfrac{\text{Shear stress}}{\text{Shear strain rate}}$

Barus Law: $\eta = \eta_o \exp(\xi p)$

Viscosity of SAE oils: $\eta = C_1 \exp\left(\dfrac{C_2}{1.8t_c + 127}\right)$

Film Parameter: $\Lambda = \dfrac{h_{\min}}{\left(R_{qa}^2 + R_{qb}^2\right)^{\frac{1}{2}}}$

Coefficient of Friction: $\mu = \dfrac{F}{W}$

Archard wear law: $v = k_1\dfrac{WL}{3H}$

Recommended Readings

Bhushan, B. (2002) *Introduction to Tribology*. John Wiley & Sons.

Bhushan, B., ed. (2001) *Modern Tribology Handbook*. CRC Press.

Dowson, D. (1998) *History of Tribology*, 2nd ed., Professional Engineering Publishing.

Hamrock, B.J., Schmid, S.R., and Jacobson, B.O. (2004) *Fundamentals of Fluid Film Lubrication*, 2nd ed., Marcel-Dekker.

Johnson, K.L. (1985) *Contact Mechanics*, Cambridge University Press.

Ludema, K. (1996) *Friction, Wear, Lubrication: A Textbook in Tribology*, CRC Press.

Szeri, A.Z. (2011) *Fluid Film Lubrication*, 2nd ed, Cambridge University Press.

Stolarski, T.A. (2000) *Tribology in Machine Design*. Butterworth Heineman.

Williams, J. (2005) *Engineering Tribology*. Cambridge University Press.

References

Bowden, F.P., and Tabor, D. (1973) *Friction–An Introduction to Tribology*, Anchor Press.

Hamrock, B.J., and Brewe, D.E. (1983) "Simplified Solution

for Stresses and Deformations," *J. Lub. Technol.*, Vol. 105, No. 2, pp. 171–177.

Hamrock, B.J., Schmid, S.R., and Jacobson, B.O. (2004) *Fundamentals of Fluid Film Lubrication*, 2nd ed., Marcel-Dekker.

Hertz, H. (1881) "The Contact of Elastic Solids," *J. Reine Angew. Math.*, vol. 92, pp. 156–171.

Jost, H.P., *Committee on Tribology Report*, Ministry of Technology, 1966.

Kalpakjian, S., and Schmid, S.R. (2010) *Manufacturing Engineering and Technology*, 6th ed., Pearson.

Seirig, A.S., and Dandage, S. (1982) "Empirical Design Procedure for the Thermodynamic Behavior of Journal Bearings," *J. Lubr. Tech.*, v. 104, pp. 135–148.

Questions

8.1 What is tribology?

8.2 Describe the difference between conformal and nonconformal surfaces.

8.3 What is an elliptical contact?

8.4 What is the difference between rectangular contact and line contact?

8.5 Under what conditions is the ellipticity ratio less than one?

8.6 What are the Hertz Approximations?

8.7 List the regimes of lubrication.

8.8 What is a babbitt?

8.9 What is elastohydrodynamic lubrication? How is it different from hydrodynamic lubrication?

8.10 List the differences between hard and soft EHL.

8.11 In boundary lubrication, what is the value of the film parameter?

8.12 What are the two measures of surface roughness?

8.13 Explain why running-in is useful.

8.14 What is the coefficient of friction?

8.15 What are the laws of friction?

8.16 What is adhesive wear?

8.17 What is the wear coefficient?

8.18 How does spalling occur?

8.19 What polymers are useful for sleeve bearings?

8.20 Define the term "wear."

Qualitative Problems

8.21 Give five examples of Hertzian contact.

8.22 Give five examples of elastohydrodynamic lubrication.

8.23 Without using equations, define *viscosity*.

8.24 Give three examples where (a) friction and (b) wear are useful. Explain your answers.

8.25 List the similarities and differences between abrasive and adhesive wear.

8.26 Explain why wear tests can have large variations in results.

8.27 Can the film thickness ever become negative? Explain your answer.

8.28 Estimate the wear rate for a pencil on paper.

8.29 When you write on paper with a ball point pen, a thick ink film is deposited on the paper regardless of the force used in writing. Explain why.

8.30 Determine which of the following contact geometries is conformal and which is nonconformal:

(a) Meshing gear teeth

(b) Ball and inner race of a ball bearing

(c) Journal bearing

(d) Railway wheel and rail contact

(e) Car making contact with the road

(f) Egg and egg cup

(g) Human knee

8.31 Describe three applications for each of the four lubrication regimes: hydrodynamic lubrication, elastohydrodynamic lubrication, boundary lubrication, and partial lubrication.

8.32 Plot the friction coefficient versus the adhesive wear coefficient using the data in Table 8.10. Comment on the results.

8.33 Show that the contact patch between two equal cylinders pressed together with their axes at right angles is approximately circular.

8.34 List the requirements of a lubricant.

8.35 It is observed that the coefficient of friction between a pad and a guideway is 0.35. To reduce friction and wet the surfaces, kerosene (a very low viscosity fluid) is applied to the interface. Instead of reducing the friction, it is now measured to be 0.38. Provide an explanation for these measurements.

Quantitative Problems

8.36 A single ball rolling in a groove has a 10-mm diameter and a 10-N normal force acting on it. The ball and the groove have a 207 GPa modulus of elasticity and a Poisson's ratio of 0.3. Assuming a 6.08-mm-radius groove in a semi-infinite steel block, determine the following:

(a) Contact zone dimensions. *Ans.* $D_y = 0.3056$ mm, $D_x = 0.1016$ mm.

(b) Maximum elastic deformation. *Ans.* $\delta_{\max} = 0.7045\ \mu$m.

(c) Maximum pressure. *Ans.* 0.6163 GPa.

8.37 A solid cylinder rolls with a load against the inside of an outer cylindrical race. The solid cylinder radius is 20 mm and the race internal radius is 150 mm. The race and the roller have the same axial length. What is the radius of a geometrically equivalent cylinder near a plane? The cylinder is made of silicon nitride ($E = 314$ GPa, $\nu = 0.26$) and the race is made of stainless steel ($E = 193$ GPa, $\nu = 0.30$). If the normal applied load per unit width is 10,000 N/m, determine

(a) Contact semiwidth *Ans.* $b^* = 46.48\ \mu$m.

(b) Maximum surface stress *Ans.* $p_{\max} = 137.0$ MPa.

(c) Maximum elastic deflection *Ans.* $\delta_{\max} = 0.3321\ \mu$m.

Also, indicate what these values would be if the silicon nitride cylinder were replaced with a stainless steel cylinder.

8.38 A 100-mm diameter shaft has a 20-mm diameter ball rolling around the outside. Find the maximum contact stress, the maximum deflection, and the contact dimensions if the ball load is 500 N. The ball is made of silicon nitride ($E = 314$ GPa, $\nu = 0.26$), and the shaft is made of steel ($E = 206$ GPa, $\nu = 0.30$). Also, determine these values if both ball and shaft are made of steel. *Ans.* For SiN on steel, $D_y = 6.1911 \times 10^{-4}$ m, $D_x = 5.513 \times 10^{-4}$ m, $p_{\max} = 2.798$ GPa.

8.39 The ball-outer-race contact of a ball bearing has a 17-mm ball diameter, an 8.84-mm outer-race groove radius, and a 44.52-mm radius from the bearing axis to the bottom of the groove. The load on the most highly loaded ball is 10,000 N. Calculate the dimensions of the contact ellipse and the maximum deformation at the center of the contact. The race and the ball are made of steel. *Ans.* $D_y = 6.935$ mm, $D_x = 0.9973$ mm.

8.40 A hydrodynamic journal bearing is loaded with a normal load W and is rotating with a surface velocity u_b. Find how much higher rotational speed the bearing needs to maintain the same minimum film thickness if the load W is doubled.

8.41 A hydrodynamic slider bearing lubricated with a mineral oil runs at 4000 rpm. Find how much thinner the oil film will be if the load is increased by a factor of 3. How much must the speed be increased to compensate for the higher load while keeping the oil film thickness constant?

8.42 A hydrodynamic bearing operates with a film parameter of 6. Calculate how much lower the speed has to be to decrease the film parameter to 3.

8.43 A machined steel surface has roughness peaks and valleys. The roughness wavelength is such that 400 peaks and 400 valleys are found during one roughness measurement. Find how much the arithmetic average surface roughness $R_a = 0.2\ \mu$m will change if one of the surface roughness peaks and one of the valleys increases from 0.25 to 2.5 μm. *Ans.* 0.2056 μm.

8.44 A surface with a triangular sawtooth roughness pattern has a peak-to-valley height of 4 μm. Find the R_a and R_q values. *Ans.* $R_a = 1.00\ \mu$m.

8.45 A precision ball bearing has a race root-mean-square surface roughness of 0.07 μm and a ball root-mean-square surface roughness of 0.02 μm. Changing the roughness of the components may give a higher Λ value. Determine which components to smooth if it costs equally as much to halve the roughness of the race as it costs to halve the roughness of the balls. Note that $h_{\min} = 0.2\ \mu$m.

8.46 The minimum film thickness in a particular application is 10 μm. Assume that two surface roughnesses of the two ground surfaces being lubricated are identical. What lubrication regime would you expect the application to be operating in? Also, what is the maximum surface roughness allowed to achieve hydrodynamic lubrication? What manufacturing operation would you recommend to achieve this roughness?

8.47 Two equally rough surfaces with rms surface roughness R_q are lubricated and loaded together. The oil film thickness is such that $\Lambda = 3$. Find the film parameter if one of the surfaces is polished so that $R_q \rightarrow 0$ (i.e., the surface becomes absolutely smooth). *Ans.* $\Lambda = 2.83$.

8.48 Two lubricated surfaces have rms surface roughnesses R_q of 0.23 and 0.04 μm, respectively. By using a new honing machine either surface can be made twice as smooth. For good lubrication, which surface roughness is it most important to decrease?

8.49 The absolute viscosity of a fluid at atmospheric conditions is 5.00×10^{-3} kgf-s/m^2. Give this absolute viscosity in

(a) Reyn. *Ans.* 7.110 μreyn.

(b) Poise. *Ans.* 0.4904 P.

(c) Pound force-seconds per square inch.

(d) Newton-seconds per square meter.

8.50 Given a fluid with an absolute viscosity η between two 1-m^2 surfaces spaced 1 mm apart, find how fast the surfaces will move relative to each other if a 10-N force is applied in the direction of the surfaces when η is

(a) 0.001 N·s/m^2 (water). *Ans.* u_b=10 m/s.

(b) 0.100 N·s/m^2 (a thin oil at room temperature).

(c) 10.0 N·s/m^2 (syrup; cold oil). *Ans.* $u_b = 1$ mm/s.

(d) 10^8 N·s/m^2 (asphalt).

8.51 A polymer box stands on a slope. The box weighs 10 kg and the slope angle is 25°. To pull the box down the slope, a force $F = 5$ N is required. The friction force is proportional to the load raised to 0.75. Find the additional weight needed in the box to make it slide down the slope. *Ans.* 5.60 kg.

8.52 A moose suddenly jumps out onto a dry asphalt road 80 m in front of a car running at 108 km/hr. The maximum coefficient of friction between the rubber tires and the road is 1.0 when they just start to slide and 0.8 when the locked wheels slide along the road. It takes 1 s for the driver to apply the brakes after the moose jumps onto the road. Find whether the car can stop in time before it hits the moose

(a) With locked wheels sliding along the road.

(b) With an automatic braking system (ABS) that keeps the sliding speed so low that maximum friction is maintained.

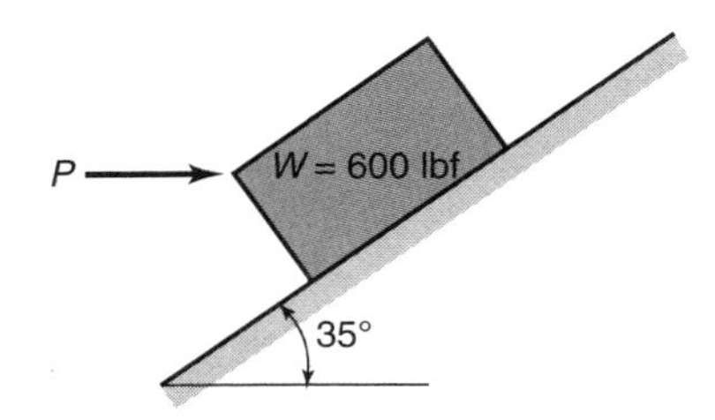

Sketch a, for Problem 8.53

8.53 Given a block on an incline as shown in Sketch a, find the force P required

(a) To prevent motion downward. *Ans.* 229.8 lb.

(b) To cause motion upward. *Ans.* 691.1 lb.

The coefficient of friction is 0.25. Draw a free-body diagram for both situations showing the forces involved.

8.54 Three equal blocks of plastic are carrying the weight of a steel slider. At first each block carries one-third of the weight, but the central block is made of polytetrafluoroethylene (PTFE) and the two outer blocks are made of polyethylene, so they have different adhesive wear constants (PTFE, $k_1 = 2 \times 10^{-5}$; polyethylene, $k_1 = 2 \times 10^{-8}$). Determine how the load redistributes between the blocks if the hardnesses of the plastics are assumed to be the same.

8.55 Given the plastic blocks in Problem 8.54 and sharing the load so that the wear is equal on each block, find the coefficient of friction.

8.56 A 50-kg copper piece is placed on a flat copper surface as shown in Sketch b. Assuming that copper has a hardness of 275 MPa, calculate the shear stress, the abrasive wear volume, and the adhesive wear volume. *Ans.* $\tau = 312$ MPa, $k_2 = 0.0669$.

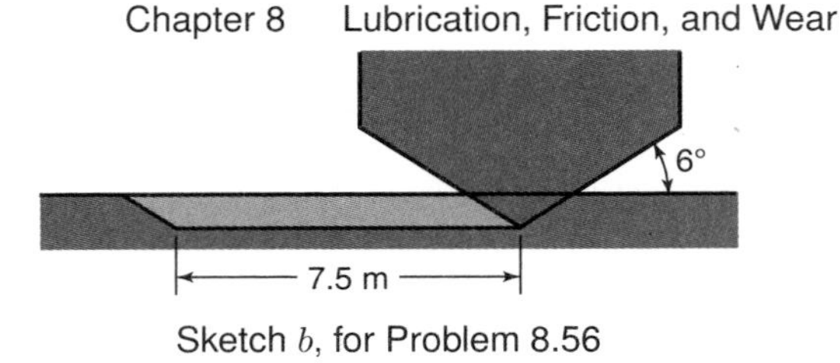

Sketch b, for Problem 8.56

Design and Projects

8.57 Design a test method for evaluating fatigue wear of rolling element bearings.

8.58 Perform a literature search to determine if there is any relationship between friction and wear.

8.59 Construct a Design Procedure that outlines the steps you would follow in order to (a) increase and (b) decrease friction when desired.

8.60 Case Study 8.1 described tribological issues in total joint replacements. A new development involves *resurfacing implants*. Consult with the technical literature and manufacturer's documentation and list the advantages of resurfacing, and write a short paper on the associated tribology concerns.

8.61 A bushing is to be selected for an application where the radial load is 350 N, the speed is 200 rpm, the mating surface is stainless steel, and a life of 3000 hours is needed. The shaft diameter is 25 mm. Select a material, and then specify the thickness and width of the bushing. Assume the maximum temperature encountered in operation is 60°C.

8.62 Consider Eq. (8.10), and explain why the pressure distribution in an elastic solid cannot be described by

$$p_H = p_{\max}\left[1 - \left(\frac{2x}{D_x}\right)^3 - \left(\frac{2y}{D_y}\right)^3\right]^{\frac{1}{2}}$$

Part II — Machine Elements

Outline

Chapter 9

Columns

Columns from the Alhambra in Granada, Spain. *Source:* Shutterstock.

And as imagination bodies forth the forms of things unknown,
The poet's pen turns them to shapes
And gives to airy nothingness a local habitation and a name.
William Shakespeare, *A Midsummer Night's Dream*

Contents

Examples

Case Study

Design Procedure

Compression members are widely used in machinery of all types. A column is a slender compression member that deforms laterally, or buckles, before stresses reach the yield strength of the material. In buckling, loads below a critical value can be supported, but once the critical load is exceeded, large deformations result. This chapter begins with a discussion of equilibrium regimes, in order to introduce nomenclature and concepts relevant to buckling. The classic derivation of Euler is then applied to elastically deforming columns, and so-called Euler buckling is described. For less slender columns, Johnson's buckling criterion is better suited to predict buckling loads. The American Institute of Steel Construction requirements and design methodology for compression members are then described and demonstrated. Finally, eccentric columns are investigated, where the applied load does not act through the column's centroid.

Machine elements in this chapter: Columns of all types; connecting rods, supports.
Typical applications: Lattice frameworks such as booms; machine frames; support structures; jacks and other actuators.
Competing machine elements: Compression springs (Chapter 17); in a general sense, beams (Chapters 4 and 5), power screws (Chapter 16).

Symbols

A	cross sectional area, m^2
C_c	slenderness ratio
C_1, C_2	integration constants
c	distance from neutral axis to outer fiber of column, m
d	diameter, m
E	modulus of elasticity, Pa
E_t	tangent modulus, Pa
e	eccentricity, m
g	gravitational acceleration, 9.807 m/s^2
h	height, m
I	area moment of inertia, m^4
l	length, m
l_e	effective length of column, m
M	moment, N-m
M'	statically equivalent moment, N-m
m_a	mass, kg
n	an integer
n_σ	stress reduction factor in AISC criteria
P	force, N
r_g	radius of gyration, m
S_y	yield strength, Pa
t	time, s
x	length dimension of column, m
y	transverse dimension or deflection of column, m
θ	position angle, deg
ρ	density, kg/m^3
σ	normal stress, Pa
σ_{all}	allowable normal stress, Pa

Subscripts

cr	critical
E	Euler
i	inner
J	Johnson
o	outer
T	tangency point

9.1 Introduction

The basic understanding of loads, stresses, and deformations obtained in the preceding chapters is related here to columns. A **column** is straight and long (relative to its cross section) and is subjected to compressive, axial loads. The reason for a special consideration of columns is that failures due to yielding, determined from Eq. (4.22), and due to deformation, determined from Eq. (4.23), are not correct in predicting failures of long columns. Because of their slender shape, columns tend to deform laterally upon loading; and if the deflection becomes larger than their respective critical values, they fail catastrophically. This situation is known as **buckling**, which can be defined as a sudden large deformation of a structure due to a slight increase of the applied load, under which the structure had exhibited little, if any, deformation before this increase.

This chapter first describes the meaning of elastic stability and end conditions. It then establishes the various failure criteria for concentrically loaded columns and describes the nature of the instability that can occur, thus predicting when buckling will take place. It also establishes the failure criteria for eccentrically loaded columns, so that proper design of concentric and eccentric columns is ensured.

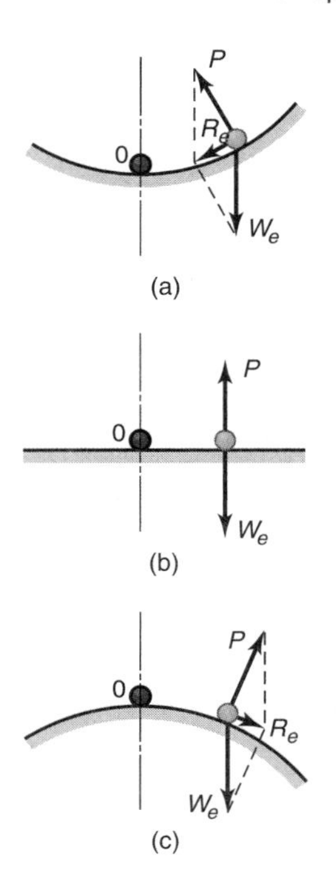

Figure 9.1: Depiction of equilibrium regimes. (a) Stable; (b) neutral; (c) unstable.

9.2 Equilibrium Regimes

To understand why columns buckle, it is first necessary to understand the equilibrium regimes. An important question is: When an equilibrium position is disturbed slightly, does the component tend to return to the equilibrium position or does it tend to depart even farther? To visualize what is happening, consider Fig. 9.1, which shows the three equilibrium regimes: stable, neutral, and unstable.

9.2.1 Stable Equilibrium

Figure 9.1a illustrates stable equilibrium. Assume that the surfaces are frictionless and that the sphere has a light weight. The forces on the sphere (gravity and normal surface reaction) are in balance whenever the surface is horizontal. The balanced position is indicated by a zero in the figure. Figure 9.1a shows the sphere displaced slightly from its equilibrium position. The forces on it no longer balance, but the resultant imbalance is a restoring force (i.e., gravity is accelerating the sphere back toward the equilibrium position). Such a situation is called *stable equilibrium*.

9.2.2 Neutral Equilibrium

Figure 9.1b considers the sphere in neutral equilibrium. After the sphere has been slightly displaced from the equilibrium position, it is still in equilibrium at the displaced position, and there is neither a tendency to return to the previous position nor to move to some other position. Equilibrium is always satisfied. Because the surface is flat, the sphere does not move after being placed in another position; this is *neutral equilibrium*.

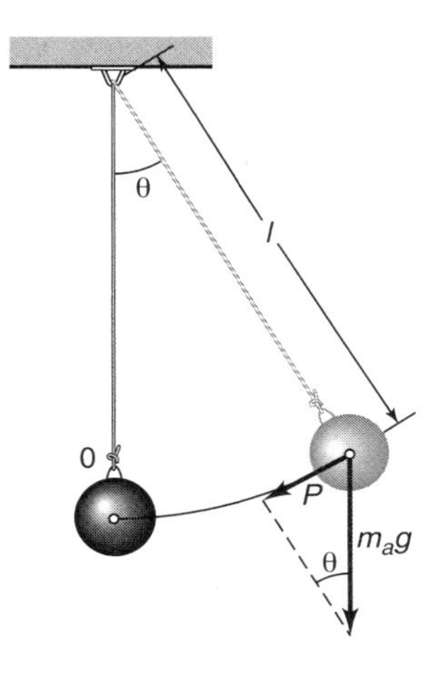

Figure 9.2: Pendulum used in Example 9.1.

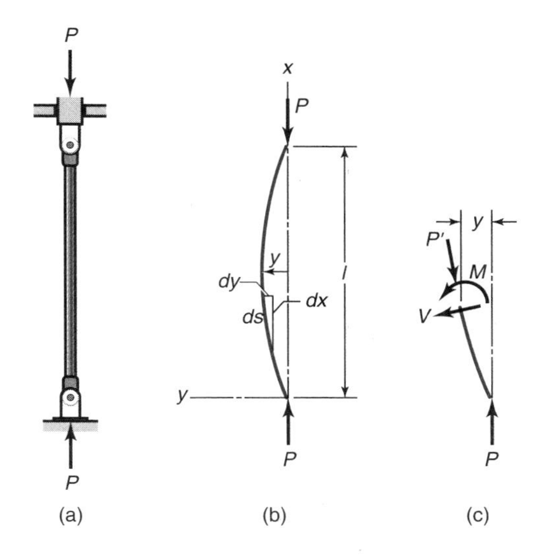

Figure 9.3: Column with pinned ends. (a) Assembly; (b) deformation shape; (c) load acting.

9.2.3 Unstable Equilibrium

Figure 9.1c shows the case of unstable equilibrium, which is the opposite situation from that presented in Fig. 9.1a. When the sphere is displaced from its equilibrium position (either to the right or the left), the resultant imbalance is a disturbing force (i.e., it accelerates the sphere away from its equilibrium position). Such a situation is called *unstable equilibrium*. Gravity and convex surfaces cause the sphere to move farther from the balanced position.

Generalizing from this example, unstable equilibrium occurs if, for small displacements or *perturbations* from the equilibrium position, the disturbing forces tend to accelerate the part away from the equilibrium position. Columns in compression display unstable equilibrium and are unreliable and hazardous when overloaded; a small displacement can cause a catastrophic change in the configuration of a column. Thus, as the load on a column increases, a critical load is reached where unstable equilibrium occurs and the column will not return to its straight configuration. The load cannot be increased beyond this value unless the column is laterally restrained. Thus, for long, slender columns, a critical buckling load occurs. This critical buckling load (when divided by the cross sectional area) gives a critical buckling stress. For columns, this critical buckling stress is much lower than the yield strength of the material. Thus, columns will generally fail from buckling long before they fail from yielding.

The shape as well as the load establishes when buckling will occur. A column may be viewed as a straight bar with a large slenderness ratio, l/r, (typically 100) subjected to axial compression. Buckling occurs in such a column when it is loaded to a critical load and marked changes in deformation occur that do not result from the material yielding.

Example 9.1: Equilibrium Regimes

Given: A simple pendulum (Fig. 9.2) in which a ball is hung by a thin wire and is acted on by gravitational acceleration.

Find: Is the system's equilibrium neutral, stable, or unstable?

Solution: The force restoring the ball to the center position ($\theta = 0$) is

$$P = m_a g \sin\theta. \qquad (a)$$

From simple harmonic motion

$$P = -m_a l \frac{d^2\theta}{dt^2}. \qquad (b)$$

Combining Eqs. (a) and (b) gives

$$\frac{d^2\theta}{dt^2} = -\frac{g}{l}\sin\theta. \qquad (c)$$

For any angle θ, the angular acceleration $d^2\theta/dt^2$ has the opposite sign from θ for $-\pi \leq \theta \leq \pi$. Thus, the ball will always return to the center position ($\theta = 0$), implying that stable equilibrium prevails.

9.3 Concentrically Loaded Columns

9.3.1 Linear-Elastic Materials

Figure 9.3a shows a concentrically loaded column with pinned ends; thus, the ends are kept in position but are free to rotate. Assume that the column is initially straight and that the load is concentric; that is, it acts through the centroid as depicted in Fig. 9.3b. Figure 9.3c shows a free-body diagram of the loads acting on the column.

Equation (5.3) relates the moment and deflection as

$$M = EI\frac{d^2y}{dx^2}. \qquad (5.3)$$

Equilibrium of the section cut from the bar requires that $M = -Py$. Substituting this into Eq. (5.3) gives

$$\frac{d^2y}{dx^2} + \frac{P}{EI}y = 0. \qquad (9.1)$$

This is a homogeneous, second-order, linear differential equation with constant coefficients. The general solution of this equation is

$$y = C_1 \sin\left(x\sqrt{\frac{P}{EI}}\right) + C_2 \cos\left(x\sqrt{\frac{P}{EI}}\right), \tag{9.2}$$

where C_1 and C_2 are integration constants. The boundary conditions are

1. $y = 0$ at $x = 0$, resulting in $C_2 = 0$
2. $y = 0$ at $x = l$, so that

$$C_1 \sin\left(l\sqrt{\frac{P}{EI}}\right) = 0. \tag{9.3}$$

However, C_1 cannot be equal to zero; otherwise, a trivial solution of $y = 0$ results and the column will always remain straight, which is contrary to experience. The other possibility of satisfying Eq. (9.3) is for

$$\sin\left(l\sqrt{\frac{P}{EI}}\right) = 0, \tag{9.4}$$

which is satisfied when

$$l\sqrt{\frac{P}{EI}} = n\pi, \tag{9.5}$$

or, solving for P,

$$P = \frac{n^2\pi^2 EI}{l^2}, \tag{9.6}$$

where n is an integer (1, 2, ...). The smallest value of P is obtained for $n = 1$. Thus, the critical load, P_{cr}, for a column with pinned ends is

$$P_{\text{cr}} = \frac{\pi^2 EI}{l^2}. \tag{9.7}$$

This load is sometimes called the **Euler load**, named after the Swiss mathematician Leonhard Euler, who originally solved this problem in 1757.

An interesting aspect of Eq. (9.7) is that the critical load is independent of the material's strength, rather, it depends only on the column's dimensions (expressed in I and l) and the material's modulus of elasticity, E. For this reason, as far as elastic buckling is concerned, long columns made, for example, of high-strength steel offer no advantage over those made of lower strength steel, since both steels have approximately the same modulus of elasticity.

Another interesting aspect of Eq. (9.7) is that the critical load capacity of a column will increase as the moment of inertia of the cross sectional area increases. Thus, columns are designed so that most of their cross sectional area is located as far as possible from the section's principal centroidal axes. This implies that a hollow tube is preferred over a solid section.

It is also important to realize that a column will buckle about the principal axis of the cross section having the least moment of inertia (the weakest axis). For example, a column having a rectangular cross section, such as a meter stick, as shown in Fig. 9.4, will buckle about the x-axis and not the y-axis.

Substituting Eq. (9.5) and $C_2 = 0$ into Eq. (9.2) gives the buckling shape as

$$y = C_1 \sin\left(\frac{\pi x}{l}\right). \tag{9.8}$$

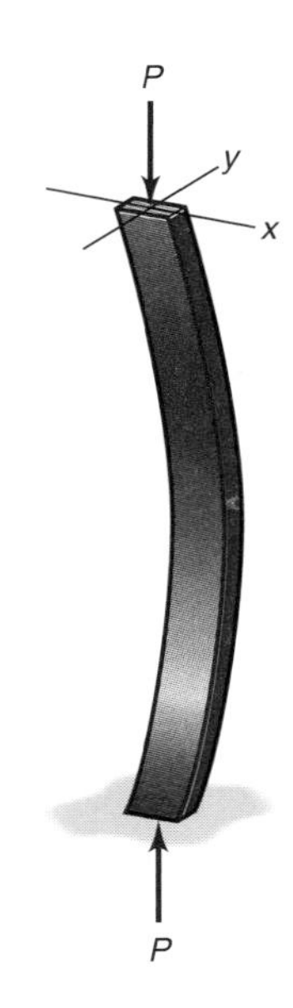

Figure 9.4: Buckling of rectangular section.

When $x = l/2$, $y = y_{\max}$ and $C_1 = y_{\max}$. Therefore,

$$y = y_{\max} \sin\left(\frac{\pi x}{l}\right). \tag{9.9}$$

Thus, the buckling shape varies sinusoidally, with zero at the ends and a maximum at half-length.

Engineers are often interested in defining the critical stress of a column. The radius of gyration, r_g, given in Eq. (4.14), is substituted into Eq. (9.7) giving the critical stress for the Euler equation as

$$(\sigma_{\text{cr}})_E = \frac{P_{\text{cr}}}{A} = \frac{\pi^2 E}{(l/r_g)^2}. \tag{9.10}$$

Also, the critical stress, σ_{cr}, is an average stress in the column just before the column buckles. This stress results in elastic strains and is therefore less than or equal to the material's yield strength.

9.3.2 Tangent Modulus

The preceding section determined the stress at which a column will buckle for a linear elastic material. As presented in Chapter 3, this is the case for most materials below the proportional limit. Above the proportional limit and up to the yield point the material may still be elastic, but it will not behave linearly. The effect is that the elastic modulus at the buckling stress in Eq. (9.10) may be significantly lower than one would expect from published values of elastic modulus. Equation (9.10) is often modified as

$$\sigma_{\text{cr}} = \frac{\pi^2 E_t}{(l/r_g)^2}, \tag{9.11}$$

where E_t is the **tangent modulus**, defined as the elastic modulus at the stress level in the column.

Equation (9.11) is called the **tangent modulus** or the **Essenger equation**. It is extremely difficult to apply in design, since the tangent modulus is rarely well quantified. However, it is mentioned here because of the importance of the tangent-modulus effect.

Table 9.1: Effective length for common column end conditions.

End condition description	Both ends pinned	One end pinned, one end fixed	Both ends fixed	One end fixed, one end free
Illustration of end condition	$l = l_e$	$l_e = 0.7l$	$l_e = 0.5l$	$l_e = 2l$
Theoretical effective column length	$l_e = l$	$l_e = 0.7l$	$l_e = 0.5l$	$l_e = 2l$
Approximated end conditions[a]	$l_e = l$	$l_e = 0.8l$	$l_e = 0.65l$	$l_e = 2.1l$

[a] American Institute of Steel Construction, *Manual of Steel Construction*, 14th ed., [2011].

9.4 End Conditions

The Euler equation was developed above for columns with pinned ends, but other end constraints are possible. Table 9.1 shows four common end conditions. The Euler equation can be used for these circumstances by recognizing that end conditions change the effective length of the column. The critical load and stress given in Eqs. (9.7) and (9.10), respectively, are modified by replacing l with the effective length l_e for the corresponding end condition. Substituting l_e for l in Eqs. (9.7) and (9.10) gives

$$(P_{\text{cr}})_E = \frac{\pi^2 EI}{l_e^2}, \tag{9.12}$$

$$(\sigma_{\text{cr}})_E = \frac{(P_{\text{cr}})_E}{A} = \frac{\pi^2 E}{(l_e/r_g)^2}. \tag{9.13}$$

These equations for the critical load and stress for the Euler criterion (see Section 9.5) are valid for any end condition. Often, end conditions are not easily classified. Table 9.1 also provides recommendations from the American Institute of Steel Construction (AISC) where "ideal conditions are approximated."

Example 9.2: Material Effect on Buckling Load

Given: A column has a square tubular cross section. The wall thickness is 10 mm and the column length is 12 m. The column is axially loaded in compression and has pinned ends. The mass of the tubular column must not exceed 200 kg.

Find: Determine which metal in Table 3.1 gives the highest buckling load; also, calculate the critical load.

Solution: The area of a square tubular cross section with outside dimension h is

$$A = h^2 - (h - 0.02)^2. \tag{a}$$

The mass of the column is

$$m_a = \rho A l = \rho l \left[h^2 - (h - 0.02)^2\right]. \tag{b}$$

The area moment of inertia for a square tube is

$$I = \frac{h^4 - (h - 0.02)^4}{12} = \frac{m_a\left[h^2 + (h - 0.02)^2\right]}{12\rho l}. \tag{c}$$

Substituting Eq. (c) into Eq. (9.7) gives the critical load as

$$P_{\text{cr}} = \frac{\pi^2 E m_a}{12\rho l^3}\left[h^2 + (h - 0.02)^2\right]. \tag{d}$$

Thus, the largest buckling load would result from the largest value of E/ρ. From Table 3.1, observe that magnesium would give the largest E/ρ, with E = 45 GPa and ρ = 1740 kg/m^3. In Eq. (*b*), given that m_a = 200 kg, l = 12 m, and ρ = 1740 kg/m^3,

$$\frac{m_a}{\rho l} = 0.04(h - 0.01);$$

$$\frac{200}{(1740)(12)(0.04)} + 0.01 = h,$$

or $h = 0.249$ m = 249 mm. From Eq. (*d*) the critical load is

$$\begin{aligned} P_{\rm cr} &= \frac{\pi^2 (45 \times 10^9)(200)}{12(1740)(12)^3}\left[(0.249)^2 + (0.229)^2\right] \\ &= 2.82 \times 10^5 \\ &= 282 \text{ kN.} \end{aligned}$$

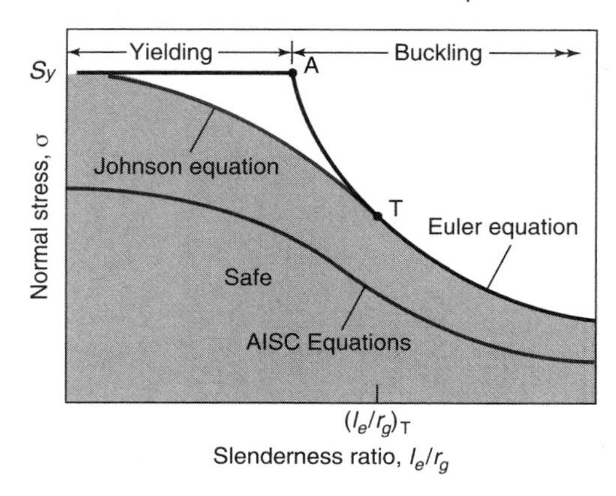

Figure 9.5: Normal stress as function of slenderness ratio obtained by Euler, Johnson, and AISC equations, as well as yield strength.

9.5 Slenderness Ratio

As discussed above, the Euler equation holds for linear elastic materials, but this is not accurate near the yield point. The American Institute of Steel Construction [2011] assumes that the proportional limit of a material exists at one-half the yield strength, or that the allowable stress for elastic buckling is

$$\sigma_{\rm all} = 0.5 S_y, \tag{9.14}$$

where S_y is the yield strength as given in Appendix A for a variety of materials. Substituting Eq. (9.14) into Eq. (9.13) gives the **critical slenderness ratio**, C_c, from Euler's formula as

$$C_c = \left(\frac{l_e}{r_g}\right)_E = \sqrt{\frac{2E\pi^2}{S_y}}. \tag{9.15}$$

C_c is called the critical slenderness ratio since it defines the applicability limit of Euler buckling. Below C_c, the tangent modulus effect will result in lower buckling stresses than predicted by the Euler equation, and a different buckling criterion must be used.

9.6 Johnson's Buckling Criterion

Figure 9.5 shows curves for normal stress as a function of slenderness ratio. Note the abrupt change in the Euler curve as it approaches the yield strength (see point A in Fig. 9.5), indicating a transition from buckling to compressive yielding or fracture. Because of the tangent modulus effect, the transition is not this abrupt, and the failure criterion needs to be modified at this location. Perhaps the most widely used modification is the **Johnson equation**:

$$(\sigma_{\rm cr})_J = \frac{(P_{\rm cr})_J}{A} = S_y - \frac{S_y^2}{4\pi^2 E}\left(\frac{l_e}{r_g}\right)^2. \tag{9.16}$$

Figure 9.5 shows the tangency point T, where the Euler and Johnson equations give the same prediction for buckling stress. This tangency point distinguishes between intermediate columns (Johnson range) and long columns (Euler range). To determine the value of $(l_e/r_g)_T$, Eqs. (9.13) and (9.16) are equated to yield

$$\frac{\pi^2 E}{(l_e/r_g)_T^2} = S_y - \frac{S_y^2}{4\pi^2 E}\left(\frac{l_e}{r_g}\right)_T^2;$$

$$\left(\frac{l_e}{r_g}\right)_T^4 - \frac{4\pi^2 E}{S_y}\left(\frac{l_e}{r_g}\right)_T^2 + \frac{4\pi^4 E^2}{S_y^2} = 0. \tag{9.17}$$

Solving for $(l_e/r_g)_T$ gives

$$\left(\frac{l_e}{r_g}\right)_T = \sqrt{\frac{2\pi^2 E}{S_y}} = \sqrt{\frac{\pi^2 EA}{P_{\rm cr}}}. \tag{9.18}$$

Design Procedure 9.1: Selecting a Buckling Equation

Often, one needs to design a column to support a given load, and constraints such as material, column length, etc, may be specified. It is often not clear beforehand whether a column will buckle according to the Euler or Johnson equations. The following procedure is useful in these cases:

1. Assume that the column buckles according to Eq. (9.7), or that it encounters Euler buckling. Use this equation to determine the column's geometry.
2. Calculate the critical slenderness ratio from Eq. (9.15) and (9.18), namely
$$C_c = \sqrt{2E\pi^2/S_y} = \sqrt{\pi^2 EA/P_{\rm cr}}.$$
3. Calculate the slenderness ratio of the column from $C = l_e/r_g$, where r_g is given by Eq. (4.14) as $r_g = \sqrt{I/A}$.
4. If $C \geq C_c$, then the column is indeed described by the Euler equation and the cross section calculated in step 1 is applicable.
5. If $C < C_c$, the Johnson equation [Eq. (9.16)] must be used to determine the column geometry.

Note that an equally useful design rule can be derived where the Johnson equation is assumed to be correct, and the Euler equation is used otherwise.

Example 9.3: Cross Section Effect on Buckling Load

Given: A column with one end fixed and the other free is to be made of aluminum alloy 2014. The column's cross sectional area is 600 mm^2, and its length is 2.5 m.

Find: The critical column buckling load for the following shapes:

(*a*) A solid round bar

(*b*) A cylindrical tube with a 50-mm outer diameter

(*c*) A square tube with a 50-mm outer dimension

(*d*) A square bar

Solution: From Table A.5 for aluminum alloy 2014, E = 72 GPa and S_y = 97 MPa (the O condition – annealed – is assumed). From Table 9.1 for one end fixed and one end free, the theoretical effective column length is $l_e = 2l = 5$ m. Note that the approximated end conditions could be used to obtain a conservative result, but it will be assumed that the end conditions are well characterized, so that the theoretical effective lengths can be used. Consider the solid round bar. The cross sectional area is

$$A = \frac{\pi d^2}{4}$$

or

$$d = \sqrt{\frac{4A}{\pi}} = \sqrt{\frac{4(600)}{\pi}} = 27.64 \text{ mm}.$$

The area moment of inertia for this circular section is

$$I = \frac{\pi d^4}{64} = \frac{\pi(27.64)^4}{64} = 28,650 \text{ mm}^4.$$

The radius of gyration is

$$r_g = \sqrt{\frac{I}{A}} = \sqrt{\frac{28,650}{600}} = 6.910 \text{ mm}.$$

Therefore, the slenderness ratio is

$$\frac{l_e}{r_g} = \frac{5000}{6.91} = 723.6.$$

From Eq. (9.18),

$$\left(\frac{l_e}{r_g}\right)_T = \sqrt{\frac{2\pi^2 E}{S_y}} = \sqrt{\frac{2\pi^2(72\times10^9)}{97\times10^6}} = 121.$$

Since $l_e/r_g > (l_e/r_g)_T$, the Euler formula applies and Eq. (9.12) gives the critical load at the onset of buckling as

$$P_{\text{cr}} = \frac{\pi^2 EI}{l_e^2} = \frac{\pi^2(72\times10^9)(28,650\times10^{-12})}{(5.00)^2} = 814 \text{ N}.$$

At this point, the remaining geometries are analyzed in the same fashion. It can be shown that all of the cross sections given result in large slenderness ratios, so that Euler buckling is applicable, and the buckling load is estimated from Eq. (9.7). The summary of cross section dimensions and buckling loads are presented in Table 9.2 and Fig. 9.6.

Table 9.2: Summary of results for Example 9.3.

Description	r_g (mm)	I (mm^4)	C_c	P_{cr} (N)
Solid round bar	6.91	28,650	723.6	814
Cylindrical tube	16.27	1.588×10^5	307	4514
Square tube	19.15	2.20×10^5	261	6250
Solid square bar	7.071	3.00×10^4	707	853

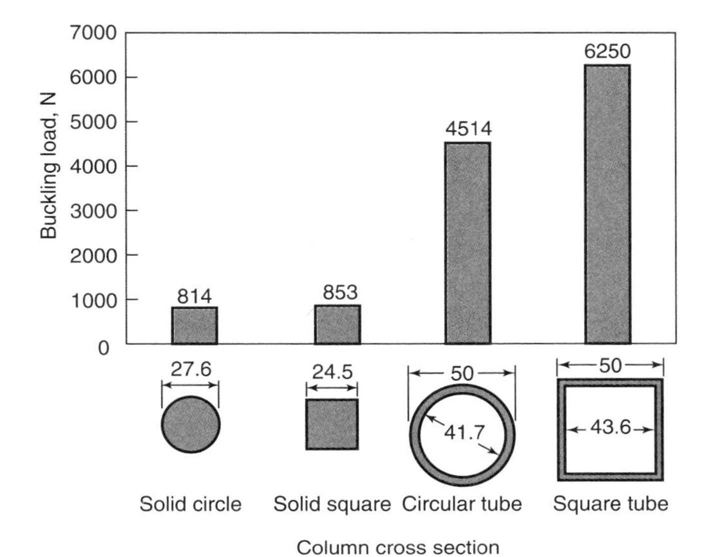

Figure 9.6: cross sectional areas, drawn to scale, from the results of Example 9.3, as well as critical buckling load for each cross sectional area.

Example 9.4: Euler and Johnson Buckling

Given: A column with one end fixed and the other end pinned is made of a low-carbon steel with S_y = 295 MPa. The column's cross section is rectangular with h = 15 mm and b = 40 mm.

Find: The buckling load for the following three lengths:

(*a*) 0.15 m

(*b*) 0.75 m

(*c*) 1.20 m

Solution: From Table 3.1 for low-carbon steel, E = 207 GPa. The cross sectional area is

$$A = bh = (0.040)(0.015) = 0.00060 \text{ m}^2.$$

For a rectangular cross section, the area moment of inertia is

$$I = \frac{bh^3}{12} = \frac{(0.040)(0.015)^3}{12} = 1.125\times10^{-8} \text{ m}^4.$$

The radius of gyration is

$$r_g = \sqrt{\frac{I}{A}} = \sqrt{\frac{1.125 \times 10^{-8}}{0.00060}} = 0.00433 \text{ m} = 4.33 \text{ mm}.$$

(*a*) From Table 9.1, for one end pinned and the other end fixed and l = 0.15 m,

$$l_e = 0.7l = 0.7(0.15) = 0.105 \text{ m}.$$

Therefore,

$$\frac{l_e}{r_g} = \frac{0.105}{0.00433} = 24.2.$$

From Eq. (9.18),

$$\left(\frac{l_e}{r_g}\right)_T = \sqrt{\frac{2\pi^2 E}{S_y}} = \sqrt{\frac{2\pi^2(207 \times 10^9)}{295 \times 10^6}} = 117.7.$$

Since $l_e/r_g < (l_e/r_g)_T$, Eq. (9.16) (the Johnson equation) should be used. Note first that

$$\begin{aligned} \frac{1}{E}\left(\frac{S_y}{2\pi}\frac{l_e}{r_g}\right) &= \frac{1}{207 \times 10^9}\left(\frac{(295 \times 10^6)(24.2)}{2\pi}\right) \\ &= 6.237 \times 10^6. \end{aligned}$$

Therefore,

$$\begin{aligned} P_{\text{cr}} &= A\left[S_y - \frac{1}{E}\left(\frac{S_y}{2\pi}\frac{l_e}{r_g}\right)^2\right] \\ &= 0.00060\left[(295 \times 10^6) - 6.237 \times 10^6\right] \\ &= 173 \text{ kN}. \end{aligned}$$

(*b*) From Table 9.1, for one end pinned and one end fixed and l = 0.75 m,

$$l_e = 0.70l = 0.7(0.75) = 0.525 \text{ m};$$

$$\frac{l_e}{r_g} = \frac{0.525}{0.00433} = 121.$$

Since $l_e/r_g > (l_e/r_g)_T$, the Euler equation should be used. From Eq. (9.12),

$$\begin{aligned} P_{\text{cr}} &= \frac{\pi^2 EI}{l_e^2} \\ &= \frac{\pi^2(207 \times 10^9)(1.125 \times 10^{-8})}{(0.525)^2} \end{aligned}$$

or $P_{\text{cr}} = 83.4$ kN. Note that the slenderness ratio of this column is close to the critical slenderness ratio. Therefore, it is not surprising that the result from the Johnson equation is also 83.4 kN.

(*c*) The Euler equation is definitely valid for l = 1.2 m, so that

$$l_e = 0.7(1.2) = 0.840 \text{ m}.$$

The critical load is

$$\begin{aligned} P_{\text{cr}} &= \frac{\pi^2 EI}{l_e^2} \\ &= \frac{\pi^2(207 \times 10^9)(1.125 \times 10^{-8})}{(0.840)^2} \\ &= 32.6 \text{ kN}. \end{aligned}$$

9.7 AISC Criteria

The American Institute of Steel Construction [2011] has produced design guidelines for elastic stability conditions. Although the guidelines are intended for steel, the wide use of steel in compression members makes the guidelines especially useful.

As shown by Eq. (9.11), the elastic modulus based on the linear-elastic portion of a stress-strain curve may lead to erroneous results. However, tangent-modulus data are not readily available in the technical literature and are difficult to obtain experimentally. The Johnson equation is needed for such circumstances. However, long columns are more difficult to design because they are extremely susceptible to defects in straightness or to eccentricity in loading. Therefore, a weighted reduction in the allowable stress is prescribed. The **AISC equations** correct for reductions in elastic modulus as the column stress exceeds the proportional limit of the material and use a sliding safety factor. For elastic buckling, the allowable stress is

$$\sigma_{\text{all}} = \frac{12\pi^2 E}{23\,(l_e/r_g)^2}, \tag{9.19}$$

and for inelastic buckling,

$$\sigma_{\text{all}} = \frac{\left\{1 - \left[\frac{(l_e/r_g)^2}{2C_c^2}\right]\right\} S_y}{n_\sigma}, \tag{9.20}$$

where C_c is the slenderness ratio for Euler buckling defined by Eq. (9.15) and n_σ is the reduction in allowable stress given by

$$n_\sigma = \frac{5}{3} + \frac{3\,(l_e/r_g)}{8C_c} - \frac{(l_e/r_g)^3}{8C_c^3}. \tag{9.21}$$

Note that n_σ is not a safety factor but a mandatory reduction in a material's allowable stress. Figure 9.5 compares the AISC as well as the Euler and Johnson equations. The American Association of State Highway and Transportation Officials (AASHTO) uses equations identical to Eqs. (9.19) and (9.20) but requires a constant stress reduction n_σ = 2.12 for both elastic and inelastic buckling.

The AISC Criteria are not intended to predict the buckling load of a given column, but instead provide a design framework for steel columns. Better predictions of buckling load are obtained from the Euler or Johnson equations [Eqs. (9.12) and (9.16)], depending on the slenderness ratio as discussed in Design Procedure 9.1.

9.8 Eccentrically Loaded Columns

The applied load is not always applied through the centroid; the distance between the load and column axes is called the **eccentricity** and is designated by e. While loads can be eccentric, the load is usually parallel with the column axis, and this situation will be analyzed in this section. Just as for concentrically loaded columns (Section 9.3) the analysis of eccentricity is restricted at first to columns with pinned ends. Figure 9.7a shows a pinned column subjected to a compressive force acting at a distance e from the centerline of the undeformed column. This loading is statically equivalent to the axial load and bending moment $M' = -Pe$ shown in Fig. 9.7b. As when considering concentrically loaded columns, small deflections and linear-elastic material behavior are assumed.

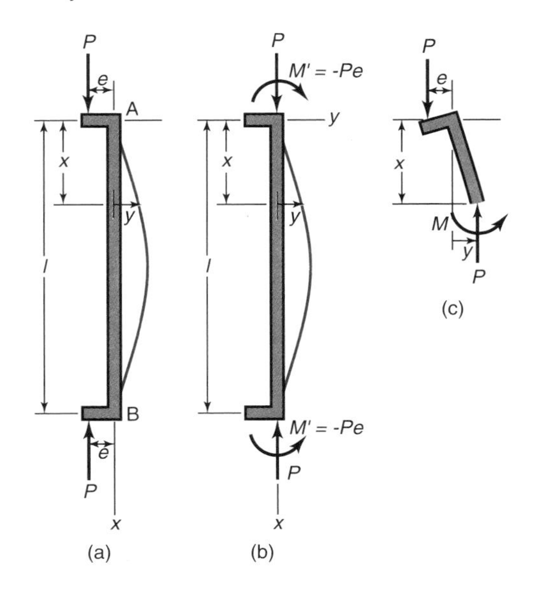

Figure 9.7: Eccentrically loaded column. (a) Eccentricity; (b) statically equivalent bending moment; (c) free-body diagram through arbitrary section.

From a free-body diagram of an arbitrary section shown in Fig. 9.7c, the internal moment in the column is

$$M = -P(e + y). \tag{9.22}$$

The differential equation for the deflection curve is obtained from Eqs. (5.3) and (9.22), or

$$\frac{d^2y}{dx^2} + \frac{Py}{EI} = -\frac{Pe}{EI} = \text{Constant}. \tag{9.23}$$

Note that if the eccentricity is zero, Eqs. (9.23) and (9.1) are identical. The general solution to Eq. (9.23) is

$$y = C_1 \sin\left(x\sqrt{\frac{P}{EI}}\right) + C_2 \cos\left(x\sqrt{\frac{P}{EI}}\right) - e. \tag{9.24}$$

The boundary conditions are

1. at $x = 0$, $y = 0$, so that $C_2 = e$.
2. at $x = l/2$, $dy/dx = 0$.

Taking the derivative of Eq. (9.24) gives

$$\frac{dy}{dx} = C_1\sqrt{\frac{P}{EI}} \cos\left(x\sqrt{\frac{P}{EI}}\right) - e\sqrt{\frac{P}{EI}} \sin\left(x\sqrt{\frac{P}{EI}}\right). \tag{9.25}$$

Using Eq. (9.25) and the second boundary condition gives

$$C_1 = e\tan\left(\frac{l}{2}\sqrt{\frac{P}{EI}}\right). \tag{9.26}$$

Substituting $C_2 = e$ and Eq. (9.26) into Eq. (9.24) gives

$$y = e\left[\tan\left(\frac{l}{2}\sqrt{\frac{P}{EI}}\right)\sin\left(x\sqrt{\frac{P}{EI}}\right) + \cos\left(x\sqrt{\frac{P}{EI}}\right) - 1\right].$$

The maximum deflection occurs at $x = l/2$, so that

$$\begin{aligned} y_{\max} &= e\left[\frac{\sin^2\left(\frac{l}{2}\sqrt{\frac{P}{EI}}\right)}{\cos\left(\frac{l}{2}\sqrt{\frac{P}{EI}}\right)} + \cos\left(\frac{l}{2}\sqrt{\frac{P}{EI}}\right) - 1\right] \\ &= e\left[\sec\left(\frac{l}{2}\sqrt{\frac{P}{EI}}\right) - 1\right]. \end{aligned} \tag{9.27}$$

The maximum stress on the column is caused by the axial load and the moment. The maximum moment occurs at the column's midheight and has a magnitude of

$$M_{\max} = |P(e + y_{\max})|;$$

$$M_{\max} = Pe\sec\left(\frac{l}{2}\sqrt{\frac{P}{EI}}\right). \tag{9.28}$$

The maximum stress in the column is compressive and is

$$\sigma_{\max} = \frac{P}{A} + \frac{M_{\max}c}{I}.$$

Making use of Eq. (9.28) gives

$$\sigma_{\max} = \frac{P}{A} + \frac{Pec}{I}\sec\left(\frac{l}{2}\sqrt{\frac{P}{EI}}\right). \tag{9.29}$$

Since the radius of gyration is $r_g^2 = I/A$, Eq. (9.29) becomes

$$\sigma_{\max} = \frac{P}{A}\left[1 + \frac{ec}{r_g^2}\sec\left(\frac{l}{2r_g}\sqrt{\frac{P}{EA}}\right)\right], \tag{9.30}$$

where

- P = critical load where buckling will occur in eccentrically loaded column, N
- A = cross sectional area of column, m^2
- e = eccentricity of load measured from neutral axis of column's cross sectional area to load's line of action, m
- c = distance from neutral axis to outer fiber of column, m
- r_g = radius of gyration, m
- l = length before load is applied, m
- E = elastic modulus of column material, Pa

For end conditions other than pinned, the length is replaced with the effective length (using Table 9.1), and Eqs. (9.28) and (9.30) become

$$y_{\max} = e\left[\sec\left(\frac{l_e}{2}\sqrt{\frac{P}{EI}}\right) - 1\right], \tag{9.31}$$

$$\sigma_{\max} = \frac{P}{A}\left[1 + \frac{ec}{r_g^2}\sec\left(\frac{l_e}{2r_g}\sqrt{\frac{P}{EA}}\right)\right]. \tag{9.32}$$

Equation (9.31) is known as the **secant equation** and the parameter ec/r_g^2 is called the **eccentricity ratio**. Note from Eq. (9.32) that it is not convenient to calculate the load explicitly.

Figure 9.8 shows the effect of slenderness ratio on normal stress for an eccentrically loaded column. These results are for structural grade steel where the modulus of elasticity is 207 GPa and the yield strength is 250 MPa. Note that

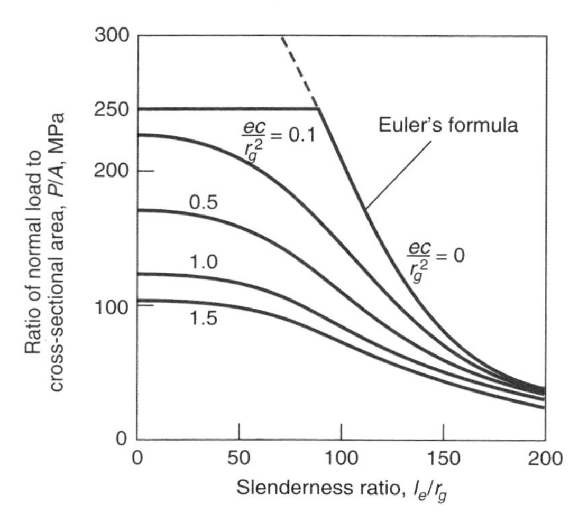

Figure 9.8: Stress variation with slenderness ratio for five eccentricity ratios, for structural steel with E = 207 GPa and S_y = 250 MPa.

as e approaches 0, σ approaches P/A, where P is the critical column load as established by the Euler formula [Eq. (9.12)].

The curves in Fig. 9.8 indicate that differences in eccentricity ratio have a significant effect on column stress when the slenderness ratio is small. On the other hand, columns with large slenderness ratios tend to fail at the Euler (or critical) load, regardless of the eccentricity ratio.

If $\sigma_{\text{max}} = \sigma_{\text{all}} = S_y/2$, Eq. (9.32) becomes

$$P = \frac{S_y A/2}{1 + \frac{ec}{r_g^2} \sec\left(\frac{l_e}{2r_g}\sqrt{\frac{P}{EA}}\right)} = f(P). \qquad (9.33)$$

Determining the load that results in buckling for eccentric columns cannot be done in closed form, and numerical or iterative methods are needed. For example, an initial guess for P is the value obtained from the concentric loading condition or

$$P = P_{\text{cr}}. \qquad (9.34)$$

Using Eq. (9.34) in Eq. (9.33) gives a new value of P. This iterative process is continued until a desired accuracy is achieved.

Example 9.5: Eccentric Loading

Given: A 36-in.-long hollow 2014 aluminum alloy tube (S_y = 14 ski) with a 3.0-in. outside diameter, a 0.03-in. wall thickness, and with both ends pinned.

Find: The critical buckling load for

(*a*) Concentric loading

(*b*) Eccentric loading, with an eccentricity of 0.15 in.

Solution:

(*a*) The inside diameter of the column is

$$d_i = d_o - 2t_h = 3.0 - 2(0.03) = 2.94 \text{ in.}$$

The cross sectional area is

$$A = \frac{\pi}{4}\left(d_o^2 - d_i^2\right) = \frac{\pi}{4}\left[(3)^2 - (2.94)^2\right] = 0.280 \text{ in.}^2$$

The area moment of inertia is

$$I = \frac{\pi}{64}\left(d_o^4 - d_i^4\right) = \frac{\pi}{64}\left[(3)^4 - (2.94)^4\right] = 0.309 \text{ in.}^4$$

The radius of gyration is

$$r_g = \sqrt{\frac{I}{A}} = \sqrt{\frac{0.309}{0.280}} = 1.050 \text{ in.}$$

From Table 9.1 for a column with both ends pinned the effective length is equal to the actual length ($l = l_e$). From Table 3.1 for 2014 aluminum alloy, E = 10.5 Mpsi. From Eq. (9.18),

$$\left(\frac{l_e}{r_g}\right)_T = \sqrt{\frac{2\pi^2 E}{S_y}} = \sqrt{\frac{2\pi^2\left(10.5 \times 10^6\right)}{14 \times 10^3}} = 121.7,$$

and

$$\frac{l_e}{r_g} = \frac{36}{1.05} = 34.29.$$

Since $l_e/r_g < (l_e/r_g)_T$, the Johnson formula [Eq. (9.16)] should be used:

$$\begin{aligned} P_{\text{cr}} &= AS_y\left[1 - \frac{S_y}{4\pi^2 E}\left(\frac{l_e}{r_g}\right)^2\right] \\ &= (0.28)\left(14 \times 10^3\right) \\ &\quad \times \left[1 - \frac{\left(14 \times 10^3\right)(34.29)^2}{4\pi^2\left(10.5 \times 10^6\right)}\right], \end{aligned}$$

or P_{cr} = 3764 lb.

(*b*) The distance from the neutral axis to the outer fiber is $c = d_o/2 = 1.5$ in. The eccentricity ratio is

$$\frac{ce}{r_g^2} = \frac{(1.5)(0.15)}{(1.05)^2} = 0.2041;$$

$$P = P_{\text{cr}} = 3764 \text{ lb.}$$

From Eq. (9.33), and noting that $S_y A/2 = 1960$ lb,

$$P_1 = \frac{1960}{1 + 0.2041 \sec\left[\frac{34.29}{2}\sqrt{\frac{3764}{\left(10.5 \times 10^6\right)(0.280)}}\right]},$$

or P_1 = 1569 lb. A second iteration leads to

$$P_2 = \frac{1960}{1 + 0.2041 \sec\left[\frac{34.29}{2}\sqrt{\frac{1569}{\left(10.5 \times 10^6\right)(0.280)}}\right]},$$

or P_2 = 1628 lb. Further iterations do not result in more accurate solutions. The critical load for the eccentric loading is 1628 lb and that for the concentric loading is 3764 lb.

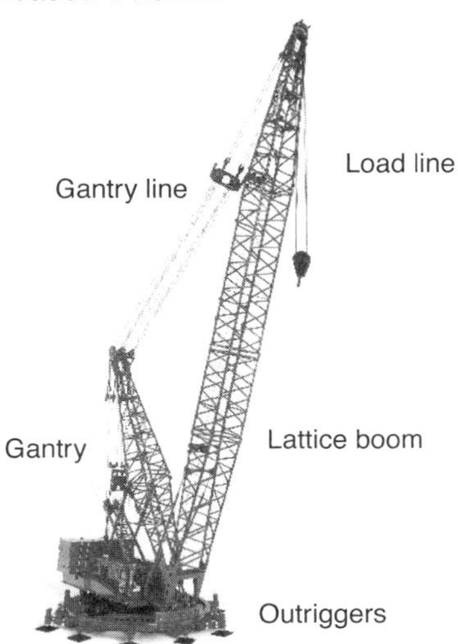

Figure 9.9: Schematic illustration of a Manitowoc 4100W RINGER crane, showing the lattice boom, outriggers, and load and gantry lines. *Source:* Courtesy of Manitowoc Co., Inc.

Figure 9.10: Illustration of a boom segment. A typical box boom consists of larger tubes or angle channels at the corners, supported by smaller braces or cords along the length. Boom segments are available in lengths ranging from 3-12 meters. *Source:* Courtesy of Manitowoc Co., Inc.

Case Study 9.1: Design Considerations for Lattice-Boom Cranes

Lattice-boom cranes are extremely common for material handling, especially in the shipping and construction industries. A schematic of a lattice-boom crane is shown in Fig. 9.9. As opposed to telescoping boom cranes, hydraulic powered cranes, or gantry type cranes, lattice-boom cranes are intended for applications where loads are very high and long boom lengths are essential. Some of the design considerations for lattice-boom cranes are:

1. The booms are constructed from segments, such as the one shown in Fig. 9.10. The segment consists of a rectangular box defined at the corners by large channels, or cords with smaller cross-bracing spaced as shown. Boom segments are usually welded fabrications (see Section 16.6) and are joined together using prestressed threaded fasteners or pins (Sections 16.4 and 16.5). A typical boom segment is 3 to 12 meters long, so a crane with a long boom requires multiple segments to be assembled. It is not feasible to produce booms of long length directly, since manufacture and transport of the booms would be overly burdensome.

2. The load is supported by the *load line,* and the *gantry lines* set the angle of the boom. Note that the boom is constructed in a horizontal position and then needs to be raised by the gantry. The weight of the boom causes a bending stress to be applied to the boom when it is lifted off the ground; this stress must be supported by the cords.

3. As can be shown from statics, the applied loads in the gantry and load lines result in pure compressive loading on the boom. Deviation from pure compressive loading occurs when the load line is not vertical, such as due to wind loadings. Crane manufacturers will specify environmental conditions (such as maximum wind speed) that should not be exceeded. However, if the load line deviates from the vertical, a horizontal force is applied to the tip of the boom, resulting in a bending moment that must be supported.

4. At low boom angles, boom buckling is not likely. This is because the large moment arm presented by the boom results in tipping of the crane should loads become excessive. Crane tipover is a serious concern, and a load chart provided by the manufacturer must be heeded to prevent such accidents. It should be noted that the crane's capacity depends on boom angle, length, and orientation with respect to the crane's axis. Special care must be taken to ensure that the crane's outriggers are properly deployed and that the ground beneath the outriggers will not collapse. (Shoring or repositioning of the crane are common corrective measures if ground strength is suspect.)

5. At high boom angles, the capacity of the crane is limited by the buckling strength of the boom. The lattice structure ensures that the cords will buckle in a high mode [see Eq. (9.6)], which can also be interpreted as a short effective length.

Buckling phenomena are well understood, and mathematical tools are available to confirm designs are adequate. Regardless, there are still occasional boom collapses, but almost always these involve misuse of cranes. Some examples are:

- In one accident, a crane boom collapsed during transport of a ship hull that was under construction. A large gantry crane normally used for this task was being repaired, so three lattice-boom cranes were used to move the hull. Clearly, the cranes cannot move while their outriggers are deployed, so the user would hook up the load lines, have the cranes raise the hull off the ground and then swing the booms to the side and lower the hull to the ground. After repositioning the cranes and resetting the outriggers, the procedure would be repeated. At some point, one of the lattice-booms buckled. A reconstruction of the accident showed that the loads that would cause boom buckling required a side load of around 20% of the vertical load — a situation that exceeded the side load that would be seen in a Class 1 hurricane!

- Another lattice-boom buckling failure involved a crane that was hooked up to an object that exceeded the crane's lift capability. That is, the load was so high that the hoist motor could not generate sufficient torque to raise the load. As a result, the crane operator attached the load to the load line at a high boom angle and applied a tension to the line until the hoist motor stalled, and then applied the hoist line brake. (The brake has a significantly higher capacity than the hoist motor.) It was only at this point that the operator lowered the outriggers. Since each outrigger can lift the entire crane and its maximum load, and the outriggers raise the crane a few inches above the ground, this had the effect of lifting the load. Not surprisingly, the boom buckled when the crane user rotated the crane superstructure.

Normally, lattice-booms are extremely reliable and of great utility. Extreme misuse is usually involved when such a boom buckles.

9.9 Summary

In an attempt to better understand column buckling, three equilibrium regimes were studied. Stable, neutral, and unstable equilibrium were explained by observing what happens to a sphere or cylinder on concave, flat, and convex surfaces. It was concluded that when columns are in unstable equilibrium, they can buckle. A small displacement can cause a catastrophic change in the configuration of the column. As the load on a column is increased, a critical load is reached where unstable equilibrium will occur and the column will not return to its straight configuration. Thus, for long, slender columns there is a critical buckling load. The column's critical buckling stress was also determined to be much less than the yield strength of the material. End conditions were found to affect the critical buckling load. The effective length was used to handle four types of end condition. The Euler and Johnson buckling criteria were developed for concentrically loaded columns. Eccentrically loaded columns were also studied and the secant formulas were derived.

Key Words

AISC equations estimations of allowable stress for prevention of buckling in structures (corrections for capacity as column stress exceeds proportional limit)

buckling sudden large deformation of a structure due to slight increase of applied load

column straight and long (relative to cross sectional dimension) member subjected to compressive axial loads

eccentricity ratio measure of how far a load is applied from a cross section's centroid

Essenger equation an expression that incorporates the tangent modulus into buckling load calculations

Euler load a statement of the critical load of an elastic column

Johnson equation an equation used to determine the critical load for inelastic buckling

secant equation an expression that allows calculation of column deflection due to eccentric loading

slenderness ratio measure of column slenderness

tangent modulus elastic modulus at a given stress level

tangent-modulus equation same as Essenger equation

Summary of Equations

Euler buckling equation: $P_{\text{cr}} = \dfrac{n^2\pi^2 EI}{l^2}$

Critical stress in Euler buckling: $(\sigma_{\text{cr}})_E = \dfrac{P_{\text{cr}}}{A} = \dfrac{\pi^2 E}{(l/r_g)^2}$

Inelastic buckling (Essenger equation): $\sigma_{\text{cr}} = \dfrac{\pi^2 E_t}{(l/r_g)^2}$

Critical slenderness ratio: $C_c = \left(\dfrac{l_e}{r_g}\right)_E = \sqrt{\dfrac{2E\pi^2}{S_y}}$

Johnson equation: $(\sigma_{\text{cr}})_J = \dfrac{(P_{\text{cr}})_J}{A} = S_y - \dfrac{S_y^2}{4\pi^2 E}\left(\dfrac{l_e}{r_g}\right)^2$

AISC buckling equations:

$$\text{Elastic buckling: } \sigma_{\text{all}} = \frac{12\pi^2 E}{23\,(l_e/r_g)^2}$$

$$\text{Inelastic buckling: } \sigma_{\text{all}} = \frac{\left\{1 - \left[\dfrac{(l_e/r_g)^2}{2C_c^2}\right]\right\} S_y}{n_\sigma},$$

$$\text{where } n_\sigma = \frac{5}{3} + \frac{3\,(l_e/r_g)}{8C_c} - \frac{(l_e/r_g)^3}{8C_c^3}$$

Secant equation: $y_{\max} = e\left[\sec\left(\dfrac{l_e}{2}\sqrt{\dfrac{P}{EI}}\right) - 1\right]$

Recommended Readings

Budynas, R.G., and Nisbett, J.K. (2011), *Shigley's Mechanical Engineering Design*, 9th ed., McGraw-Hill.

Juvinall, R.C., and Marshek, K.M. (2012) *Fundamentals of Machine Component Design*, 5th ed., Wiley.

Ketter, R.L., Lee, G.C., and Prawel, S.P. (1979) *Structural Analysis and Design*, McGraw-Hill.

Manual of Steel Construction (2011) 14th ed., American Institute of Steel Construction.

Norton, R.L. (2011) *Machine Design*, 4th ed., Prentice Hall.

Popov, E.P. (1999) *Engineering Mechanics of Solids*, 2nd ed., Prentice-Hall.

Timoshenko, S.P., and Goodier, J.N. (1970) *Theory of Elasticity*, McGraw-Hill.

Ugural, A.C., and Fenster, S.K. (2003) *Advanced Strength and Applied Elasticity*, 4th ed., Prentice-Hall.

Willems, A. (2009) *Structural Analysis in Theory and Practice*, Elsevier.

Reference

American Institute of Steel Construction (AISC) (2011) *Manual of Steel Construction*, 14th ed.

Questions

9.1 What is a column?

9.2 Define the terms neutral, stable, and unstable equilibrium.

9.3 What is the Euler equation?

9.4 Describe the tangent modulus using a stress strain curve that is typical for a carbon steel.

9.5 What is the slenderness ratio of a column?

9.6 Explain why the Euler equation does not contain the material's yield strength.

9.7 What is the Johnson equation?

9.8 What are the AISC criteria for buckling?

9.9 Define the eccentricity ratio.

9.10 What is the secant equation?

Qualitative Problems

9.11 A person rides a bike on a flat level road. Is this a neutral, stable, or unstable equilibrium position? Explain.

9.12 A golf ball is placed

(a) On top of a small hill

(b) On a horizontal flat plane

(c) In a shallow groove

Identify the type of equilibrium for each condition. Justify your answer.

9.13 List the assumptions necessary for Euler buckling to take place.

9.14 Review Table 9.1 and explain the reasons that the theoretical effective column length and approximated end conditions have the values listed.

9.15 Martensitic steels can be produced that have yield strengths as high as 1800 MPa, but they are difficult to form into structural shapes and are more expensive than other steels. List the advantages of using such materials in compression members.

9.16 Refer to Eq. (9.20) and determine if the reduction in allowable stress is larger or smaller for slender columns compared to more compact columns. Explain why you think this trend is justified.

9.17 Review Design Procedure 9.1. If the first step assumed the Johnson equation is appropriate, explain how the remainder of the Design Procedure would need to be modified.

Quantitative Problems

9.18 A column has pinned ends and is axially loaded in compression. The length is 6 m and the weight is 2310 N, but the form of the cross section can be changed. The column is made of steel (annealed AISI 1040). Find the Euler buckling load for

(a) A solid circular section. *Ans.* $P_{cr} = 113.0$ kN.

(b) A solid square section. *Ans.* $P_{cr} = 118.2$ kN.

(c) A circular tube with outer diameter of 100 mm. *Ans.* $P_{cr} = 241.8$ kN.

(d) A square tube with outside dimension of 100 mm. *Ans.* $P_{cr} = 354.7$ kN.

9.19 A column with both ends pinned is 3 m long and has a tubular section with an outer diameter of 30 mm and wall thickness of 5 mm. Find which material given in Tables A.1 through A.4 produces the highest buckling load. Also, give the buckling loads for AISI 1080 steel (quenched and tempered at 800°C), aluminum alloy 2014, and molybdenum. *Ans.* For AISI 1080, $P_{cr} = 7245$ N.

9.20 Determine the critical stresses for the four column cross sections considered in Problem 9.18 if the column length is 10 m and constructed from AISI 1080 steel, quenched and tempered at 800° C. *Ans.* For a solid square cross section, $\sigma_{cr} = 8.511$ MPa.

9.21 A column is axially loaded in compression. The ends were specified to be fixed, but because of a manufacturing error they had

(a) One end fixed and the other pinned

(b) Both ends pinned

Find how much the critical elastic buckling load is decreased because of the errors. Also, calculate the buckling loads for the 4-m-long column made of annealed AISI 1040 steel having a solid, square cross section with 30-mm sides for the three end conditions. Assume theoretical effective column length. *Ans.* For fixed-fixed, $P_{cr} = 11.26$ kN.

9.22 An elastic column has one end pinned and the other end fixed in a bushing so that the values given in Table 9.1 do not apply. Instead, the effective length is $l_e = 0.83l$. The cross section of the column is a circular tube with an outer diameter of 80 mm and 3.0-mm wall thickness.; the column is 10 m long. Calculate the elastic buckling load if the column is made of

(a) AISI 1080 steel. *Ans.* 15.97 kN.

(b) Polycarbonate. *Ans.* 183 N.

9.23 An elastic AISI 1020 (quenched and tempered at 870°C) steel column has both ends pinned. It is 12.5 m long and has a square tubular cross section with an outside dimension of 160 mm and 4-mm wall thickness. Its compressive axial load is 100 kN.

(a) Determine the safety factors guarding against buckling and yielding. *Ans.* $n_{s,\text{buckling}} = 1.325$, $n_{s,\text{yielding}} = 7.36$.

(b) If the ends are changed to fixed and the material is changed to aluminum alloy 2014, calculate the safety factors guarding against buckling and yielding.

9.24 A beam has both ends mounted in stiff rubber bushings giving bending moments in the beam ends proportional to the angular displacements at the beam ends. Calculate the effective beam length if the angular spring constant at the ends is $\partial M/\partial\theta = 10^5$ N-m/radian in all directions. The beam is 3 m long and has a solid circular cross section of AISI 1080 steel (quenched and tempered at 800°C) with a 24-mm diameter. *Ans.* 1.534 m.

9.25 Two solid circular columns are made of different materials, steel and aluminum. The cross sectional areas are the same and the moduli of elasticity are 207 GPa for the steel and 72 GPa for the aluminum. Find the ratio of the critical buckling lengths for the columns, assuming that the same buckling load is applied to both columns. *Ans.* $l_{es}/l_{ea} = 1.696$.

9.26 A television mast shown in Sketch a consists of a circular tube with an outer diameter d_o and a wall thickness t_h. Calculate how long the distance l between the anchoring points for the guy wires can be if the mast should deform plastically rather than buckle.

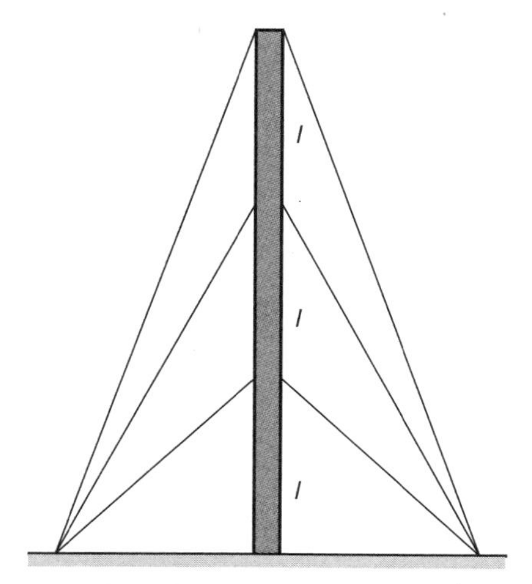

Sketch a, for Problem 9.26

9.27 A 3-m-long column of square, solid cross section with both ends fixed must sustain a critical load of 3×10^5 N. The material is steel with a modulus of elasticity of 207 GPa and a yield stress of 700 MPa. Determine the following:

(a) What minimum dimensions of the cross section are permitted without having failure? Should the Euler or the Johnson equation be used? *Ans.* $w = 0.045$ m.

(b) If the critical load is increased by two orders of magnitude to $P_{cr} = 3 \times 10^7$ N, what minimum dimensions of the cross section are permitted without failure? Also, indicate whether the Euler or the Johnson equation should be used.

9.28 A solid, round column with a length of 5 m and a diameter of 50 mm is fixed at one end and is free at the other end. The material's yield strength is 300 MPa and its modulus of elasticity is 207 GPa. Assuming concentric loading of the column, determine the following:

(a) The critical load. *Ans.* $P_{cr} = 6.268$ kN.

(b) The critical load if the free end is also fixed (i.e., both ends are now fixed). *Ans.* $P_{cr} = 534$ kN.

9.29 A circular cross section bar with a diameter of 2 in. and a length of 50 in. is axially loaded. The bar is made of medium-carbon steel. Both ends are pinned. Determine

(a) Whether the Johnson or the Euler formula should be used.

(b) The critical load. *Ans.* 91.18 kip.

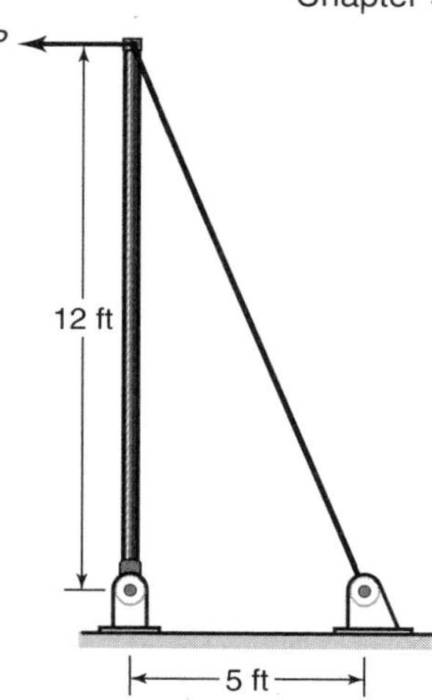

Sketch b, for Problem 9.30

9.30 A low-carbon-steel pipe (AISI 1020, quenched and tempered at 870°C), as shown in Sketch b, has an outer diameter of 2 in. and a thickness of 0.25 in. If the pipe is held in place by a guy wire, determine the largest horizontal force P that can be applied without causing the pipe to buckle. Assume that the ends of the pipe are pin connected. *Ans.* 3.193 kip.

9.31 A 20-mm-diameter, medium-carbon steel (annealed AISI 1040) rod is loaded as a column with pinned ends. If the critical load is 100 kN, how long can the rod be and still carry the following percentages of the critical load: (*a*) 90%, (*b*) 50%, and (*c*) 2%? Also, indicate whether the Johnson or the Euler equation should be used. *Ans.* (*a*) 0.3256 m; (*b*) 0.5663 m; (*c*) 2.832 m.

9.32 Most types of steel have similar moduli of elasticity but can have widely different yield strength properties depending on the alloy and the heat treatment. For a column having a solid, circular cross section with a diameter of 25 mm, the yield strengths for three steels are 300, 600, and 1000 MPa. Find for each steel the critical column length where buckling would be described by the Euler equation. *Ans.* For $S_y = 300$ MPa, $l_e = 0.7294$ m.

9.33 For a fixed cross sectional area of a tubular steel column, find the geometry that would give the highest buckling load. The minimum column wall thickness is 5 mm.

9.34 A rectangular cross section tube is axially loaded in compression. The tube's outside dimensions are 70 and 90 mm, it is 4 m long with both ends pinned; the wall thickness is 10 mm. Determine which of the steels in the table in Appendix A.2 should be used so that the Euler equation is applicable.

9.35 The column considered in Problem 9.34 is made of annealed AISI 1020 steel, quenched and tempered at 870°C. Find the critical buckling load. *Ans.* 360 kN.

9.36 A 2.5-m-long column with one end fixed and the other end free is made of aluminum alloy 2014-O and has a solid, round cross section. Determine the diameter of the column for the following loads

(a) $P = 800$ kN. *Ans.* $d = 0.1547$ m.

(b) $P = 1500$ kN.

9.37 A 1-m-long column with a 60-mm × 100-mm rectangular cross section and made of aluminum alloy 2014-O is subjected to a compressive axial load. Determine the critical buckling load

(a) If both ends are fixed. *Ans.* 565 kN.

(b) If one end is fixed and the other is free. *Ans.* 316 kN.

(c) If the load is applied eccentrically at a distance of 10 mm for case b. *Ans.* 109.6 kN.

9.38 A column with pinned ends and solid rectangular cross section with dimensions of 35 mm × 60 mm is made of annealed AISI 1040 steel. The 4-m-long column is loaded with an eccentricity of 15 mm. Find the elastic deflection if a 100-N compressive load is applied; also, calculate how large a load can be applied without permanent deformation occurring. *Ans.* $y_{\text{max}} = 68\ \mu\text{m}$, $P = 21.94$ kN.

9.39 The column considered in Problem 9.38 is subjected to a varying load in the range of 0 to 24,000 N. Calculate and plot the deformation as a function of the axial load.

Synthesis and Design

9.40 Obtain an aluminum beverage can and carefully cut the can in half and measure the wall thickness and can diameter. Calculate the critical buckling load for the can and indicate if this is greater than or less than your weight. Try to step onto another empty can. Does the can buckle? Explain why or why not. Assume the yield strength is 180 MPa and the elastic modulus is 69 GPa.

9.41 Using a compression test machine, compress bars of different lengths to provide experimental data for Table 9.1. It may be prudent to calculate required column lengths for a given cross section to make sure that elastic and inelastic buckling regimes are investigated.

9.42 Derive an expression for the weight of a column with a solid circular cross section as a function of material properties, length, and applied load assuming Euler buckling. What combination of material properties lead to the lightest weight column?

9.43 Obtain an expression for the height of a cylindrical column that will buckle under its own weight (assume Euler buckling). Compare the height that can be achieved for the materials on the inside front cover.

9.44 In *double modulus theory*, it is recognized that during buckling the tensile side of the column's cross section should use the tangent modulus of elasticity, whereas on the compressive side it may be better to use the original elastic modulus. Model the cross section of a material with two moduli, E_t and E, and show that the buckling load is

$$P_{\text{cr}} = \frac{\pi^2 E_r I}{L^2}$$

where E_r is the *reduced modulus* given by $E_r = E_t\frac{I_1}{I} + E\frac{I_2}{I}$ and I_1 and I_2 are the fractions of the moment of inertia on the tensile and compressive sides of the column, respectively.

9.45 Conduct a literature search and obtain design rules for buckling of aluminum that are analogous to Eqs. (9.19) through (9.21).

9.46 A fishing pole uses eyelets to guide the fishing line. Obtain a fishing pole and demonstrate that the pole is stiffer if all of the eyelets are used compared to when only the one at the top of the pole is used. Explain why.

9.47 In Case Study 9.1, it was explained that lattice booms are intended to buckle in a high buckling mode. Make a careful sketch of the buckling mode for a lattice boom section both with and without side load.

Chapter 10

Stresses and Deformations in Cylinders

A common beverage can. Along with food containers, these are the most common pressure vessels. *Source:* Shutterstock.

In all things, success depends on previous preparation. And without such preparation there is sure to be failure.

Confucius, *Analects*

Contents

Examples

Design Procedure

Case Study

This chapter investigates the stresses and deformations that occur in internally or externally pressurized cylinders. This basic geometry sees many practical applications, including pressure vessels such as gas storage cylinders, food and beverage cans and bottles, fuel tanks, hydraulic actuators and gun barrels; press and shrink fits; hydraulic and pneumatic tubing used for delivery of pressurized fluid; and pumps and motors. The chapter begins with a discussion of tolerances, allowances, and fits, which are essential for understanding the demanding design constraints that are needed for press and shrink fits. Classes of fit and recommended tolerances for different applications are summarized. The membrane stresses for thin-walled pressure vessels are stated, and the full linear elastic solution for thick-walled pressure vessels is derived and simplified for common special cases. This theory is then applied to press and shrink fits, two very common methods of assembly. The required assembly force for press fits and the allowable torque that can be achieved for both press and shrink fits are derived.

Machine elements in this chapter: Pressure vessels, press and shrink fits, hydraulic lines and pipelines.
Typical applications: Liquid and gas storage, assembly of parts onto shafts, actuators such as hydraulic cylinders, gun barrels.
Competing machine elements: Keyways, setscrews, splines and couplings (Chapter 11), power screws (Chapter 16), welding (Chapter 16).

Symbols

A	cross-sectional area, m^2
C_1, C_2	integration constants
c	radial clearance, m
d	diameter, m
E	modulus of elasticity, Pa
l	length, m
n_s	safety factor
P	force, N
P_b	body force per volume, N/m^3
p	pressure, Pa
r	radius, m
S_y	yield strength, Pa
T	torque, N-m
t_h	thickness, m
t_l	tolerance, m
Δt_m	temperature change, °C
x, y, z	Cartesian coordinates, m
α	cone angle, deg
β	coefficient of linear thermal expansion, (°C)$^{-1}$
δ	interference or displacement, m
ϵ	strain
θ	circumferential direction, rad
μ	coefficient of friction
ν	Poisson's ratio
ρ	density, kg/m^3
σ	normal stress, Pa
ω	angular velocity, rad/s

Subscripts

a	axial
c	circumferential
e	von Mises
f	fit
h	hub
i	inner or internal
o	outer or external
r	radial
s	shaft
t	tangential
t_m	temperature change
θ	circumferential direction

10.1 Introduction

Just as the last chapter dealt with a specific loading and the unique stresses and strains acting on columns, this chapter deals with internally and/or externally pressurized cylinders. Many important engineering applications rely on this condition.

The material in this chapter is important for understanding machine elements presented later in the text. Section 10.2 discusses tolerances and fits, a concept that is essential for design of mating parts. Section 10.3 addresses thin- and thick-walled cylinders and the stresses that develop due to internal and external pressurization, and Section 10.4 includes rotational effects. Using the theoretical approaches developed, press and shrink fits are covered in Sections 10.5 and 10.6, so that the casual reader may wish to proceed directly to those sections.

Stresses and deformations of thin-walled, thick-walled, internally pressurized, externally pressurized, and rotating cylinders are considered, as well as press and shrink fits. The material developed in this chapter is important to shafting (Chapter 11) and to a number of machine elements, such as rolling-element bearings (Chapter 13) and hydrodynamic bearings (Chapter 12).

10.2 Tolerances and Fits

A number of definitions about tolerancing are important in design:

1. **Tolerance**, t_l, is the maximum variation in the size of a part. Two types of tolerance are:
   (a) **Bilateral tolerance** — A part is permitted to vary both above and below the nominal size, such as 1.5 ± 0.003.
   (b) **Unilateral tolerance** — A part is permitted to vary either above or below the nominal size, but not both, such as $1.5^{+0.000}_{-0.003}$.
2. **Nominal diameter**, d, is the approximate size; allowances and tolerances are applied with respect to the nominal dimension.
3. **Allowance**, a, is the difference between the nominal diameters of mating parts. Allowance is used when the mating parts have a measurable positive gap between them.
4. **Interference**, δ, is the actual difference in the size of mating parts. Interference is used when the mating parts have an overlap, and can be considered the inverse of allowance.

Note that the tolerances of the shaft, t_{ls}, and the hub, t_{lh}, may be different.

Because it is impossible or impractical from a manufacturing perspective to make parts to an exact size, tolerances are used to control the variation between mating parts. Two examples illustrate the importance of tolerancing:

1. In hydrodynamic bearings (Chapter 12) a critical part of the design is the specification of the radial clearance between the journal and the bearing. The typical value is on the order of 0.02 mm. However, variations in the journal's outside diameter and the bearing's inside diameter cause larger or smaller clearances. Such variations must be accounted for in analyzing bearing performance. Too small a clearance could cause failure; too large a clearance would reduce the precision of the machine and adversely affect the lubrication. Thus, tolerancing and accuracy of the dimensions can have a significant effect on the performance of hydrodynamic bearings.
2. Rolling-element bearings (Chapter 13) are often designed to be installed on a shaft (Chapter 11) with an interference fit. The inside diameter of the bearing inner race is smaller than the outside diameter of the shaft where the bearing is to be seated. A significant force is required to press the bearing onto the shaft, thus imposing significant stresses on both the shaft and the bearing. Specification of the shaft diameter relative to the bearing bore size is important to ensure that failure due to overstressing the members does not occur. Proper tolerancing of the members will greatly contribute to successful design.

Table 10.1: Classes of fit.

Class	Type	Applications
1 (Loose)	Clearance	Where accuracy is not essential, such as in building and mining equipment
2 (Free)	Clearance	In rotating journals with speeds of 600 rpm or greater, such as in engines and some automotive parts
3 (Medium)	Clearance	In rotating journals with speeds under 600 rpm, such as in precision machine tools and precise automotive parts
4 (Snug)	Clearance	Where small clearance is permissible and where mating parts are not intended to move freely under load
5 (Wringing)	Interference	Where light tapping with a hammer is necessary to assemble the parts
6 (Tight)	Interference	In semipermanent assemblies suitable for drive or shrink fits on light sections
7 (Medium)	Interference	Where considerable pressure is required for assembly and for shrink fits of medium sections; suitable for press fits on generator and motor armatures and for automotive wheels
8 (Shrink)	Interference	Where considerable bonding between surfaces is required, such as locomotive wheels and heavy crankshaft disks of large engines

Table 10.1 describes the eight classes of fit, the type of fit, and their applications. Note that for classes 1 to 4 the members are interchangeable but for classes 5 to 8 the members are not interchangeable. Having established classes of fit, the next task will be to indicate the tolerance applicable for a specific class. Tables 10.2 and 10.3 give the tolerances for various classes of fit in inches and millimeters, respectively. In these tables the nominal diameter is designated by d.

Having established the allowance, interference, and tolerance, the next step is to establish the maximum and minimum diameters of the hub and shaft. Table 10.4 gives these dimensions for the two types of fit (clearance and interference). Tables 10.2 through 10.4 thus establish the upper and lower limits of the shaft and hub diameters for the eight classes of fit.

Example 10.1: Interference and Class of Fit

Given: A medium-force interference fit is applied to a shaft with a nominal diameter of 75 mm.

Find: Determine the interference, the hole and shaft tolerances, and the hub and shaft diameters.

Solution: From Table 10.1, the class of fit is 7. From Table 10.3 for a class 7 fit,

$$\delta = 0.0005d = 0.0005(75) = 0.0375 \text{ mm},$$

$$t_{lh} = t_{ls} = 0.0052d^{1/3} = 0.0052(3)^{1/3} = 0.0219 \text{ mm}.$$

From Table 10.4 for interference fit, the hub diameter is

$$d_{h,\max} = d + t_{lh} = 75 + 0.0219 = 75.0219 \text{ mm};$$

$$d_{h,\min} = d = 75 \text{ mm}.$$

and the shaft diameter is

$$\begin{aligned} d_{s,\max} &= d + \delta + t_{ls} \\ &= 75 + 0.0375 + 0.0219 \\ &= 75.059 \text{ mm}, \\ d_{s,\min} &= d + \delta \\ &= 75 + 0.0219 \\ &= 75.0219 \text{ mm}. \end{aligned}$$

10.3 Pressurization Effects

This section presents thin-walled and thick-walled cylinders with internal and external pressurization. The material in this section is used for a wide range of applications, some of which are

1. Pressure vessels loaded internally and/or externally
2. Storage tanks
3. Gun barrels
4. Hydraulic and pneumatic cylinders
5. Transmission pipelines
6. Machine elements, such as rolling-element bearing races or gears pressed onto shafts

A distinction must be made between **thin-walled cylinders** and **thick-walled cylinders**. In this book, when a cylinder's inner diameter, d_i, is 40 times larger than its thickness t_h, it will be considered as thin-walled. For smaller ratios of cylinder inner diameter to thickness, thick-wall analysis should be used. This criterion is somewhat arbitrary and depends on the desired accuracy in stress analysis, but is a good guideline. Mathematically expressing the above gives

$$\frac{d_i}{t_h} > 40 \quad \rightarrow \quad \text{thin-walled cylinders,} \tag{10.1}$$

$$\frac{d_i}{t_h} < 40 \quad \rightarrow \quad \text{thick-walled cylinders.} \tag{10.2}$$

10.3.1 Thin-Walled Cylinders

Figure 10.1a shows a thin-walled cylinder subjected to internal pressure, p_i. It is assumed that the stress distribution is uniform throughout the thickness. The radial stress is small relative to the circumferential stress because $t_h/d_i \ll 1$. Thus, a small element can be considered to be in plane stress with the principal stresses shown in Fig. 10.1b.

Figure 10.2, the front view of the cylinder shown in Fig. 10.1, shows the forces acting on a small element due to the internal pressure. This element also has a length dl in the thickness direction. Applying force equilibrium in the radial direction gives

$$p_i r_i \, d\theta \, dl = 2\sigma_{\theta,\text{avg}} \sin\left(\frac{d\theta}{2}\right) t_h \, dl.$$

Since $d\theta/2$ is small, $\sin(d\theta/2) \approx d\theta/2$ and thus

$$\sigma_{\theta,\text{avg}} = \frac{p_i r_i}{t_h}. \tag{10.3}$$

Table 10.2: Recommended tolerances in *inches* for classes of fit.

Class	Allowance, a	Interference, δ	Hub tolerance, t_{lh}	Shaft tolerance, t_{ls}
1	$0.0025d^{2/3}$	—	$0.0025d^{1/3}$	$0.0025d^{1/3}$
2	$0.0014d^{2/3}$	—	$0.0013d^{1/3}$	$0.0013d^{1/3}$
3	$0.0009d^{2/3}$	—	$0.0008d^{1/3}$	$0.0008d^{1/3}$
4	0.000	—	$0.0006d^{1/3}$	$0.0004d^{1/3}$
5	—	0.000	$0.0006d^{1/3}$	$0.0004d^{1/3}$
6	—	$0.00025d$	$0.0006d^{1/3}$	$0.0006d^{1/3}$
7	—	$0.0005d$	$0.0006d^{1/3}$	$0.0006d^{1/3}$
8	—	$0.0010d$	$0.0006d^{1/3}$	$0.0006d^{1/3}$

Table 10.3: Recommended tolerances in *millimeters* for classes of fit.

Class	Allowance, a	Interference, δ	Hub tolerance, t_{lh}	Shaft tolerance, t_{ls}
1	$0.0073d^{2/3}$	—	$0.0216d^{1/3}$	$0.0216d^{1/3}$
2	$0.0041d^{2/3}$	—	$0.0112d^{1/3}$	$0.0112d^{1/3}$
3	$0.0026d^{2/3}$	—	$0.0069d^{1/3}$	$0.0069d^{1/3}$
4	0.000	—	$0.0052d^{1/3}$	$0.0035d^{1/3}$
5	—	0.000	$0.0052d^{1/3}$	$0.0035d^{1/3}$
6	—	$0.00025d$	$0.0052d^{1/3}$	$0.0052d^{1/3}$
7	—	$0.0005d$	$0.0052d^{1/3}$	$0.0052d^{1/3}$
8	—	$0.0010d$	$0.0052d^{1/3}$	$0.0052d^{1/3}$

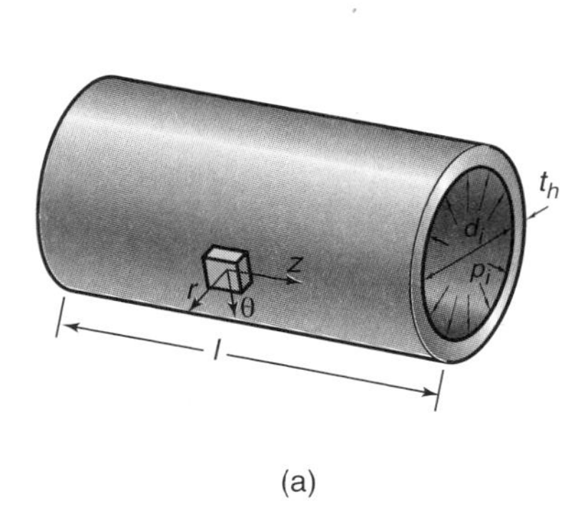

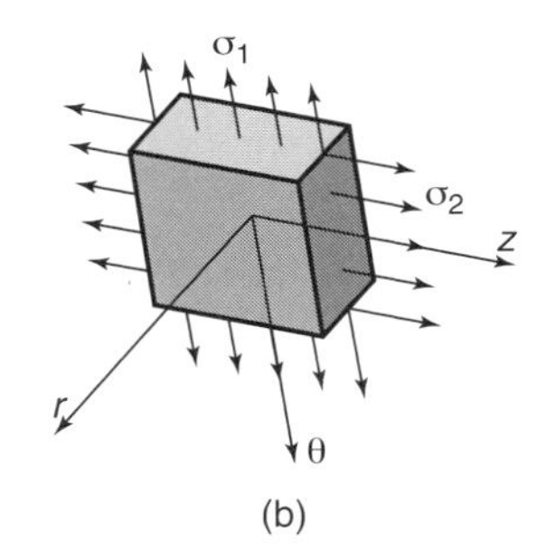

Figure 10.1: Internally pressurized thin-walled cylinder. (a) Stress element on cylinder; (b) stresses acting on element.

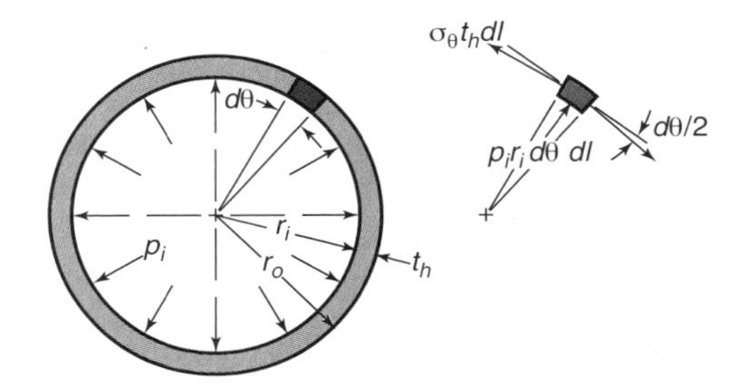

Figure 10.2: Front view of internally pressurized, thin-walled cylinder.

The θ component of stress is called the tangential stress or **hoop stress**; the term originates from the manufacture of wooden barrels by shrinking a hot iron ring, or hoop, around closely fitting planks to achieve a seal. Note, for a thin-walled pressure vessel, that $r_i \approx r$, so that

$$\sigma_\theta = \frac{p_i r}{t_h}. \tag{10.4}$$

The area subjected to axial stress is

$$A = \pi(r_o^2 - r_i^2) = 2\pi r_{\text{avg}} t_h. \tag{10.5}$$

Thus, the average axial tensile stress is

$$\sigma_{z,\text{avg}} = \frac{p_i r_i^2}{r_o^2 - r_i^2} = \frac{p_i r_i^2}{2 r_{\text{avg}} t_h}.$$

But since $r_i \approx r$,

$$\sigma_z = \frac{p_i r}{2t_h}. \tag{10.6}$$

The stress in the radial direction is, from equilibrium,

$$\sigma_r = p. \tag{10.7}$$

Table 10.4: Maximum and minimum diameters of shaft and hub for two types of fit.

Type of fit	Hub diameter Maximum $d_{h,\max}$	Hub diameter Minimum $d_{h,\min}$	Shaft diameter Maximum $d_{s,\max}$	Shaft diameter Minimum $d_{s,\min}$
Clearance	$d + t_{lh}$	d	$d - a$	$d - a - t_{ls}$
Interference	$d + t_{lh}$	d	$d + \delta + t_{ls}$	$d + \delta$

However, the pressure is usually much smaller than σ_θ or σ_z, so $\sigma_r = 0$ is a reasonable approximation. Note that the circumferential (hoop) stress is twice the axial stress. In summary, the principal stresses for thin-walled cylinders are:

$$\sigma_1 = \sigma_\theta = \frac{p_i r}{t_h}, \tag{10.8}$$

$$\sigma_2 = \sigma_z = \frac{p_i r}{2t_h}, \tag{10.9}$$

$$\sigma_3 = \sigma_r = 0. \tag{10.10}$$

Example 10.2: Thin-Walled Cylinder

Given: A 100-mm-inner-diameter cylinder made of a material with a yield strength of 400 MPa is subjected to an internal pressure of 2 MPa. Use a safety factor of 3 and assume that thin-wall analysis is adequate.

Find: Determine the wall thickness required to prevent yielding, based on

(*a*) Maximum-shear-stress theory (MSST)

(*b*) Distortion-energy theory (DET)

Solution: From Eqs. (10.8) through (10.10),

$$\sigma_1 = \sigma_\theta = \frac{p_i r}{t_h} = \frac{(2 \times 10^6)(0.050)}{t_h},$$

$$\sigma_2 = \sigma_z = \frac{p_i r}{2t_h} = \frac{(2 \times 10^6)(0.050)}{2t_h},$$

$$\sigma_3 = \sigma_r = 0.$$

Because $\sigma_1 > \sigma_2 > \sigma_3$, the principal stresses are ordered properly according to Eq. (2.18).

(*a*) From Eq. (6.8), yielding occurs according to the MSST when

$$\sigma_1 - \sigma_3 = \frac{S_y}{n_s}.$$

Substituting for σ_1 and σ_3 and solving for the thickness,

$$t_h = \frac{p_i r n_s}{S_y} = \frac{(2 \times 10^6)(0.050)(3)}{400 \times 10^6} = 0.75 \text{ mm}.$$

(*b*) Note in this case that $\sigma_2 = \sigma_1/2$. The von Mises stress for a biaxial stress state is obtained from Eq. (6.12) as:

$$\sigma_e = \left(\sigma_1^2 + \sigma_2^2 - \sigma_1\sigma_2\right)^{1/2} = \left(\frac{3\sigma_1^2}{4}\right)^{1/2} = \sqrt{\frac{3}{4}}\sigma_1.$$

Substituting for σ_1,

$$\sigma_e = (0.8660)\frac{(2 \times 10^6)(0.050)}{t_h} = \frac{(8.66 \times 10^4)}{t_h}.$$

From Eq. (6.11), yielding occurs when

$$\sigma_e = \frac{S_y}{n_s}.$$

Thus, yielding occurs when

$$t_h > \frac{(8.66 \times 10^4)\, n_s}{S_y} = \frac{(8.66 \times 10^4)\,(3)}{400 \times 10^6} = 0.650 \text{ mm}.$$

10.3.2 Thick-Walled Cylinders

It was indicated in Section 10.3.1 that if the cylinder walls are thin, the circumferential or hoop stress, σ_θ, can be assumed to be uniform throughout the wall thickness. This assumption cannot be made for thick-walled cylinders.

Figure 10.3a shows a radially loaded, thick-walled cylinder subjected to internal pressure, p_i, and external pressure, p_o. A stress element in the thick-walled cylinder is shown in Fig. 10.3b. Because the body and the loading are symmetrical about the axis, shear stresses in the circumferential and radial directions are not present, and only normal stresses σ_θ and σ_r act on the stress element. The loading is two-dimensional; therefore, only plane stresses are involved. If an axial loading is superimposed [Eq. (10.6) developed in Section 10.3.1], the third principal stress is merely changed from zero to σ_z. Recall that Eq. (10.6) is valid regardless of cylinder thickness.

Figure 10.4 shows a cylindrical polar element before and after deformation. The radial and circumferential (hoop) displacements are given by δ_r and δ_θ, respectively. The radial strain is given by:

$$\epsilon_r = \frac{\partial \delta_r}{\partial r} = \frac{\delta_r + \frac{\partial \delta_r}{\partial r}dr - \delta_r}{dr}. \tag{10.11}$$

The circumferential strain is associated with the displacement to the new radius $r + \delta_r$, or

$$\epsilon_\theta = \frac{(r + \delta_r)d\theta - r d\theta}{r d\theta} = \frac{\delta_r}{r}. \tag{10.12}$$

The circumferential variations shown in Fig. 10.4 will be ignored due to circular symmetry. Note that asymmetric cross sections, stress concentrations, or local yielding can cause such distortions, but these complications are assumed to be absent from the cylinder being analyzed. From Hooke's law, the stress-strain relationship for the biaxial stress state [see Eq. (B.58)], while making use of Eqs. (10.11) and (10.12), gives

$$\epsilon_r = \frac{\partial \delta_r}{\partial r} = \frac{1}{E}(\sigma_r - \nu\sigma_\theta), \tag{10.13}$$

$$\epsilon_\theta = \frac{\delta_r}{r} = \frac{1}{E}(\sigma_\theta - \nu\sigma_r). \tag{10.14}$$

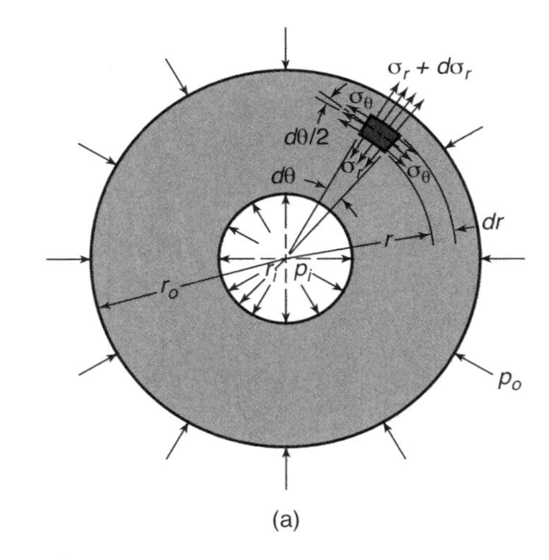

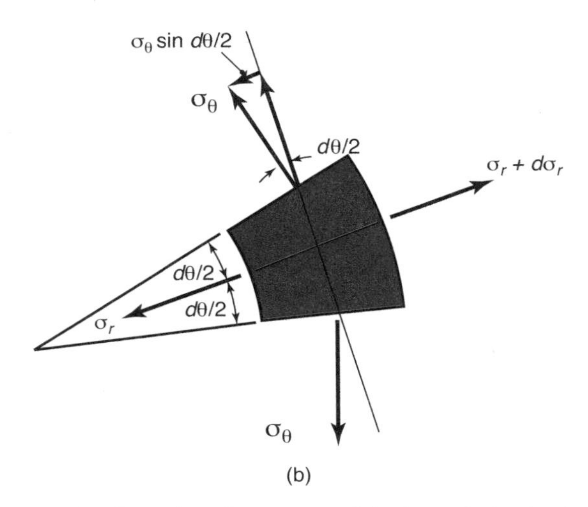

Figure 10.3: Complete front view of thick-walled cylinder internally and externally pressurized. (a) With stresses acting on cylinder; (b) detail of stresses acting on element.

The expressions for the radial and circumferential stresses are obtained from force equilibrium of the element. Figure 10.3b gives details of the stresses. Summing the forces in the radial direction gives

$$(\sigma_r + d\sigma_r)(r + dr)\, d\theta\, dz - \sigma_r r\, d\theta\, dz - 2\sigma_\theta \sin\left(\frac{d\theta}{2}\right) dr\, dz = 0. \tag{10.15}$$

For small $d\theta$, $\sin(d\theta/2) \approx d\theta/2$, and Eq. (10.15) reduces to

$$\sigma_\theta = r\frac{d\sigma_r}{dr} + \sigma_r. \tag{10.16}$$

Equations (10.13), (10.14), and (10.16) are three equations with three unknowns δ_r, σ_r, and σ_θ. Substituting Eq. (10.16) into Eqs. (10.13) and (10.14) and then differentiating Eq. (10.14) with respect to r while equating Eqs. (10.13) and (10.14) gives

$$0 = 3\frac{d\sigma_r}{dr} + r\frac{d^2\sigma_r}{dr^2}. \tag{10.17}$$

Equation (10.17) can be rewritten as

$$0 = 2\frac{d\sigma_r}{dr} + \frac{d}{dr}\left(r\frac{d\sigma_r}{dr}\right). \tag{10.18}$$

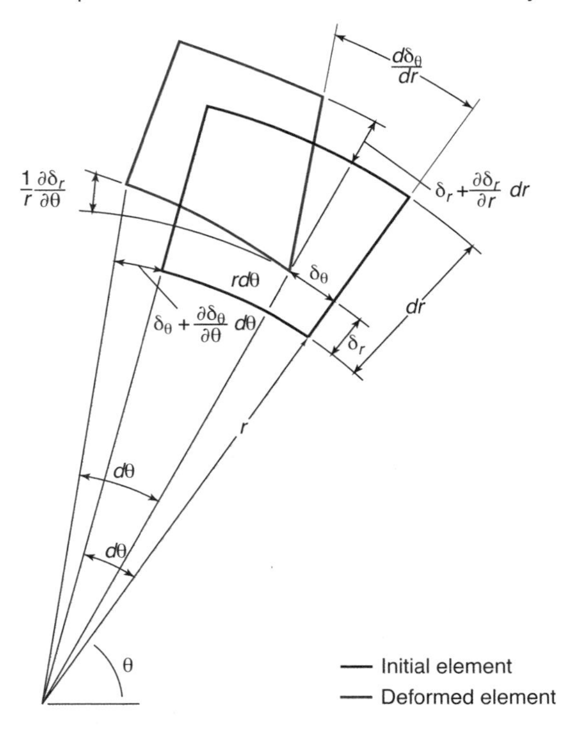

Figure 10.4: Cylindrical coordinate stress element before and after deformation.

Integrating gives

$$0 = 2\sigma_r + r\frac{d\sigma_r}{dr} + C_1.$$

This equation can be rewritten as

$$0 = \frac{d}{dr}\left(r^2\sigma_r\right) + C_1 r.$$

Integrating gives

$$\sigma_r = -\frac{C_1}{2} - \frac{C_2}{r^2}. \tag{10.19}$$

The boundary conditions for thick-walled cylinders pressurized both internally and externally are

1. $\sigma_r = -p_i$ at $r = r_i$
2. $\sigma_r = -p_o$ at $r = r_o$

Note the sign convention: a positive pressure results in a compressive (negative) stress; therefore, a negative sign appears in the boundary conditions. Applying the boundary conditions gives

$$\sigma_r = \frac{p_i r_i^2 - p_o r_o^2 + (p_o - p_i)\left(\frac{r_o r_i}{r}\right)^2}{r_o^2 - r_i^2}, \tag{10.20}$$

$$\frac{d\sigma_r}{dr} = -\frac{2(p_o - p_i)(r_o r_i)^2}{r^3(r_o^2 - r_i^2)}. \tag{10.21}$$

Substituting Eqs. (10.20) and (10.21) into Eq. (10.16) gives

$$\sigma_\theta = \frac{p_i r_i^2 - p_o r_o^2 - \left(\frac{r_i r_o}{r}\right)^2 (p_o - p_i)}{r_o^2 - r_i^2}. \tag{10.22}$$

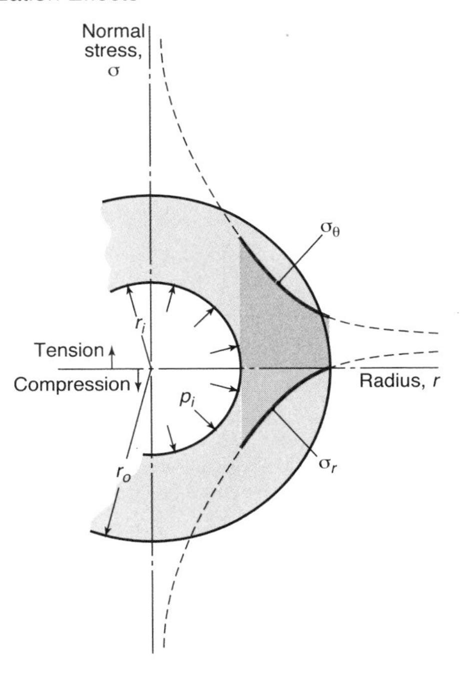

Figure 10.5: Internally pressurized, thick-walled cylinder showing circumferential (hoop) and radial stress for various radii.

Equations (10.20) and (10.22) are valuable for the general case of a thick-walled cylinder subjected to internal and external pressurization. Some special cases yield less complicated expressions that are useful for machinery applications.

Internally Pressurized

If, as in many applications, the outer pressure, p_o, is zero, Eqs. (10.20) and (10.22) reduce to

$$\sigma_r = \frac{p_i r_i^2 \left(1 - \frac{r_o^2}{r^2}\right)}{r_o^2 - r_i^2}, \tag{10.23}$$

$$\sigma_\theta = \frac{p_i r_i^2 \left(1 + \frac{r_o^2}{r^2}\right)}{r_o^2 - r_i^2}. \tag{10.24}$$

Figure 10.5 shows the radial and circumferential (hoop) stresses in an internally pressurized cylinder. The circumferential stress is tensile and the radial stress is compressive. Further, the maximum stresses occur at $r = r_i$ and are

$$\sigma_{r_{\max}} = -p_i, \tag{10.25}$$

$$\sigma_{\theta_{\max}} = p_i \left(\frac{r_o^2 + r_i^2}{r_o^2 - r_i^2}\right). \tag{10.26}$$

From Eq. (10.14), the circumferential strains for internal pressurization as evaluated at the location of maximum stress are

$$\epsilon_\theta = \frac{\delta_r}{r_i} = \frac{p_i}{E}\left(\frac{r_o^2 + r_i^2}{r_o^2 - r_i^2} + \nu\right), \tag{10.27}$$

where ν is Poisson's ratio. The radial displacement is outward and is given by

$$\delta_r = \frac{p_i r_i}{E}\left(\frac{r_o^2 + r_i^2}{r_o^2 - r_i^2} + \nu\right). \tag{10.28}$$

Example 10.3: Internally Pressurized Thick-Walled Cylinder

Given: A hydraulic cylinder can be pressurized to 100 MPa. The inner radius of the cylinder is 100 mm and the wall thickness is 35 mm.

Find: The tangential and axial stresses in the cylinder wall at the inner diameter for both thin- and thick-wall analysis.

Solution: From the thin-wall theory given in Eqs. (10.8) and (10.9), the following can be written:

$$\sigma_1 = \frac{p_i r}{t_h} = \frac{(100 \times 10^6)(0.1)}{0.035} = 285.7 \text{ MPa},$$

$$\sigma_2 = \frac{p_i r}{2t_h} = 142.9 \text{ MPa}.$$

From thick-wall theory with internal pressurization, Eq. (10.26) gives

$$\sigma_{\theta,\max} = p_i \left(\frac{r_o^2 + r_i^2}{r_o^2 - r_i^2}\right) = (100 \times 10^6)\left(\frac{0.135^2 + 0.10^2}{0.135^2 - 0.10^2}\right),$$

which is solved as $\sigma_{\theta,\max} = 343.2$ MPa. Assuming the axial stress is uniform throughout the wall,

$$\sigma_z = \frac{\pi r_i^2 p_i}{\pi(r_o^2 - r_i^2)} = \frac{(0.10)^2(100 \times 10^6)}{0.135^2 - 0.10^2} = 121.6 \text{ MPa}.$$

Note in this case there is a substantial difference between the thick- and thin-wall approaches.

Externally Pressurized

If the internal pressure is zero and the cylinder is externally pressurized, Eqs. (10.20) and (10.22) reduce to

$$\sigma_r = \frac{p_o r_o^2}{r_o^2 - r_i^2}\left(\frac{r_i^2}{r^2} - 1\right), \tag{10.29}$$

$$\sigma_\theta = -\frac{p_o r_o^2}{r_o^2 - r_i^2}\left(\frac{r_i^2}{r^2} + 1\right). \tag{10.30}$$

Figure 10.6 shows the radial and circumferential stresses in an externally pressurized cylinder. Note that both stresses are compressive. Furthermore, the maximum circumferential stress occurs at $r = r_i$ and the maximum radial stress occurs at $r = r_o$. These expressions are

$$\sigma_{r,\max} = -p_o, \tag{10.31}$$

$$\sigma_{\theta,\max} = -\frac{2r_o^2 p_o}{r_o^2 - r_i^2}. \tag{10.32}$$

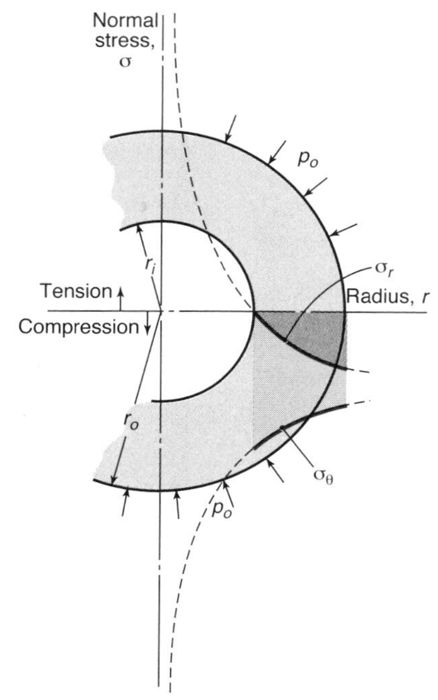

Figure 10.6: Externally pressurized, thick-walled cylinder showing circumferential (hoop) and radial stress for various radii.

Design Procedure 10.1: Stress Analysis of Thick-Walled Cylinders

A common design problem is to determine the largest permissible external and/or internal pressure to which a cylinder can be subjected without failure. Axial stresses, if present, are negligibly small. The following design procedure is useful for such circumstances:

1. For internal pressurization, both the radial and circumferential stresses are largest at the inner radius. The von Mises stress for this plane stress case can be shown to be

$$\sigma_e = p_i \sqrt{\frac{3r_o^4 + r_i^4}{(r_o^2 - r_i^2)^2}},$$

so that the allowable internal pressure is, from Eq. (6.8),

$$p_i = \frac{S_y}{n_s} \frac{r_o^2 - r_i^2}{\sqrt{3r_o^4 + r_i^4}}.$$

2. For external pressurization, it can be shown that the larger von Mises stress occurs at the inner radius, with the stresses of $\sigma_r = 0$ and σ_θ given by Eq. (10.32). This yields an expression of allowable external pressure of:

$$p_o = \frac{S_y}{n_s} \frac{r_o^2 - r_i^2}{2r_o^2}.$$

3. For combined internal and external pressurization, Eqs. (10.20) and (10.22) need to be substituted into a failure criterion from Ch. 6, such as the DET given for plane stress in Eqs. (6.10) and (6.11).

Example 10.4: Design of a Thick Walled Cylinder

Given: A thick-walled cylinder with 0.3-m internal diameter and 0.4-m external diameter has a maximum circumferential (hoop) stress of 250 MPa. The material has a Poisson's ratio of 0.3 and a modulus of elasticity of 207 GPa.

Find: Determine the following:

(*a*) For internal pressurization ($p_o = 0$) the maximum pressure to which the cylinder may be subjected.

(*b*) For external pressurization ($p_i = 0$) the maximum pressure to which the cylinder may be subjected.

(*c*) The radial displacement of a point on the inner surface for the situation presented in part a.

Solution:

(*a*) From Eq. (10.26) for internal pressurization,

$$p_i = \frac{\sigma_{\theta\max}\left(r_o^2 - r_i^2\right)}{r_o^2 + r_i^2} = \frac{\left(250 \times 10^6\right)\left(0.2^2 - 0.15^2\right)}{0.2^2 + 0.15^2},$$

which results in $p_i = 70$ MPa.

(*b*) From Eq. (10.32) for external pressurization,

$$p_o = -\frac{\sigma_{\theta\max}(r_o^2 - r_i^2)}{2r_o^2} = \frac{\left(250 \times 10^6\right)\left(0.2^2 - 0.15^2\right)}{2(0.2)^2},$$

which is solved as $p_o = 54.69$ MPa.

(*c*) The radial displacement for internally pressurized cylinders is obtained from Eq. (10.28) as

$$\begin{aligned}\delta_{r,\max} &= \frac{p_i r_i}{E}\left(\frac{r_o^2 + r_i^2}{r_o^2 - r_i^2} + \nu\right) \\ &= \frac{\left(70 \times 10^6\right)(0.15)}{\left(207 \times 10^9\right)}\left(\frac{0.2^2 + 0.15^2}{0.2^2 - 0.15^2} + 0.3\right) \\ &= 1.96 \times 10^{-4}\ \text{m} = 0.196\ \text{mm}.\end{aligned}$$

10.4 Rotational Effects

Rotating cylinders are encountered in a number of machine elements such as flywheels, gears, pulleys, and sprockets. This section considers rotation of the cylinder while assuming no pressurization ($p_i = p_o = 0$).

If a body force is included when considering the stresses acting on the elements shown in Fig. 10.3, Eq. (10.16) becomes

$$\frac{\sigma_\theta - \sigma_r}{r} - \frac{d\sigma_r}{dr} - P_b = 0, \qquad (10.33)$$

where P_b is the body force per volume. Here, the body force is the rotating inertia force that acts radially and is given by

$$P_b = r\omega^2\rho, \qquad (10.34)$$

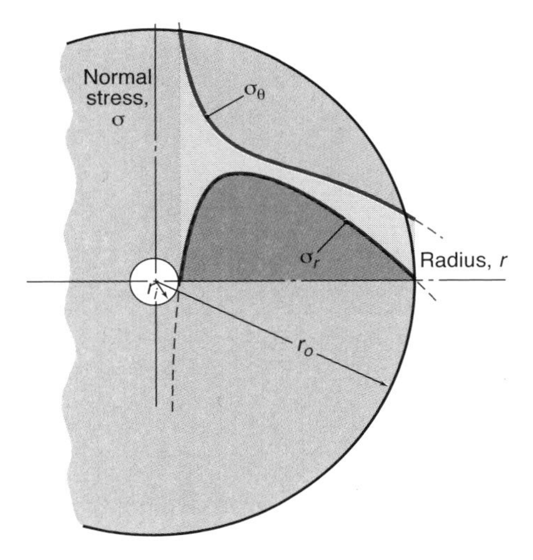

Figure 10.7: Stresses in rotating cylinder with central hole and no pressurization.

where

ω = angular velocity, rad/s
ρ = mass density, kg/m^3
r = radius of cylinder, m

Two special cases are considered: a cylinder with a central hole and a solid cylinder. The resulting equations are presented in this section since the derivation is similar to that presented in Section 10.3.

10.4.1 Cylinder with Central Hole

The circumferential (hoop) and radial stresses for a rotating cylinder with a central hole but neglecting pressurization are

$$\sigma_\theta = \frac{3+\nu}{8}\rho\omega^2\left[r_i^2 + r_o^2 + \frac{r_i^2 r_o^2}{r^2} - \frac{1+3\nu}{3+\nu}r^2\right], \quad (10.35)$$

$$\sigma_r = \frac{3+\nu}{8}\rho\omega^2\left(r_i^2 + r_o^2 - \frac{r_i^2 r_o^2}{r^2} - r^2\right). \quad (10.36)$$

Figure 10.7 shows the variation of stress with radius. Note that both stresses are tensile. The radial stress is zero at the inner and outer radii, with a maximum occurring between r_i and r_o. The circumferential stress is a maximum at $r = r_i$ and exceeds the radial stress for any value of radius. The maximum circumferential stress is

$$\sigma_{\theta,\max} = \frac{3+\nu}{4}\rho\omega^2\left[r_o^2 + \frac{r_i^2(1-\nu)}{3+\nu}\right]. \quad (10.37)$$

The maximum radial stress can be obtained by differentiating Eq. (10.36) with respect to r and setting the result equal to zero. Thus,

$$\sigma_{r,\max} = \frac{3+\nu}{8}\rho\omega^2\left(r_i - r_o\right)^2. \quad (10.38)$$

The location of maximum stress is

$$r = \sqrt{r_i r_o}. \quad (10.39)$$

Example 10.5: Rotating Press Fit

Given: Two concentric AISI 1040 steel tubes are press fit together at 110 MPa (see Section 10.5). The nominal sizes of the tubes are 100-mm outer diameter and 80-mm inner diameter and 80-mm outer diameter and 60-mm inner diameter, respectively.

Find: How fast does the combined tube have to rotate to decrease the press-fit pressure to zero?

Solution: By assuming linear elastic behavior, the press-fit pressure of 110 MPa has to be compensated for by an equally large radial stress at the fit radius. Equation (10.36) thus gives

$$\begin{aligned}\sigma_r &= \frac{3+\nu}{8}\rho\omega^2\left(r_i^2 + r_o^2 - \frac{r_i^2 r_o^2}{r^2} - r^2\right)\\ &= \frac{(3+0.3)}{8}(7850)\omega^2\\ &\quad\times\left[(0.03)^2 + (0.05)^2 - \frac{(0.03)^2(0.05)^2}{(0.04)^2} - (0.04)^2\right]\\ &= 110 \text{ MPa}.\end{aligned}$$

This results in

$$\omega = 9288 \text{ rad/s} = 88,700 \text{ rpm}.$$

Thus, at 88,700 rpm the press-fit pressure becomes zero. This very high rotational speed clearly shows that press fits of tubes do not loosen due to inertial effects in such machine elements. However, their torque transmission capability may be reduced at higher speeds.

10.4.2 Solid Cylinder

Setting $r_i = 0$ in Eqs. (10.35) and (10.36) gives the circumferential and radial stresses when considering rotation of a solid cylinder but neglecting pressurization effects. The results are

$$\sigma_\theta = \frac{3+\nu}{8}\rho\omega^2\left[r_o^2 - \frac{r^2(1+3\nu)}{3+\nu}\right], \quad (10.40)$$

$$\sigma_r = \frac{3+\nu}{8}\rho\omega^2\left(r_o^2 - r^2\right). \quad (10.41)$$

Figure 10.8 shows the stress distribution for both stress components as a function of radius for a solid cylinder. Note that both stresses are tensile, with the maximum occurring at $r = 0$ and both having the same value at that location. The maximum stress is

$$\sigma_{\theta,\max} = \sigma_{r,\max} = \frac{3+\nu}{8}\rho(r_o\omega)^2. \quad (10.42)$$

Some interesting observations can be made in comparing Eqs. (10.37) and (10.38) with Eq. (10.42). As r_i becomes very small, Eqs. (10.38) and (10.42) are in complete agreement. However, as r_i approaches 0, Eqs. (10.37) and (10.42) differ by a factor of two with respect to circumferential stress. This difference is due to the radial stress in the inner portion of the solid cylinder, which decreases the circumferential stress.

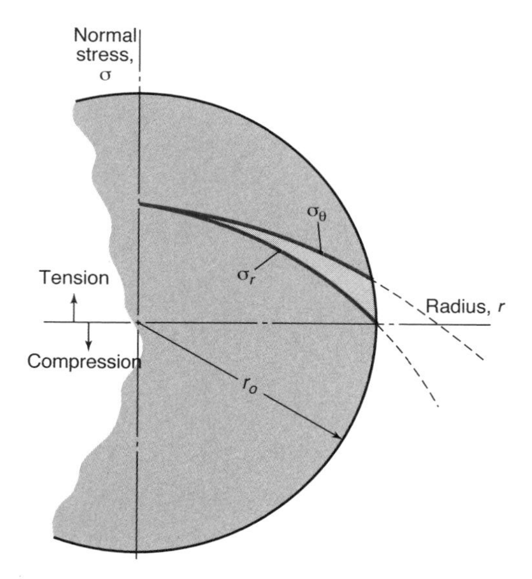

Figure 10.8: Stresses in rotating solid cylinder with no pressurization.

Example 10.6: Rotating Cylinder

Given: A cylindrical flywheel is press fitted on a solid shaft. Both are made of AISI 1080 steel, so that $\rho = 7850$ kg/m^3. The press-fit pressure is 185 MPa, the shaft diameter is 100 mm, and the outside diameter of the flywheel is 550 mm.

Find: The shaft speed when the press-fit pressure is eliminated by centrifugal effects.

Solution: When the radial stress due to centrifugal acceleration at the shaft surface is equal to the original press-fit pressure, the flywheel will start to separate from the shaft. From Table 3.1 and Eq. (10.41),

$$\begin{aligned}\sigma_r &= \frac{(3+\nu)\rho\omega^2\left(r_o^2 - r^2\right)}{8} \\ &= \frac{(3+0.3)(7850)\omega^2\left(0.275^2 - 0.050^2\right)}{8} \\ &= 185 \times 10^6 \text{ Pa.}\end{aligned}$$

Solving for angular velocity,

$$\omega = 883.9 \text{ rad/s} = 8441 \text{ rpm.}$$

Thus, at 8441 rpm the press-fit pressure becomes zero. This speed is high, but not unreasonable for a flywheel mounted shaft; thus, press fits can fail in this manner for overspeeding shafts with flywheels, and their torque transmission can be compromised at even lower speeds.

10.5 Press Fits

In a **press fit**, the pressure, p_f, is caused by the radial interference between the shaft and the hub. This pressure increases the radius of the hole and decreases the radius of the shaft. Section 10.2 described shaft and hub dimensions in terms of tolerance, which results in specific fits. This section focuses on the stress and strain found in press fits and uses equations developed in Section 10.3.2 for thick-walled cylinders.

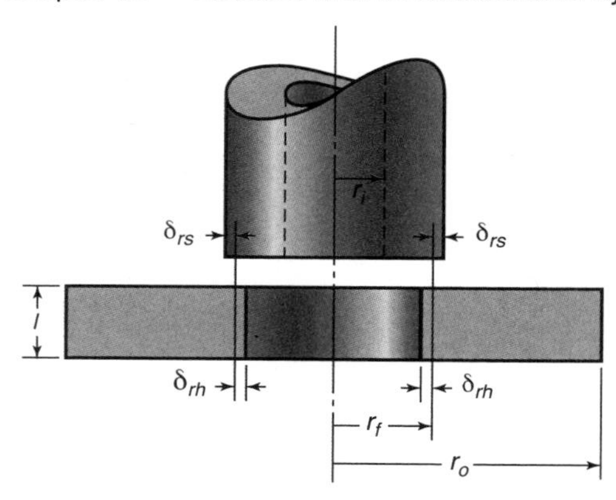

Figure 10.9: Side view showing interference in press fit of hollow shaft to hub.

Figure 10.9 shows a side view of interference in a press fit; there is a radial displacement of the hub, δ_{rh}, and a radial displacement of the shaft, δ_{rs}. Figure 10.10 shows the front view of an interference fit. In Fig. 10.10a, the cylinders are assembled with an interference fit; in Fig. 10.10b, the hub and shaft are disassembled and the dimensions of each are clearly shown. This figure also shows that the interference pressure acts internally for the hub and externally for the shaft. The shaft is shown as a tube in order to present the most general case.

10.5.1 Hub

Using Eq. (10.14), the hub displacement is

$$\delta_{rh} = \frac{r_f}{E_h}\left(\sigma_\theta - \nu_h \sigma_r\right), \tag{10.43}$$

where E_h is the elastic modulus and ν_h is the Poisson's ratio of the hub material, respectively. For internally pressurized, thick-walled cylinders and from Eqs. (10.23) and (10.24), the radial and circumferential stresses for the hub while letting $p_i = p_f$, $r = r_f$, and $r_i = r_f$, are

$$\sigma_r = \frac{p_f r_f^2\left(1 - \dfrac{r_o^2}{r_f^2}\right)}{r_o^2 - r_f^2} = -p_f, \tag{10.44}$$

and

$$\sigma_\theta = \frac{p_f r_f^2\left(1 + \dfrac{r_o^2}{r_f^2}\right)}{r_o^2 - r_f^2} = \frac{p_f\left(r_o^2 + r_f^2\right)}{r_o^2 - r_f^2}. \tag{10.45}$$

Substituting Eqs. (10.44) and (10.45) into Eq. (10.43) gives

$$\delta_{rh} = \frac{r_f p_f}{E_h}\left(\frac{r_o^2 + r_f^2}{r_o^2 - r_f^2} + \nu_h\right). \tag{10.46}$$

Since δ_{rh} is positive, the radial displacement of the hub is outward.

10.5.2 Shaft

Using Eq. (10.14), the displacement of the shaft is

$$\delta_{rs} = \frac{r_f}{E_s}\left(\sigma_\theta - \nu_s \sigma_r\right), \tag{10.47}$$

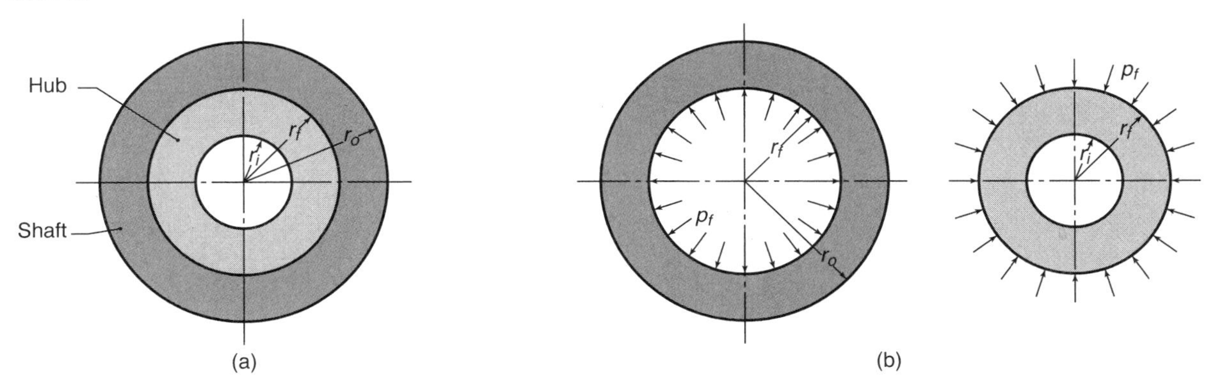

Figure 10.10: Front view showing (a) cylinder assembled with an interference fit and (b) hub and hollow shaft disassembled (also showing interference pressure).

where E_s is the elastic modulus and ν_s is Poisson's ratio of shaft material, respectively. The circumferential and radial stresses for externally pressurized, thick-walled cylinders can be obtained from Eqs. (10.29) and (10.30) by letting $p_o = p_f$, $r_o = r_f$, and $r = r_f$, yielding

$$\sigma_r = \frac{p_f r_f^2}{r_f^2 - r_i^2}\left(\frac{r_i^2}{r_f^2} - 1\right) = -p_f \tag{10.48}$$

and

$$\sigma_\theta = -\frac{p_f r_f^2}{r_f^2 - r_i^2}\left(\frac{r_i^2}{r_f^2} + 1\right) = -\frac{p_f\left(r_f^2 + r_i^2\right)}{r_f^2 - r_i^2}. \tag{10.49}$$

Substituting Eqs. (10.48) and (10.49) into Eq. (10.47) gives

$$\delta_{rs} = -\frac{r_f p_f}{E_s}\left(\frac{r_f^2 + r_i^2}{r_f^2 - r_i^2} - \nu_s\right). \tag{10.50}$$

Because the first term in parentheses in Eq. (10.50) is greater than unity and the Poisson's ratio is less than or equal to 0.5, δ_{rs} is negative; hence, shaft displacement is directed inward toward the center of the shaft.

10.5.3 Interference Fit

The total radial displacement in an interference fit is shown in Fig. 10.9. Recall that outward deflection (expansion of the inside diameter of the hub) is positive and inward deflection (reduction of the outside diameter of the shaft) is negative. Thus, the total radial interference is

$$\begin{aligned}\delta_r &= \delta_{rh} - \delta_{rs} \\ &= r_f p_f\left[\frac{r_o^2 + r_f^2}{E_h\left(r_o^2 - r_f^2\right)} + \frac{\nu_h}{E_h} + \frac{r_f^2 + r_i^2}{E_s\left(r_f^2 - r_i^2\right)} - \frac{\nu_s}{E_s}\right].\end{aligned} \tag{10.51}$$

If the shaft and the hub are made of the same material, $E = E_s = E_h$ and $\nu = \nu_s = \nu_h$, so that Eq. (10.51) reduces to

$$\delta_r = \frac{2r_f^3 p_f\left(r_o^2 - r_i^2\right)}{E\left(r_o^2 - r_f^2\right)\left(r_f^2 - r_i^2\right)}. \tag{10.52}$$

Furthermore, if the shaft is solid rather than hollow, $r_i = 0$ and Eq. (10.52) further reduces to

$$\delta_r = \frac{2r_f p_f r_o^2}{E\left(r_o^2 - r_f^2\right)}. \tag{10.53}$$

From these equations, it can be seen that if the radial displacement is known, the interference pressure can be readily obtained or vice-versa.

10.5.4 Force and Torque

The maximum force to assemble a press fit depends on the thickness and length of the outer member, the difference in diameters of the mating shaft and hub, and the coefficient of friction, μ. The maximum shear stress is

$$\tau_{\max} = p_f\mu = \frac{P_{\max}}{A} = \frac{P_{\max}}{2\pi r_f l}. \tag{10.54}$$

The torque that can be transmitted by the press fit is

$$T = P_{\max} r_f = 2\pi\mu r_f^2 l p_f. \tag{10.55}$$

It should be noted that the axial and circumferential stresses are related to the maximum stress by

$$\tau_a^2 + \tau_c^2 = \tau_{\max}^2,$$

where

$$\tau_a = \frac{P_a}{2\pi r_f l} = \text{axial stress}$$

$$\tau_c = \frac{P_c}{2\pi r_f l} = \text{circumferential stress}$$

Thus, the assembly and disassembly force given by Eq. (10.54) can be reduced by twisting or rotating the outer member while applying an axial force.

Example 10.7: Interference Fit

Given: A 150-mm-diameter steel shaft is to have a press fit with a 300-mm-outside-diameter cast iron hub. The hub is 25 mm long and defines the contact length. The maximum circumferential stress is to be 30 MPa. The moduli of elasticity are 207 GPa for steel and 100 GPa for cast iron. The Poisson's ratio for both steel and cast iron is 0.3 and the coefficient of friction for the two materials is 0.12. That is,

$$r_f = 75\text{ mm}, \qquad r_i = 0,$$

$$r_o = 150\text{ mm}, \qquad E_s = 207\text{ GPa},$$

$$E_h = 100\text{ GPa}, \qquad \nu_s = \nu_h = 0.3,$$

$$\mu = 0.12, \qquad l = 25\text{ mm}, \qquad \sigma_{\theta,\max} = 30\text{ MPa}.$$

Find:

(*a*) The interference
(*b*) The axial force required to press the hub on the shaft
(*c*) The torque that this press fit can transmit

Solution:

(*a*) From Eq. (10.45), the interference pressure is

$$p_f = \frac{\sigma_{\theta,\max}\left(r_o^2 - r_f^2\right)}{r_o^2 + r_f^2} = \frac{\left(30 \times 10^6\right)\left(0.150^2 - 0.075^2\right)}{0.150^2 + 0.075^2} = 18 \text{ MPa}.$$

From Eq. (10.51), the maximum permissible radial interference is

$$\delta_r = r_f p_f \left[\frac{r_o^2 + r_f^2}{E_h\left(r_o^2 - r_f^2\right)} + \frac{\nu_h}{E_h} + \frac{r_f^2 + r_i^2}{E_s\left(r_f^2 - r_i^2\right)} - \frac{\nu_s}{E_s}\right],$$

which is solved as $\delta_r = 31.1\ \mu$m.

(*b*) From Eq. (10.54), the force required for the press fit is

$$P_{\max} = 2\pi\mu r_f l p_f = (2)(\pi)(0.12)(0.075)(0.025)\left(18 \times 10^6\right) = 25.4 \text{ kN}.$$

(*c*) From Eq. (10.55), the torque is

$$T = P_{\max} r_f = \left(25.4 \times 10^3\right)(0.075) = 1910 \text{ N-m}.$$

Example 10.8: Disassembly of Interference Fit

Given: A wheel hub is press fit onto a 105-mm-diameter solid shaft. The coefficient of friction is 0.11 and the hub and shaft material is AISI 1080 steel. The hub's outer diameter is 160 mm and its width is 120 mm. The radial interference between the shaft and the hub is 65 μm (the shaft diameter is 130 μm larger than the inside diameter of the hub).

Find: The axial force necessary to dismount the hub.

Solution: Equation (10.53) gives the relationship between radial displacement and pressure as:

$$\delta_r = \frac{2 r_f p_f r_o^2}{E\left(r_o^2 - r_f^2\right)},$$

or

$$\left(65 \times 10^{-6}\right) = \frac{(2)(0.0525)p_f(0.080)^2}{\left(207 \times 10^9\right)\left[(0.080)^2 - (0.0525)^2\right]}.$$

This is solved as $p_f = 72.96$ MPa. The axial force necessary to dismount the hub is

$$P = \mu p_f A = (0.11)\left(72.96 \times 10^6\right)\pi(0.105)(0.120),$$

or $P = 317.7$ kN.

10.6 Shrink Fits

In producing a **shrink fit**, it is common to heat the outer component (hub) in order to expand it beyond the interference, and then slip it over the inner component (shaft); cooling then contracts the outer component. The temperature change produces thermal strain, even in the absence of stress. Although thermal strain is not exactly linear with temperature change, for temperature changes of 100° or 200°C, a linear approximation is reasonable. According to this linear relationship, the temperature difference necessary for the outer component to obtain the required expansion over the undeformed solid shaft is

$$\Delta t_m = \frac{\delta_r}{\beta r_f}, \tag{10.56}$$

where β is the coefficient of thermal expansion (see Table 3.1 and Fig. 3.16). Equation (10.56) can be expressed in terms of radial strain as

$$\epsilon_r = \frac{\delta_r}{r_f} = \beta \Delta t_m. \tag{10.57}$$

The deformation is

$$\delta_r = \epsilon_r r_f = \beta \Delta t_m r_f. \tag{10.58}$$

These equations are valid not only for shrink fits of shaft and hub but also for a wide range of thermal problems.

The strain due to a temperature change may be added algebraically to a local strain by using the *principle of superposition* (see Section 5.4). Therefore, the normal strain due to normal load and temperature effects is

$$\epsilon = \epsilon_\sigma + \epsilon_{t_m}, \tag{10.59}$$

where ϵ_σ is the strain due to normal stress and ϵ_{t_m} is the strain due to temperature change. Thus, the general (triaxial stress state) stress-strain relationship given by Eq. (B.54) can incorporate thermal strain as follows:

$$\epsilon_x = \frac{1}{E}\left[\sigma_x - \nu\left(\sigma_y + \sigma_z\right)\right] + \beta \Delta t_m, \tag{10.60}$$

$$\epsilon_y = \frac{1}{E}\left[\sigma_y - \nu\left(\sigma_z + \sigma_x\right)\right] + \beta \Delta t_m, \tag{10.61}$$

$$\epsilon_z = \frac{1}{E}\left[\sigma_z - \nu\left(\sigma_x + \sigma_y\right)\right] + \beta \Delta t_m. \tag{10.62}$$

Example 10.9: Shrink Fit

Given: A 250-mm-long steel tube, with a cross-sectional area of 5×10^{-4} m^2, expands in length by 0.2 mm from a stress-free condition at 25°C when the tube is heated to 250°C and subjected to a compressive force from its supports.

Find: The load and stress acting on the steel tube.

Solution: From Table 3.1 for steel, $\beta = 11 \times 10^{-6}$ °C^{-1}, and from Eq. (10.57),

$$\epsilon = \beta \Delta t_m = \left(11 \times 10^{-6}\right)(225) = 2.475 \times 10^{-3}.$$

Therefore, from Eq. (2.31),

$$\Delta l = l\epsilon = (0.25)\left(2.47 \times 10^{-3}\right) = 0.619 \text{ mm}.$$

Because the measured expansion was only 0.20 mm, the constraint due to compressive normal loading must apply a force sufficient to deflect the tube axially by

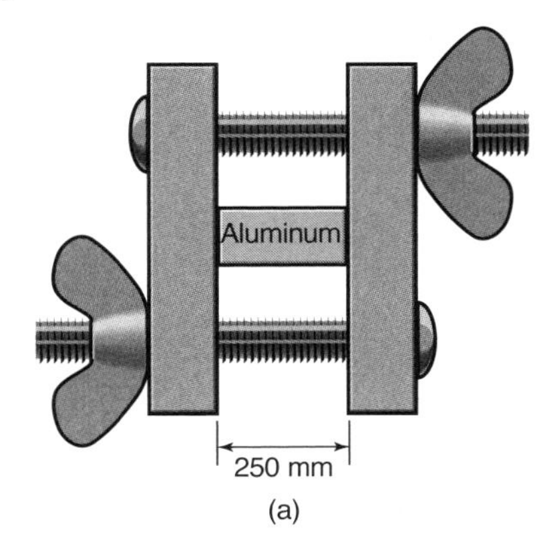

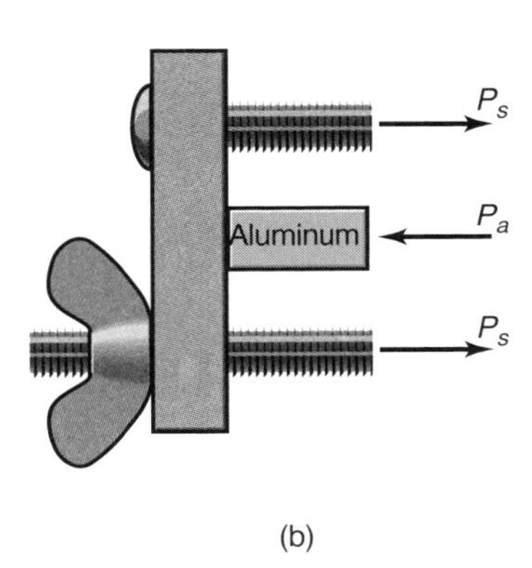

Figure 10.11: (a) Block placed between two rigid jaws of a clamp, and (b) associated forces.

$$\delta = 0.619 - 0.2 = 0.419 \text{ mm}.$$

From Table 3.1, the modulus of elasticity is 207 GPa. Therefore, from Eq. (4.26),

$$P = \frac{AE\delta}{l} = \frac{(5 \times 10^{-4})(207 \times 10^9)(0.419 \times 10^{-3})}{0.25},$$

or $P = 173$ kN. This is the compressive, normal, axial load being exerted on the steel tube. The axial stress is

$$\sigma = -\frac{P}{A} = -\frac{173 \times 10^3}{5 \times 10^{-4}} = -347 \text{ MPa}.$$

Example 10.10: Thermal Stresses

Given: A block of aluminum alloy is placed between two rigid jaws of a clamp, and the jaws are tightened to a snug state. The temperature of the entire assembly is raised by 250°C in an oven. The cross-sectional areas are 65 mm^2 for the block and 160 mm^2 for the stainless steel screws.

Find: The stresses induced in the screws and the block.

Solution: Figure 10.11 shows the block-and-screw assembly and the forces acting on these components. From force equilibrium,

$$P_a = 2P_s, \qquad (a)$$

where subscript a refers to the aluminum block and subscript s refers to the stainless steel screws. Compatibility requires that the length changes of the block and the screws be the same, or

$$\delta_a = \delta_s. \qquad (b)$$

Thermal expansion will induce an axial force as shown in Fig. 10.11b. The displacements of the block and screws are

$$\delta_a = \beta_a l \Delta t_m - \frac{P_a l}{E_a A_a}, \qquad (c)$$

$$\delta_s = \beta_s l \Delta t_m + \frac{P_s l}{E_s A_s}. \qquad (d)$$

Substituting Eqs. (a), (c), and (d) into Eq. (b) gives

$$P_s = \frac{\Delta t_m \left(\beta_a - \beta_s\right)}{\dfrac{1}{E_s A_s} + \dfrac{2}{E_a A_a}}. \qquad (e)$$

From Table 3.1, $E_a = 70$ GPa, $E_s = 193$ GPa, $\beta_a = 24 \times 10^{-6}\,(°\text{C})^{-1}$, and $\beta_s = 17 \times 10^{-6}\,(°\text{C})^{-1}$. It is given that $A_a = 65$ mm^2 and $A_s = 160$ mm^2; substituting these values into Eq. (e) gives the force acting on each screw as

$$P_s = \frac{(250)(24 - 17)(10^{-6})}{\dfrac{1}{(193 \times 10^9)(160 \times 10^{-6})} + \dfrac{2}{(70 \times 10^9)(65 \times 10^{-6})}},$$

or $P_s = 3708$ N. The force acting on the aluminum block is

$$P_a = 2P_s = 7416 \text{ N}.$$

The axial stresses of the block and screw are

$$\sigma_a = \frac{P_a}{A_a} = -\frac{7416}{(65 \times 10^{-6})} = -114.1 \text{ MPa},$$

$$\sigma_s = \frac{P_s}{A_s} = \frac{3708}{(160 \times 10^{-6})} = 23.18 \text{ MPa}.$$

Note that the stress acting on the aluminum block is compressive, and that acting on the screws is tensile.

Case Study: Design of a Shot Sleeve for a Die Casting Machine

Die casting is a common and important manufacturing process. A wide variety of products are produced through die casting, such as personal computer and camera frames, automotive structural components, fasteners, toy cars, and the like. In die casting, molten metal is placed in a shot sleeve and injected into a metal die outfitted with cooling lines to extract the heat and cause the cast metal to solidify quickly. A schematic illustration of a die casting machine is shown in Figure 10.12. In order for the process to be economically viable, a number of characteristics should be noted:

- The tooling is very expensive, and therefore a fairly large production run is required to justify the use of this process. Therefore, the cycle time (and cooling time) must be short in order to achieve required production runs.

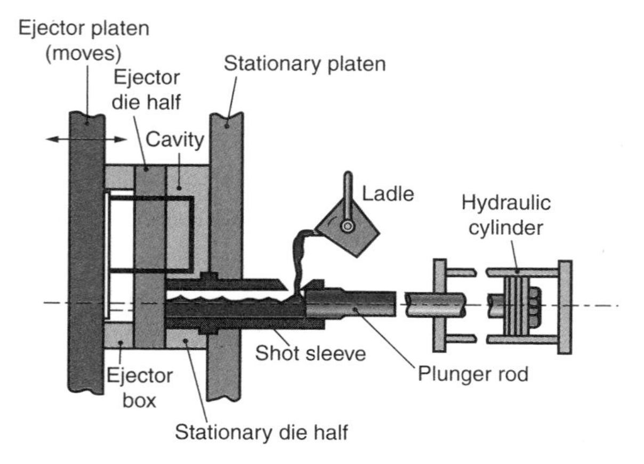

Figure 10.12: Schematic illustration of a die casting machine. *Source:* Kalpakjian and Schmid [2010].

- Recognizing that the cooling time should be very short, it is essential that the molten metal be injected into the die under high pressure and velocity to ensure that the mold is filled completely before local solidification results in a blocked channel, causing underfill. Injection pressures can be as high as 70 MPa (10 ksi), although 15 to 20 MPa (2 to 3 ksi) is more typical.
- The molten metal is transferred to the die casting machine from a furnace and held temporarily in the *shot sleeve*. The shot sleeve must have sufficient volume to fill the die cavity.
- When the metal is injected into the cavity, a high pressure is generated in the shot sleeve through the compressive action of a hydraulic cylinder.
- The die obviously must have a higher melting temperature than the metal being cast; this usually limits the tooling to die steels, and the workpiece metal to nonferrous alloys such as aluminum, magnesium, or copper alloys.

There are two basic kinds of die casting: the *hot chamber* process where the dies are maintained at a high temperature to aid in filling of the cavity, and the *cold chamber* process, where no provisions are provided to heat the die, and in fact the dies can be cooled through forced coolant flow. In the hot chamber process, the injection pressures can be as high as 35 MPa, while in the cold chamber process, pressures of 70 MPa are typical, but they can be as high as 150 MPa.

A thin-walled tube could be designed to contain such pressures (see Section 10.3.1), but the radius would be very small so that sufficient metal volume for reasonably sized parts would be unavailable. Instead, shot sleeves are designed as thick-walled cylindrical pressure vessels. A material typically used for this application is H13 steel, which has an ultimate strength of 768 MPa at room temperature and 650 MPa at casting temperatures for non-ferrous metals.

A machine may have a useful life of many decades, with a shot every few seconds. In designing such a shot sleeve, the fatigue strength of the H13 steel needs to be estimated using the procedures of Ch. 7.

The loading in this case consists of hoop and radial stresses only — significant axial stresses are not developed in the sleeve. On one end, the hydraulic cylinder applies a force that compresses the fluid and on the nozzle end the cylinder is mounted so that it does not see such loads.

Another major concern with shot sleeves is wear (Section 8.9), usually due to erosion of the sleeve material or abrasion of the plunger rod rubbing against the cylinder. For this reason, shot sleeves are often produced with a cylinder liner (see Design Procedure 8.1) that can be replaced periodically.

10.7 Summary

In this chapter, it was shown that fits must be specified to ensure the proper mating assembly of a shaft and a hub because it is impossible to manufacture these parts with exactly the desired dimensions. A system was devised to tolerate small dimensional variations of the mating shaft and hub without sacrificing their proper functioning.

Pressurization effects on cylinders were considered. Thin-wall and thick-wall analyses were described for both internal and external pressurization. These effects are important in a large number of applications, ranging from pressure vessels to gun barrels. A major assumption made in the thin-wall analysis was that the circumferential (hoop) stress is uniform throughout the wall thickness. This assumption is not valid for thick-wall analysis. Ranges of diameter to thickness were discussed for thin- and thick-wall analyses. For thick-wall situations, radial and circumferential variations with radius for both internal and external pressurization were shown. For internal pressurization, the maximum stress occurred at the inner radius for both radial and circumferential components, with the radial stress being compressive and the circumferential stress tensile. For external pressurization, the maximum radial stress occurred at the outer radius and the maximum circumferential stress occurred at the inner radius, with both stress components being compressive.

Rotational effects while assuming no pressurization of cylinders were also considered. Rotational effects are important in such machine elements as flywheels, gears, and pulleys. The rotating inertial force was considered in establishing the radial and circumferential stresses for both a cylinder with a central hole and a solid cylinder. The chapter ended with consideration of press and shrink fits of a shaft and a hub. The interference, axial force, and torque were developed for these situations.

Key Words

allowance difference between nominal diameters of mating parts

bilateral tolerance variation above and below nominal size

hoop stress circumferential stress in pressure vessel

interference difference in size of mating parts

nominal diameter approximate size of element

press fit connections where interfacial pressure is due to interference between mating parts and assembly is accomplished by elastic deformation due to large forces

shrink fit connections where interfacial pressure is due to interference between mating parts and assembly is accomplished by heating outer component

thick-walled cylinder cylinder whose ratio of diameter to thickness is less than 40

thin-walled cylinder cylinder where radial stress is negligible, approximately true for diameter-to-thickness ratios greater than 40

tolerance maximum variation in part size

unilateral tolerance variation above or below nominal size, but not both

Summary of Equations

Thin-Walled Cylinders

Hoop stress: $\sigma_{\theta,\text{avg}} = \dfrac{p_i r_i}{t_h}$

Longitudinal stress: $\sigma_z = \dfrac{p_i r}{2t_h}$

Thick-Walled Cylinders

General:

Radial strain: $\epsilon_r = \dfrac{\partial \delta_r}{\partial r} = \dfrac{1}{E}(\sigma_r - \nu\sigma_\theta)$

Circumferential strain: $\epsilon_\theta = \dfrac{\delta_r}{r} = \dfrac{1}{E}(\sigma_\theta - \nu\sigma_r)$

Radial stress: $\sigma_r = \dfrac{p_i r_i^2 - p_o r_o^2 + (p_o - p_i)\left(\dfrac{r_o r_i}{r}\right)^2}{r_o^2 - r_i^2}$

Circumferential stress:

$$\sigma_\theta = \frac{p_i r_i^2 - p_o r_o^2 - \left(\dfrac{r_i r_o}{r}\right)^2 (p_o - p_i)}{r_o^2 - r_i^2}$$

Internally Pressurized:

Maximum radial stress (at $r = r_i$): $\sigma_{r\max} = -p_i$

Maximum circumferential stress (at $r = r_i$):

$$\sigma_{\theta_{\max}} = p_i \left(\frac{r_o^2 + r_i^2}{r_o^2 - r_i^2}\right)$$

Circumferential strain (at $r = r_i$):

$$\epsilon_\theta = \frac{\delta_r}{r_i} = \frac{p_i}{E}\left(\frac{r_o^2 + r_i^2}{r_o^2 - r_i^2} + \nu\right)$$

Radial displacement: $\delta_r = \dfrac{p_i r_i}{E}\left(\dfrac{r_o^2 + r_i^2}{r_o^2 - r_i^2} + \nu\right)$

Externally Pressurized:

Maximum radial stress (at $r = r_o$): $\sigma_{r,\max} = -p_o$

Maximum circumferential stress (at $r = r_i$):

$$\sigma_{\theta,\max} = -\frac{2r_o^2 p_o}{r_o^2 - r_i^2}$$

Circumferential strain (at $r = r_i$):

$$\epsilon_\theta = \frac{\delta_r}{r_i} = \frac{p_i}{E}\left(\frac{r_o^2 + r_i^2}{r_o^2 - r_i^2} + \nu\right)$$

Radial displacement: $\delta_r = \dfrac{p_i r_i}{E}\left(\dfrac{r_o^2 + r_i^2}{r_o^2 - r_i^2} + \nu\right)$

Press Fits

Hub:

Radial stress: $\sigma_r = \dfrac{p_f r_f^2 \left(1 - \dfrac{r_o^2}{r_f^2}\right)}{r_o^2 - r_f^2} = -p_f$

Circumferential stress:

$$\sigma_\theta = \frac{p_f r_f^2 \left(1 + \dfrac{r_o^2}{r_f^2}\right)}{r_o^2 - r_f^2} = \frac{p_f\left(r_o^2 + r_f^2\right)}{r_o^2 - r_f^2}$$

Radial displacement:

$$\delta_{rh} = \frac{r_f p_f}{E_h}\left(\frac{r_o^2 + r_f^2}{r_o^2 - r_f^2} + \nu_h\right)$$

Shaft:

Radial stress: $\sigma_r = -p_f$

Circumferential stress: $\sigma_\theta = -\dfrac{p_f\left(r_f^2 + r_i^2\right)}{r_f^2 - r_i^2}$

Radial displacement:

$$\delta_{rs} = -\frac{r_f p_f}{E_s}\left(\frac{r_f^2 + r_i^2}{r_f^2 - r_i^2} - \nu_s\right)$$

Interference:

General:

$$\delta_r = \delta_{rh} - \delta_{rs} = r_f p_f \left[\frac{r_o^2 + r_f^2}{E_h\left(r_o^2 - r_f^2\right)} + \frac{\nu_h}{E_h} + \frac{r_f^2 + r_i^2}{E_s\left(r_f^2 - r_i^2\right)} - \frac{\nu_s}{E_s}\right]$$

For $E_s = E_h = E$ and $\nu_s = \nu_h = \nu$:

$$\delta_r = \frac{2r_f^3 p_f\left(r_o^2 - r_i^2\right)}{E\left(r_o^2 - r_f^2\right)\left(r_f^2 - r_i^2\right)}$$

For $E_s = E_h = E$ and $\nu_s = \nu_h = \nu$ and $r_i = 0$:

$$\delta_r = \frac{2r_f p_f r_o^2}{E\left(r_o^2 - r_f^2\right)}$$

Assembly force: $P_{\max} = 2\pi\mu r_f l p_f$

Torque: $T = P_{\max} r_f = 2\pi\mu r_f^2 l p_f$

Shrink Fits: $\delta_r = \beta\Delta t_m r_f$

Recommended Readings

Beer, F.P., Johnson, E.R., DeWolf, J., and Mazurek, D. (2011) *Mechanics of Materials*, 6th ed., McGraw-Hill.

Budynas, R.G., and Nisbett, J.K. (2011), *Shigley's Mechanical Engineering Design*, 9th ed., McGraw-Hill.

Chuse, R., and Carson, B.E. (1993) *Pressure Vessels*, 7th ed., McGraw-Hill.

Craig, R.R. (2001) *Mechanics of Materials*, 2nd ed., Wiley.

Harvey, J.F. (1991) *Theory and Design of Pressure Vessels*, 2nd ed., Van Nostrand Reinhold.

Hibbeler, R.C. (2010) *Mechanics of Materials*, 8th ed, Prentice-Hall.

Megyesy, E.F. (2008) *Pressure Vessel Handbook*, 14th ed., Pressure Vessel Handbook Publishing, Inc.

Mott, R.L. (2014) *Machine Elements in Mechanical Design*, 5th ed., Pearson.

Riley, W.F., Sturges, L.D., and Morris, D.H. (2006) *Mechanics of Materials*, 6th ed, Wiley.

Timoshenko, S., and Goodier, J. (1970) *Theory of Elasticity*, McGraw-Hill.

Ugural, A.C. and Fenster, S.K. (2011) *Advanced Mechanics of Materials and Applied Elasticity*, 5th ed, Prentice-Hall.

Reference

Kalpakjian, S., and Schmid, S.R. (2010) *Manufacturing Engineering and Technology*, 6th ed., Pearson.

Questions

10.1 What is the difference between tolerance and allowance?

10.2 Define bilateral and unilateral tolerance.

10.3 What is the difference between interference and allowance?

10.4 When is a cylinder considered to have a thin wall versus a thick wall?

10.5 What are the names of the principal stresses in cylindrical coordinates?

10.6 What are the principal stresses for a thin-walled cylinder under internal pressurization?

10.7 What is a hoop stress? Why is it so called?

10.8 What effect does rotation have on the stresses associated with internal pressurization of a cylinder?

10.9 What is a press fit? A shrink fit?

10.10 Define "hub" and "shaft."

Qualitative Problems

10.11 Give five examples of thin-walled and thick-walled pressure vessels.

10.12 Give two examples of internally pressurized and externally pressurized cylinders.

10.13 Is the radial stress compressive when the pressurization is internal or external?

10.14 Why is the coefficient of thermal expansion important? Would you select a material with a high or low coefficient of thermal expansion for a hub? Explain.

10.15 Explain the different classes of fit.

10.16 Review Table 10.2 and explain the exponents on the diameter for calculation of recommended allowance and tolerance.

10.17 What would happen if the hub's yield strength is exceeded during assembly of a press fit? What interference fit would result in this case?

10.18 Can the yield strength be exceeded in a shrink fit? Why or why not?

10.19 Using the derivation in Section 10.5, write the expressions for the force and pressure encountered by a nail when it is hammered into wood. Explain why pilot holes are drilled into wood for larger nails or harder wood.

10.20 List the concerns you would have in designing shrink or press fits with composite materials while using the theory developed in Sections 10.5 and 10.6.

Quantitative Problems

10.21 A journal bearing is to be manufactured with optimum geometry for minimum power loss for a given load and speed. The relative clearance $c/r = 0.001$; the journal diameter is 100 mm. Find the accuracy to which the bearing parts have to be manufactured so that there is no more than ±5% error in the relative clearance.

10.22 A press fit between a solid steel shaft and a steel housing is dimensioned to be of Class 7. By mistake the shaft is ground at 22°C higher temperature than originally anticipated, so that when the shaft cools, the diameter is slightly too small. The shaft material is AISI 1040 steel. What is the class of fit between the shaft and the housing because of this mistake? Also, if the grinding temperature were 50°C higher rather than 22°C higher, what is the class of fit? *Ans.* At 22°C, Class 6; at 50°C, Class 4.

10.23 A cylinder with a 0.30-m inner diameter and a 0.40-m outer diameter is internally pressurized to 100 MPa. Determine the maximum shear stress at the outer surface of the cylinder. *Ans.* $\tau_{\max} = 128.5$ MPa.

10.24 A rubber balloon has the shape of a cylinder with spherical ends. At low pressure the cylindrical part is 300 mm long with a diameter of 20 mm. The rubber material has constant thickness and is linearly elastic. How long will the cylindrical part of the balloon be when it is inflated to a diameter of 100 mm? Assume that the rubber's modulus of elasticity is constant and that the Poisson's ratio is 0.5.

10.25 A thin-walled cylinder containing pressurized gas is fixed by its two ends between rigid walls. Obtain an expression for the wall reactions in terms of cylinder length l, thickness t_h, radius r, and internal pressure p_i. *Ans.* $P = \pi p_i r^2(1 - 2\nu)$.

10.26 A pressurized cylinder has an internal pressure of 1 MPa, a thickness of 8 mm, a length of 4 m, and a diameter of 1.00 m. The cylinder is made of AISI 1080 steel. What are the hoop and axial strains upon pressurization? What is the volume increase of the cylinder due to the internal pressure? *Ans.* $\epsilon_z = 60.39$ μm/m, $\epsilon_\theta = 256.6$ μm/m.

10.27 A 1-m-diameter cylindrical container with two hemispherical ends is used to transport gas. The internal pressure is 5 MPa and the safety factor is 2.5. Using the MSST determine the container thickness. Assume that the tangential strains of the cylinder and the sphere are equal, since they are welded joints. The material is such that $E = 200$ GPa, $S_y = 430$ MPa, and $\nu = 0.3$. *Ans.* Wall: $t_h = 14.53$ mm; cap: $t_h = 7.267$ mm.

10.28 A solid cylindrical shaft made of AISI 1020 steel (quenched and tempered at 870°C) rotates at a speed that produces a safety factor of 2.5 against the stress causing yielding. To instrument the shaft, a small hole is drilled in its center for electric wires. At the same time the material is changed to AISI 1080 steel that has been quenched and tempered at 800°C. Find the safety factor against yielding of the new shaft. *Ans.* $n_s = 1.59$.

10.29 A flywheel is mounted on a tubular shaft. The shaft's inner diameter is 25 mm and its outer diameter is 50 mm. The flywheel is a cylindrical disk with an inner diameter of 50 mm, an outer diameter of 300 mm, and a thickness of 35 mm. Both the shaft and the flywheel are made of

AISI 1080 steel. The flywheel is used to store energy, so that the rotational acceleration is proportional to the angular speed ω. The angular acceleration $\partial\omega/\partial t$ is to be held to within $\pm 0.2\omega$. Calculate the shaft speed ω at which the flywheel starts to slide on the shaft if the press-fit pressure at $\omega = 0$ is 127 MPa and the coefficient of friction between the shaft and the flywheel is 0.13. *Ans.* $N = 14,980$ rpm.

10.30 Assume that the flywheel and the shaft given in Problem 10.29 are made of aluminum alloy 2014 instead of steel. Find the speed at which the flywheel will start to slide if the coefficient of friction is 0.14 and the press-fit pressure is 30 MPa. *Ans.* $N = 12,210$ rpm.

10.31 The flywheel and the shaft given in Problem 10.29 are axially loaded by a force of 60,000 N. Calculate the shaft speed at which the flywheel starts to slide on the shaft. *Ans.* $N = 8606$ rpm.

10.32 A 6-in.-diameter solid steel shaft is to have a press fit with a 12-in.-outer-diameter by 5-in.-long hub made of cast iron. The maximum allowable hoop stress is 5000 psi. The moduli of elasticity are 30×10^6 psi for steel and 20×10^6 psi for cast iron. Poisson's ratio for steel and cast iron is 0.3. The coefficient of friction for both steel and cast iron is 0.11. Determine the following:

(a) Total radial interference. *Ans.* $\delta_r = 0.001095$ in.

(b) Axial force required to press the hub on the shaft. *Ans.* $F_{\max} = 31.1$ kip.

(c) Torque transmitted with this fit. *Ans.* $T = 93.3$ kip-in.

10.33 A flat, 0.5-m-outer-diameter, 0.1-m-inner-diameter, 0.08-m-thick steel disk shown in Sketch a is shrink fit onto a shaft. If the assembly is to transmit a torque of 75 kN-m, determine the fit pressure and the total radial interference. The coefficient of friction is 0.25. *Ans.* $p_f = 238.7$ MPa, $\delta_r = 0.1201$ mm.

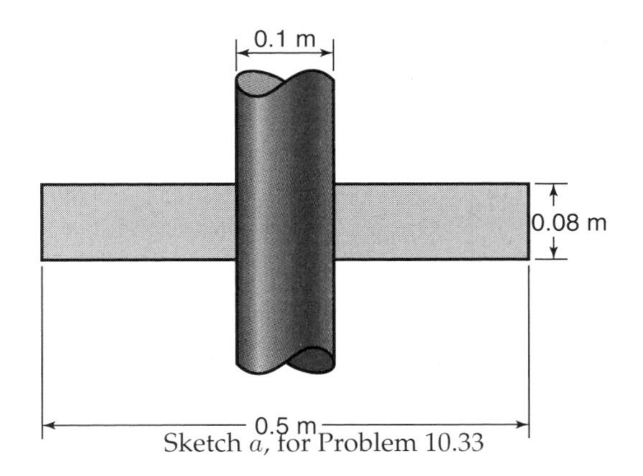

Sketch a, for Problem 10.33

10.34 A flat, 0.1-m-outer-diameter, 0.05-m-inner-diameter, 0.100-m-thick steel disk is shrink fit onto a shaft with a 0.02-m inner diameter. If the assembly is to transmit a torque of 12 kN-m, determine the fit pressure and the total radial interference. The coefficient of friction is 0.25. *Ans.* $p_f = 122.2$ MPa, $\delta_r = 44.9$ mum.

10.35 A supported cantilevered beam is loaded as shown in Sketch b. If at the time of assembly the load was zero and the beam was horizontal, determine the stress in the round rod after the load is applied and the temperature is lowered by 70°C. The beam moment of inertia is $I_{\text{beam}} = 9.8 \times 10^7$ mm^4. The beam and rod material are both high-carbon steel. *Ans.* $\sigma = 132$ MPa.

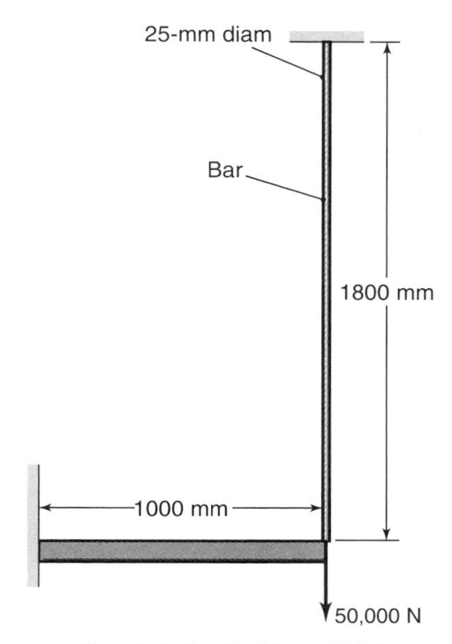

Sketch b, for Problem 10.35

10.36 A 10-in.-long tube (with properties $E = 30 \times 10^6$ psi and $\beta = 7 \times 10^{-6}$/°F) having a cross-sectional area of 1 in.^{2} is installed with fixed ends so that it is stress-free at 70°F. In operation the tube is heated throughout to a uniform 480°F. Measurements indicate that the fixed ends separate by 0.008 in. What loads are exerted on the ends of the tube, and what are the resultant stresses? *Ans.* $\sigma = -62.1$ ksi.

10.37 A solid square bar is constrained between two fixed supports as shown in Sketch c. The square bar has a cross-section of 5 by 5 in. and is 2 ft long. The bar just fits between fixed supports at the initial temperature of 70°F. If the temperature is raised to 120°F, determine the average thermal stress developed in the bar. Assume that $E = 30 \times 10^6$ psi and $\beta = 6.5 \times 10^{-6}$/°F. *Ans.* $\sigma = -9.750$ ksi.

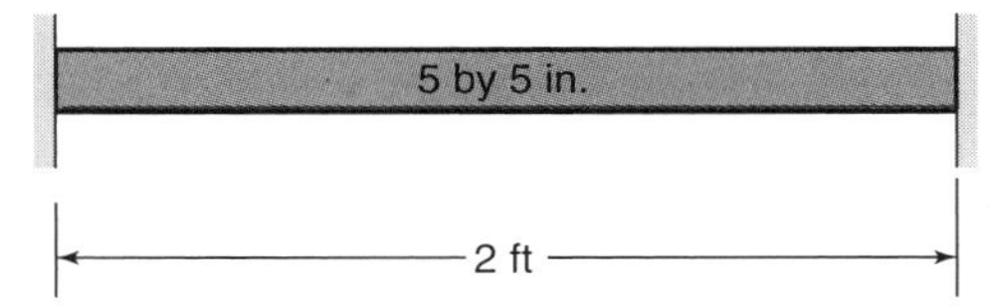

Sketch c, for Problem 10.37

10.38 Two stiff shafts are connected by a thin-walled elastic tube to a press-fit connection as shown in Sketch d. The contact pressure between the shafts and the tube is p. The coefficient of friction is μ. Calculate the maximum torque $T_{\max}$ that can be transmitted through the press fit. Describe and calculate what happens if the torque decreases from $T_{\max}$ to $\theta T_{\max}$ where $0 < \theta < 1$.

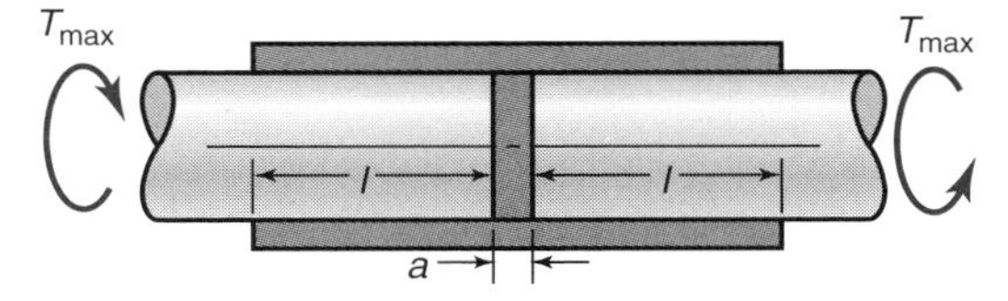

Sketch d, for Problem 10.38

10.39 A 50-mm-diameter steel shaft and a 25-mm-long cylindrical bushing of the same material with an outer diameter of 80 mm have been incorrectly shrink fit together and have to be dismounted. What axial force is needed for this operation if the diametral interference is 50 μm and the coefficient of friction is 0.2? *Ans.* $P_{\max} = 49.5$ kN.

10.40 A bushing press fit on a shaft shown in Sketch *e* is going to be dismounted. What axial force P_a is needed to dismount the bushing if, at the same time, the bushing transmits a torque $T = 500$ N-m? The diametral interference is $\delta = 30$ μm, the coefficient of friction is $\mu = 0.3$, and the modulus of elasticity is $E = 210$ GPa. *Ans.* $P_a = 144.0$ kN.

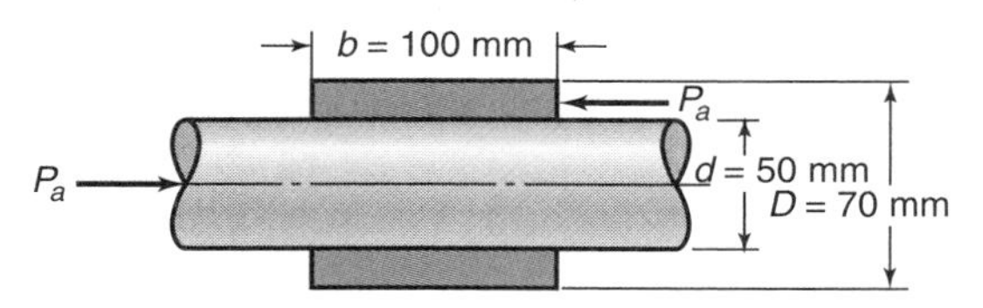

Sketch *e*, for Problem 10.40

10.41 To help in dismounting the wheel in Example 10.8, both an axial force and a torque are applied at the shaft-flywheel junction. How large a torque is needed to decrease the axial force to 40 kN when the wheel is dismounted? *Ans.* $T = 16.5$ kNm.

10.42 Two AISI 1040 steel cylinders are press fit on each other. The inner diameter is 200 mm, the common diameter is 300 mm, and the outer diameter is 400 mm. The radial interference of the two cylinders is 0.1 mm.

(a) Draw the radial and tangential stress distributions due to the press fit.

(b) If the internal pressure is 207 MPa and the external pressure is 50 MPa, determine the radial and tangential stress distributions due only to these pressures.

(c) Superimpose the stress distributions from (a) and (b) to obtain the total radial and tangential stress distribution.

10.43 A thick-walled cylinder is placed freely inside another thick-walled cylinder. What pressure is induced on the surfaces between the two cylinders by internally pressurizing the inner cylinder? *Hint:* Equate the radial displacement of the outer wall of the inner cylinder with that of the inner wall of the outer cylinder.

10.44 A 0.5-m-outer-diameter, 0.1-m-inner-diameter, 0.1-m-thick flat disk is shrink fit onto a solid shaft. Both the shaft and the disk are made of high-carbon steel. The assembly transmits 20 MW at 2000 rpm. Calculate the minimum temperature to which the disk must be heated for this shrink fit. The coefficient of friction is 0.25. *Ans.* $\Delta t_m = 222.5°$C.

10.45 Two shafts are connected by a shrink-fit bushing with an outer diameter of 120 mm. The diameter of each shaft is 80 mm and each shaft is 2 m long. Before the shrink-fit bushing was mounted, the diametral interference was $\delta = 80$ μm. The bushing and the shaft are made of steel with a modulus of elasticity $E = 210$ GPa. Find the allowable temperature increase without slip in the press fit if a power of 200 kW is transmitted and the far ends of the shafts cannot move axially. The coefficient of thermal expansion is $11.5 \times 10^{-6}/°$C, and the coefficient of friction is 0.1. The axial length of the bushing press fit on each shaft is 80 mm. The rotational speed is 1500 rpm. *Ans.* $\Delta t_m = 63.29°$C.

10.46 A 15-mm-thick, 100-mm-wide ring is shrunk onto a 100-mm-diameter shaft. The diametral interference is 75 μm. Find the surface pressure in the shrink fit and the maximum torque that can be transmitted if the coefficient of friction $\mu = 0.10$. The modulus of elasticity $E = 210$ GPa. *Ans.* $p_f = 32.15$ MPa, $T = 5050$ Nm.

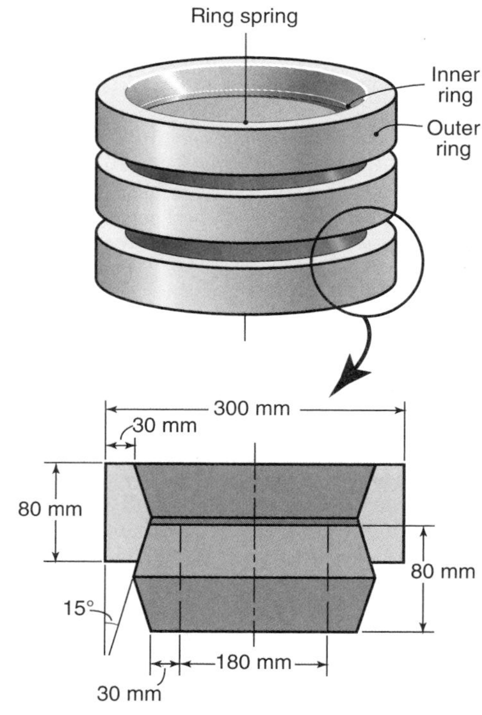

Sketch *f*, for Problem 10.47

10.47 A railway car buffer has a spring consisting of 11 outer rings and 11 inner rings where one of each ring type is a half-ring at the end of the spring. On one occasion a railway car with a total mass of 10,000 kg rolls until it hits a rigid stop. The force stopping the car is equal in the outer and the inner rings. The speed of the car just before the stop is 18 km/hr. Find the compression of the outer and inner rings and the stresses in the rings if the coefficient of friction $\mu = 0.25$ and the modulus of elasticity $E = 210$ GPa. The ring dimensions are shown in Sketch *f*. When a spring of this type is compressed, the spring rate k (i.e., force divided by deformation) is

$$k = \frac{P}{\delta} = \frac{\pi E \tan\alpha \tan(\alpha+\gamma)}{n\left(\dfrac{r_y}{A_y} + \dfrac{r_i}{A_i}\right)}$$

where

α = cone angle for spring
γ = friction angle
$\tan\gamma = \mu$ = coefficient of friction

r_y, r_i = radii to surface center of gravity
A_y, A_i = cross-sectional areas of rings

10.48 A railway car weighing 20 tons has in each end two buffers of the type described in Problem 10.47. The data are $A_y = A_i = 200$ mm^2, $r_y = 70$ mm, $r_i = 60$ mm, n = 20, $\alpha = 14°$, $\gamma = 7°$, and $E = 206$ GPa. The car hits a rigid stop with a speed of 1 m/s. Find the maximum force in each buffer and the energy absorbed. At what speed will the car bounce back?

10.49 For the situation in Example 10.4, find the maximum pressures if the yield stress is 250 MPa, not the allowable circumferential stress. *Ans.* $p_i = 60.0$ MPa, $\delta_{r,\max} = 0.168$ mm.

Synthesis and Design

10.50 Show that the principal stresses for a thin-walled, spherical pressure vessel are given by $\sigma_1 = \sigma_2 = \dfrac{pr}{2t}$.

10.51 Derive the expressions in Design Procedure 10.1.

10.52 Vitrified grinding wheels use a brittle but strong ceramic bond between abrasive particles. Serious accidents have been known to occur when a grinding wheel is damaged but still used, or when it is oversped. Considering the material in Section 10.4 and an appropriate failure theory from Chapter 6, develop an expression to predict grinding wheel failure.

10.53 Design an experiment to measure the interference pressure in press fits and shrink fits.

10.54 Consider the situation where a hub is assembled onto a shaft using an interference fit. If an adhesive is also inserted inside the hub-shaft combination, will this increase the assembly force? Explain your answer.

Chapter 11

Shafting and Associated Parts

A selection of overhead valve camshafts for automotive engines. *Source:* Courtesy of AVL Schrick.

When a man has a vision, he cannot get the power from the vision until he has performed it on the Earth for the people to see.
Black Elk, Oglala Sioux visionary

Contents

Examples

Design Procedures

This chapter focuses on shafts and related machine elements. Shafts are an essential component of most machines, and are mainly used for power transmission. Shafts are designed to transmit torque and support bending moments and axial loads; they also must be designed so that they do not deflect excessively. This includes dynamic instability as well as static and fatigue loadings. Shafts utilize a number of machine elements to provide functionality. For example, when a torque needs to be transmitted between a shaft and a machine element, that element can be mounted onto the shaft with a key or spline. If a machine element needs to be located at a certain location on a shaft axis, such as against a shoulder, then a retaining ring, set screw, or pin is often used. This chapter also discusses flywheels, which are used to store energy and provide smooth, jerk-free motion. Finally, a wide assortment of coupling types are described. Couplings are used to connect two shafts, and different coupling designs are able to accommodate misalignment and damping of vibration.

Machine elements described in this chapter: Shafts, keys, splines, set screws, retaining rings, flywheels, couplings.
Typical applications: *Shafts:* widespread use to transmit power to a point of operation, automotive crankshafts, cam shafts. *Keys, splines, and set screws:* used to transmit torque to another machine element's hub, such as for gears, pulleys, wire rope drums, mixer agitators, and flywheels. *Retaining rings:* used to fix the axial location of a component, such as rolling element bearing races, pulleys, wheels, bearing sleeves, and gears. *Flywheels:* automotive engines, crushing machinery, milling machinery, and machine tools. *Couplings:* used to connect two shafts, most commonly between the power source or motor shaft and the drive shaft.
Competing machine elements: *Shafts:* gear drives (Chapters 14 and 15), belt and wire rope drives (Chapter 19). *Keys, splines, set screws, retaining rings:* weldments, threaded retainers and adhesive joints (Chapter 16), press and shrink fits (Chapter 10). *Flywheels:* fluid couplings, large gears. *Couplings:* clutches (Chapter 18) and gear drives (Chapters 14 and 15).

Symbols

A	area, m^2
$\tilde{A}$	constant defined in Eq. (11.26)
$\tilde{B}$	constant defined in Eq. (11.27)
C_1	integration constant
C_f	coefficient of fluctuation, Eq. (11.80)
C_t	ring correction factor
c	distance from neutral axis to outer fiber, m
d	diameter, m
d_m	mean spline diameter, m
d_s	set screw diameter, m
E	modulus of elasticity, Pa
g	gravitational acceleration, 9.807 m/s^2
h	height, m
I	area moment of inertia, m^4
I_m	mass moment of inertia, kg-m^2
J	polar area moment of inertia, m^4
K_c	stress concentration factor
K_e	kinetic energy, N-m
K_f	fatigue stress concentration factor
k	spring rate, N/m
k_f	surface finish factor
k_r	reliability factor
k_s	size factor
l	length, m
l_s	spline length, m
M	moment, N-m
M_f	performance index, J/kg
m_a	mass, kg
n	number of teeth
n_s	safety factor
P	normal force, N
P_t	retaining force, N
p	pressure, Pa
p_f	interference pressure, Pa
q_n	notch sensitivity factor
r	radius, m
S_e	modified endurance limit, Pa
S'_e	endurance limit, Pa
S_{se}	shear modified endurance limit, Pa
S_{sy}	shear yield strength, Pa
S_u	ultimate strength, Pa
S_{ut}	ultimate tensile strength, Pa
S_y	yield strength, Pa
T	torque, N-m
T_l	load torque, N-m
T_m	mean torque, N-m
t	time, s
t_h	thickness, m
U	potential energy, N-m
u	velocity, m/s
W	load, N
w	width, m
x, y, z	Cartesian coordinates, m
δ	deflection, m
θ	cylindrical polar coordinate, deg
$\theta_{\omega_{\max}}$	location within a cycle where speed is maximum, deg
$\theta_{\omega_{\min}}$	location within a cycle where speed is minimum, deg
ν	Poisson's ratio
ρ	density, kg/m^3
σ	normal stress, Pa
σ_e	critical stress using distortion-energy theory, Pa
σ_ϕ	normal stress acting on oblique plane, Pa
τ	shear stress, Pa
τ_ϕ	shear stress acting on oblique plane, Pa
ϕ	oblique angle, deg
ω	angular speed, rad/s
ω_ϕ	fluctuation speed, rad/s

Subscripts

a	alternating
c	compression
i	inner
m	mean
o	outer
r	radial
s	shear
θ	circumferential
ω	speed
1,2,3	principal axes

11.1 Introduction

This chapter begins by discussing the design of shafts, making extensive use of the material from Sections 2.8 through 2.12 and Ch. 4 to develop stresses. The failure theories presented in Section 6.7 are used for static failure prediction in Section 11.2; the material in Ch. 7 is used to develop design rules for fatigue of shafts in Section 11.3. Here, combinations of loading are presented, whereas previously each type of loading was considered independently. It is important that this material be understood before proceeding with this chapter. The critical speed of rotating shafts is discussed in Section 11.5. The dynamics and the first critical speed are important, since the rotating shaft becomes dynamically unstable and large vibrations are likely to develop.

Keys, pins, and splines are used to attach devices to a shaft, and are discussed in Section 11.6. These devices use friction or mechanical interference to transmit a torque. Axial position of parts on a shaft can be done with retaining rings, cotter pins, or a number of similar devices. The design of flywheels and couplings are considered in Sections 11.8 and 11.9. Flywheels are valuable energy storage devices that also provide smooth operation. Couplings are used when two shafts need to be connected and are available in a wide variety of forms, the most common of which are presented.

11.2 Design of Shafts for Static Loading

A **shaft** is a rotating or stationary member usually having a circular cross-section much smaller in diameter than in length, and used for power transmission. Machine elements such as gears, pulleys, cams, flywheels, cranks, sprockets, and rolling-element bearings are mounted on shafts, and as such require a well-designed shaft as a prerequisite to their proper function. The loading on the shaft can include combinations of bending (almost always fluctuating); torsion (may or may not be fluctuating); shock; or axial, normal, or transverse forces. All of these types of loading were considered in Chapter 4. Some of the main considerations in designing a shaft are strength, using yield or fatigue (or both) as a criterion; deflection; or the dynamics established by the critical speeds. In general, the shaft diameter will be the variable used to satisfy the design, although in many practical applications the shaft may not have a constant diameter.

Shaft design must consider both static and fatigue failure possibilities. While at first it may appear that considering fatigue only would result in conservative designs, this is not always the case. For example, it is common that a rotating shaft sees predominantly uniform stresses, but on rare occasions encounters a much more significant stress cycle. Cumulative damage as discussed in Section 7.9 can be considered, with Miner's rule as stated in Eq. (7.24) applied to design the shaft. However, the high loading may be a rare event, such as is caused by a machine malfunction or improper operation. Thus, it may occur only extremely rarely, if at all, so that its contribution to fatigue crack growth may be minor.

Even with an overload or malfunction, there is usually some limit to the stresses that are applied to the shaft. For example, the loads can be controlled by using keys or pins (Section 11.6) or slip clutches (Section 18.10). In such circumstances, it is important to make sure that the shaft does not fail statically, especially since the shaft is usually the most difficult component to service or replace. Not surprisingly, shafts are often designed with very large safety factors.

A number of different loading conditions are considered here. Typically, the designer must establish either the minimum shaft diameter to successfully support applied loads or the safety factor for a specific design. This can be done using the approaches in Chapters 2 and 4, but the problem is so common that simplified solutions for each case are presented below.

11.2.1 Bending Moment and Torsion

Bending moments exerted on a shaft produce a maximum stress, from Eq. (4.45), of

$$\sigma_x = \frac{Mc}{I}. \tag{11.1}$$

Similarly, the shear stress due to an applied torque is, from Eq. (4.33),

$$\tau_{xy} = \frac{Tc}{J}, \tag{11.2}$$

where, for a circular cross section,

$$c = \frac{d}{2} \qquad I = \frac{\pi d^4}{64} \qquad \text{and} \qquad J = \frac{\pi d^4}{32}. \tag{11.3}$$

Substituting Eq. (11.3) into Eqs. (11.1) and (11.2) gives

$$\sigma_x = \frac{64Md}{2\pi d^4} = \frac{32M}{\pi d^3}, \tag{11.4}$$

$$\tau_{xy} = \frac{Td/2}{\pi d^4/32} = \frac{16T}{\pi d^3}. \tag{11.5}$$

Note that since $\sigma_y = 0$, these stresses result in a plane stress loading. Therefore, from Eq. (2.16),

$$\sigma_1, \sigma_2 = \frac{\sigma_x}{2} \pm \sqrt{\left(\frac{\sigma_x}{2}\right)^2 + \tau_{xy}^2}. \tag{11.6}$$

Substituting Eqs. (11.4) and (11.5) into Eq. (11.6) gives

$$\begin{aligned} \sigma_1, \sigma_2 &= \frac{16M}{\pi d^3} \pm \sqrt{\left(\frac{16M}{\pi d^3}\right)^2 + \left(\frac{16T}{\pi d^3}\right)^2} \\ &= \frac{16}{\pi d^3}\left[M \pm \sqrt{M^2+T^2}\right]. \end{aligned} \tag{11.7}$$

From Eq. (2.19), the principal shear stresses are

$$\tau_1, \tau_2 = \pm\sqrt{\tau_{xy}^2 + \left(\frac{\sigma_x}{2}\right)^2}. \tag{11.8}$$

Substituting Eqs. (11.4) and (11.5) into Eq. (11.8) gives

$$\tau_1, \tau_2 = \pm\frac{16}{\pi d^3}\sqrt{M^2+T^2}. \tag{11.9}$$

Distortion-Energy Theory

As shown in Section 6.7.1 and by Eqs. (6.11) and (6.12), the Distortion-Energy Theory (DET) predicts failure if the von Mises stress satisfies the following condition:

$$\sigma_e = \left(\sigma_1^2 + \sigma_2^2 - \sigma_1\sigma_2\right)^{1/2} = \frac{S_y}{n_s}, \tag{11.10}$$

where S_y is the yield strength of shaft material and n_s is the safety factor. Substituting Eq. (11.7) into Eq. (11.10), the DET predicts failure if

$$\frac{16}{\pi d^3}\left(4M^2 + 3T^2\right)^{1/2} = \frac{S_y}{n_s}. \tag{11.11}$$

Thus, the DET predicts the smallest diameter where failure will occur as

$$d = \left(\frac{32n_s}{\pi S_y}\sqrt{M^2 + \frac{3}{4}T^2}\right)^{1/3}. \tag{11.12}$$

If the shaft diameter is known and the safety factor is desired, Eq. (11.12) becomes

$$n_s = \frac{\pi d^3 S_y}{32\sqrt{M^2 + \frac{3}{4}T^2}}. \tag{11.13}$$

Maximum-Shear-Stress Theory

As shown in Section 6.7.1 and by Eq. (6.8), the Maximum-Shear-Stress Theory (MSST) predicts failure for a plane or biaxial stress state ($\sigma_3 = 0$) if

$$|\sigma_1 - \sigma_2| = \frac{S_y}{n_s}. \tag{11.14}$$

Equation (11.7) gives

$$\frac{32\sqrt{M^2+T^2}}{\pi d^3} = \frac{S_y}{n_s}. \tag{11.15}$$

Thus, the MSST predicts the smallest diameter where failure will occur as

$$d = \left(\frac{32n_s}{\pi S_y}\sqrt{M^2+T^2}\right)^{1/3}. \tag{11.16}$$

If the shaft diameter is known and the safety factor is an desired, Eq. (11.16) becomes

$$n_s = \frac{\pi d^3 S_y}{32\sqrt{M^2+T^2}}. \tag{11.17}$$

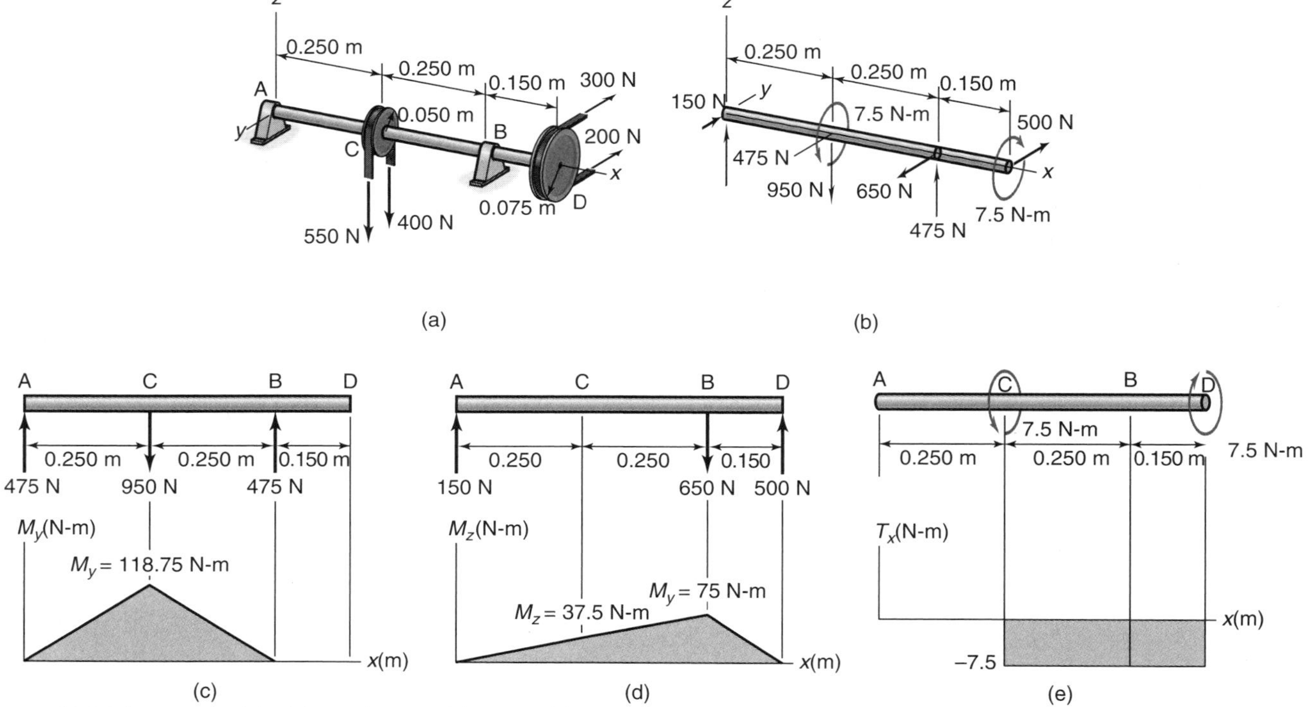

Figure 11.1: Figures used for Example 11.1. (a) Assembly drawing; (b) free-body diagram; (c) moment diagram in x-z plane; (d) moment diagram in x-y plane; (e) torque diagram.

Example 11.1: Static Design of a Shaft

Given: A shaft with mounted belt drives has tensile forces applied as shown in Fig. 11.1a and frictionless journal bearings at locations A and B. The yield strength of the shaft material is 500 MPa.

Find: Determine the smallest safe shaft diameter by using both the DET and the MSST for a safety factor of 2.0. Also, provide a free-body diagram as well as moment and torque diagrams.

Solution: A free-body diagram is shown in Fig. 11.1b; a moment diagram in the x-y plane, in Fig. 11.1c; and a moment diagram in the x-z plane in Fig. 11.1d. These have been constructed using the approach described in Section 2.8. From the moment diagrams, the maximum moment is

$$M_{\max} = \sqrt{(118.75)^2 + (37.5)^2} = 124.5 \text{ N-m}.$$

Figure 11.1e gives the torque diagram. Using the DET, the smallest safe diameter is given by Eq. (11.12) as

$$\begin{aligned} d &= \left(\frac{32n_s}{\pi S_y}\sqrt{M^2 + \frac{3}{4}T^2}\right)^{1/3} \\ &= \left\{\frac{32(2)}{\pi(500\times 10^6)}\left[124.5^2 + \frac{3}{4}(7.5)^2\right]^{1/2}\right\}^{1/3} \\ &= 17.2 \text{ mm}. \end{aligned}$$

Using the MSST as given in Eq. (11.16) gives

$$\begin{aligned} d &= \left(\frac{32n_s}{\pi S_y}\sqrt{M^2 + T^2}\right)^{1/3} \\ &= \left\{\frac{32(2)}{\pi(500\times 10^6)}\left[124.5^2 + 7.5^2\right]^{1/2}\right\}^{1/3} \\ &= 17.2 \text{ mm}. \end{aligned}$$

Since the torque is small relative to the moment, little difference exists between the DET and MSST predictions. This is not normally the case, although the results are usually close. Note that it is good design practice to specify a diameter that is rounded up to a convenient integer dimension. In this case, a diameter of 20 mm would be a good option.

11.2.2 Bending, Torsion, and Axial Loading

If, in addition to bending and torsion, an axial load is present, the normal stress is similar to Eq. (11.4) and is given by:

$$\sigma_x = \frac{32M}{\pi d^3} + \frac{4P}{\pi d^2}. \tag{11.18}$$

The shear stress is still expressed by Eq. (11.5); and the principal normal stresses, by Eq. (11.6). Substituting Eqs. (11.18) and (11.5) into Eq. (11.6) gives

$$\begin{aligned} \sigma_1, \sigma_2 &= \frac{16M}{\pi d^3} + \frac{2P}{\pi d^2} \pm \sqrt{\left(\frac{16M}{\pi d^3} + \frac{2P}{\pi d^2}\right)^2 + \left(\frac{16T}{\pi d^3}\right)^2} \\ &= \frac{2}{\pi d^3}\left[8M + Pd \pm \sqrt{(8M + Pd)^2 + (8T)^2}\right]. \end{aligned} \tag{11.19}$$

Substituting Eqs. (11.18) and (11.5) into Eq. (11.8) gives the principal shear stresses as

$$\tau_1, \tau_2 = \pm \frac{2}{\pi d^3}\sqrt{(8M + Pd)^2 + (8T)^2}. \qquad (11.20)$$

Distortion-Energy Theory

Substituting Eq. (11.19) into Eq. (11.10) shows that the DET predicts failure if

$$\frac{4}{\pi d^3}\sqrt{(8M + Pd)^2 + 48T^2} = \frac{S_y}{n_s}. \qquad (11.21)$$

This equation is more complicated than Eq. (11.16), and an explicit expression for the diameter cannot be obtained. Numerical solutions of Eq. (11.21) are relatively easy to obtain, however.

Maximum-Shear-Stress Theory

Substituting Eq. (11.19) into Eq. (11.14) shows that the MSST predicts failure if

$$\frac{4}{\pi d^3}\sqrt{(8M + Pd)^2 + 64T^2} = \frac{S_y}{n_s}. \qquad (11.22)$$

Again, when an axial loading is included, an explicit expression for the diameter cannot be obtained.

11.3 Fatigue Design of Shafts

In cyclic loading, the stresses vary throughout a cycle and do not remain constant as in static loading. In this section, a general analysis is presented for the fluctuating normal and shear stresses for ductile materials, and appropriate equations are then given for brittle materials. Significant effort is expended in deriving the ultimate expressions for diameter and safety factor, as this makes simplifying assumptions readily apparent. However, the casual reader may wish to proceed to the end of this section where useful design expressions are summarized.

11.3.1 Ductile Materials

Figure 11.2 shows the normal and shear stresses acting on a shaft. In Fig. 11.2a, the stresses act on a rectangular element, and in Fig. 11.2b, they act on an oblique plane at an angle ϕ. The normal stresses are denoted by σ and the shear stresses by τ. Subscript a designates alternating and subscript m designates mean or steady stress. Also, K_f designates the fatigue stress concentration factor due to normal loading, and K_{fs} designates the fatigue concentration factor due to shear loading. On the rectangular element in Fig. 11.2a, the normal stress is $\sigma = \sigma_m \pm K_f\sigma_a$ and the shear stress is $\tau = \tau_m \pm K_{fs}\tau_a$. This uses the approach presented in Section 7.7 that applies the stress concentration factor to the alternating stress and not the mean stress. This approximation has certain implications that will be discussed below.

The largest stress occurs when σ_a and τ_a are in phase, or when the frequency of one is an integer multiple of the frequency of the other. Summing the forces tangent to the diagonal gives

$$\begin{aligned} 0 \;=\; & -\tau_\phi A + (\tau_m + K_{fs}\tau_a)A\cos\phi\cos\phi \\ & -(\tau_m + K_{fs}\tau_a)A\sin\phi\sin\phi \\ & +(\sigma_m + K_f\sigma_a)A\cos\phi\sin\phi. \end{aligned}$$

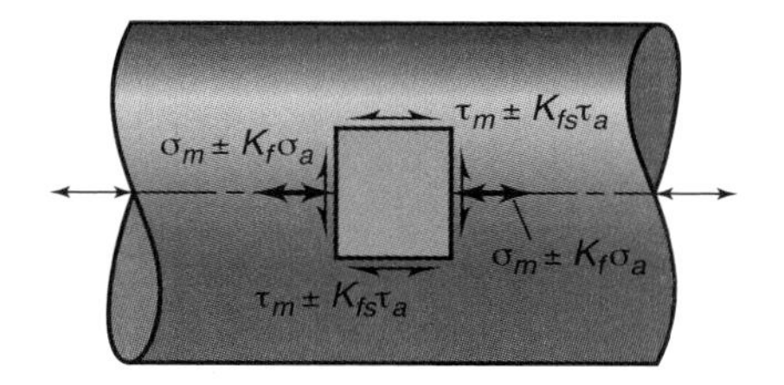

(a)

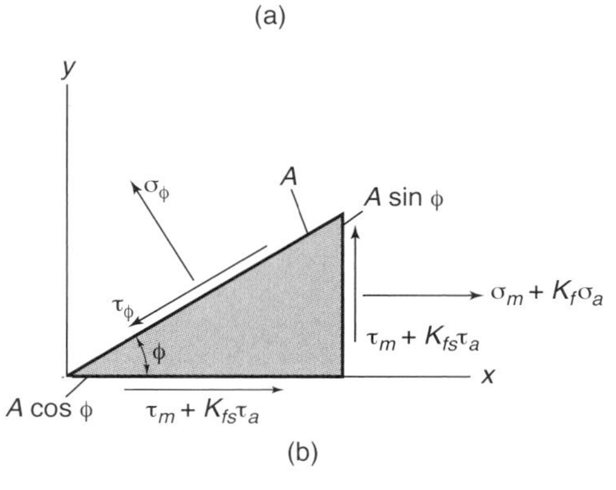

(b)

Figure 11.2: Fluctuating normal and shear stresses acting on shaft. (a) Stresses acting on rectangular element; (b) stresses acting on oblique plane at angle ϕ.

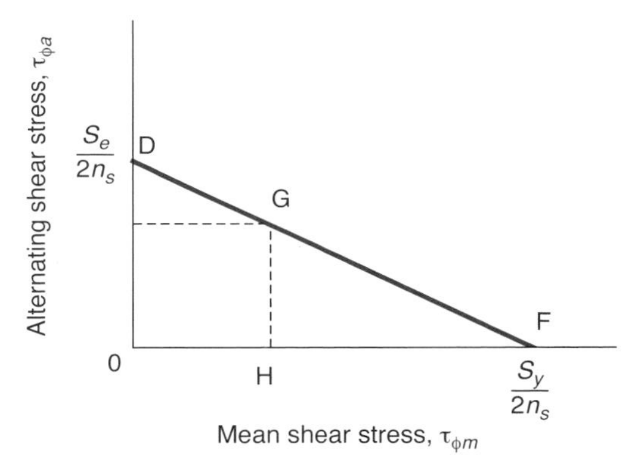

Figure 11.3: Soderberg line for shear stress.

Making use of double angle relations simplifies this expression to

$$\tau_\phi = (\tau_m + K_{fs}\tau_a)\cos 2\phi + \frac{1}{2}(\sigma_m + K_f\sigma_a)\sin 2\phi.$$

Separating the mean and alternating components of stress gives the stress acting on the oblique plane as

$$\begin{aligned} \tau_\phi &= \tau_{\phi m} + \tau_{\phi a} \\ &= \left(\frac{\sigma_m}{2}\sin 2\phi + \tau_m\cos 2\phi\right) \\ &\quad + \left(\frac{K_f\sigma_a}{2}\sin 2\phi + K_{fs}\tau_a\cos 2\phi\right). \qquad (11.23) \end{aligned}$$

Recall the Soderberg line in Fig. 7.16 for tensile loading. For shear loading, the end points of the Soderberg line are $S_{se} = S_e/2n_s$ and $S_{sy} = S_y/2n_s$. Figure 11.3 shows the Soderberg line for shear stress. From the proportional triangles GHF and D0F of Fig. 11.3,

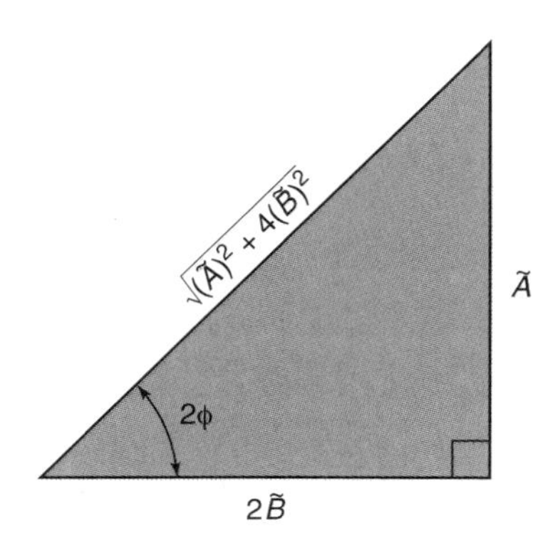

Figure 11.4: Illustration of relationship given in Eq. (11.28).

$$\frac{\text{HF}}{\text{0F}} = \frac{\text{HG}}{\text{0D}}, \qquad \text{or} \qquad \frac{S_y/2n_s - \tau_{\phi m}}{S_y/2n_s} = \frac{\tau_{\phi a}}{S_e/2n_s},$$

so that

$$\frac{1}{n_s} = \frac{\tau_{\phi a}}{S_e/2} + \frac{\tau_{\phi m}}{S_y/2}. \tag{11.24}$$

Substituting the expressions for $\tau_{\phi a}$ and $\tau_{\phi m}$ into Eq. (11.24) gives

$$\begin{aligned} \frac{1}{n_s} &= \frac{\dfrac{K_f\sigma_a}{2}\sin 2\phi + K_{fs}\tau_a \cos 2\phi}{S_e/2} \\ &+ \frac{\left(\dfrac{\sigma_m}{2}\sin 2\phi + \tau_m \cos 2\phi\right)}{S_y/2} \\ &= \tilde{A}\sin 2\phi + 2\tilde{B}\cos 2\phi, \end{aligned} \tag{11.25}$$

where

$$\tilde{A} = \frac{\sigma_m}{S_y} + \frac{K_f\sigma_a}{S_e}, \tag{11.26}$$

$$\tilde{B} = \frac{\tau_m}{S_y} + \frac{K_{fs}\tau_a}{S_e}. \tag{11.27}$$

The stress combination that produces the smallest safety factor is desired, since this corresponds to a maximum-stress situation. The minimum value of n_s corresponds to a maximum value of $1/n_s$. Differentiating $1/n_s$ in Eq. (11.25) and equating the result to zero gives

$$\frac{d}{d\phi}\left(\frac{1}{n_s}\right) = 2\tilde{A}\cos 2\phi - 4\tilde{B}\sin 2\phi = 0.$$

Therefore,

$$\frac{\sin 2\phi}{\cos 2\phi} = \tan 2\phi = \frac{\tilde{A}}{2\tilde{B}}. \tag{11.28}$$

This relationship is illustrated in Fig. 11.4, which shows that

$$\sin 2\phi = \frac{\tilde{A}}{\sqrt{\left(\tilde{A}\right)^2 + 4\left(\tilde{B}\right)^2}}$$

and

$$\cos 2\phi = \frac{2\tilde{B}}{\sqrt{\left(\tilde{A}\right)^2 + 4\left(\tilde{B}\right)^2}}. \tag{11.29}$$

Substituting these into Eq. (11.25) gives

$$\begin{aligned} \frac{1}{n_s} &= \frac{\left(\tilde{A}\right)^2}{\sqrt{\left(\tilde{A}\right)^2 + 4\left(\tilde{B}\right)^2}} + \frac{4\left(\tilde{B}\right)^2}{\sqrt{\left(\tilde{A}\right)^2 + 4\left(\tilde{B}\right)^2}} \\ &= \sqrt{\left(\tilde{A}\right)^2 + 4\left(\tilde{B}\right)^2}. \end{aligned}$$

Substituting Eqs. (11.26) and (11.27) gives

$$\frac{1}{n_s} = \sqrt{\left(\frac{\sigma_m}{S_y} + \frac{K_f\sigma_a}{S_e}\right)^2 + 4\left(\frac{\tau_m}{S_y} + \frac{K_{fs}\tau_a}{S_e}\right)^2},$$

$$\frac{S_y}{n_s} = \sqrt{\left(\sigma_m + \frac{S_y}{S_e}K_f\sigma_a\right)^2 + 4\left(\tau_m + \frac{S_y}{S_e}K_{fs}\tau_a\right)^2}. \tag{11.30}$$

Setting $\sigma_y = 0$, $\sigma_x = \sigma$, and $\tau_{xy} = \tau$ in Eq. (2.19) for biaxial stresses, the maximum shear stress is

$$\tau_{\max} = \sqrt{\left(\frac{\sigma}{2}\right)^2 + \tau^2}, \tag{11.31}$$

and the safety factor is

$$n_s = \frac{S_y/2}{\tau_{\max}} = \frac{S_y/2}{\sqrt{(\sigma/2)^2 + \tau^2}} = \frac{S_y}{\sqrt{\sigma^2 + 4\tau^2}},$$

$$\frac{S_y}{n_s} = \sqrt{\sigma^2 + 4\tau^2}. \tag{11.32}$$

Equations (11.32) and (11.30) have the same form, and

$$\sigma = \sigma_m + \frac{S_y}{S_e}K_f\sigma_a \qquad \text{and} \qquad \tau = \tau_m + \frac{S_y}{S_e}K_{fs}\tau_a.$$

Note that the normal and shear stresses each contain a steady and an alternating component, the latter weighted for the effect of fatigue and stress concentration.

By making use of Eqs. (11.4) and (11.5), Eq. (11.30) becomes

$$n_s = \frac{\pi d^3 S_y}{32\sqrt{\left(M_m + \dfrac{S_y}{S_e}K_f M_a\right)^2 + \left(T_m + \dfrac{S_y}{S_e}K_{fs}T_a\right)^2}}. \tag{11.33}$$

If the smallest safe diameter for a specified safety factor is desired, Eq. (11.33) can be rewritten as

$$d = \left[\frac{32 n_s}{\pi S_y}\sqrt{\left(M_m + \frac{S_y}{S_e}K_f M_a\right)^2 + \left(T_m + \frac{S_y}{S_e}K_{fs}T_a\right)^2}\right]^{1/3} \tag{11.34}$$

Equations (11.33) and (11.34) represent the general form of a shaft design equation using the Soderberg line and MSST. Note in Eq. (11.34) that S_y, S_y/S_e, K_f, and K_{fs} depend on the shaft diameter d. Thus, a numerical or iterative approach is needed to solve for the required diameter.

Peterson [1974] modified Eq. (11.30) by changing the coefficient of the shear stress term from 4 to 3, such that the DET is satisfied and gives

$$\frac{S_y}{n_s} = \sqrt{\left(\sigma_m + \frac{S_y}{S_e}K_f\sigma_a\right)^2 + 3\left(\tau_m + \frac{S_y}{S_e}K_{fs}\tau_a\right)^2}. \tag{11.35}$$

By making use of Eqs. (11.4) and (11.5), Eq. (11.35) becomes

$$n_s = \frac{\pi d^3 S_y}{32\sqrt{\left(M_m + \frac{S_y}{S_e}K_f M_a\right)^2 + \frac{3}{4}\left(T_m + \frac{S_y}{S_e}K_{fs}T_a\right)^2}}. \quad (11.36)$$

The smallest safe diameter corresponding to a specific safety factor can then be expressed as

$$d^3 = \frac{32 n_s}{\pi S_y}\sqrt{\left(M_m + \frac{S_y}{S_e}K_f M_a\right)^2 + \frac{3}{4}\left(T_m + \frac{S_y}{S_e}K_{fs}T_a\right)^2} \quad (11.37)$$

The distinction between Eqs. (11.33) and (11.34) and Eqs. (11.36) and (11.37) needs to be recognized. Equations (11.33) and (11.34) assume that the MSST is valid; Eqs. (11.36) and (11.37) assume that the DET is valid. All four equations are general equations applicable to ductile materials.

Example 11.2: Fatigue Design of a Shaft

Given: When a rear-wheel-drive car accelerates around a bend at high speeds, the drive shafts are subjected to both bending and torsion. The acceleration torque, T, is reasonably constant at 400 N-m while the bending moment is varying due to cornering and is expressed in newton-meters as

$$M = 250 + 800 \sin \omega t.$$

Thus, the mean and alternating moments are $M_m = 250$ N-m and $M_a = 800$ N-m. Assume there is no notch that can produce a stress concentration. The reliability must be 99% and the safety factor is 4.5. The shaft is forged from high-carbon steel, so that it has equivalent mechanical properties as AISI 1080 steel that has been quenched and tempered at 800°C.

Find: The shaft diameter using the MSST.

Solution: From Eq. (7.7) and Table A.2 the bending endurance limit for AISI 1080 steel is

$$S_e' = 0.5S_u = 0.5\,(615) = 307.5 \text{ MPa}.$$

From Fig. 7.11, the surface finish factor for an as-forged surface at $S_{ut} = 615$ MPa is $k_f = 0.42$. To evaluate the size factor, the shaft diameter needs to be chosen, and this value can be modified later if necessary. From Eq. (7.20), and assuming $d = 30$ mm,

$$k_s = 1.248d^{-0.112} = 1.248(30)^{-0.112} = 0.853.$$

From Table 7.4, for 99% probability of survival, the reliability factor is 0.82. Substituting this value into Eq. (7.18) gives the endurance limit as

$$S_e = k_f k_s k_r S_e' = (0.42)(0.853)(0.82)\left(307.5 \times 10^6\right),$$

or $S_e = 90.19$ MPa. Substituting into Eq. (11.34) with $T_a = 0$ gives

$$d^3 = \left\{\frac{32 n_s}{\pi S_y}\sqrt{\left(M_m + \frac{S_y}{S_e}K_f M_a\right)^2 + T_m^2}\right\}^{1/3}$$

$$= \frac{32(4.5)}{\pi\,(380 \times 10^6)} \times \sqrt{\left(250 + \frac{(380 \times 10^6)}{(90.19 \times 10^6)}(1)(800)\right)^2 + (400)^2}.$$

which is solved as $d = 0.0760$ m $= 76.0$ mm.
Note that this value is very different from the assumed shaft diameter of 30 mm, so at least one more iteration is required. Use the value of 76.0
mm as the new assumed value for diameter. From Eq. (7.20),

$$k_s = 1.248d^{-0.112} = 1.248(76.0)^{-0.112} = 0.768.$$

Therefore,

$$S_e = (0.42)(0.768)(0.82)(307.5 \times 10^6) = 81.33 \text{ MPa}.$$

From Eq. (11.34),

$$d^3 = \frac{32(4.5)}{\pi(380 \times 10^6)} \times \sqrt{\left[250 + \frac{(380 \times 10^6)}{(81.33 \times 10^6)}(1)(800)\right]^2 + (400)^2},$$

or $d = 0.0785$ m. Note that the size factor was calculated based on a diameter of 76.0 mm, and the updated solution is 78.5 mm. Since these are very close, no further iterations are deemed necessary. A diameter of 78.5 mm is an awkward design specification; a reasonable dimension to specify for the shaft would be 80 mm or even larger, depending on such factors as stock availability and cost.

Example 11.3: Fatigue Design of a Shaft Under Combined Loading

Given: The shaft made of AISI 1080 high-carbon steel (quenched and tempered at 800°C) shown in Fig. 11.5 is subjected to completely reversed bending and steady torsion. A standard needle bearing (see Fig. 13.1c) is to be placed on diameter d_2 and this surface will therefore be ground to form a good seat for the bearing. The remainder of the shaft will be machined. The groove between the sections ensures that the large diameter section is not damaged by the grinding operation, and is called a *grinding relief.*

Assume that standard needle bearing bore sizes are in 5-mm increments in the range 15 to 50 mm. Design the shaft so that the relative sizes are approximately (within 1 mm) $d_2 = 0.75d_3$ and $d_1 = 0.65d_3$. At this location, the loading involves completely reversed bending of 70 N-m, and steady torsion of 45 N-m. Design the shaft for infinite life.

Find: Determine the diameter d_2 that results in a safety factor of at least 5.0.

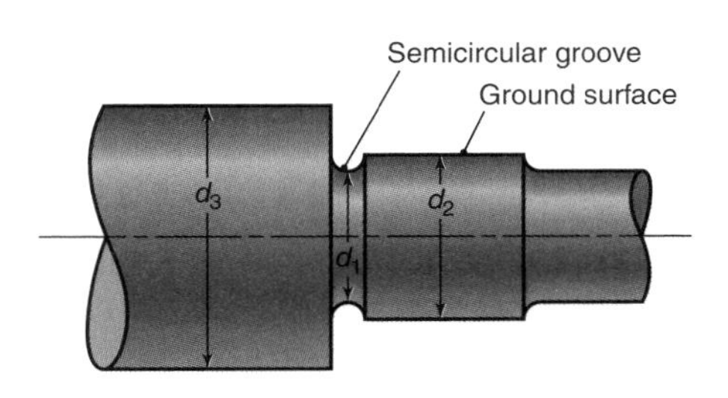

Figure 11.5: Section of shaft in Example 11.3.

Solution: The stress concentration factor can be obtained from the geometry using the ratios given. Not that some variation will occur due to rounding of the dimensions, but this has only a minor effect on the stress concentration. Therefore,

$$\frac{d_2}{d_1} = \frac{0.75d_3}{0.65d_3} = 1.154.$$

Since the grinding relief groove is semicircular,

$$\frac{r}{d} = \frac{(d_2 - d_1)/2}{d_1} = \frac{1}{2}\left(\frac{d_2}{d_1} - 1\right) = 0.0769.$$

Therefore, from Fig. 6.6b, $K_c = 2.05$. From Table A2 for this steel, $S_u = 615$ MPa and $S_y = 380$ MPa. From Eq. (7.7) for bending,

$$S_e' = 0.5S_u = 307 \text{ MPa}.$$

For the shaft, one surface is ground, but the remainder is machined. An inspection of Fig. 11.5 suggests that the ground surface has no stress concentrations and has a larger diameter than the region of grinding relief. Therefore, failure is most likely at the relief, and a surface finish correction factor will be calculated based on a machined surface. Therefore, from Eq. (7.19),

$$k_f = eS_u^f = (4.51)(615)^{-0.265} = 0.822.$$

No yield criterion has been specified, so the MSST will be used. The problem states that the diameter d_2 will be a seat for a bearing, and such bearings are available in 5-mm increments. Therefore, d_2 will be arbitrarily assigned a value of 20 mm, and the safety factor will be calculated and compared to the required value using Eq. (11.33). If the safety factor is not sufficient, then d_2 will be increased until a sufficiently high safety factor results. An alternative approach is to derive an expression for diameter based on Eq. (11.34), and then obtain a numerical solution using a mathematics software package. Either approach is valid and will produce the same results.

If d_2 is 20 mm, then d_3 is $d_2/0.75 = 26.67$ mm, which is rounded up to $d_3 = 27$ mm. Similarly, $d_1 = 0.65d_3 = 17.55$ mm, so that d_1 will be assigned a value of 18 mm. The size factor is obtained from Eq. (7.20) as

$$k_s = 1.248d^{-0.112} = 1.248(27)^{-0.112} = 0.863.$$

The selection of d_3 for use in calculating the size factor should be discussed. No detailed information is given regarding the manufacture of the shaft. It is reasonable to assume that the shaft was machined from extruded bar stock slightly larger than the d_3 dimension, thus justifying the approach in this solution. However, if the shaft were forged, roll forged, or swaged, and then machined to the final dimensions, it would be reasonable to use d_1 to obtain the size factor. Also note that k_f and k_s are coincidentally equal in this case.

No other correction factors apply to this problem, so that the modified endurance limit is obtained from Eq. (7.18):

$$S_e = k_f k_s S_e' = (0.822)(0.863)(307.5) = 218.1 \text{ MPa}.$$

If $d_1 = 18$ and $d_2 = 20$, then the notch radius is $r = 1$ mm for a semicircular groove. Therefore, from Fig. 7.10 for $S_u = 615$ MPa, the notch sensitivity factor is around $q_n = 0.7$. From Eq. (7.17),

$$K_f = 1 + (K_c - 1)q_n = 1 + (2.05 - 1)(0.7) = 1.735.$$

For completely reversed bending, $M_m = 0$ and $M_a = 70$ N-m, and for steady torsion, $T_a = 0$ and $T_m = 45$ N-m. Therefore, from Eq. (11.33),

$$n_s = \frac{\pi(0.020)^3(380 \times 10^6)}{32\sqrt{\left[\left(\frac{380}{218.1}\right)(1.735)(70)\right]^2 + 45^2}} = 1.38.$$

This safety factor is too low, since a minimum safety factor of 5.0 was prescribed. Thus, the diameter d_2 is increased, and the procedure is repeated. Table 11.1 summarizes the results for a number of values of d_2. Therefore, the value of $d_2 = 35$ mm, $d_1 = 31$ mm, and $d_3 = 47$ mm are used to design the shaft.

Table 11.1: Summary of results for Example 11.3.

d_2 (mm)	d_1 (mm)	d_3 (mm)	n_s
20	18	27	1.32
25	22	34	2.50
30	26	40	4.26
35	31	47	6.65

11.3.2 Brittle Materials

Although shafts are usually cold-worked metals that are machined to final desired dimensions, there are applications where castings, which are often brittle materials, are used as shafts. As discussed in Ch. 6, this requires a slightly different analysis approach than for ductile materials.

For brittle materials, the forces in Fig. 11.2b are assumed to be *normal* rather than tangent to the diagonal. Also, the design line for any failure theory relevant to brittle materials (see Section 6.7.2) extends from S_e/n_s to S_u/n_s instead of from $S_e/2n_s$ to $S_y/2n_s$ as was true for the ductile materials. Following procedures similar to those used in obtaining Eq. (11.30) gives

$$\frac{2S_u}{n_s} = K_c\left(\sigma_m + \frac{S_u}{S_e}\sigma_a\right) + \sqrt{K_c^2\left(\sigma_m + \frac{S_u}{S_e}\sigma_a\right)^2 + 4K_{cs}^2\left(\tau_m + \frac{S_u}{S_e}\tau_a\right)^2}. \tag{11.38}$$

where K_c is the theoretical stress concentration factor. By making use of Eqs. (11.4) and (11.5), Eq. (11.38) can be written as

$$n_s = \frac{\pi d^3 S_u/16}{K_c\Psi + \sqrt{K_c^2\Psi^2 + K_{cs}^2\left(T_m + \frac{S_u}{S_e}T_a\right)^2}}, \tag{11.39}$$

where Ψ is given by

$$\Psi = M_m + \frac{S_u}{S_e}M_a. \tag{11.40}$$

If the minimum safe diameter of the shaft is desired for a specific safety factor,

$$d=\left\{\frac{16n_s}{\pi S_u}\left[K_c\Psi+\sqrt{K_c^2\Psi^2+K_{cs}^2\left(T_m+\frac{S_u}{S_e}T_a\right)^2}\right]\right\}^{1/3} \tag{11.41}$$

The important difference in the equations developed above for the safety factor and the smallest safe diameter is that Eqs. (11.33) and (11.34) are applicable for ductile materials while assuming the MSST, Eqs. (11.36) and (11.37) are also applicable for ductile materials but while assuming the DET, and Eqs. (11.39) and (11.41) are applicable for brittle materials. Note the major differences between the equations developed for brittle and ductile materials. For brittle materials [Eqs. (11.39) and (11.41)] the stress concentration factor K_c and the ultimate stress S_u are used, whereas for ductile materials [Eqs. (11.33), (11.34), (11.36), and (11.37)] the fatigue stress concentration factor K_f and the yield stress S_y are used.

11.4 Additional Shaft Design Considerations

Sections 11.2 and 11.3 described in detail the design approach for sizing or analyzing a shaft from a stress standpoint. Rotating shafts are very likely to encounter sufficient stress cycles to necessitate design based on an endurance limit (see Section 7.8). Therefore, the same concerns discussed in Ch. 7 hold for shafts. It needs to be recognized that the design approach in Ch. 7 is empirical in nature, but must be verified through experiments. Shafts are often spared from extensive test programs because of their inherently high safety factors, the reasons for which are discussed below.

A common cyclic stress variation that occurs in practical applications is reversed bending and steady torsion. From Section 7.3, note that reversed bending implies that $\sigma_m=0$, or $M_m=0$. Also, steady torsion implies that $\tau_a=0$, or $T_a=0$. Thus, reduced forms of Eqs. (11.33) and (11.34), (11.36) and (11.37), or (11.39) and (11.41) can be readily determined. For example, for such a loading, Eqs. (11.33) and (11.34) become

$$n_s=\frac{\pi d^3 S_y}{32\sqrt{\left(\frac{S_y}{S_e}K_f M_a\right)^2+T_m^2}}, \tag{11.42}$$

$$d=\left[\frac{32n_s}{\pi S_y}\sqrt{\left(\frac{S_y}{S_e}K_f M_a\right)^2+T_m^2}\right]^{1/3}. \tag{11.43}$$

Recall that Eqs. (11.42) and (11.43) are for the MSST and Soderberg line; similar expressions can be obtained for other criteria.

As mentioned previously, shafts usually display fairly large safety factors compared to other machine elements. There are a number of reasons for this, including:

1. Shafts are usually in difficult-to-access locations, and have many machine elements mounted onto them. Replacing a shaft requires significant time merely for exposure; removing machine elements, replacing the shaft, and remounting the machine elements (and aligning them) also requires significant time. Recognizing this, designers commonly assign large safety factors to avoid high costs associated with failure and replacement of shafts.
2. Shafts themselves are usually quite expensive, and protection of the shaft is one of the main reasons that keys or pins (Section 11.6) or slip clutches (Section 18.10) are used.
3. Deflection is a major concern, and often shaft size is specified to meet a deflection requirement, leading to low stress levels. Deflection includes lateral deflection of the shaft from bending moments (Ch. 5), as well as torsional deflection (see Section 4.4).
4. Certain machine elements, such as gears or connecting rods, require that the shaft provide load support with minimal deflection. For this reason, it is not unusual to place bearings immediately adjacent to such machine elements. Thus, the spans and bending moments encountered in practice problems and examples are not reflective of well-supported shafts in practice. However, it must be recognized that providing additional bearings is not a straightforward approach, and requires careful alignment and adjustment in order to evenly distribute loads.

Design Procedure 11.1: Shafts

The general procedure for shaft design is as follows:

1. Develop a free-body diagram by replacing the various machine elements mounted on the shaft by their statically equivalent load or torque components. To illustrate this, Fig. 11.6a shows two gears exerting forces on a shaft, and Fig. 11.6b then shows a free-body diagram of the shaft.
2. Draw a bending moment diagram in the x-y and x-z planes as shown in Fig. 11.6c and d. The resultant internal moment at any section along the shaft may be expressed as

$$M_x=\sqrt{M_{xy}^2+M_{xz}^2}. \tag{11.44}$$

3. Analyze the shaft based on lateral deflection due to bending using the approach in Ch. 5. If deflection is a design constraint, select a diameter that results in acceptable deflection, or else relocate supports to reduce the bending moments encountered.
4. Develop a torque diagram as shown in Fig. 11.6e. Torque developed from one power-transmitting element must balance that from other power-transmitting elements.
5. Analyze the shaft based on deflection due to torsion (Section 4.4). If torsional deflection is a design constraint, select a diameter that results in acceptable deflection.
6. Evaluate the suitability of the shaft from a stress standpoint:

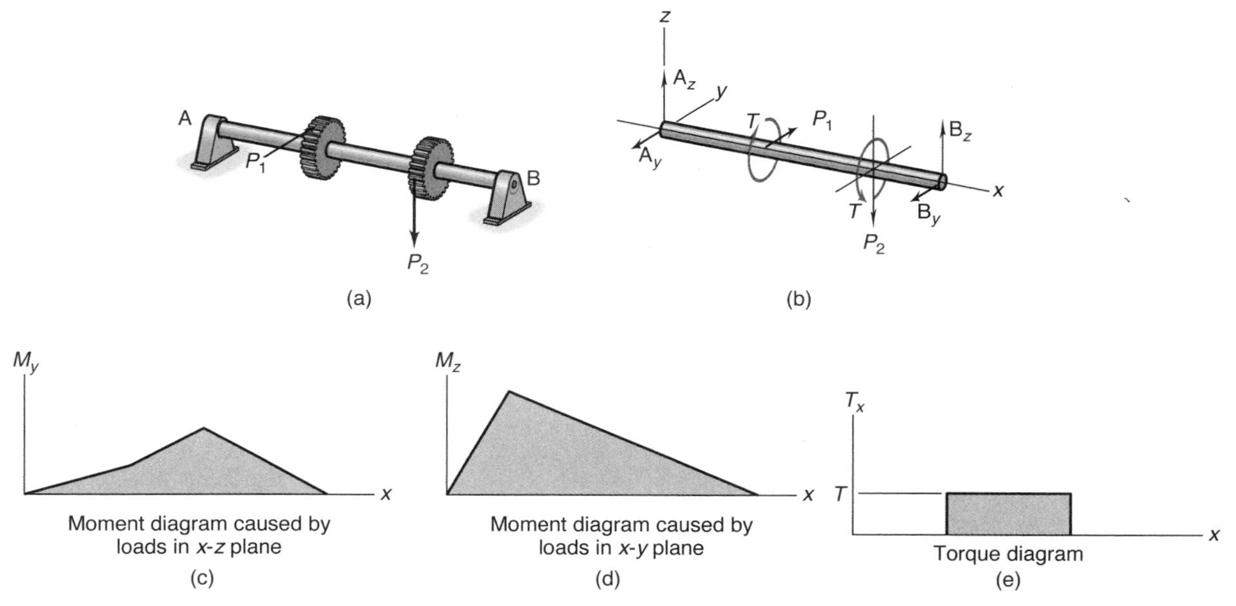

Figure 11.6: Shaft assembly. (a) Shaft with two bearings at A and B and two gears with resulting forces P_1 and P_2; (b) free-body diagram of torque and forces resulting from assembly drawing; (c) moment diagram in x-z plane; (d) moment diagram in x-y planes; (e) torque diagram.

(a) Establish the location of the critical cross-section, or the x location where the torque and moment are the largest.

(b) For ductile materials, use the MSST or the DET covered in Section 6.7.1.

(c) For brittle materials, use the maximum-normal-stress theory (MNST), the internal friction theory (IFT), or the modified Mohr theory (MMT) (see Section 6.7.2).

7. Use keyways, set screws or pins (see Section 11.6), or slip clutches (Section 18.10) where appropriate to protect the shaft.

8. Compare the critical speed of the shaft (Section 11.5) to the operating conditions. Change shaft diameter or supports to avoid critical speeds if necessary.

11.5 Critical Speed of Rotating Shafts

All rotating shafts deflect during operation. The magnitude of the deflection depends on the stiffness of the shaft and its supports, the total mass of the shaft and its attached parts, and the amount of system damping. The **critical speed** of a rotating shaft, sometimes called the **natural frequency**, is the speed at which the rotating shaft becomes dynamically unstable and large deflections associated with vibration are likely to develop. For any shaft there are an infinite number of critical speeds, but only the lowest (first) and occasionally the second are generally of interest to designers. The others are usually so high as to be well out of the operating range of shaft speed. This text considers only the first critical speed of the shaft. Two approximate methods of finding the first critical speed (or lowest natural frequency) of a system are given in this section, one attributed to Rayleigh and the other to Dunkerley.

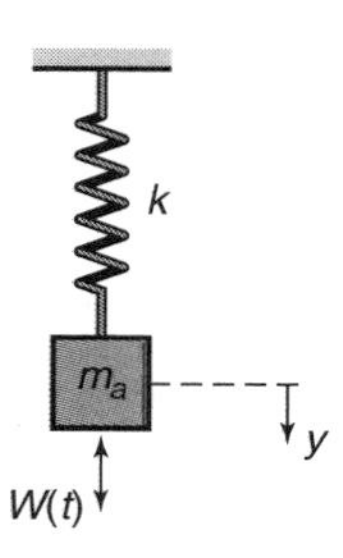

Figure 11.7: Simple single-mass system.

11.5.1 Single-Mass System

The **first critical speed** (or **lowest natural frequency**) can be obtained by observing the rate of interchange between the kinetic (energy of motion) and potential (energy of position) energies of the system during its cyclic motion. A single mass on a shaft can be represented by the simple spring and mass shown in Fig. 11.7. The dashed line indicates the static equilibrium position. The potential energy of the system is

$$U = \int_0^{\delta} (m_a g + k\delta)\, d\delta - m_a g \delta,$$

where

m_a = mass, kg
g = gravitational acceleration, 9.807 m/s^2
k = spring rate, N/m
δ = deflection, m

Integrating gives

$$U = \frac{1}{2}k\delta^2. \tag{11.45}$$

The kinetic energy of the system with the mass moving with a velocity of $\dot{\delta}$ is

$$K_e = \frac{1}{2} m_a \left(\dot{\delta}\right)^2. \tag{11.46}$$

Observe the following about Eqs. (11.45) and (11.46):

1. As the mass passes through the static equilibrium position, the potential energy is zero and the kinetic energy is at a maximum and equal to the total mechanical energy of the system.
2. When the mass is at the position of maximum displacement and is on the verge of changing direction, its velocity is zero. At this point the potential energy is at a maximum and is equal to the total mechanical energy of the system.

The total mechanical energy is the sum of the potential and kinetic energies and is constant at any time. Therefore,

$$\frac{d}{dt}\left(U + K_e\right) = 0. \tag{11.47}$$

Substituting Eqs. (11.45) and (11.46) into Eq. (11.47) gives

$$\frac{d}{dt}\left[\frac{1}{2}k\delta^2 + \frac{1}{2}m_a\left(\dot{\delta}\right)^2\right] = k\delta\dot{\delta} + m_a\dot{\delta}\ddot{\delta} = 0.$$

Factoring $\dot{\delta}$ leads to

$$\dot{\delta}\left(m_a\ddot{\delta} + k\delta\right) = 0, \tag{11.48}$$

$$\ddot{\delta} + \omega^2\delta = 0,$$

where

$$\omega = \sqrt{k/m_a},\ \text{rad/s}. \tag{11.49}$$

The general solution to this differential equation is

$$\delta = C_1 \sin(\omega t + \phi), \tag{11.50}$$

where C_1 is an integration constant. The first critical speed (or lowest natural frequency) is ω. Substituting Eq. (11.50) into Eqs. (11.45) and (11.46) gives

$$U = \frac{k}{2} C_1^2 \sin^2\left(\omega t + \phi\right), \tag{11.51}$$

$$K_e = \frac{m_a}{2} C_1^2 \omega^2 \cos^2\left(\omega t + \phi\right). \tag{11.52}$$

Note from Eq. (11.49) that, for static deflection, if $k = W/\delta$ and $m_a = W/g$, then

$$\omega = \sqrt{\frac{k}{m_a}} = \sqrt{\frac{W/y}{W/g}} = \sqrt{\frac{g}{\delta}}. \tag{11.53}$$

11.5.2 Multiple-Mass System

From Eq. (11.46), the kinetic energy for n masses is

$$K_e = \frac{1}{2}m_{a1}\left(\dot{\delta}_1\right)^2 + \frac{1}{2}m_{a2}\left(\dot{\delta}_2\right)^2 + \cdots + \frac{1}{2}m_{an}\left(\dot{\delta}_n\right)^2. \tag{11.54}$$

If the deflection is represented by Eq. (11.50), then $y_{\max} = C_1$. Also, $\dot{y}_{\max} = C_1\omega = y_{\max}\omega$. Therefore, the maximum kinetic energy is

$$K_{e,\max} = \frac{\omega^2}{2}\sum m_{an}\left(\delta_{n,\max}\right)^2. \tag{11.55}$$

From Eq. (11.45), the potential energy for n masses is

$$U = \frac{1}{2}k_1\delta_1^2 + \frac{1}{2}k_2\delta_2^2 + \cdots + \frac{1}{2}k_n\delta_n^2, \tag{11.56}$$

and the maximum potential energy is

$$U_{\max} = \frac{1}{2}k_1\left(\delta_{1,\max}\right)^2 + \frac{1}{2}k_2\left(\delta_{2,\max}\right)^2 + \cdots + \frac{1}{2}k_n\left(\delta_{n,\max}\right)^2. \tag{11.57}$$

The **Rayleigh method** assumes that $K_{e,\max} = U_{\max}$ or

$$\frac{1}{2}\omega^2 \sum_{i=1,\ldots,n} m_{ai}\left(\delta_{i,\max}\right)^2 = \frac{1}{2}\sum_{i=1,\ldots,n} k_i\left(\delta_{i,\max}\right)^2.$$

Solving for angular velocity,

$$\omega^2 = \frac{\sum_{i=1,\ldots,n} k_i\left(\delta_{i,\max}\right)^2}{\sum_{i=1,\ldots,n} m_{ai}\left(\delta_{i,\max}\right)^2}, \tag{11.58}$$

but

$$k_i = \frac{W_i}{\delta_{i,\max}} \quad \text{and} \quad m_{ai} = \frac{W_i}{g}. \tag{11.59}$$

where W_i is the ith weight placed on the shaft and g is gravitation acceleration, 9.807 m/s^2. Substituting Eq. (11.59) into Eq. (11.58) gives

$$\omega_{\text{cr}} = \sqrt{\frac{g\sum_{i=1,\ldots,n} W_i\delta_{i,\max}}{\sum_{i=1,\ldots,n} W_i\delta_{i,\max}^2}}. \tag{11.60}$$

This is the first critical speed (first natural frequency) of a multiple-mass system when using the Rayleigh method. Equation (11.60) is known as the **Rayleigh equation**. Because the actual displacements are larger than the static displacements used in Eq. (11.60), the energies in both the denominator and the numerator will be underestimated by the Rayleigh formulation. However, the error in the underestimate will be larger in the denominator, since it involves the square of the approximated displacements. Thus, Eq. (11.60) *overestimates* (provides an upper bound on) the first critical speed.

The **Dunkerley equation** is another approximation to the first critical speed of a multiple-mass system; it is given as

$$\frac{1}{\omega_{\text{cr}}^2} = \frac{1}{\omega_1^2} + \frac{1}{\omega_2^2} + \cdots + \frac{1}{\omega_n^2}, \tag{11.61}$$

where

ω_1 = critical speed if only mass 1 exists
ω_2 = critical speed if only mass 2 exists
ω_n = critical speed if only the nth mass exists

Recall from Eq. (11.53) that $\omega_i = \sqrt{g/\delta_i}$.

The Dunkerley equation *underestimates* (provides a lower bound on) the first critical speed. The major difference between the Rayleigh and Dunkerley equations is in the deflections. In the Rayleigh equation, the deflection at a specific mass location takes into account the deflections due to all the masses acting on the system; in the Dunkerley equation, the deflection is due only to the individual mass being evaluated.

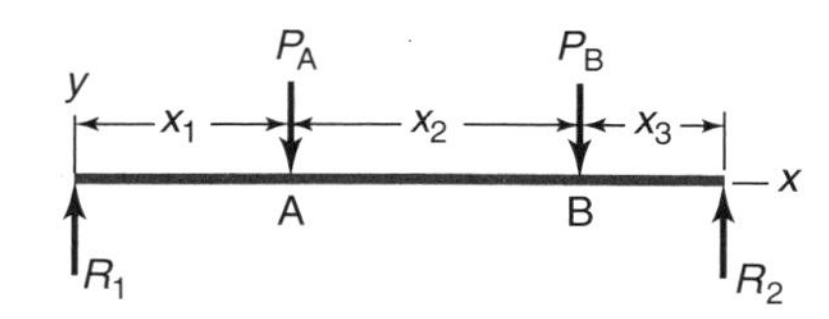

Figure 11.8: Simply supported shaft arrangement for Example 11.4.

Example 11.4: Critical Shaft Speed

Given: Figure 11.8 shows a simply supported shaft arrangement. A solid shaft of 50 mm diameter made of AISI 1020 low-carbon steel is used. The following are given: $x_1 = 0.750$ m, $x_2 = 1.000$ m, $x_3 = 0.500$ m, $P_A = 300$ N, and $P_B = 500$ N.

Find: Determine the first critical speed by using

(*a*) The Rayleigh method

(*b*) The Dunkerley method

Solution: From Table 5.1b, for simply-supported ends, the deflections are given by:

For $0 \le x \le a$

$$\delta_y = \frac{Pbx}{6lEI}(l^2 - x^2 - b^2). \qquad (a)$$

For $a \le x \le l$

$$\delta_y = \frac{Pa(l-x)}{6lEI}(2lx - a^2 - x^2). \qquad (b)$$

From Table 3.1, the modulus of elasticity for carbon steel is 207 GPa. For a solid, round shaft, the area moment of inertia is

$$I = \frac{\pi d^4}{64} = \frac{\pi (0.050)^4}{64} = 3.068 \times 10^{-7} \text{ m}^4.$$

The deflection at location A due to load P_A from Eq. (*a*) and $a = x_1 = x = 0.750$ m and $b = x_2 + x_3 = 1.5$ m is

$$\begin{aligned} \delta_{AA} &= \frac{P_A bx}{6lEI}\left(l^2 - x^2 - b^2\right) \\ &= -\frac{(300)(1.5)(0.75)\left(2.25^2 - 0.75^2 - 1.5^2\right)}{6(2.25)\left(207 \times 10^9\right)\left(3.068 \times 10^{-7}\right)} \\ &= -0.8857 \text{ mm.} \end{aligned}$$

Note that in Fig. 11.8, the y-direction is upward; thus, P_A and δ_{AA} are negative. Also, the first subscript in δ_{AA} designates the location where the deflection occurs, and the second subscript designates the loading that contributes to the deflection. The deflection at location A due to load P_B from Eq. (*a*) and $a = x_1 + x_2$, $x = x_1 = 0.750$ m, and $b = 0.5$ m is

$$\begin{aligned} \delta_{AB} &= \frac{P_B bx}{6lEI}\left(l^2 - x^2 - b^2\right) \\ &= -\frac{(500)(0.5)(0.75)\left(2.25^2 - 0.75^2 - 0.5^2\right)}{6(2.25)\left(207 \times 10^9\right)\left(3.068 \times 10^{-7}\right)} \\ &= -0.9295 \text{ mm.} \end{aligned}$$

The total deflection at location A is

$$\delta_A = \delta_{AA} + \delta_{AB} = -0.8857 - 0.9295 = -1.815 \text{ mm.}$$

The deflection at location B due to load P_B from Eq. (*a*) and $x = a = x_1 + x_2 = 1.75$ m and $b = 0.5$ m is

$$\begin{aligned} \delta_{BB} &= \frac{P_B bx}{6lEI}\left(l^2 - x^2 - b^2\right) \\ &= -\frac{(500)(0.5)(1.75)\left(2.25^2 - 1.75^2 - 0.5^2\right)}{6(2.25)\left(207 \times 10^9\right)\left(3.068 \times 10^{-7}\right)} \\ &= -0.8930 \text{ mm.} \end{aligned}$$

The deflection at location B due to load P_A from Eq. (*b*) and $a = x_1 = 0.75$ m, $b = x_2 + x_3 = 1.5$ m, and $x = x_1 + x_2 = 1.75$ m is

$$\begin{aligned} \delta_{BA} &= \frac{P_A a(l-x)}{6lEI}\left(2lx - a^2 - x^2\right) \\ &= -\frac{(300)(0.75)(2.25 - 2.75)}{6(2.25)\left(207 \times 10^9\right)\left(3.068 \times 10^{-7}\right)} \\ &\quad \times \left[2(2.25)(1.75) - 0.75^2 - 1.75^2\right], \end{aligned}$$

or $\delta_{BA} = -0.5577$ mm. Thus, the total deflection at location B is

$$\delta_B = \delta_{BA} + \delta_{BB} = -0.5577 - 0.8930 = -1.451 \text{ mm.}$$

(*a*) Using the Rayleigh method, Eq. (11.60) gives the first critical speed as

$$\begin{aligned} \omega_{\text{cr}} &= \sqrt{\frac{g(P_A\delta_A + P_B\delta_B)}{P_A\delta_A^2 + P_B\delta_B^2}} \\ &= \sqrt{\frac{(9.81)\left[(300)(0.001815) + (500)(0.001451)\right]}{(300)(0.001815)^2 + (500)(0.001451)^2}}. \end{aligned}$$

or $\omega_{\text{cr}} = 78.13$ rad/s $= 746$ rpm.

(*b*) Using the Dunkerley method, Eq. (11.61) gives

$$\frac{1}{\omega_{\text{cr}}^2} = \frac{1}{\omega_{\text{cr,A}}^2} + \frac{1}{\omega_{\text{cr,B}}^2}.$$

where

$$\omega_{\text{cr,A}} = \sqrt{\frac{g}{\delta_{AA}}} = \sqrt{\frac{9.81}{0.0008857}} = 105.2 \text{ rad/s} = 1005 \text{ rpm},$$

$$\omega_{\text{cr,B}} = \sqrt{\frac{g}{\delta_{BB}}} = \sqrt{\frac{9.81}{0.0008930}} = 104.8 \text{ rad/s} = 1001 \text{ rpm}.$$

Therefore, the critical speed is

$$\frac{1}{\omega_{\text{cr}}^2} = \frac{1}{\omega_{\text{cr,A}}^2} + \frac{1}{\omega_{\text{cr,B}}^2} = \frac{1}{1005^2} + \frac{1}{1001^2},$$

which is solved as $\omega_{\text{cr}} = 709$ rpm.

In summary, the Rayleigh equation gives $\omega_{\text{cr}} = 746$ rpm, which overestimates the first critical speed; the Dunkerley equation gives $\omega_{\text{cr}} = 709$ rpm, which underestimates the first critical speed. Therefore, the actual first critical speed is between 709 and 746 rpm, and the shaft design should avoid this range of operation.

Table 11.2: Dimensions of selected square plain parallel stock keys.

Shaft diameter, in.	Key width, in.	Distance from keyseat to opposite side of shaft, in.
0.500	0.125	0.430
0.625	0.1875	0.517
0.750	0.1875	0.644
0.875	0.1875	0.771
1.000	0.25	0.859
1.125	0.25	0.956
1.250	0.25	1.112
1.375	0.3125	1.201
1.500	0.375	1.289
1.675	0.375	1.416
1.750	0.375	1.542
1.875	0.50	1.591
2.000	0.50	1.718
2.250	0.50	1.972
2.500	0.625	2.148
2.750	0.625	2.402
3.000	0.75	2.577
3.250	0.75	2.831
3.500	0.875	3.007
3.750	0.875	3.261
4.000	1.00	3.437
4.500	1.00	3.944
5.000	1.25	4.296
6.000	1.50	5.155

11.6 Keys, Roll Pins, Splines, and Set Screws

A variety of machine elements, such as gears, pulleys, and cams, are mounted on rotating shafts. The portion of the mounted member in contact with the shaft is called the **hub**. This section describes methods of attaching the hub to the shaft. Press and shrink fits considered in Sections 10.5 and 10.6 are other options for attachment, as are using threaded hubs, or by welding or adhesively bonding the hub to the shaft (Ch. 16).

There is a great variety of **keys** and **pins**, as shown in Fig. 11.9. The simplest and most commonly used is the **parallel key**, which has a constant cross-section across its length. A **tapered key** has a constant width, but the height varies with a taper of 0.125 in. per 12 in. (0.597°), and is driven into a tapered slot on the hub until it locks. A **Gib head** can be included on the key to facilitate removal. A **Woodruff key** has a constant width and a semicircular profile, and fits into a semicircular key seat machined onto the shaft. A Woodruff key is light-duty because it has a deep seat in the shaft, which can compromise strength, but it aligns itself readily against the hub. Selected standard key sizes are summarized in Tables 11.2 through 11.4. Keys are usually manufactured from low-carbon steel, but can be heat treated to obtain higher strengths. However, this is rare, as ductility in the key material is usually important.

Roll pins or **grooved pins** are alternatives to keys, as shown in Fig. 11.9i and 11.9j. Roll and grooved pins are used for light-duty service, and remain in place because of elastic recovery of the pin in the hole, so that it bears against the hole and is restrained by friction.

Consider the behavior of a simple square key, as shown in Fig. 11.9a. The main purpose of the key is to prevent rotation between the shaft and the connected machine element through which torque is being transmitted. The purpose of using the key is to transmit the *full* torque.

A key is also intended as a safety system. Most machines have an operating speed and torque that define the required size of the key. However, in the event of a drastic increase in the load conditions, the key will shear before the shaft or machine element (gear, cam, pulley, etc.) fails. Since keys are inexpensive and can be quickly replaced, designers use them to protect more expensive machinery components.

Note in Fig. 11.9 that if $h = w$, the key is square, a special case of a flat key. Keys should fit tightly so that key rotation is not possible. In some applications, set screws may be required to restrict motion of the key.

Keyways result in stress concentrations in shafts. Figure 11.10 provides experimental measurements of stress concentrations associated with fitted keyways having a semicircular end. These stress concentrations should be applied to the shaft; any stress concentrations encountered by the key are generally incorporated into the key strength. It should further be noted that the data in Fig. 11.10 is conservative; actual stress concentrations associated with keyways are strongly dependent on the shaft diameter, keyway fillet radius, manufacturing tolerances, and key geometry. Stress concentration factors above 5 for bending have been reported for square keys near the key end. Fortunately, even with such a high stress concentration factor, failure for standard designs almost always occurs in the key instead of the shaft. Fatigue may be a concern, for which the stress concentration factors in Fig. 11.10 are considered reasonable.

Design Procedure 11.2: Keys

Keys are designed to fail when an applied torque exceeds a critical value. Failure is due to shear or compressive bearing stresses, both of which are considered here. Depending on the design motivation, it is reasonable to design based on yield strength or ultimate strength. For example, if it is desired to design a keyway that will not plastically deform under a certain loading, then use of yield strength is proper. However, if it is desired to predict the maximum torque that can be supported by a key before failure, then ultimate strength is more appropriate. This Design Procedure arbitrarily uses yield strength, but could be easily adapted for determining the maximum torque that can be generated by a key. Similarly, fatigue considerations using the approaches of Ch. 7 could be applied; however, since a time-varying torque is not the more typically encountered loading, static design is usually sufficient.

1. Failure due to shear. The shear force applied to the key is given by:

$$P = \frac{T}{d/2} = \frac{2T}{d}. \tag{11.62}$$

The shear stress at yielding, recalling that $S_{sy} = S_y/2$ (see Section 2.12), is

$$\tau = \frac{S_y}{2n_s} = \frac{P}{A_s}. \tag{11.63}$$

Therefore, the torque that causes plastic deformation of a key is obtained from Eqs. (11.62) and (11.63) as

$$T = \frac{S_y A_s d}{4n_s}. \tag{11.64}$$

The area for a key as shown in Fig. 11.9a–e is $A_s = wl$. For a round key (Fig. 11.9f), $A_s = dl$. For a Woodruff key, see Table 11.4.

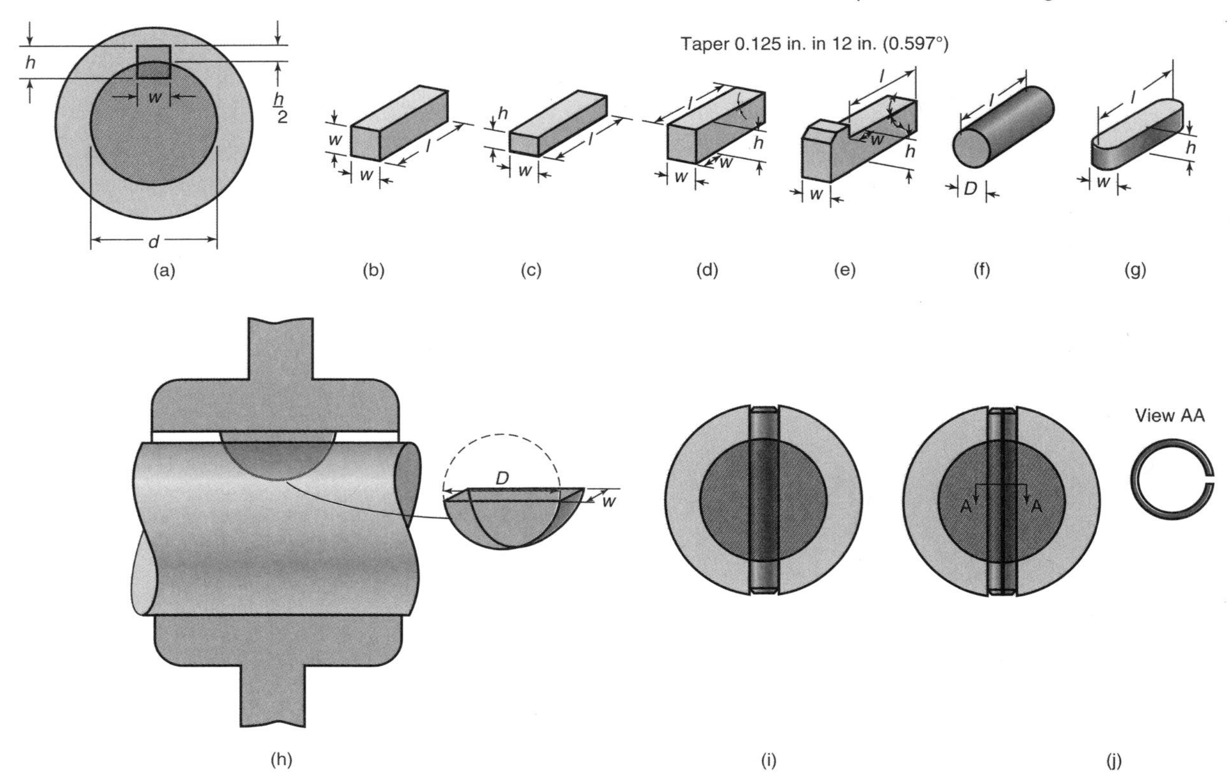

Figure 11.9: Illustration of keys and pins. (a) Dimensions of shaft with keyway in shaft and hub; (b) square parallel key; (c) flat parallel key; (d) tapered key; (e) tapered key with Gib head, or Gib-head key. The Gib head assists in key removal; (f) round key; (g) profile key; (h) Woodruff key with illustration of mounting; (i) pin, which is often grooved. The pin is slightly larger than the hole so that friction holds the pin in place; (j) roll pin. Elastic deformation of the pin in the smaller hole leads to friction forces that keep the pin in place.

2. Failure due to compressive or bearing stress. This failure mode is most commonly encountered with flat keys. The bearing stress for a key as shown in Fig. 11.9a–e is given by

$$\sigma = \frac{P}{A_c} = \frac{2T}{dlh/2} = \frac{4T}{dlh}. \tag{11.65}$$

Therefore, the torque that results in excessive bearing stress is:

$$T = \frac{S_y dlh}{4n_s}. \tag{11.66}$$

Not surprisingly, Eq. (11.66) is independent of the width, w.

Example 11.5: Key Design

Given: A 1-in.-diameter shaft and mating hub are both made of high-carbon steel. A standard square key made of AISI 1020 low-carbon steel with a yield strength of $S_y = 43$ ksi is to be selected for this application, using $n_s = 2.0$.

Find:

(*a*) The required key length to transmit 5000 in.-lb.

(*b*) The torque that can be transmitted without plastic deformation by a 1.000-in.-long key.

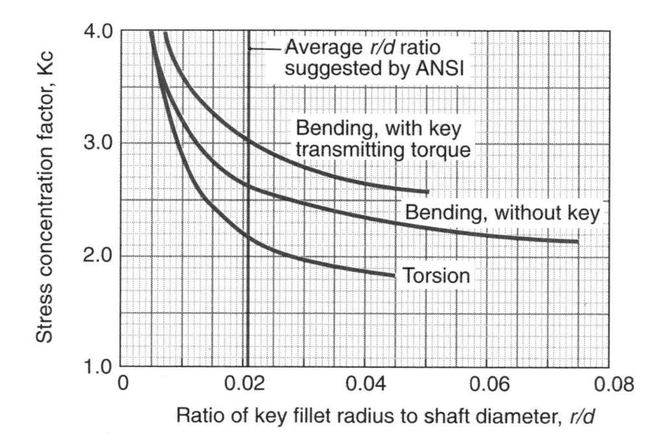

Figure 11.10: Stress concentrations for a standard profile key, where $w = h = d/4$ (see Fig. 11.9g).

Table 11.3: Dimensions of square and flat taper stock keys.

Shaft diameter, in.	Square type Width w, in.	Square type Height[a] h, in.	Flat type Width w, in.	Flat type Height[a] h, in.	Available lengths, l Minimum, in.	Maximum, in.	Available increments, in.
0.5000–0.5625	0.1250	0.1250	0.1250	0.09375	0.50	2.00	025
0.6250–0.8750	0.1875	0.1875	0.1875	0.1250	0.75	3.00	0.375
0.9375–1.2500	0.2500	0.2500	0.2500	0.1875	1.00	4.00	0.50
1.3125–1.3750	0.3125	0.3125	0.3125	0.2500	1.25	5.25	0.625
1.4375–1.7500	0.3750	0.3750	0.3750	0.2500	1.50	6.00	0.75
1.8125–2.2500	0.5000	0.5000	0.5000	0.3750	2.00	8.00	1.00
2.3125–2.7500	0.6250	0.6250	0.6250	0.4375	2.50	10.00	1.25
2.8750–3.2500	0.7500	0.7500	0.7500	0.5000	3.00	12.00	1.50
3.3750–3.7500	0.8750	0.8750	0.8750	0.6250	3.50	14.00	1.75
3.8750–4.5000	1.0000	1.0000	1.0000	0.7500	4.00	16.00	2.00
4.7500–5.5000	1.2500	1.2500	1.2500	0.8750	5.00	20.00	2.50
5.7500–6.0000	1.5000	1.5000	1.5000	1.0000	6.00	24.00	3.00

[a] Measured at a distance equal to the key width, w, from the large end.

Table 11.4: Dimensions of selected Woodruff keys.

Key no.	Suggested shaft sizes, in.	Nominal key size[a], $w \times l$, in.	Height of key, h, in.	Shearing Area, in.^{2}
204	0.3125–0.3750	0.062 × 0.500	0.203	0.030
305	0.4375–0.5000	0.094 × 0.625	0.250	0.052
405	0.6875–0.7500	0.125 × 0.625	0.250	0.072
506	0.8125–0.9375	0.156 × 0.750	0.313	0.109
507	0.8750–0.9375	0.156 × 0.875	0.375	0.129
608	1.0000–1.1875	0.188 × 1.000	0.438	0.178
807	1.2500–1.3125	0.250 × 0.875	0.375	0.198
809	1.2500–1.7500	0.250 × 1.125	0.484	0.262
810	1.2500–1.7500	0.250 × 1.250	0.547	0.296
812	1.2500–1.7500	0.250 × 1.500	0.641	0.356
1012	1.8125–2.5000	0.312 × 1.500	0.641	0.438
1212	1.8750–2.5000	0.375 × 1.500	0.641	0.517

[a] The key extends into the hub a distance of w /2.

(*c*) The torque that can be transmitted by a Woodruff key without shearing.

Solution: From Table 11.2, the width of a square key for use with a 1-in. diameter shaft is 0.25 in. Since it is a square key, $h = 0.25$ in.

(*a*) Considering shearing of the key, Eq. (11.64) can be solved for the length to yield:

$$l = \frac{4Tn_s}{S_y wd} = \frac{4(5000)(2)}{(43,000)(0.25)(1)} = 3.72 \text{ in.}$$

Similarly, for bearing stress, Eq. (11.66) gives

$$l = \frac{4Tn_s}{S_y dh} = \frac{4(5000)(2)}{(43,000)(1)(0.25)} = 3.72 \text{ in.}$$

Not surprisingly, the length is the same since this is a square key.

(*b*) For a key with a length of 1.000 in., the torque is given by Eq. (11.64) as

$$T = \frac{S_y wld}{4n_s} = \frac{(43,000)(0.25)(1)(1)}{4(2)} = 1340 \text{ in.-lb.}$$

(*c*) For a Woodruff key, Table 11.4 suggests that a key 0.188 × 1.000 in. be used; this key has a shear area of $A_s = 0.178$ in^2. Therefore, from Eq. (11.64),

$$T = \frac{S_y A_s d}{4n_s} = \frac{(43,000)(0.178)(1)}{4(2)} = 957 \text{ in.-lb.}$$

Splines are toothed sections that mate in a conformal toothed hub. While, in theory, almost any non-circular shape can serve as a spline when mated in a hub, spline profiles are standardized by the International Standards Organization, and are either *parallel-sided* or *involute*. Parallel-sided splines (Fig. 11.11) have a square cross-section with 4, 6, 10, or 16 teeth, and can be thought of as an arrangement of multiple keys. Involute splines use a geometry similar to gears, except that the pressure angle is 30° and the depth of the tooth is generally one-half that of an involute gear tooth (see Section 14.3). Involute splines are stronger than parallel-sided splines because stress concentrations are lower and better surface quality can be achieved in manufacturing. These splines have 6 to 50 teeth, and can utilize a flat root or fillet root profile (see Fig. 11.12). Splines are manufactured using similar approaches as for spur gears (see Section 14.8), although internal splines are commonly machined onto a hub by broaching, wherein a constant or slightly tapered cross section is produced by a series of progressively larger cutters. Note that splines allow for axial motion of the hub on the shaft, which is beneficial for disassembly, but may require a retaining ring (Section 11.7) or similar device to secure the

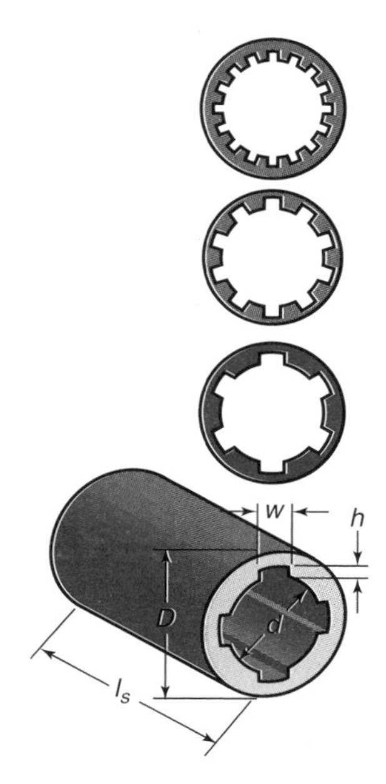

Figure 11.11: Standard parallel-sided splines hubs, showing (from top down) 16, 10, 6, and 4 teeth.

hub during operation.

Splines are produced directly from the shaft material, not from a softer metal as with a keyway, and so splines should be recognized as being suitable for transfer of large torques. The shear area of a spline is

$$A_s = \frac{\pi d_m l_s}{2}, \tag{11.67}$$

where $d_m = (d_o + d)/2$ is the mean diameter of the spline, d is the root diameter (see Fig. 11.12), and l_s is the spline width. The Society of Automotive Engineers (SAE) recommends a spline width of

$$l_s = \frac{d^3\left(1 - \frac{d_i^4}{d^4}\right)}{d_m^2}, \tag{11.68}$$

where d_i is the inner diameter in the case of a tubular shaft ($d_i = 0$ for a solid shaft). SAE further notes that only around one-fourth of the shear area is stressed because of the unavoidable presence of manufacturing tolerances and/or bending loads. Therefore, the torque that can be withstood by a spline based on shear is given by

$$T = \tau_{\text{all}} \frac{d_m}{2}\left(\frac{A_s}{4}\right) = \tau_{\text{all}}\left(\frac{d_m}{2}\right)\left(\pi \frac{d_m}{2}\frac{l_s}{4}\right) = \frac{\pi \tau_{\text{all}} d_m^2 l_s}{16}, \tag{11.69}$$

where l_s is given by Eq. (11.68).

In addition to shearing, failure due to bearing stress is also a possibility, especially for parallel-sided splines. According to SAE, the torque that can be applied before bearing failure is

$$T = \frac{pnhd_m l_s}{8}, \tag{11.70}$$

where p is the allowable bearing pressure, n is the number of teeth, h is the tooth depth (see Fig. 11.12), d_m is the mean diameter, and l_s is given by Eq. (11.68). Note that the Eq. (11.70) also incorporates the SAE finding that essentially one-fourth of the contact length transfers the torque.

Table 11.5: Holding force generated by set screws. *Source:* After Oberg et al. [2000].

Screw diameter, in.	Holding force, lb
0.25	100
0.375	250
0.50	500
0.75	1300
1.0	2500

Set screws are threaded fasteners that are driven radially through a threaded hole in the hub. The point (which can be of a number of different geometries or materials) bears against the shaft, and torque is transmitted by frictional resistance of the set screw against the shaft and the shaft against the hub. It should be noted that set screws can only develop moderate forces, as summarized in Table 11.5. Alternatively, the holding force of a set screw can be approximated as

$$P_t = 2500 d_s^{2.31}, \tag{11.71}$$

where P_t is in pounds and d_s is the set screw diameter in inches. A suitable set screw diameter as a function of shaft diameter, d, is approximated by the empirical relationship:

$$d_s = \frac{d}{8} + \frac{5}{16}, \tag{11.72}$$

where d and d_s are in inches.

11.7 Retaining Rings and Pins

Keyways, set screws, and splines are used in situations where a torque needs to be transferred between a machine element and the shaft on which it is mounted. Some machine elements, such as rolling element bearings, compression rollers, or idler wheels need to be located on a shaft but do not need to transfer a torque. If axial (thrust) loads are limited, **retaining rings** are an economical and viable approach for locating such machine elements on shafts. Retaining rings, such as those shown in Fig. 11.13, are available in a wide variety of sizes and shapes.

Retaining rings are assembled onto a shaft by expanding or reducing their diameter depending on whether they will be located on an inner or outer groove. This is usually accomplished through the use of a special hand tool, although some retaining rings are snapped in place on a groove or use a spring-loaded fit that relies on friction (so-called *push-on* designs, since they may be located without the use of tools). Recall that a retaining ring groove, if required, causes a stress concentration, as shown in Fig. 6.7.

Table 11.6 summarizes properties of selected retaining rings. It should be noted that Table 11.6 is highly abstracted; there are many more types and sizes of these rings, and forces in excess of 100 kN can be developed. A retaining ring can fail due to shear or bearing failure of the shaft groove. The holding force that can be generated by shear of the ring is given by

$$P_t = \frac{\pi C_t d t_h S_{sy}}{n_s}, \tag{11.73}$$

while the holding force that can be withstood before bearing failure of the shaft groove is

$$P_t = \frac{\pi C_t d t_g S_{ut}}{n_s}, \tag{11.74}$$

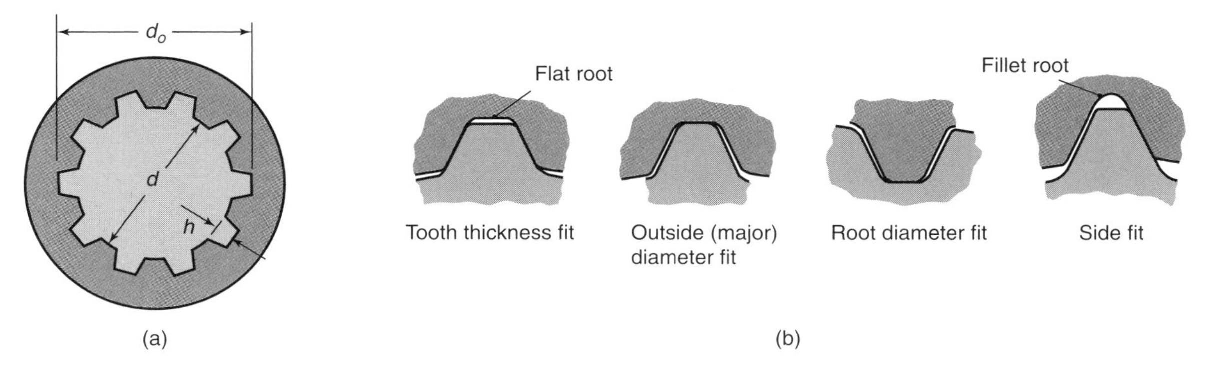

Figure 11.12: Involute splines. (a) Side view of a 10-toothed spline and hub; (b) types of fit between spline and hub.

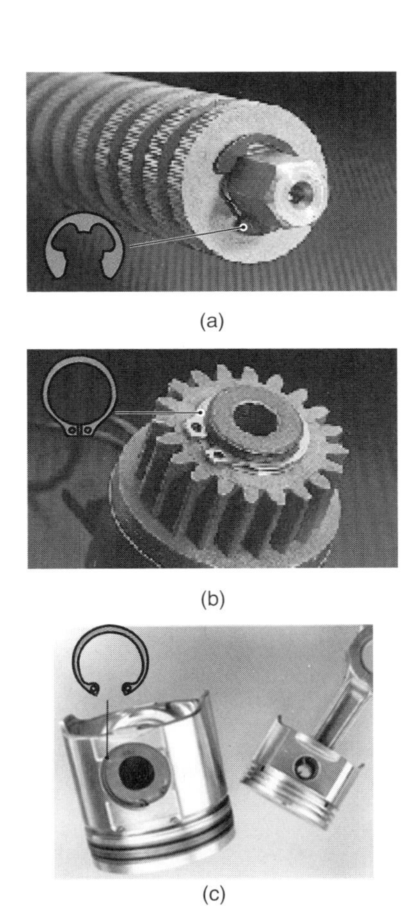

Figure 11.13: Examples of retaining rings. (a) An ME type snap ring, used to hold a series of cutters in place on a paper shredder shaft; (b) a general purpose MSH ring for locating a spur gear; (c) an internal retaining ring for securing a main piston bearing sleeve. *Source:* Courtesy of Rotor Clip Company, Inc.

where

C_t = ring correction factor, as given in Table 11.6
d = shaft diameter
t_h = retaining ring thickness
t_g = groove depth
S_{sy} = shear strength of ring material
S_{ut} = tensile strength of shaft material
n_s = safety factor

There are a number of additional factors that can affect holding force of retaining rings, including:

- Chamfers and radii are commonly used on exterior shaft corners or grooves, in order to assist manufacture, assembly, and provide for proper seating. However, such design features can change the loading state of a retaining ring from pure shear to bending, and can significantly reduce the holding force.
- Impact loads and vibration can significantly reduce the holding force of some retaining rings. It should be noted that if there is clearance between the retaining ring and the part, then vibration often results in repeated impact.
- If the machine element is allowed to rotate, friction can lead to unseating of the retaining ring.

Cotter pins are also used to secure items onto a shaft, but develop much lower forces than carbon steel retaining rings. Cotter pins are metal fasteners with two prongs that are inserted through a hole; the prongs are then bent (plastically deformed) to secure them and hold a part in place. Alternatively, some cotter pins have geometric features that serve to anchor them. Cotter pins are available in a wide variety of materials and geometry, some of which are illustrated in Fig. 11.14.

11.8 Flywheels

Large variations in angular acceleration can cause large oscillations in torque, and vice-versa. For example, consider a mechanical power press that is used to produce coins from disk-shaped workpieces. The force exerted by the press is very low until the punch contacts the workpiece, and then the force becomes very high. After the coin is formed, perhaps in less than 1 mm of travel, the punch returns to the top dead center position. In such a circumstance, the peak torque can be used to select a motor, but this would be overly large and uneconomical. Clearly, the average torque over the cycle is

Table 11.6: Properties of selected internal and external retaining rings. Ring nomenclature is manufacturer-specific; equivalent designs have other designations depending on supplier. *Source:* Courtesy of Rotor Clip Company, Inc.

Type	C_t^a	Shaft or hub diameter, mm	Groove diameter, mm	Ring width, mm	Notes
External rings					
MSH	1.00	5	4.75	0.50	General purpose shoulder ring. Axial assembly required. Can accommodate up to 254.0-mm shaft diameter.
		15	14.15	1.00	
		25	23.50	1.20	
		50	47.20	1.75	
		75	71.20	2.55	
ME	0.33	1	0.72	0.32	Popular snap-in-place retaining ring. Allows radial assembly. Can accommodate up to 34.9-mm shafts.
		5	3.90	0.70	
		10	8.00	1.00	
		20	16.00	1.40	
		25	20.00	1.40	
MCR	1.30	12	11.34	1.00	Spiral shaft ring. Available up to 280-mm shaft diameter. Suitable for higher holding forces.
		25	23.88	1.17	
		50	47.70	1.73	
		75	71.54	2.62	
		100	95.40	3.05	
LC	0.75	12.7	11.79	0.991	Two piece interlocking design, dynamically balanced when assembled. Suitable for high speeds. Allows radial assembly.
		25.4	22.15	1.422	
		50.8	46.43	2.184	
		76.2	70.97	2.616	
DTX	—	5	4.90	0.40	Self-locking push-on type. Suitable only for small loads (< 300 N). Axial assembly required.
		10	9.85	0.6	
		20	19.75	1.0	
		25	24.75	1.0	
		45	44.75	1.5	
Internal rings					
MHO	1.2	8	8.40	0.50	General purpose housing ring. Axial assembly required. Can accommodate up to 254.0-mm shaft diameter.
		12	12.65	0.70	
		25	26.60	1.20	
		50	53.10	1.75	
		75	79.70	2.55	
DTI	—	8	8.10	0.40	Self-locking push-on type. For small loads only (< 200 N). Axial assembly required.
		15	15.10	0.5	
		25	25.20	1.0	
		35	35.20	1.0	
		50	50.20	1.0	

[a] See Eqs. (11.73) and (11.74).

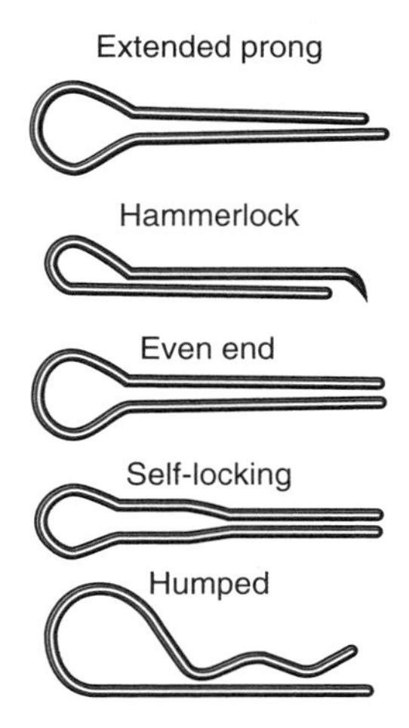

Figure 11.14: Illustration of common cotter pins. Cotter pins are often plastically deformed after insertion into a through-hole. An exception is the humped design, where the prong is inserted into a hole and the "hump" clips onto the round shaft. The hammerlock design is intended for use with castle nuts (see Table 16.4).

much lower than the peak torque, a circumstance that is usually the case. To smooth out the velocity changes, and reduce the size of the motor needed, a **flywheel** is often attached to a shaft. The use of a flywheel will allow the following to occur:

1. Reduced amplitude of speed fluctuation
2. Reduced maximum torque required
3. Energy stored and released when needed during a cycle

A flywheel with driving (mean) torque T_m and load torque T_l is shown in Fig. 11.15. The flywheel in this figure is a flat circular disk, but it can also be a large gear. A motor supplies a torque T_m, which for design purposes should be as constant as possible. The load torque T_l, for such applications as crushing machinery, punch presses, and machine tools, varies considerably.

11.8.1 Dynamics

The kinetic energy in a rotating system like that shown in Fig. 11.15 is

$$K_e = \frac{I_m \omega^2}{2}, \tag{11.75}$$

Figure 11.15: Flywheel with driving (mean) torque T_m and load torque T_l.

where I_m is the mass moment of inertia and ω is the angular velocity. The mass moments of inertia for six solids are given in Table 4.2, and mass moment of inertia is discussed in Section 4.2.6. From Newton's second law of motion as applied to the flywheel shown in Fig. 11.15,

$$-T_l + T_m = I_m \frac{d\omega}{dt}. \tag{11.76}$$

The design motor torque should be equivalent to the average torque, or $T_m = T_{\text{avg}}$. But

$$\frac{d\omega}{dt} = \frac{d\omega}{dt}\frac{d\theta}{d\theta} = \frac{d\theta}{dt}\frac{d\omega}{d\theta} = \omega\frac{d\omega}{d\theta}. \tag{11.77}$$

Substituting Eq. (11.77) into Eq. (11.76) gives

$$-T_l + T_{\text{avg}} = I_m \omega \frac{d\omega}{d\theta},$$

$$-\left(T_l - T_{\text{avg}}\right) d\theta = I_m \omega \, d\omega. \tag{11.78}$$

This equation can be written in terms of a definite integral as

$$-\int_{\theta_{\omega_{\min}}}^{\theta_{\omega_{\max}}} \left(T_l - T_{\text{avg}}\right) d\theta = \int_{\omega_{\min}}^{\omega_{\max}} I_m \omega \, d\omega$$

or

$$-\int_{\theta_{\omega_{\min}}}^{\theta_{\omega_{\max}}} \left(T_l - T_{\text{avg}}\right) d\theta = \frac{I_m}{2}\left(\omega_{\max}^2 - \omega_{\min}^2\right). \tag{11.79}$$

The left side of Eq. (11.79) represents the change in kinetic energy between the maximum and minimum shaft speeds and is equal to the area under the torque-angle diagram between the extreme values of speed. The far right side of Eq. (11.79) describes the change in energy stored in the flywheel. The only way to extract energy from the flywheel is to slow it down; adding energy will speed it up. Often it is a design constraint to achieve a desired speed variation $(\omega_{\max} - \omega_{\min})$, which can be accomplished by providing a flywheel with a sufficiently large mass moment of inertia, I_m.

The location of minimum angular speed, $\theta_{\omega_{\min}}$, within a cycle of operation occurs after the maximum positive energy has been delivered from the motor. The location of maximum angular speed $\theta_{\omega_{\max}}$ within a cycle of operation occurs after the maximum negative energy has been returned to the load, the point where the ratio of the energy summation to the area in the torque pulse is the largest negative value.

11.8.2 Flywheel Sizing

The size of a flywheel required to absorb the energy with an acceptable change in speed is a common design goal. The change in shaft speed during a cycle of operation is called the **fluctuation speed**, ω_f, and is expressed as

$$\omega_f = \omega_{\max} - \omega_{\min}.$$

The **coefficient of fluctuation** is defined as

$$C_f = \frac{\omega_{\max} - \omega_{\min}}{\omega_{\text{avg}}} = \frac{2(\omega_{\max} - \omega_{\min})}{\omega_{\max} + \omega_{\min}}. \tag{11.80}$$

Table 11.7 gives typical coefficients of fluctuation, C_f, for various types of equipment. Small values of C_f require relatively large flywheels. A larger flywheel will add more cost and weight to the system, factors that have to be considered along with the smoothness of the operation desired.

Table 11.7: Coefficient of fluctuation for various types of equipment.

Type of equipment	Coefficient of fluctuation, C_f
Crushing machinery	0.200
Electrical machinery	0.003
Electrical machinery, direct driven	0.002
Engines with belt transmissions	0.030
Flour milling machinery	0.020
Gear wheel transmission	0.020
Hammering machinery	0.200
Machine tools	0.030
Paper-making machinery	0.025
Pumping machinery	0.030-0.050
Shearing machinery	0.030-0.050
Spinning machinery	0.010-0.020
Textile machinery	0.025

The far right side of Eq. (11.79) describes the kinetic energy fluctuation of the flywheel and can be rewritten as

$$K_e = \frac{I_m}{2}\left(\omega_{\max} + \omega_{\min}\right)\left(\omega_{\max} - \omega_{\min}\right). \tag{11.81}$$

Substituting Eq. (11.80) into Eq. (11.81) yields

$$K_e = I_m \omega_{\text{avg}}^2 C_f, \tag{11.82}$$

$$I_m = \frac{K_e}{C_f \omega_{\text{avg}}^2}. \tag{11.83}$$

Given the desired coefficient of fluctuation for a specific application, obtaining the change in kinetic energy, K_e, from the integration of the torque curve, and knowing the average angular velocity, the required mass moment of inertia, I_m, can be determined. By knowing I_m, the dimensions (i.e., the thickness and diameter) of the flywheel can be determined.

The most efficient flywheel design is obtained by maximizing I_m for the minimum volume of flywheel material used. Ideally, this design would take the shape of a ring supported by a thin disk, but thin cross-sections can be difficult to manufacture. Efficient flywheels have their masses concentrated at the largest radius (or in the rim) and their hubs supported on spokes, like bicycle wheels. This configuration places the greatest part of the mass at the largest radius possible and minimizes the weight for a given mass moment of inertia while making manufacturing problems tractable.

Design Procedure 11.3: Flywheels

Flywheels are used to provide for smooth operation when the applied loading is not smooth. The design procedure for sizing a flywheel is as follows:

1. Select a material for the flywheel. Table 11.8 summarizes some common materials used for flywheels.
2. Select a coefficient of fluctuation, C_f, either from Table 11.7 or from design requirements.
3. Either from design constraints or analysis of the application, the load as a function of time needs to be determined. It is then useful to plot the load torque, T_l, versus position, θ, for one cycle.
4. Determine the average load, $T_{l,\text{avg}}$, over one cycle.
5. Find the locations $\theta_{\omega_{\max}}$ and $\theta_{\omega_{\min}}$. Often this is the most difficult task, and may require consideration of multiple intervals. The maximum velocity occurs after the interval where the energy contribution to the system is a maximum. Thus, the goal is to find the maximum value of the following expression over any interval in the load cycle:
$$\int_{\theta_{\omega_{\min}}}^{\theta_{\omega_{\max}}} \left(T_{\text{avg}} - T_l\right) d\theta.$$
6. Determine kinetic energy by integrating the torque curve as in Eq. (11.79).
7. Determine ω_{avg} from Eq. (11.80).
8. Determine I_m from Eq. (11.83).
9. Find the dimensions of the flywheel. Table 4.2 provides useful expressions for I_m for common shapes.

Example 11.6: Flywheel Design

Given: The output, or load torque, of a flywheel used in a punch press for each revolution of the shaft is 12 N-m from zero to π and from $3\pi/2$ to 2π and 144 N-m from π to $3\pi/2$. The coefficient of fluctuation is 0.05 about an average speed of 600 rpm. Assume that the flywheel's solid disk is made of low-carbon steel of constant 25 mm thickness.

Find: Determine the following:

(*a*) The average load or output torque

(*b*) The locations $\theta_{\omega_{\max}}$ and $\theta_{\omega_{\min}}$

(*c*) The energy fluctuation required

(*d*) The outside diameter of the flywheel

Solution:

(*a*) By using the load or output torque variation for one cycle from Fig. 11.16, the average load torque is

$$2\pi T_{l,\text{avg}} = \pi(12) + \frac{\pi}{2}(144) + \frac{\pi}{2}(12);$$

$$T_{l,\text{avg}} = 6 + 36 + 3 = 45 \text{ N-m}.$$

(*b*) From Fig. 11.16,

$$\theta_{\omega_{\max}} = \pi \quad \text{and} \quad \theta_{\omega_{\min}} = \frac{3\pi}{2}.$$

(*c*) From the left side of Eq. (11.79), the kinetic energy for one cycle is

$$\begin{aligned} K_e &= -\int_{\theta_{\omega_{\min}}}^{\theta_{\omega_{\max}}} \left(T_l - T_{\text{avg}}\right) d\theta \\ &= (144 - 45)\,\frac{\pi}{2} \\ &= 155.5 \text{ N-m}. \end{aligned}$$

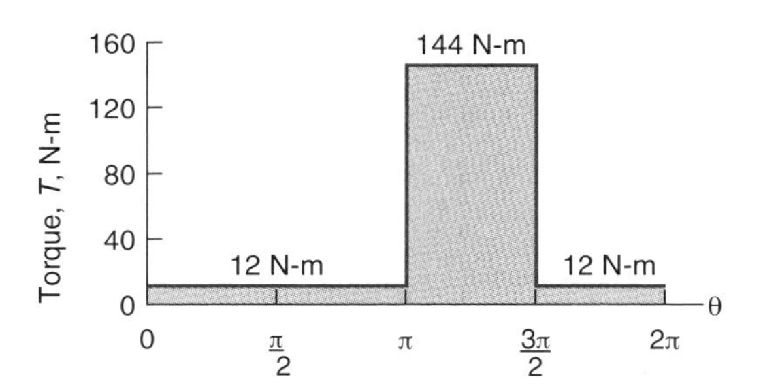

Figure 11.16: Load or output torque variation for one cycle used in Example 11.6.

(*d*) The average angular speed can be expressed as

$$\omega_{\text{avg}} = 600\frac{\text{rev}}{\text{min}}\left(\frac{1\text{ min}}{60\text{ s}}\right)\left(\frac{2\pi}{1\text{ rev}}\right) = 62.83\text{ rad/s}.$$

From Eq. (11.83), the mass moment of inertia is

$$I_m = \frac{K_e}{C_f\omega_{\text{avg}}^2} = \frac{155.5}{(0.05)(62.83)^2} = 0.7879\text{ kg-m}^2. \quad (a)$$

From Table 4.2 (also given on the inside back cover) the mass moment of inertia for a solid round disk is

$$I_m = \frac{m_a d^2}{8} = \frac{\pi d^2 t_h \rho}{4}\left(\frac{d^2}{8}\right) = \frac{\pi}{32}\rho t_h d^4.$$

From Table 3.1, ρ = 7860 kg/m^3 for low-carbon steel. Therefore,

$$I_m = \frac{\pi}{32}(7860)(0.025)d^4 = 19.29d^4. \quad (b)$$

Equating Eqs. (*a*) and (*b*) gives

$$d^4 = \frac{0.7879}{19.29} = 0.0408\text{ m}^4,$$

which is solved as $d = 0.4495$ m, or around 450 mm.

11.8.3 Stresses

The stresses in a flywheel are the sum of the rotational effects and the stresses associated with internal pressurization. Concepts introduced in Chapter 10 will be used here, in particular Section 10.4 for rotational effects and Section 10.3.2 for internal pressurization due to a press or shrink fit. The flywheel is assumed to be a disk with a central hole and constant thickness. The stresses in the flywheel due to rotational and press-fit effects can be written as

$$\sigma_\theta = \sigma_{\theta\omega} + \sigma_{\theta p}, \quad (11.84)$$

$$\sigma_r = \sigma_{r\omega} + \sigma_{rp}. \quad (11.85)$$

Making use of Eqs. (10.35) and (10.36) and Eqs. (10.23) and (10.24), respectively, and taking care to use consistent nomenclature, gives the tangential (hoop) and radial stresses as

$$\sigma_\theta = \frac{3+\nu}{8}\rho\omega^2\left(r_i^2 + r_o^2 + \frac{r_i^2 r_o^2}{r^2} - \frac{1+3\nu}{3+\nu}r^2\right) + \frac{p_f r_f^2\left(1+\frac{r_o^2}{r^2}\right)}{r_o^2 - r_f^2}, \quad (11.86)$$

$$\sigma_r = \frac{3+\nu}{8}\rho\omega^2\left(r_i^2 + r_o^2 - \frac{r_i^2 r_o^2}{r^2} - r^2\right) + \frac{p_f r_f^2\left(1-\frac{r_o^2}{r^2}\right)}{r_o^2 - r_f^2}. \quad (11.87)$$

If the shaft is solid, then the stress at the inner flywheel radius simplifies to

$$\sigma_\theta = \frac{3+\nu}{8}\rho\omega^2\left(r_o^2 - \frac{1+3\nu}{3+\nu}r_f^2\right) + \frac{p_f\left(r_f^2 + r_o^2\right)}{r_o^2 - r_f^2}, \quad (11.88)$$

$$\sigma_r = \frac{3+\nu}{8}\rho\omega^2\left(r_o^2 - r_f^2\right) - p_f. \quad (11.89)$$

Note that σ_θ and σ_r are principal stresses since there is no shear stress associated with them. Usually, σ_θ is larger than σ_r, implying that the maximum principal stress equals the tangential (hoop) stress ($\sigma_1 = \sigma_\theta$). Also, the circumferential stress is largest at $r = r_i$ for the flywheel. Thus, for *brittle* materials the design of the flywheel should be such that

$$\sigma_1 < S_u/n_s. \quad (11.90)$$

For *ductile* materials, a multiaxial failure theory, such as the DET, should be used. If $\sigma_1 = \sigma_\theta$, $\sigma_2 = \sigma_r$, and $\sigma_z = 0$, from Eq. (6.12),

$$\sigma_e = \sqrt{\sigma_\theta^2 + \sigma_r^2 - \sigma_\theta\sigma_r}.$$

From Eq. (6.11), the DET predicts failure if

$$\sigma_e \geq \frac{S_y}{n_s}.$$

Example 11.7: Flywheel Stresses

Given: A flywheel made of low-carbon steel has an outside radius of 150 mm and an inside radius of 25 mm. The flywheel is to be assembled (press fit) onto a shaft. The radial interference between the flywheel and shaft is 50 μm, and the shaft will operate at a speed of 5000 rpm.

Find: Determine the following:

(*a*) The circumferential and radial stresses on the flywheel inner radius.

(*b*) The speed at which the flywheel will break loose from the shaft.

Solution:

(*a*) For steel, Table 3.1 gives E = 207 GPa, ν = 0.3, and $\rho = 7860$ kg/m^3. Solving Eq. (10.53) for the interference pressure yields

$$p_f = \frac{\delta_r E\left(r_o^2 - r_i^2\right)}{2r_f r_o^2} = \frac{\left(50\times10^{-6}\right)\left(207\times10^9\right)\left(0.150^2 - 0.025^2\right)}{2(0.025)\left(0.150^2\right)} = 201.2\text{ MPa}.$$

Table 11.8: Materials for flywheels.

Material	Performance index, M_f kJ/kg	Comment
Ceramics	200–2000 (compression only)	Brittle and weak in tension. Use is usually discouraged.
Reinforced polymers:		
Ceramic fiber	200–500	The best performance; a good choice.
Graphite fiber	100–400	Almost as good as CFRP and cheaper; an excellent choice.
Beryllium	300	Good but expensive, difficult to work, and toxic.
High-strength steel	100–200	All about equal in performance; steel and Al-alloys less expensive than Mg and Ti alloys.
High-strength aluminum alloys	100–200	
High-strength magnesium alloys	100–200	
Titanium alloys	100–200	
Lead alloys	3	High density makes these a good (and traditional) selection when performance is velocity limited, not strength limited.
Cast iron	8–10	

Note that $\omega = 5000$ rpm $= 523.6$ rad/s. For this case, the flywheel and shaft are both the same material, and the shaft is solid, so that the circumferential stress can be obtained from Eq. (11.88) as

$$\sigma_\theta = \left[\frac{3+0.3}{8}(7860)(523.6)^2\right] \times \left[0.15^2 - \frac{1+3(0.3)}{3+0.3}(0.025)^2\right] + \frac{(201.2\times10^6)(0.15^2+0.025^2)}{0.15^2-0.025^2},$$

or $\sigma_\theta = 233$ MPa. Similarly, the radial stress is obtained from Eq. (11.89) as

$$\sigma_r = \frac{3+0.3}{8}(7860)(523.6)^2\left(0.15^2-0.025^2\right) - 201.2\times10^6.$$

or $\sigma_r = -181.8$ MPa. The negative stress indicates that σ_r is compressive at the speed of 5000 rpm.

(*b*) The flywheel breaks free when the radial stress at the inner radius is reduced to zero at an elevated speed. From Eq. (11.89),

$$\sigma_r = 0 = \frac{3+0.3}{8}(7860)\omega^2\left(0.15^2-0.025^2\right) - 201.2\times10^6.$$

Solving for the angular velocity yields $\omega = 1684$ rad/s $= 16,080$ rpm. Thus, if the shaft speed exceeded this value, the flywheel would not be held onto the shaft. Note that the ability of the flywheel to transmit a torque would be compromised at far lower speeds, as suggested by Eq. (10.55).

11.8.4 Flywheel Materials

The main purpose of a flywheel is to store energy and then use it efficiently. Small flywheels, the type found in small appliances, are often made of lead, but steel and iron are preferred for health reasons. These flywheels operate at relatively low speeds so that a high density is the sole physical property that justifies these materials. Recently, flywheels have been proposed for vehicle power storage and regenerative braking systems; high-strength steels and composites are of interest for these vehicular flywheels because they are more efficient at high speeds. Thus, a great diversity of materials is being used for flywheels. The obvious question is: What material is best suited for demanding high-speed flywheel applications?

Recall from Eqs. (11.86) and (11.87) that the speed effect is more significant than the pressurization effect. Thus, the maximum stress in a flywheel is

$$\sigma_{\max} = \left(\frac{3+\nu}{8}\right)\rho\omega^2 f(r). \tag{11.91}$$

From Eq. (11.75), the kinetic energy is

$$K_e = \frac{1}{2}I_m\omega^2. \tag{11.92}$$

The mass moment of inertia, I_m, from Table 4.2 can be represented as

$$I_m = m_a g(r), \tag{11.93}$$

so that

$$K_e = \frac{1}{2}m_a g(r)\omega^2. \tag{11.94}$$

Assume that $f(r)\omega^2 \approx g(r)\omega^2$. From Eqs. (11.94) and (11.91)

$$\frac{K_e}{m_a} \propto \frac{\sigma_{\max}}{\rho}.$$

Thus, the best materials for a flywheel, namely those that store the most energy per mass, are those with high values of the performance index given by

$$M_f = \frac{\sigma_{\max}}{\rho}. \tag{11.95}$$

Note that the units for M_f are joules per kilogram (J/kg). Also, recall that a joule (J) is 1 newton-meter (N-m).

Table 11.8 gives performance indices for various materials. From this table observe that the best materials are ceramics, beryllium, and composites. Lead and cast iron do not perform as efficiently. This seems counter intuitive, mainly because lead and cast iron are suitable for low-speed applications where optimum performance is not critical.

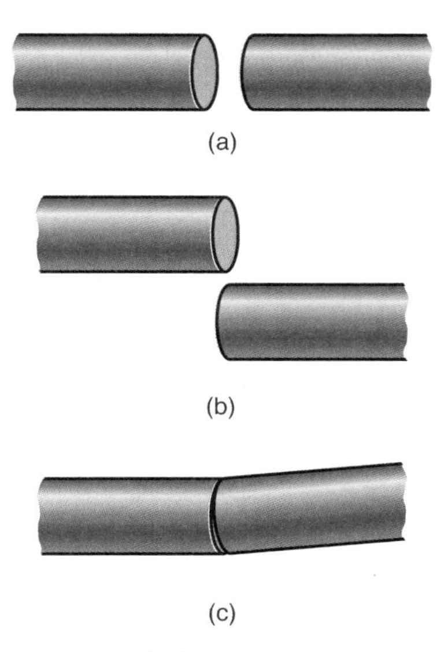

Figure 11.17: Types of shaft misalignment. (a) In-line; (b) parallel; (c) angular.

11.9 Couplings

Couplings are used to connect two shafts. This is a common need, for example, when connecting a motor to an input shaft, clutch, or transmission. If two shafts are aligned, then designing and installing a coupling is not difficult, and rigid couplings can be used. If there is some misalignment, either parallel, in-line, or angular (see Fig. 11.17), then the coupling needs to accommodate the misalignment and a *flexible coupling* may be required. Table 11.9 summarizes the most common types of couplings and their ability to accommodate misalignment, although it should be recognized that there are many variants within each classification.

Design of a coupling involves working closely with suppliers, and matching design requirements to coupling characteristics. Unless noted, the couplings discussed are available for a wide variety of torques and shaft sizes, and the design challenge is to match a coupling type with design constraints.

Whenever possible, shafts that need to be coupled should be arranged without misalignment of any form; this allows direct application of a rigid coupling. If misalignment is unavoidable, or if shock isolation is needed, Table 11.9 is a useful reference. In addition, it is not unusual to have high startup torques (compared to operating conditions). Fortunately, most electric motors have the largest torque available at zero speed. However, if this torque is insufficient to start motion, then compliant couplings may be an attractive alternative to the use of a speed reducer and its associated couplings.

11.9.1 Rigid Couplings

Examples of rigid couplings are shown in Fig. 11.18. Rigid couplings are commonly applied because of their low cost and high reliability. However, they do require close alignment and do not damp vibrations or provide electrical isolation between shafts. They have few size restrictions, although compression sleeve couplings have limited utility for large diameters; the mechanics of such couplings are discussed in Section 10.5. Rigid couplings have no backlash, making them useful for precision machines and servomechanisms. Their main downside is the inability to tolerate angular or paral-

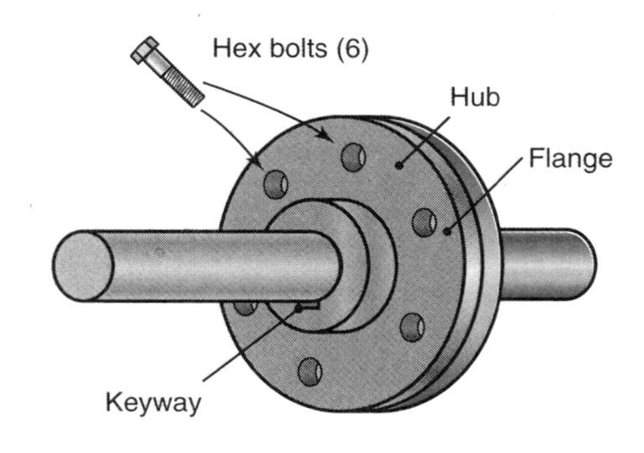

(a)

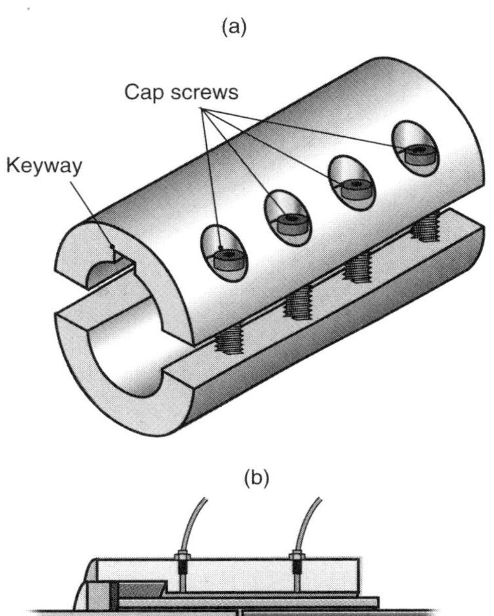

(b)

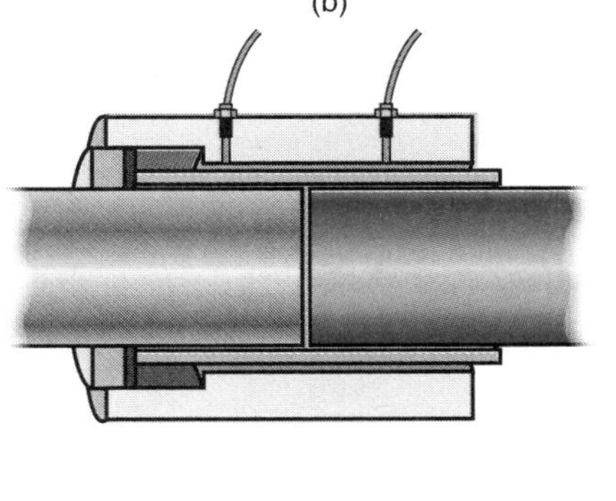

(c)

Figure 11.18: Types of rigid shaft couplings. (a) Flange coupling, consisting of a hub and flange and with a keyway for torque transmission; (b) two-piece sleeve couplings (one-piece designs are also available); (c) compression sleeve couplings, configured with hydraulic lines for installing the shaft; the hydraulic lines are not needed during the coupling's service life.

Table 11.9: General characteristics of shaft couplings.

Type	Misalignment tolerance[a]			Damping ability[a]	Notes
	In-line	Angular	Parallel		
Rigid couplings					
Flange	3	0	0	0	Useful when frequent uncoupling is required. Flanges must be installed in pairs and moved in pairs to maintain hole indexing. These couplings are usually keyed or splined to each shaft.
Sleeve	2	0	0	0	Similar to rigid flange coupling but easier to remove and install. Requires high-strength bolts and tight tolerances on sleeves and shafts.
Compression sleeve	2	0	0	0	Relies upon friction between shaft and coupling. Not practical for large diameter shafts or high torque applications.
Flexible couplings					
Elastomeric	1	1	0–2	4	Angular misalignment up to 3° and parallel misalignment up to 7.0 mm may be tolerated, depending on shaft diameter. Electrical isolation can be provided. Reduced torsional capacity compared to rigid coupling. Jaw designs can result in significant backlash.
Flexible metal	3	3	1	0–2	Can achieve higher torques for a given diameter and can withstand higher temperatures than elastomeric couplings.
Bellows	1	4	3	0	Lower capacity than rigid couplings, can be susceptible to fatigue failure. Tolerates moderate misalignment with no backlash.
Gear	4	2	0–2	0	Suitable for large torque applications. Parallel misalignment is tolerated only if space permits a pair of gear couplings with a spacer between them.
Helical	2	5	1	0	Suitable only for low torque applications, best able to accommodate angular misalignment.
Schmidt	0	0	5	1	Intended for parallel misalignment. Usually restricted to low power applications.
Universal	1	5	3–5	0	Popular rigid joint, ideal for high misalignment. Lubrication required, although can incorporate sealed bearings at higher cost.
Fluid	2	0	0	5	Intended for vibration and shock isolation. Can be provided with controlled filling to provide for soft start capability.

[a] 0 = none, 5 = high

lel misalignment, although limited in-line misalignment can be accounted for during assembly. Rigid couplings have essentially no energy loss, which can be a significant advantage compared to flexible couplings.

Rigid flange couplings, shown in Fig. 11.18a, consist of two hub and flange assemblies that are connected by bolts through the flanges. The hubs are keyed or splined (Section 11.6) onto their shafts. Special care must be taken with respect to the bolts on flange couplings. Details of bolt design are discussed in Ch. 16, but in general, it is important to make sure that the holes on the flanges are aligned carefully to ensure all of the bolts in the bolt circle are evenly loaded. This usually requires marking the holes to allow for proper realignment if the coupling is disassembled or moved to a new shaft. Tightening is usually performed in a set sequence of opposing bolts in order to ensure even load distribution.

Sleeve couplings (Fig. 11.18b) are also very popular, and can depend on a clamp or compression fit for assembly onto shafts. Some light to medium duty applications use keys for torque transmission and set screws to locate the sleeve. Clamp type designs can be one- or two-piece designs; the latter are especially useful for easy disassembly. **Compression sleeve couplings**, as shown in Fig. 11.18c, use the wedging action of a split conical section to provide a friction grip on the two shafts. Compression sleeve couplings accommodate higher torques, but require more expensive machining and have a larger diameter than equivalent capacity rigid sleeve couplings. As with rigid flange couplings, bolts must be tightened carefully and in the correct order.

11.9.2 Flexible Couplings

Flexible couplings are needed to accommodate misalignment between shafts, and different designs are intended for the different types of misalignment that may occur. Flexible couplings are inherently more complex than rigid couplings, and therefore are invariably more expensive for a given shaft diameter or torque capability. However, flexible couplings often provide the additional benefit of damping vibrations and/or isolating the shafts electrically, as well as reducing shock loads. Some flexible couplings have significant backlash, however, which can be a concern for precision applications.

Elastomeric couplings transmit torque through a hard rubber component, and are able to accommodate some misalignment, but their main advantage is the adsorption of shock and vibration between two shafts. There are many types of elastomeric couplings, but common types are **jaw** and **flexible disk** couplings. A jaw coupling uses a rubber spacer, or *spider* between two jaws as shown in Fig. 11.19. The stiffness of the spider material can be tuned for specific applications and controlled backlash. Flexible disk couplings use a rubber or metal disc separating two flanges.

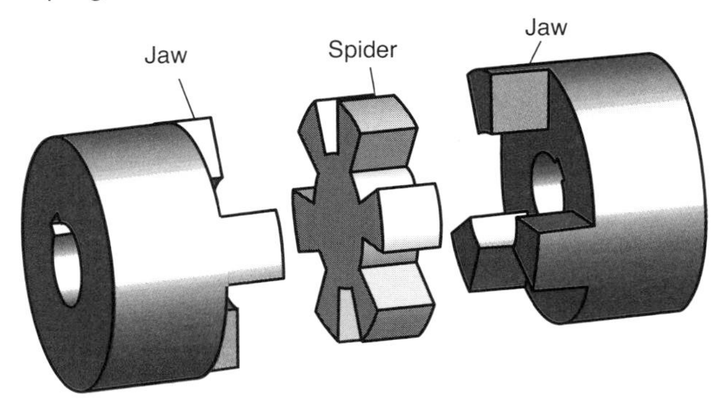

Figure 11.19: Schematic illustration of a jaw coupling.

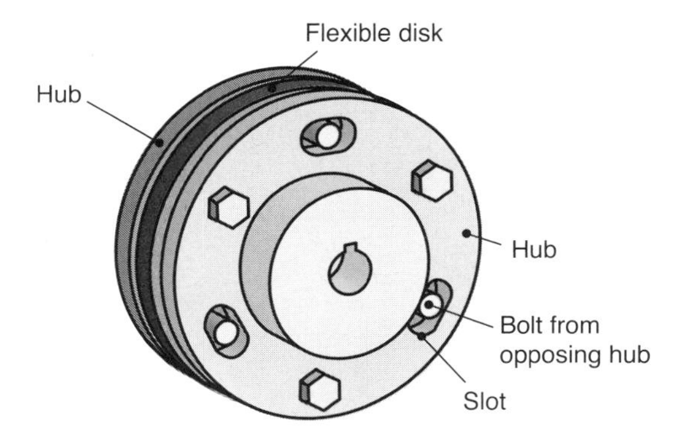

Figure 11.20: (a) An elastomeric disk coupling. Each hub is connected to the flexible disk with threaded fasteners that extend into slots on the opposing hub, allowing angular displacement between the hubs when a torque is applied. This reduces impact loads and increases damping.

The flexible disk coupling shown in Fig. 11.20 uses bolts to connect the flanges to the disk; in this case, alternating bolts connect each hub to the disk. That is, if six bolts are present in the disk, three connect it to each hub and are located in slots on the opposing side. This feature ensures that the coupling will still function (although its damping ability will be compromised) in the case of disk failure. Both jaw and flexible disk couplings are well-balanced and are suitable for high speed applications.

There are a number of **flexible metal coupling** designs, the simplest of which will be very similar to Fig. 11.20, but with a metal disk instead of a rubber one. Such couplings will be less compliant and not display the damping characteristics of elastomeric couplings, but are able to transmit much larger torques. A unique design of a flexible metal coupling is a **Falk coupling**, shown in Fig. 11.21. Falk couplings consist of a spring that is threaded through axial slots; the slots are tapered so that some deformation of the spring can occur. Under light loads, the slot grid bears near the outer edge of the hub teeth, providing for a long span, spring flexure and protection against shock. As the load increases, the distance between contact points on the hub teeth is shortened, although a span still exists.

High angular misalignment can be tolerated by **bellows** (Fig. 11.22) or **helical couplings** (Fig. 11.23). Their compliance does allow for limited in-line or parallel misalignment as well, and these couplings have no backlash. The main

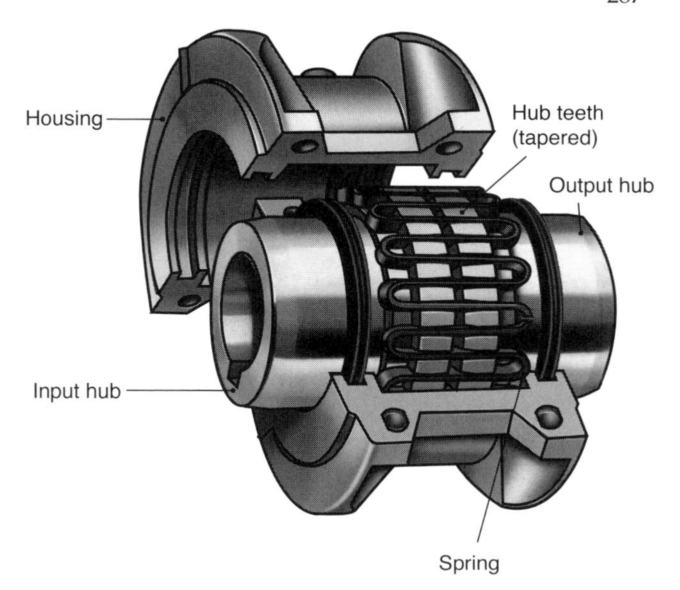

Figure 11.21: Schematic illustration of a Falk coupling, consisting of a metal spring threaded through a toothed hub. *Source:* Courtesy of Rexnord, Inc.

drawback is lower torque capacity than other couplings for a given shaft size. Bellows couplings are susceptible to fatigue failure.

When parallel misalignment occurs, **linkage** or **Schmidt couplings** can be used, as long as the shaft diameter is below 50 mm or so. Schmidt couplings, shown in Fig. 11.24, consist of three disks connected by six links. These links allow for very large parallel offsets, and some designs allow for limited angular and in-line misalignment as well. A unique feature of Schmidt couplings is that the parallel offset can be changed dynamically without removing and remounting the coupling. These couplings are useful for torque transmission at moderate speeds. The main disadvantage of these couplings is the need for proper lubrication, since the links are supported by journal or needle bearings.

Universal couplings, also known as *universal joints*, *U-joints*, or *Hooke couplings*, consist of two hinges that are axially oriented at 90° to each other, and connected by a cross shaft, as shown in Fig. 11.25. Universal joints allow for very high angular misalignment. They can also accommodate parallel misalignment if there is sufficient axial clearance. A peculiarity of universal joints is that if an input side has a uniform velocity, the output shaft will have a variable speed, which can result in excessive vibration and wear. A solution is to utilize two U-joints separated by a short spacer, so that the variations in speed are phased and a uniform output velocity can be achieved. This assembly, properly called a *double Cardan joint* or *CV joint*, is commonly employed in automotive drive axles.

Fluid couplings, as shown in Fig. 11.26, are intended for applications where significant shock loads exist, such as conveyors, crushing machines, and industrial scale mixers. Fluid couplings are also widely applied in automobiles, where they are the torque converter in automatic transmissions. Fluid couplings consist of an impeller and rotor, both enclosed in a casing. The impeller is attached to the input shaft; rotation causes hydraulic fluid to be pumped toward the rotor. If there is a velocity difference or *slip*, then the pumped fluid applies a torque to the rotor, which is directly connected to

Figure 11.22: A bellows coupling. *Source:* Courtesy of Lovejoy, Inc.

Figure 11.23: A helical coupling. *Source:* Courtesy of Lovejoy, Inc.

Figure 11.24: A Schmidt coupling. *Source:* Courtesy of Zero-Max, Inc.

Figure 11.25: A universal coupling. *Source:* Courtesy of Lovejoy, Inc.

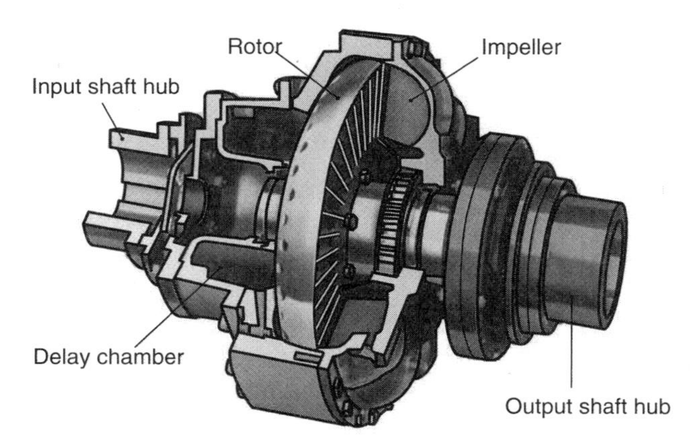

Figure 11.26: Schematic illustration of a fluid coupling. Note that a portion of the impeller and delay chamber have been removed for clarity. *Source:* Courtesy of Rexnord, Inc.

an output shaft. Thus, energy is transmitted through the hydrodynamic action of the fluid, and no direct contact between the two shafts occurs. The torque as a function of slip can be modified through use of a stator design in torque converters, or through delay chambers as shown.

Gear couplings, shown in Fig. 11.27, consist of two geared hubs connected by a sleeve with two internally geared sections. The in-line misalignment that can be tolerated depends on the width of the internal gear teeth. Further, some angular misalignment can be accommodated through the use of contoured teeth. These couplings have high torque capacity, but must be well lubricated to prevent wear.

11.10 Summary

A shaft is usually a circular cross-section rotating member that has such power-transmitting elements as gears, pulleys, flywheels, sprockets, and rolling-element bearings mounted on it. The loading on the shaft can be one or various combinations of bending, torsion, or axial or transverse shear. Furthermore, these types of loading can be either static or cyclic. A design procedure was presented that established the appropriate shaft diameter for specific conditions. If the shaft diameter is known, the safety factor (or the smallest diameter where failure first occurs) is often an important consideration. Three failure prediction theories considering important combinations of loading were presented. These theories are by no means all-inclusive but should provide the essential

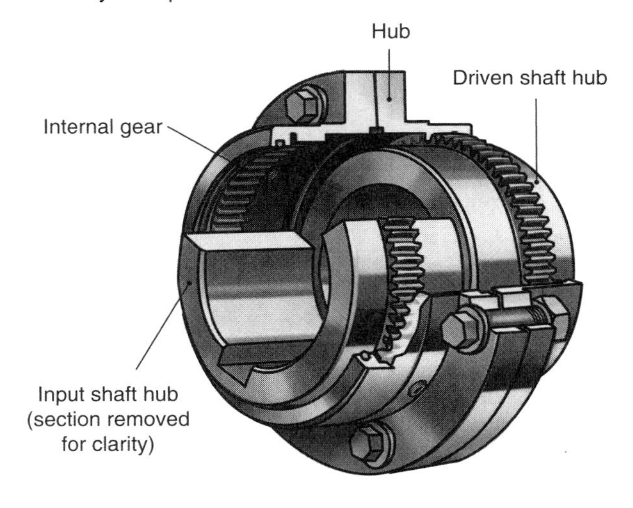

Figure 11.27: A gear coupling. *Source:* Courtesy of Rexnord, Inc.

understanding from which any other considerations can be easily obtained.

Shaft dynamics and in particular the first critical speed are important to design, since the rotating shaft becomes dynamically unstable and large vibrations are likely to develop at this speed. Both the Rayleigh and the Dunkerley equations for determining the first critical speed of a multiple-mass system were presented. The Rayleigh overestimates and the Dunkerley underestimates the exact solution, thus providing a range of operating speeds that the design should avoid.

A number of machine elements were presented for mounting parts onto shafts. These include design options that transmit a torque, such as keys, pins, and splines, as well as devices that only fix position on the shaft axis, such as retaining rings and cotter pins.

Flywheels were also discussed. In some applications, large variations in acceleration occur that can cause large oscillations in torque. The peak torque can be so high as to require an overly large motor. A flywheel is often used to smooth out the velocity changes and stabilize the back-and-forth energy flow of rotating equipment. A procedure for designing a flywheel was presented as well as flywheel dynamics, sizing, stresses, and material selection.

In the last section, some of the many types of shaft couplings were presented. Connecting two shafts is very common, such as when a motor is connected to a power transmitting shaft. The couplings all have their advantages and disadvantages, mostly associated with their ability to accommodate in-line, parallel, and/or angular misalignment. Another important capability of a coupling is damping of vibration and shock loads.

Key Words

coefficient of fluctuation dimensionless speed range, $\dfrac{\omega_{\text{max}} - \omega_{\text{min}}}{\omega_{\text{avg}}}$

critical speed speed at which a rotating shaft becomes dynamically unstable

coupling a device used to connect two shafts; many forms are available

Dunkerley equation relation for first critical speed that underestimates frequency

first critical speed lowest frequency at which dynamic instability occurs

flywheel element that stores energy through rotational inertia

hub portion of member mounted onto shaft that directly contacts shaft

key element that transmits power from shaft to hub

lowest natural frequency same as first critical speed

natural frequency same as critical speed

Rayleigh equation relation for first critical speed that overestimates frequency

retaining ring elastically deformable stamped shape that usually seats in a flat groove to axially restrain a part on a shaft

shaft rotating or stationary member, usually of circular cross-section with small diameter relative to length and used to transmit power through such elements as gears, sprockets, pulleys, and cams

spline toothed profile for transmitting power from a shaft to a hub

Summary of Equations

Shaft design

Static loading, DET: $d = \left(\dfrac{32n_s}{\pi S_y} \sqrt{M^2 + \dfrac{3}{4}T^2} \right)^{1/3}$,

$$n_s = \frac{\pi d^3 S_y}{32\sqrt{M^2 + \frac{3}{4}T^2}}$$

Static loading, MSST: $d = \left(\dfrac{32n_s}{\pi S_y} \sqrt{M^2 + T^2} \right)^{1/3}$,

$$n_s = \frac{\pi d^3 S_y}{32\sqrt{M^2 + T^2}}$$

Static loading, DET, with axial loading:

$$\frac{4}{\pi d^3}\sqrt{(8M + Pd)^2 + 48T^2} = \frac{S_y}{n_s}$$

Static loading, MSST, with axial loading:

$$\frac{4}{\pi d^3}\sqrt{(8M + Pd)^2 + 64T^2} = \frac{S_y}{n_s}$$

Fatigue loading, DET and Soderberg:

$$d^3 = \frac{32n_s}{\pi S_y}\sqrt{(\Psi)^2 + \frac{3}{4}\left(T_m + \frac{S_y}{S_e}K_{fs}T_a\right)^2}$$

$$n_s = \frac{\pi d^3 S_y}{32\sqrt{(\Psi)^2 + \frac{3}{4}\left(T_m + \frac{S_y}{S_e}K_{fs}T_a\right)^2}}$$

where $\Psi = M_m + \dfrac{S_y}{S_e}K_f M_a$.

Fatigue loading, MSST and Soderberg:

$$d^3 = \frac{32n_s}{\pi S_y}\sqrt{(\Psi)^2 + \left(T_m + \frac{S_y}{S_e}K_{fs}T_a\right)^2}$$

$$n_s = \frac{\pi d^3 S_y}{32\sqrt{(\Psi)^2 + \left(T_m + \frac{S_y}{S_e}K_{fs}T_a\right)^2}}$$

Critical speed, single mass: $\omega = \sqrt{\frac{g}{\delta}}$

Critical speed, Rayleigh equation:

$$\omega_{\text{cr}} = \sqrt{\frac{g\sum_{i=1,\ldots,n} W_i \delta_{i,\max}}{\sum_{i=1,\ldots,n} W_i \delta_{i,\max}^2}}$$

Critical speed, Dunkerley equation:

$$\frac{1}{\omega_{\text{cr}}^2} = \frac{1}{\omega_1^2} + \frac{1}{\omega_2^2} + \cdots + \frac{1}{\omega_n^2}$$

Keys

Failure due to shear: $T = \frac{S_y A_s d}{4n_s}$

Failure due to bearing stress: $T = \frac{S_y dlh}{4n_s}$

Splines

Recommended spline width: $l_s = \frac{d^3\left(1 - \frac{d_i^4}{d^4}\right)}{d_m^2}$

Failure due to shear: $T = \frac{\pi \tau_{\text{all}} d_m^2 l_s}{16}$

Failure due to bearing stress: $T = \frac{pnhd_m l_s}{8}$

Set screws

Holding force: $P_t = 2500d^{2.31}$

Approximate size: $d_s = \frac{d}{8} + \frac{5}{16}$

Retaining rings

Failure due to ring shear: $P_t = \frac{\pi C_t d t_h S_{sy}}{n_s}$

Failure due to bearing failure: $P_t = \frac{\pi C_t d t_g S_{ut}}{n_s}$

Flywheels

Coefficient of fluctuation:

$$C_f = \frac{\omega_{\max} - \omega_{\min}}{\omega_{\text{avg}}} = \frac{2(\omega_{\max} - \omega_{\min})}{\omega_{\max} + \omega_{\min}}$$

Kinetic energy fluctuation: $K_e = I_m \omega_{\text{avg}}^2 C_f$

Recommended Readings

Budynas, R.G., and Nisbett, J.K. (2011), *Shigley's Mechanical Engineering Design*, 9th ed., McGraw-Hill.

Juvinall, R.C., and Marshek, K.M. (2012) *Fundamentals of Machine Component Design*, 5th ed., Wiley.

Mott, R. L. (2014) *Machine Elements in Mechanical Design*, 5th ed., Pearson.

Neale, M., Needham, P., and Horrell, R. (2005) *Couplings and Shaft Alignment*, Wiley.

Norton, R.L. (2011) *Machine Design*, 4th ed., Prentice Hall.

Oberg, E., and Jones, F.D. (2000) *Machinery's Handbook*, 26th ed., Industrial Press.

Piotrowski, J. (1995) *Shaft Alignment Handbook*, 2nd ed., Marcel Dekker.

References

Oberg, E., et al. (2000) *Machinery's Handbook*, 26th ed., Industrial Press.

Peterson, R.E. (1974) *Stress Concentration Factors*, Wiley.

Questions

11.1 What is a shaft? How is it different from a beam?

11.2 Why is it important to consider both static and fatigue failure when designing shafts?

11.3 What is the consequence of including a thrust (axial) force on the shaft design equations?

11.4 What is meant by the critical speed of a shaft?

11.5 What is the Rayleigh equation? The Dunkerley equation?

11.6 What is the difference between a key and a pin?

11.7 What is a Gib head?

11.8 Is there an advantage to using a Woodruff key?

11.9 What should be harder, the hub or the key?

11.10 What is a flywheel?

11.11 What is the coefficient of fluctuation?

11.12 What is a rigid coupling? A flexible coupling?

11.13 What are retaining rings used for? What kind of retaining rings are there?

11.14 What is a spline?

11.15 What kind of shaft misalignments are there?

Qualitative Problems

11.16 List reasons that shafts are designed with high safety factors.

11.17 What machine elements can be used to limit the torque encountered by a shaft during an overload or other malfunction?

11.18 Explain why rotating shafts vibrate when a concentrated mass is mounted on the shaft.

11.19 Describe the manufacturing processes that are used to produce shafts, and why.

11.20 Review Fig. 11.9 and explain the conditions when you would use one type of key compared to others.

11.21 List the similarities and differences between set screws and keys.

11.22 Plot the holding force for set screws predicted by Table 11.5 and Eqs. (11.71) and (11.72). Explain the reasons for any differences between the two predictions.

11.23 Under what conditions would you use a key instead of a spline? When would you use a spline instead of a key?

11.24 When would you advise using a press fit instead of a key?

11.25 Review Table 11.6 and list two applications for each type of retaining ring shown.

11.26 Explain the consequences of driving an automobile without a flywheel.

11.27 What consequences occur if a flywheel is operated at an excessively high speed?

11.28 What material characteristics are desirable for a flywheel?

11.29 List the advantages and disadvantages of (a) rigid couplings; (b) helical couplings; (c) Schmidt couplings; (d) fluid couplings.

11.30 For a circumstance where significant parallel shaft misalignment exists, which couplings would you consider? What coupling would be more suitable if the space available for a coupling is limited? Explain.

Quantitative Problems

11.31 A 50-mm-diameter shaft is machined from AISI 1080 carbon steel (quenched and tenpered at 800°C), and because of its rotation, encounters a bending moment that varies from −20 to 55 N-m and a constant torque of 25 N-m. The shaft has a 5-mm hole drilled (machined) into it so that a roll pin can be inserted at that location. For a reliability of 99.5%, determine the safety factor for the shaft. Would the safety factor be larger or smaller if a retaining ring was used instead of the roll pin? Justify your answer. *Ans.* $n_s = 23.6$.

11.32 A shaft assembly shown in Sketch *a* is driven by a flat belt at location A and drives a flat belt at location B. The drive belt pulley diameter is 300 mm; the driven belt pulley diameter is 500 mm. The distance between sheaves is 800 mm, and the distance from each sheave to the nearest bearing is 200 mm. The belts are horizontal and load the shaft in opposite directions. Determine the size of the shaft and the types of steel that should be used. Assume a safety factor of 10. *Ans.* For AISI 1080 steel, $d = 35$ mm.

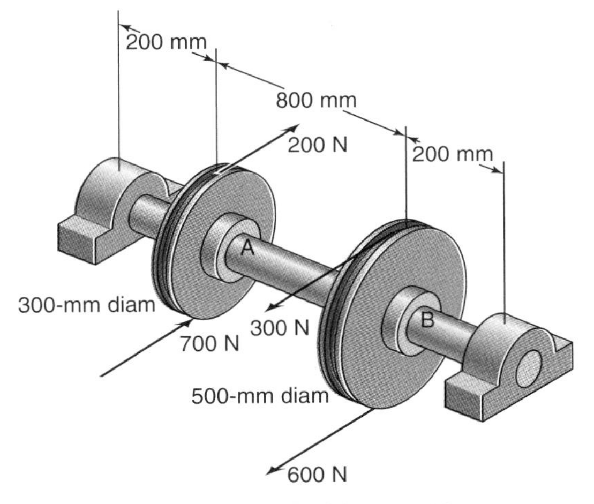

Sketch *a*, for Problem 11.32

11.33 The gears in the shaft assembly shown in Fig. 11.1a transmit 100 kW of power and rotate at 4000 rpm. Gear wheel 1 is loaded against another gear such that the force P_1 acts in a 45° direction at a radius of 80 mm from the shaft center. The force P_2 acts vertically downward at a radius of 110 mm from the shaft center. The distance from bearing A to gear 1 is 100 mm, that from gear 1 to gear 2 is 85 mm, and that from gear 2 to bearing B is 50 mm. Perform the following:

(a) Draw a free-body diagram with forces acting on the shaft when bearings A and B transmit only radial forces. *Ans.* $A_y = -750$ N, $A_z = 1212$ N.

(b) Find values of force components as well as the resultant force at locations A and B.

(c) Find the transmitted torque. *Ans.* $T = 238.7$ N-m.

(d) Draw a bending moment diagram in the x-y and x-z planes along with a torque diagram; also, indicate the maximum bending moment and the maximum torque.

(e) Find the safety factor according to the distortion-energy theory (DET) and the maximum-shear-stress theory (MSST) if the shaft has a diameter of 35 mm and is made of high-carbon steel (AISI 1080, quenched and tempered at 800°C). *Ans.* $n_{s,MSST} = 5.75$, $n_{s,DET} = 5.73$.

11.34 The shaft assembly given in Problem 11.32 has an extra loading from thermal expansion of the shaft. The bearings are assumed to be rigid in the axial direction, so that when the shaft heats up it cannot elongate but instead compressive stress builds up. Determine the thermal stress and find the safety factor by using the DET if the shaft heats up by

(a) 5°C. *Ans.* $n_s = 5.69$.

(b) 15°C. *Ans.* $n_s = 4.53$.

11.35 Given the shaft assembly in Problem 11.32 but calculating as if AISI 1080 steel, quenched and tempered at 800°C, was used and is considered to be brittle (using the MNST), find the safety factor n_s for fatigue by using the information in Problem 11.32. *Ans.* $n_{s,MNST} = 5.57$.

11.36 Gears 3 and 4 act on the shaft shown in Sketch *b*. The resultant gear force, $P_A = 600$ lb, acts at an angle of 20° from the y-axis. The yield stress for the shaft, which is made of cold-drawn steel, is 71,000 psi and the ultimate stress is 85,000 psi. The shaft is solid and of constant diameter. The safety factor is 3.0. Assume the DET throughout. Also, for fatigue loading conditions assume completely reversed bending with a bending moment amplitude equal to that used for static conditions. The alternating torque is zero. Determine the safe shaft diameter due to static and fatigue loading, rounding your answer to the next highest quarter-inch. Show shear and moment diagrams in the various planes. *Ans.* $d_{\text{static}} = 2.0$ in., $d_{\text{fatigue}} = 2.50$ in.

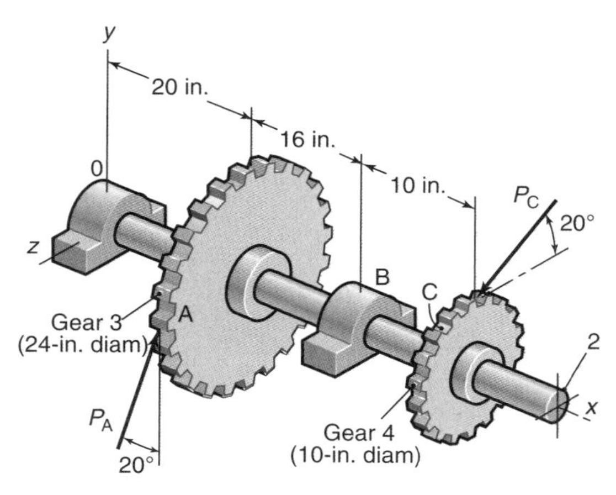

Sketch b, for Problem 11.36

11.37 Derive Eq. (11.38). Start by showing the stresses acting on an oblique plane at angle ϕ. The MNST should be used, implying that the critical stress line extends from S_e/n_s to S_u/n_s.

11.38 The shaft shown in Sketch c rotates at 1000 rpm and transfers 6 kW of power from input gear A to output gears B and C. The spur gears A and C have pressure angles of 20°. The helical gear has a pressure angle of 20° and a helix angle of 30° and transfers 70% of the input power. All important surfaces are ground. All dimensions are in mm. The shaft is made of annealed carbon steel with $S_{ut} = 636$ MPa and $S_y = 365$ MPa.

(a) Draw a free-body diagram as well as the shear and moment diagrams of the shaft.

(b) Which bearing should support the thrust load and why? *Ans.* The left bearing.

(c) Determine the minimum shaft diameter for a safety factor of 3.0 and 99% reliability. *Ans.* $d = 20$ mm.

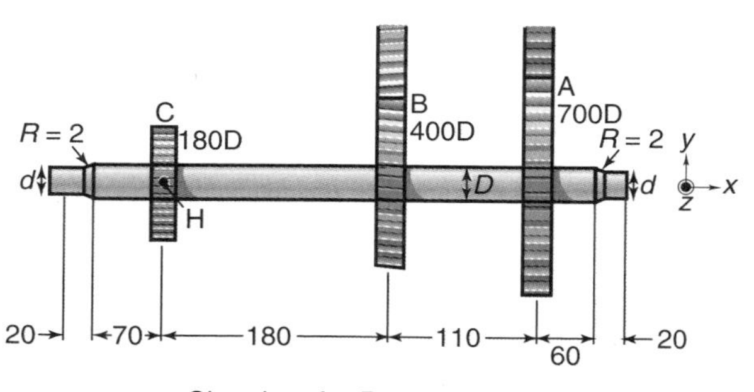

Sketch c, for Problem 11.38

11.39 In Problem 11.38 if the shaft diameter is 30 mm and it is made of AISI 1030 steel, quenched and tempered at 800°C, what is the safety factor while assuming 90% reliability? *Ans.* $n_s = 4.09$.

11.40 The shaft shown in Sketch d supports two gears. The shaft is made from high carbon steel (quenched and tempered AISI 1080), and is to be designed with a safety factor of 2.0. The gears transmit a constant torque caused by $P_A = 2000$ N acting vertically as shown. The shaft has a constant machined cross-section.

(a) What is the reaction force on gear C? *Ans.* $P_c = 4000$ N.

(b) What is the critical location in the shaft? *Ans.* At bearing B.

(c) Using the Soderberg line, obtain the required shaft diameter. (Note: Ignore all endurance limit modification factors except the surface finish factor.) *Ans.* $d = 50$ mm.

(d) One of the gears has been attached with a keyway, the other with a shrink fit. Which gear was attached with a keyway, and why?

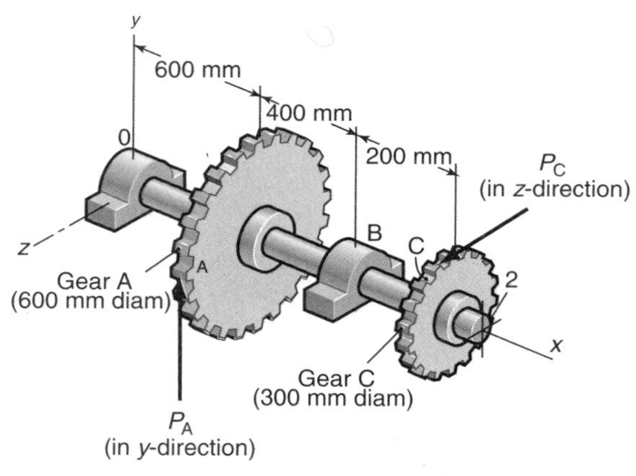

Sketch d, for Problem 11.40

11.41 The rotor shown in Sketch e has a stiff bearing on the left. Find the critical speed when the shaft is made of steel with $E = 207$ GPa. *Ans.* $N = 3018$ rpm.

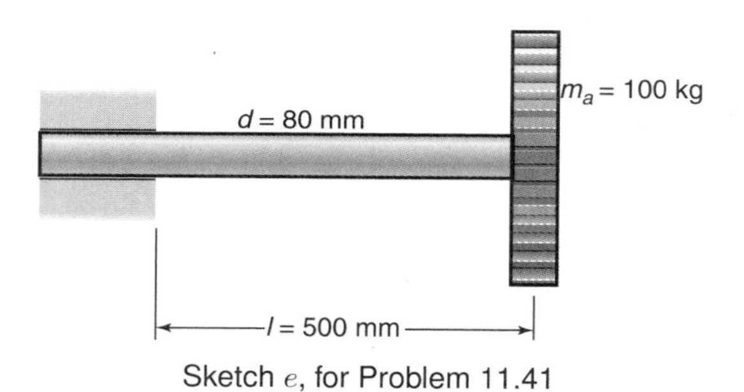

Sketch e, for Problem 11.41

11.42 Determine the critical speed in bending for the shaft assembly shown in Sketch f. The modulus of elasticity of the shaft $E = 207$ GPa, its length $l = 350$ mm, its diameter $d = 10$ mm, and the rotor mass $m_a = 2.3$ kg. *Ans.* $\omega = 250$ rad/s.

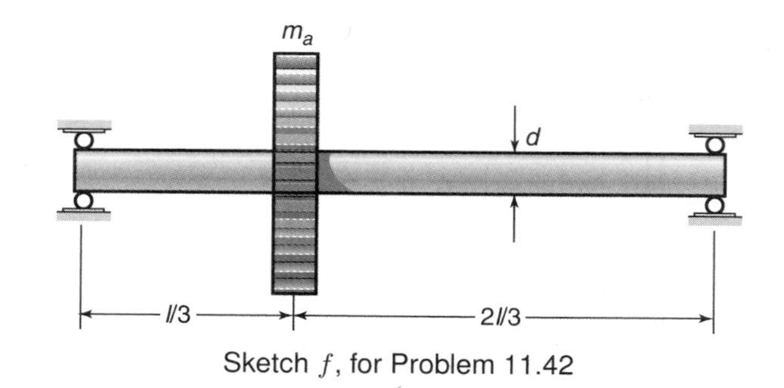

Sketch f, for Problem 11.42

11.43 Calculate the diameter of the shaft in the assembly shown in Sketch g so that the first critical speed is 9000 rpm. The shaft is made of steel with E = 207 GPa. The distance a = 300 mm and the mass m_a = 100 kg; the mass of the shaft is neglected. Use both the Rayleigh and Dunkerley methods. *Ans.* d_R = 131 mm, d_D = 133 mm.

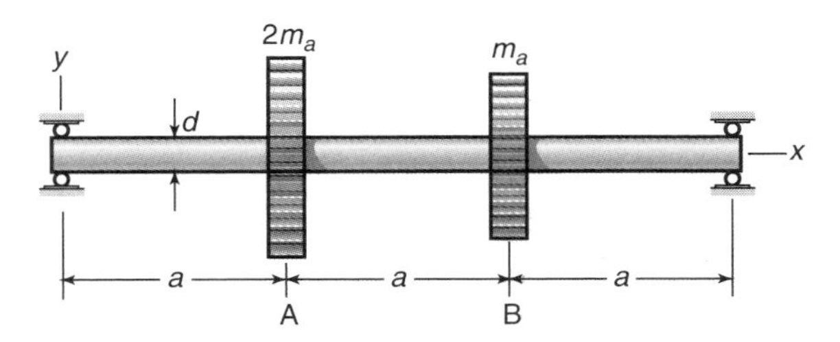

Sketch g, for Problem 11.43

11.44 The simply-supported shaft shown in Sketch h has two weights on it. Neglecting the shaft weight and using the Rayleigh method for the stainless steel ($E = 26\times10^6$ psi) shaft, determine the safe diameter to ensure that the first critical speed is no less than 3600 rpm. *Ans.* d_R = 2.23 in., d_D = 2.26 in.

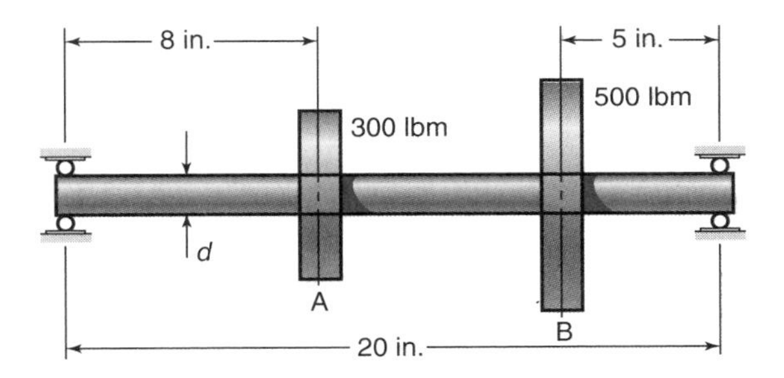

Sketch h, for Problems 11.44 and 11.50

11.45 Determine the first critical speed by the Dunkerley and Rayleigh methods for the steel shaft shown in Sketch i. Neglect the shaft mass. The area moment of inertia is $I = \pi r^4/4$, where r = shaft radius. The method of superposition may be used with the following given:

$$\delta = \frac{P}{6EI}\left[\frac{bx^3}{l} - \langle x-a\rangle^3 - \frac{xb\left(l^2-b^2\right)}{l}\right]$$

Ans. N_R = 708 rpm, N_D = 673 rpm.

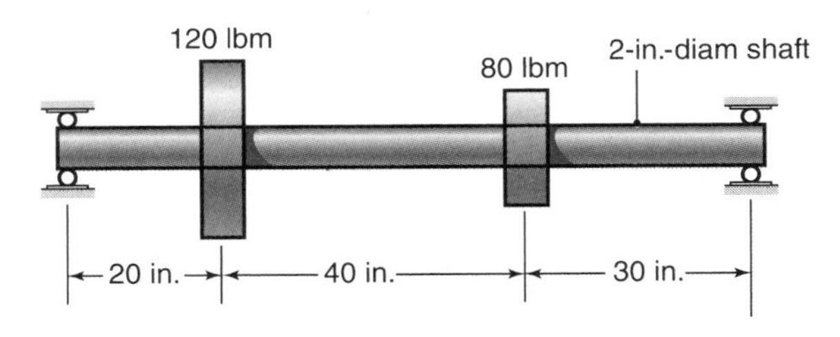

Sketch i, for Problem 11.45

11.46 Calculate the diameter d of the shaft in Sketch g so that the lowest critical speed in bending becomes 9000 rpm. The modulus of elasticity E = 107 GPa, the distance a = 300 mm, and the mass m_a = 100 kg. Neglect the mass of the shaft. Use the Rayleigh and Dunkerley methods. *Ans.* d_R = 155 mm, d_D = 156 mm.

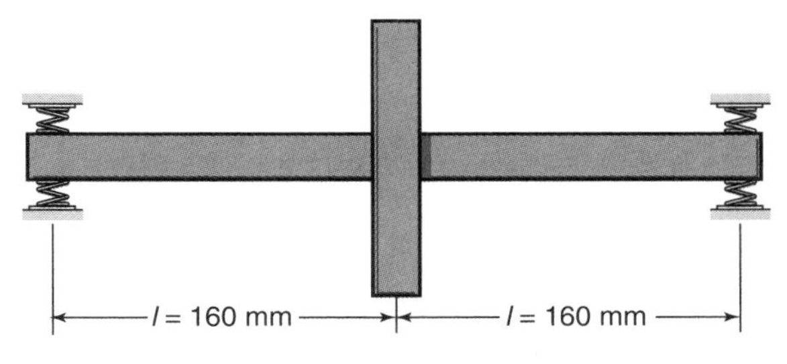

Sketch j, for Problem 11.46

11.47 The shaft from Problem 11.43 transmits 200 kW at 3000 rpm. Design the shaft diameter so that the angular deflection between the two gears is 0.75°. At this diameter, what is the critical speed of the shaft? *Ans.* d = 31.1 mm, ω_{cr} = 504 rpm.

11.48 The rotor in Sketch k has a shaft diameter of 32 mm, and the disk mass is 170 kg. Calculate the critical speed if the spring rates are 1668 and 3335 N/mm, respectively, and are the same in all directions. The elastic modulus E = 206 GPa. For a shaft without springs the influence coefficient $\alpha_{11} = l^3/(6EI)$. *Ans.* N = 1360 rpm.

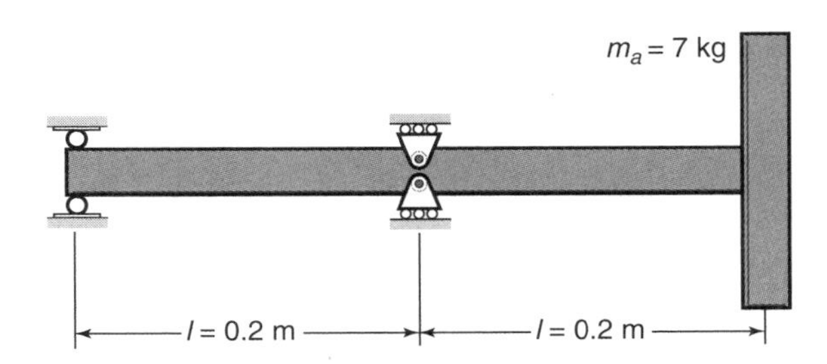

Sketch k, for Problems 11.48 and 11.49

11.49 Calculate the critical speed for a rotor shown in Sketch k that has two moment-free bearings. The shaft has a diameter of 20 mm and is made of steel with E = 207 GPa. *Ans.* N = 1993 rpm.

11.50 The deflection at the center of the stainless steel shaft ($E = 29\times10^6$ psi) shown in Sketch h is equal to 0.075 in. Find the diameter of the solid circular shaft and the first critical speed using both the Rayleigh and Dunkerly methods. *Ans.* d = 1.0 in., ω_R = 75.5 rad/s.

11.51 A flywheel has a hub made of aluminum alloy 2014-O. The hub is connected to a 20-mm-diameter annealed AISI 1040 steel shaft with a flat tapered key made of AISI 1020 steel, quenched and tempered at 870°C. Dimension the key so that a torque of 20 N-m can be transmitted with a safety factor of 4. *Ans.* w = 4.52 mm, h = 12.2 mm.

11.52 A jaw crusher is used to crush iron ore down to the particle size needed for iron ore pellet production. The pellets are later used in a blast furnace to make steel. If the ore pieces fed into the crusher are too large, the crusher is protected by a torque-limiting key made of copper. The torque on the 400-mm-diameter shaft connecting the flywheel with the jaw mechanism should never go above 700 kN-m. Dimension a copper square tapered key so that the shaft is not damaged when an oversized ore particle comes into the crusher.

11.53 Given the sketch of the Woodruff key in Figure 11.9, show that the shear area is given by

$$A_s = 2w\sqrt{c(D-c)}$$

where c is the diametral clearance and D is the key diameter.

11.54 A punch press is to be driven by a constant torque electric motor that operates at 1200 rpm. A flywheel of constant thickness is cut from 2 in. thick steel plate is to be used to ensure smooth operation. The punch press torque steps from 0 to 10,000 ft-lb, remaining constant for the first 45° of camshaft rotation, up to 60,000 ft-lb for the next 45°, and back to 0 for the remainder of the cycle. What is the minimum diameter of the flywheel? *Ans.* $d = 10.25$ in.

11.55 The output torque of a flywheel for each revolution of a shaft is 10 N-m from 0 to π and 120 N-m from π to 2π. The coefficient of fluctuation is 0.04 at an average speed of 2000 rpm. Assume that the flywheel disk is an annealed AISI 1040 steel plate of 25-mm constant thickness. Determine the following:

(a) Average load torque. *Ans.* 65 Nm.

(b) Locations $\theta_{\omega_{\max}}$ and $\theta_{\omega_{\min}}$. *Ans.* $\theta_{\omega_{\max}} = \pi$

(c) Energy required. *Ans.* $K_e = 172.8$ Nm.

(d) Outside diameter of flywheel. *Ans.* $d = 270$ mm.

11.56 The output torque of a flywheel for each revolution of the shaft is 200 in.-lb from 0 to $\pi/3$, 1600 in.-lb from $\pi/3$ to $2\pi/3$, 400 in.-lb from $2\pi/3$ to π, 900 in.-lb from π to $5\pi/3$, and 200 in.-lb from $5\pi/3$ to 2π. Input torque is assumed to be constant. The average speed is 860 rpm and the coefficient of fluctuation is 0.10. Find the diameter of the flywheel if it is to be cut from a 1-in.-thick steel plate. *Ans.* $d = 12$ in.

11.57 A 20-mm-thick flywheel is made of aluminum alloy 2014-O and runs at 9000 rpm in a racing car motor. What is the flywheel diameter if the aluminum alloy 2014 is stressed to a quarter of its yield strength at 9000 rpm? To decrease the flywheel outer diameter, higher density exotic materials are investigated. The flywheel is machined from a solid metal plate with no central hole, and the thickness cannot be larger than 20 mm. Which of the following is the best material that can be substituted for aluminum alloy 2014 but with the same safety factor?

(a) pure copper

(b) cartridge brass

(c) annealed 5% phosphor bronze

(d) magnesium alloy ZK61A

(e) annealed Ti-6Al-4V

11.58 The aluminum alloy flywheel considered in Problem 11.57 has a mass moment of inertia of 0.0025 kg-m^2. By mistake the motor is accelerated to 7000 rpm in neutral gear and the throttle sticks in the fully open position. The only way to stop the motor is to disconnect the electric lead from the spark plug, and it takes 6 s before the motor is motionless. Neglecting the inertia of all movable parts in the motor except the flywheel, calculate the internal friction moment in the motor and the (mechanical part of the) friction losses in the motor at 5000 rpm. The friction moment is assumed to be constant at all speeds. *Ans.* $T_f = -0.305$ N-m.

11.59 A one-cylinder ignition bulb motor to an old fishing boat has a flywheel that gives the motor a coefficient of fluctuation of 25% when it idles at 200 rpm. The mass moment of inertia for the flywheel is 1.9 kg-m^2. Determine the coefficient of fluctuation at 500 rpm if the compression stroke consumes equally large energy at all speeds. Also, calculate the mass moment of inertia needed to obtain a 20% coefficient of fluctuation at 500 rpm. *Ans.* $C_f = 0.040$, $I_m = 0.38$ kg-m^2.

11.60 A flywheel for a city bus drive should store as much energy as possible for a given flywheel weight. The diameter must be less than 1.5 m, and the mass must be smaller than 250 kg. Find which material from Tables A.1 and A.2 gives the highest possible stored energy for a safety factor of 4 assuming that the flywheel has constant thickness.

11.61 A flywheel on an AISI 1080 (Q&T at 800°C) steel shaft is oscillating due to disturbances from a combustion engine. The engine is a four-stroke engine with six cylinders, so that the torque disturbances occur three times per revolution. The flywheel shaft has a diameter of 20 mm and is 1 m long, and the flywheel moment of inertia is 0.5 kg-m^2. Find the engine speed at which the large torsional vibrations begin to appear. *Ans.* $N = 160$ rpm.

11.62 An external retaining ring is to be stamped from quenched and tempered AISI 1080 steel and will be designed to secure a rolling element bearing against a shaft shoulder. The shaft is produced from annealed AISI 1040 steel and has a diameter of 25 mm. Calculate the restraining force that can be developed for MSH, ME, and MCR rings using a safety factor of 2. Also estimate the theoretical stress concentration factor that results in the shaft from the retaining ring groove if the radius of the groove corner radius is 0.5 mm. *Ans.* $P_{\text{MSH}} = 4.78$ kN.

Design and Projects

11.63 A front wheel for a riding lawnmower needs to be attached to a 20 mm diameter axle. The wheel has an integral hub that will be grease lubricated and can slide axially and rotate on the shaft (it is a steering wheel, not a drive wheel, so it only needs to idle). When the lawnmower turns, a force of up to 500 N can be applied in the axial direction. List design alternatives for securing the wheel to the axle. From your list, select the three most attractive designs for further analysis, and prescribe dimensions and restraining forces for these alternatives. Write a summary of your design analysis and select the approach you think is best suited for this application.

11.64 *Hydrodynamic braking* is used as an auxiliary brake in vehicles to prevent high speeds when undergoing a large change in elevation, such as with heavy vehicles on mountain roads. Explain how a fluid coupling can be used in this manner, and describe how braking torque will vary with speed. Suggest how to dissipate the heat generated from this auxiliary brake.

11.65 Derive equivalent expressions to Eqs. (11.36) and (11.37) using the Goodman criterion and the MSST. Using a mathematics package, write a program that provides the diameter or safety factor if the modified Goodman criterion is used.

11.66 A hammer mill will be used to shred sugar cane to allow efficient extraction of sugar. Cane is fed into the hammer mill through a hopper, and impacted by cast iron hammers powered by a 8000 kW motor at 4000 rpm, so that around 10,000 tons of cane can be processed daily. When cane is struck by a hammer, it shatters, and the pulp is further reduced by repeated impacts until it is discharged. What kind of coupling would you use to connect the power source to the hammer mill? You may reference the technical literature or search the Internet for any further information you may require. Write a one-page memorandum outlining your design recommendation.

11.67 Would you recommend using a flywheel for the sugar cane shredder described in Problem 11.66? Explain.

11.68 Review Fig. 11.21, and derive an expression for the torsional stiffness of the Falk coupling from beam theory.

Chapter 12

Hydrodynamic and Hydrostatic Bearings

Components of a hydrodynamic journal bearing. *Source:* Courtesy of Kingsbury, Inc.

Contents

Examples

Design Procedures

Case Study

Getting wisdom is the most important thing you can do! And whatever else you do, get good judgment.

Proverbs 4:7

This chapter focuses on bearings that are conformal in their geometry, with a lubricant interspersed between the surfaces. There are two basic types of bearings considered in this chapter: hydrodynamic bearings, where the pressure generated in a lubricant film is caused by viscosity and bearing motion, and hydrostatic, where the pressure is produced by some external (pumping) means. Such bearings have the typical characteristics of very low friction, high efficiency and reliability, and very good load support. The main drawbacks to these bearings are their larger size compared to rolling element bearings, high startup friction, and the need to consider thermal effects on the lubricant. Wear is very low, and generally occurs only during startup. Building upon the information in Chapter 8, this chapter begins with an in-depth analysis of the fluid mechanics involved in such bearings by presenting the Reynolds equation and investigating all of its terms. Application of the Reynolds equation to thrust bearings is then followed by presentation of numerical results. The unique geometry, boundary conditions, and pressure profile in journal bearings are then discussed, and numerical solutions and design charts are presented and demonstrated. Finally, squeeze film and hydrostatic bearings are analyzed.

Machine elements described in this chapter: Hydrodynamic thrust bearings, hydrodynamic journal bearings, hydrostatic bearings.
Typical applications: Automotive main bearings, thrust bearings for turbines, ship propeller support bearings, machine spindles.
Competing machine elements: Rolling element bearings (Chapter 13).

Symbols

a_b	bearing pad load coefficient
A	area, m^2
A_b	bearing area, m^2
A_r	recess area, m^2
A_s	sill area, m^2
B_j	bearing number for journal slider bearing, given in Eq. (12.81)
B_t	bearing number for thrust slider bearing, given in Eq. (12.66)
C_p	specific heat of lubricant at constant pressure, J/(kg-°C)
C_s	volumetric specific heat, J/(m^3-°C)
C_1-C_4	integration constants
c	radial clearance in journal bearing, m
d	diameter, m
e	eccentricity of journal bearing, m
F	force, N
g	acceleration due to gravity, 9.807 m/s^2
H	dimensionless film thickness, h/s_h
h	film thickness, m
h_b	bearing pad power coefficient
h_p	power, W
l	length, m
N_a	rotational speed, rps
N_p	number of pads
n_s	step location
P	dimensionless pressure,
$P_{\max}$	dimensionless maximum film
p	pressure, Pa
p_l	lift pressure, Pa
p_r	recess pressure, Pa
p_s	supply pressure, Pa
Q	dimensionless volumetric flow rate
q	volumetric flow rate, m^3/s
q_b	bearing pad flow coefficient
q_s	side-leakage volumetric flow rate, m^3/s
q_x	volumetric flow rate in sliding direction, m^3/s
q_r'	radial flow rate per unit width, m^2/s
q_x'	volumetric flow rate per unit width in sliding direction, m^2/s
q_y'	volumetric flow rate per unit width in transverse direction, m^2/s
R_a	centerline average surface roughness, m
r	radius, m
r_b	radius of journal bearing, m
s_h	shoulder height, m
T	torque, N-m
t	time, s
Δt	time change, s
t_m	temperature, °C
Δt_m	temperature change, °C
u	fluid velocity in sliding direction, m/s
$\bar{u}$	average velocity in sliding direction, m/s
V	volume, m^3
v	fluid velocity in transverse direction, m/s
W_r	radial load, N
W_r^*	radial load per unit area, Pa
W_t	total load, N
W_z'	normal load per unit width, N/m
$\bar{W}_z$	dimensionless normal load
w	squeeze velocity, m/s
w_t	width (in side-leakage direction) of bearing, m
X	dimensionless coordinate, x/l
x	Cartesian coordinate in direction of sliding, m
y	Cartesian coordinate in side-leakage direction, m
z	Cartesian coordinate across film, m
α	inclination angle of fixed-incline slider bearing, deg
ϵ	eccentricity ratio, e/c
η	absolute viscosity, Pa-s
η_o	absolute viscosity at $p = 0$ and constant temperature, Pa-s
λ	length-to-width ratio, l/w_t
λ_j	diameter-to-width ratio, $2r/w_t$
μ	coefficient of sliding friction
ρ	density, kg/m^3
τ	shear stress, Pa
Φ	attitude angle, deg
ϕ	cylindrical polar coordinate
$\phi_{\max}$	location of maximum pressure, deg
ϕ_o	location of terminating pressure, deg
ω	angular velocity, $2\pi N_a$, rad/s

Subscripts

a	solid a (upper surface)
b	solid b (lower surface)
i	inlet; inner
m	mean; location of maximum pressure
o	outlet; outer
x, y, z	Cartesian coordinates

12.1 Introduction

Design is exciting because it involves a blending of multiple disciplines. While other machine elements demonstrate the integration of disciplines such as solid mechanics, materials science, and manufacturing, hydrodynamic bearings present a clear example requiring incorporation of fluid mechanics and (to a lesser extent) heat transfer in design. These bearings have great importance, as they are widely used in the transportation and energy industries, and often used elsewhere. Because of their importance, they have received considerable attention for over a century, and a good mathematical framework exists to guide design. As such, a strong mathematical foundation will be presented in this chapter to provide an in-depth understanding of hydrodynamic and hydrostatic bearings, but the more casual reader may wish to proceed directly to Section 12.3 for exposure to simplified design procedures based on existing numerical solutions.

The characterization of **hydrodynamic lubrication** was presented in Section 8.7.4 and will not be repeated here. For the bearings considered in this chapter, it will be assumed that **fluid film lubrication** occurs, or that the lubricated surfaces are completely separated by a fluid film. This fluid film can be generated by viscous effects and bearing motion (hydrodynamic bearings) or by supplying the lubricant under pressure (hydrostatic bearings). Also, the bearings are assumed to have conformal surfaces (see Section 8.3). These three topics (characterization of hydrodynamic lubrication, fluid film lubrication, and conformal surfaces), which were covered in Ch. 8, should be understood before proceeding with this chapter. Indeed, many equations, figures, and tables from Ch. 8 will be referred to here.

The earliest known hydrodynamic bearings were used on vehicle axles by the ancient Egyptians and thus predate written history. Such fluid lubricated bearings have been used ever since, with significant empirical knowledge developed over time. However, a modern understanding of the science and technology of hydrodynamic bearings begins with the classical experiments of Tower [1883], who detected the existence of a film from pressure measurements within the lubricant, and of Petrov [1883], who reached the same conclusion from friction measurements. Indeed, a thick, pressurized lubricant film forms readily in such bearings and allows load transfer without any contact from opposing bearing surfaces. The work of Tower and Petrov was closely followed by Reynolds' celebrated analytical paper [1886] in which he used a reduced form of the Navier-Stokes equations in association with the continuity equation to generate a second-order differential equation for the pressure in the narrow, converging gap (or conjunction) between bearing surfaces. This pressure enables a load to be transmitted between the surfaces and, since it predicts a thick film of lubricant, the load is transferred with extremely low friction. In such a situation, the physical properties of the lubricant, notably the dynamic viscosity, dictate the behavior in the conjunction.

Hydrodynamic and hydrostatic lubrication have been extensively studied because of the fundamental importance of these subjects, and many design variants exist. This chapter presents the most important aspects of these bearings and thus provides a framework that is useful for most design problems. It should be realized that this chapter is an introduction to such bearings, and specific and more detailed information can be found in Hamrock et al. [2004] or manufacturers' literature.

12.2 The Reynolds Equation

The full Navier-Stokes equation covered in most fluid mechanics texts contains inertia, body force, pressure, and viscous terms. These equations are sufficiently complicated to prohibit analytical solutions to most practical problems. However, there are a number of cases where fluid inertia and body forces can be safely ignored, so that the pressure and viscous terms dominate. Fluid film lubrication problems belong to this class.

The differential equation governing the pressure distribution in fluid film lubrication is the **Reynolds equation**, first derived in a remarkable paper by Reynolds [1886]. This paper not only contained the basic differential equation of fluid film lubrication, but also directly compared his theoretical predictions with the experimental results obtained earlier by Tower [1883].

12.2.1 Derivation of the Reynolds Equation

There are numerous approaches for deriving the Reynolds equation, but the approach shown here is the most straightforward. It involves simplifying the Navier-Stokes equations using order analysis, which basically identifies characteristics of bearings that allow unimportant terms in the Navier-Stokes equation to be identified and ignored. The velocity terms in the Navier-Stokes equations are then manipulated to allow substitution in the continuity equation. Simplification of the result leads to the general form of the Reynolds equation.

The Navier-Stokes equations can be applied to obtain detailed derivations of the fluid pressure and flow as related to the geometry, viscosity, speed, etc., but significant simplifications can be obtained if the discussion is restricted to the viscous flow seen in bearings. From an order-of-magnitude analysis, the general Navier-Stokes equations can be considerably reduced. If $h_{\min}$ is the minimum film thickness and is in the z-direction, if l is the length and is in the x-direction (the direction of motion), and if w_t is the width and is in the y-direction, then in hydrodynamic and hydrostatic bearing situations, l and w_t are several (usually at least three) orders of magnitude larger than the minimum film thickness. Thus, if terms of order $(h/l)^2$, $(h/w_t)^2$, h/l, and h/w_t and smaller terms are neglected and only first-order terms are considered, the Navier-Stokes equations reduce to

$$\frac{\partial p}{\partial x} = \frac{\partial}{\partial z}\left(\eta \frac{\partial u}{\partial z}\right), \tag{12.1}$$

$$\frac{\partial p}{\partial y} = \frac{\partial}{\partial z}\left(\eta \frac{\partial v}{\partial z}\right), \tag{12.2}$$

and

$$\frac{\partial p}{\partial z} = 0, \tag{12.3}$$

where

η = absolute viscosity, Pa-s
u = fluid velocity in x-direction, m/s
v = fluid velocity in y-direction, m/s
p = pressure, Pa

Equation (12.3) does not mean that pressure is a constant, but that it does not change across a film; it can vary with respect to other variables such as the x- and y-directions and time. Equations (12.1) and (12.2) can be directly integrated to give the general expressions for the velocity gradients:

$$\frac{\partial u}{\partial z} = \frac{z}{\eta}\frac{\partial p}{\partial x} + \frac{C_1}{\eta}, \tag{12.4}$$

$$\frac{\partial v}{\partial z} = \frac{z}{\eta}\frac{\partial p}{\partial y} + \frac{C_3}{\eta}, \tag{12.5}$$

where C_1 and C_3 are integration constants.

The lubricant viscosity might change considerably across the thin film (z-direction) as a result of temperature variations that arise in some bearing problems. In this case, progress toward developing a simple Reynolds equation is considerably complicated. An approach that is satisfactory in most fluid film applications is to treat η as the average viscosity across the film. This approach, which does not restrict variation in the x- and y-directions, is pursued here.

With η representing the average viscosity across the film, integrating Eqs. (12.4) and (12.5) gives the velocity components as

$$u = \frac{z^2}{2\eta}\frac{\partial p}{\partial x} + C_1\frac{z}{\eta} + C_2 \tag{12.6}$$

and

$$v = \frac{z^2}{2\eta}\frac{\partial p}{\partial y} + C_3\frac{z}{\eta} + C_4. \tag{12.7}$$

If zero slip at the fluid-solid interface is assumed, the boundary values for the velocity are

1. At $z = 0$ (i.e., the lower surface), $u = u_b$ and $v = v_b$.
2. At $z = h$ (the upper surface), $u = u_a$ and $v = v_a$.

The subscripts a and b refer to conditions at the upper and lower surfaces, respectively. Thus, u_a, v_a, and w_a refer to the velocity components of the upper surface in the x, y, and z-directions, respectively, and u_b, v_b, and w_b refer to the velocity components of the lower surface in the same directions.

With the boundary conditions applied to Eqs. (12.6) and (12.7), the velocity components are

$$u = -z\left(\frac{h-z}{2\eta}\right)\frac{\partial p}{\partial x} + \frac{u_b(h-z)}{h} + \frac{u_a z}{h} \tag{12.8}$$

and

$$v = -z\left(\frac{h-z}{2\eta}\right)\frac{\partial p}{\partial y} + \frac{v_b(h-z)}{h} + \frac{v_a z}{h}. \quad (12.9)$$

Consider now the volume flow rates per unit width in the x- and y-directions, which are defined as

$$q_x' = \int_0^h u\, dz, \quad (12.10)$$

$$q_y' = \int_0^h v\, dz. \quad (12.11)$$

Substituting Eqs. (12.8) and (12.9) into these equations and simplifying gives

$$q_x' = -\frac{h^3}{12\eta}\frac{\partial p}{\partial x} + \frac{h(u_a+u_b)}{2}, \quad (12.12)$$

$$q_y' = -\frac{h^3}{12\eta}\frac{\partial p}{\partial y} + \frac{h(v_a+v_b)}{2}. \quad (12.13)$$

The first term on the right side of these equations represents the pressure flow and the second term the velocity flow. Often, the pressure and velocity terms in the Reynolds equations are referred to as **Poiseuille** and **Couette** terms, respectively, as described below.

The continuity equation can be expressed as

$$\frac{\partial \rho}{\partial t} + \frac{\partial}{\partial x}(\rho u) + \frac{\partial}{\partial y}(\rho v) + \frac{\partial}{\partial z}(\rho w) = 0, \quad (12.14)$$

where ρ is the density and w is the squeeze velocity. The density, ρ, in Eq. (12.14) and the absolute viscosity, η, in Eqs. (12.8) and (12.9) are functions of x and y, but not of z. It is convenient to express Eq. (12.14) in integral form as

$$\int_0^h \left[\frac{\partial \rho}{\partial t} + \frac{\partial}{\partial x}(\rho u) + \frac{\partial}{\partial y}(\rho v) + \frac{\partial}{\partial z}(\rho w)\right] dz = 0. \quad (12.15)$$

A general rule of integration is that

$$\int_0^h \frac{\partial}{\partial x}\left[f(x,y,z)\right] dz = -f(x,y,h)\frac{\partial h}{\partial x} + \frac{\partial}{\partial x}\left[\int_0^h f(x,y,z)\, dz\right] \quad (12.16)$$

Thus, the u-component of Eq. (12.15) becomes

$$\begin{aligned}\int_0^h \frac{\partial}{\partial x}(\rho u)\, dz &= -(\rho u)_{z=h}\frac{\partial h}{\partial x} + \frac{\partial}{\partial x}\left(\int_0^h \rho u\, dz\right) \\ &= -\rho u_a \frac{\partial h}{\partial x} + \frac{\partial}{\partial x}\left(\rho\int_0^h u\, dz\right). \quad (12.17)\end{aligned}$$

Similarly, for the v-component,

$$\int_0^h \frac{\partial}{\partial y}(\rho v)\, dz = -\rho v_a \frac{\partial h}{\partial y} + \frac{\partial}{\partial y}\left(\rho\int_0^h v\, dz\right). \quad (12.18)$$

The w-component term can be integrated directly to give

$$\int_0^h \frac{\partial}{\partial z}(\rho w)\, dz = \rho(w_a - w_b). \quad (12.19)$$

After substituting Eqs. (12.17) through (12.19) into Eq. (12.15), the integrated continuity equation becomes

$$\begin{aligned}0 \;=\; & h\frac{\partial \rho}{\partial t} - \rho u_a \frac{\partial h}{\partial x} + \frac{\partial}{\partial x}\left(\rho \int_0^h u\, dz\right) \\ & -\rho v_a \frac{\partial h}{\partial y} + \frac{\partial}{\partial y}\left(\rho\int_0^h v\, dz\right) + \rho(w_a - w_b).\end{aligned} \quad (12.20)$$

The integrals in this equation represent the volume flow per unit width (q_x' and q_y') given in Eqs. (12.12) and (12.13). Introducing these flow rate expressions into Eq. (12.20) yields the general Reynolds equation:

$$\begin{aligned}0 \;=\; & \frac{\partial}{\partial x}\left(-\frac{\rho h^3}{12\eta}\frac{\partial p}{\partial x}\right) + \frac{\partial}{\partial y}\left(-\frac{\rho h^3}{12\eta}\frac{\partial p}{\partial y}\right) \\ & + \frac{\partial}{\partial x}\left[\frac{\rho h(u_a+u_b)}{2}\right] + \frac{\partial}{\partial y}\left[\frac{\rho h(v_a+v_b)}{2}\right] \\ & + \rho(w_a - w_b) - \rho u_a\frac{\partial h}{\partial x} - \rho v_a \frac{\partial h}{\partial y} + h\frac{\partial \rho}{\partial t}. \quad (12.21)\end{aligned}$$

Example 12.1: Lubricant Leakage Around a Pump Piston

Given: A pump piston without rings pumps an oil with a viscosity of 0.143 Pa-s at a pressure differential of 20 MPa. The piston moves concentrically in the cylinder so that the radial clearance is constant. The piston is 100 mm in diameter and 80 mm long and the radial clearance is 50 μm. The piston moves with a speed of 0.2 m/s.

Find: Determine the leakage past the piston when

(*a*) The piston pumps the oil (like a pump)

(*b*) The oil drives the piston (like a hydraulic cylinder)

Solution: The pressure gradient is

$$\frac{\partial p}{\partial x} = \frac{\Delta p}{\Delta x} = \frac{20\times 10^6}{0.080} = 2.5\times 10^8\ \text{N/m}^3.$$

(*a*) When the piston pumps the oil, the cylinder wall passes the piston from high pressure to low pressure. From Eq. (12.12), the pressure flow and the motion flow together are

$$\begin{aligned}q &= \pi d q_x' = \pi d\left[\frac{h^3}{12\eta}\frac{\partial p}{\partial x} + \frac{h(u_a+u_b)}{2}\right] \\ &= \pi(0.100) \\ &\quad \times \left[\frac{(50\times 10^{-6})^3(2.5\times 10^8)}{(12)(0.143)} + \frac{(50\times 10^{-6})(0.2)}{2}\right] \\ &= 7.292\times 10^{-6}\ \text{m}^3/\text{s}.\end{aligned}$$

(*b*) When the oil drives the piston, the pressure flow is opposite the flow from the motion

$$\begin{aligned} q &= \pi d\left[\frac{h^3}{12\eta}\frac{dp}{dx} - \frac{h(u_a+u_b)}{2}\right] \\ &= \pi(0.100) \\ &\quad \times\left[\frac{\left(50\times10^{-6}\right)^3\left(2.5\times10^8\right)}{12(0.143)} - \frac{\left(50\times10^{-6}\right)(0.2)}{2}\right] \\ &= 4.150\times10^{-6}\ \text{m}^3/\text{s}. \end{aligned}$$

Thus, when the piston pumps the oil, the leakage is 7.292 cm^3/s; and when the oil drives the piston, the leakage is 4.150 cm^3/s. Therefore, the cylinder wall motion pulls oil into the high-pressure region.

12.2.2 Physical Significance of Terms in the Reynolds Equation

The Reynolds equation given by Eq. (12.21) is a significant simplification over the continuity and three Navier-Stokes equations. However, this form of the Reynolds equation is usually more complicated than necessary for particular applications. Each term of the Reynolds equation can, at times, be important, and have been given formal names to clarify discussion. The first two terms in Eq. (12.21) are the Poiseuille terms and describe the net flow rates due to pressure gradients within the lubricated area; the third and fourth terms are the Couette terms and describe the net entraining flow rates due to surface velocities. The fifth to seventh terms describe the net flow rates due to a squeezing motion, and the last term describes the net flow rate due to local expansion. The x-direction defines the direction of sliding motion, so that flow in the y-direction is referred to as *side leakage*. For very wide bearings, the side leakage can be negligible. Thus, the flows or *actions* can be considered, without any loss of generality, by eliminating the side-leakage terms ($\partial/\partial y$) in Eq. (12.21). The one-dimensional Reynolds equation, with proper names of terms identified, is as follows:

$$\begin{aligned} \underbrace{\frac{\partial}{\partial x}\left(\frac{\rho h^3}{12\eta}\frac{\partial p}{\partial x}\right)}_{\text{Poiseuille}} &= \underbrace{\rho\left(w_a - w_b - u_a\frac{\partial h}{\partial x}\right)}_{\text{Squeeze}} \\ &+ \underbrace{h\frac{\partial \rho}{\partial t}}_{\text{Local expansion}} \\ &+ \underbrace{\frac{\partial}{\partial x}\left[\frac{\rho h\left(u_a+u_b\right)}{2}\right]}_{\text{Couette}} \end{aligned}$$

$$\overbrace{\underbrace{\frac{h\left(u_a+u_b\right)}{2}\frac{\partial \rho}{\partial x}}_{\text{Density wedge}} \quad \underbrace{\frac{\rho h}{2}\frac{\partial}{\partial x}\left(u_a+u_b\right)}_{\text{Stretch}} \quad \underbrace{\frac{\rho\left(u_a+u_b\right)}{2}\frac{\partial h}{\partial x}}_{\text{Physical wedge}}} \tag{12.22}$$

As can be seen in Eq. (12.22), the Couette terms lead to three distinct actions. The physical significance of each term within the Reynolds equation is now discussed in detail.

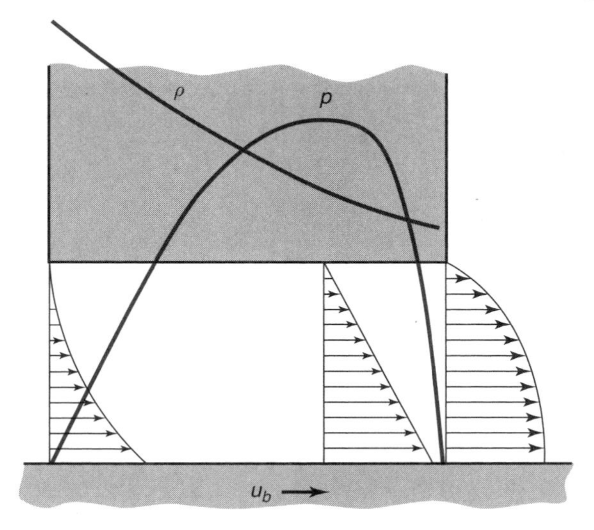

Figure 12.1: Density wedge mechanism.

Density Wedge Term, $\dfrac{h(u_a+u_b)}{2}\dfrac{\partial \rho}{\partial x}$

Consider a bearing where all terms in the Reynolds equation vanish except for the density wedge and Poiseuille terms. Figure 12.1 illustrates such a case, where a stationary bearing pad is parallel to an opposing moving surface. The remaining terms of the Reynolds equation for this bearing are:

$$\frac{\partial}{\partial x}\left(\frac{\rho h^3}{12\eta}\frac{\partial p}{\partial x}\right) = \frac{h(u_a+u_b)}{2}\frac{\partial \rho}{\partial x}.$$

Evaluating the integral and solving for pressure would result in a profile such as in Fig. 12.1.

Note from Fig. 12.1 that the density must decrease in the sliding direction if positive pressures are to be generated. This effect could be induced by raising the lubricant temperature as it flows through the bearing. While theoretically possible, the amount of heat required to make this effect significant is not generally present. For this reason, the density wedge mechanism is not important in most bearings.

Also, note the velocity profiles shown in Fig. 12.1. If the lubricant density decreases in the sliding direction, and the no slip boundary conditions are to be maintained, a changing velocity profile along the length of the bearing must exist as shown in order to satisfy mass continuity.

Stretch Term, $\dfrac{\rho h}{2}\dfrac{\partial(u_a+u_b)}{\partial x}$

As was done for the density wedge discussion, Fig. 12.2 shows a condition where the stretch term and the Poiseuille terms are all that remain in the Reynolds equation, or

$$\frac{\partial}{\partial x}\left(\frac{\rho h^3}{12\eta}\frac{\partial p}{\partial x}\right) = \frac{\rho h}{2}\frac{\partial}{\partial x}(u_a+u_b).$$

Stretch action concerns the rate at which the surface velocity changes in the sliding direction. This effect is produced if the bounding solids are elastic and the extent to which the surfaces are stretched varies through the bearing. For positive pressures to be developed, the surface velocities have to decrease in the sliding direction, as shown in Fig. 12.2. Stretch action is not encountered in conventional bearings, but a stretch action that can cause lubricant film breakdown can occur in some manufacturing operations such as metal

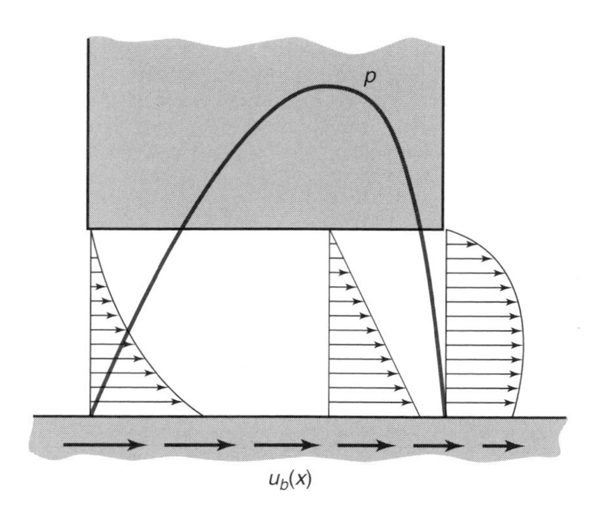

Figure 12.2: Stretch mechanism.

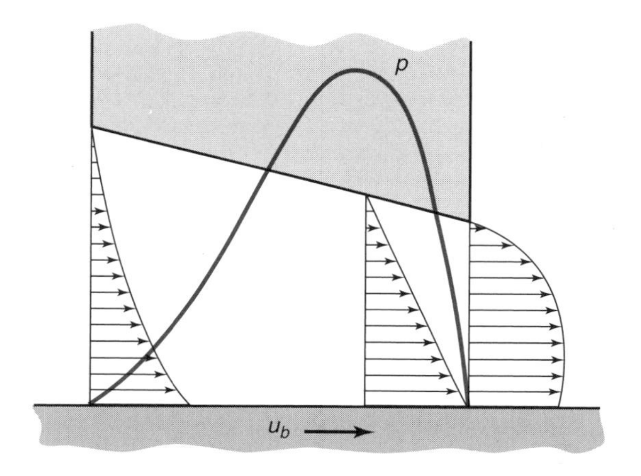

Figure 12.3: Physical wedge mechanism.

rolling or extrusion, where the workpiece increases in speed during the process.

Physical Wedge Term, $\dfrac{\rho(u_a + u_b)}{2}\dfrac{\partial h}{\partial x}$

Physical wedge action is very important in bearings and is the best known mechanism for generating pressure. Figure 12.3 illustrates this action for a plane slider and a stationary bearing pad. For this type of bearing, the Reynolds equation reduces to

$$\frac{\partial}{\partial x}\left(\frac{\rho h^3}{12\eta}\frac{\partial p}{\partial x}\right) = \frac{\rho u_b}{2}\frac{\partial h}{\partial x}.$$

Note that for a positive load-carrying capacity, the thickness of the lubricant film must decrease in the sliding direction. This is quite common, and this equation is the starting point for analyzing many bearing geometries.

Consider the velocity profile in the bearing. At each of the three sections considered, the Couette volume flow rate is proportional to the area of the triangle of height h and base u. Since h varies along the bearing, there is a different Couette flow rate at each section, and flow continuity can be achieved only if the profiles shown are achieved. This, in effect, is a superimposition of a Poiseuille flow.

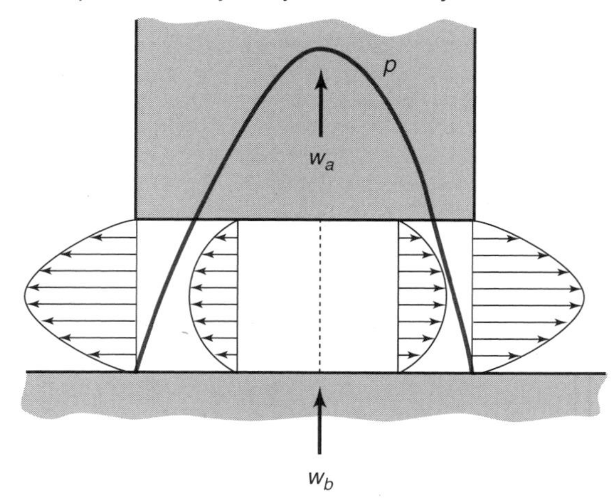

Figure 12.4: Normal squeeze mechanism.

Normal Squeeze Term, $\rho(w_a - w_b)$

The normal squeeze term is present in the case of two surfaces that are approaching each other, as shown in Fig. 12.4. The Reynolds equation for this situation is

$$\frac{\partial}{\partial x}\left(\frac{\rho h^3}{12\eta}\frac{\partial p}{\partial x}\right) = \rho(w_a - w_b).$$

Normal squeeze action provides a valuable cushioning effect when bearing surfaces tend to be pressed together. Positive pressures will be generated when the film thickness is diminishing. The physical wedge and normal squeeze actions are the two major pressure-generating mechanisms in hydrodynamic or self-acting fluid film bearings. In the absence of sliding, the squeeze effect arises directly from the difference in normal velocities $(w_a - w_b)$, as illustrated in Fig. 12.4. The normal squeeze effect is important in bearings such as main piston bearings in automobile engines and in artificial joints, for example.

Translation Squeeze Term, $-\rho u_a \dfrac{\partial h}{\partial x}$

Translation squeeze action results from the motion of inclined surfaces. The local film thickness may be reduced by the sliding of the inclined bearing surface, as shown in Fig. 12.5. If the lower surface is not moving ($u_b = 0$), the Reynolds equation is

$$\frac{\partial}{\partial x}\left(\frac{\rho h^3}{12\eta}\frac{\partial p}{\partial x}\right) = -\rho u_a\frac{\partial h}{\partial x}.$$

In this case, the pressure profile is moving over the space covered by the fixed coordinate system, the pressure at any fixed point being a function of time. Translation squeeze and physical wedge are often confused, and both may be present within a bearing. The physical wedge mechanism is a valuable method of generating a lubricant film; the translation squeeze term can be thought of as a correction. For example, consider the case where the bearing in Fig. 12.5 has $u_a = u_b = u$. For such a condition, the Reynolds equation becomes

$$\frac{\partial}{\partial x}\left(\frac{\rho h^3}{12\eta}\frac{\partial p}{\partial x}\right) = \frac{\rho(u+u)}{2}\frac{\partial h}{\partial x} - \rho u\frac{\partial h}{\partial x} = 0.$$

After application of boundary conditions, such a bearing would have zero pressure throughout, which should be supported by intuition; merely translating a bearing does not generate a lubricant film.

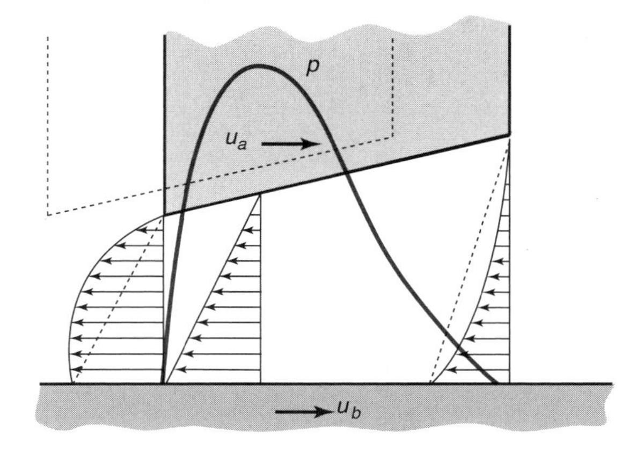

Figure 12.5: Translation squeeze mechanism. Note that the velocity u_b is negative as shown, so that the pressure developed is positive.

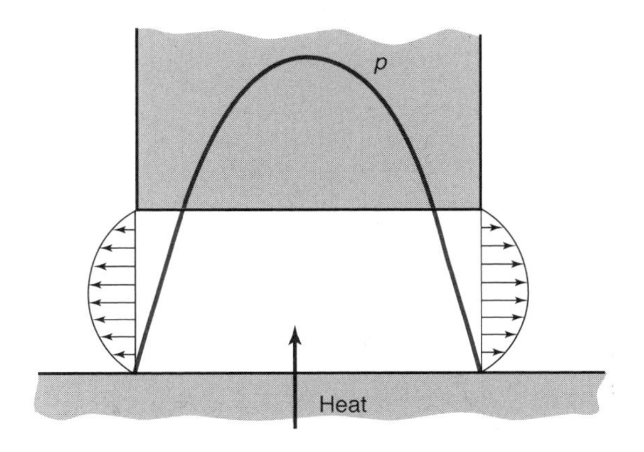

Figure 12.6: Local expansion mechanism.

Local Expansion Term, $h\frac{\partial \rho}{\partial t}$

The local time rate of density change, usually attributed to temperature rise in the lubricant, governs the local expansion term. The pressure-generating mechanism can be visualized by considering the thermal expansion of the lubricant contained between stationary bearing surfaces, as shown in Fig. 12.6. The Reynolds equation for this circumstance is given by

$$\frac{\partial}{\partial x}\left(\frac{\rho h^3}{12\eta}\frac{\partial p}{\partial x}\right) = h\frac{\partial \rho}{\partial t}.$$

If heat is supplied to the lubricant, it will expand and the excess volume will have to be expelled from the space between the bearing surfaces. In the absence of surface velocities, the excess lubricant volume must be expelled by a pressure (Poiseuille) flow action. Pressures are thus generated in the lubricant, and for a positive load-carrying capacity, $\partial\rho/\partial t$ must be negative (i.e., the volume of a given mass of lubricant must increase). Local expansion, which is a transient mechanism of pressure generation, is generally of no significance in bearing analysis, but can explain certain phenomena such as lubrication by an evaporating solid or liquid.

12.2.3 Standard Reduced Forms

For only tangential motion, where $w_a = u_a\partial h/\partial x$ and $w_b = 0$, the Reynolds equation [Eq. (12.21)] becomes

$$\frac{\partial}{\partial x}\left(\frac{\rho h^3}{\eta}\frac{\partial p}{\partial x}\right) + \frac{\partial}{\partial y}\left(\frac{\rho h^3}{\eta}\frac{\partial p}{\partial y}\right) = 12\bar{u}\frac{\partial(\rho h)}{\partial x}, \qquad (12.23)$$

where

$$\bar{u} = \frac{u_a + u_b}{2} = \text{Constant.}$$

This equation is applicable to elastohydrodynamic lubrication. For hydrodynamic lubrication, the fluid properties (density and viscosity) do not vary significantly throughout the bearing and may be considered to be constant. Thus, the corresponding Reynolds equation is

$$\frac{\partial}{\partial x}\left(h^3\frac{\partial p}{\partial x}\right) + \frac{\partial}{\partial y}\left(h^3\frac{\partial p}{\partial y}\right) = 12\bar{u}\eta_o\frac{\partial h}{\partial x}. \qquad (12.24)$$

Equation (12.23) not only allows the fluid properties to vary in the x- and y-directions but also permits the bearing surfaces to be of finite length in the y-direction. Side leakage, or flow in the y-direction, is associated with the second term in Eqs. (12.23) and (12.24). If the pressure in the lubricant film has to be considered as a function of x and y, Eq. (12.23) can rarely be solved analytically.

12.3 Thrust Slider Bearings

The surfaces of **thrust bearings** (see Fig. 12.9) are perpendicular to the axis of rotation. A hydrodynamically lubricated slider bearing develops load-carrying capacity by virtue of the relative motion of the two surfaces separated by a fluid film. A slider bearing may be viewed as a bearing that develops a positive pressure due to the physical wedge mechanism presented in Section 12.2.2. Thrust slider bearings are considered in this section and journal slider bearings are in Section 12.4. The processes occurring in a bearing with fluid film lubrication can be better understood by considering qualitatively the mechanisms whereby pressure is generated.

12.3.1 Mechanism of Pressure Development

An understanding of the development of load-supporting pressures in hydrodynamic bearings can be gleaned by considering, from a purely physical point of view, the conditions of geometry and motion required to develop pressure. An understanding of the physical situation can make the mathematics of hydrodynamic lubrication much more meaningful. By considering only what must happen to maintain continuity of flow, much of what the mathematical equations tell us later in this chapter can be deduced.

Figure 12.7 shows velocity profiles for two plane surfaces separated by a constant lubricant film. The plates are wide so that side-leakage flow (into and out of the paper) can be neglected. The upper plate is moving with a velocity u_a, and the bottom plate is held stationary. No slip occurs at the surfaces; that is, the lubricant "sticks" to the surface and develops the velocity profile as shown. The velocity varies uniformly from zero at surface AB to u_a at surface A′B′, thus implying that the rate of shear du/dz throughout the oil film is constant. Continuity and incompressibility require that the volume of fluid flowing across section AA′ per unit time has to be equal to that flowing across section BB′. The flow crossing the two boundaries results only from velocity gradients, and since the

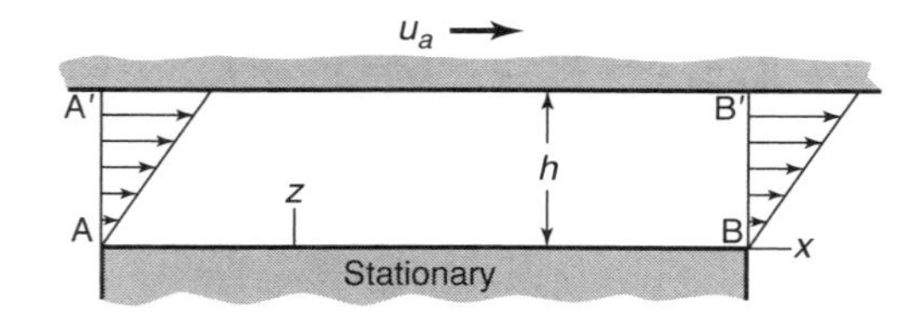

Figure 12.7: Velocity profiles in a parallel-surface slider bearing.

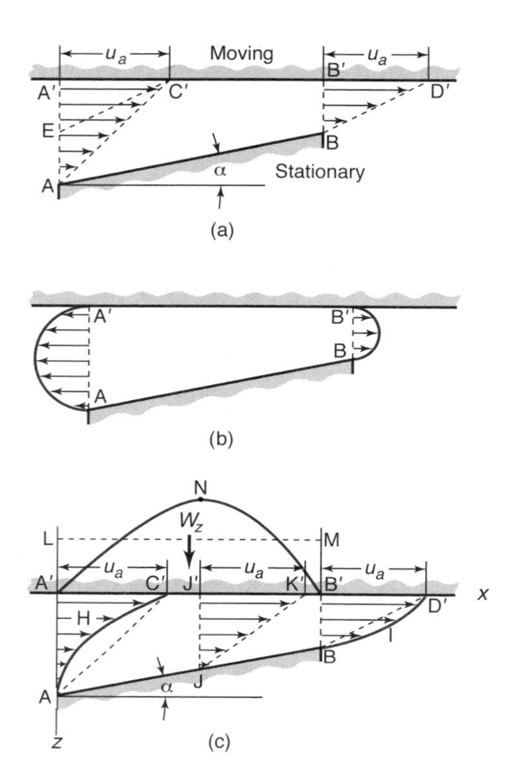

Figure 12.8: Flow within a fixed-incline slider bearing. (a) Couette flow; (b) Poiseuille flow; (c) resulting velocity profile.

gradients are equal, the flow continuity requirement is satisfied without any pressure buildup within the film. Because the ability of a lubricating film to support a load depends on the pressure buildup in the film, a slider bearing with parallel surfaces is not able to support a load. If any load is applied to the surface AB, the lubricant will be squeezed out and the bearing will operate under conditions of boundary lubrication; that is, the load will be directly transferred by surface asperities (see Section 8.7.2).

Consider now the case of two nonparallel plates as shown in Fig. 12.8a. Again, the plates are wide in the direction perpendicular to the motion, so that lubricant flow in this (side) direction is negligibly small. The volume of lubricant that the surface A′B′ tends to carry into the space between the surfaces AB and A′B′ through section AA′ during unit time is the area of triangle AC′A′. The volume of lubricant that the surface tends to discharge from the space through section BB′ during the same time is the area of triangle BD′B′. Because the distance AA′ is greater than the distance BB′, the volume AC′A′ is greater than the volume BD′B′ by the volume AEC′. From flow continuity, the actual volume of lubricant carried into the space must equal the volume discharged.

It can be surmised that there will be a pressure buildup in the lubricating film until flow continuity is satisfied. Figure 12.8b shows the velocity profiles due to Poiseuille flow. The flow is outward from both the leading and trailing edges of the bearing because flow is always from a region of higher pressure to a region of lower pressure. The pressure flow at boundary AA′ opposes the velocity flow, but the pressure flow at BB′ is in the same direction as the velocity flow.

Figure 12.8c shows the results of superimposing Couette and Poiseuille flows. The form of the velocity distribution curves obtained in this way must satisfy the condition that the flow rate through section AA′ equals the flow rate through section BB′. Therefore, the area AHC′A′ must equal the area BID′B′. The area between the straight, dashed line AC′ and the curve AHC′ in section AA′ and the area between the dashed line BD′ and the curve BID′ represent the pressure-induced flow through these areas.

The pressure is maximum in section JJ′, somewhere between sections AA′ and BB′. There is no Poiseuille flow contribution in section JJ′, since the pressure gradient is zero at this location, and all the flow is Couette. Flow continuity is satisfied in that triangle JK′J′ is equal to the areas AHC′A′ and BID′B′.

12.3.2 General Thrust Slider Bearing Theory

Solutions of the Reynolds equation presented in Section 12.2 for actual bearing configurations are usually obtained in approximate numerical form; analytical solutions are possible only for the simplest problems. However, by restricting the flow to two dimensions (say the x-z plane) analytical solutions for many common bearings become available. The quantitative value of these solutions is limited, since space limitations restrict bearing width and thus flow in the third dimension, known as **side leakage**, plays an important part in bearing performance. The two-dimensional solutions have a definite value because they provide a good deal of information about the general characteristics of bearings, information that leads to a clear physical picture and intuition about their performance. The focus of this text is on design and thus this topic will be only briefly discussed, but a detailed discussion can be found in Hamrock et al. [2004]. Following this brief discussion, the remainder of the chapter will describe the relevant equations used for finite-width bearings, present numerical solutions, and the useful design charts that result.

Although side-leakage effects are considered in the results shown in this section, a simplification is to neglect the pressure and temperature effects on the lubricant viscosity and density. Lubricant viscosity is particularly sensitive to temperature (as demonstrated in Table 8.6 and Fig. 8.13), and since the heat generated in hydrodynamic bearings is often considerable, the limitation imposed by this simplification is at once apparent. The temperature rise within the film can be calculated if it is assumed that all the heat produced by viscous shearing is carried away by the lubricant, known as the **adiabatic assumption**. The temperature rise can be accounted for and reasonable results obtained for hydrodynamic bearings. For the more demanding lubrication condition of rolling element bearings (Ch. 13), this approach cannot be safely attempted.

For finite-width bearings having hydrodynamic pressure generated within the lubricant film due to the physical wedge mechanism (see Section 12.2.2), some flow will take place in the y-direction. Assuming the heat produced in the bearing is carried away by the lubricant, the viscosity and density can be considered constant within the x-y plane. The appropriate Reynolds equation from Eq. (12.24) when the ve-

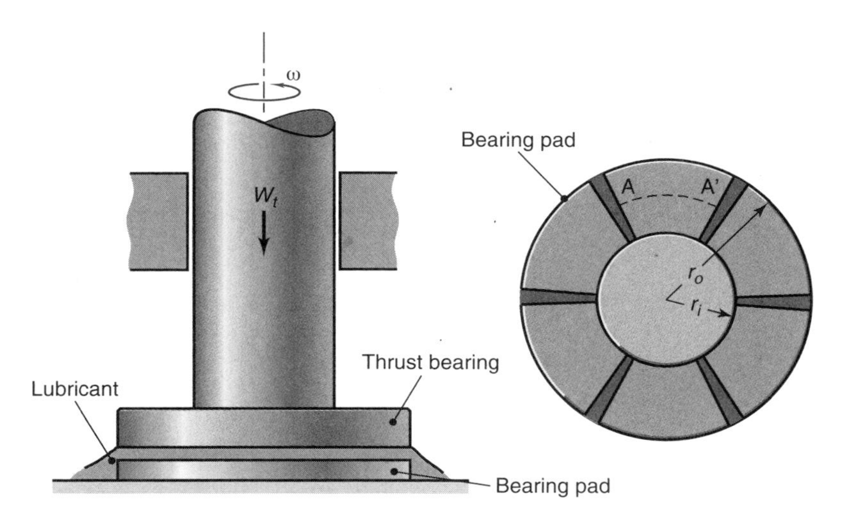

Figure 12.9: Thrust slider bearing geometry.

locity of the top surface is zero ($u_a = 0$) is

$$\frac{\partial}{\partial x}\left(h^3\frac{\partial p}{\partial x}\right)+\frac{\partial}{\partial y}\left(h^3\frac{\partial p}{\partial y}\right)=6\eta_o u_b\frac{\partial h}{\partial x}. \tag{12.25}$$

Thus, for a specific film shape $h(x,y)$, the pressure throughout the bearing can be obtained numerically. Knowing the pressure from the Reynolds equation and integrating over the x-y plane produces the load-carrying capacity.

12.3.3 Hydrodynamic Thrust Bearings - Neglecting Side Leakage

The loads carried by rotary machinery typically have components that act in the direction of the shaft's axis of rotation. These thrust loads are frequently carried by self-acting or hydrodynamic bearings of the form shown in Fig. 12.9. A thrust plate attached to, or forming part of, the rotating shaft is separated from the sector-shaped bearing pads by a lubricant film. The load-carrying capacity of the bearing arises entirely from the pressures generated by the geometry of the thrust plate over the bearing pads. The x-direction (direction of motion) is along a mean arc AA′, as shown in Fig. 12.9. The transverse direction (side-leakage direction) would be perpendicular to arc AA′. All pads are identical and separated by deep lubrication grooves, implying that atmospheric or ambient pressure exists completely around the pad. Thus, the total thrust load-carrying capacity is simply the number of pads multiplied by the load-carrying capacity of an individual pad.

This section will consider three different slider bearings while neglecting side leakage: parallel slider, fixed-incline slider, and parallel-step slider. These three slider bearings will have three different film shapes across arc AA′ in Fig. 12.9.

If side leakage is neglected, Eq. (12.25) reduces to

$$\frac{\partial}{\partial x}\left(h^3\frac{\partial p}{\partial x}\right)=6\eta_o u_b\frac{\partial h}{\partial x}. \tag{12.26}$$

This, then, is the Reynolds equation used for the three types of slider bearing being considered.

Parallel-Surface Slider Bearing

The film thickness for the parallel-surface slider bearing is constant across the arc length AA′ shown in Fig. 12.9. Therefore, Eq. (12.26) becomes

$$\frac{d}{dx}\left(\frac{dp}{dx}\right)=0.$$

Integrating twice gives

$$p=C_1x+C_2,$$

where C_1 and C_2 are integration constants. The boundary conditions are:

1. $p=0$ at $x=0$, so that $C_2=0$
2. $p=0$ at $x=l$, which leads to $C_1=0$

Therefore, the pressure is zero throughout the bearing. The normal load-carrying capacity of the bearing is the integral of the pressure across the length, and since the pressure is zero across the length, the normal applied load would also be zero. This is consistent with the observations in Section 12.3.1, but now confirmed mathematically.

Fixed-Incline Slider Bearing

The film shape expression for the fixed-incline slider bearing is

$$h=h_o+s_h\left(1-\frac{x}{l}\right). \tag{12.27}$$

Note that when $x=0$, $h=h_o+s_h$; and when $x=l$, $h=h_o$.

Equation (12.26) can be integrated to give

$$\frac{dp}{dx}=\frac{6\eta_o u_b}{h^2}+\frac{C_1}{h^3},$$

where C_1 is an integration constant. The boundary condition to be applied is

$$\text{at } h=h_m,\ \frac{dp}{dx}=0, \tag{12.28}$$

which implies that $C_1=-6h_o u_b h_m$, or

$$\frac{dp}{dx}=\frac{6\eta_o u_b}{h^3}(h-h_m). \tag{12.29}$$

This equation can be written in dimensionless form by defining

$$P=\frac{ps_h^2}{\eta_o u_b l},\qquad H=\frac{h}{s_h},\qquad H_m=\frac{h_m}{s_h},$$

$$H_o = \frac{h_o}{s_h}, \qquad X = \frac{x}{l}, \tag{12.30}$$

so that Eq. (12.29) becomes

$$\frac{dP}{dX} = 6\left(\frac{H - H_m}{H^3}\right), \tag{12.31}$$

where

$$H = \frac{h}{s_h} = H_o + 1 - X, \tag{12.32}$$

and therefore

$$\frac{dH}{dX} = -1 \text{ or } dH = -dX. \tag{12.33}$$

Note that Eqs. (12.32) and (12.33) are valid only for a fixed-incline bearing. Integrating Eq. (12.31) gives

$$P = 6\int\left(\frac{1}{H^2} - \frac{H_m}{H^3}\right) dX = -6\int\left(\frac{1}{H^2} - \frac{H_m}{H^3}\right) dH.$$

Therefore,

$$P = 6\left(\frac{1}{H} - \frac{H_m}{2H^2}\right) + C_2. \tag{12.34}$$

The boundary conditions are:

1. $P = 0$ when $X = 0$, which is the location where $H = H_o + 1$
2. $P = 0$ when $X = 1$, or where $H = H_o$

Making use of these boundary conditions gives

$$H_m = \frac{2H_o(1 + H_o)}{1 + 2H_o}, \tag{12.35}$$

$$C_2 = -\frac{6}{1 + 2H_o}. \tag{12.36}$$

Substituting Eqs. (12.35) and (12.36) into Eq. (12.34) gives

$$P = \frac{6X(1 - X)}{(H_o + 1 - X)^2(1 + 2H_o)}. \tag{12.37}$$

The normal applied load per unit width is

$$W_z' = \frac{W_z}{w_t} = \int_0^l p\,dx = \frac{\eta_o u_b l^2}{s_h^2}\int_0^1 P\,dX.$$

The dimensionless normal load is

$$\bar{W}_z = \frac{W_z s_h^2}{\eta_o u_b l^2 w_t} = \int_0^1 P\,dX. \tag{12.38}$$

Recalling that $dH = -dX$, and from Eq. (12.32) note that if $X = 0$, then $H = H_o + 1$ and if $X = 1$, then $H = H_o$. Thus, Eq. (12.38) becomes

$$\bar{W}_z = \frac{W_z s_h^2}{\eta_o u_b l^2 w_t} = \int_{H_o+1}^{H_o} -P\,dH. \tag{12.39}$$

Substituting Eqs. (12.34) and (12.36) into Eq. (12.39) gives

$$\bar{W}_z = \frac{W_z s_h^2}{\eta_o u_b l^2 w_t} = 6\ln\left(\frac{H_o + 1}{H_o}\right) - \frac{12}{1 + 2H_o}. \tag{12.40}$$

Parallel-Step Slider Bearing

For the parallel-step slider bearing, the film thickness in each of the two regions is constant, implying that the appropriate Reynolds equation is

$$\frac{d^2p}{dx^2} = 0.$$

Integrating gives

$$\frac{dp}{dx} = \text{constant}.$$

However, since the film thickness between regions differs, the pressure gradients also differ. Thus, there will be no discontinuity at the step.

$$p_{\max} = n_s l\left(\frac{dp}{dx}\right)_i = -(l - n_s l)\left(\frac{dp}{dx}\right)_o. \tag{12.41}$$

The inlet and outlet flow rates at the step must be the same, or $q'_{xo} = q'_{xi}$. Making use of Eq. (12.12) gives

$$-\frac{h_o^3}{12\eta_o}\left(\frac{dp}{dx}\right)_o + \frac{u_b h_o}{2} = -\frac{(h_o + s_h)^3}{12\eta_o}\left(\frac{dp}{dx}\right)_i + \frac{u_b(h_o + s_h)}{2} \tag{12.42}$$

Equations (12.41) and (12.42) represent a pair of simultaneous equations with unknowns $(dp/dx)_o$ and $(dp/dx)_i$. The solutions are

$$\left(\frac{dp}{dx}\right)_i = \frac{6\eta_o u_b(1 - n_s)s_h}{(1 - n_s)(h_o + s_h)^3 + n_s h_o^3}, \tag{12.43}$$

$$\left(\frac{dp}{dx}\right)_o = -\frac{6\eta_o u_b n_s s_h}{(1 - n_s)(h_o + s_h)^3 + n_s h_o^3}. \tag{12.44}$$

The maximum pressure at the step can be found by substituting Eqs. (12.43) and (12.44) into Eq. (12.41) to give

$$p_{\max} = \frac{6\eta_o u_b l n_s(1 - n_s)s_h}{(1 - n_s)(h_o + s_h)^3 + n_s h_o^3} \tag{12.45}$$

or, in dimensionless form using Eq. (12.30),

$$P_{\max} = \frac{p_{\max} s_h^2}{\eta_o u_b l} = \frac{6n_s(1 - n_s)}{(1 - n_s)(H_o + 1)^3 + n_s H_o^3}. \tag{12.46}$$

The pressures in the inlet and outlet regions are, for $0 \le X \le n_s$,

$$P_i = \frac{XP_{\max}}{n_s} = \frac{6X(1 - n_s)}{(1 - n_s)(H_o + 1)^3 + n_s H_o^3}, \tag{12.47}$$

and for $n_s \le X \le 1$,

$$P_o = \frac{(1 - X)P_{\max}}{1 - n_s} = \frac{6(1 - X)n_s}{(1 - n_s)(H_o + 1)^3 + n_s H_o^3}. \tag{12.48}$$

The normal applied load per unit width is

$$\frac{W_z}{w_t} = \frac{p_{\max} l}{2} = \frac{3\eta_o u_b l^2 n_s(1 - n_s)s_h}{(1 - n_s)(h_o + s_h)^3 + n_s h_o^3}. \tag{12.49}$$

The dimensionless normal applied load is

$$\bar{W}_z = \frac{W_z}{\eta_o u_b w_t}\left(\frac{s_h}{l}\right)^2 = \frac{3n_s(1 - n_s)}{(1 - n_s)(H_o + 1)^3 + n_s H_o^3} = \frac{P_{\max}}{2}. \tag{12.50}$$

The largest $\bar{W}_z$ is obtained when $n_s = 0.7182$ and $H_o = 0.5820$.

Example 12.2: Pressure in Thrust Bearings

Given: A fixed-incline self-acting thrust bearing has a pad width much larger than the length. The viscosity of the lubricant is 0.01 Ns/m^2, the sliding velocity is 10 m/s, the pad length is 0.3 m, the minimum film thickness is 15 μm, and the inlet film thickness is twice the outlet film thickness.

Find: The magnitude and location of the maximum pressure in the bearing.

Solution: If the width of the pad is much larger than the length, side leakage can be neglected and Eqs. (12.35) and (12.37) apply. Also, the pressure is a maximum when $dP/dX = 0$, $H = H_m$, and $X = X_m$. It is also given that $H_o = 1$. From Eq. (12.35),

$$H_m = \frac{2H_o(1+H_o)}{1+2H_o} = \frac{(2)(2)}{3} = \frac{4}{3}.$$

From Eq. (12.32),

$$X_m = H_o + 1 - H_m = 1 + 1 - \frac{4}{3} = \frac{2}{3},$$

and from Eq. (12.37),

$$\begin{aligned} P_m &= \frac{6X_m(1-X_m)}{(H_o+1-X_m)^2(1+2H_o)} \\ &= \frac{6\left(\frac{2}{3}\right)\left(1-\frac{2}{3}\right)}{\left(1+1-\frac{2}{3}\right)^2(1+2)} \\ &= \frac{\left(\frac{4}{3}\right)}{\left(\frac{4}{3}\right)^2(3)} = \frac{1}{4}. \end{aligned}$$

Thus,

$$P_m = \frac{p_m s_h^2}{\eta_o u_b l} = \frac{1}{4}, \qquad \text{or} \qquad p_m = \frac{\eta_o u_b l}{4 s_h^2}.$$

Given that $\eta_o = 0.01$ Ns/m^2, $u_b = 10$ m/s, $l = 0.3$ m, and $s_h = h_o = 15$ μm,

$$p_m = \frac{(0.01)(10)(0.3)}{4(15)^2(10^{-12})} = 33.3 \text{ MPa}.$$

12.3.4 Operating and Performance Parameters

Section 12.3.3 neglected side-leakage effects in the design of slider bearings; these effects will now be considered. The following general definitions, related to the bearing geometry shown in Fig. 12.10, and relationships are employed: The forces acting on the bearing surfaces of an individual pad can be considered in two groups. The loads, which act in the direction normal to the surface, yield normal loads that can be resolved into components W_x and W_z. The viscous surface stresses, which act in the direction tangent to the surface, yield shear forces on the solids that have components F in the x-direction. The shear forces in the z-direction are negligible. Once the pressure is obtained from the Reynolds equation [Eq. (12.25)], the force components acting on the bearing surfaces are:

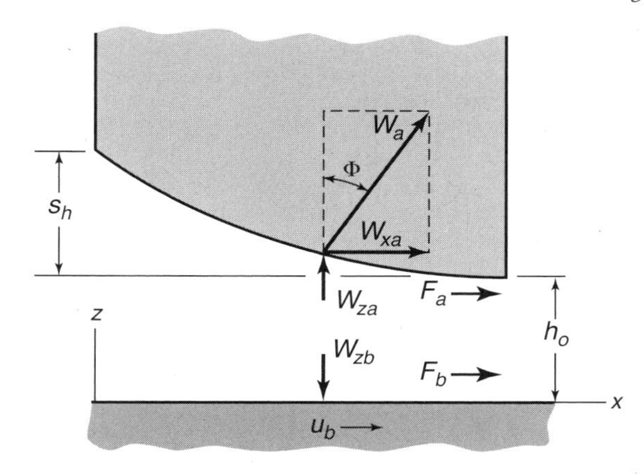

Figure 12.10: Force components and oil film geometry in a hydrodynamically lubricated thrust slider bearing.

$$W_{za} = W_{zb} = \int_0^{w_t}\int_0^{l} p\,dx\,dy, \tag{12.51}$$

$$W_{xb} = 0, \tag{12.52}$$

$$\begin{aligned} W_{xa} &= -\int_0^{w_t}\int_{h_o+s_h}^{h_o} p\,dh\,dy = -\int_0^{w_t}\int_0^{l} p\frac{dh}{dx}\,dx\,dy \\ &= -\int_0^{w_t}(ph)_o^l\,dy + \int_0^{w_t}\int_0^{l} h\frac{dp}{dx}\,dx\,dy \\ &= \int_0^{w_t}\int_0^{l} h\frac{dp}{dx}\,dx\,dy, \end{aligned} \tag{12.53}$$

$$W_b = \sqrt{W_{zb}^2 + W_{xb}^2} = W_{zb}, \tag{12.54}$$

$$W_a = \sqrt{W_{za}^2 + W_{xa}^2}, \tag{12.55}$$

$$\Phi = \tan^{-1}\left(\frac{W_{xa}}{W_{za}}\right). \tag{12.56}$$

The shear force acting on solid b is

$$F_b = \int_0^{w_t}\int_0^{l} (\tau_{zx})_{z=0}\,dx\,dy, \tag{12.57}$$

but

$$(\tau_{zx})_{z=0} = \left(\eta\frac{du}{dz}\right)_{z=0}. \tag{12.58}$$

Substituting Eq. (12.8) into Eq. (12.58) gives

$$(\tau_{zx})_{z=0} = -\frac{h}{2}\frac{dp}{dx} - \frac{\eta u_b}{h}. \tag{12.59}$$

Substituting Eq. (12.59) into Eq. (12.57) gives

$$F_b = \int_0^{w_t}\int_0^{l}\left(-\frac{h}{2}\frac{dp}{dx} - \frac{\eta u_b}{h}\right)dx\,dy.$$

Making use of Eq. (12.53) gives

$$F_b = -\frac{W_{xa}}{2} - \int_0^{w_t}\int_0^{l} \frac{\eta u_b}{h}\, dx\, dy. \quad (12.60)$$

Similarly, the shear force acting on solid a is

$$F_a = \int_0^{w_t}\int_0^{l} (\tau_{zx})_{z=h}\, dx\, dy = -\frac{W_{xa}}{2} + \int_0^{w_t}\int_0^{l} \frac{\eta u_b}{h}\, dx\, dy. \quad (12.61)$$

Note from Fig. 12.10 that

$$F_b - F_a + W_{xa} = 0, \quad (12.62)$$

$$W_{zb} - W_{za} = 0. \quad (12.63)$$

These equations represent the condition of static equilibrium.

The viscous stresses generated by the shearing of the lubricant film give rise to a resisting force of magnitude $-F_b$ on the moving surface. The rate of working against the viscous stresses, or the power loss, for one pad is

$$h_p = -F_b u_b. \quad (12.64)$$

The work done against the viscous stresses appears as heat within the lubricant. Some of this heat may be transferred to the surroundings by conduction, or it may be convected from the clearance space by lubricant flow.

The bulk temperature rise of the lubricant for the case in which all the heat is carried away by convection is known as the **adiabatic temperature rise**. The temperature increase can be calculated by equating the rate of heat generated within the lubricant to the rate of heat transferred by convection, or

$$h_p = \rho q_x C_p (\Delta t_m),$$

so that the adiabatic temperature rise may be expressed as

$$\Delta t_m = \frac{h_p}{\rho q_x C_p}, \quad (12.65)$$

where ρ is the density, q_x is volume flow rate in x-direction (direction of motion), and C_p is the lubricant's specific heat. Equations (12.51) through (12.65) are among the relevant equations used in thrust bearing analysis. Recall that the pressure-generating mechanism is the physical wedge, as discussed in Section 12.2.2. The general thrust analysis will now be used to design fixed-incline bearings. A detailed numerical evaluation will not be made of the Reynolds equation and its load and friction force components, since the emphasis of this text is on design. If the reader is interested, two references are recommended. The original source is Raimondi and Boyd [1955]; a more recent text is Hamrock et al. [2004].

The operating parameters affecting the pressure generation, film thickness, load, and friction components are:

1. **Bearing number** for a thrust slider bearing

$$B_t = \left(\frac{\eta_o u_b w_t}{W_z}\right)\left(\frac{l}{s_h}\right)^2. \quad (12.66)$$

2. **Length-to-width ratio**

$$\lambda = \frac{l}{w_t}. \quad (12.67)$$

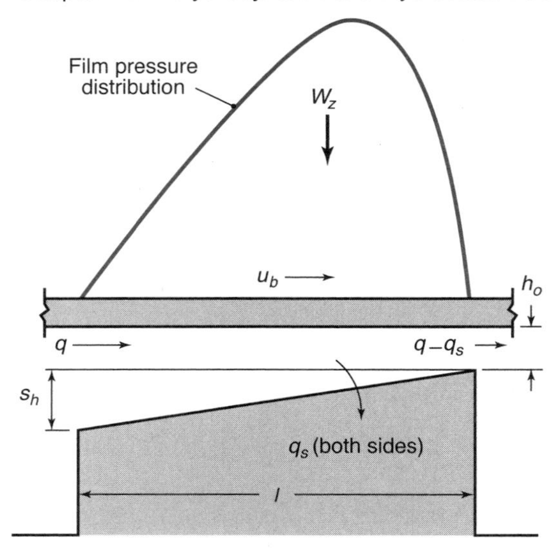

Figure 12.11: Side view of fixed-incline slider bearing.

3. Location of step from inlet, n_s.

For a fixed-incline bearing, only the first two operating parameters affect the pressure generation, film thickness, load, and friction force components. The operating parameters are viewed as the parameters given in the design of a hydrodynamic thrust bearing. The performance parameters that will result from choosing a given set of operating parameters are:

1. Outlet (minimum) film thickness, h_o
2. Temperature rise due to lubricant shearing, Δt_m
3. Power loss, h_p
4. Coefficient of sliding friction, μ
5. Circumferential and side-leakage volumetric flow rates, q and q_s

These performance parameters are needed to evaluate the design of a hydrodynamic thrust slider bearing.

12.3.5 Fixed-Incline Slider Bearing

The simplest form of fixed-incline bearing provides only straight-line motion and consists of a flat surface sliding over a fixed pad or land having a profile similar to that shown in Fig. 12.11. The fixed-incline bearing operation depends on the lubricant being drawn into a wedge-shaped space and, thus, producing pressure that counteracts the load and prevents contact between the sliding parts. Since the wedge action takes place only when the sliding surface moves in the direction in which the lubricant film converges, the fixed-incline bearing (Fig. 12.11) can carry load only in this direction. If reversibility is desired, a combination of two or more pads with their surfaces sloped in opposite directions is required, or else a pivoting pad that changes inclination is needed. Fixed-incline pads are used in multiples as in the thrust bearing shown in Fig. 12.12.

Figures 12.13 through 12.16 present solutions of the fixed-incline slider bearing based on numerical analysis. Design Procedure 12.1 outlines their effective use in analysis and

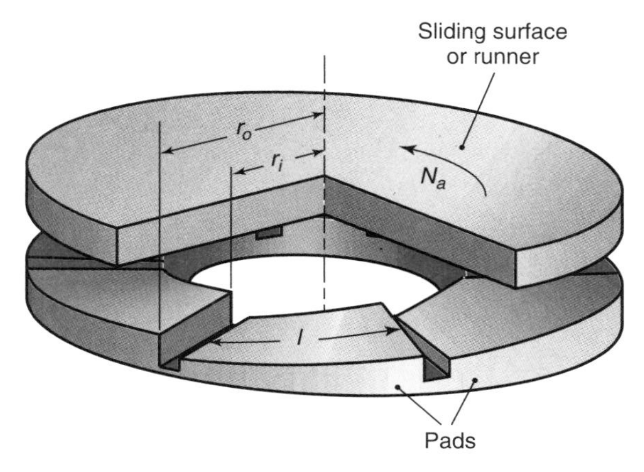

Figure 12.12: Configuration of multiple fixed-incline thrust slider bearing.

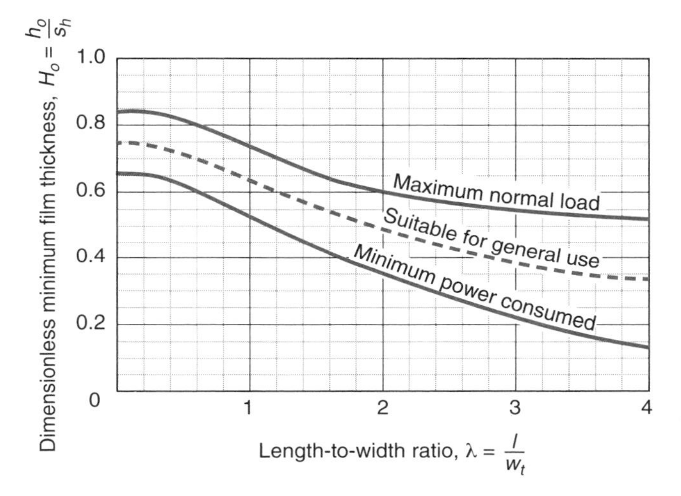

Figure 12.13: Chart for determining minimum film thickness corresponding to maximum load or minimum power-loss for various pad proportions in fixed-incline bearings. *Source:* Adapted from Raimondi and Boyd [1955].

design. The availability of such numerical solutions greatly simplifies the design of hydrodynamic bearings.

Design Procedure 12.1: Fixed-Incline Thrust Bearings

1. Choose a pad length-to-width ratio; a square pad ($\lambda = 1$) is generally considered to give good performance. If it is known whether maximum load or minimum power loss is more important in a particular application, the outlet film thickness ratio H_o can be determined from Fig. 12.13. If neither condition is a constraint, a value between these extremes is generally advisable.

2. Once λ and H_o are known, Fig. 12.14 can be used to obtain the bearing number B_t.

3. From Fig. 12.15, determine the temperature rise due to viscous shear-induced heating for a given λ and B_t. The volumetric specific heat $C_s = \rho C_p$, which is the dimensionless temperature rise parameter, is relatively constant for mineral oils and is approximately 1.36×10^6 N/(m$^2\cdot$°C).

4. Determine the lubricant temperature. The mean temperature can be expressed as

$$t_m = t_{mi} + \frac{\Delta t_m}{2}, \tag{12.68}$$

where t_{mi} is the inlet temperature and is usually known beforehand. Once the mean temperature, t_m, is known, it can be used in Fig. 8.13 to determine the viscosity of SAE oils, or Fig. 8.13 or Table 8.8 can be used.

5. Use Eqs. (12.32) and (12.66) to determine the outlet (minimum) film thickness, h_o, as

$$h_o = H_o l \sqrt{\frac{\eta_o u_b w_t}{W_z B_t}}. \tag{12.69}$$

Once the outlet film thickness is known, the shoulder height, s_h, can be directly obtained from $s_h = h_o/H_o$. If the outlet film thickness is specified and either the velocity, u_b, or the normal applied load, W_z, is not known, Eq. (12.69) can be rewritten to establish u_b or W_z.

6. Check Table 12.1 to see if the outlet (minimum) film thickness is sufficient for the surface finish as manufactured. If the result from Eq. (12.69) is greater than the recommendations in Table 12.1, go to step 7. Otherwise, consider one or both of the following steps:

   (a) Increase the bearing speed.

   (b) Decrease the load, improve the surface finish, choose a more viscous lubricant, or reduce the inlet temperature. Upon making this change, return to step 3.

7. Evaluate the remaining performance parameters. Once an adequate minimum film thickness and a proper lubricant temperature have been determined, the performance parameters can be evaluated. Specifically, the power loss, the coefficient of friction, and the total and side flows can be determined from Fig. 12.16.

Example 12.3: Analysis of a Fixed-Incline Thrust Bearing

Given: A fixed-incline slider thrust bearing has the following operating parameters: $W_z = 16$ kN, $u_b = 30$ m/s, $w_t = l = 75$ mm, SAE 10 oil, and $t_{mi} = 40$°C.

Find: For the maximum load condition, determine the following performance parameters: s_h, h_o, Δt_m, μ, h_p, q, and q_s.

Solution: From Fig 12.13, for $\lambda = l/w_t = 75/75 = 1$ and maximum normal load, the outlet film thickness ratio H_o = 0.73. Therefore, from Fig. 12.14, the bearing number B_t = 8.0, and from Fig. 12.15,

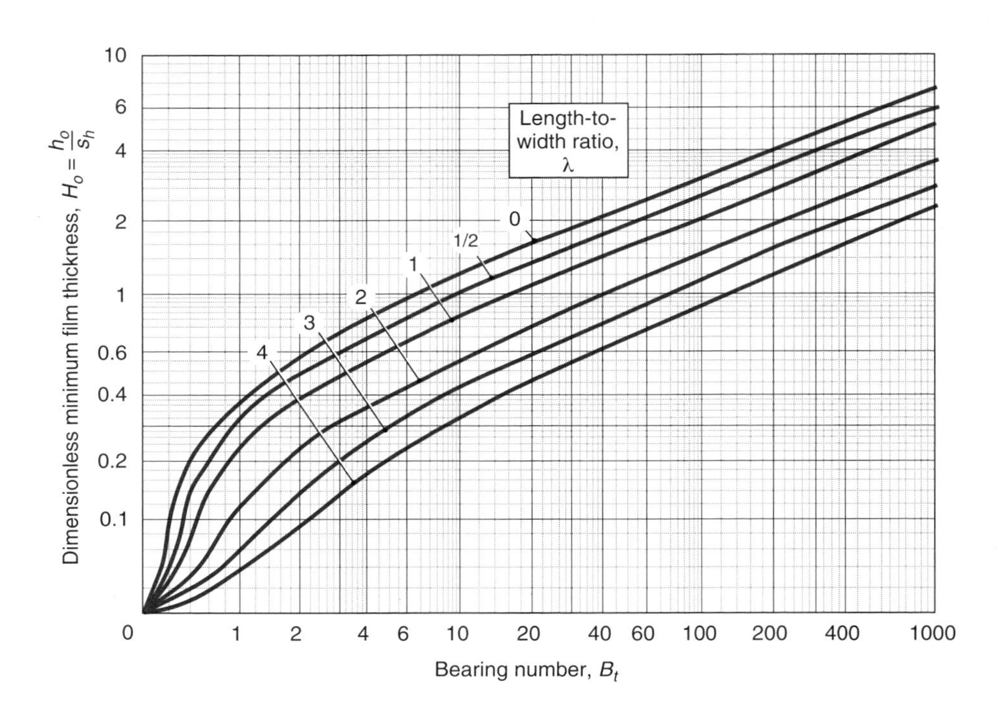

Figure 12.14: Chart for determining minimum film thickness for fixed-incline thrust bearings. *Source:* Adapted from Raimondi and Boyd [1955].

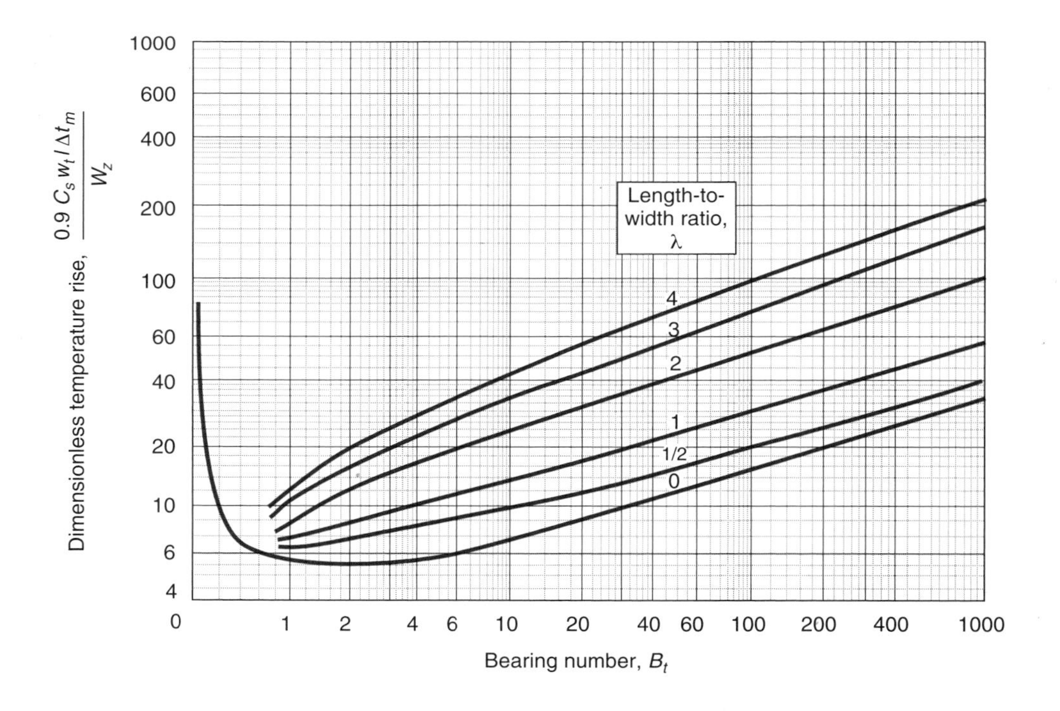

Figure 12.15: Chart for determining dimensionless temperature rise due to viscous shear heating of lubricant in fixed-incline thrust bearings. *Source:* Adapted from Raimondi and Boyd [1955].

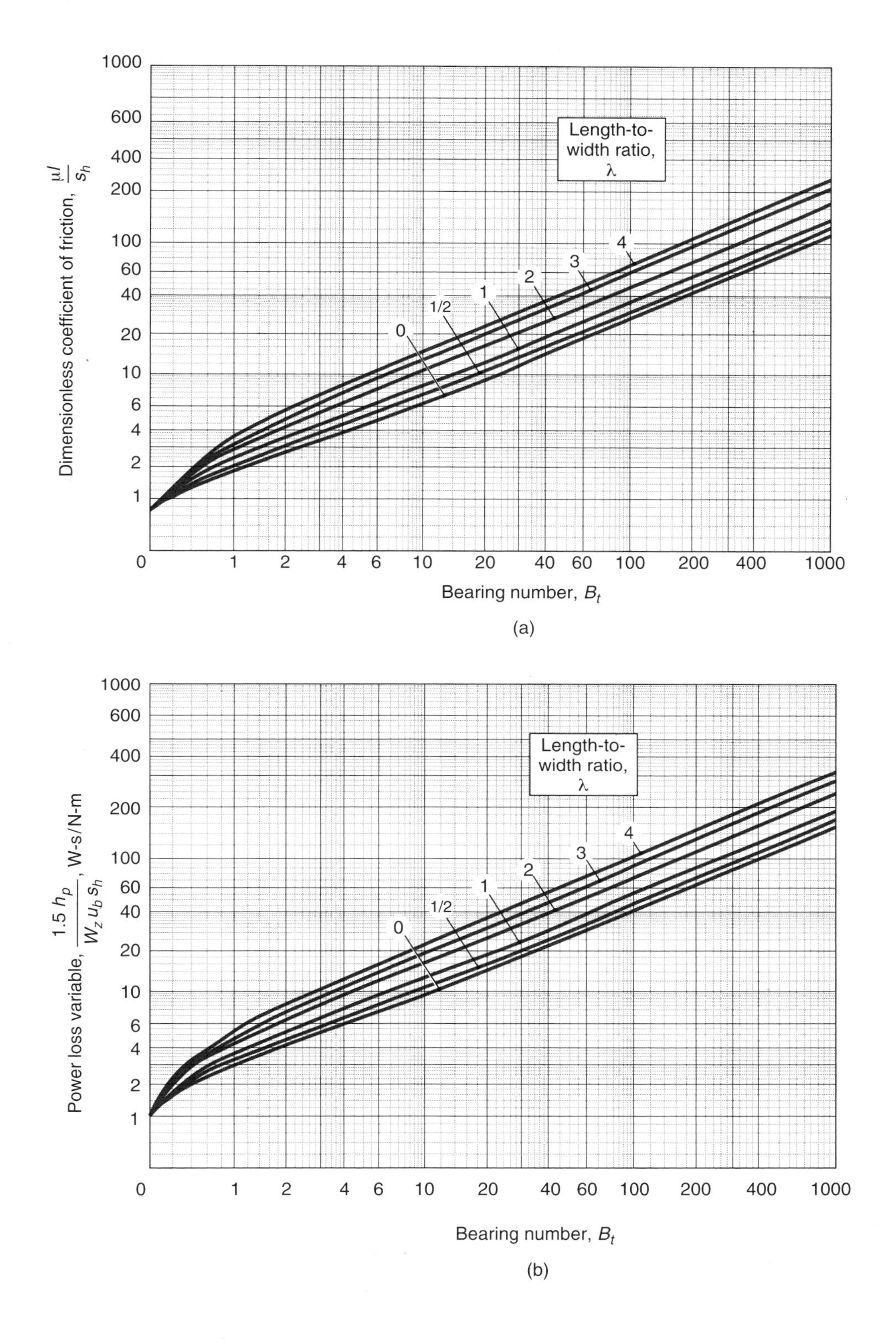

Figure 12.16: Chart for determining performance parameters of fixed-incline thrust bearings. (a) Friction coefficient; (b) power loss.

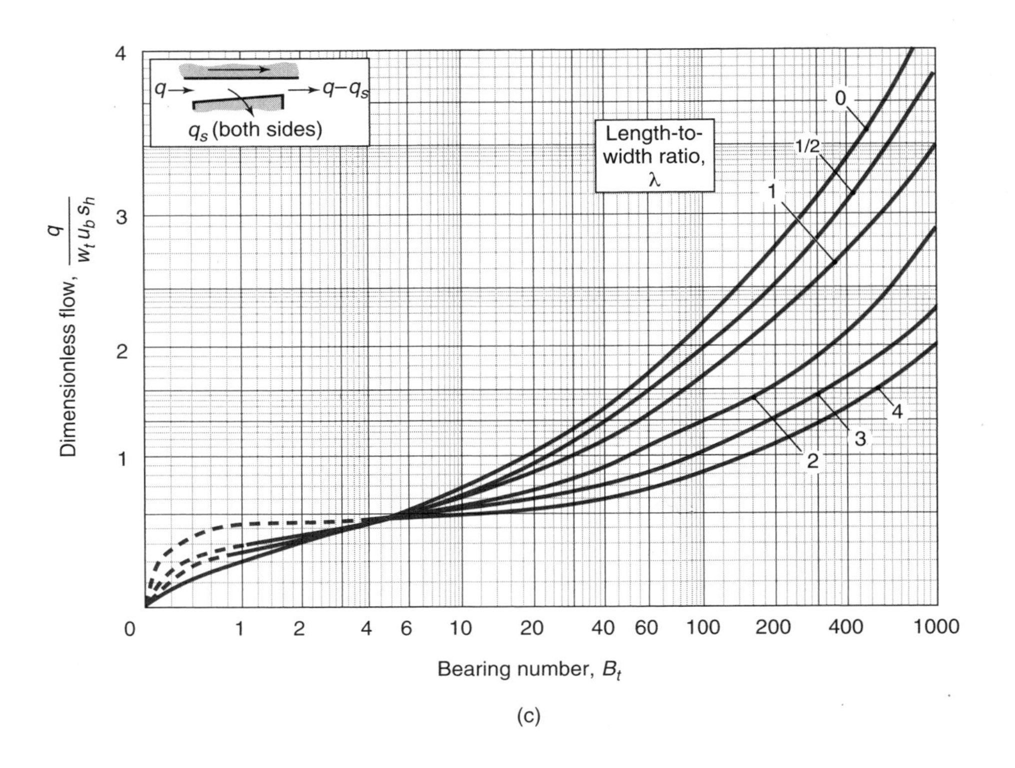

(c)

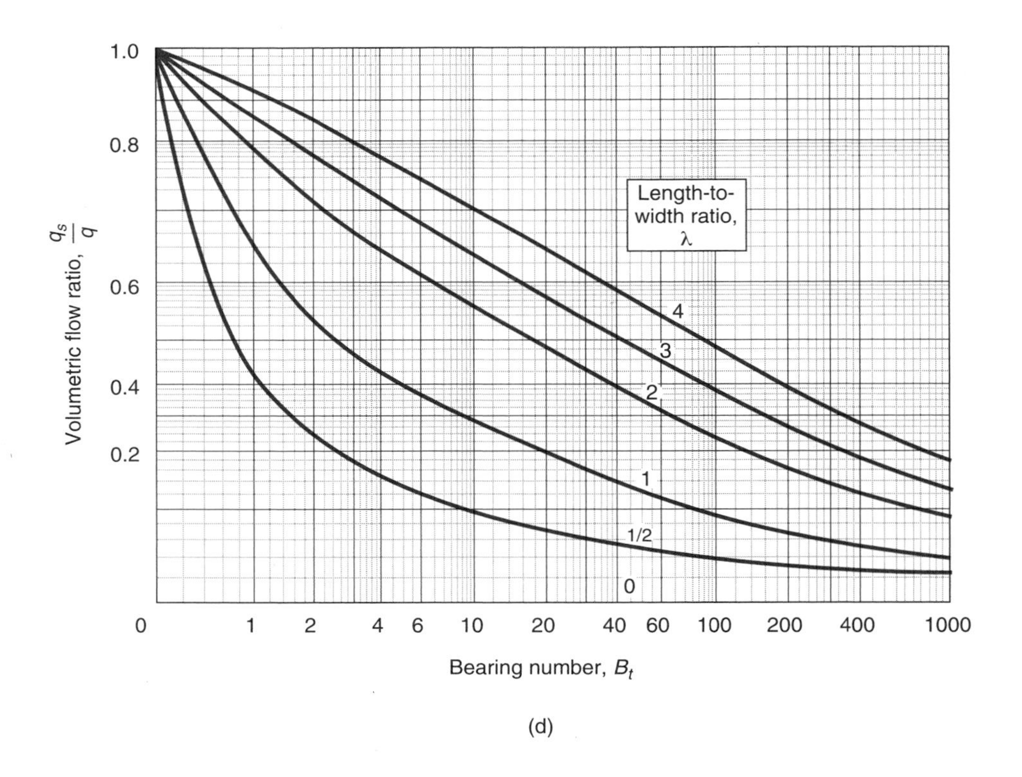

(d)

Figure 12.16: (continued) (c) Lubricant flow; (d) lubricant side flow. *Source:* Adapted from Raimondi and Boyd [1955].

Table 12.1: Allowable outlet (minimum) film thickness for a given surface finish.

Surface finish (centerline average), R_a μm	μin.	Description of surface	Examples of manufacturing methods	Approximate relative costs	Allowable outlet (minimum) film thickness[a], h_o μm	μin.
0.1–0.2	4–8	Mirror-like surface without tool marks; close tolerances	Grind, lap, and superfinish	17–20	2.5	100
0.2–0.4	8–16	Smooth surface without scratches; close tolerances	Grind and lap	17–20	6.2	250
0.4–0.8	16–32	Smooth surfaces; close tolerances	Grind, file, and lap	10	12.5	500
0.8–1.6	32–63	Accurate bearing surface withouttool marks	Grind, precision mill, and file	7	25	1000
1.6–3.2	63–125	Smooth surface without objectionable tool marks, moderate tolerances	Shape, mill, grind, and turn	5	50	2000

[a] The values of film thickness are given only for guidance. They indicate the film thickness required to avoid metal-to-metal contact under clean oil conditions with no misalignment. It may be necessary to take a larger film thickness than that indicated (e.g., to obtain an acceptable temperature rise).

$$\frac{0.9\rho C_p l w_t \Delta t_m}{W_z} = 13.$$

Substituting the known parameters and noting the fact that

$$C_s = \rho C_p = 1.36 \times 10^6 \text{ N/(m}^2\text{-}^\circ\text{C)},$$

$$\begin{aligned} \Delta t_m &= \frac{13W_z}{0.9\rho C_p l w_t} \\ &= \frac{13\left(16 \times 10^3\right)}{(0.9)\left(1.36 \times 10^6\right)(0.075)(0.075)} \\ &= 30.21^\circ\text{C}. \end{aligned}$$

From Eq. (12.68),

$$t_m = t_{mi} + \frac{\Delta t_m}{2} = 40 + \frac{30.21}{2} = 55.11^\circ\text{C}.$$

From Fig. 8.13a for SAE 10 oil at 55°C, the absolute viscosity is found to be approximately 0.016 N-s/m^2. Using Eq. (12.69) gives

$$\begin{aligned} h_o &= H_o l \sqrt{\frac{\eta_o u_b w_t}{W_z B_t}} \\ &= 0.73(0.075)\sqrt{\frac{(0.016)(30)(0.075)}{\left(16 \times 10^3\right)(8)}} \\ &= 29.0\ \mu\text{m}, \end{aligned}$$

$$s_h = \frac{h_o}{H_o} = \frac{29.0 \times 10^{-6}}{0.73} = 39.7\ \mu\text{m}.$$

Referring to Table 12.1, all but the last row of surface finishing processes will ensure that $h_o > (h_o)_{\text{all}}$. Thus, as long as the surface finish is less than 1.6 μm or so, the design is adequate. The performance parameters for $\lambda = 1$ and $B_t = 8.0$ are as follows:

(*a*) From Fig. 12.16a,

$$\frac{\mu l}{s_h} = 8.5, \quad \text{or} \quad \mu = \frac{(8.5)\left(39.7 \times 10^{-6}\right)}{0.075} = 0.0045.$$

(*b*) From Fig 12.16b,

$$\frac{1.5 h_p l}{W_z u_b s_h} = 11,$$

$$\begin{aligned} h_p &= \frac{11 W_z u_b s_h}{1.5 l} = \frac{11\left(16 \times 10^3\right)(30)\left(39.7 \times 10^{-6}\right)}{1.5(0.075)} \\ &= 1.863 \times 10^3 \text{ W} = 1.863 \text{ kW}. \end{aligned}$$

(*c*) From Fig. 12.16c,

$$\frac{q}{w_t u_b s_h} = 0.58;$$

$$\begin{aligned} q &= 0.58 w_t u_b s_h = 0.58(0.075)(30)(39.7)(10^{-6}) \\ &= 5.181 \times 10^{-5} \text{ m}^3/\text{s}. \end{aligned}$$

(*d*) $q_s = (0.3)q = 1.554 \times 10^{-5}$ m^3/s.

12.3.6 Thrust Slider Bearing Geometry

Thus far, this chapter has dealt with the performance of an individual pad of a thrust bearing. Normally, a number of identical pads are assembled in a thrust bearing as shown, for example, in Figs. 12.9 and 12.12. The length, width, speed, and load of an individual pad can be related to the geometry of a thrust bearing by the following formulas:

$$w_t = r_o - r_i, \tag{12.70}$$

$$l = \frac{r_o + r_i}{2}\left(\frac{2\pi}{N_p} - \frac{\pi}{36}\right), \tag{12.71}$$

$$u_b = \frac{(r_o + r_i)\omega}{2}, \tag{12.72}$$

$$W_t = N_p W_z \tag{12.73}$$

where N_p is the number of identical pads placed in the thrust bearing (usually between 3 and 20). The $\pi/36$ term in Eq. (12.71) accounts for the feed grooves between the pads. These are deep grooves that ensure that ambient pressure is maintained between the pads. Also, W_t in Eq. (12.73) is the total thrust load on the bearing.

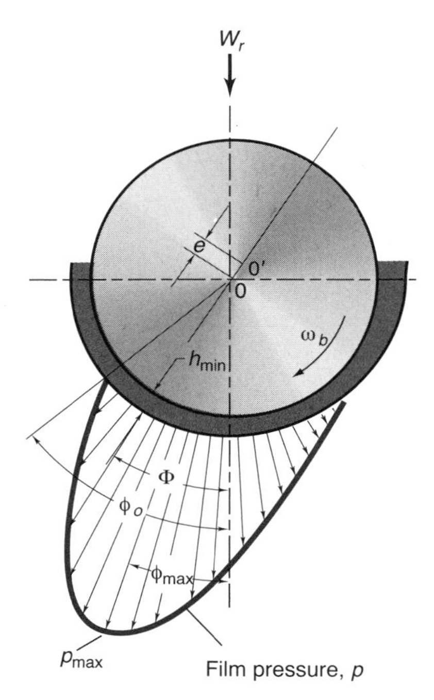

Figure 12.17: Pressure distribution around a journal bearing.

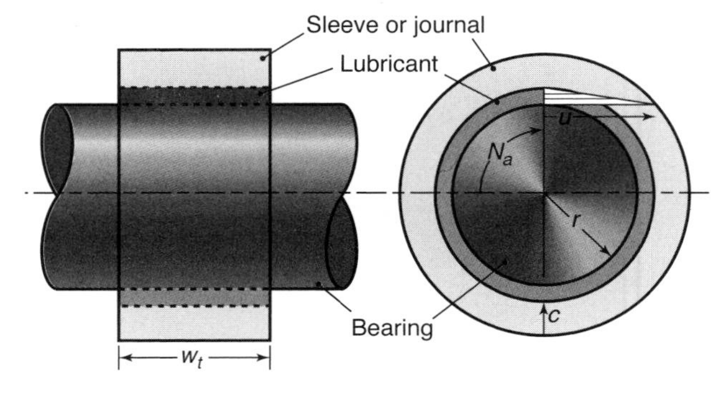

Figure 12.18: Concentric journal bearing.

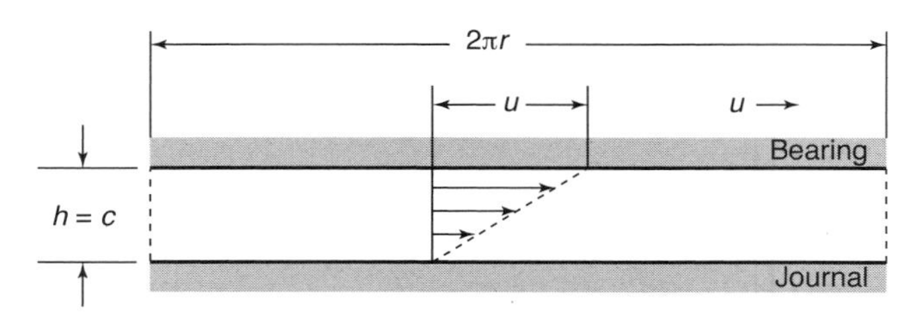

Figure 12.19: Developed journal and bearing surfaces for a concentric journal bearing.

12.4 Journal Slider Bearings

Section 12.3 dealt with slider bearing pads as used in thrust bearings. The surfaces of thrust bearings are perpendicular to the axis of rotation, as shown in Fig. 12.9. This section deals with **journal bearings**, where the bearing surfaces are parallel to the axis of rotation, as shown in Fig. 12.17. Journal bearings are used to support shafts and to carry radial loads with very low power loss and extremely low wear. The journal bearing can be represented by a plain cylindrical sleeve (bushing) wrapped around the journal (shaft) but can adopt a variety of forms. The lubricant is supplied at some convenient location in the bearing through a hole or a groove. If the bearing extends around the full 360° of the journal, it is described as a *full* journal bearing; if the wrap angle is less than 360°, the term *partial* journal bearing is used.

12.4.1 Petrov's Equation

Petrov's equation applies Newton's postulate presented in Section 8.6.1 to a full journal running concentrically with the bearing (Fig. 12.18). As shown in Section 12.4.3, the journal will run concentrically with the bearing only when any one of the following conditions prevails: (1) the radial load acting on the bearing is zero, (2) the viscosity of the lubricant is infinite, or (3) the speed of the journal is infinite. None of these conditions are very practical, though. However, if the load is small enough, if the viscosity is sufficiently high, or if the journal has a sufficiently high speed, the eccentricity of the journal relative to the bearing may be so small that the oil film around the journal can be considered to have uniform thickness.

Since the oil film thickness in a journal bearing is always low relative to the bearing radius, the curvature of the bearing surface may be ignored. It is therefore useful to consider the film as an unwrapped body having a thickness equal to the radial clearance, a length equal to $2\pi r$, and width w_t.

Assume that the viscosity throughout the oil film is constant. In Fig. 12.19, the bottom surface is stationary and the top surface is moving with constant velocity, u. Further assume that no slip exists at the interface between the lubricant and the solid surfaces.

Making use of Newton's postulate, as expressed in Eq. (8.23), gives the friction force in a concentric journal bearing as

$$F = \eta_o A \frac{u}{c} = \eta_o 2\pi r w_t \frac{2\pi r N_a}{c} = \frac{4\pi^2 \eta_o r^2 w_t N_a}{c}, \qquad (12.74)$$

where

c = radial clearance, m
r = radius of journal, m
w_t = width of journal, m
N_a = rotational speed, rps
η_o = viscosity at $p = 0$ and constant temperature

The coefficient of friction for a concentric journal bearing is thus

$$\mu = \frac{F}{W_r} = \frac{4\pi^2 \eta_o r^2 w_t N_a}{c W_r}, \qquad (12.75)$$

where W_r is the normal applied load, N. The friction torque for a concentric journal bearing can then be written as

$$T = Fr = \frac{4\pi^2 \eta_o r^3 w_t N_a}{c} = \frac{2\pi \eta_o r^3 w_t \omega}{c}, \qquad (12.76)$$

where ω is the angular velocity ($2\pi N_a$) in rad/s. Equation (12.76) is generally called **Petrov's equation** (after N. Petrov, who suggested a similar equation for torque in his paper published in 1883). This equation is notable because it was the first to give a reasonable estimate of frictional torque in lightly loaded journal bearings.

The power loss is the product of the torque and the angular velocity, and for a concentric (lightly loaded) journal bearing it can be expressed as

$$h_p = \frac{8\pi^3 \eta_o r^3 w_t N_a^2}{c} = \frac{2\pi \eta_o r^3 w_t \omega^2}{c}. \qquad (12.77)$$

12.4.2 Journal Slider Bearing Operation

Journal bearings (Fig. 12.17) rely on shaft sliding motion to generate the load-supporting pressures in the lubricant film. The shaft does not normally run concentric with the bearing as was assumed in Petrov's equation [Eq. (12.76)]. The displacement of the shaft center relative to the bearing center is known as the **eccentricity** (designated by e in Fig. 12.17). The shaft's eccentric position within the bearing clearance is influenced by the load that it carries. The amount of eccentricity adjusts itself until the load is balanced by the pressure generated in the converging lubricating film. The line drawn through the shaft center and the bearing center is called the **line of centers**. The physical wedge pressure-generating mechanism mentioned in Section 12.2.2 and used for thrust slider bearings in Section 12.3.5 is also valid in this section.

The pressure generated and therefore the load-carrying capacity of the bearing depend on the shaft eccentricity, the angular velocity, the effective viscosity of the lubricant, and the bearing dimensions and clearance. The load and the angular velocity are usually specified and the minimum shaft radius is often predetermined. To complete the design, it is usually necessary to calculate the bearing width and clearance and to choose a suitable lubricant if one is not already specified.

12.4.3 Operating and Performance Parameters

The Reynolds equation used for thrust slider bearings in Eq. (12.25) is modified for a journal slider bearing with $x = r_b\phi$ and $\bar{u} = u_b/2 = r_b\omega/2$:

$$\frac{\partial}{\partial \phi}\left(h^3 \frac{\partial p}{\partial \phi}\right) + r_b^2 \frac{\partial}{\partial y}\left(h^3 \frac{\partial p}{\partial y}\right) = 6\eta_o \omega r_b^2 \frac{\partial h}{\partial \phi}. \qquad (12.78)$$

The film thickness around the journal is expressed as

$$h = c\left(1 + \epsilon \cos\phi\right), \qquad (12.79)$$

where $\epsilon = e/c$ is the eccentricity ratio. Substituting Eq. (12.79) into Eq. (12.78) gives

$$\frac{\partial}{\partial \phi}\left(h^3 \frac{\partial p}{\partial \phi}\right) + r_b^2 h^3 \frac{\partial^2 p}{\partial y^2} = -6\eta_o \omega r_b^2 e \sin\phi. \qquad (12.80)$$

Analytical solutions to Eq. (12.80) are not normally available and numerical methods are thus needed. Equation (12.80) is often solved by using a relaxation method in which the first step is to replace the derivatives in the equation by finite difference approximations. The lubrication zone is covered by a mesh, and the numerical method relies on the fact that a function can be represented, with sufficient accuracy, over a small range by a quadratic expression. The Reynolds boundary condition which requires that $p = 0$ and $dp/dx = 0$ at $\phi = \phi^*$ (the outlet boundary) is used. The results obtained by using this numerical method are presented in this section.

The *operating* parameters for hydrodynamic journal bearings are

1. Bearing number for journal bearings (also called the **Sommerfeld number**)

$$B_j = \left(\frac{\eta_o \omega r_b w_t}{\pi W_r}\right)\left(\frac{r_b}{c}\right)^2. \qquad (12.81)$$

2. Diameter-to-width ratio

$$\lambda_j = \frac{2r_b}{w_t}. \qquad (12.82)$$

3. Angular extent of journal (full or partial).

The *performance* parameters for hydrodynamic journal bearings are

1. Eccentricity, e
2. Location of minimum film thickness, sometimes called attitude angle, Φ
3. Coefficient of sliding friction, μ
4. Total and side-leakage volumetric flow rates, q and q_s
5. Angle of maximum pressure, $\phi_{\max}$
6. Location of terminating pressure, ϕ_o
7. Temperature rise due to lubricant shearing, Δt_m

The parameters Φ, $\phi_{\max}$, and ϕ_o are described in Fig. 12.17, which gives the pressure distribution around a journal bearing. Note from this figure that if the bearing is concentric (e = 0), the film shape around the journal is uniform and equal to c, and no fluid film pressure is developed. At heavy loads, which is at the other extreme, the journal is forced downward and the limiting position is reached when $h_{\min}$ = 0 and $e = c$ (i.e., the journal is in contact with the bearing).

Temperature rise due to lubricant shearing will be considered here as was done in Section 12.3.4 for thrust bearings. In Eq. (12.80), the lubricant viscosity corresponds to the viscosity when $p = 0$ but can vary as a function of temperature. Since work is done as the fluid is being sheared, the temperature of the lubricant is higher when it leaves the conjunction than it is on entry. In Ch. 8 (Figs. 8.12 and 8.13 as well as Table 8.6), it was shown that oil viscosity drops off significantly with rising temperature. This decrease is accommodated by using a mean of the inlet and outlet temperatures:

$$t_m = t_{mi} + \frac{\Delta t_m}{2}, \qquad (12.83)$$

where t_{mi} is the inlet temperature and Δt_m is the lubricant temperature rise from the inlet to the outlet. The viscosity used in the bearing number, B_j, and other performance parameters is at the mean temperature t_m. The temperature rise of the lubricant from inlet to outlet Δt_m can then be determined from the performance charts provided in this section.

12.4.4 Numerical Solution

Having defined the operating and performance parameters, the design procedure for a hydrodynamic journal bearing can now be presented. The results are for a full journal bearing; those for a *partial* journal bearing can be obtained from Raimondi and Boyd [1958].

Most design problems are underconstrained, and thus one or more dimensions have to be specified before analysis can commence. In such a circumstance, the data in Table 12.2 can help in prescribing a candidate design, which can then be modified if the performance is unsatisfactory. For example, given the load and shaft diameter, a bearing width can be determined through use of typical average pressures from Table 12.2. This table can also be useful in confirming the results of analysis.

Figure 12.20 shows the effect of the bearing number, B_j, on the minimum film thickness for four diameter-to-width ratios. The following relationship should be observed:

$$h_{\min} = c - e. \qquad (12.84)$$

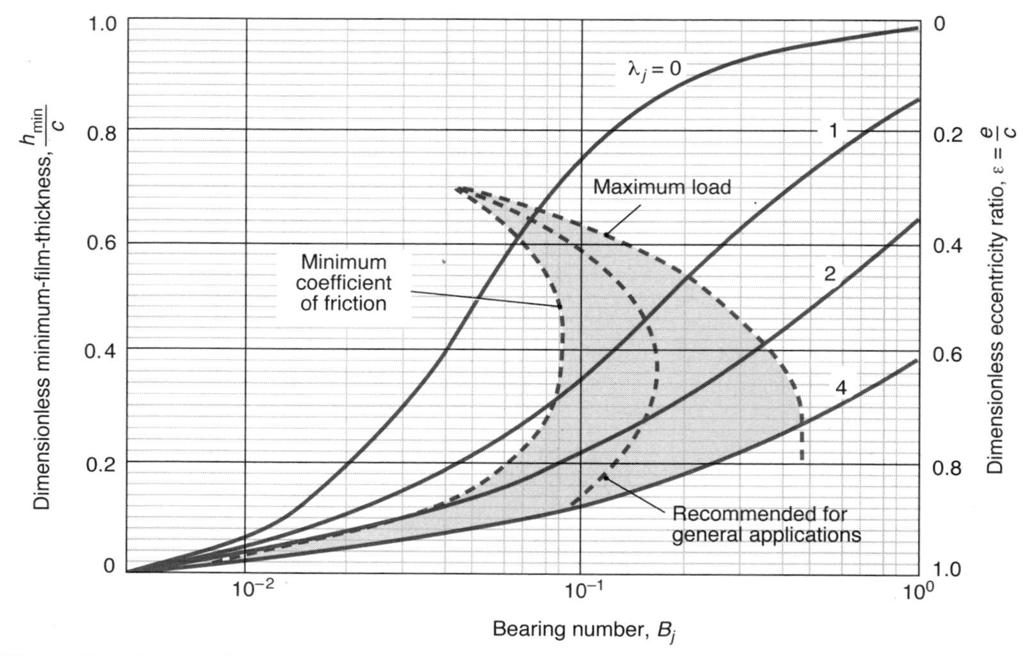

Figure 12.20: Effect of bearing number on minimum film thickness for four diameter-to-width ratios. The shaded area is the most common operating range for well-designed journal bearings. *Source:* Adapted from Raimondi and Boyd [1958].

Table 12.2: Typical radial load per area W_r^* for journal bearings.

Application	Average radial load per unit area, W_r'	
	psi	MPa
Automotive engines:		
Main bearings	600–750	4–5
Connecting rod bearing	1700–2300	10–15
Diesel engines:		
Main bearings	900–1700	6–12
Connecting rod bearing	1150–2300	8–15
Electric motors	120–250	0.8–1.5
Steam turbines	150–300	1.0–2.0
Gear reducers	120–250	0.8–1.5
Centrifugal pumps	100–180	0.6–1.2
Air compressors:		
Main bearings	140–280	1–2
Crankpin	280–500	2–4
Centrifugal pumps	100–180	0.6–1.2

In dimensionless form

$$H_{\min} = \frac{h_{\min}}{c} = 1 - \epsilon, \tag{12.85}$$

where

$$\epsilon = \frac{e}{c} = \text{eccentricity ratio.} \tag{12.86}$$

The bearing number for journal bearings is expressed in Eq. (12.81); in a given design this number is affected by:

1. Absolute lubricant viscosity, η_o
2. Angular shaft speed, ω
3. Radial load, W_r
4. Radial clearance, c
5. Journal dimensions, r_b and w_t

All of these parameters affect the design of the journal bearing.

In Fig. 12.20, a recommended operating eccentricity ratio, or minimum film thickness, is indicated as well as a preferred operating area. The left boundary of the shaded zone defines the optimum eccentricity ratio for a minimum coefficient of friction, and the right boundary, the optimum eccentricity ratio for maximum load. The recommended operating eccentricity is midway between these two boundaries, although good performance can, in general, also be achieved throughout the shaded area.

Figure 12.21 shows the effect of the bearing number on the attitude angle, Φ, defined as the angle between the load direction and a line drawn through the centers of the bearing and journal (see Fig. 12.17) for four values of λ_j. This angle establishes where the minimum and maximum film thicknesses are located within the bearing.

Figure 12.22 shows the effect of the bearing number on the coefficient of friction for four values of λ_j. The effect is small for a complete range of dimensionless load parameters.

Figure 12.23 shows the effect of bearing number on the dimensionless volumetric flow rate $Q = 2\pi q / r_b c w_t \omega$ for four values of λ_j. The dimensionless volumetric flow rate, Q, that is pumped into the converging space by the rotating journal can be obtained from this figure. Of the volume q of oil pumped by the rotating journal, an amount q_s flows out the ends and hence is called *side-leakage volumetric flow*. This side leakage can be computed from the volumetric flow ratio q_s/q of Fig. 12.24.

Figure 12.25 illustrates the maximum pressure developed in a journal bearing; note that the maximum film pressure is made dimensionless by normalizing with the load per unit area. The maximum pressure as well as its location are shown in Fig. 12.17. Figure 12.26 shows the effect of bearing number on the location of the terminating and maximum pressures for four values of λ_j.

The lubricant temperature rise, in °C, from the inlet to the outlet can be obtained from Shigley and Mitchell [1983]

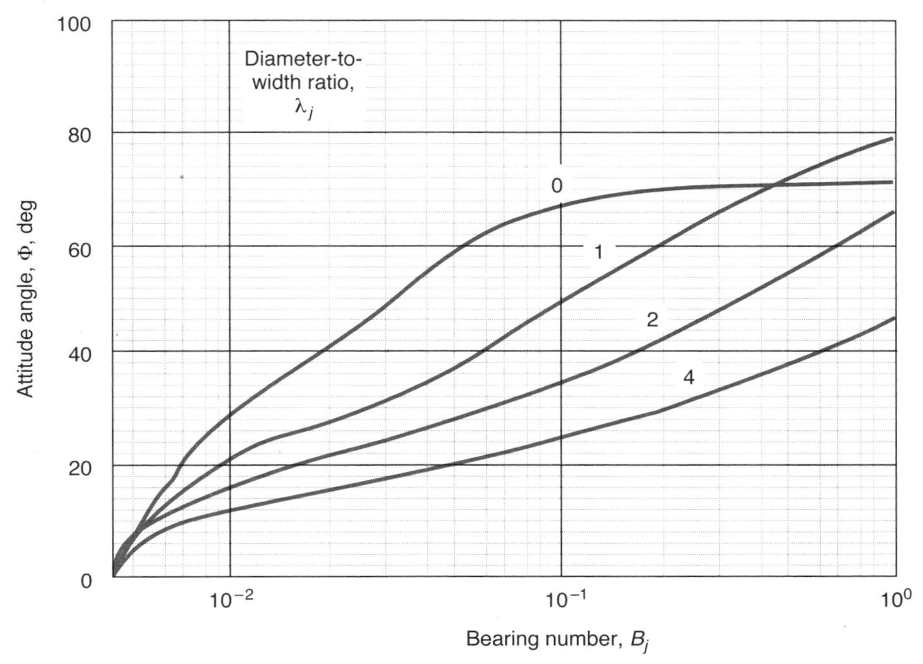

Figure 12.21: Effect of bearing number on attitude angle for four diameter-to-width ratios. *Source:* Adapted from Raimondi and Boyd [1958].

as

$$\Delta t_m = \frac{8.3 W_r^* (r_b/c)\mu}{Q(1-0.5 q_s/q)}, \tag{12.87}$$

where $W_r^* = W_r/2r_b w_t$ is in MPa. Thus, the temperature rise can be directly obtained by substituting the values of $r_b\mu/c$ obtained from Fig. 12.28, Q from Fig. 12.29, and q_s/q from Fig. 12.24 into Eq. (12.87). The temperature rise in °F is given by

$$\Delta t_m = \frac{0.103 W_r^* (r_b/c)\mu}{Q(1-0.5 q_s/q)}, \tag{12.88}$$

where

$$W_r^* = \frac{W_r}{2r_b w_t}, \tag{12.89}$$

and W_r^* is in pounds per square inch. Once the viscosity is known, the bearing number can be calculated and then the performance parameters can be obtained from Figs. 12.20 through 12.26 and Eqs. (12.89).

The results presented thus far have been for λ_j of 0, 1, 2, and 4. If λ_j has some other value, the following formula can be used to establish the performance parameters:

$$\begin{aligned} y &= \left(\frac{\lambda_j - 2}{\lambda_j}\right)\left[-\frac{(\lambda_j-1)(\lambda_j-4)}{8} y_o\right] \\ &+ (\lambda_j - 2)\left[\frac{(\lambda_j-4)y_1}{3} - \frac{(\lambda_j-1)(\lambda_j+2)y_2}{4}\right] \\ &+ (\lambda_j - 2)\left[\frac{(\lambda_j-1)y_4}{24}\right]. \end{aligned} \tag{12.90}$$

where y is any one of the performance parameters ($H_{\min}, \Phi, r_b\mu/c, Q, q_s/q, p_{\max}, \phi_o$, or $\phi_{\max}$), and where the subscript on y is the λ_j value. For example, y_1 is equivalent to y evaluated at $\lambda_j = 1$. It should be noted that reasonable results are generally obtained through careful interpolation on the proper chart; in cases where higher accuracy is needed, numerical simulations are often performed — see Hamrock et al. [2004] for details. Regardless, Eq. (12.90) ensures accuracy for any diameter-to-width ratio.

Design Procedure 12.2: Journal Bearings

1. Usually, design problems are underconstrained. Diameter is usually predetermined from shaft size, and load and speed are design requirements. Table 12.2 can be used to select a diameter-to-width ratio if it has not been specified. Otherwise, Fig. 12.20 can be used to obtain the bearing number, B_j. The dimensionless film thickness, $h_{\min}/c$, can also be obtained from Fig. 12.20. Using a film thickness from Table 12.1 allows calculation of the radial clearance, c. Alternatively, a known value of c results in an estimate of film thickness that can be compared to Table 12.1 to evaluate the design and specify journal bearing and sleeve surface roughness and manufacturing approach.

   At this point, the required lubricant viscosity can be determined to produce the required bearing number. A review of Fig. 8.13 allows selection of a lubricant and operating temperature to produce this viscosity.

2. Performance parameters for the journal bearing can then be obtained from Figs. 12.21 through 12.26.

3. The temperature rise within the journal bearing can then be obtained from Eq. (12.87). Temperature rise will indicate the adequacy of the design; the temperatures in Fig. 8.13 should in general not be exceeded within the bearing or else the lubricant may experience premature loss of viscosity and thermal degradation. Moderate temperature rises can provide useful information for design of lubricant cooling systems, if needed.

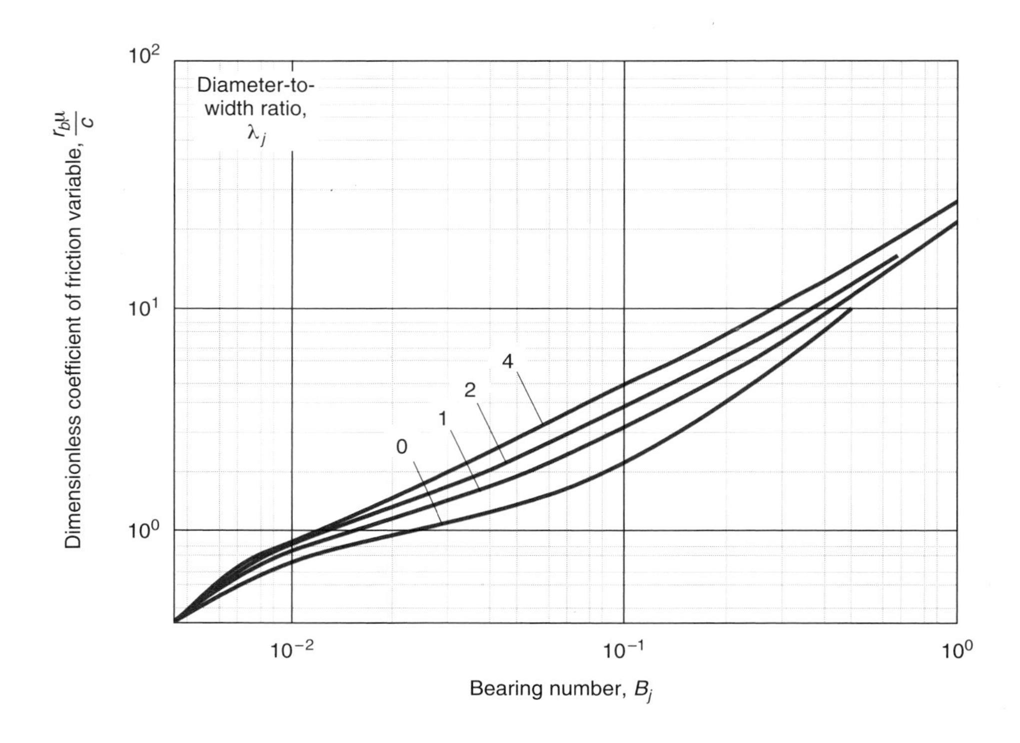

Figure 12.22: Effect of bearing number on coefficient of friction for four diameter-to-width ratios. *Source:* Adapted from Raimondi and Boyd [1958].

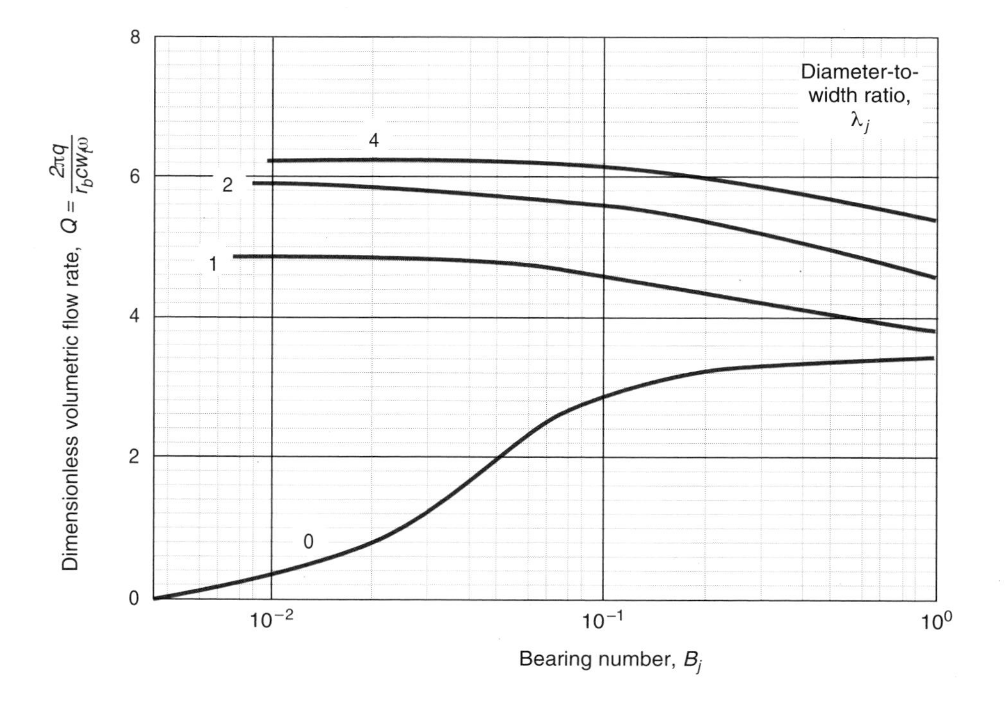

Figure 12.23: Effect of bearing number on dimensionless volumetric flow rate for four diameter-to-width ratios. *Source:* Adapted from Raimondi and Boyd [1958].

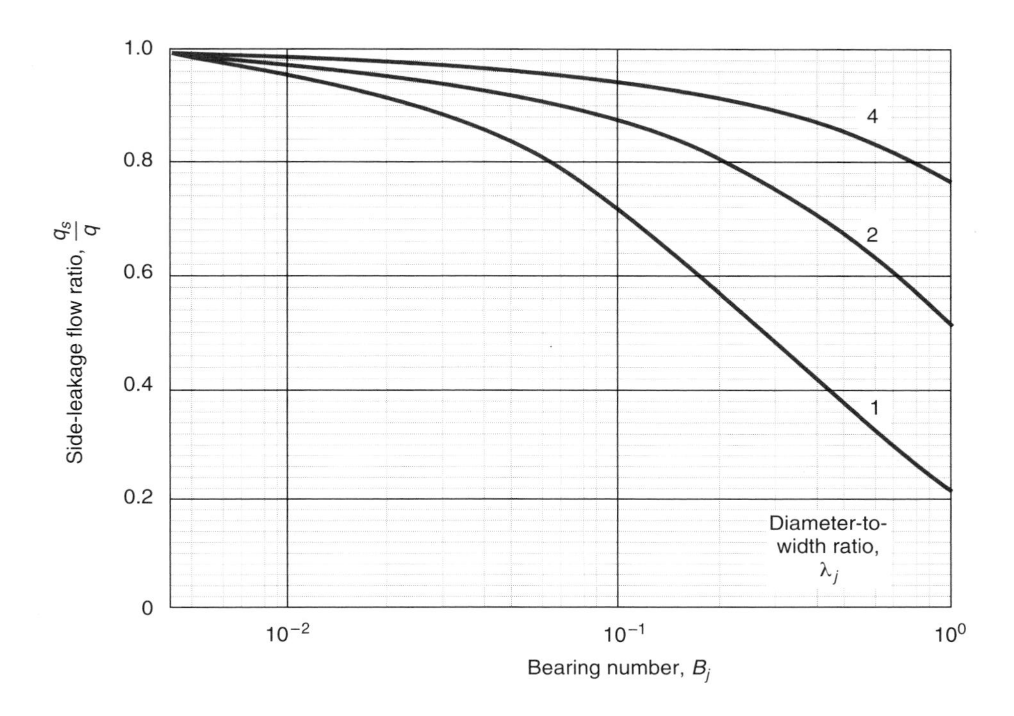

Figure 12.24: Effect of bearing number on side-leakage flow ratio for four diameter-to-width ratios. *Source:* Adapted from Raimondi and Boyd [1958].

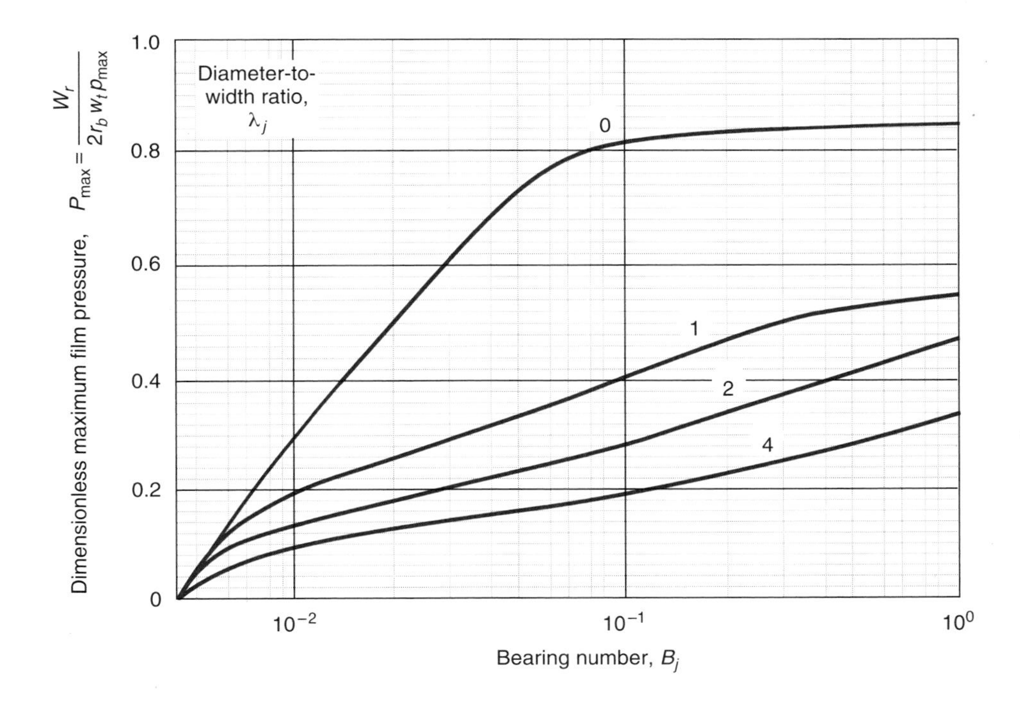

Figure 12.25: Effect of bearing number on dimensionless maximum film pressure for four diameter-to-width ratios. *Source:* Adapted from Raimondi and Boyd [1958].

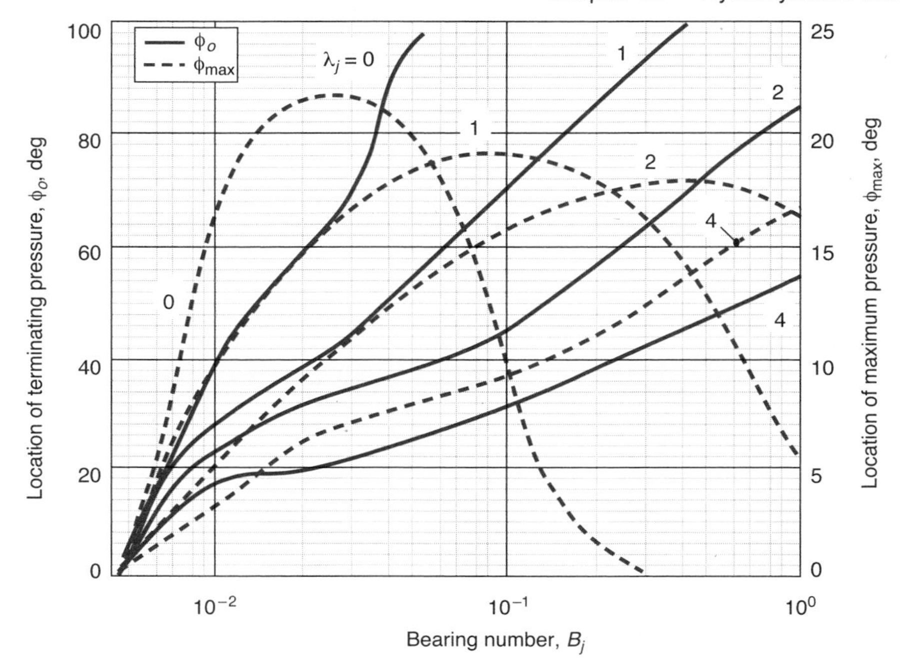

Figure 12.26: Effect of bearing number on location of terminating and maximum pressures for four diameter-to-width ratios. *Source:* Adapted from Raimondi and Boyd [1958].

4. In some cases, a bearing is specified and its performance is to be evaluated. In this case, the lubricant viscosity is unknown, since the temperature rise and mean temperature are not known beforehand. The approach in this case is to assume a temperature rise, obtain performance parameters from Figs. 12.21 through 12.26, and calculate a temperature rise. If this temperature rise is close to that assumed, then this represents the operating condition of the journal bearing. If the assumed and calculated values differ by more than 10°C or so, a revised estimate can be made and the procedure repeated. Journal bearings are well-behaved, and a solution usually is obtained after one or two iterations.

Example 12.4: Analysis of a Journal Bearing

Given: A full journal bearing uses SAE 60 oil with an inlet temperature of 40°C, N_a = 30 rev/s, W_r = 2200 N, r_b = 20 mm, and w_t = 40 mm.

Find: Establish the operating and performance parameters for this bearing while designing for maximum load.

Solution: The angular speed can be expressed as

$$\omega = 2\pi N_a = 2\pi(30) = 60\pi \text{ rad/s}.$$

The diameter-to-width ratio is

$$\lambda_j = \frac{2r_b}{w_t} = \frac{2(2)}{4} = 1.$$

From Fig. 12.20 for $\lambda_j = 1$ and designing for maximum load,

$$B_j = 0.2, \qquad \frac{h_{\min}}{c} = 0.53, \qquad \text{and} \qquad \epsilon = 0.47. \qquad (a)$$

For B_j = 0.2 and λ_j = 1, from Figs. 12.22 through 12.24,

$$\frac{r_b\mu}{c} = 4.9, \qquad Q = 4.3, \qquad \text{and} \qquad \frac{q_s}{q} = 0.6. \qquad (b)$$

From Eq. (12.89) the radial load per area is

$$W_r^* = \frac{W_r}{2r_b w_t} = \frac{2200}{2(2)(4)(10^{-4})} \text{ Pa} = 1.375 \text{ MPa}. \qquad (c)$$

The lubricant temperature rise obtained by using Eq. (12.87) and the results from Eqs. (*b*) and (*c*) is

$$\Delta t_m = \frac{8.3W_r^*(r_b/c)\mu}{Q(1-0.5q_s/q)} = \frac{8.3(1.375)(4.9)}{4.3[1-(0.5)(0.6)]} = 18.58°\text{C}.$$

From Eq. (12.83), the mean temperature of the lubricant is

$$t_m = t_{mi} + \frac{\Delta t_m}{2} = 40 + \frac{18.58}{2} = 49.29°\text{C}.$$

From Fig. 8.13a for SAE 60 oil at 49.3°C, the absolute viscosity is

$$2.5 \times 10^{-5} \text{ reyn} = 1.70 \times 10^2 \text{ centipoise} = 0.170 \text{ N-s/m}^2.$$

From Eq. (12.81), the radial clearance can be expressed as

$$c = r_b\sqrt{\frac{\eta_o\omega r_b w_t}{\pi W_r B_j}} = (0.02)\sqrt{\frac{(0.170)60\pi(0.020)(0.040)}{\pi(2200)(0.2)}}$$

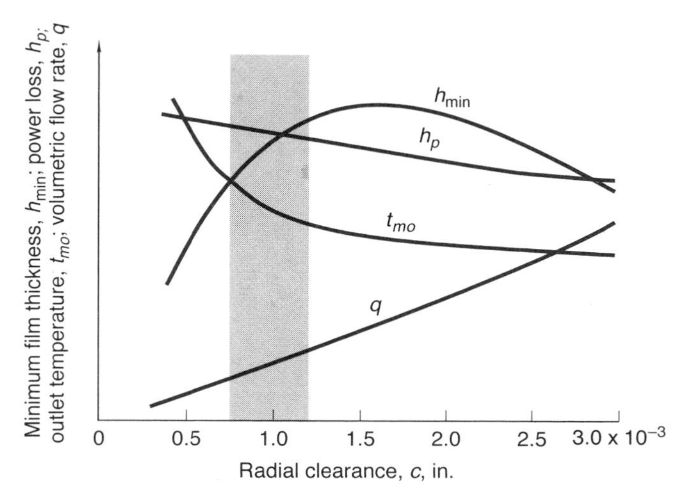

Figure 12.27: Effect of radial clearance on some performance parameters for a particular case.

or $c = 86.1 \times 10^{-6}$ m. The coefficient of friction from Eq. (b) is

$$\mu = \frac{4.9c}{r_b} = \frac{(4.9)(0.0861)(10^{-3})}{0.02} = 0.021.$$

The circumferential volumetric flow rate is

$$\begin{aligned} q &= \frac{Qr_b c w_t \omega}{2\pi} = \frac{(4.3)(0.020)\left(0.0861 \times 10^{-3}\right)(0.040)60\pi}{2\pi} \\ &= 8.89 \times 10^{-6}\ \mathrm{m^3/s}, \end{aligned}$$

$$q_s = 7.288 \times 10^{-6}\ \mathrm{m^3/s}.$$

From Fig. 12.21 for $B_j = 0.2$ and $\lambda_j = 1$, the attitude angle is 61°. From Fig. 12.25 for $B_j = 0.2$ and $\lambda_j = 1$, the dimensionless maximum pressure is $P_{\max} = 0.46$. The maximum pressure is thus

$$p_{\max} = \frac{W_r}{2r_b w_t P_{\max}} = \frac{2200}{2(0.020)(0.040)(0.46)} = 2.989\ \mathrm{MPa}.$$

From Fig. 12.26 for $B_j = 0.2$ and $\lambda_j = 1$, the location of the maximum pressure from the applied load is 18° and the location of the terminating pressure from the applied load is 86°.

12.4.5 Optimization Techniques

The radial clearance, c, is the most difficult of the performance parameters to control accurately during manufacturing, and it may increase because of wear. Figure 12.27 shows the performance of a particular bearing for a range of radial clearances. If the clearance is too low, the temperature will be too high and the minimum film thickness too low. High temperature may cause the bearing to fail by fatigue, or cause the lubricant to degrade. If the oil film is too thin, dirt particles may not pass through without scoring, or may embed themselves in the bearing. In either event, there will be excessive wear and friction, resulting in high temperatures and possible seizure. A large clearance will permit dirt particles to pass through and also permit a large flow of oil, lowering the temperature and lengthening bearing life. However, if the clearance becomes too large, the bearing becomes noisy and the minimum film thickness begins to decrease again.

Figure 12.27 shows the best compromise, when both the production tolerance and the future wear on the bearing are considered, to be a clearance range slightly to the left top of the minimum-film-thickness curve. Future wear will move the operating point to the right, thus increasing the film thickness and decreasing the operating temperature.

12.5 Squeeze Film Bearings

In Sections 12.3 and 12.4, hydrodynamic slider bearings were considered, where the positive bearing pressure is due to the physical wedge mechanism covered in Section 12.2.2. In this section, the focus is on the development of positive pressure in a fluid between two surfaces moving toward each other. Because a finite time is required to squeeze the fluid out of the gap, this action provides a useful cushioning effect in bearings. The reverse effect, which occurs when the surfaces are moving apart, can lead to cavitation in the fluid film.

The concept of **squeeze film bearings** and the associated pressure-generated mechanism was introduced in Section 12.2.2. For squeeze film bearings, a relationship needs to be developed between load and normal velocity at any instant. The time required for the separation of the surfaces to change by a specified amount can be determined by a single integration with respect to time.

The appropriate Reynolds equation for squeeze film bearings is

$$\frac{\partial}{\partial x}\left(h^3\frac{\partial p}{\partial x}\right) + \frac{\partial}{\partial y}\left(h^3\frac{\partial p}{\partial y}\right) = 12\eta_o\frac{\partial h}{\partial t} = -12\eta_o w, \quad (12.91)$$

where w is the squeeze velocity.

12.5.1 Parallel-Surface Squeeze Film Thrust Bearing

Figure 12.28 shows a simple parallel-surface squeeze film bearing. When side leakage [flow in the y-direction (the direction out from the paper)] is small, the second term in Eq. (12.91) is neglected. This assumption is valid for bearings whose width, w_t, is much larger than their length, l. Thus, if side leakage is neglected and a parallel film is assumed, Eq. (12.91) reduces to

$$\frac{\partial^2 p}{\partial x^2} = -\frac{12\eta_o w}{h_o^3}. \quad (12.92)$$

In Fig. 12.28, the coordinate system origin is at the midpoint of the bearing to take advantage of symmetry. Integrating Eq. (12.92) twice gives the pressure as

$$p = -\frac{6\eta_o w}{h_o^3}x^2 + C_1 x + C_2, \quad (12.93)$$

where C_1 and C_2 are integration constants. The boundary conditions are that $p = 0$ when $x = \pm l/2$. Making use of these boundary conditions gives

$$p = \frac{3\eta_o w}{2h_o^3}\left(l^2 - 4x^2\right). \quad (12.94)$$

The normal load-carrying capacity per unit width is

$$W_z' = \int_{-l/2}^{l/2} p\,dx = \frac{3\eta_o w}{2h_o^3}\int_{-l/2}^{l/2}\left(l^2 - 4x^2\right)dx = \frac{\eta_o l^3 w}{h_o^3}. \quad (12.95)$$

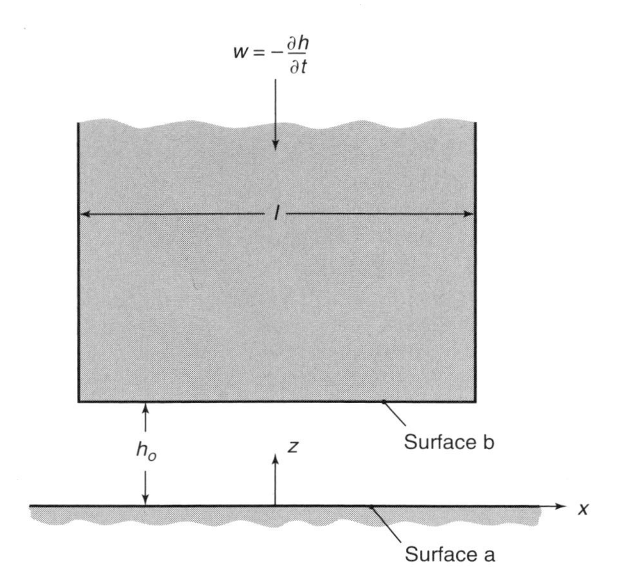

Figure 12.28: Parallel-surface squeeze film bearing.

For static loads, Eq. (12.95) can be used to determine the time for the gap between the parallel surfaces to be reduced by a given amount. The squeeze velocity w can be expressed as $-\partial h/\partial t$. Since the film thickness does not vary with x but varies with time, Eq. (12.95) becomes

$$W_z' = -\frac{\eta_o l^3}{h_o^3}\frac{\partial h_o}{\partial t}. \tag{12.96}$$

Rearranging terms and integrating,

$$-\frac{W_z'}{\eta_o l^3}\int_{t_1}^{t_2} dt = \int_{h_{o,1}}^{h_{o,2}} \frac{dh_o}{h_o^3};$$

$$\Delta t = t_2 - t_1 = \frac{\eta_o l^3}{2W_z'}\left(\frac{1}{h_{o,2}^2} - \frac{1}{h_{o,1}^2}\right). \tag{12.97}$$

The final outlet film thickness, $h_{o,2}$, can then be expressed in terms of the initial film thickness $h_{o,1}$ and the time interval, Δt, as

$$h_{o,2} = \frac{h_{o,1}}{\left[1 + \left(2W_z'\Delta t h_{o,1}^2/\eta_o l^3\right)\right]^{1/2}}. \tag{12.98}$$

Example 12.5: Film Thickness in Squeeze Film Bearings

Given: Two parallel plates are 0.025 m long and very wide. The lubricant separating the plates is initially 25 μm thick and has a viscosity of 0.5 Pa-s. The load per unit width is 20,000 N/m.

Find: Calculate the time required to reduce the film thickness to (a) 2.5 μm, (b) 0.25 μm, and (c) zero.

Solution: Making use of Eq. (12.97) gives

$$\begin{aligned}\Delta t &= \frac{\eta_o l^3}{2W_z'}\left(\frac{1}{h_{o,2}^2} - \frac{1}{h_{o,1}^2}\right) = \frac{(0.5)(0.025)^3}{2(20,000)}\left[\frac{1}{h_{o,2}^2} - \frac{1}{h_{o,1}^2}\right] \\ &= \left(2.95 \times 10^{-10}\right)\left[\frac{1}{h_{o,2}^2} - \frac{1}{(25 \times 10^{-6})^2}\right].\end{aligned}$$

Therefore, it can be noted that

1. For $h_o = 2.5\ \mu$m,

$$\Delta t = 1.95 \times 10^{-10}(0.16 - 0.0016)(10^{12}) = 30.9 \text{ s}.$$

2. For $h_o = 0.25\ \mu$m,

$$\begin{aligned}\Delta t &= 1.95 \times 10^{-10}\left[\frac{1}{(0.25 \times 10^{-6})^2} - \frac{1}{(25 \times 10^{-6})^2}\right] \\ &= 3120 \text{ s} = 52 \text{ min}.\end{aligned}$$

3. For $h_o = 0$,

$$\Delta t = 1.95 \times 10^{-10}\left[\frac{1}{0} - \frac{1}{(25 \times 10^{-6})^2}\right] = \infty.$$

Thus implying that, theoretically, the lubricant will never be squeezed out of the space between the parallel plates.

12.5.2 General Comments Regarding Squeeze Film Bearings

For a thrust bearing subjected to a squeeze velocity, a parallel film shape produces the largest load-carrying capacity of all possible film shapes. In contrast, for a thrust slider bearing, a parallel film should produce no positive pressure and thus no load-carrying capacity. Recall that a slider bearing uses a physical wedge mechanism (see Section 12.2.2) to produce a positive pressure. Squeeze film bearings can be applied to a number of bearing configurations other than the parallel-surface thrust bearing. Hamrock et al. [2004] applied squeeze films to a number of conformal and nonconformal surfaces, such as journal bearings, parallel circular plates, and an infinitely long cylinder near a plate, and serves as a good reference for more detailed information.

The second major observation about normal squeeze film bearings is that as the bearing surfaces move toward each other, the viscous fluid shows ever greater reluctance to be squeezed out the sides of the bearing. The tenacity of the squeeze film is remarkable, and the survival of many modern bearings depends on this phenomenon.

The third feature of squeeze film bearings is that a small approach velocity will provide an extremely large load-carrying capacity. The reason is mainly that in Eq. (12.95) the normal applied load per unit width is inversely proportional to the film thickness raised to the third power and directly proportional to the squeeze velocity.

12.6 Hydrostatic Bearings

Slider bearings, considered in Sections 12.3 and 12.4, use a physical wedge pressure-generating mechanism to develop pressure within the bearing. In addition to having low frictional drag and hence low power loss, such bearings have the great advantage that they are basically simple and therefore are inexpensive, reliable, and require little attention. Slider bearings have, however, some important limitations:

1. If the design speed is low, it may not be possible to generate sufficient hydrodynamic pressure.

2. Fluid film lubrication may break down during starting, direction reversal, and stopping.

P_r = pressure difference

3. In a journal slider bearing (Section 12.4), the shaft runs eccentrically and the bearing location varies with load, thus implying low stiffness.

In **hydrostatic** (also called **externally pressurized**) **lubricated bearings**, the bearing surfaces are separated by a fluid film maintained by a pressure source outside the bearing. Hydrostatic bearings avoid disadvantages 1 and 2 and reduce the variation of bearing location with load mentioned in disadvantage 3. The characteristics of hydrostatically lubricated bearings are:

1. Extremely low friction
2. Extremely high load-carrying capacity at low speeds
3. High positional accuracy in high-speed, light-load applications
4. A lubrication system more complicated than that for slider bearings

Therefore, hydrostatically lubricated bearings are used when the requirements are demanding, as in large telescopes and radar tracking units, where very heavy loads and low speeds are used; or in machine tools and gyroscopes, where extremely high speeds, light loads, and inviscid lubricants are used.

Figure 12.29 shows how a fluid film forms in a hydrostatically lubricated bearing system. In a simple bearing system under no pressure (Fig. 12.29a) the runner, under the influence of load W_z, is seated on the pad. As the source pressure builds up (Fig. 12.29b), the pressure in the pad recess also increases. The recess pressure is built up to a point (Fig. 12.29c) where the pressure on the runner over an area equal to the pad recess area is just sufficient to lift the load. This is commonly called the *lift pressure*, p_l. Just after the runner separates from the bearing pad (Fig. 12.29d) the recess pressure is less than that required to lift the bearing runner ($p_r < p_l$). After lift, flow commences through the system. Thus, a pressure drop exists between the pressure source and the bearing (across the restrictor) and from the recess to the bearing outlet. If the load is increased on the bearing (Fig. 12.29e), the film thickness will decrease and the recess pressure will rise until the integrated pressure across the land equals the load. If the load is then reduced to less than the original (Fig. 12.29f), the film thickness will increase to some higher value and the recess pressure will decrease accordingly. The maximum load that can be supported by the pad will be theoretically reached when the recess pressure is equal to the source pressure. If a load greater than this is applied, the bearing will seat and remain seated until the load is reduced and can again be supported by the supply pressure.

The load-carrying capacity of a bearing pad, regardless of its shape or size, can be expressed in a general form as

$$W_z = a_b A_p p_r, \tag{12.99}$$

where a_b is the dimensionless bearing pad load coefficient and A_p is the total projected pad area. The amount of lubricant flow across a pad and through the bearing clearance is

$$q = q_b \frac{W_z}{A_p} \frac{h_o^3}{\eta_o}, \tag{12.100}$$

where q_b is the dimensionless bearing pad flow coefficient. The pumping power required by the hydrostatic pad can be evaluated from the product of recess pressure and flow.

Assuming that the angular speed is zero, so that the power loss due to viscous dissipation is zero, the power needed to pump fluid is given by.

$$h_p = p_r q = h_b \left(\frac{W_z}{A_p}\right)^2 \frac{h_o^3}{\eta_o}, \tag{12.101}$$

where $h_b = q_b/a_b$ is the dimensionless bearing pad power coefficient.

The designer of hydrostatic bearings is primarily concerned with the three dimensionless bearing coefficients (a_b, q_b, and h_b). Values of any two of these coefficients suffice to determine the third. Bearing coefficients are dimensionless quantities that relate performance characteristics of load, flow, and power to physical parameters. The bearing coefficients for several types of common bearing pads will be considered here. Details of the derivations can be found in Hamrock et al. [2004].

12.6.1 Circular Step Bearing Pad

Consider the radial flow hydrostatic thrust bearing with a circular step bearing pad shown in Fig. 12.30a. The purpose of the hydrostatic bearing is to support thrust loads with very low friction and good stiffness. The vertical positioning of the opposing surface is influenced by the oil film thickness and its variation with load. To minimize the variation for different loads, a step is produced in the bearing as shown in Fig. 12.30a. The pressure in the step is constant, since s_h/h_o is typically a large number, usually greater than 10. Thus, $p = p_r$ through the step or for $0 < r < r_i$.

For a circular step bearing pad, the three dimensionless bearing coefficients are given by

$$a_b = \frac{1 - (r_i/r_o)^2}{2\ln(r_o/r_i)}, \tag{12.102}$$

$$q_b = \frac{\pi}{3\left[1 - (r_i/r_o)^2\right]}, \tag{12.103}$$

$$h_b = \frac{2\pi \ln(r_o/r_i)}{3\left[1 - (r_i/r_o)^2\right]^2}. \tag{12.104}$$

The total projected pad area is

$$A_p = \pi r_o^2. \tag{12.105}$$

The bearing pad load coefficient, a_b, varies from zero for extremely small recesses to unity for bearings having large recesses with respect to pad dimensions. In a sense, a_b is a measure of how efficiently the bearing uses the recess pressure to support the applied load. The bearing pad flow coefficient, q_b, varies from unity for relatively small recesses to a value approaching infinity for bearings with extremely large recesses. Physically, as the recess becomes larger with respect to the bearing, the hydraulic resistance to fluid flow decreases and thus flow increases. Also, the power coefficient, h_b, approaches infinity for extremely small recesses, decreases to a minimum as the recess size increases, and then approaches infinity again for extremely large recesses. For a circular step thrust bearing, the minimum value of h_b occurs at $r_i/r_o = 0.53$.

12.6.2 Annular Thrust Bearing

Figure 12.30b shows an annular thrust bearing with four different radii used to define the recess and the sills. The lubricant flows from the annular recess over the inner and outer

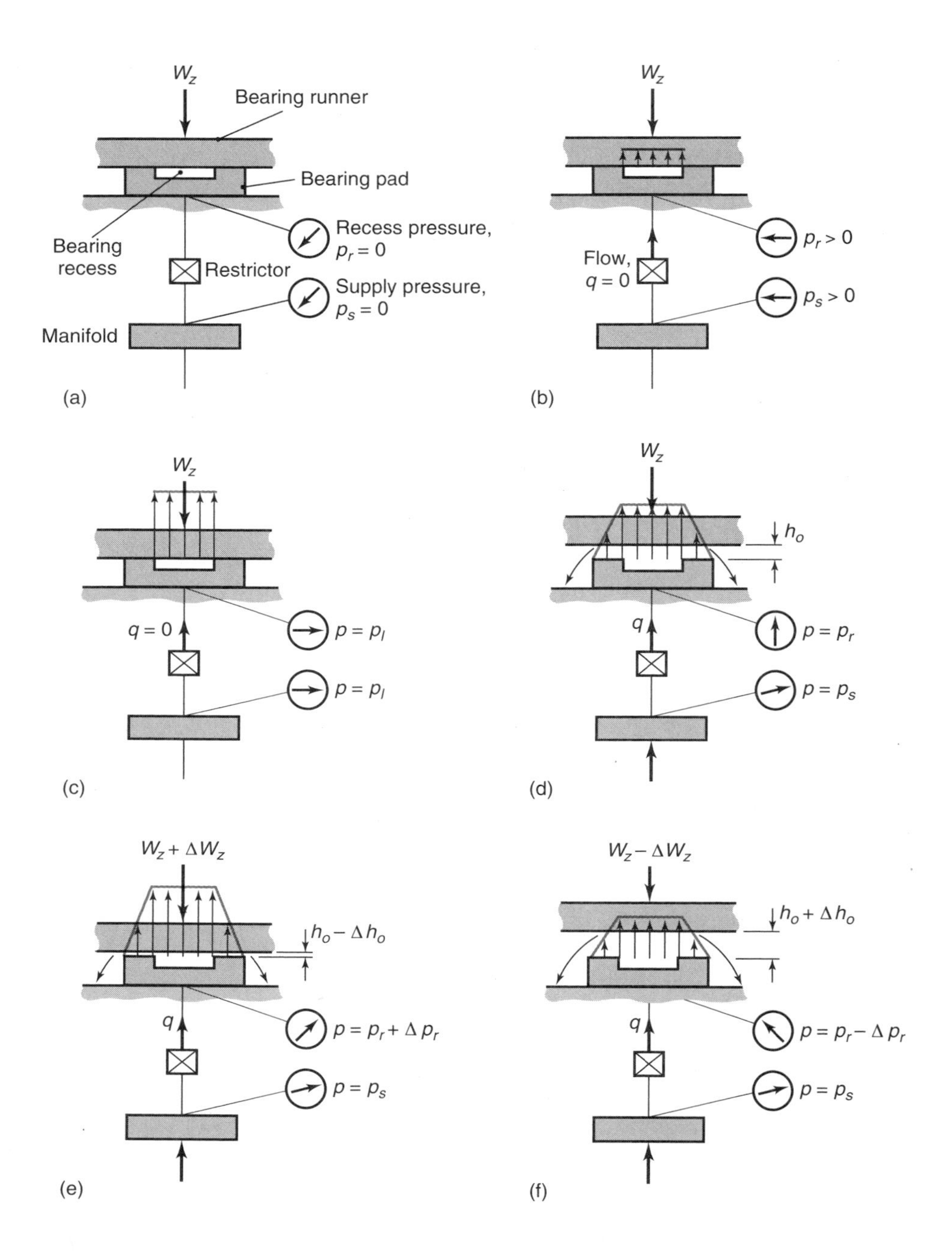

Figure 12.29: Formation of fluid film in hydrostatic bearing system. (a) Pump off; (b) pressure build up; (c) pressure times recess area equals normal applied load; (d) bearing operating; (e) increased load; (f) decreased load.

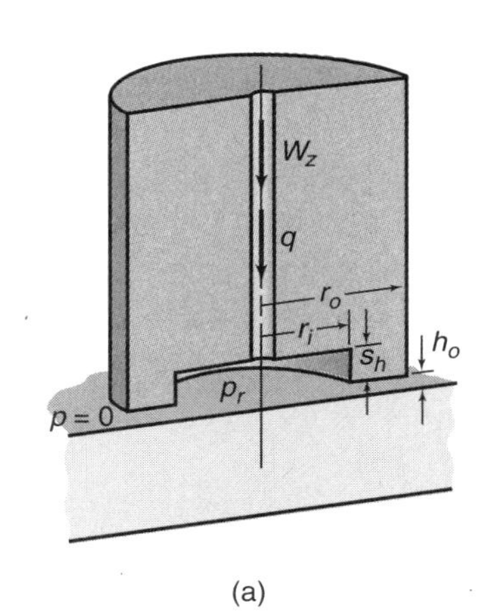

(a)

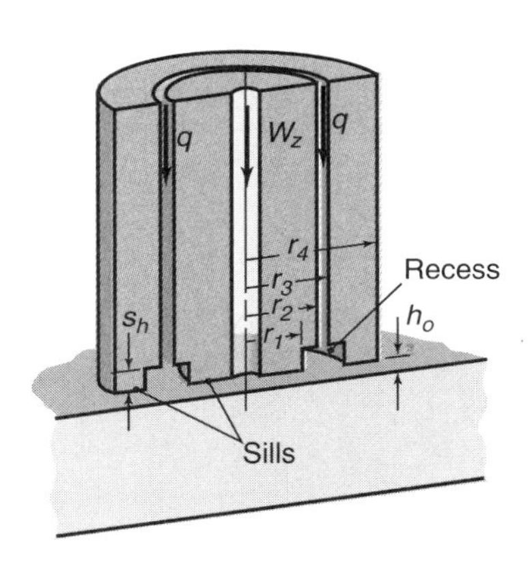

(b)

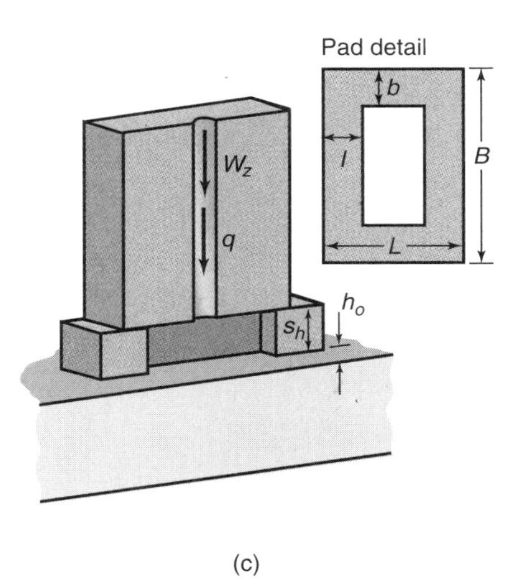

(c)

Figure 12.30: Hydrostatic thrust bearing configurations. (a) Radial-flow with circular step pad; (b) radial flow with annular pad; (c) rectangular pad.

sills. For this bearing, the following expressions are obtained for the pad coefficients:

$$a_b = \frac{1}{2\left(r_4^2 - r_1^2\right)}\left[\frac{r_4^2 - r_3^2}{\ln(r_4/r_3)} - \frac{r_2^2 - r_1^2}{\ln(r_2/r_1)},\right] \tag{12.106}$$

$$q_b = \frac{\pi}{6a_b}\left[\frac{1}{\ln(r_4/r_3)} + \frac{1}{\ln(r_2/r_1)},\right] \tag{12.107}$$

$$h_b = \frac{q_b}{a_b}. \tag{12.108}$$

The projected pad area is

$$A_p = \pi\left(r_4^2 - r_1^2\right). \tag{12.109}$$

12.6.3 Rectangular Sectors

If the pressure drop across the sill of a rectangular sector is linear, the pad coefficients can be calculated. Figure 12.30c shows a rectangular sector along with the linear pressure distribution. Assuming that the four corners are not contributing,

$$a_b = 1 - \frac{l}{L} - \frac{b}{B}, \tag{12.110}$$

$$q_b = \frac{1}{6a_b}\left(\frac{B - 2b}{l} + \frac{L - 2l}{b}\right). \tag{12.111}$$

Also, Eq. (12.108) still holds if Eqs. (12.110) and (12.111) are used for the load and flow coefficients, respectively. The areas of the bearing, the recess, and the sill are, respectively, given as

$$A_r = (L - 2l)(B - 2b), \tag{12.112}$$

$$A_s = A_b - A_r, \tag{12.113}$$

$$A_b = LB. \tag{12.114}$$

The hydrostatic bearings considered in this book have been limited to flat thrust-loaded bearings. Design information about conical, spherical, and cylindrical pads can be obtained from Rippel [1963]. The same approach used for the flat thrust-loaded bearings is used to obtain the pad coefficients for the more complex geometries.

Example 12.6: Analysis of Hydrostatic Bearings

Given: A flat, circular step hydrostatic thrust bearing is used to carry a load in a large milling machine. The purpose of the hydrostatic bearing is to position the workpiece accurately without any friction. The vertical positioning of the workpiece is influenced by the oil film thickness and its variation with load. The step outer radius is r_o = 100 mm. The oil viscosity is 0.025 N-s/m^2 and the smallest value of h_o allowed is 50 μm, and a step height of 5 mm is used to ensure a uniform reservoir pressure. The oil flow through the bearing is constant at 0.1×10^{-4} m^3/s.

Find: Determine r_i and the variation in film thickness if the load varies from 5 to 20 kN. Also, determine the largest reservoir pressure and the required power in order to specify a pump for the bearing.

Solution: Substituting Eq. (12.103) into Eq. (12.100) yields an expression for flow through the bearing as a function of load:

$$q = \frac{\pi W_z h_o^3}{3\left[1-(r_i/r_o)^2\right](\pi r_o^2)\eta_o} = \frac{W_z h_o^3}{3\eta_o\left(r_o^2 - r_i^2\right)}.$$

Solving for W_z,

$$W_z = \frac{3q\eta_o(r_o^2 - r_i^2)}{h_o^3}.$$

The minimum film thickness occurs at the maximum load; substituting $h_o = 50\ \mu$m and $W_z = 20$ kN, as well as the given values for q and r_o,

$$20 \times 10^3 = \frac{3(0.1\times10^{-4})(0.025)(0.100^2 - r_i^2)}{(50\times10^{-6})^3},$$

which is solved as $r_i = 0.082$ m. Using this radius and the load of 5000 N results in

$$5000 = \frac{3(0.1\times10^{-4})(0.025)(0.100^2 - 0.082^2)}{h_o^3},$$

hence $h_o = 78.9\ \mu$m. Therefore, the film thickness variation over this load range is $\Delta h_o = 78.9 - 50 = 28.9\ \mu$m. Equation (12.99) will be used to obtain the largest required reservoir pressure, but first the bearing pad load coefficient must be evaluated from Eq. (12.102) as

$$a_b = \frac{1-(r_i/r_o)^2}{2\ln(r_o/r_i)} = \frac{1-(0.082/0.1)^2}{2\ln(0.1/0.082)} = 0.825.$$

Therefore, solving Eq. (12.99) for the reservoir pressure, and recognizing the maximum pressure is needed for the 20 kN load,

$$p_r = \frac{W_z}{a_b A_p} = \frac{W_z}{a_b \pi r_o^2} = \frac{20,000}{(0.825)\pi(0.1)^2} = 771,000\ \text{Pa},$$

or $p_r = 771$ kPa. From Eq. (12.104),

$$h_p = p_r q = (771,000)(0.1\times10^{-4}) = 7.7\ \text{W}.$$

Case Study 12.1: Hydrodynamic Bearings in an Automotive Engine

Automobiles, motorcycles, buses, and most other vehicles utilize internal combustion engines as their power source. These engines are efficient and robust, using sophisticated designs and advanced tribology. This case study will select a number of components of automotive engines that are hydrodynamically lubricated using the methods described in this chapter. As a reference, Fig. 12.31 depicts an engine and the particular components to be addressed.

Main Bearings

The **main bearings** are journal bearings that support the crankshaft. While in theory a crankshaft needs only two main bearings, one at each end of the crankshaft, in practice such bearings are usually located immediately adjacent to the cams to minimize crankshaft deflection. Thus, there is usually one more main bearing than there are pistons. The crankshaft is in one of the most difficult-to-access locations of the engine. It is therefore imperative that the crankshaft and its bearings be designed so that maintenance will not be needed over the lifetime of the automobile. This requires that the bearings be designed to operate in the full-film lubrication regime.

The four-stroke loading cycle imparts a complicated dynamic load on the main bearing. Thus, in addition to the steady-state analysis using the approach described in Section 12.4, design of an automotive main bearing requires incorporation of squeeze terms in the Reynolds equation (Section 12.2.2) and requires dedicated simulation software to evaluate design robustness. Such squeeze films are in fact quite helpful, as their inherent damping effects reduce the peak dynamic loads to the crankshaft and supporting structure.

The crankshafts themselves usually serve as the bearing, and are ground and polished to obtain the required surface finish of approximately 0.2 μm (see Table 12.1). The journal is a replaceable pad that is attached to the engine block. Lubricant flow through the bearing is essential for cooling the bearing, suggesting that large radial clearances between the journal and bearing are advisable. However, too large of a clearance reduces the squeeze film effect (Section 12.2.2), so that smaller films result; large clearances also result in greater engine noise. Modern main bearings use radial clearances on the order of 25 μm, consistent with Fig. 12.27.

Thick lubricant films are essential for cooling, but are also beneficial for additional reasons. The shear within a bearing, especially on startup, can be very high; with a thin film, the shear is especially intense. It has been noted that for the conditions in hydrodynamic bearings, the lubricant may encounter shear localization that can cause molecules to degrade and result in a decreased viscosity. Also, contaminants from the combustion process or wear particles from engine components can cause considerable damage unless they can be flushed from the bearing and subsequently filtered.

Connecting Rod

The **connecting rod** is the machine element that transfers force from the piston to the crankshaft; a detailed view of a connecting rod is shown in Fig. 12.31. The bearing between the connecting rod and the crankshaft is the *connecting rod bearing*, and usually includes a soft metal insert.

The bearing between the connecting rod and piston is the *wrist pin*, also called the *gudgeon pin bearing*, and usually consists of a brass sleeve inserted into the connecting rod (called a fully-floating bearing), or a needle bearing (Table 13.6) press fit into the connecting rod end that supports the wrist pin (a semi-floating bearing). The connecting rod is not rigidly fixed at the piston or the crankshaft; as the crankshaft rotates, the relative motion seen by the connecting rod and wrist pin bearings is reciprocating sliding.

Both the connecting rod and wrist pin bearings encounter unsteady loads, with the largest load applied during the combustion stroke. The load during the intake, compression, and exhaust strokes is significantly smaller. Both the wrist pin and connecting rod bearing generate hydrodynamic films by Couette mechanisms (Section 12.2.2) during low load strokes, and depend on squeeze film effects during the combustion stroke to maintain the lubricant film. As a result, the connecting rod bearings are often characterized as squeeze film bearings.

Piston Rings

A **piston compression ring** (Fig. 12.31) is an open-ended ring that sits in a groove on the piston, and is used to establish a seal in the combustion chamber or cylinder. This has the dual goals of preventing oil contamination of the combustion chamber (and its inherent environmental drawbacks) and also prevents contamination of the oil by fuel and combustion products. However, some leakage can still occur, and is another reason for the need for periodic lubricant replacement. Worn piston rings are a major cause for excessive combustion of the lubricant in automotive engines.

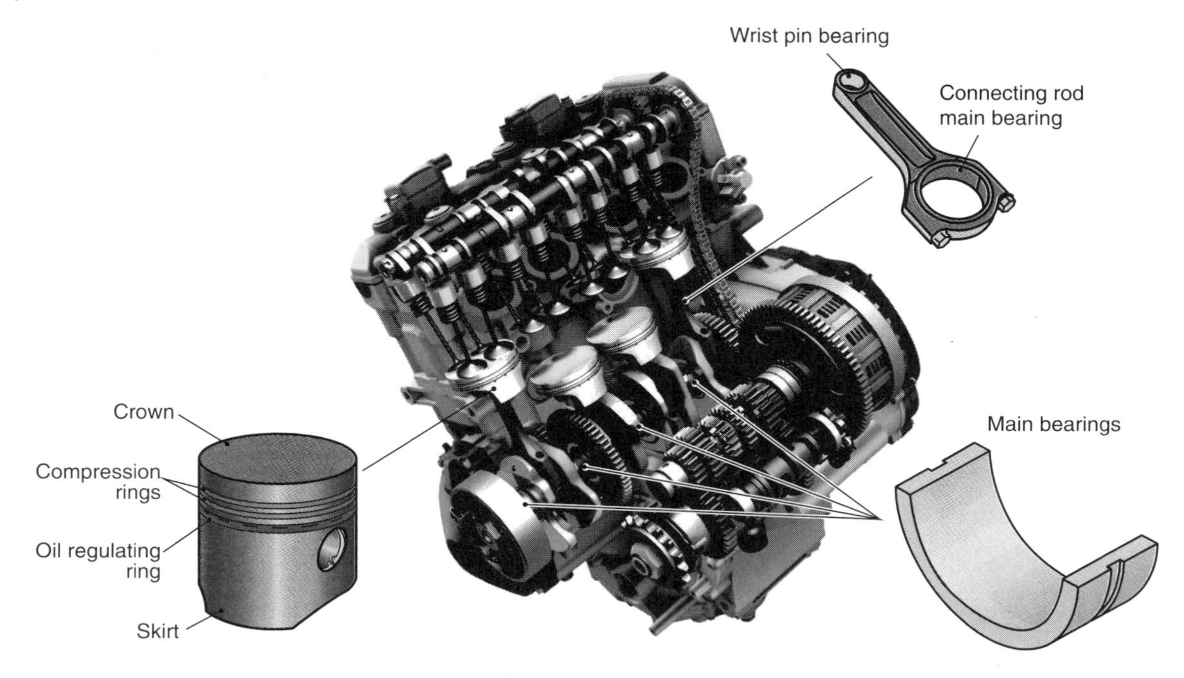

Figure 12.31: Illustration of a Suzuki motorcycle engine (2012 GSX-R1000) with selected hydrodynamically lubricated bearings. *Source:* Courtesy of Suzuki Motor of America, Inc.

As the piston travels in the cylinder, the piston ring slides against the cylinder wall or its liner. Generally a piston will use three rings; two to establish compression and one to control the supply of oil to the liner. The velocity of the piston varies with its location in the cylinder, and is hydrodynamic along most of its travel. However, since the velocity reverses direction at top and bottom dead centers of the stroke, it is at these locations that mixed or boundary lubrication can occur. For this reason, piston rings are made of wear resistant materials or are coated, often with a nitrided steel or a physical vapor deposited ceramic.

Hydrodynamic simulations of compression rings usually apply slider bearing theory without side flow (Section 12.3), but are complicated by the complex loadings due to the variation in cylinder pressure. The profile of the compression ring is usually assumed to be linear or parabolic. Piston rings, especially the second compression ring, can encounter starvation, a phenomenon where the oil supply is limited and has an adverse effect on film thickness.

12.7 Summary

The chapter began with the derivation of the Reynolds equation by coupling the Navier-Stokes equations with the continuity equation. The Reynolds equation contains Poiseuille, physical wedge, stretch, local expansion, and normal and translation squeeze terms. Each of these terms describes a specific type of physical motion, and the physical significance of each term was explained. Standard forms of the Reynolds equation were also discussed.

Design information was given for thrust and journal bearings. Results were presented for fixed-incline thrust slider bearings from numerical evaluations of the Reynolds equation. A procedure was outlined to assist in designing these bearings. The procedure provided an optimum pad configuration and described performance parameters such as normal applied load, coefficient of friction, power loss, and lubricant flow through the bearing. Similar design information was given for a plain journal bearing. Nonplain journal configurations were also considered. It was found that bearing designs with more converging and less diverging film thicknesses suppress system instabilities.

In a normal squeeze film bearing, a positive pressure was found to occur in a fluid contained between two surfaces when the surfaces are moving toward each other. A finite time is required to squeeze the fluid out of the gap, and this action provides a useful cushioning effect in bearings. The reverse effect, which occurs when the surfaces are moving apart, can lead to cavitation in fluid films. It was found for normal squeeze film bearings that a parallel film shape produces the largest normal load-carrying capacity of all possible film shapes. In contrast, for slider bearings the parallel film was shown to produce no positive pressure and therefore no load-carrying capacity. It was also found that for the normal squeeze action, as the bearing surfaces move toward each other, the viscous fluid shows great reluctance to be squeezed out the sides of the bearing. The tenacity of a squeeze film is remarkable and the survival of many modern bearings depends on this phenomenon. It was found that a relatively small approach velocity will provide an extremely large load-carrying capacity.

Hydrostatic bearings were briefly considered and were found to offer certain operating advantages over other types of bearings. Probably the most useful characteristics of such bearings are their high load-carrying capacity and inherently low friction at any speed.

Key Words

adiabatic temperature rise temperature rise in lubricant if all heat is carried away through convection

bearing number dimensionless operating parameter, specific to each type of bearing, that is important for determining bearing performance

Couette flow velocity-driven fluid flow

density wedge term dealing with density changes in bearing lubricant

eccentricity displacement of shaft center relative to bearing center

externally pressurized bearings same as hydrostatic bearings

fluid film lubrication lubrication condition where two surfaces transmitting a load are separated by a pressurized fluid film

gas-lubricated film bearings bearings where lubricating fluid is a gas

hydrodynamic lubrication lubrication activated by motion of bearing surfaces

hydrostatic lubricated bearings slider bearings where lubricant is provided at elevated pressure

journal bearings bearings whose surfaces are parallel to axis of rotation

line of centers line containing bearing and shaft centers

local expansion term involving time rate of density change

normal squeeze term dealing with approach of two bearing surfaces toward one another

Petrov's equation equation for frictional torque of concentric journal bearings.

physical wedge term dealing with rate of channel convergence

Poiseuille flow pressure-driven fluid flow

preload factor fractional reduction of film clearance when pads are brought in

Reynolds equation differential equation governing pressure distribution and film thickness in bearings

Sommerfeld number bearing characteristic number for a journal bearing

squeeze film bearings bearing where one dominant film-generating mechanism is squeeze effect (see terms in Reynolds equation)

stretch term term dealing with rate of velocity change in sliding direction

thrust bearings bearings whose surfaces are perpendicular to axis of rotation

translation squeeze term dealing with translation of two surfaces

Summary of Equations

Reynolds Equation:

$$\underbrace{\frac{\partial}{\partial x}\left(\frac{\rho h^3}{12\eta}\frac{\partial p}{\partial x}\right)}_{\text{Poiseuille}} = \underbrace{\rho\left(w_a - w_b - u_a\frac{\partial h}{\partial x}\right)}_{\text{Squeeze}} + \underbrace{h\frac{\partial \rho}{\partial t}}_{\text{Local expansion}} + \underbrace{\frac{\partial}{\partial x}\left[\frac{\rho h\left(u_a+u_b\right)}{2}\right]}_{\text{Couette}}$$

$$\overbrace{\underbrace{\frac{h\left(u_a+u_b\right)}{2}\frac{\partial \rho}{\partial x}}_{\text{Density wedge}} \quad \underbrace{\frac{\rho h}{2}\frac{\partial}{\partial x}\left(u_a+u_b\right)}_{\text{Stretch}} \quad \underbrace{\frac{\rho\left(u_a+u_b\right)}{2}\frac{\partial h}{\partial x}}_{\text{Physical wedge}}}$$

Thrust Slider Bearings:

Dimensionless variables:

$$P = \frac{p s_h^2}{\eta_o u_b l} \quad H = \frac{h}{s_h} \quad H_m = \frac{h_m}{s_h}$$

$$H_o = \frac{h_o}{s_h} \quad X = \frac{x}{l}$$

Adiabatic temperature rise: $\Delta t_m = \dfrac{h_p}{\rho q_x C_p}$

Bearing number: $B_t = \left(\dfrac{\eta_o u_b w_t}{W_z}\right)\left(\dfrac{l}{s_h}\right)^2$

Length-to-width ratio: $\lambda = \dfrac{l}{w_t}$

Journal Slider Bearings:

Petrov's Law: $T = Fr = \dfrac{4\pi^2\eta_o r^3 w_t N_a}{c} = \dfrac{2\pi\eta_o r^3 w_t \omega}{c}$

Bearing number or Sommerfeld number:

$$B_j = \left(\frac{\eta_o \omega r_b w_t}{\pi W_r}\right)\left(\frac{r_b}{c}\right)^2$$

Diameter-to-width ratio: $\lambda_j = \dfrac{2r_b}{w_t}$

Eccentricity ratio: $\epsilon = \dfrac{e}{c}$

Adiabatic temperature rise: $\Delta t_m = \dfrac{8.3W_r^*(r_b/c)\mu}{Q\left(1-0.5q_s/q\right)}$

Average pressure: $W_r^* = \dfrac{W_r}{2r_b w_t}$

Squeeze Film Bearings:

Film thickness as a function of time:

$$h_{o,2} = \frac{h_{o,1}}{\left[1+\left(2W_z'\Delta t h_{o,1}^2/\eta_o l^3\right)\right]^{1/2}}$$

Hydrostatic Bearings:

Load carrying capacity: $W_z = a_b A_p p_r$

Lubricant flow: $q = q_b \dfrac{W_z}{A_p}\dfrac{h_o^3}{\eta_o}$

Power loss: $h_p = p_r q = h_b\left(\frac{W_z}{A_p}\right)^2 \frac{h_o^3}{\eta_o}$

Circular step:

$$a_b = \frac{1-(r_i/r_o)^2}{2\ln(r_o/r_i)}$$

$$q_b = \frac{\pi}{3\left[1-(r_i/r_o)^2\right]}$$

$$h_b = \frac{2\pi\ln(r_o/r_i)}{3\left[1-(r_i/r_o)^2\right]^2}$$

Annular:

$$a_b = \frac{1}{2\left(r_4^2 - r_1^2\right)}\left[\frac{r_4^2 - r_3^2}{\ln(r_4/r_3)} - \frac{r_2^2 - r_1^2}{\ln(r_2/r_1)}\right]$$

$$q_b = \frac{\pi}{6a_b}\left[\frac{1}{\ln(r_4/r_3)} + \frac{1}{\ln(r_2/r_1)}\right]$$

$$h_b = \frac{q_b}{a_b}$$

Rectangular:

$$a_b = 1 - \frac{l}{L} - \frac{b}{B}$$

$$q_b = \frac{1}{6a_b}\left(\frac{B-2b}{l} + \frac{L-2l}{b}\right)$$

$$h_b = \frac{q_b}{a_b}$$

Recommended Readings

Bhushan, B. (2002) *Introduction to Tribology*, Wiley.

Budynas, R.G., and Nisbett, J.K. (2011), *Shigley's Mechanical Engineering Design*, 9th ed., McGraw-Hill.

Halling, J. (ed.) (1978) *Principles of Tribology*, Macmillan.

Hamrock, B.J., Schmid, S.R., and Jacobson, B.O. (2004) *Fundamentals of Fluid Film Lubrication*, 2nd ed., Marcell Dekker.

Juvinall, R.C., and Marshek, K.M. (2012) *Fundamentals of Machine Component Design*, 5th ed., Wiley.

Mott, R.L. (2014) *Machine Elements in Mechanical Design*, 5th ed., Pearson.

Norton, R.L. (2011) *Machine Design*, 4th ed., Prentice Hall.

Radzimovski, E.I. (1959) *Lubrication of Bearings*, Ronald Press.

Szeri, A.Z. (ed.) (1980) *Tribology: Friction, Lubrication and Wear*, Hemisphere Publishing Co.

References

Hamrock, B.J., Schmid, S.R., and Jacobson, B.O. (2004) *Fundamentals of Fluid Film Lubrication*, 2nd ed., Marcell-Dekker.

Petrov, N.P. (1883) Friction in Machines and the Effect of the Lubricant, *Inzh. Zh. St. Petersburg*, vol. 1, pp. 71–140; vol. 2, pp. 227–279; vol. 3, pp. 377–463; vol. 4, pp. 535–564.

Raimondi, A.A., and Boyd, J. (1955) Applying Bearing Theory to the Analysis and Design of Pad-Type Bearings, *ASME Trans.*, vol. 77, no. 3, pp. 287–309.

Raimondi, A.A., and Boyd, J. (1958) A Solution for the Finite Journal Bearing and Its Application to Analysis and Design-I, -II, -III, *ASLE Trans.*, vol. 1, no. 1, I–pp. 159–174; II–pp. 175–193; III–pp. 194–209.

Reynolds, O. (1886) On the Theory of Lubrication and Its Application to Mr. Beauchamp Tower's Experiments, Including an Experimental Determination of the Viscosity of Olive Oil, *Philos. Trans. R. Soc.*, vol. 177, pp. 157–234.

Rippel, H.C. (1963) *Cast Bronze Hydrostatic Bearing Design Manual*, 2nd ed., Cast Bronze Bearing Institute, Inc.

Shigley J.E., and Mitchell, L.D. (1983) *Mechanical Engineering Design*, 4th ed., McGraw-Hill.

Tower, B. (1883) First Report on Friction Experiments (Friction of Lubricated Bearings), *Proc. Inst. Mech. Eng. (London)*, pp. 632–659.

Questions

12.1 Explain the importance of the Reynolds equation, especially as relates to hydrodynamic bearings.

12.2 What are the functions of a lubricant?

12.3 What physical properties of a lubricant are important for the bearings described in this chapter?

12.4 Without the use of equations, define viscosity.

12.5 Give examples of hydrodynamic bearings.

12.6 Which term of the Reynolds equation contains the pressure? Which term contains the velocity?

12.7 What terms of the Reynolds equation are useful for squeeze film bearings?

12.8 What is side flow?

12.9 Define a thrust bearing. Is a journal bearing a thrust bearing? Explain.

12.10 Describe the significance of Petrov's equation.

12.11 What is eccentricity as relates to journal bearings?

12.12 What is the eccentricity ratio?

12.13 What is the attitude angle for journal bearings?

12.14 What is the Sommerfeld number?

12.15 What is a hydrostatic bearing?

12.16 What are the characteristics of hydrostatic bearings?

12.17 Give three examples of squeeze film bearings.

12.18 List the operating parameters for hydrodynamic bearings.

12.19 List the performance parameters for hydrodynamic bearings.

12.20 Why is heat dissipation important for hydrodynamic bearings?

Qualitative Problems

12.21 Explain the mechanism by which heat is generated in a hydrodynamic bearing.

12.22 List the similarities and differences between hydrodynamic thrust and journal bearings.

12.23 Why are the bearings described in this chapter especially long-lived?

12.24 In Chapter 8, it was noted that thick film lubrication holds when the ratio of film thickness to composite surface roughness is 10 or more. Give reasons that Table 12.1 requires much larger film thickness/surface roughness ratios.

12.25 It was stated in the text that the effect of bearing number on the coefficient of friction is small for a wide range of load parameters. Explain why this is so.

12.26 What are the similarities and differences between squeeze film and hydrostatic bearings?

12.27 Carefully sketch the pressure pad for a hydrostatic journal bearing.

12.28 Review Fig. 12.17. Considering the pressure distribution and the reaction forces that result, sketch the deflection (movement) of the bearing as load is increased.

12.29 Review Fig. 12.17 and explain if there is an advantage to having a full sleeve compared to a one-half sleeve journal bearing.

12.30 Using Table 12.1, plot the allowable minimum film thickness as a function of surface roughness. Comment on your observations, especially regarding active regimes of lubrication as discussed in Section 8.4.

12.31 Can air be used as the lubricant in hydrodynamic or hydrostatic bearings? Explain.

12.32 List and explain the desirable properties of a lubricant for a hydrodynamic bearing. Compare this to the list of desirable properties of a fluid used for hydrostatic bearings.

12.33 Explain the physical mechanisms that cause friction in hydrodynamic bearings, and explain how they are related to the normal force.

12.34 Using Table 12.1, plot the relative cost as a function of roughness for bearing surfaces. Comment on your observations.

12.35 Derive Eqs. (12.4) and (12.5).

Quantitative Problems

12.36 Derive Eq. (12.12).

12.37 To decrease the leakage past a seal of the type discussed in Example 12.1, the sealing gap is tapered and the film shape is

$$h(x) = h_o + x\frac{dh}{dx} = h_o + kx$$

Because of the roughness of the surfaces, the outlet (minimum) oil film thickness h_o has to be at least 10 μm. Find the optimum film thickness to minimize the total leakage past the 10-mm-wide seal during a full work cycle (back and forth) of the piston when the fluid viscosity is 0.075 Pa-s, the piston speed is 0.8 m/s in both directions, the stroke length is 85 mm, the sealing pressure is 1 MPa, and the piston diameter is 50 mm.

12.38 The oil viscosity is 0.005 Pa-s in a very wide slider bearing with an exponential oil film shape:

$$h = (h_o + s_h)\, e^{-x/l}$$

Find the pressure distribution and calculate the load-carrying capacity per unit width when the sliding speed is 10 m/s if $h_o = 60$ μm and the bearing is 0.10 m long in the direction of motion. *Ans.* $W_z/w_t = 22.53$ kN/m.

12.39 A flat strip of metal emerges from an oil bath having viscosity η_o and inlet pressure p_i with velocity u_b on passing through a slot of the form shown in Sketch a. In the initial convergent region (region 1) of the slot the film thickness decreases linearly from $h_o + s_h$ to h_o over the length $n_s l$ on each side of the strip. In the final section of the slot, the film on each side has a constant thickness h_o over the length $l(1 - n_s)$. The width of the flat strip is large relative to its length, so that it can be considered infinitely wide. Find the pressure distribution and volume flow rate in the x-direction.

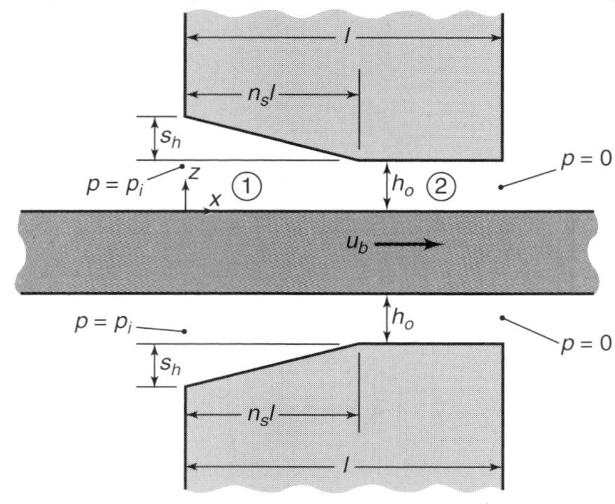

Sketch a, for Problem 12.39

12.40 A fixed-incline slider bearing is optimized to give minimum power loss. The pad length l equals the width w_t and the bearing carries the load $W_z = 5$ kN. The sliding speed $u_b = 15$ m/s. Find the minimum film thickness for the bearing if $\eta = 0.050$ N-s/m^2, $\lambda = w_t/l = 1$, and $l = 0.070$ m. *Ans.* $h_o = 62.43$ μm.

12.41 The flat slider bearing in Problem 12.40 is split into two equal halves, each with width $w_t = 0.035$ m. Find the viscosity needed at the running temperature to obtain the same minimum film thickness as in Problem 12.40. The load is equally split between the two halves, with 2.5 kN each. The bearing halves should both be optimized for minimum relative power loss. *Ans.* $\eta = 0.122$ Ns/m^2.

12.42 A very wide slider bearing lubricated with an oil having a viscosity of 0.075 N-s/m^2 consists of two parts, each with a constant film thickness (see Sketch b). Find the pressure distribution in the oil and the position and size of the bearing load per unit width of the bearing. All dimensions are in millimeters. *Ans.* $W_z/w_t = 2.068$ MN/m.

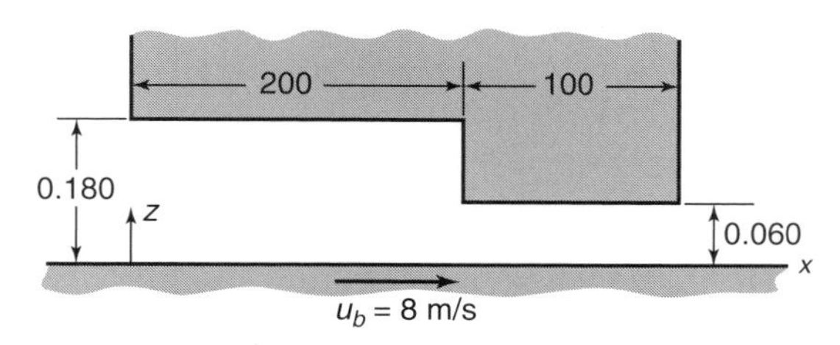

Sketch b, for Problem 12.42

12.43 Given the infinitely wide slider bearing shown in Sketch c, find the pressure distribution by using the Reynolds equation.

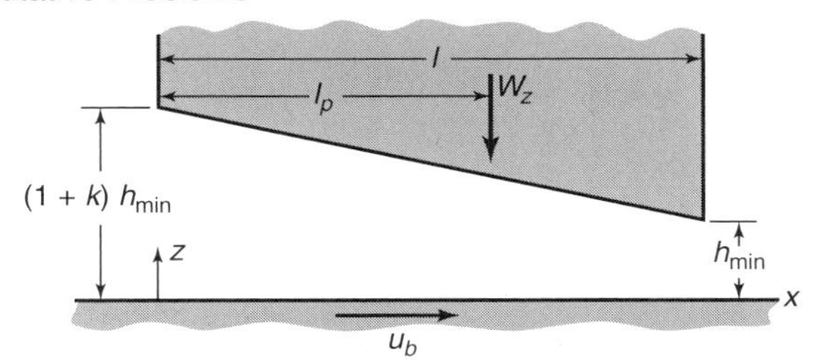

Sketch *c*, for Problem 12.43

12.44 A thrust bearing carrying the weight of a water turbine for a hydroelectric power plant is lubricated with water. The design should be optimized to obtain minimum power loss when the roughness of the bearing surfaces is $R_a = 7\ \mu$m, indicating that the bearing minimum film thickness should be 100 μm. The load on the bearing from the weight of the turbine and the water is 1 MN. The viscosity of water is 0.001 Pa-s and the bearing length-to-width ratio is 1, which gives minimum power loss for four pads circumferentially around the bearing with a mean radius equal to the bearing pad width. Determine the bearing dimensions and calculate the coefficient of friction at a rotating speed of 100 rpm. *Ans.* $w_t = 1.40$ m, $\mu = 0.00069$.

12.45 For a fixed-incline-pad thrust bearing with a total normal load of 10,000 lb, $r_o = 4$ in, $r_i = 2$ in, $N_a = 30$ r/s, $l/b = 1$, SAE 10 oil, and $t_i = 100°$F, determine the following: N_p, s_h, h_o, Δt_m, μ, h_p, q, and q_s. *Ans.* $N_p = 9$, $h_o = 270\ \mu$in., $q = 0.118$ in^3/2.

12.46 A propeller shaft thrust bearing for an oil tanker is a fixed-incline slider bearing with 12 pads. Each pad is 300 mm wide and 300 mm long and has a mean radius of 800 mm. The shaft speed is 220 rpm. Find the optimum lubricant viscosity to get minimum power loss. Calculate the coefficient of friction and the power loss. The lubricant inlet temperature is 63°C, the heat capacity of the oil is $C_s = 1.72 \times 10^6$ J/(m^3-°C), and the propeller thrust load is 1.3 MN. *Ans.* $\mu = 0.00167$, $h_p = 40.0$ kW.

12.47 For a fixed-incline-pad slider thrust bearing with $w_z = 3600$ lb, $u_b = 1200$ in/s, $l = 3$ in, $w_t = 3$ in, and SAE 10 oil with inlet temperature of 40°C, determine the following for a maximum normal load: s_h, h_o, Δt_m, μ, h_p, q, and q_s. *Ans.* $s_h = 8.756 \times 10^{-4}$ in., $h_o = 1200\ \mu$in., $q = 1.89$ in^3/s.

12.48 Assume that the shaft designed in Problem 11.36 (with $d = 2.25$ in.) rotates at 2000 rpm and will be supported by two identical journal bearings. Determine the radial clearance, coefficient of friction, and lubricant flow through the more critical bearing if the bearing surfaces are ground to a surface roughness of 12 μin. and have a diameter-to-width ratio of 2. What lubricant viscosity is needed for a general-purpose bearing? What multigrade SAE oil would you recommend for this application? *Ans.* $c \geq 893\ \mu$in, $\mu = 0.00436$, $\Delta t_m = 116°$F.

12.49 A full journal bearing is 25 mm wide. The shaft has a diameter of 50 mm. The load is 1000 N and the journal speed is 1000 rev/min. Based on maximum load carrying ability, determine the clearance that should be machined into the sleeve, the minimum film thickness, the power loss and the flow and side flow through the bearing if the operating temperature is 50°C and SAE 30 oil is used as the lubricant. *Ans.* $c = 37.0\ \mu$m, $h_{min} = 17.7\ \mu$m, $h_p = 26.0$ W.

12.50 A plain journal bearing has a diameter of 2 in. and a width of 1 in. The full journal bearing is to operate at a speed of 2000 rpm and carry a load of 750 lb. If SAE 20 oil at an inlet temperature of 100°F is to be used, determine the following:

(a) The radial clearance for optimum load-carrying capacity, the temperature rise, and the mean temperature. *Ans.* $c = 6.9 \times 10^{-4}$ in., $\Delta t_m = 95°$F.

(b) The performance parameters: coefficient of friction, flow rate, side flow, and attitude angle. *Ans.* $\mu = 0.0055$, $q = 0.120$ in^3/s.

(c) The kinematic viscosity of the oil at the mean temperature if the oil density is 0.89 g/cm^3.

12.51 An oil pump is designed like a journal bearing with a constant oil film thickness as shown in Sketch *d*. The wrap angle is 320°, the film thickness is 0.1 mm, and the oil viscosity is 0.022 N-s/m^2. The shaft diameter $2r = 60$ mm and the length of the pump in the axial direction is 50 mm. Find the volume pumped per unit time as a function of the resisting pressure at the rotational speed $N_a = 1500$ rpm.

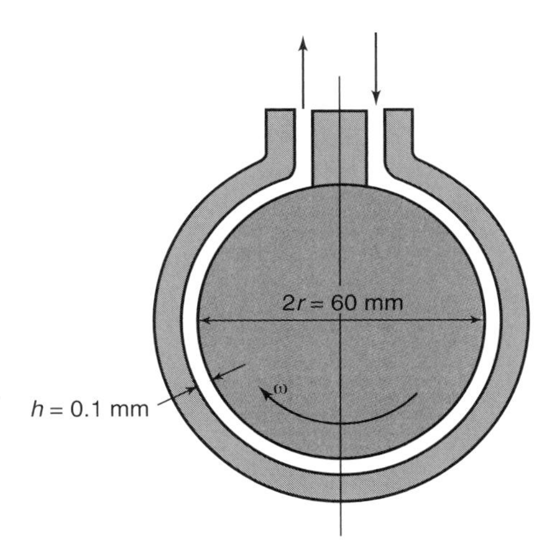

Sketch *d*, for Problem 12.51

12.52 A full journal bearing is used in a power plant generator in the United States; it has four poles and delivers electricity with 60-Hz frequency. The rotor has a 25,000 kg mass and is carried equally by two journal slider bearings. The shaft diameter is 300 mm and each bearing is 300 mm wide. The bearing and shaft surfaces are ground to a surface finish $R_a = 1\ \mu$m. Dimension the bearings for minimum power loss and calculate the coefficient of friction, the side-leakage ratio, and the location of the terminating pressure to find where the oil feed groove can be placed. To ensure that the bearings are dynamically stable, choose an eccentricity ratio of 0.82. Calculate the power loss in the bearing. *Ans.* $\mu = 0.00195$, $c = 139\ \mu$m, $h_p = 6.76$ kW.

12.53 The power plant generator considered in Problem 12.52 is rebuilt and moved to Europe. The rebuilt generator has two poles and the frequency of the electricity is 50 Hz. Redimension the bearings for minimum power loss and calculate the coefficient of friction, the side-leakage ratio, and the location of the terminating pressure. Let the eccentricity ratio be 0.82 to make sure that the bearings are dynamically stable. Calculate the power loss. *Ans.* $h_p = 13.5$ kW.

12.54 A motor speed of 110 rpm is used in a gearbox to provide a propeller speed of 275 rpm. The propeller shaft diameter is 780 mm, and the shaft is directly coupled to the outgoing shaft of the gearbox. The combined influence of the propeller shaft weight and the gear forces applies a 1.1-MN load on the journal bearing. Dimension the journal bearing for a diameter-to-width ratio of 2, and find the viscosity and radial clearance needed to give a minimum power loss; also, calculate the power loss in the bearing. The journal bearing diameter is 780 mm and the surface roughness $R_a = 3\ \mu$m. *Ans.* $\eta = 0.039$ Ns/m^2, $c = 500\ \mu$m, $h_p = 24$ kW.

12.55 A steam turbine rotor for a small power plant in a paper factory has two journal bearings, each carrying half the weight of a 24,000-N rotor. The rotor speed is 3000 rpm, the bearing diameters are both 120 mm, and the bearings are 120 mm wide. Design the bearing with surface roughness $R_a = 0.6$ mm and minimum power loss. Calculate the coefficient of friction, the attitude angle, the oil flow rate, and the side-leakage flow rate. Choose an eccentricity ratio of 0.82 to avoid dynamic instability. *Ans.* $\mu = 0.00231$, $q = 1.20 \times 10^{-4}$ m^3/s.

12.56 A full journal bearing is used in an air compressor to support a 5.0-kN radial load at a speed of 2300 rpm. The diameter-to-width ratio is 2 and SAE 50 oil is used with an inlet temperature of 100°C. Determine the minimum bearing dimensions and the radial clearance for both maximum load and minimum coefficient of friction. The radial load per area is 1.5 MPa. *Ans.* $w_t = 0.0408$ m, for maximum load, $c = 31.7\ \mu$m.

12.57 A shaft rotates at 4000 rpm and is supported by a journal bearing lubricated with SAE 30 oil at an inlet temperature of 25°C. The bearing has to support a 1.5-kN load, has a 40 mm diameter and 10 mm width. It operates at a condition which gives the largest possible load support, and the surfaces each have a roughness of 0.1 μm. Determine the film parameter, the temperature rise in the lubricant and the coefficient of friction for this bearing. *Ans.* $\Lambda = 17.07$, $\mu = 0.019$.

12.58 A 3.0-kN radial load is applied to a 50-mm-diameter shaft rotating at 1500 rpm. A journal bearing is used to carry the radial load, and has a diameter-to-width ratio of 1 and is lubricated with SAE 20 oil with an inlet temperature of 35°C. Determine the following:

(a) Mean temperature and temperature rise in the bearing. *Ans.* $\Delta t_m = 12.7$°C.

(b) Minimum film thickness and its location. *Ans.* $h_{\min} = 26\ \mu$m, $\phi = 55°$.

(c) Maximum pressure and its location. *Ans.* $p_{\max} = 2.91$ MPa, $\phi_{\max} = 76°$.

(d) Total and side-leakage flow rates. *Ans.* $q = 8.12 \times 10^{-6}$ m^3/s.

12.59 A 4000-N radial load is applied to a 50-mm-diameter shaft rotating at 2000 rpm. A journal bearing is to be designed to support the radial load. The journal has a diameter-to-width ratio of 1 and is lubricated with SAE 30 oil with an inlet temperature of 35°C. It is to be designed for minimum friction. Determine the following:

(a) The required radial clearance. *Ans.* $c = 104\ \mu$m.

(b) Mean temperature and temperature rise in the bearing. *Ans.* $\Delta t_m \approx 10$°C.

(c) Minimum film thickness. *Ans.* $h_{\min} = 36.4\ \mu$m.

(d) Total and side flow leakage rates.

12.60 A journal bearing 100 mm in diameter and 100 mm long has a radial clearance of 0.05 mm. It rotates at 2000 rpm and is lubricated with SAE 10 oil at 100°C.

(a) Estimate the power loss and the friction torque using the Petrov equation. Calculate the coefficient of friction if the applied load is 1000 N. *Ans.* $h_p = 209.4$ W, $\mu = 0.00138$.

(b) Calculate the clearance, the power loss, friction torque, and coefficient of friction if the applied load is 1000 N using the charts in Section 12.4.4. *Ans.* $h_p = 143$ W, $\mu = 0.0137$.

12.61 A 50-mm-diameter shaft rotating at 5000 rpm is to be supported by a journal bearing. The bearing is lubricated by SAE 30 oil with a mean temperature of 40°C. The bearing supports a load of 10 kN. Using a diameter-to-width ratio of 2.0, obtain the following:

(a) The required radial clearance for maximum load carrying capability. *Ans.* $c = 37.4\ \mu$m.

(b) The coefficient of friction for the bearing. *Ans.* $\mu = 0.0135$.

(c) The volume flow rate of oil through the bearing. *Ans.* $q = 9.55 \times 10^{-6}$ m^3/s.

(d) The side leakage of oil in the bearing.

12.62 A 45-kN radial load is applied to a 100-mm-diameter shaft rotating at 1000 rpm. A journal bearing is to be designed to support the radial load. The journal has a diameter-to-width ratio of 1 and is lubricated with SAE 30 oil at a mean temperature of 90°C. It is to be designed for maximum load support. Determine the following:

(a) The required radial clearance. *Ans.* $c = 20.4\ \mu$m.

(b) Coefficient of friction. *Ans.* $\mu = 0.0020$.

(c) Minimum film thickness. *Ans.* $h_{\min} = 11.01\ \mu$m.

(d) Total and side flow leakage rates.

12.63 A shaft is rotating at 1500 rpm and is supported by two journal bearings at the two ends of the shaft. The bearings are lubricated with SAE 40 oil with an inlet temperature of 25°C. A 5-kN load is applied 0.5 m from the left bearing where the total shaft length is 2.5 m. The bearing width is 25 mm, the diameter is 50 mm, and the radial clearance is 0.0315 mm. Determine

(a) Temperature rise. *Ans.* $\Delta t_m = 50$°C.

(b) Minimum film thickness. *Ans.* $h_{\min} = 10\ \mu$m.

(c) Maximum pressure. *Ans.* $p_{\max} = 9.7$ MPa.

(d) Side-leakage flow rate.

(e) Bearing power loss. *Ans.* $h_p = 119.4$ W.

12.64 A plain journal bearing has a diameter of 2 in. and a length of 1 in. The full journal bearing is to operate at a speed of 2000 r/min and carries a load of 500 lb. If SAE 10 oil at an inlet temperature of 110°F is to be used, find the radial clearance for optimum load-carrying capacity. Describe the surface finish that would be sufficient from a lubrication standpoint. Also, determine the temperature rise, coefficient of friction, flow rate, side flow rate, and attitude angle. *Ans.* $\Delta t_m = 78.6$°F, $\mu = 0.00778$.

12.65 The infinitely wide slider bearing shown in Fig. 12.28 has parallel surfaces, squeeze velocity across the film of w, oil film thickness h, and bearing length l. Find the bearing damping constant $B = -\partial W'_z/\partial w$ where W'_z is the bearing load per unit width. Use the Reynolds equation for squeeze motion. *Ans.* $B = \eta_o l^3/h_o^3$.

12.66 The circular hydrostatic thrust bearing shown in Sketch e has inner radius r_i and outer radius $r_o = 3r_i$, a constant oil film thickness h, and a constant oil flow rate q. The bearing load is P and the oil viscosity is η. Find the bearing stiffness coefficient $c_s = -\partial P/\partial h$. *Ans.* $c_s = 72\eta_o q r_i^2/h^4$

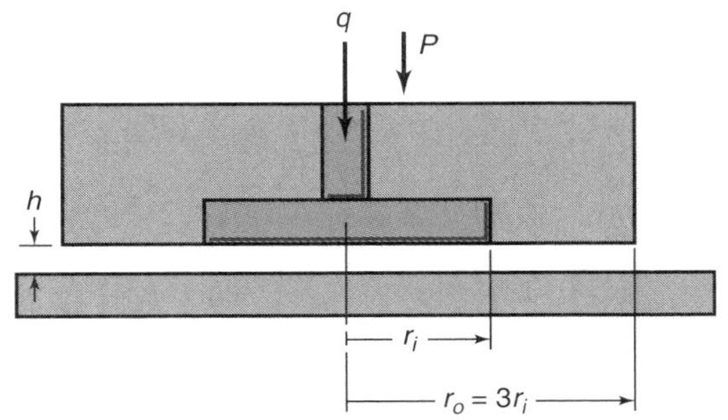

Sketch e, for Problem 12.66

12.67 A circular hydrostatic thrust bearing, shown in Sketch f, has a constant oil film thickness h. A central lubricant reservoir, whose height is much larger than the oil film thickness, is fed by a liquid with pressure p_r. Find the pump power and required flow rate if the bearing carries load P. It is given that $r_i = 27$ mm, $\eta = 0.00452$ N-s/m^2, $h = 120$ μm, $r_o = 200$ mm, $P = 3.08 \times 10^4$ N. *Ans.* $h_p = 100$ W.

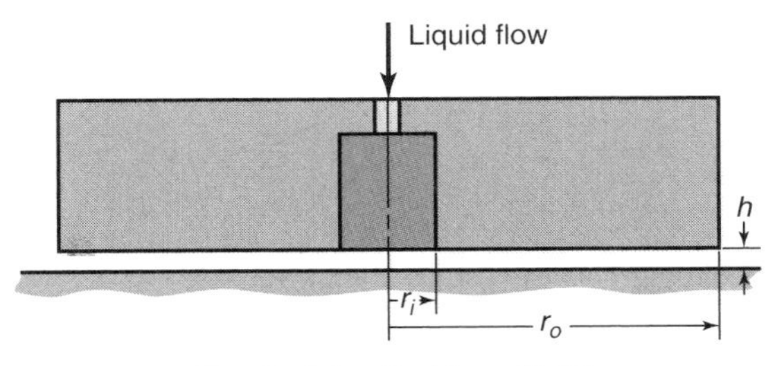

Sketch f, for Problem 12.67

12.68 A flat, circular step hydrostatic air bearing has an outer diameter of 160 mm and a 5-mm-deep recess from the 50-mm diameter to the bearing center. The air pressure in the recess is 2 MPa, and the pressure around the outside of the bearing is 0.1 MPa. Determine the pressure distribution in the bearing if 0.3 kg of air is pumped through the bearing per second. The air density is 1 kg/m^3 at atmospheric pressure and the air viscosity is 18.2×10^{-6} Pa-s at 22°C. Also, calculate the air film thickness, load support, and required pump power. *Ans.* $h_o = 185$ μm, $h_p = 597$ kW.

Design and Projects

12.69 Find the value of n_s that produces the largest pressure in a parallel step slider bearing while ignoring side flow.

12.70 Obtain, plot, and compare the dimensionless pressure distribution and load support for $H_o = 0.25$ for the following slider bearings:

(a) A fixed-incline slider bearing.

(b) A parallel step slider bearing where $n_s = 0.5$.

(c) A parallel step slider bearing where $n_s = 0.7182$.

(d) An exponential oil film shape given by $h(x) = (h_o + s_h)\, e^{-x/l}$

Comment on the results.

12.71 Another form of hydrodynamic bearing is the *foil bearing*. Conduct an Internet search and summarize the characteristics and design advantages of foil bearings.

12.72 A sculptural display uses a 300-mm-diameter granite sphere suspended on a pressurized water bed. Design the pad and pump to support the sphere so that the minimum film thickness is 50 μm.

12.73 Review the technical literature and identify the terms of the Reynolds equation that are applicable to analysis of an artificial human hip joint.

12.74 Lubricants with a higher viscosity generate higher film thicknesses in hydrodynamic bearings such as the main bearing in internal combustion engines. Since a thick lubricant film is thought to be beneficial from a wear prevention standpoint, why are more viscous lubricants than SAE 10W-30 or 10W-40 not generally used for automobiles?

12.75 The equation of state for a perfect gas is

$$p = \rho \bar{R} t_m$$

where $\bar{R}$ is the gas constant and t_m is the absolute temperature. If gas viscosity can be taken as a constant, derive the Reynolds equation suitable for gas-lubricated bearings.

12.76 It was stated in the text that a thrust bearing can use pivoted pads to allow the direction of rotation to reverse. Carefully sketch a journal bearing that uses pivoted pads in the same fashion. Is there any benefit to having pivoted pads in a journal bearing? Explain.

12.77 A squeeze film damper is used to stop the motion and damp the kinetic energy of an overhead crane. The damper consists of two flat, parallel steel plates 2 m in the side-leakage direction and 0.1 m long separated by a 1-mm-thick layer of oil with viscosity 0.1 Pa-s. When the overhead crane hits the damper, the oil is pressed out from between the plates and the oil layer becomes thinner. Calculate the damping force at the first moment of impact and the instantaneous retardation if the speed at impact is 0.2 m/s and the crane mass is 2000 kg. Also, calculate the damping force at the first moment of impact if the crane carries a 20,000 N load.

12.78 Design a journal bearing to support a 2000-lb load at 1000 rpm. Specify the bearing diameter, length, radial clearance, surface finish, and lubricant.

12.79 A generator (total weight of 500 kN) in a power plant consists of a large rotor with its weight supported by two journal bearings. Each bearing has a radius of 225 mm, a radial clearance of 0.325 mm, and a sleeve width of 350 mm. The lubricant used is SAE 40, and measurements at the inlet of each bearing indicate a flow rate of 22.4 m^3/hr and a temperature of 41°C. At the outlet, the temperature is 70.1°C. The shaft of the generator rotates at 3600 rpm. The generator has been in service for many decades and is now being refurbished. Analyze the available data for consistency and suggest improvements to the journal bearing design. In addition, estimate the heat that must be extracted from the lubricant to maintain the measured operating temperatures.

Chapter 13

Rolling-Element Bearings

An assortment of rolling-element bearings. *Source:* Courtesy of The Timken Company.

Since there is no model in nature for guiding wheels on axles or axle journals, man faced a great task in designing bearings – a task which has not lost its importance and attraction to this day.

Rolling Bearings and Their Contribution to the Progress of Technology [1986].

Contents

Examples

Design Procedures

Case Study

This chapter focuses on rolling-element bearings of all types. Rolling-element bearings consist of balls, spherical profiles, cylinders, or tapered cylinders that rotate between two races, with their spacing usually maintained by a separator or cage. These bearings are non-conformal in their contact. Significant experience has been developed in using these bearings, and they are widely applied in all industries. Rolling-element bearings have the typical characteristics of low startup friction, high efficiency and reliability, and good load support. These bearings are very compact, and some arrangements can support significant thrust loads in addition to radial loads. Such bearings have long lives with little drop in performance. Fatigue wear is a main consideration, as is proper lubrication. This chapter discusses the ratings of rolling element bearings, but only after developing the theory of rolling element behavior performance, including estimation of forces, speeds, lubricant film thickness, and the nature of fatigue wear. The use of bearing load ratings and complicated variable loading is also discussed.

Machine elements in this chapter: Deep groove, angular contact, and thrust ball bearings; cylindrical and spherical rolling-element bearings, needle bearings, and tapered rolling-element bearings.
Typical applications: Vehicle axles, gearboxes, machine tools, thrust bearings for turbines, general support for any rotating shaft or component.
Competing machine elements: Hydrodynamic and hydrostatic bearings (Chapter 12).

Symbols

A	area, m^2
$\tilde{A}$	constant
B	total conformity ratio
b_w	bearing width, m
b^*	contact semiwidth, m
C	dynamic load rating, N
$\bar{C}$	specific dynamic capacity or load rating of bearing, N
C_o	static load rating, N
c_d	diametral clearance, m
c_e	free endplay, m
c_r	distance between race curvature centers, m
D_x	diameter of contact ellipse along x-axis, m
D_y	diameter of contact ellipse along y-axis, m
$\bar{D}$	material factor
d	diameter of rolling element, m
d_a	outer diameter of outer bearing race, m
d_b	inner diameter of inner bearing race, m
d_e	pitch diameter, m
E'	effective elastic modulus
$\bar{E}$	metallurgical processing factor
$\mathcal{E}$	complete elliptic integral of second kind
$\bar{F}_l$	lubrication factor
$\mathcal{F}$	complete elliptic integral of first kind
G	dimensionless materials parameter, $\xi E'$
$\bar{G}$	speed effect factor
H	dimensionless film thickness
$\bar{H}_m$	misalignment factor
h	film thickness, m
h_o	central film thickness, m
K_1	constant defined in Eq. (13.34)
$K_{1.5}$	constant defined in Eq. (13.33)
k_e	ellipticity parameter, D_y/D_x
L_A	adjusted life, millions of cycles
$\tilde{L}$	fatigue life, millions of cycles
$\tilde{L}_{10}$	fatigue life for survival probability of 0.90, millions of cycles
l	length of rolling element, m
l_e	length of roller land, m
l_l	roller effective length, m
l_t	roller length, m
l_v	length dimension in stressed volume, m
$\tilde{M}$	probability of failure
m_k	load-life exponent
N_b	rotational speed, rpm
n	number of rolling elements per row
n_s	static safety factor
P	equivalent load, N
P_a	axial component of load, N
P_z	normal applied load, N
P_z'	normal applied load per unit width, N/m
P_r	radial component of load, N
P_t	system thrust load, N
P_o	equivalent static load, N
$\bar{P}$	bearing equivalent load, N
p	pressure, Pa
R	curvature sum, m
R_q	root-mean-square (rms) surface roughness, m
R_r	race conformity, r/d
R_x, R_y	effective radii in x- and y-directions, respectively, m
r	radius, m
r_c	corner radius, m
r_r	crown radius, m
$\tilde{S}$	probability of survival
s_h	shoulder height, m
U	dimensionless speed parameter, $\dfrac{\eta_o \tilde{u}}{E' R_x}$
u	velocity in x-direction, m/s
$\tilde{u}$	mean surface velocity in x-direction, $(u_a + u_b)/2$, m/s
$\tilde{V}_i$	elemental bearing volume, m^3
W	dimensionless load parameter, $\dfrac{P}{E' R_x^2}$
W'	dimensionless load per width, $\dfrac{P_z'}{E' R_x}$
W_t	total load, N
w_e	equivalent bearing load, N
w_o	equivalent static bearing load, N
w_t	total thrust load of bearing, N
w_z	bearing load, N
w_ψ	load at ψ, N
X	radial factor
X_o	radial factor for static loading
Y	thrust factor
Y_o	thrust factor for static loading
x	Cartesian coordinate in direction of sliding, m
y	Cartesian coordinate in side-leakage direction, m
Z_w	constant defined in Eq. (13.46)
z	Cartesian coordinate in direction of film, m
z_o	depth of maximum shear stress, m
α_r	radius ratio, R_y/R_x
β	contact angle, deg
β_f	free contact angle, deg
δ	elastic deformation, m
δ_a	axial deflection, m
δ_m	elastic deformation, m
$\delta_{\max}$	interference, or total elastic compression on load line, m
δ_t	deflection due to thrust load, m
δ_φ	elastic compression of ball, m
η	absolute viscosity, Pa-s
η_o	absolute viscosity at $p = 0$ and constant temperature, Pa-s
θ_s	angle used to define shoulder height, deg
Λ	dimensionless film parameter, $\dfrac{h_{\min}}{\sqrt{R_{qa}^2 + R_{qb}^2}}$
ν	Poisson's ratio
ξ	pressure-viscosity coefficient, m^2/N
ρ	density of lubricant, kg/m^3
ρ_o	density at $p = 0$, kg/m^3
τ_o	shear stress, Pa
ϕ_s	angle locating ball-spin vector, deg
ψ	angle to load line, deg
ψ_l	angular extent of bearing loading, deg
ω	angular velocity, rad/s
ω_b	angular velocity of surface b or of rolling element about its own axis, rad/s
ω_c	angular velocity of separator or of ball set, rad/s
ω_r	angular velocity of race, rad/s

Subscripts

a	solid a
all	allowable
b	solid b
i	inner; inlet
o	outer; outlet
x, y, z	coordinates
0	static loading

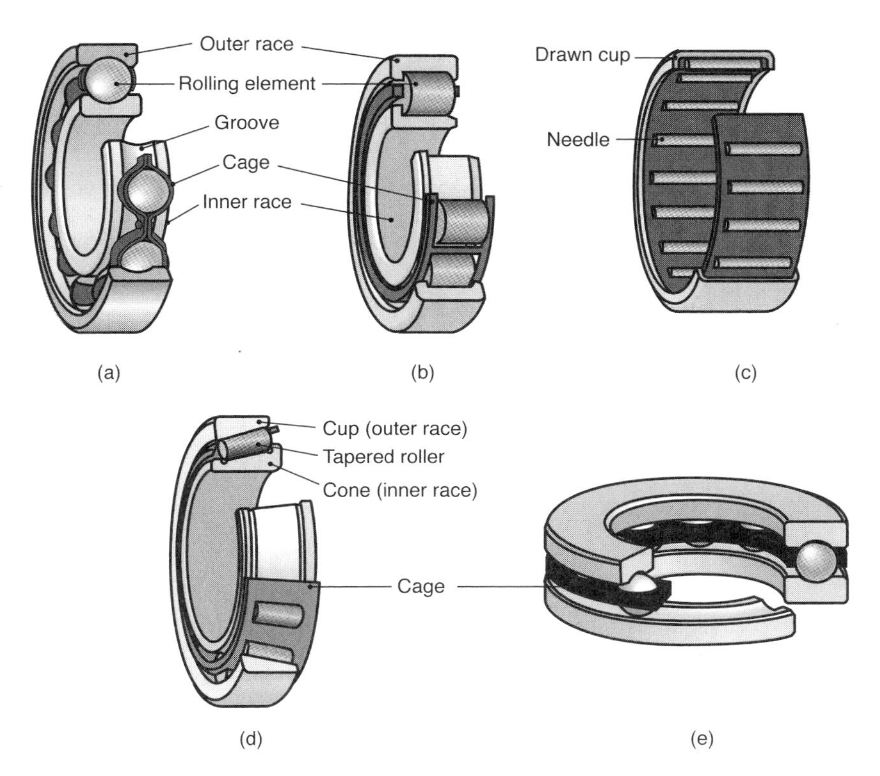

Figure 13.1: Illustration of selected single row rolling-element bearings. (a) Deep groove ball bearing; (b) cylindrical roller bearing; (c) needle bearing with drawn cup and no inner race. In such a circumstance, the needles roll on a ground and hardened shaft. (d) Tapered roller bearing; (e) thrust ball bearing.

13.1 Introduction

Chapter 12 considered hydrodynamic bearings, which involve conformal surfaces in sliding motion. This chapter considers **rolling-element bearings**, which are precise, yet simple, machine elements of great utility where the surfaces are nonconformal and the motion is primarily rolling (Fig. 13.1). Section 13.2 provides a brief introduction to the history of rolling-element bearings. Subsequent sections describe the types of rolling-element bearing and their geometry and kinematics. Selection of bearings according to static and dynamic load ratings are discussed, along with lubrication of such bearings and other factors that can affect rolling-element bearing life. The organization of this chapter is such that basic characteristics are covered in Section 13.3 followed by a more detailed discussion of the geometry (Section 13.4) and kinematics (Section 13.5) in bearings, and the role of separators (Section 13.6). The chapter then develops the theory and design practice for selecting highly loaded but unlubricated rolling-element bearings in Section 13.7; and loaded and lubricated rolling-element bearings in Sections 13.8 and 13.9.

When a designer needs to select and analyze a rolling-element bearing, the problem of interest is often constrained to a great extent. For example, if a shaft has been designed to produce a desired low deflection (see Section 5.2) or high fatigue life (Section 11.3), its diameter may be relatively large, so that all bearings that fit the shaft may have excess capacity for the required load support. Certainly, a shoulder will be machined into the shaft in order to mount the bearing (with specific dimensions given in manufacturer's literature). However, selecting a bearing with a capacity that exactly matches the applied load may require machining a very small diameter in the shaft to fit the bearing bore, and may not be practical. In many cases, the bearing selected for the shaft geometry will be sufficient; that is, the shaft deflection criterion determines the shaft dimension, which in turn determines the bearing geometry.

For more highly loaded bearings, the designer may wish to consider different bearing types as discussed in Section 13.3, and will use the approach described in Sections 13.7 and 13.9 to determine the required bearing load rating. Sections 13.4 through 13.6 are important for a detailed understanding of rolling element bearings and to aide design of specialty bearings or machinery where the manufacturer's literature may not have directly applicable products. These sections may be of less interest to the casual reader, but it should be recognized that the kinematics and mechanics described are incorporated into commercial rolling-element bearings.

13.2 Historical Overview

The purpose of a bearing is to provide relative positioning and rotational freedom while transmitting a load between two structures, usually a shaft and a housing. The basic form and concept of the rolling-element bearing are simple. If loads are to be transmitted between surfaces in relative motion, the action can be facilitated most effectively if rolling elements are interposed between the sliding members. The frictional resistance encountered in sliding is then largely replaced by the much smaller resistance associated with rolling, although the arrangement inevitably involves high stresses in the restricted contact regions.

The modern precision rolling-element bearing, a product of exacting technology and sophisticated science, has been very effective in reducing friction and wear in a wide range of machinery. The rapid development of numerous forms of rolling-element bearings in the twentieth century is well known and documented. However, the origins of these vital machine elements can be traced to long before there was a large industrial demand for such devices, and certainly long before adequate machine tools for their effective manufacture existed in large quantities. A complete history of rolling-element bearings is given in Hamrock and Dowson [1981] and in Dowson [1998]; only a brief overview is presented here.

The influence of general technological progress on the development of rolling-element bearings, particularly those used in moving heavy stone building blocks and carvings and in road vehicles, precision instruments, water-raising equipment, and windmills is discussed in Hamrock and Dowson [1981]. The history of rolling-element bearings is somewhat subject to interpretation of archeological evidence, but some early and isolated examples include wheel bearings for a Celtic cart found in Denmark that date to 300 to 100 B.C., provocative objects that resemble bearing races found in a remote Chinese village that date to around 200 B.C., and the famous Lake Nemi bearings that were components of a rotating turntable on a Roman ship circa 40 A.D. The concept of rolling-element bearings emerged in Roman times, faded from memory during the Middle Ages, was revived during the Renaissance, developed steadily in the seventeenth and eighteenth centuries for various applications, and was firmly established for individual road carriage bearings during the Industrial Revolution. Toward the end of the nineteenth century, the great merit of ball bearings for bicycles promoted interest in the manufacture of accurate steel balls. Initially, the balls were turned from bars on special lathes, with individual general machine manufacturing companies making their own bearings. Growing demand for both ball and roller bearings encouraged the formation of specialized bearing manufacturing companies at the turn of the century and thus laid the foundations of a great industry. The advent of precision grinding techniques and the availability of improved materials did much to confirm the future of the new industry.

The essential features of most forms of the modern rolling-element bearing were therefore established by the second half of the nineteenth century, but it was the formation of specialist precision-manufacturing companies in the early years of the twentieth century that finally established the rolling-element bearing as a most valuable, high-quality, readily available machine component. The availability of ball and roller bearings in standard sizes has had a large impact on machine design ever since. Such bearings still provide a challenging field for further research and development, and many engineers and scientists are currently engaged in various research projects in this area. In many cases new materials or enlightened design concepts have extended the life and range of application of rolling-element bearings. In other respects, much remains to be done in explaining the extraordinary operating characteristics of these bearings, which have served our technology so well for over a century.

13.3 Bearing Types and Selection

Ball and roller bearings are available to the designer in a great variety of designs and size ranges. Figure 13.1 and Table 13.1 illustrate some of the more widely used bearing types; numerous types of specialty bearings are also available. Size ranges are generally given in metric units because traditionally, most rolling-element bearings have been manufactured to metric dimensions, predating the efforts toward a metric standard.

Table 13.1 lists approximate relative load-carrying capacities, both radial and thrust, and approximate tolerances to misalignment. Table 13.1 also provides some guidelines to help selection of bearing types for different applications.

Rolling-element bearings are an assembly of several parts: an **inner race**, an **outer race**, a set of balls or rollers, and a **cage** or **separator**, as shown in Fig. 13.1. The cage or separator maintains even spacing of the rolling elements. A cageless bearing, in which the annulus is packed with the maximum number of rolling elements, is called a **full-complement bearing**. Full-complement bearings have high load-carrying capacity but lower speed limits than bearings equipped with cages. **Tapered-roller bearings** are an assembly of a **cup**, a **cone**, a set of tapered rollers, and a cage.

Although rolling-element bearings will function well without a lubricant and are sometimes operated that way, it is often advantageous to provide a lubricant film to extend their life. Since lubricant supply is difficult with rolling-element bearing geometry, bearings are often packed with thick greases and then sealed so that the bearing operates under lubrication. The seals are a critical component, both to keep the lubricant inside the bearing pack and also to prevent the lubricant from being washed away (e.g., as in an automobile axle when the car is driven through a puddle of water). Although seal quality is often a critical issue in bearing performance, this subject is not investigated in depth in this chapter. It should be recognized, however, that rolling-element bearings usually operate under elastohydrodynamic lubrication, often with thick greases as the fluid, and that proper sealing and lubrication is essential for long service life.

13.3.1 Ball Bearings

Ball bearings are used in greater quantity than any other type of rolling-element bearing. For an application where the load is primarily radial with some thrust load present, one of the types in the first two rows of Table 13.1 can be chosen, among others. A **Conrad**, or **deep-groove, bearing** (Fig. 13.1a) has a ball complement limited by the number that can be packed into the annulus between the inner and outer **races** with the inner race resting against the inside diameter of the outer race. A stamped and riveted two-piece **cage** or a machined two-piece cage is almost always used in a Conrad bearing. The only exception is a one-piece cage with open-sided pockets that is snapped into place. A filled-notch bearing most often has both inner and outer races notched so that a ball complement limited only by the annular space between the races can be used. It has low thrust capacity because of the filling notch.

The **self-aligning internal bearing** (Table 13.1) has an outer-race ball path ground in a spherical shape so that it can accept high levels of misalignment. The **self-aligning external bearing** shown has a multipiece outer race with a spherical inner race. It too can accept high misalignment and has higher capacity than the self-aligning internal bearing. However, the self-aligning external bearing is somewhat less self-aligning than its internal counterpart because of friction in the multipiece outer race.

Representative **angular-contact ball bearings** are also illustrated in Table 13.1. An angular-contact ball bearing has a two-shouldered ball groove in one race and a single-shouldered ball groove in the other race. Thus, it can support only a unidirectional thrust load. The cutaway shoulder allows bearing assembly by snapping the inner race over the ball set after it is positioned in the cage and outer race. It also permits use of a one-piece, machined, race-piloted cage that can be balanced for high-speed operation. Typical contact an-

Table 13.1: Selection guide for rolling-element bearings.

Bearing type	Examples	Approximate maximum diameter, mm	Radial capacity	Axial capacity	Combined load	Moment load	Speed	Stiffness	Quiet running	Misalignment compensation	Allowable axial displacement	Typical applications
Radial ball												
Deep groove	Conrad, Filling notch	1000	2	2	2	1[a]	2	2	4	1	0	Textiles, power tools, pumps, gearboxes
Self-aligning	Internal, External	120	2	1	1	0	4	1	3	4	0	Fans, paper making machinery
Angular contact	Single row	320	2	2	3	1[a]	3	2	3	1	0	Pumps, compressors, centrifuges
Duplex	Back-to-back, Tandem	320	3	2	3	2	2	3	2	0	0	Pumps, compressors, centrifuges
Two-directional	Split-ring	110	1	3	2	2[a]	3	2	2	0	0	Compressors
Thrust ball	Flat race, Grooved race	1000	0	2	0	0	1	2	1	1	0	Plastic extruder tools, crane hooks
Cylindrical roller	Separable inner ring, Separable outer ring	500	3	0	0	0	3	3	3	1	4	Traction motors, electric motors, gearboxes
Full compliment	Separable inner ring, Separable outer ring	120	4	1	2	0	1	4	1	1	2	Elevators, gearboxes
Double row		1250	4	1	2	2	1	4	1	0	2	Cranes, rolling mills, wire rope sheaves
Spherical roller	Single row	320	4	2	4	0	2	3	2	4	0	Fans, gearboxes, crushers, vibrating screens
Toroidal (CARB)		1120	4	0	0	0	2	3	2	4	4	Papermaking machines, gearboxes, fans, electric motors
Needle	Drawn cup, open end, Grease retained	185	3	0	0	0	2	3	2	0	4	Planetary gearboxes, alternators
Taper roller – single row		360	3	3	4	1	2	3	2	1	0	Gearboxes, cone crushers
Taper roller – double row	Back-to-back, Tandem	130	4	3	4	2	2	4	2	1	0	Gearboxes, rail car axles

0: Unsuitable, 1: Poor, 2: Fair, 3: Good, 4: Excellent.
[a] Dual-row only, otherwise 0.

gles vary from 15 to 40°.

Angular-contact ball bearings are often used in **duplex pairs** mounted either back-to-back or face-to-face. Duplex bearing pairs are manufactured so that they "preload" each other when clamped together in the housing and on the shaft. Preloading provides stiffer shaft support and helps prevent bearing skidding at light loads. A duplex pair can support bidirectional thrust load. The back-to-back arrangement offers more resistance to moment or overturning loads than does the face-to-face arrangement.

Where thrust loads exceed the load-carrying capacity of a single bearing, two bearings can be used in tandem, with each bearing supporting part of the thrust load. Three or more bearings are occasionally used in tandem, but this is discouraged because of the difficulty in achieving effective load sharing; even slight differences in operating temperature will cause an uneven distribution of load sharing.

The two-directional **split-ring bearing** offers several advantages. The split ring (usually the inner race) has its ball groove ground as a circular arc with a shim between the ring halves. The shim is then removed when the bearing is assembled so that the split-ring ball groove has the shape of a gothic arch. This shape reduces the axial play for a given radial play and results in more accurate axial positioning of the shaft. The bearing can support bidirectional thrust loads but must not be operated for prolonged times at predominantly radial loads. This restriction results in three-point, ball-race contact and relatively high frictional losses. As with the conventional angular-contact bearing, a one-piece, precision-machined cage is used.

Thrust ball bearings (90° contact angle), shown in Fig. 13.1e and Table 13.1, are used almost exclusively for machinery with vertically oriented shafts. The flat-race bearing allows eccentricity of the fixed and rotating members; an additional bearing must be used for radial positioning. It has low thrust load-carrying capacity because of its small ball-race contact and consequent high Hertzian stress. Grooved-race bearings have higher thrust load-carrying capacity and can support low-magnitude radial loads. All the pure-thrust ball bearings have modest speed capability because of the 90° contact angle and the consequent high level of ball spinning and frictional losses.

13.3.2 Roller Bearings

Cylindrical roller bearings (Fig. 13.1b and Table 13.1) provide purely radial load support in most applications. There are numerous forms of such bearings. Cylindrical roller bearings have moderately high radial load-carrying capacity as well as high-speed capability, exceeding those of either spherical or tapered-roller bearings. A commonly used bearing combination for supporting a high-speed rotor is an angular-contact ball bearing (or a duplex pair) and a cylindrical roller bearing. As explained in Section 13.4, the rollers in cylindrical roller bearings are seldom pure cylinders. They are usually crowned, or made slightly barrel shaped, to relieve stress concentrations on the roller ends if any misalignment of the shaft and housing is present. Cylindrical roller bearings may be equipped with one- or two-piece cages. Full-complement bearings can be used for greater load-carrying capacity but at a significant sacrifice in speed capability.

Spherical roller bearings (Table 13.1) are made as either single- or double-row bearings. The more popular bearing design (convex) uses barrel-shaped rollers. An alternative design (concave) employs hourglass-shaped rollers. Spherical roller bearings combine extremely high radial load-carrying capacity with modest thrust load-carrying capacity (with the exception of the thrust type) and excellent tolerance to misalignment. They find widespread use in heavy-duty rolling mill and industrial gear drives. Also included in Table 13.1 are **toroidal bearings**, also called *CARB bearings* (Compact Aligning Roller Bearing), which use long and slightly crowned rollers. Toroidal bearings combine the self-aligning benefits of spherical bearings, space saving profiles of needle bearings, and ability to compensate for axial misalignment of cylindrical bearings.

Tapered-roller bearings (Fig. 13.1d) are also made as single- or double-row bearings with combinations of one- or two-piece cups and cones. A four-row bearing assembly with two- or three-piece cups and cones is also available. Bearings are made with either a standard angle for applications in which moderate thrust loads are present or with a steep angle for high thrust capacity. Standard and special cages are available to suit the application requirements. Single-row tapered-roller bearings must be used in pairs because a radially loaded bearing generates a thrust reaction that must be taken up by a second bearing. Note from geometry that a tapered roller bearing can support a thrust load in only one direction, namely the direction that causes compressive stress between the roller and inner race or cone. A thrust load in the opposite direction would unload the race and could not be supported. Tapered-roller bearings are normally set up with spacers designed so that they operate with some internal play.

Needle roller bearings (Fig. 13.1c and Table 13.1) are characterized by compactness in the radial direction and are frequently used without an inner race. In this case the shaft is hardened and ground to serve as the inner race. Drawn cups, both open and closed end, are frequently used for grease retention. Drawn cups are thin-walled and require substantial support from the housing. Heavy-duty needle roller bearings have relatively rigid races and are more akin to cylindrical roller bearings with long width-to-diameter-ratio rollers.

13.3.3 Bearing Selection

Many types of specialty bearings are available other than those discussed here. Aircraft bearings for control systems, thin-section bearings, and fractured-ring bearings are some of the other bearing types among the many manufactured. Complete coverage of all bearing types is beyond the scope of this chapter, but information can be found in Hamrock and Anderson [1983] or directly from the bearing manufacturers. Most modern manufacturers maintain high-quality websites with design guides and specific bearing data.

Moment loads are difficult to support in a bearing, although duplex or double row bearings have limited moment-supporting ability. When large moments are present, one alternative is to use two bearings that are spaced apart by a short distance. This approach allows the development of moment support while causing only radial loads on the bearings.

The selection of a type of rolling-element bearing for a particular design application is often the most challenging problem, as analysis subsequently proceeds in a logical and ordered manner. Table 13.1 gives some general guidelines to help the designer select a type of rolling-element bearing for a particular application, but it should be realized that most design problems can accommodate different bearings, and the most economical and robust design will involve evaluation of competing options and communication with the vendor/supplier. Tapered roller bearings have higher fatigue lives, in general, for a given bore than ball or cylindrical roller bearings. Angular-contact ball bearings and cylindrical roller bearings are generally considered to have the highest speed capabilities. Speed limits of roller bearings are discussed in conjunction with lubrication methods. The lubrication system employed has as great an influence on bearing limiting speed as the bearing design. Grease is easier to maintain, but has a lower speed capability than oil. Additional considerations are the space available for the bearing, required bore, and level of thrust load support needed.

Example 13.1: Selection of Ball Bearings

Given: The required load-carrying capacity and type of motion desired are two important factors in the selection of rolling-element bearing type. Depending on the geometry of the races, either radial, axial, or moment loads can be transmitted through a ball bearing.

Find: Describe which rolling-element bearing type should be used for

(*a*) Pure radial load with high stiffness

(*b*) Pure axial load

(*c*) Bending moment load

(*d*) High radial stiffness

(*e*) High axial stiffness

Solution: Table 13.1 is especially useful to identify bearings for the particular loading cases. Note that the statement restricts the problem to rolling-element bearings, so only this type will be considered.

(*a*) For pure radial loading, the obvious selections are full compliment cylindrical bearings or double row taper roller bearings. Other constraints may provide differentiation between the designs, such as a comparison of costs.

(*b*) For pure axial loading, the force transmitted through the balls should be as close as possible to the bearing axis. This location gives the smallest possible internal force acting through the balls. Taper roller bearings and two-directional angular contact bearings are good selections for this loading.

(*c*) For bending moment loading, a bearing must support loads in both the axial and radial directions. As can be seen from Table 13.1, none of the rolling-element bearings are particularly suitable, and a hydrodynamic bearing (Chapter 12) may be a better option. However, double bearings have some load support, so that a double deep-groove (duplex) and double row cylindrical roller bearings can be used. Another option presents itself if the mounting can be changed: If two bearings are mounted far enough apart that their center distance is about the same as the bearing outer diameter, any type of bearing shown in Table 13.1 can be used. The bearings then support radial loads and develop moment support because of their spacing.

(*d*) If a high radial stiffness is desired, full compliment or double row cylindrical bearings, or double row taper roller bearings are most suitable. Note that the full compliment cylindrical rolling-element bearing will have the highest stiffness because it has more rollers, but it is speed limited.

(*e*) High axial stiffness is not discussed in Table 13.1, so the selection will require a study of the bearing types. The taper roller bearing, with at least two rows, will have the stiffest design.

13.4 Geometry

The operating characteristics of a rolling-element bearing depend greatly on its diametral clearance; this clearance varies for the different types of bearing discussed in the preceding section. The principal geometrical relationships governing the operation of unloaded rolling-element bearings are developed in this section. This information will be of interest when stress, deflection, load-carrying capacity, and life are considered later in the chapter.

13.4.1 Ball Bearings

As described below, the important geometric characteristics of ball bearings are pitch diameter, clearance, race conformity, contact angle, endplay, shoulder height, and curvature sum and difference.

Pitch Diameter and Clearance

The cross-section through a radial single-row ball bearing shown in Fig. 13.2 depicts the radial clearance and various diameters. The **pitch diameter**, d_e, is the mean of the inner- and outer-race contact diameters and is given by

$$d_e = d_i + \frac{d_o - d_i}{2},$$

or

$$d_e = \frac{d_o + d_i}{2}. \tag{13.1}$$

Also, from Fig. 13.2 the diametral clearance, c_d, can be written as

$$c_d = d_o - d_i - 2d. \tag{13.2}$$

Diametral clearance may, therefore, be thought of as the maximum distance that one race can move diametrically with respect to the other when a small force is applied and both races lie in the same plane. Although diametral clearance is generally used in connection with single-row radial bearings, Eq. (13.2) is also applicable to angular-contact bearings.

Race Conformity

Race conformity is a measure of the geometrical conformity of the race and the ball in a plane passing through the bearing axis, which is a line passing through the center of the bearing perpendicular to its plane and transverse to the race. Figure 13.3 is a cross-section of a ball bearing showing race conformity, expressed as

$$R_r = \frac{r}{d}. \tag{13.3}$$

In perfect conformity, the race radius is equal to the ball radius, thus R_r is equal to $1/2$. The closer the race conforms to the ball, the greater the frictional heat within the contact. On the other hand, open-race curvature and reduced geometrical conformity, which both reduce friction, also increase the maximum contact stresses and consequently reduce the bearing fatigue life. For this reason, most modern ball bearings

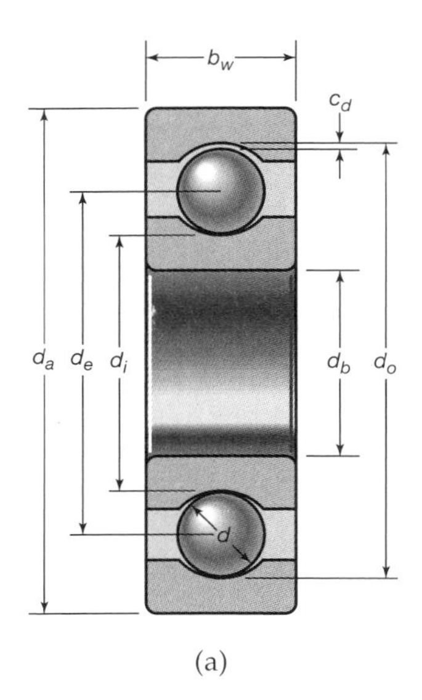

(a)

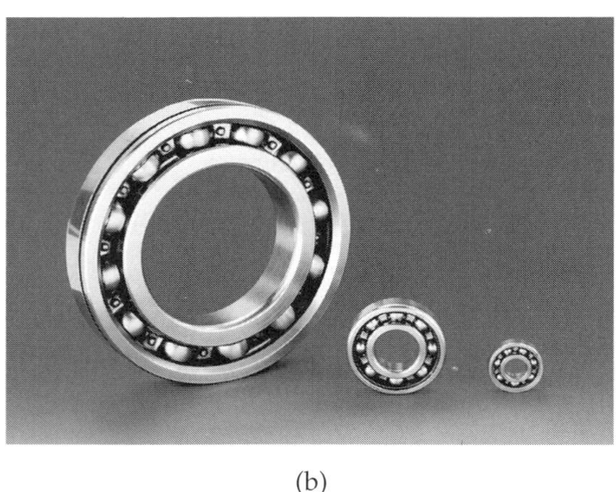

(b)

Figure 13.2: (a) Cross-section through radial single-row ball bearings; (b) examples of radial single-row ball bearings. *Source:* (b) Courtesy of SKF USA, Inc.

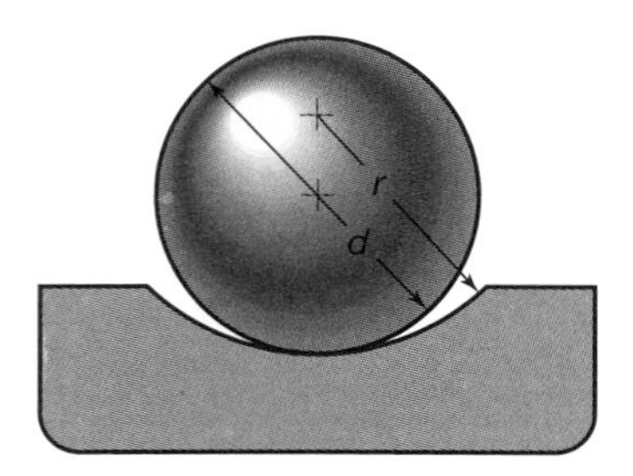

Figure 13.3: Cross-section of ball and outer race, showing race conformity.

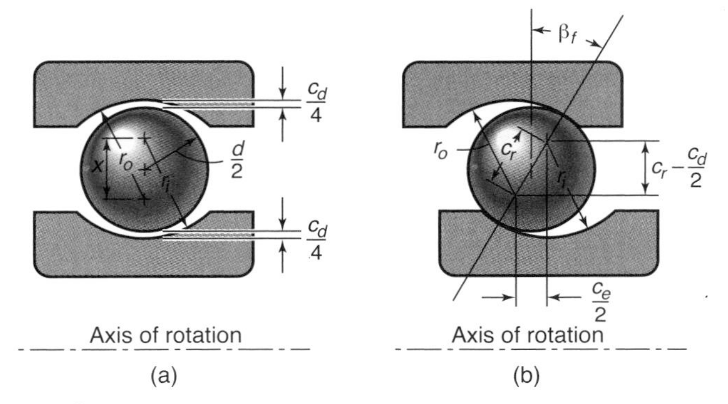

Figure 13.4: Cross-section of radial bearing, showing ball-race contact due to axial shift of inner and outer races. (a) Initial position; (b) shifted position.

have race conformity ratios in the range $0.51 \leq R_r \leq 0.54$, with $R_r = 0.52$ being the most common. The race conformity ratio for the outer race is usually made slightly larger than that for the inner race to compensate for the closer conformity in the plane of the bearing between the outer race and the ball than between the inner race and the ball. This larger ratio tends to equalize the contact stresses at the inner- and outer-race contacts. The difference in race conformities does not normally exceed 0.02.

Contact Angle

Radial bearings have some axial play since they are generally designed to have a diametral clearance, as shown in Fig. 13.4. Axial play implies a free contact angle different from zero in a mounted bearing. **Angular-contact bearings** are specifically designed to operate under thrust loads. The clearance built into the unloaded bearing, along with the race conformity ratio, determines the bearing free contact angle. Figure 13.4 shows a radial bearing with contact due to the axial shift of the inner and outer races when no measurable force is applied.

Before the free contact angle is discussed, it is important to define the distance between the centers of curvature of the two races in line with the center of the ball in Fig. 13.4a and b. This distance — denoted by x in Fig. 13.4a and by c_r in Fig. 13.4b — depends on race radius and ball diameter. When quantities referred to the inner and outer races are denoted by the subscripts i and o, respectively, Fig. 13.4b shows that

$$d = r_o - c_r + r_i,$$

or

$$c_r = r_o + r_i - d. \tag{13.4}$$

The **free contact angle**, β_f, shown in Fig. 13.4b, is defined as the angle made by a line through the points where the ball contacts both races and a plane perpendicular to the bearing axis of rotation when nonmeasurable force is applied. Note that the centers of curvature of both the outer and inner races lie on the line defining the free contact angle. From Fig. 13.4b, the expression for the free contact angle can be written as

$$\cos\beta_f = 1 - \frac{c_d}{2c_r}. \tag{13.5}$$

Equations (13.2) and (13.4) can be used to write Eq. (13.5) as

$$\beta_f = \cos^{-1}\frac{r_o + r_i - \dfrac{d_o - d_i}{2}}{r_o + r_i - d}. \tag{13.6}$$

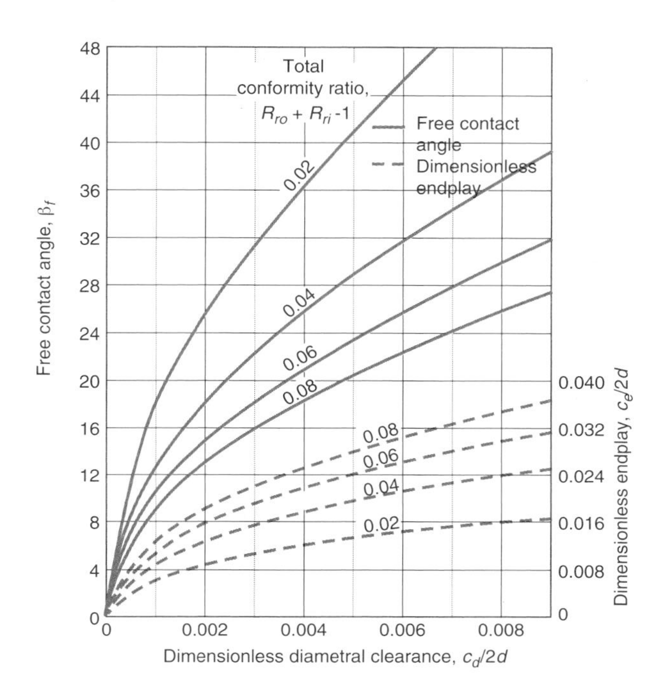

Figure 13.5: Free contact angle and endplay as function of $c_d/2d$.

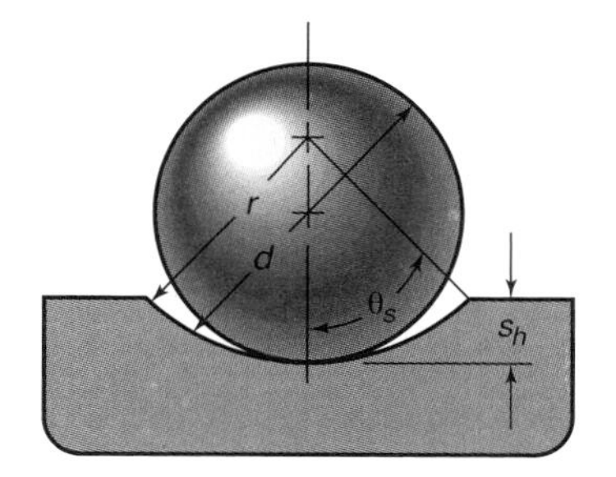

Figure 13.6: Shoulder height in a ball bearing.

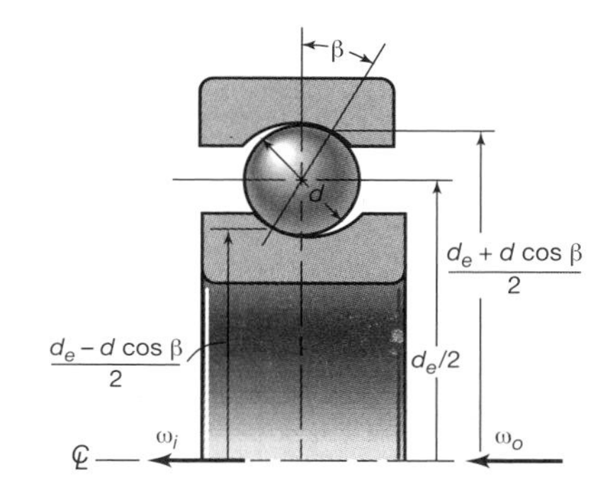

Figure 13.7: Cross-section of ball bearing.

Equation (13.6) shows that if the size of the balls is increased and everything else remains constant, the free contact angle decreases. Similarly, if the ball size is decreased, the free contact angle increases.

From Eq. (13.5), the diametral clearance, c_d, can be written as

$$c_d = 2c_r(1 - \cos\beta_f). \tag{13.7}$$

This is an alternative definition of the diametral clearance given in Eq. (13.2).

Endplay

Free **endplay**, c_e, is the maximum axial movement of the inner race with respect to the outer race when both are coaxially centered and no measurable force is applied. Free endplay depends on total curvature and contact angle, as shown in Fig. 13.4, and can be written as

$$c_e = 2c_r \sin\beta_f. \tag{13.8}$$

Figure 13.5 shows the variation of free contact angle and endplay with the diametral clearance ratio $c_d/2d$ for four values of the total conformity ratio normally found in single-row ball bearings. Eliminating β_f in Eqs. (13.7) and (13.8) enables the following relationships between free endplay and diametral clearance to be established:

$$c_d = 2c_r - \left[(2c_r)^2 - c_e^2\right]^{1/2},$$

$$c_e = \sqrt{4c_r c_d - c_d^2}.$$

Shoulder Height

Shoulder height, or **race depth**, is the depth of the race groove measured from the shoulder to the bottom of the groove and is denoted by s_h in Fig. 13.6. From this figure, the equation defining the shoulder height can be written as

$$s_h = r\left(1 - \cos\theta_s\right). \tag{13.9}$$

The maximum possible diametral clearance for complete retention of the ball-race contact within the race under zero thrust load is given by the condition $(\beta_f)_{\max} = \theta_s$. Making use of Eqs. (13.7) and (13.9) gives

$$\left(c_d\right)_{\max} = \frac{2c_r s_h}{r}. \tag{13.10}$$

Curvature Sum and Difference

The undeformed geometry of contacting solids in a ball bearing can be represented by two ellipsoids as discussed in Section 8.4.1. Figure 13.7 shows a cross-section of a ball bearing operating at a contact angle, β. Equivalent radii of curvature for both inner- and outer-race contacts in, and normal to, the direction of rolling can be calculated from this figure. The radii of curvature for the *ball–inner-race contact* are

$$r_{ax} = r_{ay} = \frac{d}{2}, \tag{13.11}$$

$$r_{bx} = \frac{d_e - d\cos\beta}{2\cos\beta}, \tag{13.12}$$

$$r_{by} - R_{ri}d = -r_i. \tag{13.13}$$

The radii of curvature for the ball–outer-race contact are

$$r_{ax} = r_{ay} = \frac{d}{2}, \tag{13.14}$$

$$r_{bx} = -\frac{d_e + d\cos\beta}{2\cos\beta}, \tag{13.15}$$

$$r_{by} = -R_{ro}d = -r_o. \tag{13.16}$$

In Eqs. (13.12) and (13.15), β is used instead of β_f because these equations are also valid when a load is applied to the contact. By setting $\beta = 0$, Eqs. (13.11) to (13.16) are equally valid for radial ball bearings. For thrust ball bearings $r_{bx} = \infty$ and the radii are defined as given in the preceding equations.

From the preceding radius-of-curvature expressions and Eqs. (8.6) and (8.7), for the *ball–inner-race* contact

$$R_{xi} = \frac{d(d_e - d\cos\beta)}{2d_e}, \tag{13.17}$$

$$R_{yi} = \frac{R_{ri}d}{2R_{ri} - 1}, \tag{13.18}$$

and for the *ball–outer-race* contact

$$R_{xo} = \frac{d(d_e + d\cos\beta)}{2d_e}, \tag{13.19}$$

$$R_{yo} = \frac{R_{ro}d}{2R_{ro} - 1}. \tag{13.20}$$

Substituting these equations into Eqs. (8.5) and (8.6) enables the curvature sum and difference to be obtained. Equations (13.17) through (13.20) are also extremely useful for analyzing the contact stress and deformations using the approach presented in Section 8.4.

Example 13.2: Bearing Geometry

Given: A single-row, deep-groove ball bearing, shown in Fig. 13.2a, has a pitch diameter of 100 mm, an inner-race diameter of 80 mm, and a ball diameter of 19.9 mm. The outer-race radius is 10.5 mm, and the inner-race radius is 10.2 mm. The shoulder height angle is 30°.

Find:

(*a*) Diametral clearance

(*b*) Inner- and outer-race conformities

(*c*) Free contact angle and endplay

(*d*) Shoulder heights for the inner and outer races

(*e*) Ellipticity parameter and the curvature sum of the contacts

Solution:

(*a*) Using Eqs. (13.1) and (13.2) gives the diametral clearance as

$$\begin{aligned} c_d &= d_o - d_i - 2d \\ &= 2d_e - 2d_i - 2d \\ &= 200 - 160 - 2(19.9) \\ &= 0.2 \text{ mm}. \end{aligned}$$

(*b*) Using Eq. (13.3) gives the conformities of the outer and inner races as:

$$R_{ro} = \frac{r_o}{d} = \frac{10.5}{19.9} = 0.528,$$

$$R_{ri} = \frac{r_i}{d} = \frac{10.2}{19.9} = 0.513.$$

(*c*) From Eq. (13.5), the free contact angle is

$$\cos\beta_f = 1 - \frac{c_d}{2c_r} = 1 - \frac{0.2}{2(0.816)} = 0.877,$$

which is solved as $\beta_f = 28.7°$. From Eq. (13.8), the endplay is

(*d*) From Eq. (13.9), the shoulder heights are

$$s_{hi} = r_i(1 - \cos\theta_s) = 10.2(1 - \cos 30°) = 1.367 \text{ mm},$$

$$s_{ho} = r_o(1 - \cos\theta_s) = 10.5(1 - \cos 30°) = 1.407 \text{ mm}.$$

From Eq. (13.10), the minimum shoulder height for complete retention of the ball-race contact is

$$(s_{hi})_{\min} = \frac{c_d r_i}{2c_r} = \frac{(0.2)(10.2)}{2(0.816)} = 1.25 \text{ mm},$$

$$(s_{ho})_{\min} = \frac{c_d r_o}{2c_r} = \frac{(0.2)(10.5)}{2(0.816)} = 1.29 \text{ mm}.$$

Note that the shoulder heights given are suitable for retaining the balls.

(*e*) *Ball–inner-race contact*

From Eqs. (13.11) through (13.13),

$$r_{ax} = r_{ay} = \frac{d}{2} = \frac{19.9}{2} = 9.95 \text{ mm},$$

$$\begin{aligned} r_{bx} &= \frac{d_e - d\cos\beta}{2\cos\beta} \\ &= \frac{100 - (19.9)\cos 28.7°}{2\cos 28.7°} \\ &= 47.1 \text{ mm}, \end{aligned}$$

$$r_{by} = -R_{ri}d = -r_i = -10.2 \text{ mm}.$$

The curvature sums in the x- and y-directions are given by Eqs. (8.6) and (8.7) as

$$\frac{1}{R_x} = \frac{1}{r_{ax}} + \frac{1}{r_{bx}} = \frac{1}{9.95} + \frac{1}{47.1},$$

$$\frac{1}{R_y} = \frac{1}{r_{ay}} + \frac{1}{r_{by}} = \frac{1}{9.95} - \frac{1}{10.2} = 0.00246 \text{ mm}^{-1}.$$

These equations result in $R_x = 8.215$ mm and $R_y = 406.0$ mm. From Eq. (8.5), the curvature sum is

$$\frac{1}{R} = \frac{1}{R_x} + \frac{1}{R_y} = 0.122 + 0.00246 = 0.1245 \text{ mm}^{-1}$$

or $R = 8.052$ mm. From Eqs. (8.8) and (8.12) the radius ratio and the ellipticity parameter are

$$\alpha_{ri} = \frac{R_y}{R_x} = \frac{406.0}{8.215} = 49.42,$$

$$k_{ei} = \alpha_{ri}^{2/\pi} = (49.42)^{2/\pi} = 12.0.$$

Ball–outer-race contact This approach uses the same equations as for the ball–inner-race contact but the radii are:

$$r_{ax} = r_{ay} = \frac{d}{2} = 9.95 \text{ mm}$$

and

$$\begin{aligned} r_{bx} &= -\frac{d_e + d\cos\beta}{2\cos\beta} \\ &= -\frac{100 + 19.9\cos(28.7^\circ)}{2\cos(28.7^\circ)} \\ &= -66.9 \text{ mm}, \\ r_{by} &= -r_o = -10.5 \text{ mm}. \end{aligned}$$

The curvature sums in the x- and y-directions are

$$\frac{1}{R_x} = \frac{1}{r_{ax}} + \frac{1}{r_{bx}} = \frac{1}{9.95} - \frac{1}{66.95} = 0.0856 \text{ mm}^{-1},$$

$$\frac{1}{R_y} = \frac{1}{r_{ay}} + \frac{1}{r_{by}} = \frac{1}{9.95} - \frac{1}{10.5} = 0.0053 \text{ mm}^{-1},$$

resulting in $R_x = 11.69$ mm and $R_y = 190.0$ mm. The curvature sum is

$$\frac{1}{R} = \frac{1}{R_x} + \frac{1}{R_y} = 0.0856 + 0.0053 = 0.0909 \text{ mm}^{-1},$$

so that $R = 11.00$ mm. The radius ratio and the ellipticity parameter are

$$\alpha_{ro} = \frac{R_y}{R_x} = \frac{190.0}{11.69} = 16.25,$$

$$k_{eo} = \alpha_r^{2/\pi} = (16.25)^{2/\pi} = 5.90.$$

13.4.2 Roller Bearings

The equations developed for the pitch diameter, d_e, and diametral clearance, c_d, for ball bearings [Eqs. (13.1) and (13.2), respectively] are directly applicable to roller bearings.

Crowning

High stresses at the edges of the rollers in cylindrical roller bearings are usually prevented by machining a **crown** into the rollers, either fully or partially as shown in Fig. 13.8, where the crown curvature is greatly exaggerated for clarity. Crowning the rollers also protects the bearing against the effects of slight misalignment. For cylindrical rollers, $r_r/d \approx 100$. In contrast, for rollers in spherical roller bearings, as shown in Fig. 13.8a, $r_r/d \approx 4$. Observe in Fig. 13.8 that the roller effective length l_l is the length presumed to be in contact with the races under loading. Generally, the roller effective length can be written as

$$l_l = l_t - 2r_c,$$

where r_c is the roller corner radius or the grinding undercut, whichever is larger.

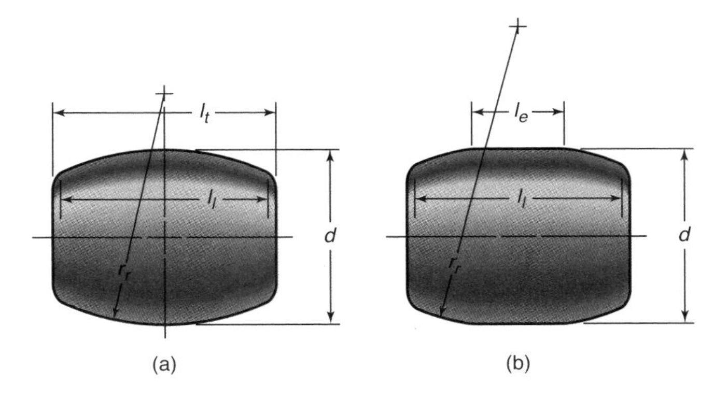

Figure 13.8: (a) Spherical roller (fully crowned); (b) cylindrical roller; (c) section of a toroidal roller bearing (CARB). *Source:* Courtesy of SKF USA, Inc.

Race Conformity

Race conformity also applies to roller bearings, and is a measure of the geometrical conformity of the race and the roller. Figure 13.9 shows a cross-section of a spherical roller bearing. From this figure the race conformity can be written as

$$R_r = \frac{r}{2r_r}.$$

In this equation, if R_r and r are subscripted with i or o, the race conformity values are for the inner- or outer-race contacts, respectively.

Free Endplay and Contact Angle

Cylindrical roller bearings have a contact angle of zero and may support thrust load only by virtue of axial flanges. Tapered-roller bearings must be subjected to a thrust load or the inner and outer races (the cone and the cup) will not remain assembled; therefore, tapered-roller bearings do not exhibit free diametral play. However, radial spherical roller bearings are normally assembled with free diametral play and hence exhibit free endplay. The diametral play, c_d, for a spherical roller bearing is the same as that obtained for ball bearings as expressed in Eq. (13.2).

Curvature Sum and Difference

The same procedure used for ball bearings will be used for defining the curvature sum and difference for roller bearings.

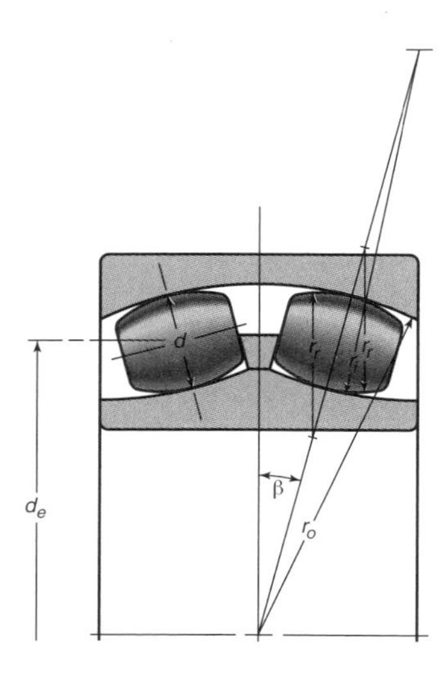

Figure 13.9: Geometry of spherical roller bearing.

For spherical roller bearings (Fig. 13.8) the radii of curvature for the *roller–inner-race contact* can be written as

$$r_{ax} = \frac{d}{2},$$

$$r_{ay} = \frac{r_i}{2R_{ri}} = r_r,$$

$$r_{bx} = \frac{d_e - d\cos\beta}{2\cos\beta},$$

$$r_{by} = -2R_{ri}r_r = -r_i.$$

The radii of curvature for the *roller–outer-race contact* can be written as

$$r_{ax} = \frac{d}{2},$$

$$r_{ay} = \frac{r_o}{2R_{ro}} = r_r,$$

$$r_{bx} = -\frac{d_e + d\cos\beta}{2\cos\beta},$$

$$r_{by} = -2R_{ro}r_r = -r_o.$$

Once the radii of curvature for the respective contact conditions are known, the curvature sum and difference can be obtained directly from Eqs. (8.6) and (8.7). Furthermore, the radius-of-curvature expressions for spherical roller bearings can be written for the *roller–inner-race contact* as

$$R_{xi} = \frac{d(d_e - d\cos\beta)}{2d_e}, \tag{13.21}$$

$$R_{yi} = \frac{r_i r_r}{r_i - r_r}, \tag{13.22}$$

and for the *roller–outer-race contact* as

$$R_{xo} = \frac{d(d_e + d\cos\beta)}{2d_e}, \tag{13.23}$$

$$R_{yo} = \frac{r_o r_r}{r_o - r_r}. \tag{13.24}$$

Substituting these equations into Eqs. (8.6) and (8.7) enables the curvature sum and difference to be obtained.

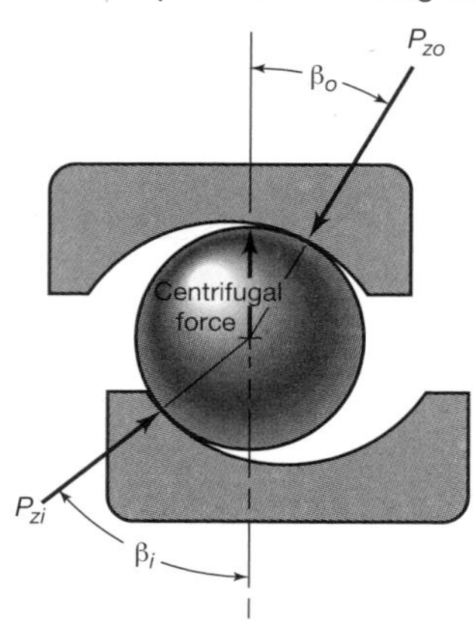

Figure 13.10: Contact angles in ball bearing at appreciable speeds.

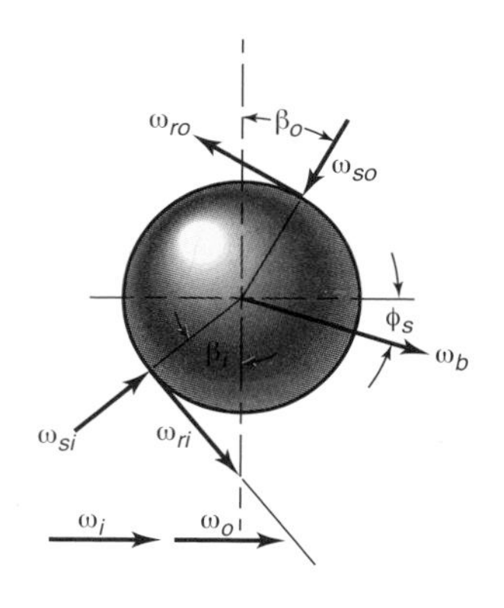

Figure 13.11: Angular velocities of ball.

13.5 Kinematics

The relative motions of the separator, the balls or rollers, and the races of rolling-element bearings are important to understand their performance. The relative velocities in a ball bearing are somewhat more complex than those in roller bearings, the latter being analogous to the specialized case of a zero- or fixed-contact-angle ball bearing. For that reason, the ball bearing is used as an example here to develop approximate expressions for relative velocities. These expressions are useful for rapid and accurate calculation of elastohydrodynamic film thickness, which can be used together with surface roughnesses to calculate the lubrication life factor as discussed in Section 13.8.

When a ball bearing operates at high speeds, the centrifugal force acting on the ball creates a divergency of the inner- and outer-race contact angles, as shown in Fig. 13.10, in order to maintain force equilibrium on the ball. For the most general case of rolling and spinning at both inner- and outer-race contacts, the rolling and spinning velocities of the ball are as shown in Fig. 13.11.

Jones [1964] developed the equations for ball and separator angular velocity for all combinations of inner- and outer-race rotation. Without introducing additional relationships to describe the elastohydrodynamic conditions at both ball-race contacts, however, the ball-spin axis orientation angle ϕ_s cannot be obtained. As mentioned, this requires a lengthy numerical solution except for the two extreme cases of outer- or inner-race control. Figure 13.12 illustrates these cases.

Race control assumes that pure rolling occurs at the controlling race, with all the ball spin occurring at the other race contact. The orientation of the ball rotation axis can then be easily determined from bearing geometry. Race control probably occurs only in dry bearings or dry-film-lubricated bearings, where Coulomb friction conditions exist in the ball-race contact ellipses. The spin-resisting moment will always be greater at one of the race contacts. Pure rolling will occur at the race contact with the higher magnitude spin-resisting moment, usually the inner race at low speeds and the outer race at high speeds.

In oil-lubricated bearings in which elastohydrodynamic films exist in both ball-race contacts, rolling with spin occurs at both contacts. Therefore, precise ball motions can only be determined numerically. The situation can be approximated with a reasonable degree of accuracy, however, by assuming that the ball rolling axis is normal to the line drawn through the centers of the two ball-race contacts (Fig. 13.12b).

The angular velocity of the separator or ball set ω_c about the shaft axis can be shown to be

$$\begin{aligned}\omega_c &= \frac{(v_i+v_o)/2}{d_e/2} \\ &= \frac{1}{2}\left[\omega_i\left(1-\frac{d\cos\beta}{d_e}\right)+\omega_o\left(1+\frac{d\cos\beta}{d_e}\right)\right],\end{aligned} \tag{13.25}$$

where v_i and v_o are the linear velocities at the inner and outer contacts. The angular velocity of a ball about its own axis, ω_b, assuming no spin, is

$$\begin{aligned}\omega_b &= \frac{v_i-v_o}{d_e/2} \\ &= \frac{d_e}{2d}\left[\omega_i\left(1-\frac{d\cos\beta}{d_e}\right)-\omega_o\left(1+\frac{d\cos\beta}{d_e}\right)\right].\end{aligned} \tag{13.26}$$

It is convenient for calculating the velocities at the ball-race contacts, which are required for calculating elastohydrodynamic film thicknesses, to use a coordinate system that rotates at ω_c. This system fixes the ball-race contacts relative to the observer. In the rotating coordinate system the angular velocities of the inner and outer races become

$$\omega_{ri} = \omega_i - \omega_c = \frac{\omega_i-\omega_o}{2}\left(1+\frac{d\cos\beta}{d_e}\right),$$

$$\omega_{ro} = \omega_o - \omega_c = \frac{\omega_o-\omega_i}{2}\left(1-\frac{d\cos\beta}{d_e}\right).$$

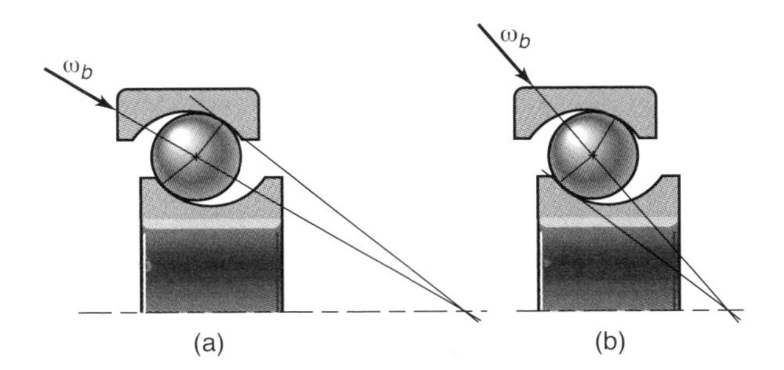

Figure 13.12: Ball-spin axis orientations for (a) outer-race control and (b) inner-race control.

The surface velocities entering the *ball–inner-race contact* for pure rolling are

$$u_{ai} = u_{bi} = \frac{\omega_{ri}\,(d_e - d\cos\beta)}{2}, \tag{13.27}$$

$$u_{ai} = u_{bi} = \frac{d_e\,(\omega_i-\omega_o)}{4}\left(1-\frac{d^2\cos^2\beta}{d_e^2}\right). \tag{13.28}$$

and those at the *ball–outer-race contact* are

$$u_{ao} = u_{bo} = \omega_{ro}\frac{d_e + d\cos\beta}{2}, \tag{13.29}$$

$$u_{ao} = u_{bo} = \frac{d_e(\omega_o-\omega_i)}{4}\left(1-\frac{d^2\cos^2\beta}{d_e^2}\right). \tag{13.30}$$

Thus,

$$|u_{ai}| = |u_{ao}.|$$

For a cylindrical roller bearing, $\beta = 0$, and Eqs. (13.25) to (13.30) become

$$\begin{aligned}\omega_c &= \frac{1}{2}\left[\omega_i\left(1-\frac{d}{d_e}\right)+\omega_o\left(1+\frac{d}{d_e}\right)\right] \\ \omega_b &= \frac{d_e}{2d}\left[\omega_i\left(1-\frac{d}{d_e}\right)+\omega_o\left(1+\frac{d}{d_e}\right)\right] \\ u_{ai} = u_{bi} &= \frac{d_e\,(\omega_i-\omega_o)}{4}\left(1-\frac{d^2}{d_e^2}\right) \\ u_{ao} = u_{bo} &= \frac{d_e(\omega_o-\omega_i)}{4}\left(1-\frac{d^2}{d_e^2}\right)\end{aligned} \tag{13.31}$$

where d is the roller diameter.

Equations directly analogous to those for a ball bearing can be used for a tapered-roller bearing if d is the average diameter of the tapered roller, d_e is the diameter at which the geometric center of the rollers is located, and β is the contact angle, as shown in Fig. 13.13.

Example 13.3: Kinematics of Rolling-Element Bearings

Given: As shown in Fig. 13.12, the motion of the balls can be guided by the inner-race contact or by the outer-race contact or by both. This is dependent on both the surface roughness and the surface geometry of angular-contact bearings. At very high speeds, the centrifugal forces acting on the balls change both the contact angle of the outer race and the friction forces between the ball and the outer race. The outer-race contact angle is 20° and the inner-race contact angle is 31°. The ball diameter is 14 mm and the pitch diameter is 134 mm.

Find: The angular speed of the balls and the separator in an angular-contact ball bearing at an inner-race speed of 10,000 rpm when the balls are

(*a*) Outer-race guided

(*b*) Inner-race guided

Solution:

(*a*) From Eq. (13.25), the separator angular speed when the balls are outer-race guided is

(a)

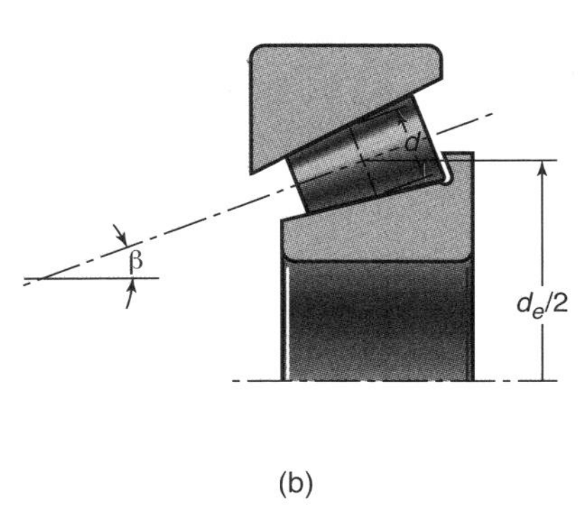

(b)

Figure 13.13: Tapered-roller bearing, (a) Tapered roller bearing with outer race removed to show rolling elements. *Source:* Courtesy of The Timken Company; (b) simplified geometry for tapered-roller bearing.

$$\begin{aligned}\omega_{co} &= \frac{1}{2}\left[\omega_i\left(1-\frac{d\cos\beta_o}{d_e}\right)+\omega_o\left(1+\frac{d\cos\beta_o}{d_e}\right)\right]\\ &= \frac{(10,000)\left(\frac{2\pi}{60}\right)}{2}\left[1-\frac{(14)\cos 20^\circ}{134}\right]\\ &= 472.2\text{ rad/s} = 4509\text{ rpm}.\end{aligned}$$

From Eq. (13.26), the ball speed when the balls are outer-race guided is

$$\begin{aligned}\omega_{bo} &= \frac{d_e}{2d}\left[\omega_i\left(1-\frac{d\cos\beta_o}{d_e}\right)\right]\\ &= \frac{(134.0)(10,000)\left(\frac{2\pi}{60}\right)}{2(14.0)}\left(1-\frac{14.0\cos 20^\circ}{134.0}\right)\\ &= 4520\text{ rad/s} = 43,160\text{ rpm}.\end{aligned}$$

(*b*) From Eq. (13.25),

$$\begin{aligned}\omega_{ci} &= \frac{1}{2}\left[\omega_i\left(1-\frac{d\cos\beta_i}{d_e}\right)\right]\\ &= \frac{(10,000)\left(\frac{2\pi}{60}\right)}{2}\left(1-\frac{14.0\cos 31^\circ}{134}\right)\\ &= 476.7\text{ rad/s} = 4552\text{ rpm}.\end{aligned}$$

From Eq. (13.26),

$$\omega_{bi} = \frac{134.0}{2(14)}\left[(10,000)\frac{2\pi}{60}\left(1-\frac{14\cos 31^\circ}{134}\right)\right]$$

which is solved as $\omega_{bi} = 43,570$ rpm.

13.6 Separators

Ball and roller bearing **separators** (sometimes called **cages** or **retainers**) are bearing components that, although they never carry load, can exert a vital influence on bearing efficiency. In a bearing without a separator, the rolling elements contact each other during operation and experience severe sliding and friction. The primary functions of a separator are to maintain the proper distance between the rolling elements and to ensure proper load distribution and balance within the bearing. Another function of the separator is to maintain control of the rolling elements so as to produce the least possible friction through sliding contact. Furthermore, a separator is necessary for several types of bearing to prevent the rolling elements from falling out of the bearing during handling. Most separator troubles occur from improper mounting, misaligned bearings, or improper (inadequate or excessive) clearance in the rolling-element pocket.

A **full compliment** bearing is one where the maximum number of rolling elements are placed between the bearing races. Clearly, there is therefore no space for a separator, and these bearings will have high load-carrying capability but will not be suitable for high-speed applications.

The materials used for separators vary according to the type of bearing and the application. In ball bearings and some sizes of roller bearing, the most common type of separator is made from two strips of carbon steel that are pressed and riveted together (Fig. 13.1a). Called *ribbon separators*, they are inexpensive and are entirely suitable for many applications; they are also lightweight and usually require little space.

The design and construction of angular-contact ball bearings allow the use of a one-piece separator. The simplicity and inherent strength of one-piece separators permit their fabrication from many desirable materials. Reinforced phenolic and bronze are the two most commonly used materials. Bronze separators offer strength and low-friction characteristics and can be operated at temperatures up to 230°C (450°F). Machined, silver-plated, ferrous-alloy separators are used in many demanding applications. Because reinforced phenolic separators combine the advantages of light weight, high strength, and nongalling properties, they are used for such high-speed applications as gyroscope bearings. Lightness and strength are particularly desirable in high-speed bearings, since the stresses increase with speed but may be greatly minimized by reducing separator weight. A limitation of phenolic separators, however, is that they have an allowable maximum temperature of about 135°C (275°F).

13.7 Static Load Distribution

Since a simple analytical expression for the deformation in terms of load was obtained in Section 8.4.1, it is possible to consider how the bearing load is distributed among the elements. Most rolling-element bearing applications involve steady-state rotation of either the inner or outer race. Rarely, both races are allowed to rotate. However, the rotational speeds are usually not so high as to cause ball or roller centrifugal forces or gyroscopic moments of significant magnitudes, and these are usually ignored. In this section, the load deflection relationships for ball and roller bearings are given, along with radial and thrust load distributions of statically loaded rolling elements.

13.7.1 Load Deflection Relationships

For an elliptical conjunction, the load deflection relationship given in Eq. (8.15) can be written as

$$w_z = K_{1.5}\delta^{3/2}, \tag{13.32}$$

where

$$K_{1.5} = \pi k_e E' \left(\frac{2\mathcal{E}R}{9\mathcal{F}^3}\right)^{1/3}. \tag{13.33}$$

Similarly, for a rectangular conjunction, from Eq. (8.19)

$$w_z = K_1 \delta_m,$$

where

$$K_1 = \frac{\frac{\pi}{2} l E'}{2\ln\left(\frac{4R_x}{b^*}\right) - 1}, \tag{13.34}$$

and l is the length of rolling element. In general then,

$$w_z = K_j \delta_m^j, \tag{13.35}$$

in which j is 1.5 for ball bearings and 1.0 for roller bearings. The total normal approach between two races separated by a rolling element is the sum of the deformations under load between the rolling element and both races. Therefore,

$$\delta_m = \delta_{mo} + \delta_{mi}, \tag{13.36}$$

where

$$\delta_{mo} = \left[\frac{w_z}{(K_j)_o}\right]^{\frac{1}{j}}, \tag{13.37}$$

$$\delta_{mi} = \left[\frac{w_z}{(K_j)_i}\right]^{\frac{1}{j}}. \tag{13.38}$$

Substituting Eqs. (13.36), (13.37), and (13.38) into Eq. (13.35) gives

$$K_j = \frac{1}{\left\{[1/(K_j)_o]^{1/j} + [1/(K_j)_i]^{1/j}\right\}^j}. \tag{13.39}$$

Recall that $(K_j)_o$ and $(K_j)_i$ are defined by Eqs. (13.33) and (13.34) for an elliptical and a rectangular conjunction, respectively. These equations show that $(K_j)_o$ and $(K_j)_i$ are functions only of the contact geometry and the material properties. The radial and thrust load analyses are presented in the following two sections and are directly applicable for radially loaded ball and roller bearings and thrust-loaded ball bearings.

13.7.2 Radially Loaded Ball and Roller Bearings

Figure 13.14 shows a radially loaded rolling-element bearing with radial clearance $c_d/2$. In the concentric arrangement (Fig. 13.14a), a uniform radial clearance between the rolling element and the races of $c_d/2$ is evident. The application of an arbitrarily small radial load to the shaft causes the inner race to move a distance $c_d/2$ before contact is made between a rolling element located on the load line and the inner and outer races. At any angle there will still be a radial clearance c that, if c_d is small relative to the radius of the groove, can be expressed with adequate accuracy by

$$c = (1 - \cos\psi)\frac{c_d}{2}.$$

On the load line when $\psi = 0°$, the clearance is zero; but when $\psi = 90°$, the clearance retains its initial value of $c_d/2$.

The application of further load will elastically deform the balls and eliminate clearance around an arc $2\psi_l$. If the interference or total elastic compression on the load line is $\delta_{\max}$, the corresponding elastic compression of the ball δ_ψ along a radius at angle ψ to the load line is given by

$$\delta_\psi = (\delta_{\max}\cos\psi - c) = \left(\delta_{\max} + \frac{c_d}{2}\right)\cos\psi - \frac{c_d}{2}.$$

It is clear from Fig. 13.14c that $\delta_{\max} + c_d/2$ represents the total relative radial displacement of the inner and outer races. Hence,

$$\delta_\psi = \delta_m \cos\psi - \frac{c_d}{2}. \tag{13.40}$$

The relationship between load and elastic compression along the radius at angle ψ to the load vector is given by Eq. (13.35) as

$$w_\psi = K_j \delta_\psi^j.$$

Substituting Eq. (13.40) gives

$$w_\psi = K_j\left(\delta_m\cos\psi - \frac{c_d}{2}\right)^j.$$

For static equilibrium, the applied load must equal the sum of the components of the rolling-element loads parallel to the direction of the applied load, or

$$w_z = \sum w_\psi \cos\psi.$$

Therefore,

$$w_z = K_j \sum \left(\delta_m \cos\psi - \frac{c_d}{2}\right)^j \cos\psi. \tag{13.41}$$

The angular extent of the bearing arc in which the rolling elements are loaded ($2\psi_l$) is obtained by setting the term in parentheses in Eq. (13.41) equal to zero and solving for ψ:

$$\psi_l = \cos^{-1}\left(\frac{c_d}{2\delta_m}\right).$$

The summation in Eq. (13.41) applies only to the angular extent of the loaded region. This equation can be written for a roller bearing as

$$w_z = \left(\psi_l - \frac{c_d}{2\delta_m}\sin\psi_l\right)\frac{nK_1\delta_m}{2\pi}, \tag{13.42}$$

and similarly in integral form for a ball bearing as

$$w_z = \frac{n}{(w_z)_{\max}} K_{1.5}\delta_m^{3/2} \int_0^{\psi_l}\left(\cos\psi - \frac{c_d}{2\delta}\right)^{3/2}\cos\psi\, d\psi.$$

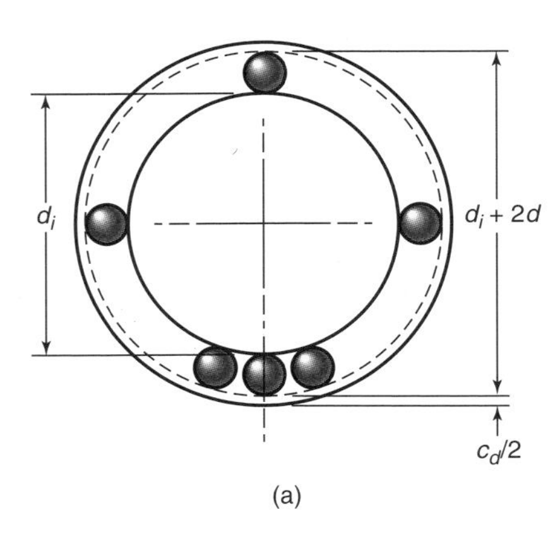

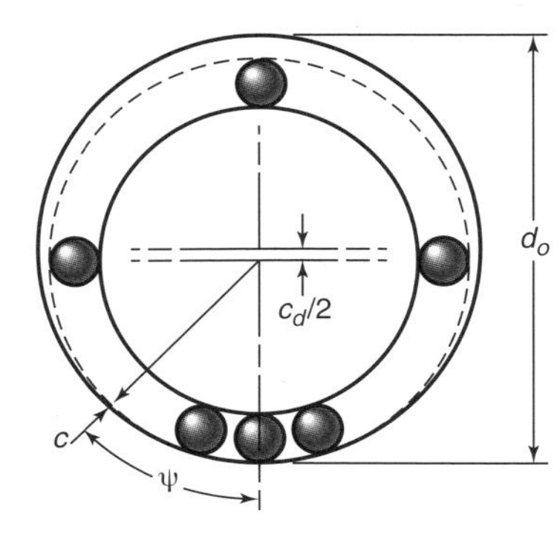

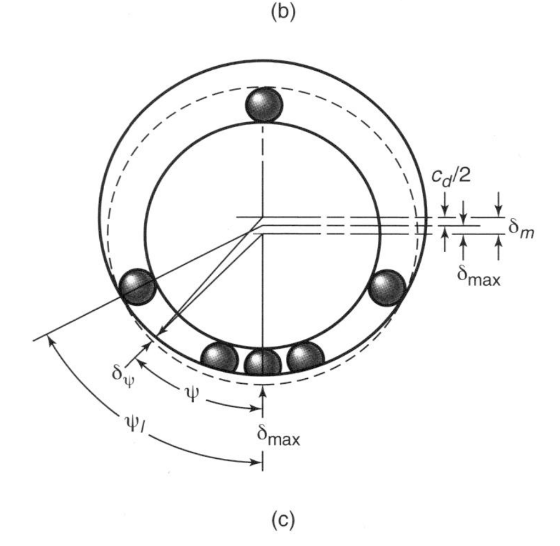

Figure 13.14: Radially loaded rolling-element bearing. (a) Concentric arrangement; (b) initial contact; (c) interference.

It has been noted that the following approximation is useful:

$$\int_0^{\psi_l}\left(\cos\psi - \frac{c_d}{2\delta}\right)^{3/2}\cos\psi\,d\psi = 2.491\left\{\left[1+\left(\frac{c_d/2\delta_m - 1}{1.23}\right)^2\right]^{1/2} - 1\right\},$$

which fits the exact numerical solution to within 2% for a complete range of $c_d/2\delta_m$.

The load carried by the most heavily loaded rolling element is obtained by substituting $\psi = 0°$ in Eq. (13.41) and dropping the summation sign to yield:

$$(w_z)_{\max} = K_j\delta_m^j\left(1 - \frac{c_d}{2\delta_m}\right)^j. \tag{13.43}$$

Recognizing that $j = 1$ for a roller bearing, and substituting for $K_1\delta_m$ in Eq. (13.42) gives

$$w_z = \frac{\left(\psi_l - \dfrac{c_d}{2\delta_m}\sin\psi_l\right)\dfrac{n(w_z)_{\max}}{2\pi}}{1 - c_d/2\delta_m}, \tag{13.44}$$

and, similarly, for a ball bearing

$$w_z = \frac{n(w_z)_{\max}}{Z_w}, \tag{13.45}$$

where

$$Z_w = \frac{\pi\left(1 - c_d/2\delta_m\right)^{3/2}}{2.491\left\{\left[1+\left(\dfrac{1 - c_d/2\delta_m}{1.23}\right)^2\right]^{1/2} - 1\right\}}. \tag{13.46}$$

For roller bearings when the diametral clearance c_d is zero, Eq. (13.44) gives

$$w_z = \frac{n\,(w_z)_{\max}}{4}. \tag{13.47}$$

For ball bearings when the diametral clearance c_d is zero, the value of Z_w in Eq. (13.45) becomes 4.37. This is the value derived by Stribeck [1901] for ball bearings of zero diametral clearance. The approach used by Stribeck was to evaluate the finite summation for various numbers of balls. He then stated the celebrated Stribeck equation for static load-carrying capacity by substituting the more conservative value of 5 for the theoretical value of 4.37:

$$w_z = \frac{n\,(w_z)_{\max}}{5}. \tag{13.48}$$

In using Eq. (13.48), it should be remembered that Z_w was considered to be a constant and that the effects of clearance and applied load on load distribution were not taken into account. However, these effects were considered in obtaining Eq. (13.45).

13.7.3 Thrust-Loaded Ball Bearings

The **static thrust load-carrying capacity** of a ball bearing may be defined as the maximum thrust load that the bearing can endure before the contact ellipse approaches a race shoulder, as shown in Fig. 13.15, or as the load at which the allowable mean compressive stress is reached, whichever is smaller.

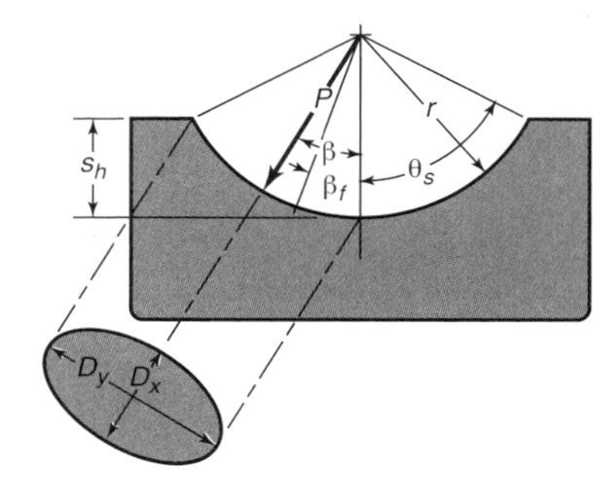

Figure 13.15: Contact ellipse in bearing race under load.

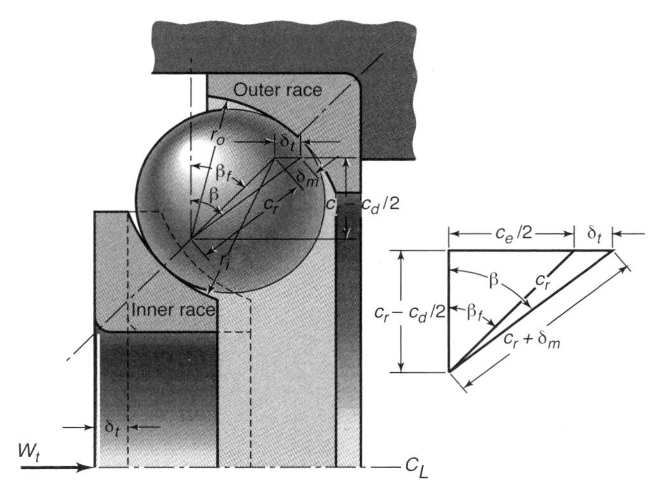

Figure 13.16: Angular-contact ball bearing under thrust load.

Both the limiting shoulder height and the mean compressive stress must be calculated to find the static thrust load-carrying capacity.

Each ball is subjected to an identical thrust component W_t/n, where W_t is the total thrust load. The initial free contact angle prior to the application of a thrust load is denoted by β_f. Under load, the normal ball thrust load, W_t, acts at the contact angle β and is written as

$$W = \frac{W_t}{n \sin \beta}. \tag{13.49}$$

Figure 13.16 shows a cross-section through an angular-contact bearing under a thrust load, W_t. From this figure, the contact angle after the thrust load has been applied can be written as

$$\beta = \cos^{-1}\left(\frac{c_r - c_d/2}{c_r + \delta_m}\right). \tag{13.50}$$

The initial free contact angle was given in Eq. (13.5). Using that equation and rearranging terms in Eq. (13.50) give, solely from the geometry of Fig. 13.16,

$$\delta_m = c_r\left(\frac{\cos \beta_f}{\cos \beta} - 1\right), \tag{13.51}$$

$$\delta_m = \delta_{mo} + \delta_{mi},$$

$$\delta_m = \left[\frac{W}{(K_j)_o}\right]^{\frac{1}{j}} + \left[\frac{W}{(K_j)_i}\right]^{\frac{1}{j}},$$

$$K_{1.5} = \frac{1}{\left\{\left[\frac{(4.5\mathcal{F}_o^3)^{1/2}}{\pi k_{eo} E'_o (R_o \mathcal{E}_o)^{1/2}}\right]^{2/3} + \left[\frac{(4.5\mathcal{F}_i^3)^{1/2}}{\pi k_{ei} E'_i (R_i \mathcal{E}_i)^{1/2}}\right]^{2/3}\right\}^{1.5}} \tag{13.52}$$

$$W = K_{1.5} c_r^{3/2}\left(\frac{\cos \beta_f}{\cos \beta} - 1\right)^{3/2} \tag{13.53}$$

where Eq. (13.33) for $K_{1.5}$ is replaced by Eq. (13.52) and k, $\mathcal{E}$, and $\mathcal{F}$ are given in Table 8.1.

From Eqs. (13.49) and (13.53),

$$\frac{W_t}{n \sin \beta} = W, \tag{13.54}$$

$$\frac{W_t}{n K_{1.5} c_r^{3/2}} = \sin \beta \left(\frac{\cos \beta_f}{\cos \beta} - 1\right)^{3/2} \tag{13.55}$$

This equation can be numerically solved by the Newton-Raphson method. The iterative equation to be satisfied is

$$\beta' - \beta = \frac{\dfrac{W_t}{n K_{1.5} c_r^{3/2}} - \sin \beta \, (\beta^*)^{3/2}}{\cos \beta \, (\beta^*)^{3/2} + \dfrac{3}{2} \cos \beta_f \tan^2 \beta \, (\beta^*)^{1/2}}, \tag{13.56}$$

where

$$\beta^* = \frac{\cos \beta_f}{\cos \beta} - 1.$$

In this equation convergence is satisfied when $\beta' - \beta$ becomes essentially zero.

When a thrust load is applied, the shoulder height limits the axial deformation, which can occur before the pressure-contact ellipse reaches the shoulder. As long as the following inequality is satisfied, the contact ellipse will not exceed this limit:

$$\theta_s > \beta + \sin^{-1}\left(\frac{D_y}{R_r d}\right).$$

From Fig. 13.5 and Eq. (13.9) the angle θ_s used to define the shoulder height can be written as

$$\theta_s = \cos^{-1}\left(1 - \frac{s_h}{R_r d}\right).$$

From Fig. 13.16, the axial deflection δ_t corresponding to a thrust load can be written as

$$\delta_t = (c_r + \delta_m) \sin \beta - c_r \sin \beta_f. \tag{13.57}$$

Substituting Eq. (13.51) into Eq. (13.57) gives

$$\delta_t = \frac{c_r \sin(\beta - \beta_f)}{\cos \beta}. \tag{13.58}$$

Note that once β has been determined from Eq. (13.55) and β_f from Eq. (13.5), the relationship for δ_t can be easily evaluated.

13.7.4 Preloading

The use of angular-contact bearings as duplex pairs preloaded against each other is discussed in Section 13.3. As shown in Table 13.1, duplex bearing pairs are used in either back-to-back or face-to-face arrangements. Such bearings are usually preloaded against each other by providing what is called *stickout* in the manufacture of the bearing. Figure 13.17 illustrates stickout for a bearing pair used in a back-to-back arrangement. The magnitude of the stickout and the bearing

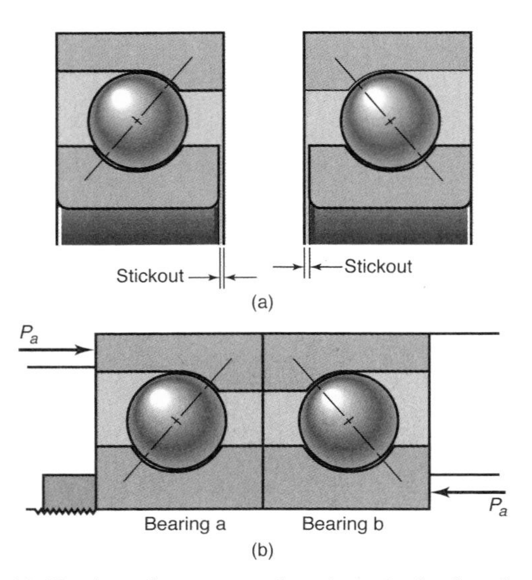

Figure 13.17: Angular contact bearings in back-to-back arrangement, shown (a) individually as manufactured and (b) as mounted with preload.

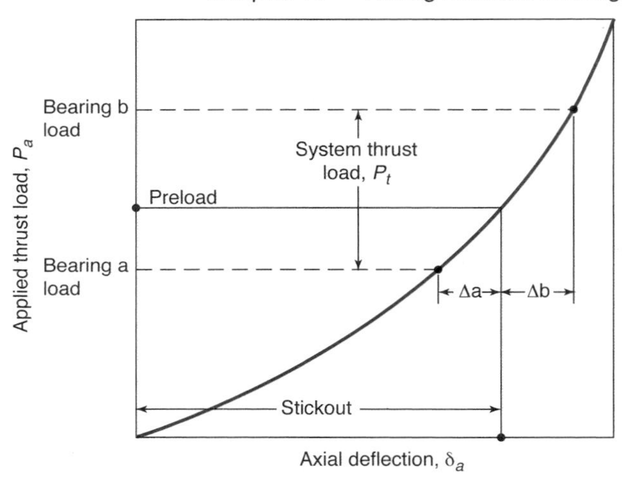

Figure 13.18: Thrust load-axial deflection curve for typical ball bearing.

design determine the level of preload on each bearing when the bearings are clamped together as in Fig. 13.17b. The magnitude of preload and the load deflection characteristics for a given bearing pair can be calculated by using Eqs. (13.5), (13.32), (13.49), and (13.55).

Figure 13.18 shows the relationship of initial preload, system load, and final load for two bearings, a and b. The load deflection curve follows the relationship $\delta_m = KP^{2/3}$. When a system thrust load, P_t, is imposed on the bearing pairs, the magnitude of load on bearing b increases while that on bearing a decreases until the difference equals the system load. The physical situation demands that the change in each bearing deflection be the same ($\Delta a = \Delta b$ in Fig. 13.18). The increments in bearing load, however, are not the same. This is important because it always requires a system thrust load far greater than twice the preload before one bearing becomes unloaded. Prevention of bearing unloading, which can result in skidding and early failure, is an objective of preloading.

13.7.5 Static Load Rating

The **static load rating** for rolling-element bearings is the maximum load that can be supported without plastic deformation. That is, below the static load rating, bearing components are loaded elastically, and are able to withstand their loading subject to the fatigue considerations in Section 13.9.

The static load rating, C_o, can be expressed as

$$C_o = n_s w_o, \tag{13.59}$$

where n_s is the safety factor and w_o is the equivalent static bearing load.

Table 13.2 gives the static load rating for selected single-row, deep-groove ball bearings. This table gives the overall dimensions of the bearing d_b, d_a, and b_w, both static and dynamic load ratings, the allowable load limit, and the bearing designation. Note that the static load rating is often lower than the dynamic load rating, and this may seem counterintuitive. If the static load rating is exceeded, then the maximum contact stresses will exceed the material flow stress. However, the volume encountering such a high stress is bounded by elastic regions; since bearing rolling elements and races are produced from stiff materials, the permanent deformation can be low and the life that can be attained can still be substantial. Bearing life as a function of load will be discussed in Section 13.9.

Table 13.2 also shows two important **speed ratings** for rolling element bearings. The *reference speed* reflects the maximum angular inner ring velocity that can be attained based on thermal criteria. If this speed is exceeded, then lubricant breakdown or interrupted supply can occur, which could significantly decrease life. This reference speed can be exceeded if a means for heat removal has been considered. The *limiting speed* is based on cage strength and bearing stability; this can be exceeded with specially designed bearings, cages, and seals. For deep groove bearings, it is not unusual for the reference speed to be higher than the limiting speed; this is attributable to their low friction and associated low heat generation. In any case, it is recommended that the lower of the speed ratings be used; non-standard bearings are available for other conditions.

The allowable load limit may be viewed as the maximum bearing load where fatigue failure will not occur. Thus, if the design load were equivalent to the allowable load limit, the safety factor would be 1 for infinite life. Also, the speed ratings in this table correspond to the maximum speed the bearing should experience before the bearing design must be altered to handle the high temperature developed in the bearing.

Table 13.3 gives similar information as Table 13.2 but for angular contact ball bearings; Table 13.4 is for single-row, cylindrical roller bearings, Table 13.5 is for toroidal (CARB) bearings, Table 13.6 presents selected needle bearings, and Table 13.7 gives the information for single-row, tapered roller bearings. In this text, consideration is limited to these types of bearings. These tables are highly abstracted, and many more sizes and types are available, as can be seen from readily available catalogs and manufacturers' websites.

13.7.6 Equivalent Static Load

A load, P, acting on a rolling-element bearing at an angle, α_p (Fig. 13.19), is a combined load, since it contains both a radial and an axial component. The radial component is $P_r = P\cos\alpha_p$ and the axial component is $P_a = P\sin\alpha_p$. The two components P_r and P_a combine to form the **equivalent static**

Table 13.2: Selected single-row, deep-groove ball bearings.

Principal dimensions d_b	d_a	b_w	Basic load ratings Dynamic C	Static C_o	Speed ratings Reference	Limiting	Mass	Designation
mm in.	mm in.	mm in.	N lb	N lb	rpm	rpm	kg lbm	—
15 0.5906	**32** 1.2598	**8** 0.3510	**5850** 1315	**2850** 641	**50,000**	**32,000**	**0.025** 0.055	**16002**
	32 1.2598	**8** 0.3543	**5850** 1315	**2850** 641	**50,000**	**32,000**	**0.030** 0.066	**6002**
	35 1.3780	**11** 0.4331	**8060** 1810	**3750** 843	**43,000**	**28,000**	**0.045** 0.099	**6202**
	35 1.3780	**13** 0.5118	**11,900** 2675	**5400** 1210	**38,000**	**24,000**	**0.082** 0.18	**6302**
20 0.7874	**42** 16535	**8** 0.3150	**7280** 1640	**4050** 910	**38,000**	**24,000**	**0.050** 0.11	**16004**
	42 1.6535	**12** 0.4724	**9950** 2240	**5000** 1120	**38,000**	**24,000**	**0.090** 0.15	**6004**
	47 1.8504	**14** 0.5512	**13,500** 3030	**6550** 1470	**32,000**	**20,000**	**0.11** 0.15	**6204**
	52 2.0472	**15** 0.5906	**16,800** 3780	**7800** 1750	**30,000**	**19,000**	**0.14** 0.31	**6304**
	72 2.8346	**19** 0.7480	**43,600** 9800	**23,600** 3370	**18,000**	**11,000**	**0.40** 0.88	**6406**
25 0.9843	**47** 1.8504	**12** 0.4724	**11,900** 2680	**6550** 1470	**32,000**	**20,000**	**0.080** 0.18	**6005**
	52 2.0472	**15** 0.5906	**14,800** 3330	**7800** 1750	**28,000**	**18,000**	**0.13** 0.29	**6205**
	62 2.4409	**17** 0.6693	**23,400** 5260	**11,600** 2610	**24,000**	**16,000**	**0.23** 0.51	**6305**
	80 3.1496	**21** 0.8268	**35,800** 8050	**19,300** 4340	**20,000**	**13,000**	**0.53** 0.51	**6405**
30 1.1811	**55** 2.1654	**15** 0.5118	**13,800** 3100	**8300** 1870	**28,000**	**17,000**	**0.12** 0.26	**6006**
	62 2.4409	**16** 0.6299	**20,300** 4560	**11,200** 2520	**24,000**	**15,000**	**0.20** 0.44	**6206**
	72 2.8346	**19** 0.7480	**29,600** 6650	**16,000** 3600	**20,000**	**13,000**	**0.35** 0.77	**6306**
	90 3.5433	**23** 0.9055	**43,600** 9800	**23,600** 5310	**18,000**	**11,000**	**0.74** 1.65	**6406**
35 1.3780	**62** 2.4409	**14** 0.5512	**16,800** 3780	**10,200** 2290	**24,000**	**15,000**	**0.16** 0.35	**6007**
	72 2.8346	**17** 0.6693	**27,000** 6070	**15,300** 3440	**20,000**	**13,000**	**0.29** 0.64	**6207**
	80 3.1496	**21** 0.8268	**35,100** 7890	**19,000** 4270	**19,000**	**12,000**	**0.46** 1.00	**6307**
	100 3.9370	**25** 0.9843	**55,300** 1240	**31,000** 6970	**16,000**	**10,000**	**0.95** 2.10	**6407**
40 1.5748	**68** 2.6672	**15** 0.5906	**17,800** 4000	**11,600** 2610	**22,000**	**14,000**	**0.19** 0.42	**6008**
	80 3.1496	**18** 0.7087	**32,500** 7310	**19,000** 4270	**18,000**	**11,000**	**0.37** 0.82	**6208**
	90 3.5433	**23** 0.9055	**42,300** 9510	**24,000** 5400	**17,000**	**11,000**	**0.63** 1.40	**6308**
	110 4.3307	**27** 1.0630	**63,700** 14,320	**36,500** 8210	**14,000**	**9000**	**1.25** 2.75	**6408**
45 1.7717	**75** 2,9528	**16** 0.6299	**22,100** 4970	**14,600** 3280	**20,000**	**12,000**	**0.25** 0.55	**6009**
	85 3.3465	**19** 0.7480	**35,100** 7890	**21,600** 4860	**17,000**	**11,000**	**0.41** 0.90	**6209**
	100 3.9370	**25** 0.9843	**55,300** 12,430	**31,500** 7080	**15,000**	**9500**	**0.83** 1.85	**6309**
	120 4.7244	**29** 1.1417	**76,100** 17,110	**45,000** 10,100	**13,000**	**8500**	**0.55** 3.40	**6409**
50 1.9685	**80** 3.1496	**16** 0.6299	**22,900** 5148	**16,000** 3600	**18,000**	**11,000**	**0.26** 0.57	**6010**
	90 3.5433	**20** 0.7874	**37,100** 8340	**23,200** 5220	**15,000**	**10,000**	**0.46** 1.00	**6210**
	110 4.3307	**27** 1.0630	**65,000** 14,610	**38,000** 8540	**13,000**	**8500**	**1.05** 2.30	**6310**
	130 5.1181	**31** 1.2205	**87,100** 19,580	**52,000** 11,700	**12,000**	**7500**	**1.90** 4.20	**6410**

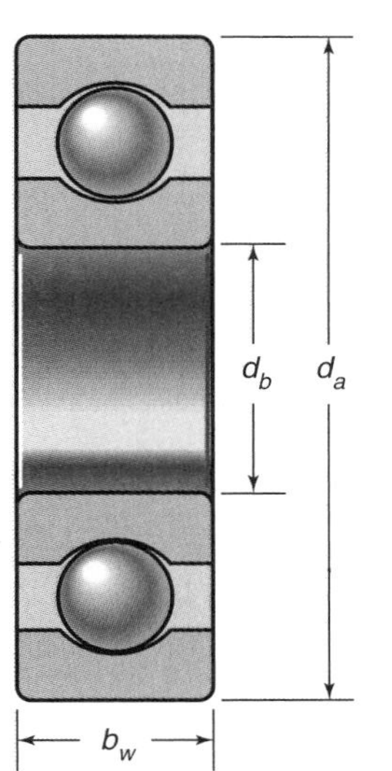

Table 13.3: Selected single-row, angular contact roller bearings. These bearings are often mounted in pairs (see Fig. 13.17).

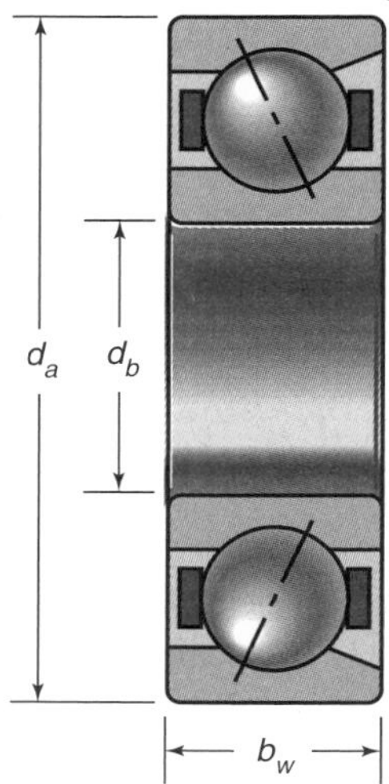

Principal dimensions			Basic load ratings		Speed ratings		Mass	Designation
			Dynamic	Static	Reference	Limiting		
d_b	d_a	b_w	C	C_o				
mm in.	mm in.	mm in.	N lb	N lb	rpm	rpm	kg lbm	
10 0.3937	**30** 1.1811	**9** 0.3543	**7020** 1580	**3350** 750	**30,000**	**30,000**	**0.03** 0.07	**7200 BE**
12 0.4724	**32** 1.2598	**10** 0.3937	**7610** 1710	**3800** 850	**26,000**	**26,000**	**0.04** 0.08	**7201 BE**
	37 1.4567	**12** 0.4724	**10,600** 2400	**5000** 1120	**24,000**	**24,000**	**0.06** 0.10	**7301 BE**
15 0.5906	**35** 1.3780	**11** 0.4331	**8840** 1990	**4800** 1080	**24,000**	**24,000**	**0.05** 0.10	**7202 BE**
	42 1.6535	**13** 0.5118	**13,000** 2900	**6700** 1510	**20,000**	**20,000**	**0.08** 0.20	**7302 BE**
20 0.7874	**47** 1.8504	**14** 0.5512	**13,300** 2990	**8300** 1870	**18,000**	**19,000**	**0.11** 0.24	**7204 BE**
	52 2.0472	**15** 0.5906	**19,000** 4300	**10,400** 2340	**18,000**	**18,000**	**0.15** 0.30	**7304 BE**
25 0.9843	**52** 2.0472	**15** 0.5906	**15,600** 3510	**10,200** 2290	**17,000**	**17,000**	**0.14** 0.31	**7205 BE**
	62 2.4409	**17** 0.6693	**26,500** 6000	**15,600** 3510	**15,000**	**15,000**	**0.24** 0.50	**7305 BE**
	80 3.1496	**21** 0.8268	**39,700** 8900	**23,600** 5300	**11,000**	**11,000**	**0.61** 1.30	**7405 B**
30 1.1811	**62** 2.4409	**16** 0.6299	**24,000** 5400	**25,600** 3510	**14,000**	**14,000**	**0.21** 0.46	**7206 BE**
	72 2.8346	**19** 0.7480	**35,500** 8000	**21,200** 4770	**13,000**	**13,000**	**0.37** 0.80	**7306 BE**
	90 3.5433	**23** 0.9055	**47,500** 10,700	**29,000** 6500	**10,000**	**10,000**	**0.85** 1.90	**7406 B**
40 1.5748	**80** 3.1496	**18** 0.7087	**36,500** 8210	**26,000** 5840	**11,000**	**11,000**	**0.39** 0.86	**7208 BE**
	90 3.5433	**23** 0.9055	**50,000** 11,200	**33,500** 7530	**10,000**	**10,000**	**0.68** 1.50	**7308 BE**
	110 4.3307	**27** 1.0630	**70,200** 15,800	**45,500** 10,200	**8000**	**8000**	**1.40** 3.10	**7408 B**
50 1.9685	**90** 3.5433	**20** 0.7874	**40,000** 8990	**30,500** 6860	**9000**	**9000**	**0.51** 1.12	**7210 BE**
	110 4.3307	**27** 1.0630	**75,000** 16,900	**51,000** 11,500	**8000**	**8000**	**1.16** 2.60	**7310 BE**
	130 5.1181	**31** 1.2205	**95,600** 21,500	**64,000** 14,400	**6300**	**6700**	**2.25** 5.00	**7410 B**
75 2.9528	**130** 5.1181	**25** 0.9843	**70,200** 15,800	**64,000** 14,400	**5600**	**6000**	**1.29** 2.80	**7215 BE**
	160 6.2992	**37** 1.4567	**132,000** 29,700	**106,000** 23,800	**5300**	**5300**	**3.26** 7.20	**7315 BE**
	190 7.4903	**45** 1.7717	**168,000** 37,800	**140,000** 31,500	**4300**	**4500**	**6.85** 15.10	**7415 B**
100 3.9370	**180** 7.0866	**34** 1.3386	**135,000** 30,300	**122,000** 27,400	**4000**	**4300**	**3.61** 8.00	**7220 BE**
	215 8.4646	**47** 1.8504	**216,000** 48,600	**190,000** 42,700	**4000**	**4000**	**8.00** 17.60	**7320 BE**
	265 10.433	**60** 2.3622	**276,000** 62,000	**275,000** 61,800	**3200**	**3200**	**15.50** 34.20	**7420**
150 5.9055	**270** 10.630	**45** 1.7717	**216,000** 48,600	**224,000** 50,400	**2600**	**2800**	**10.80** 23.80	**7230 B**
	320 12.598	**65** 2.5591	**332,000** 74,600	**365,000** 82,100	**2400**	**2400**	**25.00** 55.10	**7330B**
190 7.4803	**340** 13.386	**55** 2.1654	**307,000** 69,000	**355,000** 79,800	**2000**	**2200**	**21.90** 48.30	**7238 B**
	400 15.748	**78** 3.0709	**442,000** 99,400	**560,000** 125,900	**1900**	**1900**	**48.30** 106.50	**7338 B**
240 9.4488	**440** 17.3228	**72** 2.8346	**364,000** 81,800	**540,000** 121,400	**1600**	**1700**	**49.00** 108.00	**7248 B**

Table 13.4: Selected single-row, cylindrical roller bearings.

Principal dimensions			Basic load ratings		Speed ratings		Mass	Designation
			Dynamic	Static	Reference	Limiting		
d_b	d_a	b_w	C	C_o				
mm	mm	mm	N	N			kg	
in.	in.	in.	lb	lb	rpm	rpm	lbm	
15	**35**	**11**	**12,500**	**10,200**	**22,000**	**26,000**	**0.047**	**NU 202 ECP**
0.5906	1.3780	0.4331	2810	2290			0.10	
	42	**13**	**17,200**	**14,300**	**19,000**	**22,000**	**0.068**	**NJ 203 ECP**
	1.6535	0.5118	3870	3210			0.15	
20	**47**	**14**	**25,100**	**22,000**	**16,000**	**19,000**	**0.11**	**NU 204 ECP**
0.7874	1,8504	0.5512	5640	4950			0.24	
	52	**15**	**35,500**	**26,000**	**15,000**	**18,000**	**0.17**	**NU 304 ECP**
	2.0472	0.5906	7980	5850			0.37	
25	**52**	**15**	**28,600**	**27,000**	**14,000**	**16,000**	**0.14**	**NU 205 ECP**
0.9843	2.0472	0.5906	6430	6070			0.31	
	62	**17**	**46,500**	**36,500**	**12,000**	**15,000**	**0.28**	**NU 305 ECP**
	2.4409	0.6693	10,400	8210			0.62	
30	**62**	**16**	**44,000**	**36,500**	**13,000**	**14,000**	**0.22**	**NUP 206 ECP**
1.811	2.4409	0.6299	9890	8210			0.48	
	72	**19**	**58,500**	**48,000**	**11,000**	**12,000**	**0.40**	**NU 306 ECP**
	2.8346	0.7480	13,150	10,800			0.88	
35	**72**	**17**	**56,000**	**48,000**	**11,000**	**12,000**	**0.30**	**NU 207 ECP**
1.3780	2.8346	0.693	12,600	10,800			0.66	
	80	**21**	**75,000**	**63,000**	**9500**	**11,000**	**0.54**	**NU 307 ECP**
	3.1496	0.8268	16,900	14,200			1.19	
40	**80**	**18**	**62,000**	**53,000**	**9500**	**11,000**	**0.42**	**NU 208 ECP**
1.5748	3.1496	0.7087	13,900	11,900			0.92	
	90	**23**	**93,000**	**78,000**	**8000**	**9500**	**0.73**	**NU 308 ECP**
	3.5433	0.9055	20,900	17,500			1.61	
45	**85**	**19**	**69,500**	**64,000**	**9000**	**9500**	**0.48**	**NU 209 ECP**
1.7717	3.3465	0.7480	15,600	14,400			1.06	
	100	**25**	**112,000**	**100,000**	**7500**	**8500**	**1.0**	**NU 309 ECP**
	3.9370	0.9843	25,200	22,500			2.20	
50	**90**	**20**	**73,500**	**69,500**	**8500**	**9000**	**0.49**	**NU 210 ECP**
1.9685	3.5433	0.7874	16,500	15,600			1.08	
	110	**27**	**127,000**	**112,000**	**6700**	**8000**	**1.15**	**NU 310 ECP**
	4.3307	1.0630	28,600	25,200			2.55	
60	**110**	**22**	**108,000**	**102,000**	**6700**	**7500**	**0.86**	**NU 212 ECP**
2.3622	4.3307	0.8661	24,300	22,900			1.89	
	130	**31**	**173,000**	**160,000**	**5600**	**6700**	**1.80**	**NU 312 ECP**
	5.1181	1.2205	38,900	36,000			4.00	
75	**130**	**25**	**150,000**	**156,000**	**5600**	**6000**	**1.25**	**NU 215 ECP**
2.9528	5.1181	0.9843	33,700	35,100			2.80	
	160	**37**	**280,000**	**265,000**	**4500**	**5300**	**3.30**	**NU 315 ECP**
	6.2992	1.4567	62,900	59,600			7.30	
100	**180**	**34**	**285,000**	**305,000**	**4000**	**4500**	**3.45**	**NU 220 ECP**
3.9370	7.0866	1.3386	64,100	68,600			7.60	
	215	**47**	**450,000**	**440,000**	**3200**	**3800**	**7.80**	**NU 320 ECP**
	8.4646	1.8504	101,200	98,900			17.2	
120	**215**	**40**	**390,000**	**430,000**	**3400**	**3600**	**5.75**	**NU 224 ECP**
4.7244	8.4646	1.5748	87,700	96,700			12.6	
	260	**55**	**610,000**	**620,000**	**2800**	**3200**	**13.3**	**NU 324 ECP**
	10.236	2.1654	137,100	139,400			29.30	
150	**270**	**45**	**510,000**	**600,000**	**2600**	**2800**	**10.6**	**NU 230 ECM**
5.9055	10.630	1.7717	114,600	134,900			23.3	
	320	**65**	**900,000**	**965,000**	**2200**	**3400**	**27.5**	**NU 330 ECM**
	12.598	2.5591	202,300	216,900			60.5	
190	**340**	**55**	**800,000**	**965,000**	**2000**	**2200**	**24.0**	**NU 238 ECM**
7.4803	13.386	2.1654	179,800	216,900			52.8	
	400	**78**	**1,140,000**	**1,500,000**	**1500**	**200**	**50.00**	**NU 338 ECM**
	15.748	3.0709	256,300	337,200			110.2	

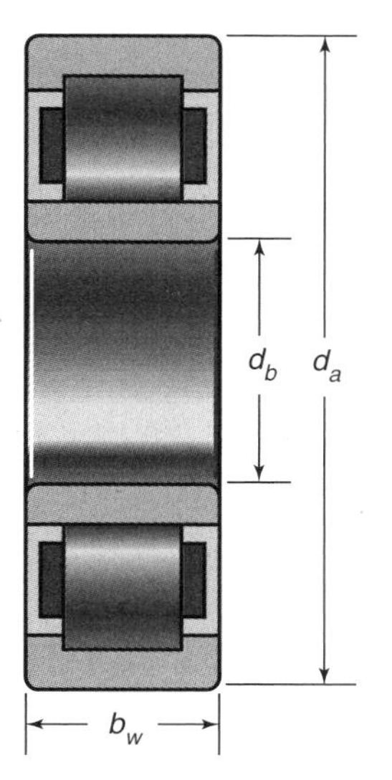

Table 13.5: Selected single-row, toroidal (CARB) roller bearing.

Principal dimensions			Basic load ratings		Speed ratings		Mass	Designation
			Dynamic	Static	Reference	Limiting		
d_b	d_a	b_w	C	C_o				
mm in.	mm in.	mm in.	N lb	N lb	rpm	rpm	kg lbm	—
25 0.9843	**52** 2.0472	**18** 0.7078	**44,000** 9900	**40,000** 9000	**13,000**	**18,000**	**0.17** 0.40	**C2205 TN9**
			50,000 11,200	**48,000** 10,800	—	**7000**	**0.18** 0.40	**C2205 V**[a]
30 1.1811	**62** 2.4409	**20** 0.7874	**69,500** 15,600	**62,000** 13,900	**11,000**	**15,000**	**0.27** 0.60	**C2206 TN9**
			76,500 17,200	**71,000** 16,000	—	**6000**	**0.29** 0.64	**C2206 V**[a]
	55 2.1654	**45** 1.7717	**134,000** 30,100	**180,000** 40,500	—	**3000**	**0.50** 1.10	**C 6006 V**[a]
40 1.5748	**80** 3.1496	**23** 0.9055	**90,000** 20,200	**86,500** 19,400	**8000**	**11,000**	**0.50** 1.10	**C2208 TN9**
			102,000 22,900	**104,000** 23,400	—	**4500**	**0.53** 1.17	**C2208 V**[a]
	62 2,4409	**22** 0.8661	**76,500** 17,200	**100,000** 22,500	—	**4300**	**0.25** 0.60	**C 4908 V**[a]
		30 1.1811	**104,000** 23,400	**143,000** 32,100	—	**3400**	**0.35** 0.80	**C 5908 V**[a]
50 1.9685	**90** 3.5433	**23** 0.9055	**98,000** 22,000	**100,000** 22,500	**7000**	**9500**	**0.59** 1.30	**C2210 TN9**
			114,000 25,600	**122,000** 27,400	—	**3800**	**0.62** 0.80	**C2210 V**[a]
	72 2.8346	**22** 0.8661	**86,500** 19,400	**120,000** 28,100	—	**3600**	**0.29** 0.60	**C 4910 V**[a]
		30 1.1811	**118,000** 26,500	**180,000** 40,500	—	**2800**	**0.42** 0.90	**C 5910 V**[a]
60 2.3622	**110** 4.3307	**28** 1.1024	**143,000** 32,100	**156,000** 35,100	**5600**	**7500**	**1.10** 2.40	**C 2212 TN9**
	110 4.3307	**28** 1.1024	**166,000** 37,300	**190,000** 42,700	—	**2800**	**1.15** 2.50	**C 2212 V**[a]
	85 3.3465	**25** 0.9843	**112,000** 25,200	**170,000** 38,200	—	**3000**	**0.46** 1.00	**C 4912 V**[a]
		34 1.3386	**150,000** 35,100	**240,000** 58,400	—	**2400**	**0.64** 1.50	**C 5912 V**[a]
75 2.9528	**130** 5.1181	**31** 1.2205	**196,000** 44,100	**208,000** 46,800	**4800**	**6700**	**1.6** 3.50	**C2215**
			220,000 49,500	**240,000** 54,000	—	**2200**	**1.65** 3.60	**C2215 V**[a]
	105 4.1339	**30** 1.1811	**166,000** 37,300	**255,000** 57,300	—	**2400**	**0.81** 1.80	**C 4015 V**[a]
		40 1.5748	**204,000** 45,900	**325,000** 73,100	—	**1900**	**1.10** 2.40	**C 5915 V**[a]
100 3.9370	**180** 7.0866	**46** 1.8110	**415,000** 93,300	**465,000** 104,500	**3600**	**4800**	**4.85** 10.70	**C 2220**
	165 6.4961	**52** 2.0472	**415,000** 93,300	**540,000** 121,400	**3200**	**4300**	**4.40** 9.70	**C 3120**
	140 5.5118	**54** 2.1260	**375,000** 84,300	**640,000** 143,900	—	**1400**	**2.70** 6.00	**C 5920 V**[a]
150 5.9055	**270** 10.630	**73** 2.8740	**980,000** 220,300	**1,220,000** 274,300	**2400**	**3200**	**17.5** 38.60	**C2230**
	250 9.8425	**80** 3.150	**880,000** 197,800	**1,290,000** 290,000	**2000**	**2800**	**15.00** 33.10	**C 3130**
	225 8.8538	**75** 2.9528	**780,000** 175,300	**1,320,000** 296,700	—	**750**	**10.5** 23.10	**C 4030 V**[a]
200 7.8740	**340** 13.3858	**112** 4.4094	**1,560,000** 350,700	**2,320,000** 521,500	**1500**	**2000**	**40.00** 88.20	**C 3140**
	310 12.205	**109** 4.2913	**1,630,000** 366,400	**2,650,000** 595,700	—	**260**	**30.50** 67.20	**C 4060 V**[a]
500 19.685	**830** 32.677	**264** 10.394	**7,500,000** 1,686,000	**12,700,000** 2,855,000	**530**	**750**	**550.00** 1212	**C31/500 M**
1000 39.370	**1580** 62.205	**462** 18.189	**22,800,000** 5,125,400	**45,500,000** 10,228,400	**220**	**300**	**3470** 7650	**C 31/1000 MB**

[a] Full-complement bearing.

Table 13.6: Selected single-row needle roller bearings. Without an inner ring, the load ratings can vary greatly depending on quality of shaft surface preparation, and are expressed for ground and hardened shafts.

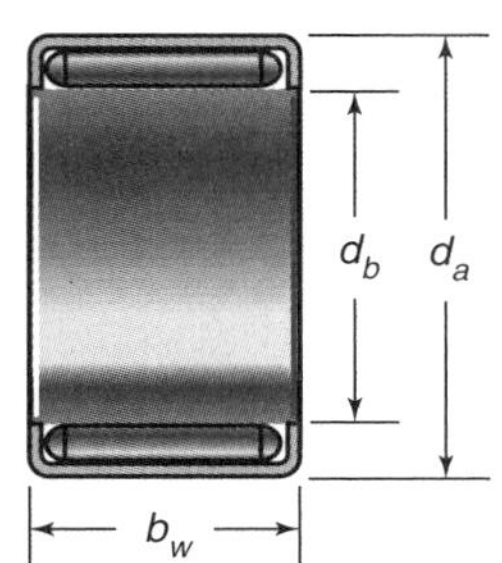

Principal dimensions			Basic load ratings		Speed ratings		Mass	Designation
			Dynamic	Static	Reference	Limiting		
d_b	d_a	b_w	C	C_o				
mm in.	mm in.	mm in.	N lb	N lb	rpm	rpm	g lb	
3 0.1181	7 0.2559	6 0.2362	1230 277	880 198	30,000	43,000	1.00 0.0025	HK 0306 TN[a]
5 0.1969	9 0.3543	9 0.3543	2380 535	2080 468	22,000	34,000	2.00 0.0043	HK 0509[a]
	10 0.3937	10 0.3937	2920 700	2700 700	36,000	40,000	10 0.01	NK 5/12 TN[a]
	15 0.5906	12 0.4724	3800 900	4250 1000	32,000	36,000	10 0.03	NKI 5/12 TN
7 0.2756	11 0.4331	9 0.3543	3030 681	3050 686	18,000	28,000	2.60 0.0056	HK 0709[a]
	14 0.5512	12 0.4724	3580 900	3570 900	32,000	36,000	10 0.02	NK 7/12 TN[a]
	17 0.6693	12 0.4724	4570 1100	5700 1300	28,000	32,000	10 0.03	NKI 7/12 TN
10 0.3937	14 0.5512	10 0.3937	4290 964	5300 1190	16000	24,000	3.00 0.0.875	HK 1010[a]
	22 0.8661	13 0.5118	8800 2000	10,400 2400	24,000	28,000	0.02 0.05	NA 4900
	17 0.6693	16 0.6299	5940 1400	8000 1800	28,000	32,000	10 0.03	NK 10/16 TN[a]
	22 0.8661	16 0.6299	10,200 2300	12,500 2900	24,000	28,000	30 0.06	NKI 10/16
12 0.4724	16 0.6299	10 0.3937	4840 1090	6400 1440	14,000	20,000	4.60 0.16	HK 1210[a]
	24 0.9449	13 0.5118	9900 2300	12,200 2800	22,000	26,000	30 0.06	NA 4901
	19 0.7480	16 0.6299	9130 2100	12,000 2700	26,000	30,000	20 0.04	NK 12/16[a]
	24 0.9449	16 0.6299	11,700 2700	15,300 3500	22,000	26,000	30 0.07	NKI 12/16
15 0.5906	21 0.8268	12 0.4724	7650 1720	9500 2140	11000	17000	11.0 0.024	HK 1512[a]
	28 1.1024	13 0.5118	11,200 2600	15,300 3500	19,000	22,000	30.0 0.07	NA 4902
	23 0.9055	16 0.6299	11,000 2500	14,000 3200	24,000	26,000	20 0.05	NK 15/16[a]
	27 1.0630	16 0.6299	13,400 3100	19,000 4300	20,000	24,000	40 0.09	NKI 15/16
20 0.7874	26 1.0236	10 0.3937	6160 1390	8500 1910	9000	14,000	12.00 0.0375	HK 2010[a]
	37 1.4567	17 0.6693	21,600 4900	28,000 6300	15,000	17,000	80 0.20	NA 4904
	28 1.1024	16 0.6299	13,200 3000	19,300 4400	19,000	22,000	30 0.06	NK 20/16[a]
	32 1.2598	16 0.6299	15,400 3500	24,500 5600	16,000	19,000	50 0.10	NKI 20/16
25 0.9843	32 1.2598	12 0.4724	10,500 2360	15,300 3440	7500	11,000	21 0.046	HK 2512[a]
	42 1.6535	17 0.6693	24,200 5500	34,500 7800	13,000	15,000	90 0.20	NA 4905
	33 1.2992	20 0.7874	19,000 4300	32,500 7400	16,000	18,000	40 0.09	NK 25/20[a]
	38 1.4961	20 0.7874	22,000 5000	36,500 8300	14,000	15,000	80 0.20	NKI 25/20
35 1.3780	42 1.6535	20 0.4724	22,900 2810	46,500 4860	5600	8000	28 0.98	HK 3512[a]
	45 1.7717	20 0.7874	26,400 5600	45,000 10,200	11,000	13,000	70 0.20	NK 35/20[a]
	50 1.9685	20 0.7874	26,400 6000	51,000 11,500	10,000	11,000	130 0.30	NKI 35/20
50 1.9685	58 2.835	20 0.7874	29,200 6560	63,000 14,200	4000	5600	72 0.16	HK 5020[a]
	72 2.8346	22 0.8661	47,300 10,700	85,000 19,200	7000	8000	270 0.60	NA 4910
100 3.9370	140 5.5118	40 1.5748	125,000 28,100	280,000 63,000	3400	4000	1900 4.20	NA 4920

[a] Open end, bearings roll against shaft surface.

Table 13.7: Selected single-row, tapered roller bearings. *Source:* Courtesy of The Timken Company.

Principal dimensions			Basic load ratings Dynamic	Static	Contact angle	Mass	Designation
d_b	d_a	b_w	C	C_o	β		
mm in.	mm in.	mm in.	N lb	N lb	deg.	kg lbm	
15 0.5906	**42** 1.6535	**14.25** 0.5610	**26,000** 5850	**22,200** 4980	**10.7**	**0.11** 0.22	**30302**
17 0.6693	**47** 1.8504	**15.25** 0.6004	**32,700** 7360	**28,400** 4460	**10.7**	**0.14** 0.29	**30303**
20 0.7874	**52** 2.0472	**16.25** 0.6398	**38,500** 8660	**34,500** 7760	**11.3**	**0.17** 0.37	**30304**
	52 2.0472	**22.5** 0.8858	**55,200** 12,400	**55,000** 12,400	**11.3**	**0.24** 0.53	**32304**
25 0.9843	**52** 2.0472	**16.25** 0.6398	**36,90000** 8300	**38,300** 8620	**11.3**	**0.15** 0.34	**30205**
	62 2.4409	**25.25** 0.9941	**72,400** 16,300	**72,400** 16,300	**11.3**	**0.37** 0.81	**32305**
30 1.1811	**72** 2.8346	**20.75** 0.8169	**67,700** 15,200	**65,300** 14,700	**11.9**	**0.39** 0.88	**30306**
	72 2.8346	**28.75** 1.1319	**87,600** 19,700	**89,800** 20,200	**11.9**	**0.56** 1.23	**32306**
35 1.3780	**72** 3.1496	**24.25** 1.2894	**74,900** 16,800	**82,300** 18,500	**11.9**	**0.44** 0.96	**32207**
	80 3.1496	**22.75** 0.8957	**87,200** 19,600	**86,100** 19,300	**11.9**	**0.53** 1.16	**30307**
40 1.5748	**90** 3.5433	**25.25** 0.9941	**117,000** 26,300	**102,000** 23,000	**12.9**	**0.73** 1.61	**30308**
	90 3.5433	**35.25** 1.3878	**157,000** 35,300	**160,000** 36,100	**12.9**	**1.1** 2.43	**32308-B**
45 1.7717	**100** 3.9370	**27.25** 1.0728	**129,000** 29,000	**139,000** 31,300	**12.9**	**1.01** 2.22	**30309**
	100 3.9370	**38.25** 1.5059	**189,000** 42,500	**187,000** 42,100	**12.9**	**1.42** 3.13	**32309-B**
50 1.9685	**110** 4.3307	**29.25** 1.1516	**142,000** 31,900	**150,000** 33,800	**12.9**	**1.25** 2.77	**30310**
	110 4.3307	**42.25** 1.6634	**187,000** 42,000	**211,000** 47,500	**12.9**	**1.83** 4.03	**32310**
60 2.3622	**130** 5.1171	**33.5** 1.3189	**203,000** 45,700	**188,000** 42,200	**12.9**	**1.96** 4.32	**30312**
	130 5.1181	**48.5** 1.9094	**264,000** 59,400	**310,000** 69,800	**12.9**	**2.89** 6.35	**32312**
75 2.9528	**115** 4.5276	**31** 1.2205	**187,000** 42,000	**239,000** 53,700	**11.2**	**1.15** 2.54	**33015**
	160 6.2992	**40** 1.5748	**248,000** 55,800	**278,000** 62,500	**12.9**	**3.46** 7.63	**30315**
100 3.9370	**150** 5.9055	**39** 1.5354	**251,000** 56,300	**393,000** 88,300	**10.8**	**2.36** 5.22	**33020**
	180 7.0866	**49** 1.9291	**368,000** 82,700	**478,000** 107,000	**15.6**	**4.92** 10.84	**32220**
120 4.7244	**165** 6.4961	**29** 1.1417	**172,000** 38,800	**317,000** 71,200	**13.1**	**1.78** 3.92	**32924**
	215 8.4646	**43.5** 1.7126	**396,000** 89,100	**508,000** 114,000	**16.2**	**6.24** 13.75	**30224**
150 5.9055	**210** 8.2677	**38** 1.4961	**324,000** 72,800	**573,000** 129,000	**12.3**	**3.99** 8.82	**32930**
	270 10.630	**49** 1.9291	**565,000** 127,000	**735,000** 165,000	**16.2**	**11.03** 24.29	**30230**
170 6.6929	**230** 9.0551	**38** 1.4961	**355,000** 79,800	**652,000** 146,000	**14.3**	**4.40** 9.71	**32934**
200 7.874	**280** 11.024	**51** 2.0079	**561,000** 126,000	**1,050,000** 235,000	**14.7**	**9.45** 20.84	**32940**
220 8.6614	**300** 11.811	**51** 2.0079	**561,000** 126,000	**1,090,000** 245,000	**15.8**	**9.90** 21.83	**32944**
	400 15.798	**72** 2.8346	**1,260,000** 283,000	**1,560,000** 350,000	**15.6**	**35.25** 77.71	**30244**
280 11.024	**380** 14.961	**63.5** 2.5000	**850,000** 191,000	**1,780,000** 401,000	**16.0**	**19.81** 43.69	**32956**
320 12.598	**480** 18.898	**100** 3.9370	**1,800,000** 406,000	**3,420,000** 768,000	**17.0**	**59.62** 131.44	**32064X**
380 14.173	**480** 18.898	**76** 2.9921	**1,250,000** 281,000	**2,780,000** 624,000	**17.0**	**45.22** 79.84	**32972**

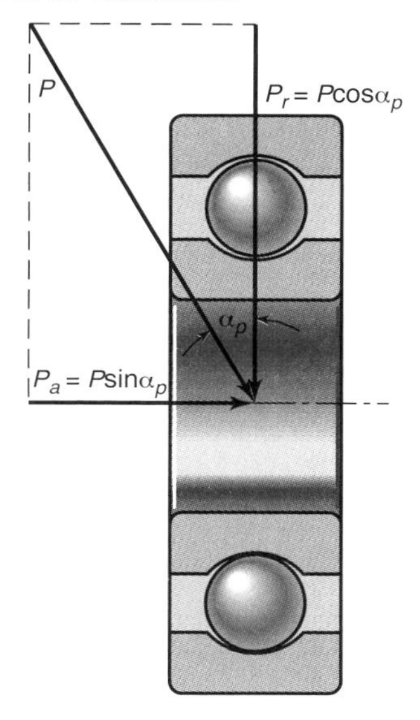

Figure 13.19: Combined load acting on a radial deep-grove ball bearing.

load, P_o, which can be expressed as

$$P_o = X_o P_r + Y_o P_a, \tag{13.60}$$

where X_o is the radial factor and Y_o is the thrust factor, both of which can be obtained from Table 13.8. Note also that in Table 13.8, β is the contact angle.

Example 13.4 Equivalent Load for a Ball Bearing

Given: A single-row, angular-contact ball bearing is loaded with a static load having an axial component of 1000 N and a radial component of 500 N. The static basic load rating is 1500 N.

Find: The contact angle between 20° and 40° that produces the largest static safety factor.

Solution: From Table 13.8, note that $X_o = 0.5$ for all values of β for a radial angular contact ball bearing. Therefore, from Eq. (13.60),

$$P_o = X_o P_r + Y_o P_a = (0.5)(500) + Y_o(1000),$$

or

$$P_o = 250 + 1000Y_o.$$

The safety factor is the largest when P_o is the smallest, which corresponds to small values of Y_o. Reviewing Table 13.8, this implies that the contact angle should be 40°, thus producing Y_o = 0.26. Therefore,

$$P_o = 250 + (0.26)(1000) = 510 \text{ N}.$$

13.8 Elastohydrodynamic Lubrication

Nonconformal surfaces, such as those in rolling-element bearings, are lubricated elastohydrodynamically as presented in Section 8.7.5. There it was shown that elastohydrodynamic lubrication is a form of fluid film lubrication where elastic deformation of the bearing surfaces becomes significant. It is usually associated with highly stressed machine components, such as rolling-element bearings.

13.8.1 Relevant Equations

The relevant equations used in elastohydrodynamic lubrication are:

- *Lubrication equation* (Reynolds equation) first developed in Chapter 12:

$$\frac{\partial}{\partial x}\left(\frac{\rho h^3}{\eta}\frac{\partial p}{\partial x}\right) + \frac{\partial}{\partial y}\left(\frac{\rho h^3}{\eta}\frac{\partial p}{\partial y}\right) = 12\tilde{u}\frac{\partial}{\partial x}(\rho h), \tag{12.23}$$

where

$$\tilde{u} = \frac{u_a + u_b}{2}.$$

- *Viscosity variation*, also known as the Barus law:

$$\eta = \eta_o e^{\xi p}, \tag{13.61}$$

where η_o is the coefficient of absolute or dynamic viscosity at atmospheric pressure and ξ is the pressure-viscosity coefficient of the fluid.

- *Density variation:* where, for mineral oils:

$$\rho = \rho_o\left(1 + \frac{0.6p}{1 + 1.7p}\right), \tag{13.62}$$

where ρ_o is the density at atmospheric conditions and p is in gigapascals.

- *Elasticity equation:*

$$\delta = \frac{2}{E'}\int\int_A \frac{p(x,y)\,dx\,dy}{\sqrt{(x-x_1)^2 + (y-y_1)^2}}, \tag{13.63}$$

where

$$E' = \frac{2}{\dfrac{1-\nu_a^2}{E_a} + \dfrac{1-\nu_b^2}{E_b}}. \tag{13.64}$$

- *Film thickness equation:*

$$h = h_o + \frac{x^2}{2R_x} + \frac{y^2}{2R_y} + \delta(x,y), \tag{13.65}$$

where

$$\frac{1}{R_x} = \frac{1}{r_{ax}} + \frac{1}{r_{bx}}, \tag{8.6}$$

$$\frac{1}{R_y} = \frac{1}{r_{ay}} + \frac{1}{r_{by}}. \tag{8.7}$$

The different radii are expressed in Eqs. (13.11) to (13.16).

Table 13.8: Radial factor X_o and thrust factor Y_o for statically stressed radial bearings.

Bearing type		Single row X_o	Single row Y_o	Double row X_o	Double row Y_o
Radial deep-groove ball		0.6	0.5	0.6	0.5
Radial angular-contact ball	β = 20°	0.5	0.42	1	0.84
	β = 25°	0.5	0.38	1	0.76
	β = 30°	0.5	0.33	1	0.66
	β = 35°	0.5	0.29	1	0.58
	β = 40°	0.5	0.26	1	0.52
Radial self-aligning ball		0.5	$0.22 \cot\beta$	1	$0.44 \cot\beta$
Radial spherical roller		0.5	$0.22 \cot\beta$	1	$0.44 \cot\beta$
Radial tapered roller		0.5	$0.22 \cot\beta$	1	$0.44 \cot\beta$

The elastohydrodynamic lubrication solution therefore requires calculating the pressure distribution within the conjunction, at the same time allowing for the effects that this pressure will have on the properties of the fluid and on the geometry of the elastic solids. The solution will also provide the shape of the lubricant film, particularly the minimum clearance between the solids. Hamrock et al. [2004] present the complete elastohydrodynamic lubrication theory.

13.8.2 Dimensionless Grouping

The variables resulting from the elastohydrodynamic lubrication theory are:

E' = effective elastic modulus, Pa
P = normal applied load, N
h = film thickness, m
R_x = effective radius in x- (motion) direction, m
R_y = effective radius in y- (transverse) direction, m
$\tilde{u}$ = mean surface velocity in x-direction, m/s
ξ = pressure-viscosity coefficient of fluid, m^2/N
η_o = atmospheric viscosity, N-s/m^2

From these variables, the following five dimensionless groupings can be established:

1. *Dimensionless film thickness:*

$$H = \frac{h}{R_x} \tag{13.66}$$

2. *Ellipticity parameter:*

$$k_e = \frac{D_y}{D_x} = \left(\frac{R_y}{R_x}\right)^{2/\pi} \tag{8.24}$$

3. *Dimensionless load parameter:*

$$W = \frac{P}{E' R_x^2} \tag{13.67}$$

4. *Dimensionless speed parameter:*

$$U = \frac{\eta_o \tilde{u}}{E' R_x} \tag{13.68}$$

5. *Dimensionless materials parameter:*

$$G = \xi E' \tag{13.69}$$

The dimensionless minimum film thickness can thus be written as a function of the other four parameters, or $H = f(k_e, U, W, G)$. The most important practical aspect of elastohydrodynamic lubrication theory is the determination of the minimum film thickness within a conjunction. That is, maintaining a fluid film thickness of adequate magnitude is extremely important to the operation of nonconformal machine elements such as rolling-element bearings. Specifically, elastohydrodynamic film thickness influences fatigue life, as discussed in Section 13.9.

13.8.3 Minimum-Film-Thickness Formula

By numerically obtaining the pressure profile and film thickness, the influence of the ellipticity parameter and the dimensionless speed, load, and materials parameters on minimum film thickness has been investigated. The resulting minimum-film-thickness formula is given as:

$$H_{\min} = 3.63 U^{0.68} G^{0.49} W^{-0.073} \left(1 - e^{-0.68 k_e}\right). \tag{13.70}$$

In this equation the most dominant exponent occurs on the speed parameter, and the exponent on the load parameter is very small and negative. The materials parameter also carries a significant exponent, although the range of this variable in engineering situations is limited.

Similarly, for rectangular contacts the dimensionless minimum-film-thickness formula is

$$H_{\min,r} = \frac{h_{\min}}{R_x} = 1.714 U^{0.694} G^{0.568} \left(W'\right)^{-0.128} \tag{13.71}$$

where

$$W' = \frac{P_z'}{E' R_x} \tag{13.72}$$

where P_z' is the normal applied load per unit width. Equations (13.70) and (13.71) are known as the **Hamrock-Dowson equations** and have wide applicability to elastohydrodynamically lubricated bodies such as rolling element bearings, gears, cams, etc.

Design Procedure 13.1: Evaluation of Ball Bearing Lubricant Film Thickness

1. It will be assumed that the radial load, number of bearings, bearing materials, lubricant properties, and geometry of the bearings are known. Of these, the lubricant properties, especially the pressure exponent of viscosity, are most likely to be unknown or will have the largest uncertainty. Regardless, these properties should be available from the lubricant supplier.
2. For ball bearings, Eq. (13.48) allows calculation of the load on an individual ball. Note that this is a conservative estimate; Eqs. (13.45) and (13.46) can be combined if necessary.
3. The velocities for the ball–inner-race and ball–outer-race contacts can be obtained from Eqs. (13.28) and (13.30). If β is not specified, it can be obtained from Eq. (13.6).
4. R_x and R_y can be calculated from Eqs. (8.6) and (8.7). Equation (13.64) then allows calculation of E'.
5. The parameters k_e, W, U, and G are then calculated from Eqs. (8.12) and (13.67) through (13.69).
6. Equation (13.70) then allows calculation of the minimum film thickness.
7. The adequacy of the lubricant film can be evaluated with respect to the film parameter, Λ, given by Eq. (13.27). A reasonable design goal is to require the bearing to operate in a partial film lubrication regime (see Section 8.7.3). This would require $\Lambda = 3$. Alternatively, Fig. 13.25 can be used to assess the adequacy of the lubricant film and modify the catalog ratings for the bearing according to Eq. (13.87).

Example 13.5: Oil Film Thickness in a Rolling Element Bearing

Given: A deep-groove ball bearing has 12 balls radially loaded by 12,000 N. The groove and ball radii are matched to give an ellipticity ratio of 10. The lubricant viscosity is 0.075 N-s/m^2, and its pressure-viscosity coefficient is 1.6×10^{-8} m^2/N. The rotational speed is 1500 rpm, the mean surface velocity is 3 m/s, and the effective radius in the x-direction is 8 mm. The modulus of elasticity for the ball and races is 207 GPa and Poisson's ratio is 0.3.

Find: The minimum oil film thickness in the bearing.

Solution: The Stribeck equation [Eq. (13.48)] gives the load acting on the most heavily loaded ball as

$$(w_z)_{\max} = \frac{5W_z}{n} = \frac{5(12,000)}{12} = 5000 \text{ N}.$$

The effective elastic modulus is

$$E' = \frac{E}{1-\nu^2} = \frac{207 \times 10^9}{1-(0.3)^2} = 227 \text{ GPa}.$$

The dimensionless load, speed, and materials parameters from Eqs. (13.67), (13.68), and (13.69), respectively, can be expressed as

$$U = \frac{\eta_o \tilde{u}}{E' R_x} = \frac{(0.075)(3)}{(227 \times 10^9)(0.008)} = 1.236 \times 10^{-10},$$

$$W = \frac{P}{E' R_x^2} = \frac{5000}{(227 \times 10^9)(0.008)^2} = 3.442 \times 10^{-4},$$

$$G = \xi E' = (1.6 \times 10^{-8})(227 \times 10^9) = 3632.$$

Equation (13.70) gives the dimensionless minimum-film-thickness as

$$\begin{aligned} H_{\min} &= \frac{h_{\min}}{R_x} = 3.63 U^{0.68} G^{0.49} W^{-0.073} \left(1 - e^{-0.68 k_e}\right) \\ &= 3.63 \left(1.236 \times 10^{-10}\right)^{0.68} (3632)^{0.49} \\ &\quad \times \left(3.442 \times 10^{-4}\right)^{-0.073} \left(1 - e^{-0.68(10)}\right) \\ &= 6.596 \times 10^{-5}. \end{aligned}$$

The dimensional minimum film thickness is

$$h_{\min} = H_{\min} R_x = \left(6.596 \times 10^{-5}\right)(0.008) = 0.53\ \mu\text{m}.$$

Thus, the minimum oil film thickness in the deep-groove ball bearing considered in this example is 0.53 μm.

Example 13.6: Effect of Race Conformity on Film Thickness

Given: It is suspected that a ball bearing lubricant film is too thin. The geometries are such that the ellipticity parameter is 2. By decreasing the race radius of curvature in the axial direction, the balls fit more tightly with the races and thus the ellipticity parameter increases from 2 to 10.

Find: How much thicker the oil film becomes in changing the ellipticity parameter from 2 to 10. Assume that all the other operating parameters remain constant.

Solution: From Eq. (13.70),

$$\frac{(H_{\min})_{k_e=10}}{(H_{\min})_{k_e=2}} = \frac{1 - e^{-0.68(10)}}{1 - e^{-0.68(2)}} = 1.344.$$

The film thickness will increase by 34.4% when the ellipticity parameter is increased from 2 to 10.

13.9 Fatigue Life

13.9.1 Contact Fatigue Theory

Rolling fatigue is a material failure caused by the application of repeated stresses to a small volume of material. It is a unique failure type—essentially a process of seeking out the weakest point at which the first failure will occur. A typical fatigue spall is shown in Fig. 13.20. On a microscopic scale there will be a large range in material strength or resistance to fatigue because of inhomogeneities in the material. Because bearing materials are complex alloys, they are not homogeneous or equally resistant to failure at all points. Therefore, the fatigue process can be expected to be one in which a group

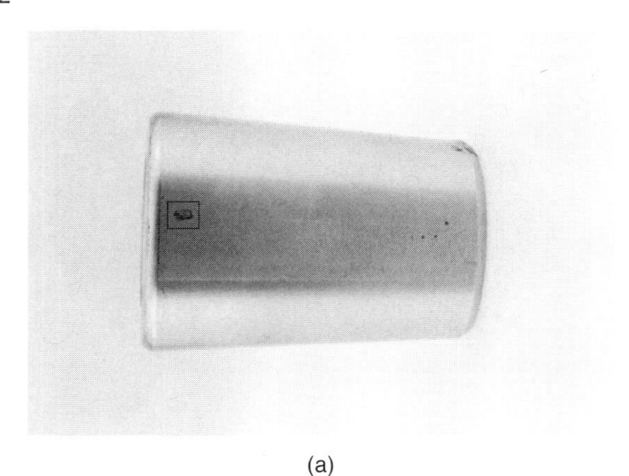

(a)

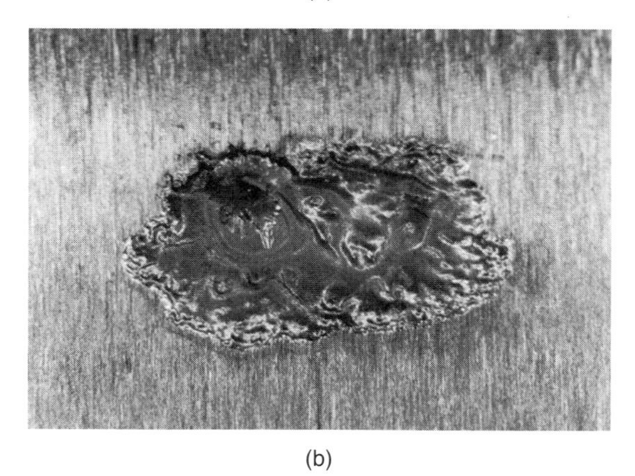

(b)

Figure 13.20: Typical fatigue spall. (a) Spall on tapered roller bearing. (b) Detail of fatigue spall. *Source:* Courtesy of The Timken Company.

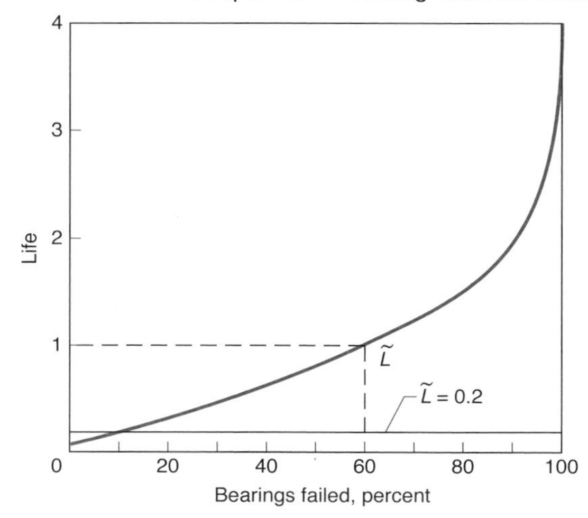

Figure 13.21: Distribution of bearing fatigue failures.

of seemingly identical specimens exhibit wide variations in failure time when stressed in the same way. For this reason it is necessary to treat the fatigue process statistically.

Predicting the life of a bearing under a specific load requires the following essential pieces of information:

1. An accurate, quantitative estimate of life dispersion or scatter
2. The life at a given survival rate or reliability level

This translates into an expression for the *load-carrying capacity*, or the ability of the bearing to endure a given load for a stipulated number of stress cycles or revolutions. If a group of supposedly identical bearings is tested at a specific load and speed, the distribution in bearing life shown in Fig. 13.21 typically occurs.

13.9.2 Weibull Distribution

Weibull [1949] postulated that the fatigue lives of a homogeneous group of rolling-element bearings are dispersed according to the following relation:

$$\ln \ln \frac{1}{\tilde{S}} = c_2 \ln \frac{\tilde{L}}{\tilde{A}}, \tag{13.73}$$

where $\tilde{S}$ is the probability of survival, $\tilde{L}$ is the fatigue life, and c_2 and $\tilde{A}$ are constants. The Weibull distribution results from a statistical theory of strength based on probability theory, where the dependence of strength on volume is explained by the dispersion in material strength. This is the "weakest link" theory. This section will derive Eq. (13.73) and explain how this expression can be used to predict bearing life.

Figure 13.22 considers a volume being stressed that is broken up into m similar volumes; $\tilde{M}$ represents the probability of failure, and $\tilde{S}$ is the probability of survival. For the entire volume

$$\tilde{S} = \tilde{S}_1 \tilde{S}_2 \tilde{S}_3 \ldots \tilde{S}_m.$$

Then,

$$\begin{aligned} 1 - \tilde{M} &= \left(1 - \tilde{M}_1\right)\left(1 - \tilde{M}_2\right)\left(1 - \tilde{M}_3\right)\ldots\left(1 - \tilde{M}_m\right) \\ &= \prod_{i=1}^{m}\left(1 - \tilde{M}_i\right), \end{aligned}$$

so that

$$\tilde{S} = \prod_{i=1}^{m}\left(1 - \tilde{M}_i\right). \tag{13.74}$$

The probability of a crack starting in the ith volume is

$$\tilde{M}_i = f(x)\tilde{V}_i, \tag{13.75}$$

where $f(x)$ is a function of the stress level, the number of stress cycles, and the depth into the material where the maximum stress occurs and $\tilde{V}_i$ is the elementary volume. Therefore, substituting Eq. (13.75) into Eq. (13.74) gives

$$\tilde{S} = \prod_{i=1}^{m}\left[1 - f(x)\tilde{V}_i\right].$$

Taking the natural log of both sides,

$$\ln \tilde{S} = \sum_{i=1}^{m} \ln\left[1 - f(x)\tilde{V}_i\right]. \tag{13.76}$$

$S_1 = 1 - M_1$ $S_2 = 1 - M_2$ $S_3 = 1 - M_3$ ··· $S_m = 1 - M_m$

Figure 13.22: Representation of m similar stressed volumes.

Now, if $f(x)\tilde{V}_i \ll 1$, then $\ln\left[1 - f(x)\tilde{V}_i\right] = -f(x)\tilde{V}_i$ and $\ln \tilde{S} = -\sum_{i=1}^{m} f(x)\tilde{V}_i$. Letting the incremental volume become very small,

$$\sum_{i=1}^{\infty} f(x)\tilde{V}_i = \int f(x)\, d\tilde{V} = \tilde{f}(x)\tilde{V}, \tag{13.77}$$

where $\tilde{f}(x)$ is the volume-average value of $f(x)$.

Lundberg and Palmgren [1947] assumed that $f(x)$ could be expressed as a power function of shear stress, τ_o, number of stress cycles, $\tilde{J}$, and depth of the maximum shear stress, z_o:

$$f(x) = \frac{\tau_o^{c_1}\tilde{J}^{c_2}}{z_o^{c_3}}. \tag{13.78}$$

They also chose the stressed volume as $\tilde{V} = D_y z_o l_v$. Substituting Eqs. (13.77) and (13.78) into Eq. (13.76) gives

$$\ln \tilde{S} = -\frac{\tau_o^{c_1}\tilde{J}^{c_2}D_y l_v}{z_o^{c_3-1}}, \tag{13.79}$$

$$\ln \frac{1}{\tilde{S}} = \frac{\tau_o^{c_1}\tilde{J}^{c_2}D_y l_v}{z_o^{c_3-1}}. \tag{13.80}$$

For a specific bearing and load (e.g., stress) τ_o, D_y, l_v, and z_o are all constant so that

$$\ln \frac{1}{\tilde{S}} = \tilde{J}^{c_2}.$$

The number of stress cycles, $\tilde{J}$, is proportional to the fatigue life based on the number of inner race revolutions, $\tilde{L}$, so that

$$\ln \frac{1}{\tilde{S}} = \left(\frac{\tilde{L}}{\tilde{A}}\right)^{c_2}.$$

Therefore, while preserving the nomenclature of Eq. (13.73),

$$\ln\ln \frac{1}{\tilde{S}} = c_2 \ln \frac{\tilde{L}}{\tilde{A}}. \tag{13.81}$$

This is the **Weibull distribution**, which relates probability of survival and life. It has two principal functions. First, bearing fatigue lives plot as a straight line on Weibull coordinates (log-log versus log), so that the life at any reliability level can be determined. Of most interest are the $\tilde{L}_{10}$ life ($\tilde{S}$ = 0.9) and the $\tilde{L}_{50}$ life ($\tilde{S}$ = 0.5). Bearing load ratings are based on the $\tilde{L}_{10}$ life. The $\tilde{L}_{10}$ life is calculated from the load on the bearing and the bearing dynamic capacity, or by the load rating given in manufacturers' catalogs and engineering journals, by using the equation

$$\tilde{L}_{10} = \left(\frac{\bar{C}}{P}\right)^{m_k} \tag{13.82}$$

where
- $\bar{C}$ = specific dynamic capacity or basic dynamic load rating, N
- P = equivalent bearing load, N
- m_k = load-life exponent; 3 for elliptical contacts and 10/3 for rectangular contacts

The life at any other reliability is strongly dependent on the sustained quality of lubricant. For very clean lubricants that are carefully filtered, the life can be estimated from Fig. 13.23. For other conditions, the life will be lower, and will depend

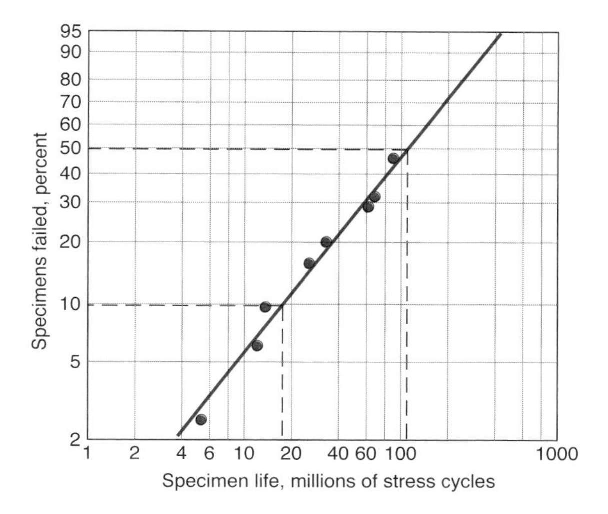

Figure 13.23: Typical Weibull plot of bearing fatigue failures for clean and filtered lubricants.

on the ability to avoid contamination. ISO [2007] outlines a detailed approach for determining life as a function of reliability, bearing size and geometry, film parameter Λ, cleanliness, and filtration effectiveness. The Timken company offers the following simpler relation for typical lubrication conditions:

$$\frac{\tilde{L}_R}{\tilde{L}_{10}} = 4.48\left(\ln \frac{100}{R}\right)^{2/3} \tag{13.83}$$

where
- R = desired reliability
- $\tilde{L}_R$ = life at a reliability of R percent
- $\tilde{L}_{10}$ = life at 90% reliability

Note that $\tilde{L}_{10}$ in Eq. (13.83) is bearing life expressed in million of revolutions. The bearing life in hours is

$$\left(\tilde{L}\right)_{\text{hour}} = \frac{10^6 \tilde{L}_{10}}{60 N_b} \tag{13.84}$$

where N_b is the rotational speed (rev/min).

13.9.3 Dynamic Load Rating

The Lundberg-Palmgren theory, on which bearing ratings are based, is expressed by Eq. (13.80). The exponents in this equation are determined experimentally from the dispersion of bearing lives and the dependence of life on load, geometry, and bearing size. As a standard of reference, all bearing load ratings are expressed in terms of the dynamic load rating, which, by definition, is the load that a bearing can carry for 1 million inner-race revolutions with a 90% chance of survival.

Factors on which specific dynamic capacity and bearing life depend are

1. Size of rolling element
2. Number of rolling elements per row
3. Number of rows of rolling elements
4. Conformity between rolling elements and races
5. Contact angle under load

6. Material properties

7. Lubricant properties

8. Operating temperature

9. Operating speed

Only factors 1 to 5 are incorporated in bearing dynamic load ratings developed from the Lundberg-Palmgren theory [1947, 1952].

Tables 13.2 through 13.7 give the dynamic load ratings for a variety of bearing types. The dynamic load ratings are based on the material and manufacturing techniques used for producing standard production bearings. They apply to loads that are constant in both magnitude and direction; for radial bearings these loads are radial, and for thrust bearings they are axial loads that act centrally.

13.9.4 Equivalent Dynamic Load

Equation (13.80), used for calculating the fatigue life, assumes a purely radial load in radial bearings and a purely axial load in thrust bearings, of constant direction and magnitude. In many rolling-element bearing applications, the load does not satisfy these conditions, since the force acts on the bearing obliquely or changes in magnitude. In such cases, a constant radial or axial force must be determined for the rating life calculation, representing, with respect to the rating life, an equivalent stress.

The **equivalent load** is expressed as

$$P = XP_r + YP_a, \tag{13.85}$$

where

X = radial factor
Y = thrust factor
P_r = radial load component, N
P_a = axial load component, N

Table 13.9 shows values of X and Y for various radial bearing types at respective contact angles; also, C_o is the static load rating given in Section 13.7.5. The radial and thrust factors for various thrust bearing types are given in Table 13.10.

Design Procedure 13.2: Design of Rolling Element Bearings Based on Dynamic Load Rating

1. It will be assumed that the radial and thrust/axial loads are known, as well as the required reliability.

2. Selection of the bearing type is important; Table 13.1 and Example 13.1 address this issue and serve as useful guides. Ultimately, selection of a type of bearing depends on the loading condition, speed, space available, and, perhaps most importantly, cost. It will be assumed that a bearing type can be selected, or a number of candidates selected for further analysis and comparison.

3. Some dimensions of the bearing may be restricted by geometry considerations not associated with the bearing. For example, a shaft may have its diameter specified by deflection considerations; a smaller bearing may be used by machining a shoulder into the shaft per manufacturer's recommendations. However, excessively small diameters with shoulders transitioning to large diameters introduce large stress concentrations (see Section 6.2). Alternatively, the bearing width or outer diameter may be constrained. Such considerations may further reduce the number of suitable bearing candidates.

4. Equivalent static load can be obtained from Eq. (13.60) using factors X_o and Y_o from Table 13.8. Bearings with a static load rating lower than the equivalent static load can be eliminated from consideration. Any other bearing can be considered.

5. A bearing for consideration can be selected from Tables 13.2 through 13.7, or from a manufacturer's catalog or website.

6. Equivalent dynamic load can be obtained from Eq. (13.85), using factors X and Y determined according to the bearing type:

   (a) For deep groove or angular contact ball bearings, the factors X and Y are obtained from Table 13.9. This requires calculation of P_a/C_o to obtain an estimate of e, then comparison of P_a/P_r with e to obtain the factors.

   (b) For thrust bearings, the factors are obtained in similar fashion from Table 13.10.

   (c) For tapered roller bearings, the factors can be obtained from Table 13.9 using the contact angle β from Table 13.7.

7. The $\tilde{L}_{10}$ life can be obtained from the bearing's dynamic load rating and equivalent dynamic load according to Eq. (13.82). The $\tilde{L}_{10}$ that results is in millions of revolutions.

8. Life at any other reliability can be estimated from Fig. 13.23 or from Eq. (13.83).

13.9.5 Life Adjustment Factors

This section presents the factors affecting the fatigue life of bearings that were not taken into account in the Lundberg-Palmgren theory. It is assumed that the various environmental or bearing design factors are multiplicative in their effects on bearing life. The following equation results:

$$L_A = \bar{D}\bar{E}\bar{F}_l\bar{G}\bar{H}_m\tilde{L}_{10} \tag{13.86}$$

$$L_A = \bar{D}\bar{E}\bar{F}_l\bar{G}\bar{H}_m\left(\frac{\bar{C}}{\bar{P}}\right)^{m_k} \tag{13.87}$$

where

$\bar{D}$ = material factor
$\bar{E}$ = metallurgical processing factor
$\bar{F}_l$ = lubrication factor
$\bar{G}$ = speed effect factor
$\bar{H}_m$ = misalignment factor

Table 13.9: Capacity formulae for rectangular and elliptical conjunctions for radial and angular bearings.

Bearing type		e	Single-row bearings				Double-row bearings			
			$\frac{P_a}{P_r}$ e		$\frac{P_a}{P_r} > e$		$\frac{P_a}{P_r}$ e		$\frac{P_a}{P_r} > e$	
			X	Y	X	Y	X	Y	X	Y
Deep-groove	$P_a/C_o = 0.025$	0.22	1	0	0.56	2.0				
ball bearings	$P_a/C_o = 0.04$	0.24	1	0	0.56	1.8				
	$P_a/C_o = 0.07$	0.27	1	0	0.56	1.6				
	$P_a/C_o = 0.13$	0.31	1	0	0.56	1.4				
	$P_a/C_o = 0.25$	0.37	1	0	0.56	1.2				
	$P_a/C_o = 0.50$	0.44	1	0	0.56	1				
Angular-	$\beta = 20°$	0.57	1	0	0.43	1	1	1.09	0.70	1.63
contact ball	$\beta = 25°$	0.68	1	0	0.41	0.87	1	0.92	0.67	1.41
bearings	$\beta = 30°$	0.80	1	0	0.39	0.76	1	0.78	0.63	1.24
	$\beta = 35°$	0.95	1	0	0.37	0.66	1	0.66	0.60	1.07
	$\beta = 40°$	1.14	1	0	0.35	0.57	1	0.55	0.57	0.93
	$\beta = 45°$	1.33	1	0	0.33	0.50	1	0.47	0.54	0.81
Self-aligning ball bearings		$1.5 \tan\beta$					1	$0.42 \cot\beta$	0.65	$0.65 \cot\beta$
Spherical roller bearings		$1.5 \tan\beta$					1	$0.45 \cot\beta$	0.67	$0.67 \cot\beta$
Tapered-roller bearings		$1.5 \tan\beta$	1	0	0.40	$0.4 \cot\beta$	1	$0.42 \cot\beta$	0.67	$0.67 \cot\beta$

Table 13.10: Radial factor X and thrust factor Y for thrust bearings.

Bearing type		e	Single acting		Double acting			
			$\frac{P_a}{P_r} > e$		$\frac{P_a}{P_r}$ e		$\frac{P_a}{P_r} > e$	
			X	Y	X	Y	X	Y
Thrust ball	$\beta = 45°$	1.25	0.66	1	1.18	0.59	0.66	1
	$\beta = 60°$	2.17	0.92	1	1.90	0.55	0.92	1
	$\beta = 75°$	4.67	1.66	1	3.89	0.52	1.66	1
Spherical roller thrust		$1.5 \tan\beta$	$\tan\beta$	1	$1.5 \tan\beta$	0.67	$\tan\beta$	1
Tapered roller		$1.5 \tan\beta$	$\tan\beta$	1	$1.5 \tan\beta$	0.67	$\tan\beta$	1

$\bar{P}$ = bearing equivalent load, N
m_k = load-life exponent; 3 for ball bearings or 10/3 for roller bearings
$\bar{C}$ = specific dynamic capacity, N

Factors $\bar{D}$, $\bar{E}$, and $\bar{F}_l$ are reviewed briefly here. Refer to Bamberger [1971] for a complete discussion of all five life adjustment factors.

Material Factors

For over a century, AISI 52100 steel has been the preferred material for rolling-element bearings. In fact, the specific dynamic capacity as defined by the Anti-Friction Bearing Manufacturers Association (AFBMA) in 1949 is based on an air-melted 52100 steel, hardened to at least Rockwell C58. Since that time, better control of air-melting processes and the introduction of vacuum-remelting processes have resulted in more homogeneous steels with fewer impurities. Such steels have extended rolling-element bearing fatigue lives to several times the catalog life; life extensions of three to eight times are not uncommon. Other steels, such as AISI M–1 and AISI M–50, chosen for their higher temperature capabilities and resistance to corrosion, also have shown greater resistance to fatigue pitting when vacuum-melting techniques are employed. Case-hardened materials, such as AISI 4620, AISI 4118, and AISI 8620, used primarily for rolling-element bearings, have the advantage of a tough, ductile steel core with a hard, fatigue-resistant surface.

The recommended material factors $\bar{D}$ for various through-hardened alloys processed by air melting are shown in Table 13.11. Insufficient definitive life data were found for case-hardened materials to recommend values of $\bar{D}$ for them. Refer to the bearing manufacturer for the choice of a specific case-hardened material.

Table 13.11: Material factors for through-hardened bearing materials.

Material	Material factor, $\bar{D}$
52100	2.0
M-1	0.6
M-2	0.6
M-10	2.0
M-50	2.0
T-1	0.6
Halmo	2.0
M-42	0.2
WB 49	0.6
440C	0.6-0.8

The processing variables considered in the development of the metallurgical processing factor $\bar{E}$ include melting practice (air and vacuum melting) and metalworking (thermomechanical working). Thermomechanical working of M–50 has also been shown to lengthen life, but in a practical sense, it is costly and still not fully developed as a processing technique. Bamberger [1971] recommends an $\bar{E}$ of 3 for consumable-electrode-vacuum-melted materials.

Lubrication Factor

Until approximately 1960, the role of the lubricant between surfaces in rolling contact was not fully appreciated; metal-to-metal contact was presumed to occur in all applications, with attendant required boundary lubrication. The development of elastohydrodynamic lubrication theory showed that lubricant films with thicknesses on the order of micrometers and tenths of micrometers can occur in rolling contact. Since surface finishes are of the same order of magnitude as the lubricant film thicknesses, the significance of rolling-element

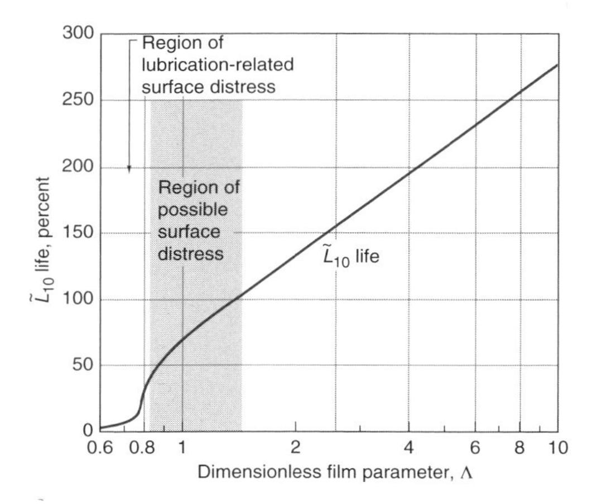

Figure 13.24: Group fatigue life $\tilde{L}_{10}$ as function of dimensionless film parameter.

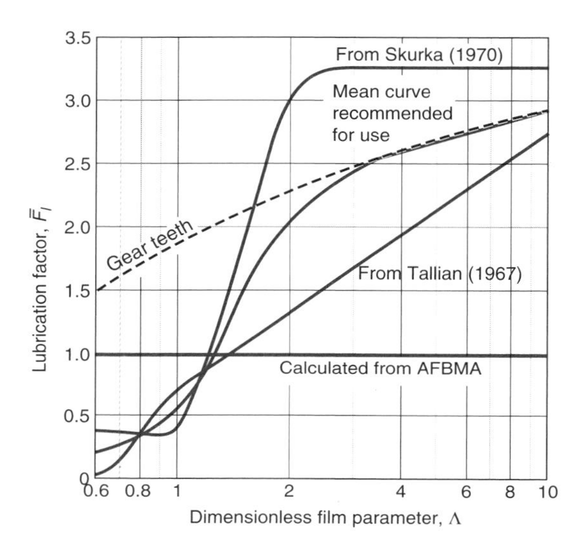

Figure 13.25: Lubrication factor as a function of dimensionless film parameter.

bearing surface roughnesses to bearing performance became apparent. Figure 13.24 shows calculated $\tilde{L}_{10}$ life as a function of the dimensionless film parameter Λ, which was introduced in Chapter 8, where

$$\Lambda = \frac{h_{\min}}{\left(R_{qa}^2 + R_{qb}^2\right)^{1/2}}. \qquad (13.88)$$

Figure 13.25 presents a curve of the recommended $\bar{F}_l$ as a function of Λ. A mean of the curves presented in Tallian [1967] for ball bearings and in Skurka [1970] for roller bearings is recommended. A formula for calculating the minimum film thickness $H_{\min}$ is given in Eq. (13.70).

Example 13.7: Fatigue Life of a Rolling Element Bearing

Given: A deep-groove 6304 ball bearing is radially loaded with 4000 N and axially loaded with 2000 N. The rotational speed is 3000 rpm.

Find: The number of hours the bearing will survive with 90% probability.

Solution: From Table 13.2 for a 6304 deep-groove ball bearing, the static load rating is 7800 N and the dynamic load is 15,900 N. Therefore,

$$\frac{P_a}{C_o} = \frac{2000}{7800} = 0.256.$$

From Table 13.9, for P_a/C_o = 0.256 for a deep-groove ball bearing the constant e is 0.37. Since $P_a/P_r = 0.5 > 0.37$, the radial and thrust factors are X = 0.56 and Y = 1.2, respectively. From Eq. (13.85), the equivalent bearing load is

$$P = XP_r + YP_a = 0.56(4000) + 1.2(2000) = 4640 \text{ N}.$$

The rotational speed is N_b = 3000 rpm. From Eqs. (13.82) and (13.84) the bearing life is

$$\begin{aligned} \left(\tilde{L}_{10}\right)_{\text{hour}} &= \frac{10^6}{60N_b}\left(\frac{\bar{C}}{P}\right)^3 \\ &= \frac{10^6}{(60)(3000)}\left(\frac{15,900}{4640}\right)^3 \\ &= 223.5 \text{ hr}. \end{aligned}$$

The bearing has a 90% probability of surviving for 223.5 hr.

Example 13.8: Effect of Reliability on Life

Given: A machine manufacturer has dimensioned the roller bearings in a paper-making machine containing 400 bearings such that the $\tilde{L}_{10}$ for each bearing is 40 years.

Find: Use Fig. 13.23 to calculate the $\tilde{L}_{50}$ life in years. Also, determine the mean time between failures in the steady state for an old paper-making machine where bearings are replaced whenever they fail.

Solution: From Fig. 13.23, the $\tilde{L}_{50}$ life is about 110 million stress cycles while the $\tilde{L}_{10}$ life is 17 million stress cycles. The ratio is

$$\frac{\tilde{L}_{50}}{\tilde{L}_{10}} = \frac{110}{17} = 6.47.$$

The $\tilde{L}_{50}$ life is thus 6.47 times the $\tilde{L}_{10}$ life, which in this example is 40 years. Therefore,

$$\tilde{L}_{50} = (6.47)(40) = 258.8 \text{ years}.$$

As there are 400 independent bearings in the machine, the mean time between failures for the steady state, that is, when the failures are evenly distributed over time, is

$$\frac{\tilde{L}_{50}}{400} = \frac{258.8}{400} = 0.647 \text{ year} = 236 \text{ days}.$$

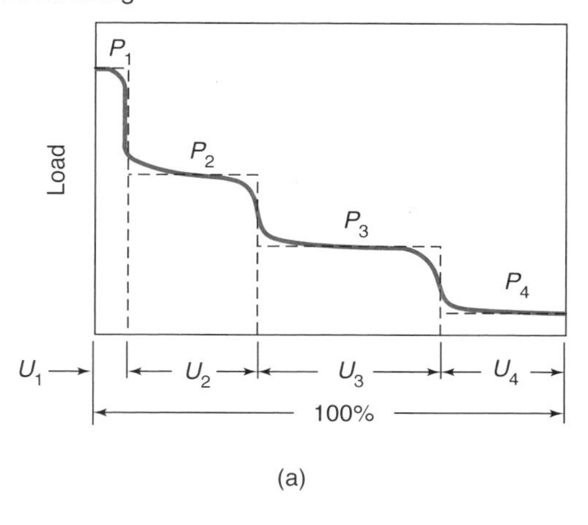

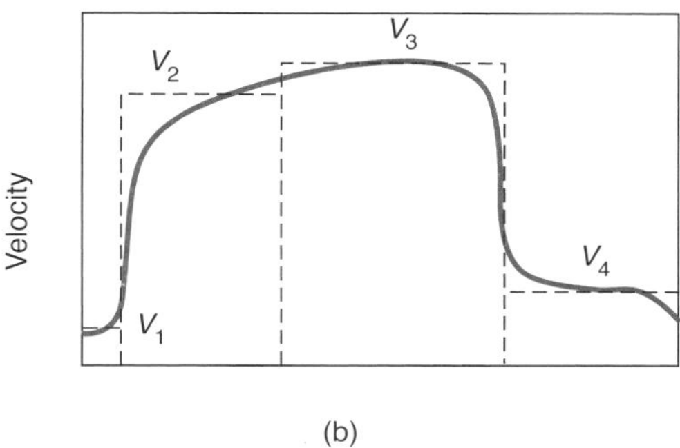

Figure 13.26: Example of duty cycles for variable operating conditions. (a) Load and (b) velocity duty cycles.

13.10 Variable Loading

It is quite common for bearings to encounter varying duty cycles over their lives, where the load magnitude, duration, and direction can change or fluctuate. The usual approach for such a problem is to consider damage to accumulate and then apply a fractional life rule such as Miner's Rule as discussed in Section 7.9. For example, Figure 13.26 shows a hypothetical variable loading, where each load and velocity shown represents a portion of the bearing's operating life. Constructing such a figure usually is very difficult, as bearings in service have very complex duty cycles; consider the range of forces encountered by the wheel bearings of an automobile as it runs over small or large potholes, for example. For every portion of the duty cycle, the life fraction is defined by

$$U_i = \frac{N_i}{N}, \tag{13.89}$$

where N_i is the the number of revolutions under a load P_i and N is the total life cycle of the application. Note that $U_1 + U_2 + U_3 + \ldots = 1$. If the life for a loading is obtained from Eq. (13.82) as $L_{10,i}$, then the life of the bearing can be estimated as

$$L_{10} = \frac{1}{\dfrac{U_1}{L_{10,1}} + \dfrac{U_2}{L_{10,2}} + \dfrac{U_3}{L_{10,3}} + \cdots}. \tag{13.90}$$

Note that Eq. (13.90) assumes that the order between the different running conditions does not change the total life for the bearing. In reality, higher loads are much more severe in terms of fatigue wear and damage accumulation, and as a result this equation should be used with caution. It should also be noted that life depends greatly on the quality of the lubricant supply; a bearing with a clean and filtered lubricant throughout its service will have a life as described by Eq. (13.82). On the other hand, a bearing that encounters third-body wear particles (see Section 8.9) that are not removed promptly, or that uses an old or contaminated lubricant (such as one infiltrated by water or other chemicals) will have much lower life than predicted by Eq. (13.90).

Case Study 13.1: Bearing Design for Windmills

Wind energy has received considerable attention in recent years as a green and renewable alternative to fossil fuels. Windmills produce energy from wind by using large blades or vanes to rotate a shaft; the shaft speed is accelerated in a gearbox and then drives a generator to produce electricity. Global wind capacity exceeded 230 gigawatts in 2011, and projections suggest that this capacity will increase to almost 2000 gigawatts by 2020. Further, many countries and states have mandated that windpower and other forms of renewable energy produce at least 25% of electricity by 2025.

A typical windmill for producing electricity is shown in Fig. 13.27. Wind causes the rotors to rotate; windpower is transferred by the rotors to the hub and the main shaft. While some new direct drive designs eliminate the need for a gearbox, generally the shaft speed is increased with a transmission and the high speed shaft is then used to power a generator. The largest windmill can produce 10 megawatts, but most wind turbines produce around 4 to 5 megawatts and have rotors that exceed 60 m in length. The rotors often weigh more than 100 tons, requiring significant load support on the rotor shaft. Typical rotor bearings have inside diameters of 1.5 m and weigh over 26 kN, and since the windmills are often located at sea where salt water exposure is possible, the bearings utilize advanced seals and lubrication monitoring and delivery systems.

Reliability of windmills is a major design concern, since access of the windmill for service is difficult and compromises power generation. For this reason, the many bearings needed are very carefully designed, and L_{10} lives of 200,000 hours are not uncommon. Proper bearing design is a key factor in reducing maintenance costs and increasing wind turbine robustness.

Duty Cycle

The duty cycle considered in design can have a significant influence on the size and geometry of the main shaft bearings. Additionally, over-simplification of the duty cycle can result in less robust designs or oversized bearings with significant adverse costs. Duty cycles are carefully generated from design programs that model the wind turbine system. A 5-s portion of such a duty cycle is shown in Fig. 13.28. The data is then sorted into bins and time durations in each bin are summed to characterize the duty cycle consistent with the approach in Section 13.10. However, it should be noted that some bearing manufacturers will use computational tools to incorporate hundreds of conditions in the duty cycle, and will generate load histograms accordingly. This allows generation of a so-called independent duty cycle.

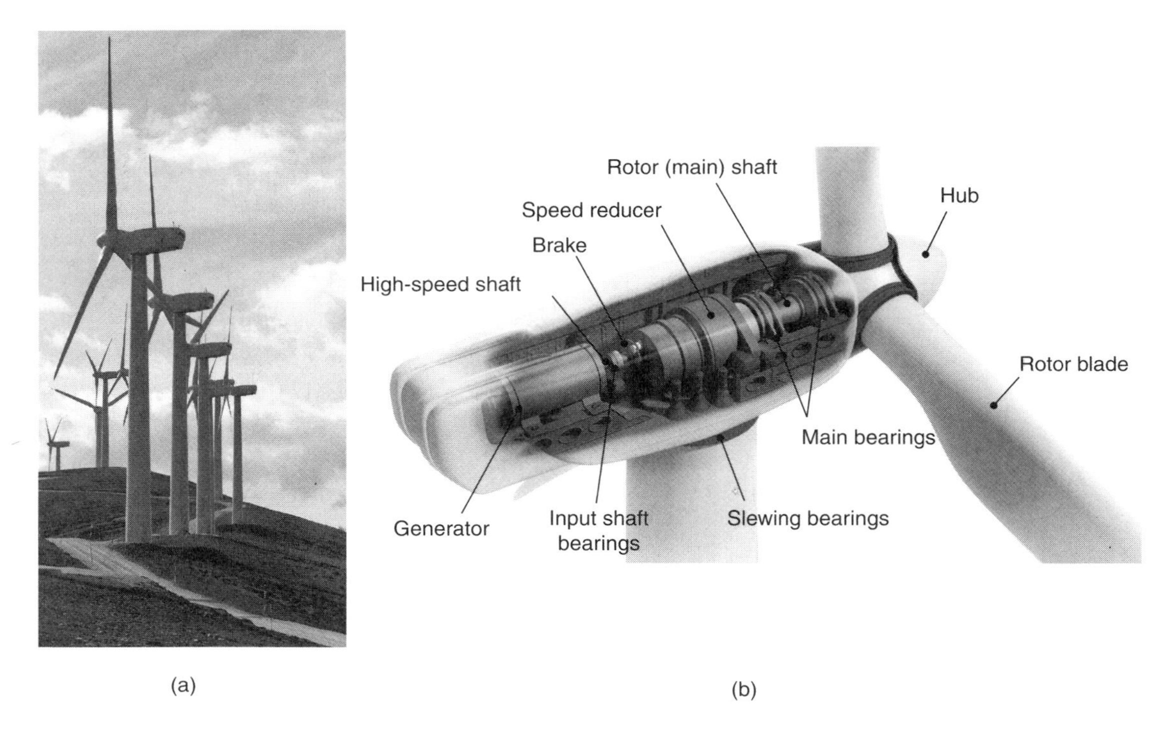

Figure 13.27: (a) Modern wind farm installation, showing multiple wind powered turbines. (b) Schematic illustration of a modern windmill and rolling-element bearings. *Source:* Courtesy of SKF USA, Inc.

A dependent duty cycle has been found to be useful for windmill design. In such a duty cycle, loads and durations are prioritized according to their importance for affecting bearing life. Bins are sized methodically for the speed and load based on their importance. For example, Timken ranks the following variables in order of importance on bearing life in windmills:

1. Speed of the main shaft, mainly due to its importance in generation of a lubricant film [see Eq. (13.70)].
2. Pitch moment, M_y (see Fig. 13.28).
3. Yaw moment, M_z.
4. Radial load, F_z.
5. Axial load, F_x.
6. Radial load, F_y.

Bins are sized according to their importance, with much more attention paid to the important load cases. Resultant duty cycles can then be constructed and used for bearing size calculation.

Bearing Life Calculations

Recognizing the complicated duty cycle, large loads, and significant speed variations involved in windmills, selection of a dynamic load rating as described in Section 13.9.3 is merely the first step in design. In fact, bearing manufacturers usually maintain in-house software to produce sophisticated calculations that can incorporate environmental effects influencing bearing life. All of the effects in Eq. (13.86) are considered, but only a brief overview is contained here for the most important factors.

1. **Lubrication.** It should be noted that all windmill bearings are carefully sealed and lubricated to prevent corrosion and promote long bearing life (see Fig. 13.25). Grease is commonly used for the lower speed mainshaft bearings, since it will not migrate or leak as easily as oil, and will exclude contaminants more effectively. Since this shaft is lower speed than the generator shaft, grease is a suitable lubricant at this location.

   The grease is formulated with a high viscosity lubricant to maintain film strength, using synthetic base oils for improved lubrication over a wide temperature range. Further, rust, oxidation, and corrosion inhibitors are included in the formulation to promote bearing life.

   Modern bearings are designed with features that promote removal of used grease and multi-port injection, as well as active pumping. This feature is especially important for low temperature operation.

   On the other hand, generator shafts are much higher speed and are, in general, oil lubricated, and incorporate filtration systems to maintain lubricant quality. Seals are incorporated on all such bearings to contain the lubricant, but also to prevent contamination, especially by water. Lubricant film thicknesses are calculated using Eq. (13.70) and then compared to surface roughness using the film parameter, Λ, as given in Eq. (13.88). The lubrication factor, $\bar{F}_l$, is obtained from Fig. 13.25.

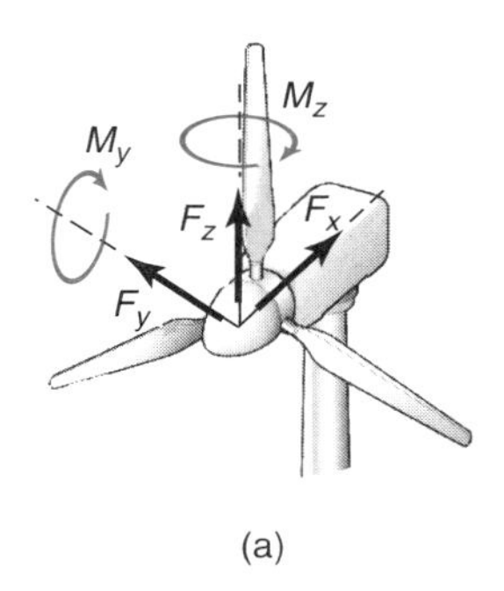

(a)

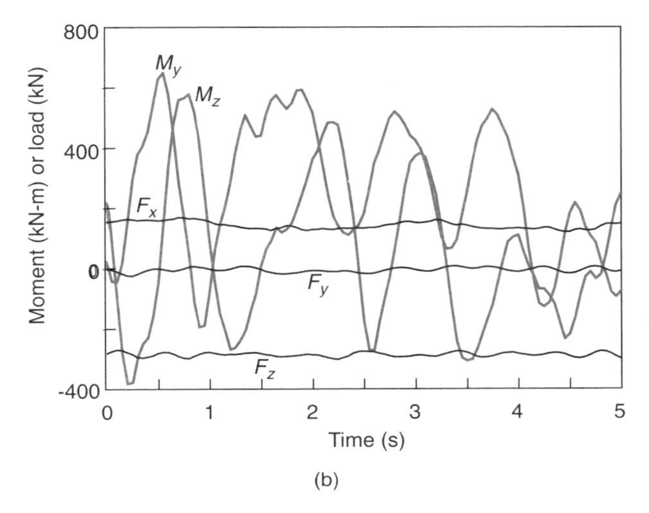

(b)

Figure 13.28: Portion of a loading for a windmill, used to develop a complicated duty cycle. (a) Illustration of rotor to demonstrate nomenclature; (b) 5-s loading excerpt. *Source:* Courtesy of The Timken Company.

2. **Misalignment.** Bearing life can be negatively affected by excessive shaft misalignment, which is attributable to high loads and pitch and yaw moments. Misalignment causes increased edge stress in roller bearings, and could cause early damage and spalling (see Fig. 13.20). Roller bearings and tapered roller bearings can be produced with special profiles to alleviate edge stresses for particular conditions, signifying the importance of an accurate load cycle and the prioritization of loadings listed above. The effect of misalignment on bearing life is accommodated through the misalignment factor, $\bar{H}_m$, where $\bar{H}_m$ is generally 1 for a misalignment of 0.0005 radians. The exact formula for $\bar{H}_m$ is supplier specific and is generally considered proprietary. Clearly, $\bar{H}_m$ is lower than 1 for higher misalignments, but it also is greater than 1 for lower levels of misalignment, indicating the importance of good shaft and support design.

3. **Thermal Effects.** Thermal effects can impact bearing life in a number of ways. Clearly, the viscosity of the lubricant will be affected by temperature (see Fig. 8.13), and this can adversely affect the lubricant film thickness in the bearings. Also, when dissimilar materials are used in construction, mismatches in thermal expansion coefficients can lead to thermal stresses that can have far-reaching effects.

   Radial expansion obviously results in increased load on bearings, but this will also increase thrust loads for tapered roller bearings. Axial expansion of shafts similarly results in increased loading.

4. **Wear modes.** The statistical framework for bearing life described in Section 13.9 is applicable for fatigue wear, but the demanding conditions in windmills can lead to other forms of wear that can lead to premature bearing failure. These include:

   - Scoring, where the end of a roller contacts an improperly lubricated flange.
   - Scuffing, where insufficient friction causes rollers to slide over the races, leading to increased heat generation, lubricant breakdown, and ultimately adhesion of material at the contact. Scuffing is usually attributed to low bearing preload or high clearance, high speeds, and/or light loads.
   - Micropitting can occur. This is different than the macropitting associated with fatigue spalls in that the pits are very small and associated with high stresses in contact zones and insufficient lubricant films. In cases of failed electrical isolation, micropits can also be caused by electrical discharge.
   - Plastic deformation, often referred to as brinelling, occurs when rollers do not rotate but instead oscillate back and forth in the thrust direction.

5. **Reliability.** Reliability is communicated by the windmill manufacturer to the bearing supplier, and as such can vary somewhat. Some windmill manufacturers specify an L_{10} life of 150,000 hours, while others use up to 200,000 hours.

Summary

Windmills are increasingly common and important for satisfying societal energy needs without contributing to carbon emissions. As has been shown, the bearings used in windmills operate in a demanding environment, with unsteady and complicated loadings being the norm. There are many windmill designs, and different types of bearings (angular contact, cylindrical or tapered roller, etc.) have been used in the same location by different manufacturers. A robust design requires sophisticated analysis using the approaches discussed in this chapter, which is essential for low-maintenance operation and reliable energy generation.

13.11 Summary

Rolling-element bearings are precise, yet simple, machine elements of great utility. This chapter drew together the current understanding of rolling-element bearings and attempted to present it in a concise manner. The detailed precise calculations of bearing performance were only summarized and appropriate references were given. The history of rolling-element bearings was briefly reviewed, and subsequent sections described the types of rolling-element bearing, their geometry, and their kinematics. Having defined ball bearing operation under unloaded and unlubricated conditions, the chapter then focused on static loading of rolling-element bearings. Most rolling-element bearing applications involve steady-state rotation of either the inner or outer race or both; however, the rotational speeds are usually not so high as to cause ball or roller centrifugal forces or gyroscopic moments of significant magnitudes; thus, these were neglected. Radial, thrust, and preloaded bearings that are statically loaded were considered in the second major thrust of the chapter, which was on loaded but unlubricated rolling-element bearings.

The last major thrust of the chapter dealt with loaded and lubricated rolling-element bearings. Topics covered were fatigue life and dynamic analyses. The chapter concluded by applying the knowledge of this and previous chapters to roller and ball bearing applications. The use of the elastohydrodynamic lubrication film thickness was integrated with the rolling-element bearing theory developed in this chapter. It was found that the most critical conjunctions of both ball and roller bearings occurred between the rolling elements and the inner races.

Key Words

angular-contact ball bearings bearings with a two-shouldered groove in one race and a single-shouldered groove in other

angular-contact bearings bearings that have clearance built into unloaded bearing, which allows operation under high thrust loads

ball bearings rolling-element bearings using spheres as rolling elements

cage fitting or spacer to keep proper distance between balls in bearing track

Conrad bearing same as deep-groove bearing

crown curvature machined into rollers to eliminate high edge stresses

cylindrical roller bearings bearings using cylinders as rolling elements

deep-groove bearing ball bearing with race containing pronounced groove for rolling elements

duplex pairs sets of two angular-contact ball bearings that preload each other upon assembly to shaft

dynamic load rating load which results in a bearing life of one million cycles

endplay maximum axial movement of inner race with respect to outer race under small forces

equivalent static load load equivalent to resultant load when considering thrust and radial components

free contact angle angle made by line through points where ball contacts both races and plane perpendicular to bearing axis rotation under low loads

needle roller bearings bearings using many small diameter but relatively long cylinders as rolling elements

pitch diameter mean of inner- and outer-race contact diameters

races grooves within bearing rings for rolling elements to roll in

race control condition where pure rolling occurs at controlling race

race depth same as shoulder height

retainers same as cage

rolling-element bearings machinery elements where surfaces are nonconformal and motion is primarily rolling

self-aligning bearings bearings with one race having spherical shape to allow for large misalignment

separators same as cage

shoulder height depth of race groove

spherical roller bearings same as self-aligning bearings

split-ring bearing bearing that has one race made from two halves allowing for accurate axial positioning of shafts

static load rating maximum load before plastic (rather than elastic) deformation occurs within bearing

static thrust load-carrying capacity maximum thrust load that bearing can endure before contact ellipse approaches race shoulder

tapered roller bearings bearings using tapered rollers as rolling elements.

thrust ball bearings bearings with race grooves arranged to support large axial or thrust forces

Weibull distribution relationship between survival and life, applied in this chapter to bearings

Summary of Equations

Bearing geometry:

Pitch diameter: $d_e = d_i + \dfrac{d_o - d_i}{2} = \dfrac{d_o + d_i}{2}$

Diametral clearance: $c_d = d_o - d_i - 2d$

Race conformity: $R_r = \dfrac{r}{d}$

Free contact angle: $\beta_f = \cos^{-1} \dfrac{r_o + r_i - \dfrac{d_o - d_i}{2}}{r_o + r_i - d}$

Free endplay: $c_e = 2c_r \sin\beta_f = \sqrt{4c_r c_d - c_d^2}$

Bearing kinematics:

Ball-inner race contact:

$$u_{ai} = u_{bi} = \frac{d_e\left(\omega_i - \omega_o\right)}{4}\left(1 - \frac{d^2\cos^2\beta}{d_e^2}\right)$$

Ball-outer race contact:

$$u_{ai} = u_{bi} = \frac{d_e\left(\omega_o - \omega_i\right)}{4}\left(1 - \frac{d^2\cos^2\beta}{d_e^2}\right)$$

Cylindrical bearing:

$$u_{ai} = u_{bi} = \frac{d_e\left(\omega_i - \omega_o\right)}{4}\left(1 - \frac{d^2}{d_e^2}\right)$$

$$u_{ao} = u_{bo} = \frac{d_e\left(\omega_o - \omega_i\right)}{4}\left(1 - \frac{d^2}{d_e^2}\right)$$

Load-deflection relationship:

$w_z = K_j \delta_m^j$

where $j = 1$ for cylindrical bearings and

$$K_1 = \frac{\dfrac{\pi}{2} l E'}{2\ln\left(\dfrac{4R_x}{b^*}\right) - 1}$$

and $j = 1.5$ for cylindrical bearings and

$$K_{1.5} = \pi k_e E' \left(\frac{2\mathcal{E}R}{9\mathcal{F}^3} \right)^{1/3}$$

Maximum load on rolling element:

$$w_z = \frac{n\,(w_z)_{\max}}{Z_w}, \text{ where}$$

$$Z_w = \frac{\pi\,(1 - c_d/2\delta_m)^{3/2}}{2.491 \left\{ \left[1 + \left(\frac{1 - c_d/2\delta_m}{1.23} \right)^2 \right]^{1/2} - 1 \right\}}$$

Stribeck equation: $w_z = \dfrac{n(w_z)_{\max}}{5}$

Minimum film-thickness formulae (Hamrock-Dowson equations):

Elliptical contacts:

$$H_{\min} = 3.63 U^{0.68} G^{0.49} W^{-0.073} \left(1 - e^{-0.68 k_e}\right)$$

Line contacts:

$$H_{\min,r} = \frac{h_{\min}}{R_x} = 1.714 U^{0.694} G^{0.568} \left(W'\right)^{-0.128}$$

Bearing life:

Basic load rating: $\tilde{L}_{10} = \left(\dfrac{\bar{C}}{P} \right)^{m_k}$

Equivalent load: $P = XP_r + YP_a$

Life adjustment factors: $L_A = \bar{D}\bar{E}\bar{F}_l\bar{G}\bar{H}_m\tilde{L}_{10}$

Variable loading: $L_{10} = \dfrac{1}{\dfrac{U_1}{L_{10,1}} + \dfrac{U_2}{L_{10,2}} + \dfrac{U_3}{L_{10,3}} + \cdots}$

Recommended Readings

Budynas, R.G., and Nisbett, J.K. (2011), *Shigley's Mechanical Engineering Design*, 9th ed., McGraw-Hill.

Halling, J. (ed.) (1978) *Principles of Tribology*, Macmillan.

Hamrock, B.J., Schmid, S.R., and Jacobson, B.O. (2004) *Fundamentals of Fluid Film Lubrication*, 2nd ed., Marcell Dekker.

Hamrock, B.J., and Dowson, D. (1981) *Ball Bearing Lubrication–The Elastohydrodynamics of Elliptical Contacts*, Wiley.

Harris, T. A., and Kotzales, M.N. (2007) *Essential Concepts of Bearing Analysis*, 5th ed., Taylor & Francis.

Harris, T. A., and Kotzales, M.N. (2007) *Advanced Concepts of Bearing Analysis*, 5th ed., Taylor & Francis.

Johnson, K.L. (1985) *Contact Mechanics*, Cambridge University Press.

Juvinall, R.C., and Marshek, K.M. (2012) *Fundamentals of Machine Component Design*, 5th ed., Wiley.

Mott, R. L. (2014) *Machine Elements in Mechanical Design*, 5th ed., Pearson.

Norton, R.L. (2011) *Machine Design*, 4th ed., Prentice Hall.

References

Bamberger, E.N. (1971) *Life Adjustment Factors for Ball and Roller Bearings–An Engineering Design Guide*, American Society for Mechanical Engineers.

Dowson, D., (1998) *History of Tribology*, 2nd ed., Wiley.

Hamrock, B.J., and Anderson, W.J. (1983) *Rolling-Element Bearings*, NASA Reference Publication 1105.

Hamrock, B.J., and Dowson, D. (1981) *Ball Bearing Lubrication–The Elastohydrodynamics of Elliptical Contacts*, Wiley-Interscience.

Hamrock, B.J., Schmid, S.R., and Jacobson, B.O. (2004) *Fundamentals of Fluid Film Lubrication*, 2nd ed., Marcell Dekker.

Jones, A.B. (1964) The Mathematical Theory of Rolling Element Bearings, *Mechanical Design and Systems Handbook*, H.A. Rothbart (ed.), McGraw-Hill, pp. 13–1 to 13–76.

Lundberg, G., and Palmgren, A. (1947) Dynamic Capacity of Rolling Bearings, *Acta Polytech. Mech. Eng. Sci.*, vol. 1, no. 3. (Also, Ingeniörs Vetenskaps Akadamien Handlingar, no. 196.)

Lundberg, G., and Palmgren, A. (1952) Dynamic Capacity of Roller Bearing, *Acta Polytech. Mech. Eng. Sci.*, vol. 2, no. 4. (Also, Ingenirs Vetenskaps Akadamien Handlingar, no. 210.)

Rolling Bearings — Dynamic Load Ratings and Rating Life, ISO 281:2007, International Standards Organization.

SKF Catalog (1991), SKF USA, Inc.

Skurka, J.C. (1970) Elastohydrodynamic Lubrication of Roller Bearings, *J. Lubr. Technol.*, vol. 92, no. 2, pp. 281–291.

Stribeck, R. (1901) Kugellager für beliebige Belastungen, *Z. Ver. Deutsch. Ing.*, vol. 45, pp. 73–125.

Tallian, T.E. (1967) On Competing Failure Modes in Rolling Contact, *Trans. ASLE*, vol. 10, pp. 418–439.

The Tapered Roller Bearing Guide (2010), Timken, Inc.

Weibull, W. (1949) A Statistical Representation of Fatigue Failures in Solids, *Transactions of the Royal Institute of Technology, Stockholm*, no. 27.

Questions

13.1 What are the components of a deep groove rolling-element bearing?

13.2 What are the advantages of ball bearings?

13.3 What are the advantages of cylindrical rollers?

13.4 What is a full compliment bearing?

13.5 What is a thrust load?

13.6 What are the functions of a separator or retainer?

13.7 What is shoulder height? Why is it important?

13.8 Do needle bearings need an inner race?

13.9 Define elastohydrodynamic lubrication.

13.10 When does a rolling element see the highest load?

13.11 Define contact angle.

13.12 Sketch the contact patch in a preloaded angular contact ball bearing.

13.13 What is a spall?

13.14 What is the static load rating?

13.15 What is the dynamic load rating?

13.16 Explain why bearing lives are expressed in terms of an $\tilde{L}_{10}$ life.

13.17 What is Lundberg-Palmgren theory?

13.18 What is a toroidal bearing?

Qualitative Problems

13.19 Referring only to Fig. 13.1, explain why cylindrical bearings cannot support a thrust load.

13.20 Rolling element bearings are often referred to as *anti-friction bearings*. Explain why.

13.21 Explain why duplex angular contact bearings are preloaded.

13.22 Would you expect Eq. (13.31) to hold for planetary gear systems described in Section 14.7.4? Explain.

13.23 List the advantages and disadvantages of back-to-back and face-to-back arrangements for angular contact ball bearings.

13.24 Explain the difference between static and dynamic load rating.

13.25 List bearings that (a) can and (b) cannot support thrust loads.

13.26 Compare the advantages and disadvantages of using grease or oil to lubricate a rolling-element bearing.

13.27 Repeat Example 13.1 for the case where a ball bearing cannot be used.

13.28 Sketch the contact patch in a tapered rolling-element bearing. Is it a rectangle? Explain.

13.29 Are there any conditions when a rolling-element bearing should be operated without a lubricant? Explain.

13.30 List the functions of a lubricant in a rolling-element bearing.

13.31 List the similarities and differences between toroidal and needle bearings.

13.32 Describe the mechanisms through which a spall is produced.

13.33 In cylindrical roller bearings the load is carried by rollers and the contacts between rollers and races are rectangular. The bearing macrogeometry determines which types of load can be carried by the bearing: radial, axial, or moment loads. Find which type of load can be carried by the different types of bearing shown in Tables 13.3 to 13.7.

13.34 Too high a power loss can be a problem in extremely high-speed applications of angular-contact ball bearings. A large portion of that power loss comes from the churning of the oil when the rolling balls displace the oil and splash it around the bearing. It is therefore important not to collect too much oil in the rolling tracks. For a duplex arrangement with the lubricating oil passing through both bearings for cooling, find if a back-to-back or face-to-face arrangement is preferred from a low-power-loss point of view.

13.35 Plot the dynamic load rating as a function of bearing diameter for the 6200 series deep-groove ball bearings (i.e., bearings 6202, 6204, 6205, etc.), NU 200 and NU300 cylindrical roller bearings, and 30000 series tapered roller bearings. Comment on your observations.

13.36 It was stated in Section 13.4.1 that race conformity usually lies between 0.51 and 0.54. Explain the consequences of using a race conformity that is (a) lower and (b) higher than this range.

13.37 Using Eqs. (13.66) through (13.69), express Eq. (13.70) in terms of dimensional variables. Compare the exponents and rank the importance of the variables.

13.38 List the advantages and disadvantages in using silicon carbide balls in a deep-groove rolling element bearing.

Quantitative Problems

13.39 A deep-groove ball bearing with its dimensions is shown in Sketch a. All dimensions are in millimeters. Also, $n = 10$, $d = 21/32$ in. $= 16.67$ mm. Find the maximum ball force and stress as well as the relative displacement of the race. The static radial load capacity of the bearing is 36,920 N. Assume the ball and race are made of steel. *Ans.* $(w_z)_{\max} = 18,460$ N, $\delta = 0.1198$ mm.

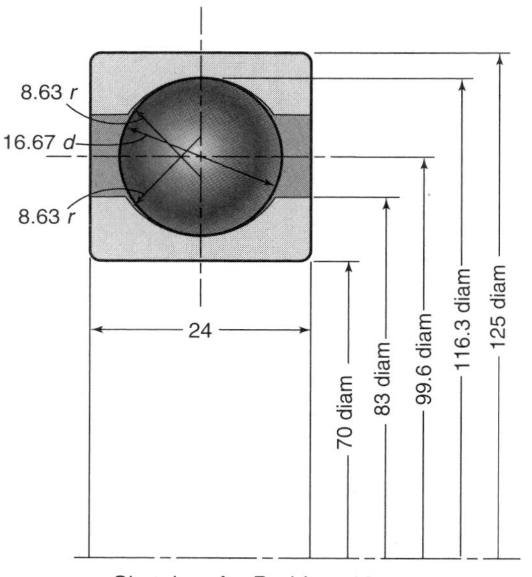

Sketch a, for Problem 13.39

13.40 In determining how large the contact areas are in a rolling-element bearing under a certain load, the geometry of each bearing part must be known. For a deep-groove ball bearing, the width-to-length ratio of the contact area is given by the race conformity, as expressed in Eq. (13.3). Calculate the race conformity for both the inner- and outer-race contacts when the ball diameter is 17 mm and the radii of curvature in the axial direction are 8.840 mm for the inner race and 9.180 mm for the outer race. Also, calculate the free contact angle and the endplay when the diametral clearance is 150 μm. *Ans.* $R_{ri} = 0.52$, $R_{ro} = 0.54$, $\beta_f = 22.11°$, $c_e = 0.7678$ mm.

13.41 A shaft will serve as a mating surface for a full-compliment set of needle rollers. If the rollers will have a diameter of 8 mm, what diameter should be ground into the shaft? If the bearings will support a load of 8 kN, what is the maximum force exerted on a roller? *Ans.* $d = 81.49$ mm, $(w_z)_{\max} = 1000$ N.

13.42 A separator for a cylindrical roller bearing is made of nylon 66 with an ultimate strength of 80 MPa at 20°C. At the maximum service temperature of 260°C the ultimate strength falls to 1 MPa. The cage material can withstand a bearing speed of 16,000 rpm at 50°C. What speed can it withstand at 150°C if the ultimate strength varies linearly with the temperature? *Ans.* $N = 11,660$ rpm.

13.43 A deep-groove ball bearing has a nylon 66 snap cage that is snapped in between the balls after they have been mounted. If the arms separating the balls have a square 3-mm by 3-mm cross-section and are 12 mm long, calculate how fast a 100-mm-diameter cage can rotate without overstressing the ball-separating arms. The maximum allowable bending stress is 20 MPa. *Ans.* $N = 14,910$ rpm.

13.44 Sketch b shows a schematic of a *Sendzimir* rolling mill, commonly used to roll thin foils. The small diameter work roller is needed for thin foils to avoid *roll flattening*, while the backup rollers are required to prevent the work roll from bending. In the metal rolling operation shown, the titanium carbide ($E = 462$ GPa, $\nu = 0.30$) work roll with a 50 mm diameter is supported by a steel backing roll with a 150 mm diameter. The 4-m wide aluminum foil requires 200 kN to roll at a speed of 2 m/s. Estimate the following:

(a) Determine the force per width between the work roll and the first backup roller. Assume the angle defined by the lines of contact between work and backup rolls is 90°. *Ans.* $P' = 70.71$ kN/m.

(b) Estimate the lubricant film thickness that develops between the backup roller and the work roll. Assume there is no slip, and a metal rolling lubricant floods all rollers. Use $\eta = 0.075$ Ns/m^2 and $\xi = 3.10 \times 10^{-8}$ m^2/N. *Ans.* $h_{\min} = 1.14$ μm.

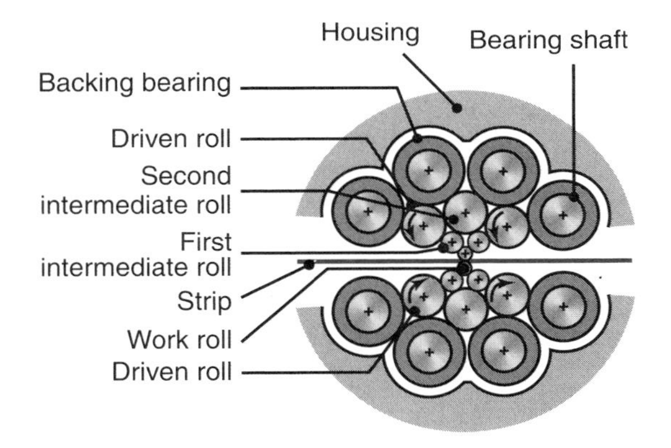

Sketch b, for Problem 13.44

13.45 Using the results from Problem 13.39, determine the minimum film thickness and the film parameter for the ball–inner-race contact and the ball–outer-race contact. The inner race rotates at 20,000 rpm and the outer race is stationary. The absolute viscosity at $p = 0$ and bearing effective temperature is 0.04 Pa-s. Also, the viscosity-pressure coefficient is 2.3×10^{-8} m^2/N. The surface roughnesses are $R_q = 0.0625$ μm for the balls and $R_q = 0.175$ μm for the races. *Ans.* For the ball–inner race, $h_{\min} = 2.394$ μm, $\Lambda = 12.88°$. For the ball–outer race, $h_{\min} = 2.777$ μm, $\Lambda = 14.94°$.

13.46 A 0.5-in.-diameter ball is loaded against a plane surface with 20 lb. The other parameters are

rms surface finish of ball	2.5 μin.
rms surface finish of plane	7.5 μin.
Absolute viscosity at $p = 0$ and bearing effective temperature	6×10^{-6} lb-s/in.^{2}
Viscosity-pressure coefficient	1.6×10^{-4} in.^{2}/lb
Modulus of elasticity for ball and plane	3×10^7 psi
Poisson's ratio for ball and plane	0.3
Film parameter	3.0

Determine the following:

(a) Minimum film thickness. *Ans.* $h_{\min} = 23.72$ μin.

(b) Mean speed necessary to achieve the above film thickness. *Ans.* $u = 426.0$ in./s.

(c) Contact dimensions. *Ans.* $D_y = 0.01221$ in.

(d) Maximum pressure. *Ans.* $\sigma_{\max} = 256$ ksi.

(e) Maximum deformation. *Ans.* $\delta_{\max} = 149.1$ μin.

13.47 A deep-groove ball bearing has steel races and silicon nitride balls. It is lubricated with a mineral oil that has a viscosity of 0.026 N-s/m^2 at the application temperature. The pressure-viscosity coefficient of the oil is 2×10^{-8} m^2/N. The ball diameter is 17 mm, and the radius of curvature in the axial direction is 8.84 mm for both races. The radii in the rolling direction are 30 mm for the inner race and 47 mm for the outer race. Calculate the minimum oil film thickness at the inner-race contact when the ball load is 20,000 N and the rolling speed is 5 m/s. What should the radius of curvature in the axial direction be for the outer race to have the same film thickness as that of the inner race? *Ans.* $h_{\min} = 0.3295$ μm, $r_{by} = 20.15$ mm.

13.48 For the bearing considered in Problem 13.47, the silicon nitride balls are changed to steel balls. Calculate how large the load can be and still maintain the same oil film thickness as obtained in Problem 13.47. *Ans.* $P = 25,880$ N.

13.49 A steel roller, shown in Sketch c, is used for rolling steel sheets that have an ultimate strength of 400 MPa. The 1-m-long roller has a diameter of 20 cm. What load per unit width will cause plastic deformation of the sheets? *Ans.* $w' = 442$ kN/m.

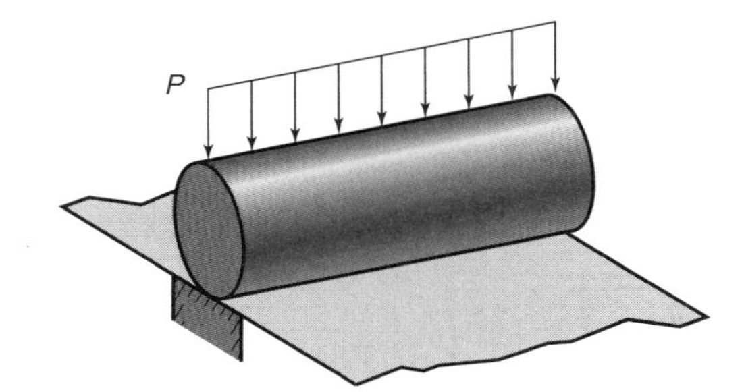

Sketch c, for Problem 13.49

13.50 Sketch d shows a gear-driven squeeze roll that, along with an idler roll, is used to produce laminated wood products. The rolls are designed to exert a normal force of 30 lb/in., and a pull of 24 lb/in. on the material. The roll speed is 300 rpm, and a life of 60,000 hours is desired. Select a pair of deep groove ball bearings; use the same size at both locations and a reliability of 0.95, and use as small a bore as possible.

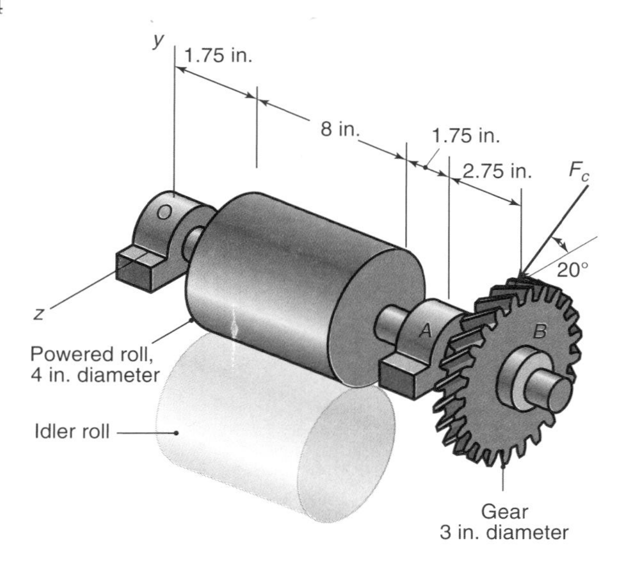

Sketch d, for Problem 13.50

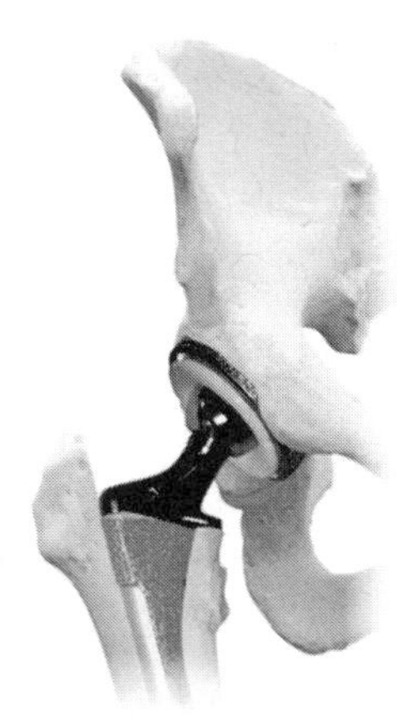

Sketch e, for Problem 13.52 (Courtesy Zimmer, Inc.)

13.51 A cylindrical roller bearing is used for the axle of a locomotive wheel, and has $d_i = 64$ mm, $d_o = 96$ mm, with nine $d = 16$ mm rollers with an effective axial length of $l_l = 16$ mm. The axle supports 10,800 N and rotates at 52.4 rad/s. The lubricant used has an absolute viscosity of 0.100 Pa-s at ambient pressure and a viscosity-pressure coefficient of 2.2×10^{-8} m^2/N. Assume the bearings and races are constructed from steel. Determine the minimum film thickness for this bearing.

13.52 An artificial hip (see Sketch e) inserted in the body is lubricated by the fluid in the joint, known as the *synovial fluid*. The following conditions are typical for a total hip replacement:

Equivalent radius, $R_x = R_y$	1 m
Viscosity of synovial fluid, η_o	0.00210 Pa-s
Pressure exponent of viscosity, ξ	2.75×10^{-8} m^2/N
Mean entraining velocity, $\tilde{u}$	0.075 m/s
Applied load, P_z	4500 N

Metal-on-metal implants use a cobalt-chrome-molybdenum (Co-Cr-Mo) alloy with $E = 230$ GPa and $\nu = 0.30$, while metal-on-polymer implants use Co-Cr-Mo on polyethylene ($E = 0.78$ GPa, $\nu = 0.46$). Perform the following:

(a) If the metal surfaces are polished to a surface roughness of 0.020 μm, estimate the film thickness and film parameter for metal-on-metal total hip replacements. *Ans.* $\Lambda = 0.8131$.

(b) Ceramic-on-ceramic implants are made from Al_2O_3, with $E = 300$ GPa, $\nu = 0.26$, and $R_q = 0.005$ μm. Calculate the film parameter for ceramic-on-ceramic implants. *Ans.* $\Lambda = 3.083$.

(c) Comment on the lubrication effectiveness for these options.

13.53 A shaft with rolling bearings and loads is shown in Sketch f. The bearings are mounted such that both the axial and radial forces caused by the external loads are absorbed. For bearing NU205ECP the geometry is cylindrical with no flange on the inner race, so that only radial load can be absorbed. The full axial thrust load P_a is thus taken up by the deep-groove ball bearing 6205. Determine the $\tilde{L}_{10}$ lives of each of the two bearings while assuming length dimensions are in meters. *Ans.* $\left(\tilde{L}_{10}\right)_{\text{NU205ECP}} = 71,500$ million revolutions, $\left(\tilde{L}_{10}\right)_{6205} = 405$ million revolutions.

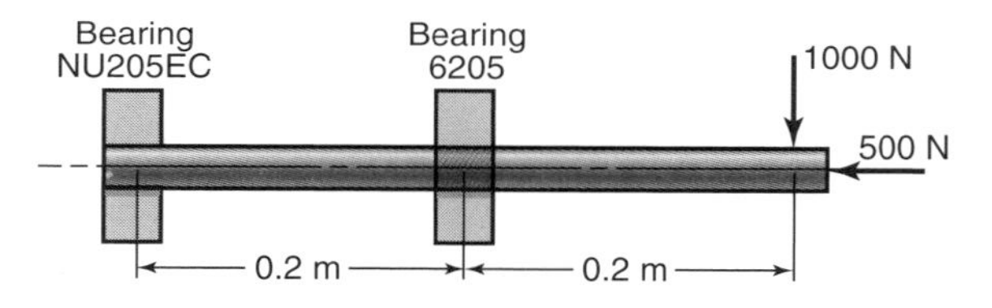

Sketch f, for Problem 13.53

13.54 Given the shaft and bearings in Problem 13.53 but with the bearing positions shifted so the 6205 deep-groove ball bearing is in the left end of the shaft and bearing NU205ECP in the middle, find the $\tilde{L}_{10}$ lives for the two bearings in their new positions. Which of the two positions would be preferred and why? *Ans.* $\left(\tilde{L}_{10}\right)_{\text{NU205ECP}} = 7098$ million revolutions, $\left(\tilde{L}_{10}\right)_{6205} = 1289$ million revolutions.

13.55 A bearing arrangement for a shaft in a gearbox is shown in Sketch g. Bearing A is a cylindrical roller bearing (NU202ECP) with $d_b = 15$ mm, $d_a = 35$ mm, $b_w = 11$ mm, and $C = 12,500$ N. Bearing B is a deep-groove ball bearing (6309) with $d_b = 45$ mm, $d_a = 100$ mm, $b_w = 25$ mm, $C = 55,300$ N, and C_o = 31,500 N. Determine which bearing is likely to fail first and what the $\tilde{L}_{10}$ is for that bearing. Length dimensions are in millimeters. *Ans.* Bearing B fails at $\tilde{L}_{10} = 1799$ million revolutions.

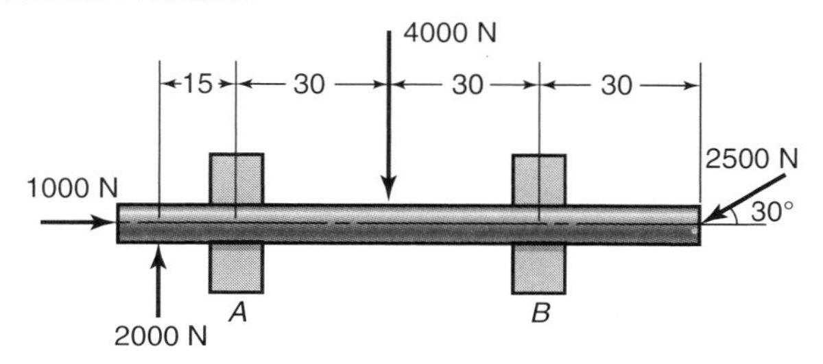

Sketch g, for Problem 13.55

13.56 The two bearings in Problem 13.55 have the same composite surface roughness of the inner-race, rolling-element contact, $\left(R_{qa}^2 + R_{qb}^2\right)^{0.5} = 0.10\ \mu$m. For bearing NU202ECP at A, the outer radius of the inner race is 9.65 mm, the roller diameter is 5.7 mm, and inner radius of the outer race is 15.35 mm. For bearing 6309 at B, the ball diameter is 17 mm, the outer radius of the inner race is 27.4 mm, and the inner radius of the outer race is 44.4 mm. The race conformity for the outer ring is 0.53 while for the inner ring it is 0.52. The lubricant used has the absolute viscosity $\eta_o = 0.1$ N-s/m^2 and the pressure-viscosity coefficient $\xi = 2 \times 10^{-8}$ m^2/N. The bearing rotational speed is 1500 rpm. The cylindrical roller bearing has 13 rollers and the ball bearing has 10 balls. The ball bearing is made of AISI 440C (E = 200 GPa), and the roller bearing is made of AISI 52100 (E = 207 GPa); both materials have ν = 0.3. Find the adjusted bearing lives $\tilde{L}_{10,\mathrm{adj}}$ for the two bearings. *Ans.* NU202ECP: $\tilde{L}_{10,\mathrm{adj}} = 1.3\tilde{L}_{10}$, 6309: $\tilde{L}_{10,\mathrm{adj}} = 2.3\tilde{L}_{10}$.

13.57 Calculate the film thickness for a steel cylindrical roller of radius 15 mm rolling against a steel inner bearing race with a radius of 50 mm if the maximum Herz pressure is 1 GPa, and the rolling speed is 4 m/s. Consider the following lubricants:

(a) Normal paraffin, $\eta = 0.331$ Ns/m^2, $\xi = 4.74 \times 10^{-7}$ m^2/N. *Ans.* $h_{\min} = 19.3\ \mu$m.

(b) Thick petroleum oil, $\eta = 7.08$ Ns/m^2, $\xi = 3.41 \times 10^{-8}$ m^2/N. *Ans.* $h_{\min} = 36.3\ \mu$m.

13.58 Given the shaft arrangement in Fig. 11.1, with the applied forces indicated, bearing B is a cylindrical roller bearing of type NU304ECP with $C = 35,500$ N, and bearing A is a deep-groove ball bearing of type 6404 ($C = 30,700$ N, $C_o = 15,000$ N). Find the $\tilde{L}_{10}$ life for each bearing and judge if they are strong enough to be used at 15,000 rpm for 20 years. *Ans.* $\left(\tilde{L}_{10}\right)_{\mathrm{NU304ECP}} = 3.03 \times 10^5$ revs, $\left(\tilde{L}_{10}\right)_{6404} = 2.343 \times 10^{11}$ revs.

13.59 Given the shaft arrangement in Problem 13.58 but with the belt drive at D changed to a worm gear giving the same radial force but also an additional axial force of 800 N directed from D to A, find the $\tilde{L}_{10}$ lives of the bearings and judge if they are strong enough to last 20 years at 15,000 rpm. *Ans.* $\left(\tilde{L}_{10}\right)_{6404} = 6.572 \times 10^9$ revs.

13.60 A flywheel has deep-groove ball bearings located close to the rotating disk. The disk mass is 50 kg, and it is mounted out of balance, so that the center of mass is situated at a radius 1 mm from the bearing centerline. The bearings are two 6305 deep-groove ball bearings with $C = 23,400$ N. Neglecting the shaft mass, find the nominal bearing $\tilde{L}_{10}$ in hours for each bearing at

(a) 500 rpm. *Ans.* $\tilde{L} = 4.13 \times 10^5$ million revolutions (12.3 million hours).

(b) 2500 rpm. *Ans.* $\tilde{L} = 1699 \times 10^5$ million revolutions (10,000 hours).

(c) 12,500 rpm. *Ans.* $\tilde{L} = 0.143 \times 10^5$ million revolutions (11 minutes).

13.61 A ball bearing manufacturer wants to make sure that its bearings provide long service lives on different components. Extra polishing operation can be applied to

(a) the ring surfaces

(b) the ball surfaces

(c) all contact surfaces

The polishing decreases the roughness by one-half the original roughness, which for the races is 0.08 μm and for the balls is 0.04 μm. Find how much the Λ value changes when either a, b, or c is polished.

13.62 Select a pair of radial bearings for the shaft shown in Problem 11.36 if the shaft rotates at 1000 rpm. Consider the case where the shaft is operated during one work shift (8 hours per day, 5 days per week, 50 weeks per year) for 2 years, with a reliability of 98% and very clean lubricant is maintained throughout the bearing life. *Ans.* $\bar{C}_{\min} = 11,960$ lb.

13.63 A 6305 ball bearing is subjected to a steady 5000-N radial load and a 2000-N thrust load, and uses a very clean lubricant throughout its life. If the inner race angular velocity is 500 rpm, find:

(a) The equivalent radial load. *Ans.* $P = 5600$ N.

(b) The L_{10} life. *Ans.* $L_{10} = 73.0$ million revolutions.

(c) The L_{50} life. *Ans.* $L_{50} = 256$ million revolutions.

13.64 The shaft shown in Sketch h transfers power between the two pulleys. The tension on the slack side (right pulley) is 30% of that on the tight side. The shaft rotates at 900 rpm and is supported uniformly by a radial ball bearing at points 0 and B. Select a pair of radial ball bearings with 99% reliability and 40,000 hr of life. Assume Eq. (13.83) can be used to account for lubricant cleanliness. All length dimensions are in millimeters. *Ans.* $\bar{C}_{\min} = 42,400$ N.

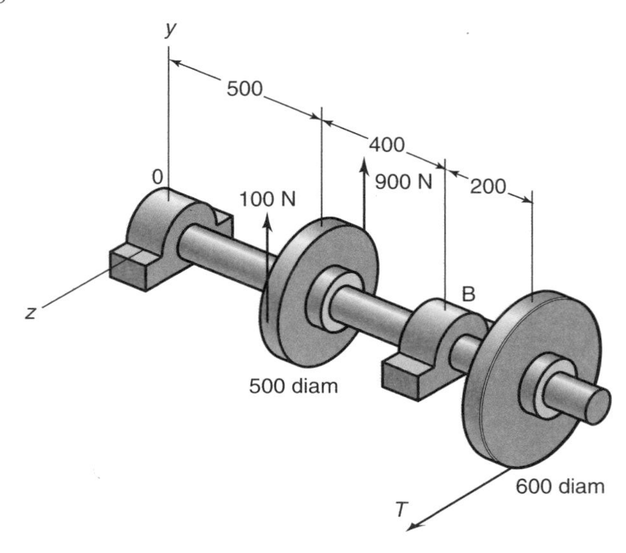

Sketch h, for Problem 13.64

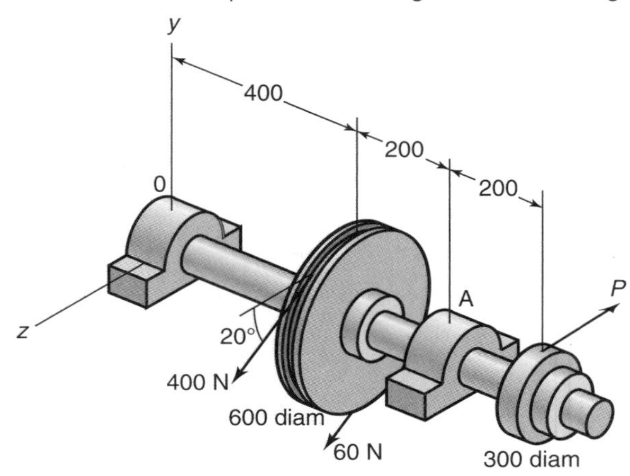

Sketch j, for Problem 13.66

13.65 In Sketch i the tension on the slack side of the left pulley is 20% of that on the tight side. The shaft rotates at 1000 rpm. Select a pair of deep-groove roller bearings to support the shaft for 99% reliability and a life of 20,000 hr. Assume Eq. (13.83) can be used to account for lubricant cleanliness. All length dimensions are in millimeters.

13.67 The power-transmitting system shown in Sketch k consists of a helical gear, a bevel gear, and a shaft that rotates at 600 rpm and is supported by two roller bearings. The load on the bevel gear is $-0.5P\bar{i} - 0.41P\bar{j} + 0.44P\bar{k}$. The left bearing supports the axial load on the shaft. Select identical single row tapered roller bearings for 36,000-hr life at 98% reliability. All length dimensions are in millimeters. Assume that the full thrust load can be applied to either bearing to make the bearings more robust and accommodating of installation errors. *Ans.* $\bar{C}_{\min} = 121,400$ N.

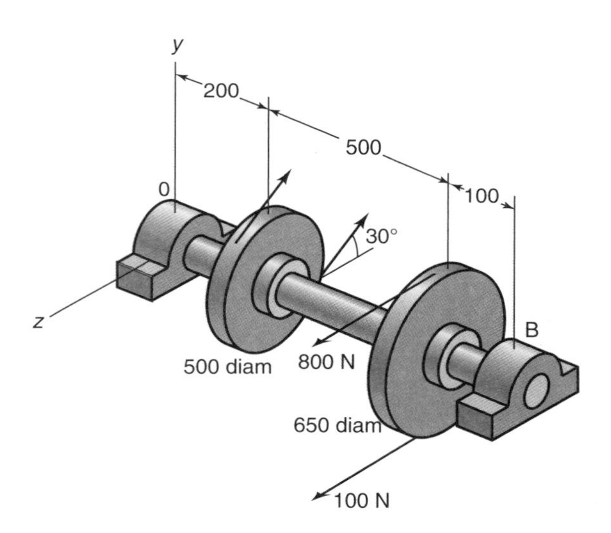

Sketch i, for Problem 13.65

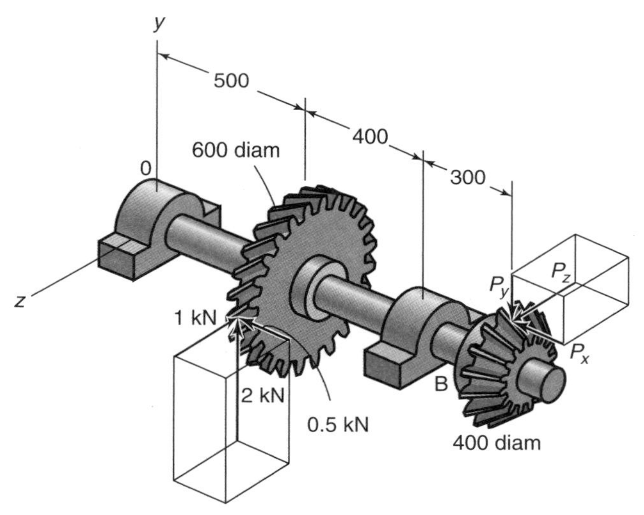

Sketch k, for Problem 13.67

13.66 In Sketch j the shaft rotates at 1100 rpm and transfers power with light shock (increases load by 20%) from a 600-mm-diameter pulley to a 300-mm-diameter sprocket. Select a deep-groove bearing for support at 0 and a cylindrical roller bearing for support at A. Assume the lubricant will be carefully monitored, maintained, and replaced to avoid bearing exposure to contaminated lubricant. The bearings should have a life of 24,000 hr at 95% reliability. All length dimensions are in millimeters.

13.68 A cylindrical roller bearing is used to support an 8-kN load for 1200 hr at 600 rpm. Calculate the load-carrying capacity of this bearing for 95% reliability. What size bearing would you recommend? *Ans.* $\bar{C} = 30,000$ N.

13.69 A deep-groove ball bearing is used to support a 6-kN load for 1000 hr at 1200 rpm. For 99% reliability, what should be the load-carrying capacity of the bearing? What size bearing should be chosen? *Ans.* $\bar{C} = 50,900$ N.

Design and Projects

13.70 During operation, a roller running in a groove develops a film thickness, but also pushes fluid to the side. Surface tension can cause this fluid to replenish, but if the bearing is operating at high speed, subsequent bearings may operate in a starved condition. Consult the technical literature and list the important factors to consider regarding starvation.

13.71 Construct a Design Procedure for evaluating lubricant film thickness for cylindrical roller bearings, similar to Design Procedure 13.1.

13.72 Sketch l shows a double bearing design intended to reduce rolling friction. Describe the motion of the rolling elements and intermediate race for this bearing. List the advantages and disadvantages of such a rolling element bearing design.

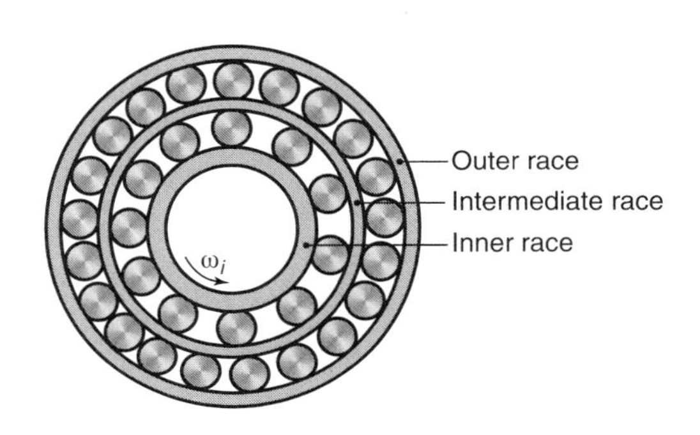

Sketch l, for Problem 13.72

13.73 A shaft has the bearing reactions and axial load as shown in Sketch m. A single row cylindrical bearing and a single row deep-groove bearing are to be used to support the shaft. Indicate the location where the deep-groove and cylindrical bearings should be placed. Select bearings for this application if the desired $\tilde{L}_{10}$ is 20,000 hours at 450 rpm.

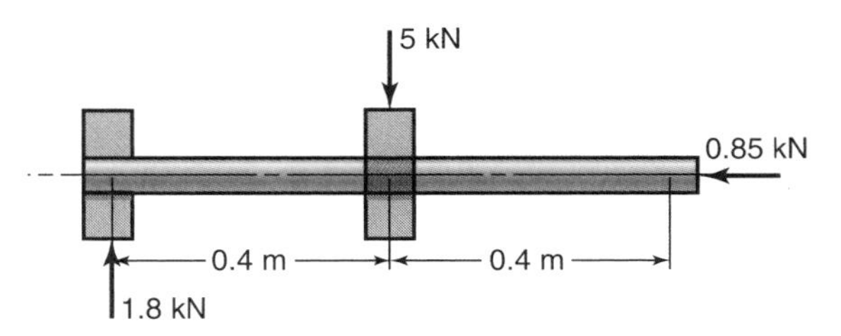

Sketch m, for Problem 13.73

13.74 For a cylindrical roller bearing the separator has the form of a cylinder with outer diameter of 80 mm, inner diameter of 74 mm, and width of 16 mm. Fourteen 13-mm by 13-mm rectangular pockets with depth of 3 mm are evenly distributed around the circumference of the separator. In each pocket is a 12.5-mm by 12.5-mm cylindrical roller. The inner-race outer diameter is 65 mm and the outer-race inner diameter is 90 mm. How fast can the bearing inner race rotate before the polyester separator fails? Use a safety factor of 10, a density of 1040 kg/m^3, and ultimate strength of 60 MPa for polyester. *Ans.* $N = 25,790$ rpm.

13.75 For the cylindrical roller bearing in Problem 13.74, calculate the sliding speed at the contact between the rollers and the separator if the inner race rotates at 5,000 rpm.

Chapter 14

General Gear Theory; Spur Gears

An assortment of gears. *Source:* Courtesy of Quality Transmission Components.

We are what we repeatedly do. Excellence, therefore, is not an act, but a habit.

Aristotle

Gears are very popular power transmission devices, combining the attributes of high efficiency, long and maintenance-free service life, reasonable manufacturing cost, and relatively large torque capability in a compact size. This chapter introduces gear concepts, including geometry and kinematics, and shows the different gear types. Spur gears, the emphasis of this chapter, are toothed wheels whose geometry does not change through their thickness. These are the easiest gears to manufacture and are very commonly used to transmit torque and power between parallel shafts. Spur gear geometry is discussed, including the importance of the involute profile. The manufacturing processes and materials used for gears are described, as well as their impact on gear performance. Two common failure modes of gears, namely gear tooth bending and surface pitting, are considered. The discussion of bending begins with the mechanics of the Lewis equation, and the standard design approach promulgated by the American Gear Manufacturers Association for bending; the similar approach for contact stress is then described. The chapter ends with a discussion of effective gear lubrication, which is essential for trouble-free performance, as well as a procedure for design synthesis of gears.

Machine elements in this chapter: Spur gears, racks, splines.
Typical applications: Power transmissions, speed reducers, gear pumps. Common machine element for torque transmission between shafts.
Competing machine elements: Helical, bevel, and worm gears (Chapter 15), belts and pulleys (Chapter 19), roller chains and sprockets (Chapter 19).

Contents

Examples

Design Procedures

Case Study

Symbols

a	addendum, m
A'	variable used in Eq. (14.47)
b	dedendum, m
b^*	Hertzian half-width, m
b_w	face width, m
b_l	backlash, m
C_e	mesh alignment correction factor
C_H	hardness ratio
C_{ma}	mesh alignment factor
C_{mc}	lead correction factor
C_{pf}	pinion proportion factor
C_{pm}	pinion proportion modifier
C_r	contact ratio
c	clearance, m
c_d	center distance, m
d	pitch diameter, m
d_r	root or fillet diameter, m
E	modulus of elasticity, Pa
E'	effective modulus of elasticity, Pa
G	dimensionless materials parameter
g_r	gear ratio
H	dimensionless film thickness
HB	Brinell hardness, kg/mm^2
h	lubricant film thickness, m
h_k	working depth, m
h_p	power, W
h_t	total depth, m
I	area moment of inertia, m^4
K_b	rim thickness factor
K_m	load distribution factor
K_o	overload factor
K_r	reliability factor
K_s	size factor
K_t	temperature factor
K_v	dynamic factor
L_{ab}	length of line of action, m
l	moment arm of tangential load, m
M_B	backup ratio
m	module, mm
N	number of teeth
N_a	rotational speed, rpm
n_s	safety factor
P_z	normal load, N
p_b	base pitch, m
p_c	circular pitch, m
p_d	diametral pitch, m^{-1}
p_H	maximum Hertzian contact pressure, Pa
Q_v	quality index
R_x	curvature sum in x-direction, m
r	pitch radius, m
S	Strength, Pa
T	torque, N-m
t	tooth thickness, m
t_h	theoretical circular tooth thickness measured on pitch circle, m
t_{ha}	actual circular tooth thickness measured on pitch circle, m
t_R	rim thickness, m
U	dimensionless speed parameter
u	velocity, m/s
$\tilde{u}$	mean velocity, m/s
v_t	pitch-line velocity, m/s
W	load, N
W'	dimensionless load parameter
W_r	radial load, N
W_t	tangential load, N
w'	load per unit width, N/m
Y	Lewis form factor
Y_j	geometry factor
Y_N	stress cycle factor for bending
Z	angular velocity ratio
Z_N	stress cycle factor for contact stress
η_o	atmospheric viscosity, Pa-s
Λ	film parameter
ξ	pressure-viscosity coefficient, m^2/N
σ	stress, Pa
ϕ	pressure angle, deg
ω	angular speed, rad/s

Subscripts

a	actual; axial
b	bending; base
c	contact
g	gear
o	outside
p	pinion
s	sliding
t	total

14.1 Introduction

A **gear** may be thought of as a toothed wheel that, when meshed with another smaller-in-diameter toothed wheel (the **pinion**), will transmit rotational motion from one shaft to another. The primary function of a gear is to transfer power from one shaft to another while maintaining a definite ratio between the shaft angular velocities. The teeth of a driving gear push on the driven gear teeth, exerting a force with a component tangent to the gear periphery. Thus, a torque is transmitted, and because the gear is rotating, power is transferred. Gears are the most rugged and durable torque transmission devices. Their power transmission efficiency can be as high as 98%, but gears are usually more costly than other power transmitting machine elements, such as chain or belt drives.

This chapter presents the methods, recommendations and terminology defined by the American Gear Manufacturers Association (AGMA), which promulgates extensive standards for gear design. However, there are occasions when the approach has been simplified. For example, AGMA follows a practice of assigning a different symbol to a variable depending on whether it is in U.S. customary or metric units. Also, some of the strength of materials terminology has been brought into agreement with the nomenclature consistent throughout this text. This will be discussed further in Section 14.8.

14.2 Types of Gears

This brief discussion explains the various gear classes and is not meant to be all-inclusive. More detailed discussion can be found in Drago [1992]. Gears can be divided into three major classes: parallel-axis gears, nonparallel but coplanar gears, and nonparallel and noncoplanar gears. This section describes gears in each of these classes. The remainder of the chapter will emphasize spur gears; helical, bevel, and worm gears will be considered in Ch. 15.

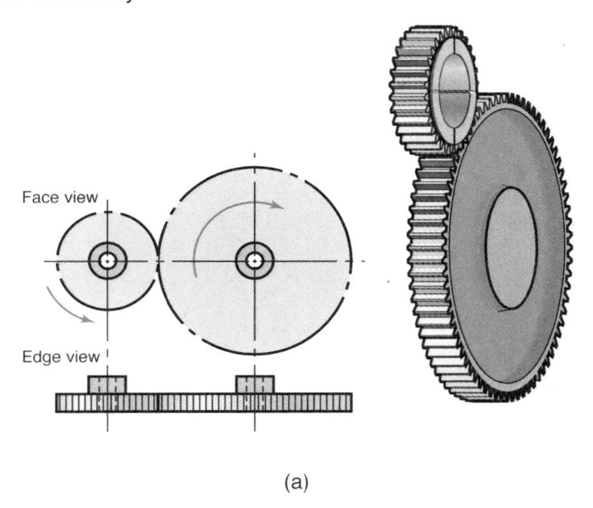

(a)

(b)

Figure 14.1: Spur gear drive. (a) Schematic illustration of meshing spur gears; (b) a collection of spur gears. *Source:* Courtesy of Boston Gear, an Altra Industrial Motion Company.

14.2.1 Parallel-Axis Gears

Parallel-axis gears are the simplest and most popular type of gear. They connect parallel shafts; have higher power transmission capacity and high efficiency. Spur and helical gears are two of the primary gears of this type.

Spur Gears

Spur gears are parallel-axis gears with teeth parallel to their axis. Figure 14.1a shows a spur gear drive with such teeth on the outside of a cylinder; they can also be located on the inner surface of a cylinder. Spur gears are the simplest and the most common type of gear, as well as the easiest to manufacture.

Helical Gears

Figure 14.2a shows a **helical gear** drive, with gear teeth cut on a spiral that wraps around a cylinder. Helical teeth enter the meshing zone progressively and, therefore, have a smoother action than spur gear teeth. Helical gears also tend to be quieter. Another positive feature of helical gears (relative to spur gears) is that the transmitted load is somewhat larger, thus implying that helical gears will last longer for the same load. Also, a smaller helical gear can transmit the same load as a larger spur gear.

A disadvantage of helical gears (relative to spur gears) is that they produce an additional end thrust along the shaft axis that is not present with spur gears. This end thrust may require an additional component, such as a thrust collar, ball bearings, or tapered-roller bearings. Another disadvantage is that helical gears have slightly lower efficiency than equivalent spur gears. Although the total load-carrying capacity is larger for helical gears, the load is distributed normally and axially, whereas for a spur gear all load is distributed normally.

14.2.2 Nonparallel, Coplanar Gears

Some gears have nonparallel axes (unlike spur and helical gears, which have parallel axes) and are coplanar (like spur and helical gears). **Bevel**, **straight**, **Zerol**, **spiral**, and **hypoid gears** are all in the nonparallel, coplanar class. The common feature of these gears is the redirection of power around a corner, as might be required, for example, when connecting a horizontally mounted engine to the vertically mounted rotor shaft on a helicopter. Figure 14.3 shows a bevel gear drive with straight teeth and illustrates the gears in this classification. Observe that the axes of the gear drive are nonparallel but coplanar.

14.2.3 Nonparallel, Noncoplanar Gears

Nonparallel, noncoplanar gears are more complex in both geometry and manufacturing than the two previous gear classifications. As a result, these gears are more expensive than the other gears discussed.

Figure 14.4 depicts a **worm gear** drive with cylindrical teeth that illustrates this class of gear. Note that the axes are nonparallel and noncoplanar. These gears can provide considerably higher reduction ratios than coplanar or simple crossed-axis gear sets, but their load-carrying capacity is low, their contact pressure is extremely high, and their wear rate is high. Thus, worm gear drives are used mainly for light-load applications.

14.3 Gear Geometry

Figure 14.5 shows the basic spur gear geometry, and Fig. 14.6 shows a detail of gear tooth nomenclature.

14.3.1 Center Distance, Circular Pitch and Diametral Pitch

The **center distance** between the rotating gear axes can be expressed as

$$c_d = \frac{d_p + d_g}{2}, \tag{14.1}$$

where d_p is the pitch diameter of the pinion and d_g is the pitch diameter of the gear, as illustrated in Fig. 14.5. The **circular pitch**, p_c, is the distance measured on the pitch circle from one point on one tooth to a corresponding point on the adjacent tooth. Mathematically, this can be expressed as

$$p_c = \frac{\pi d}{N} = \frac{\pi d_p}{N_p} = \frac{\pi d_g}{N_g}, \tag{14.2}$$

where N_p is the number of pinion teeth and N_g is the number of gear teeth. The **gear ratio**, g_r, is the ratio of gear to pinion

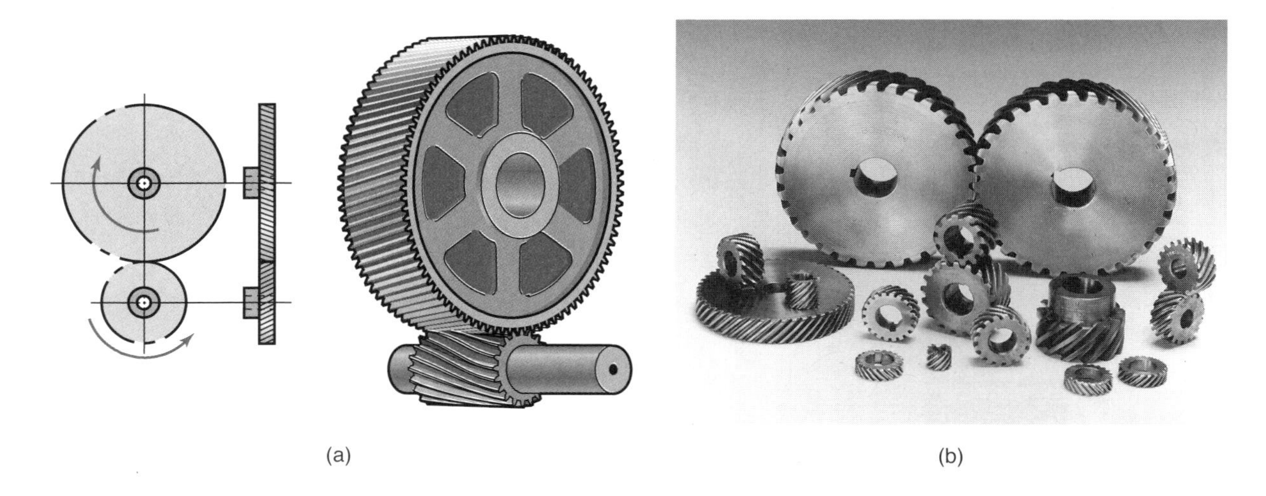

Figure 14.2: Helical gear drive. (a) Schematic illustration of meshing helical gears; (b) a collection of helical gears. *Source:* Courtesy of Boston Gear, an Altra Industrial Motion Company.

Figure 14.3: Bevel gear drive. (a) Schematic illustration of meshing bevel gears with straight teeth; (b) a collection of bevel gears. *Source:* Courtesy of Boston Gear, an Altra Industrial Motion Company.

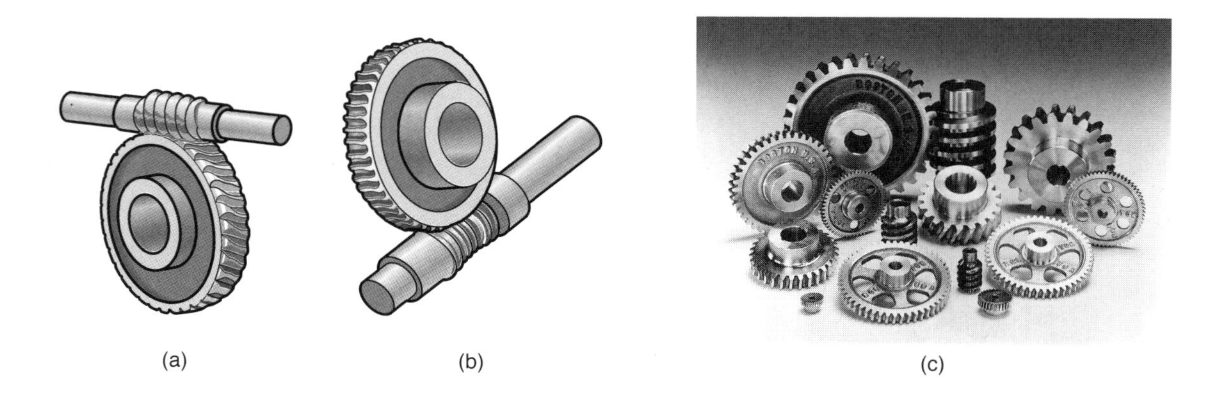

Figure 14.4: Worm gear drive. (a) Cylindrical teeth; (b) double enveloping; (c) a collection of worm gears. *Source:* Courtesy of Boston Gear, an Altra Industrial Motion Company.

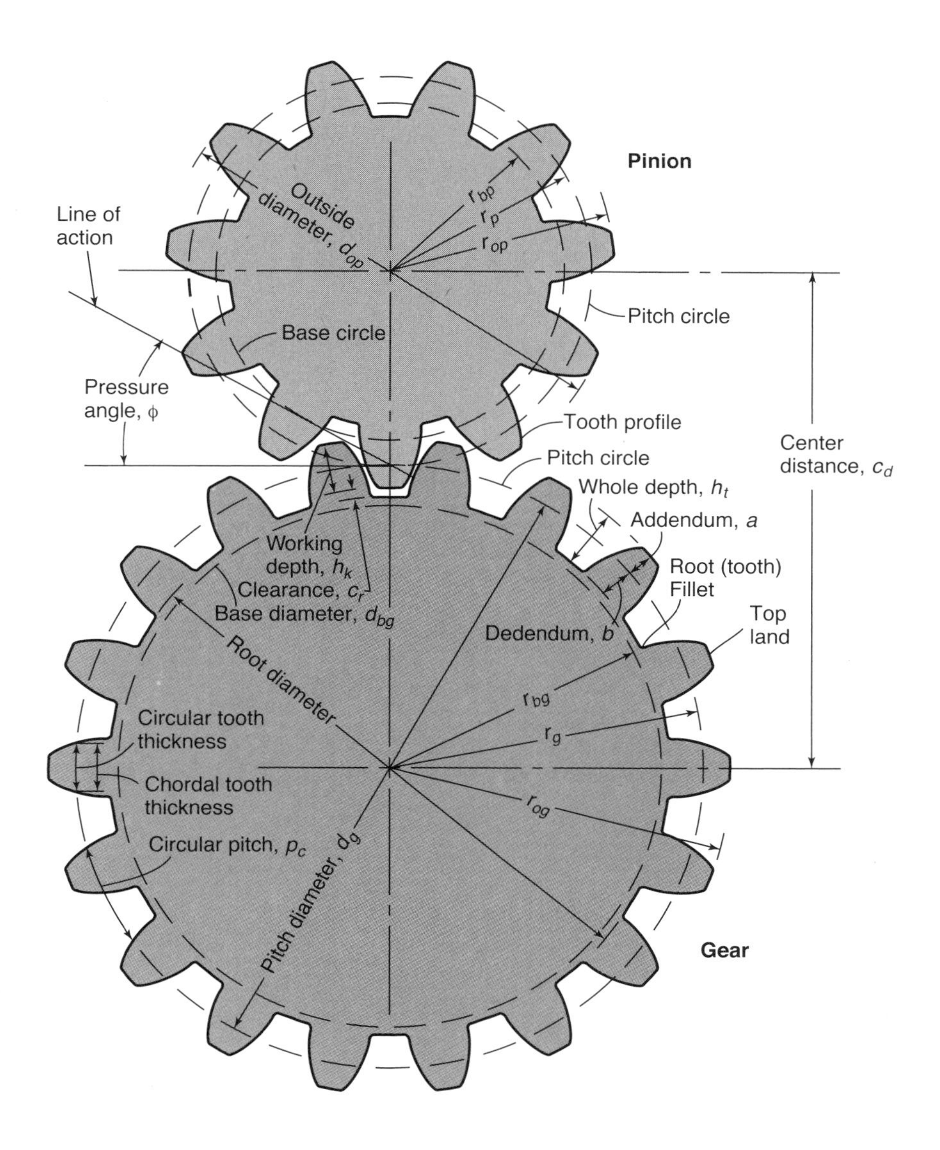

Figure 14.5: Basic spur gear geometry.

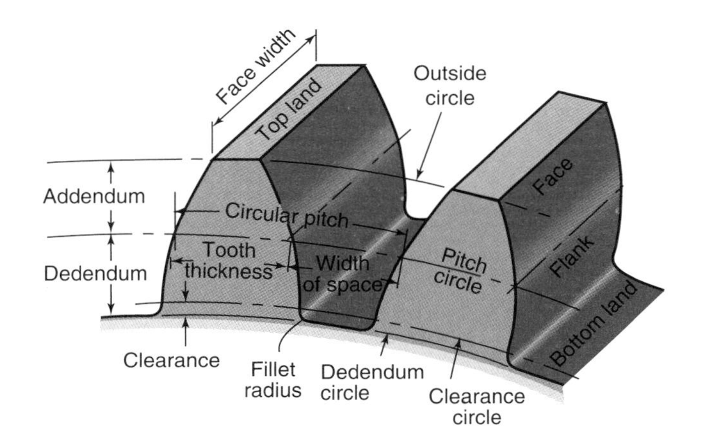

Figure 14.6: Nomenclature of gear teeth.

diameter; from Eq. (14.2),

$$g_r = \frac{d_g}{d_p} = \left(\frac{p_c N_g}{\pi}\right)\left(\frac{\pi}{p_c N_p}\right) = \frac{N_g}{N_p}. \tag{14.3}$$

Substituting Eqs. (14.2) and (14.3) into Eq. (14.1) gives

$$c_d = \frac{p_c}{2\pi}\left(N_p + N_g\right) = \frac{p_c N_p}{2\pi}\left(1 + \frac{N_g}{N_p}\right) = \frac{p_c N_p}{2\pi}\left(1 + g_r\right). \tag{14.4}$$

The **diametral pitch**, p_d, is the number of teeth in the gear per pitch diameter. For two gears to mesh, they must have the same diametral pitch. The expression for diametral pitch is

$$p_d = \frac{N_p}{d_p} = \frac{N_g}{d_g}. \tag{14.5}$$

Figure 14.7 shows standard diametral pitches and compares them with tooth size. The smaller the tooth, the larger the diametral pitch. Table 14.1 lists preferred diametral pitches for four tooth classes. Coarse and medium-coarse gears are used for power transmission applications; fine and ultrafine gears are used for instruments and control systems. The pitch

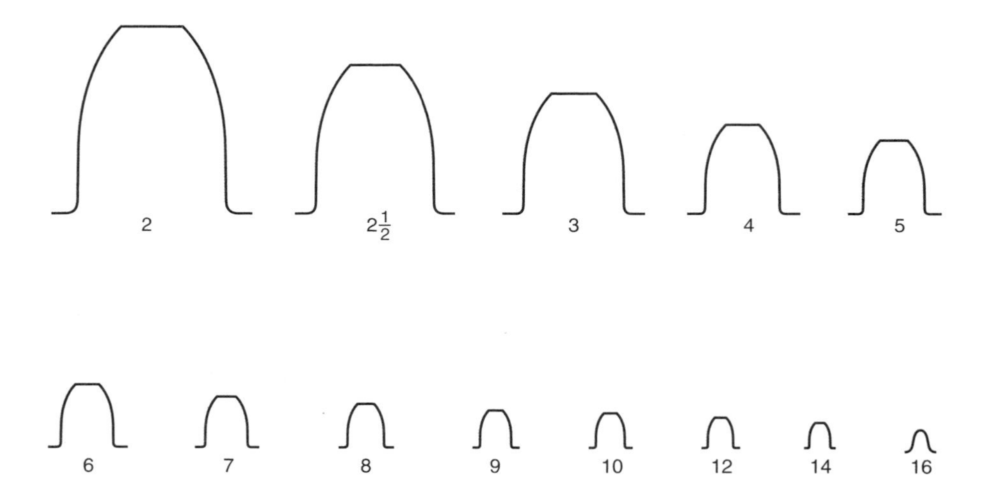

Figure 14.7: Standard diametral pitches compared with tooth size. Full size is assumed.

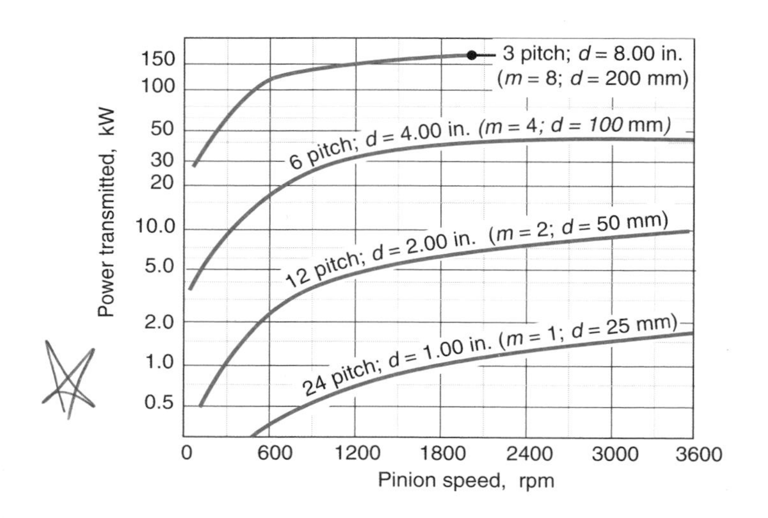

Figure 14.8: Transmitted power as a function of pinion speed for a number of diametral pitches. For all cases shown, $g_r = 4$, $N_p = 24$, $K_o = 1.0$, $\phi = 20°$. *Source:* Adapted from Mott [2003].

can also be selected from power requirements, as shown in Fig. 14.8.

Comparing Eqs. (14.2) and (14.5) yields the following:

$$p_c = \frac{\pi}{p_d}. \tag{14.6}$$

Thus, the circular pitch is inversely proportional to the diametral pitch.

The **module**, m, is the ratio of the pitch diameter to the number of teeth and is only expressed in the SI unit of millimeters as

$$m = \frac{d_g}{N_g} = \frac{d_p}{N_p}. \tag{14.7}$$

From Eqs. (14.2) and (14.7)

$$p_c = \pi m. \tag{14.8}$$

Table 14.1: Preferred diametral pitches for four tooth classes.

Class	Diametral pitch, p_d, in^{-1}
Coarse	1/2, 1, 2, 4, 6, 8, 10
Medium coarse	12, 14, 16, 18
Fine	20, 24, 32, 48, 64 72, 80, 96, 120, 128
Ultrafine	150, 180, 200

Also, from Eqs. (14.6) and (14.8)

$$m = \frac{1}{p_d}. \tag{14.9}$$

Example 14.1: Gear Geometry

Given: A 20-tooth pinion with a module of 5 mm meshes with a 63-tooth gear.

Find: The center distance, the circular pitch, and the gear ratio.

Solution: From Eq. (14.7),

$$d_p = mN_p = (5)(20) = 100 \text{ mm},$$

$$d_g = mN_p = (5)(63) = 315 \text{ mm}.$$

From Eq. (14.1), the center distance is

$$c_d = \frac{d_p + d_g}{2} = \frac{100 + 315}{2} = 207.5 \text{ mm}.$$

From Eq. (14.8), the circular pitch is

$$p_c = \pi m = \pi(5) = 15.71 \text{ mm}.$$

From Eq. (14.3), the gear ratio is

$$g_r = \frac{d_g}{d_p} = \frac{N_g}{N_p} = \frac{315}{100} = \frac{63}{20} = 3.150.$$

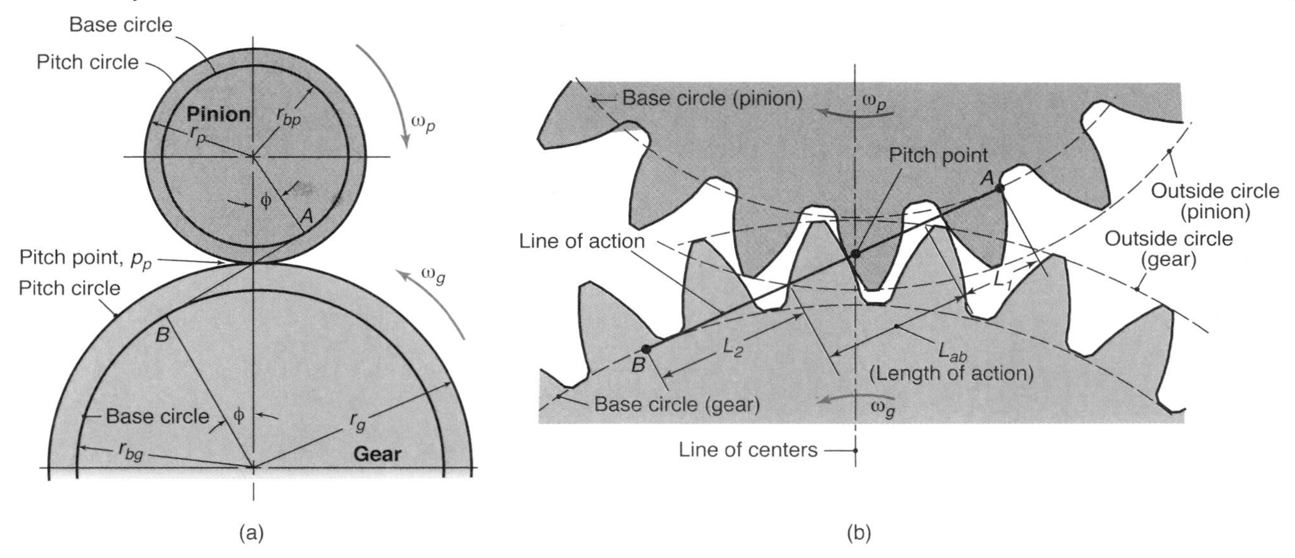

Figure 14.9: (a) Pitch and base circles for pinion and gear as well as line of action and pressure angle. (b) Detail of active profile, showing detail of line of action and length of action, L_{ab}.

14.3.2 Addendum, Dedendum, and Clearance

From Figs. 14.5 and 14.6 the **addendum**, a, is the distance from the top land to the pitch circle, and the **dedendum**, b, is the radial distance from the bottom land to the pitch circle. The **clearance** shown in Fig. 14.6 is the amount by which the dedendum exceeds the addendum. A clearance is needed to prevent the end of one gear tooth from contacting the bottom of the mating gear. Table 14.2 shows values of addendum, dedendum, and clearance for a spur gear with a 20° pressure angle and a full-depth involute. Both coarse-pitch and fine-pitch formulations are given. Note from Table 14.2 that the clearance can be expressed as:

$$c = b - a. \tag{14.10}$$

Once the addendum, dedendum, and clearance are known, a number of other dependent parameters can be obtained. Some of the more important are:

1. Outside diameter:
$$d_o = d + 2a \tag{14.11}$$
2. Root diameter:
$$d_r = d - 2b \tag{14.12}$$
3. Total depth:
$$h_t = a + b \tag{14.13}$$
4. Working depth:
$$h_k = a + a = 2a \tag{14.14}$$

These parameters can also be expressed in terms of diametral pitch and number of teeth rather than pitch diameter by using Eq. (14.5). For example, the outside diameter given in Eq. (14.11) can be expressed, by making use of the formulas given in Table 14.2 and Eq. (14.5), as

$$d_o = \frac{N}{p_d} + 2\left(\frac{1}{p_d}\right) = \frac{N+2}{p_d}. \tag{14.15}$$

It should be noted that all formulas given in this section are equally applicable for the gear and the pinion.

14.3.3 Line of Action, Pressure Angle, and Gear Involute

The purpose of meshing gear teeth is to provide constant relative motion between engaging gears. To achieve this tooth action, the common normal of the curves of the two meshing gear teeth must pass through a common point, called the **pitch point**, which also then defines the **pitch circles** of the mating gears.

Figure 14.9 shows the pitch and base circles for both the pinion and the gear. The **base circle** is the circle from which the involute profile is described, and is tangent to the **line of action**, which is the straight line where tooth contact occurs. The line of action passes through the pitch point and is tangent to the base circles (at points A and B in Fig. 14.9). The pressure angle, ϕ, is the angle between the line of centers and another line perpendicular to the line of action and going through the center of the gear or pinion. From Fig. 14.9, observe that the radius of the base circle is

$$r_b = r\cos\phi,$$

so that

$$r_{bp} = r_p \cos\phi$$

and

$$r_{bg} = r_g \cos\phi. \tag{14.16}$$

The base pitch, p_b, is the distance from a point on one tooth to the corresponding point on the adjacent tooth measured around the base circle. Recall from Eq. (14.6) that the circular pitch, p_c, is measured on the pitch circle. From Eq. (14.2), the base pitch is

$$p_b = p_c\cos\phi = \frac{\pi d}{N}\cos\phi = \frac{\pi d_p}{N_p}\cos\phi = \frac{\pi d_g}{N_g}\cos\phi. \tag{14.17}$$

The shape of the gear tooth is obtained from the involute curve shown in Fig. 14.10. Involute curves are very common for gears, and have a number of advantages including smooth rolling motion and quiet operation. Also, gears made from the involute curve have at least one pair of teeth in contact with each other. Design Procedure 14.1 provides a procedure for constructing involute curves.

Table 14.2: Formulas for addendum, dedendum and clearance ($\phi = 20°$; full-depth involute).

Parameter	Symbol	Coarse pitch ($p_d < 20$ in.$^{-1}$)	Fine pitch (p_d 20 in.$^{-1}$)	Metric module system
Addendum	a	$1/p_d$	$1/p_d$	$1.00\,m$
Dedendum	b	$1.25/p_d$	$1.200/p_d + 0.002$	$1.25\,m$
Clearance	c	$0.25/p_d$	$0.200/p_d + 0.002$	$0.25\,m$

Design Procedure 14.1: Constructing the Involute Curve

1. Divide the base circle into a number of equal distances, thus constructing $A_o, A_1, A_2, \ldots$
2. Beginning at A_1, construct the straight line A_1B_1, perpendicular with $0A_1$, and likewise beginning at A_2 and A_3.
3. Along A_1B_1, lay off the distance A_1A_o, thus establishing C_1. Along A_2B_2, lay off twice A_1A_o, thus establishing C_2, etc.
4. Establish the involute curve by using points $A_o, C_1, C_2, C_3, \ldots$
5. The involute curve extends from the outside circle to the clearance circle; a fillet transition is placed between the dedendum and clearance circles.

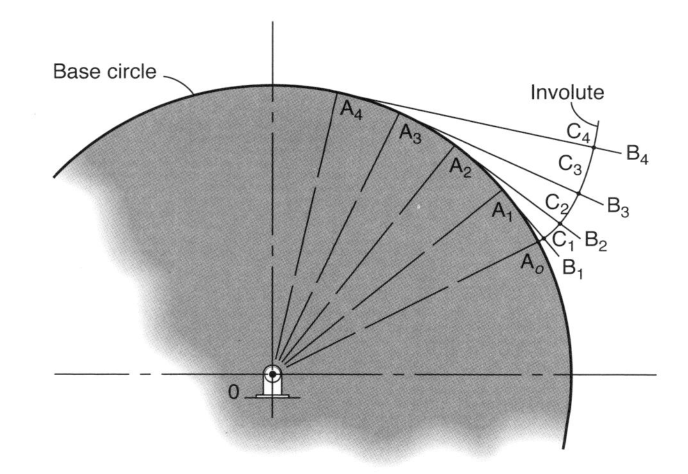

Figure 14.10: Construction of involute curve.

Example 14.2: Gear Geometry and Pressure Angle

Given: A gear with 40 teeth meshes with a 16-tooth pinion. The gear has a module of 2 mm and a pressure angle of 20°.

Find:

(*a*) Determine the circular pitch, the center distance, and the base diameter for the pinion and the gear.

(*b*) If the center distance is increased by 1 mm, find the pitch diameters for the gear and pinion and the pressure angle.

Solution:

(*a*) From Eq. (14.8), the circular pitch is $p_c = \pi m = \pi(2) = 6.283$ mm. From Eq. (14.7), the pitch diameters of the gear and pinion are

$$d_g = mN_g = (2)(40) = 80 \text{ mm},$$

$$d_p = mN_p = (2)(16) = 32 \text{ mm}.$$

From Eq. (14.1), the center distance is

$$c_d = \frac{d_p + d_g}{2} = \frac{32 + 80}{2} = 56 \text{ mm}.$$

From Eq. (14.16), the radii of the base circles for the pinion and the gear are

$$r_{bp} = \frac{d_p}{2}\cos\phi = \frac{32}{2}\cos 20° = 15.04 \text{ mm},$$

$$r_{bg} = \frac{d_g}{2}\cos\phi = \frac{80}{2}\cos 20° = 37.59 \text{ mm}.$$

(*b*) If the center distance is increased by 1 mm, $c_d = 57$ mm, and this value corresponds to a 1.8% increase in the center distance. But

$$c_d = \frac{d_p + d_g}{2},$$

so that

$$d_p + d_g = 2(57) = 114 \text{ mm}. \qquad (a)$$

From Eq. (14.5),

$$\frac{N_p}{d_p} = \frac{N_g}{d_g}.$$

Therefore,

$$\frac{16}{d_p} = \frac{40}{d_g}. \qquad (b)$$

Thus, there are two equations and two unknowns. From Eq. (*a*),

$$d_p = 114 - d_g. \qquad (c)$$

Substituting Eq. (*c*) into Eq. (*b*) gives

$$\frac{16}{114 - d_g} = \frac{40}{d_g},$$

which is solved as $d_g = 81.43$ mm. Substituting this value into Eq. (*a*) results in $d_p = 32.57$ mm. From Eq. (14.5), the diametral pitch is

$$p_d = \frac{N_p}{d_p} = \frac{16}{32.57} = 0.4912 \text{ mm}^{-1}.$$

From Eq. (14.6) the circular pitch is

$$p_c = \frac{\pi}{p_d} = \frac{\pi}{0.4912} = 6.396 \text{ mm}.$$

Changing the center distance does not affect the base circle of the pinion or the gear. Thus, from Eq. (14.16),

$$\phi = \cos^{-1}\left(\frac{r_{bp}}{r_p}\right) = \cos^{-1}\left(\frac{15.04}{32.57/2}\right) = 22.55^\circ.$$

Thus, a 1.8% increase in center distance results in a 2.55° increase in pressure angle.

14.4 Gear Ratio

For meshing gears, the rotational motion and the power must be transmitted from the driving gear to the driven gear with a smooth and uniform positive motion and with minor frictional power loss. The **fundamental law of gearing** states that the common normal to the tooth profile at the point of contact must always pass through a fixed point, called the **pitch point**, in order to maintain a constant velocity ratio of the two meshing gear teeth. The pitch point is shown in Fig. 14.9. Since the velocity at this point must be the same for both the gear and the pinion,

$$r_p \omega_p = r_g \omega_g. \tag{14.18}$$

Making use of Eq. (14.3) gives the gear ratio as

$$g_r = \frac{\omega_p}{\omega_g} = \frac{d_g}{d_p} = \frac{r_g}{r_p} = \frac{N_g}{N_p}. \tag{14.19}$$

Example 14.3: Speed on Line of Action

Given: When two involute gears work properly, the contact point moves along the line of action. Let N_p = 20, N_g = 38, and ϕ = 20°. The module is 2 mm and the pinion speed is 3600 rpm.

Find: Using Fig. 14.9 as a reference, obtain the gear velocity at the initial contact point and at the pitch point.

Solution: From Eq. (14.7), $d_p = mN_p = (2)(20) = 40$ mm. For the initial contact point, it can be seen from Fig. 14.9a that

$$\begin{aligned} \omega_p r_{bp} &= \omega_p \left(\frac{d_p}{2}\right) \cos 20^\circ \\ &= \left(\frac{3600}{60}\right) 2\pi \left(\frac{40}{2}\right) \cos 20^\circ \\ &= 7.085 \text{ m/s}. \end{aligned}$$

At the pitch point,

$$\omega_p r_p = \left(\frac{3600}{60}\right) 2\pi \left(\frac{40}{2}\right) = 7.540 \text{ m/s}.$$

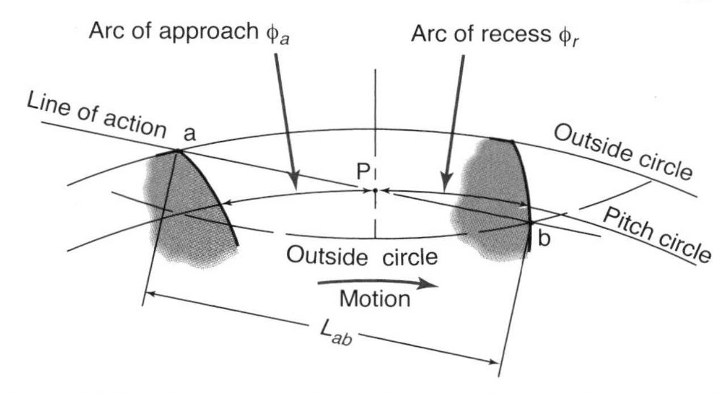

Figure 14.11: Illustration of arc of approach, arc of recess, and length of action.

14.5 Contact Ratio and Gear Velocity

To ensure smooth, continuous tooth action, as one pair of teeth ceases contact, a succeeding pair of teeth must already have come into contact. The **contact ratio** defines the average number of teeth in contact at any time, and is the ratio of the length of action to the base pitch. It is also the number of angular pitches that a tooth rotates from the beginning to the end of contact. Recall from Fig. 14.9 that the line of action is a line through the pitch point that is tangent to the base circles of the pinion and gear. The **length of action** is the length of the line of action where tooth contact occurs. Figures 14.9 and 14.11 illustrate the important parameters in defining the length of action and contact ratio. The length of action is defined by the region where the outer circles intersect the line of action, which can be calculated from the gear and pinion geometry as

$$L_{ab} = \sqrt{r_{op}^2 - r_{bp}^2} + \sqrt{r_{og}^2 - r_{bg}^2} - c_d \sin\phi. \tag{14.20}$$

The contact ratio is defined by

$$C_r = \frac{L_{ab}}{p_b}. \tag{14.21}$$

Applying Eq. (14.17), the contact ratio can be expressed as

$$\begin{aligned} C_r &= \frac{L_{ab}}{p_c \cos\phi} \\ &= \frac{1}{p_c \cos\phi}\left[\sqrt{r_{op}^2 - r_{bp}^2} + \sqrt{r_{og}^2 - r_{bg}^2}\right] - \frac{c_d \tan\phi}{p_c}. \end{aligned} \tag{14.22}$$

If the contact ratio is 1.0, then there will be a new pair of teeth coming in contact just as the old pair is leaving contact, and there will be exactly one pair of teeth in contact at all times. This may seem acceptable, but leads to poor performance, since loads will be applied at tooth tips, and there will be additional vibration, noise, and backlash. It is good practice to maintain a contact ratio of 1.2 or greater. Most spur gearsets have contact ratios between 1.4 and 2.0.

A contact ratio between 1 and 2 means that part of the time two pairs of gear teeth are in contact and during the remaining time one pair is in contact. A contact ratio between 2 and 3 means that part of the time three pairs of teeth are in contact. This is a rare situation with spur gear sets, but is typical for helical and worm gears. The contact ratio is important for performance and analysis; for example, if two pairs of teeth share the transmitted load, the stresses are lower.

The length of action of *externally* meshing gears is obtained from Eq. (14.20). For *internally* meshing gears, the following modifications are needed:

$$r_{og} = r_g - a; \qquad c_d = r_2 - r_1 = r_g - r_p;$$

$$L_{ab} = \sqrt{r_{op}^2 - r_{bp}^2} - \sqrt{r_{og}^2 - r_{bg}^2} + c_d \sin\phi. \qquad (14.23)$$

The kinematics of gears is such that three velocities for the pinion and gear are important.

1. *Velocity of pinion and gear:*

$$u_{bp} = \omega_p r_{bp} = \omega_p r_p \cos\phi \qquad (14.24)$$

$$u_{bg} = \omega_g r_{bg} = \omega_g r_g \cos\phi \qquad (14.25)$$

For no slip between meshing gears,

$$\omega_p r_p = \omega_g r_g \qquad (14.26)$$

2. *Sliding velocities in the contact*

The sliding speed is the difference in speed perpendicular to the line of action for the two gears. At the pitch point, the sliding speed is zero. The sliding speed increases on both sides of the pitch point and reaches a maximum at the far ends of the line of action (points a and b in Fig. 14.11). The sliding velocity at points a and b can be obtained from

$$u_{sa} = (L_{ab} + L_2)\omega_g - L_1\omega_p \qquad (14.27)$$

$$u_{sb} = (L_{ab} + L_1)\omega_p - L_2\omega_g \qquad (14.28)$$

where L_1 and L_2 are defined in Fig. 14.9b on the line of action. The sliding velocity of the meshing gear set is the larger of the value u_{sa} or u_{sb}. Values obtained for u_{bp} and u_{bg} in Eqs. (14.24) and (14.25) are typically two to three times larger than the values obtained for u_{sa} and u_{sb} in Eqs. (14.27) and (14.28).

3. *Lubrication velocities*

For the pinion and gear, the lubrication velocities are

$$u_p = \omega_p r_p \sin\phi \qquad (14.29)$$

$$u_g = \omega_g r_g \sin\phi \qquad (14.30)$$

Example 14.4: Gear Stubbing

Given: In a gearbox, the pinion has 20 teeth and the gear has 67 teeth. The module is 2.5 mm. The manufacturer needs to reduce the size of the gearbox but does not want to change the size of the gear teeth. The original gears are standard with an addendum of $1/p_d$ and a pressure angle of 20°. The approach to be used is to decrease the size of the gearbox by *stubbing* (reducing the addendum) on one of the wheels.

Find: Which wheel should be stubbed and what percentage smaller would the gearbox be due to the stubbing? Use a minimum contact ratio of 1.1 for the stubbed gears.

Solution: From Eq. (14.8), the circular pitch is

$$p_c = \pi m = \pi(2.5) = 7.854 \text{ mm}.$$

From Eqs. (14.1) and (14.7), the center distance is

$$\begin{aligned} c_d &= \frac{d_p + d_g}{2} \\ &= \frac{mN_p + mN_g}{2} \\ &= \frac{(2.5)(20) + (2.5)(67)}{2} \\ &= 108.75 \text{ mm}. \end{aligned}$$

From Eq. (14.7), the pitch circle radii for the pinion and gear are

$$r_p = \frac{d_p}{2} = \frac{mN_p}{2} = \frac{(2.5)(20)}{2} = 15 \text{ mm},$$

$$r_g = \frac{d_g}{2} = \frac{mN_g}{2} = \frac{(2.5)(67)}{2} = 83.75 \text{ mm}.$$

From Eq. (14.16), the base circle radii for the pinion and gear are

$$r_{bp} = r_p \cos\phi = (15)\cos 20^\circ = 14.10 \text{ mm},$$

$$r_{bg} = r_g \cos\phi = (83.75)\cos 20^\circ = 78.70 \text{ mm}.$$

The outside radii for the pinion and gear are

$$r_{op} = r_p + a,$$

$$r_{og} = r_g + a.$$

(*a*) *Standard gears:* For standard gears, the addendum is $a = 1.00m = 2.5$ mm (see Table 14.2). Therefore,

$$r_{op} = r_p + a = 15 + 2.5 = 17.5 \text{ mm}, \qquad (a)$$

$$r_{og} = r_g + a = 83.75 + 2.5 = 86.25 \text{ mm}, \qquad (b)$$

and

$$r_{op} + r_{og} = 103.75 \text{ mm}.$$

(*b*) *Pinion stubbed:* If the pinion is stubbed, the outer radius of the pinion is

$$r_{op} = r_p + xa, \qquad (c)$$

where x is a constant between 0 and 1. Note that if x = 1 a standard pinion exists and if x = 0 the pinion is stubbed to the pitch circle. Either the pinion or gear is stubbed but not both. Also, it was given that the smallest contact ratio should be at least 1.1. Thus, using Eqs. (14.22), (*b*), and (*c*),

$$\begin{aligned} 1.1 &= \frac{\sqrt{(15 + 2.5x)^2 - 14.10^2} + \sqrt{86.25^2 - 78.70^2}}{2.5\pi\cos 20^\circ} \\ &\quad - \frac{108.75\tan 20^\circ}{2.5\pi} \end{aligned}$$

Solving yields x = 0.9198. The outside radius for the pinion is

$$r_{op} = 15 + 2.5(0.9198) = 17.30 \text{ mm}.$$

The sum of the outside radii is

$$r_{op} + r_{og} = 17.30 + 86.25 = 103.55 \text{ mm}.$$

Thus, the percent reduction of the gearbox size if the pinion is stubbed is

$$\frac{(108.75 - 103.55)}{108.75} \times 100\% = 4.78\%.$$

(*c*) *Gear stubbed:* If the gear is stubbed, the addendum radius of the gear is

$$r_{og} = r_g + ya, \qquad (d)$$

where y is a constant between 0 and 1. Using Eqs. (14.22), (*d*), and (*a*) while setting the contact ratio to 1.1 gives

$$1.1 = \frac{\sqrt{17.5^2 - 14.1^2} + \sqrt{(83.75 + 2.5y)^2 - 78.70^2}}{2.5\pi \cos 20°} - \frac{108.75 \tan 20°}{2.5\pi}.$$

Numerically solving for y gives $y = 0.9443$. The outside radius for the gear is

$$r_{og} = 83.75 + 2.5(0.9443) = 86.11 \text{ mm}.$$

The sum of the outside radii is

$$r_{op} + r_{og} = 17.5 + 86.11 = 103.6 \text{ mm}.$$

Thus, the percent reduction of the gearbox size if the gear is stubbed is

$$\frac{(108.75 - 103.6)}{108.75} \times 100\% = 4.73\%.$$

Therefore, the pinion should be stubbed because it will result in a larger (4.78%) reduction of the gearbox size. Note that the dimension change is modest; stubbing does not gain a significantly more compact design, although the change would be larger with coarser teeth.

14.6 Tooth Thickness and Backlash

The **circular tooth thickness**, t_h, is the arc length, measured on the pitch circle, from one side of the tooth to the other. Theoretically, the circular tooth thickness is one-half of the circular pitch, or

$$t_h = \frac{p_c}{2} = \frac{\pi}{2p_d}. \qquad (14.31)$$

For perfect meshing of pinion and gear the tooth thickness measured on the pitch circle must be exactly one-half of the circular pitch. Because of unavoidable machining inaccuracies it is necessary to cut the teeth slightly thinner to provide some clearance so that the gears will not bind but will mesh together smoothly. This clearance, measured on the pitch circle, is called **backlash**. Figure 14.12 illustrates backlash in meshing gears.

To obtain the amount of backlash desired, it is necessary to decrease the tooth thickness given in Eq. (14.31). This decrease must be slightly greater than the desired backlash owing to errors in manufacturing and assembly. Since the amount of the decrease in tooth thickness depends on machining accuracy, the allowance for a specified backlash varies according to manufacturing conditions. Table 14.3 gives some suggestions on backlash, but these may need to be higher in some applications, such as when thermal expansion is a significant concern.

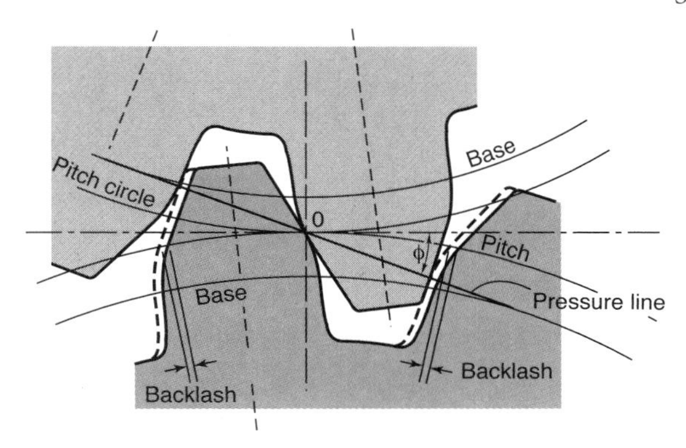

Figure 14.12: Illustration of backlash in gears.

Table 14.3: Recommended minimum backlash for coarse-pitched gears.

Diametral pitch p_d, in.$^{-1}$	Center distance, c_d, in. 2	4	8	16	32
18	0.005	0.006	—	—	—
12	0.006	0.007	0.009	—	—
8	0.007	0.008	0.010	0.014	—
5	—	0.010	0.012	0.016	—
3	—	0.014	0.016	0.020	0.028
2	—	—	0.021	0.025	0.033
1.25	—	—	—	0.034	0.042

To determine backlash in gear trains (see Section 14.7), it is necessary to sum the backlash of each mated gear pair. However, to obtain the total backlash for a series of meshed gears, it is necessary to take into account the gear ratio of each pair of mating gears relative to a chosen reference shaft in the gear train. This is beyond the scope of this text, but more details can be found in Mechalec [1962] or Townsend [1992].

In addition to reducing tooth thickness, backlash can occur if the operating center distance is greater than that obtained from Eqs. (14.31) and (14.1). Table 14.3 gives recommended minimum backlash for coarse-pitch gears. Thus, when the backlash, b_l, from Table 14.3 and the theoretical circular tooth thickness, t_h, from Eq. (14.31) are known, the actual circular tooth thickness is

$$t_{ha} = t_h - b_l. \qquad (14.32)$$

Example 14.5: Tooth Thickness, Backlash, and Contact Ratio

Given: The two situations given in Example 14.2.

Find: Determine the tooth thickness and contact ratio, and estimate the recommended backlash.

Solution:

(*a*) For the original center distance, the tooth thickness is, from Eq. (14.31),

$$t_h = p_c/2 = 6.238/2 = 3.119 \text{ mm}.$$

In this case, the diametral pitch is 0.4912 mm^{-1}, which is nearly 1/2 in^{-1}, and the center distance is 56 mm $\approx$ 2.2 in. From Table 14.3, the interpolated recommended backlash is around $b_l = 0.006$ in., or 0.152 mm. Thus, the actual tooth thickness, from Eq. (14.32), is

$$t_{ha} = t_h - b_l = 3.119 - 0.152 = 2.967 \text{ mm}.$$

From Table 14.2 the addendum for both the pinion and gear is

$$a = 1.00m = 1.00(2 \text{ mm}) = 2 \text{ mm}.$$

The outside radii for the pinion and gear are

$$r_{op} = r_p + a = 16 + 2 = 18 \text{ in}.$$

$$r_{og} = r_g + a = 40 + 2 = 42 \text{ in}.$$

By making use of the above and Eq. (14.22), the contact ratio can be expressed as

$$C_r = \frac{\sqrt{18^2 - 15.04^2} + \sqrt{42^2 - 37.59^2}}{6.238 \cos 20^\circ} - \frac{56 \tan 20^\circ}{6.238},$$

or $C_r = 1.61$. This implies that part of the time two pairs of teeth are in contact and during the remaining time one pair is in contact.

(*b*) When the center distance is increased by 1 mm, the tooth thickness is obtained from Eq. (14.31) as

$$t_h = \frac{p_c}{2} = \frac{6.3936}{2} = 3.1968.$$

The actual tooth thickness is

$$t_{ha} = t_h - b_l = 3.1968 - 0.152 = 3.045 \text{ mm}.$$

The base and outside radii for the pinion and gear are

$$r_{bp} = \frac{d_p}{2}\cos\phi = \frac{34.57}{2}\cos 22.55^\circ = 15.96 \text{ mm}$$

$$r_{op} = r_p + a = 17.285 + 2 = 19.285 \text{ mm}$$

$$r_{bg} = \frac{d_g}{2}\cos\phi = \frac{85.43}{2}\cos 22.55^\circ = 39.45 \text{ mm}$$

$$r_{og} = r_g + a = 42.715 + 2 = 44.715 \text{ mm}$$

$$C_r = \frac{\sqrt{19.285^2 - 15.96^2} + \sqrt{44.715^2 - 39.45^2}}{6.3936 \cos 22.55^\circ} - \frac{57 \tan 22.55^\circ}{6.3936} = 1.69.$$

Thus, increasing the center distance increases the pressure angle. The change in the contact angle is very small.

14.7 Gear Trains

Gear trains consist of multiple gears. They are used to obtain a desired angular velocity of an output shaft and to transfer power from an input shaft. The angular velocity ratio between input and output gears is constant. Gear trains are needed when the gear ratio is large (around 6 or so for spur gears) or when a single input shaft provides power for multiple output shafts.

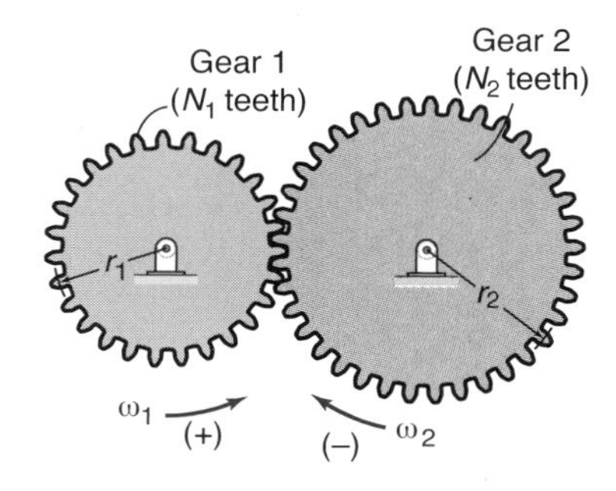

Figure 14.13: Externally meshing spur gears.

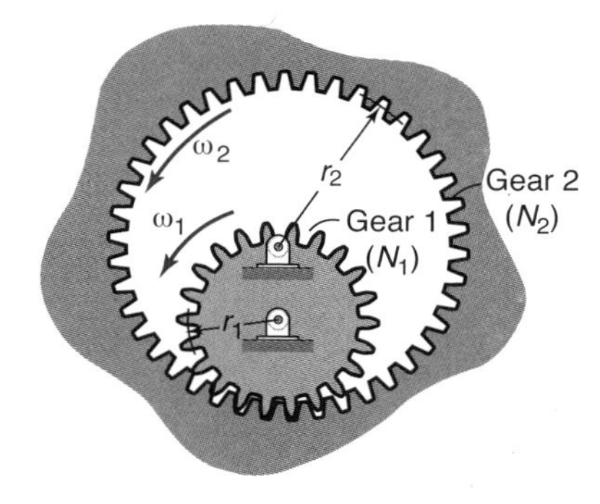

Figure 14.14: Internally meshing spur gears.

14.7.1 Single Gear Mesh

Figure 14.13 shows *externally* meshing spur gears with fixed centers. Using Eq. (14.19) and the sign convention that positive means the same direction and negative means the opposite direction of rotation gives the angular velocity ratio for the situation in Fig. 14.13 as

$$Z_{21} = \frac{\omega_2}{\omega_1} = -\frac{N_1}{N_2}. \tag{14.33}$$

Externally meshing gear teeth are extremely common, and every externally meshing gear in a train will cause the direction of rotation to change. The center distance for externally meshing spur gears is

$$c_d = r_1 + r_2. \tag{14.34}$$

Figure 14.14 shows *internally* meshing spur gears. Note that both gears are moving in the same direction and thus the angular velocity ratio is positive and is given by

$$Z_{21} = \frac{\omega_2}{\omega_1} = \frac{N_1}{N_2}. \tag{14.35}$$

14.7.2 Simple Spur Gear Trains

A simple gear train, as shown in Fig. 14.15, has only one gear mounted on each shaft. If the ith gear (i = 1,2,3,...,n) has N_i teeth and rotates with angular velocity ω_i (with a counterclockwise velocity taken as positive), the angular velocity ratio or **train value**, Z_{ji}, of the jth gear relative to the ith gear is

Figure 14.15: Simple gear train.

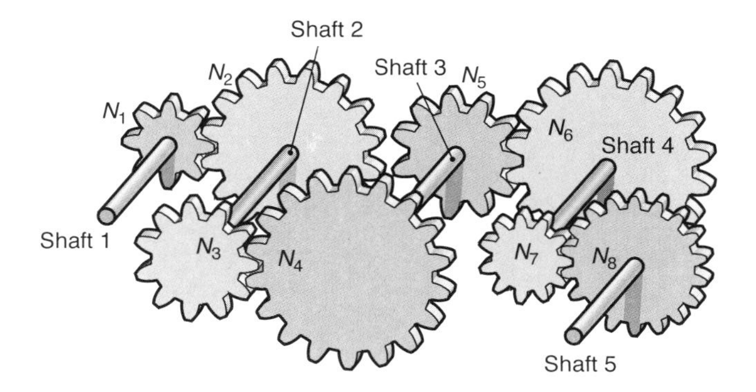

Figure 14.16: Compound gear train.

$$Z_{ji} = \frac{\omega_j}{\omega_i} = \pm\frac{N_i}{N_j}. \tag{14.36}$$

A positive sign is used in the above equation when $j - i$ is even, and a negative sign is used when $j - i$ is odd. The reason for this is that in Fig. 14.15 all the gears are external, implying that the directions of rotation of adjacent shafts are opposite.

14.7.3 Compound Spur Gear Trains

In a compound gear train (Fig. 14.16) at least one shaft carries two or more gears. In this type of train all gears are keyed (see Section 11.6) or otherwise attached to their respective shafts, so that the angular velocities of all gears are equal to that of the shaft on which they are mounted. The angular velocities of adjacent shafts are governed by the gear ratio of the associated mesh.

Let ω_i represent the angular velocity of the ith shaft. For example, ω_3 is the angular velocity of the shaft connecting gears 4 and 5. Using Fig. 14.16 while going from left to right, with counterclockwise taken as positive and clockwise negative gives

$$\frac{\omega_2}{\omega_1} = -\frac{N_1}{N_2} \qquad \frac{\omega_3}{\omega_2} = -\frac{N_3}{N_4} \qquad \frac{\omega_4}{\omega_3} = -\frac{N_5}{N_6} \qquad \frac{\omega_5}{\omega_4} = -\frac{N_7}{N_8}$$

Hence, the angular velocity ratio Z_{51} of shaft 5 relative to shaft 1 for the train in Fig. 14.16 is given by

$$Z_{51} = \frac{\omega_5}{\omega_1}, \tag{14.37}$$

where

$$\begin{aligned}\frac{\omega_5}{\omega_1} &= \left(\frac{\omega_5}{\omega_4}\right)\left(\frac{\omega_4}{\omega_3}\right)\left(\frac{\omega_3}{\omega_2}\right)\left(\frac{\omega_2}{\omega_1}\right)\\ &= \left(-\frac{N_7}{N_8}\right)\left(-\frac{N_5}{N_6}\right)\left(-\frac{N_3}{N_4}\right)\left(-\frac{N_1}{N_2}\right).\end{aligned}$$

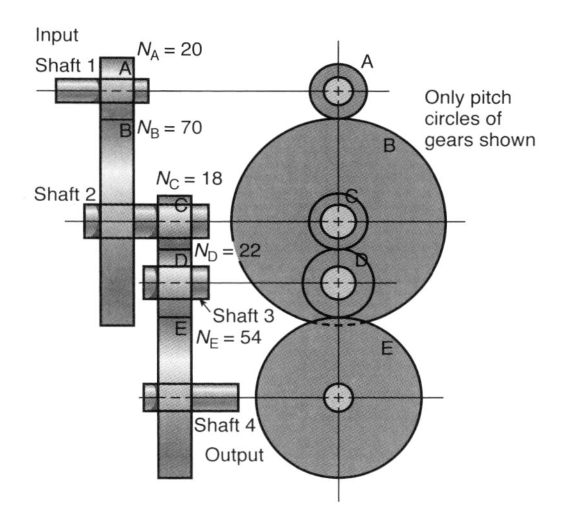

Figure 14.17: Gear train used in Example 14.6.

For the compound gear train shown in Fig. 14.16,

$$Z_{51} = \frac{N_1 N_3 N_5 N_7}{N_2 N_4 N_6 N_8}. \tag{14.38}$$

For compound trains, this systematic inspection of gear velocities and directions is essential.

Example 14.6: Gear Train

Given: The gear train shown in Fig. 14.17.

Find: Determine the angular velocity ratio for the gear train. If shaft 1 rotates clockwise at 1750 rpm, determine the speed and direction of shaft 4.

Solution: From Fig. 14.17, using an angular velocity of $\omega_1 = -1750$ rpm (negative because it is clockwise), it can be seen that shaft 2 will rotate counterclockwise, as can be seen from Eq. (14.36). Therefore, shaft 3 will rotate clockwise, and shaft 4 counterclockwise. Therefore,

$$\begin{aligned}Z_{14} &= \left(-\frac{N_B}{N_A}\right)\left(-\frac{N_D}{N_C}\right)\left(-\frac{N_E}{N_D}\right)\\ &= -\frac{N_B N_E}{N_A N_C} = -\frac{(70)(54)}{(20)(18)} = -10.5.\end{aligned}$$

Gear D is an *idler*, since it has no effect on the magnitude of the angular velocity ratio, but it does cause a direction reversal. Therefore, if gear D were not present, gear E would rotate in the opposite direction, but the angular velocity ratio would have the same numerical value but opposite sign. The output speed is

$$Z_{14} = \frac{\omega_A}{\omega_E} = -\frac{N_B N_E}{N_A N_C} = -10.5.$$

Therefore,

$$\omega_E = \frac{\omega_A}{Z_{14}} = \frac{1750}{-10.5} = -166.7 \text{ rpm}.$$

Recall that the negative sign implies that the direction is counterclockwise, or in the opposite direction as the input shaft.

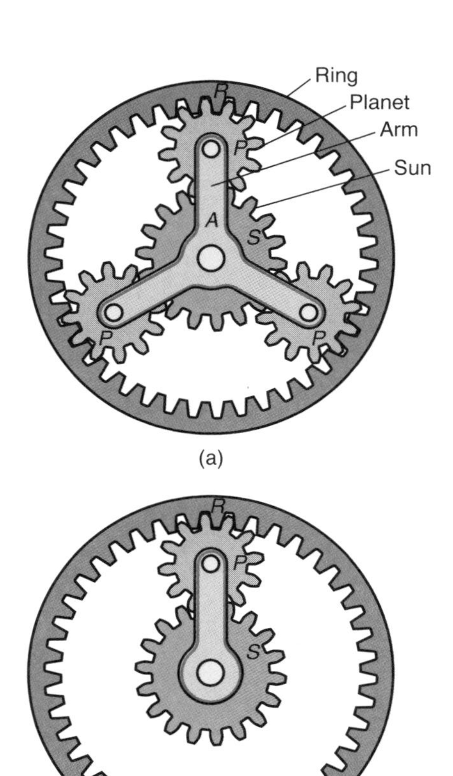

Figure 14.18: Illustration of planetary gear train. (a) With three planets; (b) with one planet (for analysis only).

14.7.4 Planetary Gear Trains

Planetary gear trains are very common, have significant flexibility in their application, and give high gear ratios in compact designs. A planetary or *epicyclic* gear train consists of a **sun gear**, a **ring gear**, a **planet carrier** or **arm**, and one or more **planet gears**. The planetary gear train shown in Fig. 14.18a depicts an arm with three planets, which is a common arrangement. However, the kinematics of gear trains can often be conceptually simplified by recognizing that the system of Fig. 14.18b gives the same angular velocities as in Fig. 14.18a, and can be used for analysis of gear trains. It should be noted that the additional planets are desirable to increase stability, and reduce gear deformations and tooth stresses.

Planetary gear trains are two degree-of-freedom trains; that is, two inputs are needed to obtain a predictable output. For example, the sun can be the input (driven by a motor or other power source), the ring can be bolted to a machine frame (so that its velocity is constrained as zero) and the planet arm can be the output (driving other devices). In this case, the input and output have parallel axes, and this arrangement is popular for power transmission devices. Another mode of operation can be achieved by driving the sun gear, fixing the ring, and using a planet gear as an output; this results in a rotating shaft that itself articulates about the input axis. This application of planetary gear trains is very common in mixers and agitators of all kinds.

Another use of a planetary gear train is in an automotive differential transmission. In this situation, the sun is an input (from the drive shaft), and the ring and armature are outputs. Normally, these outputs are frictionally coupled and are equal, but they are not rigidly constrained. Thus, one wheel can rotate independently of the other, a beneficial arrangement for slippery road surfaces, smooth ride, and low wheel wear.

Planetary gear trains can be analyzed through the application of the following equations:

$$\frac{\omega_{\text{ring}} - \omega_{\text{arm}}}{\omega_{\text{sun}} - \omega_{\text{arm}}} = -\frac{N_{\text{sun}}}{N_{\text{ring}}}, \tag{14.39}$$

$$\frac{\omega_{\text{planet}} - \omega_{\text{arm}}}{\omega_{\text{sun}} - \omega_{\text{arm}}} = -\frac{N_{\text{sun}}}{N_{\text{planet}}}, \tag{14.40}$$

$$N_{\text{ring}} = N_{\text{sun}} + 2N_{\text{planet}}. \tag{14.41}$$

The angular velocity ratio is more difficult to define, since any of these gears can be the first or last gear in the planetary gear train. The angular velocity ratio can be written as:

$$Z_p = \frac{\omega_L - \omega_A}{\omega_F - \omega_A}, \tag{14.42}$$

where

Z_p = angular velocity ratio for a planetary gear train
ω_L = angular velocity for the last gear in the planetary gear train (output gear)
ω_A = angular velocity of the arm
ω_F = angular velocity for the first gear in the planetary gear train (driven gear)

Note that if the arm is the output or input gear, then the angular velocity ratio can be calculated from Eq. (14.37) or (14.38).

Example 14.7: Planetary Gear Train

Given: A planetary gear train is used as part of a transmission for a forklift truck. The sun gear is driven by the engine at 3000 rpm, the ring is bolted to the machine frame, and the armature is connected to the track drive system. The sun has 16 teeth, and there are three planets, each with 34 teeth.

Find: Determine the angular velocities of the armature and the planets, and the angular velocity ratio for the gear train.

Solution: The number of teeth in the ring is obtained from Eq. (14.41):

$$N_{\text{ring}} = N_{\text{sun}} + 2N_{\text{planet}} = 16 + 2(34) = 84 \text{ teeth}.$$

Therefore, the angular velocity of the armature can be calculated from Eq. (14.39) as:

$$\frac{\omega_{\text{ring}} - \omega_{\text{arm}}}{\omega_{\text{sun}} - \omega_{\text{arm}}} = -\frac{N_{\text{sun}}}{N_{\text{ring}}},$$

or

$$\frac{0 - \omega_{\text{arm}}}{314 - \omega_{\text{arm}}} = -\frac{16}{84}.$$

which is solved as $\omega_{\text{arm}} = 50.24$ rad/s $= 480$ rpm. Since this is a positive value, the armature rotates in the same direction as the sun gear. The angular velocity of the planet can now be calculated from Eq. (14.40) as

$$\frac{\omega_{\text{planet}} - \omega_{\text{arm}}}{\omega_{\text{sun}} - \omega_{\text{arm}}} = -\frac{N_{\text{sun}}}{N_{\text{planet}}},$$

or

$$\frac{\omega_{\text{planet}} - 50.24}{314 - 50.24} = -\frac{16}{34}.$$

or $\omega_{\text{planet}} = -73.88$ rad/s $= -705.5$ rpm. The negative sign indicates that the planet moves in the opposite direction as the sun. The angular velocity ratio is calculated from Eq. (14.37) as

$$Z_p = \frac{\omega_{\text{arm}}}{\omega_{\text{sun}}} = \frac{480}{3000} = 0.16.$$

14.8 Gear Manufacture and Quality

14.8.1 Quality Index

Consider the shape of the spur gear shown in Fig. 14.1 with the tooth profile in Fig. 14.6. There are many manufacturing processes that can be used to produce gears, including casting, forging, powder metallurgy approaches, and machining. The manufacturing process used to create the gear has many far-reaching effects on long-term gear performance. Accurate production of the involute tooth profile and minimization of deviation from this profile are essential for the efficient operation of the gear. In addition, a gear with a smoother surface (see Section 8.2) will run quieter and with a larger film parameter, Λ, than a rougher gear. Finally, the strength of the gear teeth will be strongly influenced by the manufacturing process. Clearly, the manufacturing processes used to produce a gear have far-reaching design impact. This section is intended to introduce some of the concepts of gear manufacture; additional information can be found in Kalpakjian and Schmid [2010].

The AGMA has defined a quality index, Q_v, which is related to the pitch variation of a gearset. Table 14.4 lists typical quality index values for different applications and pitch-line velocities. The manufacturing process used to produce the gear has the largest effect on the quality index, but it should be recognized that there is a cost associated with higher gear quality, as shown in Fig. 14.19. From a design standpoint, selection of a quality index serves to constrain the manufacturing options for producing the gear. Also, note from Fig. 14.19 that each manufacturing approach can produce a range of quality indexes, depending on process parameters and control.

The **quality index**, Q_v, is defined by

$$Q_v = 0.5048 \ln N_i - 1.144 \ln P_d - 2.852 \ln(V_{PA} \times 10^4) + 14.71$$

in U.S. customary units, and

$$Q_v = 0.5048 \ln N_i + 1.144 \ln m - 2.852 \ln(V_{PA} \times 10^4) + 13.664 \qquad (14.43)$$

in SI units, where

N_i = number of pinion or gear teeth
P_d = normal diametral pitch, in.$^{-1}$
V_{PA} = absolute pitch variation, in. (or cm)
m = module, mm

It is important to use the proper units in Eq. (14.43) to obtain meaningful values for the quality index. The quality index is an empirical variable used to relate gear tolerance to performance. There is significant scatter in most gear performance evaluations, and the quality index should be recognized as an imperfect but valuable quantity. Indeed, the AGMA recommendation is to round the value of Q_v to the next lower integer, demonstrating the inexact nature of its definition. Calculations involving a quality index will therefore be approximate but conservative.

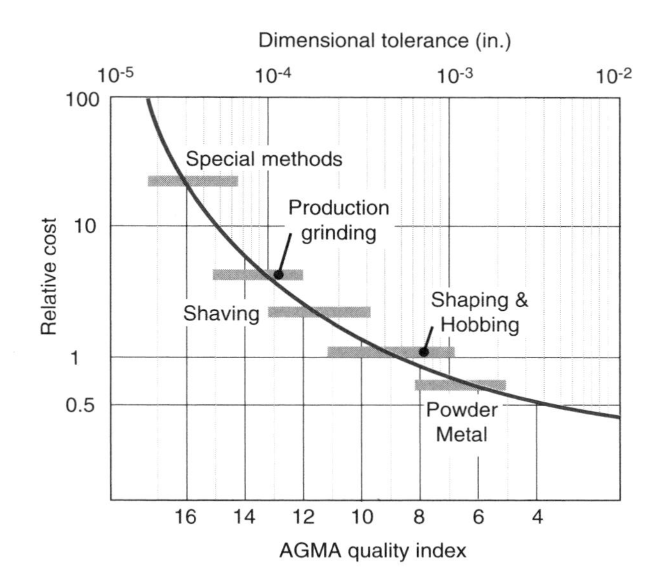

Figure 14.19: Gear cost as a function of gear quality. Note that the powder metallurgy approaches of pressing and sintering and metal injection molding can produce gears up to a quality index of 8 without additional machining. Recent research has suggested that similar quality levels can be achieved from cold forging as well.

Recent editions of the AGMA standard have allowed for an alternative measure of gear quality associated with the *transmission accuracy level*, A_v. The method of estimating A_v and Q_v are very similar, but A_v has low values for high quality gears and high values for lower quality gears. Based on gear design equations, A_v and Q_v are fundamentally related by

$$A_v = 17 - Q_v. \qquad (14.44)$$

Table 14.4: Quality index Q_v for various applications.

Application		Quality index, Q_v
Cement mixer drum driver		3–5
Cement kiln		5–6
Steel mill drives		5–6
Corn pickers		5–7
Punch press		5–7
Mining conveyor		5–7
Clothes washing machine		8–10
Printing press		9–11
Automotive transmission		10–11
Marine propulsion drive		10–12
Aircraft engine drive		10–13
Gyroscope		12–14
Pitch velocity ft/min	m/s	**Quality index, Q_v**
0–800	0–4	6–8
800–2000	4–10	8–10
2000–4000	10–20	10–12
> 4000	> 20	12–14

14.8.2 Gear Manufacture

Metal gears are manufactured through a variety of processes, including casting, forging, extrusion, drawing, thread rolling, powder metallurgy (usually compaction or metal injection molding followed by sintering), and sheet metal blanking. Non-metallic gears can be made by injection molding, compression molding, or casting. These processes yield quality factors that are low, typically below 5. However, for applications where low cost gears are needed that do not transmit much power, such as toys, small appliances, etc., these gears can have excellent performance at low cost. Some polymers can have excellent performance, although a larger addendum may be required to increase the contact ratio.

A better surface finish and more accurate profile can be achieved by machining a gear profile on a blank produced by these processes. There are two basic methods of producing a gear profile by machining: *form cutting* and *generating*.

Form Cutting

In **form cutting**, as shown in Fig. 14.20, the cutting tool has the shape of the space between the gear teeth. The gear-tooth shape is reproduced by cutting the gear blank around its periphery. The cutter travels axially along the length of the gear tooth at the appropriate depth to produce the gear tooth profile. After each tooth is cut, the cutter is withdrawn, the gear blank is rotated (indexed), and the cutter proceeds to cut another tooth. The process continues until all teeth are cut.

Form cutting is a relatively simple process and can be used for cutting gear teeth with various profiles; however, it is a relatively slow operation, and some types of machines require skilled labor. Consequently, it is suitable only for low-quantity production. Machines with semiautomatic features can be used economically for form cutting on a limited production basis.

Broaching can also be used to produce gear teeth and is particularly useful for internal teeth. The process is rapid and produces fine surface finish with high dimensional accuracy. However, because broaches are expensive and a separate broach is required for each gear size, this method is suitable mainly for high-quantity production.

(a)

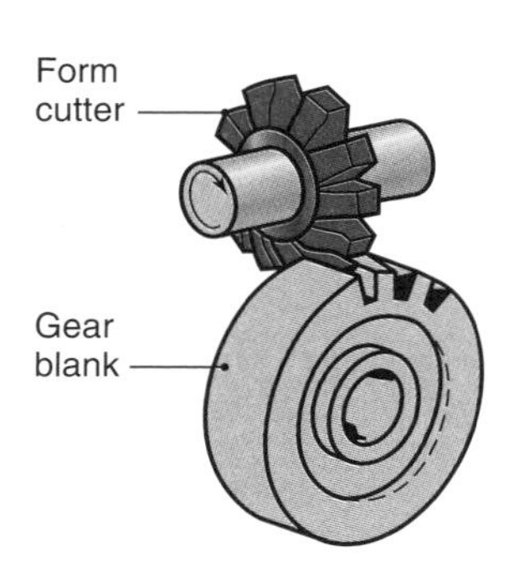

(b)

(c)

Figure 14.20: Form cutting of gear teeth. (a) A form cutter. Notice that the tooth profile is defined by the cutter profile; (b) schematic illustration of the form cutting process; (c) form cutting of teeth on a bevel gear. *Source:* (a) and (b) Kalpakjian and Schmid [2010]; (c) courtesy of Schafer Gear Works, Inc.

Generating

In gear **generating**, the tool may be either a pinion-shaped cutter (Fig. 14.21), a rack-shaped straight cutter, or a *hob*. The pinion-shaped cutter can be considered as one of the gears in a conjugate pair, and the other as the gear blank. The cutter has an axis parallel to that of the gear blank and rotates slowly with the blank at the same pitch-circle velocity, with an additional axial reciprocating motion. Cutting may take place either at the downstroke or upstroke of the machine. The process can be used for low-quantity as well as high-quantity production.

On a *rack shaper*, the generating tool is a segment of a rack which reciprocates parallel to the axis of the gear blank. Because it is not practical to have more than 6 to 12 teeth on a rack cutter, the cutter must be disengaged at suitable intervals and returned to the starting point, while the gear blank remains fixed.

A gear-cutting hob, shown in Fig. 14.22 is basically a worm or screw that has been made into a gear-generating tool by machining a series of longitudinal slots or gashes into it to produce the cutting teeth. When hobbing a spur gear, the angle between the hob and gear blank axes is 90° minus the lead angle at the hob threads. All motions in hobbing are rotary, and the hob and gear blank rotate continuously as in

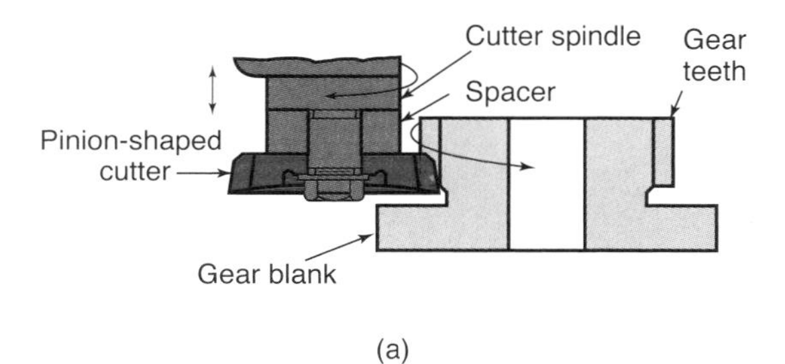

(a)

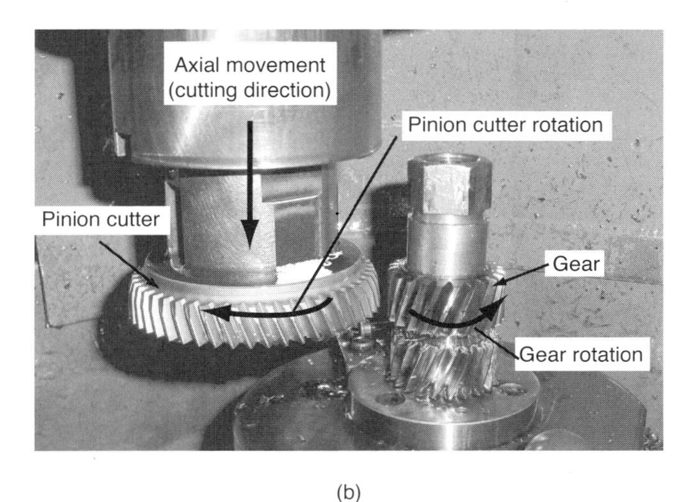

(b)

Figure 14.21: Production of gear teeth with a pinion-shaped cutter. (a) Schematic illustration of the process. *Source:* From Kalpakjian and Schmid [2010]; (b) photograph of the process with gear and cutter motions indicated. *Source:* Courtesy of Schafer Gear Works, Inc.

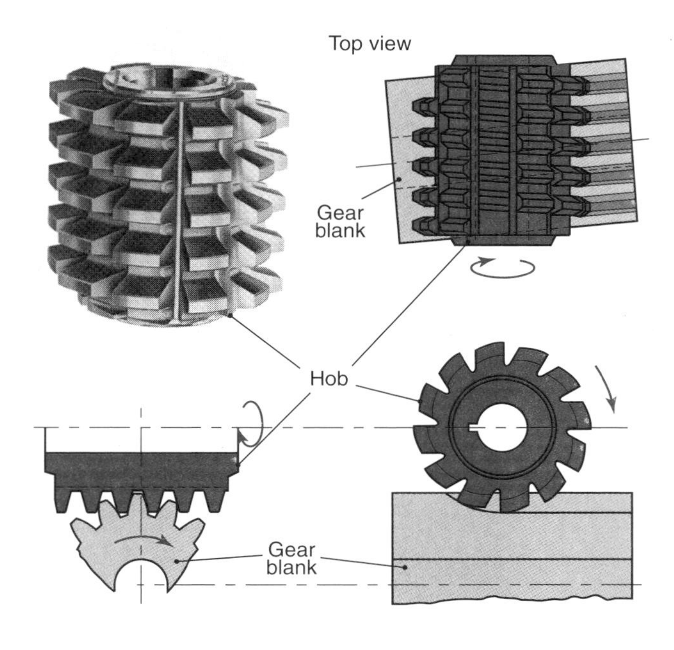

(a)

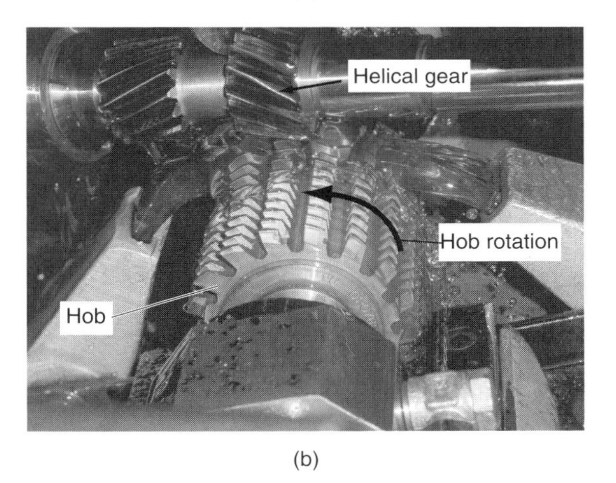

(b)

Figure 14.22: Production of gears through the hobbing process. (a) A hob, along with a schematic illustration of the process. *Source:* Kalpakjian and Schmid [2010]; (b) production of a worm gear through hobbing. *Source:* Courtesy of Schafer Gear Works, Inc.

two meshing gears until all teeth are cut.

Gear-Finishing Processes

As produced by any of the processes described above, the surface finish and dimensional accuracy of gear teeth may not be sufficiently accurate for demanding applications. Several finishing processes are available for improving the quality of gears produced by form cutting or generating.

In gear **shaving**, a cutter is made in the exact shape of the finished tooth profile, and removes small amounts of metal from the machined gear blank. Although the tools are expensive and special machines are necessary, shaving is rapid and is the most commonly used process for gear finishing. It produces gear teeth with good surface finish and tooth profile accuracy. Shaved gears may be subsequently heat treated and then ground for improved hardness, wear resistance, and tooth profile accuracy.

For highest dimensional accuracy, surface finish, tooth spacing, and form, gear teeth may be **ground**, **honed**, and **lapped**. Grinding is the most common operation, and is depicted in Fig. 14.23. There are several types of grinders for gears, with the single index form grinder being the most commonly available. In form grinding, the grinding wheel is in the identical shape of the tooth spacing.

In honing, the tool is a plastic gear impregnated with fine abrasive particles. The process requires ground gears as a workpiece, and is used to improve surface finish. To further improve the surface finish, ground gear teeth are lapped us-

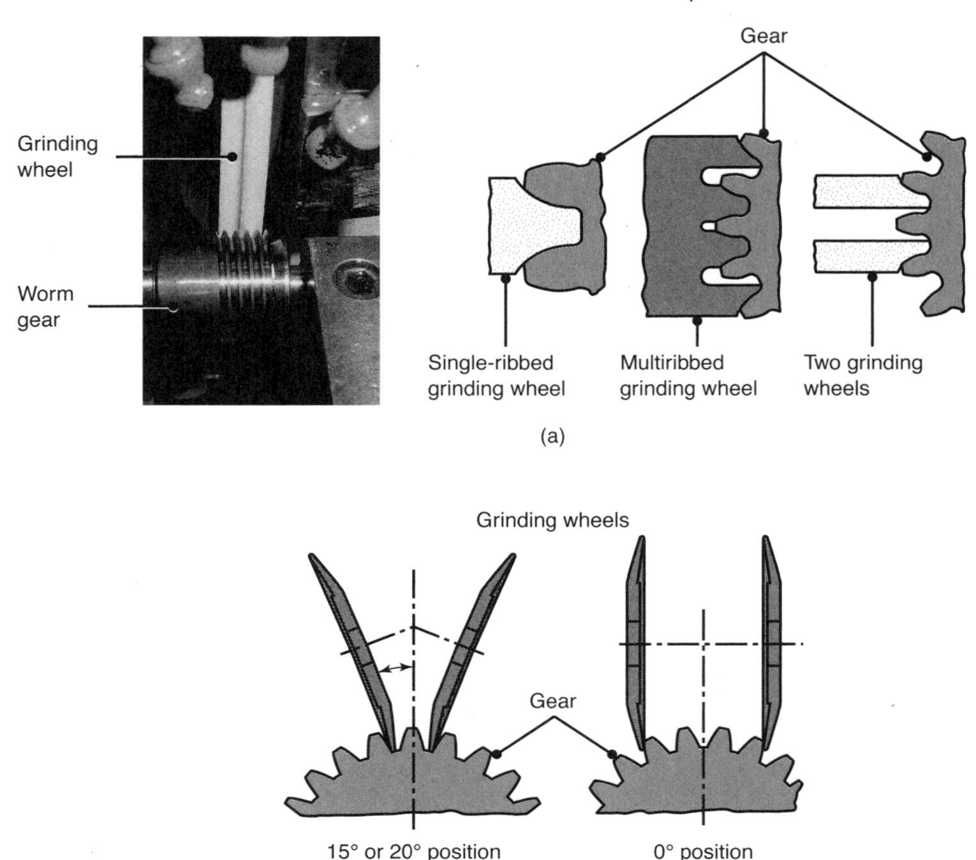

Figure 14.23: Finishing gears by grinding: (a) form grinding with shaped grinding wheels; (b) grinding by generating, using two wheels. *Source:* Kalpakjian and Schmid [2010].

ing abrasive compounds with either a gear-shaped lapping tool (made of cast iron or bronze) or a pair of mating gears that are run together. Although production rates are lower and costs are higher, these finishing operations are particularly suitable for producing hardened gears of very high quality, long life, and quiet operation.

It should be noted that **burnishing** and **shot peening** are common finishing processes for gear teeth. Burnishing, introduced in the 1960s, is basically a surface plastic deformation process using a special hardened gear-shaped burnishing die that subjects the tooth surfaces to a surface-rolling action (gear rolling). Shot peening causes surface deformation through the repeated impact of beads or shot. Cold working of tooth surfaces induces surface compressive residual stresses on the gear teeth, thus improving their fatigue life (see Section 7.8.5). However, neither burnishing nor shot peening improve gear-tooth accuracy or the quality index, Q_v.

14.9 Gear Materials

14.9.1 Allowable Stresses

The American Gear Manufacturers Association (AGMA) has promulgated a number of industry standards for the design of gears, and includes recommendations for the allowable loading of gear materials. The variety of materials used in gears is extensive. Recall that Ch. 3 focused on materials; this section addresses only a sampling of materials specific to gearing. More information about different materials, different manufacturing processes, and different coatings applied to gear teeth can be found in AGMA [2004].

Section 14.10 explains that a gear tooth acts like a cantilevered beam in resisting the force exerted on it by the mating tooth. The location of highest tensile bending stress is at the root of the tooth, where the involute curve bends with the fillet. AGMA [2004] presents a set of *allowable stress numbers* for various steels as a function of Brinell hardnesses. The AGMA uses the term "bending stress number" and "contact stress number" instead of "allowable stress" or "strength" as is the convention in this text. The terminology is intended to emphasize that the reported material properties are suitable only for gear applications. While recognizing the importance of restricting data from this chapter to gears, the nomenclature of strength, not stress number, as a material property will be continued. These strengths are shown in Fig. 14.24 and Table 14.5 for selected common gear materials. The grades differ in degree of control of the microstructure, alloy composition, cleanliness, prior heat treatment, nondestructive testing performed, core hardness values, and other factors. Grade 2 materials are more tightly controlled than Grade 1 materials.

A second, independent form of gear failure is caused by pitting and spalling of the tooth surfaces, usually near the pitch line, where the contact stresses are extremely high and loading is repetitive. The nonconformal contact of the meshing gear teeth produces high contact stresses and eventual

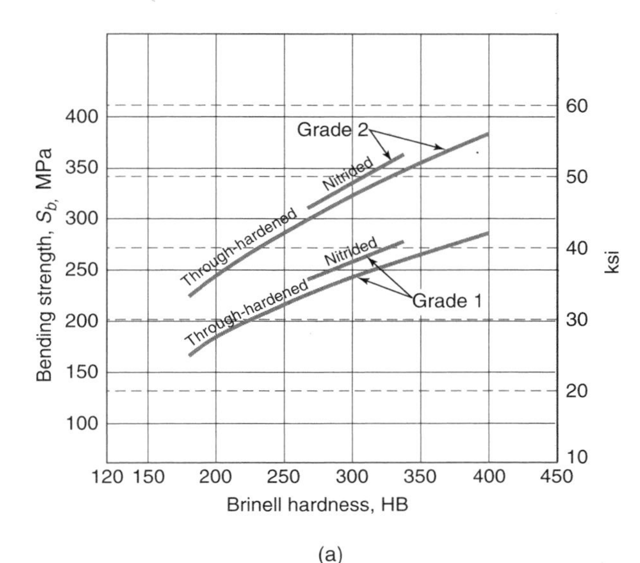

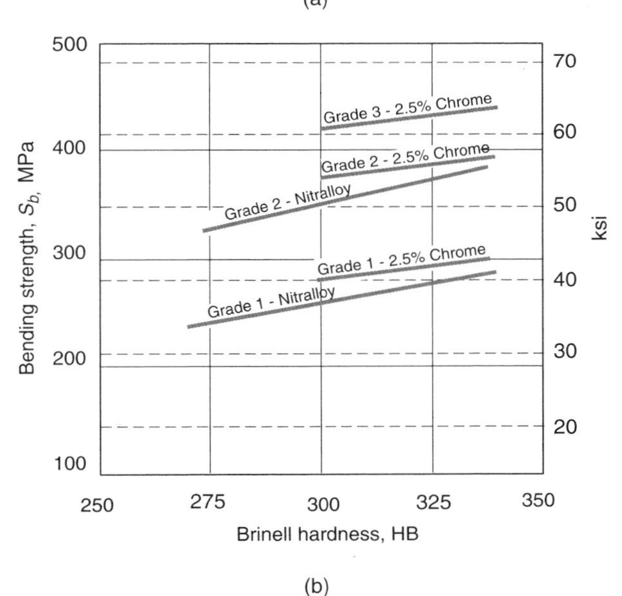

Figure 14.24: Effect of Brinell hardness on allowable bending stress for steel gears. (a) Through-hardened steels; (b) flame or induction hardened nitriding steels. Note that Brinell hardness refers to case hardness for these gears. *Source:* ANSI/AGMA Standard 2101-D04 [2004].

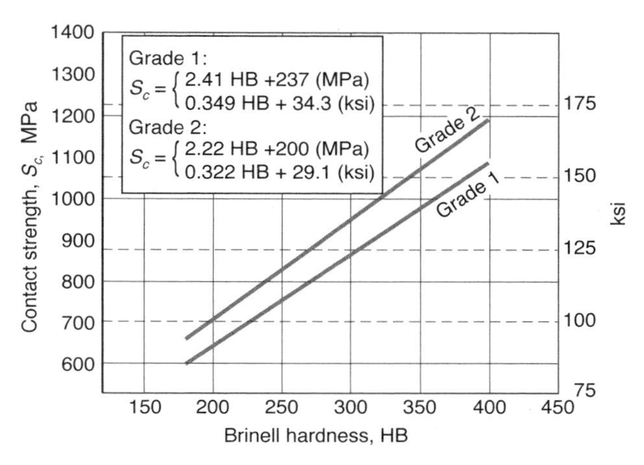

Figure 14.25: Effect of Brinell hardness on allowable contact stress number for two grades of through-hardened steel. *Source:* ANSI/AGMA Standard 2101-D04 [2004].

failure due to fatigue caused by the cyclic variation of stresses (see Section 8.9.3). The contact strength is shown in Fig. 14.25 for through-hardened steel and listed in Table 14.5 for some other materials.

14.9.2 Strength Modification Factors

AGMA recommends modifying the material strength under certain circumstances. This results in an allowable bending stress given by

$$\sigma_{b,\text{all}} = \frac{S_b}{n_s}\frac{Y_N}{K_t K_r}, \tag{14.45}$$

where

S_b = bending strength, MPa (or lb/in^2)
n_s = safety factor
Y_N = the stress cycle factor for bending
K_t = the temperature factor
K_r = the reliability factor

The allowable contact stress must also be modified under certain conditions. AGMA gives the modified allowable contact stress as:

$$\sigma_{c,\text{all}} = \frac{S_c}{n_s}\frac{Z_N C_H}{K_t K_r}, \tag{14.46}$$

where

S_c = contact strength, MPa (or lb/in.^{2})
n_s = safety factor
Z_N = stress cycle factor for contact stress
K_t = temperature factor
K_r = reliability factor
C_H = hardness ratio factor

It should be noted that the AGMA uses different variable labels depending on the unit system, but this complication has been ignored here.

Stress Cycle Factors Y_N and Z_N

The strengths discussed in Section 14.9.1 are based on 10^7 load cycles. If the allowable stresses are to be calculated for a different life, then the stress-cycle factors, Y_N for bending or Z_N for contact stress, are required. This is better defined for stress-cycle factors less than 10^7 load cycles. For more stress

Table 14.5: Bending and contact strength for selected gear materials. *Source:* Adapted from ANSI/AGMA 2101-D04 [2004] and MPIF Standard 35 [2009].

Material designation	Grade	Typical hardness[a]	Bending strength, S_t		Contact strength, S_c	
			lb/in^2	MPa	ksi	MPa
Steel						
Through-hardened	1	—	See Fig. 14.24a		See Fig. 14.25	
Through-hardened[b]	1	180–400 HB	0.0773 HB + 12.8	0.533 HB + 88.3	0.349 HB + 34.3	2.41 HB + 237
	2	180–400 HB	0.102 HB + 16.4	0.703 HB + 11.3	0.322 HB + 29.1	2.22 HB + 200
	2	—	See Fig. 14.24a		See Fig. 14.25	
Carburized and hardened	1	55–64 HRC	55.0	380	180.0	1240
	2	58–64 HRC	65.0[c]	450[c]	225.0	1550
	3	58–64 HRC	75.0	515	275.0	1895
Nitrided and through-hardened[b]	1	83.5 HR15N	0.0823 HB +12.15	0.568 HB + 83.8	150,000	1035
	2	—	0.1086 HB + 15.89	0.749 HB + 110	163,000	1125
Nitralloy 135M and Nitralloy N, nitrided[b]	1	87.5 HR15N	0.0862 HB + 12.73	0.594 HB + 87.76	170,000	1170
	2	87.5 HR15N	0.1138 HB + 16.65	0.784 HB + 114.81	183,000	1260
2.5% Chrome, nitrided[b]	1	87.5 HR15N	0.1052 HB + 9.28	0.7255 HB + 63.89	155,000	1070
	2	87.5 HR15N	0.1052 HB + 22.28	0.7255 HB + 153.63	172,000	1185
	3	87.5 HR15N	0.1052 HB + 29.28	0.7255 HB + 201.81	189,000	1305
Cast Iron						
ASTM A48 gray cast iron, as-cast	Class 20	—	5.00	34.5	50.0–60.0	345–415
	Class 30	174 HB	8.50	59	65.0–75.0	450–520
	Class 40	201 HB	13.0	90	75.0–85.0	520–585
ASTM A536 ductile (nodular) iron	60-40-18	140 HB	22.0–33.	150–230	77.0–92.0	530–635
	80-55-06	179 HB	22.0–33.0	150–230	77.0–92.0	530–635
	100-70-03	229 HB	27.0–40.0	185–275	92.0–112.0	635–770
	120-90-02	269 HB	31.0–44.0	215–305	103.0–126.0	710–870
Bronze						
$S_{ut} > 40,000$ psi ($S_{ut} > 275$ GPa)			5.70	39.5	30.0	205
$S_{ut} > 90,000$ psi ($S_{ut} > 620$ GPa)			23.6	165	65.0	450
Powder Metal						
FL-4405, $\rho = 7.30$ g/cm^3		80 HRB	49.0	340	282.0	1945
FLN2-4405, $\rho = 7.35$ g/cm^3		90 HRB	60.0	410	180.0	1240
FLC-4608, $\rho = 7.30$ g/cm^3		65 HRB	95.72	660	210.0	1450
FN-0205, $\rho = 7.10$ g/cm^3		69 HRB	30.0	210	180.0	1240

[a] Hardness refers to case hardness unless through-hardened.
[b] See Figs. 14.24 and 14.25.
[c] 70,000 psi (485 MPa) may be used if bainite and microcracks are limited to Grade 3 levels.

cycles, there is significant variation in the lives of gears, and a good correlation is difficult to obtain. For example, the life of gears with respect to pitting will be strongly affected by the quality of lubricant and the lubrication regime in which the gear operates, and gear life is more difficult to accurately predict. This will be discussed further in Section 14.12.

Figure 14.26 shows the stress cycle correction factors for bending and contact stress considerations.

Temperature Factor

While the AGMA prescribes the use of a temperature factor, the only recommendation is that the temperature factor, K_t, be taken as unity if the temperature does not exceed 120°C. For higher temperatures, K_t greater than one is needed to allow for the effect of temperature on oil film and material properties. No specific recommendations are given; experimental evaluation is therefore required.

Reliability Factor

The allowable stresses were based upon a statistical probability of 1% failure at 10^7 cycles, or 99% reliability. For other reliabilities, the data in Table 14.6 should be used to obtain K_r.

Table 14.6: Reliability factor, K_r. *Source:* From ANSI/AGMA 2101-D04 [2004].

Probability of survival, percent	Reliability factor[a] K_r
50	0.70[b]
90	0.85[b]
99	1.00
99.9	1.25
99.99	1.50

[a] Based on surface pitting. If tooth breakage is considered a greater hazard, a larger value may be required.
[b] At this value, plastic flow may occur rather than pitting.

Hardness Ratio Factor

Calculation of contact stresses was discussed in Ch. 8, and further details of such calculations specific to gears are discussed in Section 14.12. A common result from such analyses is that the contact stresses are much larger than the yield stress of the gear material, indicating that plastic deformation takes place at the contact location. Gross plastic deformation does not occur, because the plastic zone is bounded by very stiff elastic regions that restrict metal flow.

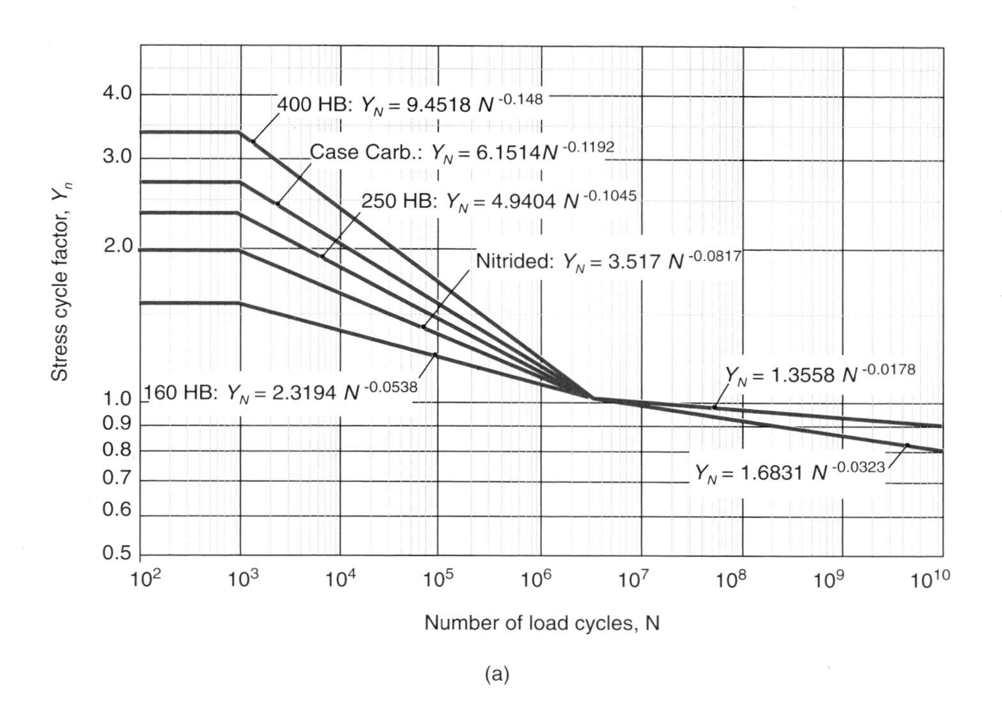

(a)

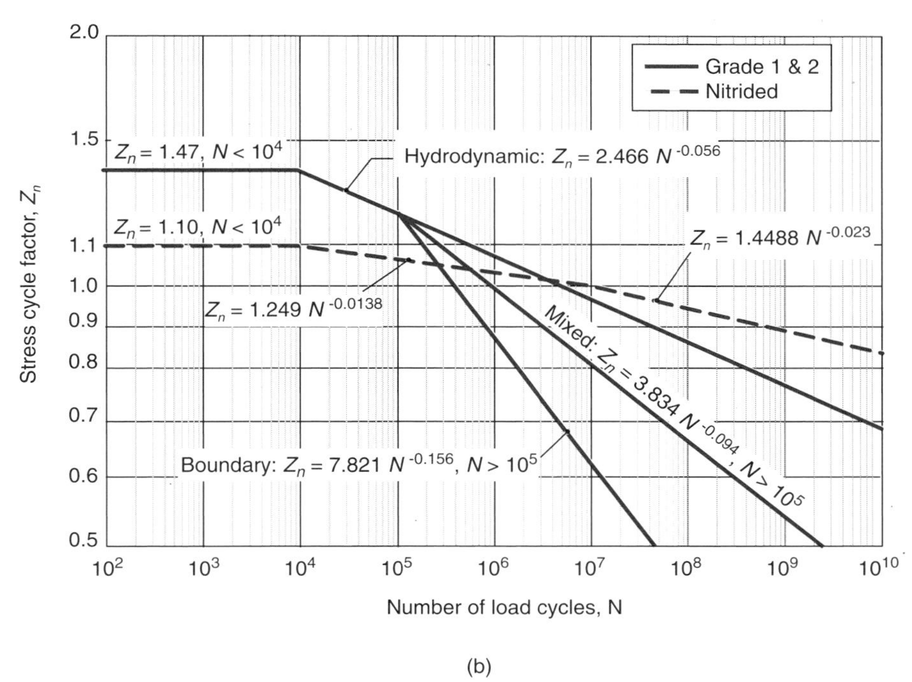

(b)

Figure 14.26: Stress cycle factor. (a) Bending strength stress cycle factor Y_N; (b) Pitting resistance stress cycle factor Z_N. *Source:* Based on ANSI/AGMA Standard 2101-D04 [2004].

Furthermore, since the pinion has fewer teeth than the gear, it will be subjected to a larger number of contact stress loadings due to its geometry. There is some justification to using a harder pinion than gear, in order to obtain a uniform safety factor for contact stress and pitting resistance failures. However, a harder pinion results in higher stresses on the gear. Therefore, a correction factor C_H is defined by AGMA, and is applied to the gear only. The hardness-ratio factor is defined by

$$C_H = 1.0 + A'(g_r - 1.0), \tag{14.47}$$

where g_r is the gear ratio given by Eq. (14.3) and A' is defined by

$$A' = \begin{cases} 0 & \dfrac{\text{HB}_P}{\text{HB}_G} < 1.2 \\ (0.00898)\left(\dfrac{\text{HB}_P}{\text{HB}_G}\right) - 0.00829 & 1.2 \le \dfrac{\text{HB}_P}{\text{HB}_G} \le 1.7 \\ 0.00698 & \dfrac{\text{HB}_P}{\text{HB}_G} > 1.7 \end{cases} \tag{14.48}$$

where HB_P and HB_G are the Brinell hardnesses of the pinion and gear, respectively.

Example 14.8: Bending and Contact Stress

Given: A Grade 1 steel gear with 54 teeth is through-hardened to a Brinell hardness of 250. It mates with a 19-tooth pinion made from Grade 1 steel through-hardened to 325 HB. It is expected that the gear will encounter 100 million load cycles over its life. It is enclosed and lubricated and designed to operate in hydrodynamic lubrication.

Find: Determine the allowable bending and contact stresses assuming a safety factor of 4.0 for bending stress and 3 for contact stress, and a reliability of 99.9%.

Solution:

(*a*) *Allowable bending stress.* Equation (14.45) will yield the allowable bending stress, but first the material's bending strength and correction factors need to be obtained. From Table 14.5, the bending strength is

$$\begin{aligned} S_b &= 0.533\,\text{HB} + 88.3 \\ &= 0.533(250) + 88.3 \\ &= 222\text{ MPa}. \end{aligned}$$

From Fig. 14.26a, the stress cycle factor can be estimated as

$$Y_n = 1.6831N^{-0.0323} = 1.6831\left(10^8\right)^{-0.0323} = 0.928.$$

Note that the lower of the two curves was chosen from Fig. 14.26a to produce a conservative result. From Table 14.6, the reliability factor is K_r=1.25. There is no information given about temperature, so it will be assumed that K_t = 1.0. Therefore,

$$\sigma_{b,\text{all}} = \frac{S_b}{n_s}\frac{Y_N}{K_t K_r} = \frac{(222)}{4}\frac{0.928}{(1)(1.25)} = 41.2\text{ MPa}.$$

(*b*) *Allowable contact stress.* Equation (14.46) will be used for allowable contact stress, but first note from Table 14.5,

$$S_c = 241\,\text{HB} + 237 = 2.41(250) + 237 = 840\text{ MPa}.$$

From Fig. 14.26b, using the curve for hydrodynamic lubrication,

$$Z_n = 2.466N^{-0.056} = 2.466\left(10^8\right)^{-0.056} = 0.879.$$

Note that HB_p/HB_g = 325/250=1.3, so that from Eq. (14.48),

$$A' = (8.898 \times 10^{-3})(1.3) - 8.29 \times 10^{-3} = 0.00338.$$

Therefore, the hardness-ratio factor is given by Eq. (14.47) as

$$C_H = 1.0 + A'(g_r - 1.0) = 1.0 + (0.00338)\left(\frac{54}{19} - 1\right),$$

or C_H = 1.0062. As before, K_r = 1.25 and K_t = 1.0. Therefore, from Eq. (14.46),

$$\sigma_{c,\text{all}} = \frac{S_c}{n_s}\frac{Z_N C_H}{K_t K_r} = \frac{840}{3}\frac{(0.879)(1.0062)}{(1)(1.25)} = 198\text{ MPa}.$$

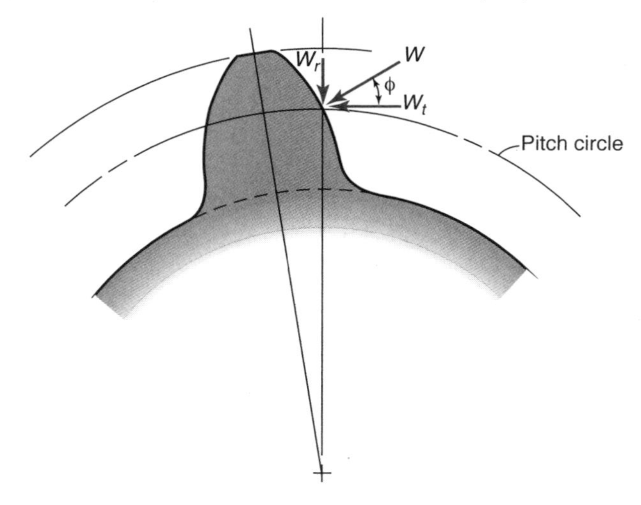

Figure 14.27: Loads acting on an individual gear tooth.

14.10 Loads Acting on a Gear Tooth

Gears transfer power by the driving teeth exerting a load on the driven teeth. Figure 14.27 shows the loads acting on an individual spur gear tooth. The tangential (W_t), radial (W_r), and normal (W) loads are shown, along with the pressure angle and the pitch circle. The **tangential load**, W_t, also called the *transmitted force*, can be obtained from Eq. (4.37) and the definition of horsepower as

$$W_t = \frac{126,050 h_p}{dN_a}, \tag{14.49}$$

where

h_p = transmitted power, hp
d = pitch diameter, in.
N_a = gear rotational speed, rpm

Equation (14.49) is valid only for English units. The comparable equation to Eq. (14.49) in SI units can be obtained directly

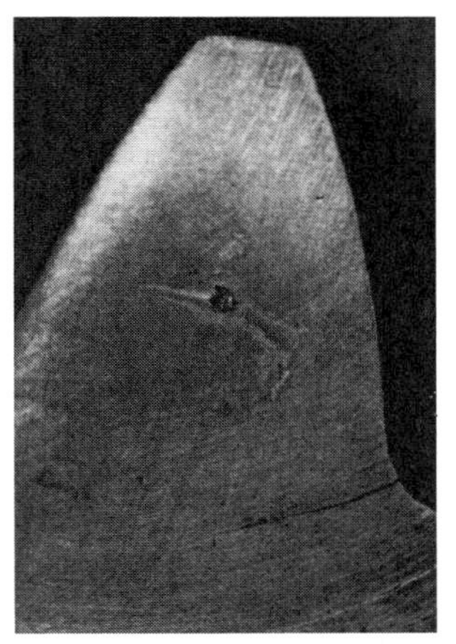

Figure 14.28: A crack that has developed at the root of a gear tooth due to excessive bending stresses. *Source:* Courtesy of the American Gear Manufacturers Association.

from Eq. (4.37):

$$W_t = \frac{h_p}{u} = \frac{60h_p}{\pi d N_a}$$

The normal load W and the radial load W_r can be computed from the following:

$$W = \frac{W_t}{\cos\phi}, \tag{14.50}$$

$$W_r = W_t \tan\phi. \tag{14.51}$$

The normal load is used in determining contact stress, contact dimensions, and lubricant film thickness.

14.11 Bending Stresses in Gear Teeth

Gear failure can occur from tooth breakage, which results when the design stress due to bending is greater than the allowable stress. Figure 14.28 shows a crack that developed at the root of a gear tooth due to excessive bending stresses. Figure 14.29a illustrates a tooth with the forces and dimensions shown, and Fig. 14.29b shows a cantilevered beam simulating the forces and dimensions acting on the gear tooth. Note that the load components shown in Fig. 14.27 have been moved to the tooth tip in Fig. 14.29. Using the bending of a cantilevered beam to simulate the stresses acting on a gear tooth is only approximate at best, since the situations present in a long, thin, uniform beam loaded by concentrated forces at the ends are quite different from those found in a gear tooth.

From Eq. (4.45), the magnitude of the maximum bending stress can be written as

$$\sigma_b = \frac{Mc}{I}. \tag{14.52}$$

From similar triangles

$$\tan\alpha = \frac{x}{t/2} = \frac{t/2}{l};$$

$$l = \frac{t^2}{4x}. \tag{14.53}$$

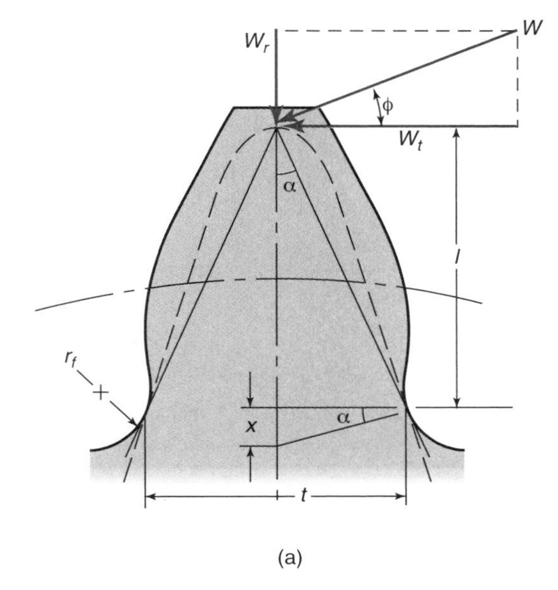

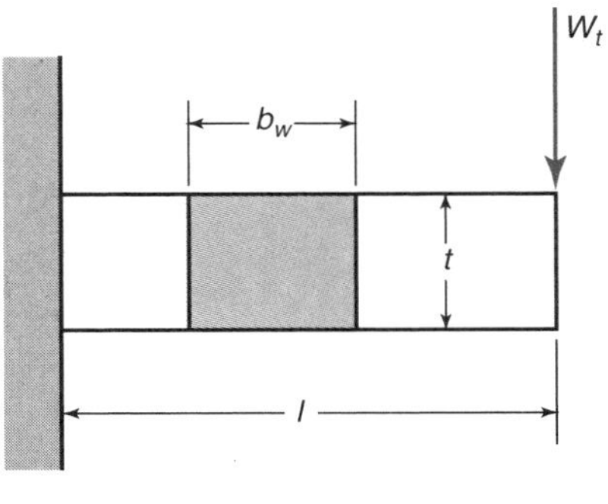

Figure 14.29: Loads and length dimensions used in determining tooth bending stress. (a) Tooth; (b) cantilevered beam.

For a rectangular section with base, b_w, and height, t, the area moment of inertia is, from Table 4.1,

$$I = \frac{bh^3}{12} = \frac{b_w t^3}{12}.$$

If $M = W_t l$ and $c = t/2$, Eq. (14.53) can be substituted into Eq. (4.45) to obtain

$$\sigma_b = \frac{W_t l(t/2)}{b_w t^3/12} = \frac{6W_t l}{b_w t^2}. \tag{14.54}$$

where b_w is the face width of the gear. Substituting Eq. (14.53) into Eq. (14.54) gives

$$\sigma_b = \frac{3W_t}{2b_w x} = \frac{3W_t p_d}{2b_w p_d x} = \frac{W_t p_d}{b_w Y}, \tag{14.55}$$

or

$$Y = \frac{2xp_d}{3}, \tag{14.56}$$

where p_d is the diametral pitch. Lewis [1892] was the first to use this approach, and thus Eq. (14.55) is known as the **Lewis equation**, and Y is called the **Lewis form factor**. This approach has had universal acceptance and illustrates the important parameters in gear failures due to excessive bending stresses.

Table 14.7: Lewis form factor for various numbers of teeth (pressure angle, 14.5°; full-depth involute).

Number of teeth	Lewis form factor	Number of teeth	Lewis form factor
10	0.176	34	0.325
11	0.192	36	0.329
12	0.210	38	0.332
13	0.223	40	0.336
14	0.236	45	0.340
15	0.245	50	0.346
16	0.256	55	0.352
17	0.264	60	0.355
18	0.270	65	0.358
19	0.277	70	0.360
20	0.283	75	0.361
22	0.292	80	0.363
24	0.302	90	0.366
26	0.308	100	0.368
28	0.314	150	0.375
30	0.318	200	0.378
32	0.322	300	0.382

Table 14.7 gives the Lewis form factor for various numbers of teeth while assuming a pressure angle of 14.5° and a full-depth involute profile. The Lewis form factor given in Eq. (14.56) is dimensionless; it is also independent of tooth size and is only a function of shape.

Equation (14.55) does not take the stress concentration at the tooth fillet into account. Introducing a geometry factor, Y_j, where $Y_j = Y/K_c$, changes the Lewis equation to

$$\sigma_b = \frac{W_t p_d}{b_w Y_j}. \tag{14.57}$$

Figure 14.30 gives values of geometry factors, Y_j, for a spur gear with a pressure angle of 20° and a full-depth involute profile. Equation (14.57) is known as the **modified Lewis equation**.

Other modifications to Eq. (14.55) are recommended in ANSI/AGMA [2004] for practical design to account for the variety of conditions that can be encountered in service. The **AGMA bending stress equation** is:

$$\sigma_b = \frac{W_t p_d K_o K_s K_m K_v K_b}{b_w Y_j}, \tag{14.58}$$

where

- W_t = transmitted load, N
- p_d = diametral pitch, m^{-1}
- K_o = overload factor
- K_s = size factor
- K_m= load distribution factor
- K_v = dynamic factor
- K_b = rim thickness factor

Each of these factors is defined below.

14.11.1 Overload Factor

The **overload factor** is used to account for load variations, vibrations, shock, speed changes, and other application-specific conditions that may result in peak loads greater than the normal transmitted load encountered during operation. In addition, the overload factor accounts for higher applied loads during startup. The overload factor is not a substitute or replacement for a safety factor, but instead accounts for normal load variations. Generally, an overload factor can only be applied after significant experience with a particular

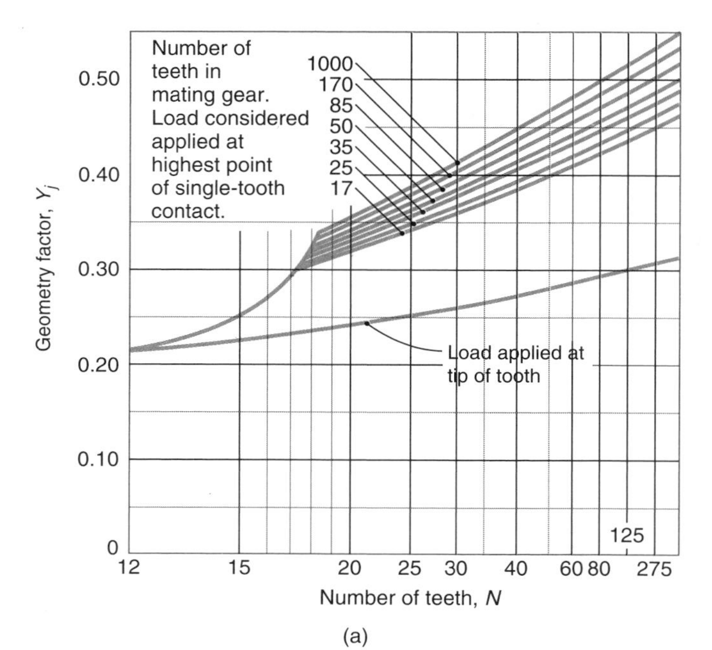

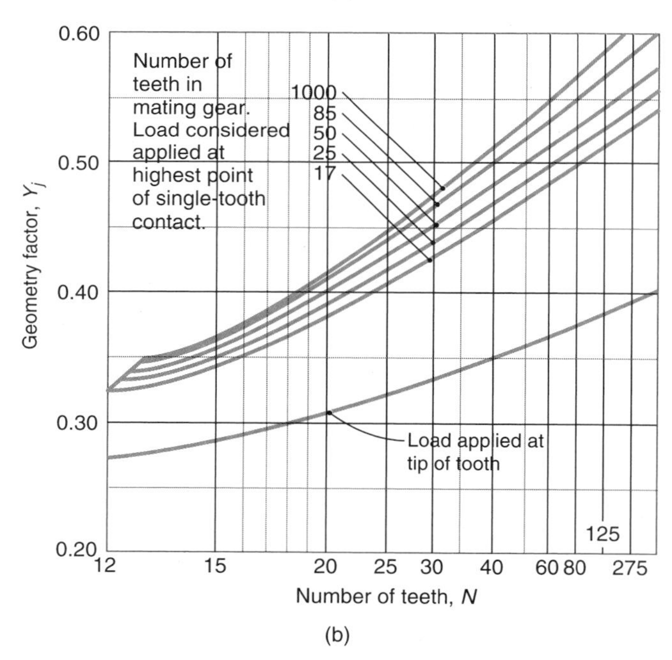

Figure 14.30: Spur gear geometry factors for full-depth involute profile. (a) $\phi = 20°$; (b) $\phi = 25°$. *Source:* AGMA Standard 908-B89 [1989].

Table 14.8: Overload factor, K_o, as function of driving power source and driven machine.

	Driven machines			
Power source	Uniform	Light shock	Moderate shock	Heavy shock
Uniform	1.00	1.25	1.50	1.75
Light shock	1.20	1.40	1.75	2.25
Moderate shock	1.30	1.70	2.00	2.75

Table 14.9: Recommended values of size factor, K_s.

Diametral pitch, p_d, in.$^{-1}$	Module, m, mm	Size factor, K_s
5	5	1.00
4	6	1.05
3	8	1.15
3	12	1.25
1.25	20	1.40

application. However, Table 14.8 gives guidance in the selection of an overload factor and incorporates both the driving power source and the driven machine. An overload factor of 1.00 would be applicable to a perfectly smooth-operating electric motor driving a perfectly smooth-operating generator through a gear speed reducer, but would still recognize that the gear would occasionally sustain momentary overloads (as in startup). Such relatively rare loadings are incorporated into strength definitions. Rougher conditions produce a value of K_o greater than 1.00. Power source classifications with typical examples are

1. *Uniform:* electric motor with constant-speed turbine

2. *Light shock:* water turbine with variable-speed drive

3. *Moderate shock:* multicylinder engine

Classification of driven machine roughness and examples are

1. *Uniform:* continuous-duty generator

2. *Light shock:* fans and low-speed centrifugal pumps, liquid agitators, variable-duty generators, uniformly loaded conveyors, rotary positive-displacement pumps

3. *Moderate shock:* high-speed centrifugal pumps, reciprocating pumps, compressors, heavy-duty conveyors, machine tool drives, concrete mixers, textile machinery, meat grinders, saws

4. *Heavy shock:* rock crushers, punch press drives, pulverizers, processing mills, tumbling barrels, wood chippers, vibrating screens, railroad car dumpers

14.11.2 Size Factor

The AGMA specifies that the **size factor**, K_s, is 1.00 for most gears. However, for gears with large teeth or large face widths, a value greater than 1.00 is recommended. Table 14.9 gives recommendations for size factors.

14.11.3 Load Distribution Factor

The **load distribution factor**, K_m, takes the non-uniform distribution of load across the tooth thickness into account. For example, if spur gear shafts are not perfectly parallel, then contact may occur only at the edges of spur gears and stresses will be much higher than if the shafts were parallel. The value of the load distribution factor is based on many variables in the design of gears, making it the most difficult factor to determine. It is defined as the ratio of the peak load intensity divided by the average or uniformly distributed load intensity.

AGMA describes a simplified approach for determining the load distribution factor, applicable to situations where

1. The face width to pinion pitch diameter ratio, b_w/d_p, is two or lower.

2. The gear elements are mounted between bearings. Overhung gears present additional complications, and require a complex theoretical approach beyond the scope of this book.

3. The face width is less than 1 m (40 in.).

4. Contact is across the full face width of the narrowest member when loaded.

For situations that do not meet these conditions, a lengthy analytical approach, discussed in AGMA 2101-D04 [2010], is applicable.

The load distribution factor can be expressed by

$$K_m = 1.0 + C_{mc}\left(C_{pf}C_{pm} + C_{ma}C_e\right), \tag{14.59}$$

where

C_{mc} = lead correction factor
C_{pf} = pinion proportion factor
C_{pm} = pinion proportion modifier
C_{ma} = mesh alignment factor
C_e = mesh alignment correction factor

These factors will be briefly described here.

Lead Correction Factor, C_{mc}

The **lead correction factor**, C_{mc}, corrects the peak load when crowning or other forms of tooth profile modification is applied. C_{mc} is given by

$$C_{mc} = \begin{cases} 1.0 & \text{for uncrowned teeth} \\ 0.8 & \text{for crowned teeth} \end{cases} \tag{14.60}$$

Pinion Proportion Factor, C_{pf}

The **pinion proportion factor** accounts for deflections under load and is shown in Fig. 14.31. If the face width and pinion pitch diameter are in millimeters, C_{pf} can be calculated according to the following:

If $b_w < 25$ mm,

$$C_{pf} = \frac{b_w}{10d_p} - 0.025;$$

for 25 mm $< b_w \leq 432$ mm,

$$C_{pf} = \frac{b_w}{10d_p} - 0.0375 + 0.000492b_w;$$

and for 432 mm $< b_w \leq 1020$ mm,

$$C_{pf} = \frac{b_w}{10d_p} - 0.1109 + 0.000815b_w - (3.53\times10^{-7})b_w^2. \tag{14.61}$$

If the face width and pinion pitch diameter are in inches, C_{pf} is given by the following:

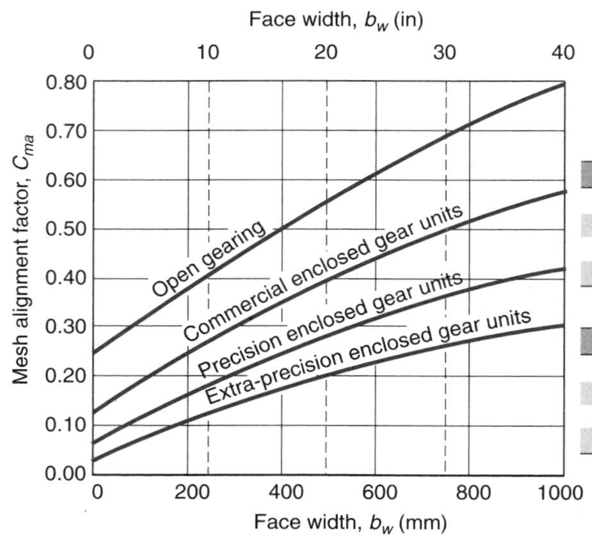

$$C_{ma} = A + Bb_w + Cb_w^2$$

If b_w is in in.:

Condition	A	B	C
Open gearing	0.247	0.0167	-0.765×10^{-4}
Commercial enclosed gears	0.12	0.0158	-1.095×10^{-4}
Precision enclosed gears	0.0675	0.0128	-0.926×10^{-4}
Extraprecision enclosed gears	0.000380	0.0102	-0.822×10^{-4}

If b_w is in mm:

Condition	A	B	C
Open gearing	0.247	6.57×10^{-4}	-1.186×10^{-7}
Commercial enclosed gears	0.127	6.22×10^{-4}	-1.69×10^{-7}
Precision enclosed gears	0.0675	5.04×10^{-4}	-1.44×10^{-7}
Extraprecision enclosed gears	0.000380	4.02×10^{-4}	-1.27×10^{-7}

Figure 14.33: Mesh alignment factor. *Source:* ANSI/AGMA Standard 2101-D04 [2004].

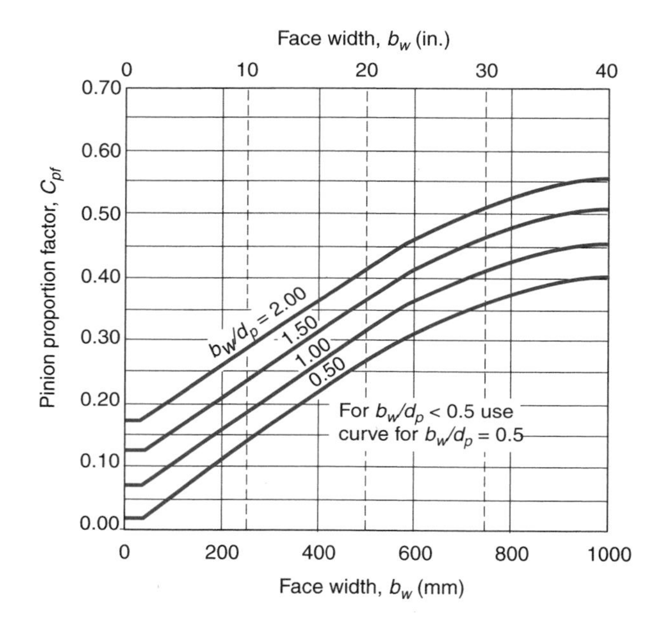

Figure 14.31: Pinion proportion factor, C_{pf}. *Source:* ANSI/AGMA Standard 2101-D04 [2004].

If $b_w \leq 1$ in.,

$$C_{pf} = \frac{b_w}{10d_p} - 0.025;$$

for 1 in. $< b_w \leq 17$ in.,

$$C_{pf} = \frac{b_w}{10d_p} - 0.0375 + 0.0125b_w;$$

and for 17 in. $< b_w \leq 40$ in.,

$$C_{pf} = \frac{b_w}{10d_p} - 0.1109 + 0.02907b_w - (2.28 \times 10^{-4})b_w^2. \quad (14.62)$$

Also, if $b_w/d_p < 0.5$, then the value of $b_w/d_p = 0.5$ should be used in Eqs. (14.61) or (14.62).

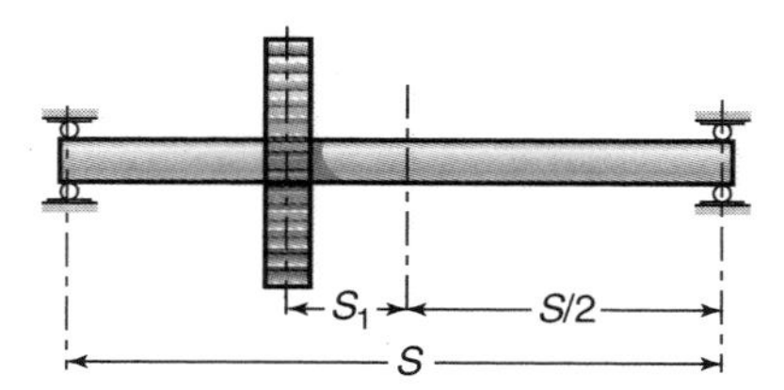

Figure 14.32: Evaluation of S and S_1. *Source:* ANSI/AGMA Standard 2101-D04 [2004].

Pinion Proportion Modifier, C_{pm}

If a gear is mounted at midspan between two equally stiff bearings, any deflections that occur will not change the load distribution across the gear tooth face. However, if the gear is mounted away from midspan, deflections of the shaft will rotate the gear and change the stress distribution across the face width. The **pinion proportion modifier,** C_{pm}, takes the location of the gear relative to the bearing centerline into account. C_{pm} is given by

$$C_{pm} = \begin{cases} 1.0, & (S_1/S) < 0.175 \\ 1.1, & (S_1/S) \geq 0.175 \end{cases} \quad (14.63)$$

where S is the bearing span and S_1 is the offset of the pinion, as shown in Fig. 14.32.

Mesh Alignment Factor, C_{ma}

The **mesh alignment factor** accounts for the misalignment of the pitch cylinder axes of rotation from causes other than elastic deformations, and is mainly associated with the quality of mounting. C_{ma} can be obtained from Fig. 14.33.

Mesh Alignment Correction Factor, C_e

When the manufacturing or assembly techniques improve the mesh alignment, the **mesh alignment correction factor,**

C_e is used. C_e is given by

$$C_e = \begin{cases} 0.80 & \text{when gearing is adjusted at assembly} \\ 0.80 & \text{when compatibility between gear teeth is improved by lapping} \\ 1.0 & \text{for all other conditions} \end{cases} \quad (14.64)$$

14.11.4 Rim Factor

Very large gears are more economically produced from rim-and-spoke designs instead of solid blanks. However, if the rim is thin, it is more likely that fracture will occur through the rim than through the tooth root. The **backup ratio** takes such failure into account and is defined as

$$m_B = \frac{t_R}{h_t}, \quad (14.65)$$

where t_R is the rim thickness. If the backup ratio is greater than 1.2, which includes the extreme case of a solid (conventional) gear, use K_b =1. Otherwise, the **rim factor** is defined by

$$K_b = -2m_B + 3.4. \quad (14.66)$$

14.11.5 Dynamic Factor

It should be noted that the American Gear Manufacturer's Association changed the definition of the **dynamic factor** in 1995; before 1995, the factor had a value less than 1.0 and it appeared in the denominator of Eq. (14.58). Equation (14.58) is consistent with the current standards and therefore K_v has a value greater than 1.0. Caution should be exercised when referring to technical literature that predates this definition change.

Further, in 2010, AGMA presented two methods for calculating the dynamic factor. The method based on quality index is presented here, but an alternative method based on transmission accuracy number, A_v is also included in the standard. A_v is useful because it is a quality measure that has the same trends as European standards in that its value decreases with increasing gear quality (the opposite is true for Q_v). This book is consistent in the approach used to obtain K_v using the quality factor, Q_v. Note that Eq. (14.44) can be used to incorporate the transmission accuracy level. However, when referring to the technical literature, care must be taken regarding calculation of K_v because of this terminology difference.

K_v accounts for the fact that the dynamic load is higher than the transmitted load, since there is some impact loading that is not accounted for in Eq. (14.57). The value of K_v depends on the accuracy of the tooth profile, the elastic properties of the tooth, and the speed with which the teeth come into contact.

Figure 14.34 gives the dynamic factors for pitch-line velocity and quality index. The pitch-line velocity, v_t, used in Fig. 14.34 is the linear velocity of a point on the pitch circle of the gear, given by

$$v_t = \pi d_p N_{ap} = \pi d_g N_{ag}, \quad (14.67)$$

where

v_t = pitch-line velocity, ft/min or m/s
N_{ap} = rotational speed of pinion, rpm or rps
N_{ag} = rotational speed of gear, rpm or rps
d_p = pitch diameter of pinion, in. or m
d_g = pitch diameter of gear, in. or m

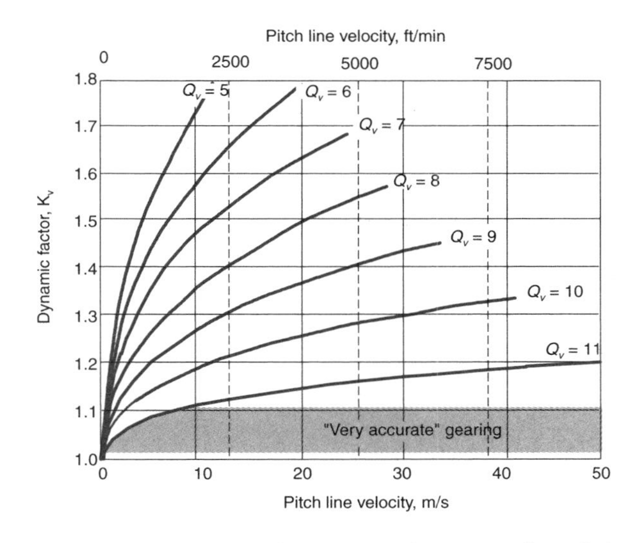

Figure 14.34: Dynamic factor as a function of pitch-line velocity and transmission accuracy level number. *Source:* ANSI/AGMA Standard 2101-D04 [2004].

The dynamic factor can be approximated for $5 \leq Q_v \leq 11$ by

$$K_v = \left(\frac{A + C\sqrt{v_t}}{A} \right)^B, \quad (14.68)$$

where

$$A = 50 + 56(1.0 - B), \quad (14.69)$$

$$B = 0.25(12 - Q_v)^{0.667}. \quad (14.70)$$

$$C = 1 \text{ for } v_t \text{ in ft/min, and } \sqrt{200} = 14.14 \text{ for } v_t \text{ in m/s}. \quad (14.71)$$

The maximum recommended pitch line velocity for a given value of Q_v is

$$v_{t,\max} = \frac{1}{C^2}\left[A + (Q_v - 3)\right]^2. \quad (14.72)$$

Design Procedure 14.2: Methods to Increase Bending Performance of Gears

If a gear does not produce a satisfactory design based on bending requirements, a design alteration may be needed. This is not always straightforward, since such alterations may help in one area and hurt in another, and may affect associated machine elements such as bearings. However, some factors that improve bending performance are the following:

1. Reduction in the load, such as by increasing contact ratio, or altering other aspects of the system.
2. Increase the center distance.
3. Apply gears with a coarser pitch.
4. Use a higher pressure angle.
5. Use a helical gear instead of a spur gear.
6. Use a carburized material.

7. Improve the gear accuracy through manufacturing process selection.
8. Select a better (stronger) gear material.
9. Use a wider effective face width.
10. Apply shot peening to the teeth.

Example 14.9: Calculation of Bending Stresses

Given: A 54-tooth, 1.25 module pinion transfers 15 kW to a 270-tooth gear. The pinion speed is 132 rad/s, and it has a 20° pressure angle, with crowned teeth, mounted between bearings in a housing that can be considered to be commercially enclosed. Use a quality index of $Q_v = 7.0$, and assume that the gears are not adjusted at assembly. The pinion's face width is 60 mm, and it has a bending strength of $S_b = 380$ MPa. Assume $K_o = K_b = Z_n = Y_n = 1.0$.

Find: Determine the safety factor based on bending stress for the pinion.

Solution: Equation (14.58) will be used to determine the stress in the gear teeth, but first the correction factors need to be determined. Note:

- From Eq. (14.7), $d_p = mN_p = (1.25)(54) = 67.5$ mm. Therefore, $r_p = 33.75$ mm = 0.03375 m.
- The transmitted load is obtained from the power:

$$W_t = \frac{h_p}{r_p \omega_p} = \frac{15,000}{(0.03375)(132)} = 3367 \text{ N}$$

- From Eq. (14.9), $p_d = 1/0.00125 = 800 \text{ m}^{-1}$
- From Table 14.9, K_s=1.0. Also, from Fig. 14.30, Y_j is approximately 0.475.
- The mesh alignment factor, K_m is obtained from Eq. (14.59), but first the following must be obtained:
  - $C_{mc} = 0.8$ from Eq. (14.60) because the teeth are crowned.
  - C_{pf} is obtained from Eq. (14.61) noting that $b_w =$ 60 mm:

$$C_{pf} = \frac{b_w}{10 d_p} - 0.0375 + 0.000492 b_w = 0.0809.$$

  - $C_{pm} = 1.0$ from Eq. (14.63), since $S_1 = 0$ for gears that are centered between bearings.
  - For commercial enclosed gears, Fig. 14.33 gives

$$C_{ma} = 0.127 + \left(6.22 \times 10^{-4}\right)(60) - \left(1.69 \times 10^{-7}\right)(60)^2 = 0.1637.$$

  - $C_e = 1.0$ from Eq. (14.64).

Therefore, Eq. (14.59) yields

$$K_m = 1.0 + (0.8)\left[(0.0809)(1.0) + (0.1637)(1.0)\right] = 1.196.$$

- The surface speed of the pinion is

$$v_t = r_p \omega_p = (0.03375)(132) = 4.455 \text{ m/s}.$$

Therefore, from Eq. (14.68), noting that

$$B = 0.25(12 - Q_v)^{0.667} = 0.25(12 - 7)^{0.667} = 0.731,$$

$$A = 50 + 56(1.0 - B) = 65.0,$$

$$C = 14.14$$

Therefore,

$$K_v = \left(\frac{A + C\sqrt{v_t}}{A}\right)^B = \left(\frac{65.0 + 14.14\sqrt{4.455}}{65.0}\right)^{0.731} = 1.32.$$

From Eq. (14.58), the bending stress is given as

$$\sigma_b = \frac{(3367)(800)(1)(1)(1.196)(1.32)(1)}{(0.060)(0.475)} = 150 \text{ MPa}.$$

Referring to Eq. (14.45), there is no information given that allows calculation of K_t or K_r, so they are assigned values of 1.0. Therefore, Eq. (14.45) becomes

$$\sigma_{b,\text{all}} = \frac{S_b}{n_s}.$$

Therefore, since $S_b = 380$ MPa, the safety factor is

$$n_s = \frac{\sigma_{t,\text{all}}}{\sigma_b} = \frac{380}{150} = 2.533.$$

14.12 Contact Stresses in Gear Teeth

In addition to bending stresses, failure can be caused by excessive contact stresses. These stresses can cause pitting, scoring, and scuffing of the gear tooth surfaces. **Pitting** is the removal of small bits of metal from the gear surface due to fatigue, leaving small holes or pits (see Section 8.9.3 and Fig. 14.35). Pitting is caused by excessive surface stress due to high normal loads, a high local temperature due to high rubbing speeds, or inadequate lubrication.

Scoring is heavy scratch patterns extending from the tooth root to the tip. It appears as if a heavily loaded tooth pair has dragged particles between sliding teeth. It can be caused by lubricant failure, incompatible materials, and overload.

Scuffing is surface damage comprising plastic material flow combined with gouges and scratches caused by loose wear particles acting as abrasives between teeth. Both scoring and scuffing are associated with welding or seizing and plastic deformation.

The contact stress is calculated by using a Hertzian analysis (see Section 8.4). The Hertzian contact stress of gear teeth is based on the analysis of two cylinders under a radial load.

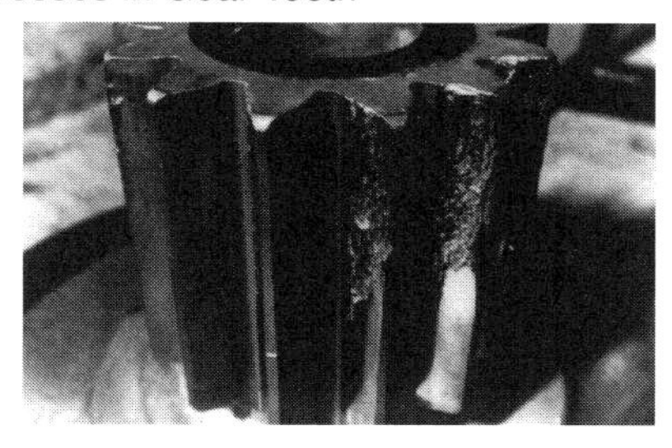

Figure 14.35: A gear showing extreme pitting or spalling. *Source:* Courtesy of the American Gear Manufacturers Association.

The radii of the cylinders are the radii of curvature of the involute tooth forms of the mating teeth at the point of contact. The load on the teeth is the normal load (see Figs. 14.27 and 14.29) and is given by Eq. (14.50) as a function of transmitted load and pressure angle.

The maximum Hertzian pressure in the contact can be written from Eq. (8.20) as

$$p_H = E'\left(\frac{W'}{2\pi}\right)^{\frac{1}{2}}. \qquad (14.73)$$

Recall from Section 8.4 that:

$$E' = \text{effective elastic modulus} = \frac{2}{\dfrac{1-\nu_a^2}{E_a}+\dfrac{1-\nu_b^2}{E_b}}, \qquad (8.16)$$

$$W' = \text{dimensionless load} = \frac{w'}{E'R_x} = \frac{W}{E'R_x b_w}. \qquad (8.18)$$

For spur gears,

$$\frac{1}{R_x} = \left(\frac{1}{r_p}+\frac{1}{r_g}\right)\frac{1}{\sin\phi} = \left(\frac{1}{d_p}+\frac{1}{d_g}\right)\frac{2}{\sin\phi}. \qquad (14.74)$$

The parameter R_x is the effective radius; for gears it is a function of the pitch radii of the pinion and the gear as well as the pressure angle. Figure 14.9 can be used to help visualize these parameters.

The contact made between meshing spur gears is rectangular, having a Hertzian half-width b^*, and the length of the contact is the face width b_w. Equation (8.17) can be used to calculate the Hertzian half-width as

$$b^* = R_x\left(\frac{8W'}{\pi}\right)^{1/2} \qquad (8.17)$$

Note that the maximum deformation in the center of the contact is given by Eq. (8.19).

Just as modification factors were used to describe the bending stress acting on the meshing gears in Eq. (14.58), there are also modification factors when dealing with contact stresses. Thus, the contact stress given in Eq. (14.73) is modified to

$$\sigma_c = E'\left(\frac{W'K_oK_sK_mK_v}{2\pi}\right)^{\frac{1}{2}} = p_H\left(K_oK_sK_mK_v\right)^{\frac{1}{2}}. \qquad (14.75)$$

The values for the overload factor K_o, the size factor K_s, the load distribution factor K_m, and the dynamic factor K_v are obtained the same way as those for the bending stress in Section 14.11. This may seem inappropriate, but the differences between bending and contact correction factors are incorporated into the material strength given in Section 14.9.2. It should also be noted that the AGMA standard [2004] also includes a surface condition factor to emphasize the importance of surface quality on pitting resistance. However, apart from admonitions that the surface condition factor is important, no numerical recommendations are given, and it has therefore been ignored in Eq. (14.75).

Design Procedure 14.3: Methods to Increase Pitting Resistance

If a gear does not produce a satisfactory design based on surface pitting requirements, a design alteration may be needed. This is not always straightforward, since such alterations may help in one area and hurt in another, and may affect associated machine elements such as bearings. However, some factors that improve pitting performance are the following:

1. Reduction in the load, such as by increasing contact ratio, or altering other aspects of the system.
2. Increase the center distance.
3. Apply gears with a finer pitch.
4. Use a higher pressure angle.
5. Use a helical gear instead of a spur gear.
6. Use a carburized material.
7. Increase the surface hardness by material selection, or by performing a surface hardening operation.
8. Improve the gear accuracy through manufacturing process selection.
9. Select a better (stronger) gear material.
10. Use a wider effective face width.
11. Increase the lubricant film thickness.

Example 14.10: Calculation of Contact Stresses

Given: Consider the pinion and gear from Example 14.9, with $S_c = 1260$ MPa and $K_B = 1.0$.

Find: Determine the safety factor based on contact stress for the pinion.

Solution: The transmitted load was calculated from the power transmitted as $W_t = 3367$ N. From Eq. (14.50), the normal load is

$$W = \frac{W_t}{\cos\phi} = \frac{3367}{\cos 20^\circ} = 3583\text{ N}.$$

Note that $d_p = 67.5$ mm and $d_g = 337.5$ mm. From Eq. (14.74),

$$\frac{1}{R_x} = \left(\frac{1}{d_p}+\frac{1}{d_g}\right)\frac{2}{\sin\phi} = \left(\frac{1}{67.5}+\frac{1}{337.5}\right)\frac{2}{\sin 20^\circ},$$

which is solved as $R_x = 0.00961$ m. Therefore,

$$W' = \frac{W}{b_w E' R_x} = \frac{3583}{(0.06)(227\times10^9)(0.00961)} = 2.731\times10^{-5}.$$

Thus, from Eq. (14.73),

$$p_H = E'\left(\frac{W'}{2\pi}\right)^{1/2} = (227\times10^9)\left(\frac{2.731\times10^{-5}}{2\pi}\right)^{1/2} = 474 \text{ MPa}.$$

From Eq. (14.75), and using the results from Example 14.9,

$$\sigma_c = p_H\left(K_o K_s K_m K_v\right)^{1/2} = (474 \text{ MPa})\left[(1.0)(1.0)(1.196)(1.32)\right]^{1/2} = 595 \text{ MPa}.$$

Therefore, the safety factor is:

$$n_s = \frac{1260}{595} = 2.11.$$

14.13 Elastohydrodynamic Film Thickness

The AGMA approach for designing gears based on tooth bending or contact stress has been described at length in this chapter. However, it has been recognized that in order to achieve their calculated capacity, gears must be adequately lubricated. In fact, AGMA states that effective lubrication is a necessary condition for applying their gear design standards.

The importance of gear lubrication is especially imperative in the design of speed reducers and gear trains. One can imagine a speed reducer that operates in three stages, designed so that the bending and contact stresses are the same on each set of gears. It would be expected that damage would be incurred by each gear, but that failure would be most likely on the highest speed gearset, since it encounters the most cycles. In practice, however, the opposite is true: the gears that operate at the lowest speed always fail first, while the high speed gears often look undamaged. This behavior can only be explained by considering the fluid film thickness developed.

The minimum-film-thickness formula for elastohydrodynamic lubrication (EHL) developed in Chapter 13 is equally applicable to gears or any other nonconformal machine element as long as the appropriate representations of the geometry, load, and speed are made. AGMA recommends that the film parameter, Λ, defined in Eq. (8.27) have a value of at least 0.4 assuming that boundary additives are included in the lubricant formulation. The relative life of a gear as a function of film parameter was included in Fig. 13.25, as well as in Fig. 14.26b, where the stress cycle factor, Z_n takes on different values depending on lubrication regime. Further information is contained in AGMA 925-A03 [2003].

Gears are usually lubricated under elastohydrodynamic lubrication, as discussed in Sections 8.7.5 and 13.8. The Hamrock-Dowson equations for hard EHL are used to determine film thickness. However, the lubrication velocities given by Eqs. (14.29) and (14.30) are used along with the contact stress calculation approach described in Section 14.12. Design Procedure 14.4 summarizes the approach needed to determine the lubricant film thickness for gears.

Design Procedure 14.4: Lubricant Film Thickness

1. It will be assumed that the power transmitted, gear and pinion angular velocity, number of teeth in pinion and gear, gear materials, lubricant properties, and geometry of the pinion and gear are known or can be determined from design constraints. Of these, the lubricant properties, especially the pressure exponent of viscosity, are most likely to be unknown or will have the largest uncertainty. Regardless, these properties should be attainable from the lubricant supplier.

2. The normal force acting on the gear teeth can be obtained from Eq. (14.50) as:

$$W = \frac{W_t}{\cos\phi}.$$

3. The lubrication velocity for the pinion and gears is obtained from Eqs. (14.29) and (14.30) as

$$\tilde{u} = \frac{u_p + u_g}{2} = \frac{\omega_p r_p \sin\phi + \omega_g r_g \sin\phi}{2}.$$

4. The effective radius is obtained from Eq. (14.74) as

$$\frac{1}{R_x} = \left(\frac{1}{r_p} + \frac{1}{r_g}\right)\frac{1}{\sin\phi}.$$

5. The effective modulus of elasticity is obtained from Eq. (8.16):

$$E' = \frac{2}{\dfrac{(1-\nu_a^2)}{E_a} + \dfrac{(1-\nu_b^2)}{E_b}}.$$

For steel-on-steel, $E' = 227.5$ GPa (32.97 Mpsi).

6. Equations (13.68), (13.69), and (13.72) yield the dimensionless load, speed, and materials parameters as:

$$W' = \frac{W}{b_w E' R_z}; \qquad U = \frac{\eta_o \tilde{u}}{E' R_x}; \qquad G = \xi E'.$$

7. The lubricant film thickness is then obtained from the Hamrock-Dowson equation for rectangular contacts given by Eq. (13.71):

$$H_{\min} = \frac{h_{\min}}{R_x} = 1.714(W')^{-0.128}U^{0.694}G^{0.568}.$$

Example 14.11: Lubricant Film Thickness

Given: A spur pinion and gear have pitch radii of 50 and 75 mm, respectively, and a pressure angle of 20°. The angular velocity of the gear is 210 rad/s, the face width is 30 mm, and the transmitted load is 2250 N. The properties of the lubricant and the gear materials are $\eta_o = 0.075$ Pa-s, $\xi = 2.2\times10^{-8}$ m^2/N, $E = 207$ GPa, and $\nu = 0.3$. The surface roughness of the gear is 0.3 μm.

Find: Calculate the minimum film thickness and the dimensionless film parameter of the meshing gear teeth.

Solution: This solution follows the steps in Design Procedure 14.4.

1. Most of the important parameters are given, except that the angular velocity of the pinion needs to be obtained from Eq. (14.18) as

$$\omega_p = \frac{r_g\omega_g}{r_p} = \frac{(0.075)(210)}{0.050} = 315 \text{ rad/s}.$$

2. The transmitted load W_t was given as 2250 N. The normal load, from Eq. (14.50), is

$$W = \frac{W_t}{\cos\phi} = \frac{2250}{\cos 20^\circ} = 2394 \text{ N}.$$

3. The lubrication velocity is then

$$\begin{aligned} \tilde{u} &= \frac{\omega_p r_p \sin\phi + \omega_g r_g \sin\phi}{2} \\ &= \frac{(315)(0.050)\sin 20^\circ + (210)(0.075)\sin 20^\circ}{2} \\ &= 5.387 \text{ m/s}. \end{aligned}$$

4. The effective radius is calculated as:

$$\begin{aligned} \frac{1}{R_x} &= \left(\frac{1}{r_p} + \frac{1}{r_g}\right)\frac{1}{\sin\phi} \\ &= \left(\frac{1}{0.050} + \frac{1}{0.075}\right)\frac{1}{\sin 20^\circ} \\ &= 97.46 \text{ m}^{-1}, \end{aligned}$$

so that $R_x = 0.01026$ m.

5. The effective elastic modulus is given by Eq. (8.16):

$$E' = \frac{2}{\dfrac{1-\nu_p^2}{E_p} + \dfrac{1-\nu_g^2}{E_g}} = \frac{\left(207 \times 10^9\right)}{1 - 0.3^2} = 227.5 \text{ GPa}.$$

6. The dimensionless load, speed, and materials parameters can then be calculated as:

$$\begin{aligned} W' &= \frac{W}{b_w E' R_x} \\ &= \frac{2394}{(0.030)\left(227.5 \times 10^9\right)\left(1.026 \times 10^{-2}\right)} \\ &= 3.419 \times 10^{-5}; \\ U &= \frac{\eta_o \tilde{u}}{E' R_x} = \frac{(0.075)(5.387)}{\left(227.5 \times 10^9\right)\left(1.026 \times 10^{-2}\right)} \\ &= 1.731 \times 10^{-10}; \end{aligned}$$

$$G = \xi E' = \left(2.2 \times 10^{-8}\right)\left(227.5 \times 10^9\right) = 5005.$$

7. The dimensionless lubricant film thickness is then obtained as:

$$\begin{aligned} H_{\min} &= \frac{h_{\min}}{R_x} = 1.714(W')^{-0.128}U^{0.694}G^{0.568} \\ &= 135.3 \times 10^{-6}. \end{aligned}$$

Therefore, since $R_x = 0.01026$, $h_{\min} = 1.388\ \mu$m. From Eq. (8.32) the dimensionless film parameter is

$$\Lambda = \frac{h_{\min}}{\left(R_{qa}^2 + R_{qb}^2\right)^{\frac{1}{2}}} = \frac{(1.388)\left(10^{-6}\right)}{\left(10^{-6}\right)\left(0.3^2 + 0.3^2\right)^{\frac{1}{2}}} = 3.272.$$

14.14 Gear Design Synthesis

The previous sections presented a mathematically robust approach for analyzing gears. Designers also must be able to perform design synthesis; that is, they must be able to establish gear characteristics based on performance criteria such as speed reduction, torque, or power transmission, etc. Design Procedure 14.5 is intended to help engineers in the synthesis of gear designs. The result is a starting point for further design analysis and optimization.

Design Procedure 14.5 presents a design approach using SI units, and is limited to fine-toothed steel gears. This design procedure is adapted from a more detailed approach [AGMA 1997].

Design Procedure 14.5: Gear Design Synthesis

1. From design requirements, determine the power transfer, total life required, and pinion and gear speed.

2. The gear ratio, g_r, must be known from design requirements. If this value is near or in excess of 6:1, a second stage is advisable.

3. If the pinion face-to-diameter ratio is not a design requirement, it can be estimated from the following equation for spur gears:

$$\frac{b_w}{d_p} = \frac{g_r}{g_r + 1}, \tag{14.76}$$

where $g_r = N_g/N_p$ is the gear ratio.

4. Estimate K_o from Table 14.8.

5. The load distribution factor, K_m, cannot be determined until knowledge of the design, manufacturing approach, and mounting is established. A rough approximation based on pinion torque can be made according to

$$K_m = 1 + \frac{b_w}{d_p}\left[0.2 + 0.0112\left(\frac{T_p K_o}{b_w/d_p}\right)^{0.33}\right]. \tag{14.77}$$

However, if the pitch diameter is known (from design requirements or selected based on experience), then the load distribution factor can be more accurately estimated from

$$K_m = 1 + \frac{b_w}{d_p}(0.2 + 0.0012 d_p). \quad (14.78)$$

Recall that the torque is related to the power by $h_p = T\omega$.

6. The dynamic factor, K_v, depends on gear speed and quality. For simplicity, a value of $K_v = 1.43$ can be assumed, which is conservative for most applications.

7. The pressure angle, ϕ, is generally chosen as 20°, since these gears are widely available. However, the pressure angle can be reduced to obtain higher contact ratios, or increased when precision and noise are not issues. Regardless, it is recommended that standard values of 14.5°, 17.5°, 22.5°, or 25° be used.

8. A geometry factor for spur gears is given by:

$$I = \frac{\sin\phi\cos\phi}{2}\frac{g_r}{g_r \pm 1}, \quad (14.79)$$

where g_r is the gear ratio. Note that the plus sign in the denominator of Eq. (14.79) applies for external gearsets, the negative sign for internal gearsets.

9. Y_j can be conservatively estimated as approximately 0.45 (see Fig. 14.30).

10. The life factors Y_n and Z_n can be estimated from Fig. 14.26.

11. Obtain the bending (σ_b) and contact (σ_c) strengths for the gear material according to Eqs. (14.45) or (14.46), respectively.

12. An estimate for the preferred number of pinion teeth is then:

$$N_p = \frac{1}{K_b}\frac{Y_j}{I}\frac{\sigma_b}{\sigma_c^2}\frac{E'}{2\pi}\frac{n_{sc}^2}{n_{sb}}. \quad (14.80)$$

For the special case of steel gears ($E' = 227$ GPa) a pressure angle of $\phi = 20°$, $K_b = 1$, $n_{sc} = n_{sb} = n_s$, and $Y_j = 0.45$, the preferred number of pinion teeth becomes

$$N_p = (101.17\text{ GPa})\frac{g_r + 1}{g_r}\frac{\sigma_b}{\sigma_c^2}n_s. \quad (14.81)$$

13. Obtain the pitting resistance constant:

$$K_c = \frac{0.3183 E' h_p K_o K_m K_v}{I\omega_p}\left(\frac{n_{sc}}{\sigma_c}\right)^2. \quad (14.82)$$

14. The pinion pitch diameter is then estimated as

$$d_p = \left(\frac{K_c}{b_w/d_p}\right)^{1/3}. \quad (14.83)$$

15. The face width can be calculated from the ratio of b_w/d_p established in step 3 if a specific value was not specified as a design requirement.

16. When the gear profile is selected from this design procedure, it is necessary to analyze the design more closely in accordance with the approaches described in Sections 14.11 and 14.12. This is clearly required since a number of approximate values were used.

17. It should also be noted that an essential part of the gear design synthesis is calculating lubricant film thickness as described in Section 14.13.

Example 14.12: Design Synthesis of a Pinion and Gear

Given: For a given pinion, the bending strength is $S_b = 380$ MPa and $S_c = 1260$ MPa. The gears transmit 15 kW at a pinion speed of 1260 rpm (132 rad/s). Assume the pressure angle is 20° and that a gear ratio of $g_r = 5.00$ is required. Assume $K_o = K_b = Z_n = Y_n = 1.0$.

Find: Estimate the number of teeth for the pinion and gear, as well as the pitch diameter and center distance for the two shafts for a safety factor of $n_{sc} = n_{sb} = 2.0$.

Solution: This solution follows the steps in Design Procedure 14.5.

1. Of the specified properties, the gear speed still needs to be obtained. From Eq. (14.19),

$$\omega_g = \frac{\omega_p}{g_r} = \frac{132}{5.00} = 26.4\text{ rad/s}.$$

2. The gear ratio of 5.00 is reasonable for a single stage of gearing, so a second stage will not be considered in the initial design synthesis.

3. The face-to-diameter ratio for the pinion is given by Eq. (14.76) as:

$$\frac{b_w}{d_p} = \frac{g_r}{g_r + 1} = \frac{5}{6} = 0.833.$$

4. The overload factor was given as $K_o = 1.0$.

5. The torque acting on the pinion is given by

$$T = \frac{h_p}{\omega_p} = \frac{15,000}{132} = 114\text{ N-m}.$$

So that the load distribution factor is estimated from Eq. (14.77) as

$$\begin{aligned} K_m &= 1 + (0.833)\left[0.2 + 0.0112\left(\frac{(114)(1.0)}{0.833}\right)^{0.333}\right] \\ &= 1.21. \end{aligned}$$

6. The dynamic factor is taken as $K_v = 1.43$.

7. The pressure angle was prescribed to be 20°, but this is also a good choice since it is a standard pressure angle that will be widely available.

8. From Eq. (14.79),

$$I = \frac{\sin\phi\cos\phi}{2}\frac{g_r}{g_r+1} = \frac{\sin 20^\circ \cos 20^\circ}{2}\frac{5}{6} = 0.134.$$

9. Y_j is assigned the vale of 0.45.

10. The life factors were assumed to be $Y_n = Z_n = 1.0$ in the problem statement. However, the approach described in Example 14.8 could be used to obtain Y_n and Z_n from the number of stress cycles and characteristics of the gear material and lubrication condition.

11. None of the correction factors in Eqs. (14.45) and (14.46) need to be considered, so that $\sigma_{b,\text{all}} = S_b = 380$ MPa and $\sigma_{c,\text{all}} = S_c = 1240$ MPa.

12. An estimate for the preferred number of teeth, from Eq. (14.80) is:

$$\begin{aligned} N_p &= \frac{1}{K_b}\frac{Y_j}{I}\frac{\sigma_b}{\sigma_c^2}\frac{E'}{2\pi}\frac{n_{sc}^2}{n_{sb}} \\ &= \frac{(0.45)\left(380\times10^6\right)\left(227\times10^9\right)(2.0)}{2\pi(0.134)\left(1240\times10^6\right)^2} \\ &= 54.0. \end{aligned}$$

13. The pitting resistance constant is obtained from Eq. (14.82):

$$K_c = \frac{(0.3183)E'h_pK_oK_mK_v}{I\omega_p}\left(\frac{n_{sc}}{\sigma_c}\right)^2,$$

which is solved as $K_c = 2.677\times10^{-4}$ m^3.

14. The estimate of the pinion diameter is, from Eq. (14.83),

$$d_p = \left(\frac{K_c}{b_w/d_p}\right)^{1/3} = \left(\frac{2.667\times10^{-4}}{0.833}\right)^{1/3} = 0.0685 \text{ m}.$$

15. The face width follows from step 3 as

$$b_w = 0.833d_p = 0.833(0.0685) = 0.057 \text{ m}.$$

Note that the pinion has a module of around 1.25 mm. To maintain the required gear ratio, the gear would use 270 teeth and have a diameter of 0.342 m.

This is a rough estimate of a starting point for design. As stated in Design Procedure 14.5, this gear should be further analyzed using the approaches described in Sections 14.11 and 14.12. This has been done in Examples 14.9 and 14.10. Note from the results of these examples that the initial pinion and gear designs have been confirmed, so that all of the design constraints have been met.

Case Study 14.1: Powder Metal Gears

The production of powder metal (PM) parts was briefly introduced in Section 3.8.1. With respect to gears, there are two principal methods of production: pressing and sintering, or metal injection molding. Additionally, any PM process can by used to produce a gear blank that is further processed by machining. Each has its own advantages, and results in gears that are ideal for their applications. This Case Study will provide an overview of powder metal gears, their processing, and their applications.

Processing

The production of powder metal gears begins with the blending of powders of different constituents, such as iron, nickel, chrome, etc. Another option is to create powders of a metal alloy, so that each powder contains the desired constituents. In order to improve part strength, powders may also be blended with small amounts of solid lubricants such as graphite or stearic acid.

For parts that will be pressed and sintered, the mixture is then fed into a die cavity and formed into a **green compact** by a mechanical press, as shown in Fig. 14.36. The green compact is very weak and brittle at this point, having strength similar to chalk. The green compact is then placed in a controlled-atmosphere furnace and sintered, so that the relatively weak bond between contacting powder particles increases due to diffusion and second phase transport. Shrinkage and some thermal distortion occurs in sintering. Details of this process can be found in Kalpakjian and Schmid [2010].

Metal injection molded gears consist of metal powders mixed with a polymer carrier, which is then injection molded at elevated temperature into a mold with a cavity that produces the desired gear shape. The molded part needs to be placed in a furnace to remove the binder polymer, then it is also sintered to develop its full strength.

Subsequent processing steps can include hardening, shaving, and roll densification. Hardening generally involves elevating the temperature in a furnace to induce a favorable phase change in the gears; this can include carburizing to also produce a compressive residual stress, and it can be combined with the sintering step (referred to as sinter hardening). Sintering is associated with part distortion, so a shaving step may be required for higher quality index applications.

In roll densification, a PM gear is meshed against a hardened gear and subjected to a high contact load. The resulting contact stress causes plastic deformation in the PM gear surface, resulting in a reduction in porosity at surface layers, as shown in Fig. 14.37. Roll densification also produces mirror-like surfaces on the gear flank face, thereby improving the film parameter for a given lubricant film thickness (see Section 14.13). The densified layer results in allowable bending stresses that are comparable to wrought materials.

One of the main advantages to powder metallurgy is that it is a net- or near-net shape process; that is, little or no machining is needed to achieve final dimensions. Machining is relatively slow and expensive; the reduction or elimination of machining is a major reason that PM gears are cost-effective, even though the raw materials are slightly more expensive than castings or forged stock.

Low Power Applications

There are a large number of applications where the power transmitted is very low, typically a small fraction of one horsepower (less than 750 watts). Typical applications include office printers, medical devices and instruments, automobile window gears, general motion control, and even hand tools.

Not only are these applications characterized by low power transmission, but they also are forgiving with respect to backlash and have relatively short lives. For these reasons,

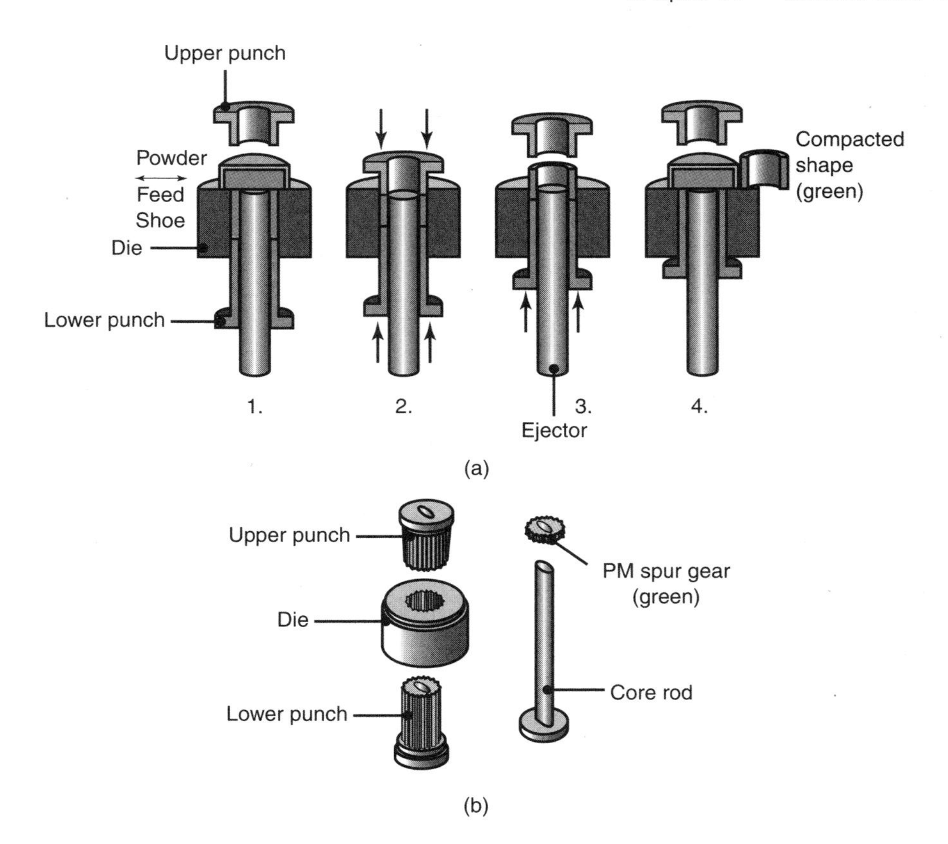

Figure 14.36: Production of gears through the powder compaction process. (a) Steps required to produce a part; (b) illustration of tooling required for a simple spur gear. *Source:* After Kalpakjian and Schmid [2010].

lower values of the quality index are acceptable, typically 4 to 6. Such gears can be produced directly from metal injection molding without any subsequent machining, making this process the most economical alternative.

Metal injection molding is used for lower quality index gears, and is suitable for producing spur gears, as well as some helical and bevel gears. The versatility of this process allows designers to combine multiple features into a single part, so that a gear can be combined with a pulley or cam, or a stepped gear can be produced (Fig. 14.38). It should be noted that MIM does require relatively high production runs because of high tooling costs.

High Power Applications

The technology associated with PM parts and materials has advanced considerably in the past 10 years, as can be seen from the increase in the use of PM gears. Generally, the pressed-and-sintered approach is used for higher power applications, as the resultant porosity is lower and thus the material strength is higher. To be suitable for PM, gear geometry has some restrictions, namely that the gear teeth cannot have an excessively fine pitch or large face widths. Generally, the face width cannot exceed 25 mm or so.

With advances in roll densification of gears discussed above, PM gears can have similar bending and contact strengths as wrought steel gears. Typical values of quality index after roll densification are between 6 and 9, but are produced more economically than machining. Figure 14.39 shows a collection of PM gears for automotive applications made by pressing with roll densification. Note that helical gears (Ch. 15) can also be produced.

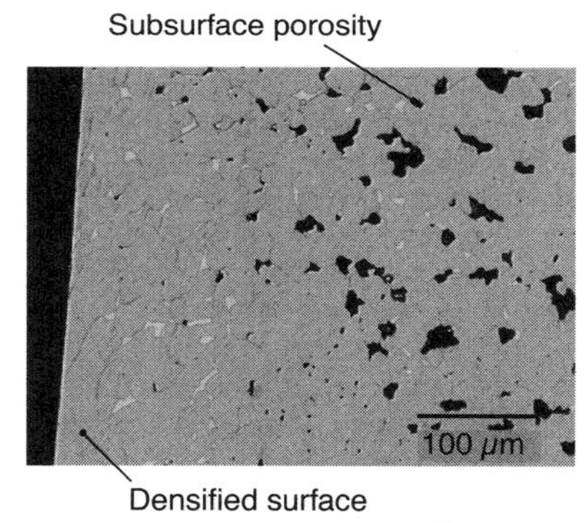

Figure 14.37: Image of a PM material after roll densification. Note the low porosity near the surface, resulting in high allowable contact stresses. *Source:* Courtesy of Capstan Atlantic Corp.

14.15 Summary

The primary function of gears is to transmit power and/or change speeds of shafts. This chapter logically developed the important methodology needed to successfully design gears. Primary emphasis was on the methodology of spur gear design. Various types of gears were discussed. Gears are divided into three major classes: parallel-axis gears, nonparallel and coplanar gears, and nonparallel and noncoplanar gears. Although spur gears were the major focus, other gear types were explained.

Figure 14.38: A stepped gear produced through powder compaction; similar stepped gears can be produced through metal injection molding. *Source:* Courtesy Perry Tool & Research, Inc.

Figure 14.39: A collection of PM gears used in automotive applications. *Source:* Courtesy of Capstan Atlantic Corp.

The geometrical factors as well as the kinematics, loads, and stresses acting in a spur gear were discussed. Gears transfer power by the driving teeth exerting a load on the driven teeth while the reacting loads act back on the teeth of the driving gear. The thrust load is the force that transmits power from the driving to the driven gear. It acts perpendicular to the axis of the shaft carrying the gear. This thrust load is used in establishing the bending stress in the tooth. The highest tensile bending stress occurs at the root of the tooth, where the involute curve blends with the fillet. The design bending stress is then compared with an allowable stress to establish whether failure will occur.

A second, independent, form of gear failure results from surface pitting and spalling, usually near the pitch line, where the contact stresses are high. Thus, failure due to fatigue occurs. The Hertzian contact pressure with modification factors was used to establish the design stress, which was then compared with an allowable stress to determine if failure due to fatigue would occur. It was found that if an adequate protective elastohydrodynamic lubrication film exists, gear life is greatly extended. Finally, a synthesis approach was developed that aides designers in the selection of gear parameters.

Key Words

backlash clearance measured on pitch circle

bevel gears nonparallel, coplanar gears with straight teeth

circular pitch distance from one tooth to corresponding point on adjacent tooth

diametral pitch number of teeth per pitch diameter

fundamental law of gearing law stating that common normal to tooth profile must pass through a fixed point, called the pitch point

gear toothed wheel that, when mated with another toothed wheel, transmits power between shafts

gear ratio ratio of number of teeth between two meshing gears

gear trains machine elements of multiple gears used to obtain desired velocity ratio

Lewis equation equation for bending stresses on gear teeth based on cantilever beam theory

modified Lewis equation Lewis equation with correction factors for gear geometry

module ratio of pitch diameter to number of teeth

pinion smaller of two meshed gears

pitch point point defined by normal to tooth profiles of meshing gears at point of contact on line of centers; see also *fundamental law of gearing*

spur gears parallel-axis gears with straight teeth

Summary of Equations

Gear Geometry

Center distance: $c_d = \dfrac{d_p + d_g}{2} = \dfrac{p_c N_p}{2\pi}(1 + g_r)$

Circular pitch: $p_c = \dfrac{\pi d}{N} = \dfrac{\pi d_p}{N_p} = \dfrac{\pi d_g}{N_g}$

Contact ratio:

$$\begin{aligned} C_r &= \frac{L_{ab}}{p_c \cos\phi} \\ &= \frac{1}{p_c \cos\phi}\left[\sqrt{r_{op}^2 - r_{bp}^2} + \sqrt{r_{og}^2 - r_{bg}^2}\right] - \frac{c_d \tan\phi}{p_c} \end{aligned}$$

Diametral pitch: $p_d = \dfrac{N_p}{d_p} = \dfrac{N_g}{d_g} = \dfrac{\pi}{p_c}$

Gear ratio: $g_r = \dfrac{d_g}{d_p} = \dfrac{N_g}{N_p} = \dfrac{\omega_p}{\omega_g}$

Module: $m = \dfrac{d_g}{N_g} = \dfrac{d_p}{N_p} = \dfrac{1}{p_d}$ (SI units only)

Velocity ratio:

Externally meshing gears: $Z_{21} = \dfrac{\omega_2}{\omega_1} = -\dfrac{N_1}{N_s}$

Internally meshing gears: $Z_{21} = \dfrac{\omega_2}{\omega_1} = \dfrac{N_1}{N_s}$

Simple train: $Z_{ji} = \dfrac{\omega_j}{\omega_i} = \pm\dfrac{N_i}{N_j}$

(positive for odd j, negative for even j)

Planetary gear trains:

Velocity ratios:

$$\frac{\omega_{\text{ring}} - \omega_{\text{arm}}}{\omega_{\text{sun}} - \omega_{\text{arm}}} = -\frac{N_{\text{sun}}}{N_{\text{ring}}}$$

$$\frac{\omega_{\text{planet}} - \omega_{\text{arm}}}{\omega_{\text{sun}} - \omega_{\text{arm}}} = -\frac{N_{\text{sun}}}{N_{\text{planet}}}$$

Numbers of teeth: $N_{\text{ring}} = N_{\text{sun}} + 2N_{\text{planet}}$

Gear Quality and Materials

Quality index:

$$Q_v = 0.5048 \ln N_i + 1.144 \ln m - 2.852 \ln(V_{PA} \times 10^4) + 13.664$$

AGMA allowable bending stress: $\sigma_{t,\text{all}} = \dfrac{S_t}{n_s} \dfrac{Y_N}{K_t K_r}$

AGMA allowable contact stress: $\sigma_{c,\text{all}} = \dfrac{S_c}{n_s} \dfrac{Z_N C_H}{K_t K_r}$

Bending Stresses in Gears

Modified Lewis equation: $\sigma_b = \dfrac{W_t p_d}{b_w Y_j}$

AGMA bending stress equation:

$$\sigma_b = \frac{W_t p_d K_o K_s K_m K_v K_b}{b_w Y_j}$$

Load distribution factor:

$K_m = 1.0 + C_{mc}\left(C_{pf} C_{pm} + C_{ma} C_e\right)$

Lead correction factor:

$C_{mc} = 1.0$ for uncrowned teeth,

$C_{mc} = 0.8$ for crowned teeth.

Pinion proportion factor:

for $b_w \leq 25$ mm

$$C_{pf} = \frac{b_w}{10 d_p} - 0.025$$

for 25 mm $< b_w \leq 432$ mm,

$$C_{pf} = \frac{b_w}{10 d_p} - 0.0375 + 0.000492 b_w$$

for 432 mm $< b_w \leq 1020$ mm,

$$C_{pf} = \frac{b_w}{10 d_p} - 0.1109 + 0.000815 b_w - (3.53 \times 10^{-7}) b_w^2$$

Pinion proportion modifier:

$$C_{pm} = \begin{cases} 1.0, & (S_1/S) < 0.175 \\ 1.1, & (S_1/S) \geq 0.175 \end{cases}$$

Dynamic factor: $K_v = \left(\dfrac{A + C\sqrt{v_t}}{A}\right)^B$

$A = 50 + 56(1.0 - B)$

$B = 0.25(12 - Q_v)^{0.667}$

$C = 1$ for v_t in ft/min or 14.14 for v_t in m/s.

Contact Stresses

Maximum Hertzian pressure: $p_H = E' \left(\dfrac{W'}{2\pi}\right)^{\frac{1}{2}}$

Effective modulus of elasticity:

$$E' = \frac{2}{\dfrac{1 - \nu_a^2}{E_a} + \dfrac{1 - \nu_b^2}{E_b}}$$

Dimensionless load: $W' = \dfrac{w'}{E' R_x}$

Effective radius:

$$\frac{1}{R_x} = \left(\frac{1}{r_p} + \frac{1}{r_g}\right) \frac{1}{\sin\phi} = \left(\frac{1}{d_p} + \frac{1}{d_g}\right) \frac{2}{\sin\phi}$$

AGMA Equation:

$$\sigma_c = E' \left(\frac{W' K_o K_s K_m K_v}{2\pi}\right)^{\frac{1}{2}} = p_H \left(K_o K_s K_m K_v\right)^{\frac{1}{2}}$$

Gear Lubrication

Normal force on teeth: $W = \dfrac{W_t}{\cos\phi}$

Lubrication velocity:

$$\tilde{u} = \frac{u_p + u_g}{2} = \frac{\omega_p r_p \sin\phi + \omega_g r_g \sin\phi}{2}$$

Effective radius: $\dfrac{1}{R_x} = \left(\dfrac{1}{r_p} + \dfrac{1}{r_g}\right) \dfrac{1}{\sin\phi}$

Effective elastic modulus: $E' = \dfrac{2}{\dfrac{(1 - \nu_a^2)}{E_a} + \dfrac{(1 - \nu_b^2)}{E_b}}$

Dimensionless parameters:

Load parameter: $W' = \dfrac{W}{b_w E' R_z}$

Speed parameter: $U = \dfrac{\eta_o \tilde{u}}{E' R_x}$

Materials parameter: $G = \xi E'$

Hamrock-Dowson Equation:

$H_{\min} = \frac{h_{\min}}{R_x} = 1.714 (W')^{-0.128} U^{0.694} G^{0.568}$

Recommended Readings

Budynas, R.G., and Nisbett, J.K. (2011), *Shigley's Mechanical Engineering Design*, 9th ed., McGraw-Hill.

Dudley, D.W. (1994) *Handbook of Practical Gear Design*, CRC Press.

Hamrock, B.J., Schmid, S.R., and Jacobson, B.O. (2004) *Fundamentals of Fluid Film Lubrication*, 2nd. ed., Marcel Dekker.

Juvinall, R.C., and Marshek, K.M. (2012) *Fundamentals of Machine Component Design*, 5th ed., Wiley.

Litvin, F.L. (2004) *Gear Geometry and Applied Theory*, 2nd ed., Cambridge University Press.

Mott, R. L. (2014) *Machine Elements in Mechanical Design*, 5th ed., Prentice Hall.

Norton, R.L. (2011) *Machine Design*, 4th ed., Prentice Hall.

Radzevich, S.P. (2011) *Gear Cutting Tools*, CRC Press.

Stokes, A. (1992) *Gear Handbook*, Butterworth Heinemann.

Townsend, D.P. (1992) *Dudley's Gear Handbook*, 2nd ed., McGraw-Hill.

References

AGMA (1989) *Geometry Factors for Determining the Pitting Resistance and Bending Strength of Spur, Helical, and Herringbone Gear Teeth*, ANSI/AGMA Standard 908–B89, American Gear Manufacturers Association.

AGMA (2003) *Effect of Lubrication on Gear Surface Distress*, AGMA Standard 925–A03, American Gear Manufacturers Association.

AGMA (1997) *A Rational Procedure for the Preliminary Design of Minimum Volume Gears*, AGMA Information Sheet 901-A92.

AGMA (2010) *Fundamental Rating Factors and Calculation Methods for Involute Spur and Helical Gear Teeth*, ANSI/AGMA Standard 2101-D04, American Gear Manufacturers Association.

Drago, R.J. (1992) Gear Types and Nomenclature, Chapter 2 in *Dudley's Gear Handbook*, D.P. Townsend, Ed., 2nd ed., McGraw-Hill.

Kalpakjian, S., and Schmid, S.R. (2010), *Manufacturing Engineering and Technology*, 6th ed., Prentice-Hall.

Lewis, W. (1892) "Investigation of the Strength of Gear Teeth," an Address to the Engineer's Club of Philadelphia, Oct. 15, 1892.

Mechalec, G.W. (1962) *Gear Handbook*, McGraw-Hill, Chapter 9.

Mott, R. L. (2003) *Machine Elements in Mechanical Design*, 4th ed., Prentice Hall.

Townsend, D.P. (1992) *Dudley's Gear Handbook*, 2nd ed., McGraw-Hill.

Questions

14.1 What is the difference between a pinion and a gear?

14.2 Define *spur*, *helical*, *worm*, and *bevel* gears.

14.3 What is the name of the line where spur gear contact occurs?

14.4 Define *addendum* and *dedendum*. Which is larger? Why?

14.5 Why is backlash necessary?

14.6 What is the circular pitch of a gear? The diametral pitch?

14.7 What is the module of a gear?

14.8 Which is largest in a planetary gear train, the ring, sun, or planet?

14.9 What is hobbing?

14.10 On what factors does the allowable bending stress depend?

14.11 What is the Lewis equation?

14.12 What factors can lead to uneven load distribution across the thickness of a spur gear tooth?

14.13 What is meant by *crowned* teeth?

14.14 What is the dynamic factor? On what variables does it depend?

14.15 What is elastohydrodynamic lubrication? How does it apply to gears?

Qualitative Problems

14.16 Explain the significance of the contact ratio.

14.17 Explain the significance of the quality index, Q_v, for gears. How is quality index related to the manufacturing process used?

14.18 What are the benefits of having a high quality factor in a gearset? What are the disadvantages?

14.19 Explain the advantages and limitations to the reliability factor defined in Table 14.6, especially for a large number of stress cycles.

14.20 The chapter discussed both failure of teeth due to bending stresses as well as surface pitting. Explain under what conditions each of these failure modes will occur for a given gear set.

14.21 Explain why proper lubrication is important for gears.

14.22 *Hunting ratio* is a ratio of numbers of gear and pinion teeth, which ensures that each tooth in the pinion will contact *every* tooth in the gear before it contacts any tooth a second time. (For example, 13 to 48 is a hunting ratio; 12 to 48 is not a hunting ratio.) Give three examples of hunting ratios. Explain the advantages of using hunting ratios in mating gears.

14.23 It seems that contact ratio does not appear in Eqs. (14.58) or (14.73). Is this true? Explain how the contact ratio influences the design approach.

14.24 Table 14.7 and Fig. 14.30 give form factors for gear teeth, but the fewest number of teeth considered is 10. With appropriate sketches, explain why fewer gear teeth are not often encountered.

14.25 Explain the significance of the quality index, Q_v.

14.26 Explain how the line of contact is defined. What is the length of contact?

14.27 Using the procedure in Design Procedure 14.1, produce a careful drawing of a gear with a 100-mm diameter and 20 teeth.

14.28 Construct a figure similar to Fig. 14.5, but for an internal contact.

14.29 One design approach for polymer gears is to use a larger addendum and dedendum than those in Table 14.2. What are the advantages and disadvantages of this approach?

14.30 Shot peening is a valuable approach for imparting a compressive residual stress on metal surfaces, but it does result in a rough surface. Under what conditions would you consider using a peened gear? Explain.

Quantitative Problems

14.31 A pinion has a pressure angle of 20°, a module of 3 mm, and 20 teeth. It is meshed with a gear having 32 teeth. The center distance between the shafts is 81 mm. Determine the gear ratio and the diametral pitch. *Ans.* $g_r = 1.60$, $p_d = 0.333\text{ mm}^{-1}$.

14.32 A gearbox has permanent shaft positions defined by the bearing mounting positions, but the gear ratio can be changed by changing the number of teeth in the pinion and gear. To achieve similar power transmission ability for different gear ratios, a manufacturer chooses to have the same module for two different gearboxes. One of the gearboxes has a pinion with 22 teeth, a gear with 68 teeth, and a center distance of 225 mm. How large is the gear module and which gear ratios are possible for a pinion with 22 or more teeth using the same module? *Ans.* $m = 5$ mm.

14.33 The blank for a normal gear with 20 teeth was made too large during manufacturing. Therefore, the addendum for the gear teeth was moved outward, and the tooth thickness at the tops of the teeth was decreased. The module for the rack tool was 3 mm. Find how large the maximum gear outside radius can be when the tool flank angle is 20° to avoid zero thickness at the gear teeth tops. *Ans.* $r_{\text{max}} = 41$ mm.

14.34 The inlet and outlet speeds for a reduction gear train are 2000 and 320 rpm, respectively. The pressure angle is designed to be 25°. The gears have a 5-mm module and their center distance is 435 mm.

(a) Determine the number of teeth for both gears, the pitch diameters, and the radii of the base, addendum, and dedendum circles. *Ans.* $N_p = 24$, $N_g = 150$, $d_p = 120$ mm, $d_g = 750$ mm.

(b) Find the change in center distance if the pressure angle changes from 25° to 29°. *Ans.* $\Delta c_d = 3.27$ mm.

14.35 The base circle radii of a gear pair are 100 and 260 mm, respectively. Their module is 8 mm and the pressure angle is 20°. Mounting inaccuracy has caused a 4-mm change in center distance. Calculate the change in the pitch radii, the center distance, and the pressure angle. *Ans.* $\Delta r_p = 1.1$ mm, $\Delta r_g = 2.90$ mm.

14.36 A gear train has a 50.3-mm circular pitch and a 25° pressure angle and meshes with a pinion having 12 teeth. Design this gear train to minimize its volume and the total number of teeth. Obtain the pitch diameters, the number of teeth, the speed ratio, the center distance, and the module of this gear train. The gearbox housing has a diameter of 620 mm. *Ans.* $d_p = 192$ mm, $g_r = 2$, $c_d = 288$ mm.

14.37 The pinion of the gear train for an endurance strength test machine (Sketch *a*) has 5 kW of power and rotates clockwise at 1200 rpm. The gears have a 12.57-mm circular pitch and a 20° pressure angle.

(a) Find the magnitude and direction of the output speed. *Ans.* $\omega = 640$ rpm.

(b) Calculate the center distance between the input and output shafts. *Ans.* $c_d = 330$ mm.

(c) Draw free-body diagrams of forces acting on gears 3 and 4.

Sketch *a*, for Problem 14.37

14.38 A standard straight gear system consists of two gears with tooth numbers $N_g = 50$ and $N_p = 30$. The pressure angle, ϕ, is 20°; the diametral pitch, p_d, is 5.08 per inch. Find the center distance, c_d, and the contact ratio, C_r. *Ans.* $c_d = 7.874$ in, $C_r = 1.702$.

14.39 Given a straight gear with diametral pitch $p_d = 6.35$ per inch, pressure angle $\phi = 20°$, $N_g = 28$, and $N_p = 20$, find the contact ratio C_r. *Ans.* $C_r = 1.594$.

14.40 Given an internal straight gear with diametral pitch $p_d = 5.08$ per inch and $N_p = 22$, find the contact ratio, C_r, if the center distance $c_d = 4.33$ in. and pressure angle is $\phi = 20°$. *Ans.* $C_r = 1.933$.

14.41 A straight gear with diametral pitch $p_d = 6.35$ per inch and face width $b_w = 1.575$ in. has a center distance of $c_d = 4.724$ in. The pinion has 25 teeth, the speed is 1000 rpm, and the pressure angle is $\phi = 20°$. Find the contact stress at the pitch point when the gear transmits 11,000 ft-lb/s. Neglect friction forces. The gear steel has modulus of elasticity $E = 30 \times 10^6$ psi and a Poisson's ratio of 0.3. *Ans.* $p_H = 76.05$ ksi.

14.42 A spur gear has $N_g = 40$, $N_p = 20$, and $\phi = 20°$. The diametral pitch is $p_d = 5.08$ per inch. Find the center distances between which the gear can work (i.e., from the backlash-free state to $C_r = 1$). *Ans.* Between 5.9055 and 6.042 in.

14.43 Given a gear with the following data: $N_g = 45$; $N_p = 20$; $p_d = 6.35$ per inch; and $\phi = 20°$, find the center distance, c_d, the contact ratio, C_r, and the tooth thickness on the pitch circle for both the pinion and the gear. *Ans.* $c_d = 5.118$ in., $C_r = 1.647$, $t_h = 0.2474$ in.

14.44 Given a gear with the center distance, c_d, find how much the pressure angle, ϕ, increases if the center distance increases by a small amount Δc_d. *Ans.* $\Delta\phi = \dfrac{\Delta c_d}{c_d \tan\phi}$.

14.45 A straight gear is free of backlash at a theoretical center distance of $c_d = a = 0.264$ m $= 10.394$ in. The pressure angle is $\phi = 20°$, $N_g = 47$, $N_p = 19$, and the coefficient of thermal expansion is $\alpha = 11 \times 10^{-6}$ (°C)$^{-1}$. Find the center distance needed to give a backlash of 0.008378 in. (0.2128 mm). What gear temperature increase is then needed relative to the housing temperature to make the gear free from backlash? *Ans.* $\Delta t = 100.6$°C.

14.46 A gear with a center distance of 8 in. is to be manufactured. Different diametral pitches can be chosen depending on the transmitted torque and speed and on the required running accuracy. Which diametral pitch should be chosen to minimize gear angular backlash if the gear ratio is 2? *Ans.* 12 per in.

14.47 A pinion having 30 teeth, a circular pitch of 0.2618 in. and rotating at 2000 rpm is to drive a gear rotating at 500 rpm. Determine the diametral pitch, number of teeth in the gear, and the center distance. *Ans.* $p_d = 14.40$ in^{-1}, $N_g = 120$, $c_d = 6.250$ in.

14.48 What is the contact ratio for two spur gears having 20° pressure angles, addendums of 12.5 mm, and a velocity ratio of 0.5? The pinion has 30 teeth. *Ans.* $C_r = 1.48$.

14.49 A drive train needs to be developed for a rear-wheel-driven car. The gear ratio in the hypoid gear from the car transmission to the rear wheels is 4.5:1. The rear wheel diameter is 550 mm. The motor can be used in the speed interval 1500 to 6000 rpm. At 6000 rpm the speed of the car should be 200 km/hr in fifth gear. In first gear it should be possible to drive as slow as 10 km/hr without slipping the clutch. Choose the gear ratios for the five gears in the gearbox.

14.50 In the train drive shown in Sketch *b*, gear 2 is the inlet of the gear train and transfers 3.0 kW of power at 2000 rpm. Each gear contact reduces the speed ratio by 5:3.

(a) Find the number of teeth for gears 3 and 4. *Ans.* $N_3 = 50$, $N_4 = 120$.

(b) Determine the rotational speed of each gear. *Ans.* $\omega_3 = \omega_4 = 1200$ rpm, $\omega_5 = 720$ rpm.

(c) Calculate the torque of each shaft, when because of frictional losses there is a 5% power loss in each matching pair. *Ans.* $T_{\text{in}} = 14.32$ N-m, $T_{\text{out}} = 35.90$ N-m.

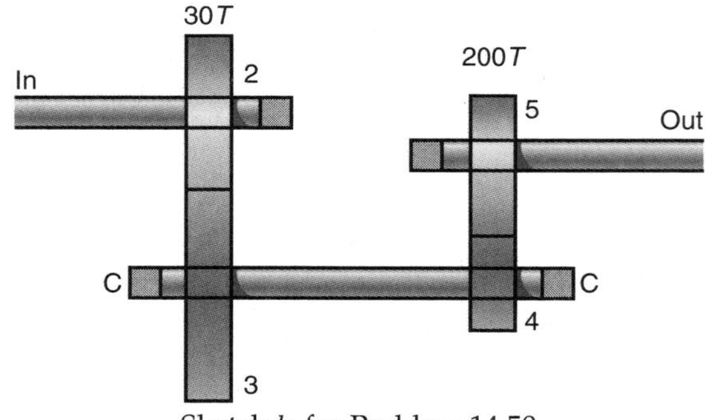

Sketch *b*, for Problem 14.50

14.51 In Problem 14.50, the module of all gears in the train drive is 8 mm and the pressure angle is 25°. Draw a free-body diagram of the forces acting on each gear in the train.

14.52 In the gear train shown in Sketch *c* the pinion rotates at 500 rpm and transfers 6.0 kW to the train. The circular pitch is 31.4 mm and the pressure angle is 25°. Find the output speed and draw a free-body diagram of the forces acting on each gear. *Ans.* $\omega_{\text{out}} = 200$ rpm.

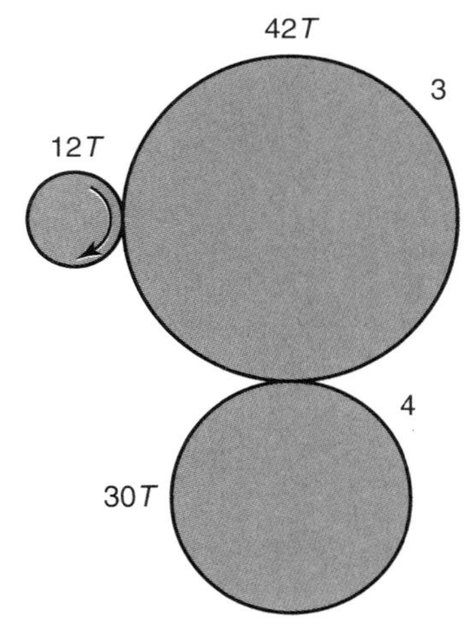

Sketch *c*, for Problem 14.52

14.53 The input shaft of the planetary gear shown in Sketch *d* has 40 kW of power and is rotating clockwise at 500 rpm. Gears 4 and 5 have a module of 4 mm. The pressure angle is 20°. The inlet and outlet shafts are coaxial. Find the following:

(a) The output speed and its direction. *Ans.* $\omega = 2200$ rpm.

(b) The number of teeth in the stationary ring gear and its pitch diameter. *Ans.* $N_r = 85$, $d_{pr} = 340$ mm.

(c) The contact loads acting on each gear. *Ans.* $W_t = 4500$ N.

Note that the sketch does not show the stationary compound ring gear.

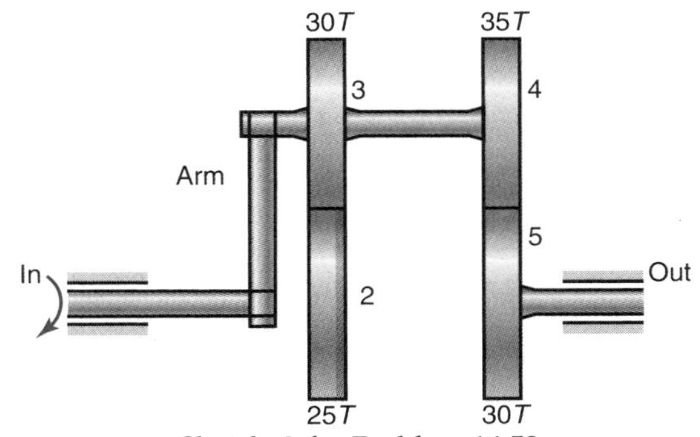

Sketch *d*, for Problem 14.53

14.54 An 18-tooth pinion rotates at 1420 rpm and transmits 52 kW of power to the gear train shown in Sketch *e*. The power of the output shaft is 35% of the input power of the train. There is 3% power loss in every gear contact. The pressure angle for all the gears is 20°. Calculate the contact forces acting on each gear. All dimensions are in millimeters. *Ans.* Gear 3: $W_t = 6.26$ kN, $W_r = 2.28$ kN.

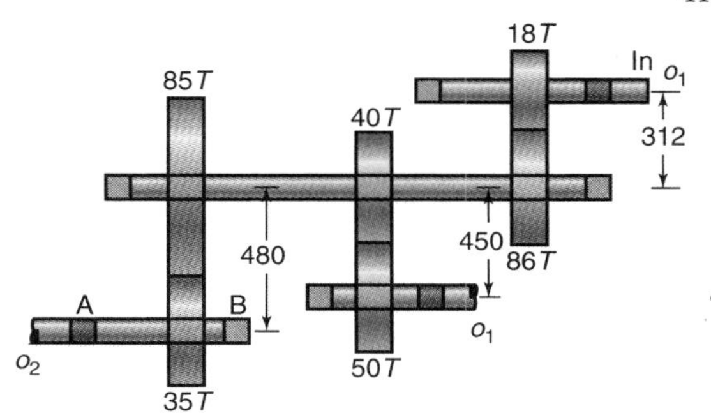

Sketch *e*, for Problem 14.54

14.55 The third gear in the gearbox of a sports car is dimensioned to have a finite life in order to save weight. When the car is used in a race, the load spectrum is totally different from the spectrum produced by normal driving. Therefore, the third gear fails by bending stresses at the gear roots and the gear teeth crack and fall off. For the load spectrum used to dimension the gearbox, bending stress failure at the gear tooth root has the same stress safety factor as surface pitting at the contact areas. The gears are made of Grade 2 steel with Brinell hardness HB = 250. By changing the material to 400 Brinell hardness, the bending stress failures disappear. Is there then any risk of surface pitting? *Ans.* No.

14.56 A speed of 520 rpm is used in a 2:1 gear reduction pair to transfer 3 kW of power for a useful life of 50 million turns with 90% reliability. A pressure angle of 20° is to be used along with a safety factor of 2. The loads are uniform with a regular mounting and hub-machined teeth. On the basis of contact stress determine the proper material, module, and number of teeth for this pair, and calculate the pitch diameters and the face width.

14.57 Determine the allowable bending and contact stresses for a Grade 1 steel through-hardened to 250 HB. Assume the desired reliability is 50% and that the pinion and gear have the same hardness, and it is expected that the gear will encounter 100,000 load cycles. *Ans.* $\sigma_{b,\text{all}} = 469.5$ MPa, $\sigma_{c,\text{all}} = 1552$ MPa.

14.58 Repeat Problem 14.57 if the gear encounters hydrodynamic lubrication and is to last ten million cycles. *Ans.* $\sigma_{b,\text{all}} = 316.6$ MPa, $\sigma_{c,\text{all}} = 1200$ MPa.

14.59 An endurance strength test machine (see Sketch *f*) consists of a motor, a gearbox, two elastic shafts, bearings, and the test gears. The gear forces are created by twisting the shafts when the gears are mounted. At one location the shafts were twisted at an angle representing one gear tooth. Find the gear force in the test gears as a function of the parameters given in the sketch. *Ans.*

$$P = \frac{G\pi^2 d^4 p_d}{16 N_p^2 l \cos\phi}.$$

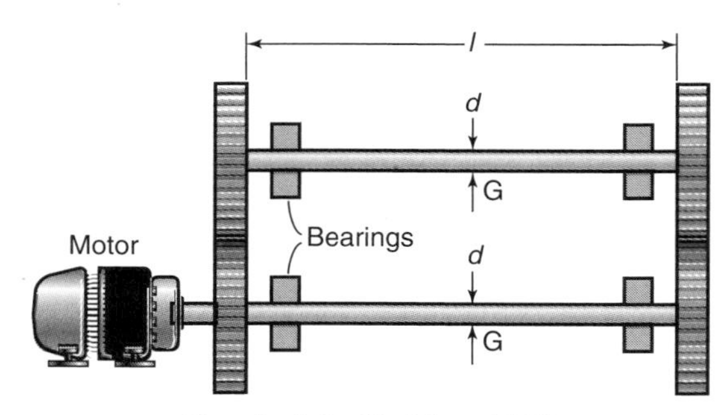

Sketch f, for Problem 14.59

14.60 For a 2:1 gear reduction the cast iron pinion is rotating at 320 rpm and transfers 5.2 kW of power. The gears are standard full depth with a 25° pressure angle. Design this gear pair for a safety factor of 2 based on bending stresses. Use Fig. 4.8 to estimate the gear module.

14.61 Electrical power sources such as motors operate at much higher speed than is needed by typical applications, and for that reason gear reducers are often needed. The quality of the reduction gear determines how large the dynamic impact forces will be in the output shaft gear. Hobbed gears of quality level 5 are used for a gear rotating at 500 rpm at a pitch diameter of 50 mm. To increase the throughput power, the speed is increased to 5000 rpm. How much higher power can be transferred through the gear at that speed? How high would the power be with extremely accurate gearing?

14.62 A 25-tooth pinion running at 1200 rpm and transferring 13 kW gives a 4:1 reduction gear pair. The gears are made of steel. The circular pitch is 15.7 mm, the face width is 49 mm, and the pressure angle is 20°. Determine the bending stress of the gears using the AGMA equation if the power source is uniform and the machine encounters moderate shock, the teeth are not crowned, and the gear is centered between supports. Use a quality factor of $Q_v = 10$ and assume that the gear will be checked at assembly to ensure proper alignment. Consider the gears to be commercial enclosed gears. *Ans.* $\sigma_t = 35.7$ MPa.

14.63 A standard commercial enclosed gearset ($Q_v = 6$) with crowned teeth and pressure angle $\phi = 20°$ and module $m = 3$ mm transfers 20 kW at a pinion rotational speed of 1500 rpm. The face width is $b_w = 15$ mm, and $N_p = 24$, $N_g = 42$, and $E = 207$ GPa for the gear steel. Find the contact stress from the AGMA equations. Assume $K_a = 1.0$. *Ans.* $\sigma_c = 1220$ MPa.

14.64 A gearbox with spur gears has a fixed shaft center distance c_d. The contact pressure between the gears at the pitch point limits the power being transmitted through the gearbox. The rotational speed of the incoming shaft is fixed. The rotational speed of the outgoing shaft varies with the gear ratio. Find the gear ratio at which the maximum power can be transmitted through the gearbox.

14.65 A standard gear has $p_d = 8.467$ per inch, $\phi = 20°$, $N_g = 40$, and $N_p = 20$. Find the contact pressure at the position where the top of the larger wheel contacts the pinion. The force is at that moment supposed to split equally between the two contacts. The transmitted power is 16,000 in-lb/s at 1500 rpm for the smaller wheel. The gear face width $b_w = 0.75$ in. and the modulus of elasticity $E = 30$ Mpsi. Neglect friction, use a quality factor of 7, assume commercial quality enclosure and mounting between bearings. *Ans.* $\sigma_c = 71.51$ ksi.

14.66 A spur gear has $m = 3$ mm, $\phi = 20°$, $N_g = 58$, and $N_p = 28$. The gear is loaded with 50 N-m of torque on the slow shaft. The gear face width is $b_w = 30$ mm and the modulus of elasticity $E = 210$ GPa with $\nu = 0.3$. Find the Hertzian contact pressure between the gears. *Ans.* $p_H = 399$ MPa.

14.67 An external spur gear has $p_d = 6.35$ per inch, $N_g = 52$, $N_p = 28$, $\phi = 20°$, pinion rotational speed of 1500 rpm, $b_w = 0.75$ in., and $E = 30$ Mpsi. Find the largest power the gear can transmit if the maximum allowable Hertzian contact pressure is $p_H = 142.1$ ksi. Ignore all correction factors except for the dynamic factor. *Ans.* $h_p = 456$ kip-in/s.

14.68 An 18-tooth pinion rotates at 1500 rpm, mates with a 72-tooth gear, and transfers 3.2 kW of power. The module is 4 mm and the pressure angle is 20°. If the gears are made of steel with a yield strength of 720 MPa, calculate the face width for a safety factor of 5 using the modified Lewis equation. Also, calculate the Hertzian contact stress between the mating teeth using this face width. *Ans.* $b_w = 3$ mm, $p_H = 859$ MPa.

14.69 A 16-tooth pinion rotates at 1000 rpm and transfers 5 kW to a 3:1 speed reduction gear pair. The circular pitch is 9.4 mm, the face width is 38 mm, and the pressure angle is 20°. Calculate the Hertzian contact stresses for this steel gear pair. Ignore correction factors. *Ans.* $p_H = 574$ MPa.

14.70 Two commercial enclosed steel gears (Grade 1 with Brinell Hardness of 300HB each, 20° pressure angle) and face width of 0.75 in. are in contact. The pinion has 12 teeth with a diametral pitch of 8 teeth per inch and the gear has 48 teeth. The pinion speed is 4000 rpm and 3100 ft-lb/s is transmitted. The gears are manufactured to a quality standard of $Q_v = 9$ and have a face width of 0.75 in. The electric power source is uniform, but the machine loading results in occasional light shock. Find the safety factor against failure due to excessive bending stresses. *Ans.* $n_s = 3.47$.

14.71 A spur gearset has 17 crowned teeth on the pinion and 51 crowned teeth on the gear. The pressure angle is 20°, the diametral pitch is 6 teeth per in. and the face width is 2 in. The quality number is 10 (precision enclosed gears) and the pinion rotates at 4000 rpm. The material is a through-hardened FL-4405 alloy powder metal.

(a) What is the center distance, gear ratio and circular pitch? *Ans.* $c_d = 5.667$ in., $g_r = 3$.

(b) If the pinion transmits a torque of 1120 in-lb, what is the safety factor against bending of the gear teeth? Use an overload factor of 1.0. *Ans.* $n_s = 4.38$.

14.72 A pair of crowned-tooth gears with pressure angles of 20° transmits 1 kW while the pinion rotates at 3450 rpm. The module is 1 mm, the pinion has 24 teeth, and the gear has 120 teeth. Both the pinion and the gear are Grade 2 steel and heat treated to 300HB. For a safety factor of 4, determine the required face width. The input power is from an electric motor, and the drive is for a small machine tool with light shock. Assume Q_v= 10 (precision enclosed gears) and $K_m = 1.0$. *Ans.* Use $b_w = 75$ mm.

14.73 Two gears (Grade 2 through-hardened steel with Brinell Hardness of 250HB each, 20° pressure angle) are in contact. The pinion has 12 crowned teeth with a diametral pitch of 8 teeth per inch and the gear has 48 crowned teeth. The pinion speed is 4000 rpm and 3100 ft-lb/s is transmitted. The gears are manufactured to a quality standard of $Q_v = 6$ and have a face width of 0.75 in. The electric power source is uniform, but the machine loading results in occasional light shock.

(a) What is the allowable bending stress and contact stress for these gears? *Ans.* $S_t = 41.9$ ksi, $S_c = 109.6$ ksi.

(b) Assume that the gears have been hardened further so that the allowable bending stress number is 50 ksi. Find the safety factor against failure due to excessive bending stresses. *Ans.* $n_s = 4.0$

14.74 Repeat part (a) of Problem 14.73 for the following circumstances, and for $n_s = 1$:

(a) The gear will see a life of 100,000 cycles with a reliability of 99.99%. $\sigma_{c,\text{all}} = 108.4$ ksi, $\sigma_{b,\text{all}} = 41.4$ ksi.

(b) The gear will see a life of 100 million cycles with a reliability of 90%, and it encounters mixed lubrication. $\sigma_{c,\text{all}} = 87.55$ ksi, $\sigma_{b,\text{all}} = 45.7$ ksi.

14.75 The gear considered in Problem 14.73 was mistakenly produced from Grade 1 steel, and was only hardened to 200 HB. What are the allowable bending and contact stresses for this case? Will the gear fail?

14.76 A vertical mixer uses an electric motor to drive the sun gear of a planetary gear train. The ring is attached to the machine frame, and the mixing attachment is attached to one of the planets. If the sun gear and the planet gears have 30 teeth, determine the angular velocity and articulating angular velocity of the mixing attachment if the sun gear is driven at 500 rpm. Repeat the problem if the sun gear has 20 teeth. *Ans.* $\omega_{\text{arm}} = 125$ rpm, $\omega_{\text{planet}} = -250$ rpm.

14.77 A planetary gear train has $N_{\text{sun}} = 30$, $N_{\text{planet}} = 60$, a diametral pitch of 6 in.$^{-1}$, a 20° pressure angle and is to transmit 6 hp. What torque can be delivered by the armature if the sun is driven and the ring is stationary? *Ans.* $T = 237.6$ in.-lb.

14.78 Design a suitable pitch and width for a pair of meshing gears that will transmit 7.5 hp. The center distance is 10 in., and the angular velocity ratio is to be 3/4. The pinion rotates at 150 rpm. Assume the pressure angle is 20°, and that the gears will be constructed from Grade 1 through-hardened steel with a Brinell hardness of 300. The power source and the driven machine are subjected to light shock, and a life of 10 million cycles and 99.99% reliability is needed. Assume $K_m = K_b = K_i = 1.0$. *Ans.* For $p_d = 6$ in^{-1}, $b_w = 1.125$ in.

14.79 Design a gear set using the approach in Design Procedure 14.5 to transmit 30 kW at a pinion speed of 2000 rpm. Assume the gear and pinion are constructed of through-hardened Grade 2 steel at 250 HB, 20° pressure angle, and that a speed reduction of 3.0 is required. The gears will see hydrodynamic lubrication and need to last 10^8 cycles at a reliability of 99.9%. *Ans.* $N_p = 67$, $N_g = 200$, $b_w = 66.4$ mm.

14.80 If the face-to-diameter ratio of the gear pair in Problem 14.79 is restricted to 0.25, what portion of the design analysis changes? By how much does the resultant center distance change? *Ans.* Increases by 73.2 mm.

14.81 An external gear has a module of 4, 52 teeth in the gear, 28 teeth in the pinion, a pressure angle of 20°, a pinion rotational speed of 1500 rpm, a face width of 20 mm, and gear and pinion material having modulus of elasticity of 206 GPa and Poisson's ratio of 0.3. The Hertzian contact pressure is 0.980 GPa. The lubricant used has an absolute viscosity of 0.35 Pa-s and pressure viscosity coefficient of 2.2×10^{-8} m^2/N. The root-mean-square surface roughness for both the pinion and gear is 0.1 μm. Determine the following:

(a) Transmitted power. *Ans.* $h_p = 54.85$ kW.

(b) Minimum film thickness. *Ans.* $h_{\min} = 2.45$ μm.

(c) The film parameter. *Ans.* $\Lambda = 17.3$.

(d) Is the film parameter adequate? Justify your answer. If it is not adequate, what changes would you suggest to improve the situation?

14.82 If the pinion and gear in Example 14.10 are precision enclosed gears with $Q_v = 10$, centered between bearings with a uniform power source and driven machine with light shock, find the safety factor if the allowable contact stress is 1.2 GPa. Assume that the teeth are properly crowned and that the gears are mounted so that they are centered between two bearings. *Ans.* $n_s = 1.99$.

14.83 Calculate the lubricant film thickness for the gears in Problem 14.82 if the lubricant used is a synthetic paraffinic oil with antiwear additives and the operating temperature is 38°. *Ans.* $h_{\min} = 1.67$ μm.

Synthesis and Design

14.84 You are a manufacturer of high-speed dental drills that use a steel planetary gear system for speed reduction. A vendor has suggested that less expensive polymer spur gears are suitable for this application. List the questions you would ask before considering the change.

14.85 You are a manufacturer of high-quality compression molded spur gears. You are scheduled to meet with a manufacturer of dental drills that use a steel planetary gear system for speed reduction. List the benefits of plastic gears, with the intent of convincing the manufacturer to become a customer.

14.86 Design a test apparatus to evaluate the pitting resistance of gears. Incorporate lubricants in the tests, and allow controlled adjustment of as many of the AGMA modification factors as possible.

14.87 Review the technical literature and prepare a two-page paper on the topic of scuffing in gears.

14.88 A *gear pump* is shown schematically in Sketch *g*. An input shaft drives a gear that mates with an idler in a housing with a small clearance to the gear. Fluid is conveyed between the gear teeth, and squeezed out on the outlet side of the gear pump. List the variables that you expect will affect the performance of a gear pump. Derive an expression for fluid flow as a function of incoming shaft speed.

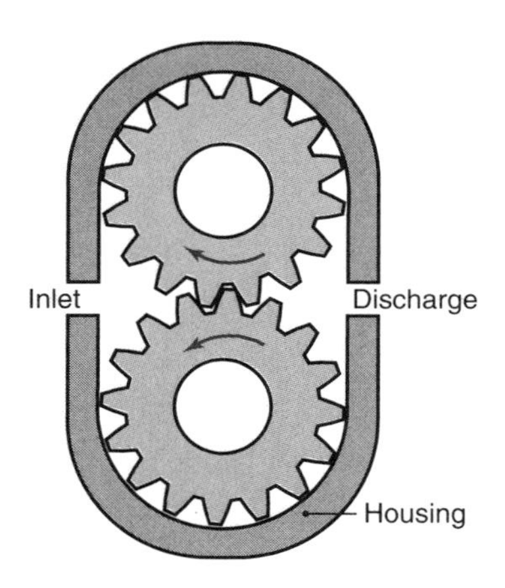

Sketch g, for use in Problem 14.88

14.89 One method of making cardboard or honeycomb composite structures is the *corrugation* process.

(a) Conduct an internet search and obtain a concise description of corrugation.

(b) Design two mating gears that will produce corrugated paper. Specify the gear geometry, including profile and centerline distance.

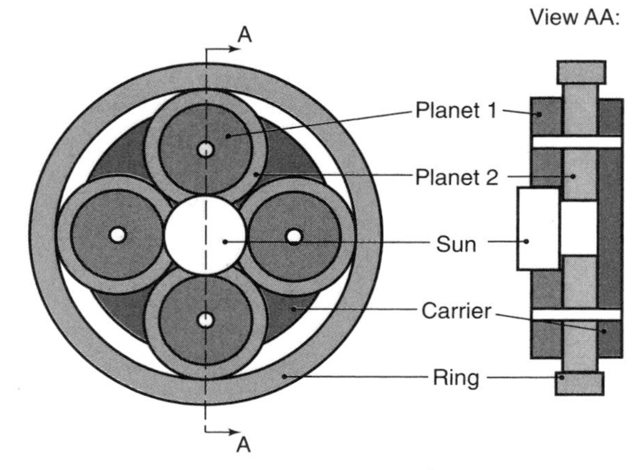

Sketch h, for use in Problem 14.91

14.90 Consider the situation where a two-stage gear reducer is to be designed with a total reduction of $g_r = 8.0$. Using a Grade 1 steel through-hardened to 250HB for both the pinion and the gear, determine the number of teeth for the two pinions and gears in the speed reducer. Assume $\phi = 20°$, $K_o = 1.0$, and that the gear needs to last for 2 million cycles with a reliability of 50%. The speed reducer needs to transmit 10 kW at an input speed of 1000 rpm using a safety factor of 3.0.

14.91 Consider the situation in Example 14.12 and design the pinion and gear for the case when $n_{sc} = 4.0$ and $n_{sb} = 2.5$.

Chapter 15

Helical, Bevel, and Worm Gears

A combined helical and worm gearset speed reducer. *Source:* Courtesy of Boston Gear, an Altra Industrial Motion Company.

Contents

Examples

Design Procedures

Case Study

Just stare at the machine. There is nothing wrong with that. Just live with it for a while. Watch it the way you watch a line when fishing and before long, as sure as you live, you'll get a little nibble, a little fact asking in a timid, humble way if you're interested in it. That's the way the world keeps on happening. Be interested in it.

Robert Pirsig, *Zen and the Art of Motorcycle Maintenance*

Chapter 14 introduced the background science of gears, including kinematics, profile geometry, gear manufacture and the use of gear trains, all applied to spur gears. This chapter extends the discussion to helical, bevel and worm gears. While their geometry and manufacturing approaches are slightly more complicated, and these gears are more expensive as a result, each of these gears displays characteristics that make them ideal for certain applications. Helical gears are applied to parallel shafts in this chapter; helical gears have higher contact ratio and smoother operation than spur gears, but use very similar design approaches. Bevel gears allow for non-parallel shafts inside a plane, and are commonly used for efficient torque transfer between two shafts. Bevel gears have a number of thread geometries, including straight, Zerol, and hypoid. The design approach presented is similar to spur and helical gears, although geometry correction factors are slightly more complicated. Worm gears use perpendicular, non-planar shafts, and are especially useful for large speed reductions in a compact design. Worm gears are different from other gear types in that they must be designed with respect to abrasive wear considerations. Effective lubrication, which is essential for all gears, is thus especially important for worm gears.

Machine elements in this chapter: Helical gears, bevel gears, worm gears.
Typical applications: Power transmissions and speed reducers. Common machine element for torque transmission from shafts.
Competing machine elements: Spur gears and trains (Ch. 14), belts and pulleys (Ch. 19), roller chains and sprockets (Ch. 19).

Symbols

a	addendum, m
b	dedendum, m
b_w	face width, m
b_{wp}	projected face width, m
C_m	gear ratio correction factor
C_r	contact ratio
C_s	materials factor
C_v	velocity correction factor
c_d	center distance, m
d	pitch diameter, m
d_r	root diameter, m
d_w	pitch diameter of worm, m
E	modulus of elasticity, Pa
E'	effective modulus of elasticity, Pa
HB	Brinell hardness
h_p	transmitted horsepower, hp
I_b	Geometry factor for contact stresses for bevel gears
K_o	overload factor
K_b	rim factor
K_i	idler factor
K_m	load distribution factor
K_{mb}	mounting factor
K_s	size factor
K_v	dynamic factor
K_x	crowning factor
L	lead of worm, m
m	module, d/N, mm
n_a	fractional part of axial contact ratio
n_r	fractional part of transverse contact ratio
N	number of teeth
N_n	equivalent number of teeth
N_v	equivalent number of teeth
N_w	number of teeth or starts in worm
p_a	axial pitch for helical gears, m
p_c	circular pitch, $\pi d/N$, m
p_{cn}	normal circular pitch, m
p_d	diametral pitch, N/d, in.$^{-1}$
p_{dn}	normal diametral pitch, in.$^{-1}$
p_H	maximum Hertzian contact pressure, Pa
q	exponent in Eq. (15.32)
r	pitch radius, m
r_c	cutter radius, m
R_x	rolling radius, m
T	torque, N-m
v_t	pitch-line velocity, m/s
W	load, N
W'	dimensionless load
W_{ag}	thrust load on worm gear, N
W_{aw}	thrust load on worm, N
W_f	friction force, N
W_r	radial load, N
W_s	separation force, N
W_t	tangential load, N
W_{tg}	tangential load on worm gear, N
W_{tw}	tangential load on worm, N
Y_a	geometry factor for a 75 tooth helical gear
Y_b	geometry factor in bending for bevel gears
Y_h	geometry factor in bending for helical gears
Y_m	geometry correction factor
Z	angular velocity ratio
α_p	pitch cone angle, deg
λ	lead angle, deg
μ	coefficient of friction
ν	Poisson's ratio
σ	stress, Pa
σ_t	bending stress of gear, Pa
ϕ	pressure angle, deg
ϕ_n	pressure angle in normal direction, deg
ψ	helix angle, deg
ψ_b	base helix angle, deg
ω	angular speed of gear, rad/s

Subscripts

a	actual; axial
b	bending
c	contact
g	gear
o	outside
p	pinion
t	total

15.1 Introduction

Chapter 14 discussed the basic geometry of gears and presented the design methodology developed by the American Gear Manufacturers Association (AGMA) for spur gears. Similar design approaches have been developed for helical, bevel, and worm gears and will be briefly summarized in this chapter.

The interested student, faced with a menu from which to select basic gear types, and then catalogs from which to select specific gears, may feel overwhelmed by the unconstrained nature of the design problem. Section 14.2 presented some fundamental considerations relating to the orientation of drive and driven shafts for a gearset. This often provides sufficient restrictions on the design that a class, that is, spur, worm, bevel, etc., can be selected and analysis can proceed to determine required gear dimensions. However, shaft orientation is often a design variable, and no gear type can be easily eliminated from consideration. Some of the advantages and disadvantages of gear types are given in Table 15.1. Further information to aid designers in gear selection is contained in Dudley [1994], but also will arise from familiarity.

15.2 Helical Gears

The material covered in Chapter 14 related primarily to spur gears. This section describes **helical gears** as shown in Fig. 15.1a. Although helical gears were introduced in Section 14.2.1, this section describes them in more detail. However, instead of repeating much of the theoretical background and nomenclature, this section builds upon the information obtained from spur gears and describes only the differences that exist between spur gears and helical gears. Many of the equations and approaches apply directly to helical gears; Design Procedures 14.2 and 14.3 are especially pertinent.

Tooth strength is improved with helical gears (relative to spur gears) because of the elongated helical wrap-around tooth base support, as shown in Fig. 15.1a. For helical gears, the contact ratio is higher owing to the axial tooth overlap. Helical gears, therefore, tend to have greater load-carrying capacity than spur gears of the same size. Spur gears, on the other hand, have a somewhat higher efficiency.

15.2.1 Helical Gear Relationships

Note from Fig. 15.1 that the teeth intersect a cylindrical surface, such as the pitch cylinder, in a helical fashion. All the relationships governing spur gears apply to helical gears with

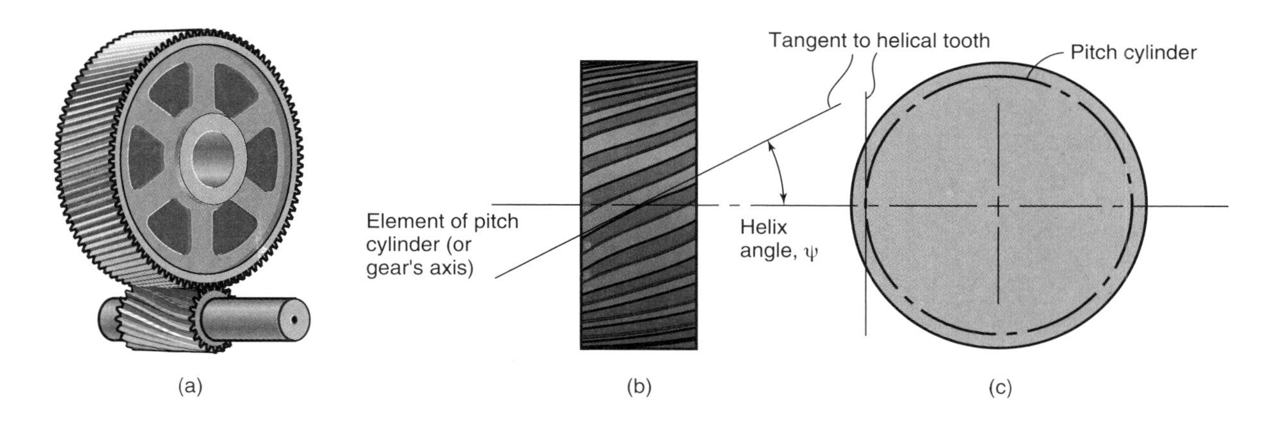

Figure 15.1: Helical gear. (a) Illustration of meshing helical gears; (b) front view; (c) side view.

Table 15.1: Design considerations for gears.

Gear type	Advantages	Disadvantages
Spur	Inexpensive, simple to design, no thrust load is developed by the gearing, wide variety of manufacturing options.	Can generate significant noise, especially at high speeds, and are usually restricted to pitch-line velocities below 20 m/s (4000 ft/min).
Helical	Useful for high-speed and high-power applications, quiet at high speeds. Often used instead of spur gears for high-speed applications.	Generate a thrust load on a single face, more expensive than spur gears.
Bevel	High efficiency (can be 98% or higher), can transfer power across nonintersecting shafts. Spiral bevel gears transmit loads evenly and are quieter than straight bevel.	Shaft alignment is critical, rolling element bearings are therefore often used with bevel gears. This limits power transfer for high-speed applications (where a journal bearing is preferable). Can be expensive.
Worm	Compact and cost-effective designs for large gear ratios. Efficiency can be as high as 90% or as low as 50%, and is lower than for other gear sets.	Wear by abrasion is of higher concern than other gear types, can be expensive. Generate very high thrust loads. Worm cannot be driven by gear; worm must drive gear.

some slight modifications to account for the axial twist of the teeth described by the **helix angle**, ψ. The helix angle varies from the base of the tooth to the outside radius. The helix angle is defined as the angle between an element of the pitch cylinder and the tangent to the helicoidal tooth at the intersection of the pitch cylinder and the tooth profile, as shown in Fig. 15.1.

15.2.2 Pitches of Helical Gears

Helical gears have two related pitches: one in the plane of rotation and the other in a plane normal to the tooth. For spur gears the pitches were described only in terms of the plane of rotation, but there is an additional axial pitch for helical gears. Figure 15.2 shows the circular and axial pitches of helical gears, which are related by the **normal circular pitch**:

$$p_{cn} = p_c \cos\psi, \tag{15.1}$$

where ψ is the helix angle. The normal diametral pitch is

$$p_{dn} = \frac{p_d}{\cos\psi}. \tag{15.2}$$

The axial pitch of a helical gear is the distance between corresponding points on adjacent teeth measured parallel to the gear's axis (see Fig. 15.2b). The axial pitch is related to the circular pitch by the following expression:

$$p_a = p_c \cot\psi = \frac{p_{cn}}{\sin\psi}. \tag{15.3}$$

15.2.3 Equivalent Number of Teeth and Pressure Angle

Helical gear teeth have an involute profile in the plane of rotation. If a cross section is taken in a plane perpendicular to a helical gear's axis, the helical teeth have a resemblance to involute teeth with the normal circular pitch, p_{cn}. However, the shape of the tooth in this cross section corresponds to the profile of a spur gear with a larger number of teeth. This **equivalent number of teeth** for helical gears is given as

$$N_n = \frac{N}{\cos^3\psi}. \tag{15.4}$$

For helical gears there is a normal pressure angle as well as the usual pressure angle in the plane of rotation, and they are related to the helix angle by

$$\tan\phi = \frac{\tan\phi_n}{\cos\psi}. \tag{15.5}$$

15.2.4 Helical Tooth Proportions

Helical gear tooth proportions follow the same standards as those for spur gears. Addendum, dedendum, whole depth, and clearance are exactly the same. A transverse cross section of a helical gear will display an involute profile, so that within this plane, the analysis of spur gears is very valuable. The inclusion of a helix angle complicates matters outside of this plane, however.

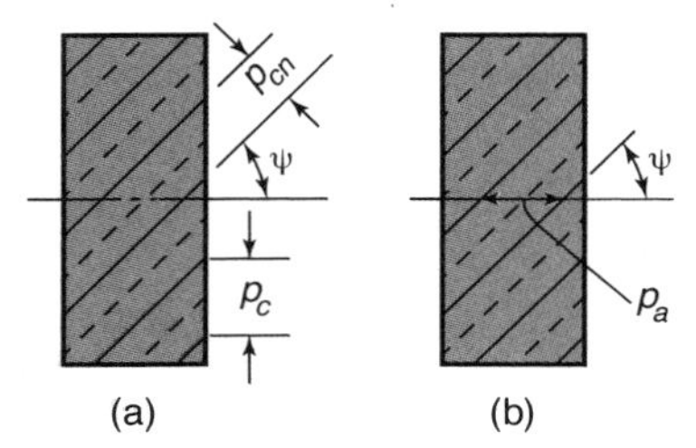

Figure 15.2: Pitches of helical gears. (a) Circular; (b) axial.

The pitch diameter can be expressed from the diametral pitch by Eq. (14.5), or from the normal diametral pitch and helix angle according to

$$d = \frac{N}{p_d} = \frac{N}{p_{dn}\cos\psi}, \tag{15.6}$$

where
d = pitch diameter, m (in.)
N = number of teeth
p_d = diametral pitch, m^{-1} (in.^{-1})
p_{dn} = normal diametral pitch, m^{-1} (in.^{-1})
ψ = helix angle, deg.

The center distance, using Eq. (14.1), is

$$c_d = \frac{d_p + d_g}{2} = \frac{N_p + N_g}{2p_{dn}\cos\psi}. \tag{15.7}$$

The contact ratio of helical gears is enhanced by the axial overlap of the teeth. The **total contact ratio** is the sum of the transverse and axial contact ratios or

$$C_{rt} = C_r + C_{ra}. \tag{15.8}$$

Recall that the **transverse contact ratio**, C_r, is given by Eq. (14.22), restated here for convenience:

$$\begin{aligned} C_r &= \frac{L_{ab}}{p_c\cos\phi} \\ &= \frac{1}{p_c\cos\phi}\left[\sqrt{r_{op}^2 - r_{bp}^2} + \sqrt{r_{og}^2 - r_{bg}^2}\right] - \frac{c_d\tan\phi}{p_c}. \end{aligned}$$

The **axial contact ratio**, C_{ra}, is given by

$$C_{ra} = \frac{b_w}{p_a} = \frac{b_w\tan\psi}{p_c} = \frac{b_w\sin\psi}{p_{cn}}. \tag{15.9}$$

where b_w is the face width. Recall that the (transverse) contact ratio of spur gears should be at least 1.4, but is rarely above 2.0 (see Section 14.5). With spur gears, this was interpreted as the average number of gear teeth that mesh along the line of contact, and it was noted that poor performance results if the contact ratio becomes too small. The total contact ratio is similarly a valuable design tool for helical gears. Conventional helical gears will have a total contact ratio above 2.0, and the total contact ratio can be very large for high helix angles. However, it should be noted that if $C_{ra} < 1$, the helical gear is referred to as a *low axial contact ratio* helical, and these gears need special treatment in their design. It should also be noted that it is possible to have helical gears that have their shafts cross. Detailed information regarding low axial contact and cross-shaft helical gears is available in AGMA [2010c].

Since the total contact ratio can be very large, there are potentially larger numbers of gear teeth in contact, but they may not be engaged along their entire length. Therefore, a concept of the **minimum contact length**, $L_{\min}$, is useful. In any mating helical gears, this quantity is the minimum total contact length that will support a load. To calculate $L_{\min}$, one must first define n_a as the fractional part of the axial contact ratio, C_{ra}, and n_r as the fractional part of the transverse contact ratio. That is, if $C_{ra} = 1.51$, then $n_a = 0.51$. $L_{\min}$ is then obtained from:

$$L_{\min} = \begin{cases} \dfrac{C_{ra}b_w - n_a n_r p_a}{\cos\psi_b} & \text{for } n_a \le 1 - n_r \\[2ex] \dfrac{C_{ra}b_w - (1-n_a)(1-n_r)p_a}{\cos\psi_b} & \text{for } n_a > 1 - n_r \end{cases} \tag{15.10}$$

where ψ_b is the base helix angle, given by

$$\cos\psi_b = \frac{\cos\psi\cos\phi_n}{\cos\phi} \tag{15.11}$$

and ϕ_n is the normal pressure angle given by Eq. (15.5).

15.2.5 Loads and Stresses

The tangential load, W_t, is the same for spur or helical gears. Recall that the tangential load is the force that transmits power from the driving to the driven gear. It acts perpendicular to the axis of the shaft carrying the gear. Thus, Eqs. (14.49) and (14.57) are equally valid for spur gears or helical gears.

The *axial*, or thrust, load in a helical gear is

$$W_a = W_t\tan\psi. \tag{15.12}$$

Note that the axial load increases as the helix angle increases. Helix angles typically range from 15 to 45°. The *radial* load is the force that acts toward the center of the gear (i.e., radially). Equation (14.51) gives the radial load as

$$W_r = W_t\tan\phi. \tag{15.13}$$

The *normal* load used to evaluate surface pitting or elastohydrodynamic lubrication is

$$W = \frac{W_t}{\cos\phi_n\cos\psi}. \tag{15.14}$$

The bending and contact stresses for helical gears are the same as those given for spur gears in Eqs. (14.58) and (14.75) with the coefficients in those equations having the same meaning but different numerical values. Chapter 14 explained the methodology for spur gears and gave a brief discussion of the difference in going from a spur gear to a helical gear.

Example 15.1: Geometry of a Helical Gear

Given: An involute gear drives a high-speed centrifuge. The speed of the centrifuge is 18,000 rpm. It is driven by a 3000-rpm electric motor through a 6:1 speedup gearbox. The pinion has 21 teeth and the gear has 126 teeth with a diametral pitch of 14 per inch. The width of the gears is 1.8 in, and the pressure angle is 20°.

Find: The gear contact ratio for

(*a*) A spur gear

(*b*) A helical gear with helix angle of 30°

Also, find the helix angle if the contact ratio is 3.

Solution:

(*a*) *Spur gear:* From Eq. (14.6), the circular pitch is

$$p_c = \frac{\pi}{p_d} = \frac{\pi}{14} = 0.224 \text{ in.}$$

From Eqs. (14.1) and (14.5), the center distance is

$$c_d = \frac{d_p + d_g}{2} = \frac{N_p + N_g}{2p_d} = \frac{21 + 126}{2(14)} = 5.25 \text{ in.}$$

From Eq. (14.5), the pitch circle radii for the pinion and gear are

$$r_p = \frac{N_p}{2p_d} = \frac{21}{2(14)} = 0.75 \text{ in.}$$

$$r_g = \frac{N_g}{2p_d} = \frac{126}{2(14)} = 4.5 \text{ in.}$$

The radii of the base circles for the pinion and gear are obtained from Eq. (14.16) as

$$r_{bp} = r_p \cos\phi = 0.75 \cos 20^\circ = 0.7048 \text{ in.}$$

$$r_{bg} = r_g \cos\phi = 4.5 \cos 20^\circ = 4.2286 \text{ in.}$$

The outside radii for the pinion and gear are

$$r_{op} = r_p + a = r_p + \frac{1}{p_d} = 0.75 + \frac{1}{14} = 0.8214 \text{ in.}$$

$$r_{og} = r_g + a = r_g + \frac{1}{p_d} = 4.5 + \frac{1}{14} = 4.5714 \text{ in.}$$

From Eq. (14.22), the contact ratio for the spur gear is

$$C_r = \frac{\sqrt{(0.8214)^2 - (0.7048)^2} + \sqrt{(4.571)^2 - (4.229)^2}}{(0.2244)\cos 20^\circ}$$

$$- \frac{5.250 \tan 20^\circ}{0.2244} = 1.712.$$

(*b*) Helical gear: The total contact ratio for a helical gear is given by Eqs. (15.8) and (15.9) as

$$C_{rt} = C_r + C_{ra} = 1.712 + \frac{(1.8)\tan 30^\circ}{0.2244} = 6.343.$$

A contact ratio of 3 can be obtained by selecting an appropriate helix angle. Using Eq. (15.9),

$$C_{ra} = \frac{b_w \tan\psi}{p_c} = 3.00 - 1.712 = 1.288.$$

This results in $\psi = 9.122^\circ$. Section 14.5 discussed the importance of the contact ratio. Clearly, a much higher contact ratio can be achieved for the helical gear than for the spur gear in this example. This illustrates that transmitted loads will be shared by more teeth with a helical gear, and this also results in smoother operation.

15.2.6 AGMA Design Approach for Helical Gears

Bending Stress

The design approach for helical and spur gears is very similar. Indeed, the ANSI/AGMA standard [2010c] uses the same equations for pitting resistance and bending strength for both spur and helical gears. Equation (14.58) still holds for bending stress in helical gears:

$$\sigma_t = \frac{W_t p_d K_o K_s K_m K_v K_b}{b_w Y_h}, \tag{15.15}$$

where

- W_t = transmitted load, N (lb)
- p_d = diametral pitch, m^{-1} (in.^{-1})
- K_o = overload factor, as given in Table 14.8
- K_s = size factor, as given in Table 14.9
- K_m = load distribution factor, as given by Eq. (14.59)
- K_v = dynamic factor, as given by Eq. (14.68) or Fig. 14.34
- K_b = rim factor, as given in Section 14.11.4
- b_w = face width of gear, m (in.)
- Y_h = geometry correction factor for bending

The geometry correction factor for bending, Y_h, not surprisingly, is different for helical gears than for spur gears, but the other correction factors are calculated as described in Ch. 14. The geometry correction factor can be obtained using an iterative numerical approach or using tabulated values as given in AGMA [1989], or it can be approximated from

$$Y_h = Y_a Y_m, \tag{15.16}$$

where Y_a is the geometry correction factor for a helical gear mating against a gear with 75 teeth and Y_m is a modifier factor to account for a different number of teeth in the mating gear. This is the simplest approach; these factors can be obtained from Figs. 15.3 and 15.4.

Contact Stress

The equation for contact stress in helical gears is the same as Eq. (14.75):

$$\sigma_c = p_H \left(K_o K_s K_m K_v\right)^{\frac{1}{2}}, \tag{15.17}$$

where

- K_o = overload factor, as given in Table 14.8
- K_s = size factor, as given in Table 14.9
- K_m = load distribution factor, as given by Eq. (14.59)
- K_v = dynamic factor, as given by Eq. (14.68) or Fig. 14.34

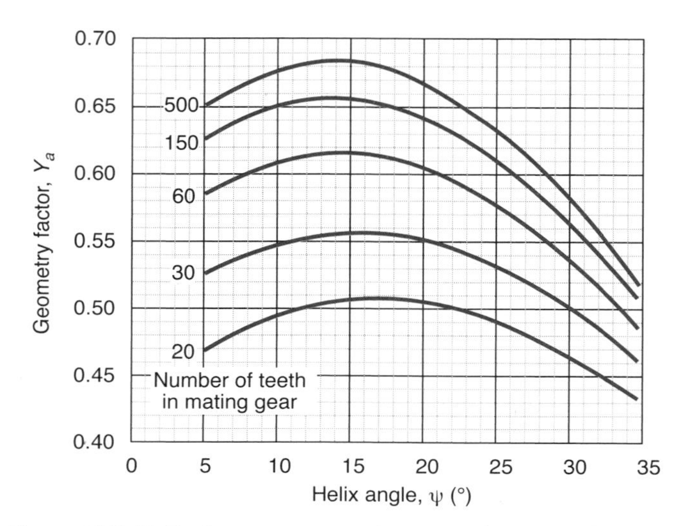

Figure 15.3: Helical gear geometry factor as a function of helix angle when mating with a 75-tooth gear. *Source:* Courtesy of the American Gear Manufacturers Association.

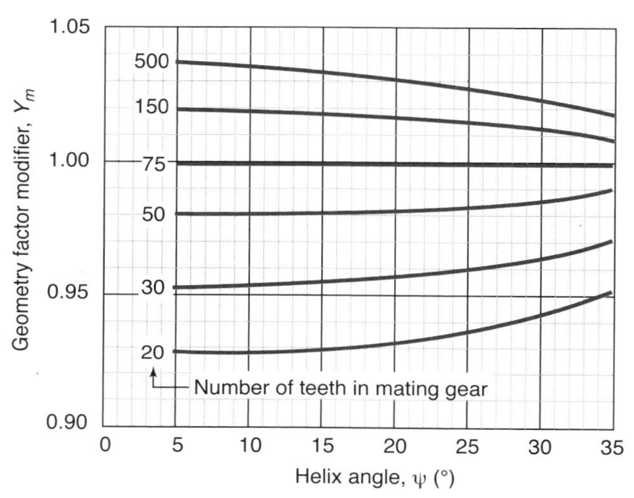

Figure 15.4: Helical gear geometry correction factor as a function of helix angle. *Source:* Courtesy of the American Gear Manufacturers Association.

The maximum Hertzian contact stress, p_H, was given in Eq. (8.20) as

$$p_H = E'\left(\frac{W'}{2\pi}\right)^{1/2}, \tag{8.20}$$

where

$$E' = \frac{2}{\dfrac{(1-\nu_a^2)}{E_a} + \dfrac{(1-\nu_b^2)}{E_b}}, \tag{8.16}$$

$$W' = \frac{w'}{E'R_x} = \frac{W}{L_{\min}E'R_x}, \tag{15.18}$$

and

$$\frac{1}{R_x} = \left(\frac{1}{r_p} + \frac{1}{r_g}\right)\frac{1}{\sin\phi} = \left(\frac{1}{d_p} + \frac{1}{d_g}\right)\frac{2}{\sin\phi}. \tag{14.74}$$

With helical gears, the dimensionless normal load, W', needs some clarification. Spur gears usually have a contact ratio between 1.4 and 2, as discussed in Section 14.5, so that only one pair of teeth will mesh at some point in the contact. Thus, the length of the contact is the face width of the gear for spur gears. Since multiple helical gear teeth can be partially engaged, the total contact length for helical gears is $L_{\min}$, as defined by Eq. (15.10).

Example 15.2: Bending Stresses in Helical Gear Teeth

Given: A 20-tooth pinion with a helix angle of $\psi = 30°$ and normal pressure angle of $\phi_n = 20°$ mates with a 60-tooth gear. At a pinion speed of 3000 rpm, a power of 5 kW is transmitted. The gears have a module of 3 and a face width of 20 mm. The allowable bending stress number for both the pinion and the gear is 200 MPa, and assume $K_m = 1.20$, $K_v = 1.40$, and $K_o = K_s = K_i = K_b = 1.0$.

Find: Determine the safety factor for the pinion based on the AGMA bending stress.

Solution: Note that $\omega_p = 3000$ rpm $= 314$ rad/s. From Eq. (14.9), note that $p_d = 1/m = 333.3\ \text{m}^{-1}$. From Eqs. (14.5) and (14.9),

$$r_p = \frac{N_p}{2p_d} = \frac{N_p m}{2} = \frac{(20)(3)}{2} = 30\ \text{mm},$$

$$r_g = \frac{N_g m}{2} = \frac{(60)(3)}{2} = 90\ \text{mm}.$$

Therefore, the transmitted force can be obtained from the power as

$$W_t = \frac{h_p}{r_p\omega_p} = \frac{5000}{(0.03)(314)} = 531\ \text{N}.$$

From Fig. 15.3, $Y_a = 0.465$ and from Fig. 15.4, $Y_m = 0.995$. Therefore, from Eq. (15.16),

$$Y_h = Y_a Y_m = (0.465)(0.995) = 0.463.$$

From Eq. (15.15), the bending stress is

$$\begin{aligned}\sigma_t &= \frac{W_t p_d K_o K_s K_m K_v K_i K_b}{b_w Y_h}\\ &= \frac{(531)(333.3)(1.20)(1.40)}{(0.020)(0.463)}\\ &= 32.1\ \text{MPa}.\end{aligned}$$

Since the allowable bending stress is 200 MPa, the safety factor is

$$n_s = \frac{200}{32.1} = 6.23.$$

Example 15.3: Analysis of Surface Pitting in a Helical Gear

Given: Consider the same mating pinion and gear as in Example 15.2, along with an allowable contact stress of 450 MPa.

Find: Determine the safety factor based on contact stress. Assume the gears are made of steel, with $E_s = 207$ GPa and $\nu_s = 0.3$.

Solution: Using the information from Example 15.2, Eq. (14.9) yields

$$p_c = \pi m = 9.42 \text{ mm}.$$

From Eq. (15.3),

$$p_a = p_c \cot \psi = 0.0163 \text{ m}.$$

From Eq. (15.5),

$$\phi = \tan^{-1} \frac{\tan \phi_n}{\cos \psi} = \tan^{-1} \frac{\tan 20^\circ}{\cos 30^\circ} = 22.8^\circ.$$

From Eq. (14.73),

$$\frac{1}{R_x} = \left(\frac{1}{r_p} + \frac{1}{r_g} \right) \frac{1}{\sin \phi} = \left(\frac{1}{30} + \frac{1}{90} \right) \frac{1}{\sin 22.8^\circ},$$

or $R_x = 8.719$ mm. From Eq. (8.16),

$$E' = \frac{E_s}{1 - \nu_s^2} = \frac{207 \times 10^9}{1 - 0.3^2} = 227 \text{ GPa}.$$

From Eq. (15.14),

$$W = \frac{W_t}{\cos \phi_n \cos \psi} = \frac{531}{\cos 20^\circ \cos 30^\circ} = 652 \text{ N}.$$

Before $L_{\min}$ can be determined, it is necessary to obtain the base and outside radii for the pinion and gear. From Eq. (14.16),

$$r_{bp} = r_p \cos \phi = (30)(\cos 22.8^\circ) = 27.7 \text{ mm},$$

$$r_{bg} = r_g \cos \phi = (90)(\cos 22.8^\circ) = 83.0 \text{ mm}.$$

From Table 14.2, $a = 1.00m = 3.00$ mm, so that from Eq. (14.11),

$$r_{op} = r_p + a = 30 + 3 = 33 \text{ mm}.$$

Similarly, $r_{og} = 93$ mm. Therefore, from Eq. (14.22),

$$\begin{aligned} C_r &= \frac{L_{ab}}{p_c \cos \phi} \\ &= \frac{1}{p_c \cos \phi} \left[\sqrt{r_{op}^2 - r_{bp}^2} + \sqrt{r_{og}^2 - r_{bg}^2} \right] - \frac{c_d \tan \phi}{p_c}, \end{aligned}$$

which results in $C_r = 1.54$. From Eq. (15.9),

$$C_{ra} = \frac{b_w \tan \psi}{p_c} = \frac{(0.020) \tan 30^\circ}{0.00942} = 1.22.$$

Recall that n_r is the fractional part of C_r, or $n_r = 0.54$. Similarly, n_a is the fractional part of C_{ra}, or $n_a = 0.22$. Note from Eq. (15.11) that

$$\cos \psi_b = \frac{\cos 30^\circ \cos 22.8^\circ}{\cos 20^\circ}$$

or $\psi_b = 31.8^\circ$. Since $n_a < 1 - n_r$, Eq. (15.10) results in

$$\begin{aligned} L_{\min} &= \frac{C_{ra} b_w - n_a n_r p_a}{\cos \psi_b} \\ &= \frac{(1.22)(0.020) - (0.22)(0.54)(0.0163)}{\cos 31.8^\circ} \\ &= 0.0264 \text{ m}. \end{aligned}$$

Therefore, W' can be calculated from Eq. (15.18) as

$$\begin{aligned} W' &= \frac{W}{L_{\min} E' R_x} = \frac{652}{(0.0264)(227 \times 10^9)(0.008719)} \\ &= 1.248 \times 10^{-5}, \end{aligned}$$

and Eq. (8.20) yields the maximum Hertzian contact pressure as

$$p_H = E' \left(\frac{W'}{2\pi} \right)^{1/2} = (227 \times 10^9) \left(\frac{1.248 \times 10^{-5}}{2\pi} \right)^{1/2},$$

or $p_H = 320$ MPa. From Eq. (15.17),

$$\begin{aligned} \sigma_c &= p_H (K_o K_s K_m K_v)^{1/2} = (320 \times 10^6) [(1.2)(1.4)]^{1/2} \\ &= 415 \text{ MPa}. \end{aligned}$$

The safety factor against pitting is

$$n_s = \frac{450}{415} = 1.08.$$

15.3 Bevel Gears

15.3.1 Types of Bevel Gears

As discussed in Section 14.2.2, **bevel gears** have nonparallel axes that lie in the same plane. Usually, bevel gears are mounted perpendicular to each other, but almost any shaft angle can be accommodated. Bevel gear blanks are conical, and a number of tooth shapes can be produced. The simplest tooth form occurs on **straight bevel gears**, where the teeth have a profile as shown in Figs. 14.3 and 15.5a. Note that since the bevel gear is a conic section, the thickness of a gear tooth is not constant.

Two meshing straight bevel gears are depicted in Fig. 15.6 with the appropriate terminology identified. The tooth profile closely approximates a spur gear involute tooth profile (see Section 14.3.3), but with more teeth than are present in the bevel gear. Thus, the **virtual number of teeth** is defined by

$$N_v = \frac{N}{\cos \alpha_p}, \tag{15.19}$$

where N_v is the virtual number of teeth, N is the actual number of teeth, and α_p is the pitch cone angle. Straight bevel gears are usually limited to pitch-line velocities of 5 m/s (1000 ft/min) or less. Alignment of the bevel gear shaft is critical, and for this reason bevel gears are usually supported by rolling element bearings instead of journal bearings. A complication for mounting bevel gears is that they must also be mounted at the right distance from the cone center.

Straight bevel gears are commonly made with 14.5°, 17.5°, or 20° pressure angles, with 20° being the most popular. Straight bevel gears are the least costly form of bevel gears, and standard sizes are widely available.

Zerol bevel gears, as shown in Fig. 15.5b, are similar to straight bevel gears, except that the tooth is curved in the lengthwise direction, but with zero spiral angle. Zerol gears can be generated on a rotary cutter and ground when necessary. Because of the tooth curvature, Zerol gears have a slight overlapping action, and are better suited than straight bevel gears for higher speed applications. Zerol gears are produced in matched sets and are more expensive than straight bevel gears. Interchangeability (as for maintenance or replacement

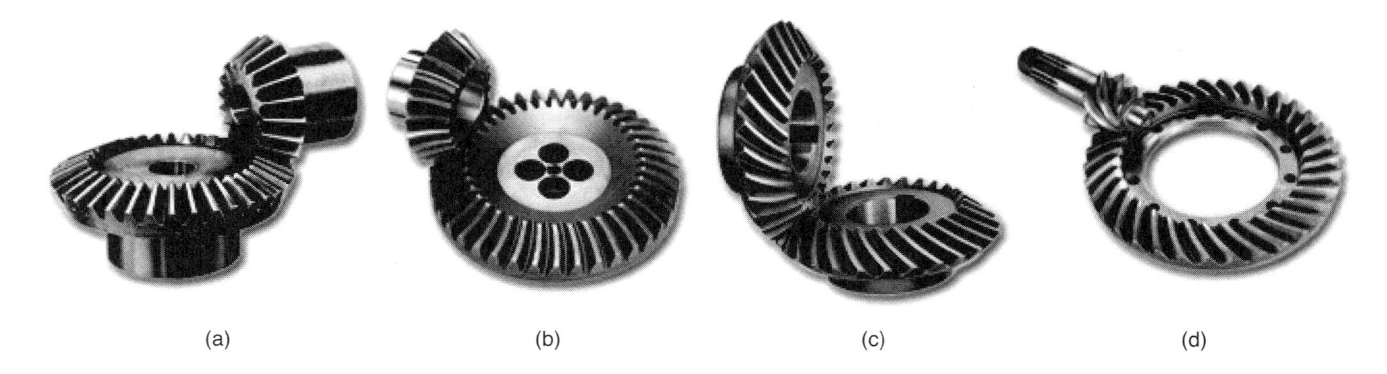

Figure 15.5: Types of bevel gears: (a) A straight bevel gear set; (b) a Zerol gear set; (c) a spiral bevel gear set; (d) a hypoid bevel gear set. *Source:* Courtesy of ATI Gear, Inc.

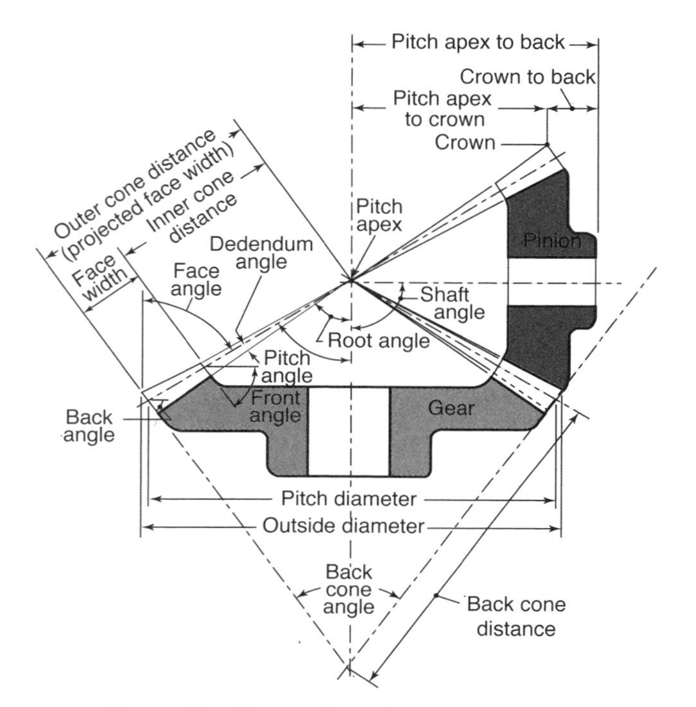

Figure 15.6: Terminology for bevel gears.

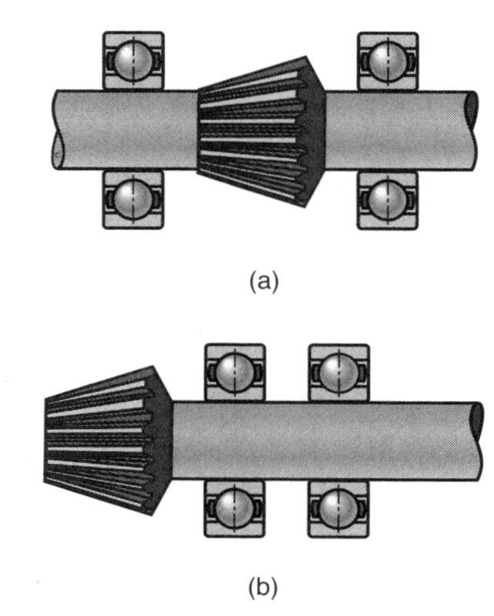

Figure 15.7: Schematic illustration of the two basic forms of gear mounting. (a) Straddle mounting, where the gear is located between bearings; (b) overhang mounting. Note that deep-groove rolling element bearings are shown, but often a bearing better suited for thrust load support is required in at least one location.

of gears) can be achieved by using master sets of gears to fit against production gears.

Spiral bevel gears are similar to Zerol bevel gears in that the teeth have a lengthwise curvature, but there is an appreciable spiral angle, as shown in Fig. 15.5c. Like Zerol gears, spiral bevel gears are produced in matched sets and are more expensive than straight bevel gears. Spiral gears have more tooth overlap than Zerol gears, and are therefore even better suited for high-speed applications. However, spiral gears generate a significantly larger thrust force compared to Zerol gears, and this thrust force must be accommodated by the mounting bearings.

A gear commonly found in automotive and industrial drives is the **hypoid gear**, shown in Fig. 15.5d, which resembles bevel gears in many respects. Hypoid gears are used on non-parallel, coplanar shafts and the parts usually taper like bevel gears. The main difference is that axes of the pinion and gear for hypoid gear sets do not intersect; the distance between the axes is called the **offset**. If the offset is large enough, the mounting shafts can pass each other and a compact straddle mounting can be used on the gearset (see Fig. 15.7). This has performance advantages, as will be discussed in Section 15.3.3. A spiral gearset can be considered as a hypoid gearset with zero offset.

Hypoid gears are different from spur, helical, and bevel gears in that their pitch diameters are not proportional to the number of teeth. The module, diametral pitch, pitch diameter, and pitch angle are used for the gear only, since the pitch for the pinion is smaller than that for the gear. Hypoid gears are produced in matched sets, and economic considerations similar to those for Zerol and spiral gears must be taken into account.

15.3.2 Bevel Gear Geometry and Forces

Figure 15.6 shows two meshing bevel gears with important nomenclature. The angular velocity ratio for bevel gears is defined as in Eq. (14.33), but if the shaft angle is 90°, the fol-

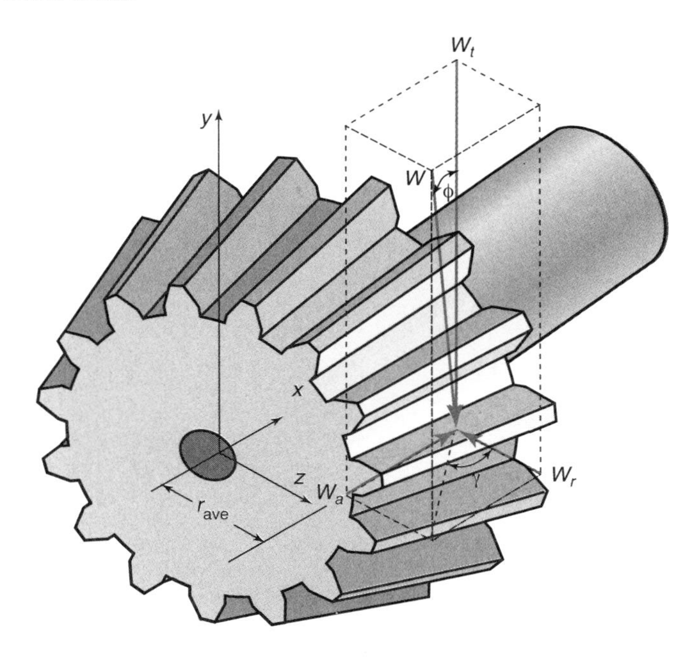

Figure 15.8: Forces acting on a bevel gear.

lowing expression can be used:

$$Z_{21} = \frac{\omega_2}{\omega_1} = \frac{N_1}{N_2} = \tan\alpha_1 = \cot\alpha_2. \tag{15.20}$$

Note that the negative sign in Eq. (14.33) has been omitted in Eq. (15.20) because it is impossible to know the sign of the angular velocity without knowledge of the mounting geometry.

Figure 15.8 shows the resultant forces acting on a straight bevel gear. The force acting on a gear tooth has three components: the transmitted load, W_t (which also results in a transmitted torque), a radial force, W_r, and an axial or thrust force, W_a. Assuming these forces act at the midpoint of the gear tooth, it can be shown from statics, for a straight bevel gear, that:

$$W = \frac{W_t}{\cos\phi} = \frac{T}{r_{\text{ave}}\cos\phi}, \tag{15.21}$$

$$W_a = W_t \tan\phi \sin\alpha, \tag{15.22}$$

$$W_r = W_t \tan\phi \cos\alpha, \tag{15.23}$$

where
- T = transmitted torque, N-m (lb-in.)
- r_{ave} = pitch radius at midpoint of tooth, m (in.)
- ϕ = pressure angle, deg
- α = pitch cone angle (see Fig. 15.6), deg

For a spiral bevel gear, the thrust and radial forces are more complicated. For a right-handed spiral rotating clockwise when viewed from its large end, or for a left-handed spiral rotating counterclockwise from its large end, the radial and thrust forces acting on the tooth are:

$$W_a = \frac{W_t}{\cos\psi}\left(\tan\phi\sin\alpha - \sin\psi\cos\alpha\right), \tag{15.24}$$

$$W_r = \frac{W_t}{\cos\psi}\left(\tan\phi\cos\alpha + \sin\psi\sin\alpha\right), \tag{15.25}$$

while for a right-handed spiral rotating counterclockwise or a left-handed spiral rotating clockwise,

$$W_a = \frac{W_t}{\cos\psi}\left(\tan\phi\sin\alpha + \sin\psi\cos\alpha\right), \tag{15.26}$$

$$W_r = \frac{W_t}{\cos\psi}\left(\tan\phi\cos\alpha - \sin\psi\sin\alpha\right), \tag{15.27}$$

where ψ is the helix angle.

15.3.3 AGMA Design Approach for Bevel Gears

The American Gear Manufacturers Association [2010b] has developed a bevel gear design methodology that is similar to that presented for spur gears in Chapter 14. As with the spur gear methodology, AGMA defines a bending or contact stress, which is then compared to an allowable strength for a candidate material. For the most part, allowable stresses for bevel and spur gears are similar for a given material, and the data presented in Section 14.9 can be used for illustrative purposes in this chapter. It should be noted that extensive material properties are available in the technical literature and from AGMA, and this data is essential for real applications. Just as in Ch. 14, the terms "stress" and "strength" are used instead of "bending stress number" and "allowable bending stress number," respectively, to maintain consistent terminology in this textbook.

The equations for bending and contact stress, using terminology consistent with this text, are:

$$\sigma_t = \frac{2T_p}{b_w d_p}\frac{p_d K_o K_v K_s K_m}{K_x Y_b}, \tag{15.28}$$

$$\sigma_c = \sqrt{\frac{T_p E'}{\pi b_w d_p^2 I_b} K_o K_v K_m K_s K_x}, \tag{15.29}$$

where
- T_p = the pinion torque, N-m (or lb-in.)
- b_w = the face width, m (in.)
- d_p = the pinion outer pitch diameter, m (in.)
- p_d = outer transverse diametral pitch, m^{-1} (in.^{-1})
- K_o = overload factor, as given in Table 14.8
- K_v = dynamic factor, as given in Fig. 14.34
- K_m= load distribution factor
- K_s = size factor
- K_x = crowning factor
- Y_b = bending strength geometry factor
- I_b = contact strength geometry factor
- E' = effective elastic modulus, given by Eq. (8.16).

The overload factor K_o and dynamic factor K_v are defined as for spur gears in Sections 14.11.1 and 14.11.5, respectively. The remaining factors will be briefly summarized here.

Load Distribution Factor, K_m

The **load distribution factor** can be calculated from

$$K_m = \begin{cases} K_{mb} + 0.0036 b_w^2 & \text{for } b_w \text{ in inches} \\ K_{mb} + \left(5.6\times10^{-6}\right) b_w^2 & \text{for } b_w \text{ in mm} \end{cases} \tag{15.30}$$

where
- K_{mb} = 1.00 for both gear and pinion straddle mounted (bearings on both sides of gear)
  - = 1.10 for only one member straddle mounted
  - = 1.25 for neither member straddle mounted

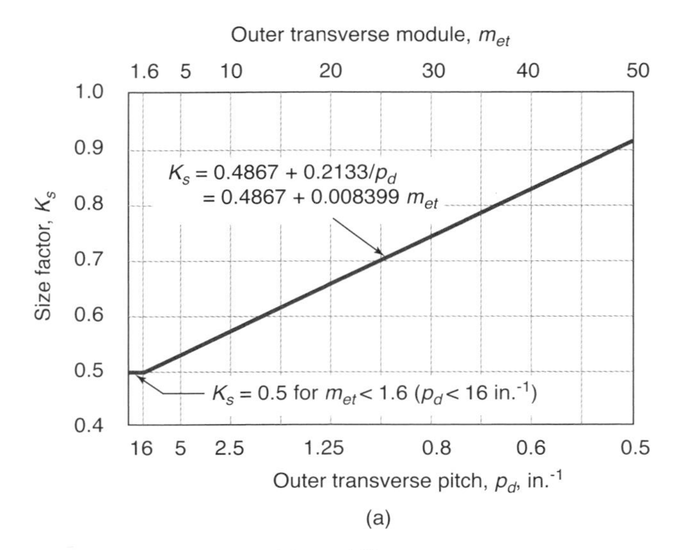

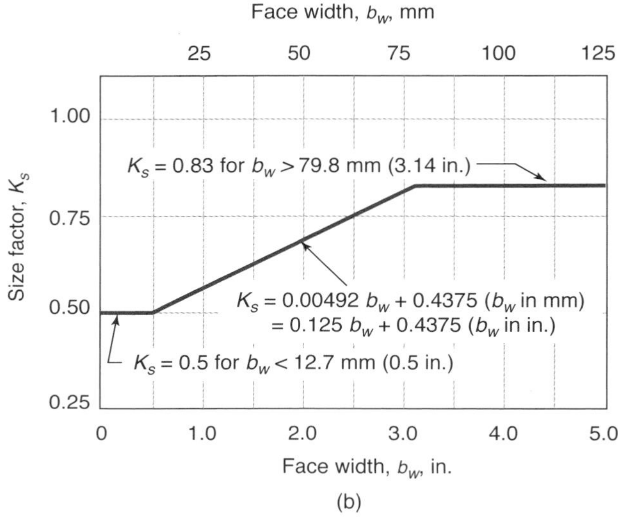

Figure 15.9: Size factor for bevel gears. (a) Size factor for bending stress; (b) size factor for contact stress or pitting resistance. *Source:* From AGMA [2010].

Straddle and overhang mounting are illustrated in Fig. 15.7.

Size Factor, K_s

The **size factor** depends on the face width for contact stress considerations, and on the outer transverse pitch or module for bending. Figure 15.9 gives the size factor for contact stress and bending circumstances.

Crowning Factor, K_x

The teeth of most bevel gears are crowned in the lengthwise direction to accommodate deflection of the mountings. Under light loads, this results in localized contact and higher stresses. On the other hand, lack of crowning will result in localized contact and higher stresses at the end of the tooth. AGMA recommends that a **crowning factor** be included in contact stress calculations according to:

$$K_x = \begin{cases} 1.5 & \text{for properly crowned teeth} \\ 2.0 & \text{(or larger) for non-crowned teeth} \end{cases} \tag{15.31}$$

For bending stress calculations, AGMA recommends that K_x be set equal to unity for straight or Zerol bevel gears. For spiral bevel gears, K_x is given by

$$K_x = 0.211\left(\frac{r_c}{A_m}\right)^q + 0.789, \tag{15.32}$$

where r_c is the cutter radius used for producing the gears, A_m is the the mean cone distance, and q is given by

$$q = \frac{0.279}{\log(\sin\psi)}. \tag{15.33}$$

Geometry Factors, Y_b and I_b

The geometry factors for bevel gears are different than for spur or helical gears. The AGMA standard provides geometry factors for straight, spiral, and Zerol gears for a number of helix and pressure angles. A selection of these correction factors is provided in Figs. 15.10 to 15.12, but more such charts and background mathematical derivations are contained in AGMA [2003].

Design Procedure 15.1: Bevel Gear Synthesis

This design procedure will assist in the selection of preliminary bevel gear geometries where the application's load, speed, and desired gear ratios are known. The discussion will be limited to spiral bevel gears, but similar approaches for straight, Zerol, or hypoid gears can be found in AGMA [2003]. The approach is also restricted to 90° shaft angles. The procedure described here is empirically based, and has been shown to produce useful designs. Departure from this procedure may be required to meet other design constraints and can still produce good gears; this procedure should be recognized as a method of obtaining a starting point for gear analysis.

1. An estimate for the required pinion diameter can be obtained from the pinion torque and gear ratio using Fig. 15.13 for surface pitting and Fig. 15.14 for bending strength. For precision finished gears (which have a cost penalty), the pinion diameter from pitting resistance can be multiplied by 0.80. From the two pinion diameter estimates, the larger value should be selected for further evaluation.

2. The pinion diameter selected in Step 1 is based on using case hardened steel with a hardness of 55 HRC, and other materials will require a modification in the pinion diameter. Table 15.2 lists material modification factors for selected materials. An updated pinion diameter can be obtained by multiplying the estimate obtained from Step 1 by the materials factor from Table 15.2.

3. Figure 15.15 provides an estimate for the number of teeth that should be machined into the pinion. Spiral bevel gears can maintain a higher contact ratio than straight or Zerol bevel gears, so departure from the recommendation in Fig. 15.15 is not uncommon. Also, note that Fig. 15.15 is for a 35° spiral angle, so that a high contact ratio can be preserved for fewer teeth (see Step 6).

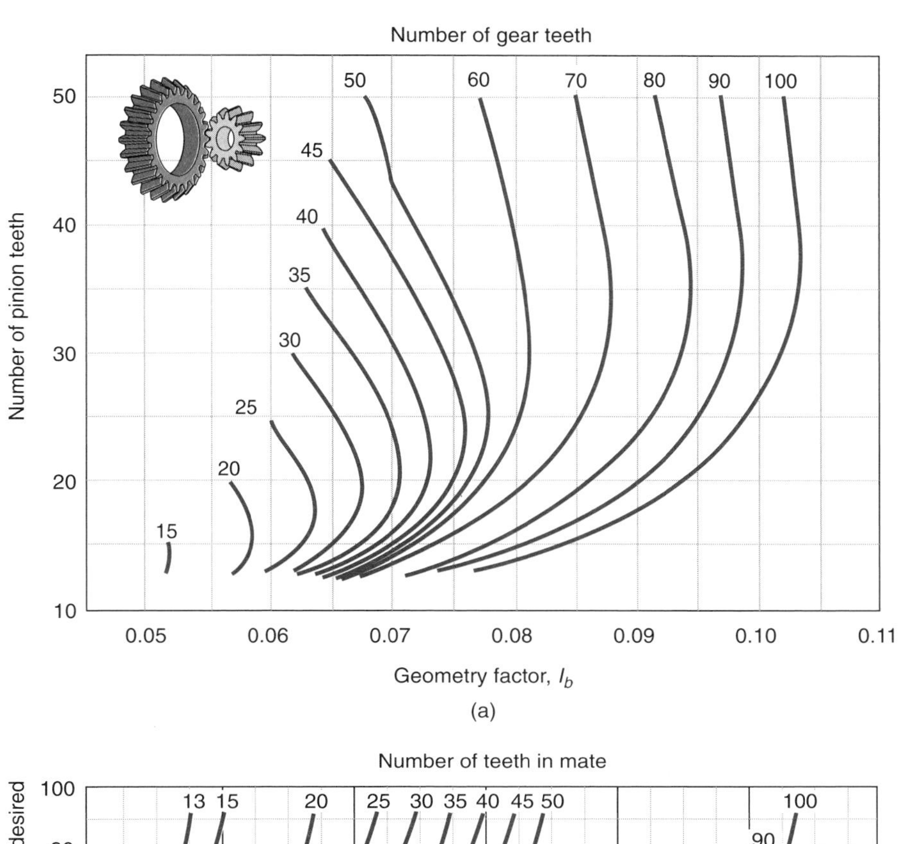

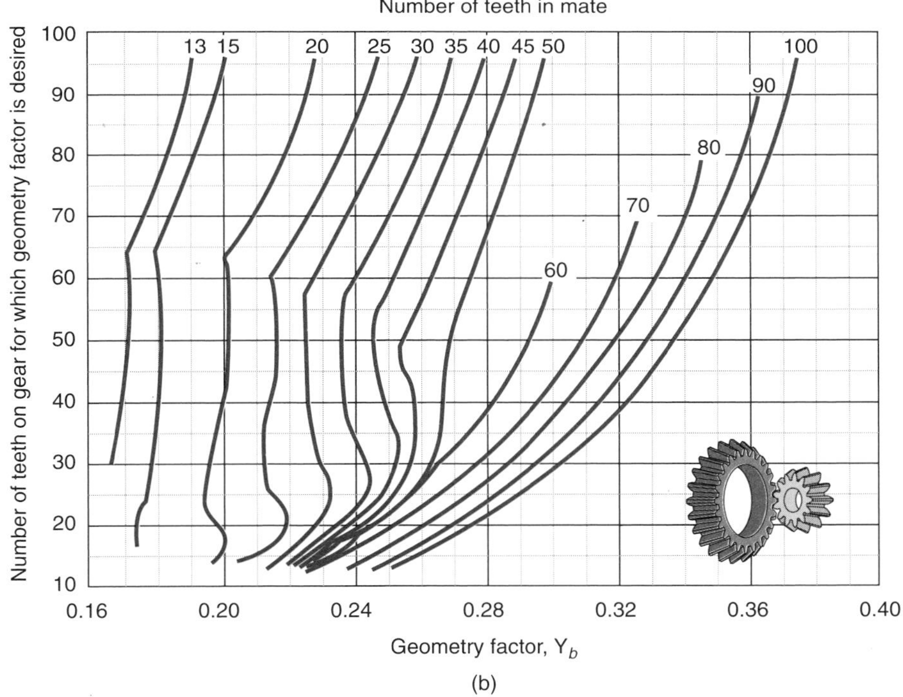

Figure 15.10: Geometry factors for straight bevel gears, with pressure angle $\phi = 20^\circ$ and shaft angle = 90°. (a) Geometry factor for contact stress I_b; (b) geometry factor for bending Y_b. *Source:* AGMA [2010b].

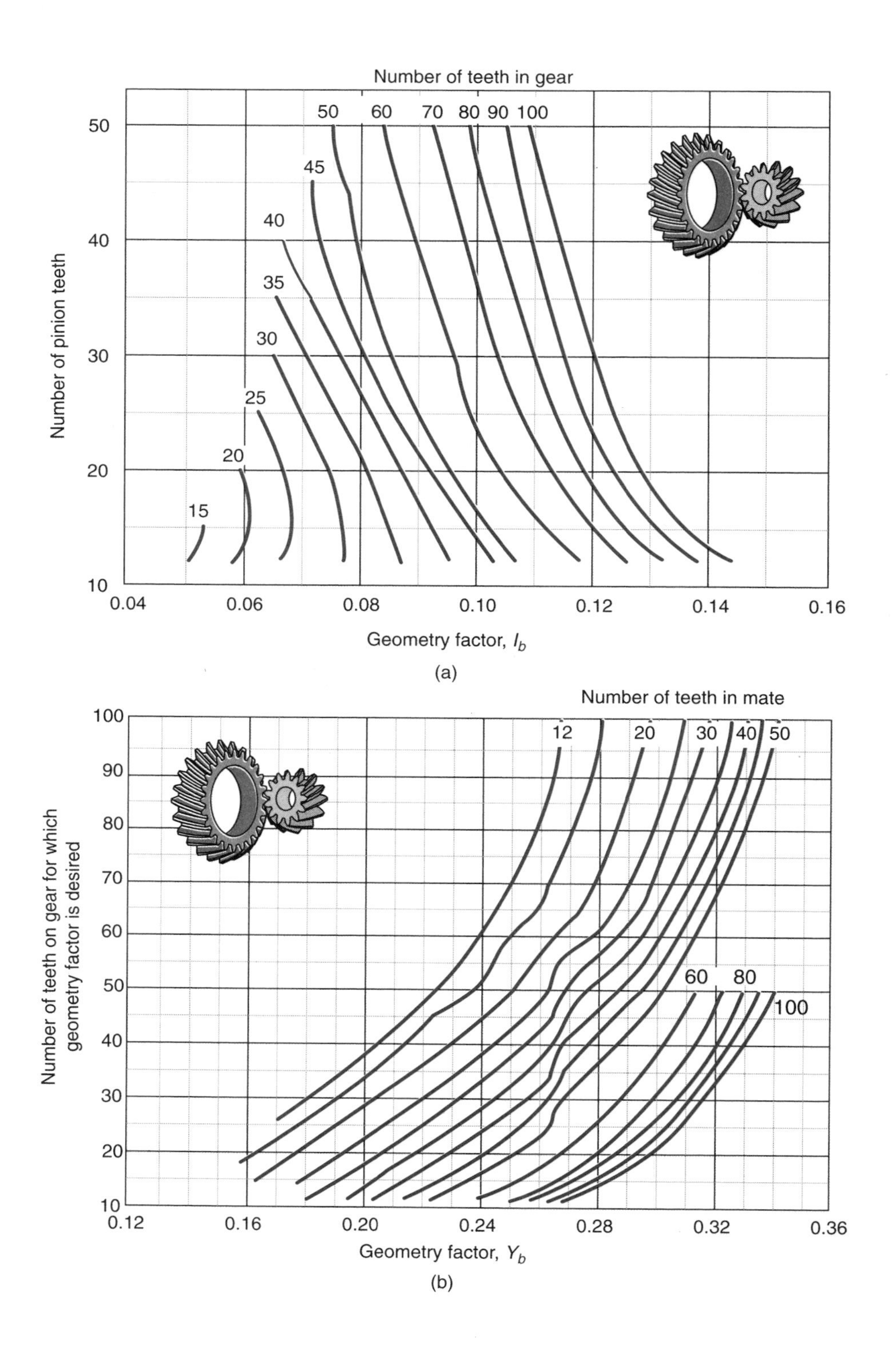

Figure 15.11: Geometry factors for spiral bevel gears, with pressure angle $\phi = 20°$, spiral angle $\psi = 25°$ and shaft angle = 90°. (a) Geometry factor for contact stress I_b; (b) geometry factor for bending Y_b. *Source:* AGMA [2010b].

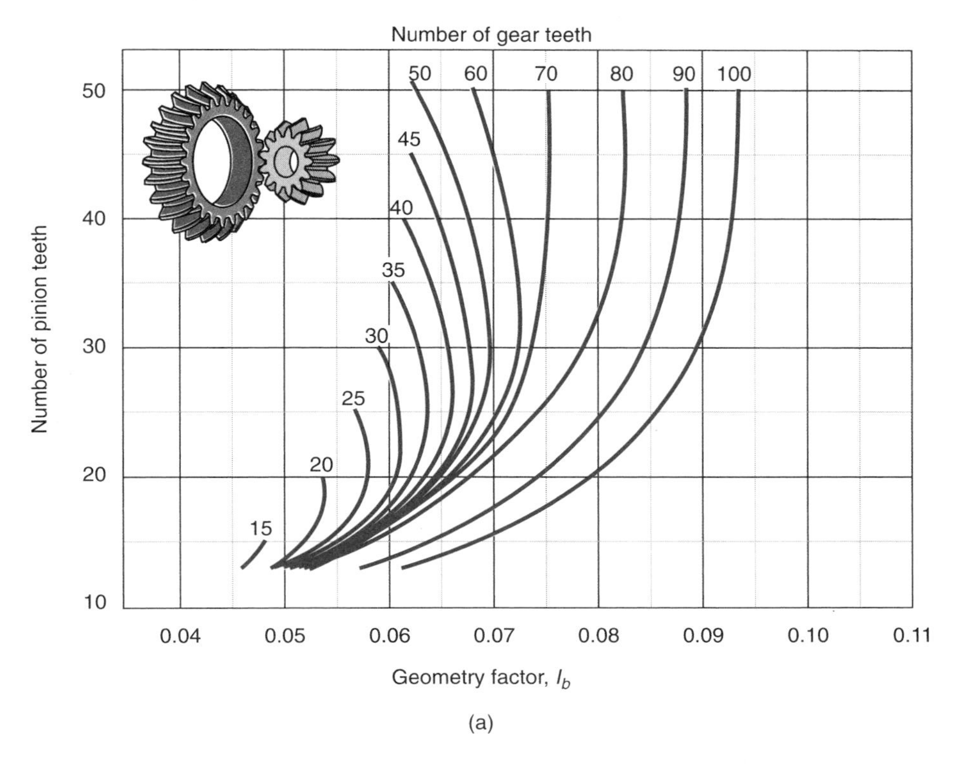

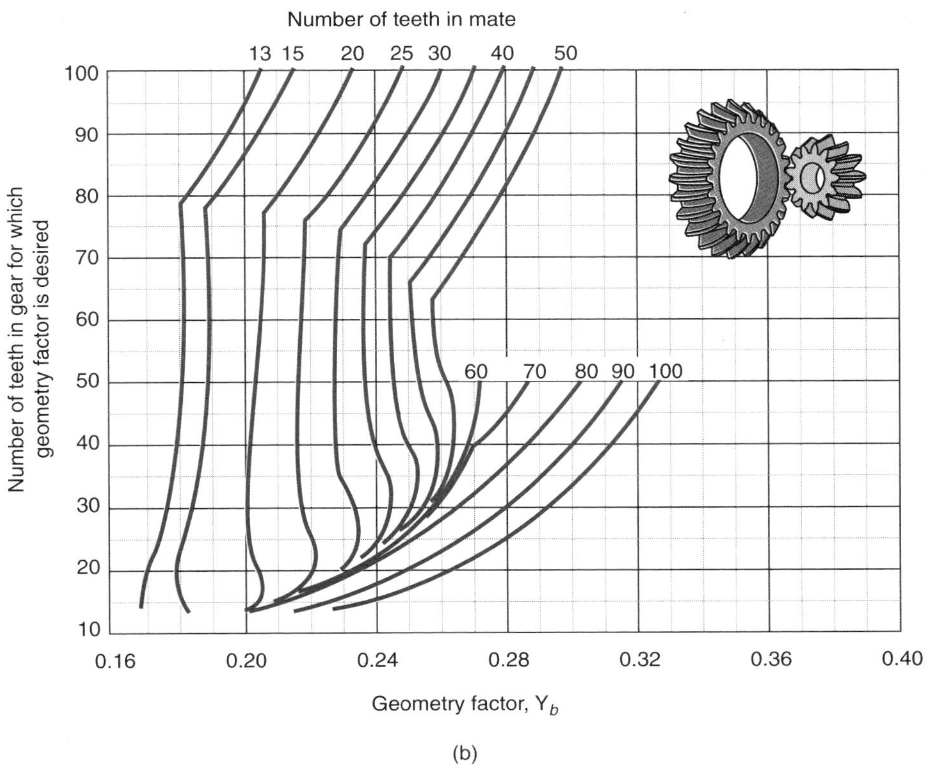

Figure 15.12: Geometry factors for Zerol bevel gears, with pressure angle $\phi = 20°$, spiral angle $\psi = 25°$ and shaft angle = 90°. (a) Geometry factor for contact stress I_b; (b) geometry factor for bending Y_b. *Source:* AGMA [2010b].

4. The outer transverse diametral pitch, p_{do}, can be obtained from Eq. (14.5) as the ratio of the number of teeth in the pinion to the pinion diameter.

5. The face width of the spiral bevel gear can be obtained from Fig. 15.16 as a function of pinion pitch diameter and gear ratio. The face width should not exceed $10/p_{do}$, however.

6. The spiral angle should be selected based on a face contact ratio of 2.0. AGMA [2003] recommends higher values for smooth and quiet operation, or high speed applications. Face contact ratios below 2.0 can be tolerated for some applications, but this is a reasonable value for preliminary design synthesis. The face contact ratio is given by

$$C_{rf} = \frac{A_o}{A_m} \frac{p_{do} b_w \tan\psi}{\pi}, \tag{15.34}$$

so that

$$\tan\psi = \frac{\pi C_{rf} A_m}{p_{do} b_w A_o}, \tag{15.35}$$

where
- A_o = outer cone distance (see Fig. 15.6).
- A_m = mean cone distance
- p_{do} = other transverse diametral pitch
- b_w = net face width
- ψ = mean spiral angle at the pitch surface.

7. The most common pressure angle for bevel gears is 20°, and is recommended for initial design synthesis. However, higher or lower pressure angles can be used. Lower pressure angles increase the contact ratio, reduce axial and separating forces, and increase the tooth slot widths. However, lower pressure angles increase the risk of undercut gear teeth and associated high stress concentrations.

The results from this Design Procedure generally are a reasonable starting point for gear design. As is usually the case, results from this approach must be modified slightly to produce a useful result. For example, if Fig. 15.15 suggests 32.2 teeth should be used, clearly one should specify 32 or 33 teeth. Also, it is good practice to use blanks that are of standard size, etc.

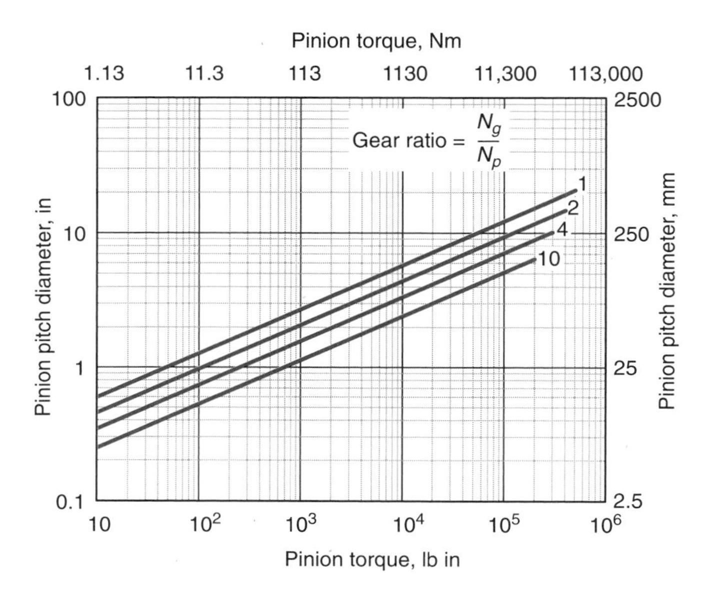

Figure 15.13: Estimated pinion pitch diameter as a function of pinion torque for a number of gear ratios, and based on pitting resistance. *Source:* AGMA [2003].

Example 15.4: Bevel Gear Design Analysis

Given: Two spiral bevel gears mesh with a shaft angle of 90°, have properly crowned teeth with a pressure angle of $\phi = 20°$, and a spiral angle of $\psi = 25°$. The face width is $b_w = 1.00$ in., the diametral pitch is $p_d = 5.60$ in.$^{-1}$, the pinion has 14 teeth, and the gear has 40 teeth. They are produced from steel with a quality factor of $Q_v = 11$, an allowable bending stress of 28,200 psi and an allowable contact stress of 200,000 psi. The pinion is overhung mounted and the gear is straddle mounted. The pinion is driven at 1750 rpm with a torque of 1440 in.-lb. The gears were produced with a cutter radius of 4.5 in., and the mean cone distance is 3.20 in. Assume the overload factor is $K_o = 1.0$.

Find: What is the factor of safety based on pinion design?

Solution: From Eq. (14.5), the diameter of the pinion is

$$d_p = \frac{N_p}{p_d} = \frac{14}{5.60} = 2.50 \text{ in.}$$

For steel, $E = 30 \times 10^6$ psi and $\nu = 0.3$, so

$$\begin{aligned} E' &= \frac{2}{\left(\frac{1-\nu_a^2}{E_a}\right) + \left(\frac{1-\nu_b^2}{E_b}\right)} \\ &= \frac{E}{1-\nu^2} \\ &= \frac{30 \times 10^6 \text{ psi}}{1-0.30^2} \\ &= 32.97 \times 10^6 \text{ psi.} \end{aligned}$$

The contact stress will be calculated from Eq. (15.29), but the assorted factors must be obtained first. From Fig. 15.11a, $I_b = 0.095$. The pitch velocity is calculated as

$$\begin{aligned} v_t &= \omega r = N_p(2\pi)\frac{d}{2} = (1750 \text{ rpm})(2\pi)\left(\frac{2.5 \text{ in.}}{2}\right) \\ &= 13,740 \text{ in./min} = 1145 \text{ ft/min.} \end{aligned}$$

Therefore, using a quality factor of $Q_v = 11$, it can be seen from Fig. 14.34 that $K_v = 1.08$. Since only one member is straddle mounted, $K_{mb} = 1.10$, and from Eq. (15.30),

$$K_m = K_{mb} + 0.0036 b_w^2 = 1.10 + (0.0036)(1.0)^2 = 1.1036.$$

From Fig. 15.9,

$$K_s = 0.125 b_w + 0.4375 = 0.5625.$$

Table 15.2: Material factor for pinion diameter estimate for selected gearset materials. *Source:* AGMA [2003].

Gear		Pinion		
Material	Hardness	Material	Hardness	Material factor
Case hardened steel	58 HRC	Case hardened steel	60 HRC	0.85
Case hardened steel	55 HRC	Case hardened steel	55 HRC	1.00
Flame hardened steel	50 HRC	Case hardened steel	55 HRC	1.05
Flame hardened steel	50 HRC	Flame hardened steel	50 HRC	1.05
Oil hardened steel	375–425 HB	Oil hardened steel	375–425 HB	1.20
Heat treated steel	210–300 HB	Case hardened steel	55 HRC	1.45
Cast iron	—	Case hardened steel	55 HRC	1.95
Cast iron	—	Flame hardened steel	55 HRC	2.00
Cast iron	—	Annealed steel	160–200 HB	2.10
Cast iron	—	Cast iron	—	3.10

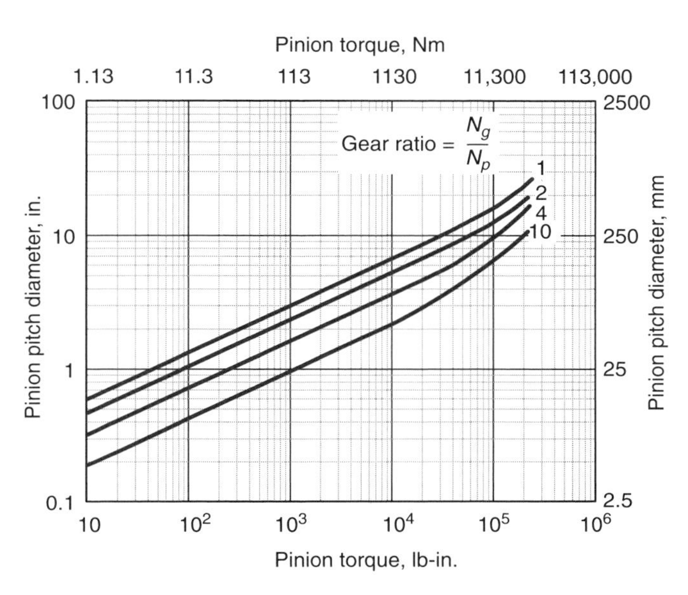

Figure 15.14: Estimated pinion pitch diameter as a function of pinion torque for a number of gear ratios, and based on bending strength. *Source:* AGMA [2003].

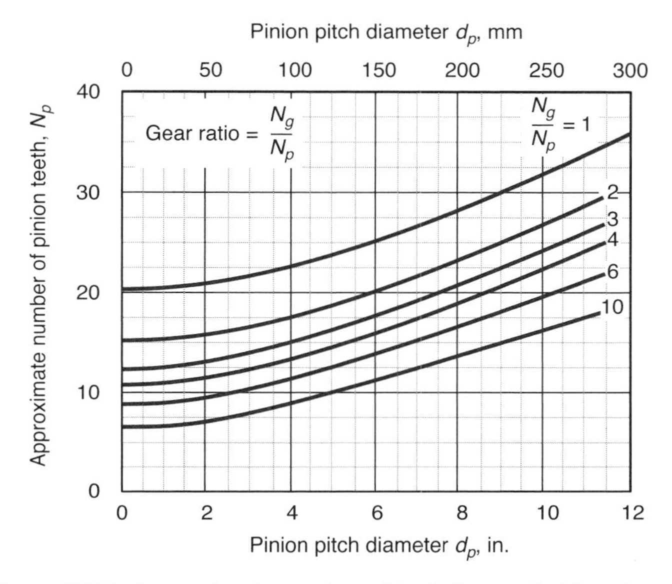

Figure 15.15: Approximate number of teeth for a spiral bevel gear as a function of pinion pitch diameter for various gear ratios. *Source:* AGMA [2003].

Note that to use these equations, b_w needed to be expressed in inches. Since the teeth are properly crowned, Eq. (15.31) suggests that $K_x = 1.5$. Therefore the contact stress is given by Eq. (15.29) as

$$\sigma_c = \sqrt{\frac{T_p E'}{\pi b_w d_p^2 I_b} K_o K_v K_m K_s K_x}$$

$$= \sqrt{\frac{(1440)(32.97 \times 10^6)}{\pi(1.0)(2.5)^2(0.095)}(1.08)(1.1036)(0.5625)(1.5)}$$

or $\sigma_c = 160.0$ ksi. Therefore, the safety factor against surface pitting failure is

$$n_s = \frac{200,000}{160,000} = 1.25.$$

The bending stress is given by Eq. (15.28). The factors obtained above can be used, except that the bending geometry factor is obtained from Fig. 15.11b as $Y_b = 0.21$. Also, K_x needs to be recalculated based on Eq. (15.32). Since this is a spiral gear,

$$q = \frac{0.279}{\log \sin \psi} = \frac{0.279}{\log \sin 25^\circ} = -0.7459.$$

Therefore, from Eq. (15.32),

$$K_x = 0.211 \left(\frac{4.5}{3.20}\right)^{-0.7459} + 0.789 = 0.9526.$$

The bending stress is

$$\sigma_t = \frac{2T_p}{b_w d_p} \frac{p_d K_o K_v K_s K_m}{K_x Y_b}$$

$$= \frac{2(1440)}{(1.0)(2.50)} \frac{(5.60)(1.0)(1.08)(0.5625)(0.9526)}{(1.0)(0.21)},$$

or $\sigma_t = 21.62$ ksi. The safety factor against bending is therefore

$$n_s = \frac{28.20}{21.62} = 1.30.$$

Therefore, surface pitting has the lower safety factor, with the value of 1.25.

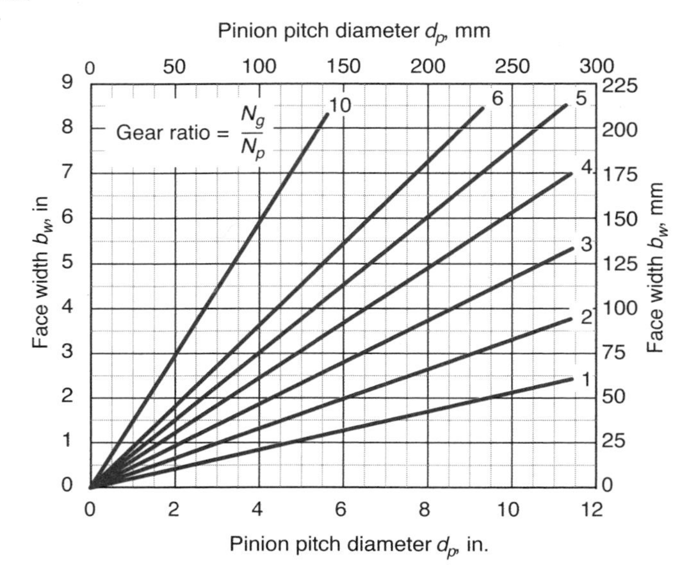

Figure 15.16: Face width of spiral bevel gears operating at a 90° shaft angle. *Source:* AGMA [2003].

15.4 Worm Gears

Worm gears, such as the set shown in Fig. 14.4, are extremely popular for situations where large speed reductions are needed. Worm gears are unique in that they cannot be back-driven; that is, the worm can drive the worm gear, but the worm gear cannot drive the worm. This is a function of the thread geometry and is referred to as **self-locking**, a topic that is explored in greater detail in Section 16.3.3. Worm gears provide very large speed reductions in simple assemblies with compact profiles, but they are less efficient than other gear types.

One of the other main differences between worm gears and other types of gears is that adhesive and abrasive wear are the primary concern with worm gears, and contact fatigue and bending stresses are much less of a concern. This suggests that effective lubrication is required for the worm, and the thread geometry will be designed to avoid point contacts and associated high contact pressures. This is accomplished by incorporating **envelopment** or **curvature** in the transverse plane, at the very least in the worm gear. Single-enveloping worm gear sets (Fig. 14.4a) have a cylindrical worm and a *throated* worm gear and generate line contact. Double-enveloping worm gear sets (Fig. 14.4b) have throated worms and worm gears and generate areas of contact. Clearly, larger forces can be transmitted with double-enveloping worm gears, but at a cost penalty associated with more complicated manufacture.

Furthermore, *wear-in* effects can occur, where initial abrasive wear causes the gear thread profiles to conform to the worm teeth, which are usually constructed of a much harder material. Thus, manufacturing tolerances, misalignment, etc, can cause initial contact to be point contact, but wear-in leads to line contact as the teeth become better conforming with time. This wear-in results in more effective lubrication and efficient transfer of loads, so that the wear rates after an initial wear-in period are much lower.

Worm gears are produced with speed ratios of 1:1 up to 360:1, although most worm gears range from 3:1 to 100:1 speed ratios. Ratios greater from 3:1 to 30:1 use single threaded worms, while larger ratios will usually use worms

Table 15.3: Suggested minimum number of worm gear teeth for customary designs. *Source:* From ANSI/AGMA [1993].

Pressure angle ϕ, deg	Minimum number of wormgear teeth
14.5	40
17.5	27
20	21
22.5	17
25	14
27.5	12
30	10

with multiple threads, or **starts**. In general, the larger the center distance between the worm and the worm gear, the larger the number of starts. The majority of worm threads are right-handed.

15.4.1 Tooth Geometry

A worm will usually have 2 to 3 teeth in contact with the worm gear at any time, as shown in Fig. 15.17. As the worm gear rotates, the contact location shifts as shown in the figure. As a result, a worm and worm gear profile must be designed to allow for efficient transfer of force along line contacts. Worm profiles are more complex than the involute tooth profiles described in Chapter 14, and the interested reader is referred to the American Gear Manufacturers Association standards [AGMA 1993] for different forms of worm gear tooth profiles. For example, recessed designs can be used to increase the number of teeth in contact.

The pressure angle for worm gears is usually restricted to the values in Table 15.3. In general, higher pressure angles are used when high worm gear tooth depths are needed, but this can result in fewer teeth in contact. In addition, the bearing loads can be higher with higher pressure angles, and the bending stresses and deflections in the worm must be considered.

AGMA recommends the minimum number of worm gear teeth as a function of pressure angle as summarized in Table 15.3. The number of threads in the worm is determined according to:

$$N_w = \frac{N_g}{Z}, \tag{15.36}$$

where N_w is the number of starts in the worm, N_g is the number of teeth in the worm gear, and Z is the required speed ratio.

The pitch of the worm is the same as for spur gears; it is the distance in the axial plane from a point on one thread to the corresponding point on the next thread. Pitches can be selected for the worm gear just as they were selected for spur gears in Ch. 14. The worm pitch must equal the circular pitch of the worm gear. The **lead** of the worm is defined as

$$L = N_w p. \tag{15.37}$$

The **lead angle** is given by

$$\lambda = \tan^{-1} \frac{L}{\pi d_{pw}}. \tag{15.38}$$

The addendum and dedendum for worms are given by

$$a = \frac{p_b}{\pi} \tag{15.39}$$

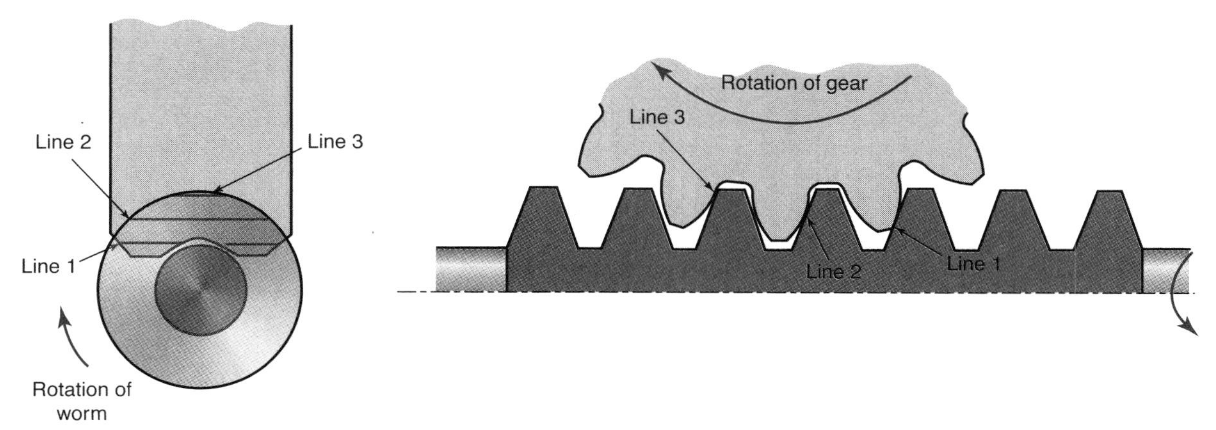

Figure 15.17: Illustration of worm contact with a worm gear, showing multiple teeth in contact.

$$b = \begin{cases} \dfrac{1.157p_b}{\pi} & \text{for } p_b > 4.06 \text{ mm } (0.160 \text{ in.}) \\ \dfrac{1.200p_b}{\pi} + 0.05 \text{ mm} & \text{for } p_b < 4.06 \text{ mm} \\ \dfrac{1.200p_b}{\pi} + 0.002 \text{ in.} & \text{for } p_b < 0.160 \text{ in.} \end{cases} \quad (15.40)$$

The outside and root diameters of the worm are given by

$$d_o = d + 2a,$$
$$d_r = d - 2b. \quad (15.41)$$

Example 15.5: Worm Gear Geometry

Given: A double-threaded worm with a pitch diameter of 3 in. meshes with a 25-tooth worm gear. The worm gear has a pitch diameter of 5 in.

Find: The speed ratio and the lead angle for the gearset.

Solution: The speed ratio is obtained from Eq. (15.36) as

$$Z = \frac{N_g}{N_w} = \frac{25}{2} = 12.5.$$

The worm gear pitch is

$$p = \frac{\pi d_g}{N_g} = \frac{\pi(5)}{25} = 0.6283 \text{ in.}$$

From Eq. (15.37),

$$L = N_w p = 2(0.6283) = 1.257 \text{ in.}$$

Therefore, the lead angle is calculated from Eq. (15.38) as:

$$\lambda = \tan^{-1}\frac{L}{\pi d_{pw}} = \tan^{-1}\frac{1.257}{\pi(3)} = 7.595^\circ.$$

15.4.2 Forces on Worm Gears

Figure 15.18 shows the forces acting on a right-handed worm thread. The nomenclature of the forces is slightly more complicated than for other gears, since the thrust force for the worm gear is equal to the tangential force for the worm and vice-versa. The tangential force for the worm gear and the thrust force for the worm are equal and are given by

$$W_{tg} = W_{aw} = \frac{2T_g}{d_m}, \quad (15.42)$$

where

W_{tg} = worm gear tangential force, N (lb)
W_{aw} = worm axial or thrust force, N (lb)
T_g = worm gear torque, N-m (in.-lb)
d_m = worm gear mean diameter, m (in.)

The gear separation force is given by

$$W_s = \frac{W_{tg}\tan\phi_n}{\cos\lambda_n}, \quad (15.43)$$

where W_s is the separation force, ϕ_n is the normal pressure angle, and λ is the worm mean lead angle. The worm gear thrust force and the worm tangential force are equal and given by:

$$W_{ag} = W_{tw} = \frac{2T_w}{d_m}, \quad (15.44)$$

where

W_{ag} = worm gear axial or thrust force, N (lb)
W_{tw} = worm tangential force, N (lb)
T_w = worm torque, N-m (in.-lb)
d_m = worm mean diameter, m (in.)

It can be shown from statics that worm and worm gear thrust loads are related by the following equation:

$$W_{tw} = W_{tg}\left(\frac{\cos\phi_n\sin\lambda + \mu\cos\lambda}{\cos\phi_n\cos\lambda - \mu\sin\lambda}\right) \quad (15.45)$$

15.4.3 AGMA Equations

A number of attempts have been made by various researchers and organizations to develop a power rating system for worm gears. This has been elusive since the long-term success of worm gears depends on many factors that are more difficult to adequately quantify than with other gear types. For example, while lubrication is always a critical concern, it is especially important with worm gears since the dominant wear mechanism is abrasive wear instead of surface pitting. The ability to remove wear particles from the lubricant, the operating temperature of the lubricant, the heat removed from the worm, the number of stops and starts, the frequency and severity of dynamic loadings, etc., are all important considerations that are somewhat addressed in current worm

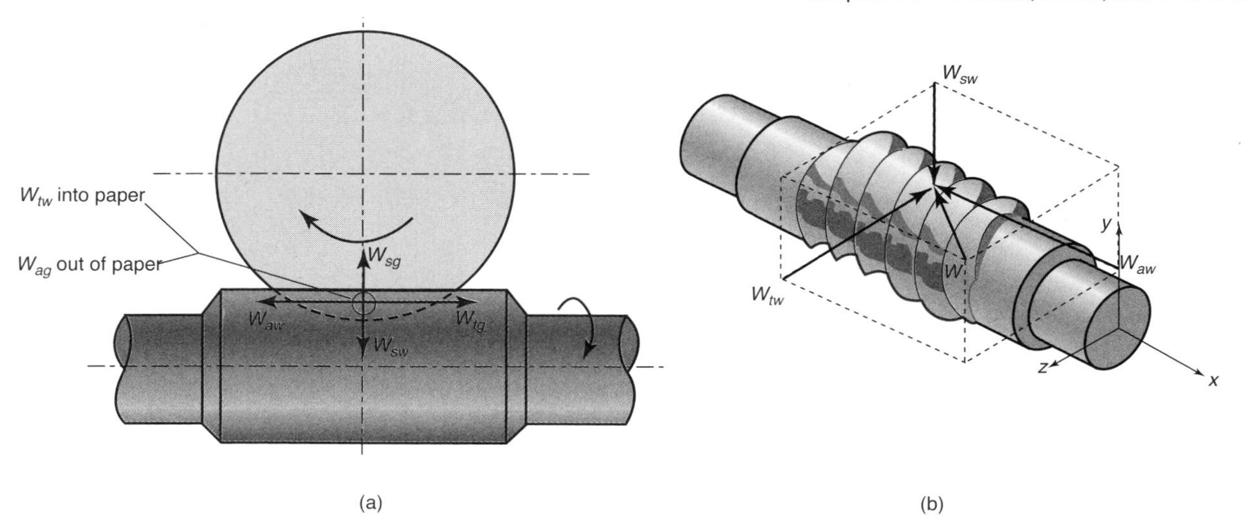

Figure 15.18: Forces acting on a worm. (a) Side view, showing forces acting on worm and worm gear. (b) Three-dimensional view of worm, showing worm forces. The worm gear has been removed for clarity.

gear rating equations, but are not robustly understood. Regardless, the AGMA equations have been demonstrated to work well for most worm gear applications.

The AGMA [1993] gives the following formula for the input power rating for a worm gearset:

$$h_{pi} = \frac{\omega W_t d_g}{2Z} + \frac{\omega W_f d_{wm}}{2\cos\lambda}, \tag{15.46}$$

where

- h_{pi} = rated input power, W (in.-lb/s)
- d_g = gear diameter, m (in.)
- Z = gear ratio, $Z = N_g/N_w$
- W_t = rated tangential load on the worm gear tooth, N (lb)
- W_f = friction force on the worm gear tooth, N (lb)
- ω = angular speed of the worm, rad/s
- d_{wm} = mean worm diameter, m (in.)
- λ = lead angle, see Eq. (15.38)

Equation (15.46) contains two terms; the first term gives the output power at the gear, while the second estimates the power loss at the tooth mesh. Recall that the worm will be constructed of a harder material than the worm gear, and the system is designed so that the worm gear conforms to the worm. Thus, the critical component for power ratings is the worm gear.

The rated tangential load for use in Eq. (15.46) to establish a gear set rating is prescribed by AGMA as:

$$W_t = \begin{cases} C_s d_{gm}^{0.8} b_w C_m C_v & \text{English units} \\ \dfrac{C_s d_{gm}^{0.8} b_w C_m C_v}{75.948} & \text{SI units} \end{cases} \tag{15.47}$$

where

- C_s = a materials factor, as given in Fig. 15.19 and Table 15.4 for bronze worm gears
- d_{gm} = mean gear diameter, mm (in.)
- b_w = face width of gear, mm (in.), not to exceed $0.67 d_{gm}$
- C_m = ratio correction factor, given by

$$C_m = \begin{cases} 0.0200\left(-Z^2 + 40Z - 76\right)^{0.5} + 0.46 & 3 \le Z < 20 \\ 0.0107\left(-Z^2 + 56Z + 5145\right)^{0.5} & 20 \le Z < 76 \\ 1.1483 - 0.00658Z & 76 \le Z \end{cases} \tag{15.48}$$

- C_v = velocity correction factor, given by

$$C_v = \begin{cases} 0.659\exp\left(-0.0011 v_t\right) & 0 < v_t \le 700\text{ ft/min} \\ 13.31 v_t^{(-0.571)} & 700 < v_t \le 3000\text{ ft/min} \\ 65.52 v_t^{(-0.774)} & 3000\text{ ft/min} < v_t \end{cases} \tag{15.49}$$

The velocity can be obtained by

$$v_t = \frac{\omega d_{wm}}{2\cos\lambda}. \tag{15.50}$$

The mean diameter has been used instead of the pitch diameter so that Eq. (15.50) is equally valid for cylindrical or enveloping worms. Note that the velocity in m/s can be multiplied by 196.85 to obtain a velocity in ft/min.

The friction force W_f in Eq. (15.46) is given by

$$W_f = \frac{\mu W_t}{\cos\lambda\cos\phi_n}, \tag{15.51}$$

where μ is the friction coefficient, λ is the lead angle at the mean worm diameter, and ϕ_n is the normal pressure angle at the mean worm diameter.

For steel worms and bronze worm gears, the AGMA offers the following expressions for coefficient of friction:

$$\mu = \begin{cases} 0.150 & v_t = 0\text{ ft/min} \\ 0.124\exp\left(-0.074 v_t^{0.645}\right) & 0 < v_t \le 10\text{ ft/min} \\ 0.103\exp\left(-0.110 v_t^{0.450}\right) + 0.012 & 10\text{ ft/min} < v_t \end{cases} \tag{15.52}$$

The rated torque for a worm gear can thus be calculated from the tangential force W_t using Eq. (15.42).

Table 15.4: Materials factor, C_s, for bronze worm gears with the worm having surface hardness of 58 HRC. *Source:* From AGMA [2010a].

Manufacturing process	Pitch diameter	Units for pitch diameter in.	Units for pitch diameter mm
Sand casting	$d \leq$ 64 mm (2.5 in.)	1000	1000
	$d \geq$ 64 mm	$1190 - 476.5 \log d$	$1859 - 476.5 \log d$
Static chill cast or forged	$d \leq$ 200 mm (8 in.)	1000	1000
	$d >$ 200 mm	$1412 - 455.8 \log d$	$2052 - 455.8 \log d$
Centrifugally cast	$d \leq$ 625 mm (25 in.)	1000	1000
	$d >$ 625 mm	$1251 - 179.8 \log d$	$1504 - 179.8 \log d$

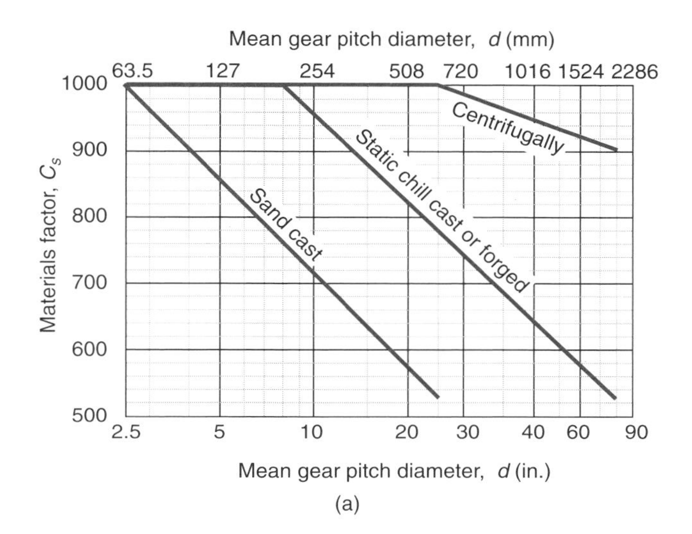

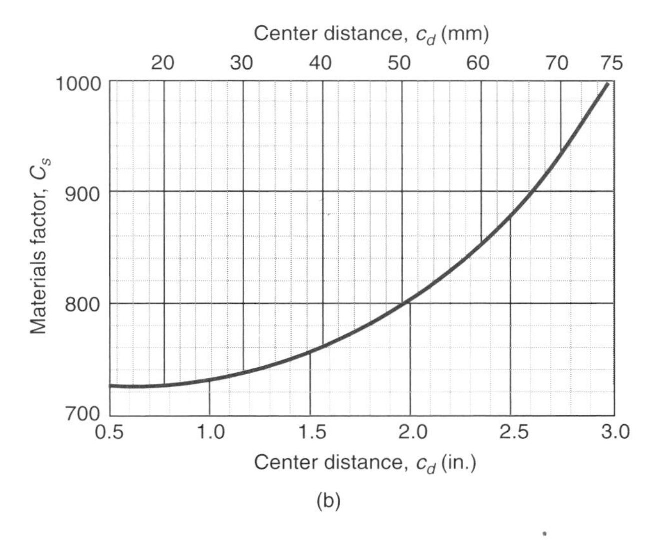

Figure 15.19: Materials parameter C_s for bronze worm gears and worm minimum surface hardness of 58 HRC. (a) Materials factor for center distances c_d greater than 76 mm (3 in.); (b) Materials factor for center distances c_d less than 76 mm (3 in.). When using part (b), the value from part (a) should be checked, and the lower value used. See also Table 15.4. *Source:* From AGMA [2010a].

Design Procedure 15.2: Worm Gear Synthesis

This design procedure closely follows ANSI/AGMA [2010a], and is a general guide for developing preliminary designs of cylindrical worm gearing. The results from this approach are, in general, good starting points for analysis, but deviation from the method and resulting values is not unreasonable. Worm gears are complicated systems, and final design attributes need to be selected by also considering the specific application, loading environment, lubrication condition, etc. This design procedure is presented here as a useful guide for preliminary worm gear layout.

The approach is restricted to coarse pitch cylindrical worm gears operated at right angles. It is recognized that high power transmission at high speeds will require fine-pitch teeth. It is assumed that the worm and worm gear speeds are known, as is the input power.

1. A pressure angle needs to be selected based on the design application. Higher pressure angles lead to higher tooth strength, but have the drawback of higher bearing reaction loads and worm bending stress, as well as resulting in fewer teeth in contact. Lower pressure angles are better suited for high speed and quiet operation. Table 15.3 lists the customary pressure angles for worm gears.

2. The minimum number of teeth in the worm gear is given in Table 15.3. More teeth can be selected if the center distance is not a design constraint.

3. The number of teeth in the worm is then obtained from Eq. (15.36).

4. The worm pitch diameter usually falls between the following ranges:

$$\frac{c_d^{0.875}}{3.0} \leq d_w \leq \frac{c_d^{0.875}}{1.6} \quad \text{English units}$$
$$\frac{c_d^{0.875}}{2.0} \leq d_w \leq \frac{c_d^{0.875}}{1.07} \quad \text{SI units} \qquad (15.53)$$

where c_d is the the center distance between the axes of the worm and worm gear.

5. The worm gear pitch diameter is calculated as

$$d_g = 2c_d - d_w. \qquad (15.54)$$

6. The axial pitch of the worm gear is

$$p_x = \frac{\pi d_g}{N_g}. \qquad (15.55)$$

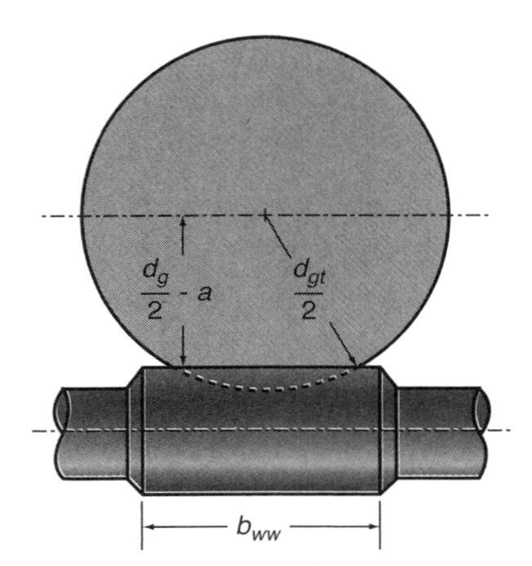

Figure 15.20: Worm face width.

7. Fig. 15.20 illustrates the required face width of the worm, which is given as

$$b_{ww} = 2\sqrt{\left(\frac{d_{gt}}{2}\right)^2 - \left(\frac{d_g}{2} - a\right)^2}. \tag{15.56}$$

A slightly larger face width should be used to allow for manufacturing and assembly tolerances. However, most worm gears have between two and three teeth in contact at all times, so using $b_{ww} = 5p_x$ gives a reasonable face width, allowing an extra axial pitch before and after contact.

8. If $p_x \geq 0.160$ in. (4.06 mm), the worm gear face width can be determined from

$$b_{wg} = \frac{2}{3}d_w,$$

and if $p_x < 0.160$ in. (4.06 mm),

$$b_{wg} = 1.125\sqrt{(d_o + 2c)^2 - (d_o - 4a)^2}, \tag{15.57}$$

where

b_{wg} = worm gear face width, m
d_o = worm gear outer diameter, given by Eq. (15.41), m
a = addendum, given by Eq. (15.39), m
c = clearance, m

Example 15.6: Worm Gear Design

Given: A worm with two starts and a face width of 25 mm meshes with a 50-tooth worm gear with a pitch diameter of 200 mm. The worm has a pitch diameter of 50 mm, and a pressure angle of 20°. The worm gear is sand cast from bronze and the worm is heat treated to a hardness of 58 HRC.

Find: Determine the rated input power for the gearset if the worm is driven at 2000 rpm.

Solution: The rated horsepower is given by Eq. (15.46), but a number of factors need to be determined before this equation can be used.

Note that the angular velocity is $\omega = 2000$ rpm $= 209.4$ rad/s and the gear ratio is

$$Z = \frac{N_g}{N_w} = \frac{50}{2} = 25.$$

The pitch of the worm gear (which is also the pitch of the worm) is

$$p = \frac{\pi d_g}{N_g} = \frac{\pi(200)}{50} = 12.57 \text{ mm}.$$

The lead is given by Eq. (15.37) as

$$L = N_w p = (2)(12.57) = 25.13 \text{ mm}.$$

Therefore, the lead angle is calculated from Eq. (15.38):

$$\lambda = \tan^{-1}\frac{L}{\pi d_{pw}} = \tan^{-1}\frac{25.13}{\pi(50)} = 9.090°.$$

Therefore, from Eq. (15.50), the pitch line velocity is calculated as

$$v_t = \frac{\omega d_{wm}}{2\cos\lambda} = \frac{(209.4)(0.050)}{2\cos 9.090°} = 5.30 \text{ m/s}.$$

From Table 15.4 for a sand cast bronze worm gear,

$$C_s = 1859 - 476.5\log d_g = 1859 - 476.5\log 200 = 762.6.$$

From Eq. (15.48),

$$\begin{aligned} C_m &= 0.0107\left(-Z^2 + 56Z + 5145\right)^{0.5} \\ &= 0.0107\left(-(25)^2 + 56(25) + 5145\right)^{0.5} \\ &= 0.8233. \end{aligned}$$

Since $v_t = 5.30$ m/s $= 1041$ ft/min, Eq. (15.49) gives

$$C_v = 13.31(v_t)^{-0.571} = 0.2515.$$

Therefore, the tangential load used to obtain a power rating for the worm gear, W_t, is given by Eq. (15.47) as

$$\begin{aligned} W_t &= \frac{C_s d_g^{0.8} b_w C_m C_v}{75.948} \\ &= \frac{(762.6)(200)^{0.8}(25)(0.8233)(0.2515)}{75.948} \\ &= 3603 \text{ N}. \end{aligned}$$

The coefficient of friction is given by Eq. (15.52) as

$$\begin{aligned} \mu &= 0.103\exp\left(-0.110 v_t^{0.45}\right) + 0.012 \\ &= 0.103\exp\left[-0.110(1041)^{0.45}\right] + 0.012 \\ &= 0.0204. \end{aligned}$$

Therefore, the friction force is given by Eq. (15.51) as

$$W_f = \frac{\mu W_t}{\cos\lambda\cos\phi} = \frac{(0.0204)(3603 \text{ N})}{\cos 9.090° \cos 20°} = 79.3 \text{ N},$$

so that the rated power, from Eq. (15.46) is:

$$\begin{aligned} h_{pi} &= \frac{\omega W_t d_g}{2Z} + \frac{\omega W_f d_{wm}}{2\cos\lambda} \\ &= \frac{(209.4)(3607)(0.201)}{2(25)} + \frac{(209.4)(79.3)(0.05)}{2\cos 9.090°} \\ &= 3457 \text{ Nm/s}. \end{aligned}$$

Or, the rated input power is roughly 3.5 kW.

Figure 15.21: The gears used to transmit power from an electric motor to the agitators of a commercial mixer. *Source:* Courtesy of Hobart, Inc.

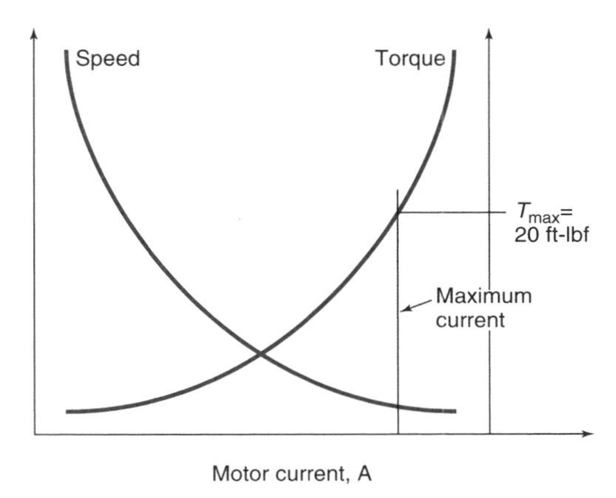

Figure 15.22: Torque and speed of motor as a function of current for the industrial mixer used in the case study.

Case Study 15.1: Gears for an Industrial Food Mixer

Given: Figure 15.21 depicts a view of the gearing present in a common food mixer of the type found in restaurants and small bakeries. Different speeds can be achieved by engaging different gears with the drive, each gear with a prescribed number of teeth. The different speeds are required to perform various operations, such as kneading, mixing, and whipping. This problem involves the specification of a bevel gear to effect mixing or kneading of dough. The motor is attached to a speed reducer whose output speed is 200 rpm and whose torque and speed curves are as shown in Fig. 15.22. Use an allowable bending stress of 40 ksi and an allowable contact stress of 350 ksi.

Find: A helical gear mating pair, as shown in the figure, so that the output shaft speed is near 90 rpm. (This speed is then the input to a planet of a planetary gear train driving the agitator.)

Solution: The torque and speed characteristics are extremely important. If the mixer has been overloaded by using improper ingredients or by overfilling, a torque high enough to stall the motor may be developed. The highest torque that can be developed is that at the maximum current provided by the power supply, or 20 ft-lb. (The power supply circuitry is designed to limit the stall current through placement of resistive elements.) It is important that a stalled motor does not lead to gear tooth breakage, since gears are a relatively expensive component and difficult to replace. Thus, the gears will be designed to survive the stall torque, even though they should ideally never be subjected to such high loading.

Given the required speed ratios, the number of teeth in the driven gears must be approximately 2.22 times the number of teeth in the driving gear. At this point, an infinite number of gear tooth combinations will serve to fulfill the design requirements, since a face width has not been specified. However, because it is difficult to manufacture gears with few teeth, the driving gear will be arbitrarily assigned 24 teeth, and the driven gear, 52 teeth. Note that this is a particular solution to a problem with many solutions.

This application is not so demanding as to require fine or ultrafine tooth classes; a coarse spacing may even be sufficient, but a medium-coarse spacing will be chosen as a balance between performance and cost. Using a diametral pitch of 12 per inch gives the gear pitch diameters as 2 and 4.32 in. for the driving and driven gears, respectively. We will use standard values of $\phi = \psi = 20°$. The torque exerted onto the drive shaft causes a load of $W_t = T/r = T/(1) = 240$ lb.

The gear and pinion must be designed based on both bending and contact stress based on Eqs. (15.15) and (15.17). Before these equations can be applied, the assorted factors need to be determined. From Table 14.8, we can reasonably assign a value of K_o=1.0, since the power source is uniform, as is the machine operation. The size factor K_s, from Table 14.9, is clearly 1.00, since the diametral pitch is larger than 5. The load distribution factor is calculated from Eq. (14.58), but first we note the following.

From Eq. (14.60) using uncrowned teeth, $C_{mc} = 1.0$. From Eq. (14.62), assuming $b_w \le 1$ in.,

$$C_{pf} = \frac{b_w}{10d_p} - 0.025.$$

Note that if the face width needs to be larger than 1 in., a different relationship for C_{pf} will be needed and this portion of the analysis will have to be repeated. From Eq. (14.63), allowing flexibility for the design of bearing supports, we take $C_{pm} = 1.1$. From Fig. 14.33,

$$C_{ma} = 0.127 + 0.0158b_w - 1.095 \times 10^{-4}b_w^2.$$

Finally, from Eq. (14.64), $C_e = 1$, so that the gears do not need to be lapped or require complicated adjustment procedures. These can be substituted into Eq. (14.60) to obtain the load distribution factor K_m as a function of face width b_w:

$$\begin{aligned} K_m &= 1.0 + \left(\frac{b_w}{10(2\text{in.})} - 0.025\right)(1.1) \\ &\quad + \left(0.127 + 0.0158b_w - 1.095 \times 10^{-4}b_w^2\right). \end{aligned}$$

The pitch line velocity is given by

$$v_t = \pi d_p N_{ap} = \pi(2\text{ in.})(200\text{ rpm}) = 1256\text{ in./min},$$

or $v_t = 105$ ft/min. From Fig. 14.34, all gears have $K_v \approx 1.0$ at this low speed, so we assign $K_v = 1$ and take $Q_v = 5$, although this may be modified at a later time with no design consequences in order to reduce noise or any other reason.

Similarly, this is not a large gear produced from a rim-and-spoke design, so we use $K_b = 1.0$. The geometry factor is obtained from Eq. (15.16) and Figs. 15.3 and 15.4 as $Y_h = 0.51$. Since σ_t was defined as 40 ksi, Eq. (15.15) gives:

$$\begin{aligned}\sigma_t &= \frac{W_t p_d K_o K_s K_m K_v K_b}{b_w Y_h} \\ &= \frac{(240)(12)}{b_w(0.51)}\left[1 + \left(\frac{b_w}{20} - 0.025\right)(1.1)\right] \\ &\quad + \frac{(240)(12)}{b_w(0.51)}\left(0.127 + 0.0158 b_w - 1.095 \times 10^{-4} b_w^2\right).\end{aligned}$$

This equation is solved numerically as $b_w = 0.15$ in., but since this is an awkward dimension, a face width of $b_w = 0.25$ in. is selected. Note that the correction factors C_{pf} and K_m as calculated above are correct for this face width.

It is now important to determine a face width for acceptable contact stress. Equation (15.17) gives the contact stress, and many of the factors calculated above can be used directly. However, some additional values are needed before solving Eq. (15.17). From Eq. (8.16), noting that for steel $E = 30 \times 10^6$ psi and $\nu = 0.30$,

$$E' = \frac{E}{1-\nu^2} = \frac{30 \times 10^6\text{ psi}}{1-(0.30)^2} = 32.96 \times 10^6\text{ psi}.$$

From Eq. (14.74),

$$\frac{1}{R_x} = \left(\frac{1}{r_p} + \frac{1}{r_g}\right)\frac{1}{\sin\phi} = \left(\frac{1}{2\text{ in.}} + \frac{1}{4.32\text{ in.}}\right)\frac{1}{\sin 20^\circ},$$

or $R_x = 0.4676$ in. The applied load is $W = W_t/cos\phi = 240/cos20^\circ = 255$ lb. Therefore, from Eq. (14.73),

$$\begin{aligned}p_H &= E'\left(\frac{W}{2\pi b_w R_x E'}\right)^{1/2} \\ &= (32.96 \times 10^6)\left(\frac{255}{2\pi b_w (32.96 \times 10^6)(0.4676)}\right)^{1/2} \\ &= 94,800(b_w)^{-1/2}.\end{aligned}$$

where p_H is in psi and b_w is in inches. Therefore, we can apply Eq. (15.17) to obtain the contact stress, which was restricted to a maximum value of 350 ksi:

$$\begin{aligned}\sigma_c &= 350,000\text{ psi} = P_H\left(K_o K_s K_m K_v\right)^{1/2} \\ &= \left(94,800(b_w)^{-1/2}\text{ psi}\right)\left((1)(1)(1.1)(1)\right)^{1/2}.\end{aligned}$$

This is solved as $b_w = 0.081$ in. Therefore, a face width of 0.25 in., based on the bending results, is selected to complete the design. This problem is typical of most design problems in that an infinite number of solutions exist. This case study merely demonstrates the design procedure used to obtain a reasonable answer, not the "right" answer.

15.5 Summary

This chapter extended the discussion of gears beyond spur gears addressed in the previous chapter to helical, bevel and worm gears. All of these gears have their advantages, with the main limitation being higher cost compared to spur gears.

Helical gears are similar to spur gears, but the tooth geometry is wound in a helix around the outside of the gear cylinder. Helical gears have the main advantage (over spur gears) of being quieter, especially at high speeds. Bevel gears are quiet and efficient, and do not require parallel shafts. Worm gear sets consist of a worm, which is similar to a helical gear with a very small helix angle, and a worm gear that is similar to a spur gear. Worm gears are commonly used for compact designs that achieve large speed reductions, and are unique among gears in that abrasive wear is the dominant wear mechanism.

The particular kinematics and dynamics of these gears was analyzed using first principles, and design methodologies standardized by the American Gear Manufacturers Association were briefly outlined. These robust design approaches allow for the selection of gears and the analysis of specific gear geometries to ensure long service lives.

Key Words

bevel gears nonparallel, coplanar gears with straight, hypoid, spiral, or Zerol teeth

circular pitch distance from one tooth to corresponding point on adjacent tooth

helical gears parallel-axis gears with teeth cut on helix that wraps around cylinder

gear toothed wheel that, when mated with another toothed wheel, transmits power between shafts

gear ratio ratio of number of teeth between two meshing gears

hypoid bevel gear A spiral gearset with zero offset

module ratio of pitch diameter to number of teeth

normal pitch number of teeth per pitch diameter in a direction normal to tooth face

pinion smaller of two meshed gears

spiral bevel gears nonparallel coplanar gears with teeth cut on a spiral

straight bevel gears nonparallel coplanar gears with straight teeth

starts the number of threads in a worm

worm gear nonparallel, noncoplanar gear with typically high reduction ratio

Zerol bevel gears similar to a straight bevel gear, but with curved teeth in the lengthwise direction and zero spiral angle

Summary of Equations

Helical Gears:

Normal circular pitch: $p_{cn} = p_c \cos\psi$

Normal diametral pitch: $p_{dn} = \dfrac{p_d}{\cos\psi}$

Axial pitch: $p_a = p_c \cot\psi = \dfrac{p_{cn}}{\sin\psi}$

Pitch diameter: $d = \dfrac{N}{p_d} = \dfrac{N}{p_{dn}\cos\psi}$

Center distance: $c_d = \dfrac{d_p + d_g}{2} = \dfrac{N_p + N_g}{2p_{dn}\cos\psi}$

Transverse contact ratio:

$$\begin{aligned} C_r &= \frac{L_{ab}}{p_c \cos\phi} \\ &= \frac{1}{p_c\cos\phi}\left[\sqrt{r_{op}^2 - r_{bp}^2} + \sqrt{r_{og}^2 - r_{bg}^2}\right] - \frac{c_d \tan\phi}{p_c} \end{aligned}$$

Axial contact ratio: $C_{ra} = \dfrac{b_w}{p_a} = \dfrac{b_w \tan\psi}{p_c} = \dfrac{b_w \sin\psi}{p_{cn}}$

Minimum contact length:

for $n_a \le 1 - n_r$,

$$L_{\min} = \frac{C_{ra} b_w - n_a n_r p_a}{\cos\psi_b}$$

for $n_a > 1 - n_r$,

$$L_{\min} = \frac{C_{ra} b_w - (1 - n_a)(1 - n_r)p_a}{\cos\psi_b}$$

Base helix angle: $\cos\psi_b = \dfrac{\cos\psi\cos\phi_n}{\cos\phi}$

Axial load: $W_z = W_t \tan\psi$

Radial load: $W_r = W_t \tan\phi$

Normal load: $W = \dfrac{W_t}{\cos\phi_n \cos\psi}$

AGMA bending stress equation:

$$\sigma_t = \frac{W_t p_d K_o K_s K_m K_v K_i K_b}{b_w Y_h}$$

AGMA contact stress equation:

$$\sigma_c = p_H (K_o K_s K_m K_v)^{1/2}$$

Maximum Hertzian contact stress:

$$p_H = E'\left(\frac{W'}{2\pi}\right)^{1/2}$$

Effective elastic modulus:

$$E' = \frac{2}{(1 - \nu_a^2)/E_a + (1 - \nu_b^2)/E_b}$$

Effective radius:

$$\frac{1}{R_x} = \left(\frac{1}{r_p} + \frac{1}{r_g}\right)\frac{1}{\sin\phi} = \left(\frac{1}{d_p} + \frac{1}{d_g}\right)\frac{2}{\sin\phi}$$

Dimensionless load: $W' = \dfrac{W}{L_{\min} E' R_x}$

Bevel Gears:

Virtual number of teeth: $N_v = \dfrac{N}{\cos\alpha_p}$

Normal force: $W = \dfrac{W_t}{\cos\phi} = \dfrac{T}{r_{ave}\cos\phi}$

Axial force: $W_a = W_t \tan\phi \sin\alpha$

Radial force: $W_r = W_t \tan\phi \cos\alpha$

AGMA bending stress equation:

$$\sigma_t = \frac{2T_p}{b_w d_p}\frac{p_d K_o K_v K_s K_m}{K_x Y_b}$$

AGMA contact stress equation:

$$\sigma_c = \sqrt{\frac{T_p E'}{\pi b_w d_p^2 I_b} K_o K_v K_m K_s K_x}$$

Load distribution factor:

$$K_m = \begin{cases} K_{mb} + 0.0036 b_w^2 & \text{for } b_w \text{ in inches} \\ K_{mb} + \left(5.6 \times 10^{-6}\right) b_w^2 & \text{for } b_w \text{ in mm} \end{cases}$$

Crowning factor for bending stress:

For straight or Zerol®: $K_x = 1$,

For spiral: $K_x = 0.211\left(\dfrac{r_c}{A_m}\right)^q + 0.789$

where $q = \dfrac{0.279}{\log(\sin\psi)}$

Crowning factor for contact stress:

$$K_x = \begin{cases} 1.5 & \text{for properly crowned teeth} \\ 2.0 & \text{(or larger) for non-crowned teeth} \end{cases}$$

Worm Gears:

Number of threads in worm: $N_w = \dfrac{N_g}{Z}$

Worm lead: $L = N_w p$

Lead angle: $\lambda = \tan^{-1}\dfrac{L}{\pi d_{pw}}$

Addendum: $a = \dfrac{p_b}{\pi}$

Dedendum:

$$b = \begin{cases} \dfrac{1.157 p_b}{\pi} & \text{for } p_b > 4.06 \text{ mm (0.160 in.)} \\ \dfrac{1.200 p_b}{\pi} + 0.05 \text{ mm} & \text{for } p_b < 4.06 \text{ mm} \\ \dfrac{1.200 p_b}{\pi} + 0.002 \text{ in.} & \text{for } p_b < 0.160 \text{ in.} \end{cases}$$

Outside and root diameters: $d_o = d + 2a$, $d_r = d - 2b$

Tangential force on worm gear or thrust force for worm:

$$W_{tg} = W_{aw} = \frac{2T}{d_m}$$

Gear separation force: $W_s = \dfrac{W_{tg}\tan\phi_n}{\cos\lambda_n}$

Worm gear thrust force or worm tangential force:

$$W_{ag} = W_{tw} = \frac{2T_w}{d_m}$$

AGMA input power rating: $h_{pi} = \dfrac{\omega W_t d_g}{2Z} + \dfrac{\omega W_f d_{wm}}{2\cos\lambda}$

Rated transmitted load:

$$W_t = \begin{cases} C_s d_{gm}^{0.8} b_w C_m C_v & \text{English units} \\ \dfrac{C_s d_{gm}^{0.8} b_w C_m C_v}{75.948} & \text{SI units} \end{cases}$$

Ratio correction factor:

$$C_m = \begin{cases} 0.0200\left(-Z^2 + 40Z - 76\right)^{0.5} + 0.46 & 3 \le Z < 20 \\ 0.0107\left(-Z^2 + 56Z + 5145\right)^{0.5} & 20 \le Z < 76 \\ 1.1483 - 0.00658Z & 76 \le Z \end{cases}$$

Velocity: $v_t = \dfrac{\omega d_{wm}}{2\cos\lambda}$

Velocity factor:

$$C_v = \begin{cases} 0.659\exp\left(-0.0011 v_t\right) & 0 < v_t \le 700 \text{ ft/min} \\ 13.31 v_t^{(-0.571)} & 700 < v_t \le 3000 \text{ ft/min} \\ 65.52 v_t^{(-0.774)} & 3000 \text{ ft/min} < v_t \end{cases}$$

Friction force: $W_f = \dfrac{\mu W_t}{\cos\lambda\cos\phi_n}$

Friction coefficient:

$$\mu = \begin{cases} 0.150 & v_t = 0 \text{ ft/min} \\ 0.124\exp\left(-0.074 v_t^{0.645}\right) & 0 < v_t \le 10 \text{ ft/min} \\ 0.103\exp\left(-0.110 v_t^{0.450}\right) + 0.012 & 10 \text{ ft/min} < v_t \end{cases}$$

Recommended Readings

Budynas, R.G., and Nisbett, J.K. (2011), *Shigley's Mechanical Engineering Design*, 9th ed., McGraw-Hill.
Crosher, W.P. (2002) *Design and Application of the Worm Gear*, ASME.
Dudley, D.W. (1994) *Handbook of Practical Gear Design*, CRC Press.
Hamrock, B.J., Schmid, S.R., and Jacobson, B.O. (2004) *Fundamentals of Fluid Film Lubrication*, 2nd ed., Marcel Dekker.
Howes, M.A., Ed. (1980) *Source Book on Gear Design Technology and Performance*, American Society for Metals.
Juvinall, R.C., and Marshek, K.M. (2012) *Fundamentals of Machine Component Design*, 5th ed., Wiley.
Litvin, F.L. (1994) *Gear Geometry and Applied Theory*, Prentice-Hall.
Mott, R.L. (2014) *Machine Elements in Mechanical Design*, 5th ed., Pearson.
Norton, R.L. (2011) *Machine Design*, 4th ed., Prentice Hall.
Stokes, A. (1992) *Gear Handbook*, Butterworth Heinemann.
Townsend, D.P. (1992) *Dudley's Gear Handbook*, 2nd ed., McGraw-Hill.

References

AGMA (1989) *Geometry Factors for Determining the Pitting Resistance and Bending Strength of Spur, Helical, and Herringbone Gear Teeth*, ANSI/AGMA Standard 908-B89, American Gear Manufacturers Association.
AGMA (1993) *Design Manual for Cylindrical Wormgearing*, ANSI/AGMA Standard 6022-C93. American Gear Manufacturing Association.
AGMA (2003) *Design Manual for Bevel Gears*, ANSI/AGMA 2005-D03, American Gear Manufacturing Association.
AGMA (2010a) *Practice for Enclosed Cylindrical Wormgear Speed Reducers and Gearmotors*, ANSI /AGMA Standard 6034-B92. American Gear Manufacturing Association.
AGMA (2010b) *Rating the Pitting Resistance and Bending Strength of Generated Straight Bevel, Zerol Bevel and Spiral Bevel Gear Teeth*, ANSI/AGMA Standard 2003-C10. American Gear Manufacturing Association.
AGMA (2010c) *Fundamental Rating Factors and Calculation Methods for Involute Spur and Helical Gear Teeth*, ANSI/AGMA Standard 2101- D04, American Gear Manufacturers Association.
AGMA (2011) *Gear Nomenclature, Definitions of Terms with Symbols*, ANSI/AGMA Standard 1012-G05, American Gear Manufacturers Association.
Dudley, D.W. (1994) *Handbook of Practical Gear Design*, McGraw-Hill.

Questions

15.1 What are the differences between spur and helical gears?

15.2 What is the difference between circular pitch and normal circular pitch for helical gears?

15.3 Define "equivalent number of teeth".

15.4 What are typical values of contact ratio for helical gears? Is this different or similar to the spur gears considered in Ch. 14?

15.5 How are helical, bevel, and worm gears manufactured?

15.6 Why do the terms K_i and K_b appear in Eq. (15.15) but not Eq. (15.17)?

15.7 Can a planetary gear train be constructed from helical gears? Explain your answer.

15.8 Explain the difference between straight bevel, spiral bevel, and Zerol bevel gears.

15.9 What is the main advantage of hypoid gears?

15.10 Are helical, bevel, and worm gears involute? Explain.

15.11 What is a straddle mount? What is an overhang mount?

15.12 What is the difference between a worm and a worm gear?

15.13 What is the lead for a worm? What is the lead angle?

15.14 What is harder, a worm or a worm gear? Why?

15.15 What is meant by the term "starts" for a worm? Do other gear types have multiple starts?

Qualitative Problems

15.16 *Miter* gears are bevel gears with a speed ratio of unity. What are the advantages and disadvantages of using miter gears?

15.17 Carefully sketch a helical gear with a helix angle of (a) $\psi = 30°$, and (b) $\psi = 10°$. What observations can you make regarding gear performance based on your sketches?

15.18 What are the similarities and differences between the concepts of equivalent number of teeth for helical gears and virtual number of teeth for bevel gears?

15.19 Without the use of equations, explain how the contact ratio is calculated for helical gears.

15.20 Note that the helix angle does not appear in Eq. (15.15). Does this mean that the bending stress in helical gears is independent of helix angle? Explain.

15.21 How does the AGMA bending stress equation change if one is analyzing a spur or a helical gear?

15.22 Carefully sketch an end view of a bevel gear with (a) straight; (b) Zerol; and (c) spiral teeth. Which of these tooth forms has a constant tooth thickness?

15.23 Review Table 11.9 and add a row for two shafts that are coupled through meshing bevel gears.

15.24 Compare Eqs. (15.15) and (15.28) and list the similarities and differences.

15.25 It has been suggested that worm gears are merely helical gears with crossed shafts. Do you agree or disagree? Explain your answer.

15.26 Explain why there is no equivalent bending or contact stress equation for worm gears.

15.27 Plot coefficient of friction versus speed for worm gears, and indicate the regimes of lubrication.

15.28 What are the implications of using fewer worm gear teeth than suggested by Table 15.3?

15.29 With suitable sketches, explain the significance of run in with respect to worm gears.

15.30 Explain the importance of effective lubrication for helical, bevel, and worm gears.

Quantitative Problems

15.31 For gear transmissions there is a direct proportionality between the local sliding speeds experienced by the contacting surfaces and the power loss. For a normal spur or helical gear, the mean sliding speed is typically 20% of the load transmission speed. The coefficient of friction in lubricated gears is typically between 0.05 and 0.1. This leads to a power loss of 1 to 2% and a power efficiency of 98 to 99%. For a hypoid gear in the rear axle of a car, the sliding speed can be 60% of the load transmission speed, and for a worm gear it can be 1200%. Estimate the power efficiencies for a hypoid gear and a worm gear. *Ans.* $e_{\text{hypoid}} = 94 - 97\%$, $w_{\text{worm}} = 45.5 - 62\%$.

15.32 A 16-tooth, commercial enclosed ($Q_v = 6$) right-handed helical pinion turns at 2000 rpm and transfers 5 kW to a 38-tooth gear. The helix angle is 30°, the pressure angle is 25°, and the normal module is 2 mm. For a safety factor of 3 for bending, calculate the pitch diameters; the normal, axial, and tangential pitch; the normal pressure angle; and the face width. The material is 2.5% chrome steel (Grade 3), hardened to 315 HB. Assume $K_o = 1.0$. *Ans.* $d_p = 27.7$ mm, $d_g = 65.8$ mm, $p_a = 9.42$ mm, $\phi_n = 22°$.

15.33 Two meshing commercial enclosed helical gears are made of steel (allowable bending stress = 60 ksi and allowable contact stress = 200 ksi), and are mounted on parallel shafts 10 in. apart. The desired velocity ratio is 0.35, with a diametral pitch of 8.47 in.$^{-1}$ and the gears have a 2-in. face width, with a 20° pressure angle and 30° helix angle. What is the maximum horsepower that can be safely transmitted at a pinion speed of 1000 rpm for bending and for surface pitting failure modes? Use $Q_v = 9$. *Ans.* For bending, $h_{p,\max} = 210$.

15.34 Two helical gears have shafts at 90° and helix angles of 45°. The speed ratio is 4:1, and the gears have a normal module of 5. Find the center distance if the pinion has 20 teeth. *Ans.* $c_d = 353.5$ mm.

15.35 A 25-tooth pinion transmits 5 hp at 8000 rpm to a 100-tooth gear. Both gears have $\phi = \psi = 30°$ and a face width of 0.50 in. The pinion and gear are constructed of Grade 2 through-hardened steel. Determine the minimum diameter and the required Brinell hardness based on bending stress. *Ans.* For Grade 2 hardened to 400 BH, $d_p = 0.388$ in.

15.36 A 25-tooth helical gear with a 20° pressure angle in the plane of rotation has the helix angle $\psi = 20°$. What is the pressure angle in the normal plane, ϕ_n, and the equivalent number of teeth, N_n? *Ans.* $N_n = 30.13$ teeth.

15.37 Two meshing helical gears on parallel shafts have 25 teeth and 43 teeth, respectively, and have the helix angle $\psi = 25°$. The circular pitch is $p_c = 6.5$ mm and $\phi = 20°$. How wide do the gears have to be to obtain a total contact ratio of 2.5? *Ans.* $b_w = 8.27$ mm.

15.38 In a helical gear set for a marine main propulsion drive (on a large ship), a desired life is 6000 hours at a pinion speed of 3000 rpm with a reliability of 99%. The power transferred is 15 kW and the gear ratio is 2.5. For the gears, $\phi = 20°$, $\psi = 30°$, $r_p = 50$ mm, $m = 5$ mm, and the required safety factor for bending and pitting resistance is 1.5. For a rim thickness factor of $K_b = 2$, and carburized and hardened Grade 1 steel with an allowable bending stress of 380 MPa and an allowable contact stress of 1250 MPa, find the face width. Assume $K_o = 1.25$, $K_m = 1.15$, and $Q_v = 11$. *Ans.* $b_w \approx 20$ mm.

15.39 A bevel gear with a 60° pitch angle and 52 teeth meshes with a pinion with a 30° pitch angle and 20 teeth. Calculate the virtual number of teeth for the gear and the pinion. *Ans.* $N_{vg} = 104$ teeth, $N_{vp} = 23.1$ teeth.

15.40 Two 20° pressure angle straight bevel gears have a diametral pitch of 4 in.$^{-1}$ at the outside radius, with a 2-in. face width. The pinion has 20 teeth, and the gear has 40. The pinion is driven at 1500 rpm. Determine the maximum horsepower that can be transmitted by the gears if the allowable contact stress is 230 ksi and the allowable bending stress is 35 ksi. Both gears are made from steel, straddle mounted, properly crowned, and have $Q_v = 11$. *Ans.* $h_{p,\max} = 266$ hp.

15.41 A 90° straight bevel gear set is needed for a 6:1 speed reduction. Determine the gear forces if the 25° pressure angle pinion has 20 teeth with a module of 3, and 850 W are transmitted at a pinion speed of 1000 rpm. *Ans.* $W_t = 271$ N, $W_a = 125$ N, $W_r = 20.9$ N.

15.42 Determine the safety factor of the gearset described in Example 15.4 based on analysis of the gear instead of the pinion. *Ans.* For contact stress, $n_s = 2.37$.

15.43 Two Zerol bevel gears with 90° shaft angles and mean cone distance of 80 mm have a face width of 20 mm, a module of 5, properly crowned teeth, and were produced with a cutter radius of 110 mm. The gears have $\phi = 20°$, $\psi = 25°$, $N_p = 30$, $N_g = 60$, and both pinion and gear are straddle mounted. If the allowable contact stress is 650 MPa and the allowable bending stress is 205 MPa, find the maximum power rating for the gear set if the pinion is driven at 1000 rpm. Assume $K_o = 1.0$ and $Q_v = 8$. *Ans.* $h_p = 89.0$ kW.

15.44 A pair of straight bevel gears are mounted on shafts that are 90° apart, and the pinion transmits 10 hp at 300 rpm. The pinion has a 6-in. outside pitch diameter, 2-in face width, 20 teeth, and has $\phi = 20°$. For a velocity ratio of 0.5, draw a free body diagram of the gears showing all forces. What is the torque produced about the gear axis?

15.45 Determine the required bending and contact strength of the gear materials for Problem 15.44.

15.46 Design a spiral gear set using straight teeth that transmits 9000 in-lb/s at an input speed of 500 rpm and a speed ratio of 2:1. Use a material factor of 1.0. *Ans.* $d_p = 4.0$ in., $N_p = 18$, $\phi = 20°$.

15.47 How much extra space is needed (in terms of volume of an enclosure for the gears) if the bevel gears in Problem 15.46 are changed to cast iron?

15.48 A worm with two starts has a lead angle of 6° and meshes with a worm gear having 50 teeth and a module of 7. Find the center distance between the shafts. *Ans.* $c_d = 241.6$ mm.

15.49 The efficiency η of a gearset is the ratio of the output to input power. Obtain an expression for the efficiency of a worm gearset based on the AGMA approach. Using the gearset of Example 15.6, plot the efficiency of the gearset as a function of velocity.

15.50 A worm with two starts has a lead angle of 8°, a pressure angle of 35°, and a mean diameter of 75 mm. The worm is driven by a 2-kW motor at 1500 rpm. The worm gear has 40 teeth. Calculate the worm and worm gear tangential, separation, and thrust forces. *Ans.* $W_{tw} = 339.5$ N, $W_s = 1200$ N, $W_{aw} = 1688$ N.

15.51 A triple-threaded worm has a lead angle of 12° and a pitch diameter of 2.281 in. Find the center distance when the worm is mated with a worm gear of 60 teeth. *Ans.* $c_d = 5.99$ in.

15.52 For a worm gear speed reducer shown in sketch a, the lead angle is 10° and the pressure angle is 25°. If the tangential worm gear force is 2000 N, find (a) the axial and radial forces on the worm and gear, and (b) the bearing loads. Assume the friction coefficient is 0.05.

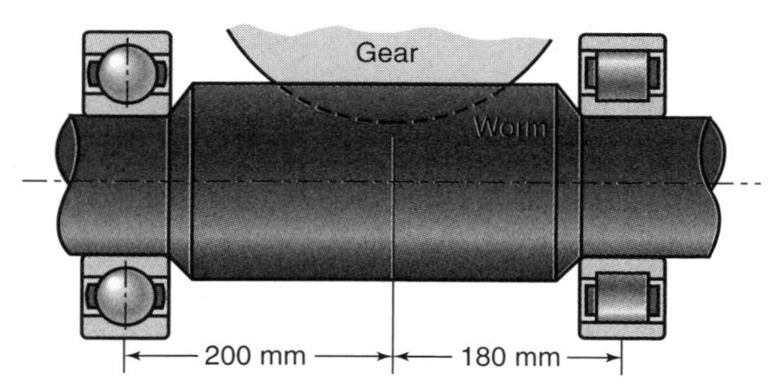

Sketch a, for Problem 15.52

15.53 Find the ratio of the output to input torque for a worm gear set in which the speed ratio is 50:1, the center distance is 250 mm, the worm has one start with a mean diameter of 100 mm, the lead angle is 5°, and the pressure angle is 14.5°. The gear mean diameter is 400 mm, and $\mu = 0.10$. What is the torque ratio if $\mu = 0$? *Ans.* $T_g/T_w = 41.5$

15.54 Find the power rating of the worm gear set described in Problem 15.53 if the worm speed is 2000 rpm and the gear was sand cast with a width of 40 mm. *Ans.* $h_p = 9407$ W

15.55 A single threaded steel worm meshes with a 42-tooth forged bronze worm gear. If the pitch diameter of the worm gear is 3 in., the center distance is 12 in., the pressure angle is 20°, and the pinion turns at 860 rpm, find the diameter and maximum input horsepower if the width is 2 in. *Ans.* $h_{pi} = 49.45$ hp.

15.56 A double-threaded worm gear has a worm pitch diameter of 50 mm, a pitch of 10 mm, and a speed reduction of 25:1. Find the lead angle, the worm gear diameter, and the center distance. *Ans.* $\lambda = 7.256°$, $c_d = 0.105$.

15.57 For the worm gear in Problem 15.56, determine the forces acting on the worm and worm gear if the transmitted torque is 100 N-m at 1000 rpm. Use a pressure angle of 25°.

15.58 Design a triple-threaded worm gearset to deliver 15 hp from a shaft rotating at 1500 rpm to another rotating at 75 rpm. Use $\phi = 25$ and $\psi = 20$, a forged bronze worm gear, and a maximum center distance of 15 in. *Ans.* $N_g = 60$, $b_{ww} = 7.33$ in., $b_{wg} = 0.933$ in.

15.59 A bakery carousel oven consists of six trays, 10 ft × 3 ft, mounted onto a wheel and rotated within a heated enclosure. It is desired to attain a rotational speed of the carousel of 2 rpm, while a 3-hp motor to be used to power the oven has an operating speed of 700 rpm. Design a worm gearset, including specification of worm and gear materials, numbers of teeth, face width, numbers of starts, and center distance. How large of an enclosure would be needed for this gearset? Note that the motor can be mounted outside the oven, so no thermal effects need to be considered.

15.60 Repeat Problem 15.59 using a helical planetary gear system in the speed reducer. Comment on the strengths and weaknesses of each design.

15.61 Design a coarse pitch worm gear set that has a worm speed of 400 rpm, a speed ratio of 40:1, and an input power of 550 ft-lb/s by following the steps in Design Procedure 15.2, and then evaluate the power rating of your worm gear set using the AGMA approach using sand cast bronze gears. Use a pressure angle of 25° and a maximum center distance of 25 in. *Ans.* $N_g = 40$, $d_w = 6$ in.

Design and Projects

15.62 The AGMA equation for contact stress is based on Hertzian contact stress, but expressed as

$$\sigma_c = C_p \left(\frac{W_t K_o K_s K_m K_v}{d_p b_w I} \right)^{1/2}$$

where C_p is an elastic coefficient defined by

$$C_p = \sqrt{\frac{1}{\pi \left[\dfrac{1-\nu_1^2}{E_1} + \dfrac{1-\nu_2^2}{E_2} \right]}}$$

Compare this to Eq. (15.17) and derive an expression for the geometry correction factor, I. List the advantages and disadvantages of using such a correction factor.

15.63 Review Design Procedure 14.5 and prepare a detailed design procedure to outline the approach for design synthesis of helical gears.

15.64 Prepare a figure equivalent to Fig. 15.17 that shows double enveloping worm gears, and obtain an expression for the face width, b_{ww}.

15.65 Design a gearset that is to have two stages involving helical gears with prescribed center distances of $c_{d1} = 7.0$ in. and $c_{d2} = 18.0$ in. Assume the application is for a smoothly operating conveyor driven by an electric motor. The overall speed reduction needs to be 20:1. Restrict your analysis to through-hardened Grade 1 steel with a Brinell hardness of 300, input pinion speed of 1750 rpm, a life of 10,000 hours, a 20° pressure angle and 15° helix angle for all gears, and a power transfer of 50 hp. Assume the rim factors are 1.0 for all gears.

15.66 For the data in Problem 15.65, design a worm gear set and compare the housing dimension to that resulting from the use of helical gears.

Chapter 16

Fasteners, Connections, and Power Screws

A collection of threaded fasteners. *Source:* Courtesy of CC Fasteners.

The simplest things are also the most extraordinary things, and only the wise can see them.

Paulo Coelho, *The Alchemist*

Contents

This chapter focuses on fasteners and connections. Since the manufacture of intricate products often requires assembly of components, engineers are thus confronted with the task of fastening various members together. A number of options exist, including threaded fasteners, rivets, welds, and adhesive joints. Each of these types is considered in this chapter. The mechanics of threaded fasteners are discussed, and applied to the design of bolted joints. Common design considerations for welds and rivets are presented. Welding is a common method of achieving permanent junctions, and takes many forms. The chapter discusses analysis of welding processes that produce beads or weld lines, spot welds, or area welds. Brazing, soldering, and adhesive bonding are discussed, and these processes are analyzed in the same fashion as welding. The chapter closes with a discussion of integrated snap fasteners, crimping, and compound fastening methods.

Machine elements in this chapter: Power screws, bolts, screws, studs, rivets, welds, adhesive joints, integrated snap fasteners.
Typical applications: Construction of all kinds, pressure vessels, flanged connections, assembly of wheels on axles, frames and lattice structures, electrical connections, covers and enclosures.
Competing machine elements: Press and shrink fits (Chapter 10), keys, splines, set screws, couplings, retaining rings (Chapter 11), clutches (Chapter 18).

Symbols

A cross-sectional area, m^2
A_g gasket area per bolt, m^2
A_i constant
A_s shear stress area, m^2
A_t tensile stress area, m^2
B_i constant
b width, m
C_k dimensionless stiffness parameter
c distance from neutral axis to outer fiber, m
d major diameter, largest possible diameter of thread, m
d_c crest diameter, largest actual diameter of thread, m
d_i smallest diameter of frustum cone, m
d_p pitch diameter, m
d_r root diameter, smallest diameter of thread, m
d_w washer diameter, m
E modulus of elasticity, Pa
e efficiency, percent
e_k elongation, m
h maximum cantilever thickness, m
h_e weld leg length, m
h_p power, W
h_t largest possible thread height, m
I area moment of inertia, m^4
I_u unit area moment of inertia, m^3
J polar area moment of inertia, m^4
J_u unit polar area moment of inertia, m^3
K_f fatigue stress concentration factor
k stiffness, N/m
k_f surface finish factor
k_m miscellaneous factor
k_r reliability factor
k_s size factor
L length, m
L_g grip length, m
L_s shank length, m
L_{se} shank effective length, m
L_t threaded length, m
L_{tot} total threaded length, m
L_{te} effective threaded length, m
L_w weld length, m
l lead, m
M bending moment, N-m
m 1, 2, 3, for single-, double-, and triple-threaded screws, respectively
N_a rotational speed, rpm
N_r number of rivets in width of member
n number of threads per inch, $in.^{-1}$
n_s safety factor
P force, N
P_i preload, N
P_n load acting on diagonal of parallelepiped, N
p pitch, in.
p_o minimum gasket seal pressure, Pa
r radius, distance from centroid of weld group to farthest point in weld, m
S_e modified endurance limit, Pa
S'_e endurance limit, Pa
S_p proof strength, Pa
S_y yield strength, Pa
S_{ut} ultimate tensile strength, Pa
T torque, N-m
T_l torque needed to lower load, N-m
T_r torque needed to raise load, N-m
t_e weld throat length, m
t_m thickness of thinnest member, m
W load; mating force; tensile force, N
x, y coordinates, m
$\bar{x}, \bar{y}$ centroid coordinates, m
Z_m section modulus, m^3
α lead angle; wedge angle of cantilever tip; angle of inclination, deg
α_f angle of frustum cone stress representation, deg
β thread angle, deg
ϵ strain
θ_n angle created by force P_n for diagonal of parallelepiped shown in Fig. 16.7a, deg
μ coefficient of friction
σ normal stress, Pa
τ shear stress, Pa
ω rotational speed, rad/s

Subscripts

a alternating
b bolt
c collar
d direct
g gasket
i preload; inner
j joint
m mean
n normal, nominal
o outer
p proof
s shear
t thread; torsional

16.1 Introduction

This chapter begins with a discussion of thread geometries, followed by the mechanics of power screws. In Section 16.4, the mechanics of power screws are applied to the analysis and design of threaded fasteners, and addresses bolt and joint stiffnesses, proper preload, bolt strength, and design for different failure modes. Section 16.5 considers rivets, which are cost-effective and reliable, but do not lend themselves to disassembly. Welding (Section 16.6) is a class of extremely versatile manufacturing processes that lead to permanent joints. Brazing and soldering are similar to welding, but do not rely on diffusion of material across a conjunction. In that manner, they are similar to adhesive bonding (Section 16.7), where the strength of a joint derives from adhesion. Finally, miscellaneous fasteners such as integrated snap fasteners, crimps, and combination joints are described in Section 16.8.

16.2 Thread Terminology, Classification, and Designation

In 1904, a fire broke out in the John E. Hurst & Co. building in Baltimore, Maryland. The fire engulfed the entire structure, and leaped from building to building until it engulfed an 80-block area. Additional firefighters were requisitioned and arrived from New York, Philadelphia, and Washington, but the reinforcements had little effect — their hoses' threads did not match those on the Baltimore fire hydrants. The fire

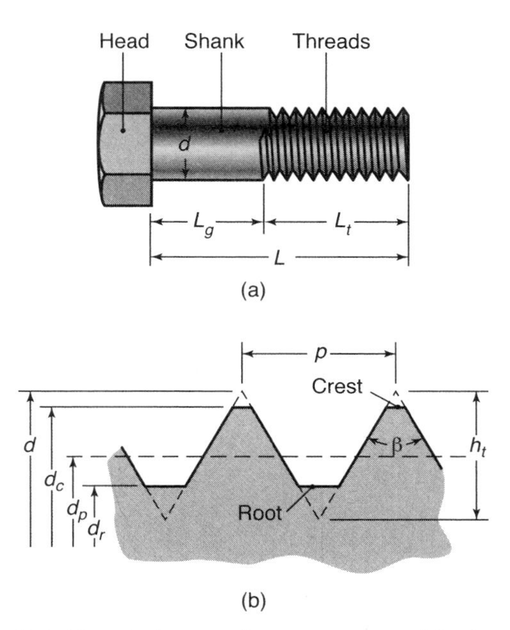

Figure 16.1: Terminology and parameters used in thread profiles. (a) View of a bolt, with parts and lengths identified; (b) detail of the thread.

eventually destroyed over 2000 buildings and burned fiercely for more than 30 hours, with smoldering fires continuing for weeks.

As a direct response to the Baltimore fire, and the recognized need for improved standardization in an increasingly industrial society, the American Engineering Standards Committee, now known as the American National Standards Institute, was founded in 1916. It was no coincidence that the first standard promulgated was one that defined the geometry of pipe threads. Standardized thread geometries are essential for proper mating of parts and allow mass production of high-quality threads. The initial standards, with only slight modification, are still relevant today.

Figure 16.1 describes the terminology and dimensions of threaded parts. Note in particular the differences between the various diameters (major, crest, pitch, and root), the largest possible thread height, h_t, and the thread angle, β. The **pitch**, p, shown in Fig. 16.1, is the distance from a point on one thread to the same point on an adjacent thread, and its unit is meters or inches. A parameter that can be used instead of pitch is **threads per inch**, n. Pitch and threads per inch are related by

$$p = \frac{1}{n}. \tag{16.1}$$

Another thread parameter is **lead**, which is the distance that the screw or bolt would advance relative to the nut in one revolution. For a single-threaded screw, $l = p$, and for a double-threaded screw, $l = 2p$, etc. Figure 16.2 shows the differences between single-, double-, and triple-threaded screws.

A number of different thread profiles can be used for a wide variety of applications. Figure 16.3 shows selected types. The Acme thread profile is used for power screws and machine tool threads. The unified (UN) is used extensively in threaded fasteners. The Acme profile has a thread angle of 29°, whereas the UN profile has a thread angle of 60°. The metric profile (M) is quite similar to the UN profile.

Figure 16.4 shows more details of the M and UN thread

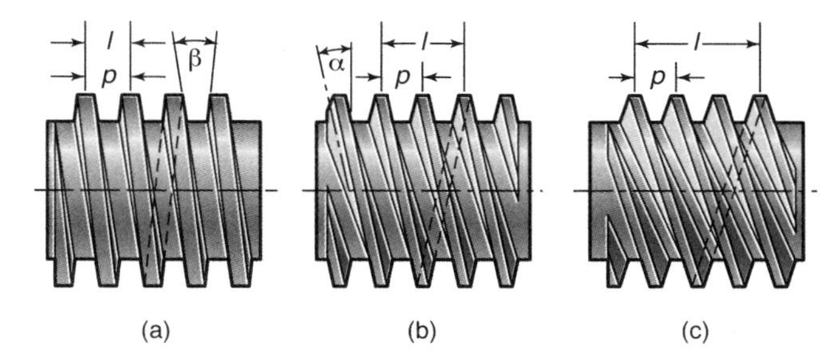

Figure 16.2: (a) Single-, (b) double-, and (c) triple-threaded screws.

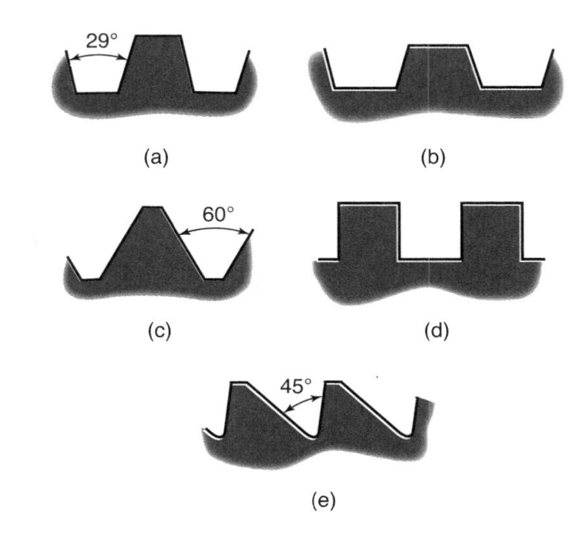

Figure 16.3: Thread profiles. (a) Acme; (b) stubbed Acme, identical to the Acme profile but with smaller thread height; (c) UN (a stubbed UN profile also exists); (d) square; (e) buttress. Buttress threads have increased shearing load capacity.

profiles than are shown in Fig. 16.3. From this figure

$$h_t = \frac{0.5p}{\tan 30^\circ} = 0.8660p. \tag{16.2}$$

Once pitch, p, or the largest possible thread height, h_t, are known, the various dimensions of the UN and M thread profiles can be obtained. It should be noted that a large number of specialty profiles are available. The most common are the UN thread profile in English units and the M profile for metric units.

The term **thread series** can be applied to threads of any size. The UN thread profile has eight thread series of constant pitch. Each thread series has the same number of threads per inch. The helix angle that the thread makes around the fastener varies with the fastener diameter. However, the thread depth is constant, regardless of the fastener diameter, because the 60° thread angle remains constant. This situation can be clearly observed from Eq. (16.2). The eight constant-pitch UN thread series are 4, 6, 8, 12, 16, 20, 28, and 32 threads per inch.

In addition to thread series, thread profiles are classified by **coarseness**, which refers to the quality and number of threads per length produced on a common diameter of the fastener. These designations after UN have the following meanings:

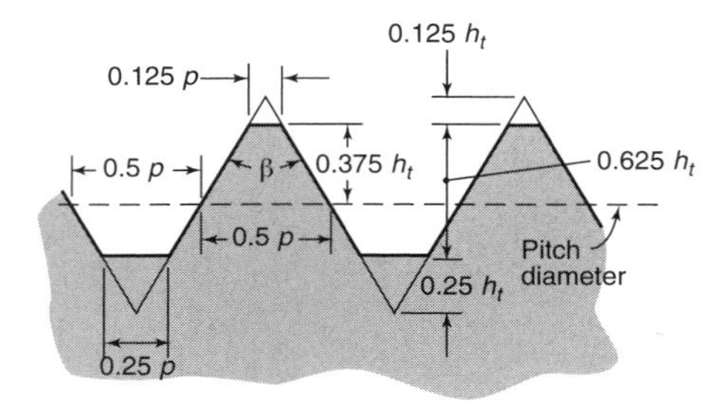

Figure 16.4: Details of M and UN thread profiles. The M profile is essentially the same as the UN profile, but is manufactured using SI dimensions.

1. C — coarse-pitch threads
2. F — fine-pitch threads
3. EF — extra-fine-pitch threads

The coarseness designation is followed by the crest diameter in inches and the number of threads per inch. Thus, UNF $1/2 \times 16$ means a UN thread profile with fine-pitch threads, a crest diameter of 1/2 in., and 16 constant-pitch threads per inch.

For metric thread profiles, the coarseness designation is usually considered only as coarse or fine, omitting the extra-fine designation. Instead of using threads per inch, the metric thread series simply uses the pitch distance between two threads measured in millimeters. For example, MF 6×1 means an M thread profile with fine-pitch threads, a crest diameter of 6 mm, and a pitch distance of 1 mm.

The classifications given above are only applicable to individual threads and do not consider how the male and female parts of the fastener fit together. That is, is the fit loose and sloppy or is it tight? In English units, the letter A designates external threads and B designates internal threads. There are three fits: 1 (the loosest fit), 2 (a normal fit), and 3 (a tight fit). Thus, for a very dirty environment a 1 fit would be specified; a 3 fit requires an extra degree of precision and a very clean environment. A 2 fit is usually designated for normal applications. Thus, the designation UNC 2×8 – 1B means a UN thread profile with coarse-pitch threads, a crest diameter of 2 in., eight constant-pitch threads per inch, and a loose fit, with the internal part of the fastener being specified.

The fit designation has more options in metric units than in English units. There G and H denote internal threads, and e, f, g, and h denote external threads. For example, G and e define the loosest fit and the greatest clearance, and H and h define zero allowance, no deviation from the basic profile. Thus, a very dirty environment would require a G and e fit; an H and h fit would require an extra degree of precision and a very clean environment.

The tolerance grade for metric threads is denoted by a number symbol. Seven tolerance grades have been established and are identified by numbers 3 to 9. Nine defines the loosest fit (the most generous tolerance) and 3 defines the tightest. Thus, MF 8×2 – G6 means an M thread profile with fine-pitch threads, a crest diameter of 8 mm, a pitch of 2 mm, and a normal fit of the internal threads.

Table 16.1 gives approximate equivalent fits of English and metric unit threads. This table does not show all possible equivalencies but does give an idea of how the two systems work.

Table 16.1: Inch and metric equivalent thread classifications.

Inch series		Metric series	
Bolts	Nuts	Bolts	Nuts
1A	1B	8g	7H
2A	2B	6g	6H
3A	3B	8h	5H

Example 16.1: Thread Designation

Given: A screw thread is going to be chosen for a nuclear power plant pressure vessel lid. The outside diameter of the screws is limited to 160 mm, and the length of the threaded part screwed into the pressure vessel wall is 220 mm. For M160 metric threads, three different pitches are preferred: 6, 4, and 3 mm. The manufacturing tolerance for the screw thread outer diameter is 0.6 mm, and the difference in thermal expansion of the pressure vessel wall relative to the screw is 0.2 mm because the screw heats more slowly than the wall during start-up. The manufacturing tolerance for the thread in the pressure vessel wall is 0.9 mm.

Find: Which pitch should be chosen?

Solution: The highest mean shear stress in the thread is equal to the force divided by π times the root diameter of the screw (if there is no play between the screw and nut) times the length of the thread. If there is play between the screw and nut, a smaller part consisting of only the tops of the threads needs to be sheared off; and if a weaker material is used in the pressure vessel than in the screws, that material will break before the screw threads. If the pitch is 3 mm, the height of the thread profile is

$$(0.625)\left(\frac{\sqrt{3}}{2}\right)(3) = 1.624 \text{ mm},$$

which is less than the combined thermal expansion and manufacturing error. For maximum safety, the largest pitch of 6 mm should be chosen.

16.3 Power Screws

A **power screw** is a device used to change angular motion into linear motion and, usually, to transmit power. More specifically, power screws are used:

1. To obtain greater mechanical advantage in order to lift weight, as in a screw type of jack for cars
2. To exert large forces, as in a trash compactor or a screw press
3. To obtain precise axial positioning, as in a micrometer screw or the lead screw of a lathe

In each of these applications, a torque is applied to the screws, thus creating an axial load.

Power screws use the Acme thread profile shown in Fig. 16.3a. Figure 16.5 shows more details of Acme threads. The thread angle is 29°, and the thread dimensions can be easily determined once the pitch is known. Table 16.2 gives the crest diameter, the number of threads per inch, and the tensile and shear stress areas for Acme power screw threads. The tensile stress area is

$$A_t = \frac{\pi}{4}\left(\frac{d_r + d_p}{2}\right)^2. \tag{16.3}$$

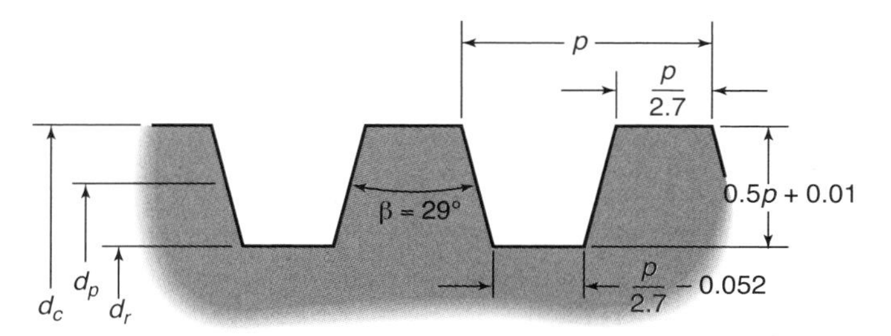

Figure 16.5: Details of Acme thread profile. (All dimensions are in inches.)

Table 16.2: Crest diameters, threads per inch, and stress areas for Acme thread.

Crest diameter, d_c, in.	Number of threads per inch, n	Tensile stress area, A_t, in.^{2}	Shear stress area, A_s, in.^{2}
1/4	16	0.02632	0.3355
5/16	14	0.04438	0.4344
3/8	12	0.06589	0.5276
7/16	12	0.09720	0.6396
1/2	10	0.1225	0.7278
5/8	8	0.1955	0.9180
3/4	6	0.2732	1.084
7/8	6	0.4003	1.313
1	5	0.5175	1.493
1 1/8	5	0.6881	1.722
1 1/4	5	0.8831	1.952
1 3/8	4	1.030	2.110
1 1/2	4	1.266	2.341
1 3/4	4	1.811	2.803
2	4	2.454	3.262
2 1/4	3	2.982	3.610
2 1/2	3	3.802	4.075
2 3/4	3	4.711	4.538
3	2	5.181	4.757
3 1/2	2	7.338	5.700
4	2	9.985	6.640
4 1/2	2	12.972	7.577
5	2	16.351	8.511

The shear stress area, A_s, is also given in Table 16.2, but a simple formulation of it is not readily available. Note that A_s represents the area in shear at approximately the pitch line of the threads for a 1.0-in. length of engagement, or the axial length over which the threads are engaged with each other. Lengths other than 1.0 in. would require that the shear stress area be modified accordingly. Because of this, the modified shear stress area differs considerably from the tensile stress area, which is based on the cross section. Table 16.2 and Fig. 16.5 should be used jointly in evaluating Acme power screws. The pitch diameter of an Acme power screw thread is:

$$d_p = d_c - 0.5p - 0.2\text{ mm} \qquad \text{(metric threads)} \qquad (16.4)$$

$$d_p = d_c - 0.5p - 0.01\text{ in.} \qquad \text{(inch series)}. \qquad (16.5)$$

16.3.1 Forces and Torque

Figure 16.6 shows a load, W, which is supported by the screw and which can be raised or lowered by rotating the screw. It is assumed that the load is mounted, such as in a collar or a nut, so that it moves proportionally with screw rotation. Also shown in Fig. 16.6 is the thread angle of an Acme power screw, which is $\beta = 29°$, and the lead angle, α. The lead angle relates the lead to the pitch circumference through the following equation:

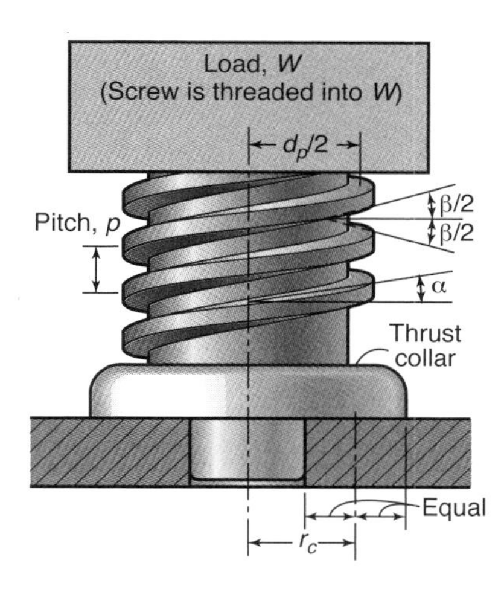

Figure 16.6: Dimensions and angles of a power screw with collar.

$$\alpha = \tan^{-1}\left(\frac{l}{\pi d_p}\right), \qquad (16.6)$$

where d_p is the pitch diameter, and the lead is given by

$$l = mp. \qquad (16.7)$$

If $m = 1$, the screw is *single-threaded*; if $m = 2$, it is *double-threaded*, etc. Figure 16.6 also shows a thrust collar, which will be discussed later in this chapter. The distance traveled in the axial direction per revolution is the lead.

Raising the Load

To determine the force, P, required to raise a load, W, it is necessary to observe the forces acting on a screw thread. The net force on the threads is represented by a single statically equivalent force, P_n, shown in Fig. 16.7a, that is normal to the thread surface. From Fig. 16.7, recall the representation of the lead angle, α, and the thread angle, β. The angle θ_n is created by the force P_n, which is the diagonal of the parallelepiped. Side ABE0 is an axial section through the bolt and is shown in Fig. 16.7b. Note from Fig. 16.7a and b that the sides DC, AB, and 0E of the parallelepiped are equal. Thus, the following must be true:

$$\sin\theta_n = \cos\theta_n \cos\alpha \tan\left(\frac{\beta}{2}\right), \qquad (16.8)$$

so that

$$\theta_n = \tan^{-1}\left[\cos\alpha\tan\left(\frac{\beta}{2}\right)\right]. \qquad (16.9)$$

In Fig. 16.7a, the projection of load P_n into plane ABE0 is inclined at an angle $\beta/2$, where β is the thread angle.

Side ACH0 lies in the plane tangent to the pitch point. The projection of P_n into this plane is inclined at the lead angle, α, calculated at the screw pitch radius and is shown in Fig. 16.7c. This figure contains not only the normal components but also the friction forces due to the screw and nut as well as the friction forces due to the collar.

Summing the vertical forces shown in Fig. 16.7c gives

$$P_n \cos\theta_n \cos\alpha = \mu P_n \sin\alpha + W. \qquad (16.10)$$

Solving for P_n,

$$P_n = \frac{W}{\cos\theta_n\cos\alpha - \mu\sin\alpha}. \qquad (16.11)$$

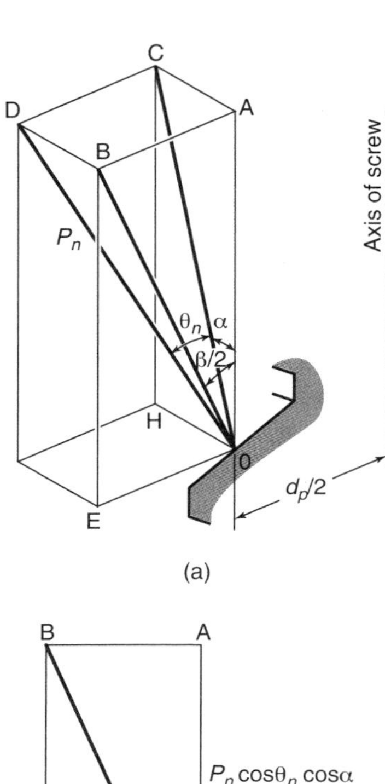

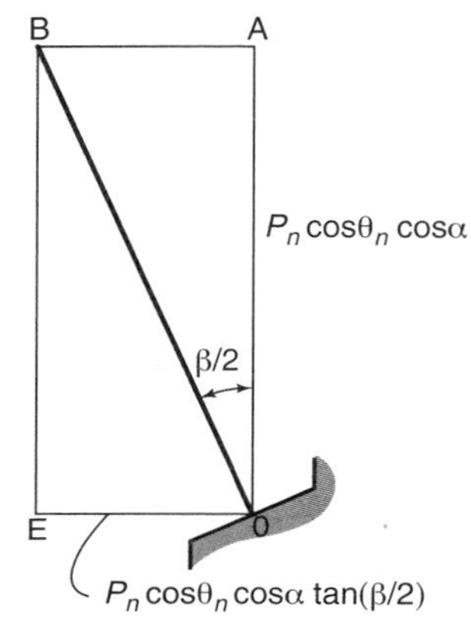

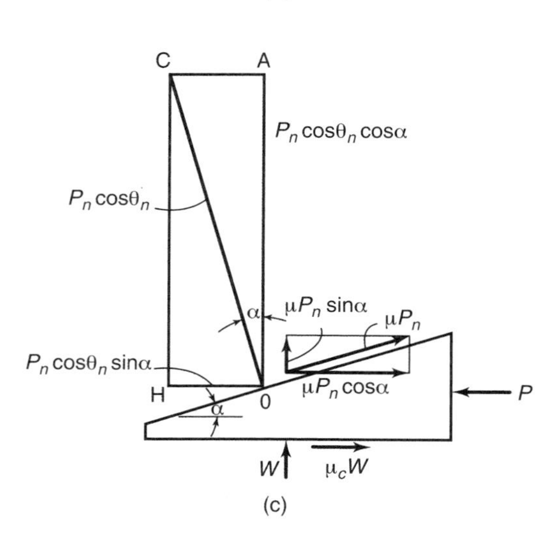

Figure 16.7: Forces acting in raising load of power screw. (a) Forces acting on parallelepiped; (b) forces acting on axial section; (c) forces acting on tangential plane.

All thread forces act at the thread pitch radius, $d_p/2$. From Fig. 16.6, the force of the collar acts at the radius of r_c, the midpoint of the collar surface. The torque required to raise the load is obtained by multiplying the *horizontal* forces by the appropriate radii. Thus,

$$T_r = P_n \left(\frac{d_p}{2}\right)(\cos\theta_n \sin\alpha + \mu \cos\alpha) + r_c \mu_c W. \tag{16.12}$$

Substituting Eq. (16.10) into Eq. (16.12) gives

$$T_r = W\left[\frac{(d_p/2)(\cos\theta_n \tan\alpha + \mu)}{\cos\theta_n - \mu\tan\alpha} + r_c\mu_c\right]. \tag{16.13}$$

Recall that μ is the coefficient of friction between the threads and μ_c is the coefficient of friction for the collar. Note that Eqs. (16.13) and (16.9) can be applied to any threaded fastener or other application, such as threaded fasteners, even though they were developed for power screws.

In typical power screw problems, the coefficients of friction μ and μ_c are given and the thread and lead angles are known. In practice, only approximate values of μ and μ_c can be obtained, leading to some uncertainty in the actual forces and torques required. If the collar consists of thrust rolling-element bearings, the collar friction is extremely low and can be neglected ($\mu_c = 0$).

Lowering the Load

Lowering the load differs from raising the load, as shown in Fig. 16.7, only in that the friction force components (Fig. 16.7c) become negative. These changes, after a summation of the vertical forces, yield

$$P_n = \frac{W}{\cos\theta_n \cos\alpha + \mu\sin\alpha}. \tag{16.14}$$

As was done previously, the torque required to lower the load is obtained by multiplying the horizontal forces by the appropriate radii. Thus,

$$T_l = P_n \left(\frac{d_p}{2}\right)(\cos\theta_n \sin\alpha - \mu \cos\alpha) - r_c \mu_c W. \tag{16.15}$$

Substituting Eq. (16.14) into Eq. (16.15) gives

$$T_l = -W\left[\frac{(d_p/2)(\mu - \cos\theta_n \tan\alpha)}{\cos\theta_n + \mu\tan\alpha} + r_c\mu_c\right]. \tag{16.16}$$

16.3.2 Power and Efficiency

Once the torque is known, the power transferred can be obtained from Eqs. (4.35) and (4.38). For metric units, the power, h_p, is given by

$$h_p = T\omega$$

where the power is in watts, T is measured in N-m, and ω is the rotational speed in rad/s.

The efficiency of a screw mechanism is the ratio of work output to work input, or

$$e = \frac{Wl}{2\pi T} \times 100\%, \tag{16.17}$$

where l is the lead and W is the load. Once the torque required to raise the load is known, the efficiency of the power screw mechanism can be calculated. Note from Eqs. (16.13) and (16.17) that if the coefficients of friction of the threads and the collar are equal to zero,

$$e = \frac{Wl}{2\pi W(d_p/2)\tan\alpha} \times 100\% = \frac{l}{2\pi(d_p/2)\tan\alpha} \times 100\%.$$

Making use of Eq. (16.6), it can be shown that if μ and μ_c can be reduced to zero, the efficiency will be 100% for small values of α.

Example 16.2: Power Screw Torque Requirements

Given: A double-threaded Acme power screw has a load of 1000 lb, a crest diameter of 1 in., a collar diameter of 1.5 in., a thread coefficient of friction of 0.16, and a collar coefficient of friction of 0.12.

Find: Determine the torque required to raise and lower the load. Also, determine the efficiency for raising the load.

Solution: From Table 16.2 for d_c = 1 in., the number of threads per inch is $n = 5$. From Eq. (16.1), $p = 0.2$ in. Since the screw is double-threaded, the lead is obtained from Eq. (16.7) as $l = 2p = 0.4$ in. Also, Eq. (16.5) gives

$$d_p = d_c - 0.5p - 0.01 = 1.0 - 0.5(0.2) - 0.01 = 0.89 \text{ in.}$$

From Eq. (16.6), the lead angle is

$$\alpha = \tan^{-1}\left(\frac{l}{\pi d_p}\right) = \tan^{-1}\left[\frac{0.4}{\pi(0.89)}\right] = 8.142^\circ.$$

Recall that, for an Acme screw, the thread angle is $\beta = 29^\circ$. Making use of Eq. (16.9) gives

$$\begin{aligned}\theta_n &= \tan^{-1}\left[\cos\alpha\tan\left(\frac{\beta}{2}\right)\right] \\ &= \tan^{-1}(\cos 8.142^\circ \tan 14.5^\circ) \\ &= 14.36^\circ.\end{aligned}$$

Note that θ_n and $\beta/2$ are nearly equal. Using Eq. (16.13) gives the torque required to raise the load as

$$\begin{aligned}T_r &= W\left[\frac{(d_p/2)(\cos\theta_n\tan\alpha + \mu)}{\cos\theta_n - \mu\tan\alpha} + r_c\mu_c\right] \\ &= 1000\left[\frac{(0.445)(\cos 14.36^\circ \tan 8.142^\circ + 0.16)}{\cos 14.36^\circ - 0.16\tan 8.142^\circ}\right] \\ &\quad +(1000)(0.75)(0.12), \\ &= 230 \text{ in-lb.}\end{aligned}$$

Using Eq. (16.16) gives the torque required to lower the load as

$$\begin{aligned}T_l &= -W\left[\frac{(d_p/2)(\mu - \cos\theta_n\tan\alpha)}{\cos\theta_n + \mu\tan\alpha} + r_c\mu_c\right] \\ &= -1000\left[\frac{(0.445)(0.16 - \cos 14.36^\circ \tan 8.142^\circ)}{\cos 14.36^\circ + 0.16\tan 8.142^\circ}\right] \\ &\quad +(1000)(0.75)(0.12), \\ &= -99.60 \text{ in-lb.}\end{aligned}$$

Note that 90.4%.of the torque is due to collar friction. As would be expected, less torque is required to lower the load than to raise the load. The minus sign indicates the opposite direction in lowering the load relative to raising the load. Using Eq. (16.17) gives the efficiency in raising the load as

$$e = \frac{Wl}{2\pi T_r} \times 100\% = \frac{(1000)(0.4)}{2\pi(230.5)} \times 100\% = 27.62\%.$$

Example 16.3: Single- and Double-Threaded Power Screws

Given: The exact situation in Example 16.2 except that a single-threaded power screw is used instead of a double-threaded one.

Find: Determine the torque required to raise and lower the load. Also, determine the efficiency for raising the load.

Solution: In this case, the lead equals the pitch ($l = p = 0.2$ in.). The lead angle is

$$\alpha = \tan^{-1}\left(\frac{l}{\pi d_p}\right) = \tan^{-1}\left(\frac{0.2}{0.89\pi}\right) = 4.091^\circ.$$

From Eq. (16.9)

$$\begin{aligned}\theta_n &= \tan^{-1}[\cos\alpha\tan(\beta/2)] \\ &= \tan^{-1}(\cos 4.091^\circ \tan 14.5^\circ) \\ &= 14.46^\circ.\end{aligned}$$

Using Eq. (16.13) gives the torque required to raise the load as

$$T_r = W\left[\frac{(d_p/2)(\cos\theta_n\tan\alpha + \mu)}{\cos\theta_n - \mu\tan\alpha} + r_c\mu_c\right].$$

Therefore,

$$\begin{aligned}T_r &= 1000\left[\frac{(0.445)(\cos 14.46^\circ \tan 4.091^\circ + 0.16)}{\cos 14.46^\circ - 0.16\tan 4.091^\circ}\right] \\ &\quad +(1000)(0.75)(0.12) \\ &= 196.6 \text{ in-lb.}\end{aligned}$$

Note that to raise the load, the single-threaded screw requires lower torque than the double-threaded screw. Using Eq. (16.17) gives the efficiency in raising the load as

$$e = \frac{Wl}{2\pi T_r} \times 100\% = \frac{(1000)(0.2)}{2\pi(196.6)} \times 100\% = 16.19\%.$$

The single-threaded screw is thus less efficient (by 11 percentage points) than the double-threaded screw considered in Example 16.2. Using Eq. (16.16) gives the torque required to lower the load as

$$\begin{aligned}T_l &= -W\left[\frac{(d_p/2)(\mu - \cos\theta_n\tan\alpha)}{\cos\theta_n + \mu\tan\alpha} + r_c\mu_c\right] \\ &= -1000\left[\frac{(0.445)(0.16 - \cos 14.46^\circ \tan 4.091^\circ)}{\cos 14.46^\circ + 0.16\tan 4.091^\circ}\right] \\ &\quad +(1000)(0.75)(0.12), \\ &= -131.2 \text{ in-lb.}\end{aligned}$$

16.3.3 Self-Locking Screws

If the screw thread is very steep (i.e., has a large lead angle), the friction force may not be able to positively support the load, and gravity will cause the load to fall. For power screws, the lead angle is usually small, and the friction force of the thread interaction is large enough to oppose the load and keep it from lowering in an uncontrolled fashion. Such a screw is called **self-locking**, a desirable characteristic to have in power screws.

If the collar uses a rolling-element bearing, or effective lubrication is achieved, the collar friction can be neglected. From Eq. (16.16), the torque required to lower the load is negative if

$$\mu - \cos\theta_n \tan\alpha \geq 0.$$

Thus, under static conditions self-locking occurs if

$$\mu > \cos\theta_n \tan\alpha = \frac{l\cos\theta_n}{\pi d_p}. \tag{16.18}$$

Note that self-locking is not assured by Eq. (16.18); for example, vibrations may cause the load to lower regardless of the screw geometry.

Example 16.4: Self-Locking

Given: The double-threaded screw of Example 16.2 and the single-threaded screw of Example 16.3.

Find: Determine the thread coefficient of friction required to ensure that self-locking occurs. Assume that rolling-element bearings have been installed at the collar.

Solution: For the double-threaded screw, where $\theta_n = 14.36°$ and $\alpha = 8.142°$, self-locking occurs if

$$\mu > \cos\theta_n \tan\alpha = \cos 14.36° \tan 8.142° = 0.1386.$$

For the single-threaded screw, where $\alpha = 4.091°$, self-locking occurs if

$$\mu > \cos\theta_n \tan\alpha = \cos 14.36° \tan 4.091° = 0.0693.$$

Example 16.5: Triple-Threaded Power Screw

Given: A triple-threaded Acme power screw has a crest diameter of 2 in. and four threads per inch. A rolling-element bearing is used at the collar, so that the collar coefficient of friction is zero. The thread coefficient of friction is 0.15.

Find: Determine the load that can be raised by a torque of 400 in-lb. Also, check to see if self-locking occurs.

Solution: The pitch and the lead are, from Eqs. (16.1) and (16.7), respectively:

$$p = \frac{1}{n} = 0.25 \text{ in.}$$

$$l = mp = 3(0.25) = 0.75 \text{ in.}$$

The pitch diameter is obtained from Eq. (16.5) as

$$d_p = d_c - 0.5p - 0.01 = 2 - 0.5(0.25) - 0.01 = 1.865 \text{ in.}$$

From Eq. (16.6), the lead angle is

$$\alpha = \tan^{-1}\left(\frac{l}{\pi d_p}\right) = \tan^{-1}\left(\frac{0.75}{\pi(1.865)}\right) = 7.295°.$$

From Eq. (16.9),

$$\begin{aligned}\theta_n &= \tan^{-1}(\cos\alpha\tan\beta/2)\\ &= \tan^{-1}(\cos 7.295° \tan 14.5°)\\ &= 14.39°.\end{aligned}$$

Using Eq. (16.13) gives the torque required to raise the load:

$$T_r = W\left[\frac{(d_p/2)(\cos\theta_n\tan\alpha + \mu)}{\cos\theta_n - \mu\tan\alpha}\right],$$

or

$$400 = W\left[\frac{(1.865/2)(\cos 14.39° \tan 7.295° + 0.15)}{\cos 14.39° - 0.15\tan 7.295°}\right].$$

Solving this expression yields $W = 1486$ lb. From Eq. (16.18), self-locking occurs if

$$\mu > \cos\theta_n\tan\alpha = \cos 14.39° \tan 7.295° = 0.1240.$$

Since the coefficient of friction is actually 0.15, self-locking occurs.

16.4 Threaded Fasteners

A **fastener** is a device used to connect or join two or more members. Many fastener types and variations are commercially available. This section considers threaded fasteners, which are widely used because of their strength, reliability, and ability to accommodate disassembly.

16.4.1 Types of Threaded Fastener

Figure 16.8 shows three types of threaded fastener: the bolt and nut, the cap screw, and the stud. Most threaded fasteners consist of a **bolt** passing through a hole in the members being joined and mating with a **nut**, as shown in Fig. 16.8a. Occasionally, the bolt mates with threads machined into one of the members rather than with a nut, as shown in Fig. 16.8b. This type is called a **cap screw**. A **stud** (Fig. 16.8c) is threaded on both ends and screwed into the threaded hole in one or both of the members being joined.

A selection of available bolt and screw types is shown in Table 16.3 along with some notes regarding their typical applications; head, washers, and nut types are shown in Table 16.4. Most high strength bolt heads (and nuts) are hexagonal, and this text considers only this type, although much of the mechanics remain applicable to other types. A more detailed coverage of bolted joints can be found in Bickford [1995].

16.4.2 Load Analysis of Bolts and Nuts

When a bolt is assembled, the nut is turned until a snug state is reached, which is the condition where the members to be joined are in contact. Further, if there was any noncompliance, such as due to a small warpage or deviation from flatness in the members, this nonconformity is eliminated by tightening the nut. Once this snug state is reached, the nut is tightened further for highly stressed bolts. The reasons for this will be discussed in the following sections, but the mechanics of bolt tightening will be reviewed here and they

Table 16.3: Common types of bolts and screws.

Illustration	Type	Description	Application notes
	Hex-head bolt	An externally threaded fastener with a trimmed hex head, often with a washer face on the bearing side.	Used in a variety of general purpose applications in different grades depending on the required loads and material being joined.
	Carriage bolt	A round head bolt with a square neck under the head and a standard thread.	Used in slots where the square neck keeps the bolt from turning when being tightened.
	Elevator bolt or belt bolt	A bolt with a wide, countersunk flat head, a shallow conical bearing surface, an integrally formed square neck under the head, and a standard thread.	Used in belting and elevator applications where head clearances must be minimal.
	Serrated flange bolt	A hex bolt with integrated washer, but wider than standard washers and incorporating serrations on the bearing surface side.	Used in applications where loosening hazard exists, such as vibration applications. The serrations grip the surface so that more torque is needed to loosen than tighten the bolt.
	Flat cap screw (slotted head shown)	A flat, countersunk screw with a flat top surface and conical bearing surface.	A common fastener for assembling joints where head clearance is critical.
	Buttonhead cap screw (socket head shown)	Dome shaped head that is wider and has a lower profile than a flat cap screw.	Designed for light fastening applications where their appearance is desired. Not recommended for high-strength applications.
	Lag screw	A screw with spaced threads, a hex head, and a gimlet point. (Can also be made with a square head.)	Used to fasten metal to wood or with expansion fittings in masonry.
	Step bolt	A plain, circular, oval head bolt with a square neck. The head diameter is about three times the bolt diameter.	Used to join resilient materials or sheet metal to supporting structures, or for joining wood since the large head will not pull through.

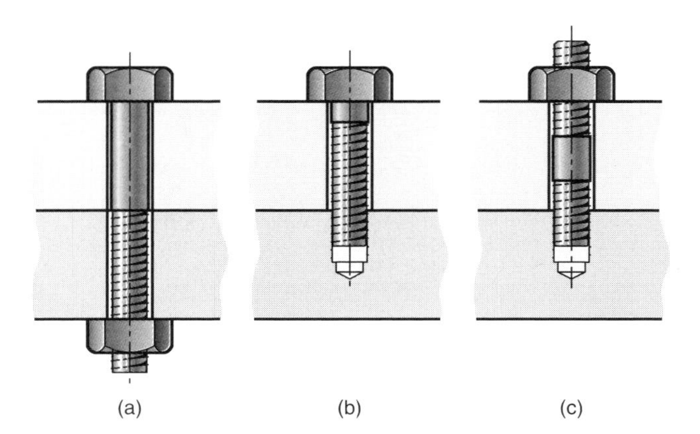

Figure 16.8: Three types of threaded fastener. (a) Bolt and nut; (b) cap screw; (c) stud.

provide an important mathematical framework for threaded fastener design.

The bolt and nut shown in Fig. 16.8a can be thought of as a spring system, as shown in Fig. 16.9. The bolt is viewed as a spring in tension with stiffness k_b. The joint, which can consist of multiple members being joined, is viewed as a spring loaded in compression due to the action of the bolt. The stiffness of the joint is denoted as k_j.

Figure 16.10 shows the forces and deflections acting on the bolt and the joint. In Fig. 16.10a, the bolt and the joint are considered to be unassembled, and the load and the deflection are shown for each. For the bolt the force is tensile and the deflection is an extension of the bolt, whereas for the joint the deflection is a contraction. Figure 16.10b considers the assembled bolt, nut, and joint. The slopes of the load-deflection lines are the same as in Fig. 16.10a, but the slopes for the bolt and joint are not only opposite in sign but have quite different values. The bolt and joint load-deflection lines intersect at a preload P_i.

Figure 16.11 shows the forces as functions of deflection for a bolt and a joint when an external load has been applied.

Table 16.4: Common nuts and washers for use with threaded fasteners.

Illustration	Type	Description	Application notes
Nuts			
	Hex nut	A six-sided internally threaded fastener. Specific dimensions are prescribed in industry standards.	The most commonly used general-purpose nut.
	Nylon insert stop	A nut with a hex profile and an integral nylon insert.	The nylon insert produces higher friction on the threads and prevents loosening due to vibration or corrosion.
	Cap nut	Similar to a hex nut with a dome top.	Used to cover exposed, dangerous bolt threads or for aesthetic reasons.
	Castle nut	A type of slotted nut.	Used for general purpose fastening and locking. A cotter pin or wire can be inserted through the slots and a hole drilled through the fastener.
	Coupling nut	A six-sided double chamfered nut.	Used to join two externally threaded parts of equal thread diameter and pitch.
	Hex jam nut	A six-sided internally threaded fastener, thinner than a normal hex nut.	Used in combination with a hex nut to keep the nut from loosening.
	K-lock or keplock nut	A hex nut preassembled with a free spinning external tool lock washer. When tightened, the teeth bite into the member to achieve locking.	A popular lock nut because of ease of use and low cost.
	Wing nut	An internally threaded nut with integral pronounced flat tabs.	Used for applications where repetitive hand tightening is required.
	Serrated nut	A hex nut with integrated washer, but wider than standard washers and incorporating serrations on the bearing surface side.	Used in applications where loosening hazard exists, such as vibration applications. The serrations grip the surface so that more torque is needed to loosen than tighten the bolt.
Washers			
	Flat washer	A circular disk with circular hole, produced in accordance with industry standards. Fender washers have larger surface area than conventional flat washer.	Designed for general-purpose mechanical and structural use.
	Belleville washer	A conical disk spring.	Used to maintain load in bolted connections.
	Split lock washer	A coiled, hardened, split circular washer with a slightly trapezoidal cross-section.	Preferred for use with hardened bearing surfaces. Applies high bolt tension per torque, resists loosening caused by vibration and corrosion.
	Tooth lock washer	A hardened circular washer with twisted teeth or prongs.	Internal teeth are preferred for aesthstics since the teeth are hidden under the bolt head. External teeth give greater locking efficiency. Combination teeth are used for oversized or out-of-round holes or for electrical connections.

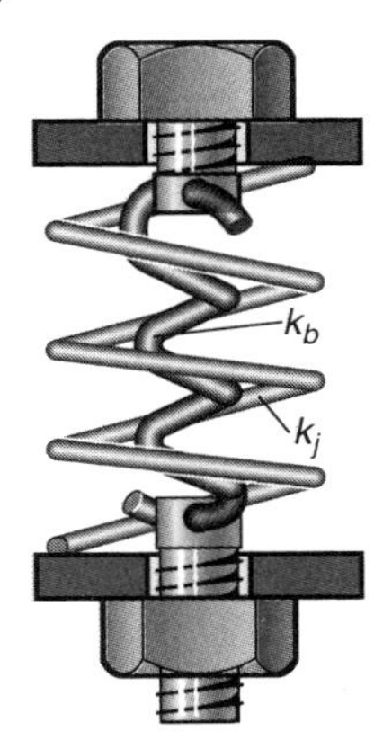

Figure 16.9: Bolt-and-nut assembly modeled as bolt-and-joint spring.

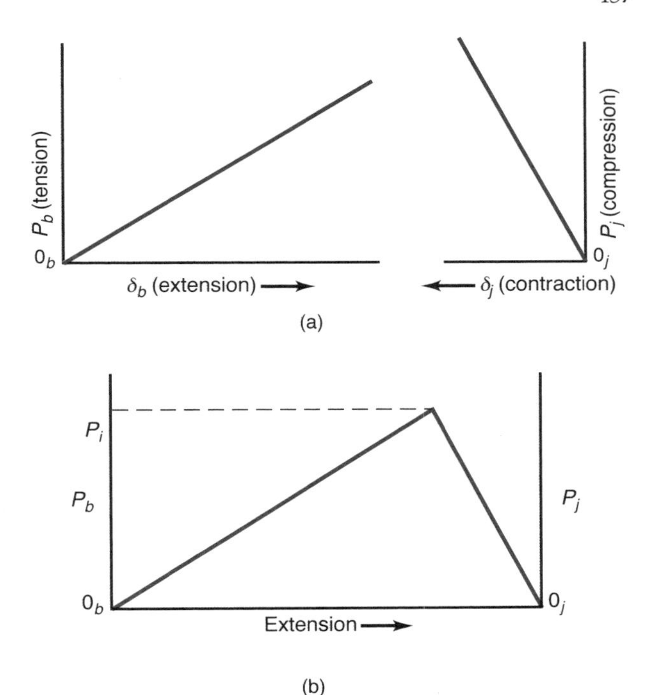

Figure 16.10: Force versus deflection of bolt and member. (a) Separated bolt and joint; (b) assembled bolt and joint.

The bolt is elongated by e_k and the contraction of the joint is reduced, as shown in Fig. 16.11. Thus, loading the bolt reduces the force on the member compared to the unloaded but preloaded state. The force on the bolt becomes $P_i + k_b e_k$, and the force on the joint becomes $P_i - k_j e_k$.

From the equilibrium of the bolt,

$$P + P_i - k_j e_k - P_i - k_b e_k = 0.$$

Therefore,

$$e_k = \frac{P}{k_b + k_j}. \tag{16.19}$$

Making use of Eq. (16.19) gives the load on the bolt as

$$P_b = P_i + k_b e_k = P_i + \frac{P k_b}{k_b + k_j} = P_i + C_k P, \tag{16.20}$$

where the dimensionless stiffness parameter is

$$C_k = \frac{k_b}{k_b + k_j}. \tag{16.21}$$

From Eq. (16.21), C_k is less than 1. Usually, C_k is between 0.15 and 0.40, although it can be slightly lower or higher.

The load on the joint is

$$P_j = P_i - k_j e_k = P_i - \frac{P k_j}{k_j + k_b} = P_i - (1 - C_k)P. \tag{16.22}$$

16.4.3 Stiffness Parameters

Recall from Eq. (4.26) that the spring rate is the normal load divided by the elastic deflection for axial loading and that the elastic deflection for axial loading is given in Eq. (4.25). Making use of these equations gives

$$k = \frac{P}{\left(\frac{PL}{AE}\right)} = \frac{AE}{L}, \tag{16.23}$$

where A is the cross-sectional area, L is the length, and E is the elastic modulus.

Bolt Stiffness

A bolt with threads can be modeled as a stepped shaft. The root diameter is used for the threaded section of the bolt, and the crest diameter is used for the unthreaded section, called the **shank** (see Fig. 16.1). The bolt is treated as two springs in series when considering the shank and the threaded section. For more complicated bolts, there may be more complicated profiles, so that the stiffness of the bolt can be expressed as

$$\frac{1}{k_b} = \frac{1}{k_{b1}} + \frac{1}{k_{b2}} + \frac{1}{k_{b3}} + \ldots \tag{16.24}$$

Figure 16.12a shows a bolt-and-nut assembly, and Fig. 16.12b shows the stepped-shaft representation of the shank and the threaded section. In Fig. 16.12b, the effective length of the shaft and the threaded section includes slight extension into the bolt head and the nut to account for their deflections. Making use of Eq. (16.23) and Fig. 16.12b gives

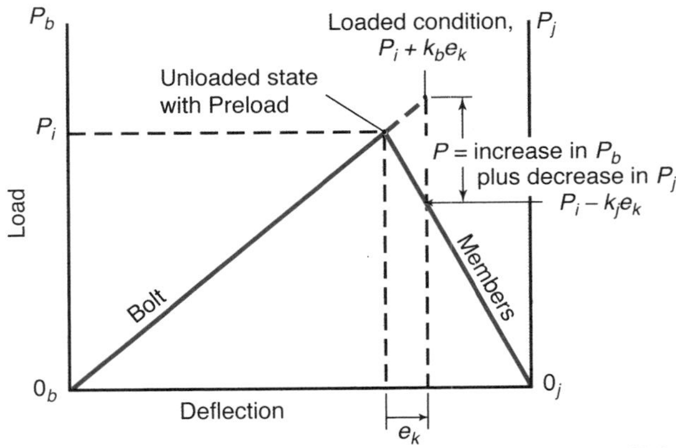

Figure 16.11: Forces versus deflection of bolt and joint when external load is applied.

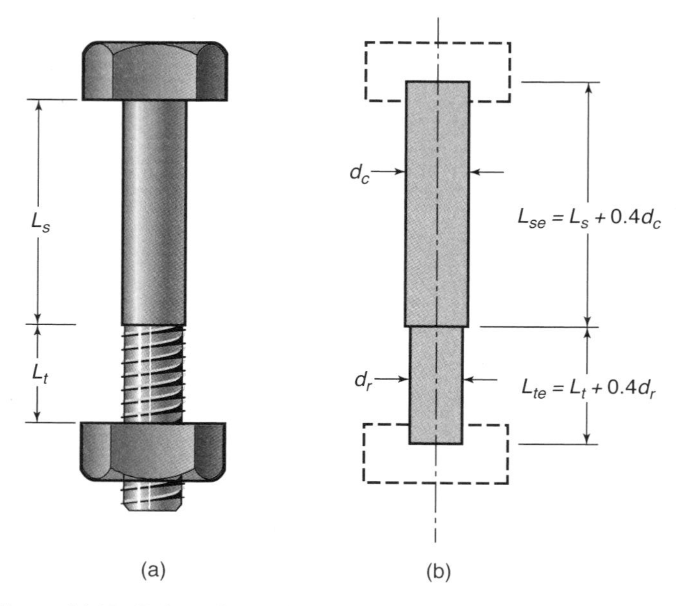

Figure 16.12: Bolt and nut. (a) Assembled; (b) stepped-shaft representation of shank and threaded section.

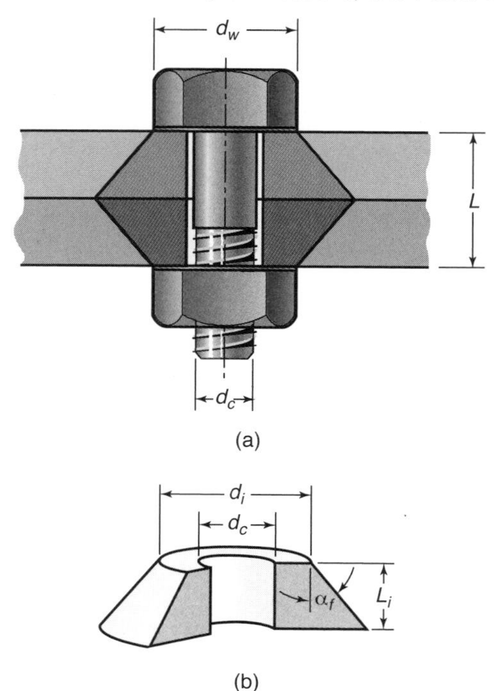

Figure 16.13: (a) Bolt-and-nut assembly with conical frustum stress representation of joint; (b) detail of frustum with important dimensions.

$$\frac{1}{k_b} = \frac{4}{\pi E}\left[\frac{L_{se}}{d_c^2} + \frac{L_{te}}{d_r^2}\right] = \frac{4}{\pi E}\left[\frac{L_s + 0.4d_c}{d_c^2} + \frac{L_t + 0.4d_r}{d_r^2}\right], \tag{16.25}$$

where d_c is the crest diameter and d_r is the root diameter. Equation (16.25) is valid for a shank having a constant crest diameter. In some designs, the shank has a more complicated profile. This situation can be easily analyzed by using Eq. (16.24) and the appropriate lengths and diameters. Note that $0.4d_c$ is added to the length only for the section closest to the bolt head.

Often, a particular application will require a special fastener, and any combination of diameter and thread length is possible. However, it is usually much more economical to use mass-produced threaded fasteners; these threads are generally of higher quality (as they may be rolled instead of cut) and they have a more repeatable strength. Usually, a specially designed and manufactured bolt or cap screw will have as little of its length threaded as practicable to maximize bolt stiffness. However, for standardized fasteners, the threaded length is given by

$$L_{\text{tot}} = \begin{cases} 2d_c + 6 & L \le 125 \\ 2d_c + 12 & 125 < L \le 200 \\ 2d_c + 25 & L > 200 \end{cases} \quad \text{(metric threads)} \tag{16.26}$$

$$L_{\text{tot}} = \begin{cases} 2d_c + 0.25 \text{ in.} & L \le 6 \text{ in.} \\ 2d_c + 0.50 \text{ in.} & L > 6 \text{ in.} \end{cases} \quad \text{(inch series)} \tag{16.27}$$

where L_{tot} is the total threaded length, L is the total bolt length, and d_c is the crest diameter.

Joint Stiffness

Determining joint stiffness is much more complicated than determining bolt stiffness. The volume of material that is compressed by the bolt is not intuitively obvious; clearly material far away from the bolt is not compressed. The effective loaded volume is often approximated as a cylinder or conic section with a central hole (i.e., a frustum). Two conical frustums symmetric about the joint midplane and each having a vertex angle of $2\alpha_f$ are often used to represent the stresses in the joint. Figure 16.13 shows the conical frustum stress representation of the joint in a bolt-and-nut assembly. Note that d_w is the diameter of the washer through which the load is transferred. Shigley and Mischke [1989] arrived at the following expression for stiffness of a frustum:

$$k_{ji} = \frac{\pi E_j d_c \tan\alpha_f}{\ln\left[\dfrac{(2L_i\tan\alpha_f + d_i - d_c)(d_i + d_c)}{(2L_i\tan\alpha_f + d_i + d_c)(d_i - d_c)}\right]}, \tag{16.28}$$

where L_i is the axial length of the frustum and d_i is the smaller frustum cone diameter, as shown in Fig. 16.13b. For the member closest to the bolt head or to the nut, $d_i = d_w$. If information is not available regarding the washer, it is reasonable to assume that $d_i = 1.5d_c$ to allow design calculations; the assumption can be evaluated when a specific washer has been selected.

Equation (16.28) gives the stiffness of a single frustum, but a joint is made up of at least two frusta. Thus, the resulting joint stiffness is obtained from

$$\frac{1}{k_j} = \frac{1}{k_{j1}} + \frac{1}{k_{j2}} + \frac{1}{k_{j3}} + \ldots \tag{16.29}$$

Wileman et al. [1991] performed a finite element analysis of bolted connections, and determined joint stiffness for a range of materials and geometry. Their expression for joint stiffness is

$$k_j = E_i d_c A_i e^{B_i d_c/L}, \tag{16.30}$$

where A_i and B_i are dimensionless constants given in Table 16.5. Note that Eq. (16.30) only holds for two members of the same material. If two plates of different material are joined,

Table 16.5: Constants used to obtain joint stiffness from Eq. (16.30). *Source:* From Wileman, et al. [1991].

Material	Poisson's ratio, ν	Elastic modulus, E, GPa	Numerical constants, A_i	B_i
Steel	0.291	206.8	0.78715	0.62873
Aluminum	0.334	71.0	0.79670	0.63816
Copper	0.326	118.6	0.79568	0.63553
Gray cast iron	0.211	100.0	0.77871	0.61616

then Eq. (16.28) must be used with Eq. (16.29) to obtain the joint stiffness.

Once the joint and bolt stiffnesses are known, the dimensionless stiffness parameter can be calculated from Eq. (16.21). Design Procedure 16.1 summarizes the approaches for determining joint stiffness.

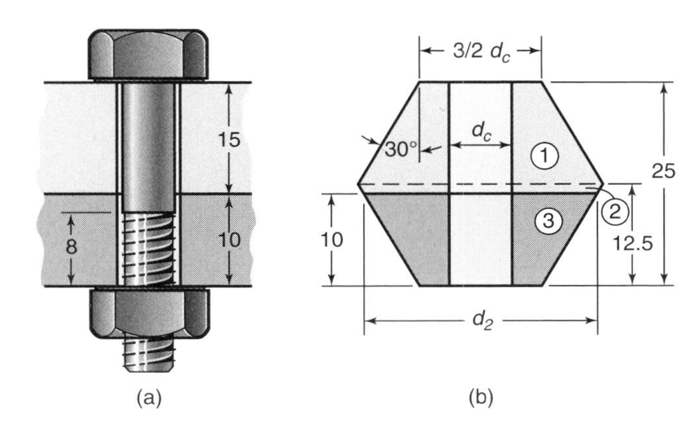

Figure 16.14: Hexagonal bolt-and-nut assembly used in Example 16.6. (a) Assembly and dimensions; (b) dimension of frusta. (All dimensions are in millimeters.)

Design Procedure 16.1: Evaluation of Joint Stiffness

1. If the members all have the same elastic modulus, then the Wileman approach can be used and Eq. (16.30) yields the member stiffness. Because of its simplicity and good results, this is the recommended approach. However, if the conical frusta approach is desired, the following simplifications can be useful:

   (a) For two frusta, with a total grip length L, the stiffness of the joint is given as:

$$k_j = \frac{\pi E_j d_c \tan\alpha_f}{2\ln\left[\frac{(L\tan\alpha_f + d_i - d_c)(d_i + d_c)}{(L\tan\alpha_f + d_i + d_c)(d_i - d_c)}\right]}. \tag{16.31}$$

   (b) For the further simplification of $d_i = d_w = 1.5d_c$ and $\alpha_f = 30°$,

$$k_j = \frac{0.9069 E_j d_c}{\ln\left[\frac{(2.887L + 2.5d_c)}{(0.5774L + 2.5d_c)}\right]}. \tag{16.32}$$

   Note that this approach neglects the contribution of the washers to joint stiffness; because washers are usually steel and very thin, they do not normally contribute significantly to joint stiffness.

2. If the joint is not constructed of members with the same stiffness, the conical frustum approach must be used. To obtain joint stiffness from conical frusta, the following steps must be taken:

   (a) Construct a careful sketch of the joint, including the bolt, washer, and members.

   (b) Select a frustum angle, α_f. Generally, good results can be obtained with $\alpha_f = 30°$. Draw the frusta by starting at the inside surface of the bolt and a diameter equal to $1.5d_c$ (to account for washers), and extending to the center of the joint using the frustum angle.

   (c) Identify the number of frusta needed to obtain joint stiffness. Frusta can extend past a joint boundary if the elastic modulus does not change; otherwise, new frusta are required at joint boundaries. Determine the dimensions of the frusta from the problem geometry.

   (d) The stiffness of a frustum can be obtained from Eq. (16.28). If $\alpha_f = 30°$, this can be simplified as:

$$k_{ji} = \frac{1.813 E_j d_c}{\ln\left[\frac{(1.15L_i + d_i - d_c)(d_i + d_c)}{(1.15L_i + d_i + d_c)(d_i - d_c)}\right]}. \tag{16.33}$$

   (e) Equation (16.29) then yields the joint stiffness.

Example 16.6: Bolt and Joint Stiffness

Given: A hexagonal bolt-and-nut assembly, shown in Fig. 16.14a, is used to join two members. The bolt and the nut are made of steel and the frustum cone angle is taken as 30°. The thread crest diameter is 14 mm and the root diameter is 12 mm. Use $d_w = 1.5d_c$, but assume washers are thin enough that their contribution to joint stiffness can be neglected.

Find: Find the bolt and joint stiffnesses as well as the dimensionless stiffness parameter. Use the modulus of elasticity and Poisson's ratio from Table 16.5 and consider the cases where:

(*a*) Both members are made of steel. Use both the conical frusta and Wileman approaches.

(*b*) The 15-mm thick member is made of aluminum and the 10-mm thick member is made of steel.

Solution: In both cases, the bolt stiffness will be needed to obtain the dimensionless stiffness parameter. From Eq. (16.25) the bolt stiffness is

$$\frac{1}{k_b} = \frac{4}{\pi\left(206.8\times10^9\right)} \times\left[\frac{(0.017)+0.4\,(0.014)}{(0.014)^2}+\frac{(0.008)+0.4\,(0.012)}{(0.012)^2}\right] = 7.954\times10^8 \text{ N/m}.$$

(*a*) *Steel joint:* Approximating the joint as two conical back-to-back frusta with $\alpha = 30^\circ$ and $d_i = 1.5d_c$, the member stiffness is given by Eq. (16.32) as

$$k_j = \frac{0.9069E_j d_c}{\ln\left[\dfrac{(2.887L_i+2.5d_c)}{(0.5774L_i+2.5d_c)}\right]} = \frac{0.9069\left(206.8\times10^9\right)(0.014)}{\ln\left[\dfrac{2.887(25)+2.5(14)}{0.5774(25)+2.5(14)}\right]} = 3.393\times10^9 \text{ N/m}.$$

Using the Wileman method, Eq. (16.30) yields

$$k_j = E_i d_c A_i e^{B_i d_c/L} = \left(206.8\times10^9\right)(0.014)(0.78715)e^{(0.62873)(14)/(25)} = 3.242\times10^9 \text{ N/m}.$$

where the values of A_i and B_i have been taken from Table 16.5. Note that the Wileman calculations are simple and quite close to the frusta approach results. The Wileman result will therefore be used for calculating joint stiffness.

The dimensionless stiffness parameter is given by Eq. (16.21) as

$$C_k = \frac{k_b}{k_b+k_j} = \frac{0.7954}{0.7954+3.242} = 0.197.$$

(*b*) *Steel-aluminum joint:* If there are two different joint materials, the Wileman approach cannot be used. Instead, the approach outlined in Design Procedure 16.1 results in the joint stiffness. First, note that a sketch of the joint has been made in Fig. 16.14b. Note also that the asymmetry of the joint requires that the members be considered as three frusta. The member stiffnesses are, using the simplified expression for $\alpha_f = 30^\circ$ given in Eq. (16.33):

$$k_{j1} = \frac{1.813E_j d_c}{\ln\left[\dfrac{(1.15L_i+d_i-d_c)(d_i+d_c)}{(1.15L_i+d_i+d_c)(d_i-d_c)}\right]} = \frac{1.813(71\times10^9)(0.014)}{\ln\left\{\dfrac{[1.15(12.5)+21-14](21+14)}{[1.15(12.5)+21+14](21-14)}\right\}}, = 4.799\times10^9 \text{ N/m}.$$

Similarly,

$$k_{j2} = \frac{1.813(71\times10^9)(0.014)}{\ln\left\{\dfrac{[1.15(2.5)+32.55-14](32.55+14)}{[1.15(2.5)+32.55+14](32.55-14)}\right\}} = 2.141\times10^{10} \text{ N/m},$$

$$k_{j3} = \frac{1.813(206.8\times10^9)(0.014)}{\ln\left\{\dfrac{[1.15(10)+21-14](21+14)}{[1.15(10)+21+14](21-14)}\right\}} = 7.632\times10^9 \text{ N/m}.$$

Substituting these values into Eq. (16.29) gives

$$\frac{1}{k_j} = \frac{1}{4.799\times10^9}+\frac{1}{2.141\times10^{10}}+\frac{1}{7.632\times10^9},$$

which results in $k_j = 2.590\times10^9$ N/m. The dimensionless stiffness parameter is given by Eq. (16.21)] as

$$C_k = \frac{k_b}{k_b+k_j} = \frac{0.7954}{0.7954+2.590} = 0.2350.$$

16.4.4 Strength

The **proof load** of a bolt is the maximum load that it can withstand without acquiring any permanent deformation. The **proof strength** is the limiting value of the stress determined by using the proof load and the tensile stress area. Although proof strength and yield strength are close for bolt materials, the yield strength is usually higher because it is based on a 0.2% permanent deformation.

Standard bolt classes that specify material, heat treatment, and minimum proof strength have been defined by organizations such as the Society of Automotive Engineers (SAE), the American Society for Testing and Materials (ASTM), and International Organization for Standardization (ISO). Table 16.6 gives the strength information for several SAE bolt grades, and Table 16.7 gives similar information for metric bolts. SAE grade numbers range from 1 to 8 and metric grade numbers from 4.6 to 12.9, with higher numbers indicating greater strength.

Tables 16.8 and 16.9 give the dimensions and the tensile stress areas for UN and M coarse and fine threads. Dimensions for common hexagonal bolts, cap screws, and nuts are given in Appendix D. The tensile stress area in Tables 16.8 and 16.9 is calculated from

$$A_t = (0.7854)(d_c - 0.9382p)^2 \quad \text{(metric threads)} \quad (16.34)$$

$$A_t = (0.7854)\left(d_c - \frac{0.9743}{n}\right)^2 \quad \text{(inch series)} \quad (16.35)$$

where n is the number of threads per inch and p is the pitch.

16.4.5 Bolt Preload — Static Loading

The preload on the bolt is an important quantity; if the preload is too low, the joint may separate in service and subject the bolt to an excessive alternating stress as discussed in Section 16.4.6. If the preload is too high, the bolt or threads may fail. Thus, special care is taken to ensure that a proper

Table 16.6: Strength of steel bolts for various sizes in inches.

SAE grade	Head marking	Range of crest diameters, in.	Ultimate tensile strength S_u, ksi	Yield strength S_y, ksi	Proof strength S_p, ksi
1		0.25 – 1.5	60	36	33
2		0.25 – 0.75 > 0.75 – 1.5	74 60	57 36	55 33
4		0.25 – 1.5	115	100	65
5		0.25 – 1 >1 – 1.5	120 105	92 81	85 74
5.2		0.25 – 1	120	92	85
7		0.25 – 1.5	133	115	105
8		0.25 – 1.5	150	130	120
8.2		0.25 – 1	150	130	120

Table 16.7: Strength of various metric series steel bolts.

Metric grade	Head marking	Range of crest diameters, mm	Ultimate tensile strength S_u, MPa	Yield strength S_y, MPa	Proof strength S_p, MPa
4.6	4.6	M5 – M36	400	240	225
4.8	4.8	M1.6 – M16	420	340[a]	310
5.8	5.8	M5 – M24	520	415[a]	380
8.8	8.8	M17 – M36	830	660	600
9.8	9.8	M1.6 – M16	900	720[a]	650
10.9	10.9	M6 – M36	1040	940	830
12.9	12.9	M1.6 – M36	1220	1100	970

[a] Yield strengths are approximate and are not included in the standard.

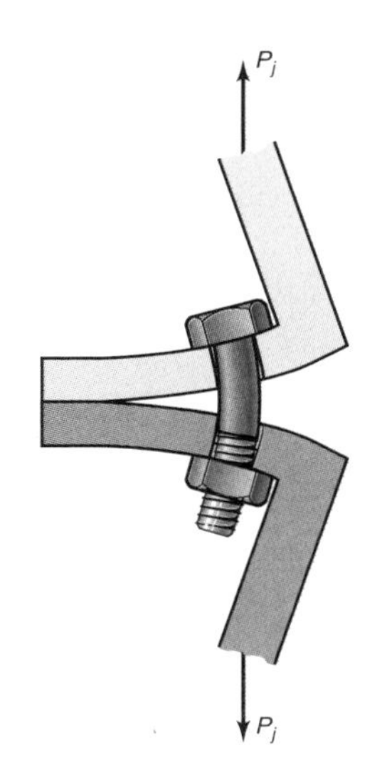

Figure 16.15: Separation of joint.

preload is achieved. Note from Eq. (16.20) that

$$\sigma_b = \frac{P_b}{A_t} = \frac{P_i}{A_t} + C_k \frac{P}{A_t}. \tag{16.36}$$

The limiting value of the bolt stress is the proof strength, which is given in Tables 16.6 and 16.7. Also, introducing the safety factor for bolt stress gives the proof strength as

$$S_p = \frac{P_i}{A_t} + \frac{P_{\max} n_s C_k}{A_t}, \tag{16.37}$$

where A_t is the tensile stress area given in Table 16.8 or 16.9 and P_i is the preload. Note that the safety factor is not applied to the preload stress; preload will be discussed shortly. Equation (16.37) can be rewritten to give the bolt failure safety factor as

$$n_{sb} = \frac{A_t S_p - P_i}{P_{\max,b} C_k}, \tag{16.38}$$

where $P_{\max,b}$ is the maximum load applied to bolt.

Equation (16.38) suggests that the safety factor is maximized by having zero preload on the bolt. For statically loaded bolts where separation is not a concern, this is certainly true. However, a preload is often applied to statically loaded bolted connections to make certain that members are tightly joined in order to provide tight fits. Further, many applications require members to avoid separation, as shown in Fig. 16.15. Separation occurs when the joint becomes fully unloaded, or $P_j = 0$. The safety factor guarding against separation is

$$n_{sj} = \frac{P_i}{P_{\max,j}(1 - C_k)}, \tag{16.39}$$

where $P_{\max,j}$ is the maximum load applied to the joint.

The amount of preload that is applied to bolts in practice is therefore a compromise between bolt overloading (where zero preload is most beneficial) and separation (where a large preload is desirable). The preload is given for reused and permanent connections as

$$P_i = \begin{cases} 0.75 P_p & \text{for reused connection} \\ 0.90 P_p & \text{for permanent connections} \end{cases} \tag{16.40}$$

Table 16.8: Dimensions and tensile stress areas for UN coarse and fine threads. Root diameter is calculated from Eq. (16.2) and Fig. 16.4.

	Coarse threads (UNC)			Fine threads (UNF)		
Crest diameter, d_c, in.	Number of threads per, inch, n	Root diameter, d_r, in.	Tensile stress area, A_t, in.^{2}	Number of threads per, inch, n	Root diameter, d_r, in.	Tensile stress area, A_t, in.^{2}
0.0600	—	—	—	80	0.04647	0.00180
0.0730	64	0.05609	0.00263	72	0.05796	0.00278
0.0860	56	0.06667	0.00370	64	0.06909	0.00394
0.0990	48	0.07645	0.00487	56	0.07967	0.00523
0.1120	40	0.08494	0.00604	48	0.08945	0.00661
0.1250	40	0.09794	0.00796	44	0.1004	0.00830
0.1380	32	0.1042	0.00909	40	0.1109	0.01015
0.1640	32	0.1302	0.0140	36	0.1339	0.01474
0.1900	24	0.1449	0.0175	32	0.1562	0.0200
0.2160	24	0.1709	0.0242	28	0.1773	0.0258
0.2500	20	0.1959	0.0318	28	0.2113	0.0364
0.3125	18	0.2523	0.0524	24	0.2674	0.0580
0.3750	16	0.3073	0.0775	24	0.3299	0.0878
0.4750	14	0.3962	0.1063	20	0.4194	0.1187
0.5000	13	0.4167	0.1419	20	0.4459	0.1599
0.5625	12	0.4723	0.182	18	0.5023	0.203
0.6250	11	0.5266	0.226	18	0.5648	0.256
0.7500	10	0.6417	0.334	16	0.6823	0.373
0.8750	9	0.7547	0.462	14	0.7977	0.509
1.000	8	0.8647	0.606	12	0.9098	0.663
1.125	7	0.9703	0.763	12	1.035	0.856
1.250	7	1.095	0.969	12	1.160	1.073
1.375	6	1.195	1.155	12	1.285	1.315
0.500	6	1.320	1.405	12	1.140	1.581
1.750	5	1.533	1.90	—	—	—
2.000	4.5	1.759	2.5	—	—	—

Table 16.9: Dimensions and tensile stress areas for metric coarse and fine threads. Root diameter is calculated from Eq. (16.2) and Fig. 16.4.

	Coarse threads (MC)			Fine threads (MF)		
Crest diameter, d_c, mm	Pitch, p, mm	Root diameter, d_r, mm	Tensile stress area, A_t, mm^2	Pitch, p, mm	Root diameter, d_r, mm	Tensile stress area, A_t, mm^2
1	0.25	0.7294	0.460	—	—	—
1.6	0.35	1.221	1.27	0.20	1.383	1.57
2	0.4	1.567	2.07	0.25	1.729	2.45
2.5	0.45	2.013	3.39	0.35	2.121	3.70
3	0.5	2.459	5.03	0.35	2.621	5.61
4	0.7	3.242	8.78	0.5	3.459	9.79
5	0.8	4.134	14.2	0.5	4.459	16.1
6	1.0	4.917	20.1	0.75	5.188	22
8	1.25	6.647	36.6	1.0	6.917	39.2
10	1.5	8.376	58.0	1.25	8.647	61.2
12	1.75	10.11	84.3	1.25	10.65	92.1
14	2.0	11.84	115	1.5	12.38	124
16	2.0	13.83	157	1.5	14.38	167
20	2.5	17.29	245	1.5	18.38	272
24	3.0	20.75	353	2.0	21.83	384
30	3.5	26.21	561	2.0	27.83	621
36	4.0	31.67	817	3.0	32.75	865
42	4.5	37.13	1121	—	—	—
48	5.0	42.59	1473	—	—	—

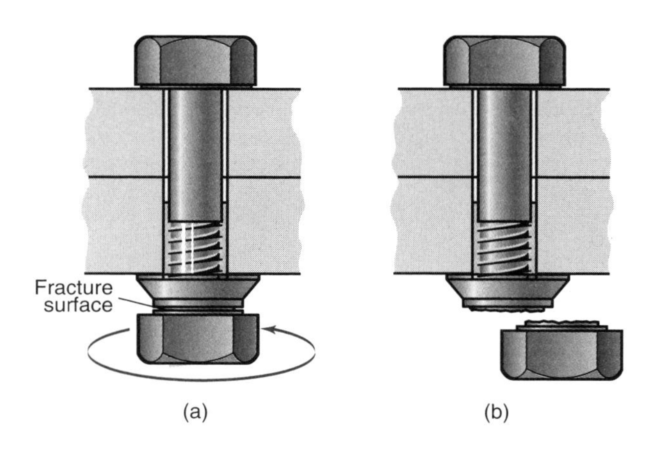

Figure 16.16: An example of a twist-off fastener. (a) During assembly, a tightening torque is applied to the nut as shown. (b) When the torque reaches a critical value, the nut shears and fractures adjacent to the hex head, leaving the remainder of the nut in place. Note that this design complicates disassembly; the use of a second hex head instead of a flared section, for example, would allow disassembly.

where P_p is the proof load, or S_pA_t. Equation (16.40) is applicable to static as well as fatigue conditions considered in Section 16.4.6.

Recall in Eq. (16.37) that the safety factor was not applied to the preload, and this can be understood from Eq. (16.40). Preloading a bolt should not cause failure during normal assembly, but if it does, the bolt is replaced and the new bolt is properly preloaded. The remaining design margin is based on the difference between the proof strength and the portion of the load carried by the bolts.

In design practice, preloads are rarely specified because preloads are extremely difficult to measure during assembly of bolted connections. Three alternatives are used:

1. Specify a torque to be applied during tightening, which is then controlled by using a torque wrench.
2. Define a number of rotations from a snug state, such as a half-turn.
3. A breakaway design is used, such as the one shown in Fig. 16.16.

Each alternative has difficulties associated with them: specifying a torque or using breakaway designs does not accommodate changes in thread friction, which clearly affect the resulting preload, and a snug state is inherently subjective. Also, breakaway designs result in higher cost and some designs require special tooling. Thus, although Eq. (16.40) can be used in design problems and for determining a pretorque or deformation, it should be recognized that bolts preload varies in practice.

Example 16.7: Comparison of Bolt Failure and Joint Separation

Given: The results of Example 16.6 for the steel-aluminum joint.

Find: Determine the maximum load for bolt-and-joint failure while assuming reused connections and a static safety factor of 2.5. Assume a 5.8 grade and coarse threads.

Solution: From Table 16.7 for 5.8 grade, S_p = 380 MPa, and from Table 16.9 for a crest diameter of 14 mm and coarse threads,

$$\begin{aligned} A_t &= (0.7854)(d_c - 0.9382p)^2 \\ &= (0.7854)\left[14 - (0.9382)(2)\right]^2 \\ &= 115.4\ \text{mm}^2. \end{aligned}$$

For reused connections, Eq. (16.40) yields

$$\begin{aligned} P_i &= 0.75P_p = 0.75A_tS_p \\ &= (0.75)\left(115.4\times10^{-6}\right)\left(380\times10^{6}\right) \\ &= 32,900\ \text{N}. \end{aligned}$$

From Eq. (16.38), the maximum bolt load with a safety factor of 2.5 is

$$P_{\max,b} = \frac{A_tS_p - P_i}{n_{sb}C_k} = \frac{(115.4)(380) - 32,900}{(2.5)(0.2350)} = 18,640\ \text{N}.$$

From Eq. (16.39), the maximum load before separation occurs is

$$P_{\max,j} = \frac{P_i}{n_{sj}(1 - C_k)} = \frac{32,900}{(2.5)(1 - 0.2350)} = 17,200\ \text{N}.$$

Thus, failure due to separation will occur before bolt failure.

16.4.6 Bolt Preload — Fatigue Considerations

The importance of preload is even greater for fatigue loaded joints than for those that are statically loaded. Figure 16.17 illustrates the force versus deflection of the bolt and joint as a function of time for a cyclic loading. From Fig. 16.17, observe that the bolt load is in tension and varies with time from the preload P_i to some maximum bolt load $P_{b,\max}$. Simultaneously, the joint encounters compressive force that varies with time from the preload P_i to the minimum joint load $P_{j,\min}$. When bolt load is a maximum, the joint load has a minimum value and vice-versa. Unless the joint separates, the deflection of the bolt and members will be equal. However, as discussed in Section 16.4.3, the members are stiffer than the bolt. That is, C_k varies from around 0.15 to 0.40. For these reasons, the amount of load variation for one cycle is much larger for the joint than for the bolt.

Figure 16.17 also shows the deflection variation with time for both the bolt and the joint. The quantitative amount is the same, but for the bolt the deflection is an increased elongation whereas for the joint the contraction is decreased.

Consider fatigue failure of the bolt, since this is more likely than fatigue of the members. Equation (16.20) gives the alternating and mean loads acting on the bolt as

$$P_{ba} = \frac{P_{b,\max} - P_{b,\min}}{2} = \frac{C_k\left(P_{\max} - P_{\min}\right)}{2} = C_kP_a \tag{16.41}$$

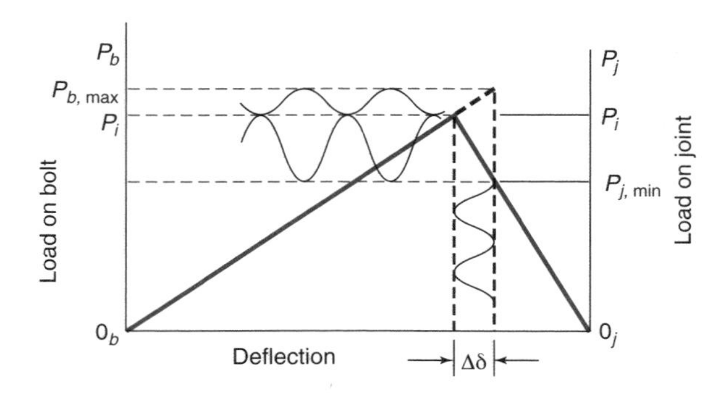

Figure 16.17: Forces versus deflection of bolt and joint as a function of time.

$$\begin{aligned} P_{bm} &= \frac{P_{b,\max} + P_{b,\min}}{2} \\ &= P_i + \frac{C_k\left(P_{\max} + P_{\min}\right)}{2} \\ &= P_i + C_k P_m. \end{aligned} \tag{16.42}$$

The alternating and mean stresses can be expressed as

$$\sigma_a = \frac{C_k P_a n_s}{A_t}, \tag{16.43}$$

$$\sigma_m = \frac{P_i + C_k P_m n_s}{A_t}. \tag{16.44}$$

Recall that the safety factor is not applied to the preload.

To derive a safety factor applicable to bolt fatigue, consider the Goodman fatigue failure criterion given in Section 7.10. Since the safety factor has already been incorporated into Eqs. (16.43) and (16.44), the corresponding Goodman equation is

$$\frac{K_f \sigma_a}{S_e} + \frac{\sigma_m}{S_{ut}} = 1. \tag{16.45}$$

Substituting Eqs. (16.43) and (16.44) into Eq. (16.45) gives

$$\frac{K_f C_k P_a n_s}{S_e A_t} + \frac{P_i + C_k P_m n_s}{A_t S_{ut}} = 1. \tag{16.46}$$

Equation (16.46) can also be expressed in terms of stresses as

$$\frac{n_s K_f C_k \sigma_a}{S_e} + \frac{\sigma_i + C_k \sigma_m n_s}{S_{ut}} = 1. \tag{16.47}$$

Solving for the safety factor gives

$$n_s = \frac{S_{ut} - \sigma_i}{C_k\left[K_f \sigma_a\left(\dfrac{S_{ut}}{S_e}\right) + \sigma_m\right]}. \tag{16.48}$$

The fatigue stress concentration factor, K_f, for threaded elements can be obtained from Table 16.10. Cutting is the simplest method of producing external threads for a specialty fastener. However, rolling external threads is preferred for mass production and provides a smoother thread finish (fewer initial cracks). It is usually safe to assume that the threads have been rolled, unless specific information is available. A generous fillet between the head and the shank reduces the fatigue stress concentration, as shown in Table 16.10. The modified endurance limit given in Eq. (7.18) needs to be used in Eq. (16.48). Note, however, that the surface finish correction factor has already been incorporated into the fatigue stress concentration factor given in Table 16.10.

Table 16.10: Fatigue stress concentration factors for threaded elements.

SAE grade	Metric grade	Rolled threads	Cut threads	Fillet
0–2	3.6–5.8	2.2	2.8	2.1
4–8	6.6–10.9	3.0	3.8	2.3

At this point, a more sophisticated discussion of preload can occur. Recall from the discussion of statically loaded bolts that a preload is desired to prevent joint separation, minimize tolerances, and produce a tight fit. A preload also reduces vibrations and sound associated with loose fits. However, the smallest safety factor is the ratio of ultimate stress to prestress (if the mean and alternating stresses are zero). Examining the values in Table 16.6 and the requirements of Eq. (16.40) leads to the conclusion that the safety factor for properly preloaded bolts cannot be greater than 2.4 and is usually below 2.0. Thus, it would appear that a preload is undesirable for statically loaded bolts, and indeed, proper preload is not as stringently specified as for fatigue applications.

For fatigue applications, a preload is especially important. The reasons can be seen in Eqs. (16.43) and (16.44) and by recalling some subtleties of fatigue analysis discussed in Chapter 7. As long as separation does not occur, the alternating stress experienced by the bolt is reduced by the dimensionless stiffness parameter C_k. The mean stress is increased by the preload. However, the failure theories in Section 7.10 (e.g., Gerber, Goodman, Soderberg, etc.) show that the alternating stress has a much greater effect on service life than the mean stress. That is, the endurance limit is usually a small fraction of the yield or ultimate strength. Thus, a large enough preload to prevent separation is critical since it reduces the value of alternating stress. For the purposes of this text, this preload can be obtained by using the relations given by Eq. (16.40).

Example 16.8: Bolt Fatigue Failure

Given: The hexagonal bolt-and-nut assembly with one steel and one aluminum member shown in Fig. 16.14a and considered in Examples 16.6 and 16.7 has an external load of 11,000 N applied cyclically in a released-tension manner (see Section 7.3).

Find: Find the safety factor guarding against fatigue failure of the bolt. Assume a survival probability of 90%, rolled threads, and a 5.8 metric grade.

Solution: For released tension with a maximum load of 11,000 N,

$$P_a = P_m = 5500 \text{ N}.$$

With the stress area of 115.4 mm^2, the stress amplitude is $\sigma_a = 47.66$ MPa. The endurance limit for axial loading from Eq. (7.6) and Table 16.7 for 5.8-metric-grade thread is

$$S_e' = 0.45 S_u = 0.45\left(520 \times 10^6\right) = 234 \text{ MPa}.$$

From Table 16.10 for 5.8-metric-grade thread and rolled threads, the fatigue stress concentration factor is $K_f = 2.2$. From Table 7.4, the reliability factor for 90% survival probability is $k_r = 0.90$.

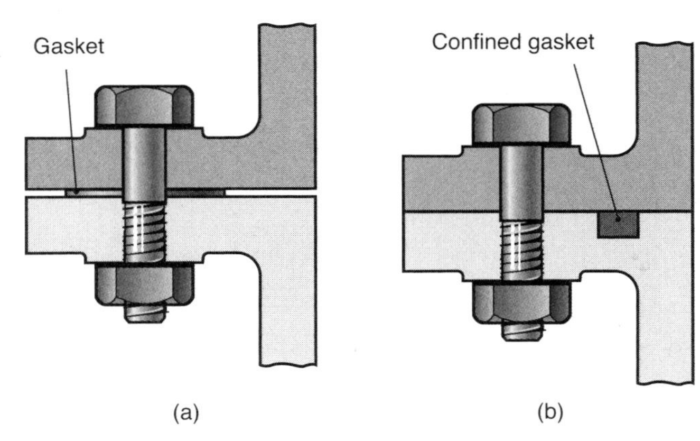

Figure 16.18: (a) Unconfined gasket. The gasket is often produced in an elastomer to match a housing profile, or can be produced from a curing polymer. The joint stiffness is dominated by the gasket in such designs. (b) Confined gasket. Such gaskets can be bonded to one member or be a separate element, and do not affect joint stiffness.

Recall that $k_f = 1$ (because k_f is incorporated into the values of K_f in Table 16.10) and $k_s = 1$ (because the bolt encounters axial loading). Thus, the modified endurance limit from Eq. (7.18) is

$$\begin{aligned} S_e &= k_f k_s k_r k_m S_e' \\ &= (1)(1)(0.9)(1)\left(234 \times 10^6\right) \\ &= 210.6 \text{ MPa.} \end{aligned}$$

From Eq. (16.34), for a crest diameter of 14 mm and coarse threads the tensile stress area is 115.4 mm^2. Recall from Example 16.6 that $C_k = 0.2350$ and from Example 16.7 that $\sigma_i = 285$ MPa. Therefore, from Eq. (16.48),

$$\begin{aligned} n_s &= \frac{S_{ut} - \sigma_i}{\sigma_a C_k \left[\dfrac{K_f S_{ut}}{S_e} + 1\right]} \\ &= \frac{\left(520 \times 10^6\right) - \left(285 \times 10^6\right)}{\left(47.66 \times 10^6\right)(0.2350)\left[2.2\left(\dfrac{520}{210.6}\right) + 1\right]} \\ &= 3.262. \end{aligned}$$

16.4.7 Gasketed Joints

Figure 16.18 shows a threaded fastener with an unconfined **gasket** joining two members. An unconfined gasket extends over the entire diameter (as shown in Fig. 16.18a) of a joint, whereas a confined gasket extends over only a part of the joint diameter. The bolts for a confined gasket joint may be designed as if the gasket is not present. However, the members still need to provide a properly sized cavity to compress the gasket and provide a seal, and this cavity should ideally be placed just outside the conical frusta zone.

An unconfined gasket is made of an elastomeric material or soft metal that conforms to the joint members. Recall that the various members of the joint can be considered as individual springs with different spring rates depending on their materials. Gaskets are soft relative to the other joint members. Since the modulus of elasticity is directly proportional to the stiffness,

$$k_g \ll k_1, k_2, \ldots \tag{16.49}$$

Furthermore,

$$\frac{1}{k_g} \gg \frac{1}{k_1}, \frac{1}{k_2}, \cdots \tag{16.50}$$

Making use of Eqs. (16.29) and (16.50) gives, for an unconfined gasket,

$$\frac{1}{k_j} = \frac{1}{k_1} + \frac{1}{k_2} + \cdots + \frac{1}{k_g} \approx \frac{1}{k_g}. \tag{16.51}$$

Thus, for unconfined gaskets, the joint stiffness is equivalent to the gasket stiffness.

The preload for an unconfined gasket must be large enough to achieve the minimum sealing pressure. Thus,

$$P_i \geq A_g p_o, \tag{16.52}$$

where A_g is the gasket area per bolt and p_o is the minimum gasket seal pressure.

16.5 Riveted Fasteners

Rivets are in wide use as fasteners in aircraft, buildings, bridges, boilers, ships, and (in general) for any framework. A rivet is a pin with a head on one end that is inserted into an aligned hole in the members to be joined. The free end, or *blind* side, is then plastically deformed to lock the rivet in place.

The advantages of rivets compared to threaded fasteners include:

1. They will not shake loose.
2. They are inexpensive, and require short assembly times.
3. They are lightweight.
4. With some designs, they can be inserted and fastened from the same side, as opposed to a bolted connection that needs access to the bolt head and nut during tightening.

However, rivets cannot provide as strong an attachment as a threaded fastener of the same diameter. Also, perhaps most significantly, rivets do not allow any disassembly of the joint.

Rivets must be spaced neither too close nor too far apart. The minimum rivet spacing, center to center, for structural steel work is usually taken as three rivet diameters. A somewhat closer spacing is used in boilers. Rivets should not be spaced too far apart or plate buckling or joint separation will occur. The maximum spacing is usually taken as 16 times the thickness of the outside plate.

Riveted and threaded fasteners in shear are treated exactly alike in design and failure analysis. Failure due to a shear force is primarily caused by the four different modes of failure shown in Fig. 16.19: bending of a member, shearing of the rivet, tensile failure of a member, and the bearing of the rivet on a member or the bearing of a member on the rivet. Each of these four modes is presented here.

1. *Bending of a member:* To avoid this failure mode, the following should be valid:

$$\frac{S_{yj}}{n_s} = \frac{PL}{2Z_m}, \tag{16.53}$$

where L is the grip length, Z_m is the section modulus of the weakest member, where $Z_m = I/c$, and S_{yj} is the yield strength of weakest member.

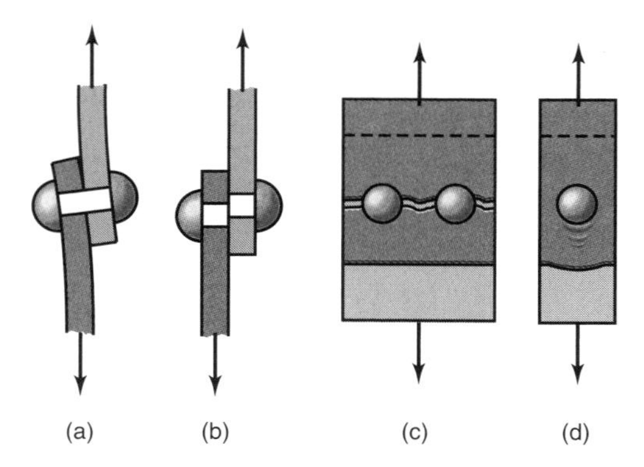

Figure 16.19: Failure modes due to shear loading of riveted fasteners. (a) Bending of member; (b) shear of rivet; (c) tensile failure of member; (d) bearing of member on rivet.

2. *Shear of rivets:* To avoid this failure,

$$\frac{S_{sy}}{n_s} = \frac{4P}{\pi d_c^2}, \tag{16.54}$$

where d_c is the crest diameter and S_{sy} is the rivet yield strength in shear. In Eq. (16.54), the diameter of the rivet rather than the diameter of the hole is used. Normally, these are equivalent, but the rivet may enlarge the hole if there is an interference fit (which is advantageous for fatigue loadings). Clearly if the hole is oversized, the rivet still fails according to its cross section regardless of hole size. Shear of rivets is a major failure mode considered in the design of threaded or riveted fasteners. This mode is also called the direct shear mode of failure.

3. *Tensile failure of member:* To avoid this failure, the following should be valid:

$$\frac{S_{yj}}{n_s} = \frac{P}{(b - N_r d_c)t_m}, \tag{16.55}$$

where b is the width of the member, N_r is the number of rivets across the member's width, and t_m is the thickness of the member. Note that it has been assumed that the members are ductile and that plastic deformation is allowed to relieve stress concentrations. If this is not the case, stress concentration factors that reflect the rivet geometry are needed.

4. *Compressive bearing failure:* To avoid this failure, the following should be valid:

$$\frac{S_{yj}}{n_s} = \frac{P}{d_c t_m}. \tag{16.56}$$

These four failure modes relate to failure of an individual rivet or an individual member. Threaded and riveted fasteners are used in groups, and the torsional shear of the group must be considered as a failure mode. The result of the shear stress acting on the rivet is the vectorial sum of the direct and torsional shear stresses:

$$\tau = \tau_d + \tau_t, \tag{16.57}$$

where τ_d is the shear stress of rivet due to shear (Section 4.6) and τ_t is the shear stress due to torsion (Section 4.4). When dealing with a group of threaded or riveted fasteners, the centroid of the group must be evaluated and equilibrium of vertical and horizontal loads must be observed.

Example 16.9: Analysis of a Riveted Joint

Given: The riveted connection shown in Fig. 16.20a fastens a beam to a column, and a load is applied as shown. Rivets A and C are 5/8 in. in diameter and rivets B and D are 7/8 in. in diameter.

Find: Determine the following:

(*a*) Centroid of the rivet assembly

(*b*) Direct and resultant shear stresses acting on each rivet

(*c*) Safety factor guarding against shear of the rivet

(*d*) Safety factor guarding against bending of the member

Assume that the shear strength of the rivet is 34 ksi and the yield strength of the member is 54 ksi.

Solution:

(*a*) The total area of the four rivets is

$$\sum A_i = \frac{\pi}{2}\left[\left(\frac{5}{8}\right)^2 + \left(\frac{7}{8}\right)^2\right] = 1.816 \text{ in.}^2$$

In Fig. 16.20a, the $x-y$ coordinate is located at the center of rivet C. The centroid of the rivet group is given by Eqs. (4.5) and (4.6) as

$$\bar{y} = \frac{\sum A_i \bar{y}_i}{\sum A_i} = \frac{0.3068(0+6) + 0.6013(0+6)}{1.816} = 3.000 \text{ in.}$$

$$\bar{x} = \frac{\sum A_i \bar{x}_i}{\sum A_i} = \frac{0.3068(0+0) + 0.6013(7+7)}{1.816} = 4.635 \text{ in.}$$

The eccentricity of the loading is

$$e_c = 8 + 7 - 4.635 = 10.36 \text{ in.}$$

Figure 16.20b shows the distances from the center of the rivets to the centroid, and Fig. 16.20c shows the resulting triangles. From these figures

$$r_A = r_C = \sqrt{3^2 + 4.635^2} = 5.521 \text{ in.}$$

$$r_B = r_D = \sqrt{3^2 + 2.365^2} = 3.820 \text{ in.}$$

(*b*) The direct shear stress acting on each of the rivets is

$$\tau_d = \frac{P}{\sum A} = \frac{1000 \text{ lb}}{1.816 \text{ in.}^2} = 550.7 \text{ psi.}$$

Figure 16.20d illustrates the direct and torsional shear stresses acting on each rivet. The statically equivalent torque acting at the centroid is

$$T = Pe_c = (1000 \text{ lb})(10.36 \text{ in.}) = 10,360 \text{ lb-in.}$$

Also, if the rivet size is small enough so that it does not contribute to the polar moment of inertia, the parallel axis theorem (see Section 4.2.3) results in

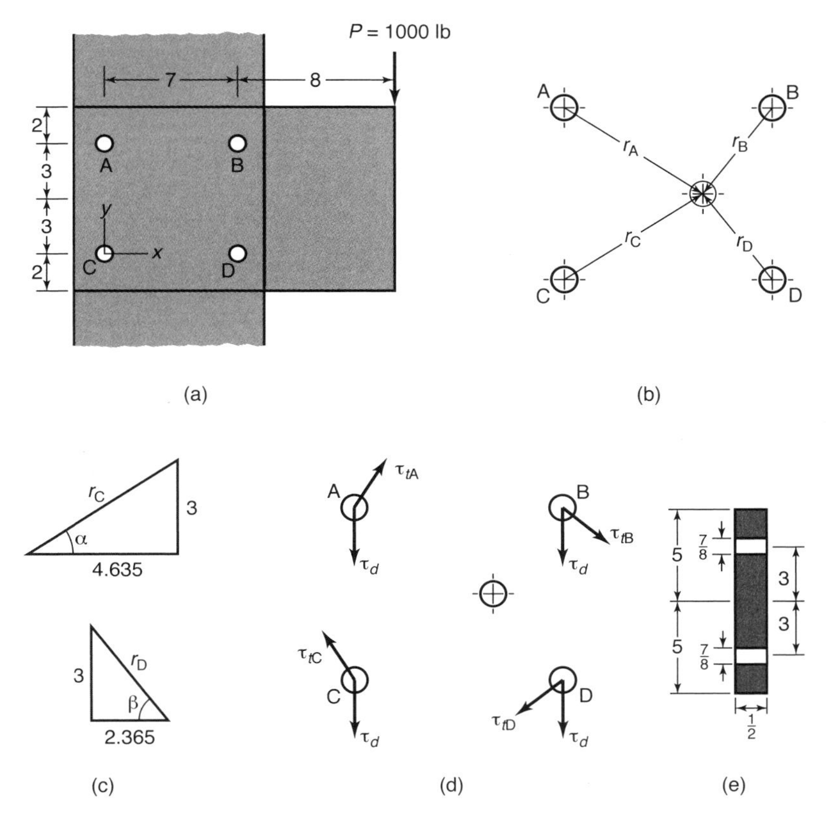

Figure 16.20: Group of riveted fasteners used in Example 16.9. (a) Assembly of rivet group; (b) radii from centroid to center of rivets; (c) resulting triangles; (d) direct and torsional shear acting on each rivet; (e) side view of member. (All dimensions are in inches.)

$$J = \sum r_j^2 A_j = 2\left(r_C^2 A_C + r_D^2 A_D\right) = 36.28 \text{ in.}^4$$

so that the shear stress due to torsional loading is, from Eq. (4.32),

$$\tau_{ti} = \frac{T r_i}{\sum r_j^2 A_j} = \frac{10,360 r_i}{36.28} = 285.6 r_i.$$

Therefore,

$$\tau_{tA} = \tau_{tC} = (285.6)(5.521) = 1577 \text{ lb/in.}^2$$

$$\tau_{tB} = \tau_{tD} = (285.6)(3.820) = 1091 \text{ lb/in.}^2$$

The x- and y-components of the torsional stresses are

$$\tau_{tAx} = \tau_{tCx} = \tau_{tA} \sin\alpha = 1577 \sin 32.91^\circ = 856.9 \text{ psi}$$

$$\tau_{tAy} = \tau_{tCy} = \tau_{tA} \cos\alpha = 1577 \cos 32.91^\circ = 1324 \text{ psi}$$

$$\tau_{tBx} = \tau_{tDx} = \tau_{tB} \sin\beta = 1091 \sin 51.75^\circ = 856.8 \text{ psi}$$

$$\tau_{tBy} = \tau_{tDy} = \tau_{tB} \cos\beta = 1091 \cos 51.75^\circ = 675.4 \text{ psi.}$$

The resultant shear stress at the four rivets, from Fig. 16.20d, is

$$\begin{aligned} \tau_A &= \tau_C = \sqrt{\left(\tau_{tAy} - \tau_d\right)^2 + \tau_{tAx}^2} \\ &= \sqrt{(1324 - 550.7)^2 + (856.8)^2} \\ &= 1154 \text{ psi,} \end{aligned}$$

$$\begin{aligned} \tau_B &= \tau_D = \sqrt{\left(\tau_{tDy} + \tau_d\right)^2 + \tau_{tDx}^2} \\ &= \sqrt{(675.4 + 550.7)^2 + (856.8)^2} \\ &= 1496 \text{ psi.} \end{aligned}$$

The critical shear stress is at rivets B and D. Once the shear stress is known, the shear force can be obtained from

$$P_B = P_D = \tau_B A = (1496)(0.6013) = 899.5 \text{ lb.}$$

(*c*) The safety factor guarding against shear of the rivets is

$$n_s = \frac{\tau_{\text{all}}}{\tau_D} = \frac{34.0}{1.496} = 22.73.$$

(*d*) Figure 16.20e shows a side view of the member at a critical section where there are two rivets across the width. The distance from the neutral axis to the outer fiber is 5 in. The bending of the member, from Fig. 16.20a, is

$$M_B = (1000)(8) = 8000 \text{ lb-in.}$$

The area moment of inertia for the critical section shown in Fig. 16.20e is

$$I = \frac{1}{12}\left(\frac{1}{2}\right)(10^3) - 2\left[\left(\frac{1}{12}\right)\left(\frac{1}{2}\right)\left(\frac{7}{8}\right)^3 + \left(\frac{1}{2}\right)\left(\frac{7}{8}\right)(3)^2\right],$$

or $I = 33.74$ in.^{4}. The bending stress at rivet B is

$$\sigma_B = \frac{M_B c}{I} = \frac{(8000)(5)}{33.74} = 1.186 \times 10^3 \text{ psi.}$$

The safety factor guarding against bending of the member is

$$n_s = \frac{\sigma_{\text{all}}}{\sigma_B} = \frac{54}{1.186} = 45.5.$$

Therefore, if failure should occur, it would occur first from shear of the rivets. However, because the safety factor guarding against shear of the rivets is 22.73, it is highly unlikely that failure will occur.

16.6 Welded, Brazed, and Soldered Joints

A **weld** is a permanent joint that results from diffusion of material across an interface, and may include a filler metal. Brazing and soldering develop an interface from the adhesion between members and a filler metal. Brazing and soldering result in non-permanent joints, since the filler metal can be melted and removed, but welds are considered permanent. The distinction between brazing and soldering is that brazing occurs above 450°C, thus soldering generally involves tin and other low melting point metals, and is very common for assembling and joining electronic devices.

There are a wide variety of welding methods, as discussed by Kalpakjian and Schmid [2010]. **Fusion welding** involves melting and coalescing materials by means of heat, usually supplied by chemical, electrical, or high energy means. Examples of fusion welding are oxyfuel gas welding, consumable- and nonconsumable-electrode *arc welding*, and *high energy beam welding*. **Solid state welding** involves joining without fusion; that is, there is no liquid (molten) phase involved. The basic categories of solid state welding are *cold, ultrasonic, friction, resistance, explosion welding,* and *diffusion bonding*.

There are many configurations of welds. Figure 16.21 shows the standard weld symbols, indicating that weld joints offer considerable design flexibility. Although welded joints are available in a wide variety of forms, the following classification of welded, brazed, and soldered joints assists in design analysis:

1. **Spot welds**, such as in resistance spot welding, where electrodes cause a localized weld to form between two metals.

2. **Line welds**, where a weld line or *bead* is produced. For example, consider shielded metal arc welding (SMAW), illustrated in Fig. 16.22. In this method, the two members to be joined are clamped together, with each of them in electrical contact with one terminal of an electrical power source. A low-voltage, high-current arc is struck with an electrode to complete the electric circuit at the joint. The distance between the electrode and the work is controlled such that the electric arc is sustained, creating temperatures high enough to melt the members at the joint. The electrode is gradually consumed in the process, supplying additional metal to the joint as it melts. A flux covering the electrode provides a shielding gas (hence, the term "shielded metal arc welding") and a chalky covering (slag), both of which serve to protect the metal from oxidation. The slag is usually chipped off once the weld has cooled sufficiently. The result of SMAW is a **weld bead** or **weld line** that is very long compared to its width.

3. **Area welds**, when a weld joint is developed over a large area. Examples include cold and explosive welding and diffusion bonding. Brazing and soldering generally result in area bonds.

Some advantages of welded joints over threaded fasteners are that they are inexpensive and there is no danger of the joint loosening. Some disadvantages of welded joints over threaded fasteners are that they produce residual stresses, they distort the shape of the member, metallurgical changes occur, and disassembly is a problem.

Weld quality depends greatly on operator skill, surface preparation and process parameters. It is expected that welds will be designed with higher safety factors than other fastening approaches, and this is due in large part to variations in weld strength.

16.6.1 Spot Welds

In resistance spot welding (RSW), the tips of two opposing solid, cylindrical electrodes touch a lap joint of two sheet metals, and resistance heating produces a spot weld (Fig. 16.23). RSW is one of the most commonly used processes in sheet-metal fabrication and automotive body panel assembly. In order to obtain a strong bond in the weld nugget, pressure is applied until the current is turned off and the weld has solidified. Accurate control and timing of the alternating electric current and of the pressure are essential in resistance welding. The resultant surface of the spot weld has a slightly discolored indentation.

Note that a spot weld develops a weld nugget that is slightly larger than the electrode. The weld nugget is generally 6 to 10 mm (0.25 to 0.375 in.) in diameter. From a stress analysis standpoint, a spot weld can be treated as a rivet as described in Section 16.5. However, since spot welds are generally used on sheet metals, the grip is short and failure of the weld nugget by bending is not likely; instead, it is probable that failure of the spot weld by shear will take place. Shear of the nugget can be analyzed by using Eq. (16.54) where the nugget diameter is used instead of the rivet diameter.

16.6.2 Line Welds

Recall that a large number of welding processes result in weld beads or lines that have a small width compared to their length. A number of different joints can be achieved, one of which is the **fillet weld**, shown in Fig. 16.24a, and will be discussed here. The approach for fillet welds can be easily extended to other weld geometries.

The thinnest section of a fillet weld is at the **throat** of the weld, at 45° from the legs. The governing stress in fillet welds

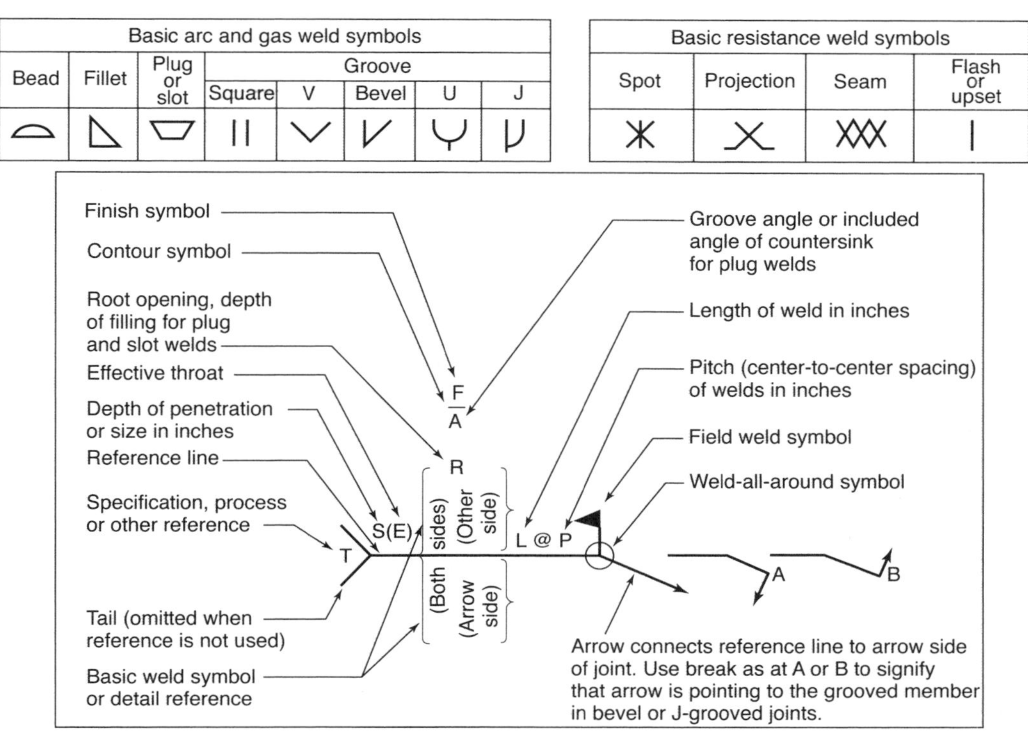

Figure 16.21: Basic weld symbols. *Source:* Kalpakjian and Schmid [2010].

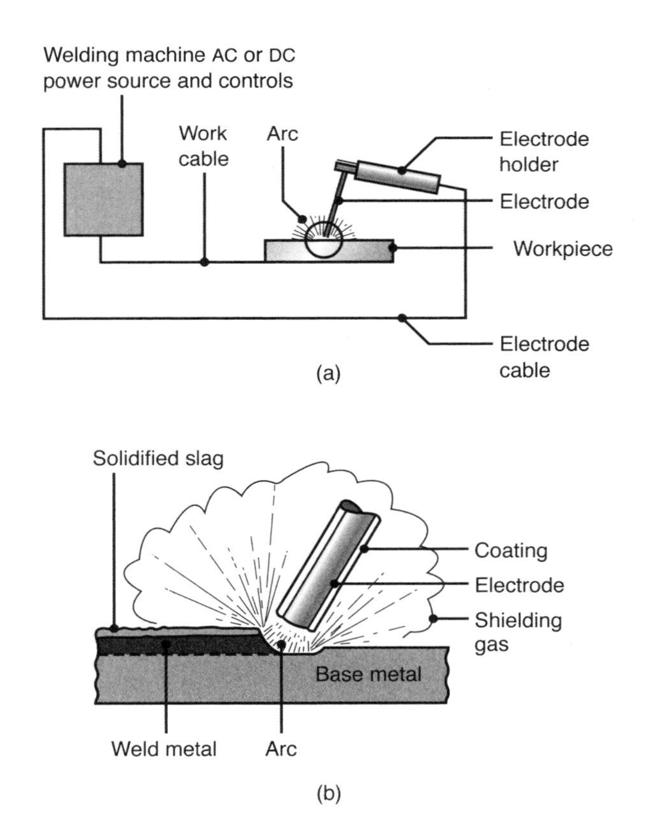

Figure 16.22: Schematic illustration of the shielded-metal arc welding (SMAW) process. About 50% of all large-scale industrial welding operations use this process. *Source:* Kalpakjian and Schmid [2010].

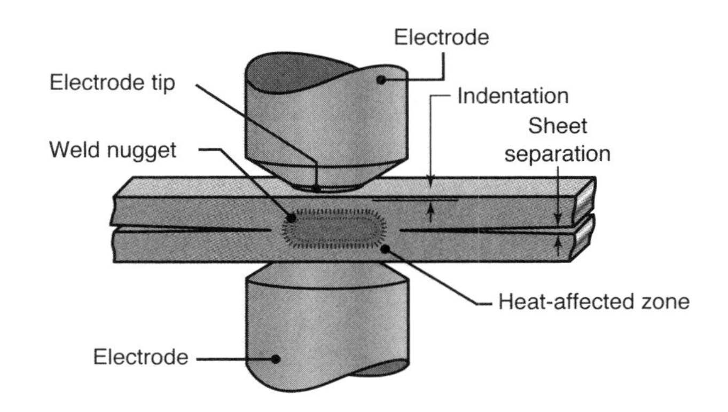

Figure 16.23: Schematic illustration of the resistance spot welding process, showing the weld nugget and indentation of the surface. *Source:* Kalpakjian and Schmid [2010].

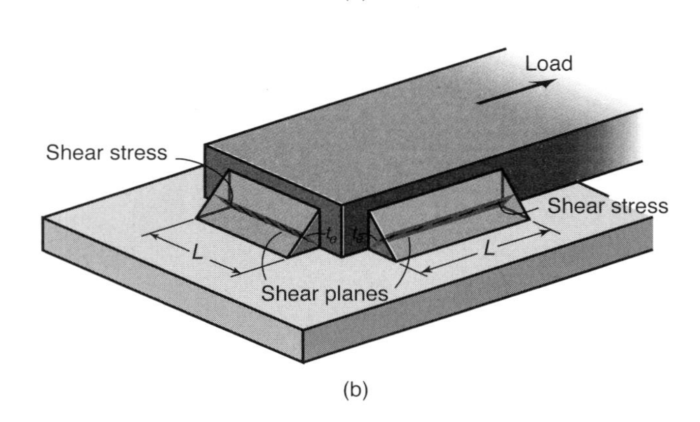

Figure 16.24: Fillet weld. (a) Cross section of weld showing throat and legs; (b) shear planes.

is shear on the throat of the weld as shown in Fig. 16.24b. Observe from this figure that in the fillet weld aligned parallel to the load, the shear stress occurs along the throat of the fillet parallel to the load. In a fillet weld aligned transverse to the load, the shear stress occurs at 45° to the load, acting transverse to the axis of the fillet.

Parallel and Transverse Loading

Fillet welds fail by shearing at the minimum section, which is at the throat of the weld, shown in Fig. 16.24a. This is true whether the weld has parallel (on the side) or transverse (at the end, as shown in Fig. 16.24b) loading. The shear stress from these types of loading is

$$\tau = \frac{P}{t_e L_w} = \frac{P}{0.707 h_e L_w} = \frac{1.414P}{h_e L_w}, \tag{16.58}$$

where t_e is the weld throat length, h_e is the weld leg length, and L_w is the weld length. Thus, to avoid failure, the following should be valid:

$$\frac{S_{sy}}{n_s} = \frac{P}{t_e L_w}. \tag{16.59}$$

Torsional Loading

For torsional loading, the resultant shear stress acting on the weld group is the vectorial sum of the direct and torsional shear stresses, as shown in Eq. (16.55) for rivets. The direct (or transverse) shear stress in the weld is

$$\tau_d = \frac{V}{A} = \frac{\text{Shear force}}{\text{Total throat area}}. \tag{16.60}$$

The torsional shear stress is

$$\tau_t = \frac{Tr}{J}, \tag{16.61}$$

where r is the distance from the centroid of the weld group to farthest point in the weld, T is the applied torque, and J is the polar area moment of inertia. The critical section for torsional loading is the throat section, as it is for parallel and transverse loading. The analysis can be simplified for line welds by using the concept of the **unit polar moment of inertia**, J_u, where

$$J = t_e J_u = 0.707 h_e J_u. \tag{16.62}$$

Note that J_u has units of length to the third power. Table 16.11 gives values of unit polar area moment of inertia for nine weld groups. Using this table simplifies analysis of torsional loading. The alternative is to calculate the polar area moment of inertia using the approaches outlined in Section 4.2.

Bending

In bending, the welded joint experiences a transverse shear stress as well as a normal stress. The direct (or transverse) shear stress is the same as that given in Eq. (16.58). The moment M produces a normal bending stress σ in the welds. It is customary to assume that the stress acts normal to the throat area. The area moment of inertia is calculated from the **unit area moment of inertia** as:

$$I = t_e I_u = 0.707 h_e I_u. \tag{16.63}$$

Table 16.11 gives values of unit area moment of inertia for nine weld groups. The normal stress due to bending is

$$\sigma = \frac{Mc}{I}, \tag{16.64}$$

where c is the distance from the neutral axis to the outer fiber. Once the shear stress and the normal stress are known, the principal normal stresses or the shear stresses can be determined from Eqs. (2.16) or (2.19), respectively. Once these principal stresses are obtained, Eq. (6.8) (the maximum-shear-stress theory) or Eq. (6.9) (the distortion-energy theory) can be applied to determine if the weld will fail.

Weld Bead Strength

The electrodes used in arc welding are identified by the letter E followed by a four-digit number, such as E6018. The first two numbers indicate the strength of the deposited material in thousands of pounds per square inch. The last digit indicates variables in the welding technique, such as current supply. The next-to-last digit indicates the welding position as, for example, flat, vertical, or overhead.

Table 16.12 lists the minimum strength properties for some electrode classes.

Example 16.10: Welded Joint

Given: A bracket is welded to a beam as shown in Fig. 16.25a. Assume a steady loading of 20 kN and weld lengths l_1 = 150 mm and l_2 = 100 mm. Assume an electrode number of E60XX and a fillet weld.

Find: Determine the minimum weld leg length for the eccentric loading shown in Fig. 16.25a based on torsion and a safety factor of 5.0.

Solution: The total weld area is

$$\begin{aligned}\sum A_i &= t_e(l_1 + l_2) = 0.7071 h_e (0.150 + 0.100) \\ &= (0.1768\ \text{m}) h_e = (0.250\ \text{m}) t_e.\end{aligned}$$

Table 16.11: Geometry of welds and parameters used when considering various types of loading. Note that failure is assumed to occur at the tensile side; this requires that the members bear against each other during welding so that compressive forces are not transferred solely through the weld bead. *Source:* Adapted from Mott [2014]

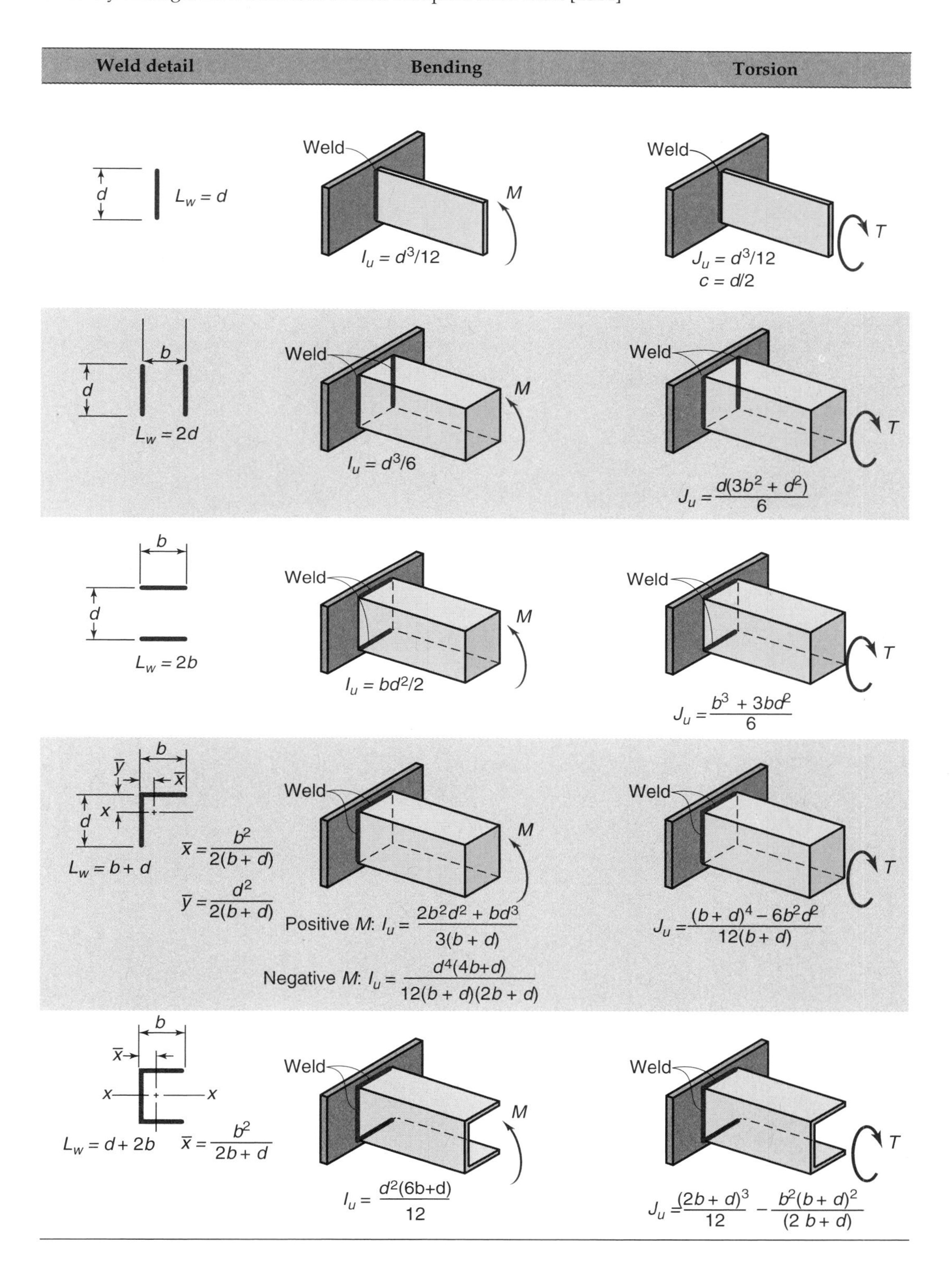

Table 16.11: *(continued)*

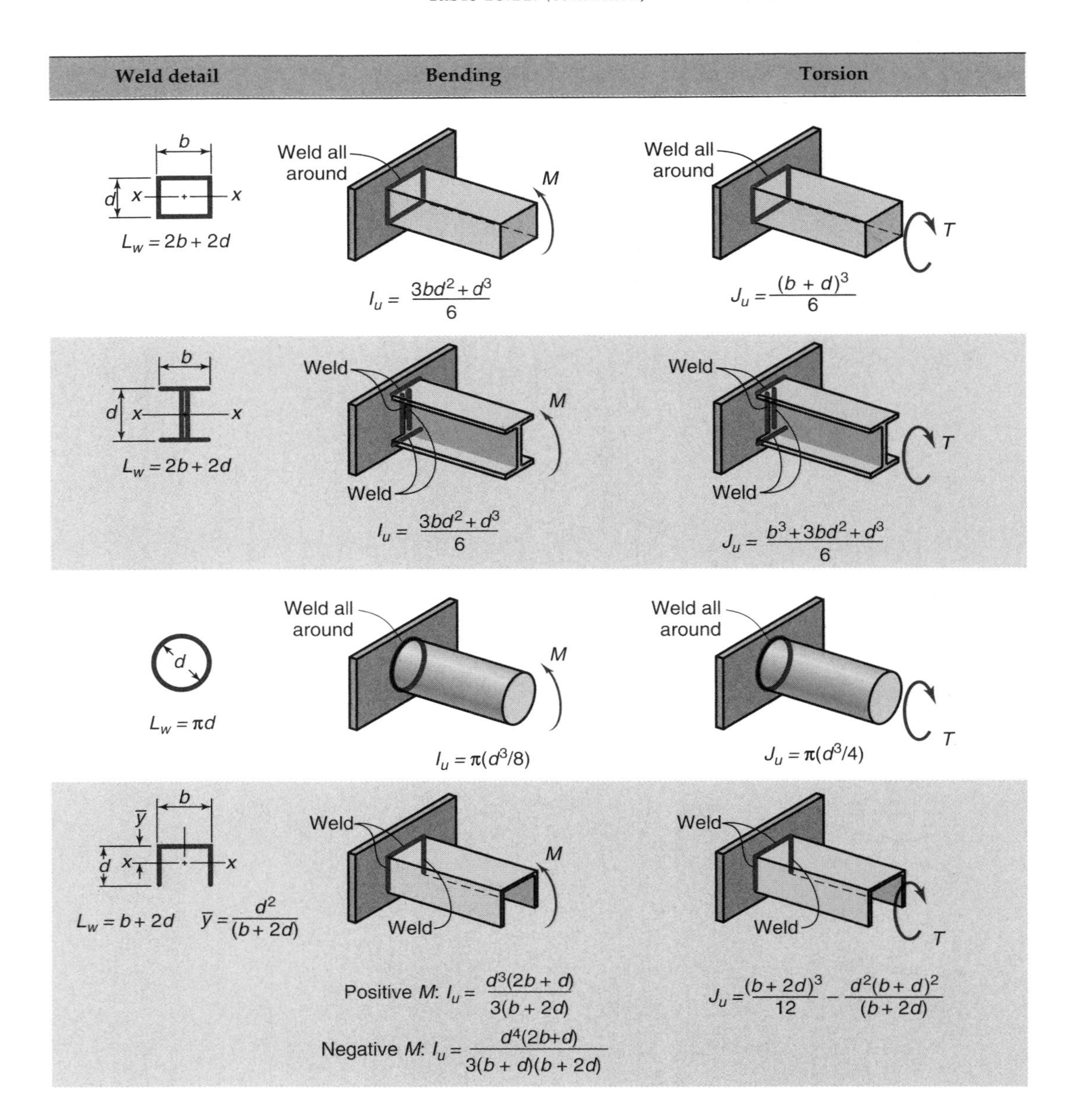

The centroid of the weld group is obtained from Eqs. (4.5) and (4.6) as

$$\bar{x} = \frac{\sum A_i x_i}{\sum A_i} = \frac{0(150)t_e + 50(100)t_e}{250t_e} = 20 \text{ mm},$$

$$\bar{y} = \frac{\sum A_i y_i}{\sum A_i} = \frac{75(150)t_e + 150(100)t_e}{250t_e} = 105 \text{ mm}.$$

The applied torque is

$$T = Pr = (20,000)(0.3 - 0.02) \text{ N-m} = 5.6 \text{ kN-m}.$$

Also,

$$\begin{aligned} J_u &= \frac{(b+d)^4 - 6b^2d^2}{12(b+d)} \\ &= \frac{(0.100 + 0.150)^4 - 6(0.100)^2(0.150)^2}{12(0.100 + 0.150)} \\ &= 8.521 \times 10^{-4} \text{ m}^3. \end{aligned}$$

The polar area moment of inertia is

$$\begin{aligned} J &= t_e J_u = 0.7071 h_e J_u = (0.7071)(h_e)\left(8.521 \times 10^{-4}\right) \\ &= \left(6.025 \times 10^{-4} \text{ m}^3\right) h_e. \end{aligned}$$

From Eq. (16.60) the direct (or transverse) shear at locations A and B shown in Fig. 16.25 is

$$\tau_{dA} = \tau_{dB} = \frac{P}{A} = \frac{20,000}{0.1768 h_e} = \frac{(1.131 \times 10^5 \text{ N/m})}{h_e}.$$

From Fig. 16.25b, the torsional shear stress components at point A are

Table 16.12: Minimum strength properties of electrode classes.

Electrode number	Ultimate tensile strength, S_{ut}, MPa (ksi)	Yield, strength, S_y, MPa (ksi)	Elongation, e_k, percent
E60XX	427 (62)	345 (50)	17–25
E70XX	482 (70)	393 (57)	22
E80XX	552 (80)	462 (67)	19
E90XX	620 (90)	531 (77)	14–17
E100XX	689 (100)	600 (87)	13–16
E120XX	827 (120)	738 (107)	14

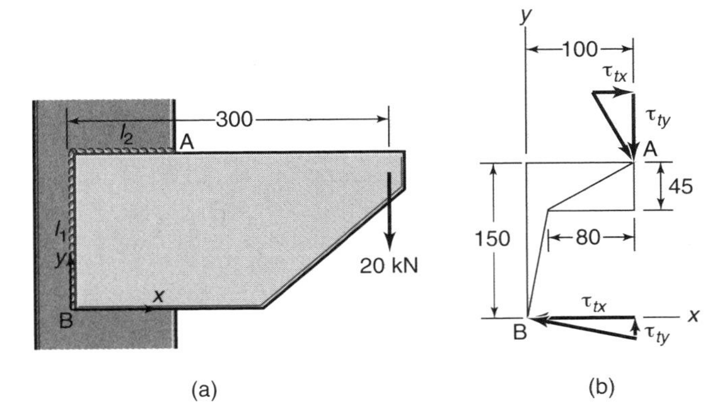

Figure 16.25: Welded bracket used in Example 16.10. (a) Dimensions, load, and coordinates; (b) torsional shear stress components at points A and B. (All dimensions are in millimeters.).

$$\tau_{tAx} = \frac{T(0.045)}{J} = \frac{(5600)(0.045)}{(6.025\times10^{-4})\,h_e} = \frac{(4.183\times10^5\text{ N/m})}{h_e},$$

$$\tau_{tAy} = \frac{T(0.080)}{J} = \frac{(5600)(0.080)}{(6.025\times10^{-4})\,h_e} = \frac{(7.436\times10^5\text{ N/m})}{h_e}.$$

Therefore, the x-component of the shear stress at location A is

$$\tau_{Ax} = \tau_{tAx} = \frac{(4.183\times10^5\text{ N/m})}{h_e}.$$

In the y-direction, the shear stress is due to both direct shear and torsion, so that:

$$\tau_{Ay} = \tau_{dA} + \tau_{tAy} = \frac{1.131\times10^5}{h_e} + \frac{7.436\times10^5}{h_e} = \frac{(8.567\times10^5\text{ N/m})}{h_e},$$

and the combined shear stress is

$$\tau_A = \sqrt{\tau_{Ax}^2+\tau_{Ay}^2} = \frac{1}{h_e}\sqrt{(4.183\times10^5)^2+(8.567\times10^5)^2} = \frac{(9.534\times10^5\text{ N/m})}{h_e}.$$

Similarly, at point B

$$\tau_{Bx} = \frac{T(0.105)}{J} = \frac{(5600)(0.105)}{(6.025\times10^{-4})\,h_e} = \frac{(9.759\times10^5\text{ N/m})}{h_e},$$

$$\tau_{tBy} = \frac{T(0.020)}{J} = \frac{(5600)(0.020)}{(6.025\times10^{-4})\,h_e} = \frac{(1.859\times10^5\text{ N/m})}{h_e}.$$

Note from Fig. 16.24b that the direct shear stress and torsional shear stress are in opposite directions; therefore,

$$\tau_{By} = \tau_{dB} - \tau_{tBy} = \frac{1.859\times10^5}{h_e} - \frac{1.131\times10^5}{h_e} = \frac{(7.280\times10^4\text{ N/m})}{h_e},$$

so that

$$\tau_B = \sqrt{\tau_{Bx}^2+\tau_{By}^2} = \frac{1}{h_e}\sqrt{(9.759\times10^5)^2+(7.280\times10^4)^2} = \frac{(9.786\times10^5\text{ N/m})}{h_e}.$$

Because the shear stress is larger at point B, this point is selected for design. From Table 16.12, the yield strength is 50 ksi (345 MPa) for electrode E60XX. Therefore,

$$n_s = \frac{\tau_{\text{all}}}{\tau_B} = \frac{345\times10^6}{\left(\frac{9.786\times10^5}{h_e}\right)} = \left(352.5\text{ m}^{-1}\right)h_e.$$

For a safety factor of 5.0, h_e is found to be 14.18 mm.

16.6.3 Area Welds

Many welding operations produce a weld joint over an area, such as cold rolling, diffusion bonding, or friction welding; this is also the situation for brazed and soldered joints. From a design analysis standpoint, these joints can be treated like adhesively bonded joints as discussed in Section 16.7, but the allowable stresses over the interfaces are accordingly larger. However, it must be recognized that these welds require good control over process parameter, and are especially sensitive to surface preparation.

16.6.4 The Weld Joint

Three distinct zones can be identified in a typical weld joint, as shown in Fig. 16.26:

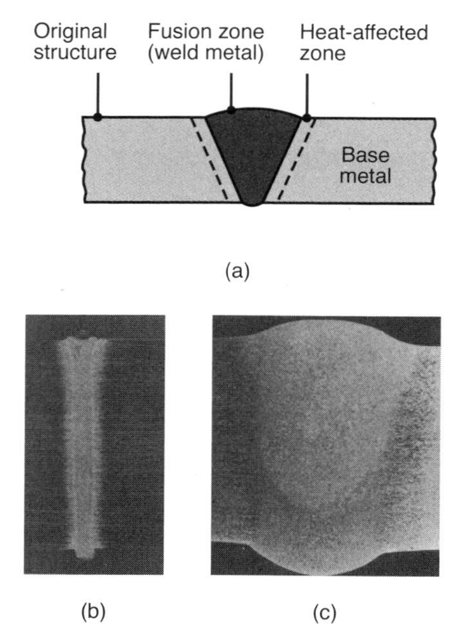

Figure 16.26: (a) Characteristics of a typical fusion-weld zone in oxyfuel-gas and arc welding.*Source:* Kalpakjian and Schmid [2010] (b)-(c) Comparison of the welds produced in (b) electron beam welding and (c) tungsten-arc welding. *Source:* Courtesy of the American Welding Society.

1. Base metal
2. Heat-affected zone
3. Weld metal

The geometry, metallurgy, and physical and mechanical properties of the second and third zones strongly depend on the type of metals joined, the particular joining process, the filler metals used (if any), and welding process variables.

Heat-Affected Zone

The portions of the base metal that are far away from the heat source are not exposed to high temperatures, and therefore do not undergo any microstructural changes. The chemistry of the **heat-affected zone** (HAZ) is identical to the base metal, but it has a different microstructure because it has been temporarily subjected to elevated temperatures during welding.

The strength and hardness of the heat-affected zone depend partly on how the original strength and hardness of the base metal was developed prior to the welding. For example, consider a metal that was cold worked, as by rolling, and is then welded. The metal far away from the weld will retain its original microstructure, but closer to the weld, it will recrystallize, with larger grains the closer the metal is to the weld. Consequently, the metal will be softer and have lower strength adjacent to the weld and will be weakest inside the heat-affected zone.

The size and magnitude of the HAZ is an important factor in understanding weld failure. Note from Fig. 16.26b and c that different processes can produce larger or smaller beads and heat-affected zones. Generally, the smaller the HAZ, the less pronounced a reduction in mechanical properties occurs in the HAZ. Thus, electron beam welding is generally preferable to arc welding; a process without any HAZ (such as cold welding) is even better.

The effects of the HAZ for joints made from dissimilar metals and for alloys strengthened by other methods are so complex as to be beyond the scope of this book. Details can be found in the more advanced references listed in the Recommended Readings at the end of this chapter.

Weld Quality

As a result of a history of thermal cycling and its attendant microstructural changes, a welded joint may develop various discontinuities. Welding discontinuities also can be caused by an inadequate or careless application of proper welding technologies or by poor operator training. The major discontinuities that affect weld quality are:

- **Porosity.** Porosity in welds is caused by gases released during melting of the weld area but trapped during solidification, chemical reactions during welding, or other contaminants.

- **Slag inclusions.** Slag inclusions are compounds such as oxides, fluxes, and electrode-coating materials that are trapped in the weld zone. If shielding gases are not effective during welding, contamination from the environment also may contribute to such inclusions. Welding conditions also are important; with control of welding process parameters, the molten slag will float to the surface of the molten weld metal and thus will not become entrapped.

- **Incomplete fusion and penetration.** Incomplete fusion (lack of fusion) produces poor weld beads, such as those shown in Fig. 16.27. Such defects are generally associated with poor welding practices.

- **Cracks.** Cracks may occur in various locations and directions in the weld area. Typical types of cracks are longitudinal, transverse, crater, underbead, and toe cracks (Fig. 16.28). Cracks in welding are usually associated with thermal stresses, either because of high thermal gradients in the weld zone, or restriction of contraction due to rigid clamping.

- **Surface damage.** Some of the metal may spatter during welding and be deposited as small droplets on adjacent surfaces. In arc-welding processes, the electrode inadvertently may touch the parts being welded at places other than the weld zone (arc strikes). Such surface discontinuities may be objectionable for reasons of appearance or of subsequent use of the welded part. If severe, these discontinuities may adversely affect the properties of the welded structure, particularly for notch-sensitive metals.

- **Residual stresses.** Because of localized heating and cooling during welding, the expansion and contraction of the weld area causes residual stresses in the workpiece. Residual stresses usually place the weld and the HAZ in a state of residual tension, which is harmful from a fatigue standpoint. Many welded structures will use cold-worked materials (such as extruded or roll-formed shapes), and these are relatively strong and fatigue-resistant. The weld itself may have porosity, which can act as a stress riser and aid fatigue-crack growth, or there could be other cracks that can grow in fatigue. In general, the HAZ is less fatigue-resistant than the base metal.

Therefore, it is not surprising that welds in high stress applications need to have the tensile residual stress relieved by

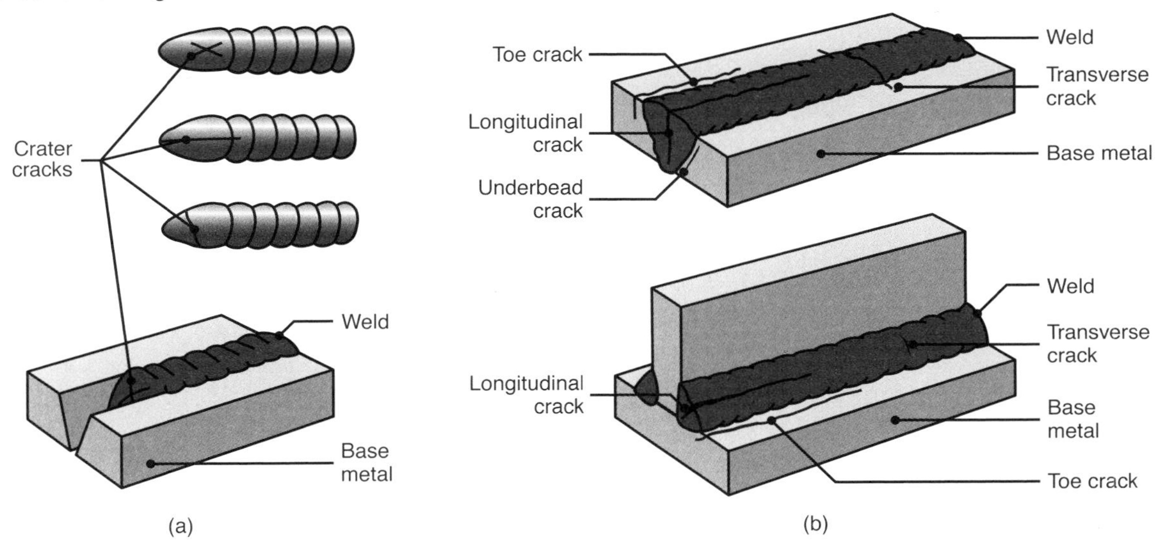

Figure 16.28: Types of cracks developed in welded joints. The cracks are usually caused by thermal stresses. *Source:* From Kalpakjian and Schmid [2010].

processes such as shot blasting, peening, hammering, or surface rolling of the weld-bead area. These techniques reduce or eliminate porosity and induce compressive residual stresses.

16.6.5 Fatigue Strength of Welds

When welded members are placed into an environment where cyclic loadings are encountered, the joint generally fails in the HAZ before the welded members. Since the electrode material contains a large amount of alloying elements, it is relatively strong, and it is not clear why the weld strength is suspect. However, because a stress concentration is associated with every weld, the stresses are highest in the immediate vicinity of the weld. (This also explains the general design rule that welds should be moved away from geometric stress concentrations whenever possible.) Further, cracks rarely propagate through the weld material, but rather fatigue failure arises by crack propagation through the HAZ of the welded material, as discussed above.

The technical literature has numerous warnings against the use of welds in fatigue applications because of these shortcomings. However, since welded connections are routinely placed in situations where fatigue is possible, such admonitions provide no assistance to machine designers. Regardless, it should be recognized that it is difficult to design welds for fatigue applications.

Recognizing the difficulties associated with determining the actual magnitude of the stress concentration associated with welds, Shigley and Mischke [1989] recommend the fatigue strength reduction factors shown in Table 16.13. These factors should be applied to the member's wrought material strength (regardless of the manufacturing history prior to welding) as well as to the weld material strength. Recognizing that weld quality can be improved by shot blasting, peening, etc., the recommendations in Table 16.13 may be conservative.

Table 16.13: Fatigue strength reduction factors for welds *Source:* From Shigley and Mischke [1989].

Type of weld	Fatigue stress concentration factor, K_f
Reinforced butt weld	1.2
Tow of transverse fillet weld	1.5
End of parallel fillet weld	2.7
T-butt joint with sharp corners	2.0

16.7 Adhesive Bonding

Adhesive bonding is the process of joining materials chemically through the formation of interatomic or intermolecular bonds, or adhesion. Adhesive bonding can be used to join a wide variety of materials (metal to metal, metal to ceramic, metal to polymer, etc.) for both structural and nonstructural uses. Table 16.14 summarizes the main characteristics of selected structural adhesives.

Some of the advantages of adhesive bonding over other fastening techniques covered earlier in this chapter are

1. Uniform stress distribution with resultant increased life
2. Reduced weight
3. Improved fatigue resistance
4. Ability to join thick or thin materials
5. Ability to join dissimilar materials
6. The adhesive results in leakproof joints
7. Vibration-damping and insulation properties
8. Economic advantages associated with cost of adhesive and ease of assembly

Some of the limitations of adhesive bonding are

1. Possible need for extensive surface preparation
2. Service temperatures are more limited compared to welds, brazes, or solders
3. Tendency to creep under sustained load
4. Questionable long-term durability

Table 16.14: Typical properties and characteristics of common structural adhesives.

	Epoxy	Polyurethane	Modified acrylic	Cyanoacrylate	Anaerobic
Impact resistance	Poor	Excellent	Good	Poor	Fair
Tension-shear strength, MPa	15–22	12–20	20–30	18.9	17.5
ksi	2.2–3.2	1.7–2.9	2.9–4.3	2.7	2.5
Peel strength, N/m[a]	< 523	14,000	5250	< 525	1750
lb/in.	3	80	30	3	10
Service temperature range, °C	–55–120	–40–90	–70–120	–55–80	–55–150
°F	–70–250	–250–175	–100–250	–70–175	–70–300
Solvent resistance	Excellent	Good	Good	Good	Excellent
Moisture resistance	Good–excellent	Fair	Good	Poor	Good
Odor	Mild	Mild	Strong	Moderate	Mild
Toxicity	Moderate	Moderate	Moderate	Low	Low
Flammability	Low	Low	High	Low	Low

[a] Peel strength varies widely depending on surface preparation and quality.

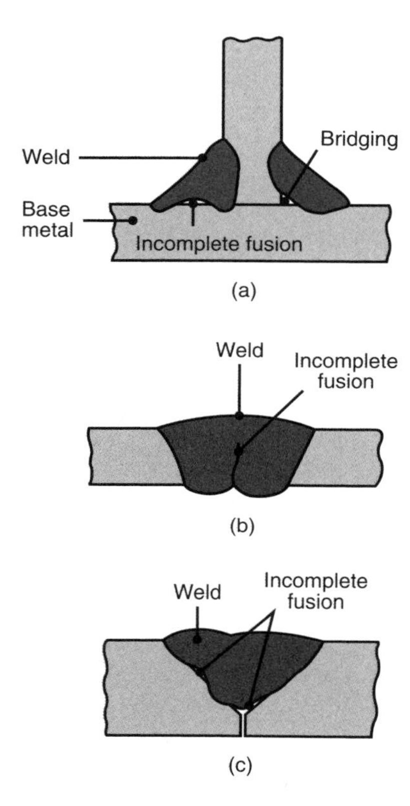

Figure 16.27: (a)-(c) Examples of various discontinuities in fusion welds. *Source:* Kalpakjian and Schmid [2010].

Some common adhesive joints are shown in Fig. 16.29. It should be recognized that there is a general goal of increasing the adhesively bonded area with such joints, so that strap joints are in general stronger than lap joints, which are in turn stronger than scarf joints, which are stronger than butt joints. If the adhesion for a lap joint is uniform over the lapped surfaces, as shown in Fig. 16.29c, the average shear stress is

$$\tau_{\text{avg}} = \frac{P}{A} = \frac{P}{bL}. \tag{16.65}$$

In practice, the shear stress would not necessarily be constant over the area and would be highest at the edges and lowest in the center. For a typical lap joint configuration, without any

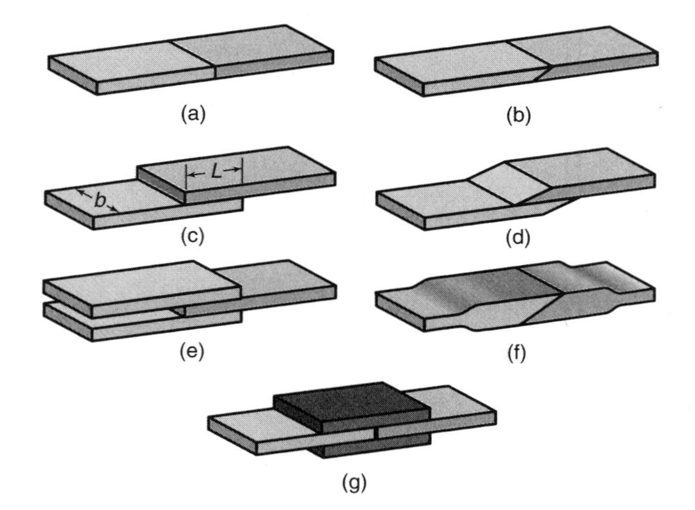

Figure 16.29: Examples of adhesively bonded joints. (a) Butt; (b) scarf; (c) lap; (d) bevel; (e) double lap; (f) increased thickness; (g) strap.

bevels (Fig. 16.29c),

$$\tau_{\max} = 2\tau_{\text{avg}} = \frac{2P}{bL}. \tag{16.66}$$

The stress distribution factor varies with material and aspect ratio, b/L, but Eq. (16.66) is reasonable for $L = b$.

For a scarf joint adhesively bonded under axial load, as shown in Fig. 16.30a, while assuming a uniform stress distribution

$$\sigma_x = \frac{P}{bt_m/\sin\theta} = \frac{P\sin\theta}{bt_m}, \tag{16.67}$$

where t_m is the thickness of the thinnest member. The normal and shear stress components can be written as

$$\sigma_n = \sigma_x \sin\theta = \frac{P}{bt_m}\sin^2\theta, \tag{16.68}$$

$$\tau = \sigma_x \cos\theta = \frac{P}{bt_m}\sin\theta\cos\theta = \frac{P}{2bt_m}\sin 2\theta. \tag{16.69}$$

For *bending*, shown in Fig. 16.30b, the stress distribution along the scarf surface (A-A) is

$$\sigma_n = \frac{Mc}{I},$$

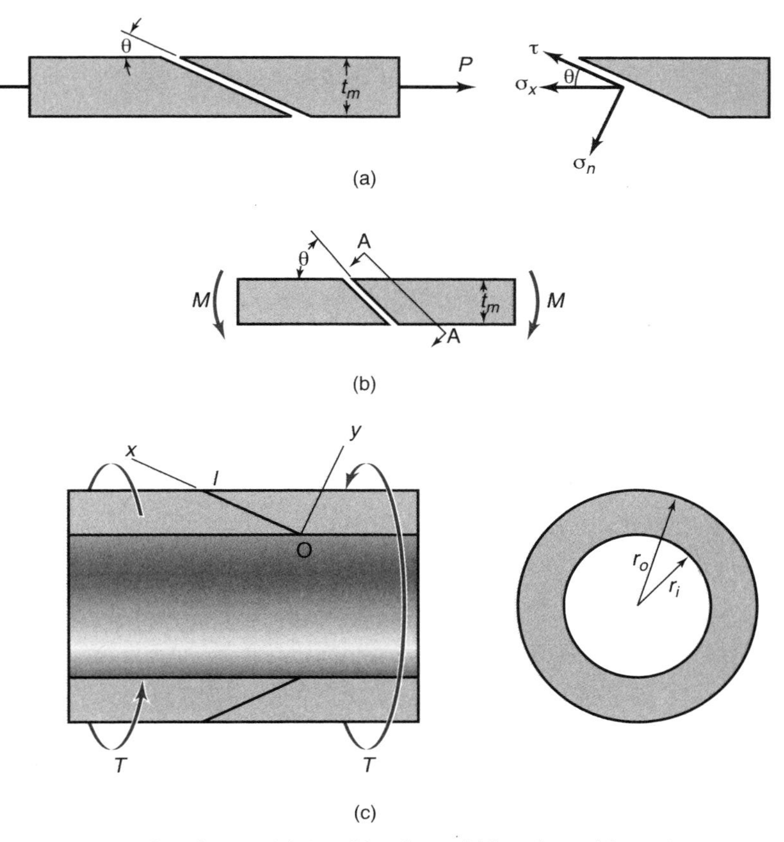

Figure 16.30: Scarf joint. (a) Axial loading; (b) bending; (c) torsion.

so that

$$\sigma_n = \frac{M(t_m/2\sin\theta)}{\dfrac{b}{12}\left(\dfrac{t_m}{\sin\theta}\right)^3} = \frac{6M\sin^2\theta}{bt_m^2}. \tag{16.70}$$

The shear stress along the scarf surface due to *bending* is

$$\tau = \frac{\sigma_n}{\tan\theta} = \frac{6M}{bt_m^2}\sin^2\theta\frac{\cos\theta}{\sin\theta} = \frac{6M}{bt_m^2}\sin\theta\cos\theta = \frac{3M}{bt_m^2}\sin 2\theta. \tag{16.71}$$

For *torsional* loading, shown in Fig. 16.30c,

$$\tau = \frac{Tr}{J},$$

where

$$J = \int r^2 dA = 2\pi\int_0^l r^3\,dx,$$

but

$$dx = \frac{dr}{\sin\theta},$$

so that

$$J = \frac{2\pi}{\sin\theta}\int_{r_i}^{r_o} r^3\,dr = \frac{\pi}{2\sin\theta}\left(r_o^4 - r_i^4\right).$$

Therefore, the shear stress is

$$\tau = \frac{2Tr\sin\theta}{\pi\left(r_o^4 - r_i^4\right)}. \tag{16.72}$$

Also, $\sigma_n = 0$.

Example 16.11: Adhesive Scarf Joint

Given: A scarf joint in a hollow, round shaft rotates at 1200 rpm (125.7 rad/s). The shaft has an inner- to outer-radius ratio of 0.8, and the allowable shear stress is 30 MPa. The scarf joint has a 30° orientation with respect to the shaft axis.

Find: The required outer radius as well as the contact area needed to transmit 150 kW, using a safety factor of 2.0.

Solution: From Eq. (4.38), the torque is

$$T = \frac{h_p}{\omega} = \frac{150\times 10^3}{125.7} = 1194\text{ N-m}.$$

From Eq. (16.72), when $r = r_o$

$$\tau_{\max} = \frac{2Tr_o\sin\theta}{\pi\left(r_o^4 - r_i^4\right)} = \frac{2T\sin\theta}{\pi r_o^3\left[1-\left(\dfrac{r_i}{r_o}\right)^4\right]} = \frac{\tau_{\text{all}}}{n_s}.$$

Solving for the outer radius,

$$\begin{aligned}(r_o)_{\text{cr}} &= \left[\frac{2Tn_s\sin\theta}{\pi\tau_{\text{all}}\left(1-\left(\dfrac{r_i}{r_o}\right)^4\right)}\right]^{\frac{1}{3}} \\ &= \left[\frac{2(1194)(2)\sin 30^\circ}{\pi(30)(10^6)(1-0.8^4)}\right]^{\frac{1}{3}}\end{aligned}$$

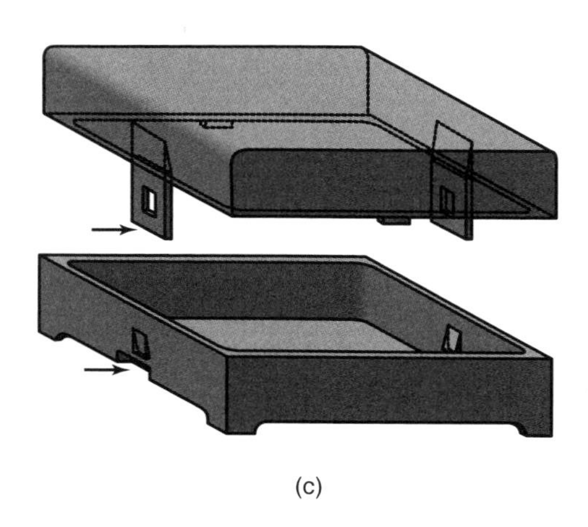

Figure 16.31: Common examples of integrated fasteners. (a) Module with four cantilever lugs; (b) cover with two cantilever and two rigid lugs; (c) separable snap joints for chassis cover.

or $(r_o)_{\text{cr}} = 0.035$ m or 3.5 cm. The normal area is

$$A_n = \pi\left(r_o^2 - r_i^2\right) = \pi r_o^2\left(1 - 0.8^2\right) = \pi 3.5^2\left(1 - 0.8^2\right),$$

so that $A_n = 13.85$ cm^2. The scarf area is

$$A = \frac{A_n}{\sin\theta} = \frac{13.85}{\sin 30^\circ} = 27.71 \text{ cm}^2.$$

16.8 Integrated Snap Fasteners

Snap fasteners are extremely popular when integrated into machine components because they greatly simplify assembly. Figure 16.31 shows common integrated snap fasteners. Integrated snap fasteners are commonly used with plastic housings or components, since they can be molded directly into the elements. This is one of the most powerful mechanisms for simplifying assemblies and reducing component and manufacturing costs.

Analyzing snap fasteners is complicated by the highly nonlinear elastic behavior of the polymers used. While there are many forms of integrated snap fasteners, this section emphasizes a cantilever design as depicted in Fig. 16.32.

The maximum deflection for the cantilever shown in Fig. 16.32 is given by:

$$y = A\frac{\epsilon l^2}{h}. \tag{16.73}$$

where A is a constant for given cantilever shape, as given in Fig. 16.33, ϵ is the allowable strain for the polymer, l is the

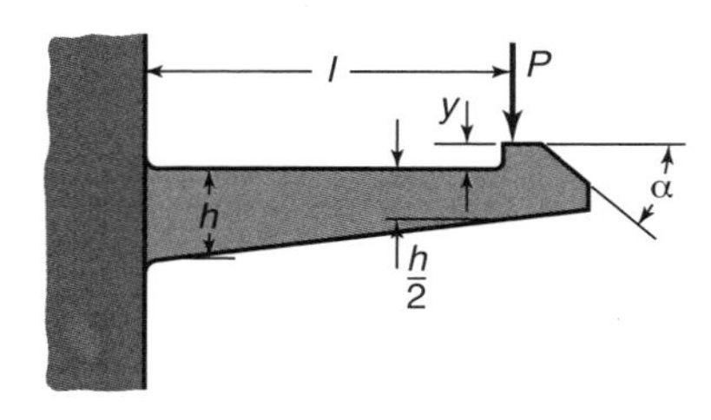

Figure 16.32: Cantilever snap joint.

	Cross section	
	Rectangle	Trapezoid
Constant cross section	0.67	$\frac{a+b}{2a+b}$
Tapered thickness	1.09	$1.64\,\frac{a+b}{2a+b}$
Tapered width	0.86	$1.28\,\frac{a+b}{2a+b}$

Figure 16.33: Shape constant, A, used to obtain deflection of snap fastener cantilevers.

cantilever length, and h is the maximum cantilever thickness. For a rectangular cross section, the force required to obtain this deformation is

$$P = \frac{bh^2}{6}\frac{E_s\epsilon}{l}, \tag{16.74}$$

where b is the cantilever width and E_s is the secant modulus of elasticity. During assembly, however, the mating force that must be overcome is directed along the cantilever axis and also includes frictional effects. The mating force is given by

$$W = P\frac{\mu + \tan\alpha}{1 - \mu\tan\alpha}, \tag{16.75}$$

where μ is the coefficient of friction and α is the wedge angle of the cantilever tip. Table 16.15 gives some typical friction coefficients for polymers used in integrated snap fasteners.

Table 16.15: Coefficients of friction for common snap fastener polymers.

Material	Coefficient of friction On steel	On self-mated polymer
Polytetrafluoroethylene (PTFE or *teflon*)	0.12-0.22	—
Polyethylene (rigid)	0.20–0.25	0.40–0.50
Polyethylene (flexible)	0.55–0.60	0.66–0.72
Polypropylene	0.25–0.30	0.38–0.45
Polymethylmethacrylate (PMMA)	0.50–0.60	0.60–0.72
Acrylonitrile-butadiene-styrene (ABS)	0.50–0.65	0.60–0.78
Polyvinylchloride (PVC)	0.55–0.60	0.55–0.60
Polystyrene	0.40–0.50	0.48–0.60
Polycarbonate	0.45–0.55	0.54–0.66

Design Procedure 16.2: Design of Integrated Snap Fasteners

The following considerations should be followed when designing cantilever-type integrated snap fasteners:

1. $\frac{l}{h} \geq 5$. If this is not the case, deformations will not be restricted to the cantilever, and the beam stiffness can be much larger than expected.
2. A secant modulus must be used because the polymers are usually highly nonlinear. The secant modulus is the slope of the straight line drawn from zero strain to ϵ on the stress-strain curve.
3. $\frac{l}{b} \geq 5$. If the beam is wider, it approaches a plate configuration and will be much stiffer due to the Poisson effect.
4. Snaps should be located away from stress risers such as sharp corners and manufacturing complications such as mold gates or weld lines.

Example 16.12: Integrated Snap Fastener

Given: A rectangular cross section cantilever snap-fit must be designed with a constant decrease in thickness from h at the root to $h/2$ at the end of the hook. The undercut dimension serves as the maximum deformation needed for assembly and is decided based on disassembly force requirements. The material is polycarbonate on polycarbonate, length l = 19 mm, width b = 9.5 mm, angle of inclination α = 30°, undercut y = 2.4 mm, secant modulus of elasticity E_s = 1.815 GPa.

Find:

(*a*) The thickness, h, at which full deflection causes a strain of 2%.

(*b*) The associated force, P, that causes this deflection.

(*c*) The resultant mating force, W.

Figure 16.34: Photograph of a jet engine highlighting the blisk. (Shutterstock)

Solution:

(*a*) From Fig. 16.33, $A = 1.09$. Therefore, from Eq. (16.73), and a maximum strain of $\epsilon = 2\% = 0.02$,

$$y = 1.09\frac{\epsilon l^2}{h}.$$

Solving for h,

$$h = 1.09\frac{\epsilon l^2}{y} = 1.09\frac{(0.02)(0.019)^2}{0.0024} = 0.00328 \text{ m}.$$

(*b*) From Eq. (16.74),

$$\begin{aligned} P &= \frac{bh^2}{6}\frac{E_s\epsilon}{l} \\ &= \frac{(0.0095)(0.00328)^2}{6}\frac{(1.815 \times 10^9)(0.02)}{0.019} \\ &= 32.5 \text{ N}. \end{aligned}$$

(*c*) For polycarbonate from Table 16.15, an average coefficient of friction is $\mu = 0.60$. Using this value and the given value of $\alpha = 30°$ in Eq. (16.75) yield the mating force of

$$\begin{aligned} W &= P\frac{\mu + \tan\alpha}{1 - \mu\tan\alpha} \\ &= (32.5 \text{ N})\left(\frac{0.60 + \tan 30°}{1 - 0.60\tan 30°}\right) \\ &= 58.54 \text{ N}. \end{aligned}$$

Case Study 16.1: Blanes and Blisks

Titanium alloy bladed vanes (or *blanes*) and bladed disks (*blisks*) are integral components of modern jet engines. Figure 16.34 shows a typical jet engine with casings removed to highlight the blisks; Fig. 16.35 shows details of a typical blisk. Note that there are many blades mounted in close proximity to each other, and that very strict tolerances must be maintained for operating efficiency. Further, the environment in a jet engine is very demanding; temperatures can easily exceed

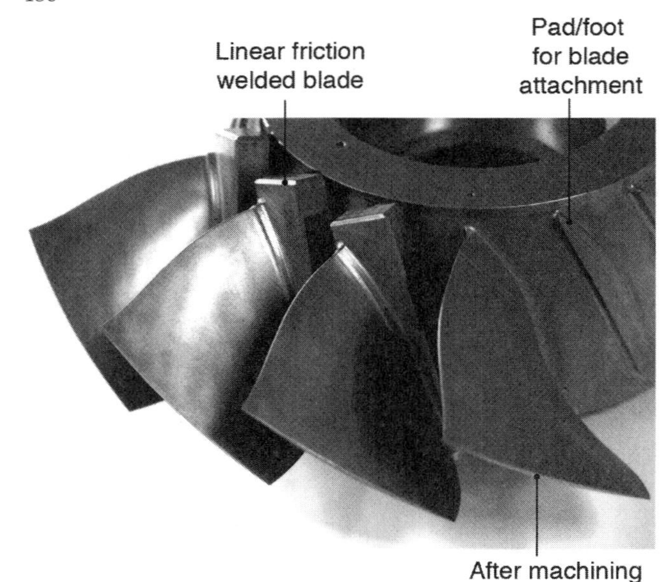

Figure 16.35: Detailed view of blades that have been attached to a compressor disk through linear friction welding. During linear friction welding, the parts encounter plastic deformation; the block shown is removed by machining. *Source:* Courtesy of ACB Presses UK.

1000°C, and loadings are unsteady, so that fatigue is an issue.

Blanes and blisks traditionally required skilled machinists to attach blades to a central hub using mechanical fasteners. However, this approach was time-consuming and expensive, and the quality of the product was difficult to control. Beginning in the 1990s, laser welding started to be used to fasten blades to disks, and significant economic and performance improvements resulted. However, blade failures in the heat affected zone still occurred.

Linear friction welding has the advantages of producing a high strength, compact weld without a heat affected zone. It is also a process that is suitable for the demanding materials involved in aircraft jet engines. Blanes and blisks started appearing in aerospace applications in 2001, and have seen steadily increasing use ever since. Linear friction welding involves reciprocating sliding motion under controlled pressure; as the temperature increases, the load is also increased, eventually causing plastic deformation at the interface between parts to be joined. This deformation removes surface oxides and other defects from the joint. When the desired deformation is achieved, the relative motion between parts is stopped, resulting in a strong diffusion-based joint. Since the part cools fairly quickly, the joint is cold worked and has an advantageous microstructure for fatigue resistance.

As can be seen in Fig. 16.35, the blades are produced with a relatively large block, and the disks have a prepared shoe or pad for the blades. After welding, the block needs to be removed by machining resulting in the high-quality blisks required in modern aircraft. The blades are attached with a more fatigue-resistant weld zone without a heat affected zone, and the blanes and blisks are more reliable as a result.

16.9 Summary

Members can be joined by using a number of different techniques. This chapter investigated threaded, riveted, welded, and adhesive joining of members. Power screws were also covered. A power screw is a device used to change angular motion into linear motion and to transmit power. A general analysis of the forces and torques required to raise and lower a load was presented. It was assumed that an Acme thread is specified for power screws. Collar friction was also considered. Power and efficiency of power screws, including the effects of self-locking, were discussed.

The types and classes of thread were given for threaded fasteners. Pretension plays a major role in the successful design of threaded connections. Analysis of the bolted joint required determination of the bolt and joint stiffness. Different approaches for joint stiffness approximation are available, all of which define a stressed volume in the compressed members. Once the preload and external loading is known, the design stresses imposed by the threaded fastener can be determined. The allowable stress and safety factor for threaded fasteners were obtained by considering the proof stress and proof load. Both static and dynamic considerations were made and gasketed joints were covered.

Riveted and threaded fasteners in shear are treated alike in design and failure analysis. Four modes of failure were presented: bending of a member, shear of a rivet, tensile failure of a member, and compressive bearing failure.

The design of welded fasteners was also considered. Welding processes are available in a wide variety of types, some of which produce beads, others spot welds, and others that join two materials along an area. The types of defects encountered in welded connections and the importance of the heat affected zone were discussed. Each type of weld was analyzed. Fillet welds were analyzed using unit moment of inertia and unit polar moment of inertia, which have been derived for common weld geometries. For axial loading when the fillet weld is aligned parallel to the load, the shear stress occurs along the throat of the fillet parallel to the load. The load-and-stress analysis occurs along the critical section, which is the plane created by the throat dimension extending the length of the weld. Failure could be predicted for various types of loading while also considering the strength of the weld. Fatigue of welds was discussed, with a recognition that fatigue failures commonly occur in the heat affected zone adjacent to weld fillets.

The last two fastener methods to be considered were adhesive bonding and snap fasteners. Bonded joints, like those produced by area welding operations like friction welding, rely on adhesive forces to develop joint strength. The common types of adhesively bonded joints were described, and the common lap and scarf joints were analyzed. In addition, integrated snap fasteners were examined.

Key Words

adhesive bonding process of joining materials chemically through formation of interatomic or intermolecular bonds

bolt externally threaded fastener intended to be used with nut

cap screw externally threaded fastener intended to be used with internally threaded hole

coarseness quality and threads per inch of thread profiles on bolt, nut, or screw

gasket compliant member intended to provide seal

heat affected zone (HAZ) the region adjacent to a weld fillet where the substrate microstructure has been affected by temperature

lead distance that screw would advance in one revolution

nut internally threaded mating member for bolt

pitch distance from point on one thread to same point on adjacent thread

power screws power transmission device using mechanical advantage of threads to apply large loads

proof load maximum load that bolt can withstand without acquiring permanent set

proof strength limiting value of stress determined by proof load and tensile stress area

rivets fasteners that function through mechanical interference, usually by upsetting one end of rivet extending outside of free hole

self-locking power screw where thread friction is high enough to prevent loads from lowering in absence of externally applied torque

snap fasteners integrated fasteners that operate through elastic deformation and recovery of part of structure after insertion into proper retainer

stud externally threaded member with threads on both ends in lieu of cap

thread series standardized thread profile, either UN (unified) or M (metric)

throat thinnest section of weld

weld junction formed through diffusion of two materials to be joined, combined with optional filler material

Summary of Equations

Thread Geometry:

Pitch: $p = 1/n$

M and UN profiles: $\beta = 60°$, $h_t = \dfrac{0.5p}{\tan 30°}$

Acme profile: $\beta = 29°$, $h_t = 0.5p + 0.01$ in.

Lead angle: $\alpha = \tan^{-1}\left(\dfrac{l}{\pi d_p}\right)$

Thread length (metric threads):

$$L_{\text{tot}} = \begin{cases} 2d_c + 6 & L \le 125 \\ 2d_c + 12 & 125 < L \le 200 \\ 2d_c + 25 & L > 200 \end{cases}$$

Tensile stress area:

UN profiles: $A_t = (0.7854)\left(d_c - \dfrac{0.9743}{n}\right)^2$

M profiles: $A_t = (0.7854)(d_c - 0.9382p)^2$

Power Screws:

Raising the load:

$$T_r = W\left[\frac{(d_p/2)(\cos\theta_n \tan\alpha + \mu)}{\cos\theta_n - \mu\tan\alpha} + r_c\mu_c\right]$$

Lowering the load:

$$T_l = -W\left[\frac{(d_p/2)(\mu - \cos\theta_n \tan\alpha)}{\cos\theta_n + \mu\tan\alpha} + r_c\mu_c\right]$$

Efficiency: $e = \dfrac{100Wl}{2\pi T} = \dfrac{(100)l}{2\pi(d_p/2)\tan\alpha}$

Self-locking if $\mu > \dfrac{l\cos\theta_n}{\pi d_p}$

Bolted Connections:

Stiffness or joint parameter: $C_k = \dfrac{k_b}{k_b + k_f}$

Bolt stiffness: $\dfrac{1}{k_b} = \dfrac{4}{\pi E}\left[\dfrac{L_s + 0.4d_c}{d_c^2} + \dfrac{L_t + 0.4d_r}{d_r^2}\right]$

Joint stiffness using frusta:

$$k_{ji} = \frac{\pi E_j d_c \tan\alpha_f}{\ln\left[\dfrac{(2L_i\tan\alpha_f + d_i - d_c)(d_i + d_c)}{(2L_i\tan\alpha_f + d_i + d_c)(d_i - d_c)}\right]}$$

where $\dfrac{1}{k_j} = \dfrac{1}{k_{j1}} + \dfrac{1}{k_{j2}} + \dfrac{1}{k_{j3}} + \dots$

Joint stiffness by Wileman method: $k_j = E_i d_c A_i e^{B_i d_c/L}$

Failure modes:

Bolt overload: $n_{sb} = \dfrac{A_t S_p - P_i}{P_{\max,b} C_k}$

Joint separation: $n_{sj} = \dfrac{P_i}{P_{\max,j}(1 - C_k)}$

Fatigue: $n_s = \dfrac{S_{ut} - \sigma_i}{C_k\left[K_f\sigma_a\left(\dfrac{S_{ut}}{S_e}\right) + \sigma_m\right]}$

Rivets:

Bending of a member: $\sigma = \dfrac{PL_g}{2Z_m} < 0.6\,(S_y)_j$

Shear of rivets: $\tau = \dfrac{4P}{\pi d_c^2} < S_{sy}$

Tensile failure of member (without stress concentration factor):

$$\sigma = \frac{P}{(b - N - rd_c)t_m} < (S_y)_j$$

Compressive bearing failure: $\sigma = \dfrac{P}{d_c t_m} < 0.9\,(S_y)_j$

Note: also can apply to bolted connections.

Welds:

Parallel and transverse loading: $\tau = \dfrac{P}{t_c L_w} < (S_{sy})_{\text{weld}}$

Torsional loading: $\tau_t = \dfrac{Tr}{J}$, where $J = t_e J_u$

Bending: $\sigma = \dfrac{Mc}{I}$, where $I = t_c I_u L_w$

Integrated Snap Fasteners:

Maximum deflection: $y = A\dfrac{\epsilon l^2}{h}$

Force to achieve maximum deflection: $P = \dfrac{bh^2}{6}\dfrac{E_s\epsilon}{l}$

Assembly force: $W = P\dfrac{\mu + \tan\alpha}{1 - \mu\tan\alpha}$

Recommended Readings

Adams, R.D. (2005) *Adhesive Bonding: Science, Technology and Applications*, CRC Press.

ASM Handbook Volume 6: *Welding, Brazing, and Soldering* (1993) American Society for Metals.

AWS Welding Handbook, 9th ed., Volume 1, *Welding Science and Technology* (2011) American Welding Society.

Bickford, J.H., and Nassar, S. (1998) *Handbook of Bolts and Bolted Joints*. Marcel Dekker.

Budynas, R.G., and Nisbett, J.K. (2011), *Shigley's Mechanical Engineering Design*, 9th ed., McGraw-Hill.

Chapman, W.W. (2004) *Guide to Threads, Threading and Threaded Fasteners*, Hanser Gardner.

Jacobson, D.M. (2005) *Principles of Brazing*, American Society for Metals.

Juvinall, R.C., and Marshek, K.M. (2012) *Fundamentals of Machine Component Design*, 5th ed., Wiley.

Mott, R. L. (2014) *Machine Elements in Mechanical Design*, 5th ed., Pearson.

Norton, R.L. (2011) *Machine Design*, 4th ed., Prentice Hall.
Parmly, R.O. (1996) *Standard Handbook of Fastening and Joining*, 3rd ed., McGraw-Hill.
Petrie, E. (2006) *Handbook of Adhesives and Sealants*, 2nd ed., McGraw-Hill.

References

Bickford, J.H. (1995) *An Introduction to the Design and Behavior of Bolted Joints*, 3rd ed., Marcel Dekker.
Kalpakjian, S., and Schmid, S.R. (2010) *Manufacturing Engineering and Technology*, 6th ed. Prentice Hall.
Mott, R. L. (2014) *Machine Elements in Mechanical Design*, 5th ed., Pearson.
Shigley, J.E., and Mischke, C.R. (1989) *Mechanical Engineering Design*, 5th ed., McGraw-Hill.
Wileman, J., Choudhury, M., and Green, I. (1991) "Computation of Member Stiffness in Bolted Connections," *J. Machine Design*, ASME, vol. 113, pp. 432–437.

Questions

16.1 Define *pitch* as it relates to threads. Explain with an appropriate sketch why the pitch of a thread must match that of a mating nut.

16.2 What is the difference between a UNF and a UNC thread?

16.3 What is a power screw? What are power screws used for?

16.4 Is the lead the same as the pitch? Why or why not?

16.5 What is the difference between a bolt, a screw and a stud?

16.6 What is the significance of the joint parameter, C_k?

16.7 Define the term *proof strength*. How is this different from the *grade* of a bolt?

16.8 Why do rolled threads result in a lower stress concentration than cut threads (see Table 16.10)? Why does the stress concentration depend on bolt grade?

16.9 What is a gasket? When is a gasket used? Does a gasket always reduce the stiffness of the bolted members? Explain.

16.10 What is the difference between a rivet and a bolt? What are the advantages of rivets?

16.11 Describe the common types of discontinuities in welded joints.

16.12 What is the difference between brazing and soldering?

16.13 What is a HAZ? Why is it important?

16.14 Is there a HAZ in brazing or soldering? Explain.

16.15 What is a *scarf* joint? How is it different than a *lap* joint?

16.16 What are the advantages of integrated snap fasteners?

16.17 What are the advantages of adhesives?

16.18 Why are breakaway bolts used?

16.19 Do all welding processes involve high temperatures and melting? Explain.

16.20 Explain how a rivet can be assembled from the blind side.

Qualitative Problems

16.21 Can the pitch of a thread vary along a fastener's length? Explain your answer.

16.22 Without the use of equations, explain the reasons that the torque required to raise a load on a power screw is larger than the torque needed to lower a load.

16.23 Define *self-locking*. Give two applications where self-locking threads are advantageous.

16.24 Review Table 16.3 and sketch the mating hole required for full functionality for the following bolt types: (a) hex-head bolt; (b) carriage bolt; (c) flat cap screw.

16.25 Describe the advantages of using a castle nut as shown in Table 16.4.

16.26 Explain why it is difficult to precisely attain desired level of preload in threaded fasteners.

16.27 Is a high preload advantageous for fatigue applications of threaded connections? Is a high preload desirable for static loading? Explain your answers.

16.28 Give three applications where adhesive bonding is the best joining method.

16.29 Rate lap, butt and scarf joints in terms of joint strength. Explain your answers.

16.30 List the advantages and disadvantages of welding.

16.31 What are the advantages of electron-beam and laser-beam welding as compared to arc welding?

16.32 List and explain the factors that affect the strength of weld beads.

16.33 Explain why weld beads are often shot blasted or peened.

16.34 Is a high preload advisable for bolted connections that are loaded statically? What if the load is cyclic? Explain.

16.35 List the defects that can occur in fusion welding.

16.36 Sketch the Mohr's circle diagram for a bolt while it is being tightened. Sketch the Mohr's circle diagram for a tightened bolt that is loaded in (a) tension; (b) shear.

16.37 Make a sketch of the following bolts: (a) $\frac{3}{8} - -16$ UNC - $2\frac{1}{2}$; (b) M24 $\times$ 3 – 75; (c) a UNC bolt with crest diameter of 1 in. and length of 4 in.

16.38 Explain how you would determine the polar moment of inertia and the moment of inertia if a weldment consisted of segments of weld bead spaced by a short distance.

16.39 Is there any advantage to matching the stiffness of bolt and members? For example, is there an advantage in using an aluminum bolt to fasten aluminum members? Explain.

16.40 Explain the advantages and disadvantages of applying a lubricant to a bolt and nut before assembly.

Quantitative Problems

16.41 Compare the member stiffness of a bolted connection consisting of two 0.5-in.-thick steel members as a function of bolt crest diameter using (a) conical frusta with $\alpha_f = 45°$, (b) conical frusta with $\alpha_f = 30°$, and (c) the Wileman approach.

16.42 An Acme-threaded power screw with a crest diameter of 1.125 in. and single thread is used to raise a load of 20,000 lb. The collar mean diameter is 1.5 in. The coefficient of friction is 0.12 for both the thread and the collar. Determine the following:

(a) Pitch diameter of the screw. *Ans.* $d_p = 1.015$ in.

(b) Screw torque required to raise the load. *Ans.* $T_r = 3709$ in-lb.

(c) Maximum thread coefficient of friction allowed to prevent the screw from self-locking if collar friction is eliminated. *Ans.* $\mu = 0.0607$.

16.43 A car jack consists of a screw and a nut, so that the car is lifted by turning the screw. Calculate the torque needed to lift a load with a mass of 1000 kg if the lead of the thread is $l = 9$ mm, its pitch diameter is 22 mm, and its thread angle is 30°. The coefficient of friction is 0.10 in the threads and zero elsewhere. *Ans.* $T_r = 25.56$ Nm.

16.44 A power screw gives the axial tool motions in a numerically controlled lathe. To get high accuracy in the motions, the heating and power loss in the screw have to be low. Determine the power efficiency of the screw if the coefficient of friction is 0.08, pitch diameter is 30 mm, lead is 6 mm, and thread angle is 25°. *Ans.* $e = 43.5\%$.

16.45 To change its oil, an 18,000-lb truck is lifted a height of 5 ft by a screw jack. The power screw has Acme threads and a crest diameter of 5 in. with two threads per inch, and the lead equals the pitch. Calculate how much energy has been used to lift and lower the truck if the only friction is in the threads, where the coefficient of friction is 0.10. *Ans.* $E_r = 368$ kip-ft.

16.46 A single-threaded M32×3.5 power screw is used to raise a 10-kN load at a speed of 25 mm/s. The coefficients of friction are 0.08 for the thread and 0.12 for the collar. The collar mean diameter is 55 mm. Determine the power required. Also, determine how much power is needed to lower the load at 50 mm/s. *Ans.* $h_{pr} = 2290$ W, $h_{pl} = 3570$ W.

16.47 A double-threaded Acme power screw is used to raise a 1000-lb load. The outer diameter of the screw is 1.25 in. and the mean collar diameter is 2.0 in. The coefficients of friction are 0.13 for the thread and 0.16 for the collar. Determine the following:

(a) Required torque for raising and lowering the load. *Ans.* $T_r = 302.3$ in-lb, $T_l = 172.6$ in.-lb.

(b) Geometrical dimensions of the screw. *Ans.* $l = 0.4$ in., $d_p = 1.14$ in., $\alpha = 6.376°$, $\theta = 14.4°$.

(c) Efficiency in raising the load. *Ans.* $e = 21.06\%$.

(d) Load corresponding to the efficiency if the efficiency in raising the load is 18%. *Ans.* $W = 854.7$ lb.

16.48 Sketch a shows a stretching device for steel wires used to stabilize the mast of a sailing boat. Both front and side views are shown and all dimensions are in millimeters. A screw with square threads ($\beta = 0$), a lead and pitch of 4 mm, and an outer diameter of 20 mm is used. The screw can move axially but is prevented from rotating by flat guiding pins (side view in Sketch a). Derive an expression for the tightening torque as a function of the stretching force, P, when the coefficient of friction is 0.20 everywhere. Also, calculate the torque needed when the tightening force is 1000 N. *Ans.* For $P = 1000$ N, $T = 5.61$ N-m.

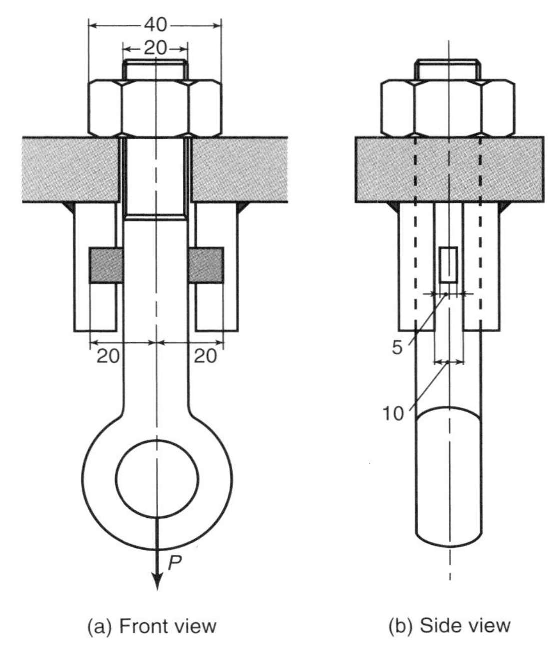

Sketch a, for Problem 16.48

16.49 A flywheel of a motorcycle is fastened by a thread manufactured directly in the center of the flywheel as shown in Sketch b. The flywheel is mounted by applying a torque T. The cone angle is γ. Calculate the tensile force W in the shaft between the contact line at N and the thread as a function of D_1, D_2, γ, and T. The lead angle is α at the mean diameter D_1. The shaft is assumed to be rigid.

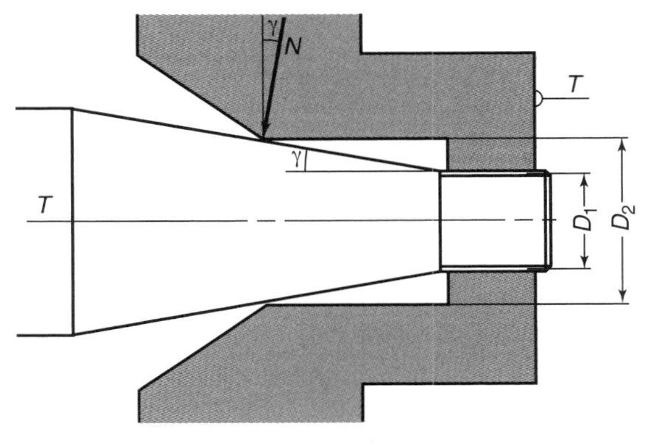

Sketch b, for Problem 16.49

16.50 A 25-kN load is raised by two Acme-threaded power screws with a minimum speed of 35 mm/s and a maximum power of 1750 W per screw. Because of space limitations, the screw diameter should not be larger than 45 mm. The coefficient of friction for both the thread and the collar is 0.09. The collar mean diameter is 65 mm. Assuming that the loads are distributed evenly on both sides, select the size of the screw to be used and calculate its efficiency.

16.51 The lead screw of a small lathe is made from a 1/2-in. crest diameter Acme threaded shaft. The lead screw has to exert a force on the lathe carriage for a number of operations, and it is powered by a belt drive from the motor. If a force of 500 lb is desired, what is the torque required if the collar diameter is twice the pitch diameter of the screw? Use $\mu = \mu_c = 0.15$. With what velocity does the lead screw move the crosshead if the lead screw is single-threaded and is driven at 250 rpm? *Ans.* $T = 58.3$ in-lb.

16.52 A screw with Acme thread can have more than one entrance to the thread per screw revolution. A single thread means that the pitch and the lead are equal, but for double and triple threads the lead is larger than the pitch. Determine the relationship between the number of threads per inch n, the pitch p, and the lead l.

16.53 A section of a bolt circle on a large coupling is shown in Sketch *c*. Each bolt is loaded by a repeated force $P = 5000$ lb. The members are steel, and all bolts have been carefully preloaded to $P_i = 25,000$ lb each. The bolt is to be an SAE Grade 5, 0.75-in. crest diameter with fine threads, (so that $d_r = 0.674$ in.) and the nut fits on this bolt has a thickness of 0.50 in. The threads have been manufactured through rolling, and use a survival probability of 90%.

(a) If hardened steel washers 0.134-in. thick are to be used under the bolt and nut, what length of bolts should be used? *Ans.* $L = 2.5$ in.

(b) Find the stiffness of the bolt, the members, and the joint constant. *Ans.* $k_b = 5.798 \times 10^6$ lb/in., $k_j = 24.25$ Mlb/in. (Wileman).

(c) What is the factor of safety guarding against a fatigue failure? *Ans.* $n_s = 4.78$

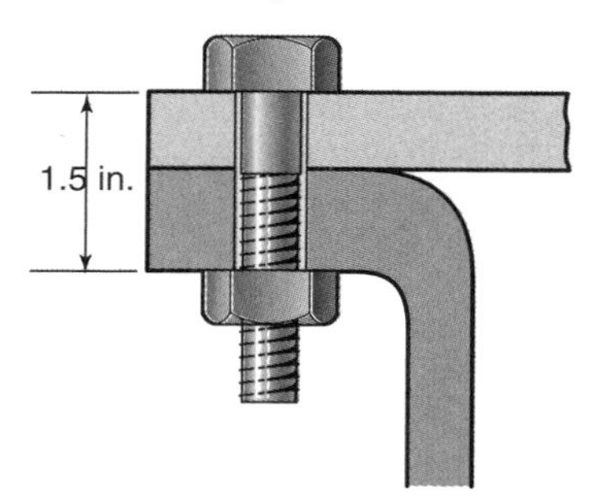

Sketch *c*, for Problem 16.53

16.54 An M12, coarse-pitch, Class-5.8 bolt with a hexagonal nut assembly is used to keep two machine parts together as shown in Sketch *d*. Determine the following:

(a) Bolt stiffness and clamped member stiffness. *Ans.* $k_b = 297.5$ MN/m, $k_j = 769.7$ MN/m (Wileman).

(b) Maximum external load that the assembly can support for a load safety factor of 2.5. *Ans.* $P_{\max,b} = 11.5$ kN.

(c) Safety factor guarding against separation of the members. *Ans*, $n_s = 3.20$.

(d) Safety factor guarding against fatigue if a repeated external load of 10 kN is applied to the assembly. *Ans.* $n_s = 2.36$.

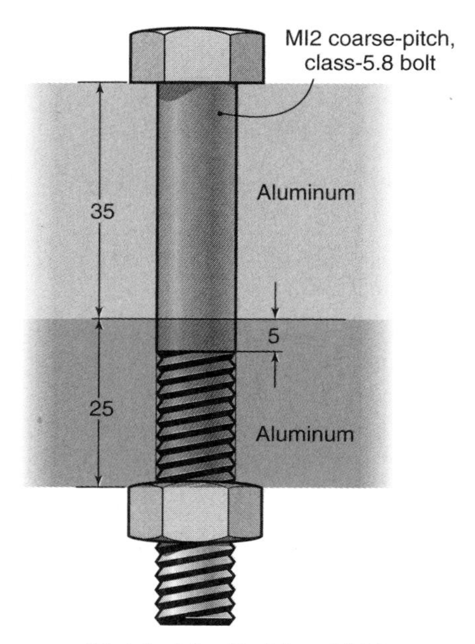

Sketch *d*, for Problem 16.54

16.55 Repeat Problem 16.54 if the 25-mm thick member is made of steel. *Ans.* $k_j = 998.7$ MN/m, $n_{sj} = 2.22$, $n_s = 2.81$.

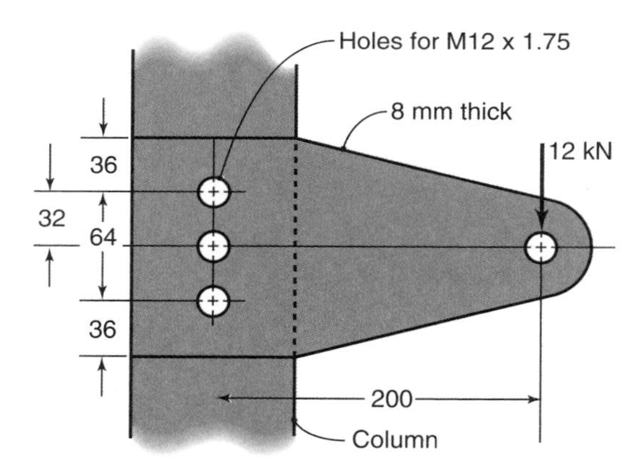

Sketch *e*, for Problem 16.56

16.56 Find the total shear load on each of the three bolts for the connection shown in Sketch *e*. Also, compute the shear stress and the bearing stress. Find the area moment of inertia for the 8-mm-thick plate on a section through the three bolt holes. *Ans.* $P_{\text{center}} = 4$ kN, $P_{\text{top}} = P_{\text{bottom}} = 37.7$ kN.

16.57 A coarse-pitch, SAE Grade-5 bolt with a hexagonal nut assembly is used to keep two machine parts together as shown in Sketch *f*. The major diameter of the bolt is 0.5 in. The bolt and the bottom member are made of carbon steel. Assume that the connection is to be reused. Length dimension is in inches. Determine the following:

(a) Length of the bolt. *Ans.* $L = 1.25$ in.

(b) Stiffnesses of the bolt and the member, assuming a washer ensures $d_i = 1.5d_c$ under the bolt and the washer. *Ans.* $k_b = 2.00$ Mlb/in, $k_j = 6.307$ Mlb/in.

(c) Safety factor guarding against separation of the members when the maximum external load is 5000 lb. *Ans.* $n_j = 2.97$.

(d) Safety factor guarding against fatigue if the repeated maximum external load is 2500 lb in a released-tension loading cycle. *Ans.* $n_s = 4.36$.

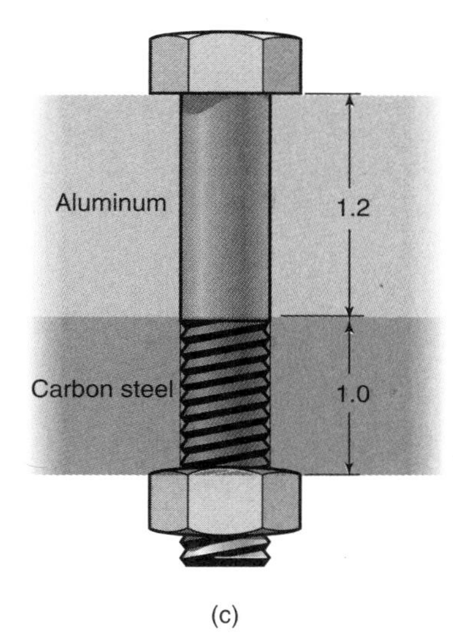

Sketch f, for Problem 16.57

16.58 An electric-motor-driven press (Sketch g) has the total press force $P = 5000$ lb. The screws are Acme type with $\beta = 29°$, $d_p = 3$ in., $p = l = 0.5$ in., and $\mu = 0.05$. The thrust bearings for the screws have $d_c = 5$ in. and $\mu_c = 0.06$. The motor speed is 2000 rpm, the total speed ratio is 75:1, and the mechanical efficiency $e = 0.95$. Calculate

(a) Press head speed. *Ans.* $v = 13.33$ in./min.

(b) Power rating needed for the motor. *Ans.* $h_p = 58.9$ in-lb/s.

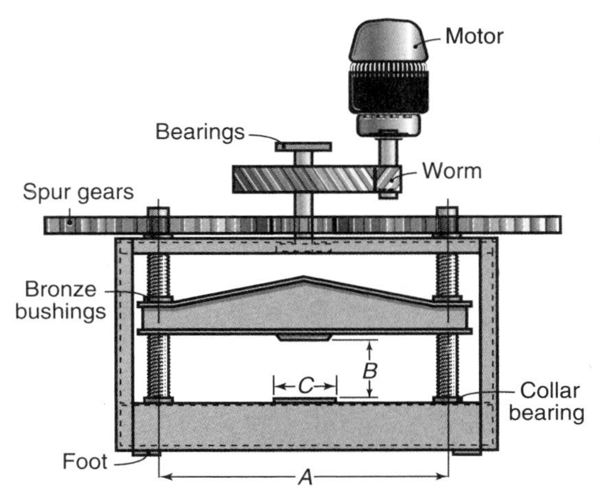

Sketch g, for Problem 16.58

16.59 A valve for high-pressure air is shown in Sketch h. The spindle has thread M12 with a pitch diameter of 10.9 mm, lead $l = 1.75$ mm, and a thread angle of 60°. Derive the relationship between torque and axial thrust force, and calculate the axial force against the seating when the applied torque is 15 N-m during tightening. The coefficient of friction is 0.15. *Ans.* $T_r = (0.00123\text{ m})P$.

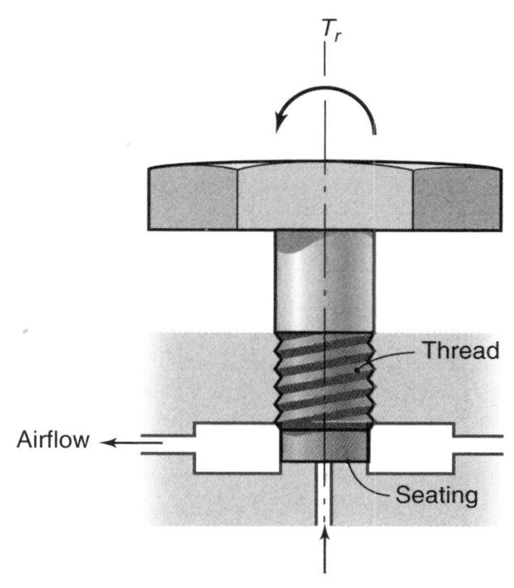

Sketch h, for Problem 16.59

16.60 Derive the expression for the power efficiency of a lead screw with a flat thread (thread angle $\beta = 0°$) and find the lead angle α that gives maximum efficiency in terms of the coefficient of friction. Also, give results if $\mu = 0.15$. *Ans.* $e_{\mu=0.15} = 74.2\%$.

16.61 A car manufacturer has problems with the cylinder head studs in a new high-power motor. After a relatively short time the studs crack just under the nuts, the soft cylinder head gasket blows out, and the motor stops. To be able to analyze the problem, the car manufacturer experimentally measures the stiffnesses of the various components. The stiffness for all bolts together is 400 N/μm, the stiffness of the gasket is 600 N/μm, and the stiffness of the cylinder head that compresses the gasket is 10,000 N/μm. By comparing the life-stress relationships with those for rolling-element bearings, the car manufacturer estimates that the stress amplitude in the screws needs to be halved to get sufficient life. How can that be done?

16.62 A pressure vessel is used as an accumulator to make it possible to use a small air compressor that works continuously. The stiffness parameter for the lid around each of the 10-mm diameter bolts is $k_j = 900$ MN/m. The shank length is 20 mm. Because the air consumption is uneven, the air pressure in the container varies between 0.2 and 0.8 MPa many times during a week. After 5 years of use one of the bolts holding down the top lid of the pressure vessel cracks off. A redesign is then made, decreasing the stress variation amplitude by 25%, to increase the life of the bolts to at least 50 years. The stress variation amplitude is decreased by lengthening the bolts and using circular tubes with the same cross-sectional area as the solid circular cross section of the bolt to transfer the compressive force from the bolt head to the lid. Calculate how long the tubes should be. *Ans.* $l = 7.089$ mm.

16.63 A loading hook of a crane is fastened to a block hanging in six steel wires. The hook and block are bolted together with four 10-mm-diameter screws prestressed to

20,000 N each. The shank length is 80 mm and the thread length is 5 mm. The stiffness of the material around each screw is 1 GN/m. One of the screws of the crane cracks due to fatigue after a couple of years of use. Will it help to change the screws to 12-mm diameter while other dimensions are unchanged if the stress variation needs to be decreased by at least 20%? *Ans.* No.

16.64 Depending on the roughness of the contacting surfaces of a bolted joint, some plastic deformation takes place on the tops of the roughness peaks when the joint is loaded. The rougher the surfaces are, the more pressure in the bolted joint is lost by plastic deformation. For a roughness profile depth of 20 μm on each of the surfaces, a plastic deformation of 6.5 μm can be expected for the two surfaces in contact. For a bolt-and-nut assembly as shown in Fig. 16.13, three sets of two surfaces are in contact. The stiffness of the two steel plates together is 700 MN/m when each is 40 mm thick. The bolt diameter is 16 mm with metric thread. The shank length is 70 mm. The bolt is prestressed to 20 kN before plastic deformation sets in. Calculate how much of the prestress is left after the asperities have deformed. *Ans.* $F_p = 14.8$ kN.

16.65 An ISO M12 $\times$ 1.75 class = 12.9 bolt is used to fasten three members as shown in Sketch i. The first member is made of cast iron, the second is low-carbon steel, and the third is aluminum. The static loading safety factor is 3.0. Dimensions are in millimeters. Determine

(a) Total bolt length, threaded length, and threaded length in the joint. *Ans.* $L = 80$ mm.

(b) Bolt-and-joint stiffness using a 30° cone. *Ans.* $k_b = 286.8$ MN/m, $k_j = 928$ MN/m.

(c) Preload for permanent connections. *Ans.* $P_i = 73.6$ kN.

(d) Maximum static load that the bolt can support. *Ans.* $P_{\max,b} = 11.54$ kN.

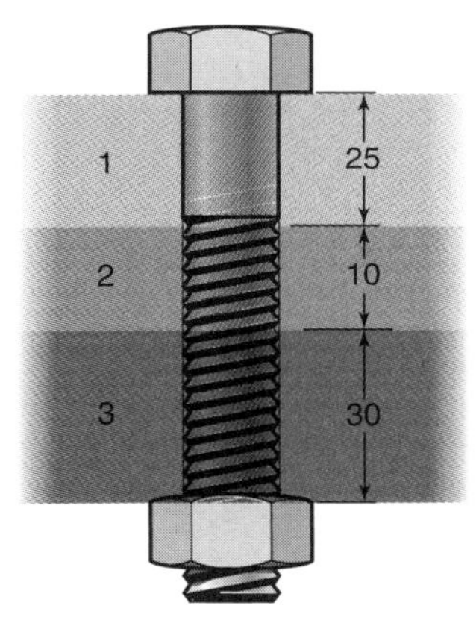

Sketch i, for Problem 16.65

16.66 A pressurized cast iron cylinder shown in Sketch j is used to hold pressurized gas at a static pressure of 10 MPa. The cylinder is joined to a low-carbon-steel cylinder head by bolted joints. Dimensions are in millimeters. Determine the required number of bolts. Use Grade 12.9 bolts, with M36 $\times$ 100 mm coarse threads, and a safety factor of 2. *Ans.* $L_t = 78$ mm, $d_r = 31.67$ mm, $n = 39$ bolts.

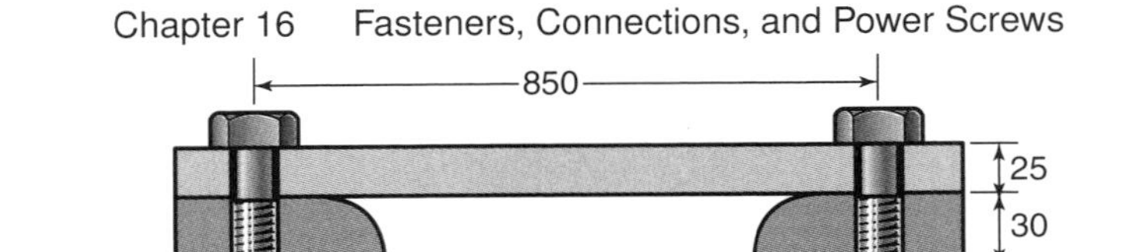

Sketch j, for Problem 16.66

16.67 In the bolted joint shown in Sketch k the first member is made of low-carbon steel, the second member is aluminum, and the third member is cast iron. Assuming that the members can be rearranged and the frustum cone angle is 45°, find the arrangement that can support the maximum load. Dimensions are in millimeters.

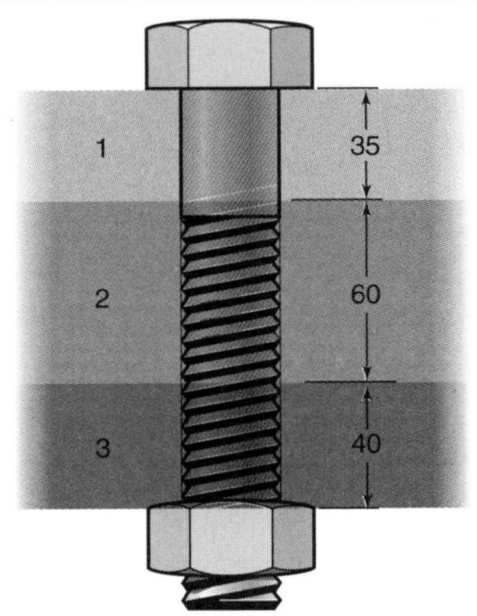

Sketch k, for Problem 16.67

16.68 The flange of a ship's propeller shaft is riveted in the radial direction against the hollow shaft. The outside diameter is 1 m and there are 180 rivets around the circumference, each with a diameter of 25 mm. The rivets are made of AISI 1020 steel (quenched and tempered at 870°C) and placed in three rows. Calculate the maximum allowable propeller torque transmitted through the rivets for a safety factor of 5. *Ans.* $T = 1.043 \times 10^6$ N-m.

16.69 A steel plate (Sketch l) is riveted to a vertical pillar. The three rivets have a 5/8-in. diameter and carry the load and moment resulting from the external load of 1950 lb. All length dimensions are in inches. The yield strengths of the materials are $S_{y,rivet} = 85,000$ psi and $S_{y,plate} = 50,000$ psi. Calculate the safety factors for

(a) Shear of rivet when $S_{sy} = 0.5S_y$. *Ans.* $n_s = 1.86$.

(b) Bearing of rivet. *Ans.* $n_s = 4.74$.

(c) Bearing of plate. *Ans.* $n_s = 2.79$.

(d) Bending of plate. *Ans.* $n_s = 4.16$.

Sketch l, for Problem 16.69

16.70 The cylinder shown in Sketch m is pressurized up to 2 MPa and is connected to the cylinder head by 32 M24 $\times$ 3 metric Grade 8.8 bolts. The bolts are evenly spaced around the perimeters of the two circles with diameters of 1.2 and 1.5 m, respectively. The cylinder is made of cast iron and its head is made of high-carbon steel. Assume that the force in each bolt is inversely related to its radial distance from the center of the cylinder head. Calculate the safety factor guarding against failure due to static loading. *Ans.* $n_s = 4.02$.

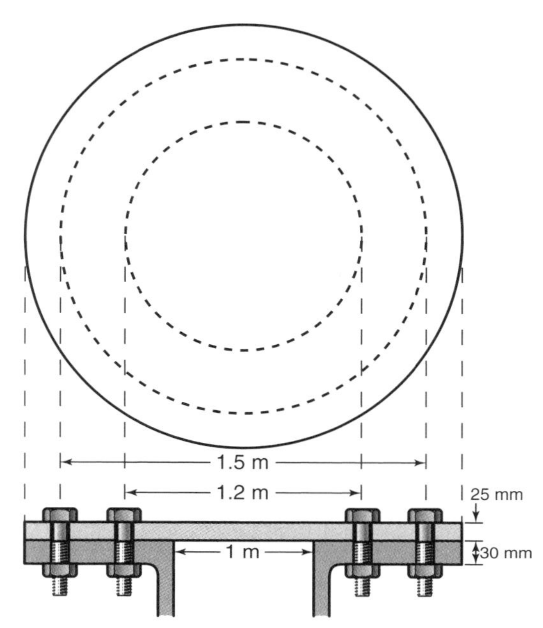

Sketch m, for Problem 16.70

Calculate the safety factors for

(a) Shear of rivet when $S_{sy} = 0.57S_y$
(b) Bearing of rivet
(c) Bearing of plate
(d) Bending of plate

16.71 An extruded beam as shown in Sketch n is welded to a thick column using shielded metal arc welding and an E60XX series electrode. Select the weld size for a safety factor of 3.0, assuming the weld is placed around the entire periphery, and indicate your recommendation using standard weld symbols. *Ans.* $h_e = 3$ mm.

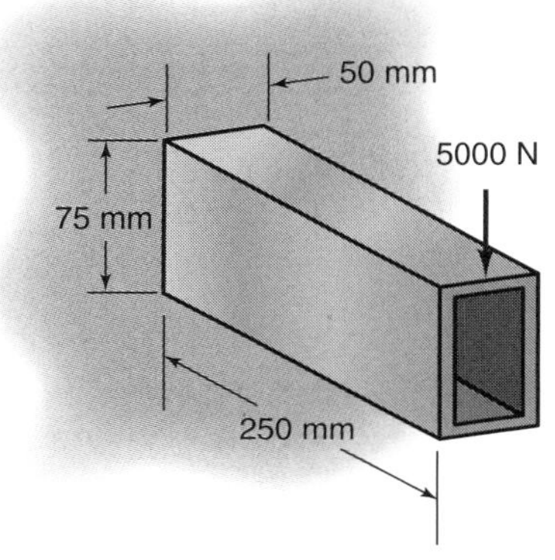

Sketch n, for Problem 16.71

16.72 For the beam of Problem 16.71, assume that instead of a weld around the entire periphery, two parallel lines will be used. Should the welds be placed on the horizontal or vertical surfaces? To compare the geometries, determine which would require a larger volume of electrode to achieve the same safety factor. Make a sketch of your recommendation using standard weld symbols. *Ans.* $h_e = 5$ mm.

16.73 The bracket shown in Sketch o is attached to a beam using 15-mm-diameter rivets, whose allowable shear stress is 500 MPa. The bracket will support a mainly vertical load, but the direction of the load could be as much as $\pm 30^\circ$ from the vertical as shown. Determine the safety factor based on shear of the rivets. *Ans.* $n_s = 1.47$.

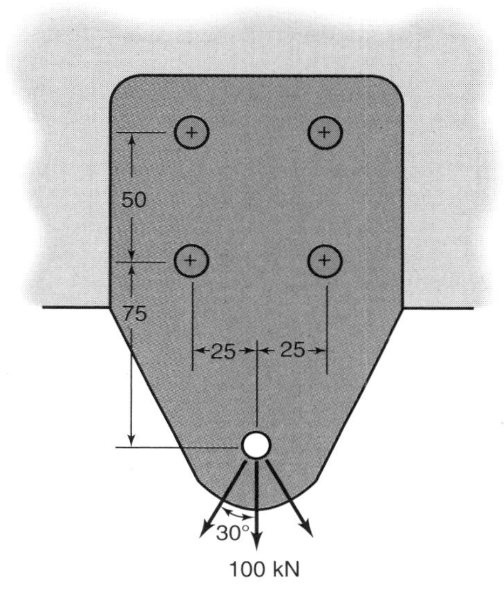

Sketch o, for Problem 16.73

16.74 A rectangular steel plate is connected with rivets to a steel beam as shown in Sketch p. Assume the steel to be low-carbon steel. The rivets have a yield strength of 600 MPa. A load of 24 kN is applied. For a safety factor of 3, calculate the diameter of the rivets. *Ans.* $d = 15$ mm.

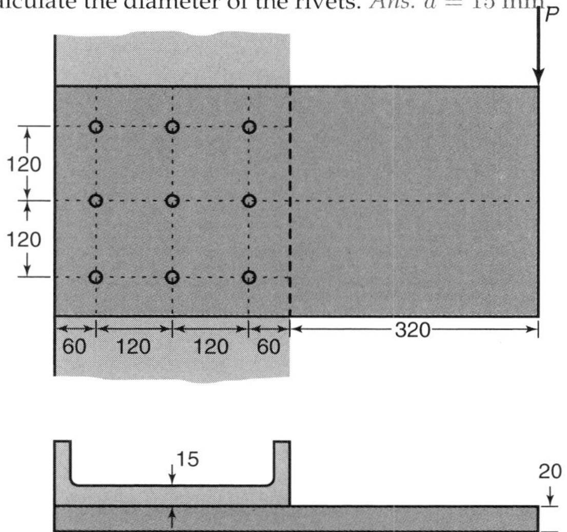

Sketch p, for Problem 16.74

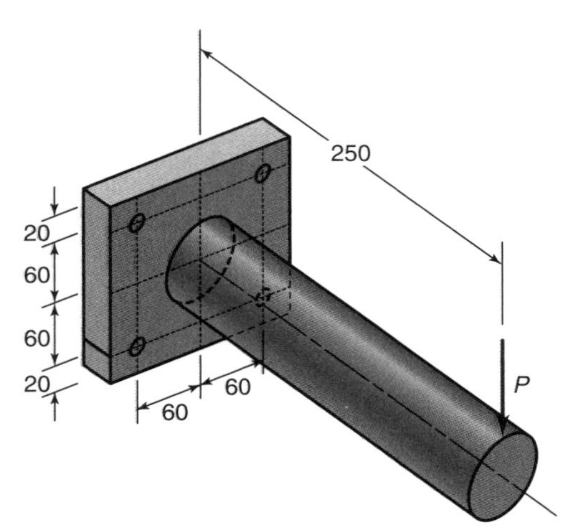

Sketch q, for Problem 16.75

16.75 Repeat Problem 16.74 but with the plate and beam shown in Sketch q. *Ans.* $d = 13$ mm.

16.76 The steel plate shown in Sketch r is welded against a wall. Length dimensions are in inches. The vertical load $W = 4000$ lb acts 6.8 in. from the left weld. Both welds are made by AWS electrode number E8000. Calculate the weld size for a safety factor of $n_s = 5.0$. *Ans.* $h_e = 0.5$ in.

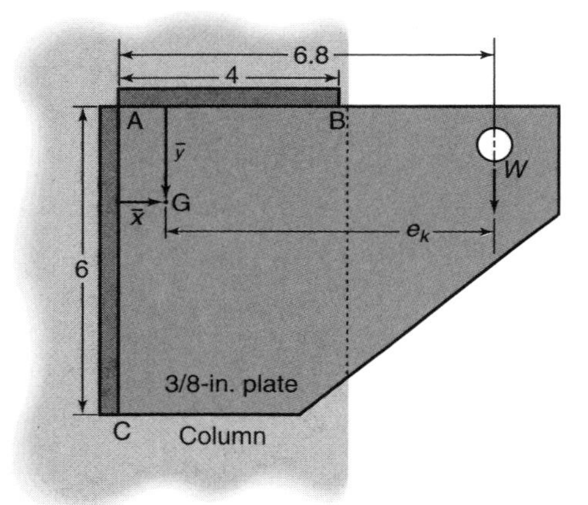

Sketch r, for Problem 16.76

16.77 Determine the weld size required if only the top (AB) portion is welded in Problem 16.76. *Ans.* $h_e = 2.5$ in.

16.78 Two medium-carbon steel (annealed AISI 1040) plates are attached by parallel-loaded fillet welds as shown in Sketch s. E60 series welding rods are used. Each of the welds is 3 in. long. What minimum leg length must be used if a load of 4000 lb is to be applied with a safety factor of 3.5? *Ans.* $h_e = 3/16$ in.

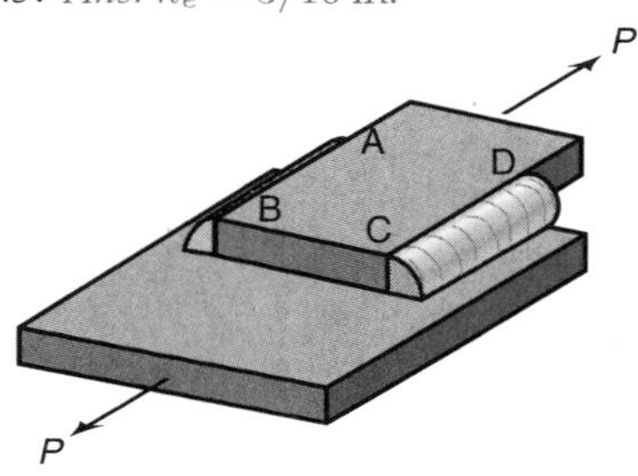

Sketch s, for Problem 16.78

16.79 The universal joint on a car axle is welded to the 60-mm-outside-diameter tube and should be able to transfer 1500 N-m of torque from the gearbox to the rear axle. Calculate how large the weld leg should be to give a safety factor of 10 if the weld metal is of class E70. *Ans.* $h_e = 3$ mm.

16.80 The steel bar shown in Sketch t is welded by an E60XX electrode to the wall. A 1000-N load is applied in the y-direction at the end of the bar. Calculate the safety factor against yielding. Also, would the safety factor change if the direction of load P is changed to the z-direction? Dimensions are in millimeters. *Ans.* $n_s = 7.14$.

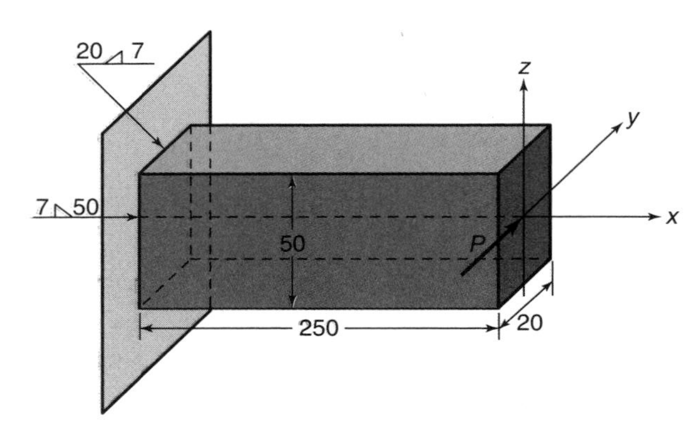

Sketch t, for Problem 16.80

16.81 The bar shown in Sketch u is welded to the wall by AWS electrodes. A 10-kN load is applied at the top of the bar. Dimensions are in millimeters. For a safety factor of 3.0 against yielding, determine the electrode number that must be used and the weld throat length, which should not exceed 10 mm.

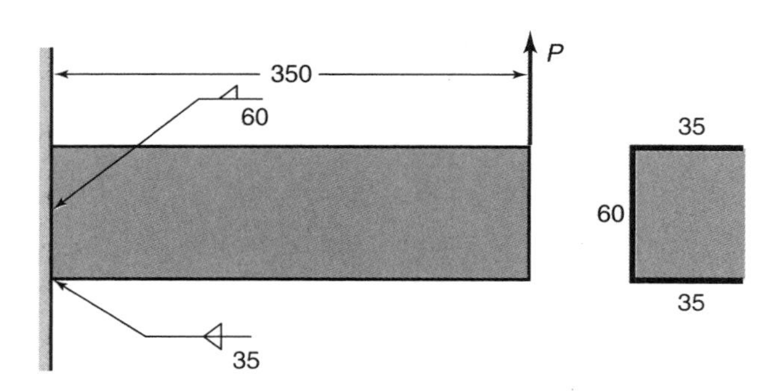

Sketch u, for Problem 16.81

16.82 The bracket shown in Sketch v is spot welded to the annealed AISI 1040 steel column using three 7-mm-diameter electrodes. If 90% of the column material shear strength is developed by the spot welds, calculate the safety factor against weld nugget shear. Dimensions are in millimeters. *Ans* $n_s = 1.09$.

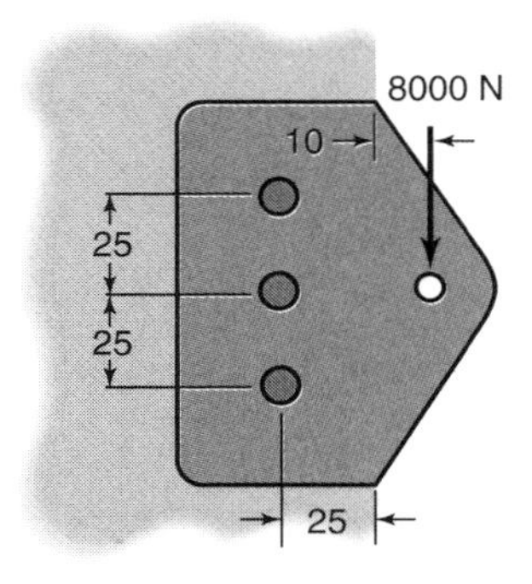

Sketch v, for Problem 16.82

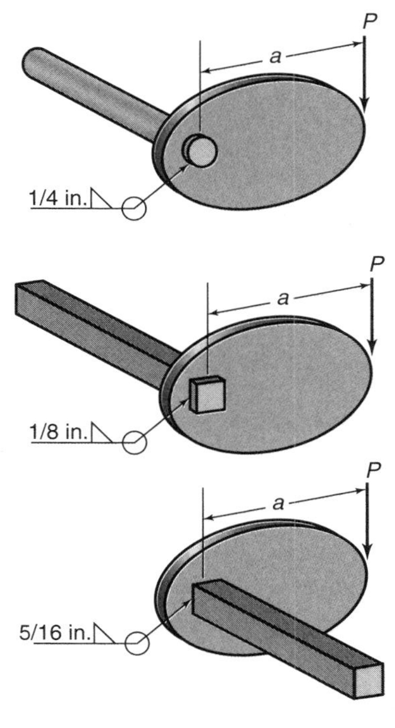

Sketch x, for Problem 16.84

16.83 AWS electrode number E100XX is used to weld a bar, shown in Sketch w, to a wall. For a safety factor of 3 against yielding, find the maximum load that can be supported. Dimensions are in millimeters. *Ans.* $P_{\max} =$ 359 N.

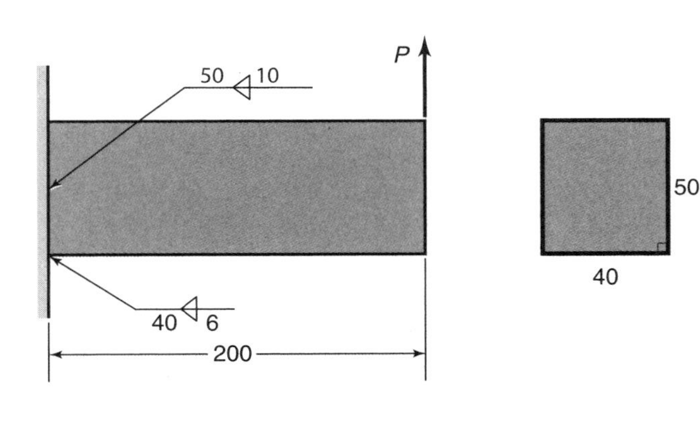

Sketch w, for Problem 16.83

16.84 A cam is to be attached to an extruded channel using arc welds. Three designs are proposed as shown in Sketch x. If E60 electrodes are used, find the factor of safety of the welded joint if $P = 200$ pounds for each case. Use $a = 15$ in. You can assume that the extrusion fits tightly with the mating hole in the cam for the first two designs so that direct shear can be neglected. The rightmost design has no mating hole, but the channel is welded to the cam surface. Design based on yielding of the weld. Assume the round channel has a diameter of 2 in., and the square channels are 2 × 2 in.

16.85 When manufacturing the fuselage of a commuter airplane, aluminum plates are glued together with lap joints. Because the elastic deformation for a single plate differs from the deformation for two plates glued together in a lap joint, the maximum shear stress in the glue is twice as high as the average shear stress. The shear strength of the glue is 20 MPa, the tensile strength of the aluminum plates is 95 MPa, and their thickness is 4.0 mm. Calculate the overlapping length needed to make the glue joint twice as strong as the aluminum plate. *Ans.* $L = 76$ mm.

16.86 The ropes holding a children's swing are glued into two plastic tubes with an inner diameter of 10 mm and a length of 100 mm. The difference in elasticity between the rope and the rope plus the plastic tube gives a maximum shear stress 2.5 times as high as the mean shear stress in the glue. The glue is an epoxy type with an ultimate shear strength of 12 MPa. How heavy can the person on the swing be without overstressing the glue if the speed of the swing at its lowest point is 6.5 m/s and the distance from the center of gravity of the person to the fastening points of the ropes is 2 m? The safety factor is 10. *Ans.* Maximum mass is 97 kg.

16.87 A fishing rod is made of carbon-fiber-reinforced plastic tube. To get optimum elastic properties along the length of the rod, and to therefore be able to make long and accurate casts, the concentrations of the fibers in the various parts of the rod have to be different. It is necessary to scarf joint the rod parts. The tensile strength of the epoxy glue joint is 10 MPa and its shear strength is 12.5 MPa. These strengths are independent of each other. Find the optimum scarf angle to make the rod as strong as possible in bending.

16.88 Refer to the simple butt and lap joints shown in Fig. 16.29. (a) Assuming the area of the butt joint is 5 mm x 20 mm and referring to the adhesive properties given in Table 16.14, estimate the minimum and maximum tensile force that this joint can withstand. (b) Estimate these forces for the lap joint assuming that the members overlap by 20 mm.

16.89 A strap joint as depicted in Fig. 16.29g is used to connect two strips of 2014-O aluminum with a thickness of 1 mm and a width of 25 mm. What overlap is needed to develop the full strength of the strip using an epoxy adhesive? *Ans.* $l = 2.2$ mm.

Synthesis and Design

16.90 A threaded shaft, when rotated inside a smooth cylinder, can be used as a pump; the geometry is as shown in Sketch y. Obtain an expression for the flow rate as a function of geometry and shaft speed. *Hint:* Consider the flow as a ribbon of fluid that is being translated by the screw motion.

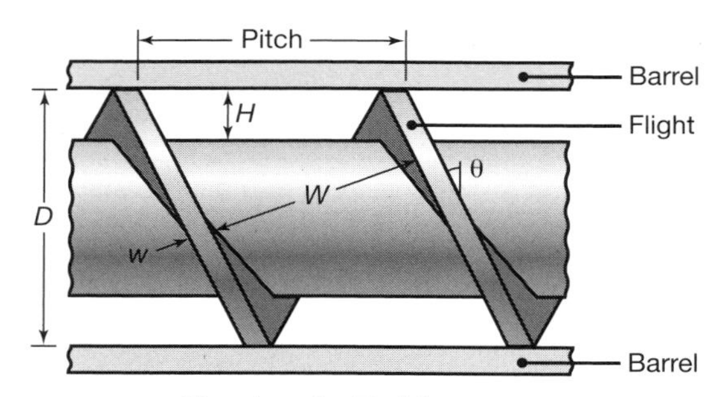

Sketch y, for Problem 16.90

16.91 Consider the situation where large diameter bolts are in service for extended periods of time and are replaced to ensure a high safety factor. It is proposed to refurbish those bolts by machining off (removing) a surface layer, and cutting new threads. Explain:

(a) whether this approach has technical merit, and

(b) any technical and ethical concerns you would have with this approach.

16.92 Design a joint to connect two 25-mm-wide, 5-mm-thick steel members. The overlap may be as much as 25 mm, and any one approach described in this chapter can be used.

16.93 For the same members in Problem 16.92, design a joint using threaded fasteners arranged in one row. Do you advise the use of one large fastener or many small fasteners? Explain.

16.94 For the same members in Problem 16.92, design a joint using a *combination* of joining techniques.

16.95 Design three twist-off fasteners, other than the one depicted in Fig. 16.16, that can be disassembled.

16.96 Alclad stock is made from 5182 aluminum alloy and has both sides coated with a thin layer of pure aluminum. The 5182 provides high strength, while the outside layers of pure aluminum provide good corrosion resistance because of their stable oxide film. Hence, Alclad is used commonly in aerospace structural applications. Investigate other common roll-bonded metals and their uses, and prepare a summary table.

16.97 Sketch z shows a schematic of a compound joint that combines rivets with adhesive bonding. List the advantages and disadvantages of such a design, and develop mathematical expressions to predict the axial force that causes failure of the joint. Consider all relevant failure modes discussed in Section 16.5.

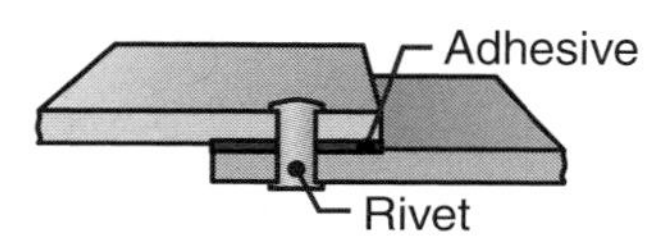

Sketch z, for Problem 16.97

16.98 For a fast-moving power screw, modify Eq. (16.13) to incorporate inertia of the load and screw.

Chapter 17

Springs

A collection of helical compression springs. *Source:* Courtesy of Danly IEM.

It must be confessed that the inventors of the mechanical arts have been much more useful to men than the inventors of syllogisms.
Voltaire

Contents

Examples

Design Procedures

Case Study

Springs are among the most common machine element, and have a wide variety of forms and functions. This chapter introduces the most common types of springs and describes their design. The chapter begins by describing the materials and properties that are needed to produce an effective spring, and discusses the unique strength characteristics that arise from spring metals that have been highly cold worked. Helical compression springs are analyzed, with geometric concerns and associated forces and stresses, and the approach is then applied to extension springs. Torsion springs are then examined, and the normal stresses that result from an applied moment are derived. Torsion springs are also unique in that the number of active coils changes as the spring deflects. Leaf springs are then summarized; these are cantilevers that are commonly applied to vehicle suspensions because of their compact designs. Gas springs are commonly used as counterbalances, but also introduce favorable damping characteristics. Belleville springs, also known as Belleville lock washers, are a special conical disk spring that has a naturally small profile, and that can be stacked in series or parallel to accentuate their performance. Finally, wave springs are discussed, which use a helical pattern with a superimposed wave; the resulting springs result in very compact designs.

Machine elements in this chapter: Helical compression and extension springs; torsion springs, leaf springs, gas springs, Belleville springs, wave springs.
Typical applications: General energy storage, vehicle suspensions, counterweights, preloading components.
Competing machine elements: Flywheels (Ch. 10), tension or compression members for supports (Ch. 5 and 10), retaining rings (Ch. 11), lock washers (Ch. 16).

Symbols

A	cross-sectional area, m^2
A_p	intercept, Pa
b	width of leaf spring, m
b_s	slope
C	spring index, D/d
$\bar{C}$	intercept
c	distance from neutral axis to outer fiber, m
D	mean coil diameter, m
D_i'	coil inside diameter after loading, m
d	wire diameter, m
E	modulus of elasticity, Pa
f_n	lowest natural frequency, Hz
G	shear modulus of elasticity, Pa
g	gravitational acceleration, 9.807 m/s^2
g_a	gap in open end of extension spring, m
h	height, m
I	area moment of inertia, m^4
J	polar area moment of inertia, m^4
K	multiple wave factor
K_b	Bergsträsser factor
K_d	transverse shear factor
K_i	spring stress concentration factor,
K_w	Wahl curvature correction factor,
K_1	defined in Eq. (17.55)
k	spring rate, N/m
k_t	spring rate considering torsional loading, N/m
k_θ	angular spring rate, Nm/rev
l	length, m
M	moment, N-m
m	slope
N	number of coils
N_a	number of active coils
N_a'	number of active coils after loading
N_w	number of waves per turn
n	number of leaves
n_s	safety factor
P	force, N
p	pitch, m; gas pressure, Pa
R	radius used in applying torque; linear arm length for bracket, m
$\bar{R}$	specific gas constant, J/kg-K
R_d	diameter ratio, D_o/D_i
r	radius, m
S_{se}	modified endurance limit, Pa
S_{se}'	endurance limit, Pa
S_{sf}	modified shear fatigue strength, Pa
S_{su}	shear ultimate strength, Pa
S_{sy}	shear yield strength, Pa
S_y	yield stress, Pa
S_{ut}	tensile ultimate stress, Pa
T	torque, N-m
t	thickness, m
t_m	temperature, K
U	stored elastic energy, N-m
ΔU	change in energy, N-m
v	volume, m^3
x	Cartesian coordinate, m
$\gamma_{\theta z}$	shear strain due to torsional loading
Δ_v	loss coefficient
δ	deflection, m
ζ	cone angle
θ	angular deflection, rad
ν	Poisson's ratio
ρ	density, kg/m^3
σ	stress, Pa
τ	shear stress, Pa

Subscripts

a	alternating
b	body
c	conical
d	transverse (or direct)
e	end
f	free (without load)
h	hook
i	installed; inside; preload
l	loop
o	operating; outside
s	solid
t	torsional; total
u	ultimate
w	wire

17.1 Introduction

A **spring** is a flexible machine element used to exert a force or a torque and, at the same time, store energy. Energy is stored in the solid that is bent, twisted, stretched, or compressed to form the spring. The energy is recoverable by the elastic return of the distorted material. Springs must have the ability to elastically withstand desired deflections. Springs frequently operate with high working stresses and with loads that are continuously varying.

Some applications of springs are

1. To store and return energy, as in a gun recoil mechanism
2. To apply and maintain a definite force, as in relief valves and governors
3. To isolate vibrations, as in automobile suspensions
4. To indicate and/or control load, as in a scale
5. To return or displace a component, as in a brake pedal or engine valve

17.2 Spring Materials

Strength is one of the most important characteristics to consider when selecting a spring material. Figure 3.21 shows a plot of elastic modulus, E, against strength, S. In this figure, strength refers to yield strength for metals and polymers, compressive crushing strength for ceramics, tear strength for elastomers, and tensile strength for composites and woods. As discussed by Ashby [2010], the normalized strength, S/E, is the best parameter for evaluating lightweight strength-based designs. From Fig. 3.21, note that engineering polymers have values of S/E in the range 0.01 to 0.1. The values for metals are 10 times lower. Even ceramics in compression are not as strong as engineering polymers, and in tension they are far weaker. Composites and woods lie on the 0.01 contour, as good as the best metals. Because of their exceptionally low elastic modulus, elastomers have higher S/E, between 0.1 and 1.0, than does any other class of material.

The *loss coefficient*, Δ_v, is the second parameter that is important in selecting a spring material. The loss coefficient measures the fractional energy dissipated in a stress-strain cycle. Figure 17.1 shows the stress-strain variation for a complete cycle. The loss coefficient is

$$\Delta_v = \frac{\Delta U}{2U}, \tag{17.1}$$

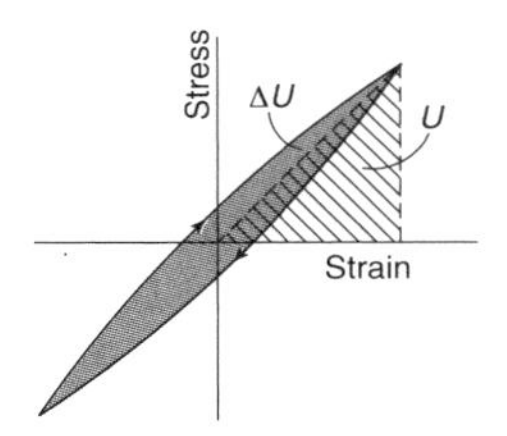

Figure 17.1: Stress-strain curve for one complete cycle.

where ΔU is the change of energy over one cycle, and U is the stored elastic energy. The loss coefficient is a dimensionless parameter. A material used for springs should have a *low* loss coefficient. Elastomers have the highest loss coefficients, and ceramics have the lowest, with a four-order-of-magnitude range between them. Obviously, ceramics are not a suitable spring material, due to their brittleness. High-carbon steels have just slightly higher loss coefficients than ceramics and are a more suitable spring material. In bulk form, elastomers are a good spring material. In a different shape, such as a cantilever or helix, the high stiffness of steel is not a detriment, and its low cost and fatigue performance make it an attractive spring material.

For these reasons, most commercial springs are produced from the group of high-strength, low-loss-coefficient materials that includes high-carbon steel; cold-rolled, precipitation-hardened stainless steel; nonferrous alloys; and a few specialized nonmetallics such as laminated fiberglass. Table 17.1 gives typical properties for common spring materials (modulus of elasticity, shear modulus of elasticity, density, and maximum service temperature) and gives additional design information.

Springs are manufactured by either hot- or cold-working processes, depending on the size of the material, the spring index, and the properties desired. In general, prehardened wire should not be used if $D/d < 4$ or if $d > 1/4$ in. Winding the spring induces residual stresses through bending, but these are normal to the direction of torsional working stresses in a coiled spring. When a spring is manufactured, it is quite frequently stress-relieved after winding by a mild thermal treatment.

The ultimate strength of a spring material varies significantly with wire size, so that the ultimate strength cannot be specified unless the wire size is known. The material and its processing also have an effect on tensile strength. Results of extensive testing show that wire strength versus wire diameter can be approximated by:

$$S_{ut} = \frac{A_p}{d^m}, \tag{17.2}$$

where A_p and m are constants. Table 17.2 gives values of A_p and m for selected spring materials. Equation (17.2) is valid only for the limited size range given in Table 17.2. Also, note that A_p in thousand pounds per square inch requires d to be in inches and that A_p in megapascals requires d to be in millimeters.

In the design of springs, the allowable stress is the torsional yield strength rather than the ultimate strength. Once the ultimate strength is known from Eq. (17.2), the shear yield stress can be expressed as

$$S_{sy} = \tau_{\text{all}} \approx 0.40 S_{ut}. \tag{17.3}$$

Note that Eq. (17.3) approximates the shear yield stress from the material's ultimate strength in tension. There normally is not a generally applicable rule that allows determination of yield strength from ultimate strength, so it should be recognized that Eq. (17.3) is applicable only to spring steels and is somewhat conservative in nature. For example, if the maximum shear stress theory (Section 6.7.1) is used, then the shear yield stress is one-half the uniaxial yield strength.

Example 17.1: Comparison of Spring Designs

Given: A spring arrangement is being considered for opening the lid of a music box. There can be either one spring behind the lid or one weaker spring on each side of the lid pushing it out after a snap mechanism is released. Except for the spring wire diameter, the spring force and the spring geometry are the same for the two options. The material chosen is music wire. For the one-spring option, the wire diameter needed is 1 mm.

Find:

(*a*) Which option is more economical if the cost of the springs is proportional to the weight of the spring material?

(*b*) Is there any risk that the two-spring option will suffer from fatigue if the one-spring option does not fatigue?

Solution:

(*a*) From Table 17.2 for music wire, $m = 0.146$ and $A_p = 2170$ MPa. Substituting these values into Eq. (17.2) gives

$$S_{ut} = \frac{2170}{1^{0.146}} = 2170 \text{ MPa}.$$

This value is the allowable stress for the single spring design. From Appendix D for a concentrated load at the end of the beam ($a = x = l$), the deflection at the free end of the beam is

$$y = \frac{Pl^3}{3EI}. \tag{a}$$

The spring has a circular cross section, so that the area moment of inertia is

$$I = \frac{\pi r^4}{4}. \tag{b}$$

Substituting Eq. (*b*) into Eq. (*a*) gives

$$y = \frac{Pl^3 4}{3E\pi r^4} = \left(\frac{4l^3}{3\pi E}\right)\left(\frac{P}{r^4}\right). \tag{c}$$

The material and the spring length are the same for options 1 and 2. Thus, if subscript 1 refers to option 1 and subscript 2 refers to option 2, from Eq. (*c*)

$$\frac{P_1}{r_1^4} = \frac{P_1/2}{r_2^4}.$$

Solving for r_2,

Table 17.1: Typical properties of common spring materials. *Source:* Adapted from Relvas [1996].

Material/Specification	Elastic modulus, E, GPa (Mpsi)	Shear modulus, G, GPa (MPsi)	Density, ρ, kg/m^3 (lbm/in^3)	Maximum service temperature, °C(°F)	Principal characteristics
High-carbon steels					
Music wire (ASTM A228)	207 (30.0)	79.3 (11.5)	7840 (0.283)	120 (248)	High strength; excellent fatigue life
Hard drawn (ASTM A227)	207 (30.0)	79.3 (11.5)	7840 (0.283)	120 (248)	General purpose use; poor fatigue life
Stainless steels					
Martensitic (AISI 410, 420)	200 (29.0)	75.8 (11.0)	7750 (0.280)	250 (482)	Unsatisfactory for subzero applications
Austenitic (AISI 301, 302)	193 (28.0)	68.9 (9.99)	7840 (0.283)	315 (600)	Good strength at moderate temperatures; low stress relaxation
Copper-based alloys					
Spring brass (ASTM B134)	110 (15.9)	41.4 (6.00)	8520 (0.308)	90 (194)	Low cost; high conductivity; poor mechanical properties
Phosphor bronze (ASTM B159)	103 (14.9)	43.4 (6.29)	8860 (0.320)	90 (194)	Ability to withstand repeated flexures; popular alloy
Beryllium copper (ASTM B197)	131 (19.0)	44.8 (6.50)	8220 (0.297)	200 (392)	High yield and fatigue strength; hardenable
Nickel-based alloys					
Inconel 600	214 (31.0)	75.8 (11.0)	8500 (0.307)	315 (600)	Good strength; high corrosion resistance
Inconel X-750	214 (31.0)	75.8 (11.0)	8250 (0.298)	600 (1110)	Precipitation hardening; for high temperatures
Ni-Span C	186 (27.0)	66.2 (9.60)	8140 (0.294)	90 (194)	Constant modulus over a wide temperature range

Table 17.2: Coefficients used in Eq. (17.2) for selected spring materials.

Material	Size range in.	Size range mm	Exponent, m	Constant, A_p ksi	Constant, A_p MPa
Music wire[a]	0.004–0.250	0.10–6.5	0.146	196	2170
Oil-tempered wire[b]	0.020–0.500	0.50–12	0.186	149	1880
Hard-drawn wire[c]	0.028–0.500	0.70–12	0.192	136	1750
Chromium vanadium[d]	0.032–0.437	0.80–12	0.167	169	2000
Chromium silicon[e]	0.063–0.375	1.6–10	0.112	202	2000
302 stainless steel	0.013–0.10	0.33–2.5	0.146	169	1867
	0.10–0.20	2.5–5	0.263	128	2065
	0.20–0.40	5–10	0.478	90	2911
Phosphor-bronze[f]	0.004–0.022	0.1–0.6	0	145	1000
	0.022–0.075	0.6–2	0.028	121	913
	0.075–0.30	2–7.5	0.064	110	932

[a] Surface is smooth and free from defects and has a bright, lustrous finish.
[b] Surface has a slight heat-treating scale that must be removed before plating.
[c] Surface is smooth and bright with no visible marks.
[d] Aircraft-quality tempered wire; can also be obtained annealed.
[e] Tempered to Rockwell C49 but may also be obtained untempered.
[f] SAE CA510, tempered to Rockwell B92-B98.

$$r_2 = \frac{r_1}{(2)^{0.25}} = 0.8409 r_1. \qquad (d)$$

The cost will change proportionally to the weight, or

$$\frac{\pi r_1^2 l\rho}{2\pi r_2^2 l\rho} = \frac{r_1^2}{2r_2^2} = \frac{r_1^2}{2(0.8409 r_1)^2} = 0.7071. \qquad (e)$$

Thus, the one-spring option costs only 0.7071 times the two-spring option and is therefore 29.3% cheaper.

(*b*) From Eq. (4.45), the bending (design) stress for one spring (option 1) is

$$\sigma_1 = \frac{Mc}{I} = \frac{Plr_1}{\frac{\pi r_1^4}{4}} = \frac{4Pl}{\pi r_1^3}. \qquad (f)$$

For two springs (option 2), the design stress is

$$\sigma_2 = \frac{4(P/2)l}{\pi r_2^3} = \frac{2Pl}{\pi(0.8409)^3 r_1^3} = \frac{\sigma_1}{2(0.8409)^3} = 0.8409\sigma_1.$$

Using Eq. (17.2) gives the allowable stress for option 2 as

$$S_{ut} = \frac{2170}{(0.8409)^{0.146}} = 2226 \text{ MPa}.$$

Thus, the allowable stress is higher for option 2 than for option 1 and the design stress is lower for option 2 than for option 1; therefore, the safety factor for option 2 is much larger than that for option 1. Recall from Eq. (1.1) that the safety factor is the allowable stress divided by the design stress. Therefore, if there is no risk of fatigue failure for option 1, there is no risk of fatigue failure for option 2.

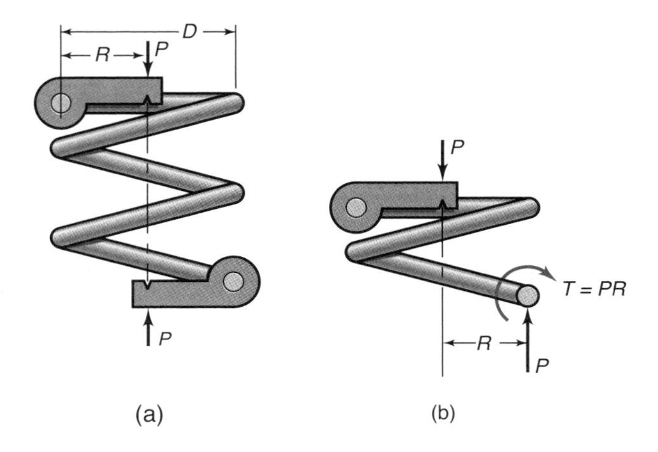

Figure 17.2: Helical coil. (a) Coiled wire showing applied force; (b) coiled wire with section showing torsional and direct (vertical) shear acting on the wire.

17.3 Helical Compression Springs

The helix is the spiral form of spring wound with constant coil diameter and uniform pitch. **Pitch** is the distance, measured parallel to the coil axis, from the center of one coil to the center of the adjacent coil. In the most common form of the **helical compression spring**, a round wire is wrapped into a cylindrical form with a constant pitch between adjacent coils.

Figure 17.2a shows a wire that has been formed into a helix of N coils with mean coil radius R. The coiled wire is in equilibrium under the action of two equal and opposite forces P. In Fig. 17.2b, a section has been taken through a wire, and the statically equivalent force and torque are shown. Note from Fig. 17.2b that the wire experiences both transverse and torsional shear. Torsional shear in a helical coil spring is the primary stress, but transverse (or direct) shear is significant, so that it cannot be neglected.

17.3.1 Torsional Shear Stress

From Eq. (4.33), the maximum torsional shear stress is, using Table 4.1:

$$\tau_{t,\max} = \frac{Tc}{J} = \frac{Td(32)}{(2)\pi d^4} = \frac{16PR}{\pi d^3} = \frac{8PD}{\pi d^3}, \tag{17.4}$$

where D is the mean coil diameter and d is the wire diameter. Figure 17.3 shows the stress distribution across the wire cross section. From Fig. 17.3a, for pure torsional stress, the shear stress is a maximum at the outer fiber of the wire and zero at the center of the wire.

17.3.2 Transverse Shear Stress

The maximum transverse (also called direct) shear stress shown in Fig. 17.3b can be expressed for a solid circular cross section as

$$\tau_{d,\max} = \frac{P}{A} = \frac{4P}{\pi d^2}. \tag{17.5}$$

In Fig. 17.3b, the maximum stress occurs at the midheight of the wire.

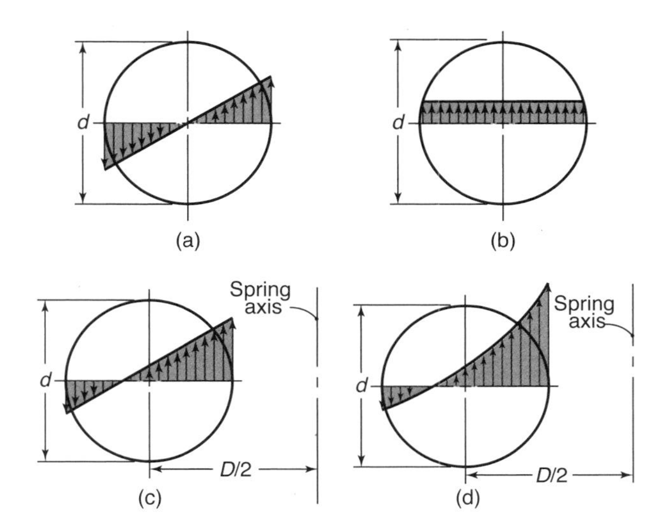

Figure 17.3: Shear stresses acting on wire and coil. (a) Pure torsional loading; (b) transverse loading; (c) torsional and transverse loading with no curvature effects; (d) torsional and transverse loading with curvature effects.

17.3.3 Combined Torsional and Transverse Shear Stress

The maximum shear stress resulting from summing the torsional and transverse shear stresses, using Eqs. (17.4) and (17.5), is

$$\tau_{\max} = \tau_{t,\max} + \tau_{d,\max} = \frac{8PD}{\pi d^3} + \frac{4P}{\pi d^2} = \frac{8DP}{\pi d^3}\left(1 + \frac{d}{2D}\right). \tag{17.6}$$

Figure 17.3c shows that the maximum shear stress occurs at the midheight of the wire and at the coil inside diameter. Curvature effects are not considered in Eq. (17.6) and are not shown in Fig. 17.3c. The **spring index**, which is a measure of coil curvature, is

$$C = \frac{D}{d}. \tag{17.7}$$

For most springs, the spring index is between 4 and 12. Equation (17.6) can be rewritten as

$$\tau_{\max} = \frac{8DK_dP}{\pi d^3}, \tag{17.8}$$

where K_d is the transverse shear factor, given by

$$K_d = \frac{C + 0.5}{C}. \tag{17.9}$$

Note from Eq. (17.6) that if the transverse shear were small relative to the torsional shear, then as it relates to Eq. (17.8), K_d would be equal to 1. Any contribution from the transverse shear term would make the transverse shear factor greater than 1.

Also, recall that the spring index C is usually between 4 and 12. If that range is used, the range of transverse shear factor is 1.0417 to 1.1667. Thus, the contribution due to transverse shear is indeed small relative to that due to torsional shear.

Recall from Section 4.5.3 that, for a curved member, the stresses can be considerably higher at the inside surface than at the outside surface. Thus, incorporating curvature can play a significant role in the spring design. Curvature effects

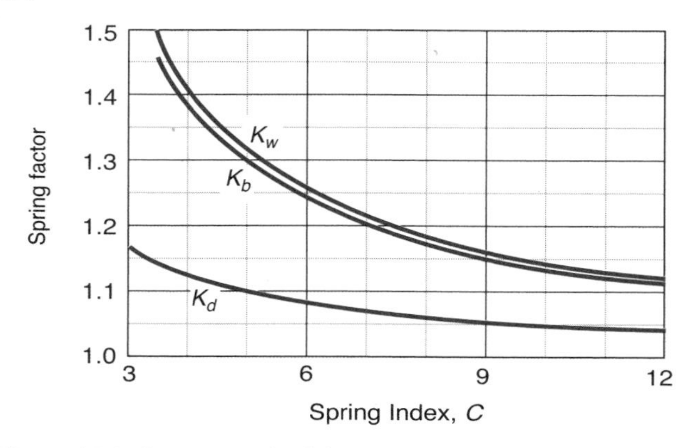

Figure 17.4: Comparison of the Wahl and Bergsträsser curvature correction factors used for helical springs. The transverse shear factor is also shown.

can be included in Eq. (17.8) by simply replacing the transverse shear factor with another factor that accounts for stress concentrations. A curvature correction factor attributed to A.M. Wahl results in the following:

$$\tau_{\max} = \frac{8DK_wP}{\pi d^3}, \tag{17.10}$$

where

$$K_w = \frac{4C-1}{4C-4} + \frac{0.615}{C}. \tag{17.11}$$

Another curvature correction factor in common use is the *Bergsträsser factor*, given by

$$K_b = \frac{4C+2}{4C-3}. \tag{17.12}$$

Selecting a curvature correction factor requires consideration of a number of factors, including the application, previous design history, and anticipated loading. In general, a spring that is to be designed for static failure without consideration of multiple cycles can often use the transverse shear factor, K_d. For more demanding applications, or when fatigue is a possibility, either K_w or K_b provides good results; they are very close but K_w is slightly more conservative (see Fig. 17.4).

Figure 17.3d shows the stress distribution when curvature effects and both torsional and transverse shear stresses are considered. The maximum stress occurs at the midheight of the wire and at the coil inside diameter. This location is where failure should first occur in the spring.

17.3.4 Deflection

From Eq. (4.27), the shear strain due to torsional loading is

$$\gamma_{\theta z} = \frac{r\theta}{l} = \frac{\delta_t}{l} = \frac{D}{2}\frac{\theta}{l}. \tag{17.13}$$

Making use of Eq. (4.31) gives

$$\delta_t = \left(\frac{D}{2}\right)\left(\frac{Tl}{JG}\right). \tag{17.14}$$

Applying this equation to a helical coil spring, using Fig. 17.2, and assuming that the wire has a circular cross section results in

$$\delta_t = \left(\frac{D}{2}\right)\left[\frac{(D/2)P(2\pi)(D/2)N_a}{G(\pi d^4/32)}\right] = \frac{8PD^3N_a}{Gd^4} = \frac{8PC^3N_a}{Gd} \tag{17.15}$$

where C is the spring index, D/d, N_a is the number of active coils, and G is the elastic shear modulus. The circumferential deflection due to torsional and transverse shear loading can be derived by using Castigliano's theorem (Section 5.6). The total strain energy from Table 5.2 is

$$U = \frac{T^2l}{2GJ} + \frac{P^2l}{2AG}. \tag{17.16}$$

The first term on the right of the equal sign corresponds to torsional loading and the second term corresponds to transverse shear for a circular cross section.

Applying Eq. (17.16) to a helical coil spring of circular cross section gives the total strain energy as

$$\begin{aligned} U &= \frac{(PD/2)^2(\pi DN_a)}{2G(\pi d^4/32)} + \frac{P^2(\pi DN_a)}{2G(\pi d^2/4)} \\ &= \frac{4P^2D^3N_a}{Gd^4} + \frac{2P^2DN_a}{Gd^2}. \end{aligned}$$

Using Castigliano's theorem, Eq. (5.30), gives

$$\delta = \frac{\partial U}{\partial P} = \frac{8PD^3N_a}{Gd^4} + \frac{4PDN_a}{Gd^2} = \frac{8PC^3N_a}{Gd}\left(1+\frac{0.5}{C^2}\right). \tag{17.17}$$

Comparing Eq. (17.15) with (17.17) shows that the second term in Eq. (17.17) is the transverse shear term and that, for a spring index in the normal range $3 \le C \le 12$, the deflection due to transverse shear is extremely small.

From Eq. (17.17), the **spring rate** is

$$k = \frac{P}{\delta} = \frac{Gd}{8C^3N_a\left(1+\dfrac{0.5}{C^2}\right)}. \tag{17.18}$$

The difference between spring index, C, and spring rate, k, is important. Spring index is a dimensionless geometric variable whereas spring rate has units of newtons per meter and incorporates material shear modulus. Also, a stiff spring has a small spring index and a large spring rate. (Recall that the spring index for most conventional springs varies between 4 and 12.) A spring with excessive deflection has a large spring index and a small spring rate.

17.3.5 End Conditions and Spring Length

Figure 17.5 shows four types of ends commonly used in compression springs. Some end coils produce an eccentric application of the load, increasing the stress on one side of the spring, while others result in more uniform loading. The importance of end conditions is self-evident; plain ends are less expensive than squared and ground ends, for example, but are not as uniformly loaded and therefore more susceptible to fatigue failure. Note that the end coils can remain flush with the element they are bearing against. Thus, a spring can have an active number of coils that is different from the total. It is difficult to identify just how many coils should be considered end coils, as this can vary with spring index, solid length, and specific manufacturing parameters. However, an average number based on experimental results is used in Table 17.3 and is useful for designers.

Figure 17.5a shows plain ends that have a noninterrupted helicoid; the spring rates for the ends are the same as if they were not cut from a longer coil. Figure 17.5b shows a plain end that has been ground. In Fig. 17.5c, a spring with plain ends that are squared (or closed) is obtained by deforming the ends to a 0° helix angle. Figure 17.5d shows squared

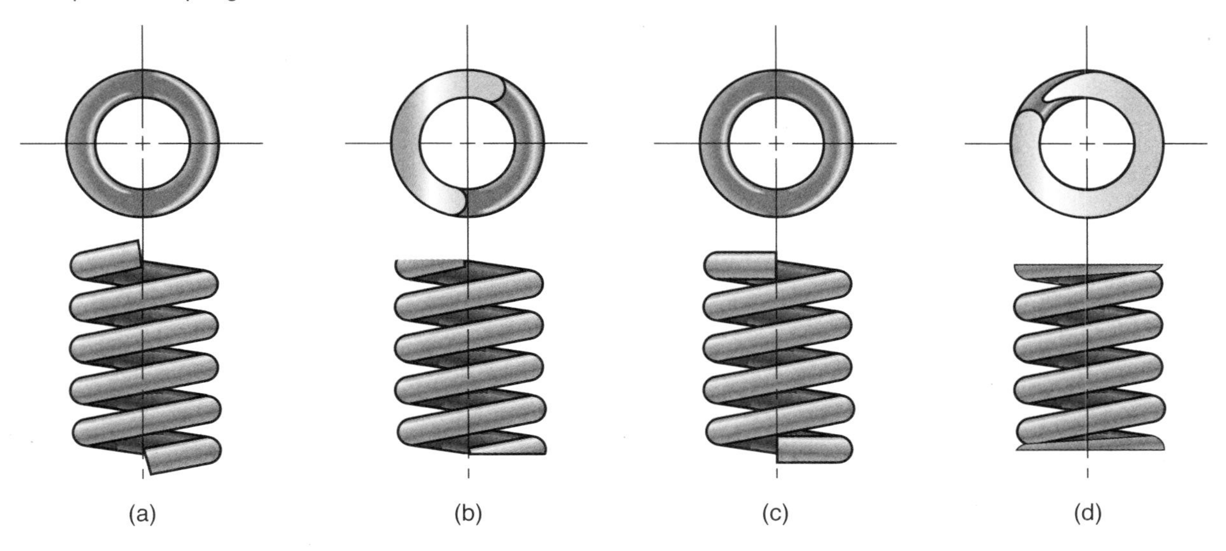

Figure 17.5: Four end types commonly used in compression springs. (a) Plain; (b) plain and ground; (c) squared; (d) squared and ground.

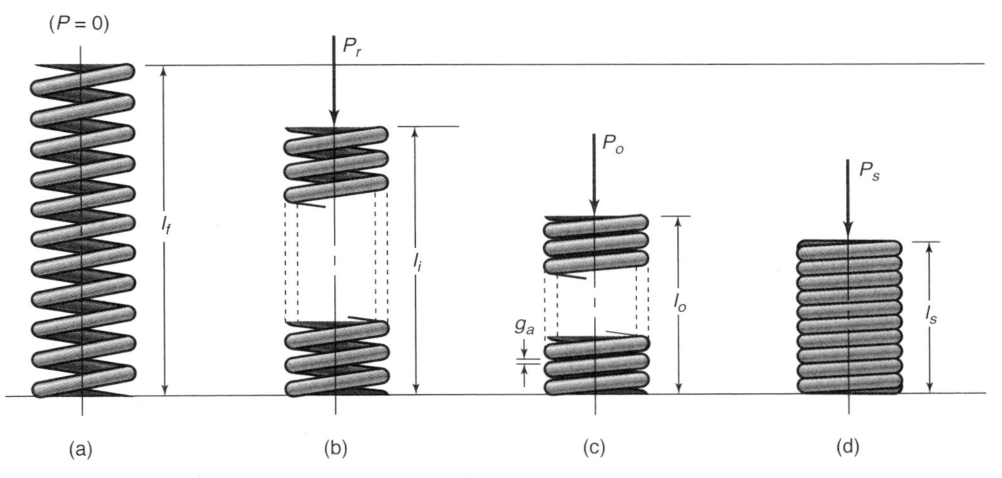

Figure 17.6: Various lengths and forces applicable to helical compression springs. (a) Unloaded; (b) under initial load; (c) under operating load; (d) under solid load.

and ground ends. A better load transfer into the spring is obtained by using ground ends, but at a higher cost.

Table 17.3 also shows useful formulas for the pitch, length, and number of coils for compression springs. Recall that the pitch is the distance measured parallel to the spring axis from the center of one coil to the center of an adjacent coil. The **solid length** is the length of the spring when all adjacent coils are in metal-to-metal contact. The **free length** is the length of the spring when no external forces are applied to it. Figure 17.6 shows the various lengths and forces applicable to helical compression springs.

Figure 17.7 shows the interdependent relationships between force, deflection, and spring length for four distinctively different positions: free, initial, operating, and solid. The deflection is zero at the top right corner and moves to the left for positive lengths. Also, the free length, l_f, is equal to the solid deflection plus the solid length, or

$$l_f = l_s + \delta_s. \tag{17.19}$$

17.3.6 Buckling and Surge

Relatively long compression springs should be checked for buckling. Figure 17.8 shows critical buckling conditions for parallel and nonparallel ends. The critical deflection where buckling will first start to occur can be determined from this figure. If buckling is a problem, it can be prevented by placing the spring in a cage or over a rod. However, the coils rubbing on these guides will take away some of the spring force and thus reduce the load delivered at the spring ends. There is also the potential for wear at the cage or rod contact points with the spring.

A longitudinal vibration that should be avoided in spring design is a **surge**, or pulse of compression, passing through the coils to the ends, where it is reflected and returned. An initial surge is sustained if the natural frequency of the spring is close to the frequency of the repeated loading. The equation for the lowest natural frequency in cycles per

Table 17.3: Useful formulas for compression springs with four end conditions.

Term	Plain	Plain and ground	Squared or closed	Squared and ground
	Type of spring end			
Number of end coils, N_e	0	1	2	2
Total number of active coils, N_a	N_t	$N_t - 1$	$N_t - 2$	$N_t - 2$
Free length, l_f	$pN_a + d$	$p(N_a + 1)$	$pN_a + 3d$	$pN_a + 2d$
Solid length, l_s	$d(N_t + 1)$	dN_t	$d(N_t + 1)$	dN_t
Pitch at free length, p	$(l_f - d)/N_a$	$l_f/(N_a + 1)$	$(l_f - 3d)/N_a$	$(l_f - 2d)/N_a$

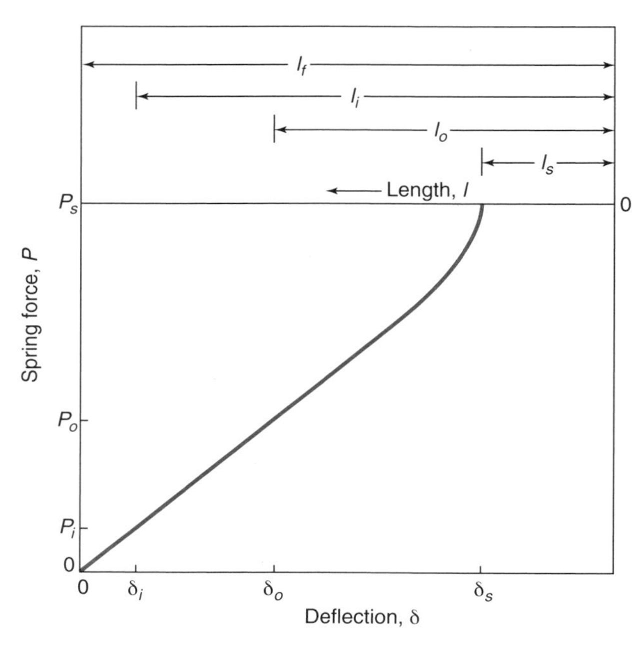

Figure 17.7: Graphical representation of deflection, force, and length for four spring positions.

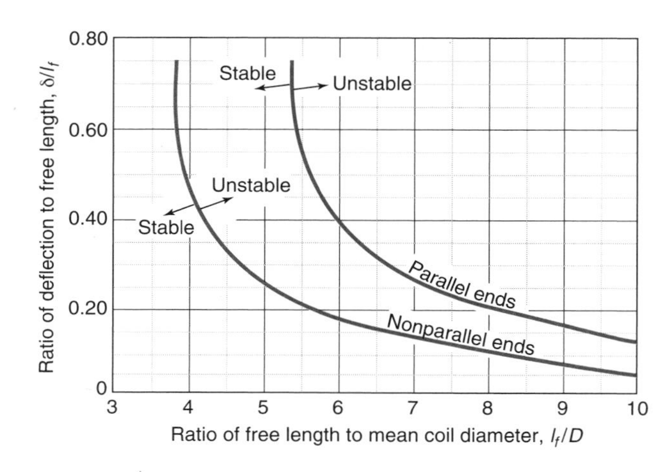

Figure 17.8: Critical buckling conditions for parallel and nonparallel ends of compression springs. *Source:* Adapted from *Engineering Guide to Spring Design*, Barnes Group, Inc. [1987].

second is

$$f_n = \frac{2}{\pi N_a} \frac{d}{D^2} \sqrt{\frac{G}{32\rho}}, \tag{17.20}$$

where G is the shear modulus of elasticity and ρ is the mass density. Vibrations may also occur at whole multiples, such as two, three, and four times the lowest frequency. The spring design should avoid these frequencies.

Example 17.2: Analysis of a Helical Spring

Given: A helical compression spring with plain ends is to be designed with a spring rate of approximately 100,000 N/m, a wire diameter of 10 mm, and a spring index of 5. The maximum allowable shear stress is 480 N/mm^2 and the shear modulus of elasticity is 80 GPa.

Find: The number of active coils, the maximum allowable static load, and the manufactured pitch so that the maximum load just compresses the spring to its solid length.

Solution: From Eq. (17.18), the number of active coils is

$$N_a = \frac{Gd}{8C^3k\left(1 + \frac{0.5}{C^2}\right)} = \frac{(80 \times 10^9)(0.010)}{8(5)^3(10^5)\left(1 + \frac{0.5}{25}\right)},$$

or $N_a = 7.843$ coils. From Eq. (17.9), the transverse shear factor is

$$K_d = \frac{C + 0.5}{C} = \frac{5 + 0.5}{5} = 1.10.$$

For $\tau_{\max} = \tau_{\text{all}} = 480\ \text{N/mm}^2$, Eq. (17.8) gives the maximum force as

$$P_{\max} = \frac{\pi d^3 \tau_{\max}}{8DK_d} = \frac{\pi(0.010)^3 (480 \times 10^6)}{(8)(0.050)(1.10)} = 3427\ \text{N}.$$

The maximum deflection just to bring the spring to solid length is

$$\delta_s = \delta_{\max} = \frac{P_{\max}}{k} = \frac{3427}{10^5} = 34.27\ \text{mm}.$$

From Table 17.3, the solid length for plain ends is

$$l_s = d(N_a + 1) = (0.010)(8.843) = 0.08843\ \text{m} = 88.43\ \text{mm}.$$

From Eq. (17.19), the free length is

$$l_f = l_s + \delta_s = 88.43 + 34.27 = 122.7\ \text{mm}.$$

From Table 17.3, the pitch is

$$p = \frac{(l_f - d)}{N_a} = \frac{(122.7 - 10)}{7.843} = 14.37\ \text{mm}.$$

Example 17.3: Buckling of a Helical Spring

Given: A compression coil spring is made of music wire with squared and ground ends. The spring is to have a spring rate of 1250 N/m. The force corresponding to the solid length is 60 N. The spring index is fixed at 10. Static loading conditions are assumed with few loading cycles, so that the transverse shear factor can be used.

Find: Find the wire diameter, mean coil diameter, free and solid lengths, and indicate whether buckling is a problem. Analyze the nominal case ($n_s = 1$) and give a design recommendation.

Solution: The tensile ultimate strength can be determined from Eq. (17.2) and Table 17.2 as

$$S_{ut} = \frac{A_p}{d^m} = \frac{2.170 \times 10^9}{d^{0.146}},$$

where S_{ut} is in pascals if d is in millimeters. From Eq. (17.3) the shear yield stress is

$$S_{sy} = 0.40 S_{ut} = \frac{868.0 \times 10^6}{d^{0.146}}. \qquad (a)$$

For static loading and given that the spring index is 10, the transverse shear factor is, from Eq. (17.9),

$$K_d = 1 + \frac{0.5}{C} = 1.05.$$

From Eq. (17.8) the maximum design shear stress is

$$\tau_{\max} = \frac{8CK_dP}{\pi d^2} = \frac{8(10)(1.05)(60)}{\pi d^2} = \frac{1604}{d^2}. \qquad (b)$$

Equating Eqs. (a) and (b) gives

$$\frac{868.0 \times 10^6}{d^{0.146}} = \frac{1604}{d^2}, \qquad (c)$$

or $d = 0.808$ mm. Normally, the diameter would be rounded to a convenient value for design purposes, but this analysis can proceed with $d = 0.808$ mm for illustration purposes. The mean coil diameter is, from Eq. (17.7),

$$D = Cd = (10)(0.838) = 83.8 \text{ mm}.$$

Since the spring rate is 1250 N/m and the force corresponding to a solid length is 60 N,

$$\delta_s = \frac{P_s}{k} = \frac{60}{1250} = 0.0480 \text{ m} = 48 \text{ mm}.$$

From Table 17.1, the shear modulus of elasticity for music wire is $G = 79.29 \times 10^9$ Pa. Therefore, from Eq. (17.18), the number of active coils is found to be

$$N_a = \frac{Gd}{8k_tC^3} = \frac{(79.29 \times 10^9)(0.000808)}{8(1250 \times 10^3)} = 6.41 \text{ coils}.$$

From Table 17.3, for squared and ground ends, the total number of coils is

$$N_t = N_a + 2 = 8.41 \text{ coils}.$$

The solid length is

$$l_s = dN_t = (0.808)(8.41) = 5.17 \text{ mm}.$$

Equation (17.19) gives the free length as

$$l_f = l_s + \delta_s = 5.17 + 48 = 53.17 \text{ mm}.$$

The pitch is

$$p = \frac{(l_f - 2d)}{N_a} = \frac{(53.17 - 1.736)}{6.64} = 7.75 \text{ mm}.$$

Making use of Fig. 17.8 when

$$\frac{l_f}{D} = \frac{53.17}{83.8} = 0.634,$$

and

$$\frac{\delta_s}{l_f} = \frac{48.00}{53.17} = 0.9027,$$

and, assuming parallel ends, shows that stability should not be a problem.

17.3.7 Cyclic Loading

Because spring loading is most often continuously fluctuating, allowance must be made in designing for fatigue and stress concentration. The material developed in Ch. 7 is applied in this section and should be understood before proceeding. Helical springs are never used as both compression and extension springs, as is readily apparent from their end coils. Furthermore, springs are assembled with a preload in addition to the working stress. Thus, fatigue is based on nonzero mean stress approaches considered in Section 7.3. The worst condition would occur if there were no preload (i.e., when $\tau_{\min} = 0$).

For cyclic loading, the Wahl or Bergsträsser curvature correction factors given in Eqs. (17.11) and (17.12) should be used instead of the transverse shear factor defined in Eq. (17.8), which is used for static conditions. The Wahl and Bergsträsser curvature correction factors can be viewed as a fatigue stress concentration factor. However, a major difference for springs in contrast to other machine elements is that for springs the curvature correction factor is applied to both the mean stress and the stress amplitude. The reason is that in its true sense, the curvature correction factors are not fatigue stress concentration factors (as considered in Section 7.7), but a way of calculating the shear stress inside the coil. Plastic deformation is to be avoided in springs, as this does not reduce stress concentration factors as is the case with other geometries, but instead tends to increase stress.

The alternating and mean forces can be expressed as

$$P_a = \frac{P_{\max} - P_{\min}}{2}, \qquad (17.21)$$

$$P_m = \frac{P_{\max} + P_{\min}}{2}. \qquad (17.22)$$

The alternating and mean stresses, from Eq. (17.10), are then

$$\tau_a = \frac{8DK_wP_a}{\pi d^3}, \qquad (17.23)$$

$$\tau_m = \frac{8DK_wP_m}{\pi d^3}. \qquad (17.24)$$

For springs, the safety factor guarding against torsional endurance limit fatigue is

$$n_s = \frac{S_{se}}{\tau_a}. \qquad (17.25)$$

Against torsional yielding it is

$$n_s = \frac{S_{sy}}{\tau_a + \tau_m}, \tag{17.26}$$

and against torsional fatigue strength (when infinite life cannot be attained — see Section 7.6) it is

$$n_s = \frac{S_{sf}}{\tau_a}. \tag{17.27}$$

where S_{sf} is the modified shear fatigue strength. Most manufacturers provide information regarding the fatigue properties of their springs. The best general purpose data for the torsional endurance limits of spring steels are those of Zimmerli [1957]. The surprising fact about these data is that size, material, and tensile strength have no effect on the endurance limits of spring steels with wire diameters under 3/8 in. (10 mm). Zimmerli's results are

$$\begin{array}{ll} S'_{se} = 310 \text{ MPa (45.0 ksi)} & \text{for unpeened springs} \\ S'_{se} = 465 \text{ MPa (67.5 ksi)} & \text{for peened springs} \end{array} \tag{17.28}$$

These results are valid for all materials in Table 17.2. Shot peening is working the surface material to cause compressive residual stresses that improve fatigue performance (see Section 7.8.5). The endurance limit given in Eq. (17.28) is corrected for all factors given in Eq. (7.18) except the reliability factor.

The shear ultimate strength, S_{su}, for spring steels can be expressed as

$$S_{su} = 0.60 S_{ut}. \tag{17.29}$$

Example 17.4: Stresses in Helical Springs

Given: A helical compression spring has 14 active coils, a free length of 1.25 in., and an outside diameter of 7/16 in. The ends of the spring are squared and ground and the end plates are constrained to be parallel. The material is music wire with a diameter of 0.042 in.

Find:

(*a*) For *static* conditions compute the spring rate, the solid length, and the stress when the spring is compressed to the solid length. Will static yielding occur before the spring is compressed to its solid length?

(*b*) For *dynamic* conditions with $P_{\min} = 0.9$ lb and $P_{\max} = 2.9$ lb, will the spring experience torsional endurance limit fatigue, torsional yielding, or torsional fatigue failure? Assume a survival probability of 90%, unpeened coils, and fatigue failure based on 50,000 cycles.

Solution:

(*a*) From Eq. (17.7), the spring index is

$$C = \frac{D}{d} = \frac{D_o - d}{d} = \frac{\frac{7}{16} - 0.042}{0.042} = 9.417.$$

From Eq. (17.9), the transverse shear factor is

$$K_d = 1 + \frac{0.5}{C} = 1 + \frac{0.5}{9.417} = 1.053.$$

Either the Wahl or Bergsträsser factors could be used, and will give very similar results. Arbitrarily selecting the Wahl factor, Eq. (17.11) yields

$$K_w = \frac{4C-1}{4C-4} + \frac{0.615}{C} = \frac{4(9.417)-1}{4(9.417)-4} + \frac{0.615}{9.417} = 1.154.$$

From Table 17.3 for squared and ground ends

$$N_t = N_a + 2 = 14 + 2 = 16 \text{ coils.}$$

The solid length is

$$l_s = dN_t = (0.042)(16) = 0.6720 \text{ in.}$$

The pitch is

$$p = \frac{l_f - 2d}{N_a} = \frac{1.25 - 0.084}{14} = 0.0833 \text{ in.}$$

Also, from Eq. (17.19), the deflection to solid length is

$$\delta_s = l_f - l_s = 1.25 - 0.6720 = 0.5780 \text{ in.}$$

From Eq. (17.2) and Table 17.2, the ultimate strength in tension is

$$S_{ut} = \frac{A_p}{d^m} = \frac{196 \times 10^3}{(0.042)^{0.146}} = 311.4 \times 10^3 \text{ psi.}$$

From Eq. (17.3), the torsional yield stress, or the allowable shear stress for static loading, is

$$\begin{aligned} \tau_{\text{all}} &= S_{sy} = 0.40 S_{ut} = (0.40)\left(311.4 \times 10^3\right) \\ &= 124.6 \times 10^3 \text{ psi.} \end{aligned}$$

Table 17.1 gives $G = 11.5 \times 10^6$ psi. Therefore, from Eq. (17.15), the force required to compress the coils to a solid length is

$$\begin{aligned} P_s &= \frac{G d \delta_s}{8 C^3 N_a} \\ &= \frac{\left(11.5 \times 10^6\right)(0.042)(0.5780)}{8(9.417)^3(14)} \\ &= 2.985 \text{ lb.} \end{aligned}$$

From Eq. (17.8), the maximum design stress is

$$\begin{aligned} \tau_{\max} &= \frac{8 D K_d P_s}{\pi d^3} = \frac{8 C K_d P_s}{\pi d^2} \\ &= \frac{8(9.417)(1.053)(2.985)}{\pi (0.042)^2} \\ &= 42,730 \text{ psi.} \end{aligned}$$

The safety factor guarding against static yielding is

$$n_s = \frac{\tau_{\text{all}}}{\tau_{\max}} = \frac{124,600}{42,730} = 2.916.$$

Thus, failure should not occur due to static yielding. Checking for buckling gives

$$\frac{\delta_s}{l_f} = \frac{0.5780}{1.25} = 0.4624.$$

$$\frac{l_f}{D} = \frac{1.25}{7/16 - 0.042} = 3.161.$$

From Fig. 17.8, buckling is not a problem.

(*b*) The alternating and mean forces are

$$P_a = \frac{P_{\max} - P_{\min}}{2} = \frac{2.9 - 0.9}{2} = 1.0 \text{ lb},$$

$$P_m = \frac{P_{\max} + P_{\min}}{2} = \frac{2.9 + 0.9}{2} = 1.9 \text{ lb}.$$

From Eq. (17.10), the alternating and mean stresses are

$$\begin{aligned} \tau_a &= \frac{8CK_wP_a}{\pi d^2} \\ &= \frac{8(9.417)(1.154)(1.0)}{\pi(0.042)^2} \\ &= 15.69 \times 10^3 \text{ psi}, \end{aligned}$$

$$\tau_m = \frac{8CK_wP_m}{\pi d^2} = \tau_a\left(\frac{P_m}{P_a}\right) = 29.81 \times 10^3 \text{ psi}.$$

The safety factor guarding against torsional yielding is

$$n_s = \frac{S_{sy}}{\tau_a + \tau_m} = \frac{124.6}{15.69 + 29.81} = 2.738.$$

Therefore, failure should not occur from torsional yielding. From Table 7.4 and Eq. (17.28), the modified endurance limit is

$$S_{se} = k_r S'_{se} = (0.9)\left(45 \times 10^3\right) = 40.5 \times 10^3 \text{ psi}.$$

The safety factor against torsional endurance limit fatigue is

$$n_s = \frac{S_{se}}{\tau_a} = \frac{40.5 \times 10^3}{15.69 \times 10^3} = 2.581.$$

Therefore, failure should not occur from torsional endurance limit fatigue. From Eq. (17.29), the modulus of rupture for spring steel is

$$S_{su} = 0.60S_{ut} = 0.60\left(311.4 \times 10^3\right) = 186.8 \times 10^3 \text{ psi}.$$

From Eq. (7.11), the slope used to calculate the fatigue strength is

$$\begin{aligned} b_s &= -\frac{1}{3}\log\left(\frac{0.72S_{su}}{S_{se}}\right) \\ &= -\frac{1}{3}\log\left[\frac{(0.72)\left(186.8 \times 10^3\right)}{40.5 \times 10^3}\right], \end{aligned}$$

or $b_s = -0.1738$. From Eq. (7.12), the intercept used to calculate the fatigue strength is

$$\bar{C} = \log\left[\frac{(0.72S_{su})^2}{S_{se}}\right] = \log\left[\frac{(0.72)^2\left(186.8 \times 10^3\right)^2}{40.5 \times 10^3}\right],$$

or $\bar{C} = 5.650$. From Eq. (7.13), the fatigue strength is

$$S_{sf} = 10^{\bar{C}}\left(N_t\right)^{b_s} = \left(10^{5.650}\right)(50,000)^{-0.1738}$$

or $S_{sf} = 68.12 \times 10^3$ psi. From Eq. (17.27), the safety factor guarding against torsional fatigue strength failure is

$$n_s = \frac{S_{sf}}{\tau_a} = \frac{68.12}{15.69} = 4.342$$

Therefore, failure should not occur under either static or dynamic conditions.

Design Procedure 17.1: Design Synthesis of Helical Springs

The approach described in Section 17.3 is very useful for design analysis purposes. However, it is often the case that a designer needs to specify parameters of a spring and obtain a reasonable design that is then suitable for analysis and optimization. The following are important considerations for synthesis of springs. The considerations are strictly applicable to helical compression springs, but will have utility elsewhere as well.

1. The application should provide some information regarding the required force and spring rate or total deflection for the spring. It is possible that the solid and free lengths are also prescribed. Usually, there is significant freedom for the designer, and not all of these quantities are known beforehand.
2. Select a spring index in the range of 4 to 12. A spring index lower than 4 will be difficult to manufacture, while a spring index higher than 12 will result in springs that are flimsy and tangle easily. Higher forces will require a smaller spring index. A value between 8 and 10 is suitable for most design applications.
3. The number of active coils should be greater than 2 in order to avoid manufacturing difficulties. The number of active coils can be estimated from a spring stiffness design constraint.
4. For initial design purposes, the solid height should be specified as a maximum dimension. Usually, applications will allow a spring to have a smaller solid height than the geometry allows, so the solid height should not be considered a strict constraint.
5. When a spring will operate in a cage or with a central rod, a clearance of roughly 10% of the spring diameter must be specified. This is also useful in compensating for a coating thickness from an electroplating process, for example.
6. At the free height, the spring has no restraining force, and therefore a spring should have at least some preload.
7. To avoid compressing a spring to its solid length, and the impact and plastic deformation that often result, a clash allowance of at least 10% of the maximum working deflection should be required before the spring is compressed solid.

8. Consider the application when designing the spring and the amount of force variation that is required. Sometimes, such as in a garage door counterbalance spring, it is useful to have the force vary significantly, because the load changes with position. For such applications, a high spring rate is useful. However, it is often the case that only small variations in force over the spring's range of motion are desired, which suggests that low spring rates are preferable. In such circumstances, a preloaded spring with a low stiffness will represent a better design.

17.4 Helical Extension Springs

There are often circumstances when a spring is needed to support tensile loads, and the ends of these springs will have coils or hooks, as appropriate. In applying the load in **helical extension springs**, these hooks, or end turns, must be designed so that the stress concentration caused by the presence of sharp bends are decreased as much as possible. In Fig. 17.9a and b the end of the extension spring is formed by merely bending up a half-loop. If the radius of the bend is small, the stress concentration at the cross section, at B in Fig. 17.9b, will be large.

The most obvious method for avoiding these severe stress concentrations is to make the mean radius of the hook, r_2, larger. Figure 17.9c shows another method of achieving this goal. Here, the hook radius is smaller, which increases the stress concentration. The stress is not as severe, however, because the coil diameter is reduced. A lower stress results because of the shorter moment arm. The largest stress occurs at cross section B shown in Fig. 17.9d, a side view of Fig. 17.9c.

Figure 17.10 shows some of the important dimensions of a helical extension spring. All the coils in the body are assumed to be active. The total number of coils depends on the design of the hooks, but generally it can be stated that

$$N_t = N_a + 1. \tag{17.30}$$

The length of the body is

$$l_b = dN_t. \tag{17.31}$$

The free length is measured from the inside of the end loops, or hooks, and is

$$l_f = l_b + l_h + l_l. \tag{17.32}$$

When extension springs are wound, they can be twisted during the manufacturing process so that a preload exists even when the spring is at its solid length. The result is that for close-wound extension springs, this initial force needs to be exceeded before any deflection occurs, and after the initial force the force-deflection curve is linear. Thus, the force is

$$P = P_i + \frac{\delta G d^4}{8 N_a D^3}, \tag{17.33}$$

where P_i is the preload. The spring rate is

$$k = \frac{P - P_i}{\delta} = \frac{d^4 G}{8 N_a D^3} = \frac{dG}{8 N_a C^3}. \tag{17.34}$$

The shear stresses for static and dynamic conditions are given by Eqs. (17.8) and (17.10), respectively. The preload stress τ_i is obtained by using Eq. (17.4).

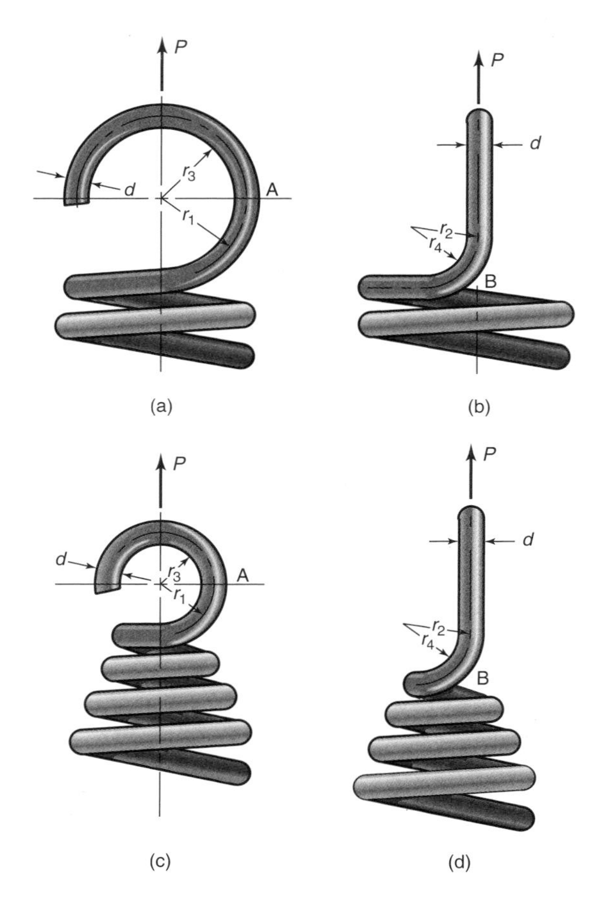

Figure 17.9: Ends for extension springs. (a) Conventional design; (b) side view of Fig. 17.9a; (c) improved design over Fig. 17.9a; (d) side view of Fig. 17.9c.

The preferred range of 4 to 12 for the spring index is equally valid for extension and compression springs. The initial tension wound into the spring, giving a preload, is described by

$$P_i = \frac{\pi \tau_i d^3}{8D} = \frac{\pi \tau_i d^2}{8C}. \tag{17.35}$$

Recommended values of τ_i depend on the spring index and are given in Fig. 17.11. Springs should be designed for midway in the preferred spring index range in Fig. 17.11.

The critical stresses in the hook occur at sections A and B, as shown in Fig. 17.9. At section A the stress is due to bending and transverse shear, and at section B the stress is due to torsion. The bending moment and transverse shear stresses acting at section A can be expressed as

$$\sigma_A = \left(\frac{Mc}{I}\right)\left(\frac{r_1}{r_3}\right) + \frac{P_A}{A} = \left(\frac{32 P_A r_1}{\pi d^3}\right)\left(\frac{r_1}{r_3}\right) + \frac{4P_A}{\pi d^2}. \tag{17.36}$$

The shear stress acting at section B is

$$\tau_B = \frac{8 P_B C}{\pi d^2}\left(\frac{r_2}{r_4}\right). \tag{17.37}$$

Radii r_1, r_2, r_3, and r_4 are given in Fig. 17.9. Recommended design practice is that $r_4 > 2d$. Values of σ_A and τ_B given in Eqs. (17.36) and (17.37) are the design stresses. These stresses are compared with the allowable stresses, taking into account

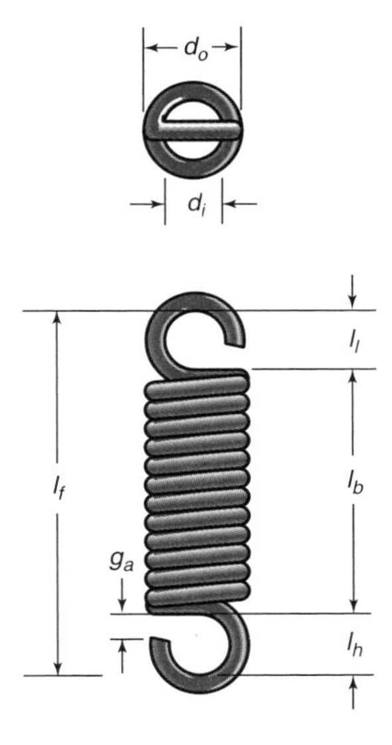

Figure 17.10: Important dimensions of a helical extension spring.

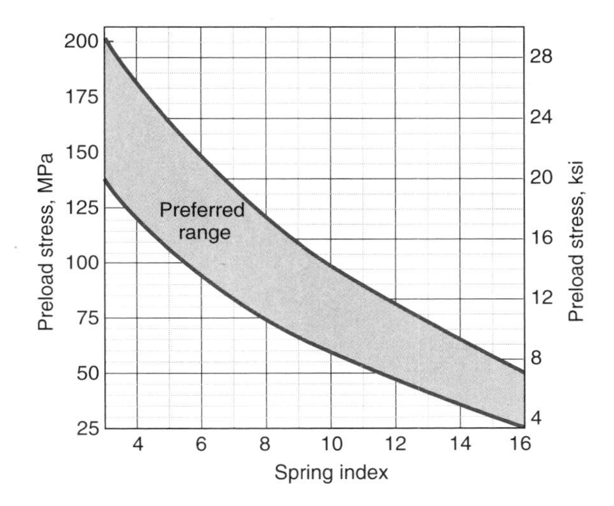

Figure 17.11: Preferred range of preload stress for various spring indexes.

the type of loading to determine if failure will occur. The allowable shear stress to be used with Eq. (17.37) is the shear yield stress given in Eq. (17.3). The allowable stress to be used with Eq. (17.36) is the yield strength in tension, given as

$$S_y = 0.60 S_{ut}. \tag{17.38}$$

Example 17.5: Analysis of a Helical Tension Spring

Given: A helical extension spring similar to that shown in Figs. 17.9a and b is made of hard-drawn wire with a mean coil diameter of 10 mm, a wire diameter of 2 mm, and 120 active coils. The hook radius is 6 mm (r_1 = 6 mm) and the bend radius is 3 mm (r_2 = 3 mm). The preload is 30 N and the free length is 264 mm.

Find: Determine the following:

(*a*) Tensile and torsional yield strength of the wire

(*b*) Initial torsional shear stress of the wire

(*c*) Spring rate

(*d*) Force required to cause the normal stress in the hook to reach the tensile yield strength

(*e*) Force required to cause the torsional stress in the hook to reach the yield stress

(*f*) Distance between hook ends if the smaller of the two forces found in parts d and e is applied

Solution:

(*a*) From Table 17.2 for hard-drawn wire, m = 0.192 and A_p = 1750 MPa. From Eq. (17.2), the ultimate strength in tension is

$$S_{ut} = \frac{A_p}{d^m} = \frac{1750 \times 10^6}{(2)^{0.192}} = 1.532 \text{ GPa}.$$

From Eq. (17.3), the torsional yield stress, which in this case is the allowable shear stress, is

$$S_{sy} = 0.40 S_{ut} = 0.40\left(1.532 \times 10^9\right) = 0.6128 \text{ GPa}.$$

From Eq. (17.38), the tensile yield strength is

$$S_y = 0.60 S_{ut} = 0.60\left(1.532 \times 10^9\right) = 0.9192 \text{ GPa}.$$

(*b*) For static loading, the transverse shear factor is given by Eq. (17.9) as

$$K_d = 1 + \frac{0.5}{C} = 1 + \frac{0.5}{5} = 1.1.$$

Using Eq. (17.35) gives

$$P_i = \frac{\pi \tau_i d^2}{8C} = \frac{\pi \tau_i (0.002)^2}{8(5)} = 30 \text{ N},$$

or $\tau_i = 95.49$ MPa.

(*c*) From Table 17.1 for hard-drawn wire, $G = 79.29 \times 10^9$ Pa. From Eq. (17.34), the spring rate is

$$k = \frac{dG}{8N_a C^3} = \frac{(0.002)\left(79.29 \times 10^9\right)}{8(120)(5)^3} = 1322 \text{ N/m}.$$

(*d*) The critical bending stress in the hook, from Eq. (17.36) and the yield stress in tension given above, is

$$\sigma_A = \left(\frac{32 P_A r_1}{\pi d^3}\right)\left(\frac{r_1}{r_3}\right) + \frac{4P_A}{\pi d^2} = S_y.$$

From Fig. 17.9a, note that $r_3 = r_1 - 1 = d/2 = 5$ mm. Therefore,

$$\frac{32 P_A (0.006)^2}{\pi (0.002)^3 (0.005)} + \frac{4P_A}{\pi (0.002)^2} = 0.9192 \times 10^9.$$

Solving for P_A gives the critical load where failure will first start to occur by normal stress in the hook as $P_A = 96.90$ N.

(*e*) The critical torsional shear stress in the hook can be determined from Eq. (17.37), or

$$\tau_B = \frac{8P_B C}{\pi d^2}\left(\frac{r_2}{r_4}\right) = S_{sy}.$$

Therefore,

$$\frac{8P_B(5)}{\pi(0.002)^2}\left(\frac{3}{2}\right) = 0.6128 \times 10^9,$$

so that $P_B = 128.3$ N. The smaller load is $P_A = 96.91$ N, which indicates that failure will first occur by normal stress in the hook.

(*f*) Using P_A in Eq. (17.34) gives the deflection as

$$\delta = \frac{P_A - P_i}{k} = \frac{96.91 - 30}{1322} = 0.05061 \text{ m} = 50.61 \text{ mm}.$$

The distance between hook ends is

$$l_f + \delta = 264 + 50.61 = 314.6 \text{ mm}.$$

Example 17.6: Helical Tension Spring Stiffness

Given: The ends of a helical extension spring are manufactured according to the improved design shown in Fig. 17.9c. The conical parts have five coils at each end with a winding diameter changing according to

$$D = D_o\left(1 - \frac{\zeta}{15\pi}\right),$$

where ζ is a coordinate. The cylindrical part of the spring has an outside diameter D_o and 20 coils.

Find: How much softer does the spring become with the conical parts at the ends?

Solution: Referring to Fig. 17.2, the torque in the spring is $T = PR = PD/2$. From Eq. (4.31), the change of angular deflection for a small length is

$$d\theta = \frac{T}{GJ}\,dl.$$

The axial compression of the conical part of the spring is

$$d\delta_c = R d\theta = \frac{RT}{GJ}dl = \frac{PR^2}{GJ}R\,d\zeta = \frac{PR^3}{GJ}\,d\zeta.$$

The total deflection of the two conical ends is

$$\begin{aligned}\delta_c &= 2\int_0^{10\pi} \frac{PR^3}{GJ}\,d\zeta \\ &= \frac{2P}{GJ}\int_0^{10\pi}\left(\frac{D_o}{2}\right)^3\left(1 - \frac{\zeta}{15\pi}\right)^3 d\zeta \\ &= \frac{75\pi}{81}\frac{PD_o^3}{GJ}.\end{aligned}$$

From Eq. (17.14), the deflection in the main part of the spring is

$$\delta = \left(\frac{D_o}{2}\right)\frac{Tl}{GJ} = \left(\frac{D_o}{2}\right)\frac{PD_o\pi D_o(20)}{2GJ} = \frac{5\pi PD_o^3}{GJ}.$$

Therefore,

$$\frac{\delta_t}{\delta} = \frac{\delta + \delta_c}{\delta} = \frac{\left(5\pi + \dfrac{75\pi}{81}\right)\dfrac{PD_o^3}{GJ}}{5\pi\dfrac{PD_o^3}{GJ}} = 1.185.$$

Thus, the spring becomes 18.5% more compliant by including the conical ends.

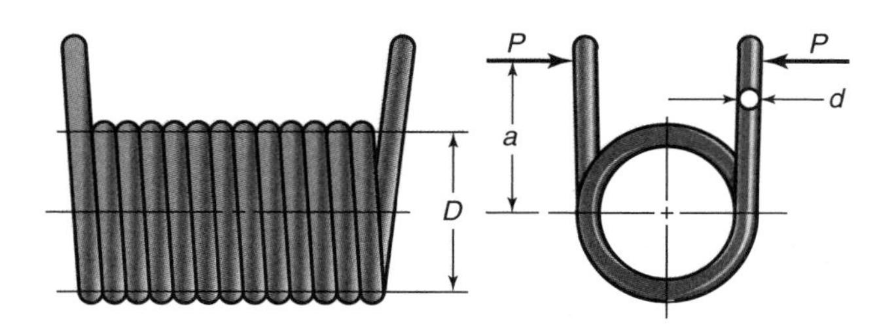

Figure 17.12: Helical torsion spring.

17.5 Helical Torsion Springs

The previous section demonstrated that the helical coil spring can be loaded either in compression or in tension. That is, helical springs can support axial loads. This section considers a spring that supports a torsional load. Figure 17.12 shows a typical example of a **helical torsion spring**. The coil ends can have a great variety of shapes to suit various applications. The coils are usually close wound like an extension spring, as shown in Fig. 17.12, but differ from extension springs in that torsion springs do not have any initial tension. Also, note from Fig. 17.12 that the ends are shaped to transmit torque rather than force as is the case for compression and tension springs. The torque is applied about the axis of the helix. The torque actually acts as a bending moment on each section of the wire, so that the primary stress in a torsional spring is a normal stress due to bending. (In contrast, in a compression or extension spring, the load produces a torsional stress in the wire.) During manufacture, the residual stresses built in during winding are in the same direction, but of opposite sign, as the working stresses that occur during use. These residual stresses are useful in making the spring stronger by opposing the working stress, provided that the load is always applied to cause the spring to wind up. Because the residual stress opposes the working stress, torsional springs are designed to operate at stress levels that equal or even exceed the yield strength of the wire.

The maximum bending stress occurs at the inner fiber of the coil, and, for wire of circular cross-section, it is given as

$$\sigma = \frac{K_i M c}{I} = \frac{32 K_i M}{\pi d^3}, \tag{17.39}$$

where

$$K_i = \frac{4C^2 - C - 1}{4C(C-1)}. \tag{17.40}$$

The angular deflection in radians is

$$\theta_{\rm rad} = \frac{Ml_w}{EI}, \tag{17.41}$$

where $l_w = \pi D N_a$ is the length of wire in the spring. The angular deflection in revolutions is

$$\theta_{\rm rev} = \frac{M\pi D N_a}{E\left(\frac{\pi d^4}{64}\right)}\left(\frac{1 \text{ rev}}{2\pi \text{ rad}}\right) = \frac{32MDN_a}{\pi E d^4} = \frac{10.19MDN_a}{Ed^4}. \tag{17.42}$$

The angular spring rate, in Nm/rev, is

$$k_\theta = \frac{M}{\theta_{\rm rev}} = \frac{Ed^4}{10.19DN_a}. \tag{17.43}$$

The number of active coils is

$$N_a = N_b + N_e, \tag{17.44}$$

where

N_b = number of coils in body

N_e = number of end coils = $\dfrac{(l_1 + l_2)}{3\pi D}$

l_1, l_2 = length of ends, m

The inside diameter of a loaded torsion spring is

$$D_i' = \frac{N_a D_i}{N_a'}, \tag{17.45}$$

where

N_a = number of active coils at no load

D_i = coil inside diameter at no load, m

N_a' = number of active coils when loaded

D_i' = coil inside diameter when loaded, m

Torsion springs are frequently used over a round bar. When a load is applied to a torsion spring, the spring winds up, causing a decrease in the inside diameter. For design purposes, the inside diameter of the spring must never become equal to the diameter of the bar; otherwise, a spring failure will occur. On the other hand, a locating device or spring-loaded clip can use this feature to secure the spring in a desired location; when load is applied to the spring, its inner diameter will increase, allowing the spring to slide freely over a bar or tube. When load is removed, it will develop a restrictive force due to the reduction in spring diameter.

Example 17.7: Analysis of a Torsion Spring

Given: A torsion spring similar to that shown in Fig. 17.12 is made of 0.055-in.-diameter music wire and has six coils in the body of the spring and straight ends. The distance a in Fig. 17.12 is 2 in. The outside diameter of the coil is 0.654 in.

Find:

(*a*) If the maximum stress is set equal to the yield strength of the wire, what is the corresponding moment?

(*b*) With load applied, what is the angular deflection in revolutions?

(*c*) When the spring is loaded, what is the resulting inside diameter?

(*d*) What size bar should be placed inside the coils?

Solution:

(*a*) The mean coil diameter is

$$D = D_o - d = 0.654 - 0.055 = 0.599 \text{ in.}$$

The coil inside diameter without a load is

$$D_i = D - d = 0.599 - 0.055 = 0.544 \text{ in.}$$

The spring index is

$$C = \frac{D}{d} = \frac{0.599}{0.055} = 10.89.$$

From Eq. (17.40),

$$K_i = \frac{4C^2 - C - 1}{4C(C-1)} = \frac{4(10.89)^2 - 10.89 - 1}{4(10.89)(10.89-1)} = 1.074.$$

From Eq. (17.2) and Table 17.2 for music wire,

$$S_{ut} = \frac{A_p}{d^m} = \frac{196 \times 10^3}{(0.055)^{0.146}} = 299.3 \times 10^3 \text{ psi.}$$

From Eq. (17.38),

$$S_y = 0.6 S_{ut} = 179.6 \times 10^3 \text{ psi.}$$

After equating the bending stress to the yield strength, Eq. (17.39) gives

$$M = \frac{\pi d^3 S_y}{32K_i} = \frac{\pi(0.055)^3\left(179.6 \times 10^3\right)}{32(1.074)} = 2.731 \text{ lb-in.}$$

(*b*) The number of active coils is

$$N_a = N_b + N_e = 6 + \frac{4}{3\pi(0.599)} = 6.709.$$

From Table 17.1, the modulus of elasticity for music wire is 30×10^6 psi. From Eq. (17.43), the angular spring rate is

$$k_\theta = \frac{Ed^4}{10.18DN_a} = \frac{\left(30 \times 10^6\right)(0.055)^4}{10.18(0.599)(6.709)} = 6.710 \text{ lb-in./rev.}$$

The angular deflection is

$$\theta_{\rm rev} = \frac{M}{k_\theta} = \frac{2.731}{6.710} = 0.4070 \text{ rev.}$$

(*c*) The number of coils when loaded is

$$N_a' = N_a + \theta_{\rm rev} = 6.709 + 0.4070 = 7.116.$$

From Eq. (17.45), the inside diameter after loading is

$$D_i' = \frac{N_a}{N_a'} D_i = \frac{6.709}{7.116}(0.544) = 0.5129 \text{ in.}$$

(*d*) Thus, if the bar inside the coils is 1/2 in. in diameter, the spring should not fail.

Example 17.8: Stress Analysis of a Torsion Spring

Given: A hand-squeeze grip training device with two 100-mm-long handles is connected to a helical torsion spring with 4.375 coils. The handles and coils are made of music wire. When actuated, the helical torsion spring has 4.5 coils and has a spring rate of 2000 N/m.

Find: Determine the spring dimensions for a safety factor of 2 against breakage.

Solution: From Eq. (17.39), the maximum bending stress is

$$\sigma = \frac{K_i M c}{I} = \frac{K_i M r}{I}. \qquad (a)$$

Using Eq. (17.41) gives the moment needed to deform the spring to 4.5 coils as

$$M = \frac{EI\theta_{\text{rad}}}{l_w} = \frac{EI(4.5 - 4.375)2\pi}{l_w} = \frac{0.25\pi EI}{l_w}. \qquad (b)$$

Substituting Eq. (*b*) into Eq. (*a*) gives

$$\sigma = \frac{\pi}{8}\frac{K_i d E}{l_w}.$$

But $l_w = \pi D N_a$, so that

$$\sigma = \frac{\pi K_i d E}{8\pi D N_a} = \frac{K_i E}{8 C N_a}.$$

For music wire, $E = 207$ GPa, and it was given that $N_a = 4.375$. Therefore, the stress is

$$\sigma = \frac{K_i}{C}\left[\frac{(207 \times 10^9)}{8(4.375)}\right] = 5.914\frac{K_i}{C}. \qquad (c)$$

To get the spring rate of 2000 N/m at the ends of the handles (see Fig. 17.12), the angular spring rate is

$$k_\theta = (2000)(0.1)(0.1) = 20 \text{ N-m/rad}.$$

From Eq. (17.43),

$$d^3 = \frac{(10.18)CN_a k_\theta}{E} = C\frac{(10.18)(4.375)(20)}{207 \times 10^9},$$

or

$$d^3 = 4.303 \times 10^{-9} C. \qquad (d)$$

As a first try, assume $C = 6$. From Eq. (17.40),

$$K_i = \frac{4C^2 - C - 1}{4C(C-1)} = \frac{4(6)^2 - 6 - 1}{4(6)(5)} = 1.142.$$

From Eq. (*d*),

$$d^3 = 25.82 \times 10^{-9},$$

$$d = 2.956 \times 10^{-3} \text{ m} = 2.956 \text{ mm}.$$

Therefore,

$$D = 6\left(2.956 \times 10^{-3}\right) = 17.73 \times 10^{-3} = 17.73 \text{ mm}.$$

From Eq. (*c*) the maximum design stress is

$$\sigma = (5.914)\frac{1.142}{6} = 1.126 \text{ GPa}. \qquad (e)$$

From Eq. (17.2), the allowable stress is

$$\sigma_{\text{all}} = S_{ut} = \frac{A_p}{d^m} = \frac{2170 \times 10^6}{(2.956)^{0.146}} = 1.852 \text{ GPa}.$$

The safety factor is

$$n_s = \frac{\sigma_{\text{all}}}{\sigma} = \frac{1.852}{1.126} = 1.645.$$

Since the safety factor is too low, try $C = 7.5$. From Eq. (17.40),

$$K_i = \frac{4C^2 - C - 1}{4C(C-1)} = \frac{4(7.5)^2 - 7.5 - 1}{4(7.5)(6.5)} = 1.110.$$

From Eq. (*d*)

$$d^3 = 4.303 \times 10^{-9}(7.5) = 32.27 \times 10^{-9},$$

$$d = 3.184 \times 10^{-3} \text{ m} = 3.184 \text{ mm}.$$

Therefore,

$$D = 7.5\left(3.184 \times 10^{-3}\right) = 23.88 \times 10^{-3} \text{ m} = 23.88 \text{ mm}.$$

From Eq. (*c*), the maximum design stress is

$$\sigma = (5.914)\frac{1.110}{7.5} = 0.8753 \text{ GPa}.$$

From Eq. (17.2), the allowable stress is

$$\sigma_{\text{all}} = S_{ut} = \frac{A_p}{d^m} = \frac{2170 \times 10^6}{(3.184)^{0.146}} = 1.832 \text{ GPa}.$$

Solving for the safety factor,

$$n_s = \frac{\sigma_{\text{all}}}{\sigma} = \frac{1.832}{0.8753} = 2.09.$$

Since $n_s > 2$, the design is adequate according to the constraint given in the problem statement.

17.6 Leaf Springs

Multiple-**leaf springs** are in wide use, especially in the railway and automotive industries, such as the truck suspension arrangement shown in Fig. 17.13. A multiple-leaf spring can be considered as a simple cantilever type, as shown schematically in Fig. 17.14b. It can also be considered as a triangular plate, as shown in Fig. 17.14a. The triangular plate is cut into n strips of width b and stacked in the graduated manner shown in Fig. 17.14b. The layers can slide over each other, as long as they are not welded, clamped, or fastened together. Some interaction between layers is desirable, however, as is discussed below.

Before analyzing a multiple-leaf spring, first consider a single-leaf, cantilevered spring with constant rectangular cross section. For bending of a straight beam, from Eq. (4.45), the magnitude of the bending stress is given by

$$\sigma = \frac{Mc}{I}. \qquad (17.46)$$

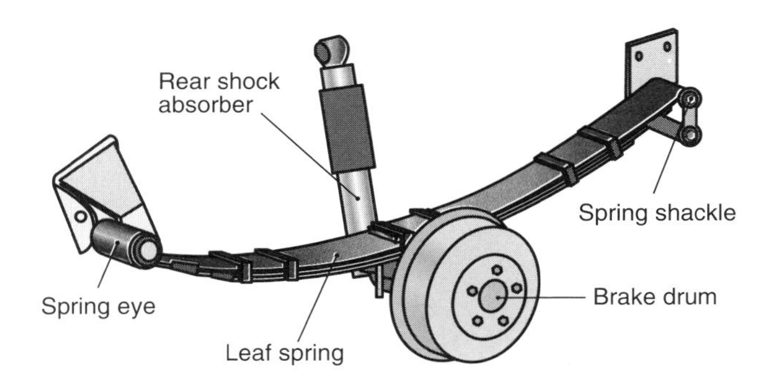

Figure 17.13: Illustration of a leaf spring used in an automotive application.

For a rectangular cross section of base, b, height, t, and moment, $M = Px$, Eq. (17.46) becomes

$$\sigma = \frac{6M}{bt^2} = \frac{6Px}{bt^2}. \tag{17.47}$$

The maximum moment occurs at $x = l$ and at the outer fiber of the cross section, or

$$\sigma_{\max} = \frac{6Pl}{bt^2} \tag{17.48}$$

From Eq. (17.47) the stress is a function of position along the beam. One method of design optimization requires that the stress be constant all along the beam. To accomplish this, either t is held constant and b is allowed to vary, or vice-versa, so that the stress is constant for any x. It is convenient from a manufacturing perspective to use a constant thickness, so that cold-rolled shapes can be used. Thus,

$$\frac{b(x)}{x} = \frac{6P}{t^2\sigma} = \text{Constant}. \tag{17.49}$$

Equation (17.49) is linear, giving the triangular shape shown in Fig. 17.14a and a constant stress for any x.

The triangular-plate spring and the equivalent multiple-leaf spring have identical stresses and deflection characteristics with two exceptions:

1. Interleaf friction provides damping in the multiple-leaf spring.
2. The multiple-leaf spring can carry a full load in only one direction.

The deflection and spring rate for the ideal leaf spring are

$$\delta = \frac{6Pl^3}{Enbt^3}, \tag{17.50}$$

$$k = \frac{P}{\delta} = \frac{Enbt^3}{6l^3}. \tag{17.51}$$

Example 17.9: Leaf Spring

Given: A 35-in.-long cantilever spring is composed of eight graduated leaves. The leaves are 1.75 in. wide. A load of 500 lb at the end of the spring causes a deflection of 3 in. The spring is made of steel with modulus of elasticity of 30×10^6 psi.

Find: Determine the thickness of the leaves and the maximum bending stress.

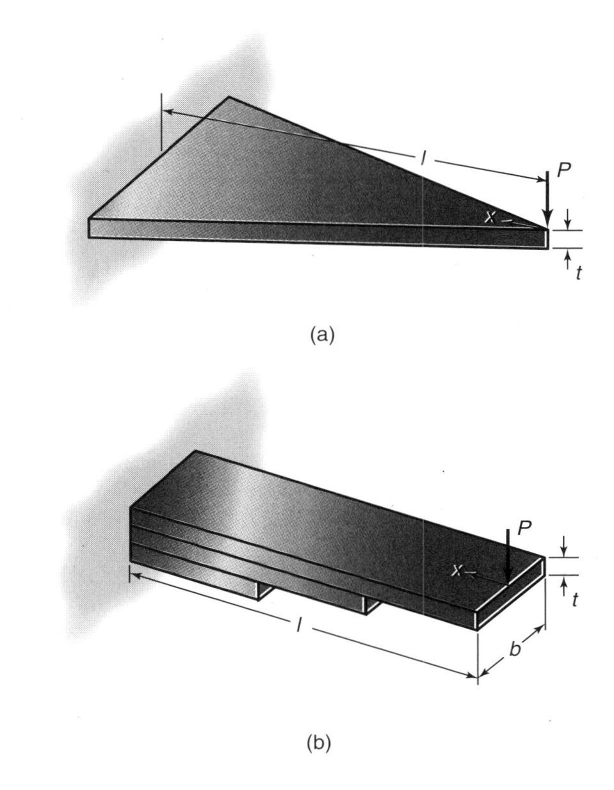

Figure 17.14: Leaf spring. (a) Triangular plate, cantilever spring; (b) equivalent multiple-leaf spring.

Solution: From Eq. (17.50),

$$\delta = \frac{6Pl^3}{Enbt^3},$$

or

$$t^3 = \frac{6Pl^3}{\delta Enb},$$

so that

$$t^3 = \frac{6(500)(35)^3}{3\,(30 \times 10^6)\,(8)(1.75)} = 0.1021 \text{ in.}^3$$

or $t = 0.4674$ in. From Eq. (17.48), the maximum bending stress is

$$\sigma_{\max} = \frac{6Pl}{nbt^2} = \frac{6(500)(35)}{(8)(1.75)(0.4674)^2} = 34,340 \text{ psi}.$$

Example 17.10: Design of a Leaf Spring

Given: A leaf spring for a locomotive wheel axle is made of spring steel with a thickness of 20 mm and an allowable bending stress of 1050 MPa. The spring has a modulus of elasticity of 207 GPa, is 1.6 m long from tip to tip, and carries a weight of 12,500 N at the middle of each leaf spring.

Find:

(*a*) The width of the spring for a safety factor of 3

(*b*) How high the locomotive has to be lifted during overhauls to unload the springs

Solution:

(*a*) From Eq. (1.1), the design stress is

$$\sigma_{\max} = \frac{\sigma_{\text{all}}}{3} = \frac{1050 \times 10^6}{3} = 0.350 \text{ GPa.} \qquad (a)$$

From Eq. (17.48), the maximum bending stress is

$$\sigma_{\max} = \frac{6Pl}{bt^2}.$$

Because P is the force applied at the tip, the load is 6250 N and the length is 0.8 m in the locomotive spring. Therefore,

$$\sigma_{\max} = \frac{6Pl}{bt^2} = \frac{6(6250)(9.807)(0.8)}{b(0.020)^2} = \frac{0.7355 \times 10^9}{b}.$$

Making use of Eq. (*a*) results in $b = 2.101$ m. Splitting into 10 leaves gives the width of the leaf spring as

$$\frac{b}{10} = 0.2101 \text{ m} = 210.1 \text{ mm}.$$

(*b*) The locomotive has to be lifted at least as high as the spring's deflection. Thus, from Eq. (17.51),

$$\delta = \frac{P}{k} = \frac{6Pl^3}{Ebt^3} = \frac{6(6250)(9.807)(0.8)^3}{(207 \times 10^9)(2.101)(0.020)^3} = 0.05412 \text{ m} = 54.1 \text{ mm}.$$

17.7 Gas Springs

Gas springs, as shown in Fig. 17.15, are very commonly applied as support struts or counterweights. These are commonly seen in applications such as automobile liftgate or hood counterbalances or support structures that allow height adjustment for office chairs and hospital beds.

A cross-section of a typical gas spring is shown in Fig. 17.15b. There are many potential mounting options, ranging from threaded shafts to incorporation of eyelets or integral brackets. The gas spring has a high-pressure (typically 10 to 20 MPa) nitrogen or air chamber that acts against the piston as shown. The piston has small diameter orifices that allow the pressurized gas to flow in order to equalize pressure on both sides of the piston. Referring to Fig. 17.15, the pressurized gas on the left side of the piston acts on the entire internal (bore) area. On the right side, the pressurized gas acts on the annular region around the rod. For slow operation, a force on the moving rod is opposed by a force

$$F = pA, \qquad (17.52)$$

where p is the initial pressurization and A is the area of the rod. Note that for rapid operation, a much higher force is needed because the pressure in the two chambers is not equal, and one of the chambers will be compressed by the sudden movement of the piston. This increased resistance to

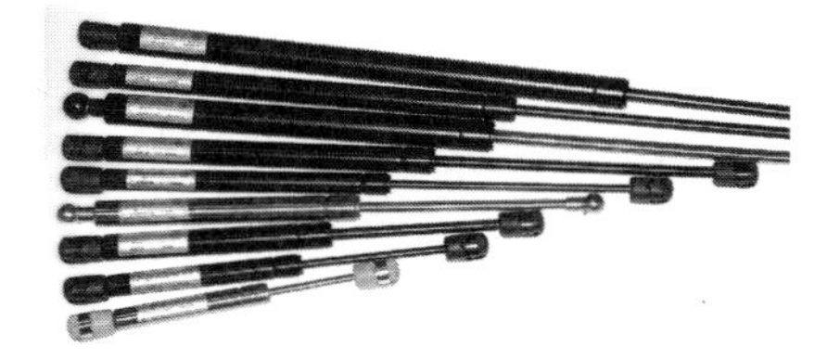

(a)

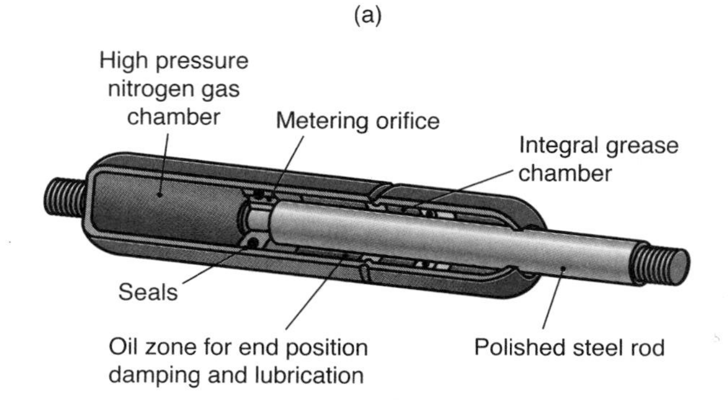

(b)

Figure 17.15: Gas springs. (a) A collection of gas springs. Note that the springs are available with a wide variety of end attachments and strut lengths. *Source:* Courtesy of Newport Engineering Associates, Inc. (b) Schematic illustration of a typical gas spring.

motion produces a damping effect, which is a fundamental characteristic of such springs. The amount of damping can be tailored to a particular application by changing the diameter and length of the orifice in the piston. A small amount of oil is added to one side of the piston to provide end position damping. This can be obtained at either the end of the extension stroke or compression stroke by changing the mounting orientation of the gas spring.

Notice in Fig. 17.15b that there is also an integral grease chamber behind the rod seals. The rod is polished, and the grease helps to prevent wear of the rod and also assists in sealing the pressurized gas. Clearly, the rod cannot be subjected to bending stresses or damage from impact or abrasion from dirt or other debris, or else the seal will be compromised, pressure will be lost, and uneven rod wear will result.

There is considerable flexibility in the design of gas springs, and a very long stroke can be achieved with minimal change in force. By changing the length of the cylinder and/or rod, long strokes can be achieved over relatively long distances. However, gas springs have a minimum length which must be accommodated by the application. Gas springs are rated at room temperature, but changes in ambient temperatures will affect the spring capacity. The **Ideal Gas Law** can be used to estimate the change in pressure and therefore the change in piston force as a function of temperature. The Ideal Gas Law using the specific gas constant is given by

$$p = \frac{\bar{R}t_m}{v}, \qquad (17.53)$$

where

p = pressure, N/m^2
$\bar{R}$ = specific gas constant, 296.8 J/kg-K for nitrogen, 286.9 J/kg-K for air
t_m = absolute temperature, K
v = specific volume, m^3/kg

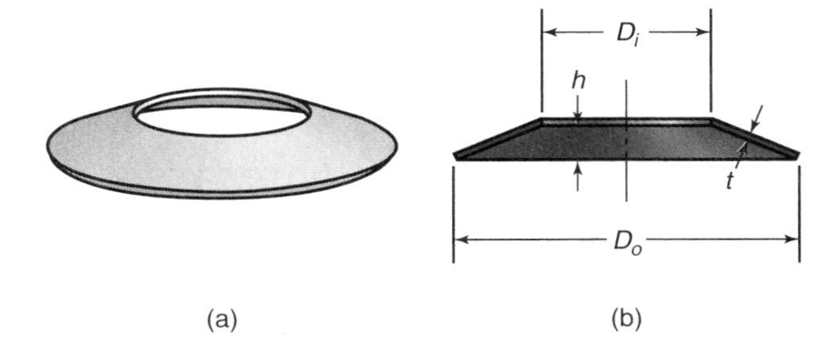

Figure 17.16: Typical Belleville spring. (a) Isometric view of Belleville spring; (b) cross section, with key dimensions identified.

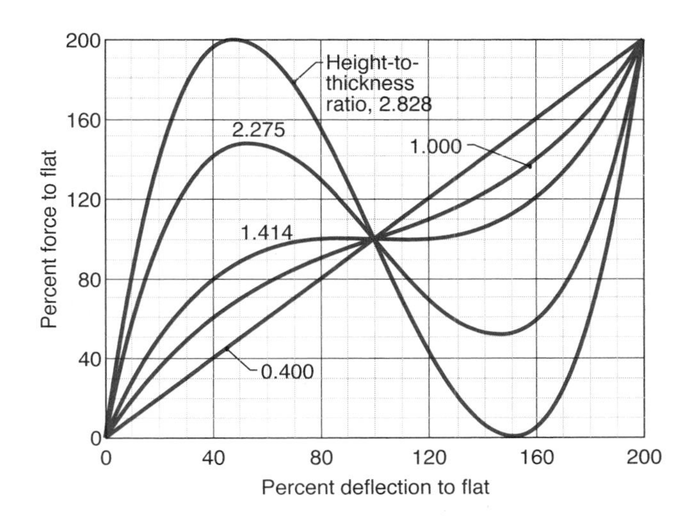

Figure 17.17: Force-deflection response of Belleville spring given by Eq. (17.54).

Equation (17.53) can be used to obtain the resisting force of a gas spring for a wide temperature range. However, most gas spring manufacturers will provide estimates, such as a 3.5% increase in force with a 10°C increase in temperature, etc., and these approximations may be sufficient for design purposes.

17.8 Belleville Springs

Belleville springs are named after their inventor, J.F. Belleville, who patented their design in 1867. Shaped like a coned disk, these springs are especially useful where large forces are desired for small spring deflections. In fact, because many lock washers used with bolts follow this principle to obtain a bolt preload (see Section 16.4.5), these springs are often referred to as Belleville washers. Typical applications include clutch plate supports, gun recoil mechanisms, and a wide variety of bolted connections.

Figure 17.16 shows the cross section of a Belleville spring. Two of the critical parameters affecting a Belleville spring are the diameter ratio, $R_d = D_o/D_i$, and the height-to-thickness ratio, h/t. From Fig. 17.17, the behavior of a Belleville spring is very complex and varies considerably with a change in h/t. For low h/t values, the spring acts almost linearly, whereas large h/t values lead to highly nonlinear behavior. A phenomenon occurring at large h/t is "snapthrough buckling." In snapthrough, the spring deflection is unstable once the maximum force is applied; the spring quickly deforms, or snaps, to the next stable position.

The force and deflection for Belleville springs are related by

$$P = \frac{4E\delta}{K_1 D_o^2 (1-\nu^2)} \left[(h-\delta)\left(h - \frac{\delta}{2}\right) t + t^3 \right] \tag{17.54}$$

where E is the elastic modulus, δ is the deflection from the unloaded state, D_o is the coil outside diameter, ν is Poisson's ratio for the material, h is the spring height, and t is the spring thickness. The factor K_1 is given by

$$K_1 = \frac{6}{\pi \ln R_d} \left[\frac{(R_d - 1)^2}{R_d^2} \right], \tag{17.55}$$

where R_d is the diameter ratio, given by

$$R_d = \frac{D_o}{D_i}. \tag{17.56}$$

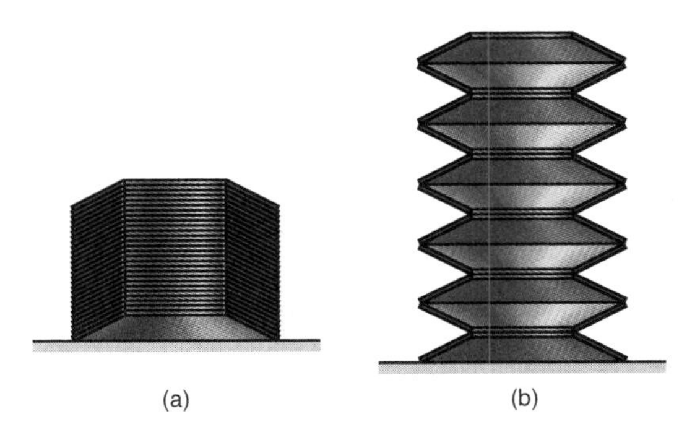

Figure 17.18: Stacking of Belleville springs. (a) In parallel; (b) in series.

The force required to totally flatten a Belleville spring is given by

$$P_{\text{flat}} = \frac{4Eht^3}{K_1 D_o^2 (1-\nu^2)}. \tag{17.57}$$

The forces associated with a Belleville spring can be multiplied by stacking them in parallel, as shown in Fig. 17.18a. The deflection associated with a given force can be increased by stacking the springs in series, as shown in Fig. 17.18b. Because such a configuration may be susceptible to buckling, a central support is necessary.

17.9 Wave Springs

Figure 17.19 shows a **wave spring**, which consists of a helical pattern with an incorporated wave. Wave springs have been used for a variety of industrial applications, such as in valves, preloading bearings, etc. Wave springs are used instead of helical compression springs when compact designs are required, as wave springs generate the same spring rates as helical springs but require up to one-half the axial space. The stress in a wave spring is given by

$$\sigma = \frac{3\pi PD}{4bt^2 N_w^2}, \tag{17.58}$$

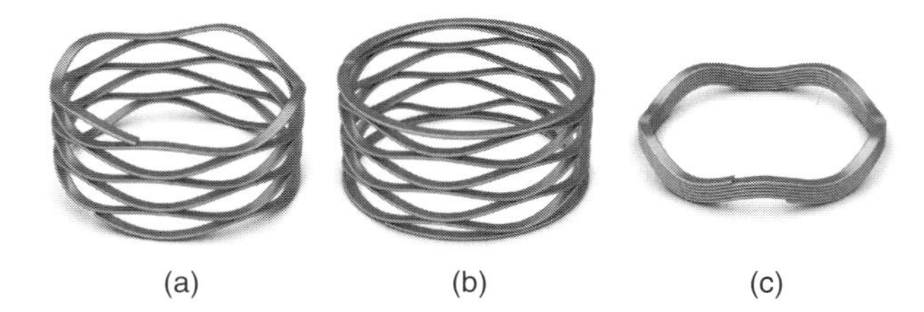

Figure 17.19: Examples of common wave spring configurations. (a) Common crest-to-crest orientation; (b) crest-to-crest orientation with shim ends; (c) nested wave springs. *Source:* Courtesy of Smalley Co.

Table 17.4: Multiple wave factor, K_w, used to calculate wave spring stiffness. *Source:* from Lee Spring, Inc.

Waves per turn, N_w	Multiple wave factor, K_w
2.0–4.0	3.88
4.5–6.5	2.9
7.0–9.5	2.3
> 9.5	2.13

where
P = applied force, N
D = mean spring diameter, m
b = width of wave, m
t = thickness of material, m
N_w= number of waves per turn.

Note that the number of waves per turn must be in 1/2 turn increments, or the wave spring will not stack properly. Wave springs are linear over a design range, but should not be compressed to their solid length. Thus, it is preferred that the spring be compressed so that the height per turn is not less than twice the material thickness. Most wave springs have between 2 and 20 turns, with a width between 3 and 10 times the material thickness. Care must be taken in prescribing the cages for wave springs, as these springs expand slightly (usually less than 10% of the spring diameter) when compressed.

The stiffness of a wave spring is given by

$$k = \frac{P}{d} = \frac{Ebt^3 N_w^4 D_o}{K_w D^3 N_a D_i}, \qquad (17.59)$$

where
E = elastic modulus, N/m^2
D_o = outer diameter, m
D_i = inner diameter, m
K_w= multiple wave factor, as given in Table 17.4
N_a = number of active turns

Wave springs have an advantage over stacked Belleville springs discussed in Section 17.8 and shown in Fig. 17.18 in that a wave spring is a single component and therefore assembly is simplified.

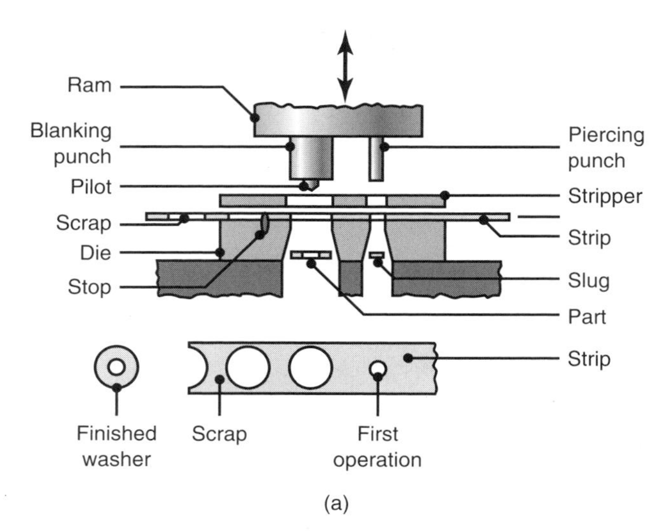

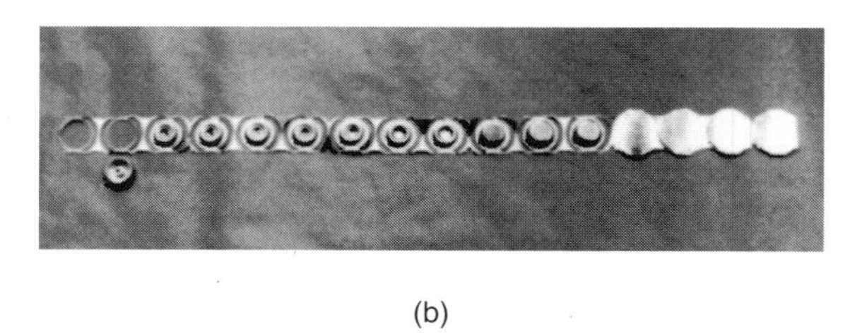

Figure 17.20: Illustration of a simple part that is produced by a progressive die. (a) Schematic illustration of the two-station die set needed to produce a washer; (b) sequence of operations to produce an aerosol can lid. *Source:* Kalpakjian and Schmid [2010]

Case Study: Springs for a Dickerman Feed Unit

Figure 17.20 illustrates a simple part that is manufactured through progressive dies. As can be seen, this small part has a number of operations performed on it, each needing specific features in the associated die. Such parts are produced economically in large numbers by using progressive or multi-staged dies, and by using a feed mechanism that advances the material a controlled and repeatable distance with every press stroke.

Figure 17.21 shows a Dickerman feed unit, which is one of the most widely used approaches for feeding material by a controlled distance. During a stroke of a mechanical press, a cam attached to the press ram or punch lowers, pushing the follower shown to the right and causing one or more springs to compress. When the ram rises, the compressed spring forces the gripping unit to the left, and since the gripping unit mechanically interferes with notched strip or sheet stock, it forces new material into the press.

Dickerman feeds are among the oldest and least expensive feeding devices in wide use. They can be mounted to feed material in any direction, on both mechanical and hydraulic presses, and the feed length can be adjusted by changing the low-cost, straight cams. Grippers can be blades (as shown) or cylinder assemblies.

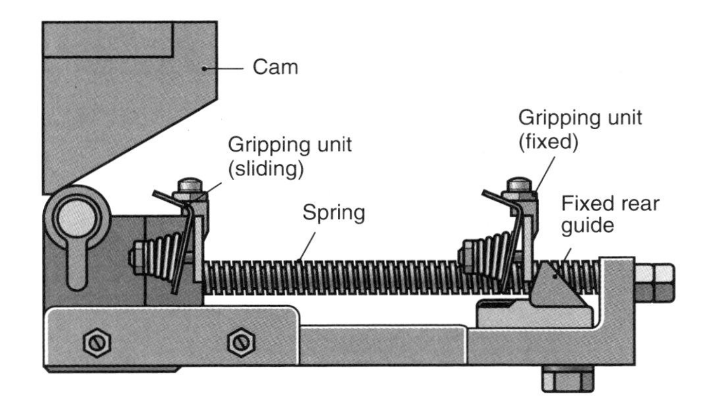

Figure 17.21: Dickerman feed unit.

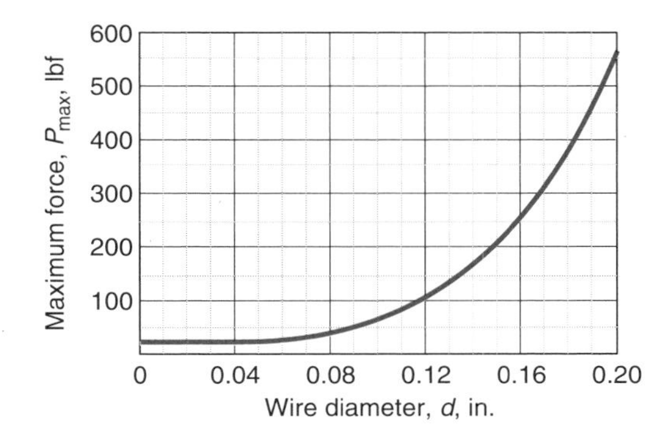

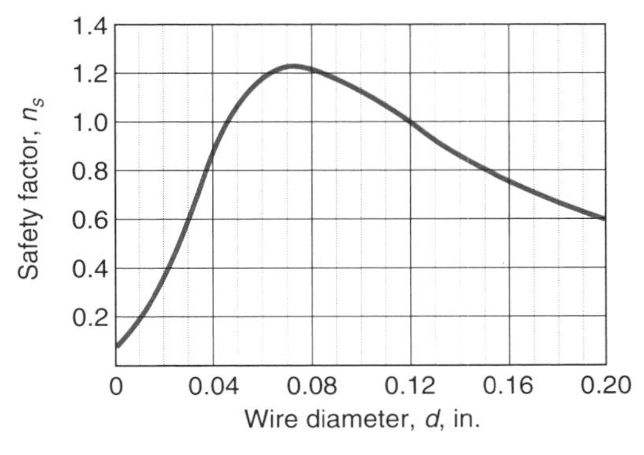

Figure 17.22: Performance of the spring in Case Study 17.1.

The case study will investigate the design alternatives in selecting a helical compression spring for a Dickerman feed unit. It is desired to use a spring that will provide at least 20 lb and have a 6-in. stroke. Buckling of the spring is a concern with this combination of load and stroke (see Section 17.3.6), so a 1/2-in.-diameter shaft will be used to guide the spring.

Not much more information is usually given before selecting a spring. Noting that the Dickerman drive can be used for different applications, it is necessary that 20 lb be provided whether the deflection is 6 in., 4 in., or a very small distance. Thus, the spring will be chosen so that at a length of 8 in., it is loaded with 20 lb of compressive force, and a relatively small stiffness will be selected while satisfying other design constraints, as discussed in Design Procedure 17.1. Another requirement on the spring is that it must have a solid length not greater than 4 in., or it will bottom out with every stroke of the press.

Economics is not a critical issue here. As long as the spring is selected with standard dimensions and made from spring steel, the cost difference between candidate springs will be low, especially when spread over the total number of parts that will be produced with the Dickerman feed. It is much more essential that a highly reliable spring be chosen because an hour of downtime (required to replace a spring and set up the Dickerman drive) costs far more than a spring. Thus, the ends will be squared and ground, as discussed in Section 17.3.5, for better load transfer into the spring. The material used will be music wire, an outstanding spring material.

No maximum force requirement is given; indeed, it appears that the maximum force is inconsequential. However, if the compressed spring contains too large a force, it could buckle the press stock, or else dynamic effects could cause too much material to feed with each stroke. An inconsequential effect is the reduction of press capacity; loads on the order of tens of pounds are not high enough to be significant.

Since mechanical presses used for progressive die work are operated unattended for long periods and typically will operate at 100 strokes per minute or faster, the Dickerman feed unit must be designed for fatigue. The important equations are (17.21) to (17.27). The solution method involves a variation of the unknown independent parameters (i.e., d, D, and N_a). All other quantities can be calculated from these parameters. However, the problem is actually much simpler, for the following reasons:

1. The more active coils in the spring, the lower the spring stiffness and hence the lower the maximum force. In addition to reducing the stress levels encountered by the spring, a large number of coils makes a more smoothly operating drive for the reasons discussed above.

2. The spring diameter is not specified. However, the guide rod in the center of the spring is 0.5 in. in diameter, so that the minimum spring diameter is 0.5 in. + d. Normally, the spring diameter would be treated as another parameter to be varied, but in this problem it has been set at 0.75 in.

   The reasons for this are twofold: First, Design Procedure 17.1 recommended a 10% clearance based on the spring diameter, which would suggest a minimum diameter around 0.55 in. However, a standard size will be more readily available, and the use of a 0.75-in.-diameter spring will lead to lower stresses and better reliability.

A design approach that combines the Goodman failure criterion (see Section 7.10.1), the use of Wahl correction factor given by Eq. (17.11), wire properties from Eq. (17.2), and the number of active coils for squared and ground ends (see Table 17.3) results in the performance prediction shown in Fig. 17.22.

From the charts, the safety factor is highest for a spring using a wire diameter of approximately 0.08 in. Thus, a wire diameter of 0.081 in. (2 mm) is chosen. The resulting spring will have 47.4 active coils and result in a maximum force (when compressed to a length of 4 in.) of 24.31 lb, with a free length of approximately 11.9 in. The safety factor for this case is 1.23.

The fact that the safety factor is small for extremely thin wires should not be surprising, since the stresses are inversely proportional to the cube of wire diameter. However, it is not immediately obvious that the safety factor would eventually increase for increasing wire diameters. The ultimate strength of the wire decreases rapidly with increasing wire diameter, so that there is in fact an optimum wire diameter. Indeed, the finding that there is an optimum wire diameter is typical.

17.10 Summary

When a machine element needs flexibility or deflection, or needs to store energy, some form of spring is often selected. Springs are used to exert forces or torques in a mechanism or to absorb the energy of suddenly applied forces. Springs frequently operate with high stresses and continuously varying forces.

This chapter provided information about spring design while considering popular types of spring designs available to engineers. Different spring materials were presented. Spring strength is a material parameter that is obviously important in spring design. The loss coefficient, which measures the fractional energy dissipated in a stress-strain cycle, is an equally important parameter.

This chapter emphasized helical compression springs. Both torsional shear stress and transverse shear stress were considered. Transverse shear is small relative to torsional shear. The maximum stress occurs at the midheight of the wire at the coil inside diameter.

The spring index is dimensionless and is the mean coil diameter divided by the wire diameter. The spring rate is the force divided by the deflection, thus having the unit of newtons per meter. The difference between these terms is important. Different conditions cause different amounts of eccentric loading that must be compensated for in the design of a compression spring. Relatively long compression springs must be checked for buckling. Also, a surge, or longitudinal vibration, should be avoided in spring design. Avoiding the natural frequency is thus recommended. Spring loading is most often continuously fluctuating, so this chapter considered the design allowance that must be made for fatigue and stress concentrations.

In helical extension springs, it was found that the hooks need to be shaped so that the stress concentration effects are decreased as much as possible. Two critical locations within a hook were analyzed. In one, the designer must consider normal stress caused by the bending moment and the transverse shear stresses; and in the other, shear stress must be considered. Both should be checked.

A spring can be designed for axial loading, either compressive or tensile, or it can be designed for torsional loading, as discovered when helical torsion springs were considered. The ends of torsion springs are shaped to transmit a torque rather than a force as for compression and tensile springs. The torque is applied about the axis of the helix and acts as a bending moment on each section of the wire.

The leaf spring, used extensively in the automobile and railway industries, was also considered. An approximate analysis considered a triangular-plate, cantilever spring, and an equivalent multiple-leaf spring. Finally, Belleville springs were considered, and their advantages explained.

Key Words

Belleville spring coned disk spring

free length length of spring when no forces are applied to it

gas spring a spring that utilizes a compressed gas and moving piston to achieve fairly uniform force over long lengths

helical compression spring most common spring, wherein round wire is wrapped into cylindrical form with constant pitch between adjacent coils and is loaded in compression

helical extension spring spring wherein round wire is wrapped into cylindrical form with constant pitch between adjacent coils and is loaded in tension

helical torsion spring helical coil spring loaded in torsion

leaf spring spring based on cantilever action

pitch distance, measured parallel to coil axis, from center of one coil to center of adjacent coil

solid length length of spring when all adjacent coils are in metal-to-metal contact

spring flexible machine element used to exert force or torque and, at the ame time, to store energy

spring index ratio of coil to wire diameter, measure of coil curvature

spring rate ratio of applied force to spring deflection

surge stress pulse that propagates along spring

wave spring a helical coil spring with a superimposed waviness

Summary of Equations

Spring Materials:

Loss coefficient: $\Delta_v = \dfrac{\Delta U}{2U}$

Wire strength: $S_{ut} = \dfrac{A_p}{d^m}$

Wire shear strength: $S_{sy} = 0.40 S_{ut}$

Fatigue strength:

$S'_{se} = 310$ MPa (45.0 ksi) for unpeened springs,

465 MPa (67.5 ksi) for peened springs

Helical Compression Springs:

Spring index: $C = \dfrac{D}{d}$

Maximum shear stress: $\tau_{\max} = \dfrac{8DP}{\pi d^3}\left(1 + \dfrac{d}{2D}\right)$

Transverse shear factor: $K_d = \dfrac{C + 0.5}{C}$

Wahl factor: $K_w = \dfrac{4C - 1}{4C - 4} + \dfrac{0.615}{C}$

Bergsträsser factor: $K_b = \dfrac{4C + 2}{4C - 3}$

Deflection: $\delta = \dfrac{8PC^3N_a}{Gd}\left(1 + \dfrac{0.5}{C^2}\right)$

Spring rate: $k = \dfrac{P}{\delta} = \dfrac{Gd}{8C^3N_a\left(1 + \dfrac{0.5}{C^2}\right)}$

Lowest natural frequency: $f_n = \dfrac{2}{\pi N_a} \dfrac{d}{D^2} \sqrt{\dfrac{G}{32\rho}}$

Extension Springs:

Spring index: $k = \dfrac{P - P_i}{\delta} = \dfrac{d^4 G}{8 N_a D^3}$

Helical Torsion Springs:

Maximum bending stress: $\sigma = \dfrac{32 K_i M}{\pi d^3}$

where $K_i = \dfrac{4C^2 - C - 1}{4C(C-1)}$

Angular spring rate: $k_\theta = \dfrac{M}{\theta_{\text{rev}}} = \dfrac{Ed^4}{10.186 D N_a}$

Spring diameter: $D_i' = \dfrac{N_a D_i}{N_a'}$

Leaf Springs:

Maximum stress: $\sigma_{\max} = \dfrac{6Pl}{bt^2}$

Deflection: $\delta = \dfrac{6Pl^3}{Enbt^3}$

Spring rate: $k = \dfrac{P}{\delta} = \dfrac{Enbt^3}{6l^3}$

Gas Springs:

Ideal gas law: $p = \dfrac{\bar{R} t_m}{v}$

Belleville Springs:

Force-deflection relation:

$$P = \frac{4E\delta}{K_1 D_o^2 (1 - \nu^2)} \left[(h - \delta) \left(h - \frac{\delta}{2} \right) t + t^3 \right]$$

where $K_1 = \dfrac{6}{\pi \ln R_d} \left[\dfrac{(R_d - 1)^2}{R_d^2} \right]$

and $R_d = \dfrac{D_o}{D_i}$

Flattening force: $P_{\text{flat}} = \dfrac{4Eht^3}{K_1 D_o^2 (1 - \nu^2)}$

Wave Springs:

Maximum stress: $\sigma = \dfrac{3\pi PD}{4bt^2 N_w^2}$

Stiffness: $k = \dfrac{P}{d} = \dfrac{Ebt^3 N_w^4 D_o}{K_w D^3 N_a D_i}$

Recommended Readings

Budynas, R.G., and Nisbett, J.K. (2011), *Shigley's Mechanical Engineering Design*, 9th ed., McGraw-Hill.

Carlson, H. (1978) *Spring Designers Handbook*, Marcel Dekker.

Juvinall, R.C., and Marshek, K.M. (2012) *Fundamentals of Machine Component Design*, 5th ed., Wiley.

Krutz, G.W., Schuelle, J.K., and Claar, P.W. (1994) *Machine Design for Mobile and Industrial Applications*, Society of Automotive Engineers.

Mott, R. L. (2014) *Machine Elements in Mechanical Design*, 5th ed., Pearson.

Norton, R.L. (2011) *Machine Design*, 4th ed., Prentice Hall.

Rothbart, H.A., Ed. (2006) *Mechanical Design Handbook*, 2nd ed., McGraw-Hill.

SAE (1982) *Manual on Design and Application of Leaf Springs*, Society of Automotive Engineers.

SAE (1996) *Spring Design Manual*, 2nd ed., Society of Automotive Engineers.

References

Ashby, M.J. (2010) *Materials Selection in Mechanical Design*, 4th ed., Butterworth-Heinemann.

Design Handbook (1987) *Engineering Guide to Spring Design*. Associated Spring Corp., Barnes Group Inc.

Kalpakjian, S., and Schmid, S.R. (2010), *Manufacturing Engineering and Technology*, 6th ed., Pearson.

Relvas, A.A. (1996) Springs, Chapter 28 in *Mechanical Design Handbook*, H.A. Rothbart, Ed., McGraw-Hill.

Zimmerli, F.P. (1957) "Human Failures in Spring Applications," *The Mainspring*, no. 17, Associated Spring Corp., Barnes Group, Inc.

Questions

17.1 What is a spring? What are springs used for?

17.2 What is music wire?

17.3 What kind of stress is developed in a compression spring? What about a torsion spring?

17.4 What is the Wahl factor?

17.5 How is buckling avoided with springs?

17.6 What is spring surge?

17.7 What is a leaf spring?

17.8 Explain why gas springs provide almost uniform forces over their entire stroke.

17.9 Why do gas springs provide damping?

17.10 What is a Belleville spring?

Qualitative Problems

17.11 What is the *loss coefficient*? Why is steel used for springs even though other materials have a superior loss coefficient?

17.12 Explain why the strength of a wire increases as the diameter decreases.

17.13 Using appropriate sketches, explain why the spring index for helical compression springs is usually between 4 and 12.

17.14 What are the advantages and disadvantages of using plain ends on helical compression springs? What about for plain and ground ends?

17.15 Review Design Procedure 17.1 and list applications where a low spring index is preferable. Create a second list for applications where a high spring index is preferable.

17.16 What is detrimental about compressing a spring to its solid length?

17.17 Explain why helical extension springs are not used to support compressive loads, and why helical compression springs are not used to support tensile loads.

17.18 Without the use of equations, explain why the diameter of torsion springs changes as the spring is loaded or unloaded.

17.19 Referring to Fig. 17.14, explain what machining operations you would apply to the triangular plate to create the equivalent multi-leaf spring. What is the base of the triangular plate?

17.20 Why do wave springs need to have multiples of 0.5 waves per turn?

Quantitative Problems

17.21 Sketch a shows a guide wire, used to deliver catheters in the human body during angioplasty operations. The compression spring at the end is used as a flexible support for the nose as it opens the artery and prevents its puncture by the guide wire. The spring has an outside diameter of 0.0250 in, a wire diameter of 0.003 in., a free length of 0.110 in. and 30 total coils, and has squared and ground ends. The wire is a stainless steel (E = 29 Mpsi, G = 11 Mpsi), but has a strength so that A_p = 100 ksi, m = 0.0474). Determine

(a) The solid length and spring rate. *Ans.* $l_s = 0.090$ in., $k = 0.350$ lb/in.

(b) The force needed to compress the spring to its solid length. *Ans.* $F = 0.0070$ lb.

(c) The safety factor against static overload if the spring is compressed to its solid length. Use the Wahl factor. *Ans.* $n_s = 3.21$.

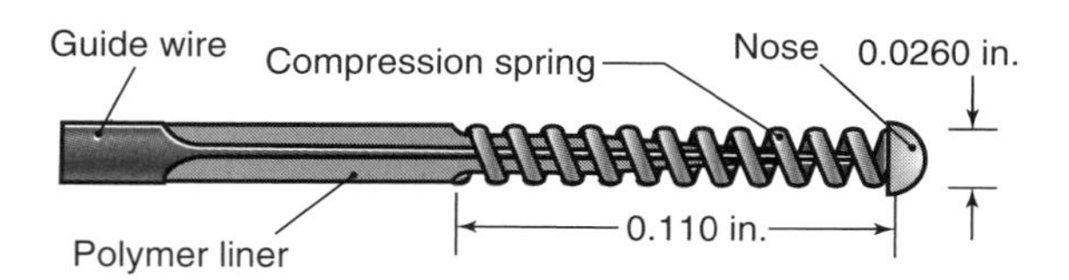

Sketch a, for Problem 17.21

17.22 A helical compression spring with square and ground ends has an outside diameter of 0.560 in., 20 total coils, a wire diameter of 0.085 in. (music wire), and a free length of 4.22 in. Calculate the spring index, the pitch, the solid length, and the shear stress in the wire when the spring is compressed to its solid length. *Ans.* $C = 5.588$, $p = 0.225$ in., $l_s = 1.700$ in., $\tau_{\max} = 250$ ksi.

17.23 An overflow valve, shown in Sketch b, has a piston diameter of 15 mm and a slit length of 5 mm. The spring has mean coil diameter D = 10 mm and wire diameter d = 2 mm. The valve should open at 1 bar (0.1 MPa) pressure and be totally open at 3 bar (0.3 MPa) pressure when the spring is fully compressed. Calculate the number of active coils, the free length, and the pitch of the spring. The shear modulus for the spring material G = 80 GPa. The spring ends are squared and ground. Determine the maximum shear stress for this geometry. *Ans.* $N_a = 22.6$, $l_f = 56.7$ mm.

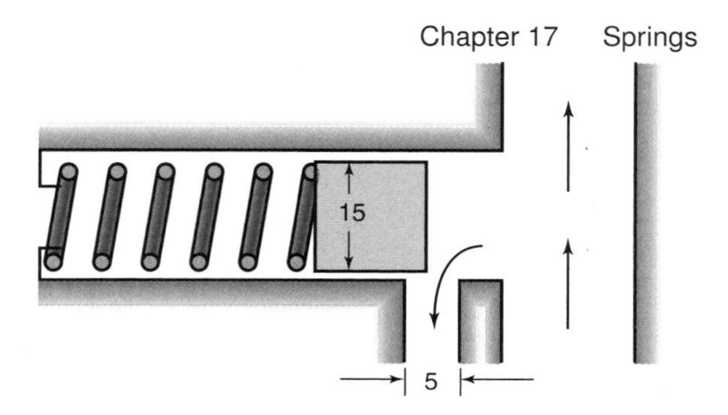

Sketch b, for Problem 17.23

17.24 A vehicle has individual wheel suspension in the form of helical springs. The free length of the spring $l_f = 360$ mm and the solid length $l_s = 160$ mm at a compressive force of 5000 N. The shear modulus G = 80 GPa. Use $D/d = 10$ and calculate the shear stress for pure torsion of the spring wire. The spring ends are squared and ground. Find N_a, p, d, and D, and $\tau_{\max}$. *Ans.* $\tau_{\max} = 370$ MPa, $N_a = 7.06$, $p = 38.5$ mm, $d = 17.65$ mm.

17.25 Two equally long cylindrical helical compression springs are placed one inside the other (see Sketch c) and loaded in compression. How should the springs be dimensioned to get the same shear stresses in both springs?

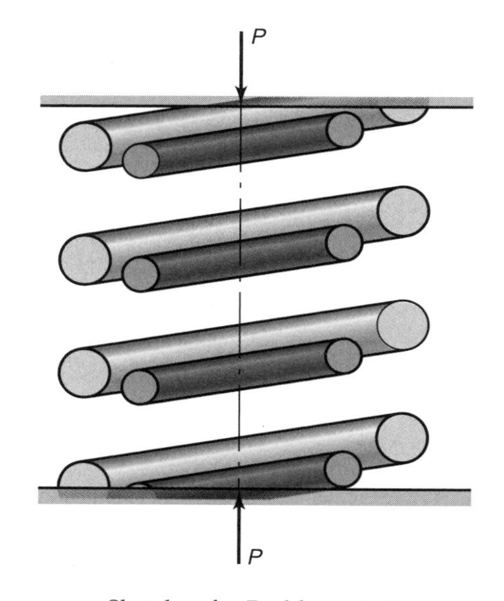

Sketch c, for Problem 17.25

17.26 A mechanism is used to press as hard as possible against a moving horizontal surface shown in Sketch d. The mechanism consists of a stiff central beam and two flexible bending springs made of circular rods with length l, diameter d, modulus of elasticity E, and allowable stress σ_{all}. Wheels are mounted on these rods and can roll over a bump with height f. Calculate the diameter of the springs so that the prestress of the wheels against the moving surface is as high as possible without plastically deforming the springs when the bump is rolled over. The deflection of a spring is shown in Sketch e. *Ans.* $= \frac{\sigma l^2}{Ef}$, $P = \frac{\pi\sigma^4 l^5}{128E^3 f^3}$.

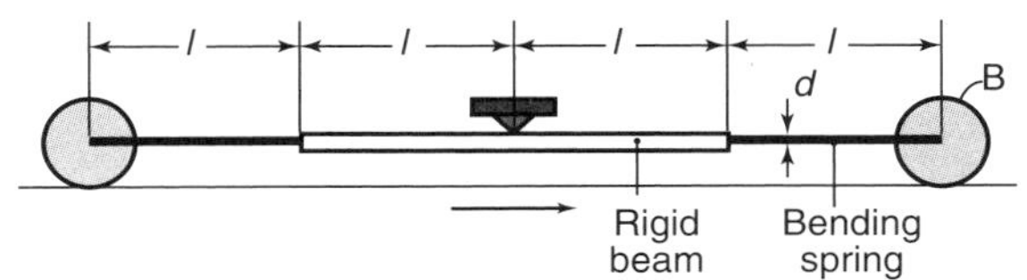

Sketch d, for Problem 17.26

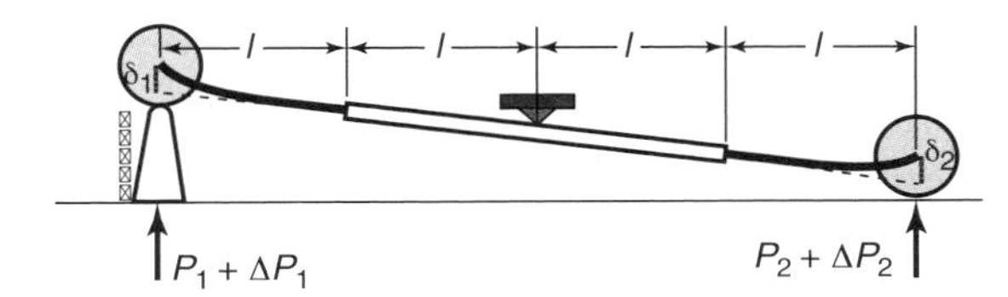

Sketch e, for Problem 17.26

17.27 A helical compression spring is used as a catapult. Calculate the maximum speed of a body weighing 5 kg being thrown by the catapult, given that $\tau_{\max} = 500$ MPa, $D = 50$ mm, $d = 8$ mm, $N_a = 20$, and $G = 80$ GPa. *Ans.* Using a Wahl factor, $v = 6.51$ m/s.

17.28 A spring is preloaded with force P_i and is then exposed to a force increase whereby the shear stress increases to a certain value τ. Choose the mean coil diameter D to maximize the energy absorption caused by the force increase. *Ans.* $D = \dfrac{\pi}{8\sqrt{3}}\dfrac{\tau d^3}{P_i}$.

17.29 A compression spring made of music wire is used for static loading. Wire diameter $d = 1.5$ mm, coil outside diameter $D_o = 12.1$ mm, and there are eight active coils. Also, assume that the spring ends are squared and ground. Find the following:

(a) Spring rate and solid length. *Ans.* $k = 5214$ N/m, $l_s = 15$ mm.

(b) Greatest load that can be applied without causing a permanent set in excess of 2%. *Ans.* $P_{\max} = 95.5$ N.

(c) The spring free length if the load determined in part (b) causes the spring to compress to a solid state. *Ans.* 32.3 mm.

(d) Whether buckling is a problem. If it is, recommend what you would change in the redesign.

17.30 A helical compression spring will be used in a pressure relief valve. When the valve is closed, the spring length is 2.0 in., and the spring force is to be 1.50 lb. The spring uses a wire diameter of $d = 0.0577$ in. and a spring diameter of $D = 0.375$ in., and has a solid length of $l_s = 1.25$ in. Use hard drawn ASTM A227 wire, squared and ground ends, and ignore buckling since the spring is in a cage. Find the force needed to compress the spring to its solid length, the number of active coils, the spring's free length, the spring rate, and the safety factor against torsional yielding. *Ans.* $F_s = 12.88$ lb, $N_a = 19.66$, $l_f = 2.0988$ in., $k = 15.18$ lb/in., $n_s = 1.19$.

17.31 Two compression springs will be used in a stamping press to open a die after it has been pushed closed by the action of the press ram. The springs will stay in a retainer, which means buckling is not an issue and also means that the springs will be allowed to be preloaded. Also, this retainer geometry requires that $D = 1$ in. It is desired to have a total deflection of 7 in., and each spring should exert 250 lb on the die when the spring is compressed and 150 lb when extended. Squared and ground ends will be used to ensure good performance. If 30 active coils are used in each spring, what is the required wire diameter and spring solid length? Use $G = 11.6$ Mpsi. *Ans.* $d = 0.131$ in., $l_s = 4.192$ in.

17.32 Design a helical compression spring made of music wire with squared and parallel ends. Assume a spring index $C = 12$, a spring rate $k = 300$ N/m, and a force to solid length of 60 N. Find the wire diameter and the mean coil diameter while assuming a safety factor $n_s = 1.5$. Also, check whether buckling is a problem. Assume steady loading. *Ans.* $d = 1.90$ mm, $D = 22.8$ mm.

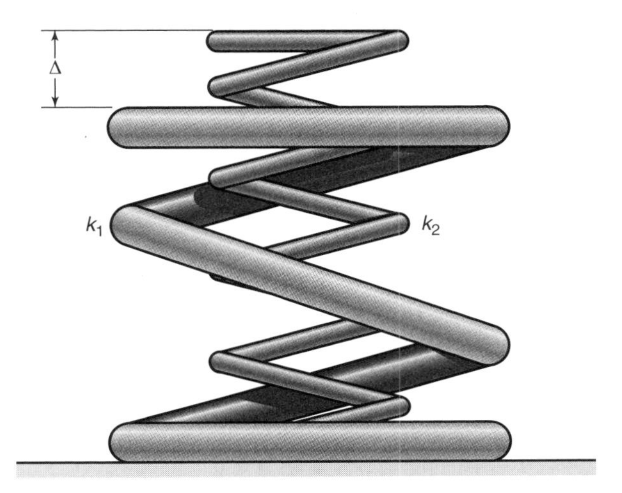

Sketch f, for Problem 17.32

17.33 Two helical springs with spring rates of k_1 and k_2 are mounted one inside the other as shown in Sketch f. The difference in unstressed length is Δ and the second spring is longer. A loose clamping device is mounted at the top of the springs so that they are deformed and become equal in length. Calculate the forces in the two springs. An external force P is the load applied thereafter to the spring. In a diagram show how the spring forces vary with P. *Ans.* $P_1 = P_o + \dfrac{k_1 P}{k_1 + k_2}$, $P_2 = P_o - \dfrac{k_2 P}{k_1 + k_2}$.

17.34 Using Eqs. (17.4) and (17.18) describe how the spring diameter can be chosen to give as low a maximum shear stress as possible.

17.35 Consider a helical compression spring with plain ends made of hard-drawn wire with a spring index of 12 and a stiffness of 0.3 kN/m. The applied load on the spring is 60 N. For a safety factor of 1.5 guarding against yielding find the spring diameter, the wire diameter, the number of coils, and the free and solid lengths of the spring. Does this spring have buckling and/or dynamic instability problems? *Ans.* $d = 2.18$ mm, $D = 26.16$ mm, $N_a = 41.5$.

17.36 An 18-mm mean diameter helical compression spring has 20 coils, has 2-mm wire diameter, and is made of chromium vanadium. Determine the following:

(a) The maximum load-carrying capacity for a safety factor of 1.5 guarding against yielding. *Ans.* $P = 78.5$ N

(b) The maximum deflection of the spring. *Ans.* $\delta_t = 0.0577m$.

(c) The free length for squared and ground ends. *Ans.* $l_f = 0.1075$ m.

17.37 A 60-mm mean diameter helical compression spring with plain ends is made of hard-drawn steel and has a wire diameter of 2.5 mm. The shear strength of the wire material is 500 MPa and the spring has 20 coils. The free length is 500 mm. Find the following:

(a) The required load needed to compress the spring to its solid length. *Ans.* $P = 40.11$ N.

(b) By applying the load found in part (a) and then unloading it, whether the spring will return to its free length.

17.38 A desk lamp has four high-carbon steel helical extension springs mounted on a mechanism that makes it possible to position it over different parts of the desk. Each spring is preloaded to 15 N, so that no deflection takes place for forces below 15 N. Above 15-N force the spring rate is 100 N/m. The mean coil diameter is 12 mm and the wire diameter is 1 mm. Calculate the torsional deflection needed on the wire during manufacturing to get the correct preload, and calculate how many coils are needed to obtain the correct spring rate. *Ans.* $\theta/l = 9.63$ rad/m, $N_a = 57.36$.

17.39 A spring balance for weighing fish needs to be dimensioned. The weighing mechanism is a sharp hook hanging in a helical extension spring. To make it easy to read the weight of the fish in the range from 0 to 10 kg, the length of the scale should be 100 mm. The spring material is music wire. For $d = 2$ mm and $C = 10$, determine the number of active coils and the safety factor. *Ans.* $N_a = 20.1$, $n_s = 1.10$.

17.40 A muscle-training device consists of two handles with three parallel 500-mm-long springs in between. The springs are tightly wound but without prestress. When the springs are fully extended to 1600 mm, the force in each spring is 100 N. The springs are made of music wire. Dimension the springs. *Ans.* For $C = 10$ and $d = 2$ mm, $n_s = 1.08$.

17.41 A 45-mm-diameter extension spring (similar to that shown in Fig. 17.9a and b) has 102 coils and is made of 4-mm-diameter music wire. The stress due to preload is equivalent to 10% of the yield shear strength. The hook radius is 5 mm and the bend radius is 2.5 mm. Determine the following:

(a) The solid length of the spring and the spring stiffness. *Ans.* $k = 273$ N/m, $l_s = 0.412$ m.

(b) The preload and the load that causes failure. *Ans.* $P_i = 39.6$ N.

17.42 The extension spring shown in Sketch g is used in a cyclic motion in turning on and off a power switch. The spring has a 15-mm outer diameter and is made of a 1.5-mm-diameter wire of hard-drawn steel. The spring has no preload. In a full stroke of the spring the force varies between 25 and 33 N. Determine the following:

(a) The number of coils, the free length, the maximum and minimum lengths during cyclic loading, and the spring rate. *Ans.* $N_a = 38.0$, $l_f = 73.5$ mm.

(b) For infinite life with 99% reliability, the safety factors guarding against static and fatigue failure. *Ans.* $n_{s,\text{static}} = 1.66$.

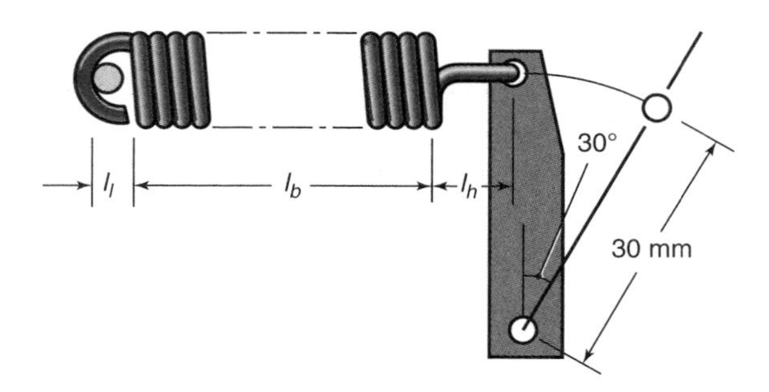

Sketch g, for Problem 17.42

17.43 Calculate the safety factor guarding against fatigue failure if the spring given in Problem 17.42 is designed for 50,000 strokes of motion with 50% reliability. *Ans.* $n_s = 11.8$.

17.44 The extension spring shown in Sketch h is used in a car braking system. The spring is made of 2-mm-diameter wire of hard drawn steel, has a mean coil diameter of 10 mm, and has a free length of 106 mm. The spring has 12 active coils and a deflection under braking force of 11 mm. Determine the following:

(a) The safety factor of the spring. *Ans.* $n_s = 1.01$.

(b) The torsional and normal stresses at the hook. *Ans.* $\tau = 577$ MPa, $\sigma = 815$ MPa.

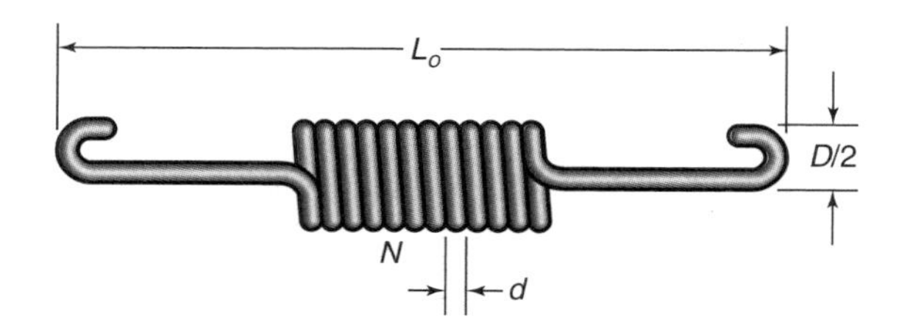

Sketch h, for Problem 17.44

17.45 A torsion rod is used as a vehicle spring. The torque on the rod is created by a force $P = 1500$ N acting on radius $R = 200$ mm. Maximum torsion angle is 30°. Calculate the diameter and length of the rod if the maximum shear stress is 500 MPa. *Ans.* $d = 14.50$ mm, $L = 0.606$ m.

17.46 A torsional spring, shown in Sketch i, consists of a steel cylinder with a rubber ring glued to it. The dimensions of the ring are $D = 45$ mm, $d = 15$ mm, and $h = 20$ mm. Calculate the torque as a function of the angular deflection. The shear modulus of elasticity for rubber is 150 N/cm^2. *Ans.* $T = (23.9 \text{ N-m/rad})\phi$.

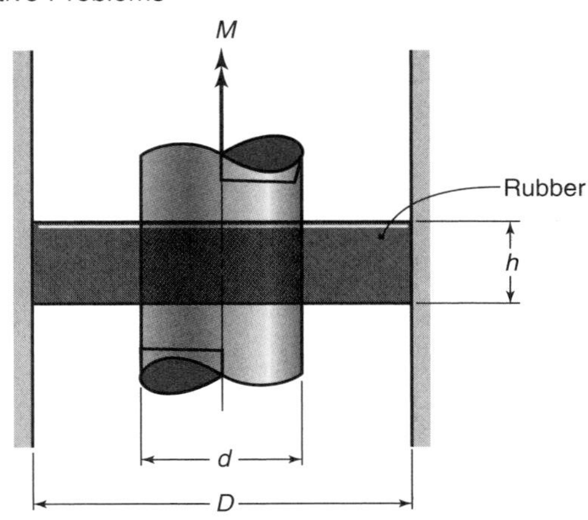

Sketch i, for Problem 17.46

17.47 To keep a sauna door shut, helical torsion springs are mounted at each of the two door hinges. The friction torque in each hinge is 0.2 N-m. It should be possible to open the door 180° without plastically deforming the spring. Dimension the springs so that a 10-kg door 700 mm wide will shut itself in 2 s. The length of the wire ends can be neglected. The air drag on the door is also neglected. The springs are manufactured from music wire and have a diameter of 2.5 mm. Use $C = 10$. *Ans.* $N_a = 10.03$.

17.48 A helical torsion spring used in a door handle is made of hard-drawn steel wire with a diameter of 3 mm. The spring has 10 active coils and a coil diameter of 21 mm. After 5 years of use the spring cracks due to fatigue. To get longer fatigue life, the stresses should be decreased by 5%, but the only available space is that used by the present spring, which has an outside diameter of 24 mm, an inside diameter of 18 mm, and a length of 30 mm. The spring rate for the new spring should be the same as for the old one. Dimension a new spring with a rectangular cross-section to satisfy the new stress requirement. *Ans.* $w = 3$ mm, $h = 2.36$ mm.

17.49 A mouse trap uses two anti-symmetric torsion springs. The wire has a diameter of 2 mm and the inside diameter of the spring in the unset position is 12 mm. Each spring has 10 turns when unset. About 50 Newtons are needed to set the trap (not to start setting, but rather the largest load needed to deflect the spring to the set position).

(a) Find the number of coils and diameter of the spring prior to assembly into the mouse trap, that is, in unstressed state. *Ans.* $N_A = 9.58$.

(b) Find the maximum stress in the spring wire when the trap is set. *Ans.* $\sigma = 3010$ MPa.

17.50 The oven door on a kitchen stove is kept shut by a helical torsion spring. The spring torque is 1 N-m when the door is shut. When the oven door is fully open, it must stay open by gravitational force. The door height is 450 mm and it weighs 4 kg. Dimension the helical torsion spring using music wire as the spring material and a wire diameter of 4.5 mm. Is it also possible to use 3-mm-diameter wire? *Ans.* For D/d=10, $N_a = 5.89$, $l_w = 0.83$ m.

17.51 A helical torsion spring, shown in Sketch j, is made from hard-drawn steel with a wire diameter of 2.2 mm and 8.5 turns. Dimensions are in millimeters.

(a) Using a safety factor of 2, find the maximum force and the corresponding angular displacement. *Ans.* $P = 12.5$ N, $\theta = 0.18$ rad.

(b) What would the coil inside diameter be when the maximum load is applied? *Ans.* $D_i = 0.0194$ m.

(c) For 100,000 loading cycles calculate the maximum moment and the corresponding angular displacement for a safety factor of 2.5 guarding against fatigue failure. *Ans.* $\theta = 0.180$ rev.

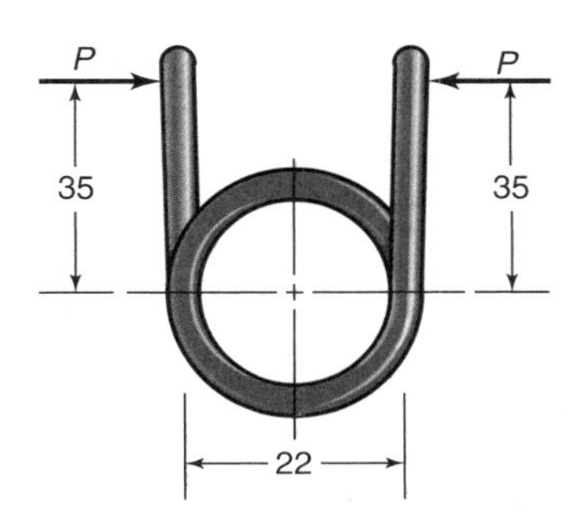

Sketch j, for Problem 17.51

17.52 The helical torsion spring shown in Sketch k has a coil outside diameter of 22 mm, 10.25 turns, and a wire diameter of 2 mm. The spring material is hard-drawn steel. What would the applied moment be if the maximum stress equaled the yield limit? Calculate the inside diameter of the loaded spring and the corresponding angular displacement. Dimensions are in millimeters. *Ans.* $D_i' = 17.29$ mm, $\theta = 0.421$ rev.

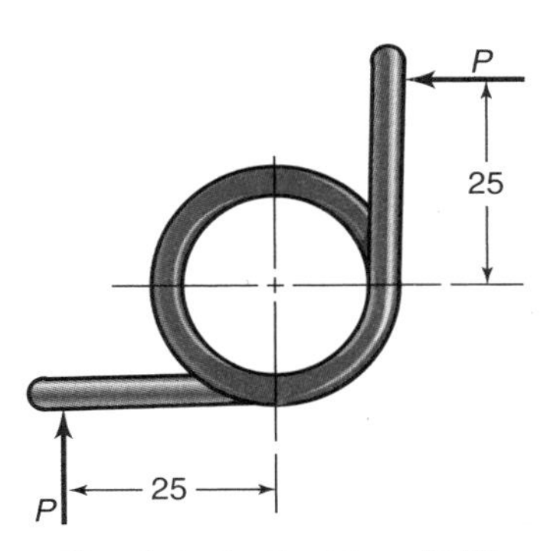

Sketch k, for Problem 17.52

17.53 A helical torsion spring made of music wire has a coil diameter of 17.5 mm and a wire diameter of 1.5 mm while supporting a 0.15-N-m moment with a 20% fluctuation. The maximum number of turns is 12, and the load is 22 mm from the center of the spring. For infinite life with 99% reliability find the safety factors guarding against static and fatigue failure. Also, determine the inside diameter of the spring when load has been applied. *Ans.* $n_{s,\text{static}} = 3.52$, $n_{s,\text{fatigue}} = 2.51$.

17.54 A fishing rod is made like an ideal leaf spring with rectangular cross sections. It is made of carbon-fiber-reinforced plastic with a 150-GPa modulus of elasticity. The thickness is constant at 8 mm and the length is 2 m. Find how large the cross section must be at the handle to carry a 0.3-kg fish by the hook without bending the top of the rod more than 200 mm. Neglect the weight of the lure. Also, calculate the bending stress. *Ans.* $b = 9.19$ mm, $\sigma = 60.0$ MPa.

17.55 A trampoline is made like a leaf spring with variable width so that the maximum bending stress in each section of the trampoline is constant. The material is glass-fiber-reinforced plastic with a modulus of elasticity of 28 GPa and a bending strength of 300 MPa. Calculate the spring rate at the tip of the trampoline and the corresponding safety factor if a swimmer weighing 80 kg jumps onto the trampoline from a height of 2 m. The active length of the trampoline is 3 m, its width is 1.2 m, and its thickness is 38 mm. *Ans.* $k = 11.38$ kN/m, $n_s = 4.24$.

17.56 The leaf spring of a truck should be able to accommodate 55-mm deflections (up and down) of the wheels from an equilibrium position when the truck is driven on a rough road. The static load in the middle of the leaf spring is 50,000 N. Assume an allowable stress of 1050 MPa, a safety factor of 3, leaf half-length of 0.8 m, a leaf thickness of 0.02 m, and that there are 10 leaf layers. Determine the width of the spring. The modulus of elasticity for the spring is 207 GPa. *Ans.* $b = 0.1714$ m.

17.57 A Belleville spring is formed from cold-rolled steel (AISI 1040) with $t = 1$ mm, $h = 2.5$ mm, $D_o/D_i = 2.0$ and $D_i = 7$ mm. Calculate the force needed to flatten the spring to $h_f = 1.5$ mm. What force is needed to fully flatten the spring? Explain your answer. *Ans.* For $h_f = 1.5$, $P = 10.1$ kN, $P_{flat} = 6.79$ kN.

17.58 A Belleville spring is to be used to control the pretension applied to a bolt by a nut. Thus, it is desired that the inner diameter be set equal to the crest diameter and that $R_d = 1.5$ so that the Belleville spring will have the same basic size as the bolt head. If the head height is to be $h = 3$ mm, what thickness is needed for a metric grade 4.6 bolt with a 12-mm crest diameter if the connection is to be reused? Is this a feasible design? *Ans.* $t = 1.08$ mm.

17.59 Consider a wave spring to be used to open a forging die after it has been pushed closed by the action of the press ram. The retainer allows for an outer diameter of 25 mm. The wave spring should support 1000 N, and assume that 4 waves per turn are to be used. If the wave spring is to be constructed from steel with a width of $b = 3$ mm and an allowable stress of 800 MPa, determine the thickness of the wave ribbon. How much will the spring compress under the 1000 N load if there are 25 turns? Compare this spring with that analyzed in Problem 17.31. *Ans.* $t = 1.16$ mm, $d = 3.163$ mm.

17.60 Given a steel wave spring with an outer diameter of 150 mm and a width of 10 mm, with a thickness of 2 mm, plot the stiffness of the spring as a function of the number of waves per turn if the spring has 10 active turns.

Design and Projects

17.61 As shown in the opening photo for the chapter, helical compression springs can have a rectangular cross section. Derive the equation for maximum stress for such a cross section. Conduct an Internet search and obtain the equivalent to the Wahl or Bergsträsser factor for such a geometry.

17.62 Create a table that lists every type of spring discussed in the chapter, the size or capacity of such springs, and two or three typical applications.

17.63 Review Fig. 17.15 and explain what design features you would incorporate to allow for modification of the gas charge pressure.

17.64 It is possible to create a compression spring by laser cutting a pattern into a tube. This can result in a wide variety of spring shapes, but consider the case when a helix is machined into the tube. Make a sketch of the resulting spring and list the geometric variables that define the spring's geometry. Explain how the stress analysis would differ, if at all, from the discussion in Section 17.3. Would there be any advantages to having a helix with more than one start (see Section 15.4)?

Chapter 18

Brakes and Clutches

A truck brake drum with cooling fins around the periphery for extended life and improved performance. *Source:* Courtesy of Webb Wheel Products, Inc.

Nothing has such power to broaden the mind as the ability to investigate systematically and truly all that comes under thy observation in life.

Marcus Aurelius, Roman Emperor

Contents

Examples

Design Procedures

Case Study

This chapter deals with clutches and brakes, two machine elements that are very similar in function and appearance. Brakes convert mechanical energy to heat, and are widely used to bring all types of machinery to rest. Examples are the thrust pad and long-shoe, internal expanding (drum) brakes on automobiles, and external expanding and pivot shoe brakes in machinery. Clutches serve to bring one shaft to the same speed as another shaft, and are available in a wide variety of sizes. Both brakes and clutches use opposing surfaces, and rely on friction to fulfill their function. Friction causes heat generation, and it is important that brakes and clutches be designed properly to ovoid overheating. For this reason, clutches for high power applications are often operated while submerged in a fluid (wet clutches). However, thermal and material considerations are paramount for effective brake and clutch designs. The chapter begins by analyzing thrust brakes and clutches, as well as the related cone clutches. Block or short-shoe brakes introduce the concept of self-energizing brake shoes, which is further examined with drum brakes and clutches. Band brakes, pivot-shoe brakes, and slip clutches are also discussed.

Machine elements in this chapter: Brakes and clutches of all kinds.
Typical applications: Automobiles, aircraft, vehicles of all kinds; shafts on any machinery; transmissions.
Competing machine elements: Shaft couplings (Ch. 11), springs (Ch. 17).

Symbols

A	area, m^2; constant
b	cone or face width, m
C	cost
C_p	specific heat, J/(kg°C)
c	constant
D	largest diameter of cone, m
d	smallest diameter of cone, m
d_1-d_{10}	distances used for brakes, m
F	friction force, N
F_1	pin reaction force, N
F_2	actuating force, N
h_p	work or energy conversion rate, W
M	moment, N-m
m_a	mass, kg
N	number of sets of disks
n_s	safety factor
P	normal force, N
p	contact pressure, Pa
Q	energy, J
p_o	uniform pressure, Pa
R	reaction force, N
r	radius, m
T	torque, N-m
$\bar{T}$	dimensionless torque, $\dfrac{T}{(2\mu P r_o)}$
t_m	temperature, °C
u	sliding velocity, m/s
W	actuating force, N
α	half-cone angle, deg
β	radius ratio, $\dfrac{r_i}{r_o}$
β_o	optimum radius ratio
γ	extent of brake pad
θ	circumferential coordinate, deg
θ_a	angle where $p = p_{\max}$, deg
θ_1	location where shoe begins, deg
θ_2	location where shoe ends, deg
μ	coefficient of friction
ϕ	wrap angle, deg
ω	angular velocity, rad/s

Subscripts

c	conduction
d	deenergizing
F	friction force
f	friction
h	convection
i	inner
m	mean
o	outer
P	normal force
p	uniform pressure
s	self-energizing, storage
w	uniform wear

18.1 Introduction

Brakes and clutches are examples of machine elements that use friction in a useful way. Clutches are required when shafts must be frequently connected and disconnected. The function of a clutch is twofold: first, to provide a gradual increase in the angular velocity of the driven shaft, so that its speed can be brought up to that of the driving shaft without shock; second, when the two shafts are rotating at the same angular velocity, to act as a coupling without slip or loss of speed in the driving shaft. A **brake** is a device used to bring a moving system to rest, to slow its speed, or to control its speed to a certain value. The function of the brake is to turn mechanical energy into heat. The design of brakes and clutches is subjected to uncertainties in the value of the coefficient of friction that must necessarily be used. Material selection topics from Sections 3.5 and 3.7, as well as friction and wear covered in Sections 8.8 and 8.9 will be used in this chapter.

Figure 18.1 illustrates selected brakes and clutches that are covered in this chapter. These include a **rim type** that has internal expanding and external contracting shoes, a **band brake**, a **thrust disk**, and a **cone disk**. This figure also shows the actuating forces being applied to each brake or clutch. Table 18.1 summarizes the types of brakes and clutches as well as some typical applications.

Brakes and clutches are similar, but different from other machinery elements in that they are tribological systems where friction is intended to be high. Therefore, much effort has been directed toward identifying and developing materials that result in simultaneous high coefficients of friction and low wear so that a reasonable combination of performance and service life can be achieved. In previous years, brake and clutch materials were asbestos-fiber-containing composites, but the wear particles associated with these materials resulted in excessive health hazards to maintenance personnel. Modern brakes and clutches use "semimetallic" materials (i.e., metals produced using powder metallurgy techniques) in the tribological interface, even though longer life could be obtained by using the older asbestos-based linings. This substitution is a good example of multidisciplinary design, in that a consideration totally outside of mechanical engineering has eliminated a class of materials from consideration.

Typical brake and clutch design also involves selecting components of sufficient size and capacity to attain reasonable service life. Many of the problems are solid mechanics oriented; the associated theory is covered in Chapters 4 to 7 and is not repeated here. Because this chapter is mainly concerned with the performance of brake and clutch systems, the focus here is on the actuating forces and resultant torques. Components of brake systems, such as springs, rivets, etc., are considered elsewhere and will not be repeated here. However, it should be noted that a brake or clutch consists of a number of components integrated into a system.

18.2 Thermal Considerations

A critical consideration in the design of brake and clutch components is temperature. Whenever brakes or clutches are activated, one high-friction material slides over another under a large normal force. The associated energy is converted into heat which always results in elevated temperatures in the lining material. While all brakes encounter wear, thermal effects can lead to accelerated wear and can compromise performance and life. Obviously, such circumstances can only be discovered through periodic inspection. Therefore, regular maintenance of brake and clutch systems is essential, and their service lives are often much lower than those of other machinery elements.

As mentioned above, thermal effects are important in braking and clutch systems. If temperatures become too high, damage to components could result, which could compromise the useful life or performance of brake and clutch systems. Thermally induced damage often takes the following forms:

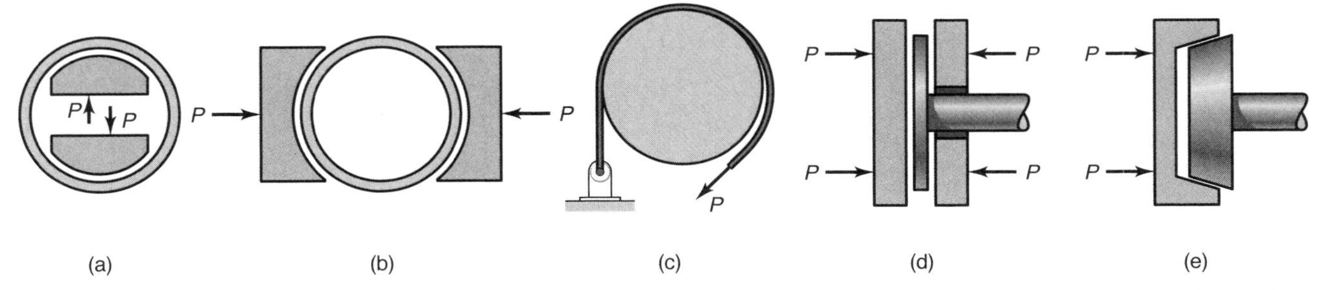

Figure 18.1: Five types of brake and clutch. (a) Internal, expanding rim type; (b) external contracting rim type; (c) band brake; (d) thrust disk; (e) cone disk.

Table 18.1: Typical applications of clutches and brakes.

Type	Application notes
Thrust pad (disc)	Extremely common and versatile arrangement; can be wet or dry; wide variety of materials including carbon-carbon composites for aircraft brakes; preferred for front axles of vehicles because of superior convective cooling; cannot self-lock.
Cone	Higher pressure and torque for the same sized clutch compared to thrust pad due to wedging action of cone; common for lower speed applications with little sliding such as washing machines or extractors, or high-performance applications such as vehicle racing.
Block or short-shoe	Available in a wide variety of configurations and capacities; commonly applied to roller coasters, industrial equipment and positioning devices.
Long-shoe (drum)	Widely applied in vehicles on rear axles; self-locking promotes "parking brake" function; economical and reliable; limited heat dissipation capability.
Pivot-shoe	Used for low-torque applications in architecture, fishing equipement; higher torque applications include hoists and cranes; difficult to properly locate pivot.
Band brakes	Simple, compact, and rugged, widely applied to chain saws, go-karts, motorcycles, and some bicycles; susceptible to chatter or grabbing.
Slip clutches	Used to prevent excessive torque transfer to machinery; available in a wide variety of sizes and capacities; applied to machinery to prevent overload, some garage door operators, cranes as an anti-two blocking device; torque is difficult to control.

- **Warped components**, such as out-of-round drums or non-planar rotors, result from excessive heating and are often associated with recovery of residual stresses. Such conditions can also result from improper machining of components during service or repair, or from mishandling or improper installation.

- **Heat checks**, shown in Fig. 18.2, are small cracks caused by repeated heating and cooling of brake surfaces, and could result in high wear rates and compromised performance. Light heat checking is a normal condition, and the small cracks form and are worn away during normal operation; the component can be machined to remove the damaged surface during service, as long as manufacturer's tolerances are maintained. Excessive heat checks can be caused by an operator using the brakes excessively, damaged brake components (such as a worn return spring or bushing) or by out-of round or warped drums or rotors.

- **Glazed lining surfaces** are associated with excessive heat that can be attributable to a number of causes, and result in lower friction and diminished performance. Unless the lining is significantly worn, the glazed surfaces can be removed with an abrasive such as emery cloth to re-establish proper performance.

- **Hard spots** in the brake surface (Fig. 18.3) are caused by highly localized heating and cooling cycles, and result in chatter or pulsation during brake action, with the ultimate result of compromised performance and life. Hard spots are often associated with warped or out-of-round components.

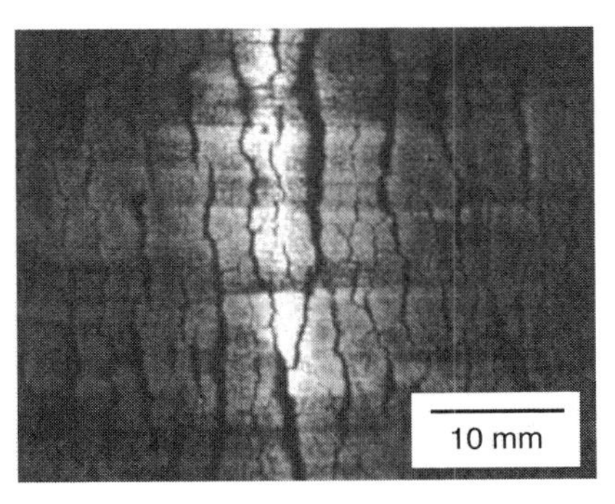

Figure 18.2: Brake drum surface showing a high level of heat checking. *Source:* Courtesy Webb Wheel Products, Inc.

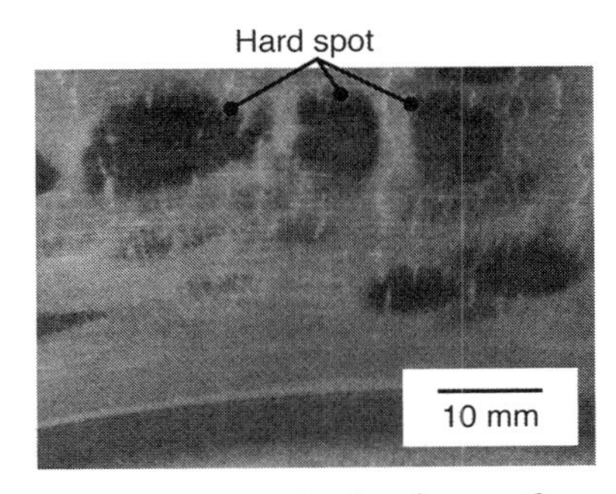

Figure 18.3: Hard spot on a brake drum. *Source:* Courtesy Webb Wheel Products, Inc.

Predicting temperatures of brake and clutch systems is extremely difficult in practice because they are operated under widely varying conditions. Neglecting radiation, the first law of thermodynamics requires that

$$Q_f = Q_c + Q_h + Q_s, \tag{18.1}$$

where

Q_f = energy input into brake or clutch system from friction between sliding elements
Q_c = heat transferred by conduction through machinery elements
Q_h = heat transferred by convection to surrounding environment
Q_s = energy stored in brake and clutch components, resulting in temperature increase

If conduction and convection are negligible, the temperature rise in the brake or clutch material is given by

$$\Delta t_m = \frac{Q_f}{C_p m_a}, \tag{18.2}$$

where C_p is the specific heat of the material, and m_a is the mass. This equation is useful for determining the instantaneous temperature rise in a brake or clutch pad, since the frictional energy is dissipated directly on the contacting surfaces and does not have time to be conducted or conveyed away. Brake pads and clutches usually have an area in contact that then moves out of contact (i.e., heat is conveyed) and can cool. The actual maximum operating temperature is therefore a complicated function of heat input and cooling rates.

The main difficulty in predicting brake system temperatures is that the heat conducted and the heat conveyed depend on the machine ambient temperature and the brake or clutch geometry, and can vary widely. In previous circumstances in this text, a worst-case analysis would be performed, which in this case quickly reduces to circumstances where brakes and clutches become obviously overheated. This result is not incorrect: most brake and clutch systems are overheated when abused and can sustain serious damage as a result. The alternative is to use such massive brake and clutch systems as to make the economic burden unbearable to responsible users. It is far more reasonable to use brake systems that require periodic maintenance and can be damaged through abuse than to incur the economic costs of surviving worst-case analyses. This differs from previous circumstances, where a worst-case analysis still resulted in a reasonable product.

Some clutches are intended to be used with a fluid (**wet clutches**) to aid in cooling the clutch. Similarly, some pads or shoes will include grooves so that air or fluid can be better entrained, and increased flow and convective heat transfer result. Predicting wet clutch temperatures is a complex problem and requires numerical, usually finite element, methods.

Obviously, the proper size of brake components is extremely difficult to determine with certainty. For the purposes of this text the values given by Juvinall and Marshek [2006] for the product of brake shoe or pad contact pressure and sliding velocity, pu, can be used to estimate component sizes (Table 18.2). As discussed in Section 18.3.2, pu is proportional to the power dissipated. Most manufacturers rely heavily on experimental verification of designs; the application of these numbers in the absence of experimental verification requires extreme caution, but is useful for evaluating designs and estimating component sizes.

Table 18.2: Product of contact pressure and sliding velocity for brakes and clutches. *Source:* Adapted from Juvinall, R.C., and Marshek, K.M. [2006].

	pu	
Operating condition	**(kPa)(m/s)**	**(psi)(ft/min)**
Continuous: poor heat dissipation	1050	30,000
Occasional: poor heat dissipation	2100	60,000
Continuous: good heat dissipation as in oil bath	3000	85,000

18.3 Thrust Pad Clutches and Brakes

A thrust disk has its axis of rotation perpendicular to the contacting surfaces, as shown in Fig. 18.1d; Fig. 18.4 illustrates the components of an automotive disk brake. Basically, a rotor or disk is attached to the vehicle's axle, and a caliper that is mounted on the automobile body contains two brake pads. The pads consist of a friction material supported by a backing plate. Pressurized brake fluid hydraulically actuates the brake cylinder, causing brake pads to bear against the rotor on opposite sides. The pressure applied determines the contacting pressure, friction, and torque, as will be shown below.

Figure 18.5 shows the various radii of the thrust disk clutch or brake. A typical design task is to obtain the axial force, P, needed to produce a certain torque, T, and the resulting contact pressure, p, and wear depth, δ. For some elemental area,

$$dA = (r\,d\theta)\,dr, \tag{18.3}$$

the normal force and the torque can be expressed as

$$dP = p\,dA = pr\,d\theta\,dr, \tag{18.4}$$

and

$$T = \int r\,dF = \int \mu r\,dP = \int\int \mu p r^2\,dr\,d\theta. \tag{18.5}$$

Only a single set of disks will be analyzed in the following sections, but the torque for a single set of disks is multiplied by N to get the torque for N sets of disks.

18.3.1 Uniform Pressure Model

For new, accurately flat, and aligned disks, the pressure will be uniform, or $p = p_o$. Substituting this into Eqs. (18.4) and (18.5) gives

$$P_p = \pi p_o\left(r_o^2 - r_i^2\right), \tag{18.6}$$

$$T_p = \frac{2\pi\mu p_o}{3}\left(r_o^3 - r_i^3\right) = \frac{2\mu P_p\left(r_o^3 - r_i^3\right)}{3\left(r_o^2 - r_i^2\right)}. \tag{18.7}$$

Thus, expressions for the normal load and torque can be determined from the uniform pressure, the geometry (r_o and r_i), and the coefficient of friction, μ.

18.3.2 Uniform Wear Model

If the mating surfaces of the clutch are sufficiently rigid, it can be assumed that uniform wear will occur. This assumption generally holds true after some initial running in. For example, consider the Archard wear law of Eq. (8.39). For a thrust disk clutch, the sliding distance per revolution is proportional to the radius; that is, the outside of the disk sees a

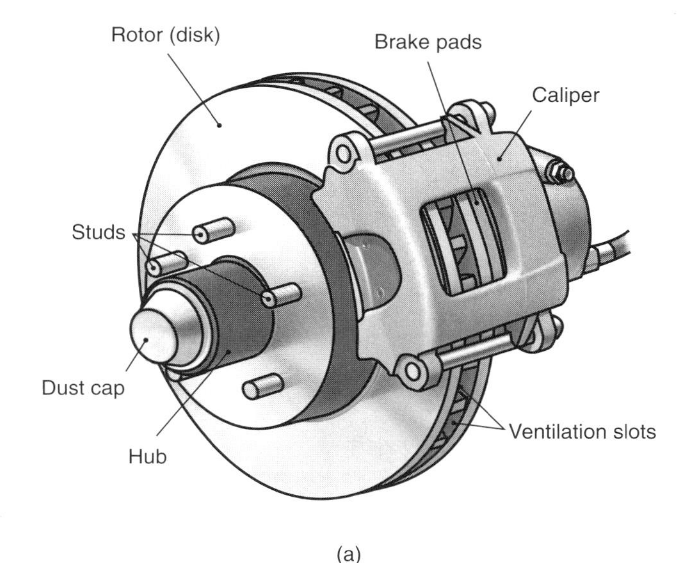

(a)

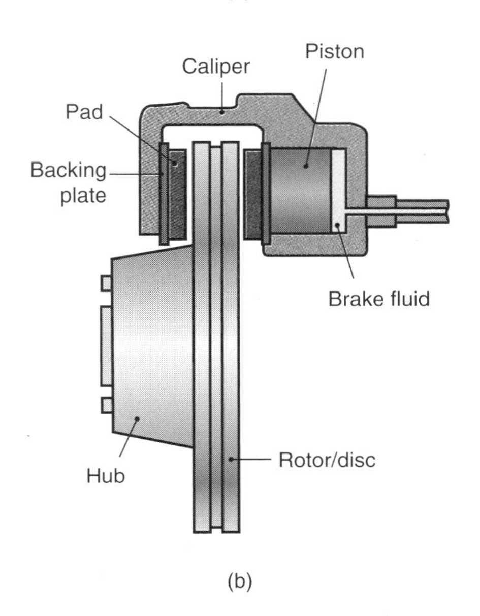

(b)

Figure 18.4: Thrust brake terminology and operation. (a) Illustration of a thrust brake, with wheel removed for clarity. Note that the caliper shown has a window to allow observation of the brake pad thickness, a feature that is not always present. (b) Section view of the disk brake, showing the caliper and brake cylinder.

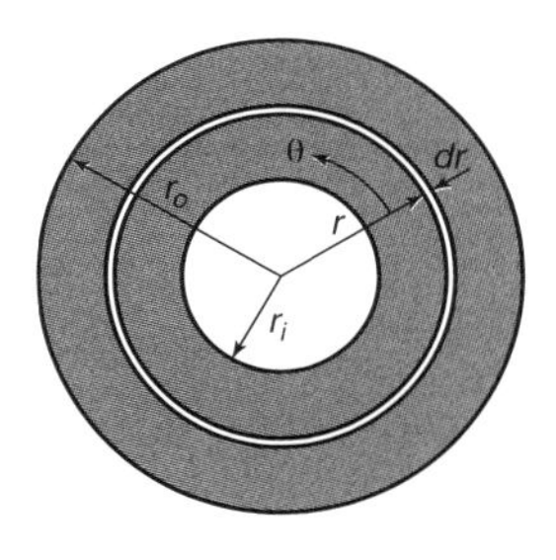

Figure 18.5: Thrust disk clutch surface with various radii.

larger sliding distance than the inner radius. If the pressure is uniform, and the hardness of a brake pad or clutch disk is constant, then more wear will occur on the outside of the disc. Of course, this results in a redistribution of pressure, so that after initial wear has taken place, uniform wear becomes possible. Thus, for the disk shown in Fig. 18.5 the wear is constant over the surface area $r_i \leq r \leq r_o$ and around the circumference of the disk.

The Archard wear law suggests that the rate of wear is proportional to the product of force and velocity, as can be seen by taking the derivative with respect to time of Eq. (8.39). Since the product of force and speed is power,

$$h_p = Fu = \mu Pu = \mu pAu, \tag{18.8}$$

where

F = friction force, N
μ = coefficient of friction
P = normal force, N
u = velocity, m/s
A = area, m^2

If the brake cylinder actuates linearly, and tolerances are reasonably tight, any initial uniform pressure will need to adjust. Where the surfaces wear the most, the pressure decreases the most. Thus, multiplying pressure and velocity will produce a constant work or energy conversion, implying that the wear should be uniform at any radius. Then, μpAu remains constant; and if μA is constant, p is inversely proportional to u, and for any radius, r,

$$p = \frac{c}{r}. \tag{18.9}$$

Substituting Eq. (18.9) into Eq. (18.4) gives

$$P_w = 2\pi c\left(r_o - r_i\right). \tag{18.10}$$

Since $p = p_{\max}$ at $r = r_i$, it follows from Eq. (18.9) that

$$c = p_{\max} r_i. \tag{18.11}$$

Substituting Eq. (18.11) into Eq. (18.10) results in

$$P_w = 2\pi p_{\max} r_i\left(r_o - r_i\right), \tag{18.12}$$

and inserting Eq. (18.9) into Eq. (18.5) produces

$$T_w = \mu c \int\int r\,dr\,d\theta = \frac{2\pi\mu c}{2}\left(r_o^2 - r_i^2\right). \tag{18.13}$$

Substituting Eq. (18.11) into Eq. (18.13) gives

$$T_w = \pi\mu r_i p_{\max}\left(r_o^2 - r_i^2\right). \tag{18.14}$$

Substituting Eq. (18.12) into Eq. (18.14) gives

$$T_w = F_w r_m = \frac{\mu P_w\left(r_o + r_i\right)}{2}. \tag{18.15}$$

By coincidence, Eq. (18.15) gives the same result as if the torque was obtained by multiplying the mean radius $r_m = (r_o + r_i)/2$ by the friction force, F.

Equations (18.7) and (18.15) can be expressed as the dimensionless torque for uniform pressure, $\bar{T}_p$, and for uniform wear, $\bar{T}_w$, by the following equations:

$$\bar{T}_w = \frac{T_w}{2\mu P_w r_o} = \frac{(1+\beta)}{4}, \tag{18.16}$$

$$\bar{T}_p = \frac{T_p}{2\mu P_p r_o} = \frac{(1-\beta^3)}{3(1-\beta^2)}, \tag{18.17}$$

Table 18.3: Representative properties of contacting materials operating dry, when rubbing against smooth cast iron or steel.

Friction material	Coefficient of friction, μ	Maximum contact pressure,[a] p_{max} psi	Maximum contact pressure,[a] p_{max} kPa	Maximum bulk temperature, $t_{m,\max}$ °F	Maximum bulk temperature, $t_{m,\max}$ °C
Molded	0.25–0.45	150–300	1030–2070	400–500	204–260
Woven	0.25–0.45	50–100	345–690	400–500	204–260
Sintered metal	0.15–0.45	150–300	1030–2070	400–1250	204–677
Cork	0.30–0.50	8–14	55–95	180	82
Wood	0.20–0.30	50–90	345–620	200	93
Cast iron; hard steel	0.15–0.25	100–250	690–1720	500	260

[a] Use of lower values will give longer life.

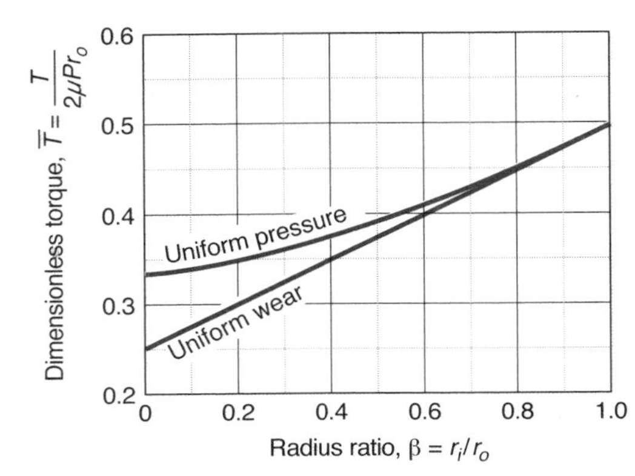

Figure 18.6: Effect of radius ratio on dimensionless torque for uniform pressure and uniform wear models.

where

$$\beta = \frac{r_i}{r_o}. \tag{18.18}$$

Figure 18.6 shows the effect of radius ratio, β, on dimensionless torque for the uniform pressure and uniform wear models. The largest difference between these models occurs at a radius ratio of zero, and the smallest difference occurs at a radius ratio of 1. Also, for the same dimensionless torque the uniform wear model requires a larger radius ratio than does the uniform pressure model. This larger radius ratio implies that a smaller area is predicted by the uniform wear model. Thus, the uniform wear model may be viewed as the safer approach, although the two approaches yield similar torque predictions. Usually a brake or clutch is analyzed using both conditions to make sure a design is robust over its life. The rationale is that, when new, the pressure can be uniform, but as the brake wears, the pressure will adjust until wear is uniform across the pad area.

Table 18.3 gives the coefficient of friction for several materials rubbing against smooth cast iron or steel under dry conditions. It also gives the maximum contact pressure and the maximum bulk temperature for these materials. Table 18.4 gives the coefficient of friction for several materials, including those in Table 18.3, rubbing against smooth cast iron or steel in oil. As would be expected, the coefficients of friction are much smaller in oil than under dry conditions.

Equations (18.6) and (18.7) for the uniform pressure model and Eqs. (18.12) and (18.15) for the uniform wear model, which are applicable for thrust disk clutches, are also applicable for thrust disk brakes provided that the disk shape is similar to that shown in Fig. 18.5. A detailed analysis of disk brakes gives equations that result in slightly larger torques than those resulting from the clutch equations. This text assumes that the brake and clutch equations are identical.

Table 18.4: Coefficient of friction for contacting materials operating in oil when rubbing against steel or cast iron.

Friction material	Coefficient of friction, μ
Molded	0.06–0.09
Woven	0.08–0.10
Sintered metal	0.05–0.08
Paper	0.10–0.14
Graphitic	0.12 (avg.)
Polymeric	0.11 (avg.)
Cork	0.15–0.25
Wood	0.12–0.16
Cast iron; hard steel	0.03–0.16

18.3.3 Partial Contact

It should be noted that many brakes do not use a full circle of contact, as shown in Fig. 18.4. A common practice is to assume that a pad can be approximated as a sector or wedge, so that it covers a fraction of the rotor or disc. This has to be accommodated in the equations for the actuating force and torque developed for the uniform pressure and uniform wear models. This can be easily accomplished by defining γ as the percentage of the rotor covered by the pad. Then, for the uniform pressure model, Eqs. (18.6) and (18.7) become

$$P_p = \gamma\pi p_o\left(r_o^2 - r_i^2\right), \tag{18.19}$$

$$T_p = \frac{2\gamma\pi\mu p_o}{3}\left(r_o^3 - r_i^3\right), \tag{18.20}$$

and for the uniform wear model, Eq. (18.12) and (18.14) become

$$P_w = 2\pi\gamma p_{\max} r_i\left(r_o - r_i\right), \tag{18.21}$$

$$T_w = \pi\gamma\mu r_i p_{\max}\left(r_o^2 - r_i^2\right). \tag{18.22}$$

Example 18.1: Optimum Size of a Thrust Disk Clutch

Given: A single set of thrust disk clutches is to be designed for use in an engine with a maximum torque of 150 N-m. A woven fiber reinforced polymer will contact steel in a dry environment. A safety factor of 1.5 is assumed in order to account for slippage at full engine torque. The outside diameter should be as small as possible.

Find: Determine the appropriate values for r_o, r_i, and P.

Solution: For a woven material in contact with steel in a dry environment, Table 18.3 gives the coefficient of friction as $\mu = 0.35$ and the maximum contact pressure as $p_{max} = 345$ kPa = 0.345 MPa. The average coefficient of friction has been used to estimate average performance; however, the smallest pressure has been selected for a long life. Making use of the above and Eq. (18.14) gives

$$\begin{aligned} r_i\left(r_o^2 - r_i^2\right) &= \frac{n_s T_w}{\pi \mu p_{\max}} \\ &= \frac{(1.5)(150)}{\pi(0.35)\left(0.345 \times 10^6\right)} \\ &= 5.931 \times 10^{-4}\ \text{m}^3. \end{aligned}$$

Solving for the outside radius, r_o,

$$r_o = \sqrt{\frac{5.931 \times 10^{-4}}{r_i} + r_i^2}. \qquad (a)$$

The minimum outside radius is obtained by taking the derivative of the outside radius with respect to the inside radius and setting it equal to zero:

$$\frac{dr_o}{dr_i} = \frac{0.5}{\sqrt{\dfrac{5.931 \times 10^{-4}}{r_i} + r_i^2}}\left(-\frac{5.931 \times 10^{-4}}{r_i^2} + 2r_i\right) = 0,$$

which is numerically solved as $r_i = 66.69$ mm. Therefore, Eq. (a) gives

$$r_o = \sqrt{\frac{5.931 \times 10^{-4}}{0.06669} + 0.06669^2} = 0.1155\ \text{m} = 115.5\ \text{mm}.$$

The radius ratio is

$$\beta = \frac{r_i}{r_o} = \frac{66.69}{115.5} = 0.5774.$$

The radius ratio required to maximize the torque capacity is the same as the radius ratio required to minimize the outside radius for a given torque capacity. Thus, the radius ratio for maximizing the torque capacity or for minimizing the outside radius is

$$\beta_o = \sqrt{\frac{1}{3}} = 0.5774.$$

From Eq. (18.15), the maximum normal force that can be applied to the clutch without exceeding the pad pressure constraint is

$$P = \frac{2n_s T_w}{\mu(r_o + r_i)} = \frac{2(1.5)(150)}{(0.35)(0.1155 + 0.06669)} = 7057\ \text{N}.$$

18.4 Cone Clutches and Brakes

Cone clutches use wedging action to increase the normal force on the lining, thus increasing the friction force and the torque. Usually the half cone angle, α, is above 4° to avoid jamming, and is usually between 6° and 8°. Figure 18.7 shows a conical surface with forces acting on an element. The area of the element and the normal force on the element are

$$dA = (r\,d\theta)\left(\frac{dr}{\sin\alpha}\right), \qquad (18.23)$$

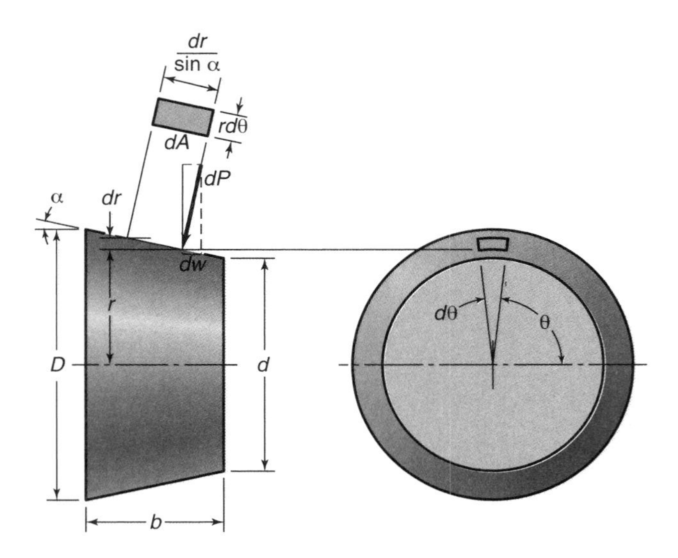

Figure 18.7: Forces acting on elements of a cone clutch.

$$dP = p\,dA. \qquad (18.24)$$

The actuating force is the thrust component, dW, of the normal force, dP, or

$$dW = dP\sin\alpha = p\,dA\sin\alpha = pr\,dr d\theta.$$

Using Eq. (18.4) gives the actuating force as

$$W = \int\int pr\,dr\,d\theta = 2\pi\int_{d/2}^{D/2} pr\,dr. \qquad (18.25)$$

Similarly, Eq. (18.5) gives the torque as

$$T = \int \mu r\,dP = \frac{2\pi}{\sin\alpha}\int_{d/2}^{D/2} \mu p r^2\,dr. \qquad (18.26)$$

18.4.1 Uniform Pressure Model

As was discussed in Section 18.3.1, the pressure for a new thrust disk clutch is assumed to be uniform over the surfaces, or $p = p_o$. Using this uniform pressure model for a cone clutch gives the actuating force as

$$W = \frac{\pi p_o}{4}\left(D^2 - d^2\right). \qquad (18.27)$$

Similarly, the torque is

$$T = \frac{2\pi p_o \mu}{3\sin\alpha}\left(\frac{1}{8}\right)\left(D^3 - d^3\right) = \frac{\pi p_o \mu}{12\sin\alpha}\left(D^3 - d^3\right). \qquad (18.28)$$

Making use of Eq. (18.27) enables Eq. (18.28) to be rewritten as

$$T = \frac{\mu W\left(D^3 - d^3\right)}{3\sin\alpha\left(D^2 - d^2\right)}. \qquad (18.29)$$

18.4.2 Uniform Wear Model

Substituting Eq. (18.9) into Eq. (18.25) gives the actuating force as

$$W = 2\pi c\int_{d/2}^{D/2} dr = \pi c\,(D - d). \qquad (18.30)$$

Similarly, substituting Eq. (18.9) into Eq. (18.26) gives the torque as

$$T = \frac{2\pi\mu c}{\sin\alpha}\int_{d/2}^{D/2} r\,dr = \frac{\pi\mu c}{4\sin\alpha}\left(D^2 - d^2\right). \qquad (18.31)$$

Making use of Eq. (18.30) enables Eq. (18.31) to be rewritten as

$$T = \frac{\mu W}{4\sin\alpha}\left(D + d\right). \qquad (18.32)$$

Example 18.2: Cone Clutch

Given: A cone clutch similar to that shown in Fig. 18.7 has the following dimensions: D = 330 mm, d = 306 mm, and b = 60 mm. The clutch uses sintered metal on steel, with a coefficient of friction of 0.26, and the torque transmitted is 200 N-m.

Find: Determine the minimum required actuating force and the associated contact pressure by using the uniform pressure and uniform wear models.

Solution:

1. *Uniform wear:* From Fig. 18.7, the half-cone angle of the cone clutch is

$$\tan\alpha = \frac{D-d}{2b} = \frac{165-153}{60} = \frac{12}{60} = 0.200,$$

$$\alpha = 11.31^\circ.$$

The pressure is a maximum when $r = d/2$. Thus, making use of Eqs. (18.31) and (18.9) gives

$$T = \frac{\pi\mu d p_{\max}}{8\sin\alpha}\left(D^2 - d^2\right),$$

and

$$\begin{aligned} p_{\max} &= \frac{8T\sin\alpha}{\pi\mu d\left(D^2 - d^2\right)} \\ &= \frac{8(200)\sin 11.31^\circ}{\pi(0.26)(0.306)(0.330^2 - 0.306^2)} \\ &= 82.25 \text{ kPa.} \end{aligned}$$

From Eq. (18.32), the actuating force can be written as

$$W = \frac{4T\sin\alpha}{\mu(D+d)} = \frac{4(200)\sin 11.31^\circ}{(0.26)(0.330+0.306)} = 948.8 \text{ N.}$$

2. *Uniform pressure:* From Eq. (18.29), the actuating force can be expressed as

$$\begin{aligned} W &= \frac{3T\sin\alpha\left(D^2 - d^2\right)}{\mu\left(D^3 - d^3\right)} \\ &= \frac{3(200)\sin 11.31^\circ\left(0.330^2 - 0.306^2\right)}{(0.26)(0.330^3 - 0.306^3)} \\ &= 948.4 \text{ N.} \end{aligned}$$

From Eq. (18.27), the uniform pressure required is

$$\begin{aligned} p_{\max} &= p_o = \frac{4W}{\pi\left(D^2 - d^2\right)} \\ &= \frac{4(948.4)}{\pi(0.330^2 - 0.306^2)} \text{ Pa} \\ &= 79.11 \text{ kPa.} \end{aligned}$$

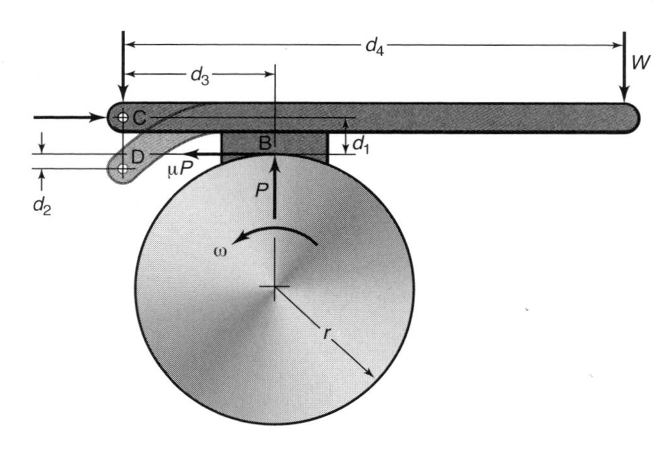

Figure 18.8: Block, or short-shoe brake, with two configurations.

18.5 Block or Short-Shoe Brakes

A **block**, or **short-shoe**, **brake** can be configured to move radially against a cylindrical drum or plate, as shown in Fig. 18.8, although other configurations exist. A normal force, P, develops a friction force, $F = \mu P$, on the drum or plate, where μ is the coefficient of friction. The actuating force, W, is also shown in Fig. 18.8 along with critical hinge pin dimensions d_1, d_2, d_3, and d_4. The normal force, P, and the friction force, μP, are the forces acting on the brake. For block or short-shoe brakes, a constant pressure is assumed over the pad surface. As long as the pad is short relative to the circumference of the drum, this assumption is reasonably accurate.

A brake is considered to be **self-energizing** if the friction moment assists the actuating moment in applying the brake. This implies that the signs of the friction and actuating moments are the same. **Deenergizing** effects occur if the friction moment counteracts the actuating moment in applying the brake. Figure 18.8 can be used to illustrate self-energizing and deenergizing effects. Note that merely changing the cylinder's direction of rotation will change a self-energizing shoe into a deenergizing shoe.

Summing the moments about the hinge at C (see Fig. 18.8) and setting equal to zero results in

$$d_4W + \mu P d_1 - d_3P = 0.$$

Since the signs of the friction and actuating moments are the same, the brake hinged at C is self-energizing. Solving for the normal force gives

$$P = \frac{d_4W}{d_3 - \mu d_1}. \qquad (18.33)$$

The braking torque at C is

$$T = Fr = \mu r P = \frac{\mu r d_4 W}{d_3 - \mu d_1}, \qquad (18.34)$$

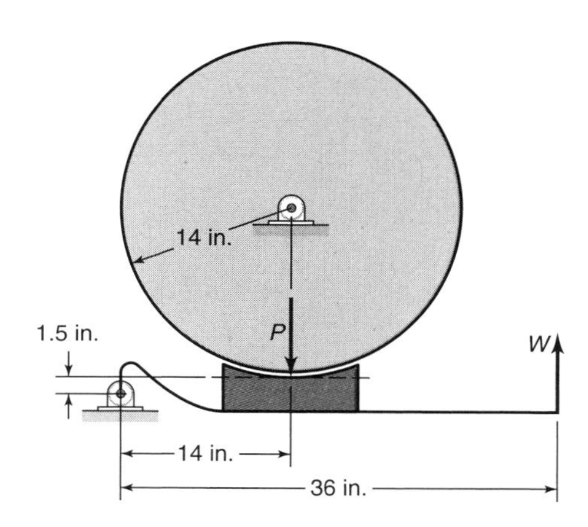

Figure 18.9: Short-shoe brake used in Example 18.3.

where r is the radius of the brake drum. Summing the moments about the hinge at D (see Fig. 18.8) and setting the sum equal to zero results in

$$-Wd_4 + \mu P d_2 + d_3 P = 0.$$

Since the signs of the friction and actuating moments are opposite, the brake hinged at D is deenergizing. Solving for the normal force gives

$$P = \frac{Wd_4}{d_3 + \mu d_2}. \tag{18.35}$$

The torque of the brake hinged at D in Fig. 18.8 is

$$T = \frac{\mu d_4 r W}{d_3 + \mu d_2}. \tag{18.36}$$

A brake is considered to be **self-locking** if the actuating force (W in Fig. 18.8) equals zero for a non-zero torque. Self-locking brake geometries are avoided, since they seize or grab, thus operating unsatisfactorily or even dangerously.

Block brakes such as shown in Fig. 18.8 reduce an angular velocity, and in vehicle applications require a high friction force to exist between a wheel and road to affect braking. This does not exist for some conveyor designs, or for vehicles such as roller coasters that ride on rails where the friction between the wheel and rail is very low. In such circumstances, a common brake uses a vehicle-mounted rectangular pad or fin that bears against a flange or fin on the moving component. This is discussed in Case Study 18.1.

With such a block brake, when a plate or fin approaches the brake system, the fins bear against the brake pads and thereby develop a normal force. Not surprisingly, the leading edge of the brake pad will wear more than other sections of the brake. Clearly, the brake does not encounter uniform pressure. However, from a tribological standpoint, there is little difference between a concentrated load and a distributed load in generating a friction force. As long as the real area of contact between the pad and the fin remains small, a linear relationship exists between the friction and the normal force. For all practical purposes, this approximation is reasonable for the entire time of brake-fin contact.

Example 18.3: Short-Shoe Brake

Given: A 14-in.-radius brake drum contacts a single short shoe, as shown in Fig. 18.9, and sustains 2000 in.-lb of torque at 500 rpm. Assume that the coefficient of friction for the drum and shoe combination is $\mu = 0.30$.

Find: Determine the following:

(*a*) Normal force acting on the shoe

(*b*) Required actuating force, W, when the drum has clockwise rotation

(*c*) Required actuating force, W, when the drum has counterclockwise rotation

(*d*) Required change in the 1.5-in. dimension (Fig. 18.9) for self-locking to occur if the other dimensions do not change

Solution:

(*a*) The torque of the brake is

$$T = rF = r\mu P,$$

where

$$P = \frac{T}{\mu r} = \frac{2000}{(0.3)(14)} = 476.2 \text{ lb},$$

and

$$\mu P = (0.3)(476.2) = 142.9 \text{ lb}.$$

(*b*) For clockwise rotation, summing the moments about the hinge pin and setting the sum equal to zero yields

$$(1.5)(142.9) + 36W - 14(476.2) = 0, \tag{a}$$

which is solved as $W = 179.2$ lb. Since the signs of the friction and actuating moments are the same, the brake is self-energizing.

(*c*) For counterclockwise rotation, summing the moments about the hinge pin and setting the sum equal to zero gives

$$(1.5)(142.9) - 36W + 14(476.2) = 0,$$

or $W = 191.1$ lb. Since the signs of the friction and actuating moments are not the same, the brake is deenergizing.

(*d*) If, in Eq. (*a*), $W = 0$,

$$x(142.9) + 36W - 14(476.2) = 0,$$

or, solving for x,

$$x = \frac{(14)(476.2)}{142.9} = 46.65 \text{ in}.$$

Therefore, self-locking will occur if the distance of 1.5 in. in Fig. 18.9 is changed to 46.65 in. Since self-locking is not a desirable effect in a brake and 1.5 in. is quite different from 46.65 in., we would not expect the brake to self-lock.

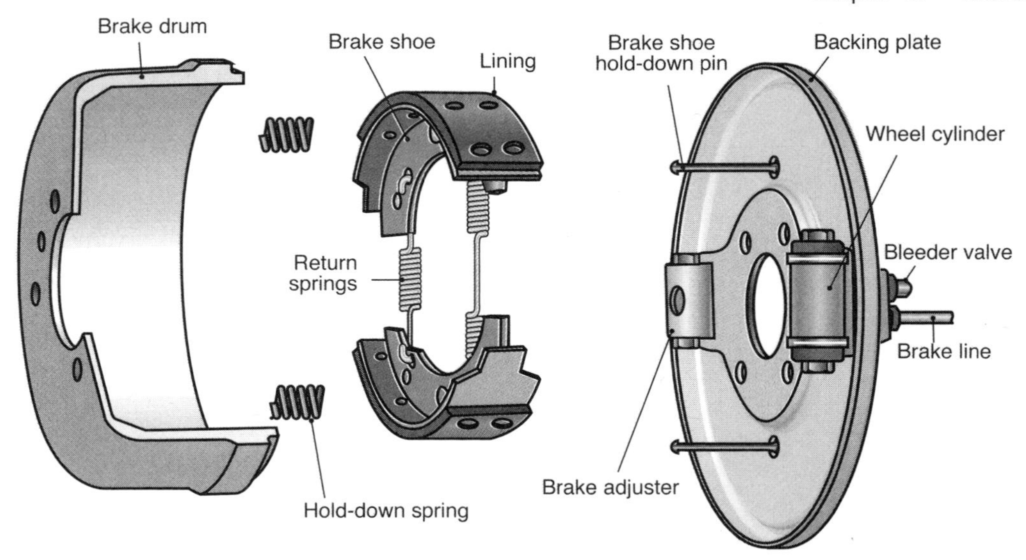

Figure 18.10: A typical automotive long-shoe, internal, expanding rim brake, commonly called a drum brake.

18.6 Long-Shoe, Internal, Expanding Rim Brakes

Figure 18.10 shows a long-shoe, internal, expanding rim brake with two **pads** or **shoes**, often referred to as **drum brakes**. They are widely used, especially in automobiles, where they are commonly applied for rear axle braking. As mentioned previously, disc brakes are commonly used for front brakes because they promote more convective heat transfer than drum designs, and the front axle encounters more air flow over the brake or disc. This can be accentuated with grooves or holes in the rotor, as shown in this chapter's opening illustration. Drum brakes are used because they have a self-energizing shoe, which is very easy to incorporate into a parking brake feature; disc brakes do not have self-energizing shoes, and when all-wheel disc brakes are used, it is common that a drum brake is built into the rear disc brakes to provide a parking brake feature.

Figure 18.11 illustrates the shoes and the drum, and defines important dimensions. The hinge pin for the right shoe is at A in Fig. 18.11. The *heel* of the pad is the region closest to the hinge pin, and the *toe* is the region closest to the actuating force, W. A major difference between a short shoe (Fig. 18.8) and a long shoe (Fig. 18.11) is that the pressure can be considered constant for a short shoe but not for a long shoe. In a long shoe, little if any pressure is applied at the heel, and the pressure increases toward the toe. This sort of pressure variation suggests that the pressure may vary sinusoidally. Thus, a relationship of the contact pressure, p, in terms of the maximum pressure, $p_{\max}$, may be written as

$$p = p_{\max}\left(\frac{\sin\theta}{\sin\theta_a}\right), \tag{18.37}$$

where θ_a is the angle where pressure is at a maximum value. Observe from Eq. (18.37) that $p = p_{\max}$ when $\theta = 90°$. If the shoe has an angular extent less than 90° (such as is the case in Fig. 18.11 if θ_2 was less than 90°), $p = p_{\max}$ when $\theta_a = \theta_2$.

Observe also in Fig. 18.11 that the distance d_6 is perpendicular to the actuating force W. Figure 18.12 shows the forces and critical dimensions of a long-shoe, internal, expanding rim brake. In Fig. 18.12, the θ-coordinate begins with a line drawn from the center of the drum to the center of the hinge pin. Also, the shoe lining does not begin at $\theta = 0°$, but at some θ_1 and extends until θ_2. At any angle, θ, of the lining, the differential normal force dP is

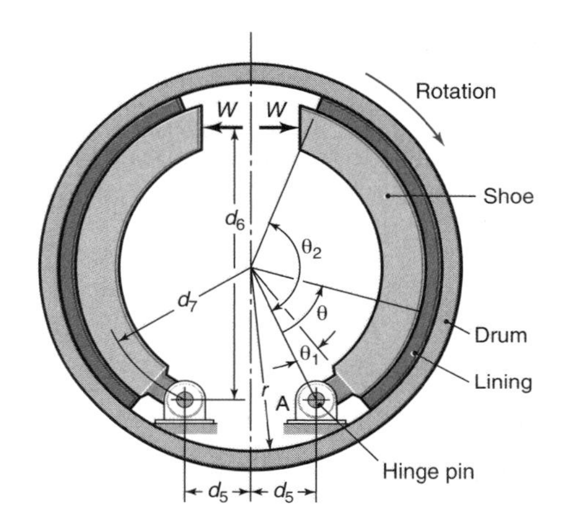

Figure 18.11: Long-shoe, internal, expanding rim brake with two shoes.

$$dP = pbr\,d\theta, \tag{18.38}$$

where b is the face width of the shoe (the distance perpendicular to the paper). Substituting Eq. (18.37) into Eq. (18.38) gives

$$dP = \frac{p_{\max} br \sin\theta\,d\theta}{\sin\theta_a}. \tag{18.39}$$

From Eq. (18.39), the normal force has a moment arm of $d_7 \sin\theta$ so that its associated moment is

$$\begin{aligned} M_P &= \int d_7 \sin\theta\,dP = \frac{d_7 br p_{\max}}{\sin\theta_a}\int_{\theta_1}^{\theta_2}\sin^2\theta\,d\theta \\ &= \frac{brd_7 p_{\max}}{4\sin\theta_1}\left[2\left(\theta_2-\theta_1\right)\frac{\pi}{180°} - \sin 2\theta_2 + \sin 2\theta_1\right], \end{aligned} \tag{18.40}$$

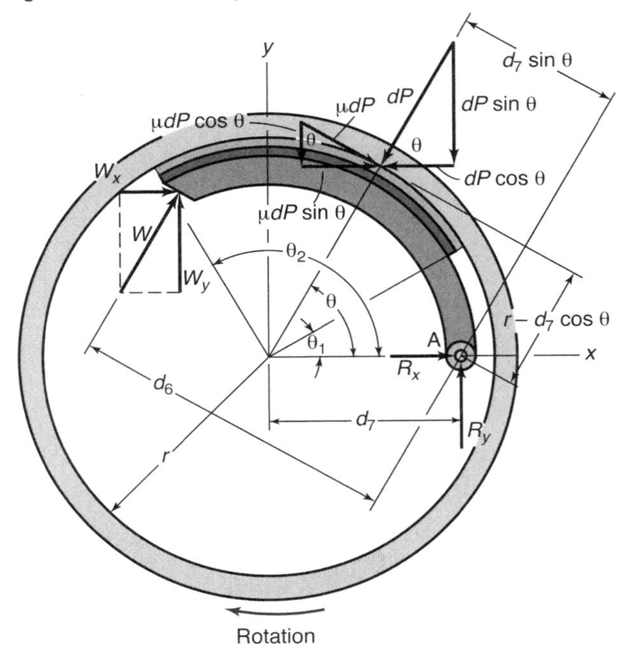

Figure 18.12: Forces and dimensions of long-shoe, internal expanding rim brake.

where θ_1 and θ_2 are in degrees. From Eq. (18.39), the friction force has a moment arm of $r - d_7 \cos\theta$, so that its moment is

$$\begin{aligned} M_F &= \int (r - d_7 \cos\theta)\, \mu\, dP \\ &= \frac{\mu p_{\max} br}{\sin\theta_1} \int_{\theta_1}^{\theta_2} (r - d_7 \cos\theta) \sin\theta \, d\theta, \end{aligned}$$

or

$$M_F = -\frac{\mu p_{\max} br}{\sin\theta_a} \left[r\left(\cos\theta_2 + \cos\theta_1\right) - \frac{d_7}{2}\left(\sin^2\theta_2 - \sin^2\theta_1\right)\right] \tag{18.41}$$

Long-shoe brakes can be designed so that all, some, or none of the shoes are self-energizing. Each type of shoe needs to be analyzed separately.

18.6.1 Self-Energizing Shoe

Setting the sum of the moments about the hinge pin equal to zero results in

$$-Wd_6 - M_F + M_P = 0. \tag{18.42}$$

Since the actuating and friction moments have the same sign in Eq. (18.42), the shoe shown in Fig. 18.12 is self-energizing. It can also be concluded just from Fig. 18.12 that the shoe is self-energizing because W_x and $\mu\, dP \sin\theta$ are in the same direction. Solving for the actuating force in Eq. (18.42) gives

$$W = \frac{M_P - M_F}{d_6}. \tag{18.43}$$

From Eqs. (8.5) and (18.39), the braking torque is

$$\begin{aligned} T &= \int r\mu\, dP \\ &= \frac{\mu p_{\max} br^2}{\sin\theta_a} \int_{\theta_1}^{\theta_2} \sin\theta\, d\theta \\ &= \frac{\mu p_{\max} br^2 \left(\cos\theta_1 - \cos\theta_2\right)}{\sin\theta_a}. \end{aligned} \tag{18.44}$$

Figure 18.12 shows the reaction forces as well as the friction force and the normal force. Summing the forces in the x-direction and setting the sum equal to zero results in

$$R_{xs} + W_x - \int \cos\theta\, dP + \int \mu \sin\theta\, dP = 0. \tag{18.45}$$

Substituting Eq. (18.39) into Eq. (18.45) gives the reaction force in the x-direction for a self-energizing shoe as

$$\begin{aligned} R_{xs} = &\ \frac{p_{\max,s} br}{\sin\theta_a} \int_{\theta_1}^{\theta_2} \sin\theta \cos\theta\, d\theta \\ &- \frac{\mu p_{\max,s} br}{\sin\theta_a} \int_{\theta_1}^{\theta_2} \sin^2\theta\, d\theta - W_x, \end{aligned}$$

or

$$\begin{aligned} R_{xs} = &-W_x + \frac{p_{\max,s} br}{4\sin\theta_a} \left\{2\left(\sin^2\theta_2 - \sin^2\theta_1\right)\right\} \\ &- \frac{\mu p_{\max} br}{4\sin\theta_a} \left[2\left(\theta_2 - \theta_1\right)\frac{\pi}{180^\circ} - \sin 2\theta_2 + \sin 2\theta_1\right], \end{aligned} \tag{18.46}$$

where θ_1 and θ_2 are in degrees. From Fig. 18.12, summing the forces in the y-direction and setting the sum equal to zero gives

$$R_{ys} + W_y - \int \mu\, dP \cos\theta - \int dP \sin\theta = 0, \tag{18.47}$$

and

$$\begin{aligned} R_{ys} = &-W_y + \frac{2\mu p_{\max,s} br}{4\sin\theta_a} \left(\sin^2\theta_2 - \sin^2\theta_1\right) \\ &+ \frac{p_{\max,s} br}{4\sin\theta_a} \left[2\left(\theta_2 - \theta_1\right)\frac{\pi}{180^\circ} - \sin 2\theta_2 + \sin 2\theta_1\right]. \end{aligned} \tag{18.48}$$

In Eqs. (18.45) to (18.48), the reference system has its origin at the center of the drum. The positive x-axis is taken to be through the hinge pin. The positive y-axis is in the direction of the shoe.

18.6.2 Deenergizing Shoe

If, in Fig. 18.12, the direction of rotation is changed from clockwise to counterclockwise, the friction forces change direction. Thus, summing the moments about the hinge pin and setting the sum equal to zero,

$$-Wd_6 + M_F + M_P = 0. \tag{18.49}$$

The only difference between Eq. (18.42) and Eq. (18.49) is the sign of the friction moment. Solving for the actuating force in Eq. (18.49) gives

$$W = \frac{M_P + M_F}{d_6}. \tag{18.50}$$

For a deenergizing shoe, the only changes from the equations derived for the self-energizing shoe are that in Eqs. (18.46) and (18.48) a sign change occurs for terms containing the coefficient of friction, μ, resulting in the following:

$$\begin{aligned} R_{xd} = &-W_x + \frac{p_{\max,d} br}{4\sin\theta_a} \left[2\left(\sin^2\theta_2 - \sin^2\theta_1\right)\right] \\ &+ \frac{\mu p_{\max,d} br}{4\sin\theta_a} \left[2\left(\theta_2 - \theta_1\right)\frac{\pi}{180^\circ} - \sin 2\theta_2 + \sin 2\theta_1\right], \end{aligned} \tag{18.51}$$

$$R_{yd} = -W_y - \frac{\mu p_{\max,d} b r}{2 \sin\theta_a}\left(\sin^2\theta_2 - \sin^2\theta_1\right) + \frac{p_{\max,d} b r}{4\sin\theta_a}\left[2\left(\theta_2-\theta_1\right)\frac{\pi}{180^\circ} - \sin 2\theta_2 + \sin 2\theta_1\right]. \quad (18.52)$$

The maximum contact pressure used in evaluating a self-energizing shoe is taken from Table 18.3. The maximum contact pressure used in evaluating a deenergizing shoe is less than that for the self-energizing shoe, since the actuating force is the same for both types of shoe.

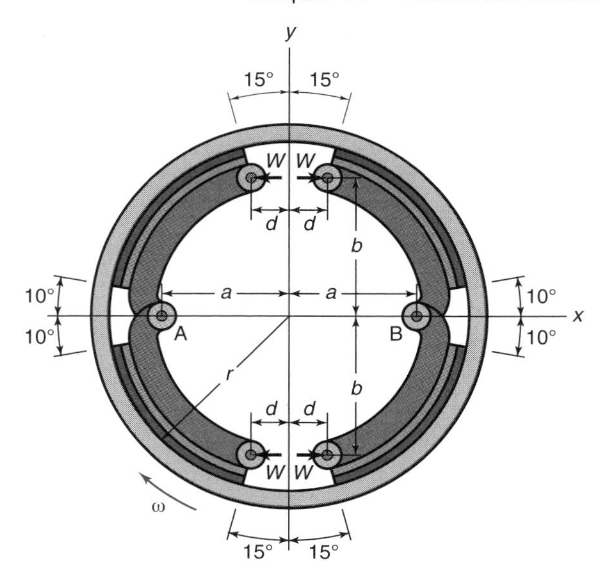

Figure 18.13: Four-long-shoe, internal expanding rim brake used in Example 18.4.

Design Procedure 18.1: Long-Shoe, Internal, Expanding Brake Analysis

It will be assumed that the application has a known velocity, and that the physical dimensions of the brake are known. This Design Procedure outlines the method used to obtain the maximum allowable brake force (which can be controlled by design of the hydraulic or pneumatic actuators) and braking torque.

1. Select a brake material. A reasonable starting point is to assume the drum is made of steel, using sintered metal lining material. Table 18.2 then allows estimation of maximum allowable contact pressure and friction coefficient. Table 18.3 also recommends a maximum pressure, but based on thermal conditions. The lower of the two contact pressures should be used for further analysis.

2. Draw a free body diagram of the brake shoes, paying special attention to the force that acts on the shoes due to friction. Identify which of the shoes, if any, are self-energizing or deenergizing. In a self-energizing shoe, the moment due to frictional force applied to the shoe will have the same sign as the moment due to the applied force. If it is not clear that a shoe is self-energizing or deenergizing, then assume the brake is self-energizing in order to be conservative regarding maximum shoe pressure. In any case, if the friction moment is close to zero, then the braking torque will be similar whether the brake was assumed to be self-energizing or deenergizing.

3. Evaluate M_P and M_F from Eqs. (18.40) and (18.41), respectively. Note that one or more terms may be unknown, but they can be treated as variables.

4. Consider the self-energizing shoe first. The self-energizing shoe will encounter a higher pressure than the deenergizing shoe, so that the limiting pressure determined above can be used to evaluate M_P and M_F.

5. Equation (18.43) can be used to determine the maximum braking force. Note that a lower braking force can be applied, but a higher braking force would exceed the allowable stress of the lining material, leading to plastic deformation or compromised brake life. If the braking force was prescribed, then Eq. (18.43) can be used to obtain the pressure in the shoe, which can be compared to the maximum allowable pressure obtained previously.

6. Equation (18.44) can be used to obtain the torque for the self-energizing shoe.

7. Equations (18.46) and (18.48) can be used to obtain the hinge pin reaction forces.

8. In most brakes, the force applied to the self-energizing and deenergizing shoes are the same. However, the maximum pressure on the deenergizing shoe will be lower than the self-energizing one. Thus, the applied force and pressure can be taken from the self-energizing shoe analysis, as this will reflect the higher stress.

9. Equation (18.50) allows calculation of the maximum pressure on the deenergizing shoe.

10. The torque can be obtained from Eq. (18.45) using the maximum pressure for the deenergizing shoe.

11. Equations (18.51) and (18.52) allow calculation of the hinge pin reaction forces.

Example 18.4: Long-Shoe Internal Brake

Given: Figure 18.13 shows four long shoes in an internal, expanding rim brake. The brake drum has a 400-mm inner diameter. Each hinge pin (A and B) supports a pair of shoes. The actuating mechanism is to be arranged to produce the same actuating force W on each shoe. The shoe face width is 75 mm. The material of the shoe and drum produces a coefficient of friction of 0.24 and a maximum contact pressure of 1 MPa. Additional dimensions for use in Fig. 18.13 are as follows: $d = 50$ mm, $b = 165$ mm, and $a = 150$ mm.

Find:

(*a*) Which shoes are self-energizing and which are deenergizing?

(*b*) What are the actuating forces and total torques for the four shoes?

(*c*) What are the hinge pin reactions as well as the resultant reaction?

Solution:

(*a*) With the drum rotating in the clockwise direction (Fig. 18.13) the top-right and bottom-left shoes have their actuating and friction moments acting in the same direction. Thus, they are self-energizing shoes. The top-left and bottom-right shoes have their actuating and friction moments acting in opposite directions. Thus, they are deenergizing shoes.

(*b*) The dimensions given in Fig. 18.13 correspond to the dimensions given in Figs. 18.11 and 18.12 as $d_5 = 50$ mm, $d_6 = 165$ mm, and $d_7 = 150$ mm. Also, since $\theta_2 < 90°$, then $\theta_2 = \theta_a$. Because the hinge pins in Fig. 18.13 are at A and B, $\theta_1 = 10°$ and $\theta_2 = \theta_a = 75°$.

Self-energizing shoes: Making use of the above and Eq. (18.40) gives the normal force moment as

$$\begin{aligned} M_{Ps} &= \frac{brd_7 p_{\max,s}}{4\sin\theta_a} \times \left[2(\theta_2-\theta_1)\frac{\pi}{180°} - \sin 2\theta_2 + \sin 2\theta_1\right] \\ &= \frac{(0.075)(0.2)(0.15)\left(1\times 10^6\right)}{4\sin 75°} \times \left[2(75-10)\frac{\pi}{180°} - \sin 150° + \sin 20°\right] \\ &= 1229 \text{ N-m.} \end{aligned}$$

From Eq. (18.41), the friction moment is

$$\begin{aligned} M_{Fs} &= -\frac{\mu p_{\max,s} br}{\sin\theta_a}\left[r(\cos\theta_2 - \cos\theta_1) + \frac{d_7}{2}\left(\sin^2\theta_2 - \sin^2\theta_1\right)\right] \\ &= 288.8 \text{ N-m.} \end{aligned}$$

From Eq. (18.43), the actuating force for both the self-energizing and deenergizing shoes is

$$W_s = W_d = \frac{M_P - M_F}{d_6} = \frac{1229 - 288.8}{0.165} = 5698 \text{ N.}$$

From Eq. (18.44), the braking torque for each self-energizing shoe is

$$\begin{aligned} T_s &= \frac{\mu p_{\max,s} br^2(\cos\theta_1 - \cos\theta_2)}{\sin\theta_a} \\ &= \frac{(0.24)\left(0.075\times 10^6\right)(0.2)^2(\cos 10° - \cos 75°)}{\sin 75°} \\ &= 541.2 \text{ N-m.} \end{aligned}$$

Deenergizing shoes: The only change in the calculation of the normal and friction moments for the deenergizing shoes is the maximum pressure. This is unknown, but using Eqs. (18.40), (18.41) and (18.50) will allow its determination. M_{Pd} and M_{Fd} are obtained as

$$\begin{aligned} M_{Pd} &= \frac{brd_7 p_{\max,d}}{4\sin\theta_a} \times \left[2(\theta_2-\theta_1)\frac{\pi}{180°} - \sin 2\theta_2 + \sin 2\theta_1\right] \\ &= \frac{(0.075)(0.2)(0.15)p_{\max,d}}{4\sin 75°} \times \left[2(75°-10°)\frac{\pi}{180°} - \sin 150° + \sin 20°\right] \\ &= 0.001229 p_{\max,d}, \end{aligned}$$

$$\begin{aligned} M_{Fd} &= \frac{\mu p_{\max,d} br}{\sin\theta_a}\left[-r(\cos\theta_2 - \cos\theta_1) - \frac{d_7}{2}\left(\sin^2\theta_2 - \sin^2\theta_1\right)\right] \\ &= 0.0002888 p_{\max,d}. \end{aligned}$$

From Eq. (18.50) the actuating load for the deenergizing shoes is

$$\begin{aligned} W_d &= W_s = 5698 \text{ N} = \frac{M_{Pd} + M_{Fd}}{d_6} \\ &= \left(\frac{0.001229 + 0.0002888}{0.165}\right) p_{\max,d}. \end{aligned}$$

Solving for the maximum pressure yields $p_{\max,d} = 0.6194$ MPa. The braking torque for the deenergizing shoes is

$$T_d = T_s\left(\frac{p_{\max,d}}{p_{\max,s}}\right) = 541.2\left(\frac{0.6194}{1.000}\right) = 335.2 \text{ N-m.}$$

The total braking torque of the four shoes, two of which are self-energizing and two of which are deenergizing, is

$$T = 2(T_s + T_d) = 2(541.2 + 335.2) = 1753 \text{ N-m.}$$

(*c*) The hinge reactions are obtained from Eqs. (18.46) and (18.48) for the self-energizing and deenergizing shoes, respectively. First, for the self-energizing shoes:

From Eq. (18.46),

$$\begin{aligned} R_{xs} &= -W_x + \frac{p_{\max,s} br}{4\sin\theta_a}\left\{2\left(\sin^2\theta_2 - \sin^2\theta_1\right)\right\} - \mu\frac{\mu p_{\max} br}{4\sin\theta_a} \times \left[2(\theta_2-\theta_1)\frac{\pi}{180°} - \sin 2\theta_2 + \sin 2\theta_1\right], \\ &= -654.6 \text{ N} \end{aligned}$$

From Eq. (18.48)

$$\begin{aligned} R_{ys} &= \left[\frac{(1\times 10^6)(0.075)(0.2)}{4\sin 75°}\right] \times \left[2(75°-10°)\frac{\pi}{180°} - \sin 150° + \sin 20° + 2(0.24)\left(\sin^2 75° - \sin^2 10°\right)\right] \\ &= 9878 \text{ N.} \end{aligned}$$

Similarly, $R_{xd} = -137.5$ N.

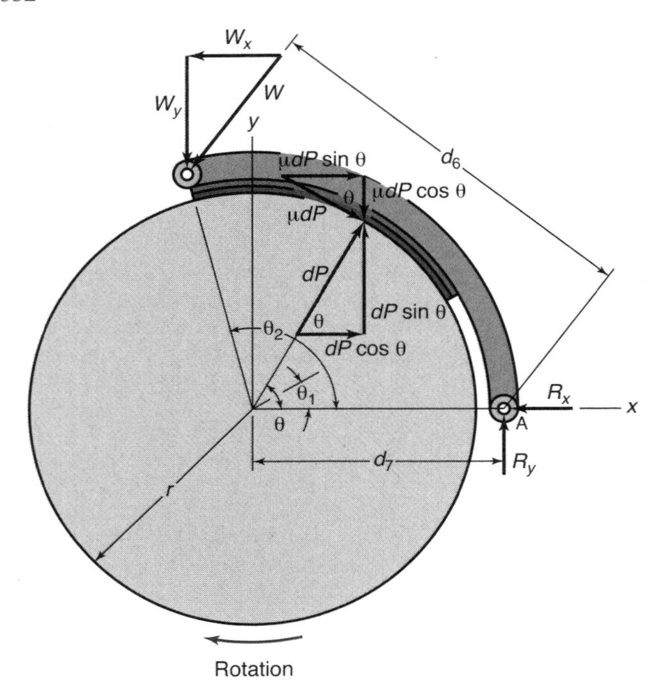

Figure 18.14: Forces and dimensions of long-shoe, external, contracting rim brake.

Deenergizing shoes: From Eq. (18.52),

$$\begin{aligned} R_{yd} &= \left[\frac{(0.6194)(1\times 10^6)(0.075)(0.2)}{4\sin 75^\circ}\right] \\ &\quad \times \left[2(75^\circ - 10^\circ)\frac{\pi}{180^\circ} - \sin 150^\circ + \sin 20^\circ\right] \\ &\quad -2(0.24)(\sin^2 75^\circ - \sin^2 10^\circ)] \\ &= 4034 \text{ N}. \end{aligned}$$

The resultant forces of the reactions in the hinge pin in the horizontal and vertical directions are

$$R_H = -654.6 - 137.5 = -792.1 \text{ N},$$

$$R_V = 9878 - 4034 = 5844 \text{ N}.$$

The resultant force at the hinge pin is

$$R = \sqrt{R_H^2 + R_V^2} = \sqrt{(-792.1)^2 + (5844)^2} = 5897 \text{ N}.$$

18.7 Long-Shoe, External, Contracting Rim Brakes

Figure 18.14 shows the forces and dimensions of a long-shoe, external, contracting rim brake. In Fig. 18.12, the brake is internal to the drum, whereas in Fig. 18.14 the brake is external to the drum. The symbols used in these figures are similar. The equations developed in Section 18.6 for internal shoe brakes are exactly the same as those for external shoe brakes as long as one properly identifies whether the brake is self-energizing or deenergizing.

The *internal* brake shoe in Fig. 18.12 was found to be *self-energizing* for clockwise rotation, since in the moment summation of Eq. (18.42), the actuating and friction moments have the same sign. The *external* brake shoe in Fig. 18.14 is *deenergizing* for clockwise rotation. Summing the moments and equating the sum to zero

$$W d_6 - M_F - M_P = 0. \tag{18.53}$$

The actuating and friction moments have opposite signs and thus the external brake shoe shown in Fig. 18.14 is deenergizing.

If, in Figs. 18.12 and 18.14, the direction of rotation were changed from clockwise to counterclockwise, the friction moments in Eqs. (18.42) and (18.53) would have opposite signs. Therefore, the internal brake shoe would be deenergizing and the external brake shoe would be self-energizing.

Example 18.5: External Long-Shoe Brake

Given: An external, long-shoe rim brake is to be cost optimized. Three lining geometries are being considered, covering the entire 90° of the shoe, covering only 45° of the central portion of the shoe, or covering only 22.5° of the central portion of the shoe. The braking torque must be the same for all three geometries, and the cost of changing any of the brake linings is half of the cost of a 22.5° lining. The cost of the lining material is proportional to the wrap angle. The wear rate is proportional to the pressure. The input parameters for the 90° lining are $d_7 = 100$ mm, $r = 80$ mm, $b = 25$ mm, $\theta_1 = 0^\circ$, $\theta_2 = 90^\circ$, $\mu = 0.27$, and $T = 125$ N-m.

Find: Which of the wrap angles (90°, 45°, or 22.5°) would be the most economical?

Solution: The braking torque is given by Eq. (18.44) and is the same for all three geometries. For the 90° wrap angle ($\theta_1 = 0^\circ$, $\theta_2 = 90^\circ$, and $\theta_a = 90^\circ$):

$$\begin{aligned} (p_{\max})_{90^\circ} &= \frac{T\sin\theta_a}{\mu b r^2(\cos\theta_1 - \cos\theta_2)} \\ &= \frac{(125)\sin 90^\circ}{(0.27)(0.025)(0.08)^2(\cos 0^\circ - \cos 90^\circ)} \\ &= 2.894\times 10^6 \text{ Pa} \\ &= 2.894 \text{ MPa}. \end{aligned}$$

For the 45° wrap angle ($\theta_1 = 22.5^\circ$, $\theta_2 = 67.5^\circ$, and $\theta_a = 67.5^\circ$):

$$\begin{aligned} (p_{\max})_{45^\circ} &= \frac{(125)\sin 67.5^\circ}{(0.27)(0.025)(0.08)^2(\cos 22.5^\circ - \cos 67.5^\circ)} \\ &= 4.940\times 10^6 \text{ Pa} \\ &= 4.940 \text{ MPa}. \end{aligned}$$

For the 22.5° wrap angle ($\theta_1 = 33.75^\circ$, $\theta_2 = 56.25^\circ$, and $\theta_a = 56.25^\circ$):

$$(p_{\max})_{22.5^\circ} = \frac{(125)\sin 56.25^\circ}{(0.27)(0.025)(0.08)^2(\cos 33.7^\circ - \cos 56.2^\circ)}$$

which is solved as 8.720 MPa.

The cost of changing the lining is C. The lining costs are $2C$ for a 22.5° lining, $4C$ for a 45° lining, and $8C$ for a 90° lining. The wear rate is proportional to the pressure, or the time it takes for the shoe to wear out is inversely proportional to the pressure. Thus, the times it takes for the shoe to wear out for the three geometries are

$$t_{90^\circ} = \frac{A}{(p_{\max})_{90^\circ}} = \frac{A}{(2.894 \times 10^6)} = 3.455 \times 10^{-7} A,$$

where A is a constant independent of geometry. Similarly,

$$t_{45^\circ} = \frac{A}{(p_{\max})_{45^\circ}} = \frac{A}{(4.940 \times 10^6)} = 2.024 \times 10^{-7} A,$$

$$t_{22.65^\circ} = \frac{A}{(p_{\max})_{22.5^\circ}} = \frac{A}{(8.720 \times 10^6)} = 1.147 \times 10^{-7} A.$$

The costs per unit time for the three geometries are:

90° wrap angle:

$$\frac{8C + C}{(3.455 \times 10^{-7})\, A} = (26.05 \times 10^6)\, \frac{C}{A}.$$

45° wrap angle:

$$\frac{4C + C}{(2.024 \times 10^{-7})\, A} = (24.70 \times 10^6)\, \frac{C}{A}.$$

22.5° wrap angle:

$$\frac{2C + C}{(1.147 \times 10^{-7})\, A} = (26.16 \times 10^6)\, \frac{C}{A}.$$

The 45°-wrap-angle shoe gives the lowest cost, 5.6% lower than the 22.5°-wrap-angle shoe and 5.2% lower than the 90°-wrap-angle shoe.

18.8 Symmetrically Loaded Pivot-Shoe Brakes

Figure 18.15 shows a symmetrically loaded **pivot-shoe brake**. The major difference between the internal and external rim brakes considered previously and the symmetrically loaded pivot-shoe brake shown in Fig. 18.15 is the pressure distribution around the shoe. Recall from Eq. (18.37) for the internal rim brake that the maximum pressure was at $\theta = 90^\circ$ and the pressure distribution from the heel to the top of the brake was sinusoidal. For the symmetrically loaded pivot-shoe brake (Fig. 18.15), the maximum pressure is at $\theta = 0^\circ$, which suggests the pressure variation is

$$p = \frac{p_{\max} \cos\theta}{\cos\theta_a} = p_{\max} \cos\theta. \qquad (18.54)$$

For any angular position from the pivot, θ, a differential normal force dP acts with a magnitude of

$$dP = pbr\, d\theta = p_{\max} br\, \cos\theta\, d\theta. \qquad (18.55)$$

The design of a symmetrically loaded pivot-shoe brake is such that the distance d_7, which is measured from the center of the drum to the pivot, is chosen so that the resulting friction moment acting on the brake shoe is zero. From Fig. 18.15,

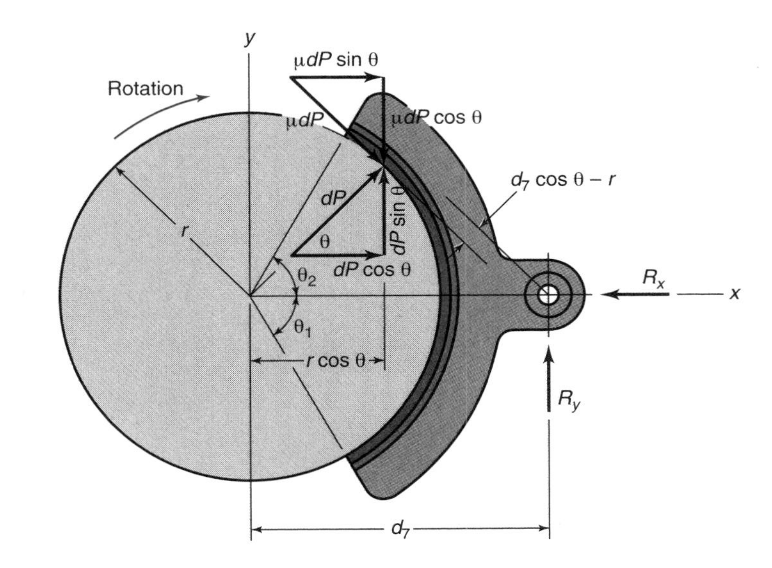

Figure 18.15: Symmetrically loaded pivot-shoe brake.

the friction moment, when set equal to zero, is

$$M_F = 2\int_0^{\theta_2} \mu\, dP\, (d_7 \cos\theta - r) = 0. \qquad (18.56)$$

Substituting Eq. (18.55) into Eq. (18.56) gives

$$2\mu p_{\max} br \int_0^{\theta_2} \left(d_7 \cos^2\theta - r\cos\theta\right) = 0.$$

This reduces to

$$d_7 = \frac{4r\sin\theta_2}{2\theta_2\left(\frac{\pi}{180^\circ}\right) + \sin 2\theta_2}. \qquad (18.57)$$

This value of d_7 produces a friction moment equal to zero ($M_F = 0$). The braking torque is

$$\begin{aligned} T &= 2\int_0^{\theta_2} r\mu\, dP \\ &= 2\mu r^2 b p_{\max} \int_0^{\theta_2} \cos\theta\, d\theta \\ &= 2\mu r^2 b p_{\max} \sin\theta_2. \end{aligned} \qquad (18.58)$$

Note from Fig. 18.15 that, for any x, the horizontal friction force components of the upper half of the shoe are equal and opposite in direction to the horizontal friction force components of the lower half of the shoe. For a fixed x, the horizontal *normal* components of both halves of the shoe are equal and in the same direction, so that the horizontal reaction force is

$$R_x = 2\int_0^{\theta_2} dP\cos\theta = \frac{p_{\max} br}{2}\left[2\theta_2\left(\frac{\pi}{180^\circ}\right) + \sin 2\theta_2\right]. \qquad (18.59)$$

Making use of Eq. (18.57) gives

$$R_x = \frac{2br^2 p_{\max}\sin\theta_2}{d_7}. \qquad (18.60)$$

For a fixed y, the vertical *friction force* components of the upper half of the shoe are equal and in the same direction as the vertical friction force components of the lower half of the shoe. For a fixed y, the vertical *normal* components of both halves of the shoe are equal and opposite in direction, so that the vertical reaction force is

$$R_y = 2\int_0^{\theta_2} \mu\, dP \cos\theta = \frac{\mu p_{\max} br}{2}\left[2\theta_2\left(\frac{\pi}{180^\circ}\right) + \sin 2\theta_2\right]. \quad (18.61)$$

Making use of Eq. (18.59) gives

$$R_y = \frac{2\mu b r^2 p_{\max}\sin\theta_2}{d_7} = \mu R_x.$$

Example 18.6: Pivot-Shoe Brake

Given: A symmetrically loaded, pivot-shoe brake has the distance d_7 shown in Fig 18.11 optimized for a 180° wrap angle. When the brake lining is worn out, it is replaced with a 90°-wrap-angle lining symmetrically positioned in the shoe. The actuating force is 11,000 N, the coefficient of friction is 0.31, the brake drum radius is 100 mm, and the brake width is 45 mm.

Find: Calculate the pressure distribution in the brake shoe and the braking torque.

Solution: The distance d_7 can be expressed from Eq. (18.57) for the 180° wrap angle as

$$\begin{aligned}(d_7)_{180^\circ} &= \frac{4r\sin\theta_2}{2\theta_2\left(\frac{\pi}{180^\circ}\right)+\sin 2\theta_2}\\ &= \frac{4(0.1)\sin 90^\circ}{2(90)\left(\frac{\pi}{180^\circ}\right)+\sin 180^\circ}\\ &= 0.1273 \text{ m}.\end{aligned}$$

For the 90°-wrap-angle, symmetrically loaded, pivot-shoe brake, the pressure distribution will be unsymmetrical. The maximum pressure, which will occur at θ_2, needs to be determined from shoe equilibrium, or

$$\int_{-\pi/4}^{\pi/4} pr\, d\theta(b)(d_7\sin\theta) - \int_{-\pi/4}^{\pi/4} \mu pr\, d\theta(b)(d_7\cos\theta - r) = 0. \quad (a)$$

If the wear rate is proportional to the pressure and the maximum pressure is at $\theta = \theta_o$, the pressure distribution is

$$p = p_{\max}\cos(\theta - \theta_o). \quad (b)$$

Substituting Eq. (b) into Eq. (a) gives

$$\begin{aligned}0 &= \int_{-\pi/4}^{\pi/4} d_7\cos(\theta-\theta_o)\sin\theta\, d\theta\\ &\quad -\int_{-\pi/4}^{\pi/4}\mu\cos(\theta-\theta_o)(d_7\cos\theta - r)\, d\theta.\end{aligned}$$

But

$$\cos(\theta-\theta_o) = \cos\theta\cos\theta_o + \sin\theta\sin\theta_o.$$

Therefore,

$$\begin{aligned}&\int_{-\pi/4}^{\pi/4} d_7\left(\cos\theta_o\cos\theta\sin\theta + \sin\theta_o\sin^2\theta\right)d\theta\\ &= \int_{-\pi/4}^{\pi/4}\mu(\cos\theta_o\cos\theta + \sin\theta_o\sin\theta)(d_7\cos\theta - r)d\theta.\end{aligned}$$

Integrating gives

$$\begin{aligned}&\frac{d_7\cos\theta_o}{2}\left[\sin^2\frac{\pi}{4} - \sin^2\left(-\frac{\pi}{4}\right)\right] + d_7\sin\theta_o\left(\frac{\pi}{4}-\frac{1}{2}\right)\\ &= \mu d_7\cos\theta_o\left(\frac{\pi}{4}+\frac{1}{2}\right)\\ &\quad +\mu d_7\sin\theta_o\left(\frac{\sin^2\frac{\pi}{4}}{2} - \frac{\sin^2\left(-\frac{\pi}{4}\right)}{2}\right)\\ &\quad -\mu r\cos\theta_o\sqrt{2} + \mu r\sin\theta_o\left(\frac{1}{\sqrt{2}}-\frac{1}{\sqrt{2}}\right).\end{aligned}$$

This reduces to

$$d_7\sin\theta_o\left(\frac{\pi}{4}-\frac{1}{2}\right) = \mu\cos\theta_o\left[d_7\left(\frac{\pi}{4}+\frac{1}{2}\right) - r\sqrt{2}\right],$$

or

$$\begin{aligned}\tan\theta_o &= \frac{\mu\left[d_7\left(\frac{\pi}{4}+\frac{1}{2}\right) - r\sqrt{2}\right]}{d_7\left(\frac{\pi}{4}-\frac{1}{2}\right)}\\ &= \frac{0.31\left[0.1273\left(\frac{\pi}{4}+\frac{1}{2}\right) - (0.100)\sqrt{2}\right]}{0.1273\left(\frac{\pi}{4}-\frac{1}{2}\right)}\\ &= 0.1895,\end{aligned}$$

or $\theta_o = 10.73^\circ$. The actuating force is

$$W = \int_{-\pi/4}^{\pi/4} pr\, d\theta b\cos\theta + \int_{-\pi/4}^{\pi/4}\mu pr\, d\theta b\sin\theta. \quad (c)$$

Substituting Eq. (b) into Eq. (c) gives

$$\begin{aligned}W &= rbp_{\max}\int_{-\pi/4}^{\pi/4}\cos(\theta-\theta_o)\cos\theta\, d\theta\\ &\quad + rbp_{\max}\int_{-\pi/4}^{\pi/4}\mu\cos(\theta-\theta_o)\sin\theta\, d\theta. \quad (d)\end{aligned}$$

But

$$\int_{-\pi/4}^{\pi/4}\cos(\theta-\theta_o)\cos\theta\, d\theta = \left(\frac{\pi}{4}+\frac{1}{2}\right)\cos\theta_o, \quad (e)$$

and

$$\int_{-\pi/4}^{\pi/4}\cos(\theta-\theta_o)\sin\theta\, d\theta = \left(\frac{\pi}{4}-\frac{1}{2}\right)\sin\theta_o. \quad (f)$$

Substituting Eqs. (e) and (f) into Eq. (d) while solving for the maximum pressure gives

$$\begin{aligned}p_{\max} &= \frac{W}{rb\left[\left(\frac{\pi}{4}+\frac{1}{2}\right)\cos\theta_o + \mu\left(\frac{\pi}{4}-\frac{1}{2}\right)\sin\theta_o\right]}\\ &= \frac{11,000}{(0.0045)\left[(1.285)\cos 10.7^\circ + (0.08835)\sin 10.7^\circ\right]}\\ &= 1.911\times 10^6 \text{ Pa} = 1.911 \text{ MPa}.\end{aligned}$$

From Eq. (*b*) the pressure distribution can be expressed as

$$p = \left(1.911 \times 10^6\right)\cos(\theta - 10.73^\circ).$$

The braking torque is

$$\begin{aligned} T &= \mu p_{\max} b r^2 \int_{-\pi/4}^{\pi/4} \cos(\theta - \theta_o)\, d\theta \\ &= \mu p_{\max} b r^2 \left[\sin\left(\frac{\pi}{4} - \theta_o\right) + \sin\left(\frac{\pi}{4} + \theta_o\right)\right] \\ &= (0.31)\left(1.911 \times 10^6\right)(0.045)(0.1)^2 \\ &\quad \times [\sin 34.27^\circ + \sin 55.73^\circ] \\ &= 370.4 \text{ N-m}. \end{aligned}$$

18.9 Band Brakes

Figure 18.16 shows the components of a **band brake**, which consists of a band wrapped partly around a drum. The brake is actuated by pulling the band tighter against the drum, as shown in Fig. 18.17a. The band is assumed to be in contact with the drum over the entire wrap angle, ϕ in Fig. 18.17a. The pin reaction force is given as F_1 and the actuating force as F_2. In Fig. 18.17a, the heel of the brake is near F_1 and the toe is near F_2. Since some friction will exist between the band and the drum, the actuating force will be less than the pin reaction force, or $F_2 < F_1$.

Figure 18.17b shows the forces acting on an element of the band. The forces are the normal force, P, and the friction force, F. Summing the forces in the radial direction while using Fig. 18.17b gives

$$(F + dF)\sin\left(\frac{d\theta}{2}\right) + F\sin\left(\frac{d\theta}{2}\right) - dP = 0;$$

$$dP = 2F\sin\left(\frac{d\theta}{2}\right) + dF\sin\left(\frac{d\theta}{2}\right).$$

Since $dF \ll F$,

$$dP = 2F\sin\left(\frac{d\theta}{2}\right).$$

Since $d\theta/2$ is small, then $\sin d\theta/2 \approx d\theta/2$. Therefore,

$$dP = F d\theta. \tag{18.62}$$

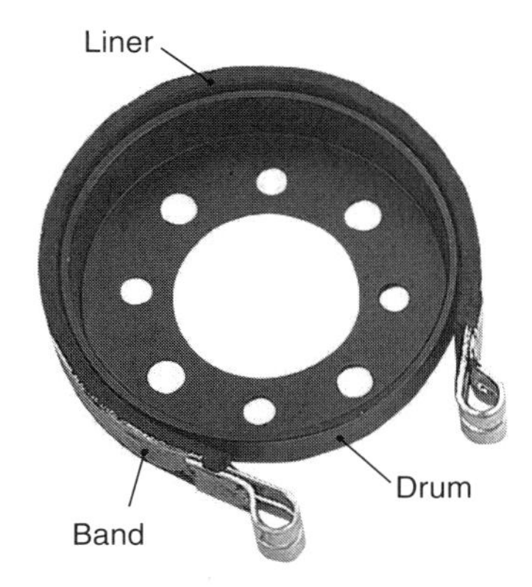

Figure 18.16: A typical band brake. *Source:* Courtesy of Northern Tool & Equipment Company, Inc.

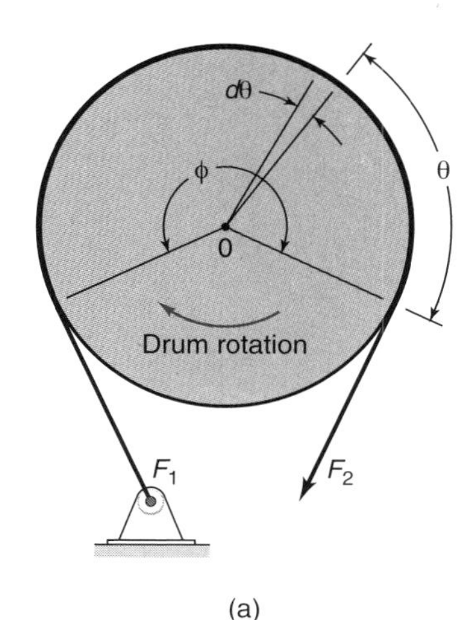

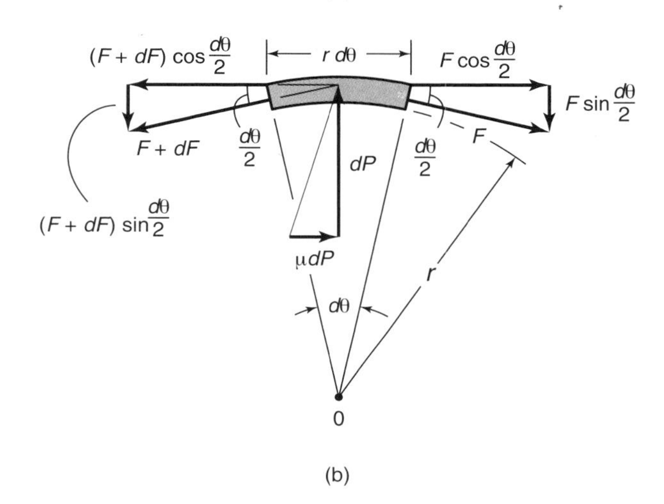

Figure 18.17: Band brake. (a) Forces acting on band; (b) forces acting on element.

Summing the forces in the *horizontal* (tangential) direction while using Fig. 18.17b gives

$$(F + dF)\cos\left(\frac{d\theta}{2}\right) - F\cos\left(\frac{d\theta}{2}\right) - \mu dP = 0;$$

$$dF\cos\left(\frac{d\theta}{2}\right) - \mu dP = 0.$$

Since $d\theta/2$ is small, then $\cos(d\theta/2) \approx 1$. Therefore,

$$dF - \mu dP = 0. \tag{18.63}$$

Substituting Eq. (18.62) into Eq. (18.63) gives

$$dF - \mu F d\theta = 0 \quad \text{or} \quad \int_{F_2}^{F_1} \frac{dF}{F} = \mu \int_0^{\phi} d\theta.$$

Integrating gives

$$\ln\left(\frac{F_1}{F_2}\right) = \frac{\mu\phi\pi}{180^\circ},$$

or

$$\frac{F_1}{F_2} = e^{\mu\phi\pi/180^\circ}, \tag{18.64}$$

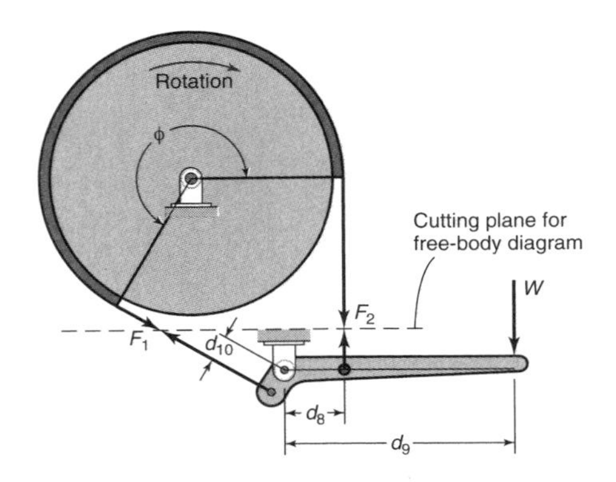

Figure 18.18: Band brake used in Example 18.7.

where ϕ is the wrap angle. The torque applied to the drum is

$$T = r(F_1 - F_2). \tag{18.65}$$

The differential normal force dP acting on the element in Fig. 18.17b, with width b (coming out of the paper) and length $r\,d\theta$, is

$$dP = pbr\,d\theta, \tag{18.66}$$

where p is the contact pressure. Substituting Eq. (18.66) into Eq. (18.62) gives

$$p = \frac{F}{br}. \tag{18.67}$$

The pressure is proportional to the tension in the band. The maximum pressure occurs at the heel, or near the pin reaction force, and has the value

$$p_{\max} = \frac{F_1}{br}. \tag{18.68}$$

Example 18.7: Band Brake

Given: The band brake shown in Fig. 18.18 has r = 4 in., b = 1 in., d_9 = 9 in., d_8 = 2 in., d_{10} = 0.5 in., $\phi = 270°$, $\mu = 0.2$, and $p_{\max}$ = 75 psi.

Find: Determine the braking torque, actuating force, and value of d_{10} when the brake force locks.

Solution: From Eq. (18.68), the pin reaction force is

$$F_1 = p_{\max} br = (75)(1)(4) = 300 \text{ lb}.$$

From Eq. (18.64), the actuating force is

$$F_2 = F_1 e^{-\mu\phi\pi/180°} = 300e^{-0.2(270)\pi/180} = 116.9 \text{ lb}.$$

From Eq. (18.65), the braking torque is

$$T = r(F_1 - F_2) = 4(300 - 116.9) = 732.4 \text{ lb-in}.$$

Summing the moments about the hinge pin and setting the sum equal to zero give

$$-d_9 W + d_8 F_2 - d_{10} F_1 = 0.$$

Solving for the actuating force W,

$$W = \frac{d_8 F_2 - d_{10} F_1}{d_9} = \frac{2(116.9) - (0.5)(300)}{9} = 9.311 \text{ lb}.$$

If $W = 0$, the brake will self-lock, therefore:

$$d_8 F_2 - d_{10} F_1 = 0,$$

$$d_{10} = \frac{d_8 F_2}{F_1} = \frac{2(116.9)}{300} = 0.7793 \text{ in}.$$

The brake will self-lock if $d_{10} \geq 0.7793$ in.

18.10 Slip Clutches

A clutch will often be used as a torque-limiting device, usually to prevent machinery damage in a malfunction or undesired event, although they can also be used to control peak torques during startup. A **slip clutch**, in a simple manifestation, consists of two surfaces held together by a constant force so that they slip when a preset level of torque is applied to them. Slip clutches come in a wide variety of sizes but are very compact. They are designed to be actuated only rarely and thus the friction elements do not need to be sized for wear. Also, slip clutches are almost always contacting disks, mainly because it is imperative to prevent the possibility of a self-energizing shoe (which would compromise the torque-limiting control).

Slip clutches are mainly used to protect machinery elements and are not relied upon to prevent personal injury. After slip clutches spend long time periods in contact, the friction surfaces can stick or weld together, requiring a larger torque to initiate slippage. This torque is usually not large enough to break a gear, for example, but can be enough of an increase to result in a serious injury. Regardless, a slip clutch is an effective torque limiter and can be used instead of shear pins or keyways with the advantage that no maintenance is required after the excessive torque condition has been corrected.

Case Study 18.1: Roller Coaster Braking System

This case study discusses some of the design decisions that are made for the braking system of a roller coaster such as shown in Fig. 18.19. Figure 18.20 is a schematic of a typical roller coaster brake. The brake systems for automobiles or other vehicles, that is, drum and disk brakes associated with wheels, will not work for a roller coaster. The friction between the steel wheel and the steel rail is extremely low, so that the cars can slide on the rails without losing much energy. This is important for roller coaster operation, as the cars are pulled up a main hill (usually by a chain drive — see Section 19.6), and then coast for the remainder of the ride.

Roller coasters have braking stations, where the cars must be brought to a stop so that passengers can get on or off. A typical brake system uses pads that bear against a flange, or fin, on the roller coaster cars. The roller coaster operator controls a pneumatic actuator that applies force through the linkage to aggressively decelerate the cars. The maximum braking force should not expose the passengers to greater than 1/2 g of deceleration, and preferably not more than 2.5 m/s^2 or so.

Typically, several brakes are used at the point of access to a roller coaster. This case study will show a four-brake system, which is not unusual. Note that the thermal effects

Figure 18.19: A typical roller coaster. (Shutterstock)

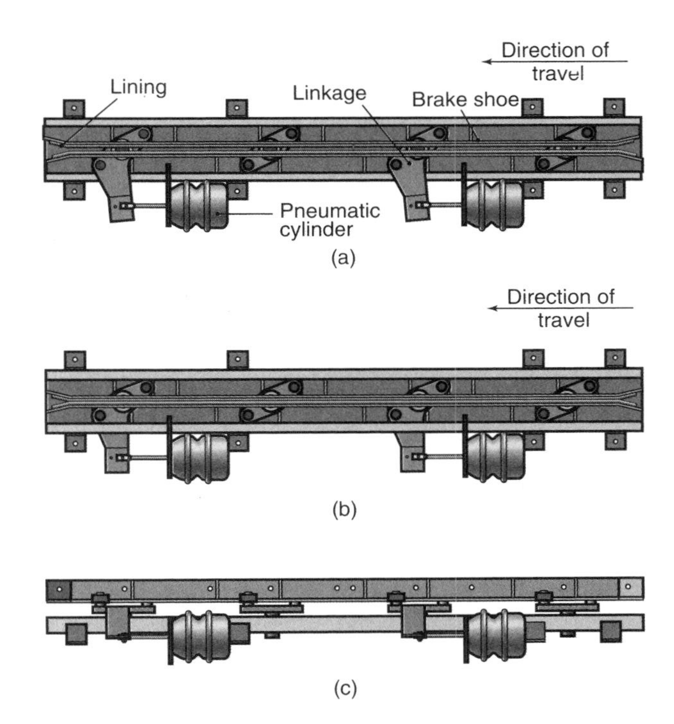

Figure 18.20: Schematic illustration of a roller coaster brake system. (a) Components of the roller coaster and shown when the brake is not engaged, as seen by the gap between the liner pads; (b) top view of an engaged brake; (c) side view of engaged brake.

that are so prominent in other brake designs are not considered here because roller coaster brakes are actuated only intermittently, so that they avoid the thermal checking and temperature problems typical of automotive disk or drum brakes.

Brake Pad–Fin Interaction

The first part of the design problem is to correlate brake system variables (actuating force, pad area, etc.) with performance (braking force developed). The coefficient of friction can be reasonably taken as at least 0.25 for sintered brass on steel (it is usually higher for these materials). Friction-reducing contaminants on the roller coaster are not a serious concern. This is mainly because the braking system is pneumatic, not hydraulic, so no leaking hydraulic fluids can contaminate the brake pads.

Figure 18.21 illustrates a three-car chain of roller coaster cars, but the manufacturer allows as many as seven cars provided that the front and rear cars are configured as shown. Many details, especially case-specific decorations, have been omitted for clarity. Each car can weigh 1900 lb and carry as many as four adults. (It is more common that two adults and two children will ride in a roller coaster car.) The 95th percentile adult male weighs 216 lb, and weight of the car and contents is calculated accordingly. The maximum incoming velocity is approximately 30 mph (44 ft/s) on the basis of the energy attainable from the first roller coaster hill.

A braking force is then determined to achieve the deceleration of 0.5 g, which is then used to obtain the actuating force. However, the brake pad arrangement shown in Fig. 18.20 develops frictional force on both sides of the fin. The normal force obtained from such a calculation represents the worst-case design load in the cars. No increase is necessary from a safety standpoint for the following reasons:

1. For most cases this friction force is more than adequate.

2. Typical installations will have a series of at least three brakes spaced in alternate bays. If deceleration can be maintained throughout the entire contact length between the cars and the brakes, extra braking capacity is inherently placed in the system.

3. Even in the event of insufficient braking (perhaps due to excessive wear of the braking elements), emergency brakes in the operator's station can stop the cars. The alternative is to generate larger braking forces with the subject braking system, leading to excessive deceleration and possible injuries to riders from jarring and jerking. The target deceleration has been chosen to provide good braking potential while minimizing the risk of such injuries. A stronger brake would actually lead to a less safe system.

Brake Actuation System

Figure 18.22 shows a detail of the brake-actuating cylinder and associated linkage. Recall from Fig. 18.20 that two cylinders are used on each brake and two unpowered stabilizing elements are used to distribute the contact forces over the whole shoe. The brake is based on a pneumatic cylinder/spring system so that the normal cylinder position is with the piston totally extended, corresponding to the brake generating maximum brake force. The reason for such a system is that many difficulties can arise with pneumatic systems: air hoses can leak, pneumatic couplings can decouple, power can fail, etc. The system is fail safe if the brakes are engaged in such a reasonably foreseeable event.

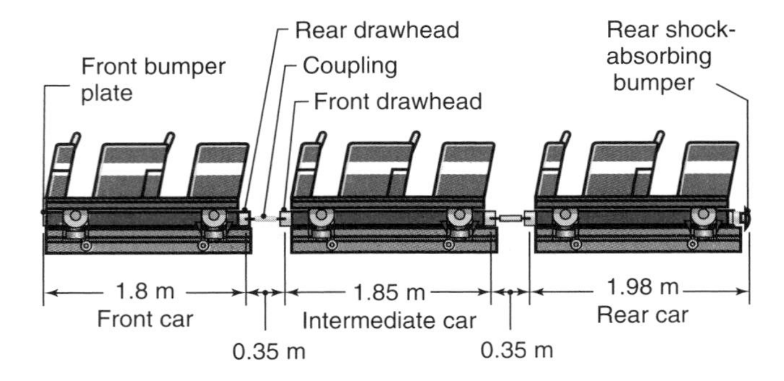

Figure 18.21: Schematic illustration of typical roller coaster cars.

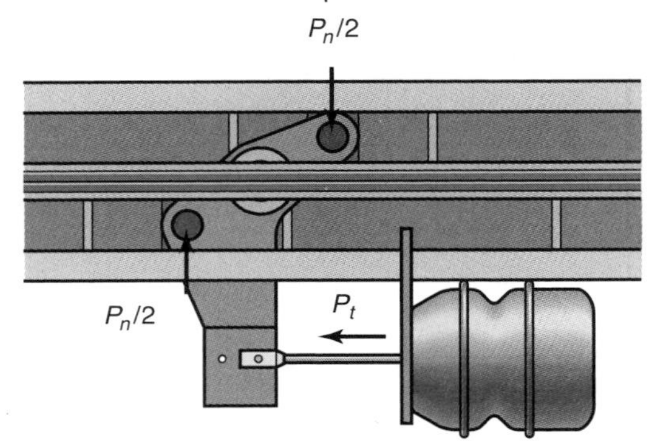

Figure 18.22: Detail of brake-actuating cylinder with forces shown.

Pneumatic brake cylinders are available in a range of capacities from a brake manufacturer, each of which has a characteristic spring force curve. The nonlinear nature of this curve is typical of pneumatic springs (as discussed in Section 17.7), but this cylinder uses a combination of air pressure and helical springs for actuation as shown in the breakaway view in Fig. 18.23.

A subtlety in sizing such brake cylinders must be explained. During assembly the piston is totally extended and may be difficult to affix to the hinge mechanism. Therefore, a piston extension is threaded for a distance on its length. The cylinder can be assembled and then the collar on the piston extension adjusted for brake actuation.

Summary

Short-shoe brakes have a long history of safe operation on roller coasters. As discussed in this case study, the brake system has a number of redundancies to make sure that failure of any one braking station does not result in catastrophic injury of the passengers.

Also, it should be noted that roller coasters have a stringent maintenance schedule, and the ride is inspected before the ride is placed in service every day. If a lining is worn, there will be an audible indication from the exposed rivets on the brake shoes "grinding" against the fin, and this is a clear indication that the pads need replacement. Any reasonably attentive maintenance staff can diagnose and repair such conditions as needed.

Source: Courtesy of Brian King, Recreation Engineering LLC.

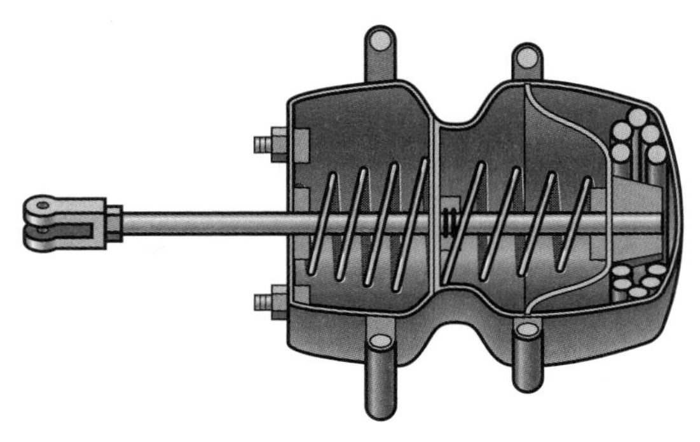

Figure 18.23: Cross-section of the actuating pneumatic cylinder, highlighting the helical springs incorporated into the design.

18.11 Summary

This chapter focused on two machine elements, clutches and brakes, that are associated with motion and have the common function of dissipating and/or transferring rotating energy. In analyzing the performance of clutches and brakes, the actuating force, the torque transmitted, and the reaction force at the hinge pin were the major focuses of this chapter. The torque transmitted is related to the actuating force, the coefficient of friction, and the geometry of the clutch or brake. This is a problem in statics, where different geometries were studied separately.

For long-shoe clutches and brakes, two theories were studied: the uniform pressure model and the uniform wear model. It was found that for the same dimensionless torque the uniform wear model requires a larger radius ratio than does the uniform pressure model for the same maximum pressure. This larger radius ratio implies that a larger area is needed for the uniform wear model. Thus, the uniform wear model was viewed as a safer approach.

Key Words

band brake brake that uses contact pressure of flexible band against outer surface of drum

brake device used to bring moving system to rest through dissipation of energy to heat by friction

clutches power transfer devices that allow coupling and decoupling of shafts

cone disk brake or clutch that uses shoes pressed against convergent surface of cone

deenergizing brake or clutch shoe where frictional moment hinders engagement

rim type brake or clutch that uses internal shoes which expand onto inner surface of drum

self-energizing brake or clutch shoe where frictional moment assists engagement

slip clutch clutch where maximum transferred torque is limited

thrust disk brake or clutch that uses flat shoes pushed against rotating disk

Summary of Equations

Heat Transfer

First Law of Thermodynamics: $Q_f = Q_c + Q_h + Q_s$

Lining temperature rise: $\Delta t_m = \dfrac{Q_f}{C_p m_a}$

Thrust (Disk) Brakes:

Uniform pressure model:

Actuating force: $P_p = \pi p_o \left(r_o^2 - r_i^2\right)$

Torque: $T_p = \frac{2\pi\mu p_o}{3}\left(r_o^3 - r_i^3\right) = \frac{2\mu P_p\left(r_o^3 - r_i^3\right)}{3\left(r_o^2 - r_i^2\right)}$

Uniform wear model:

Actuating force: $P_w = 2\pi p_{\max} r_i \left(r_o - r_i\right)$

Torque: $T_w = \pi\mu r_i p_{\max}\left(r_o^2 - r_i^2\right) = \dfrac{\mu P_w\left(r_o + r_i\right)}{2}$

Cone Clutches and Brakes:

Uniform pressure model:

Actuating force: $W = \frac{\pi p_o}{4}\left(D^2 - d^2\right)$

Torque: $T = \dfrac{\mu W\left(D^3 - d^3\right)}{3\sin\alpha\left(D^2 - d^2\right)}$

Uniform wear model:

Actuating force: $W = \pi p r \left(D - d\right)$

Torque: $T = \dfrac{\mu W}{4\sin\alpha}\left(D + d\right)$

Block (Short-Shoe) Brakes:

Normal force: $P = \dfrac{W d_4}{d_3 + \mu d_2}$

Torque: $T = \dfrac{\mu d_4 r W}{d_3 - \mu d_1}$ (energizing)

$T = \dfrac{\mu d_4 r W}{d_3 + \mu d_2}$ (deenergizing)

Long-Shoe, Internal, Expanding Rim (Drum) Brakes:

Pressure distribution: $p = p_{\max}\left(\dfrac{\sin\theta}{\sin\theta_a}\right)$

Normal force moment:

$$M_P = \frac{b r d_7 p_{\max}}{4\sin\theta_1}\left[2\left(\theta_2 - \theta_1\right)\frac{\pi}{180^\circ} - \sin 2\theta_2 + \sin 2\theta_1\right]$$

Friction force moment:

$$M_F = \frac{\mu p_{\max} b r}{\sin\theta_a} \times \left[-r\left(\cos\theta_2 - \cos\theta_1\right) - \frac{d_7}{2}\left(\sin^2\theta_2 - \sin^2\theta_1\right)\right]$$

Self-energizing shoe: $-W d_6 - M_F + M_P = 0$

Deenergizing shoe: $-W d_6 + M_F + M_P = 0$

Pivot-Shoe Brakes:

Pressure distribution: $p = p_{\max}\cos\theta$

Torque: $T = 2\mu r^2 b p_{\max}\sin\theta_2$

Band Brakes:

Forces: $\dfrac{F_1}{F_2} = e^{\mu\phi\pi/180^\circ}$

Maximum pressure: $p_{\max} = \dfrac{F_1}{br}$

Torque: $T = r(F_1 - F_2)$

Recommended Readings

Baker, A.K. (1986) *Vehicle Braking*, Pentech Press.

Breuer, B. (2008) *Brake Technology Handbook*, Society of Automotive Engineers.

Budynas, R.G., and Nisbett, J.K. (2011), *Shigley's Mechanical Engineering Design*, 9th ed., McGraw-Hill.

Crouse, W.H., and Anglin, D.L. (1983) *Automotive Brakes, Suspension and Steering*, 6th ed., McGraw-Hill.

Juvinall, R.C., and Marshek, K.M. (2012) *Fundamentals of Machine Component Design*, 5th ed., Wiley.

Krutz, G.W., Schuelle, J.K, and Claar, P.W. (1994) *Machine Design for Mobile and Industrial Applications*, Society of Automotive Engineers.

Monroe, T. (1977) *Clutch and Flywheel Handbook*, H.P. Books.

Mott, R. L. (2014) *Machine Elements in Mechanical Design*, 5th ed., Pearson.

Norton, R.L. (2011) *Machine Design*, 4th ed., Prentice Hall.

Orthwein, W.C. (1986) *Clutches and Brakes: Design and Selection*. 2nd ed., CRC Press.

Reference

Juvinall, R.C., and Marshek, K.M. (2006) *Fundamentals of Machine Component Design*, 4th ed., Wiley.

Questions

18.1 What is a clutch? How is it different from a brake?

18.2 What is the uniform pressure model? The uniform wear model?

18.3 When is a rectangular pad disk brake used?

18.4 What is a cone clutch?

18.5 What is a self-energizing shoe?

18.6 Can a short-shoe brake be self-energizing?

18.7 What is a long-shoe brake? What is the main difference between a long-shoe and a short-shoe brake?

18.8 What is the difference between an external long-shoe brake and a pivot-shoe brake?

18.9 Describe the operating mechanisms of a band brake.

18.10 What is a heat check? How are they avoided?

Qualitative Problems

18.11 List the material properties that are (*a*) necessary, and (*b*) useful for a brake or clutch lining.

18.12 List the advantages and disadvantages of cone clutches compared to thrust disc clutches.

18.13 Explain the conditions when the uniform pressure model is more appropriate than the uniform wear model. When is the uniform wear model more appropriate?

18.14 If a composite material is used as a brake or clutch lining, what preferred orientation of the reinforcement would you recommend, if any? Explain your answer.

18.15 Explain the difference between self-energizing and self-locking.

18.16 What are the similarities and differences between band and thrust disc brakes?

18.17 What are the main indications of an out-of-round brake drum?

18.18 What are the similarities and differences between heat checks and hard spots?

18.19 Without the use of equations, explain why disk brakes have more heat loss through convection than drum brakes, at least in the front axle of vehicles.

18.20 What are the reasons for using a slip clutch?

Quantitative Problems

18.21 The disk brake shown in Sketch *a* has brake pads in the form of circular sections with inner radius r, outer radius $2r$, and section angle $\pi/4$. Calculate the brake torque when the pads are applied with normal force P. The brake is worn in so that pu is constant, where p is the contact pressure and u is the sliding velocity. The coefficient of friction is μ. *Ans.* $T = 3\mu rP$.

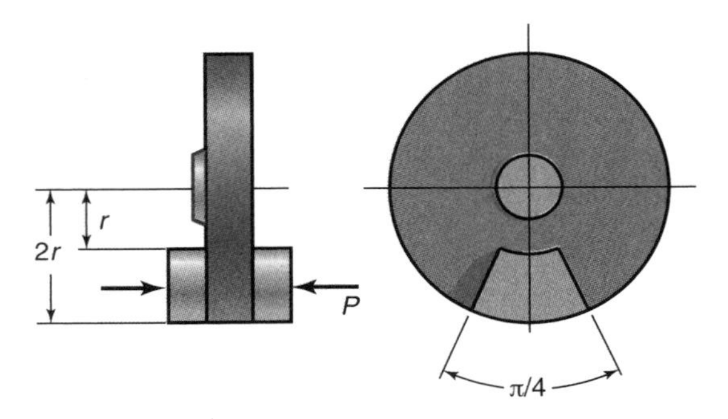

Sketch *a*, for Problem 18.21

18.22 A plate clutch has a single pair of mating friction surfaces 350 mm outer diameter by 250 mm inner diameter. The mean value of the coefficient of friction is 0.25, and the actuating force is 5kN. Find the maximum pressure and torque capacity when the clutch is new, as well as after a sufficiently long run-in period. *Ans.* $T_p = 189$ N-m, $p_{\max,p} = 0.127$ MPa, $T_w = 187.5$ N-m.

18.23 An automotive clutch with a single friction surface is to be designed with a maximum torque of 140 N-m. The materials are chosen such that $\mu = 0.35$ and $p_{\max} = 0.35$ MPa. Use safety factor $n_s = 1.3$ with respect to slippage at full engine torque and as small an outside diameter as possible. Determine appropriate values of r_o, r_i, and P by using both the uniform pressure and uniform wear models. *Ans.* $P_p = 8743$ N, $P_w = 6160$ N.

18.24 The brakes used to stop and turn a tank are built like a multiple-disk clutch with three loose disks connected through splines to the drive shaft and four flat rings connected to the frame of the tank. The brake has an outer contact diameter of 600 mm, an inner contact diameter of 300 mm, and six surface contacts. The wear of the disks is proportional to the contact pressure multiplied by the sliding distance. The coefficient of friction of the brake is 0.15, and the friction between the caterpillar and the ground is 0.25, which gives a braking torque of 12,800 N-m needed to block one caterpillar track so that it slides along the ground. Calculate the force needed to press the brake disks together to block one caterpillar track. Also, calculate the force when the brake is new. *Ans.* $P_w = 63.2$ kN, $P_p = 60.9$ kN.

18.25 A disk brake used in a printing machine is designed as shown in Sketch *b*. The brake pad is mounted on an arm that can swivel around point 0. Calculate the braking torque when the force $P = 4000$ N. The friction pad is a circular sector with the inner radius equal to half the outer radius. Also, $a = 150$ mm, $b = 50$ mm, $D = 300$ mm, and $\mu = 0.25$. The wear of the brake lining is proportional to the pressure and the sliding distance. *Ans.* $T = 157$ Nm.

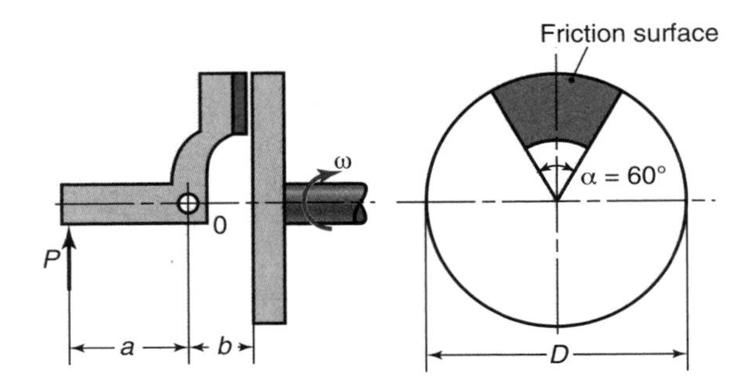

Sketch *b*, for Problem 18.25

18.26 Three thrust disk clutches, each with a pair of frictional surfaces, are mounted on a shaft. The hardened-steel clutches are identical, with an inside diameter of 100 mm and an outside diameter of 250 mm. What is the torque capacity of these clutches based on (*a*) uniform wear and (*b*) uniform pressure? *Ans.* $T_w = 2970$ N-m, $T_p = 5513$ N-m.

18.27 A pair of disk clutches has an inside diameter of 250 mm and an outside diameter of 420 mm. A normal force of 18.5 kN is applied and the coefficient of friction of the contacting surfaces is 0.25. Using the uniform wear and uniform pressure assumptions determine the maximum pressure acting on the clutches. Which of these assumptions would produce results closer to reality? *Ans.* $p_{\max,p} = 206.8$ kPa, $p_{\max,w} = 277$ kPa.

18.28 A disk clutch is made of cast iron and has a maximum torque of 210 N-m. Because of space limitations the outside diameter must be minimized. Using the uniform wear assumption and a safety factor of 1.3, determine

(a) The inner and outer radii of the clutch. *Ans.* $r_i = 0.068$ m, $r_o = 0.118$ m.

(b) The maximum actuating force needed. *Ans.* $P_w = 11.3$ kN.

18.29 A disk brake for a flywheel is designed as shown in Sketch *c*. The hydraulic pistons actuating the brake need to be placed at a radius r_p so that the brake pads wear evenly over the entire contact surface. Calculate the actuating force P and the radius so that the flywheel can be stopped within 4 s when it rotates at 1000 rpm and has a kinetic energy of 4×10^5 N-m. The input parameters are $\mu = 0.3$, $\alpha = 30°$, $r_o = 120$ mm, $r_i = 60$ mm. *Ans.* $P = 35.4$ kN, $r_p = 0.089$ m.

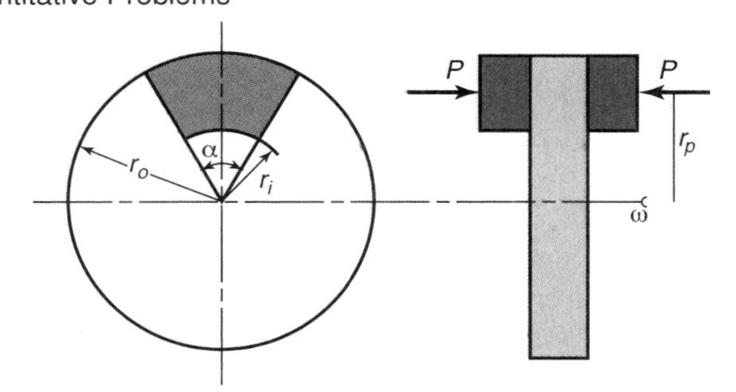

Sketch *c*, for Problem 18.29

18.30 A disk clutch produces a torque of 125 N-m and a maximum pressure of 315 kPa. The coefficient of friction of the contacting surfaces is 0.28. Assume a safety factor of 1.8 for maximum pressure and design the smallest disk clutch for the above constraints. What should the normal force be? *Ans.* $P_w = 4.39$ kN.

18.31 A cone clutch is to transmit 1200 lb-in. of torque. The half-cone angle $\alpha = 10°$, the maximum diameter of the friction surface is 12 in., and face width $b = 2$ in. For coefficient of friction $\mu = 0.20$ find the normal force P and the maximum contact pressure p by using both the uniform pressure and uniform wear models. *Ans.* $P_p = P_w = 174$ lb, $p_{\max,p} = 13.15$ psi, $p_{\max,w} = 13.5$ psi.

18.32 The synchronization clutch for the second gear of a car has a major cone diameter of 50 mm and a minor diameter of 40 mm. When the stick shift is moved to second gear, the synchronized clutch is engaged with an axial force of 100 N, and the moment of inertia of 0.005 kg-m^2 is accelerated 200 rad/s^2 in 1 s to make it possible to engage the gear. The coefficient of friction of the cone clutch is 0.10. Determine the smallest cone clutch width that still gives large enough torque. Assume the clutches are worn in. *Ans.* $b = 21.6$ mm.

18.33 A safety brake for an elevator is a self-locking cone clutch. The minor diameter is 120 mm, the width is 60 mm, and the major diameter is 130 mm. The force applying the brake comes from a prestressed spring. Calculate the spring force needed if the 2-ton (2000 kg) elevator must stop from a speed of 3 m/s in a maximum distance of 3 m while the cone clutch rotates five revolutions per meter of elevator motion. The coefficient of friction in the cone clutch is 0.20. *Ans.* $P = 4781$ N.

18.34 A cone clutch is used in a car automatic gearbox to fix the planet wheel carriers to the gearbox housing when the gear is reversing. The car weighs 1300 kg with 53% loading on the front wheels. The gear ratio from the driven front wheels to the reversing clutch is 16.3:1 (i.e., the torque on the clutch is 16.3 times lower than the torque on the wheels if all friction losses are neglected). The car wheel diameter is 550 mm, the cone clutch major diameter is 85 mm and the minor diameter is 80 mm, and the coefficients of friction are 0.3 in the clutch and 1.0 between the wheel and the ground. Dimension the width of the cone clutch so that it is not self-locking. Calculate the axial force needed when the clutch is worn in. *Ans.* $b < 0.0083$ m, $P = 2647$ N.

18.35 A cone clutch has a major diameter of 328 mm and a minor diameter of 310 mm, is 50 mm wide, and transfers 250 N-m of torque. The coefficient of friction is 0.33. Using the assumptions of uniform pressure and uniform wear, determine the actuating force and the contact pressure. *Ans.* $P_p = 854$ N, $p_{\max,p} = 101$ kPa.

18.36 The coefficient of friction of a cone clutch is 0.25. It can support a maximum pressure of 410 kPa while transferring a maximum torque of 280 N-m. The width of the clutch is 65 mm. Minimize the major diameter of the clutch. Determine the clutch dimensions and the actuating force.

18.37 A block brake is used to stop and hold a rope used to transport skiers from a valley to the top of a mountain. The distance between cars used to transport the skiers is 100 m, the length of the rope from the valley to the top of the mountain is 4 km, and the altitude difference is 1.4 km. The rope is driven by a V-groove wheel with a diameter of 2 m. The rope is stopped and held with a block brake mounted on the shaft of the V-groove wheel, shown in Sketch *d*. Neglect all friction in the different parts of the ropeway except the friction in the driving sheave, and assume that the slope of the mountain is constant. Dimension the brake for 20 passengers with each passenger's equipment weighing 100 kg, and assume that all ropeway cars descending from the top are empty of people. The direction of rotation for the drive motor is shown in the figure. Calculate the braking force W needed to hold the ropeway still if all passengers are on their way up. Do the same calculation if all passengers continue on down to the valley with the ropeway. The coefficient of friction in the brake is 0.25. *Ans.* $W_{\text{up}} = 2.197$ MN, $W_{\text{down}} = 671$ kN.

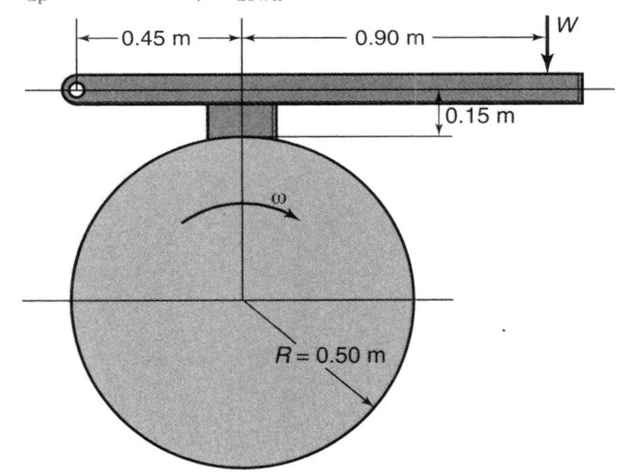

Sketch *d*, for Problem 18.37

18.38 The motion of an elevator is controlled by an electric motor and a block brake. On one side the rotating shaft of the electric motor is connected to the gearbox driving the elevator, and on the other side it is connected to the block brake. The motor has two magnetic poles and can be run on either 60- or 50-Hz electricity (3600 or 3000 rpm). When the elevator motor is driven by 50-Hz electricity, the braking distance needed to stop is 52 cm when going down and 31 cm when going up with maximum load in the elevator. To use it with 60-Hz electricity and still be able to stop exactly at the different floor levels without changing the electric switch positions, the brake force at the motor should be changed. How should it be changed for going up and for going down? The brake geometry is like that shown in Fig. 18.5 with $d_1 = 0.030$ m, $d_3 = 0.100$ m, $d_4 = 0.400$ m, $r = 0.120$ m, and $\mu = 0.20$. Only the inertia of the elevator needs to be considered, not the rotating parts associated with the drive or other components. Make the brake self-energizing when the elevator is going down.

18.39 An anchor winding is driven by an oil hydraulic motor with a short-shoe brake to stop the anchor machinery from rotating and letting out too much anchor chain when the wind moves the ship. The maximum force transmitted from the anchor through the chain is 1.1 MN at a radius of 2 m. Figure 18.9 describes the type of block brake used, which is self-energizing. Calculate the brake force W needed when the the brake dimensions are $d_1 = 0.9$ m, $d_3 = 1.0$ m, $d_4 = 6$ m, $r = 3$ m, and $\mu = 0.25$. Also, calculate the contact force between the brake shoe and the drum. *Ans.* $W = 379$ kN, $P = 2.93$ MN.

18.40 The hand brake shown in Sketch e has an average pressure of 600 kPa across the shoe and is 50 mm wide. The wheel runs at 150 rpm and the coefficient of friction is 0.25. Dimensions are in millimeters. Determine the following:

(a) If $x = 150$ mm, what should the actuating force be? *Ans.* $P_n = 2.97$ kN.

(b) What value of x causes self-locking? *Ans.* $x = 2.0$ m.

(c) What torque is transferred? *Ans.* $T = 303$ N-m.

(d) If the direction of rotation is reversed, how would the answers to parts (a) to (c) change?

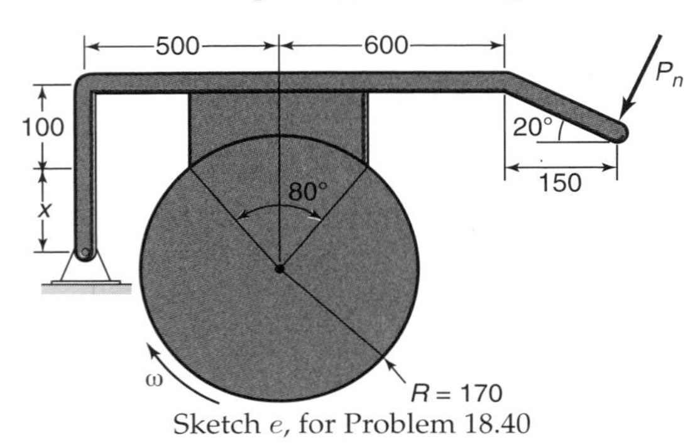

Sketch e, for Problem 18.40

18.41 The short-shoe brake shown in Sketch f has an average pressure of 1 MPa and a coefficient of friction of 0.32. The shoe is 250 mm long and 45 mm wide. The drum rotates at 310 rpm and has a diameter of 550 mm. Dimensions are in millimeters.

(a) Obtain the value of x for the self-locking condition. *Ans.* $x = 1.81$ m.

(b) Calculate the actuating force if $x = 275$ mm. *Ans.* $P = 3982$ N.

(c) Calculate the braking torque. *Ans.* $T = 3.44$ kN-m.

(d) Calculate the reaction at point A.

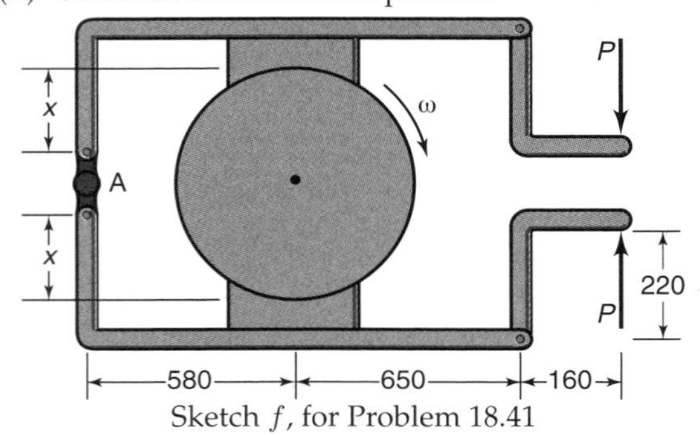

Sketch f, for Problem 18.41

18.42 The brake on the rear wheel of a car is the long-shoe internal type. The brake dimensions according to Fig. 18.7 are $\theta_1 = 10°$, $\theta_2 = 120°$, $r = 95$ mm, $d_7 = 73$ mm, $d_6 = 120$ mm, and $d_5 = 30$ mm. The brake shoe lining is 38 mm wide, and the maximum allowable contact pressure is 5 MPa. Calculate the braking torque and the fraction of the torque produced from each brake shoe when the brake force is 5000 N. Also, calculate the safety factor against contact pressure that is too high. The coefficient of friction is 0.25. *Ans.* $T_{\text{tot}} = 534$ N-m, $T_{se} = 359$ N-m.

18.43 Sketch g shows a long-shoe, internal expanding shoe brake. The inside rim diameter is 280 mm and the dimension d_7 is 90 mm. The shoes have a face width of 30 mm.

(a) Find the braking torque and the maximum pressure for each shoe if the actuating force is 1000 N, the drum rotation is counterclockwise, and the coefficient of friction is 0.30. *Ans.* $T = 218.7$ N-m.

(b) Find the braking torque if the drum rotation is clockwise.

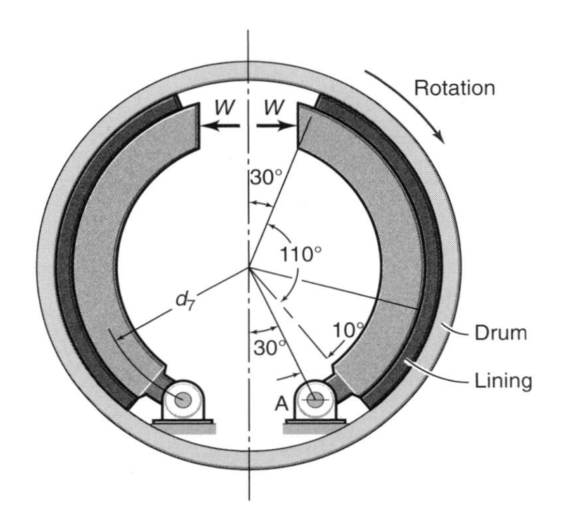

Sketch g, for Problem 18.43.

18.44 Sketch h shows a 450-mm inner diameter brake drum with four internally expanding shoes. Each of the hinge pins A and B supports a pair of shoes. The actuating mechanism is a linkage which produces the same actuating force W on each shoe as shown. The lining material is to be sintered metal. The face width of each shoe is 70 mm. Using the mid-range values of friction coefficient and contact pressure,

(a) Identify the self-energizing and de-energizing shoes and label them.

(b) Determine the maximum pressure on the deenergizing shoe. *Ans.* $p_{\max} = 1.071$ MPa.

(c) Determine the maximum force W and the braking torque T at this force. *Ans.* $T = 3430$ Nm.

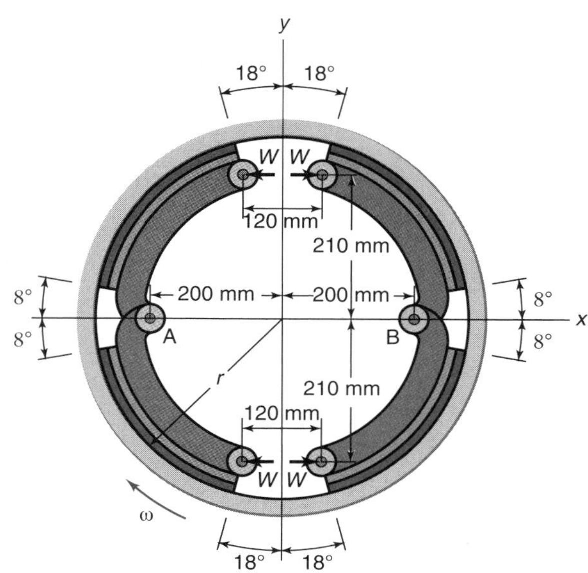

Sketch h, for Problem 18.44

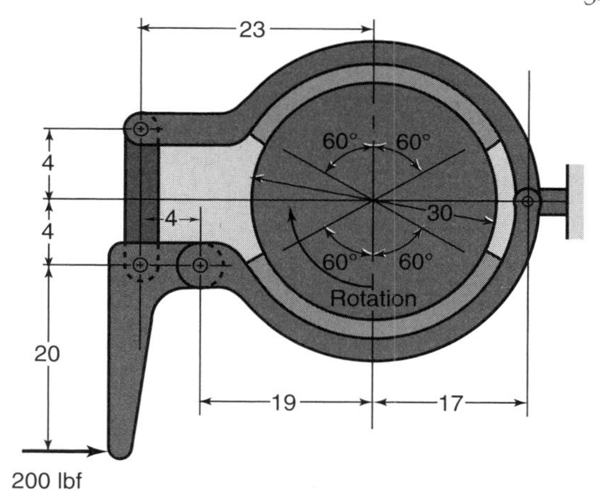

Sketch i, for Problem 18.47

18.45 The maximum volume of the long-shoe internal brake on a car is given as 10^{-3} m^3. The brake should have two equal shoes, one self-energizing and one deenergizing, so that the brake can fit on both the right and left sides of the car. Calculate the brake width and radius for maximum braking power if the space available inside the wheel is 400 mm in diameter and 100 mm wide. The brake lining material has a maximum allowed contact pressure of 4 MPa and a coefficient of friction of 0.18. Also, calculate the maximum braking torque. *Ans.* $T = 730$ N-m.

18.46 A long-shoe brake in a car is designed to give as high a braking torque as possible for a given force on the brake pedal. The ratio between the actuating force and the pedal force is given by the hydraulic area ratio between the actuating cylinder and the cylinder under the pedal. The brake shoe angles are $\theta_1 = 10°$, $\theta_2 = 170°$, and $\theta_a = 90°$. The maximum brake shoe pressure is 5 MPa, the brake shoe width is 40 mm, and the drum radius is 100 mm. Find the distance d_7 that gives the maximum braking power for a coefficient of friction of 0.2 at any pedal force. What braking torque would result if the coefficient of friction were 0.25?

18.47 An external drum brake assembly (see Sketch i) has a normal force $P = 200$ lb acting on the lever. Dimensions are in inches. Assume that coefficient of friction $\mu = 0.25$ and maximum contact pressure $p = 100$ psi. Determine the following from long-shoe calculations:

(a) Free-body diagram with the directionality of the forces acting on each component.

(b) Which shoe is self-energizing and which is deenergizing.

(c) Total braking torque. *Ans.* $T = 20.45$ kip-in.

(d) Pad width as obtained from the self-energizing shoe (deenergizing shoe width equals self-energizing shoe width). *Ans.* $b = 1.26$ in.

(e) Pressure acting on the deenergizing shoe.

18.48 A long-shoe external brake as shown in Fig. 18.15 has a pivot point such that $d_7 = 4r$, $d_6 = 2r$, $\theta_1 = 5°$, and $\theta_2 = 45°$. Find the coefficient of friction needed to make the brake self-lock if the rotation is in the direction shown in Fig. 18.15. If the shaft rotates in the opposite direction, calculate the drum radius needed to get a braking torque of 180 N-m for the actuating force of 10,000 N. *Ans.* $r = 43.32$ mm.

18.49 An external, long brake shoe is mounted on an elastic arm. When a load is applied, the arm and the brake lining bend and redistribute the pressure. Instead of the normal sine pressure distribution the pressure becomes constant along the length of the lining. For a given actuating force calculate how the brake torque changes when the pressure distribution changes from sinusoidal to a constant pressure. Also, assume that $d_7 = 110$ mm, $r = 90$ mm, $b = 40$ mm, $\theta_1 = 20°$, $\theta_2 = 160°$, $d_6 = 220$ mm, $\mu = 0.25$, and $W = 12$ kN.

18.50 A long-shoe external brake has two identical shoes coupled in series so that the peripheral force from the first shoe is directly transferred to the second shoe. No radial force is transmitted between the shoes. Each of the two shoes covers 90° of the circumference, and the brake linings cover the central 70° of each shoe, leaving 10° at each end without lining as shown in Sketch j. The actuating force is applied tangentially to the brake drum at the end of the loose shoe, 180° from the fixed hinge point of the other shoe. Calculate the braking torques for both rotational directions when $d_7 = 150$ mm, $r = 125$ mm, $b = 50$ mm, $W = 14,000$ N, and $\mu = 0.2$. Also show a free-body diagram of these forces acting on the two shoes. *Ans.* $T = 717$ N-m. If reversed, $T = 1197$ N-m.

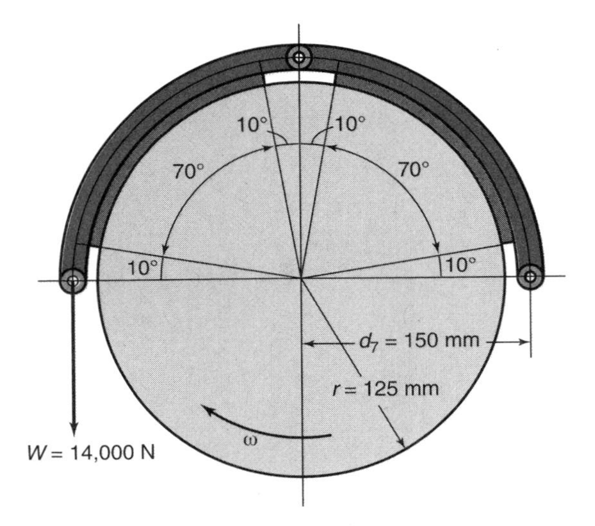

Sketch j, for Problem 18.50

18.51 A special type of brake is used in a car factory to hold the steel panels during drilling operations so that the forces from the drill bits cannot move the panels. The brake is shown in Sketch k. Calculate the braking force P_B on the steel panel when it moves to the right with the speed u_b between drilling operations. The actuating force is P_M. The brake lining is thin relative to the other dimensions. *Ans.* $P_b = \frac{9}{7}\mu P_M$.

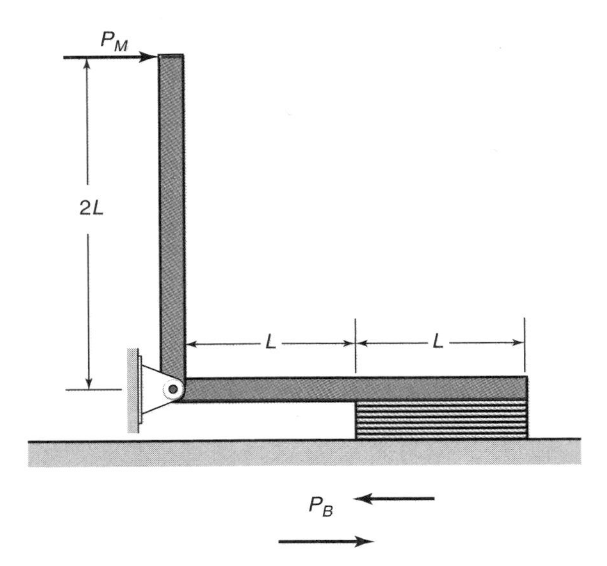

Sketch k, for Problem 18.51

18.52 Redo Problem 18.39 for long-shoe assumptions. The average contact pressure occurs at 40°. Determine the maximum contact pressure and its location. Assume that the distance x is 150 mm. What is the braking torque? Also, repeat this problem while reversing the direction of rotation. Discuss the changes in the results.

18.53 A symmetrically loaded, pivot-shoe brake has a wrap angle of 180° and the optimum distance d_7, giving a symmetrical pressure distribution. The coefficient of friction of the brake lining is 0.30. A redesign is considered that will increase the breaking torque without increasing the actuating force. The wrap angle is decreased to 80° (+40° to –40°) and the d_7 distance is decreased to still give a symmetrical pressure distribution. How much does the brake torque change? *Ans.* 15% decrease.

18.54 The band brake shown in Sketch l is activated by a compressed-air cylinder with diameter d_c. The brake cylinder is driven by air pressure $p = 0.7$ MPa. Calculate the maximum possible brake torque if the coefficient of friction between the band and the drum is 0.20. The mass force on the brake arm is neglected, $d_c = 50$ mm, $r = 200$ mm, $l_1 = 500$ mm, $l_2 = 200$ mm, and $l_3 = 500$ mm. *Ans.* $T = 719$ N-m.

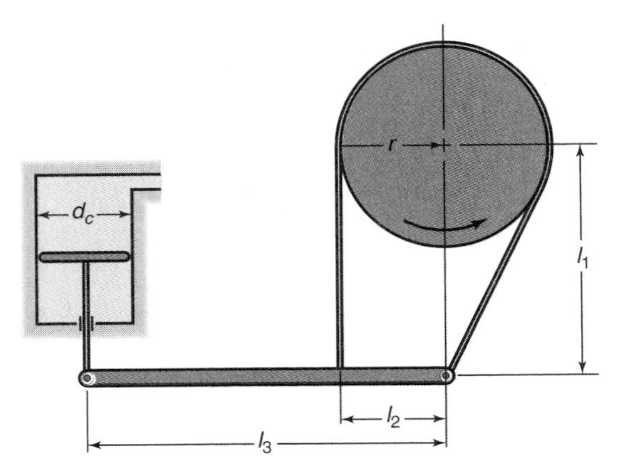

Sketch l, for Problem 18.54

18.55 The band brake shown in Sketch m has wrap angle $\phi = 225°$ and cylinder radius $r = 80$ mm. Calculate the brake torque when the lever is loaded by 100 N and coefficient of friction $\mu = 0.3$. How long is the braking time from 1200 rpm if the rotor mass moment of inertia is 2.5 kg-m^2? *Ans.* $T = 54$ N-m, $t = 5.825s$.

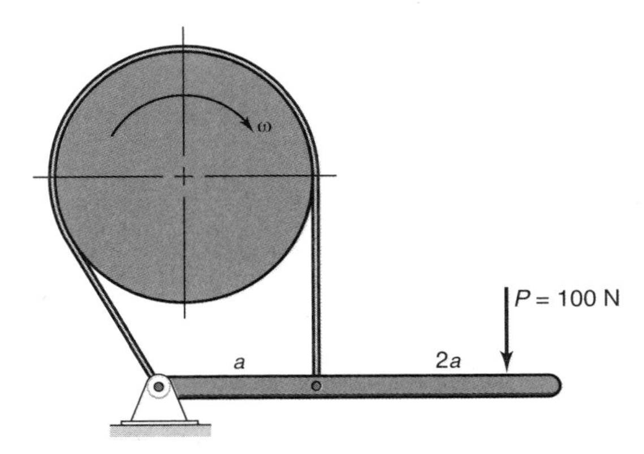

Sketch m, for Problem 18.55

18.56 The band brake shown in Sketch n is 40 mm wide and can take a maximum pressure of 1.1 MPa. All dimensions are in millimeters. The coefficient of friction is 0.3. Determine the following:

(a) The maximum allowable actuating force.

(b) The braking torque. *Ans.* $T = 1.26$ kN-m

(c) The reaction supports at 0_1 and 0_2.

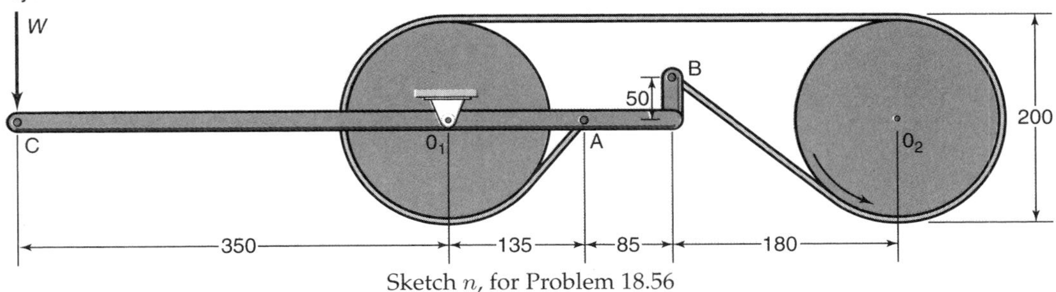

Sketch n, for Problem 18.56

(d) Whether it is possible to change the distance 0_1A in order to have self-locking. Assume point A can be anywhere on line C0_1A.

18.57 A brake (see Sketch o) consists of a drum with a brake shoe pressing against it. The drum radius is $r = 80$ mm. Calculate the brake torque when $P = 7000$ N, $\mu = 0.35$, and brake pad width $b = 40$ mm. The wear is proportional to the contact pressure times the sliding distance. *Ans.* $T = 205$ Nm.

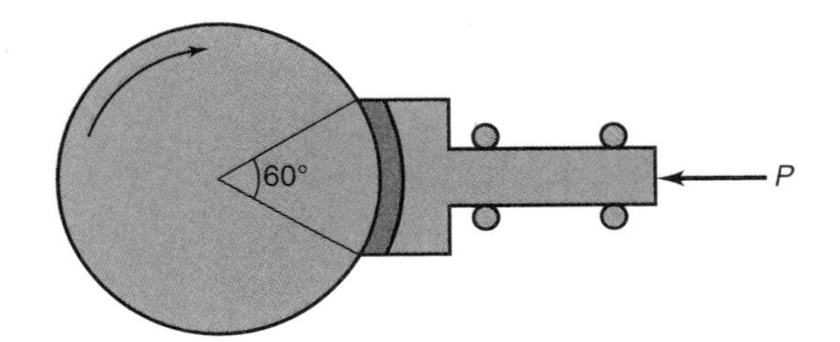

Sketch o, for Problem 18.57

18.58 For the band brake shown in Sketch p, the following conditions are given: $d = 350$ mm, $p_{\max} = 1.2$ MPa, $\mu = 0.25$, and $b = 50$ mm. All dimensions are in millimeters. Determine the following:

(a) The braking torque. *Ans.* $T = 1.26$ kN-m.

(b) The actuating force. *Ans.* $P = 2.57$ kN.

(c) The forces acting at hinge 0. *Ans.* $O_x = 0$, $O_y = 2.57$ kN.

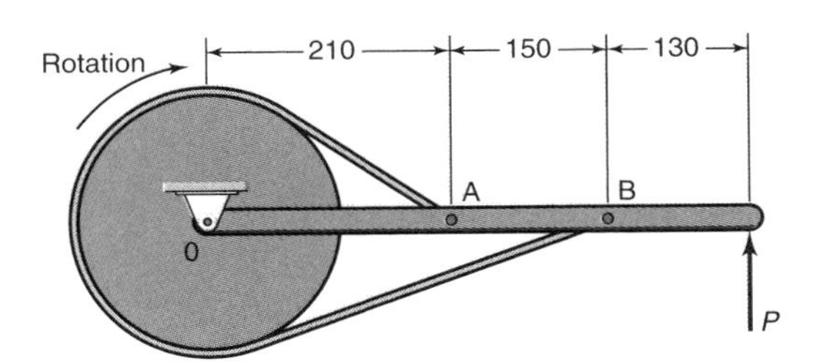

Sketch p, for Problem 18.58

18.59 The band brake shown in Sketch q has wrap angle $\phi = 215°$ and cylinder radius $r = 60$ mm. Calculate the brake torque when coefficient of friction $\mu = 0.25$. How long is the braking time from 1500 rpm if the rotor moment of inertia is $J = 2$ kg-m^2? *Ans.* $t = 20.0$ s.

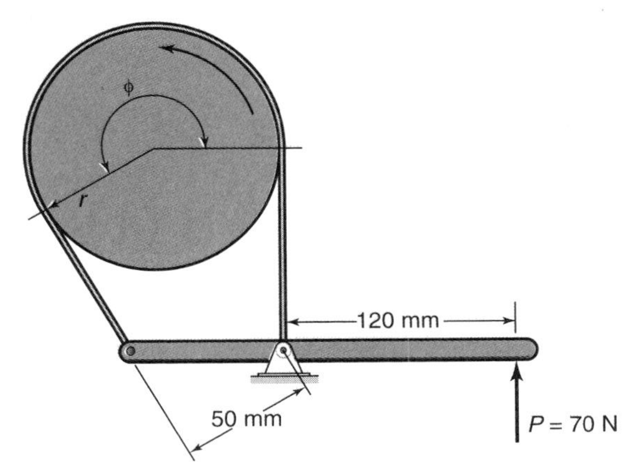

Sketch q, for Problem 18.59

Design and Projects

18.60 Many brake systems are designed so that they will have one component wear preferentially to the others. For example, a pad will wear faster than a disk in a thrust disk clutch. Explain why this feature is incorporated into brakes, instead of designing the entire system with the same intended life.

18.61 Excessive heating of brake materials leads to brake checking, warping, and out-of-roundness for brake drums. Conduct a literature search and investigate the highest temperature that exists between contacting asperities of brake systems.

18.62 Write a computer program to determine the optimum angular coverage of lining material for the pivot-shoe brake of Example 18.5.

18.63 Electric vehicles often use *regenerative braking* to recover some of the kinetic energy when vehicles are brought to rest. Write a one-page summary of the design features and disadvantages of regenerative braking.

18.64 What are anti-lock brakes? Conduct an Internet search and explain their operation.

Chapter 19

Flexible Machine Elements

A rolling chain on a sprocket. *Source:* Shutterstock.

Scientists study the world as it is, engineers create the world that never has been.

Theodore von Karman

Contents

Examples

Design Procedures

Case Study

Rubber belts, roller chains, and wire ropes are three examples of power transmission devices that are widely used in many industries. These machine elements have the common characteristic of flexibility. Rubber belts are used around pulleys, and transmit a torque based on friction associated with the belt-pullet interface. Flat belts can be constructed from soft or hard rubber, depending on the power to be transmitted, but synchronous and V-belts are advanced composite materials. Reinforced rubber belts are economically attractive and reliable machine elements that have further advantages of reducing impact stresses, but they will slip if torque becomes too large. Synchronous (timing) belts have teeth that prevent slip, and have the further advantages of low preload and associated reduced bearing loads. Roller chains are longer-lived and more robust devices than belts, but are more expensive. Roller chains and their sprockets require effective lubrication to achieve long lives. There are a number of design variables that can be modified with roller chains, and conventional and silent chains are presented in this chapter. Wire rope consists of many strands of wire carefully wound around a core; such rope is widely used to hoist loads or provide a force through a winch or equivalent device. Smaller diameter ropes are often used as cables to transmit a force in order to open a clutch or engage a brake, for example. Wire ropes are especially useful when power needs to be transferred efficiently over long distances.

Machine elements in this chapter: Flat belts, V-belts, synchronous belts, roller and silent chains, and wire ropes.
Typical applications: Power transmission from electric motors or internal combustion engines to power other devices; bicycle power transmission; cranes, hoists, and derricks; elevators.
Competing machine elements: Shaft couplings (Ch. 11), gears (Ch. 14 and 15).

Symbols

A_m	cross-sectional area of metal strand in rope, m^2
a	link plate thickness, m
a_1	service factor
a_2	multiple-strand factor in rolling chain
c	distance from neutral axis to outer fiber, m
c_d	center distance, m
D	sheave or pulley diameter, m
d	diameter, m
d_w	wire diameter, m
E	modulus of elasticity, Pa
F	force, N
F_a	force due to acceleration, N
F_f	fatigue allowable force, N
F_h	static force, N
F_i	initial tensile force, N
F_r	rope weight, N
F_t	total friction force, N
F_w	deadweight, N
f_1	overload service factor
f_2	power correction factor
g	gravitational acceleration, 9.807 m/s^2
g_r	velocity ratio
HB	Brinell hardness, kg/mm^2
h_p	power transmitted, W
h_{pb}	rated power per belt, W
h_{pr}	rated power, W
h_t	V-belt height, m
I	area moment of inertia, m^4
L	belt length, m
M	bending moment, N-m
m	mass, kg
m'	mass per unit length, kg/m
N	number of teeth in sprocket or number of belts
N_a	angular speed, rpm
n_s	safety factor
p	bearing pressure, Pa
p_{all}	allowable bearing pressure, Pa
p_t	pitch or datum, m
r	sheave or pulley radius, m
Δr	chordal rise, m
r_c	chordal radius, m
S_u	ultimate strength, Pa
t	thickness, m
u	belt velocity, m/s
w_t	belt width, m
w_z	weight, N
w'_z	weight per unit length, N/m
α	angle used to describe loss in arc of contact, deg
θ_r	angle of rotation to give chordal rise, deg
μ	coefficient of friction
σ	normal stress, Pa
σ_b	bending stress, Pa
σ_t	tensile normal stress, Pa
ϕ	wrap angle; gantry line angle, deg
ω	angular velocity, rad/s

Subscripts

1	driver pulley or sheave
2	driven pulley or sheave

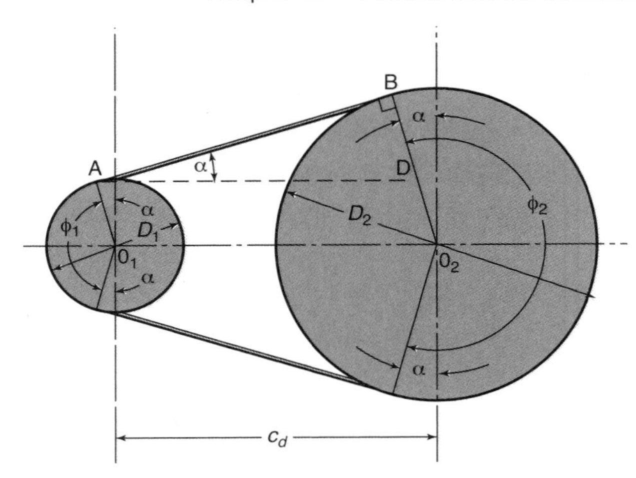

Figure 19.1: Dimensions, angles of contact, and center distance of open flat belt.

19.1 Introduction

The machine elements considered in this chapter, like the machine elements considered in Chapter 18, use friction as a useful agent to produce a high and uniform force and to transmit power. A belt, rope, or chain provides a convenient means for transferring power from one shaft to another, and are especially useful for transfer of power over long distances. Table 19.1 compares belt, chain, and gear power transmission approaches.

19.2 Flat Belts

Flat belts find considerable use in applications requiring small pulley diameters, high belt surface speeds, low noise levels, low weight, and/or low inertia. They cannot be used where absolute synchronization between pulleys must be maintained because they rely on friction for their proper functioning, so some slip is inevitable. Flat belts must be kept under tension to function, and therefore they require tensioning devices.

There are many applications where belts are used to transfer very low power levels. This book emphasizes machinery applications, where the power that is transferred is generally larger than 100 Nm/s (75 ft-lb/s). Smaller belts are readily available, and usually can be selected based solely on kinematic constraints.

19.2.1 Belt Length

Figure 19.1 shows the dimensions, angles of contact, and center distance of an open flat belt. The term "open" is used to distinguish it from the geometry of a crossed belt, where the belt forms a figure eight when viewed on end. Note that distance 0_2D is equal to $r_1 = D_1/2$, distance BD is equal to $r_2 - r_1 = (D_2 - D_1)/2$, and distance AD is the center distance c_d. Also, triangle ABD is a right triangle so that

$$\text{AB}^2 + \text{BD}^2 = \text{AD}^2,$$

so that

$$\text{AB}^2 + (r_2 - r_1)^2 = c_d^2;$$

$$\text{AB} = \sqrt{c_d^2 - (r_2 - r_1)^2}. \tag{19.1}$$

Table 19.1: Comparison of selected power transmission devices.

Constraint	Power transmission device					
	Flat belt	V-belt	Synchronous belt	Roller chain	Silent chain	Spur gear
Synchronization	1	1	1	4	4	4
Efficiency	1	1	2	4	4	4
Anti-shock	4	4	3	2	3	1
Noise/vibration	4	4	3	2	3	1
Compactness						
High speed/low load	2	3	3	1	4	3
Low speed/high load	1	1	2	4	3	2
Lubrication	None	None	None	Required	Required	Required
Bearing loads	2	1	2	4	3	4
Longevity	1	2	2	3	3	4

1, Poor; 2, Fair; 3, Good; 4, Excellent.

The length of the open flat belt can be expressed as

$$L = 2\text{AB} + r_1\phi_1\frac{\pi}{180} + r_2\phi_2\frac{\pi}{180}, \tag{19.2}$$

where ϕ is the wrap angle. The wrap angles can be expressed as

$$\phi_1 = 180^\circ - 2\alpha \qquad \text{and} \qquad \phi_2 = 180^\circ + 2\alpha. \tag{19.3}$$

Also, from right triangle ABD in Fig. 19.1, the angle used to describe the loss in arc of contact is

$$\sin\alpha = \frac{D_2 - D_1}{2c_d} \qquad \text{or} \qquad \alpha = \sin^{-1}\left(\frac{D_2 - D_1}{2c_d}\right). \tag{19.4}$$

The angle α is in degrees in order to be consistent with Eq. (19.3), and is equal to zero only if the pulleys have a 1-to-1 ratio. By substituting Eqs. (19.3) and (19.4) into Eq. (19.2), the belt length can be expressed as

$$\begin{aligned} L &= \sqrt{(2c_d)^2 - (D_2 - D_1)^2} \\ &\quad + \frac{D_1\pi}{360}\left[180^\circ - 2\sin^{-1}\left(\frac{D_2 - D_1}{2c_d}\right)\right] \\ &\quad + \frac{D_2\pi}{360}\left[180^\circ + 2\sin^{-1}\left(\frac{D_2 - D_1}{2c_d}\right)\right], \end{aligned}$$

or

$$\begin{aligned} L &= \sqrt{(2c_d)^2 - (D_2 - D_1)^2} + \frac{\pi}{2}(D_1 + D_2) \\ &\quad + \frac{\pi(D_2 - D_1)}{180}\sin^{-1}\left(\frac{D_2 - D_1}{2c_d}\right). \end{aligned} \tag{19.5}$$

19.2.2 Belt Forces

The basic equations developed for band brakes are also applicable here. From Section 18.9, the following torque and friction equations can be written:

$$\frac{F_1}{F_2} = e^{\mu\phi\pi/180^\circ}; \tag{18.64}$$

$$T = \frac{(F_1 - F_2)D_1}{2}, \tag{18.65}$$

where

- ϕ = wrap angle, deg
- μ = coefficient of friction
- F_1 = tight-side or driver force, N
- F_2 = slack-side or driven force, N

In obtaining the preceding equations, it was assumed that the coefficient of friction on the belt is uniform over the entire angle of wrap and that centrifugal forces on the belt can be neglected.

In transmitting power from one shaft to another by means of a flat belt and pulleys, the belt must have an initial tensile force, F_i. The required initial belt tension (or tensile force) depends on the elastic characteristics of the belt and friction, but can be approximated by

$$F_i = \frac{F_1 + F_2}{2}. \tag{19.6}$$

Also, from Eq. (18.65), note that when power is being transmitted, the tensile force, F_1, in the tight side exceeds the tension in the slack side, F_2.

The power transmitted is

$$h_p = (F_1 - F_2)u, \tag{19.7}$$

where u is the belt velocity. The centrifugal force acting on the belt can be expressed as

$$F_c = m'u^2 = \frac{w_z'}{g}u^2, \tag{19.8}$$

where

- m' = mass per unit length, kg/m
- u = belt velocity, m/s
- w_z' = weight per unit length, N/m

When the centrifugal force is considered, Eq. (18.64) becomes

$$\frac{F_1 - F_c}{F_2 - F_c} = e^{\mu\phi\pi/180^\circ} \tag{19.9}$$

Of course, Eq. (18.65) is unchanged when centrifugal forces are considered.

Example 19.1: Forces on a Flat Belt

Given: A flat belt is 6 in. wide and 1/3 in. thick and transmits 8250 ft-lb/s (15 hp). The center distance is 8 ft. The driving pulley is 6 in. in diameter and rotates at 2000 rpm such that the loose side of the belt is on top. The driven pulley has a diameter of 18 in. The belt material weighs 0.035 lb/in.^{3}.

Find: Determine the following:

(*a*) If μ = 0.30, determine F_1 and F_2.

(*b*) If μ is reduced to 0.20 because of oil getting on part of the pulley, what are F_1 and F_2? Would the belt slip?

(*c*) What is the belt length?

Solution:

(*a*) The belt velocity is

$$u = \frac{\pi D_1 N_1}{12} = \frac{\pi(6)(2000)}{12} = 3142 \text{ ft/min} = 52.37 \text{ ft/s}.$$

From Eq. (19.7),

$$F_1 - F_2 = \frac{h_p}{u} = \frac{8250}{52.37} = 157.5. \text{ lb} \qquad (a)$$

The weight per volume was given as

$$\frac{w_z}{L w_t t} = 0.035 \text{ lb/in.}^3$$

so that

$$\begin{aligned} \frac{w_z}{L} &= 0.035 w_t t = (0.035)(6)\left(\frac{1}{3}\right) \\ &= 0.070 \text{ lb/in.} = 0.84 \text{ lb/ft}. \end{aligned}$$

The centrifugal force acting on the belt is, from Eq. (19.8),

$$F_c = \frac{w_z}{L}\frac{u^2}{g} = \frac{(0.84)(3142)^2}{(32.2)(60)^2} = 71.54 \text{ lb}.$$

From Eq. (19.4),

$$\begin{aligned} \alpha &= \sin^{-1}\left(\frac{D_2 - D_1}{2c_d}\right) \\ &= \sin^{-1}\left(\frac{18-6}{(2)(8)(12)}\right) \\ &= 3.583^\circ. \end{aligned}$$

The wrap angle is given by Eq. (19.3) as

$$\phi = 180^\circ - 2(\alpha) = 180^\circ - 2(3.583^\circ) = 172.8^\circ. \qquad (b)$$

Making use of Eq. (19.9) gives

$$\frac{F_1 - 71.54}{F_2 - 71.54} = e^{(0.3)(172.8)\pi/180} = 2.472.$$

Therefore,

$$F_1 - 71.54 = 2.472F_2 - 176.8,$$

so that

$$F_1 = 2.472F_2 - 105.3. \qquad (c)$$

Substituting Eq. (*c*) into Eq. (*a*) gives

$$2.472F_2 - F_2 = 105.3 + 157.5;$$

$$F_2 = \frac{262.8}{1.472} = 178.5 \text{ lb}.$$

Therefore, from Eq. (*a*),

$$F_1 = 157.5 + 178.5 = 336.0 \text{ lb}.$$

From Eq. (19.6), the initial belt tension is

$$F_i = \frac{F_1 + F_2}{2} = \frac{336.0 + 178.5}{2} = 257.3 \text{ lb}.$$

$$F_1 = 1.828F_2 - (71.54)(1.828) + 71.54 = 1.828F_2 - 59.24, \qquad (d)$$

(*b*) If $\mu = 0.20$ instead of 0.30,

$$\frac{F_1 - 71.54}{F_2 - 71.54} = e^{(0.2)(172.8)\pi/180} = 1.828.$$

Since the initial belt tension is still $F_i = 257.3$ lb, substituting Eq. (*e*) into Eq. (19.6) gives

$$\frac{1.828F_2 - 59.24 + F_2}{2} = 257.3,$$

or $F_2 = 202.9$ lb. Substituting this result into Eq. (*d*) produces

$$F_1 = 1.828(202.9) - 59.24 = 311.7 \text{ lb}.$$

From Eq. (19.7) the power transmitted is

$$\begin{aligned} h_p &= (F_1 - F_2)\,u \\ &= (311.7 - 202.9)(52.37) \\ &= 5698 \text{ ft-lb/s}. \end{aligned}$$

Since this is less than the required power, the belt will slip.

(*c*) From Eq. (19.5) while making use of Eq. (*c*),

$$\begin{aligned} L &= 2\sqrt{8^2(12)^2 - (9-3)^2} \\ &\quad + \frac{3\pi}{180^\circ}(172.8^\circ) + \frac{9\pi}{180^\circ}(187.2^\circ) \\ &= 230.1 \text{ in}. \end{aligned}$$

19.2.3 Slip

Slip is detrimental for a number of reasons. It reduces belt efficiency and can cause thermal damage to the belt, leading to elongation (and resulting in lower belt tension and more slip) or chemical degradation. To eliminate slip, the initial belt tension needs to be retained. But as the belt stretches over time, some of the initial tension will be lost. One solution might be to have a very high initial tension, but this would put large loads on the pulley, shaft, and bearings, and also shorten the belt life. Some better approaches are the following:

1. Develop means of adjusting tension during operation.
2. Increase the wrap angle.
3. Change the belt material to increase the coefficient of friction.
4. Use a larger belt section.
5. Use an alternative design, such as a synchronous belt (see Section 19.3) or a chain as discussed in Section 19.6.

There are many tensioning devices involving spring-loaded pulleys or weights to apply loads to belts. Figure 19.2 illustrates one way of maintaining belt tension. The slack side of the belt is on the top, so that the sag of the belt acts to increase its wrap angle.

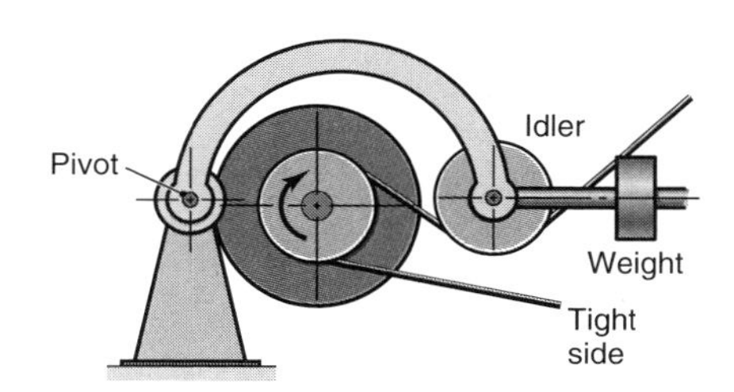

Figure 19.2: Weighted idler used to maintain desired belt tension.

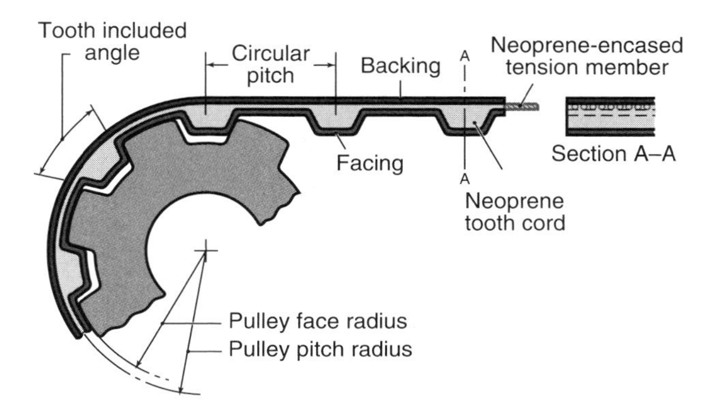

Figure 19.3: Synchronous, or timing, belt.

19.3 Synchronous Belts

Synchronous belts, or **timing belts**, are basically flat belts with a series of evenly spaced teeth on the inside circumference, thereby combining the advantages of flat belts with the excellent traction of gears and chains (see Section 19.6). A synchronous belt is shown in Fig. 19.3.

Unlike flat belts, synchronous belts do not slip or creep, and the required belt tension is low, resulting in very small bearing loads. Synchronous belts will not stretch and require no lubrication. Speed is transmitted uniformly because there is no chordal rise and fall of the pitch line as in rolling chains (Section 19.6). The equations developed for flat-belt length and torque in Section 19.2 are equally valid for synchronous belts.

Example 19.2: Optimum Pitch Diameter for a Timing Belt

Given: A timing belt is used to transfer power from a high-speed motor to a grinding wheel. The timing belt is 750 mm long and weighs 180 g. The maximum allowable force in the belt is 2000 N, and the speed of the turbine is the same as the speed of the grinding wheel, 5000 rpm.

Find: Calculate the optimum pulley pitch diameter for maximum power transfer.

Solution: If the total maximum force in the belt is F_1 and the centrifugal force is F_c, the maximum power transmitted is

$$h_p = u(F_1 - F_c).$$

Letting F_1 be equal to the maximum allowable force and making use of Eq. (19.8) for the centrifugal force gives

$$h_p = u\left(2000 - \frac{0.180}{0.750}u^2\right) = u\left(2000 - 0.24u^2\right).$$

The optimum power transmitted occurs when

$$\frac{\partial h_p}{\partial u} = 2000 - (0.24)(3)u^2 = 0.$$

Solving for the velocity yields $u = 52.70$ m/s. The pulley diameter that produces the maximum power transfer is

$$\frac{\omega D}{2} = u$$

$$D = \frac{2u}{\omega} = \frac{2(52.70)}{(5000)\left(\frac{2\pi}{60}\right)} = 0.2013 \text{ m} = 201.3 \text{ mm}.$$

19.4 V-Belts

V-belts are an extremely common power transmission device, used on applications as diverse as blowers, compressors, mixers, machine tools, etc. One or more V-belts are used to drive accessories on an automobile and transfer power from the internal-combustion engine. V-belts are made to standard lengths and with standard cross-sectional sizes, as shown in Fig. 19.4a. V-belts A through E are standard shapes that were standardized as early as the 1940s, but the more modern 3V, 5V, and 8V belts shown have higher power ratings for the same cross-sectional area. V-belts have a fiber-reinforced construction, as shown in Fig. 19.4b and c, so they should be recognized as advanced composite materials. The flexible reinforcement is often a cord, not an individual fiber, and can be made from nylon, kevlar (aramid), polyester, etc. The impregnated woven jacket shown protects the interior and provides a wear- and oil-resistant surface for the belt. The compressive side of the belt is produced from a high-strength, fatigue-resistant rubber. Because their interior tension cords are stretch and creep resistant, V-belts (unlike flat belts) do not require frequent adjustment of initial tension. The grooved pulleys that V-belts run in are called **sheaves**. They are usually made of cast iron, pressed steel, or die-cast metal. Figure 19.5 shows a V-belt in a sheave groove.

V-belts can also be used in multiples as shown in Fig. 19.4c. The obvious advantage of such a belt is that it can transmit higher power; a two-belt design can transmit twice the power of a single belt, etc. The tie band construction shown ensures that the belt maintains alignment when it is outside of the sheaves, which helps the belt run over the sheave properly.

V-belts find frequent application where synchronization between shafts is not important; that is, V-belts can slip (see Section 19.2.3). V-belts are easily installed and removed, quiet in operation (but not quite as quiet as flat belts), low in maintenance, and provide shock absorption between the driver and driven shafts. V-belt drives normally operate best at belt velocities between 7.5 and 30 m/s (1500 and 6500 ft/min).

The **velocity ratio** is similar to the gear ratio given in Eq. (14.19), or

$$g_r = \frac{N_{a1}}{N_{a2}} = \frac{r_2}{r_1}. \tag{19.10}$$

V-belts can operate satisfactorily at velocity ratios up to approximately 7 to 1. V-belts typically operate at 90 to 98% efficiency, lower than that found for flat belts.

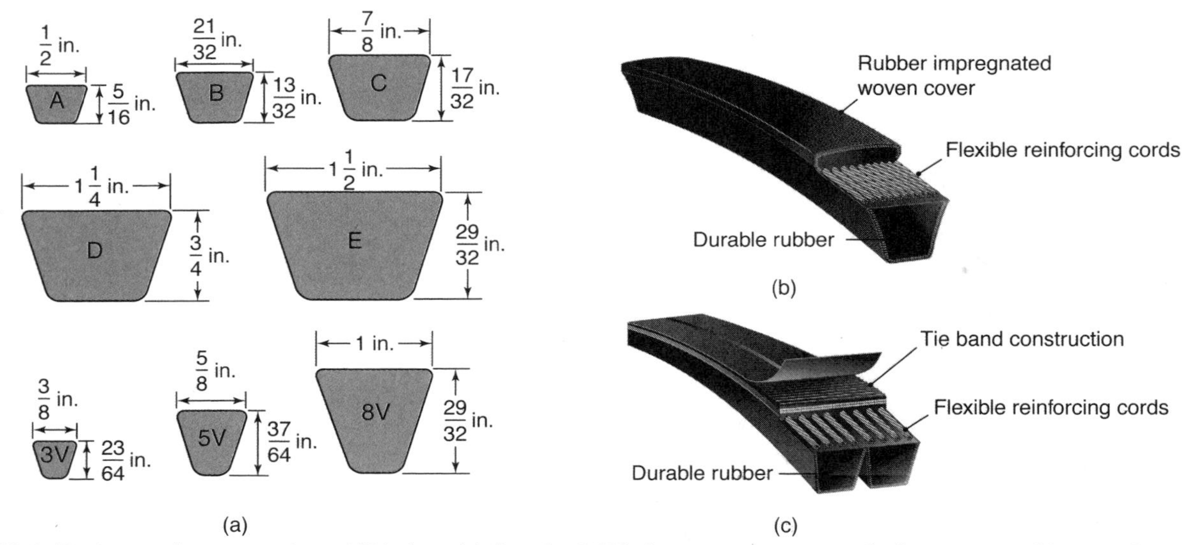

Figure 19.4: Design and construction of V-belts. (a) Standard V-belts cross sections with dimensions; (b) typical single-belt, showing reinforcing cords and wear-resistant exterior; (c) double V-belt, used for higher power transmission than single belts. Up to five belts can be combined in this fashion.

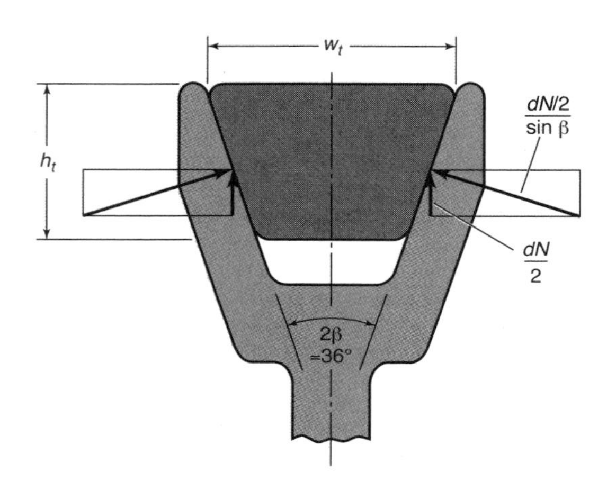

Figure 19.5: V-belt in sheave groove.

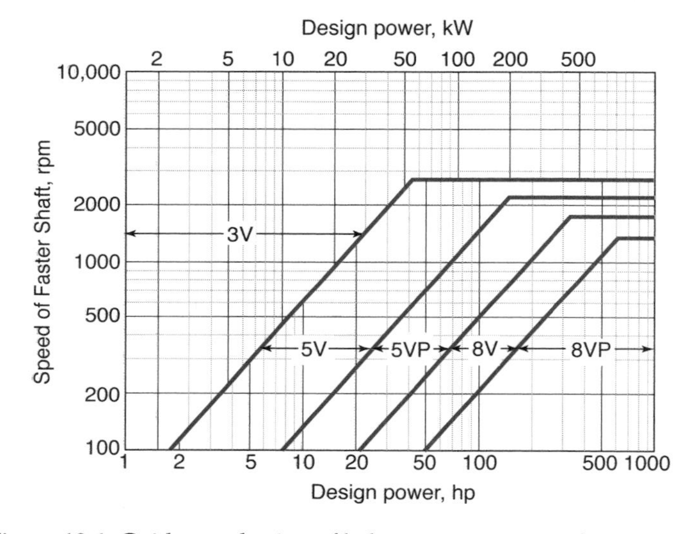

Figure 19.6: Guide to selection of belt cross section as a function of power transmitted and shaft speed.

A major advantage of V-belts over flat belts is their wedging action, which increases the normal force from dN for flat belts to $(dN/2)/\sin\beta$ (as shown in Fig. 19.5) for V-belts, where β is the sheave angle. Because the V-belt has a trapezoidal cross section, the belt rides on the side of the groove and a wedging action increases the traction. Pressure and friction forces act on the side of the belt. The force equations developed in Section 19.2 for flat belts are equally applicable for V-belts if the coefficient of friction μ is replaced with $\mu/\sin\beta$. Also, the belt length equation for flat belts given by Eq. (19.5) is applicable to V-belts.

19.4.1 Input Normal Power Rating

It is essential that the maximum possible load conditions be considered in designing a V-belt. The design power rating of the belt needs to consider the service factor, or

$$h_{pr} = f_1 h_p, \tag{19.11}$$

where f_1 is the overload service factor for various applications as given in Table 19.2, and h_p is the input power. Figure 19.6 shows the belt profiles that are commonly applied as a function of design power rating.

19.4.2 Sheave Size

The materials and construction of V-belts have improved dramatically in the past few decades. For example, a B-Section in a 7.0-in. sheave was rated at 4.2 hp in 1945, while a modern belt under the same conditions has a rating over 11 hp. This is due in large part to the transition from multiple reinforcing cords centrally located in the belt to a single tensile cord line located higher in the cross section. One subtle complication occurs with such belt designs: the pitch diameter (from which velocities are calculated) shifts upward when a modern belt replaces a conventional design. For this

Table 19.2: Typical overload service factors, f_1. *Source:* Courtesy of the Gates Corporation.

	Power source					
	Normal[a]			Demanding[b]		
	Service (hr/day)			Service (hr/day)		
Driven machine	3–5	8–10	16–24	3–5	8–10	16–24
Dispensing, display equipment, measuring equipment, office and projection equipment.	1.0	1.1	1.2	1.1	1.2	1.3
Liquid agitators, appliances, sewing machines, sweepers, light-duty conveyors, fans, light duty machine tools (drill presses, lathes, saws), woodworking equipment.	1.1	1.2	1.3	1.2	1.3	1.4
Semi-liquid agitators, centrifuges, centrifugal compressors, heavy-duty conveyors, dough mixers, generators, laundry equipment, heavy-duty machine tools (boring mills, grinders, mills, shapers), presses, shears, printing machinery, centrifugal and gear pumps.	1.1	1.2	1.4	1.2	1.3	1.5
Brick machinery, piston compressors, screw conveyors, bucket elevators, extractors, hammer mills, paper pulpers, pulverizers, piston pumps, extruders, rubber calendar mills, textile machinery.	1.2	1.3	1.5	1.4	1.5	1.6
Jaw crushers, hoists, ball mills, rod and tube mills, sawmill machinery.	1.3	1.4	1.6	1.5	1.6	1.8

[a] Includes normal torque, squirrel cage, synchronous and split phase AC motors; shunt wound DC motors; multiple cylinder internal combustion engines.

[b] Includes high torque, high slip, repulsion-induction, single phase, series wound AC motors; series wound, compound wound DC motors; single piston internal combustion engines.

Table 19.3: Recommended minimum datum diameters, in inches, of sheaves for general purpose 60-cycle electric motors. *Source:* Courtesy of the Gates Corporation.

Motor power,	Motor rpm					
hp	575	690	870	1160	1750	3450
0.5	2.5	2.5	2.2	—	—	—
0.75	3.0	2.5	2.4	2.2	—	—
1	3.0	3.0	2.4	2.4	2.2	—
1.5	3.0	3.0	2.4	2.4	2.4	2.2
2	3.8	3.0	3.0	2.4	2.4	2.4
5	4.5	4.5	3.8	3.0	3.0	2.6
10	6.0	5.2	4.6	4.4	3.8	3.0
15	6.8	6.0	5.4	4.6	4.4	3.8
20	8.2	6.8	6.0	5.4	4.6	4.4
30	10.0	9.0	6.8	6.8	5.4	—
50	11.0	10.0	9.0	8.2	6.8	—
75	14.0	13.0	10.5	10.0	9.0	—
100	18.0	15.0	12.5	11.0	10.0	—
200	22.0	22.0	22.0	—	—	—

reason, the International Standards Organization introduced the term **datum line** to differentiate pitch lines associated with new and conventional belt designs; this standard was adopted in the United States in 1988. Modern sheaves are specified based on a datum diameter; the concepts of datum diameter and pitch diameter are the same, and the sheave datum diameter can be used for velocity calculations. For standard belt cross-sections, the datum diameter is nearly the same as the outside diameter of the sheave, since this is very close to the value of the reinforcing line for most belts.

The design of a V-belt drive should use the largest possible sheaves. Unfortunately, large sheaves are usually more expensive, and there is an inherently larger center distance associated with their use compared to smaller sheaves. Small sheaves are less efficient, require larger belt preload (and associated higher load on bearings and shafts), and greatly reduce belt life because of slip and extreme flexing of the belt. Table 19.3 shows minimum recommended sheave datum diameters for general purpose electric motors.

Usually, much larger sheave datum diameters are used than suggested by Table 19.3. For some applications, it may be economically justifiable to select sheave designs based on design constraints and manufacture them to the desired dimensions. However, it is often preferable to utilize standard sheaves. Table 19.4 presents some combinations of standard sheave sizes, and also provides a power correction factor, f_2, which will modify the belt power rating as discussed in Design Procedure 19.1.

19.4.3 Design Power Rating

Table 19.5 gives the rated power for selected belt types. The belts considered are the 3V and 5V cross sections, but the data for the other cross sections shown in Fig. 19.4a are readily available in manufacturers' literature. It should be noted that Table 19.5 represents a very small sampling of available V-belts; there are many more cross-sections, and many more lengths and capacities depending on specific materials and belt construction. Table 19.5 is useful for illustration purposes, but for detailed design, a wider selection of belts should be considered, as can be readily obtained from manufacturers' web sites.

Design Procedure 19.1: V-Belt Drives

It will be assumed that a belt drive will be designed for power transmission where the shaft speeds (and hence speed ratio) and desired center distance are known. The power available can be obtained from the rating of the motor, or else it can be obtained from design requirements. Based on these quantities, this design procedure provides a methodology for selecting a cross-section of a belt, choosing sheaves and number of belts required.

1. Estimate the overload service factor from Table 19.2 and use it to obtain the required belt power rating using Eq. (19.11).

2. Select a cross section of the belt from the required belt power rating and the shaft speed using Fig. 19.6.

3. Obtain the minimum allowable sheave datum diameter from Table 19.3.

Table 19.4: Center distance and power correction factor, f_2, for standard sheaves. *Source:* Courtesy of the Gates Corporation.

Belt type	Speed ratio[a]	Datum diameter, in.		Belt length, in.						
		Small sheave	Large sheave	25	30	35	40	45	50	60
3V	1.00	2.65	2.65	8.3	10.8	13.3	15.8	18.3	20.8	25.8
		3.00	3.00	7.8	10.3	12.8	15.3	17.8	20.3	25.3
		3.35	3.35	7.2	9.7	12.2	14.7	17.2	19.7	24.7
		4.50	4.50	5.4	7.9	10.4	12.9	15.4	17.9	22.9
	1.25	2.50	3.15	8.1	10.6	13.1	15.6	18.1	20.6	25.6
		3.65	4.50	6.1	8.6	11.1	13.6	16.1	18.6	23.6
		4.50	5.60	—	7.0	9.6	12.1	14.6	17.1	22.1
		4.75	6.00	—	6.5	9.0	11.5	14.0	16.5	21.5
	1.5	2.80	4.12	7.0	9.5	12.0	14.6	17.1	19.6	24.6
		3.00	4.50	6.8	9.1	11.6	14.1	16.6	19.1	24.1
		3.35	5.00	5.9	8.4	10.9	13.4	15.9	18.4	23.4
		5.30	8.00	—	—	—	9.5	12.0	14.5	19.5
	2.0	2.65	5.30	6.1	8.7	11.2	13.7	16.2	19.1	13.1
		3.00	6.00	5.2	7.8	10.3	12.8	15.4	18.3	22.4
		4.12	8.00	—	—	7.7	10.3	12.8	15.4	23.4
		5.30	10.60	—	—	—	—	9.6	12.2	22.9
	2.5	2.65	6.50	—	7.6	10.1	12.7	15.2	17.7	22.7
		2.80	6.90	—	7.1	9.7	12.2	14.7	17.3	22.3
		3.15	8.00	—	—	8.4	11.0	13.5	16.1	21.1
		5.60	14.00	—	—	—	—	—	—	14.0
	3.0	2.65	8.00	—	6.0	8.7	11.3	13.9	16.4	21.5
		3.65	10.60	—	—	—	8.0	10.7	13.4	18.5
		4.75	14.00	—	—	—	—	—	—	14.5

Belt type	Speed ratio[a]	Datum diameter		Belt length, in.						
		Small sheave	Large sheave	60	71	80	90	100	125	150
5V	1.00	7.10	7.10	18.8	24.3	28.8	33.8	38.8	51.3	63.8
		8.00	8.00	17.4	22.9	27.4	32.4	37.4	49.9	62.4
		9.00	9.00	15.9	21.4	25.9	30.9	35.9	48.4	60.9
		12.50	12.50	—	15.9	20.4	25.4	30.4	42.9	55.4
	1.5	7.50	11.30	15.1	20.6	25.2	30.2	35.2	47.7	60.2
		9.25	14.00	—	17.1	21.6	26.6	31.7	44.2	56.7
		12.50	18.70	—	—	—	20.3	25.3	37.9	50.4
	2.0	7.10	14.00	13.0	18.6	23.2	28.2	33.2	45.8	63.3
		8.00	16.00	—	16.2	20.8	25.8	30.9	43.5	61.0
		14.00	28.00	—	—	—	—	—	28.7	46.5
	3.0	8.00	23.60	—	—	—	18.5	23.9	36.9	49.6
		7.10	21.20	—	—	16.2	21.6	26.8	39.6	52.3
		12.50	37.50	—	—	—	—	—	—	33.4

[a] Nominal speed ratio, actual value may vary slightly.

Power Correction Factor:

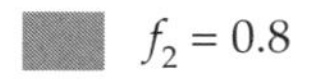

$f_2 = 0.8$

$f_2 = 0.9$

$f_2 = 1.0$

4. Locate the sheave diameter combinations in Table 19.4 that are suitable for a desired speed ratio. Disregard from consideration any candidates that are smaller than the minimum values obtained in Step 3. From the remaining candidates, select a sheave size that is consistent with space requirements.

5. From Table 19.4, locate the center distance that most closely matches design constraints, and obtain the power correction factor, f_2. Note that the belt length can be calculated from Eq. (19.5) or read directly from Table 19.4.

6. From Table 19.5, locate the proper belt cross section and center distance, and obtain the basic power rating per belt, h_1. Note that for very high speeds or small sheaves, an additional power may be required. This is usually a small amount and is neglected in this design procedure.

7. The rated power per belt is given by

$$h_{pb} = f_2 h_1. \tag{19.12}$$

8. The number of belts required can be obtained from the required power from Step 1:

$$N = \frac{h_{pr}}{h_{pb}}. \tag{19.13}$$

Example 19.3: V-Belt Design

Given: A 10-hp squirrel cage electric motor operating at a speed of 2000 rpm drives a centrifugal air compressor in normal service at 1000 rpm. It is desired to have a center distance around 15 in.

Find: Determine the proper belt cross section for this application, select pulleys, and choose the number of belts.

Solution: This approach will closely follow Design Procedure 19.1.

1. Referring to Table 19.2, note from the footnote that the squirrel cage motor represents a normal power source. Therefore, an overload service factor of $f_1 = 1.2$ is selected, assuming the pump is operated during a normal 8-hr shift. Therefore, the design power is, from Eq. (19.11),

$$h_{pr} = f_1 h_p = (1.2)(10) = 12 \text{ hp}.$$

2. For 10 hp and a shaft speed of 2000 rpm, Fig. 19.6 suggests a 3V cross-section is proper. Note, however, that the application is very close to the limit of 3V effectiveness, so a 5V belt may also provide a reasonable design solution. However, this solution will continue using a 3V belt as recommended by Fig. 19.6.

Table 19.5: Rated power in horsepower per belt for selected 3V and 5V cross sections. *Source:* Courtesy of the Gates Corporation.

Belt	Speed (rpm)	Small sheave outside diameter											
		2.65	2.80	3.00	3.15	3.35	3.65	4.12	4.50	4.75	5.00	5.30	5.60
3V	200	0.27	0.31	0.35	0.39	0.43	0.50	0.61	0.70	0.75	0.81	0.88	0.95
	400	0.49	0.55	0.64	0.71	0.80	0.93	1.14	1.30	1.41	1.52	1.64	1.77
	600	0.68	0.78	0.91	1.01	1.14	1.33	1.63	1.87	2.02	2.18	2.37	2.55
	800	0.86	0.99	1.16	1.29	1.45	1.70	2.09	2.41	2.61	2.81	3.05	3.30
	1000	1.03	1.19	1.40	1.55	1.76	2.07	2.54	2.92	3.17	3.42	3.72	4.01
	1200	1.19	1.38	1.62	1.81	2.05	2.41	2.98	3.43	3.72	4.01	4.36	4.70
	1400	1.35	1.56	1.84	2.05	2.33	2.75	3.39	3.91	4.25	4.58	4.98	5.37
	1600	1.49	1.73	2.05	2.29	2.60	3.07	3.80	4.38	4.75	5.13	5.57	6.01
	1800	1.63	1.90	2.25	2.52	2.87	3.39	4.19	4.83	5.25	5.66	6.15	6.63
	2000	1.76	2.06	2.45	2.74	3.12	3.69	4.57	5.27	5.72	6.17	6.70	7.22
	2400	2.02	2.36	2.82	3.16	3.60	4.27	5.29	6.09	6.62	7.13	7.74	8.33
	3000	2.35	2.77	3.32	3.73	4.26	5.06	6.27	7.21	7.82	8.41	9.11	9.79
	4000	2.80	3.33	4.02	4.52	5.19	6.16	7.61	8.72	9.41	10.1	10.8	11.6
	5000	3.12	3.73	4.53	5.11	5.87	6.95	8.53	9.69	10.4	–	–	–
Belt	**Speed (rpm)**	**Small sheave outside diameter**											
		7.1	7.5	8.0	8.5	9.0	9.25	9.75	10.3	11.3	12.5	14.0	16.0
5V	100	1.36	1.48	1.64	1.80	1.96	2.04	2.19	2.36	2.67	3.04	3.50	4.11
	200	2.52	2.76	3.06	3.36	3.66	3.81	4.11	4.44	5.03	5.73	6.60	7.75
	300	3.60	3.96	4.40	4.83	5.27	5.49	5.92	6.40	7.25	8.27	9.54	11.2
	400	4.63	5.10	5.67	6.24	6.81	7.09	7.66	8.27	9.39	10.7	12.3	14.5
	500	5.63	6.19	6.90	7.60	8.29	8.64	9.33	10.1	11.4	13.1	15.1	17.7
	600	6.58	7.25	8.08	8.91	9.73	10.1	10.9	11.8	13.4	15.3	17.6	20.7
	800	8.41	9.27	10.3	11.4	12.5	13.0	14.0	15.2	17.2	19.6	22.5	26.3
	1000	10.1	11.2	12.5	13.8	15.0	15.7	16.9	18.3	20.7	23.5	27.0	31.3
	1500	13.9	15.4	17.2	18.9	20.7	21.5	23.2	25.0	28.1	31.7	35.8	40.6
	2000	17.1	18.8	21.0	23.1	25.1	26.1	27.9	29.9	33.3	–	–	–

3. From Table 19.3, note that the 10-hp motor operating at 2000 rpm will require sheaves that are slightly smaller than 3.8 in.

4. In Table 19.4, there are four sheave combinations that are suitable for this application. Either the 4.12-in. or 5.30-in. smaller sheave are above the minimum sheave dimensions and will yield reasonable designs. This solution will assume that the bearing reaction forces are not as much of a concern as low center distance, so the 4.12- and 8.00-in. sheave combination will be selected.

5. For the required nominal center distance of 15 in., a 50-in. belt is most suitable. From the color code in Table 19.4, note that the application factor for this belt is $f_2 = 0.90$.

6. Referring to Table 19.5, a 4.12-in. outer diameter sheave operating at 2000 rpm has a basic power rating of $h_1 = 4.57$ hp.

7. The rated power per belt is therefore, from Eq. (19.12),

$$h_{pb} = f_2 h_1 = (0.90)(4.57) = 4.113 \text{ hp}.$$

8. The number of belts required is given by Eq. (19.13) as

$$N = \frac{h_{pr}}{h_{pb}} = \frac{12}{4.113} = 2.917.$$

Therefore, the number of belts are chosen as $N = 3$.

In summary, three 50-in.-long, 3V belts operating on sheaves of 4.12 and 8.00 in. are needed for this application. Since three belts are needed, a triple belt design (see Fig. 19.4) and associated sheaves are specified.

19.5 Wire Ropes

Wire ropes are used instead of flat belts or V-belts when power must be transmitted over long distances as in hoists, elevators, ski lifts, etc. Figure 19.7 shows cross sections of selected wire ropes. The center portion (dark section) is the **core** of the rope and is often made of hemp (a tall Asiatic herb), but can also be constructed of a polymer such as polypropylene or steel strands. The purposes of the core are to elastically support the strands and to lubricate them to prevent excessive wire wear.

The **strands** are groupings of wires placed around the core. In Fig. 19.7a there are six strands, and each strand consists of 19 wires. Wire ropes are typically designated as, for example, "$1\frac{1}{8}, 6 \times 19$ hauling rope." The $1\frac{1}{8}$ gives the wire rope diameter in inches, designated by the symbol d. The 6 designates the number of strands and 19 designates the number of wires in a strand. The wire diameter is designated by the symbol d_w. The term "hauling rope" designates the application in which the wire rope is to be used. Generally, wire ropes with more but smaller wires are more resistant to fatigue, whereas ropes with fewer but larger wires have greater abrasive wear resistance.

When a wire rope is bent and unbent around a sheave, the strands and wires slide over each other at a small length scale. The sliding velocities are low, but the pressure is high, so that wire wear is a concern. For this reason, wire rope lubrication is essential for prolonged life, and lubricant needs to be periodically replaced. Lubricants need to be matched to core materials to ensure proper wetting, and are also essential for corrosion protection.

Figure 19.8 shows two **lays** of wire rope. The **regular lay** (Fig. 19.8b) has the wire twisted in one direction to form strands and the strands twisted in the opposite direction to form the rope. Visible wires are approximately parallel to the rope axis. The major advantage of the regular lay is that the rope does not kink or untwist and is easy to handle. The **Lang lay** (Fig. 19.8a) has the wires in the strands and the strands in

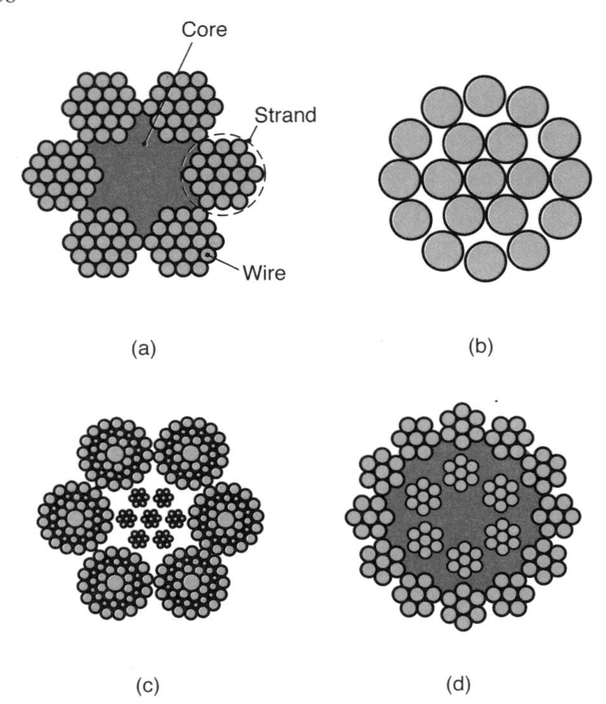

Figure 19.7: Cross sections of selected wire rope. (a) 6×19 fiber core; (b) 1×19; (c) 6×36 wire core; (d) 18×7 fiber core.

the rope twisted in the same direction. This type of lay has more resistance to abrasive wear and bending fatigue than the regular lay. Lang-lay ropes are, however, more susceptible to handling abuses, pinching in undersized grooves, and crushing when improperly wound on drums. Also, the twisting moment acting in the strand tends to unwind the strand, causing excessive rope rotation. Lang-lay ropes should therefore always be secured at the ends to prevent the rope from unlaying.

Although steel is most popular, wire rope is made of many kinds of metal, such as copper, bronze, stainless steel, and wrought iron. Table 19.6 lists some of the various ropes that are available, together with their characteristics and properties. The cross-sectional area of the metal strand in standard hoisting and haulage ropes is $A_m \approx 0.38d^2$.

19.5.1 Tensile Stress

The total force acting on the rope is

$$F_t = F_w + F_r + F_a + F_h, \tag{19.14}$$

where

- F_w = weight being supported, N
- F_r = rope weight, N
- F_a = force due to acceleration, N
- F_h = static load, often due to a *headache ball*, N

The static load or dead weight, F_h is essential for most applications to maintain tension in the wire rope and to prevent it from slipping off of a sheave. The tensile stress is

$$\sigma_t = \frac{F_t}{A_m}, \tag{19.15}$$

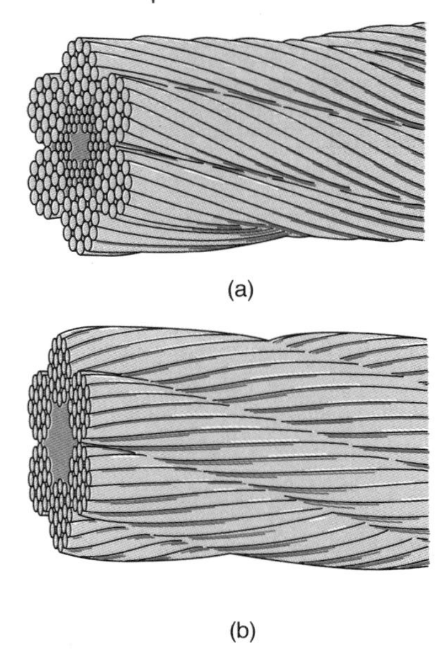

Figure 19.8: Two lays of wire rope. (a) Lang; (b) regular.

where A_m is the cross-sectional area of metal strand in standard hoisting and haulage ropes. The allowable stress is obtained from Table 19.6.

From Eq. (1.1), the safety factor is

$$n_s = \frac{\sigma_{\text{all}}}{\sigma_t}. \tag{19.16}$$

Table 19.7 gives minimum safety factors for a variety of wire rope applications. The safety factor obtained from Eq. (19.16) should be larger than the safety factor obtained from Table 19.7.

19.5.2 Bending Stress

From Eqs. (4.46) and (4.47), the bending moment applied to the wires in a rope passing over a pulley is

$$M = \frac{EI}{r} \quad \text{and} \quad M = \frac{\sigma I}{c}. \tag{19.17}$$

Equating these expressions and solving for the bending stress gives

$$\sigma_b = \frac{Ec}{r},$$

where r is the radius of curvature that the rope will experience and c is the distance from the neutral axis to the outer fiber of the wire. A rope bearing against a sheave may flatten slightly, but the radius of curvature that the rope experiences is very nearly the sheave radius $D/2$. Similarly, c can be taken as $d_w/2$, yielding

$$\sigma_b = \frac{Ed_w}{D}, \tag{19.18}$$

where d_w is the wire diameter and D is the pulley diameter. Note from Eq. (19.18) that if D/d_w is very large, the bending stress will be small. Suggested minimum sheave diameters in Table 19.6 are based on a D/d_w ratio of 400. If possible, the sheaves should be designed for a larger ratio. If the ratio D/d_w is less than 200, heavy loads will often cause a permanent set of the rope. Thus, for a safe design with very long service-free life, it is recommended that $D/d_w \geq 400$.

Table 19.6: Wire rope data. *Source:* Shigley and Mitchell [1983].

Rope	Weight per height, lb/ft	Minimum sheave diameter, in.	Rope diameter, d, in.	Material	Size of outer wires	Stiffness,[a] psi	Strength,[b] psi
6 × 7 Haulage	$1.50d^2$	$42d$	$\frac{1}{4}-1\frac{1}{2}$	Monitor steel	$d/9$	14×10^6	100×10^3
				Plow steel	$d/9$	14×10^6	88×10^3
				Mild plow steel	$d/9$	14×10^6	76×10^3
6 × 19 Standard hoisting	$1.60d^2$	$26d-34d$	$\frac{1}{4}-2\frac{3}{4}$	Monitor steel	$d/13-d/16$	12×10^6	106×10^3
				Plow steel	$d/13-d/16$	12×10^6	93×10^3
				Mild plow steel	$d/13-d/16$	12×10^6	80×10^3
6 × 37 Special flexible	$1.55d^2$	$18d$	$\frac{1}{4}-3\frac{1}{2}$	Monitor steel	$d/22$	11×10^6	100×10^3
				Plow steel	$d/22$	11×10^6	88×10^3
8 × 19 Extra flexible	$1.45d^2$	$21d-26d$	$\frac{1}{4}-1\frac{1}{2}$	Monitor steel	$d/15-d/19$	10×10^6	92×10^3
				Plow steel	$d/15-d/19$	10×10^6	80×10^3
7 × 7 Aircraft	$1.70d^2$	—	$\frac{1}{16}-\frac{3}{8}$	Corrosion-resistant steel	—	—	124×10^3
				Carbon steel	—	—	124×10^3
7 × 9 Aircraft	$1.75d^2$	—	$\frac{1}{8}-1\frac{3}{8}$	Corrosion-resistant steel	—	—	135×10^3
				Carbon steel	—	—	143×10^3
19-Wire aircraft	$2.15d^2$	—	$\frac{1}{32}-\frac{5}{16}$	Corrosion-resistant steel	—	—	165×10^3
				Carbon steel	—	—	165×10^3

[a] The stiffness is only approximate; it is affected by the loads on the rope and, in general, increases with the life of the rope.
[b] The strength is based on the nominal area of the rope. The figures given are only approximate and are based on 1-in. rope sizes and 1/4-in. aircraft cable sizes.

However, these design rules are hardly ever followed for some applications, for a number of reasons. First, consider a 2-in.-diameter wire rope, such as would typically be used in the crane or in the dragline discussed in Case Study 19.1. According to the rules just stated, the sheave on the crane would have to be 80 in., or over 6 ft, in diameter. Further, since the wire rope is wound on a drum and failure can occur in the rope adjacent to the drum, such a large diameter would also be needed for the drum. Because the required motor torque is the product of the hoist rope tension and the drum radius, a very large motor would be required for lifting relatively light loads. As a result, the entire crane would be much larger and more expensive, and other design challenges would arise. For example, ground collapse below the crane would be more common, energy consumption would increase, and it would be more difficult to move cranes to different sites.

The design rules just stated are recommendations for applications where the wire rope should attain infinite life. The economic consequences of infinite-life wire rope are usually too large to bear, and smaller pulleys are usually prescribed. To prevent failures that result in property damage or personal injury, the wire ropes are periodically examined for damage. Since a broken wire will generally be easily detected and will snag a cotton cloth dragged over the rope's surface, rope life requirements are often expressed in terms of the number of broken wires allowed per length of wire rope. For example, the American Society of Mechanical Engineers [2007] requires inspections of crane wire ropes every 6 months, and if any section has more than six broken wires within one lay (revolution of a strand) of the rope, or three in any strand within a lay, the entire wire rope must be replaced. The same standard calls for pulley-to-wire diameter ratios of 12:1, which obviously results in wire rope with finite service life. Figures 19.9 and 19.10 relate the decrease in strength and service life associated with smaller pulley diameters.

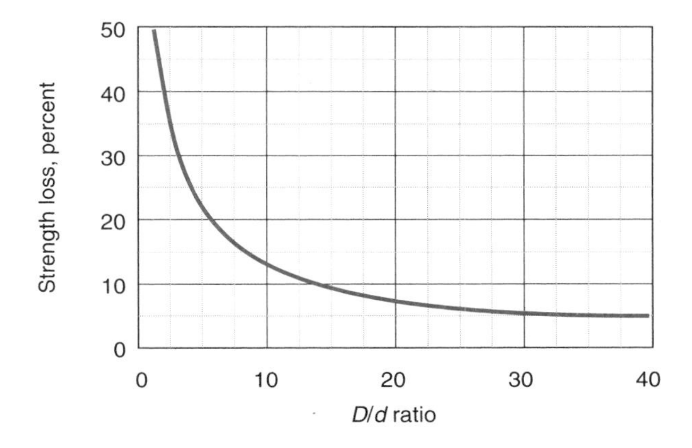

Figure 19.9: Percent strength loss in wire rope for different D/d ratios.

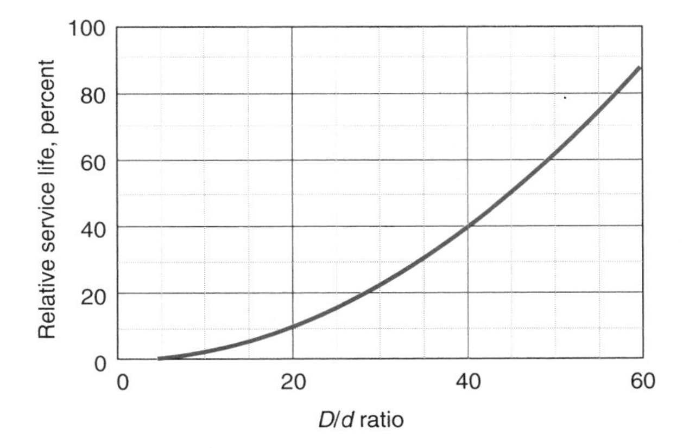

Figure 19.10: Service life for different D/d ratios.

19.5.3 Bearing Pressure

The rope stretches and rubs against the pulley, causing wear of both the rope and pulley. The amount of wear depends on

Table 19.7: Minimum safety factors for a variety of wire rope applications. Note that the use of these safety factors does not preclude a fatigue failure. *Source:* From Shigley and Mitchell [1983].

Application	Safety factor, n_s
Track cables	3.2
Guys	3.5
Mine shafts, ft	
Up to 500	8.0
1000–2000	7.0
2000–3000	6.0
Over 3000	5.0
Hoisting	5.0
Haulage	6.0
Cranes and derricks	6.0
Electric hoists	7.0
Hand elevators	5.0
Private elevators	7.5
Hand dumbwaiters	4.5
Grain elevators	7.5
Passenger elevators	
Up to 50 ft/min	7.60
50–300	9.20
300–800	11.25
800–1200	11.80
1200–1500	11.90
Freight elevators	
Up to 50 ft/min	6.65
50–300	8.20
300–800	10.00
800–1200	10.50
1200–1500	10.55
Powered dumbwaiters	
Up to 50 ft/min	4.8
50–300	6.6
300–800	8.0

the pressure on the rope in the pulley groove, or

$$p = \frac{2F_t}{dD}, \tag{19.19}$$

where d is the rope diameter. The pressure obtained from Eq. (19.19) should be less than the maximum pressure obtained from Table 19.8 for various pulley materials and types of rope.

19.5.4 Fatigue

For the rope to have a long life, the total force, F_t, must be less than the fatigue allowable force, F_f, where

$$F_f = \frac{S_u dD}{2000}. \tag{19.20}$$

The ultimate strength given by Eq. (19.20) for three materials is

$$\begin{array}{ll} \text{Monitor plow steel} & 240 \le S_u \le 280 \times 10^3 \text{ psi} \\ \text{Plow steel} & 210 \le S_u \le 240 \times 10^3 \text{ psi} \\ \text{Mild plow steel} & 180 \le S_u \le 210 \times 10^3 \text{ psi} \end{array} \tag{19.21}$$

Table 19.8: Maximum allowable bearing pressures for various sheave materials and types of rope *Source:* From Shigley and Mitchell [1983].

	Maximum bearing pressure for sheave material			
Rope	Cast iron [a]	Cast steel [b]	Chilled cast iron [c]	Manganese steel [d]
Regular lay				
6 × 7	300	550	650	1470
6 × 19	480	900	1100	2400
6 × 37	585	1075	1325	3000
8 × 19	680	1260	1550	3500
Lang lay				
6 × 7	350	600	715	1650
6 × 19	550	1000	1210	2750
6 × 37	660	1180	1450	3300

[a] For minimum HB = 125.
[b] 0.30–0.40% carbon; HB (min.) = 160.
[c] Use only with uniform surface hardness.
[d] For high speeds with balanced sheaves having ground surfaces.

Example 19.4: Analysis of Wire Rope

Given: Lighting fixtures in a large theater are to be raised and lowered by a wire rope, and need to be moved a maximum of 90 ft. The maximum load to be hoisted is 3000 lb at a velocity not exceeding 2 ft/s and an acceleration of 4 ft/s^2. Use 1-in. plow steel in the form of 6 × 19 standard hoisting ropes.

Find: Determine the safety factor while considering

(*a*) Tensile stress

(*b*) Bending stress

(*c*) Bearing pressure

(*d*) Fatigue

Solution: From Table 19.6 for 6 × 19 standard hoisting wire rope, assuming the use of only one rope,

$$F_r = 1.60d^2h_2 = 1.60(1)^2(90) = 144.0 \text{ lb/rope},$$

$$F_w = W_{\max} = 3000 \text{ lb}.$$

The force due to acceleration is

$$F_a = ma = \frac{W}{g}a = \frac{(3000+144)(4)}{32.3} = 389.3 \text{ lb}.$$

The total force on the rope is

$$F_t = F_w + F_r + F_a = 3000 + 144 + 389.3 = 3533 \text{ lb}.$$

(*a*) *Tensile stress:*

$$\sigma_t = \frac{F_t}{A_m} = \frac{3533}{0.38} = 9297 \text{ psi}.$$

From Table 19.6 for 6 × 19 standard hoisting wire rope made of plow steel, $\sigma_{\text{all}} = 93,000$ psi. The safety factor is

$$n_s = \frac{\sigma_{\text{all}}}{\sigma_t} = \frac{93,000}{9297} = 10.0.$$

From Table 19.7, the recommended safety factor for hoisting applications is 5.0. Thus, one rope is sufficient as far as the tensile stress is concerned.

(*b*) *Bending stress:* From Table 19.6, the minimum pulley diameter for 6 × 19 standard hoisting wire rope is $26d$ to $34d$. No design constraints have been given regarding the pulley, so assign the conservative value of $D = 34d = 34$ in. Also, from the same table, the wire diameter should be between $d/13$ and $d/16$. Choose $d_w = d/16 = 1/16$ in., so that

$$\frac{D}{d_w} = \frac{34}{1/16} = 544.$$

Permanent set should not be a concern, since $D/d_w \geq 400$. From Table 19.6, the modulus of elasticity is 12×10^6 psi. The bending stress is, from Eq. (19.18),

$$\sigma_b = E\frac{d_w}{D} = \frac{12 \times 10^6}{544} = 22.06 \times 10^3 \text{ psi.}$$

The safety factor due to bending is

$$n_s = \frac{\sigma_{\text{all}}}{\sigma_b} = \frac{93,000}{22,060} = 4.22.$$

This safety factor is less than the 5.0 recommended, but the safety factor is for static loading. Changing the material from plow steel to monitor steel would produce a safety factor of 4.81, closer to 5.

Note that increasing the number of ropes will not alter the results, since the rope will still be wrapped around the same sized sheave. The safety factor was not a constraint in this example, but the obvious solution is to use an even larger pulley.

(*c*) *Bearing pressure:* From Eq. (19.19),

$$p = \frac{2F_t}{dD} = \frac{2(3533)}{(1)(34)} = 207.8 \text{ psi.}$$

From Table 19.8 for a 6 × 19 Lang lay for a cast steel pulley, p_{all} = 1000 psi. The safety factor is

$$n_s = \frac{p_{\text{all}}}{p} = \frac{1000}{207.8} = 4.81.$$

(*d*) Fatigue: For monitor steel the ultimate strength is 280×10^3 psi. The allowable fatigue force from Eq. (19.20) is

$$F_f = \frac{S_u dD}{2000} = \frac{(280 \times 10^3)(1)(34)}{2000} = 4760 \text{ lb.}$$

The safety factor is

$$n_s = \frac{F_f}{F_t} = \frac{4760}{3535} = 1.35.$$

Thus, fatigue failure is the most likely failure to occur, since it produced the smallest safety factor. Using four ropes instead of one would produce a safety factor greater than 5.

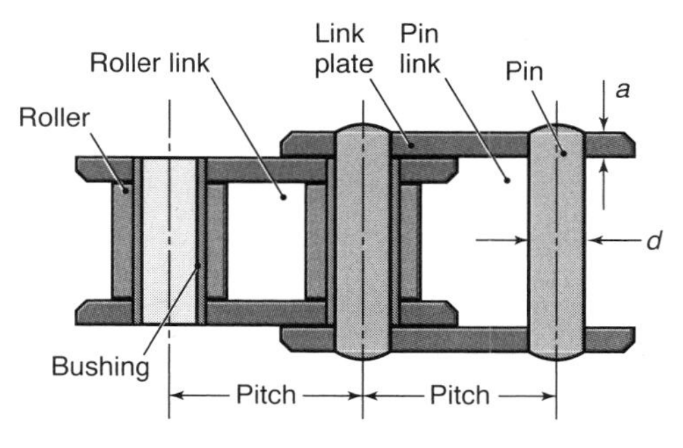

Figure 19.11: Various parts of rolling chain.

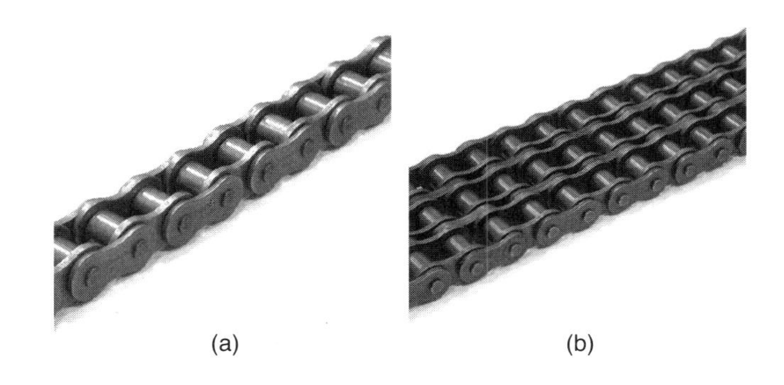

Figure 19.12: Typical rolling chain. (a) One-strand rolling chain; (b) three-strand chain.

19.6 Rolling Chains

Rolling chains are used to transmit power between two sprockets rotating in the same plane. The machine element that it most resembles is a timing belt (shown in Fig. 19.3). The major advantage of using a rolling chain compared to a belt is that rolling chains do not slip. Large center distances can be dealt with more easily with rolling chains with fewer elements and in less space than with gears. Rolling chains also have high efficiency. No initial tension is necessary and shaft loads are therefore smaller than with belt drives. The only maintenance required after careful alignment of the elements is periodic lubrication, and with proper lubrication, a long life can be attained.

19.6.1 Operation of Rolling Chains

Figure 19.11 shows the various parts of a rolling chain with pins, bushings, rollers, and link plates. The rollers turn on bushings that are press fitted to the inner link plates. The pins are prevented from rotating in the outer plates by the press-fit assembly. Table 19.9 gives dimensions for standard sizes. The size range given in Table 19.9 is large, and multiple strands can be used (see Fig. 19.12), so that chains can be used for both large and small power transmission levels. A large reduction in speed can be obtained with rolling chains if desired. The tolerances for a chain drive are larger than for gears, and the installation of a chain is relatively easy.

The minimum wrap angle of the chain on the smaller sprocket is 120°. A smaller wrap angle can be used on idler sprockets, which are used to adjust the chain slack where the center distance is not adjustable. Horizontal drives (the line connecting the axes of sprockets is parallel to the ground) are

Table 19.9: Standard sizes and strengths of rolling chains.

Chain number[a]	Pitch, p_t, in.	Roller Diameter, in.	Roller Width, in.	Pin diameter, d, in.	Link plate thickness, a, in.	Average breaking strength, S_u, lb	Weight per foot, lb
[b]25	1/4	0.130	1/8	0.0905	0.030	875	0.084
[b]35	3/8	0.200	3/16	0.141	0.050	2100	0.21
[b]41	1/2	0.306	1/4	0.141	0.050	2000	0.28
40	1/2	5/16	5/16	0.156	0.060	3700	0.41
50	5/8	2/5	3/8	0.200	0.080	6100	0.68
60	3/4	15/32	1/2	0.234	0.094	8500	1.00
80	1	5/8	5/8	0.312	0.125	14,500	1.69
100	$1\frac{1}{4}$	3/4	3/4	0.375	0.156	24,000	2.49
120	$1\frac{1}{2}$	7/8	1	0.437	0.187	34,000	3.67
140	$1\frac{3}{4}$	1	1	0.500	0.219	46,000	4.93
160	2	$1\frac{1}{8}$	$1\frac{1}{4}$	0.562	0.250	58,000	6.43
180	$2\frac{1}{4}$	$1\frac{13}{32}$	$1\frac{13}{32}$	0.687	0.281	76,000	8.70
200	$2\frac{1}{2}$	$1\frac{9}{16}$	$1\frac{1}{2}$	0.781	0.312	95,000	10.51
240	3	$1\frac{7}{8}$	$1\frac{7}{8}$	0.937	0.375	130,000	16.90

[a] The pitch can be obtained from the chain number by taking the left digits and dividing by 8.
[b] Without rollers.

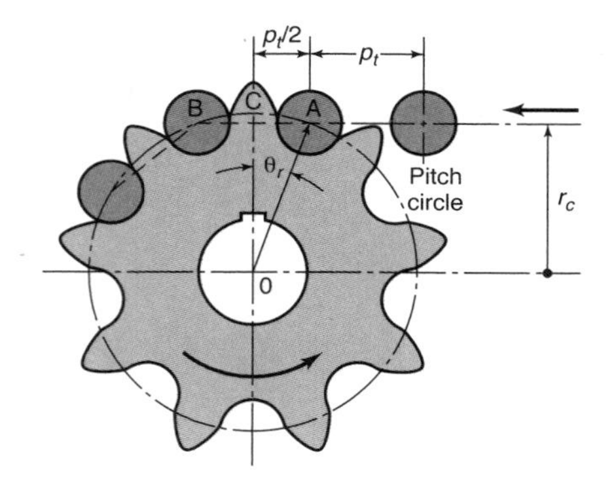

Figure 19.13: Chordal rise in rolling chains. Note that the chain link travels upwards as well as horizontally when moving from position A to position B.

recommended; vertical drives (the line connecting the axes of sprockets is perpendicular to the ground) are less desirable. Vertical drives, if used, should be used with idlers to prevent the chain from sagging and to avoid disengagement from the lower sprocket.

19.6.2 Kinematics

The velocity ratio, comparable to the gear ratio given in Eq. (14.19), is

$$g_r = \frac{d_2}{d_1} = \frac{\omega_1}{\omega_2} = \frac{N_2}{N_1}, \qquad (19.22)$$

where N is the number of teeth in the sprocket, ω is the angular velocity in rad/s, and d is the diameter. For a one-step transmission it is recommended that $g_r < 7$. Values of g_r between 7 and 10 can be used at low speed (below 650 ft/min).

19.6.3 Chordal Rise

An important factor affecting the smoothness of rolling chain drive operation, especially at high speeds, is **chordal rise**, shown in Fig. 19.13. From right triangle 0CA

$$r_c = r\cos\theta_r. \qquad (19.23)$$

The chordal rise while using Eq. (19.23) gives

$$\Delta r = r - r_c = r\left(1 - \cos\theta_r\right) = r\left[1 - \cos\left(\frac{180}{N}\right)\right], \qquad (19.24)$$

where N is the number of teeth in the sprocket. Note also from triangle 0CA

$$\sin\theta_r = \frac{p_t/2}{r} \qquad \text{or} \qquad p_t = 2r\sin\theta_r = D\sin\theta_r. \qquad (19.25)$$

19.6.4 Chain Length

The number of links is

$$\frac{L}{p_t} = \frac{2c_d}{p_t} + \frac{N_1+N_2}{2} + \frac{(N_2-N_1)^2}{4\pi^2\left(\frac{c_d}{p_t}\right)}, \qquad (19.26)$$

where c_d is the center distance between sprockets. It is normally recommended that c_d/p_t lie between 30 and 50 pitches. If the center distance per pitch is not given, the designer can fix c_d/p_t and calculate L/p_t from Eq. (19.26). The next larger even integer L/p_t should be chosen. With L/p_t as an integer, the center distance per pitch becomes

$$\frac{c_d}{p_t} = A + \sqrt{A^2 - \frac{B^2}{2}}, \qquad (19.27)$$

where

$$A = \frac{1}{4}\left(\frac{L}{p_t} - \frac{N_1+N_2}{2}\right), \qquad (19.28)$$

$$B = \frac{N_2-N_1}{2\pi}. \qquad (19.29)$$

The value of c_d/p_t obtained from Eq. (19.27) should be decreased by about 1% to provide slack in the nondriving chain strand.

The chain velocity is

$$u = N_a p_t N, \qquad (19.30)$$

Table 19.10: Service factor, a_1, for rolling chains.

	Type of input power		
Type of driven load	Internal combustion engine with hydraulic drive	Electric motor or turbine	Internal combustion engine with mechanical drive
Smooth	1.0	1.0	1.2
Moderate shock	1.2	1.3	1.4
Heavy shock	1.4	1.5	1.7

Table 19.11: Multiple-strand factor for rolling chains.

Number of strands	Multiple-strand factor, a_2
1	1.0
2	1.7
3	2.5
4	3.3
5	3.9
6	4.6

where

N_a = angular speed, rev/min

p_t = chain pitch, in.

N = number of teeth in the sprocket

19.6.5 Power Rating

The required power rating of a chain is given by

$$h_{pr} = h_p \frac{a_1}{a_2}, \tag{19.31}$$

where h_p is the power that is transmitted or input, a_1 is the service factor obtained from Table 19.10 and a_2 is the strand factor obtained from Table 19.11.

Rolling chains are available in a wide variety of sizes; Fig. 19.14 and Design Procedure 19.2 provide some guidelines for selection of rolling chains. Note that Fig. 19.14 has four scales, depending on the number of strands in the chain (see Fig. 19.12). For each type of roller chain, there is a characteristic shape to the power curve, where the power rating increases up until a certain sprocket size, and then decreases. This is attributed to a change in failure mode; at low speeds, link plate fatigue will be the dominant failure mode, while at higher speeds, wear of the pins becomes dominant.

Table 19.12 provides power ratings for selected standard-sized roller chains. Similar data can be readily obtained for other chain configurations from manufacturers' literature and web sites. The four types of lubrication given in Table 19.12 are

- Type A – Manual or drip lubrication, oil applied periodically with brush or spout can or applied between the link plate edges with a drop lubricator.
- Type B – Oil bath or oil slinger, where the oil level is maintained in a casing at a predetermined height, and the chain is immersed into the bath during at least a part of its travel.
- Type C – Oil stream lubrication, where oil is supplied by a circulating pump inside the chain loop or lower span.

Example 19.5: Power and Force in Rolling Chains

Given: A four-strand, No. 25 rolling chain transmits power from a 25-tooth driving sprocket that turns at 900 rpm. The speed ratio is 4:1.

Find: Determine the following:

(*a*) Power that can be transmitted for this drive

(*b*) Tension in the chain

(*c*) Chain length if center distance is about 10 in.

Solution:

(*a*) From Table 19.12 for a smaller sprocket of 25 teeth and a speed of 900 rpm, the power rating is 1.21 hp and bath lubrication is required. From Table 19.11 for four strands, a_2 = 3.3. Nothing is mentioned about the type of input power or drive load, so assume that a_1 = 1. From Eq. (19.31), the power that can be transmitted is

$$h_p = h_{pr}\frac{a_2}{a_1} = (1.21)\left(\frac{3.3}{1}\right) = 3.993 \approx 4.0 \text{ hp}.$$

(*b*) From Table 19.9 for No. 25 chain, the pitch is 0.25 in. The velocity can be calculated from Eq. (19.30) as

$$u = N_a p_t N = (900)(0.25)(25) = 5625 \text{ in./min}$$

or $u = 7.81$ ft/s. The power that can be transmitted is 4 hp = 2200 ft-lb/s. Therefore, the tension in the chain is

$$P = \frac{h_p}{u} = \frac{2200}{7.81} = 281.7 \text{ lb}.$$

(*c*) The number of teeth on the larger sprocket given a speed ratio of 4 is 4(25) = 100 teeth. Note that for a center distance of 10 in.,

$$\frac{c_d}{p_t} = \frac{10}{0.25} = 40,$$

which is between the 30 and 50 pitches recommended. From Eq. (19.26),

$$\frac{L}{p_t} = 2\left(\frac{c_d}{p_t}\right) + \frac{N_1+N_2}{2} + \frac{(N_2-N_1)^2}{4\pi^2\left(\frac{c_d}{p_t}\right)}$$

$$= 80 + \frac{25+100}{2} + \frac{(100-25)^2}{4\pi^2(40)} = 146.1.$$

Since it is good practice to specify the next largest even number of links, L is chosen as 148 pitches, or 37 in.

19.6.6 Silent Chain

An interesting type of chain is the **inverted tooth** or **silent chain**, shown in Fig. 19.15, and made up of stacked rows of flat, tooth shaped links that mesh with sprockets having compatible tooth spaces, much the way gear teeth mesh (see Fig. 14.9). Because of the geometry advantages, these chains are much quieter than conventional roller chains, especially at high speeds, but they are somewhat more expen-

Table 19.12: Power rating of selected standard roller chains, in horsepower.

Chain no.	No. of teeth[a]	Maximum speed of small sprocket (rpm)										
		50	100	300	500	900	1500	2100	3000	4000	5000	6000
25	11	0.03	0.06	0.19	0.30	0.53	0.87	1.20	1.69	1.38	0.99	0.75
	12	0.04	0.07	0.20	0.33	0.58	0.95	1.31	1.84	1.57	1.12	0.86
	15	0.05	0.09	0.25	0.41	0.72	1.18	1.63	2.30	2.20	1.57	1.20
	18	0.05	0.11	0.30	0.49	0.87	1.42	1.96	2.76	2.89	2.07	1.57
	20	0.06	0.12	0.34	0.55	0.97	1.58	2.18	3.07	3.38	2.42	1.84
	22	0.07	0.13	0.37	0.60	1.06	1.73	2.40	3.37	3.90	2.79	2.12
	25	0.08	0.15	0.42	0.69	1.21	1.97	2.72	3.84	4.73	3.38	2.57
	28	0.08	0.16	0.47	0.77	1.35	2.21	3.05	4.30	5.60	4.01	3.05
	30	0.09	0.18	0.50	0.82	1.45	2.37	3.27	4.60	6.07	4.45	3.38
	35	0.11	0.21	0.59	0.96	1.69	2.76	3.81	5.37	7.08	5.60	4.26
	40	0.12	0.23	0.67	1.10	1.93	3.15	4.36	6.14	8.09	6.85	4.91
	45	0.14	0.26	0.76	1.24	2.17	3.55	4.90	6.90	9.10	8.17	1.38
35	11	0.11	0.22	0.62	1.02	1.80	2.93	4.05	2.94	1.91	1.37	1.04
	12	0.12	0.24	0.68	1.11	1.96	3.20	4.42	3.35	2.17	1.56	1.18
	15	0.15	0.30	0.85	1.39	2.45	4.00	5.52	4.68	3.04	2.17	1.65
	18	0.18	0.36	1.02	1.67	2.94	4.80	6.63	6.15	3.99	2.86	2.17
	20	0.20	0.40	1.14	1.86	3.26	5.33	7.36	7.20	4.68	3.35	2.55
	22	0.22	0.44	1.25	2.04	3.59	5.86	8.10	8.31	5.40	3.86	2.94
	25	0.25	0.50	1.42	2.32	4.08	6.66	9.20	10.07	6.54	4.68	3.56
	28	0.29	0.55	1.58	260	4.57	7.46	10.31	11.93	7.75	5.55	–
	30	0.31	0.59	1.70	2.79	4.90	8.00	11.05	13.23	8.59	6.15	–
	35	0.36	0.69	1.99	3.25	5.71	9.33	12.89	16.67	10.83	0.34	–
	40	0.41	0.79	2.32	3.71	6.53	10.66	14.73	20.37	11.04	–	–
	45	0.46	0.89	2.56	4.18	7.35	11.99	16.57	23.33	3.11	–	–
50	11	0.52	1.00	2.88	4.70	8.27	7.33	4.42	2.59	1.68	1.20	0.92
	12	0.56	1.09	3.14	5.13	9.02	8.35	5.04	2.95	1.92	1.37	1.04
	15	0.70	1.37	3.93	6.41	11.28	11.67	7.05	4.13	2.68	1.92	–
	18	0.84	1.64	4.71	7.70	13.53	15.34	9.26	5.42	3.52	0.05	–
	20	0.94	1.82	5.24	8.55	15.04	17.97	10.85	6.35	4.13	–	–
	22	1.03	2.01	5.76	9.41	16.54	20.73	12.52	7.33	4.76	–	–
	25	1.17	2.28	6.55	10.69	18.79	25.11	15.16	8.88	–	–	–
	28	1.31	2.55	7.33	11.97	21.05	29.77	17.97	10.52	–	–	–
	30	1.41	2.74	7.86	12.83	22.55	33.01	19.93	11.67	–	–	–
	35	1.64	3.19	9.16	14.97	26.31	41.60	25.11	0.94	–	–	–
	40	1.88	3.65	10.47	17.10	30.07	49.11	30.68	–	–	–	–
	45	2.11	4.10	11.78	19.24	33.83	55.24	36.61	–	–	–	–
80	11	2.07	4.03	11.56	18.87	22.97	10.76	6.53	3.77	2.45	–	–
	12	2.26	4.39	12.61	20.59	26.17	12.26	7.45	4.30	2.79	–	–
	15	2.82	5.49	15.76	25.74	36.58	17.14	10.41	6.01	–	–	–
	18	3.39	6.59	18.91	30.88	48.08	22.54	13.69	7.90	–	–	–
	20	3.76	7.32	21.01	34.32	56.32	26.40	16.03	–	–	–	–
	22	4.14	8.05	23.12	37.75	64.97	30.45	18.49	–	–	–	–
	25	4.70	9.15	26.27	42.89	75.42	36.88	20.64	–	–	–	–
	28	5.27	10.25	29.42	48.04	84.47	43.72	–	–	–	–	–
	30	5.54	10.96	31.52	51.47	90.50	48.49	–	–	–	–	–
	35	6.58	12.81	36.78	60.05	105.58	61.10	–	–	–	–	–
100	11	3.96	7.71	22.14	36.15	27.46	12.87	8.29	4.51	–	–	–
	12	4.32	8.41	24.15	39.44	31.29	14.66	9.45	5.14	–	–	–
	15	5.41	10.51	30.19	49.30	43.73	20.50	13.20	–	–	–	–
	18	6.49	12.62	36.23	59.15	57.48	26.94	17.35	–	–	–	–
	20	7.21	14.02	40.25	65.73	67.32	31.55	20.32	–	–	–	–
	22	7.93	15.42	44.28	72.30	77.67	36.40	23.45	–	–	–	–
	25	9.01	17.52	50.31	82.16	94.09	44.09	–	–	–	–	–
	28	10.09	19.63	56.35	92.02	111.5	52.26	–	–	–	–	–
	30	10.81	21.03	60.38	98.59	123.7	56.57	–	–	–	–	–
	35	12.61	24.53	70.44	115.0	155.9	–	–	–	–	–	–
160	11	15.02	29.23	83.91	96.58	39.99	18.74	12.07	–	–	–	–
	12	16.39	31.88	91.54	110.0	45.57	21.36	13.76	–	–	–	–
	15	20.49	39.86	114.4	153.8	63.69	29.85	–	–	–	–	–
	18	24.59	47.83	137.3	202.2	83.72	–	–	–	–	–	–
	20	27.32	53.14	152.6	236.8	98.05	–	–	–	–	–	–
	22	30.05	58.45	167.8	273.2	113.1	–	–	–	–	–	–
	25	34.15	66.43	190.7	311.4	137.0	–	–	–	–	–	–
	28	38.24	74.40	213.6	348.8	162.4	–	–	–	–	–	–
	30	40.98	79.71	228.9	373.7	180.1	–	–	–	–	–	–
	35	47.81	93.00	267.0	436.0	112.6	–	–	–	–	–	–
200	11	27.54	53.58	153.84	115.46	47.81	22.40	–	–	–	–	–
	12	30.05	58.45	167.82	131.56	54.48	25.52	–	–	–	–	–
	15	37.56	73.07	209.78	183.86	76.13	–	–	–	–	–	–
	20	50.08	97.42	279.70	283.07	117.21	–	–	–	–	–	–
	25	62.60	121.78	349.83	395.60	163.81	–	–	–	–	–	–

[a] The use of fewer than 17 teeth is possible, but should be avoided when practical.

Required Lubrication:

- Manual/drip
- Bath or disc
- Oil stream

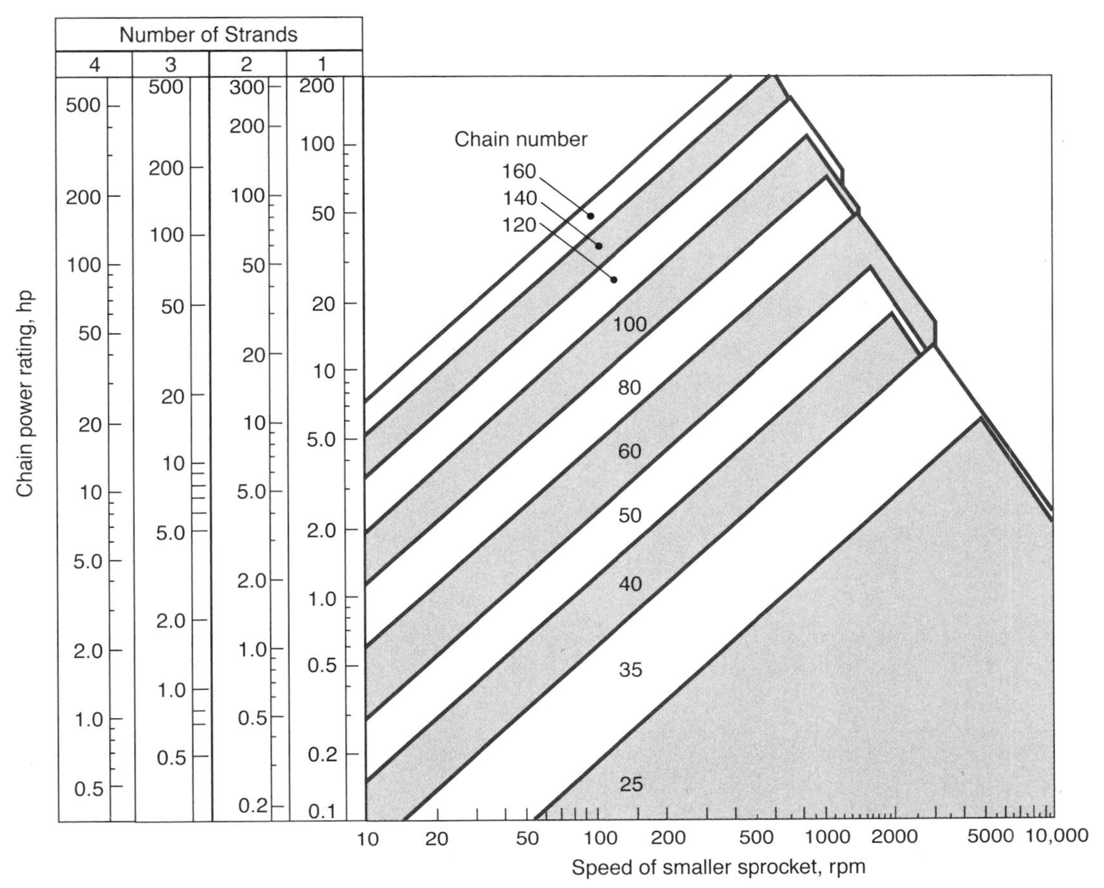

Figure 19.14: Design guideline for standard roller chains.

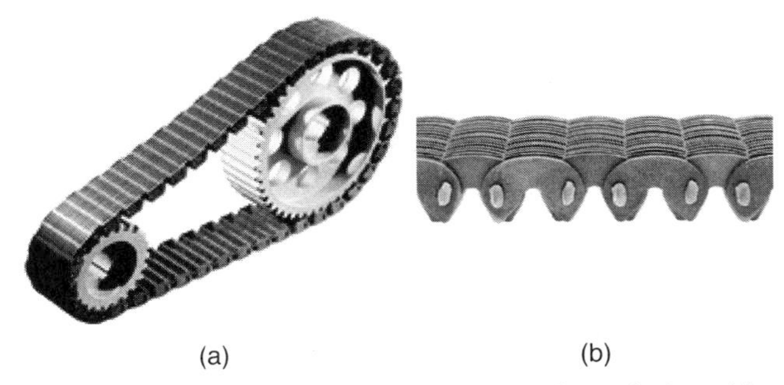

Figure 19.15: A silent chain drive. (a) Silent chain with sprockets; (b) detail of silent chain links. *Source:* Courtesy of Ramsey Products Corp.

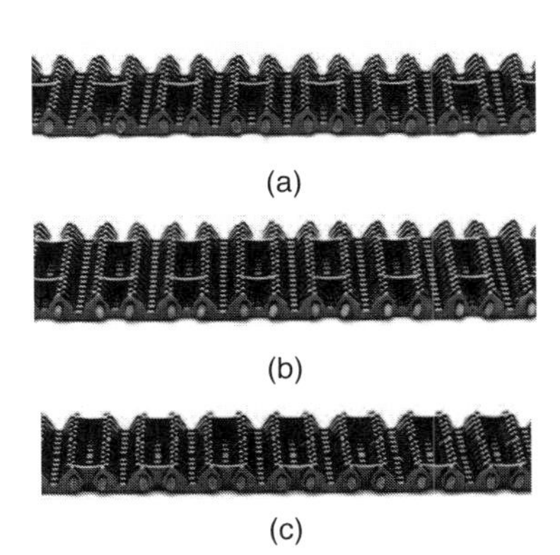

Figure 19.16: The use of guide links in silent chains. (a) One guide link in center of chain; (b) two center guide links; (c) two side guide links. *Source:* Courtesy of Ramsey Products Corp.

sive. Washers or spacers may be present in some chain constructions, especially for conveyor applications. All of these components are held together by riveted pins located in each chain joint. Although all silent chains have these basic features, there are still many different styles, designs, and configurations. The width of a silent chain can be varied according to horsepower constraints. Silent chains are available for power transmission, as considered in this section, but they can also be modified to incorporate links that provide a stable and wear resistant surface for conveyor applications.

Guide links are used to maintain tracking on sprockets, as shown in Fig. 19.16. The guide links may be located within the chain (center guide) or along the outer edges of the chain (side guide). A chain with two rows of guide links within the chain is referred to as *two center guide*. In many applications, any of these guide types will be satisfactory but it is essential that sprockets be selected with the same guide type as the chain.

Most silent chains have teeth that engage sprockets on only one side of the chain. However, duplex silent chain designs are available that have teeth on both sides and are designed for use in serpentine drives, where sprockets are driven from both sides of the chain.

Silent chains are able to transfer higher powers at higher

speeds than roller chains. Essentially, the width of the chain can be increased to accommodate any power. Because they do not have a chordal rise, there is less vibration and longer sprocket life as well. However, the main advantage of silent chains is that they are much quieter than conventional rolling chains.

Silent chains can be made in large widths and used as conveyor belts. In such cases, it is common to use a spacer between links to produce a more open belt.

Design Procedure 19.2: Design of Chain Drives

This Design Procedure presents a method for designing power transmission devices consisting of chains and sprockets. While the approach is intended for roller chains, the Design Procedure can be used for silent chains, with caveats as noted. It will be assumed that operating conditions and design constraints are adequately described. For the purposes of this Design Procedure, the power transmitted (or chain force and speed), power source, speed ratio, and loading environment need to be known, or at least be somewhat constrained.

1. Obtain the service factor from Table 19.10. Calculate the chain's required power rating from Eq. (19.31), taking $a_2 = 1.0$.
2. Select a chain size from Fig. 19.14 using the required power rating and the small sprocket speed. Note that using the fewest number of chain strands while satisfying power requirements usually results in the most economical design.
3. Obtain the strand factor, a_2, from Table 19.11.
4. The required power rating, given by Eq. (19.31), needs to be recalculated if a multiple strand chain is to be used.
5. Referring to Table 19.12, identify the column of the table that corresponds to the small sprocket's speed. Reading down from the top, find the number of teeth in the smaller sprocket that produces the required modified power rating, h'_{pr}. This is the minimum number of teeth that are required for the application. Larger sprockets can be used if desired.
6. If a multiple strand chain is being considered, record the modified power rating from Table 19.12, and multiply by the strand factor to obtain the chain's power rating.
7. Note the required lubrication method in Table 19.12. Variation from the lubrication approach may compromise chain longevity.
8. The number of teeth on the larger sprocket can be calculated from the desired velocity ratio by using Eq. (19.22).
9. If the center distance has not been prescribed, it can be estimated by recognizing that c_d/p_t should be between 30 and 50, although larger lengths can be allowed if chain guides are incorporated into the design. If the center distance exceeds space limitations, increase the number of strands or select the next largest pitch chain and return to Step 4.
10. The number of links in the chain can be calculated from Eq. (19.26), rounded up to the next highest even integer.

Example 19.6: Chain Drive Design

Given: A chain will transmit power from a 10-hp motor to a drive roller on a belt conveyor. The motor is connected to a speed reducer so that the input speed is 100 rpm, and the desired conveyor drive roller speed is 25 rpm. The conveyor has some starts and stops, so it should be considered to have moderate shock. The center distance of the shafts should be around 50 in.

Find: Select a roller chain and sprockets for this application.

Solution: This solution will closely follow Design Procedure 19.2.

1. Referring to Table 19.11, note that for an electric motor with a driven load with moderate shock, the service factor is $a_1 = 1.3$. The required power rating, assuming $a_2 = 1$ is obtained from Eq. (19.31) as

$$h_{pr} = h_p \frac{a_1}{a_2} = (10)(1.3) = 13 \text{ hp}.$$

2. Noting that the smaller sprocket will have a speed of 100 rpm, Fig. 19.14 suggests that a single strand of standard No. 100 chain can be used for this application. Note that a No. 120 chain may also lead to a reasonable design, and may be worth investigating, but this solution will continue with a No. 100 chain.
3. From Table 19.11, the strand factor is $a_2 = 1.0$.
4. A single strand chain is being analyzed, so that the power rating of 13 hp can still be used. Note that the power rating would need to be recalculated if a multiple strand chain was selected.
5. Referring to Table 19.12, for a No. 100 chain with a small sprocket speed of 100 rpm, 20 teeth are required in order to exceed the modified power rating of 13 hp.
6. Table 19.12 also indicates that a bath-type of lubrication is required, and will need to be incorporated into the chain system.
7. From Eq. (19.22), the number of teeth on the larger sprocket can be obtained as:

$$g_r = \frac{\omega_1}{\omega_2} = \frac{N_2}{N_1},$$

$$\frac{100}{25} = \frac{N_2}{20},$$

or $N_2 = 80$.

8. Note from Table 19.9 that the pitch of No. 100 chain is 1.25 in, so that $\frac{c_d}{p_t} = \frac{50}{1.25} = 40$. This is within the standard range, so no further modification (changing chain size, using more strands, etc.) is required.
9. From Eq. (19.26),

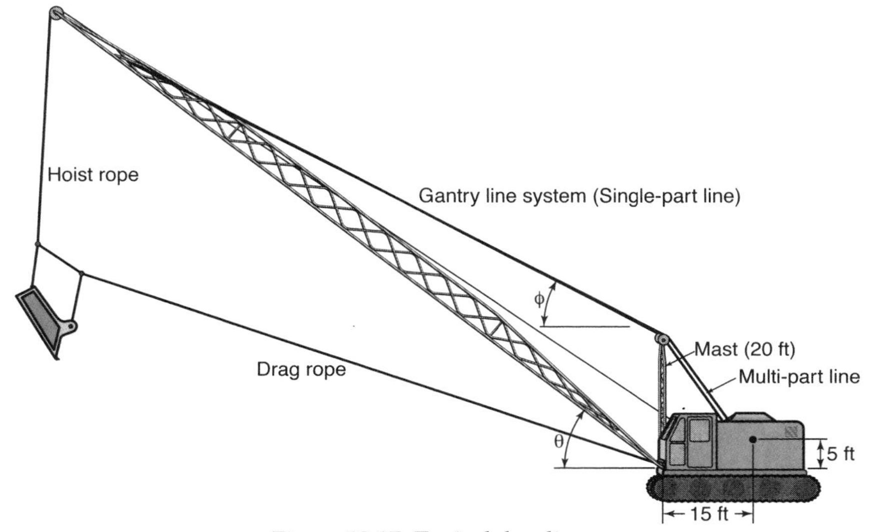

Figure 19.17: Typical dragline.

$$\begin{aligned}\frac{L}{p_t} &= \frac{2c_d}{p_t} + \frac{N_1+N_2}{2} + \frac{(N_2-N_1)^2}{4\pi^2\left(\frac{c_d}{p_t}\right)} \\ &= \frac{2(50)}{1.25} + \frac{20+80}{2} + \frac{(80-20)^2}{4\pi^2\left(\frac{50}{1.25}\right)} = 132.2.\end{aligned}$$

or $L = 165.3$. Since it is good practice to specify an even integer of links, a value of $L/p_t = 134$ is selected. This would result in an actual center distance of $c_d = 51.1$ in. as obtained from Eq. (19.27). Note that the center distance should be reduced by 1% to provide some slack, so that an actual center distance of around 50.6 in. should be used.

Case Study 19.1: Design of a Gantry for a Dragline

A dragline, often used for mining or dredging operations, uses a large bucket to remove material and transport it elsewhere, usually into a trailer or a train boxcar. The gantry line system is the portion of the dragline that supports the boom and fixes the boom angle, as shown in Fig. 19.17. The boom typically weighs over 10,000 lb, is around 100 ft long, and can be considered to have its center of gravity at its geometric center. The dragline tips if the moment from the hoist rope and the boom exceeds the moment from the dragline about the tipping point. Tipovers are to be avoided, of course, but it should be recognized that the load applied to the gantry cable cannot be larger than that resulting from a tipover condition.

Figure 19.17 shows the dimensions of a typical dragline. The gantry line lengths shown were selected so that the 20-ft mast is vertical at a boom angle of 30°. Since draglines have booms, many people confuse them with cranes. There are significant differences, however. Dragline load is limited by the size of the bucket, whereas a crane can attempt to lift extremely large loads. Further, a dragline will operate at a set boom angle for extended periods of time and will rarely, if ever, operate at very high or very low boom angles. Therefore, the range of boom angles that must be achieved is somewhat limited.

Just as with cranes, however, the gantry line system serves to eliminate bending forces on the boom. Thus, the gantry line must attach at the tip of the boom, known as the *boom point*. Attaching the gantry line to the crane's superstructure near the boom pivot point will result in extremely large forces, as can be seen from moment equilibrium. A mast is used to offset the gantry line system and obtain reasonably low forces in the gantry line.

A number of clever design features have been incorporated into draglines in the past (Fig. 19.17). The lines closest to the crane superstructure have multiple pulleys (and hence are called a *multiple-part line*) to share the load and reduce the stress on the cable. Above the multiple-part line is a single line attached by proper couplings to the boom. Given the operating characteristics of a dragline, this single line is never wound over a pulley and its size is determined by the rated cable strength and does not have to be reduced for fatigue effects (see Fig. 19.10).

Because the mast rotates as the boom angle changes, the load supported by the gantry lines will change with the boom angle. The gantry line tension is highest at a boom angle of 0°. Draglines are not operated at such low boom angles because the lifting capacity is severely limited, but analyzing a boom angle of 0° is important. Draglines are shipped in multiple parts and assembled at the construction site, so that the boom needs to be lifted, at least for the first time that the dragline is placed into service.

A number of equalizer pulleys (a multiple-part line) are used in Fig. 19.17 so that the load in the gantry line is reduced. The load in the gantry line depends on the number of pulleys or parts in the line, each of which supports an equal share of the load. The earlier discussion (Section 19.5.2) regarding sheave sizing is relevant. If the pulley and drum are too large, the motor capacity will be excessive. Per industry standard requirements, pulley and drum diameters of $12d$ are to be chosen. These diameters reduce gantry line strength by approximately 12% (Fig. 19.9), but a maintenance procedure of checking for broken wires in the wire rope must still be followed. This leads to a very safe system, since draglines rarely change their boom angle, so the gantry line does not run over pulleys often.

19.7 Summary

Belts and ropes are machine elements (like brakes and clutches) that use friction as a useful agent, in contrast to other machine elements in which friction is to be kept as low as possible. Belts, ropes, and chains provide a convenient means for transferring power from one shaft to another. This chapter discussed flat, synchronous, and V-belts. All flat belts are subject to slip, that is, relative motion between the pulley surface and the adjacent belt. For this reason, flat belts must be kept under tension to function and require tensioning devices. One major advantage of V-belts over flat belts is that the wedging action of V-belts increases the normal force from dN to $dN/\sin\beta$, where β is the sheave angle. V-belts also can transfer much higher power than a similarly sized flat belt.

Wire ropes are used when power must be transmitted over very long center distances. Wire rope is widely used in hoisting applications including cranes and elevators. The major advantage of using rolling chains over flat or V-belts is that rolling chains do not slip. Another advantage of rolling chains over belt drives is that no initial tension is necessary and thus the shaft loads are smaller. The required length, power rating, and modes of failure were considered for belts, ropes, and rolling chains.

Key Words

core center of wire rope, mainly intended to support outer strands

flat belts power transmission device that consists of loop of rectangular cross section placed under tension between pulleys

inverted tooth chain see *silent chain*

lay type of twist in wire rope (regular or Lang); distance for strand to revolve around wire rope

rolling chains power transmission device using rollers and links to form continuous loop, used with sprockets

sheaves grooved pulleys that V-belts run in

silent chain a chain that consists of specially formed links that mesh with a gear-shaped sprocket

strands grouping of wire used to construct wire rope

synchronous belt flat belt with series of evenly spaced teeth on inside circumference, intended to eliminate slip and creep

timing belt same as *synchronous belt*

V-belt power transmission device with trapezoidal cross section placed under tension between grooved sheaves

wire rope wound collection of strands

Summary of Equations

Belts:

Belt length:

$$L = \sqrt{(2c_d)^2 - (D_2 - D_1)^2} + \frac{\pi}{2}(D_1 + D_2) + \frac{\pi(D_2 - D_1)}{180}\sin^{-1}\left(\frac{D_2 - D_1}{2c_d}\right)$$

Force: $\dfrac{F_1}{F_2} = e^{\mu\phi\pi/180^\circ}$

Including centrifugal force: $\dfrac{F_1 - F_c}{F_2 - F_c} = e^{\mu\phi\pi/180^\circ}$

Torque: $T = \dfrac{(F_1 - F_2)D_1}{2}$

Velocity ratio: $g_r = \dfrac{N_{a1}}{N_{a2}} = \dfrac{r_2}{r_1}$

Design power rating: $h_{pr} = f_1 h_p$

Rated power per belt: $h_{pb} = f_2 h_1$

Number of belts required: $N = \dfrac{h_{pr}}{h_{pb}}$

Wire Rope:

Total force: $F_t = F_w + F_r + F_a + F_h$

Wire stress: $\sigma = \dfrac{Ed_w}{D}$

Bearing pressure: $p = \dfrac{2F_t}{dD}$

Fatigue: $F_f = \dfrac{S_u dD}{2000}$

Rolling Chains:

Velocity ratio: $g_r = \dfrac{d_2}{d_1} = \dfrac{\omega_1}{\omega_2} = \dfrac{N_2}{N_1}$

Chordal rise: $\Delta r = r\left[1 - \cos\left(\dfrac{180}{N}\right)\right]$

Number of links: $\dfrac{L}{p_t} = \dfrac{2c_d}{p_t} + \dfrac{N_1 + N_2}{2} + \dfrac{(N_2 - N_1)^2}{4\pi^2\left(\dfrac{c_d}{p_t}\right)}$

Center distance: $\dfrac{c_d}{p_t} = A + \sqrt{A^2 - \dfrac{B^2}{2}}$

$$A = \frac{1}{4}\left(\frac{L}{p_t} - \frac{N_1 + N_2}{2}\right)$$

$$B = \frac{N_2 - N_1}{2\pi}$$

Power rating: $h_{pr} = h_p\dfrac{a_1}{a_2}$

Recommended Readings

American Chain Association, *Chains for Power Transmission and Material Handling*, Marcel Dekker.

Budynas, R.G., and Nisbett, J.K. (2011), *Shigley's Mechanical Engineering Design*, 9th ed., McGraw-Hill.

Dickie, D.E. (1985) *Rigging Manual*, Construction Safety Association of Ontario.

Juvinall, R.C., and Marshek, K.M. (2012) *Fundamentals of Machine Component Design*, 5th ed., Wiley.

Krutz, G.W., Schuelle, J.K., and Claar, P.W. (1994) *Machine Design for Mobile and Industrial Applications*, Society of Automotive Engineers.

Mott, R. L. (2014) *Machine Elements in Mechanical Design*, 5th ed., Pearson.

Rossnagel, W.E., Higgins, L.R., and MacDonald, J.A. (1988) *Handbook of Rigging*, 4th ed., McGraw-Hill.

Shapiro, H.I. et al. (1991) *Cranes and Derricks*, 2nd ed., McGraw-Hill.

Wire Rope Users Manual (1972) Armco Steel Corp.

References

ASME (2007) B30.5 *Mobile and Locomotive Cranes*, American Society of Mechanical Engineers.

Shigley, J.E., and Mitchell, L.D. (1983), *Mechanical Engineering Design*, 4th ed., McGraw-Hill.

Questions

19.1 What is the difference between a flat belt and a V-belt?

19.2 What is a synchronous belt?

19.3 What do the terms 3V, A, E, 5V, and 8V have in common?

19.4 Define *slip* as it relates to machine elements of this chapter.

19.5 What is the difference between a strand and a wire?

19.6 Describe the difference between Lang and regular lay.

19.7 Why is a headache ball (static weight) used with wire rope?

19.8 What is larger, a rope or a strand? A strand or a wire?

19.9 What is the difference between a sheave and a sprocket?

19.10 Why is a preload needed for belts?

19.11 Should there be a preload on roller chains? Explain your answer.

19.12 What are the components of a roller chain?

19.13 What is a silent chain?

19.14 What are the consequences of having a vertical chain drive?

19.15 Should a V-belt be lubricated? Why or why not?

Qualitative Problems

19.16 Define the term *datum* and explain its importance.

19.17 What are the main advantages of V-belts over spur gears?

19.18 List the advantages and disadvantages of V-belts compared to chain drives.

19.19 List the reasons that a wire rope requires lubrication. What kind of lubricant would you recommend?

19.20 What are the similarities and differences between V-belts, wire ropes, and roller chains?

19.21 Using proper sketches, determine if there is a chordal rise with synchronous belts.

19.22 Explain why there is a minimum recommended sheave diameter for V-belts.

19.23 Explain the similarities and differences between reinforcing strands in V-belts and wire rope.

19.24 Review Figures 3.19 through 3.22 and identify materials that would be useful for wire ropes other than steel.

19.25 Plot power rating for V-belts as a function of small sheave diameter, and describe your observations.

19.26 List the characteristics and advantages of (*a*) regular lay, and (*b*) Lang lay.

19.27 Why is steel popular for wire rope?

19.28 Explain why the curves in Fig. 19.14 have their shape.

19.29 A vendor approaches you and asks to replace your steel roller chain with a wear-resistant polymer chain. What questions would you ask before considering the change?

19.30 Explain why the reinforcing cords are located near the exterior of a V-belt.

Quantitative Problems

19.31 Repeat Example 19.3 using a 5V cross-section. *Ans.* 2 belts needed, with 7.1-in. and 14-in. sheaves.

19.32 An open flat belt 8 in. wide and 0.13 in. thick connects a 16-in.-diameter pulley with a 36-in.-diameter pulley over a center distance of 15 ft. The belt speed is 3600 ft/min. The allowable preload per unit width of the belt is 100 lb/in., and the weight per volume is 0.042 lb/in.^{3}. The coefficient of friction between the belt material and the pulley is 0.8. Find the length of the belt, the maximum forces acting on the belt before failure is experienced, and the maximum power that can be transmitted. *Ans.* $L = 441$ in., $h_{p,\max} = 171$ hp.

19.33 The driving and the driven pulley in an open-flat-belt transmission each have a diameter of 160 mm. Calculate the preload needed in the belt if a power of 7 kW should be transmitted at 1000 rpm by using only half the wrap angle on each pulley (that is, $\phi = \pi/2$). The coefficient of friction is 0.20, the density of the belt material is 1500 kg/m^3, and the allowable belt preload stress is 5 MPa. Also, calculate the cross-sectional area of the belt. *Ans.* $F_i = 2740$ N.

19.34 A belt transmission is driven by a motor hinged at point 0 in Sketch *a*. The motor mass is 50 kg and its speed is 1500 rpm. Calculate the maximum transferable power for the transmission. Check, also, that the largest belt stress is lower than σ_{all} when the bending stresses are included. Belt dimensions are 100 by 4 mm, the coefficient of friction is 0.32, the modulus of elasticity is 100 MPa, the belt density is 1200 kg/m^3, and the allowable stress is 20 MPa. *Ans.* $h_p = 10.88$ kW.

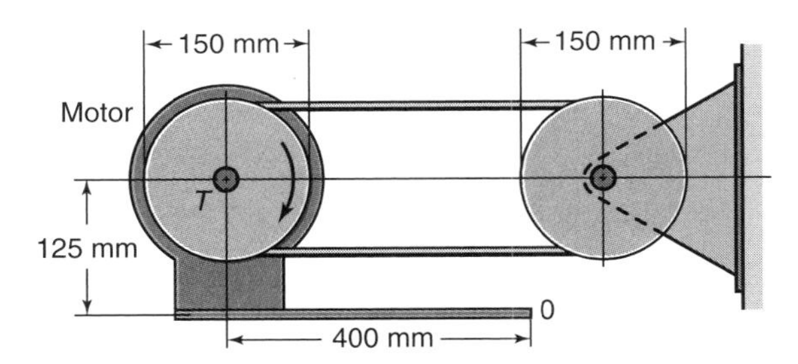

Sketch *a*, for Problem 19.34

19.35 The tension in a flat belt is given by the motor weight, as shown in Sketch *b*. The mass is 80 kg and is assumed to be concentrated at the motor shaft position. The motor speed is 1405 rpm and the pulley diameter is 400 mm. Calculate the belt width when the allowable belt stress is 6.00 MPa, the coefficient of friction is 0.5, the belt thickness is 5 mm, the modulus of elasticity is 150 MPa, and the density is 1200 kg/m^3. *Ans.* $w_t = 33.84$ mm.

Sketch *b*, for Problem 19.35

19.36 Calculate the maximum possible power transmitted when a wrap angle is used as shown in Sketch *c*. The belt speed is 5 m/s and the coefficient of friction is 0.25. *Ans.* $h_p = 585$ W.

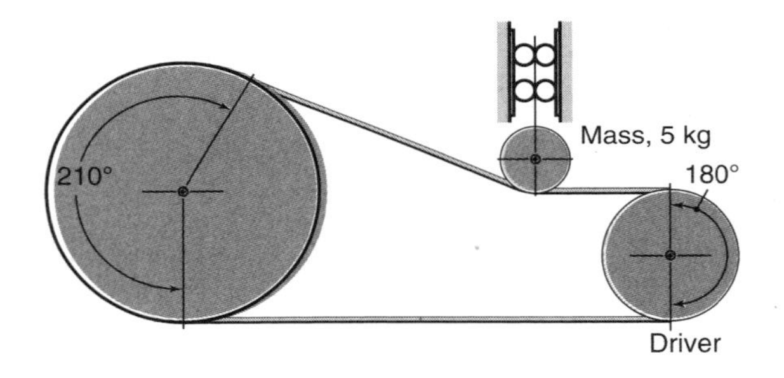

Sketch *c*, for Problem 19.26

19.37 Equation (19.9) gives the belt forces, where F_1 is the largest belt force. What is the maximum power the belt can transmit for a given value of F_1? *Hint:* Express the power as a function of speed, and then take the derivative with respect to speed and set the derivative equal to zero.

19.38 In a flat belt drive with parallel belts the motor and pulley are pivoted at point 0 as shown in Sketch *d*. Calculate the maximum possible power transmitted at 1200 rpm. The motor and pulley together have a 50-kg mass at the center of the shaft. The centrifugal forces on the belt have to be included in the analysis. The coefficient of friction is 0.2, belt area is 300 mm^2, and density is 1500 kg/m^3. *Ans.* $h_p = 5.69$ kW.

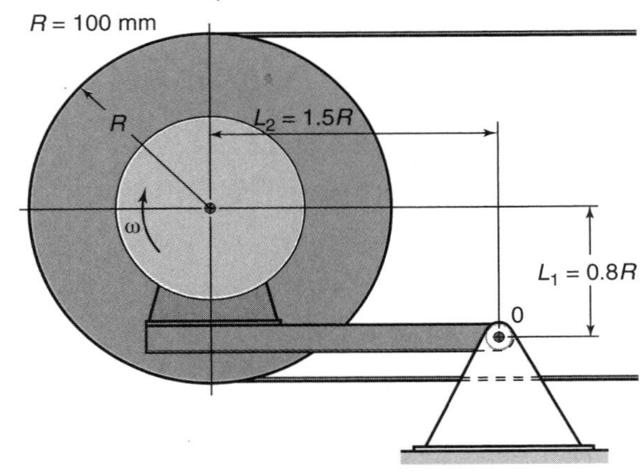

Sketch *d*, for Problem 19.38

19.39 A flat belt drive, shown in Sketch *e*, has the top left pulley driven by a motor and driving the other two pulleys. Calculate the wrap angles for the two driven pulleys when the full wrap angle is used on the driving pulley and the power taken from pulley 2 is twice the power from pulley 3. The coefficient of friction is 0.3 and $l = 4r$. *Ans.* $\alpha_3 = 37.2°$.

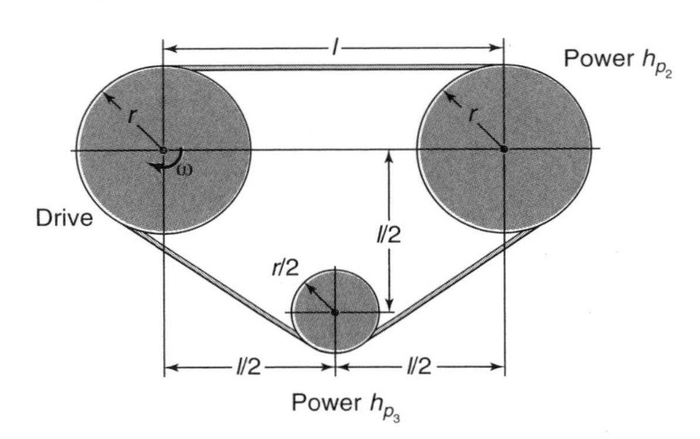

Sketch *e*, for Problem 19.39

19.40 A flat belt 6 mm thick and 60 mm wide is used in a belt transmission with a speed ratio of 1. The radius of the pulleys is 100 mm, their angular speed is 50 rad/s, the center distance is 1000 mm, the belt modulus of elasticity is 200 MPa, and the density is 1000 kg/m^3. The belt is pretensioned by increasing the center distance. Calculate this increase in center distance if the transmission should use only half the wrap angle to transmit 1 kW. Centrifugal effects should be considered. The coefficient of friction is 0.3. *Ans.* $\Delta a = 8$ mm.

19.41 A 120-mm-wide and 5-mm-thick flat belt transfers power from a 250-mm-diameter driving pulley to a 700-mm-diameter driven pulley. The belt has a mass per length of 2.1 kg/m and a coefficient of friction of 0.25. The input power of the belt drive is 60 kW at 1000 rpm. The center distance is 3.5 m. Determine

(a) The maximum tensile stress in the belt. *Ans.* $\sigma = 15.04$ MPa.

(b) The loads for each pulley on the axis. *Ans.* $F_x = 12.7$ kN.

19.42 A flat belt drive is used to transfer 100 kW of power. The diameters of the driving and driven pulleys are 300 and 850 mm, respectively, and the center distance is 2 m. The belt has a width of 200 mm, thickness of 10 mm, speed of 20 m/s, and coefficient of friction of 0.40. For a safety factor of 2.0 for static loading determine the following:

(a) The contact angles and belt length for an open configuration. *Ans.* $\phi_1 = 164.2^\circ$, $\phi_2 = 195.8^\circ$, $L = 5.84$ m.

(b) The loads for each pulley on the axis. *Ans.* $F_x = 9.56$ kN, $F_y = 687$ N.

19.43 A synchronous belt transmission is used as a timing belt for an overhead camshaft on a car motor. The belt is elastically prestressed at standstill to make sure it does not slip at high speeds due to centrifugal forces. The belt is 1100 mm long and weighs 200 g. The elastic spring constant for 1 m of belt material is 10^5 N/m. The belt prestress elongates it 2 mm. It needs to elongate 4 mm more to start slipping. Calculate the maximum allowable motor speed if the pulley on the motor shaft has a diameter of 100 mm. *Ans.* $\omega = 1095$ rad/s.

19.44 A timing belt for power transfer should be used at a velocity ratio of one-third, so that the outgoing speed should be three times as high as the incoming speed. The material for the belt reinforcement can be chosen from glass fiber, carbon fiber, and steel wire. These materials give different belt densities as well as different tensile strengths. The glass-fiber-reinforced belt has a density of 1400 kg/m^3 and an allowable stress of 300 MPa. The carbon-fiber-reinforced belt has a density of 1300 kg/m^3 and an allowable stress of 600 MPa. The steel-wire-reinforced belt has a density of 2100 kg/m^3 and an allowable stress of 400 MPa. Calculate the maximum power for each belt type if the belt speed is 30 m/s, the belt cross-section size is 50 mm^2, and the safety factor is 12. *Ans.* For glass fibers, $h_p = 37.34$ kW.

19.45 A 20-hp, 1750-rpm electric motor drives a machine through a multiple V-belt drive. The driver sheave is 3.7 in. in diameter and the wrap angle of the driver is 165°. The weight per length is 0.012 lb/in. Maximum belt preload is 150 lb, and the coefficient of friction of the belt material acting on the sheave is 0.2. How many belts should be used? Assume a sheave angle of 18°. *Ans.* Two are needed.

19.46 A V-belt drive has $r_1 = 200$ mm, $r_2 = 100$ mm, $2\beta = 36^\circ$, and $c_d = 700$ mm. The speed of the smaller sheave is 1200 rpm. The cross-sectional area of the belt is 160 mm^2 and its density $\rho = 1500$ kg/m^3. How large is the maximum possible power transmitted by six belts if each belt is prestressed to 200 N? The coefficient of friction $\mu = 0.30$. If 15 kW is transmitted, how large a part of the periphery of the smaller wheel is then active? *Ans.* $\phi = 85.0^\circ$.

19.47 A combined V-belt and flat belt drive, shown in Sketch *f*, has a speed ratio of 4. Three V-belts drive the sheave in three grooves but connect to the larger sheave like flat belts on the cylindrical surface. The center distance is six times the smaller sheave radius, which is 80 mm with a speed of 1500 rpm. Determine which sheave can transmit the largest power without slip. Determine the ratio between the powers transmitted through the two sheaves. The angle 2β is 36° and the coefficient of friction is 0.30. *Ans.* $h_{p,\text{small}} = 2.64 h_{p,\text{large}}$.

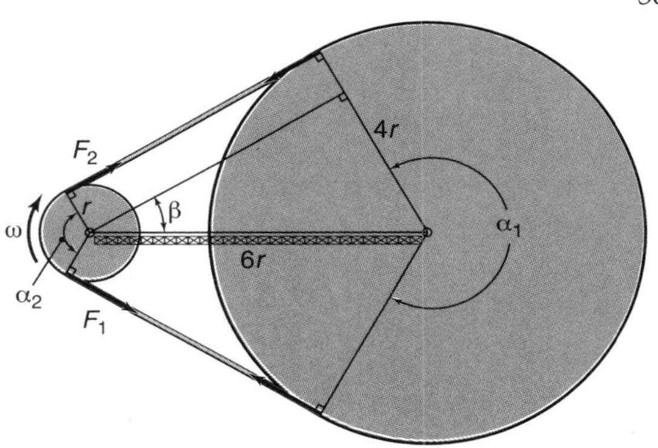

Sketch *f*, for Problem 19.47

19.48 A small lathe (consider the power source to be "normal", with a service of 8 hr/day) uses a belt drive to transmit power from a 10-hp electric motor to a spindle. The maximum motor speed is 1000 rpm, and the maximum spindle speed is 1000 rpm. Select a belt and sheaves for this application, as well as the center distance. *Ans.* 3V cross-section with four belts, 5.3-in small sheave, $c_d = 9.6$ in.

19.49 The input power to shaft A, shown in Sketch *g*, is transferred to shaft B through a pair of mating spur gears and then to shaft C through a 3*V*-section V-belt drive. The sheaves on shafts B and C have 76- and 200-mm diameters. For the maximum power the belt can transfer, determine the following:

(a) The input and output torques of the system. *Ans.* $T_B = 98.9$ N-m.

(b) The belt length for an approximate center distance of 550 mm. *Ans.* $L = 1.32$ m.

Recall that 1 hp = 746 W.

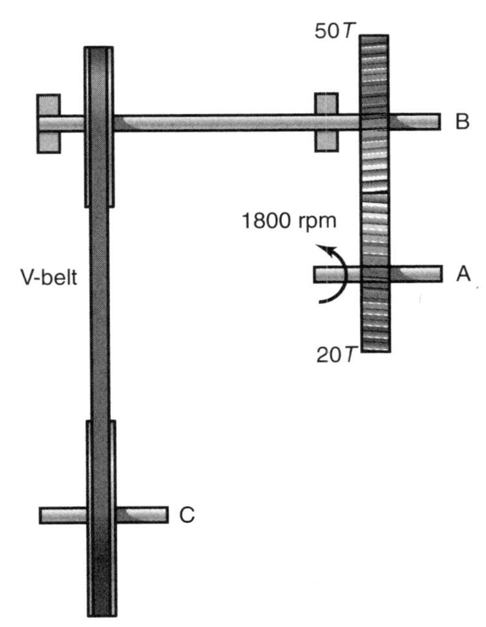

Sketch *g*, for Problem 19.49

19.50 An automobile fan is driven by an engine through a V-belt drive. The engine's sheave has an 8 in. diameter and is running at 880 rpm. The fan's sheave has a 3.15 in. diameter. The power required to move air is 1.2 hp.

Select the size of the V-belt to obtain a compact design. *Ans.* Single 3V belt, $L = 35$ in.

19.51 A hoist uses a 2-in., 6×19 monitor steel wire rope with Lang lay on cast steel sheaves. The rope is used to haul loads up to 8000 lb a distance of 400 ft.

(a) Using a maximum acceleration of 2 ft/s^2, determine the tensile and bending stresses in the rope and their corresponding safety factors. *Ans.* $\sigma_t = 7379$ psi, $\sigma_b = 28.6$ ksi.

(b) Determine the bearing pressure in the rope and the safety factor when it interacts with the sheave. Also, determine the stretch of the rope. *Ans.* $p = 187$ psi, $\delta = 2.95$ in.

(c) Determine the safety factor due to fatigue and anticipate the number of bends until failure.

19.52 A steel rope for a crane has a cross-sectional area of 31 mm^2 and an ultimate strength of 1500 N/mm^2. The rope is 12 m long and is dimensioned to carry a maximum load of 1000 kg. If that load is allowed to free-fall 1 m before the rope is tightened, how large will the stress be in the rope? The modulus of elasticity is 68 GPa. *Ans.* $\sigma = 2236$ MPa.

19.53 Two 2-cm, 6×19 plow-steel wire ropes are used to haul mining material to a depth of 150 m at a speed of 8 m/s and an acceleration of 2 m/s^2. The pulley diameter is 80 cm. Using a proper safety factor, determine the maximum hauling load for these ropes. Are two ropes enough for infinite life? *Ans.* $F_w = 3580$ lb.

19.54 Using six 16-mm, 6×19 wire ropes, an elevator is to lift a 2500-kg weight to a height of 81 m at a speed of 4 m/s and an acceleration of 1 m/s^2. The 726-mm-diameter pulley has a strength of 6.41 MPa. Determine the safety factors for tension, bending, bearing pressure, and fatigue failure. Also, calculate the maximum elongation of the ropes. *Ans.* $n_{s,\text{tension}} = 11.1$, $n_{s,\text{fatigue}} = 1.80$.

19.55 Sketch h shows a wire rope drive used on an elevator to transport dishes between floors in a restaurant. The maximum mass being transported is 300 kg and hangs by a steel wire. The line drum is braked during the downward motion by a constant moment of 800 N-m. The rotating parts have a mass moment of inertia of 13 kg-m^2. Calculate the maximum force in the wire and find when it first appears. *Ans.* $F_{\max} = 3960$ N.

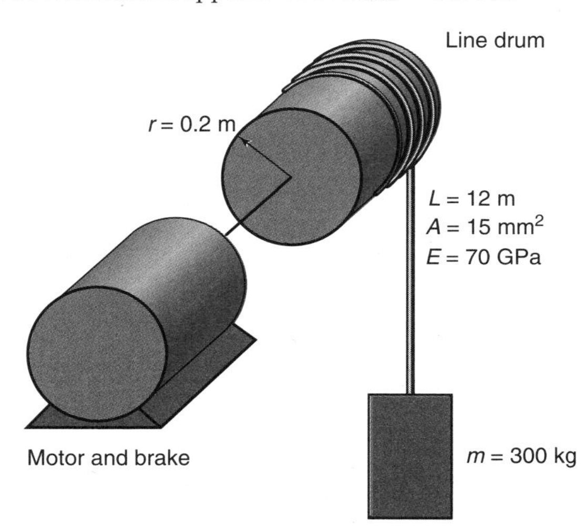

Sketch h, for Problem 19.55

19.56 End and front views of a hoisting machine are shown in Sketch i. Calculate the lifting height per drum revolution. Calculate the maximum force in the wire when the drum is instantly started with angular speed ω to lift a mass m. The free length of the wire is L and its cross-sectional area is A.

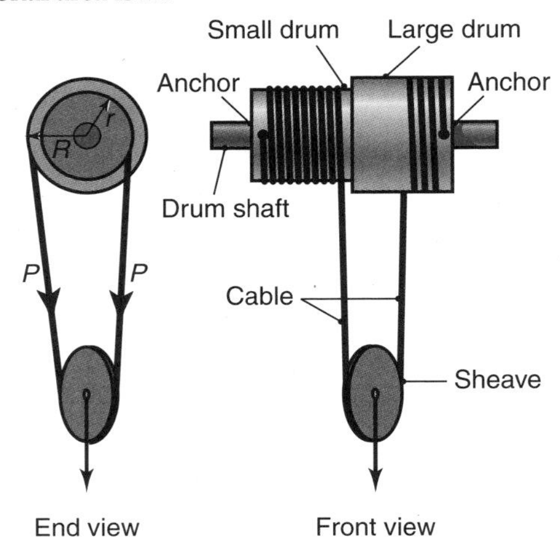

Sketch i, for Problem 19.56

19.57 A hauling unit with 25-mm, 8×19 wire rope made of fiber-core plow steel is used to raise a 45-kN load to a height of 150 m at a speed of 5 m/s and an acceleration of 2 m/s^2. Design for a minimum pulley diameter while determining the safety factors for tensile strength, bending stress, bearing pressure, and fatigue endurance. *Ans.* For $D/d = 21$, $n_{st} = 2.10$, $n_{sb} = 2.99$.

19.58 A building elevator operates at a speed of 5 m/s and 1.2-m/s^2 acceleration and is designed for 10-kN deadweight, five 70-kg passengers, and 150-kg overload. The building has twelve 4-m stories. Using the proper safety factor, design the wire rope required for the elevator. A maximum of four ropes may be used.

19.59 A three-strand, ANSI 50 roller chain is used to transfer power from a 20-tooth driving sprocket that rotates at 500 rpm to a 60-tooth driven sprocket. The input power is from an internal combustion engine, and the chain experiences moderate shock. Determine the following:

(a) The power rating. *Ans.* $h_p = 7.125$ hp per chain.

(b) The length of the chain for an approximate center distance of 550 mm. *Ans.* $L = 1760$ mm.

19.60 To transfer 100 kW of power at 400 rpm, two strands of roller chain are needed. The load characteristics are heavy shock with an internal combustion engine and mechanical drive. The driving sprocket has 12 teeth and the driven sprocket has 42 teeth. Determine the kind, length, and size of the chain for a center distance close to 50 pitches. *Ans.* For ANSI 160 chain, $L = 255$ in.

19.61 A two-strand AISI 50 roller chain system is used to transmit the power of an electric motor rotating at 500 rpm. The driver sprocket has 12 teeth and the driven sprocket has 60 teeth. Calculate the power rating per strand, the power that can be transmitted, and the chain length if $c_d = 50p_t$. *Ans.* $h_p = 6.71$ hp per strand, $h_{pr} = 5.13$ hp, $L = 85.7$ in.

19.62 A roller chain is used to transfer 12 hp of power from a 20-tooth driving sprocket running at 500 rpm to a 40-tooth sprocket. Design the chain drive for heavy shocks, an electric motor, and a single-strand chain. Also, determine the required chain length for a center distance of approximately 70 pitches. *Ans.* For ANSI 100 chain, $h_{pr} = 18$ hp, $L = 213$ in.

Design and Projects

19.63 It has been proposed to use thin metal foils moving over smooth spools as power transmission devices instead of rubber belts on sheaves. List the advantages and disadvantages of metal foils for this application and determine under what conditions you would expect metal foils would be preferable.

19.64 Assume you are investigating a failure of wire rope on a mobile crane. What would you expect to find at the fracture surfaces of the wires if

(a) the rope was overloaded;

(b) the rope was in service for too long;

(c) a kink was put in the rope when it was installed.

Make appropriate sketches and describe what you would do to determine which mode was the cause of the wire rope failure.

19.65 Explain the function of the tie layer in a multiple belt arrangement. Describe what steps you would take to determine the load on the tie layer, and how this would allow you to design the tie layer properly.

19.66 One form of a V-belt drive uses two grooved sheaves with a flat pulley in between, so that the flat side of the V-belt runs against the pulley. What concerns would you have about such an arrangement? What design considerations would you have regarding the belt?

19.67 A mobile crane is being used to help construct a dam. In this particular application, wooden molds and steel reinforcement (rebar) is located, then concrete is poured in place. After the concrete has set, the molds are attached to a crane hook, unbolted and pried off of the concrete. Often the molds will fall 5 to 10 ft because of slack in the load line, and then will be caught by the wire rope. List your concerns regarding this situation.

19.68 Wire rope is often used for high-voltage electrical power transmission. These are often called *high-tension* lines because the tension is very high to avoid excessive sag in the rope. Conduct an Internet search, and summarize the theory of operation for cable pullers that apply a tension to wire rope.

19.69 Assume you are asked to investigate a ski lift to make certain that it is in proper operating condition for the start of winter. Construct an inspection protocol outlining the necessary steps you would take regarding the wire rope.

19.70 In the footnote to Table 19.6, it states that the stiffness of wire ropes is affected by load and increases with the life of the rope. Explain why this would be the case. How would you expect the stiffness of V-belts and roller chains to change, if at all, during their service lives? Explain your answer.

Appendix A

Physical and Mechanical Properties of Materials

A.1 Mechanical Properties of Selected Alloy Steels

AISI number	Condition[a]	Yield strength, MPa (ksi)	Ultimate tensile strength, MPa (ksi)	Elongation in 50 mm, %	Reduction in area, %	Brinell hardness, HB
Chromium-Molybdenum						
4130	Annealed	361 (52)	560 (81)	28	56	156
	N 870 °C	436 (63)	670 (97)	25	59	197
	Q&T 205 °C	1460 (212)	1630 (236)	10	41	467
	Q&T 315 °C	1380 (200)	1500 (217)	11	43	430
	Q&T 425 °C	1190 (173)	1280 (186)	13	49	380
	Q&T 540 °C	910 (132)	1030 (150)	17	57	315
	Q&T 650 °C	703 (102)	814 (118)	22	64	245
4140	Annealed	417 (61)	655 (95)	26	57	200
	N 870 °C	655 (95)	1020 (148)	18	47	300
	Q&T 205 °C	1640 (238)	1770 (257)	8	38	510
	Q&T 315 °C	1430 (208)	1550 (225)	9	43	445
	Q&T 425 °C	1140 (165)	1250 (181)	13	49	370
	Q&T 540 °C	834 (121)	951 (138)	18	58	285
	Q&T 650 °C	655 (95)	758 (110)	22	63	230
Nickel-Chromium-Molybdenum						
4320	Annealed	425 (61.6)	580 (84)	29	50	160
	N 870 °C	460 (66)	790 (110)	21	40	235
4340	Annealed	470 (68)	745 (108)	22	50	220
	Q&T 315 °C	1590 (230)	1720 (250)	10	40	486
	Q&T 425 °C	1360 (198)	1470 (213)	10	44	430
	Q&T 540 °C	1080 (156)	1170 (170)	13	50	360
	Q&T 650 °C	855 (124)	965 (140)	20	60	280
8620	A	357 (52)	536 (77)	31	60	150
	N 870 °C	385 (56)	632 (93)	26	59	183
Chromium-Vanadium						
6150	Annealed	407 (59)	662 (96)	23	50	190
	N 870 °C	615 (89.1)	940 (130)	22	30	270
	Q&T 540 °C	1160 (168)	1200 (170)	15	20	350

[a] N, normalized; Q&T, quenched and tempered

A.2 Mechanical Properties of Selected Carbon Steels

AISI number	Condition[a]	Yield strength, MPa (ksi)	Ultimate tensile strength, MPa (ksi)	Elongation in 50 mm, %	Reduction in area, %	Brinell hardness, HB
1006	Hot rolled	170 (24)	300 (43)	30	55	85
	Cold drawn	280 (41)	330 (48)	20	45	95
1010	Cold drawn	305 (44.2)	365 (52.9)	20	40	105
	Hot rolled	180 (26)	325 (47.1)	28	50	95
1015	Cold drawn	325 (47.1)	385 (55.8)	18	40	111
	Hot rolled	190 (27.5)	340 (50)	28	50	100
1018	Cold drawn	370 (53)	440 (63)	15	40	126
	Hot rolled	220 (31)	400 (58)	25	50	116
1020	Q&T 870°C	295 (43)	395 (57)	37	60	100
	Cold drawn	350 (50)	420 (60)	15	40	121
	Hot rolled	205 (29.7)	380 (55)	25	50	110
1030	Annealed	317 (46)	430 (62)	30	60	130
	N 925°C	345 (50)	520 (75)	32	61	150
	Q&T 205°C	648 (94)	848 (123)	17	47	495
	Q&T 315°C	621 (90)	800 (116)	19	53	400
	Q&T 425°C	579 (84)	731 (106)	23	60	300
	Q&T 540°C	517 (75)	669 (97)	28	65	250
	Q&T 650°C	441 (64)	586 (85)	32	70	210
	Cold drawn	440 (63.8)	525 (76.1)	12	35	149
	Hot rolled	260 (37.5)	70 (68)	20	40	130
1040	Annealed	350 (51)	520 (75)	30	57	150
	N 900°C	374 (54)	590 (86)	28	55	170
	Q&T 205°C	593 (86)	779 (113)	19	48	262
	Q&T 425°C	552 (80)	758 (110)	21	54	240
	Q&T 650°C	434 (63)	634 (92)	29	65	192
	Cold drawn	490 (71)	585 (84.8)	12	35	170
	Hot rolled	290 (42)	525 (76.1)	18	40	149
1050	Annealed	365 (53)	636 (92)	24	40	190
	N 900°C	427 (62)	748 (108)	20	39	220
	Q&T 205°C	807 (117)	1120 (163)	9	27	514
	Q&T 425°C	793 (115)	1090 (158)	13	36	444
	Q&T 650°C	538 (78)	717 (104)	28	65	235
	Cold drawn	580 (84.1)	690 (100)	10	30	197
	Hot rolled	340 (49.5)	620 (90)	15	35	179
1060	Annealed	372 (54)	626 (91)	22	38	179
	N 900°C	421 (61)	776 (112)	18	37	230
	Q&T 425°C	765 (111)	1080 (156)	14	40	310
	Q&T 540°C	669 (97)	965 (140)	17	45	280
	Q&T 650°C	524 (76)	800 (116)	23	40	230
	Hot Rolled	370 (54)	680 (98)	12	30	201
1080	Q&T 800°C	380 (55)	615 (89)	25	30	255
	Hot rolled	420 (61.5)	770 (112)	10	25	229
1095	Annealed	380 (55)	658 (95)	13	21	190
	N 900°C	500 (72)	1010 (147)	9	13	293
	Q&T 315°C	813 (118)	1260 (183)	10	30	375
	Q&T 425°C	772 (112)	1210 (176)	12	32	360
	Q&T 540°C	676 (98)	1090 (158)	15	37	320
	Q&T 650°C	552 (80)	896 (130)	21	30	269
	Hot rolled	455 (65.9)	825 (119)	10	25	248
	Cold drawn	525 (76.1)	680 (98)	10	20	197

[a] N, normalized; Q&T, quenched and tempered.

A.3 Mechanical Properties of Selected Cast Irons

ASTM class	Condition[a]	Tensile yield strength, MPa (ksi)	Ultimate tensile strength, MPa (ksi)	Compressive strength, MPa (ksi)	Hardness, HB
Gray cast irons					
20	As cast	—	152 (22)	572 (83)	156
25	As cast	—	179 (26)	669 (97)	174
30	As cast	—	214 (31)	752 (109)	210
35	As cast	—	252 (36.5)	855 (124)	212
40	As cast	—	293 (42.5)	965 (140)	235
50	As cast	—	362 (52.5)	1130 (164)	262
60	As cast	—	431 (62.5)	1293 (187.5)	302
Ductile (nodular or spheroidal) cast irons					
60-40-18	Annealed	324 (47)	448 (65)	359 (52)	160
65-45-12	Annealed	331 (48)	462 (67)	365 (53)	174
80-55-06	Annealed	365 (53)	565 (82)	385 (56)	228
120-90-02	Q&T	827 (120)	965 (140)	924 (134)	325

[a] Q&T, quenched and tempered.

A.4 Mechanical Properties of Selected Stainless Steels

AISI no.	Manufacturing history[a]	Yield strength, MPa (ksi)	Ultimate tensile strength, MPa (ksi)	Elongation in 50 mm, %	Hardness
301	Annealed	205 (29.7)	515 (74.7)	40	85 HRB
	1/16-hard	310 (45.0)	620 (89.9)	40	95 HRB
	1/8-hard	380 (55.0)	690 (100)	40	20 HRC
	1/4-hard	515 (74.7)	860 (125)	25	25 HRC
	1/2-hard	758 (110)	1034 (150)	18	32 HRC
	3/4-hard	930 (135)	1205 (175)	12	37 HRC
	Full-hard	965 (140)	1276 (185)	9	41 HRC
302	Annealed	207 (30.0)	517 (75.0)	40	150 HB
	1/4-hard	517 (75.0)	860 (125)	10	310 HB
	1/2-hard	758 (110)	1034 (150)	9	320 HB
	Full-hard	965 (140)	1275 (185)	3	335 HB
304	Annealed	215 (31.2)	505 (73.2)	40	201 HB
304L	Annealed	170 (24.6)	485 (70.3)	40	201 HB
304H	Annealed	205 (29.7)	515 (74.7)	40	201 HB
316	Annealed	240 (35.0)	585 (85.0)	55	150 HB
316L	Annealed	207 (30.0)	538 (78.0)	55	145 HB
410	Annealed	275 (39.9)	485 (70.3)	20	95 HB
420	Annealed	345 (50.0)	655 (95.0)	25	241 HB
	Q&T 204°C	1360 (197)	1600 (232)	12	444 HB
	Q&T 427°C	1420 (205)	1620 (234)	10	461 HB
	Q&T 650°C	680 (98.6)	895 (130)	20	262 HB
431	Annealed	655 (95.0)	862 (125)	20	285 HB
	Q&T 204°C	1055 (153)	1345 (195)	20	388 HB
	Q&T 427°C	1080 (156)	1350 (196)	19	388 HB
	Q&T 650°C	695 (101)	960 (139)	20	277 HB
631	Cold rolled	1275 (185)	1030 (149)	6	41 HRC

[a] Q&T, quenched and tempered

A.5 Mechanical Properties of Selected Aluminum Alloys

Alloy	Temper	Yield strength, MPa (ksi)	Ultimate tensile strength, MPa (ksi)	Elongation in 50 mm, %
Wrought				
1100	O	35 (5.07)	90 (13.0)	40
	H14	120 (17.4)	125 (18.1)	25
1350	O	28 (4.0)	83 (12)	—
	H19	165 (24)	186 (27)	—
2011	T3	296 (43)	379 (55)	15
	T8	310 (45)	407 (59)	12
2014	O	97 (14)	186 (27)	18
	T4	290 (42)	427 (62)	20
	T6	414 (60)	483 (70)	13
2017	T4	276 (40)	427 (62)	22
2024	O	75 (11)	190 (27.6)	20
	T3	345 (50)	483 (70)	18
	T4	325 (47.1)	470 (68.2)	20
2219	O	76 (11)	179 (26)	20
	T87	393 (57)	476 (69)	10
3003	O	40 (5.8)	110 (16.0)	35
	H12	117 (17)	131 (19)	20
	H14	145 (21.0)	150 (21.7)	12
	H16	165 (24)	179 (26)	14
3004	O	69 (10)	179 (26)	20
	H34	186 (27)	234 (34)	6
	H38	234 (34)	276 (40)	6
3105	O	55 (8)	117 (17)	24
	H14	152 (22)	172 (25)	5
	H18	193 (28)	214 (31)	3
	H25	159 (23)	179 (23)	—
5005	H34	138 (20)	159 (23)	8
5052	O	90 (13.0)	190 (27.6)	25
	H32	186 (27)	234 (34)	62
	H34	215 (31.2)	260 (37.7)	12
	H36	234 (34)	269 (39)	10
5056	O	152 (22)	290 (42)	35
	H18	407 (59)	434 (63)	10
5083	O	145 (21)	290 (42)	22
	H321	228 (330	317 (46)	16
5086	H32	207 (30)	290 (42)	12
	H34	255 (37)	324 (47)	10
	H112	131 (19)	269 (39)	14
5454	O	117 (17)	248 (36)	22
	H32	207 (30)	276 (40)	10
	H34	241 (35)	303 (44)	10
	H112	124 (18)	248 (36)	18
6061	O	55 (8.0)	125 (18.1)	25
	T4	145 (21)	241 (35)	22
	T6	275 (40.0)	310 (45.0)	15
6063	T5	145 (21)	186 (27)	12
	T6	214 (31)	241 (35)	12
7050	T7651	490 (71)	552 (80)	—
7075	O	105 (15.2)	230 (33.3)	16
	T6	500 (72.5)	570 (82.7)	11
Cast				
319.0	T6	165 (24)	248 (36)	2.0
333.0	T5	172 (25)	234 (34)	1.0
	T6	207 (30)	289 (42)	1.5
335.0	T6	172 (25)	241 (35)	3.0
	T7	248 (36)	262 (38)	0.5

A.6 Mechanical Properties of Selected Magnesium Alloys

Alloy	Temper	Yield strength, MPa (ksi)	Ultimate tensile strength, MPa (ksi)	Elongation in 50 mm, %	Hardness
Sand and permanent-mold castings					
AM100-A	T61	150 (22)	275 (40)	1.0	69 HB
AZ63A	T6	130 (22)	275 (40)	5.0	73 HB
EZ233A	T5	110 (16)	160 (23)	2.0	50 HB
HK31A	T6	105 (15)	220 (32)	8.0	55 HB
ZC63A	T6	125 (18)	210 (30)	3.5	62 HB
ZK61A	T6	195 (28)	310 (45)	10.0	70 HB
Die castings					
AS41A	F	150 (22)	220 (32)	4.0	—
AZ91A	F	150 (22)	230 (33)	3.0	63 HB
Extrusions					
AZ31B	F	200 (29)	260 (38)	15	—
AZ61A	F	230 (33)	310 (45)	16.0	60 HB
AZ80A	T5	275 (40)	380 (55)	7	—
ZK60A	T5	305 (44)	365 (53)	11.0	88 HB
Sheet and plate					
AZ31B	H24	220 (32)	290 (42)	15.0	73 HB
HK31A	H24	200 (29)	255 (37)	9.0	68 HB
ZE10	F	163 (24)	263 (38)	16	—
ZEK199	F	308 (45)	311 (45)	19	—
ZK60A	T5	300 (44)	365 (53)	11	—

A.7 Mechanical Properties of Selected Wrought Copper and Copper Alloys

Alloy	UNS No.	Temper	Yield Strength MPa (ksi)	Ultimate Tensile Strength MPa (ksi)	Elongation in 50 mm, %	Hardness
Pure copper						
0FHC	C10200	—	69-365 (10-53)	221-455 (33-66)	55-4	—
High-copper alloys						
Beryllium-copper	C17200	Annealed	—	490 (71)	35	60 HRB
		Hardened	1050 (152)	1400 (203)	2	42 HRC
Brass						
Gilding, 5% Zn	C21000	Annealed	77 (11)	245 (36)	45	52 HRF
		Hard	350 (51)	392 (57)	5	64 HRB
Red brass, 15% Zn	C23000	Annealed	91 (13)	280 (41)	47	64 HRF
		Hard	406 (59)	434 (63)	5	73 HRB
Cartridge brass, 30% Zn	C26000	Annealed	133 (19)	357 (52)	55	72 HRF
		Hard	441 (64)	532 (77)	8	82 HRB
Muntz metal, 40% Zn	C28000	Annealed	119 (17)	378 (55)	45	80 HRF
		Hard	350 (51)	490 (71)	15	75 HRB
High lead brass, 36% Zn, 2% Pb	C35300	Annealed	119 (17)	350 (51)	52	68 HRF
		Hard	318 (46)	420 (61)	7	80 HRB
Bronze						
Phosphor bronze, 5% Sn	C51000	Annealed	175 (25)	350 (51)	55	40 HRB
		Hard	581 (84)	588 (85)	9	90 HRB
Phosphor bronze, 10% Sn	CS52400	Annealed	250 (36)	483 (70)	63	62 HRB
		Hard	658 (95)	707 (103)	16	96 HRB
Aluminum bronze	C60800	Annealed	175 (25)	420 (61)	66	49 HRB
		Cold rolled	441 (64)	700 (102)	8	94 HRB
	C63000	Extruded	414 (60)	690 (100)	15	96 HRB
		Halfhard	517 (75)	814 (118)	15	98 HRB
High-silicon bronze	C65500	Annealed	210 (31)	441 (64)	55	66 HRB
		Hard	406 (59)	658 (95)	8	95 HRB
Other alloys						
Cupronickel, 30% Ni	C71500	Annealed	126 (18)	385 (56)	36	40 HRB
		Cold rolled	553 (80)	588 (85)	3	86 HRB
Nickel Silver	C75700	Annealed	196 (28)	427 (62)	35	55 HRB
		Hard	525 (76)	595 (86)	4	89 HRB

A.8 Mechanical Properties of Selected Nickel-base Alloys at 25°C

Alloy	Yield strength, MPa (ksi)	Ultimate tensile strength, MPa (ksi)	Elongation in 50 mm, %	Hardness
Commercially pure and low-alloy nickels				
Nickel 200	148 (21.5)	462 (67)	47	109 HB
Nickel 201	103 (15)	403 (58.5)	50	129 HB
Nickel 211	240 (35)	530 (77)	40	—
Duranickel 301	862 (125)	1170 (170)	25	30 HRC
Nickel-copper alloys				
Alloy 400	240 (35)	550 (80)	40	110 HB
Alloy K-500	790 (115)	1100 (160)	20	300 HB
Monel R-405	230 (33)	525 (76)	35	—
Monel K-500	750 (108)	1050 (152)	30	—
Nickel-molybdenum and nickel-silicon alloys				
Hastelloy B				
As cast	345 (50)	586 (85)	10	93 HRB
Sheet	386 (56)	834 (121)	63	92 HRB
Hastelloy C-4	400 (58)	785 (114)	54	—
Hastelloy D	—	793 (115)	—	35 HRC
Nickel-chromium-iron alloys				
Inconel 600	310 (45)	655 (95)	40	75 HRB
Inconel 617	350 (51)	755 (110)	58	173 HB
Inconel 690	348 (50)	725 (105)	41	88 HRB
Inconel 751	976 (142)	1310 (190)	22	352 HB
Inconel 800	295 (43)	600 (87)	44	138 HB
Other alloys				
Hastelloy C-276	372 (54)	785 (114)	62	209 HB
Hastelloy G	320 (47)	690 (100)	50	79 HRB
Inconel 625	517 (75)	930 (135)	42.5	190 HB
Inconel 825	310 (45)	690 (100)	45	—

A.9 Properties of Selected Nickel-based Superalloys at 870°C (1600°F)

Alloy	Condition	Ultimate tensile strength, MPa	Yield strength, MPa	Elongation in 50 mm %
Astroloy	Wrought	770	690	25
Hastelloy X	Wrought	255	180	50
IN-100	Cast	885	695	6
IN-102	Wrought	215	200	110
Inconel 625	Wrought	285	275	125
Inconel 718	Wrought	340	330	88
MAR-M 200	Cast	840	760	4
MAR-M 432	Cast	730	605	8
René 41	Wrought	620	550	19
Udimet 700	Wrought	690	635	27
Waspaloy	Wrought	525	515	35

A.10 Mechanical Properties of Selected Zinc Alloys

Alloy	Ultimate tensile strength, MPa (ksi)	Elongation in 50 mm, %	Hardness
Die-casting alloys			
Z35541	359 (52)	7	100 HB
Z33520	283 (41)	10	82 HB
Z33531	329 (48)	7	91 HB
Z33523	283 (41)	14	76 HB
Z35635	374 (54)	8	103 HB
Z35630	404 (58)	5	100 HB
Z35840	426 (62)	2	119 HB
Wrought alloys (hot rolled)			
Z21220	150 (21)	52	43 HB
Z44330	170 (24)	50	52 HB
Z41320	221 (32)	38	61 HB

A.11 Mechanical Properties of Selected Titanium Alloys

Alloy	Condition	Temperature °C	Yield strength, MPa (ksi)	Ultimate tensile strength, MPa (ksi)	Elongation in 50 mm, %
Unalloyed grades					
99.5% Ti	Annealed	25	240 (35)	330 (48)	30
		300	95 (14)	150 (22)	—
ASTM Grade 1	Annealed	25	170 (25)	240 (35)	—
ASTM Grade 4	Annealed	25	480 (70)	550 (80)	—
Alpha and near-alpha alloys					
Ti-5Al-2.5Sn	Annealed	25	810 (117)	860 (125)	16
		300	450 (65)	565 (82)	—
Ti-8Al-1Mo-1V	Annealed	25	830 (120)	900 (130)	—
Ti-2.25Al-11Sn-5Zr-1Mo	Annealed	25	900 (130)	1000 (145)	—
Ti-6Al-2Sn-4Zr-2Mo	Annealed	25	830 (120)	900 (130)	—
Alpha-beta alloys					
Ti-6Al-4V	Annealed	25	925 (134)	1000 (145)	10
		300	650 (94.3)	725 (105)	—
	Solution + age	25	1100 (159)	1175 (170)	10
		300	900 (130)	980 (142)	—
Ti-6Al-2Sn-4Zr-6Mo	Solution + age	25	1100 (160)	1170 (170)	—
Beta alloys					
Ti-3Al-8V-6Cr-4Mo-4Zr	Annealed	25	830 (120)	900 (130)	—
Ti-15V-3Cr-3Al-3Sn	Solution + age	25	965 (140)	1000 (145)	—
Ti-10V02Fe-3Al	Annealed	25	1100 (160)	1170 (170)	—
Ti-13V-11Cr-3Al	Solution + age	25	1210 (175)	1275 (185)	8
		300	830 (121)	1100 (160)	—

A.12 Mechanical Properties of Selected Powder Metal Alloys

Material	Yield strength, MPa (ksi)	Ultimate tensile strength, MPa (ksi)	Elastic modulus, GPa (Mpsi)	Hardness	Elongation in 25 mm (%)	Density (g/cm^3)
Ferrous						
F-0008-20	170 (25)	200 (29)	85 (12)	35 HRB	< 1	5.8
F-0008-35	260 (38)	390 (57)	140 (20)	70 HRB	1	7.0
F-0008-55HT		450 (65)	115 (17)	22 HRC	< 1	6.3
F-0008-85HT		660 (98)	150 (22)	35 HRC	< 1	7.1
FC-0008-30	240 (35)	240 (35)	85 (12)	50 HRB	< 1	5.8
FC-0008-60	450 (65)	520 (75)	155 (22)	84 HRB	< 1	7.2
FC-0008-95		720 (104)	150 (22)	43 HRC	< 1	7.1
FN-0205-20	170 (25)	280 (41)	115 (17)	44 HRB	1	6.6
FN-0205-35	280 (41)	480 (70)	170 (25)	78 HRB	5	7.4
FN-0205-180HT		1280 (186)	170 (25)	78 HRB	< 1	7.4
FX-1005-40	340 (49)	530 (77)	160 (23)	82 HRB	4	7.3
FX-1005-110HT		830 (120)	160 (23)	38 HRC	< 1	7.3
Stainless steel						
SS-303N1-38	310 (45)	470 (68)	115 (17)	70 HRB	5	6.9
SS-304N1-30	260 (38)	300 (43)	105 (15)	61 HRB	< 1	6.4
SS-316N1-25	230 (33)	280 (41)	105 (15)	59 HRB	< 1	6.4
SS-316N2-38	310 (45)	480 (70)	140 (20)	65 HRB	< 1	6.9
Copper and copper alloys						
CZ-1000-9	70 (10)	120 (17)	80 (12)	65 HRH	9	7.6
CZ-1000-11	80 (12)	160 (23)	100 (14)	80 HRH	12	8.1
CZP-3002-14	110 (16)	220 (32)	90 (13)	88 HRH	16	8.0
CT-1000-13	110 (16)	150 (22)	60 (8.7)	82 HRH	4	7.2
Aluminum alloys						
Ax 123-T1	200 (29)	270 (39)	—	47 HRB	3	2.7
Ax 123-T6	390 (57)	400 (58)	—	72 HRB	< 1	2.7
Ax 231-T6	200 (29)	220 (32)	—	55 HRB	1	2.7
Ax 231-T6	310 (45)	320 (46)	—	77 HRB	< 1	2.7
Ax 431-T6	270 (39)	300 (43)	—	55 HRB	5	2.8
Ax 431-T6	440 (64)	470 (68)	—	80 HRB	2	2.8
Titanium alloys						
Ti-6Al-4V (HIP)	827 (120)	917 (133)	—	—	—	—
Superalloys						
Stellite 19	—	1035 (150)	—	49 HRC	—	< 1

A.13 Mechanical Properties of Selected Engineering Plastics at Room Temperature

Material	Ultimate tensile strength, MPa (ksi)	Elastic modulus, GPa (Mpsi)	Elongation, %	Poisson's ratio, ν
Thermoplastics:				
Acrylonitrile-butadiene-styrene (ABS)	28-55 (4.1-8.0)	1.4-2.8 (0.20-0.40)	75-5	-
ABS, reinforced	100 (14)	7.5 (1.1)	-	0.35
Acetal	55-70 (8.0-10)	1.4-3.5 (0.20-0.50)	75-25	-
Acetal, reinforced	135 (20)	10 (1.4)	-	0.35-0.40
Acrylic	40-75 (5.8-11)	1.4-3.5 (0.20-0.50)	50-5	-
Cellulosic	10-48 (1.4-7.0)	0.4-1.4 (0.06-0.20)	100-5	-
Fluorocarbon	7-48 (1.0-7.0)	0.7-2 (0.10-0.29)	300-100	0.46-0.48
Nylon	55-83 (8.0-12)	1.4-2.8 (0.20-0.40)	200-60	0.32-0.40
Nylon, reinforced	70-210 (10-30)	2-10 (0.29-1.4)	10-1	-
Polycarbonate	55-70 (8.0-10)	2.5-3 (0.36-0.43)	125-10	0.38
Polycarbonate, reinforced	110 (16)	6 (0.87)	6-4	-
Polyester	55 (8.0)	2 (0.29)	300-5	0.38
Polyester, reinforced	110-160 (16-23)	8.3-12 (1.2-1.7)	3-1	-
Polyethylene	7-40 (1.0-5.8)	0.1-1.4 (0.01-0.20)	1000-15	0.46
Polypropylene	20-35 (2.9-5.1)	0.7-1.2 (0.10-0.17)	500-10	-
Polypropylene, reinforced	40-100 (5.8-14)	3.5-6 (0.51-0.87)	4-2	-
Polystyrene	14-83 (2.0-12)	1.4-4 (0.20-0.58)	60-1	0.35
Polyvinyl chloride	7-55 (1.0-8.0)	0.014-4 (0.0020-0.58)	450-40	-
Thermosets:				
Epoxy	35-140 (5.1-20)	3.5-17 (0.51-2.5)	10-1	-
Epoxy, reinforced	70-1400 (10-200)	21-52 (3.0-7.5)	4-2	-
Phenolic	28-70 (4.0-10)	2.8-21 (0.40-3.0)	2-0	-
Polyester, unsaturated	30 (4.3)	5-9 (0.7-1.3)	1-0	-
Elastomers:				
Chloroprene (neoprene)	15-25 (2.2-3.6)	1-2 (0.14-0.29)	100-500	0.5
Natural rubber	17-25 (2.5-3.6)	1.3 (0.19)	75-650	0.5
Silicone	5-8 (0.7-1.2)	1-5 (0.14-0.72)	100-1100	0.5
Styrene-butadiene	10-25 (1.4-3.6)	2-10 (0.29-1.4)	250-700	0.5
Urethane	20-30 (2.9-4.3)	2-10 (0.29-1.4)	300-450	0.5

A.14 Mechanical Properties of Selected Ceramics at Room Temperature

Material	Symbol	Transverse rupture strength, MPa (ksi)	Compressive strength, MPa (ksi)	Elastic modulus, GPa (Mpsi)	Hardness, (HK)	Poisson's ratio, ν
Aluminum oxide	Al_2O_3	140–240 (20–35)	1000–2900 (140–420)	310–410 (45–59)	2000–3000	0.26
Cubic boron nitride	cBN	725 (105)	7000 (1015)	850 (123)	4000–5000	—
Diamond	—	1400 (200)	7000 (1015)	830–1000 (120–145)	7000–8000	—
Silica, fused	SiO_2	—	1300 (188)	70 (10)	550	0.25
Silicon carbide	SiC	100–750 (14–109)	700–3500 (101–510)	240–480 (35–70)	2100–3000	0.14
Silicon nitride	Si_3N_4	480–600 (70–87)	—	300–310 (43–45)	2000–2500	0.24
Titanium carbide	TiC	1400–1900 (200–276)	3100–3850 (450–560)	310–410 (45–60)	1800–3200	—
Tungsten carbide	WC	1030–2600 (150–380)	4100–5900 (600–860)	520–700 (75–100)	1800–2400	—
Partially stabilized zirconia	PSZ	620 (90)	—	200 (30)	1100	0.30

Note: These properties vary widely depending on the condition of the material.

A.15 Mechanical Properties of Selected Materials used in Rapid Prototyping.

Material	Tensile strength, MPa (ksi)	Elastic modulus, GPa (Mpsi)	Elongation in 50 mm (%)	Characteristics
Stereolithography				
Accura 60	68 (10)	3.10 (0.44)	5	Transparent; good general-purpose material for rapid prototyping
Somos 9920	9 (1.3)	1.81 (0.25)	15	Transparent amber; good chemical resistance; good fatigue properties; used for producing patterns in rubber molding
WaterClear Ultra	55 (8.0)	2.9 (0.41)	6–9	Optically clear resin with ABS-like properties
WaterShed 11122	53 (7.7)	2.5 (0.36)	15	Optically clear with a slight green tinge; mechanical properties similar to those of ABS; used for rapid tooling
DMX-SL 100	32 (4.6)	2.4 (0.35)	12–28	Opaque beige; good general-purpose material for rapid prototyping
Polyjet				
FC720	60 (6.6)	2.9 (0.42)	20	Transparent amber; good impact strength, good paint adsorption and machinability
FC830	50 (7.29)	2.49 (0.36)	20	White, blue, or black; good humidity resistance; suitable for general-purpose applications
FC 930	1.4 (0.20)	0.185 (0.027)	218	Semiopaque, gray, or black; highly flexible material used for prototyping of soft polymers or rubber
Fused-deposition modeling				
Polycarbonate	52 (7.6)	2.0 (0.29)	3	White; high-strength polymer suitable for rapid prototyping and general use
Ultem 9085	72 (10.5)	2.2 (0.32)	5.9	Opaque tan, high-strength FDM material, good flame, smoke and toxicity rating.
ABS-M30i	36 (5.25)	2.4 (0.35)	4	Available in multiple colors, most commonly white; a strong and durable material suitable for general use; biocompatible
PC	68 (9.9)	2.28 (0.33)	4.8	White; good combination of mechanical properties and heat resistance
Selective laser sintering				
WindForm XT	78 (11.4)	7.32 (1.1)	2.6	Opaque black polymide and carbon; produces durable heat- and chemical-resistant parts; high wear resistance.
Polyamide PA 3200GF	45 (6.6)	3.3 (0.40)	6	White; glass-filled polyamide has increased stiffness and is suitable for higher temperature applications
SOMOS 201	–	0.015 (0.002)	110	Multiple colors available; mimics mechanical properties of rubber
ST-100c	305 (44.5)	137 20.0)	10	Bronze-infiltrated steel powder
Electron-beam melting				
Ti-6Al-4V	970 (140)	120 (17.5)	12–16	Can be heat - treated by hot isostatic pressing to obtain up to 600 MPa fatigue strength

Appendix B

Stress-Strain Relationships

Symbols

C	elastic material coefficients, Pa
E	modulus of elasticity, Pa
G	shear modulus, Pa
K	bulk modulus, Pa
x, y, z	Cartesian coordinate system, m
x', y', z'	rotated Cartesian coordinate system, m
γ	shear strain
δ	deformation, m
ϵ	normal strain
λ	Lame's constant, Pa
ν	Poisson's ratio
σ	normal stress, Pa
τ	shear stress, Pa

Subscripts

x, y, z	Cartesian coordinates
x', y', z'	rotated Cartesian coordinates
$1, 2, 3$	principal axes

B.1 Introduction

This Appendix presents selected equations of elasticity, including derivations and definitions of terms that are of importance to the design of structures and machine elements. The derivations in this Appendix are much more mathematical than typical treatment elsewhere in the book, with much less effort expended to explain approaches. It is hoped that the statement and rapid derivation of equations will suffice to demonstrate their existence; the interested reader is encouraged to find far more in-depth treatment in the classic text by Timoschenko and Goodier [1970], as well as any of the other Recommended Readings at the end of the Appendix.

This appendix presents the laws of stress transformation that are used in Chapter 2, the Generalized Hooke's Law in three dimensions, defines elastic constants by their proper names, and gives generalized equations for stress and strain.

B.2 Laws of Stress Transformation

Let a new orthogonal coordinate system $0'x'y'z'$ be placed having origin P, having the z' axis coincident with the normal n to the plane, and having the x' and y' axes (which must parallel the plane with normal n) coincident with the desired shear stresses directions. Such a coordinate system is shown in Fig. B.1, where the plane ABC is imagined to pass through point P, since the tetrahedron is very small. Figure B.1 also shows the three mutually perpendicular stress components

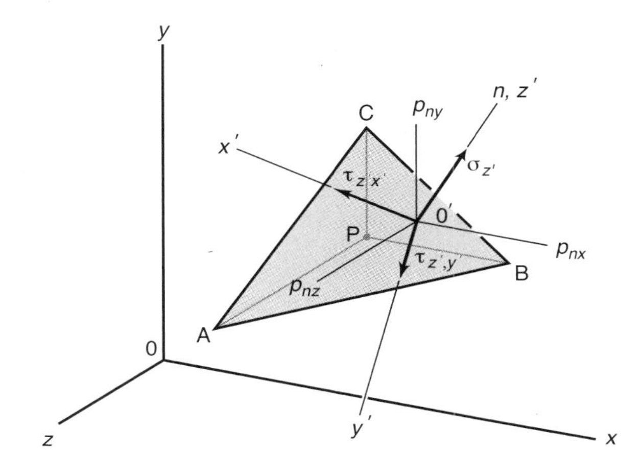

Figure B.1: Small tetrahedral element cut from body at point P. The three perpendicular stress components $\sigma_{z'}$, $\tau_{z'x'}$, and $\tau_{z'y'}$ acting on the inclined plane are shown.

$\sigma_{z'}$, $\tau_{z'x'}$, and $\tau_{z'y'}$ acting on the inclined plane. The laws of stress transformation give these stresses in terms of σ_x, σ_y, σ_z, τ_{xy}, τ_{yz}, and τ_{zx}. The previously established rules for shear stress subscripts are equally applicable here. For example, $\tau_{z'x'}$ is the shear stress directed in the x' direction and acting in a plane through P whose normal is directed in the z' direction.

Because the normal direction coincides with the z' direction,

$$\begin{aligned} p_{nz} &= \sigma_z \cos(z', z) + \tau_{xz} \cos(z', x) + \tau_{yz} \cos(z', y) \\ p_{ny} &= \tau_{zx} \cos(z', z) + \sigma_x \cos(z', x) + \tau_{yx} \cos(z', y) \\ p_{nx} &= \tau_{zy} \cos(z', z) + \tau_{xy} \cos(z', x) + \sigma_y \cos(z', y) \end{aligned} \quad \text{(B.1)}$$

where

$\cos(z', x)$ = cosine of angle between z' and x

The stress $\sigma_{z'}$ must equal the sum of the projections of p_{nz}, p_{ny}, and p_{nx} onto the z' axis,

$$\sigma_{z'} = p_{nz} \cos(z', z) + p_{nx} \cos(z', x) + p_{ny} \cos(z', y) \quad \text{(B.2)}$$

Similarly,

$$\tau_{z'x'} = p_{nz} \cos(x', z) + p_{nx} \cos(x', x) + p_{ny} \cos(x', y) \quad \text{(B.3)}$$

By substituting Eqs. (B.1) into Eqs. (B.2) and (B.3) while also making use of the fact developed earlier that the shear stresses are symmetrical (e.g., $\tau_{x'y'} = \tau_{y'x'}$), Eqs. (B.4) and (B.5) can be developed. Equations (B.6) to (B.9) can be derived in an entirely similar way by using two more tetrahedrons with inclined planes having outer normals parallel to the x' and y' axes. The result of all of the above is

$$\begin{aligned}\sigma_{z'} &= \sigma_z \cos^2(z',z) + \sigma_x \cos^2(z',x) + \sigma_y \cos^2(z',y)\\ &\quad +2\tau_{zx}\cos(z',z)\cos(z',x)\\ &\quad +2\tau_{xy}\cos(z',x)\cos(z',y)\\ &\quad +2\tau_{yz}\cos(z',y)\cos(z',z)\end{aligned} \tag{B.4}$$

$$\begin{aligned}\sigma_{x'} &= \sigma_x \cos^2(x',x) + \sigma_y \cos^2(x',y) + \sigma_z \cos^2(x',z)\\ &\quad +2\tau_{xy}\cos(x',x)\cos(x',y)\\ &\quad +2\tau_{xy}\cos(x',y)\cos(x',z)\\ &\quad +2\tau_{zx}\cos(x',z)\cos(x',x)\end{aligned} \tag{B.5}$$

$$\begin{aligned}\sigma_{y'} &= \sigma_y \cos^2(y',y) + \sigma_z \cos^2(y',z) + \sigma_x \cos^2(y',x)\\ &\quad +2\tau_{yz}\cos(y',y)\cos(y',z)\\ &\quad +2\tau_{zx}\cos(y',z)\cos(y',x)\\ &\quad +2\tau_{xy}\cos(y',x)\cos(y',y)\end{aligned} \tag{B.6}$$

$$\begin{aligned}\tau_{z'x'} &= \sigma_z\cos(z',z)\cos(x',z) + \sigma_x\cos(z',x)\cos(x',x)\\ &\quad +\sigma_y\cos(z',y)\cos(x',y)\\ &\quad +\tau_{zx}\left[\cos(z',z)\cos(x',x) + \cos(z',x)\cos(x',z)\right]\\ &\quad +\tau_{xy}\left[\cos(z',x)\cos(x',y) + \cos(z',y)\cos(x',x)\right]\\ &\quad +\tau_{yz}\left[\cos(z',y)\cos(x',z) + \cos(z',z)\cos(x',y)\right]\end{aligned} \tag{B.7}$$

$$\begin{aligned}\tau_{x'y'} &= \sigma_x\cos(x',x)\cos(y',x) + \sigma_y\cos(x',y)\cos(y',y)\\ &\quad +\sigma_z\cos(x',z)\cos(y',z)\\ &\quad +\tau_{xy}\left[\cos(x',x)\cos(y',y) + \cos(x',y)\cos(y',x)\right]\\ &\quad +\tau_{yz}\left[\cos(x',y)\cos(y',z) + \cos(x',z)\cos(y',y)\right]\\ &\quad +\tau_{zx}\left[\cos(x',z)\cos(y',x) + \cos(x',x)\cos(y',z)\right]\end{aligned} \tag{B.8}$$

$$\begin{aligned}\tau_{y'z'} &= \sigma_y\cos(y',y)\cos(z',y) + \sigma_z\cos(y',z)\cos(z',z)\\ &\quad +\sigma_x\cos(y',x)\cos(z',x)\\ &\quad +\tau_{yz}\left[\cos(y',y)\cos(z',z) + \cos(y',z)\cos(z',y)\right]\\ &\quad +\tau_{zx}\left[\cos(y',z)\cos(z',x) + \cos(y',x)\cos(z',z)\right]\\ &\quad +\tau_{xy}\left[\cos(y',x)\cos(z',y) + \cos(y',y)\cos(z',x)\right]\end{aligned} \tag{B.9}$$

It is important to recognize the meaning of these equations. For example, Eq. (B.7) can be used to find the shear stress acting in the x' direction on a surface having an outer normal directed in the z' direction if the six Cartesian stress components σ_x, σ_y, σ_z, τ_{xy}, τ_{yz}, and t_{zx} at the point and the orientation of the coordinate system $0'x'y'z'$ are known.

B.3 Laws of Strain Transformation

In the study of stresses we found that three normal stresses σ_x, σ_y, and σ_z and three shear stresses τ_{xy}, τ_{yz}, and τ_{zx} act on planes parallel to the Cartesian planes. Similarly, three normal strains ϵ_x, ϵ_y, and ϵ_z and three shear strains γ_{xy}, γ_{yz}, and γ_{zx} characterize the behavior of line segments originally parallel to the Cartesian axes. From Durelli et al. (1958) the laws of strain transformation can be written directly from the laws of stress transformation given in Eqs. (B.4) to (B.9) if the following replacements are made to these equations:

$$\begin{gathered}\sigma_x \leftarrow \epsilon_x, \quad \sigma_y \leftarrow \epsilon_y, \quad \sigma_z \leftarrow \epsilon_z, \quad 2\tau_{xy} \leftarrow \gamma_{xy},\\ 2\tau_{yz} \leftarrow \gamma_{yz}, \quad 2\tau_{zx} \leftarrow \gamma_{zx}\end{gathered} \tag{B.10}$$

B.4 Hooke's Law Generalized

What is the relationship between stresses and strains when the stress system is not in simple tension or compression as in Eq. (3.22)? The stress system can be defined by the six components $\sigma_x, \sigma_y, \sigma_z, \tau_{xy}, \tau_{yz}$, and τ_{zx}; and the strain system, by the six components ϵ_x, ϵ_y, ϵ_z, γ_{xy}, γ_{yz}, and γ_{zx}. Thus, a generalization of Hooke's law is to make each stress component a linear function of the strain components or

$$\begin{aligned}\sigma_x &= C_{11}\epsilon_x + C_{12}\epsilon_y + C_{13}\epsilon_z + C_{14}\gamma_{xy}\\ &\quad +C_{15}\gamma_{yz} + C_{16}\gamma_{zx}\\ \sigma_y &= C_{21}\epsilon_x + C_{22}\epsilon_y + C_{23}\epsilon_z + C_{24}\gamma_{xy}\\ &\quad +C_{25}\gamma_{yz} + C_{26}\gamma_{zx}\\ &\cdots\\ &\cdots\\ &\cdots\\ \tau_{zx} &= C_{61}\epsilon_x + C_{62}\epsilon_y + C_{63}\epsilon_z + C_{64}\gamma_{xy}\\ &\quad +C_{65}\gamma_{yz} + C_{66}\gamma_{zx}\end{aligned} \tag{B.11}$$

where C_{11}, C_{12}, ..., C_{66} = elastic material coefficients independent of stress or strain. Equation (B.11) is valid for many hard materials over a strain range of practical interest in designing machine elements.

The assumption of an isotropic material implies that the stress and strain components referred to a coordinate system $0x'y'z'$ of any arbitrary orientation must be related by the same elastic material coefficients $C_{11}, C_{12}, \ldots, C_{66}$. Thus, the six Eqs. (B.11) imply that the subscripts change, $x \to x'$, $y \to y'$, and $z \to z'$, while the constants C_{11}, C_{12}, ..., C_{66} remain the same. Thus,

$$\begin{aligned}\sigma_{x'} &= C_{11}\epsilon_{x'} + C_{12}\epsilon_{y'} + C_{13}\epsilon_{z'} + C_{14}\gamma_{x'y'}\\ &\quad +C_{15}\gamma_{y'z'} + C_{16}\gamma_{z'x'}\\ \sigma_{y'} &= C_{21}\epsilon_{x'} + C_{22}\epsilon_{y'} + C_{23}\epsilon_{z'} + C_{24}\gamma_{x'y'}\\ &\quad +C_{25}\gamma_{y'z'} + C_{26}\gamma_{z'x'}\\ &\cdots\\ &\cdots\\ &\cdots\\ \tau_{z'x'} &= C_{61}\epsilon_{x'} + C_{62}\epsilon_{y'} + C_{63}\epsilon_{z'} + C_{64}\gamma_{x'y'}\\ &\quad +C_{65}\gamma_{y'z'} + C_{66}\gamma_{z'x'}\end{aligned} \tag{B.12}$$

Recall that $0x'y'z'$ is an arbitrary coordinate system.

First, obtaining a new coordinate system $0x'y'z'$ by rotating $0z$ through a 180° angle results in Fig. B.2. Recall that

$$(x,y) = \text{Angle between } x \text{ and } y$$

From Fig. B.2 and the above notation

$$\begin{aligned}(x',x) &= 180^\circ, & (x',y) &= 90^\circ, & (x',z) &= 90^\circ\\ (y',y) &= 180^\circ, & (y',z) &= 90^\circ, & (y',x) &= 90^\circ\\ (z',z) &= 0^\circ, & (z',x) &= 90^\circ, & (z',y) &= 90^\circ\end{aligned} \tag{B.13}$$

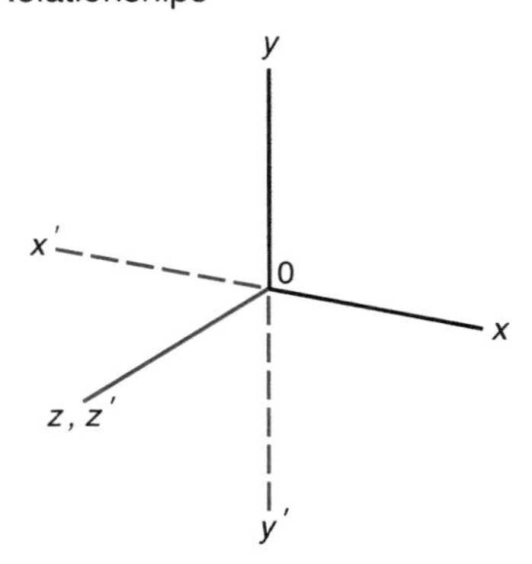

Figure B.2: Coordinate system $0x'y'z'$ obtained by rotating $0z$ by 180°.

Making use of Eq. (B.13) reduces Eqs. (B.4) to (B.9) to

$$\sigma_{z'} = \sigma_z, \quad \sigma_{x'} = \sigma_x, \quad \sigma_{y'} = \sigma_y, \quad \tau_{x',y'} = \tau_{xy},$$
$$\tau_{y'z'} = -\tau_{yz}, \quad \tau_{z'x'} = -\tau_{zx} \tag{B.14}$$

Similarly, for the strains

$$\epsilon_{z'} = \epsilon_z, \quad \epsilon_{x'} = \epsilon_x, \quad \epsilon_{y'} = \epsilon_y, \quad \gamma_{x'y'} = \gamma_{xy},$$
$$\gamma_{y'z'} = -\gamma_{yz}, \quad \gamma_{z'x'} = -\gamma_{zx} \tag{B.15}$$

Substituting Eqs. (B.14) and (B.15) into Eq. (B.12) gives

$$\begin{aligned}
\sigma_{x'} &= C_{11}\epsilon_x + C_{12}\epsilon_y + C_{13}\epsilon_z + C_{14}\gamma_{xy} \\
&\quad - C_{15}\gamma_{yz} - C_{16}\gamma_{zx} \\
\sigma_y &= C_{21}\epsilon_x + C_{22}\epsilon_y + C_{23}\epsilon_z + C_{24}\gamma_{xy} \\
&\quad - C_{25}\gamma_{yz} - C_{26}\gamma_{zx} \\
&\cdots \\
&\cdots \\
&\cdots \\
\tau_{z'x'} &= -C_{61}\epsilon_x - C_{62}\epsilon_y - C_{63}\epsilon_z - C_{64}\gamma_{xy} \\
&\quad + C_{65}\gamma_{yz} + C_{66}\gamma_{zx}
\end{aligned} \tag{B.16}$$

Comparing Eqs. (B.11) and (B.16) shows that

$$C_{15} = -C_{15} \qquad C_{16} = -C_{16}$$

implying that

$$\begin{aligned}
&C_{15} = C_{16} = C_{25} = C_{26} = C_{35} = C_{36} = C_{45} = C_{46} = 0 \\
&C_{51} = C_{52} = C_{53} = C_{54} = C_{61} = C_{62} = C_{63} = C_{64} = 0
\end{aligned} \tag{B.17}$$

Substituting Eqs. (B.17) into Eqs. (B.11) gives

$$\begin{aligned}
\sigma_x &= C_{11}\epsilon_x + C_{12}\epsilon_y + C_{13}\epsilon_z + C_{14}\gamma_{xy} \\
\sigma_y &= C_{21}\epsilon_x + C_{22}\epsilon_y + C_{23}\epsilon_z + C_{24}\gamma_{xy} \\
\sigma_z &= C_{31}\epsilon_x + C_{32}\epsilon_y + C_{33}\epsilon_z + C_{34}\gamma_{xy} \\
\tau_{xy} &= C_{41}\epsilon_x + C_{42}\epsilon_y + C_{43}\epsilon_z + C_{44}\gamma_{xy} \\
\tau_{yz} &= C_{55}\gamma_{yz} + C_{56}\gamma_{zx} \\
\tau_{zx} &= C_{65}\gamma_{yz} + C_{66}\gamma_{zx}
\end{aligned} \tag{B.18}$$

Thus, using the coordinate system of Fig. B.2 with the isotropic assumption reduces the 36 elastic material constants of Eqs. (B.11) to the 20 expressed in Eqs. (B.18).

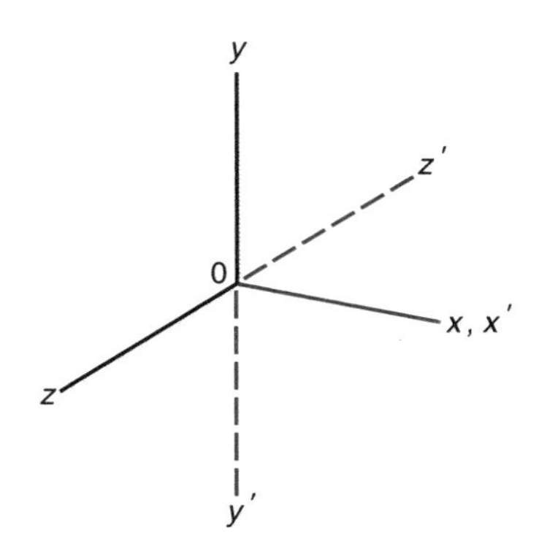

Figure B.3: Coordinate system $0x'y'z'$ obtained by rotating $0x$ by 180°.

Obtaining a new coordinate system by rotating $0x$ through a 180° angle results in Fig. B.3, giving

$$\begin{aligned}
(x', x) &= 0^\circ, & (x', y) &= 90^\circ, & (z', z) &= 90^\circ \\
(y', y) &= 180^\circ, & (y', x) &= 90^\circ, & (y', z) &= 90^\circ \\
(z', z) &= 180^\circ, & (z', x) &= 90^\circ, & (z', y) &= 90^\circ
\end{aligned} \tag{B.19}$$

Making use of Eqs. (B.19) reduces Eqs. (B.4) to (B.9) to

$$\begin{aligned}
\sigma_{z'} &= \sigma_z, & \sigma_{x'} &= \sigma_x, & \sigma_{y'} &= \sigma_y \\
\tau_{z'x'} &= -\tau_{zx}, & \tau_{x'y'} &= -\tau_{xy}, & \tau_{y'z'} &= \tau_{yz} \\
\epsilon_{x'} &= \epsilon_x, & \epsilon_{y'} &= \epsilon_y, & \epsilon_{z'} &= \epsilon_z \\
\gamma_{x'y'} &= -\gamma_{xy}, & \gamma_{y',z'} &= \gamma_{yz}, & \gamma_{z',x'} &= -\gamma_{zx}
\end{aligned} \tag{B.20}$$

The assumption of an isotropic material requires that the new coordinate system $0x'y'z'$ shown in Fig. B.3 must be related to the same elastic material constants $C_{11}, C_{12}, \ldots, C_{66}$ expressed in Eqs. (B.18), implying that

$$\begin{aligned}
\sigma_{x'} &= C_{11}\epsilon_{x'} + C_{12}\epsilon_{y'} + C_{13}\epsilon_{z'} + C_{14}\gamma_{x'y'} \\
\sigma_{y'} &= C_{21}\epsilon_{x'} + C_{22}\epsilon_{y'} + C_{23}\epsilon_{z'} + C_{24}\gamma_{x'y'} \\
\sigma_{z'} &= C_{31}\epsilon_{x'} + C_{32}\epsilon_{y'} + C_{33}\epsilon_{z'} + C_{34}\gamma_{x'y'} \\
\tau_{x'y'} &= C_{41}\epsilon_{x'} + C_{42}\epsilon_{y'} + C_{43}\epsilon_{z'} + C_{44}\gamma_{x'y'} \\
\tau_{y'z'} &= C_{55}\gamma_{y'z'} + C_{56}\gamma_{z'x'} \\
\tau_{z'x'} &= C_{65}\gamma_{y'z'} + C_{66}\gamma_{z'x'}
\end{aligned} \tag{B.21}$$

Substituting Eqs. (B.20) into Eqs. (B.21) gives

$$\begin{aligned}
\sigma_x &= C_{11}\epsilon_x + C_{12}\epsilon_y + C_{13}\epsilon_z - C_{14}\gamma_{xy} \\
\sigma_y &= C_{21}\epsilon_x + C_{22}\epsilon_y + C_{23}\epsilon_z - C_{24}\gamma_{xy} \\
\sigma_z &= C_{31}\epsilon_x + C_{32}\epsilon_y + C_{33}\epsilon_z - C_{34}\gamma_{xy} \\
-\tau_{xy} &= C_{41}\epsilon_x + C_{42}\epsilon_y + C_{43}\epsilon_z - C_{44}\gamma_{xy} \\
\tau_{yz} &= C_{55}\gamma_{yz} - C_{56}\gamma_{zx} \\
-\tau_{zx} &= C_{65}\gamma_{yz} - C_{66}\gamma_{zx}
\end{aligned} \tag{B.22}$$

Equations (B.18) and (B.22) will agree only if

$$C_{14} = C_{24} = C_{34} = C_{41} = C_{42} = C_{43} = C_{56} = C_{65} = 0 \tag{B.23}$$

Substituting Eqs. (B.13) into Eqs. (B.8) gives

$$\begin{aligned}
\sigma_x &= C_{11}\epsilon_x + C_{12}\epsilon_y + C_{13}\epsilon_z \\
\sigma_y &= C_{21}\epsilon_x + C_{22}\epsilon_y + C_{23}\epsilon_z \\
\sigma_z &= C_{31}\epsilon_x + C_{32}\epsilon_y + C_{33}\epsilon_z \\
\tau_{xy} &= C_{44}\gamma_{xy} \\
\tau_{yz} &= C_{55}\gamma_{yz} \\
\tau_{zx} &= C_{66}\gamma_{zx}
\end{aligned} \tag{B.24}$$

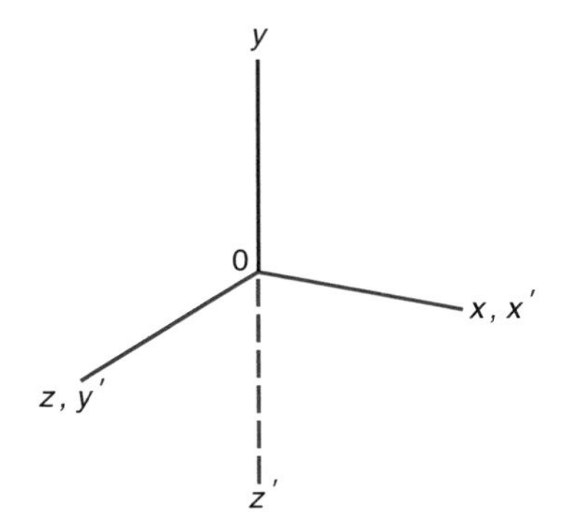

Figure B.4: Coordinate system $0x'y'z'$ obtained by rotating $0x$ by $90°$.

Thus, using the coordinate system of Fig. B.3 with the isotropic assumption reduces the 20 elastic material constants of Eqs. (B.18) to the 12 expressed in Eqs. (B.24).

Third, rotating $0x$ through a $90°$ angle results in Fig. B.4, implying that

$$\begin{array}{lll} (x',x)=0°, & (x',y)=90°, & (x',z)=90° \\ (y',y)=90°, & (y',x)=90°, & (y',z)=0° \\ (z',z)=90°, & (z',x)=90°, & (z',y)=180° \end{array} \tag{B.25}$$

Making use of Eqs. (B.25) reduces Eqs. (B.4) to (B.9) and Eq. (2.37) to

$$\begin{array}{lll} \sigma_{z'}=\sigma_y, & \sigma_{x'}=\sigma_x, & \sigma_{y'}=\sigma_z \\ \tau_{z'x'}=-\tau_{xy}, & \tau_{x'y'}=\tau_{zx}, & \tau_{y'z'}=-\tau_{yz} \\ \epsilon_{z'}=\epsilon_y, & \epsilon_{x'}=\epsilon_x, & \epsilon_{y'}=\epsilon_z \\ \gamma_{z'x'}=-\gamma_{xy}, & \gamma_{x'y'}=\gamma_{zx}, & \gamma_{y'z'}=-\gamma_{yz} \end{array} \tag{B.26}$$

The assumption of an isotropic material requires that the new coordinate system $0x'y'z'$ shown in Fig. B.4 must be related to the same elastic material constants expressed in Eqs. (B.24), implying that

$$\begin{aligned} \sigma_{x'} &= C_{11}\epsilon_{x'}+C_{12}\epsilon_{y'}+C_{13}\epsilon_{z'} \\ \sigma_{y'} &= C_{21}\epsilon_{x'}+C_{22}\epsilon_{y'}+C_{23}\epsilon_{z'} \\ \sigma_{z'} &= C_{31}\epsilon_{x'}+C_{32}\epsilon_{y'}+C_{33}\epsilon_{z'} \\ \tau_{x'y'} &= C_{44}\gamma_{x'y'} \\ \tau_{y'z'} &= C_{55}\gamma_{y'z'} \\ \tau_{z'x'} &= C_{66}\gamma_{z'x'} \end{aligned} \tag{B.27}$$

Substituting Eqs. (B.26) into Eqs. (B.27) gives

$$\begin{aligned} \sigma_x &= C_{11}\epsilon_x+C_{12}\epsilon_y+C_{13}\epsilon_z \\ \sigma_y &= C_{21}\epsilon_x+C_{22}\epsilon_y+C_{23}\epsilon_z \\ \sigma_z &= C_{31}\epsilon_x+C_{32}\epsilon_y+C_{33}\epsilon_z \\ \tau_{xy} &= C_{44}\gamma_{xy} \\ \tau_{yz} &= C_{55}\gamma_{yz} \\ \tau_{zx} &= C_{66}\gamma_{zx} \end{aligned} \tag{B.28}$$

Equations (B.24) and (B.28) will agree only if

$$C_{12}=C_{13} \qquad C_{21}=C_{31} \qquad C_{23}=C_{32}$$
$$C_{22}=C_{33} \qquad C_{44}=C_{66} \tag{B.29}$$

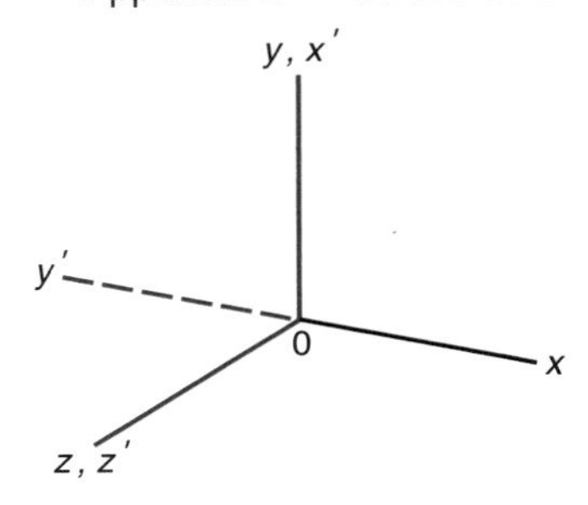

Figure B.5: Coordinate system $0x'y'z'$ obtained by rotating $0z$ by $90°$.

Substituting Eqs. (B.29) into Eqs. (B.28) gives

$$\begin{aligned} \sigma_x &= C_{11}\epsilon_x+C_{12}(\epsilon_y+\epsilon_z) \\ \sigma_y &= C_{21}\epsilon_x+C_{22}\epsilon_y+C_{23}\epsilon_z \\ \sigma_z &= C_{21}\epsilon_x+C_{22}\epsilon_z+C_{23}\epsilon_y \\ \tau_{xy} &= C_{44}\gamma_{xy} \\ \tau_{yz} &= C_{55}\gamma_{yz} \\ \tau_{zx} &= C_{66}\gamma_{zx} \end{aligned} \tag{B.30}$$

Thus, using the coordinate system of Fig. B.4 with the isotropic assumption reduces the 12 elastic material constants of Eqs. (B.24) to the 7 expressed in Eqs. (B.30).

Fourth, rotating $0z$ through a $90°$ angle results in Fig. B.5, giving

$$\begin{array}{lll} (x',x)=90°, & (x',y)=0°, & (x',z)=90° \\ (y',y)=90°, & (y',x)=180°, & (y',z)=90° \\ (z',z)=0°, & (z',x)=90°, & (z',y)=90° \end{array} \tag{B.31}$$

Making use of Eqs. (B.31) reduces Eqs. (B.4) to (B.9) and Eq. (2.37) to

$$\begin{array}{lll} \sigma_{z'}=\sigma_z, & \sigma_{x'}=\sigma_y, & \sigma_{y'}=\sigma_x \\ \tau_{z'x'}=\tau_{yz}, & \tau_{x'y'}=-\tau_{xy}, & \tau_{y'z'}=-\tau_{zx} \\ \epsilon_{z'}=\epsilon_z, & \epsilon_{x'}=\epsilon_y, & \epsilon_{y'}=\epsilon_x \\ \gamma_{z'x'}=\gamma_{yz}, & \gamma_{x'y'}=-\gamma_{xy}, & \gamma_{y'z'}=-\gamma_{zx} \end{array} \tag{B.32}$$

The assumption of an isotropic material requires that Eqs. (B.30) be written as

$$\begin{aligned} \sigma_{x'} &= C_{11}\epsilon_{x'}+C_{12}(\epsilon_{y'}+\epsilon_{z'}) \\ \sigma_{y'} &= C_{21}\epsilon_{x'}+C_{22}\epsilon_{y'}+C_{23}\epsilon_{z'} \\ \sigma_{z'} &= C_{21}\epsilon_{x'}+C_{22}\epsilon_{z'}+C_{23}\epsilon_{y'} \\ \tau_{x'y'} &= C_{44}\gamma_{x'y'} \\ \tau_{y'z'} &= C_{55}\gamma_{y'z'} \\ \tau_{z'x'} &= C_{66}\gamma_{z'x'} \end{aligned} \tag{B.33}$$

Substituting Eqs. (B.32) into Eqs. (B.33) gives

$$\begin{aligned} \sigma_x &= C_{11}\epsilon_y+C_{12}(\epsilon_x+\epsilon_z) \\ \sigma_y &= C_{21}\epsilon_y+C_{22}\epsilon_x+C_{23}\epsilon_z \\ \sigma_z &= C_{21}\epsilon_y+C_{22}\epsilon_z+C_{23}\epsilon_x \\ \tau_{xy} &= C_{44}\gamma_{xy} \\ \tau_{yz} &= C_{55}\gamma_{yz} \\ \tau_{zx} &= C_{66}\gamma_{zx} \end{aligned} \tag{B.34}$$

Equations (B.30) and (B.34) will agree only if

$$C_{22}=C_{11} \qquad C_{21}=C_{12} \qquad C_{23}=C_{12} \qquad C_{55}=C_{44} \tag{B.35}$$

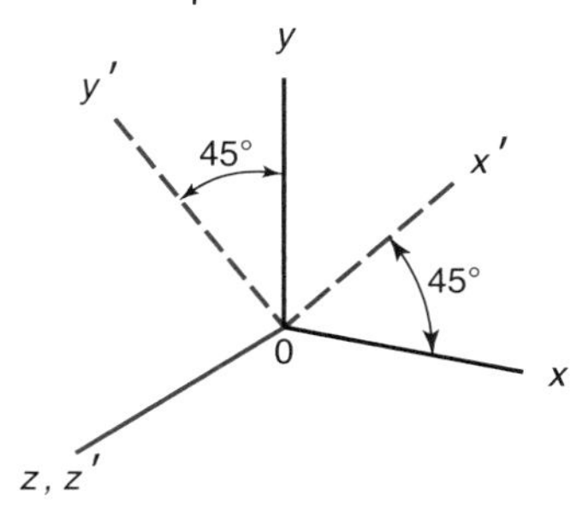

Figure B.6: Coordinate system $0x'y'z'$ obtained by rotating $0z$ by $45°$.

Substituting Eqs. (B.37) into Eqs. (B.34) gives

$$\begin{aligned}\sigma_x &= C_{11}\epsilon_x + C_{12}(\epsilon_y + \epsilon_z)\\ \sigma_y &= C_{11}\epsilon_y + C_{12}(\epsilon_x + \epsilon_z)\\ \sigma_z &= C_{11}\epsilon_z + C_{12}(\epsilon_x + \epsilon_y)\\ \tau_{xy} &= C_{44}\gamma_{xy}\\ \tau_{yz} &= C_{55}\gamma_{yz}\\ \tau_{zx} &= C_{66}\gamma_{zx}\end{aligned} \quad \text{(B.36)}$$

Thus, using the coordinate system of Fig. B.5 with the isotropic assumption reduces the seven elastic material constants of Eqs. (B.30) to the three expressed in Eqs. (B.36).

Fifth and finally, rotating $0z$ through a $45°$ angle results in Fig. B.6, implying that

$$\begin{array}{lll}(x',x)=45°, & (x',y)=45°, & (x',z)=90°\\ (y',y)=45°, & (y',x)=135°, & (y',z)=90°\\ (z',z)=0°, & (z',x)=90°, & (z',y)=90°\end{array} \quad \text{(B.37)}$$

Making use of Eqs. (B.37) reduces Eqs. (B.4) to (B.9) and Eq. (2.37) to

$$\sigma_{z'} = \sigma_z,$$

$$\sigma_{x'} = \frac{1}{2}(\sigma_x + \sigma_y) + \tau_{xy},$$

$$\sigma_{y'} = \frac{1}{2}(\sigma_x + \sigma_y) - \tau_{xy}$$

$$\tau_{z'x'} = \frac{1}{\sqrt{2}}(\tau_{zx} + \tau_{yz}),$$

$$\tau_{x'y'} = \frac{1}{2}(\sigma_y - \sigma_x),$$

$$\tau_{y'z'} = \frac{1}{\sqrt{2}}(\tau_{yz} - \tau_{zx})$$

$$\epsilon_{z'} = \epsilon_z,$$

$$\epsilon_{x'} = \frac{1}{2}(\epsilon_x + \epsilon_y + \gamma_{xy}),$$

$$\epsilon_{y'} = \frac{1}{2}(\epsilon_x + \epsilon_y - \gamma_{xy})$$

$$\gamma_{z'x'} = \frac{1}{\sqrt{2}}(\gamma_{zx} + \gamma_{yz}),$$

$$\gamma_{x'y'} = \frac{1}{2}(\epsilon_y - \epsilon_x),$$

$$\gamma_{y'z'} = \frac{1}{\sqrt{2}}(\gamma_{yz} - \gamma_{zx}) \quad \text{(B.38)}$$

The assumption of an isotropic material requires that Eqs. (B.37) be written as

$$\begin{aligned}\sigma_{x'} &= C_{11}\epsilon_{x'} + C_{12}(\epsilon_{y'} + \epsilon_{z'})\\ \sigma_{y'} &= C_{11}\epsilon_{y'} + C_{12}(\epsilon_{x'} + \epsilon_{z'})\\ \sigma_{z'} &= C_{11}\epsilon_{z'} + C_{12}(\epsilon_{x'} + \epsilon_{y'})\\ \tau_{x'y'} &= C_{44}\gamma_{x'y'}\\ \tau_{y'z'} &= C_{55}\gamma_{y'z'}\\ \tau_{z'x'} &= C_{66}\gamma_{z'x'}\end{aligned} \quad \text{(B.39)}$$

Substituting Eqs. (B.38) into the first of Eqs. (B.39) gives

$$\begin{aligned}\frac{1}{2}(\sigma_x + \sigma_y) + \tau_{xy} &= C_{11}\left(\frac{\gamma_{xy}}{2} + \frac{\epsilon_x + \epsilon_y}{2}\right)\\ &\quad + C_{12}\left(\frac{\epsilon_x + \epsilon_y}{2} - \frac{\gamma_{xy}}{2} + \epsilon_z\right)\end{aligned}$$

and substituting the expressions for σ_x and σ_y in Eqs. (B.36) into the above equation gives

$$\tau_{xy} = (C_{11} - C_{12})\frac{\gamma_{xy}}{2} \quad \text{(B.40)}$$

Comparing Eq. (B.40) with the txy expression in Eqs. (B.36) gives

$$C_{11} = C_{12} + 2C_{44} \quad \text{(B.41)}$$

Letting $C_{12} = \lambda$ and $C_{44} = G$ and making use of Eq. (B.41) gives Eqs. (B.36) as

$$\begin{aligned}\sigma_x &= (\lambda + 2G)\epsilon_x + \lambda(\epsilon_y + \epsilon_x)\\ \sigma_y &= (\lambda + 2G)\epsilon_y + \lambda(\epsilon_x + \epsilon_z)\\ \sigma_z &= (\lambda + 2G)\epsilon_z + \lambda(\epsilon_x + \epsilon_y)\\ \tau_{xy} &= G\gamma_{xy}\\ \tau_{yz} &= G\gamma_{yz}\\ \tau_{zx} &= G\gamma_{zx}\end{aligned} \quad \text{(B.42)}$$

Thus, using Fig. B.6 with the isotropic assumption reduces the number of elastic material constants from the 36 expressed in Eqs. (B.11) to the 2 expressed in Eqs. (B.42).

Equation (B.42) can be solved for strains to give

$$\begin{aligned}\epsilon_x &= \frac{(\lambda + G)\sigma_x}{G(3\lambda + 2G)} - \frac{\lambda(\sigma_y + \sigma_z)}{2G(3\lambda + 2G)}\\ \epsilon_y &= \frac{(\lambda + G)\sigma_y}{G(3\lambda + 2G)} - \frac{\lambda(\sigma_x + \sigma_z)}{2G(3\lambda + 2G)}\\ \epsilon_z &= \frac{(\lambda + G)\sigma_z}{G(3\lambda + 2G)} - \frac{\lambda(\sigma_x + \sigma_y)}{2G(3\lambda + 2G)}\\ \gamma_{xy} &= \frac{\tau_{xy}}{G}\\ \gamma_{yz} &= \frac{\tau_{yz}}{G}\\ \gamma_{zx} &= \frac{\tau_{zx}}{G}\end{aligned} \quad \text{(B.43)}$$

For isotropic but *nonhomogeneous* materials the constants λ and G are functions of the space coordinates x, y, and z and vary from point to point. For isotropic and homogeneous materials these constants are not functions of the space coordinates and do not vary from point to point. They depend only on the particular material.

If the $0x$, $0y$, and $0z$ axes are chosen along the principal axes of stress, $\tau_{xy} = \dot{\tau}_{yz} = \tau_{zx} = 0$. From Eqs. (B.43) it follows that $\gamma_{xy} = \gamma_{yz} = \gamma_{zx} = 0$. Thus, for isotropic materials the principal axes of stress and strain coincide.

B.5 Physical Significance of Elastic Material Constants

The first elastic material constant is shear modulus, or modulus of rigidity, G in pascals. The second elastic material constant λ, known as Lamé's constant, is of no particular physical significance.

In a uniaxial stress state where $\sigma_y = \sigma_z = \tau_{xy} = \tau_{yz} = \tau_{zx} = 0$ and σ_x is the applied uniaxial stress, Eqs. (B.43) give

$$\epsilon_x = \frac{(\lambda + G)\sigma_x}{G(3\lambda + 2G)} \tag{B.44}$$

$$\epsilon_y = \epsilon_x = -\frac{\lambda\sigma_x}{2G(3\lambda + 2G)} \tag{B.45}$$

Comparing Eq. (B.44) with Eq. (3.23) gives

$$\lambda = \frac{G(E - 2G)}{3G - E} \tag{B.46}$$

where E = modulus of elasticity, the third elastic material constant, covered in Sec. 3.5.2. Relating the transverse strain in Eq. (B.45) to Poisson's ratio in Eq. (3.4) gives

$$\nu = \frac{\lambda}{2(\lambda + G)} \tag{B.47}$$

or solving for λ

$$\lambda = \frac{2G\nu}{1 - 2\nu} \tag{B.48}$$

Thus, Eqs. (B.46) and (B.48) express λ in terms of two known elastic material constants. Equating Eqs. (B.46) and (B.48) gives

$$\nu = \frac{E - 2G}{2G} \tag{B.49}$$

where Poisson's ratio ν is the fourth elastic material constant. Equation (B.49) is equivalent to Eq. (3.6). The range of λ is between zero and 0.5. At $\nu = 0$ no transverse deformation, but rather longitudinal deformation, occurs. At $\nu = 0.5$ the material exhibits constant volume. The volume increase longitudinally is the same as the shrinkage in the transverse direction.

Besides the four elastic material constants G, λ, E, and ν a fifth is provided by bulk modulus K, the ratio of applied hydrostatic pressure to observed volume shrinkage per unit volume, which is in pascals (pounds per square inch). The bulk modulus is obtained from hydrostatic compression or when

$$\sigma_x = \sigma_y = \sigma - z = -p \qquad \text{for } p > 0$$

$$\tau_{xy} = \tau_{yz} = \tau_{zx} = 0$$

Substituting the above into Eqs. (B.43) gives

$$\epsilon_x = \epsilon_y = \epsilon_z = -\frac{p}{3\lambda + 2G} \tag{B.50}$$

But the total strain is

$$\epsilon = \epsilon_x + \epsilon_y + \epsilon_z \rightarrow p = -\frac{(3\lambda + 2G)\epsilon}{3} = -K\epsilon \tag{B.51}$$

where

$$K = \frac{3\lambda + 2G}{3} \tag{B.52}$$

It is possible to express any of the elastic material constants (G, λ, E, ν, and K) if two of these five constants are given. These relationships are expressed in Table B.1.

B.6 Stress-Strain Equations in Terms of Modulus of Elasticity and Poisson's Ratio

From Table B.1, λ and G can be expressed in terms of E and ν as

$$\lambda = \frac{\nu E}{(1 + \nu)(1 - 2\nu)} \qquad \text{and} \qquad G = \frac{E}{2(1 + \nu)} \tag{B.53}$$

Substituting Eqs. (B.53) into Eqs. (B.43) gives

$$\begin{aligned}
\epsilon_x &= \frac{1}{E}\left[\sigma_x - \nu(\sigma_y + \sigma_z)\right] \\
\epsilon_y &= \frac{1}{E}\left[\sigma_y - \nu(\sigma_z + \sigma_x)\right] \\
\epsilon_z &= \frac{1}{E}\left[\sigma_z - \nu(\sigma_x + \sigma_y)\right] \\
\gamma_{xy} &= \frac{2(1 + \nu)}{E}\tau_{xy} \\
\gamma_{yz} &= \frac{2(1 + \nu)}{E}\tau_{yz} \\
\gamma_{zx} &= \frac{2(1 + \nu)}{E}\tau_{zx}
\end{aligned} \tag{B.54}$$

Similarly, the stress components can be expressed in terms of strains as

$$\begin{aligned}
\sigma_x &= \frac{E}{(1 + \nu)(1 - 2\nu)}\left[(1 - \nu)\epsilon_x + \nu(\epsilon_y + \epsilon_z)\right] \\
\sigma_y &= \frac{E}{(1 + \nu)(1 - 2\nu)}\left[(1 - \nu)\epsilon_y + \nu(\epsilon_x + \epsilon_z)\right] \\
\sigma_z &= \frac{E}{(1 + \nu)(1 - 2\nu)}\left[(1 - \nu)\epsilon_z + \nu(\epsilon_x + \epsilon_y)\right] \\
\tau_{xy} &= \frac{E}{2(1 + \nu)}\gamma_{xy} \\
\tau_{yz} &= \frac{E}{2(1 + \nu)}\gamma_{yz} \\
\tau_{zx} &= \frac{E}{2(1 + \nu)}\gamma_{zx}
\end{aligned} \tag{B.55}$$

Although the stress and strain equations given in Eqs. (B.54) and (B.55), respectively, are expressed in terms of E and ν, by using Table B.1 these equations can be rewritten in terms of any two of the five elastic material constants (G, λ, E, ν, and K). The reason for using E and ν is that data for a particular material can be readily obtained for these constants as demonstrated in Tables 3.2 and 3.3. For rubber ($\nu \rightarrow 0.5$) it is more accurate to use the bulk modulus K together with the shear modulus G, since $1 - 2\nu$ appears in the denominator.

For the special case in which the x, y, and z axes are coincidental with the principal axes 1, 2, and 3 (the *triaxial stress state*), Eqs. (B.54) and (B.55) are simplified by virtue of all shear stresses and shear strains being equal to zero:

$$\begin{aligned}
\epsilon_1 &= \frac{1}{E}\left[\sigma_1 - \nu(\sigma_2 + \sigma_3)\right] \\
\epsilon_2 &= \frac{1}{E}\left[\sigma_2 - \nu(\sigma_1 + \sigma_3)\right] \\
\epsilon_3 &= \frac{1}{E}\left[\sigma_3 - \nu(\sigma_2 + \sigma_3)\right]
\end{aligned} \tag{B.56}$$

Table B.1: Relationships between elastic material constants for isotropic materials.

Constants involved	Lamé's constant, λ	Shear modulus, G	Modulus of elasticity, E	Poisson's ratio, ν	Bulk modulus K
λ G	—	—	$E = \frac{G(3\lambda+2G)}{\lambda+G}$	$\nu = \frac{\lambda}{2(\lambda+G)}$	$K = \frac{3\lambda+2G}{3}$
λ E	—	$G = \frac{A^{\dagger}+(E-3\lambda)}{4}$	—	$\nu = \frac{A-(E+\lambda)}{4\lambda}$	$K = \frac{A+(3\lambda+E)}{6}$
λ ν	—	$G = \frac{\lambda(1-2\nu)}{2\nu}$	$E = \frac{\lambda(1+\nu)(1-2\nu)}{\nu}$	—	$K = \frac{\lambda(1+\nu)}{3\nu}$
λ K	—	$G = \frac{3(K-\lambda)}{2}$	$E = \frac{9K(K-\lambda)}{3K-\lambda}$	$\nu = \frac{\lambda}{3K-\lambda}$	—
G E	$\lambda = \frac{G(2G-E)}{E-3G}$	—	—	$\nu = \frac{E-2G}{2G}$	$K = \frac{GE}{3(3G-E)}$
G ν	$\lambda = \frac{2G\nu}{1-2\nu}$	—	$E = 2G(1+\nu)$	—	$K = \frac{2G(1+\nu)}{3(1-2\nu)}$
G K	$\lambda = \frac{3K-2G}{3}$	—	$E = \frac{9KG}{3K+G}$	$\nu = \frac{3K-2G}{2(3K+G)}$	—
E ν	$\lambda = \frac{\nu E}{(1+\nu)(1-2\nu)}$	$G = \frac{E}{2(1+\nu)}$	—	—	$K = \frac{E}{3(1-2\nu)}$
E K	$\lambda = \frac{3K(3K-E)}{9K-E}$	$G = \frac{3EK}{9K-E}$	—	$\nu = \frac{3K-E}{6K}$	—
ν K	$\lambda = \frac{3K\nu}{1+\nu}$	$G = \frac{3K(1-2\nu)}{2(1+\nu)}$	$E = 3K(1-2\nu)$	—	—

$^{\dagger}A = \left[(E+\lambda)^2 + 8\lambda^2\right]^{1/2}$

$$\begin{aligned}
\sigma_1 &= \frac{E}{(1+\nu)(1-2\nu)}\left[(1-\nu)\epsilon_1 + \nu(\epsilon_2+\epsilon_3)\right]\\
\sigma_2 &= \frac{E}{(1+\nu)(1-2\nu)}\left[(1-\nu)\epsilon_2 + \nu(\epsilon_1+\epsilon_3)\right]\\
\sigma_3 &= \frac{E}{(1+\nu)(1-2\nu)}\left[(1-\nu)\epsilon_3 + \nu(\epsilon_1+\epsilon_2)\right] \quad \text{(B.57)}
\end{aligned}$$

For the commonly encountered *biaxial stress state*, one of the principal stresses (say, σ_3) is zero and Eqs. (B.56) become

$$\begin{aligned}
\epsilon_1 &= \frac{\sigma_1 - \nu\sigma_2}{E}\\
\epsilon_2 &= \frac{\sigma_2 - \sigma_1}{E}\\
\epsilon_3 &= -\frac{\nu(\sigma_1+\sigma_2)}{E} \quad \text{(B.58)}
\end{aligned}$$

For $\sigma_3 = 0$ the third of Eqs. (B.57) gives

$$\epsilon_3 = -\frac{\nu(\epsilon_1+\epsilon_2)}{1-\nu} \quad \text{(B.59)}$$

Substituting Eq. (B.59) into Eqs. (B.57) gives

$$\begin{aligned}
\sigma_1 &= \frac{E(\epsilon_1 + \nu\epsilon_2)}{1-\nu^2}\\
\sigma_2 &= \frac{E(\epsilon_2 + \nu\epsilon_1)}{1-\nu^2}\\
\sigma_3 &= 0 \quad \text{(B.60)}
\end{aligned}$$

For the *uniaxial stress state* Eqs. (B.58) and (B.60) must reduce to

$$\epsilon_1 = \frac{\sigma_1}{E} \qquad \epsilon_2 = \epsilon_3 = -\frac{\nu\sigma_1}{E} \quad \text{(B.61)}$$

$$\sigma_1 = E\epsilon_1 \qquad \sigma_2 = \sigma_3 = 0 \quad \text{(B.62)}$$

These expressions for the uniaxial, biaxial, and triaxial stress states can be expressed in tabular form as shown in Table B.2.

Recall that we chose to express the stress and strain in terms of the modulus of elasticity and Poisson's ratio. Furthermore, we are considering the special case in which the x, y, and z axes are coincidental with the principal axes 1, 2, and 3, thus implying that the shear stresses and strains are equal to zero. For this case the shear stress and strain are equal to zero.

Example B.1: Hookian Material

Given: Equations (B.42) and (B.43).

Find: Determine the modulus of elasticity E as a function of G and ν, where ν is given by Hooke's law for uniaxial tension.

$$\epsilon_x = \frac{\sigma_x}{E}$$

$$\epsilon_y = -\frac{\nu\sigma_x}{E}$$

$$\epsilon_z = -\frac{\nu\sigma_x}{E}$$

Solution: Equation (B.43) gives

$$\epsilon_x = \frac{(\lambda+G)\sigma_x}{G(3\lambda+2G)} - \frac{\lambda(\sigma_y+\sigma_z)}{2G(3\lambda+2G)} = \frac{\sigma_x}{E} - \frac{\nu(\sigma_y+\sigma_z)}{E}$$

$$\epsilon_y = \frac{(\lambda+G)\sigma_y}{G(3\lambda+2G)} - \frac{\lambda(\sigma_x+\sigma_z)}{2G(3\lambda+2G)} = \frac{\sigma_y}{E} - \frac{\nu(\sigma_x+\sigma_z)}{E}$$

$$\epsilon_z = \frac{(\lambda+G)\sigma_z}{G(3\lambda+2G)} - \frac{\lambda(\sigma_x+\sigma_y)}{2G(3\lambda+2G)} = \frac{\sigma_z}{E} - \frac{\nu(\sigma_x+\sigma_y)}{E}$$

$$\left.\begin{aligned}\frac{\lambda+G}{G(3\lambda+2G)} &= \frac{1}{E}\\ \frac{\lambda}{2G(3\lambda+2G)} &= \frac{\nu}{E}\end{aligned}\right\} \frac{\lambda}{2(\lambda+G)} = \nu$$

Table B.2: Principal stresses and strains in terms of modulus of elasticity and Poisson's ratio for uniaxial, biaxial, and triaxial stress states. (It is assumed that the x, y, and z axes are coincident with the principal axes 1, 2, and 3, thus implying that the shear stresses and strains are equal to zero.)

Type of stress	Principal strains	Principal stresses
Uniaxial	$\epsilon_1 = \frac{\sigma_1}{E}$	$\sigma_1 = E\epsilon_1$
	$\epsilon_2 = -\nu\epsilon_1$	$\sigma_2 = 0$
	$\epsilon_3 = -\nu\epsilon_1$	$\sigma_3 = 0$
Biaxial	$\epsilon_1 = \frac{\sigma_1}{E} - \frac{\nu\sigma_2}{E}$	$\sigma_1 = \frac{E(\epsilon_1 + \nu\epsilon_2)}{1-\nu^2}$
	$\epsilon_2 = \frac{\sigma_2}{E} - \frac{\nu\sigma_1}{E}$	$\sigma_2 = \frac{E(\epsilon_1 + \nu\epsilon_2)}{1-\nu^2}$
	$\epsilon_3 = -\frac{\nu\sigma_1}{E} - \frac{\nu\sigma_2}{E}$	$\sigma_3 = 0$
Triaxial	$\epsilon_1 = \frac{\sigma_1}{E} - \frac{\nu\sigma_2}{E} - \frac{\nu\sigma_3}{E}$	$\sigma_1 = \frac{E\epsilon_1(1-\nu) + \nu E(\epsilon_2 + \epsilon_3)}{1-\nu-2\nu^2}$
	$\epsilon_2 = \frac{\sigma_2}{E} - \frac{\nu\sigma_1}{E} - \frac{\nu\sigma_3}{E}$	$\sigma_2 = \frac{E\epsilon_2(1-\nu) + \nu E(\epsilon_1 + \epsilon_3)}{1-\nu-2\nu^2}$
	$\epsilon_3 = \frac{\sigma_3}{E} - \frac{\nu\sigma_1}{E} - \frac{\nu\sigma_2}{E}$	$\sigma_3 = \frac{E\epsilon_3(1-\nu) + \nu E(\epsilon_1 + \epsilon_2)}{1-\nu-2\nu^2}$

$$(\lambda + G)E = 3\lambda G + 2G^2$$

$$G(E - 2G) = \lambda(3G - E) \qquad \lambda = \frac{G(E-2G)}{3G-E}$$

$$\nu = \frac{G(E-2G)}{2(3G-E)\left[\frac{G(E-2G)}{3G-E} + G\right]}$$

$$E = (\nu + 1)2G \quad \text{and} \quad G = \frac{E}{2(1+\nu)}$$

The modulus of elasticity $E = 2G(\nu + 1)$, the shear modulus $G = E/2(1+\nu)$, and Poisson's ratio $\nu = E/2G - 1$.

B.7 Summary

Hooke's law was generalized in this appendix. That is, the linear relationship between stress and strain in the elastic range was generalized for the six components of stress: three normal stresses and three shear stresses. By using the laws of stress and strain transformation, the 36 elastic material constants were reduced to 2 through five different coordinate orientations. It was also found that, for the isotropic and homogeneous materials assumed throughout this chapter, these elastic material constants are not functions of the space coordinates and do not vary from point to point. They depend only on the particular material.

Relationships were presented between the four elastic material constants (the shear modulus or modulus of rigidity G, the modulus of elasticity E, the bulk modulus K, and Poisson's ratio ν) and Lamé's constant λ, which has no particular physical significance. It was shown how any two of the five constants can be expressed in terms of the other constants. For the special case in which the x, y, and z axes are coincidental with the principal axes, simplified equations were expressed where all shear stresses and shear strains were zero. For these situations uniaxial, biaxial, and triaxial principal stresses could be expressed. These equations serve as the foundation for the main text of this book.

References

Durelli, A.J., Phillips, E.A., and Tsao, C.H. (1958) *Introduction to the Theoretical and Experimental Analysis of Stress and Strain*, McGraw-Hill.

Timoshenko, S., and Goodier, J.N., (1970) *Theory of Elasticity*, McGraw-Hill.

Appendix C

Stress Intensity Factors for Some Common Crack Geometries

This appendix summarizes some stress intensity factors for common machine element configurations and test specimen geometries. Most of this appendix is taken from Suresh [1998], although Tada et al. [2000] was also used in the compilation.

C.1 Internally-Cracked Tension Specimen

The stress intensity factor for a center crack of length l_c in a plate of width b is given by

$$K_I = \sigma\sqrt{\frac{\pi l_c}{2}}Y \tag{C.1}$$

The geometry correction factor Y is shown in Fig. C.1. If the tensile stress is remote ($h >> b$), then the geometry factor is given by

$$Y = \sqrt{\sec\frac{\pi l_c}{2b}} \tag{C.2}$$

If an internal crack of length l_c exists a distance d from the edge of the tension member of width b, the geometry correction factor is shown in Fig. C.2.

C.2 Edge-Cracked Tension Specimen

For a single edge-cracked tension specimen of crack width l_c and width b, the stress intensity factor is given by

$$K_I = \sigma\sqrt{l_c}Y \tag{C.3}$$

The geometry factor Y is shown in Fig. C.3. If the tensile stress is remote ($h >> b$), then the geometry factor is given by

$$Y = 1.99 - 0.41\frac{l_c}{b} + 18.7\left(\frac{l_c}{b}\right)^2 - 38.48\left(\frac{l_c}{b}\right)^3 + 53.83\left(\frac{l_c}{b}\right)^4 \tag{C.4}$$

For a tensile specimen of width b with two edge cracks, each of length l_c, the geometry factor for use in Eq. (C.3) is given by

$$Y = 1.99 + 0.76\left(\frac{l_c}{b}\right) - 8.48\left(\frac{l_c}{b}\right)^2 + 27.36\left(\frac{l_c}{b}\right)^3 \tag{C.5}$$

C.3 Bending Specimen

The stress intensity factor for an edge-cracked bending specimen is given by

$$K_I = \left(\frac{6M}{w_t b^2}\right)\sqrt{\pi l_c}Y \tag{C.6}$$

where

M = applied moment, Nm (in.-lb)
w_t = beam width, m (in.)
b = beam height, m (in.)
Y = a geometry correction factor, given by

$$Y = 1.122 - 1.4\frac{l_c}{b} + 7.33\left(\frac{l_c}{b}\right)^2 - 13.08\left(\frac{l_c}{b}\right)^3 + 14.0\left(\frac{l_c}{b}\right)^4 \tag{C.7}$$

For a three-point bending specimen (see Fig. C.4), the stress intensity factor is given by

$$K_I = \left(\frac{3aF}{2w_t b^2}\right)\sqrt{\pi l_c}Y \tag{C.8}$$

Two cases are available. Using the notation of $\alpha = l_c/b$,

1. For $a/b = 2$,

$$Y = \frac{1.99 - \alpha(1-\alpha)\left(2.15 - 3.93\alpha + 2.7\alpha^2\right)}{(1+2\alpha)(1-\alpha)^{3/2}\sqrt{pi}} \tag{C.9}$$

2. For $a/b = 4$,

$$Y = 1.106 - 1.552\alpha + 7.71\alpha^2 - 13.53\alpha^3 + 14.23\alpha^4 \tag{C.10}$$

C.4 Central Hole with Crack in Tension Specimen

A common occurrence when a central hole is placed in a tension specimen is that a crack will initiate at the locations of

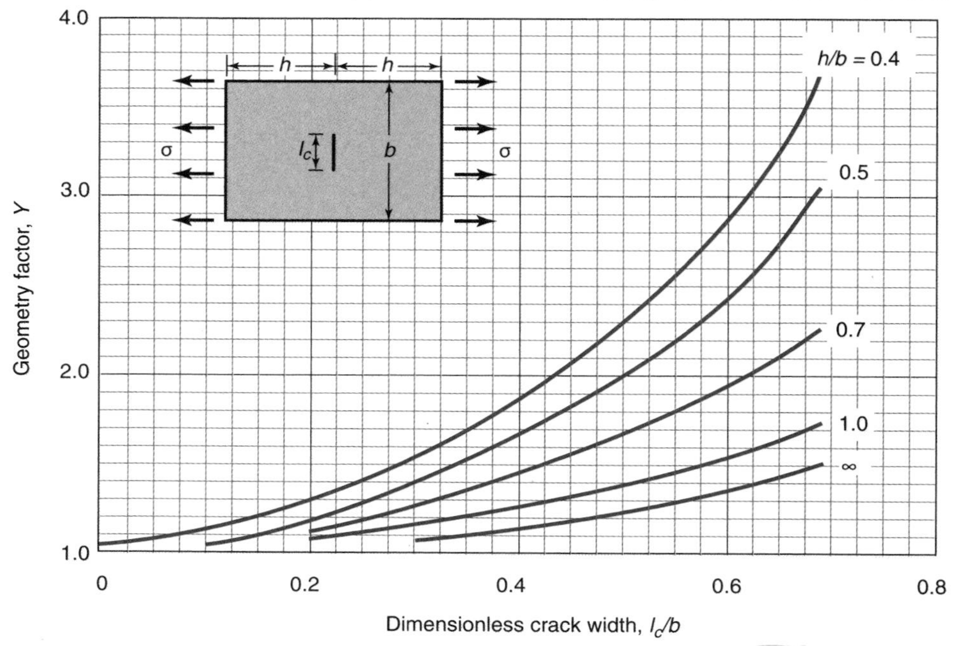

Figure C.1: Geometry factor for center-cracked tension specimen.

maximum stress. The effect on the stress state of the tension member can be quite significant, and a very large apparent crack can develop. This is treated as a center crack (the stress intensity factor is given by Eq. (C.1) but with a geometry correction factor as given in Fig. C.5.

C.5 Crack in Wall of Pressure Vessel

In the circumstance where the internally applied pressure varies with time, a crack can develop and propagate within the wall of a pressure vessel. For such occurrences, the stress concentration factor is given by Eq. (C.1) with a geometry correction factor as given in Fig. C.6.

References

Suresh, S. (1998) *Fatigue of Materials*, 2nd. ed., Cambridge University Press.

Tada, H., Paris, P.C., and Irwin, G.R. (2000), *The Stress Analysis of Cracks Handbook*, ASME Press.

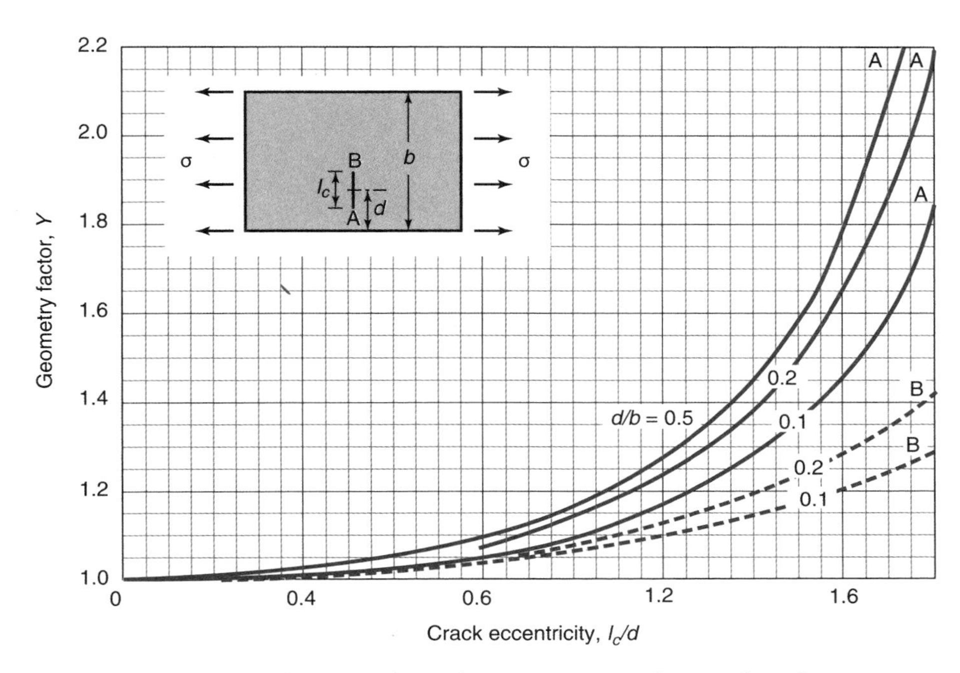

Figure C.2: Geometry factor for an non-centered internal crack.

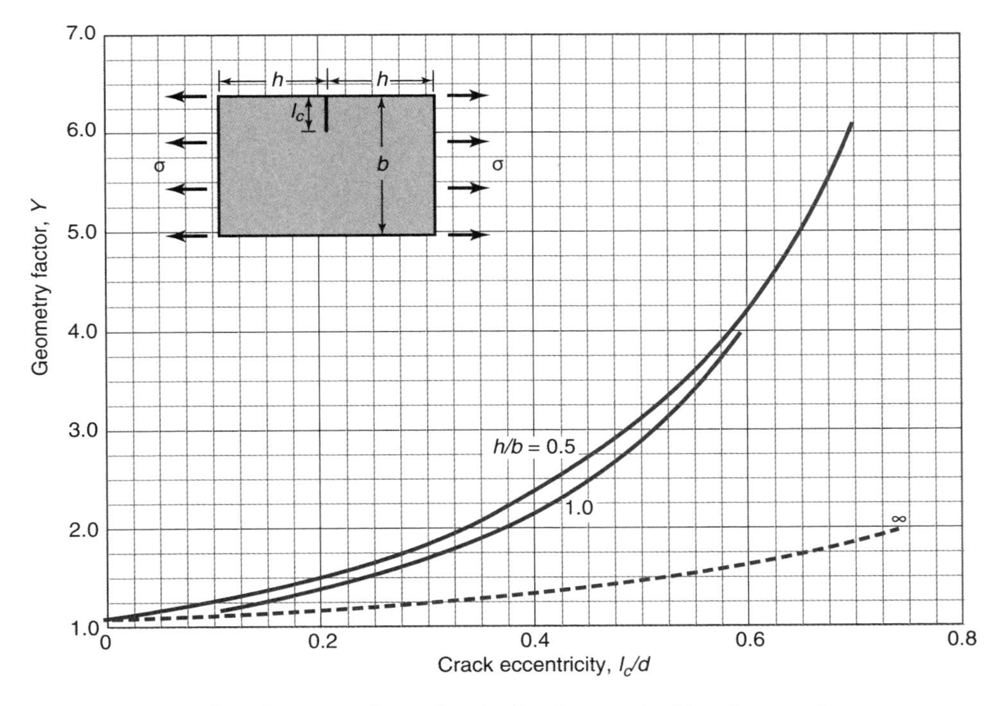

Figure C.3: Geometry factor for single edge-cracked tension specimen.

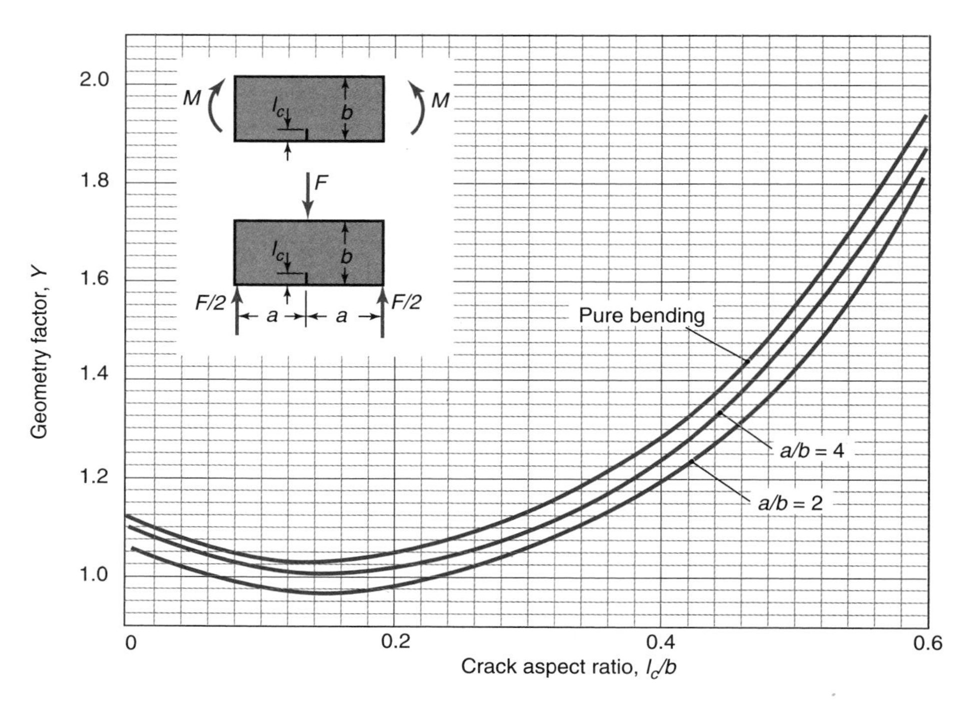

Figure C.4: Geometry factor for single edge-cracked tension specimen.

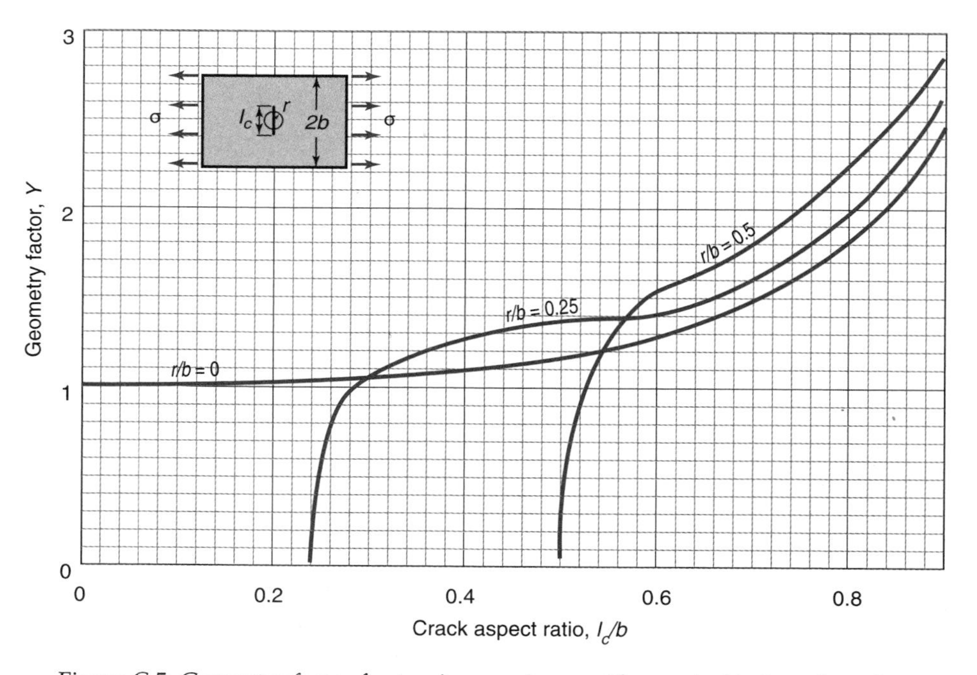

Figure C.5: Geometry factor for tension specimen with a central hole and cracks.

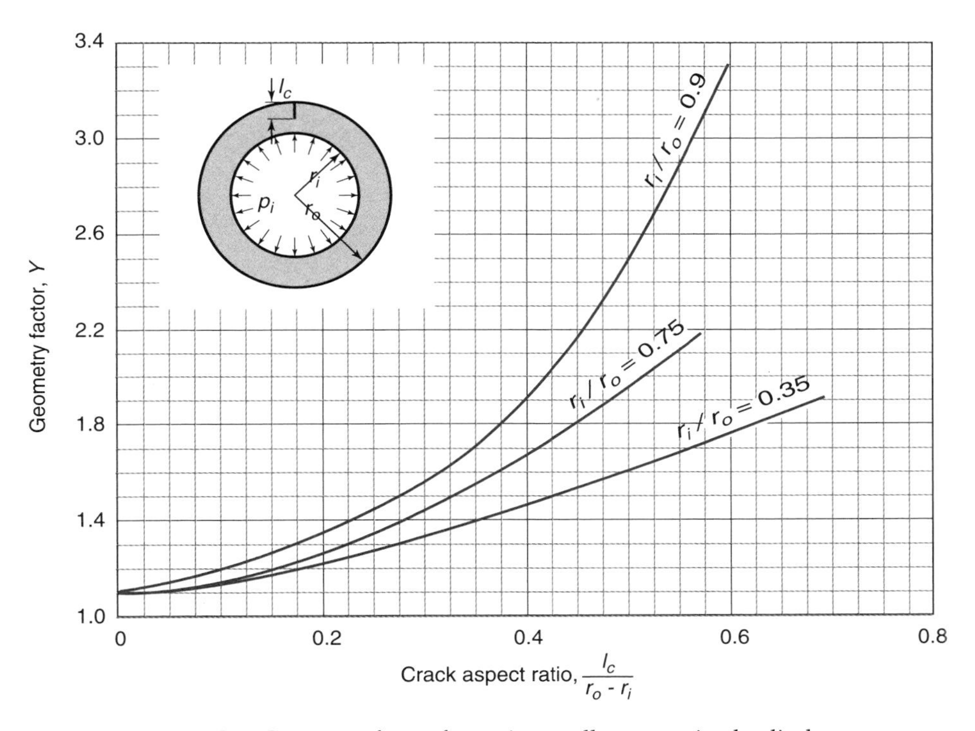

Figure C.6: Geometry factor for an internally pressurized cylinder.

Appendix D

Shear, Moment and Deflection of Selected Beams and Cantilevers

D.1 Cantilever loaded at free end

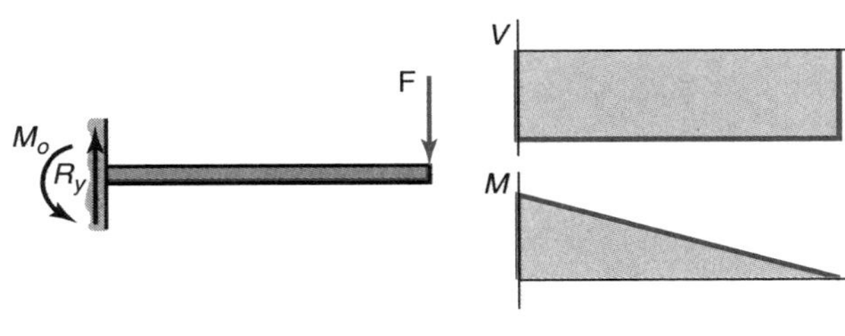

$$R_y = -V = F$$

$$M_o = F(l - x)$$

$$y = \frac{Fx^2}{6EI}(x - 3l)$$

$$y_{\max} = -\frac{Fl^3}{3EI}$$

D.2 Cantilever loaded off-end

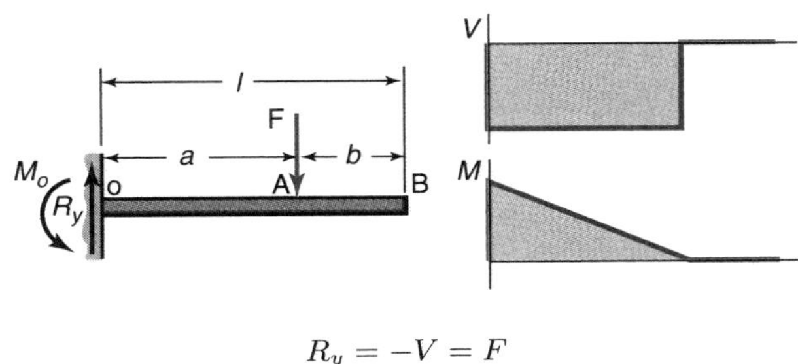

$$R_y = -V = F$$

$$M_{oA} = F(a - x)$$

$$M_{AB} = 0$$

$$M_o = Fa$$

$$y_{oA} = \frac{Fx^2}{6EI}(x - 3a)$$

$$y_{AB} = \frac{Fa^2}{6EI}(a - 3x)$$

$$y_{\max} = \frac{Fa^2}{6EI}(a - 3l)$$

D.3 Cantilever with uniform load

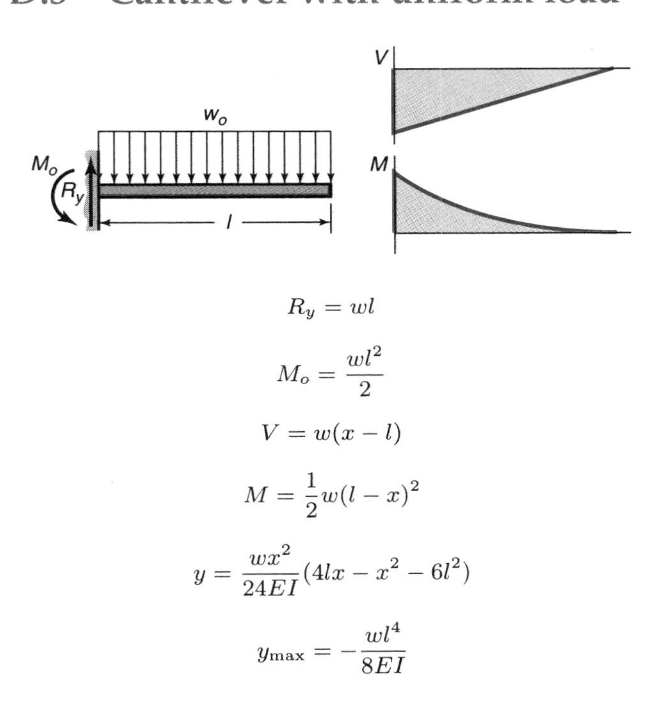

$$R_y = wl$$

$$M_o = \frac{wl^2}{2}$$

$$V = w(x - l)$$

$$M = \frac{1}{2}w(l - x)^2$$

$$y = \frac{wx^2}{24EI}(4lx - x^2 - 6l^2)$$

$$y_{\max} = -\frac{wl^4}{8EI}$$

D.4 Cantilever with moment load

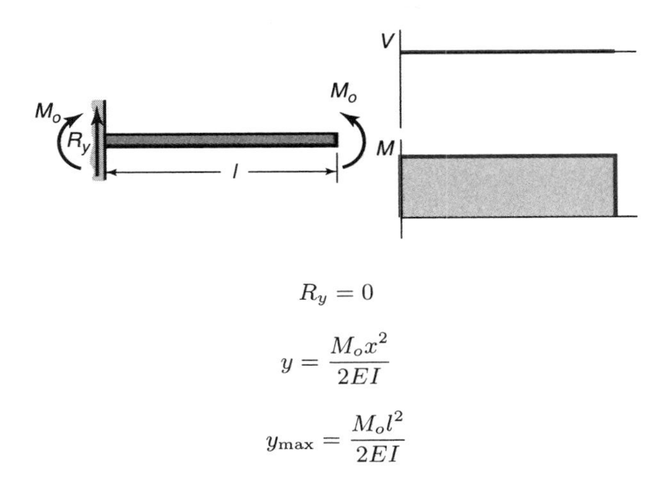

$$R_y = 0$$

$$y = \frac{M_o x^2}{2EI}$$

$$y_{\max} = \frac{M_o l^2}{2EI}$$

D.5 Simply-supported beam with point load at mid-span

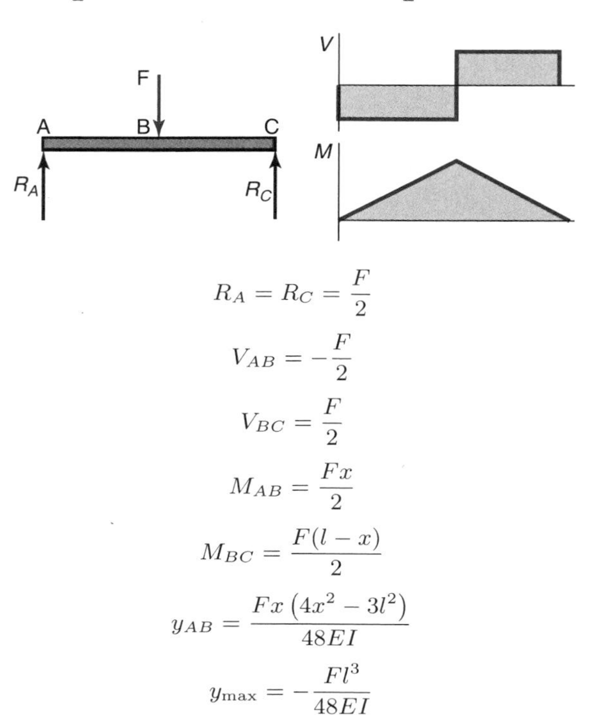

$$R_A = R_C = \frac{F}{2}$$

$$V_{AB} = -\frac{F}{2}$$

$$V_{BC} = \frac{F}{2}$$

$$M_{AB} = \frac{Fx}{2}$$

$$M_{BC} = \frac{F(l-x)}{2}$$

$$y_{AB} = \frac{Fx\left(4x^2 - 3l^2\right)}{48EI}$$

$$y_{\max} = -\frac{Fl^3}{48EI}$$

D.6 Simply-supported beam with intermediate point load

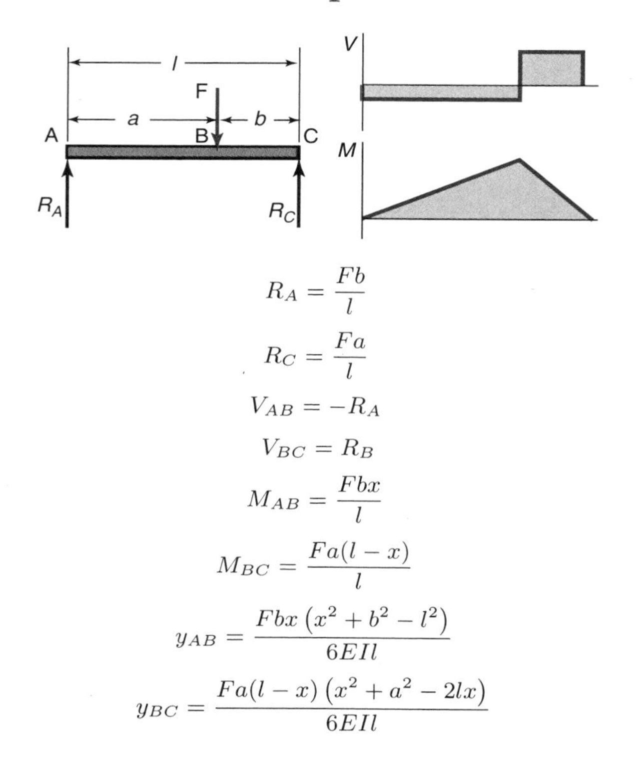

$$R_A = \frac{Fb}{l}$$

$$R_C = \frac{Fa}{l}$$

$$V_{AB} = -R_A$$

$$V_{BC} = R_B$$

$$M_{AB} = \frac{Fbx}{l}$$

$$M_{BC} = \frac{Fa(l-x)}{l}$$

$$y_{AB} = \frac{Fbx\left(x^2 + b^2 - l^2\right)}{6EIl}$$

$$y_{BC} = \frac{Fa(l-x)\left(x^2 + a^2 - 2lx\right)}{6EIl}$$

D.7 Simply-supported beam with uniform load

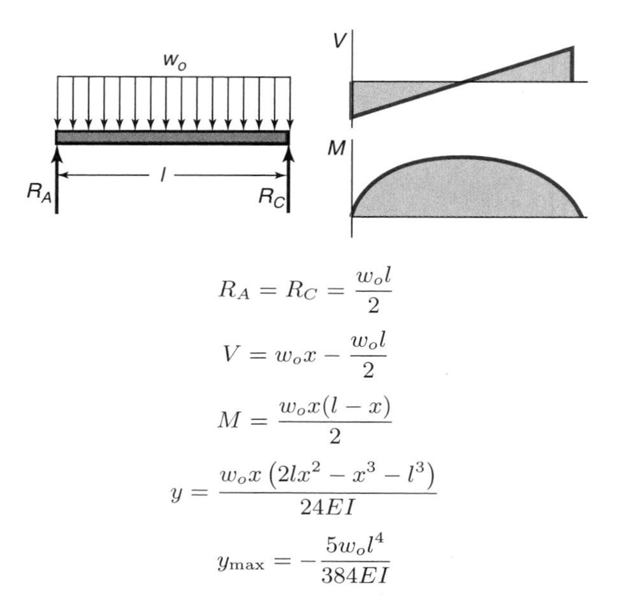

$$R_A = R_C = \frac{w_o l}{2}$$

$$V = w_o x - \frac{w_o l}{2}$$

$$M = \frac{w_o x(l-x)}{2}$$

$$y = \frac{w_o x\left(2lx^2 - x^3 - l^3\right)}{24EI}$$

$$y_{\max} = -\frac{5w_o l^4}{384EI}$$

D.8 Simply-supported beam with moment load

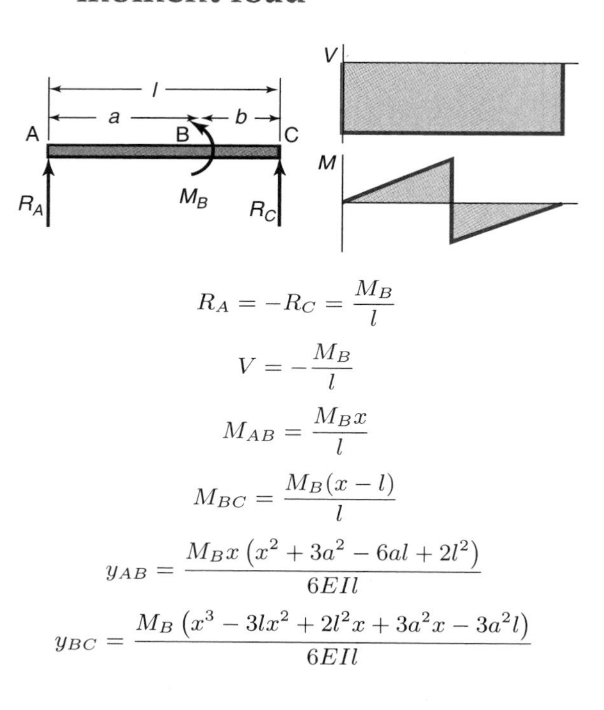

$$R_A = -R_C = \frac{M_B}{l}$$

$$V = -\frac{M_B}{l}$$

$$M_{AB} = \frac{M_B x}{l}$$

$$M_{BC} = \frac{M_B(x-l)}{l}$$

$$y_{AB} = \frac{M_B x\left(x^2 + 3a^2 - 6al + 2l^2\right)}{6EIl}$$

$$y_{BC} = \frac{M_B\left(x^3 - 3lx^2 + 2l^2x + 3a^2x - 3a^2l\right)}{6EIl}$$

D.9 Simply-supported beam with twin loads

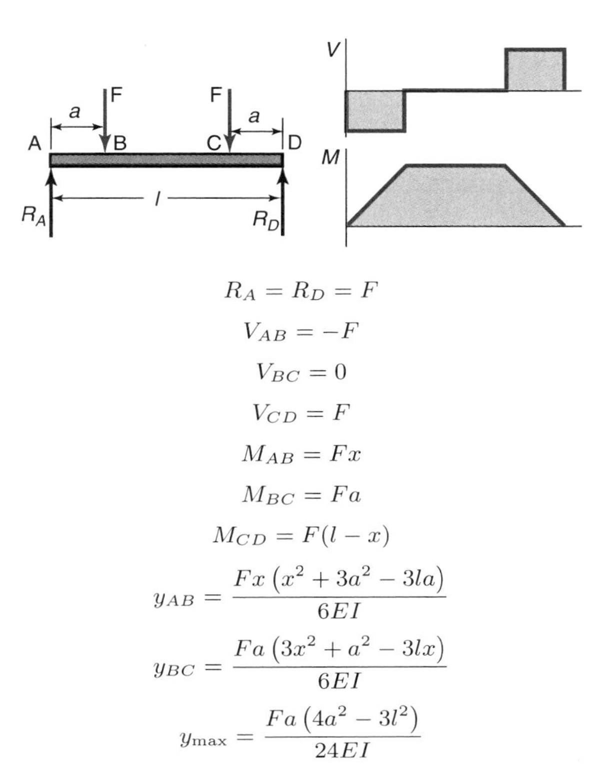

$$R_A = R_D = F$$

$$V_{AB} = -F$$

$$V_{BC} = 0$$

$$V_{CD} = F$$

$$M_{AB} = Fx$$

$$M_{BC} = Fa$$

$$M_{CD} = F(l - x)$$

$$y_{AB} = \frac{Fx\left(x^2 + 3a^2 - 3la\right)}{6EI}$$

$$y_{BC} = \frac{Fa\left(3x^2 + a^2 - 3lx\right)}{6EI}$$

$$y_{\max} = \frac{Fa\left(4a^2 - 3l^2\right)}{24EI}$$

D.10 Simply-supported beam with overhanging load

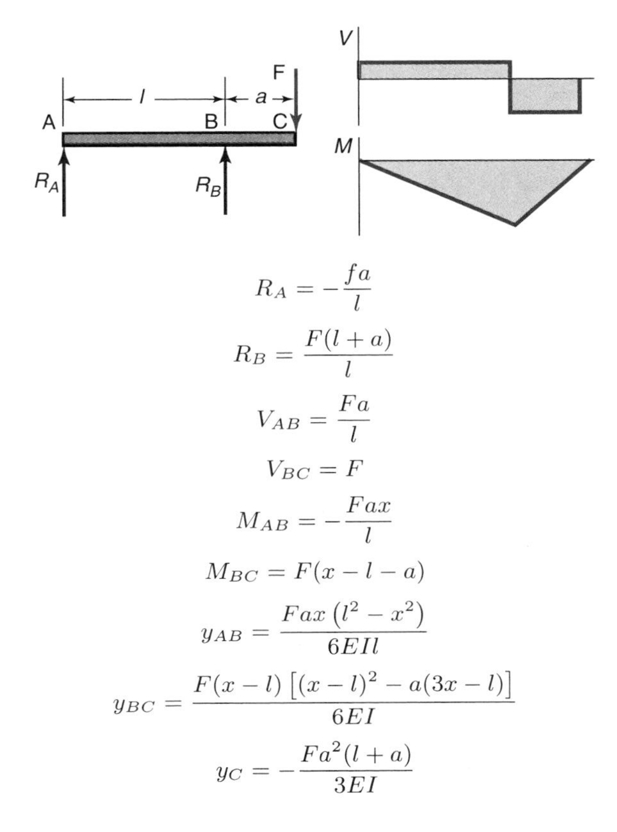

$$R_A = -\frac{fa}{l}$$

$$R_B = \frac{F(l + a)}{l}$$

$$V_{AB} = \frac{Fa}{l}$$

$$V_{BC} = F$$

$$M_{AB} = -\frac{Fax}{l}$$

$$M_{BC} = F(x - l - a)$$

$$y_{AB} = \frac{Fax\left(l^2 - x^2\right)}{6EIl}$$

$$y_{BC} = \frac{F(x - l)\left[(x - l)^2 - a(3x - l)\right]}{6EI}$$

$$y_C = -\frac{Fa^2(l + a)}{3EI}$$

D.11 Fixed-simply supported beam with point load at mid-span

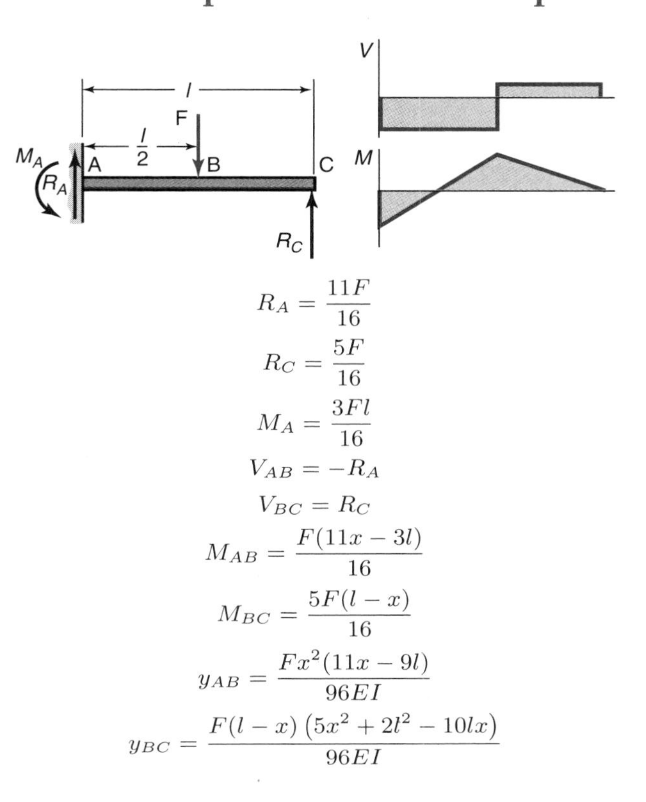

$$R_A = \frac{11F}{16}$$

$$R_C = \frac{5F}{16}$$

$$M_A = \frac{3Fl}{16}$$

$$V_{AB} = -R_A$$

$$V_{BC} = R_C$$

$$M_{AB} = \frac{F(11x - 3l)}{16}$$

$$M_{BC} = \frac{5F(l - x)}{16}$$

$$y_{AB} = \frac{Fx^2(11x - 9l)}{96EI}$$

$$y_{BC} = \frac{F(l - x)\left(5x^2 + 2l^2 - 10lx\right)}{96EI}$$

D.12 Fixed-simply supported beam with intermediate point load

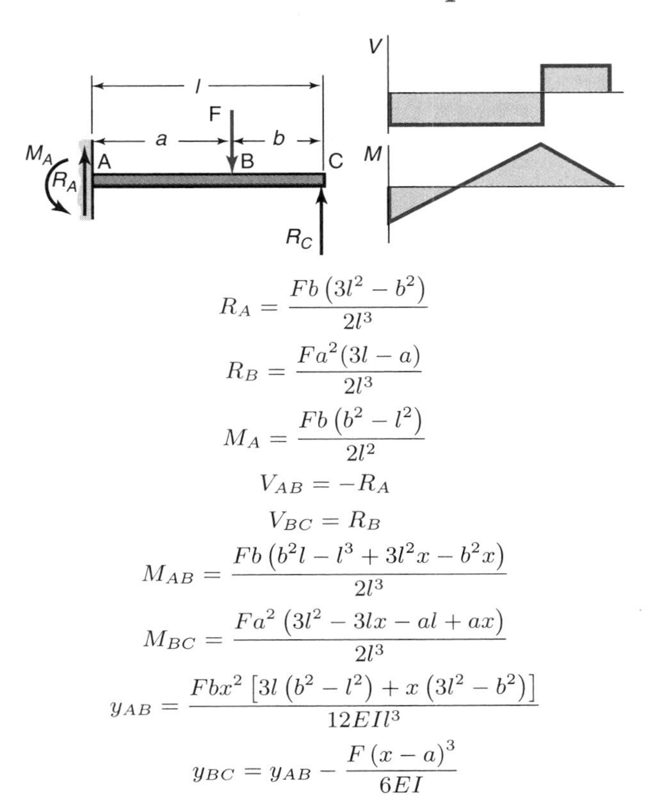

$$R_A = \frac{Fb\left(3l^2 - b^2\right)}{2l^3}$$

$$R_B = \frac{Fa^2(3l - a)}{2l^3}$$

$$M_A = \frac{Fb\left(b^2 - l^2\right)}{2l^2}$$

$$V_{AB} = -R_A$$

$$V_{BC} = R_B$$

$$M_{AB} = \frac{Fb\left(b^2 l - l^3 + 3l^2 x - b^2 x\right)}{2l^3}$$

$$M_{BC} = \frac{Fa^2\left(3l^2 - 3lx - al + ax\right)}{2l^3}$$

$$y_{AB} = \frac{Fbx^2\left[3l\left(b^2 - l^2\right) + x\left(3l^2 - b^2\right)\right]}{12EIl^3}$$

$$y_{BC} = y_{AB} - \frac{F(x - a)^3}{6EI}$$

D.13 Fixed-simply supported beam with uniform load

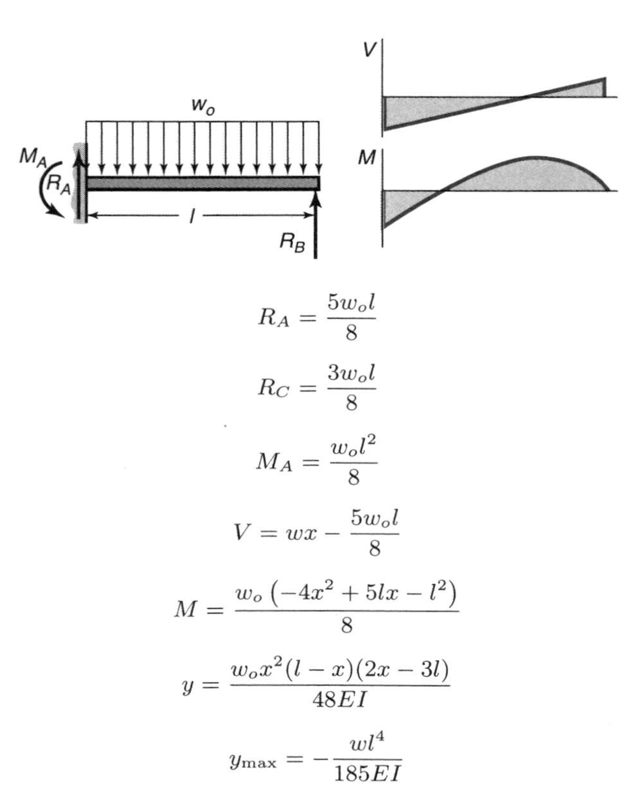

$$R_A = \frac{5w_o l}{8}$$

$$R_C = \frac{3w_o l}{8}$$

$$M_A = \frac{w_o l^2}{8}$$

$$V = wx - \frac{5w_o l}{8}$$

$$M = \frac{w_o\left(-4x^2 + 5lx - l^2\right)}{8}$$

$$y = \frac{w_o x^2(l-x)(2x-3l)}{48EI}$$

$$y_{\max} = -\frac{wl^4}{185EI}$$

D.14 Fixed-fixed beam with point load at mid-span

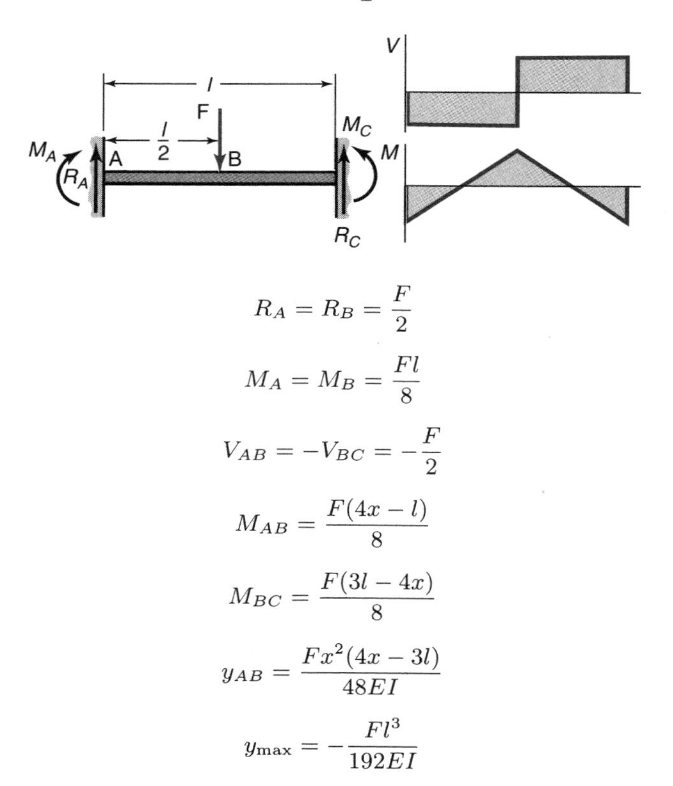

$$R_A = R_B = \frac{F}{2}$$

$$M_A = M_B = \frac{Fl}{8}$$

$$V_{AB} = -V_{BC} = -\frac{F}{2}$$

$$M_{AB} = \frac{F(4x-l)}{8}$$

$$M_{BC} = \frac{F(3l-4x)}{8}$$

$$y_{AB} = \frac{Fx^2(4x-3l)}{48EI}$$

$$y_{\max} = -\frac{Fl^3}{192EI}$$

D.15 Fixed-fixed beam with intermediate point load

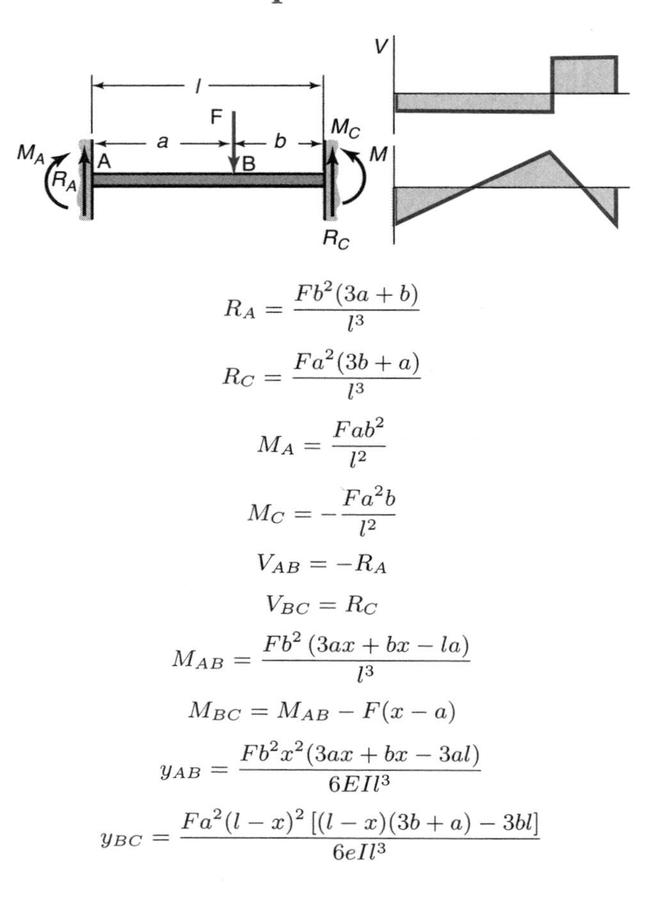

$$R_A = \frac{Fb^2(3a+b)}{l^3}$$

$$R_C = \frac{Fa^2(3b+a)}{l^3}$$

$$M_A = \frac{Fab^2}{l^2}$$

$$M_C = -\frac{Fa^2 b}{l^2}$$

$$V_{AB} = -R_A$$

$$V_{BC} = R_C$$

$$M_{AB} = \frac{Fb^2\left(3ax + bx - la\right)}{l^3}$$

$$M_{BC} = M_{AB} - F(x-a)$$

$$y_{AB} = \frac{Fb^2x^2(3ax+bx-3al)}{6EIl^3}$$

$$y_{BC} = \frac{Fa^2(l-x)^2\left[(l-x)(3b+a) - 3bl\right]}{6eIl^3}$$

D.16 Fixed-fixed beam with distributed load

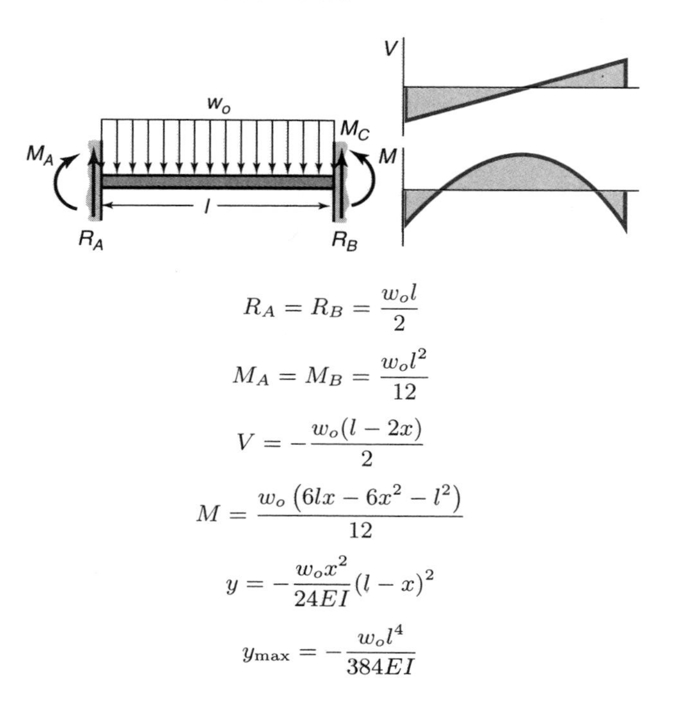

$$R_A = R_B = \frac{w_o l}{2}$$

$$M_A = M_B = \frac{w_o l^2}{12}$$

$$V = -\frac{w_o(l-2x)}{2}$$

$$M = \frac{w_o\left(6lx - 6x^2 - l^2\right)}{12}$$

$$y = -\frac{w_o x^2}{24EI}(l-x)^2$$

$$y_{\max} = -\frac{w_o l^4}{384EI}$$

Appendix E

Dimensions of Threaded Fasteners

E.1 Dimensions of Selected Bolts

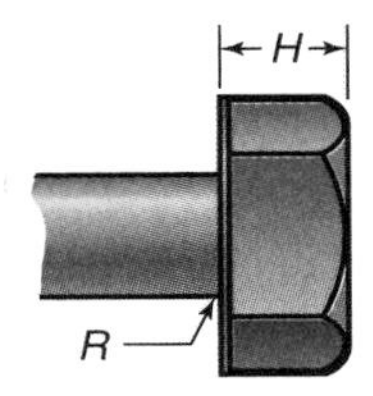

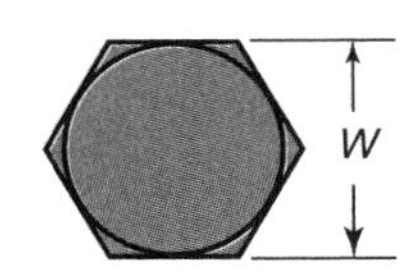

Nominal size	Square		Hexagonal head type: Regular			Heavy			Structural		
	W	H	W	H	R_{min}	W	H	R_{min}	W	H	R_{min}
Inch series – dimension in in.											
0.2500	0.3750	0.1719	0.4375	0.1719	0.010	—	—	—	—	—	—
0.3125	0.5000	0.2031	0.5000	0.2188	0.010	—	—	—	—	—	—
0.3750	0.5626	0.2500	0.5625	0.2500	0.010	—	—	—	—	—	—
0.4375	0.6250	0.2969	0.6250	0.2969	0.010	—	—	—	—	—	—
0.5000	0.7500	0.3281	0.7500	0.3438	0.010	0.8750	0.3438	0.010	0.8750	0.3125	0.009
0.6250	0.9375	0.4219	0.9375	0.4219	0.020	1.0625	0.4219	0.020	1.0625	0.3906	0.021
0.7500	1.1250	0.5000	1.1250	0.5000	0.020	1.2500	0.5000	0.020	1.2500	0.4688	0.021
1.0000	1.5000	0.6562	1.5000	0.6719	0.030	1.6250	0.6719	0.030	1.6250	0.6094	0.062
1.1250	1.6875	0.7500	1.6875	0.7500	0.030	1.8125	0.7500	0.030	1.8125	0.6875	0.062
1.2500	1.8750	0.8438	1.8750	0.8438	0.030	2.0000	0.8438	0.030	2.0000	0.7812	0.062
1.3750	2.0625	0.9062	2.0625	0.7632	0.030	2.1875	0.9062	0.030	2.1875	0.8438	0.062
1.5000	2.2500	1.0000	2.2500	1.0000	0.030	2.3750	1.0000	0.030	2.3750	0.9375	0.062
Metric series – dimensions in mm											
5	8	3.58	8	3.58	0.2	—	—	—	—	—	—
6	—	—	10	4.38	0.3	—	—	—	—	—	—
8	—	—	13	5.68	0.4	—	—	—	—	—	—
10	—	—	16	6.85	0.4	—	—	—	—	—	—
12	—	—	18	7.95	0.6	21	7.95	0.6	—	—	—
14	—	—	21	9.25	0.6	24	9.25	0.6	—	—	—
16	—	—	24	10.75	0.6	27	10.75	0.6	27	10.75	0.6
20	—	—	30	13.40	0.8	34	13.40	0.8	34	13.40	0.8
24	—	—	36	15.90	0.8	41	15.90	0.8	41	15.90	1.0
30	—	—	46	19.75	1.0	50	19.75	1.0	50	19.75	1.2
36	—	—	55	23.55	1.0	60	23.55	1.0	60	23.55	1.5

E.2 Dimensions of Hexagonal Cap Screws

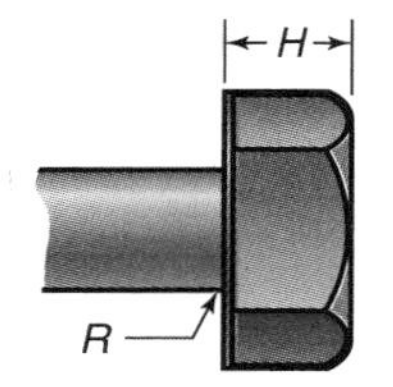

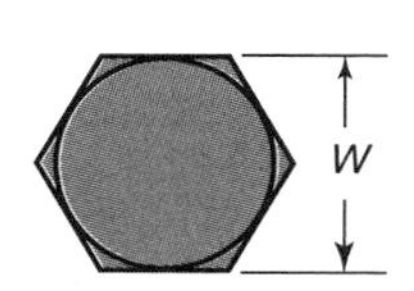

Nominal size	Fillet radius, *R*	Type of Screw: Cap, *W*	Heavy, *W*	Height, *H*
Inch series – dimensions in in.				
0.2500	0.015	0.4375	—	0.1562
0.3125	0.015	0.5000	—	0.2031
0.3750	0.015	0.5625	—	0.2344
0.4375	0.015	0.6250	—	0.2812
0.5000	0.015	0.7500	0.8750	0.3125
0.6250	0.020	0.9375	1.0625	0.3906
0.7500	0.020	1.1250	1.2500	0.4688
0.8750	0.040	1.3125	1.4375	0.5469
1.0000	0.060	1.5000	1.1250	0.6094
1.2500	0.060	1.8750	2.0000	0.7812
1.3750	0.060	2.0625	2.1875	0.8438
1.5000	0.060	2.2500	2.3750	0.9375
Metric series – dimensions in mm				
5	0.2	8	—	3.65
6	0.3	10	—	4.15
8	0.4	13	—	5.50
10	0.4	16	—	6.63
12	0.6	18	21	7.76
14	0.6	21	24	9.09
16	0.6	24	27	10.32
20	0.8	30	34	12.88
24	0.8	36	41	15.44
30	1.0	46	50	19.48
36	1.0	55	60	23.38

E.3 Dimensions of Hexagonal Nuts

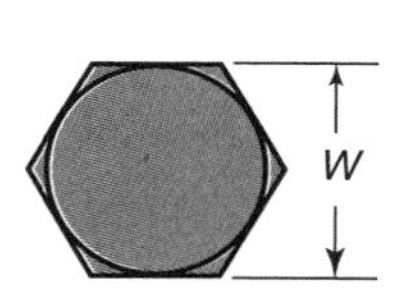

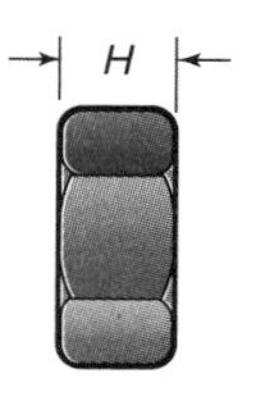

Nominal size	Width, *W*	Height, *H*: Regular hexagonal	Thick or slotted	JAM
Inch series – dimensions in in.				
0.2500	0.4375	0.2188	0.2812	0.1562
0.3125	0.5000	0.2656	0.3281	0.1875
0.3750	0.5625	0.3281	0.4062	0.2188
0.4375	0.6875	0.3750	0.4531	0.2500
0.5000	0.7500	0.4375	0.5625	0.3125
0.5625	0.8750	0.4844	0.6094	0.3125
0.6250	0.9375	0.5469	0.7188	0.3750
0.7500	1.1250	0.6406	0.8125	0.4219
0.8750	1.3125	0.7500	0.9062	0.4844
1.0000	1.5000	0.8594	1.0000	0.5469
1.1250	1.6875	0.9688	1.1562	0.6094
1.2500	1.8750	1.0625	1.2500	0.7188
1.3750	2.0625	1.1719	1.3750	0.7812
1.5000	2.2500	1.2812	1.5000	0.8438
Metric series – dimensions in mm				
5	8	4.7	5.1	2.7
6	10	5.2	5.7	3.2
8	13	6.8	7.5	4
10	16	8.4	9.3	5
12	18	10.8	12.0	6
14	21	12.8	14.1	7
16	24	14.8	16.4	8
20	30	18.0	20.3	10
24	36	21.5	23.9	12
30	46	25.6	28.6	15
36	55	31.0	34.7	18

E.4 Dimensions of American Standard Washers. (Dimensions in inches.)

Fastener size	Washer size	Inner diameter	Outer diameter	Thickness
#6	0.138	0.156	0.375	0.049
#8	0.164	0.188	0.438	0.049
#10	0.190	0.219	0.500	0.049
0.1875	0.188	0.250	0.562	0.049
#12	0.216	0.250	0.562	0.049
0.2500 N	0.250	0.281	0.625	0.065
0.2500 W	0.250	0.312	0.734	0.065
0.3125 N	0.312	0.344	0.688	0.065
0.3125 W	0.312	0.375	0.875	0.083
0.3750 N	0.375	0.406	0.812	0.065
0.3750 W	0.375	0.438	1.000	0.083
0.4375 N	0.438	0.469	0.922	0.065
0.4375 W	0.438	0.500	1.250	0.083
0.5000 N	0.500	0.531	1.062	0.095
0.5000 W	0.500	0.562	1.375	0.109
0.5625 N	0.562	0.594	1.156	0.095
0.5625 W	0.562	0.625	1.469	0.109
0.6250 N	0.625	0.656	1.312	0.095
0.6250 W	0.625	0.688	1.750	0.134
0.7500 N	0.750	0.812	1.469	0.134
0.7500 W	0.750	0.812	2.000	0.148
0.8750 N	0.875	0.938	1.750	0.134
0.8750 W	0.875	0.938	2.250	0.165
1.0000 N	1.000	1.062	2.000	0.134
1.0000 W	1.000	1.062	2.500	0.165
1.1250 N	1.125	1.250	2.250	0.134
1.1250 W	1.125	1.250	2.750	0.165
1.2500 N	1.250	1.375	2.500	0.165
1.2500 W	1.250	1.375	3.000	0.165
1.3750 N	1.375	1.500	2.750	0.165
1.3750 W	1.375	1.500	3.250	0.180
1.5000 N	1.500	1.625	3.000	0.165
1.5000 W	1.500	1.625	3.500	0.180
1.6250	1.625	1.750	3.750	0.180
1.7500	1.750	1.875	4.000	0.180
1.8750	1.875	2.000	4.250	0.180
2.0000	2.000	2.125	4.500	0.180
2.2500	2.250	2.375	4.750	0.220
2.5000	2.500	2.625	5.000	0.238
2.7500	2.750	2.875	5.250	0.259
3.0000	3.000	3.215	5.500	0.284

Note: N = narrow; W = wide.

E.5 Dimensions of Metric Washers. (Dimensions in millimeters.)

Washer Size	Minimum Inner Diameter	Maximum Outer Diameter	Maximum Thickness
1.6 N	1.95	4.00	0.70
1.6 R	1.95	5.00	0.70
1.6 W	1.95	6.00	0.90
2 N	2.50	5.00	0.90
2 R	2.50	6.00	0.90
2 W	3.50	8.00	0.90
2.5 N	3.00	6.00	0.90
2.5 R	3.00	8.00	0.90
2.5 W	3.00	10.00	1.20
3 N	3.50	7.00	0.90
3 R	3.50	10.00	1.20
3 W	3.50	12.00	1.40
3.5 N	4.00	9.00	1.20
3.5 R	4.00	10.00	1.40
3.5 W	4.00	15.00	1.75
4 N	4.70	10.00	1.20
4 R	4.70	12.00	1.40
4 W	4.70	16.00	2.30
5 N	5.50	11.00	1.40
5 R	5.50	15.00	1.75
5 W	5.50	20.00	2.30
6 N	6.65	13.00	1.75
6 R	6.65	18.80	1.75
6 W	6.65	25.40	2.30
8 N	8.90	18.80	2.30
8 R	8.90	25.40	2.30
8 W	8.90	32.00	2.80
10 N	10.85	20.00	2.30
10 R	10.85	28.00	2.80
10 W	10.85	39.00	3.50
12 N	13.30	25.40	2.80
12 R	13.30	34.00	3.50
12 W	13.30	44.00	3.50
14 N	15.25	28.00	2.80
14 R	15.25	39.00	3.50
14 W	15.25	50.00	4.00
16 N	17.25	32.00	3.50
16 R	17.25	44.00	4.00
16 W	17.25	56.00	4.60
20 N	21.80	39.00	4.00
20 R	21.80	50.00	4.60
20 W	21.80	66.00	5.10
24 N	25.60	44.00	4.60
24 R	25.60	56.00	5.10
24 W	25.60	72.00	5.60
30 N	32.40	56.00	5.10
30 R	32.40	72.00	5.60
30 W	32.40	90.00	6.40
36 N	38.30	66.00	5.60
36 R	38.30	90.00	6.40
36 W	38.30	110.00	8.50

Note: N = narrow; R = Regular; W = wide.

Index

G

H

I

J

K

L

M

N

O

P

Q

R

S

1,350

Table 4.1: Centroid, area moment of inertia, and area for common cross sections.

Cross section	Centroid	Area moment of inertia	Area
Circle 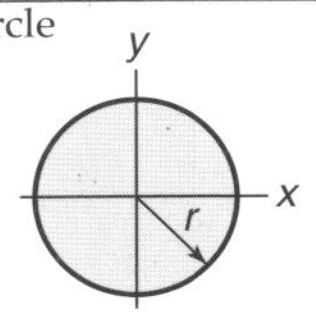	$\bar{x} = 0$ $\bar{y} = 0$	$I_x = I_{\bar{x}} = \frac{\pi}{4} r^4$ $I_y = I_{\bar{y}} = \frac{\pi}{4} r^4$ $J = \frac{\pi}{2} r^4$	$A = \pi r^2$
Hollow circle 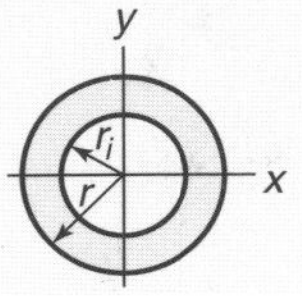	$\bar{x} = 0$ $\bar{y} = 0$	$I_x = I_{\bar{x}} = \frac{\pi}{4}\ r^4 - r_i^4$ $I_y = I_{\bar{y}} = \frac{\pi}{4}\ r^4 - r_i^4$ $J = \frac{\pi}{2}\ r^4 - r_i^4$	$A = \pi\ r^2 - r_i^2$
Triangle 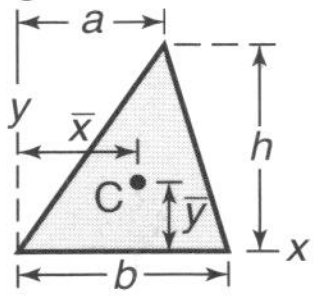	$\bar{x} = \frac{a+b}{3}$ $\bar{y} = \frac{h}{3}$	$I_x = \frac{bh^3}{12}, I_{\bar{x}} = \frac{bh^3}{36}$ $I_y = \frac{bh\left(b^2 + ab + a^2\right)}{12}$ $I_{\bar{y}} = \frac{bh\left(b^2 - ab + a^2\right)}{36}$ $\bar{J} = \frac{bh}{36}\left(b^2 + h^2 + a^2 - ab\right)$	$A = \frac{bh}{2}$
Rectangle 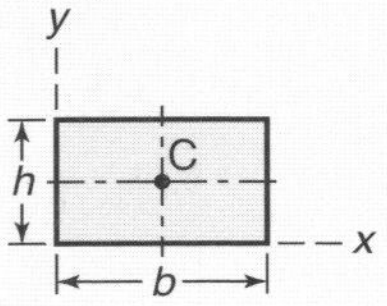	$\bar{x} = \frac{b}{2}$ $\bar{y} = \frac{h}{2}$	$I_x = \frac{bh^3}{3}, I_{\bar{x}} = \frac{bh^3}{12}$ $I_y = \frac{b^3h}{3}, I_{\bar{y}} = \frac{b^3h}{12}$ $\bar{J} = \frac{bh}{12}\ b^2 + h^2$	$A = bh$
Circular sector 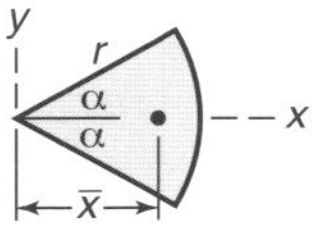	$\bar{x} = \frac{2}{3}\frac{r \sin\alpha}{\alpha}$	$I_x = \frac{r^4}{4}\left(\alpha - \frac{1}{2}\sin 2\alpha\right)$ $I_y = \frac{r^4}{4}\left(\alpha + \frac{1}{2}\sin 2\alpha\right)$ $J = \frac{1}{2} r^4 \alpha$	$A = r^2\alpha$
Quarter-circle 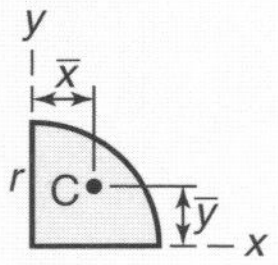	$\bar{x} = \bar{y} = \frac{4r}{3\pi}$	$I_x = I_y \frac{\pi r^4}{16}$ $I_{\bar{x}} = I_{\bar{y}} = \left(\frac{\pi}{16} - \frac{4}{9\pi}\right) r^4$ $J = \frac{\pi r^4}{8}$	$A = \frac{\pi r^2}{4}$
Elliptical quadrant 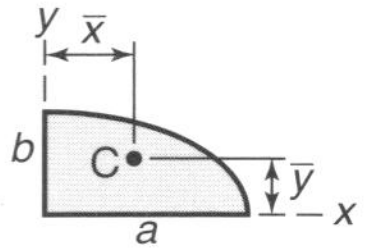	$\bar{x} = \frac{4a}{3\pi}$ $\bar{y} = \frac{4b}{3\pi}$	$I_x = \frac{\pi ab^3}{16}, I_{\bar{x}} = \left(\frac{\pi}{16} - \frac{4}{9\pi}\right) ab^3$ $I_y = \frac{\pi a^3 b}{16}, I_{\bar{y}} = \left(\frac{\pi}{16} - \frac{4}{9\pi}\right) a^3 b$ $J = \frac{\pi ab}{16}\left(a^2 + b^2\right)$	$A = \frac{\pi ab}{4}$